Advances in the Physics of Particles and Nuclei
Volume 29

Advances in the Physics of Particles and Nuclei

The series *Advances in the Physics of Particles and Nuclei* (APPN) is devoted to the archiving, in printed high-quality book format, of the comprehensive, long shelf-life reviews published in *The European Physical Journal A* and *C*. APPN will be of benefit in particular to those librarians and research groups, who have chosen to have only electronic access to these journals. Occasionally, original material in review format and refereed by the series' editorial board will also be included.

Series Editors

Douglas H. Beck
Department of Physics
University of Illinois at Urbana-Champaign
1110 West Green Street
Urbana, IL 61801-3080
USA

Dieter Haidt
DESY
Notkestraße 85
22603 Hamburg
Germany

John W. Negele
William A. Coolidge Professor of Physics
Massachusetts Institute of Technology
Center for Theoretical Physics
77 Massachusetts Ave. NE25-4079
Cambridge MA 02139
USA

Flavor in the Era of the LHC

Reports of the CERN Working Groups

Edited by

Robert Fleischer
Tobias Hurth
Michelangelo L. Mangano

Volume 29

Contributions to this Volume:

Flavor physics of leptons and dipole moments
Collider aspects of flavor physics at high Q
B, D and K decays

Robert Fleischer
CERN, PH Theory, 1211 Geneva 23, Switzerland
e-mail: robert.fleischer@cern.ch

Tobias Hurth
Michelangelo L. Mangano
CERN, TH Division, 1211 Geneva 23, Switzerland
e-mail: tobias.hurth@cern.ch
michelangelo.mangano@cern.ch

Originally published in Eur. Phys. J. C 57, no. 1-2 (2008)
© Springer-Verlag / Società Italiana di Fisica 2008

ISSN 1868-2146   e-ISSN 1861-440X
ISBN 978-3-540-95941-0   e-ISBN 978-3-540-95942-7
DOI 10.1007/978-3-540-95942-7
Springer Dordrecht Heidelberg London New York

Library of Congress Control Number: 2009926138

© Springer-Verlag Berlin Heidelberg 2009
This work is subject to copyright. All rights are reserved, whether the whole or part of the material is concerned, specifically the rights of translation, reprinting, reuse of illustrations, recitation, broad-casting, reproduction on microfilm or in any other way, and storage in data banks. Duplication of this publication or parts thereof is permitted only under the provisions of the German Copyright Law of September 9, 1965, in its current version, and permission for use must always be obtained from Springer. Violations are liable to prosecution under the German Copyright Law.
The use of general descriptive names, registered names, trademarks, etc. in this publication does not imply, even in the absence of a specific statement, that such names are exempt from the relevant protective laws and regulations and therefore free for general use.

Cover design: eStudioCalamar, Figueres/Berlin

Printed on acid-free paper

Springer is part of Springer Science+Business Media (www.springer.com)

Preface*

R. Fleischer, T. Hurth, M.L. Mangano[a]

Physics Department, CERN, 1211 Geneva, Switzerland

In the history of quantum and particle physics, discrete symmetries and their violation have played an outstanding rôle. First, the assumption of the conservation of P (parity), C (charge conjugation), CP and CPT (T denotes time reversal) helped theorists to restrict theoretical predictions, such as in Fermi's 1934 seminal paper on weak interactions. In 1957, the observation of P (and C) violation in weak interactions gave a new impact and led to the conjecture that CP was still a conserved symmetry. In 1963, one year before the surprising observation of CP violation in $K_L \to \pi^+\pi^-$ decays, the concept of quark-flavour mixing was introduced by Cabibbo. In 1973, it was discovered by Kobayashi and Maskawa that quark-flavour mixing actually allows us to accommodate CP violation in the framework of the standard model, provided the fermion content of this theory comes at least in three different replicas. Already in 1970, Glashow, Iliopoulos and Maiani introduced the charm quark to suppress the flavour-changing neutral currents, and the mass of that quark was estimated with the help of the K^0–$\bar{K}^0$ oscillation frequency by Gaillard and Lee in 1974. Moreover, the large value of the top-quark mass was first suggested by the large B_d^0–$\bar{B}_d^0$ mixing seen by ARGUS (DESY) and UA1 (CERN). In addition to the quark sector, also experimental information on the lepton sector was crucial for our understanding of the electroweak interactions, and forbidden decays, such as $\mu \to e\gamma$ have intrigued people.

Since the early days of the standard model, flavour physics has continued to progress, and flavour-changing neutral-current processes and CP-violating phenomena are still key targets of research, as they may be sensitive to the physics lying beyond the standard model. The particles of all three generations have been discovered, and non-vanishing neutrino masses have been established, leading to a rich flavour phenomenology in the lepton sector, and pointing towards new physics. The exploration of the quark-flavour sector was dominated for more than 30 years by the kaon system. In this past decade, the key player has been the B-meson system, and we also witnessed the appearance on stage of the top quark. Thanks to the e^+e^- B factories with their detectors BaBar (SLAC) and Belle (KEK), CP violation is now also firmly seen in B-meson decays, where the "golden" decay $B_d^0 \to J/\psi K_S$ shows CP-violating effects at the level of 70%. These effects can be translated into the angle β of the "unitarity triangle" (UT), which characterizes the Kobayashi–Maskawa mechanism of CP violation. Several strategies to determine the other angles of the triangle, α and γ, have been proposed and successfully applied to the B-factory data. After important first steps at the LEP experiments (CERN) and SLD (SLAC), in 2006 the CDF and D0 collaborations of the Tevatron (FNAL) collider could eventually measure the B_s^0–$\bar{B}_s^0$ oscillation frequency ΔM_s. In 2007, the B factories reported evidence for D^0–$\bar{D}^0$ mixing, which was the last missing meson–anti-meson mixing phenomenon.

So far, these results and intensive theoretical work have shown that the Kobayashi–Maskawa mechanism of CP violation is working remarkably well, thereby complementing the precision tests of the gauge sector of the standard model and, thus, also highly constraining any new-physics scenario beyond the standard model. On the other hand, neutrino oscillations and the baryon asymmetry of the universe require sources of flavour mixing and CP violation beyond what present in the SM. This demands the continued exploration of flavour phenomena, improving the current accuracy and probing new observables.

These efforts will soon be boosted with the startup of the LHC at CERN. B-decay studies will be the main theme of the LHCb experiment, while ATLAS and CMS will mostly focus on the properties of the top quark, and on the direct search of new particles, which could themselves be the mediators of new flavour and CP violating interactions. The new territory of the B-physics landscape that can be fully explored at the LHC is the B_s-meson system, which was not accessible at the e^+e^- B factories operating at the $\Upsilon(4S)$ resonance. The experimental value of ΔM_s is consistent with the standard model prediction, which suffers from lattice QCD uncertainties, and still leaves a lot of room for

---

*Introduction to the Report of the CERN Workshop "Flavour in the era of the LHC", Geneva, Switzerland, November 2005–March 2007.

[a] e-mail: michelangelo.mangano@cern.ch

CP-violating new-physics contributions to B_s^0–$\bar{B}_s^0$ mixing, which could be detected at the LHC with the help of the $B_s^0 \to J/\psi\phi$ decay. Another aspect of B_s physics is the fact that it will open various new ways to determine the angle γ of the UT. These methods make use, on the one hand, of pure "tree" decays (e.g. $B_s^0 \to D_s^\pm K^\mp$) and, on the other hand, of decays with penguin contributions (e.g. $B_s^0 \to K^+K^-$). Moreover, the $B_s^0 \to \phi\phi$ channel will allow us to shed more light on possible new-physics contributions to the CP asymmetries of various $b \to s$ penguin modes, which may be indicated by the current B-factory data for $B_d^0 \to \pi^0 K_S$, $B_d^0 \to \phi K_S$ and similar modes. Another key aspect of the LHC B-physics programme are studies of strongly suppressed rare decays, such as $B_s \to \mu^+\mu^-$, which could be highly enhanced through the impact of physics beyond the standard model.

These studies can be complemented through investigations of the extremely rare decays $K^+ \to \pi^+\nu\bar{\nu}$ and $K_L \to \pi^0\nu\bar{\nu}$. These are very clean from the theoretical point of view, but unfortunately hard to measure. Nevertheless, there is a proposal to take this challenge and to measure the former channel at the CERN SPS, and efforts to explore the latter – even more difficult decay – at J-PARC in Japan. Moreover, there are many other fascinating aspects of flavour physics, where the D-meson system is an interesting example. The recently observed D^0–$\bar{D}^0$ mixing can be accommodated in the standard model, but suffers from large theoretical uncertainties, so that new physics may actually be hiding there; it could be unambiguously detected through CP-violating effects. The programme of charm physics starting at the new BES-III τ-charm factory, extending the reach of the finishing CLEO-c, will provide, among other things, crucial ingredients for the testing and validation of lattice calculations, as well as for the measurements of B decays.

Other important flavour probes are offered by top-quark physics and flavour violation in the neutrino and charged lepton sectors; regarding the latter, the MEG experiment at PSI, searching for $\mu \to e\gamma$ decays, has started its commissioning at the end of 2007, and new explorations of $\mu \to e$ conversion on nuclei are proposed at FNAL and J-PARC. Further studies in this direction using τ decays at the LHC and at a possible future super-B factory will be important. Finally, continued searches of electric dipole moments and measurements of the anomalous magnetic moment of the muon are essential parts of the future experimental programme, are being pursued in several laboratories and with a broad variety of constantly-improving techniques.

As well known, there is a very active community dedicated to the complete and precise determination of neutrinos' masses, mixings and, possibly, CP violation. The state of the art and progress in this field are well documented in a great variety of workshop proceedings and study-group reports. For this reason, during this workshop we simply focused on the theoretical aspects of modeling neutrino mixing, and on the study of possible correlations of the values of the neutrino mixing parameters with flavour phenomena in the quark and charged-lepton sectors, as well as with the possible production of new heavy particles at the LHC.

In view of the approaching start of the LHC, there is a burning question: what is the synergy between the plenty of information following from analyses of the flavour sector with the high-Q programme of the ATLAS and CMS experiments? This topic was the centre of the five meetings of an extended workshop, which was held at CERN between November 2005 and March 2007. The goals of the workshop have been to outline and document a programme for flavour physics for the next decade, to discuss new experimental proposals, and to address the complementarity and synergy between the LHC and the flavour factories with respect to the discovery and exploration potential for new physics.

The activities have been coordinated by three working groups: (i) flavour aspects of high-Q collider physics, (ii) B, D and K decays, and (iii) flavour physics of leptons and of dipole moments. In addition to overviewing the status of theoretical progress and experimental results, several new studies have been performed. Seminal discussions took place on two proposals for an e^+e^- super-B factory, namely at KEK in Japan and near Frascati in Italy. Such a flavour factory would allow for precision experiments in quark and lepton flavour physics by accessing the B, the τ, and the charm sector. This discussion was complemented in the final meeting with the review of LHCb's upgrade plans. A miniworkshop on the prospects for future studies of dipole moments was also organized, bringing together expertise from both the experimental, the accelerator and the theoretical communities.

The following three articles document the outcome of the activity of the working groups. They confirm that flavour physics is an essential element in the future of high-energy physics. Further testing of the flavour sector of the standard model leaves still room for the detection of possible, unambiguous inconsistencies. Should new-physics particles be produced at the LHC, studies of flavour physics will play a key rôle, helping us to find the underlying new-physics scenario, to study the properties of the new-physics particles, and to detect or exclude new sources of CP violation and flavour structures.

Appendix: List of Workshop presentations

We collect here the full list of presentations that have been given during the Workshop, both in the plenary and in the parallel sessions. The slides of the presentations can be obtained from the links to the meeting agendas given on the Workshop web page, http://cern.ch/flavlhc.

Contributions to the 1st meeting, 7–10 November 2005

	Plenary Sessions
Y. Nir	Exploring BSM phenomena with B physics
M. Hazumi	Future prospects for B factories
O. Schneider	B physics prospects at the LHC
I. Bigi	Probing BSM phenomena with charm physics
G. Isidori	Exploring BSM phenomena with K physics
L. Littenberg	Future prospects for K-decay experiments
A. Romanino	Flavour phenomena in the lepton sector
T. Mori	LFV, status and prospects
Y. Semertzidis	Prospects for future measurements of muon $g-2$ and EDMs of muon, deuteron and neutron
R. Oldeman	Flavour studies and BSM searches at the Tevatron
G. Polesello	Flavour studies and BSM searches at the LHC
M. Nojiri	Flavour studies and SUSY at the LHC
G. Perez	Flavour physics beyond SUSY
D. Hitlin	The relevance of heavy flavour physics in the LHC era

	Working Group 1
T. Lari	SUSY (s)quark flavour studies with ATLAS
I. Borjanovic	SUSY (s)lepton flavour studies with ATLAS
S. Heinemeyer	Testing the NMFV MSSM with precision observables
J. Guasch	Heavy quark production by SUSY FCNC at the LHC
G. Unel	Search for isosinglet quarks with the ATLAS detector
G. Burdman	Extra dimensions flavour physics
C. Verzegnassi	Search for Supersymmetric electroweak effects in top production at the LHC
N. Castro	Study of top anomalous couplings and FCNC with the ATLAS detector
L. Benucci	Preliminary studies of top FCNC in CMS
P. Ko	$B_s \to \mu\mu$ and various SUSY scenarios

	Working Group 2
L. Cavoto	Measurements of γ and V_{ub}
T. Gershon	Unitarity triangle angles from Belle
M. Pierini	Rare charmless decays and measurements of α and β in penguins
E. Barberio	Photos
L. Fernandez	B_s^0 mass difference Δm_s and mixing phase ϕ_s at LHCb
S. Stone	Charm physics, experimental aspects
A. Soni	Null tests of the SM
P. Colangelo	Topics on nonleptonic B_s decays
S. Khalil	Probing the flavour structure in supersymmetric theories
M. Papucci	Next to minimal flavour violation
T. Iijima	Prospects for measurements of $b \to s\gamma$, $b \to s\ell\ell$, and $b \to c\tau\nu/\tau\nu$ at the super-B factory
S. Playfer	BaBar results on radiative and leptonic B decays

C. Ay	B-physics at D0
N. Nikitine	Rare dimuon decays at ATLAS
T. Speer	Sensitivity to new physics in B decays at CMS
P. Koppenburg	$B \to \mu\mu K^*$ and $B \to \ell\ell K$ at LHCb
J. Foster	Probing the flavour structure of SUSY breaking with rare B-processes: a beyond leading order analysis
Z. Ligeti	Who needs SCET in $B \to X\ell^+\ell^-$?
A. Buras	$K \to \pi\nu\bar{\nu}$, MFV and beyond It
I. Scimemi	$K \to 3\pi$, unveiling ϵ'/ϵ and π–π scattering lengths
G. Ruggiero	Future kaon program at CERN
T. Komatsubara	Future kaon program at J-PARC
A. Robert	A model-independent analysis of new physics contributions in $\Delta F = 2$ transitions
A. Stocchi	Constraining new physics with the UT fit

Working Group 3

P. Iaydjiev	The neutron EDM and CryoEDM experiments at ILL
K. Kirch	Towards a neutron EDM experiment at the PSI ultra-cold neutron source
Y. Semertzidis	The deuteron EDM at the 10^{-29} e cm level with the storage ring method
P. Paradisi	Higgs mediated LFV
B. Bajc	Seesaw in SO(10) and split SUSY
A. Ilakovac	LFV in MSSM based on the minimal SO(10) model
M. Picariello	Global analysis of neutrino data and implications for future experiments
G. Branco	Leptogenesis and low energy observables
A. Baldini	Improving the $\mu \to e\gamma$ sensitivity, MEG and beyond
J. Hosek	Dynamical generation of fermion masses by large Yukawa couplings
Y. Kuno	A high-intensity, high-luminosity muon source PRISM and search for muon to electron conversion
M. Felcini	A test of CP symmetry in positronium
G. Onderwater	Measurements of muon dipole moments
G. Colangelo	Hadronic contributions to muon $g - 2$
A. Soni	Massive neutrinos in a grounds-up approach

Joint sessions, Working Groups 1 + 3

G. Isidori	Minimal lepton flavour violation
R. Rueckl	Slepton flavour violation
A. Deandrea	Tests of R-parity violation
A. Ibarra	LFV in scenarios with stau NLSP
W. Porod	Lepton flavour and number violation at the LHC
O. Vives	Realistic models of flavour at LHC
Y. Kuno	A study on $\mu(e)$–τ conversion in DIS
G. Marchiori	Study of μ–τ conversion with high-intensity muon beams
Z. Was	Aspects of CP violation in τ production and decays

	Joint session, Working Groups 1 + 2 + 3
T. Hurth	Possible interplay between B-physics and collider physics
J.A. Aguilar-Saavedra	Signals of new quark singlets at large colliders and B-factories
S. Banerjee	Lepton flavour violation in τ decays: status and perspectives
I. Bigi	T-violating polarization in K and lepton decays
T. Shindou	Impact of B physics in the LHC era

Contributions to the 2nd meeting, 6–8 February 2006

	Working group 1
G. Unel	Update on E_6 isosinglet quark studies
S. Bejar	Higgs FCNC decays into top quark in the 2HDM model
C. Verzegnassi	A complete 1-loop MSSM calculation of tW production
J. Guasch	Single top-quark production by direct supersymmetric FCNC at the LHC
M.M. Najafabadi	tWb anomalous couplings
B. Fuks	SUSY CKM and mixed squark production

	Working group 2
J. Malcles	CKM fits in the SU(3) limit
R. Zwicky	B meson form factors from sum rules and kaon DA
S. Duerr	User's guide to lattice QCD
P. Reznicek	ATLAS $\Lambda_b \to \Lambda\mu\mu$
M. Patel	LHCb $B \to DK$ ADS method
U. Haisch	Theoretical status of $K \to \pi\nu\nu$
P. Paradisi	Higgs-mediated $K \to \pi\nu\nu$ at large $\tan\beta$
T. Feldmann	Hadronic uncertainties in two-body B decays
S. Jaeger	NLO hard spectator scattering in QCDF
M. Ciuchini	CKM bounds from $B \to K\pi\pi$
E. Baracchini	QED corrections to hadronic B decays
Z. Was	News on PHOTOS Monte Carlo: issue of systematic errors
C. Lazzeroni	LHCb $B \to DK$ Dalitz method
All	Discussion on hadronic uncertainties

	Working group 2, round table
G. Ruggiero	Kaons
F. Muheim	Status of study groups
M. Hazumi	Super B-factories/Belle

	Working group 3
A. Pilaftsis	Flavour in resonant leptogenesis, LFV and EDMs
O. Lebedev	Neutron–electron EDM correlations in SUSY
I. Masina	On power and complementarity of the experimental constraints on seesaw models
S. Albino	Strength and correlations of slepton flavour violation in SUSY seesaw models

S. King	Lepton flavour violation in non-minimal supergravity
E. Paoloni	Search for T-violation in τ decays at super-τ/B factories
A. Lusiani	Feasibility study for a fixed target $\mu \to \tau$ conversion experiment
M. Giffels	Status and plans of $\tau \to 3\mu$ at CMS
Y. Takanishi	The see-saw mechanism, neutrino Yukawa couplings, LFV decays and leptogenesis
S. Petcov	Charged LFV decays, Majorana CP-violating phases, leptogenesis and $\beta\beta0\nu$-decay
F. Deppisch	Enhanced lepton flavor violation in the inverse seesaw model
A. Ibarra	Reconstructing see-saw models from low energy data
A. Strumia	Minimal dark matter
P. Paradisi	Higgs-mediated LFV effects including all the family transitions
M. Herrero	Lepton flavour violating τ and μ decays induced by SUSY
B. Bajc	The minimal SO(10) GUT
M. Krawczyk	Large 2HDM(II) one-loop corrections in leptonic τ decays

	Joint sessions, Working group 1 + 2
A. Raklev	Search for a light stop
T. Lari	Search for light stop with ATLAS
S. Paktinat	Search for stop in CMS
G. Polesello	SUSY parameter measurement and b-physics
P. Gambino	NLO QCD corrections to $b \to s\gamma$ in the MSSM
All	Discussion for Tools

	Joint sessions, Working group 1 + 2 + 3
G. Moreau	Neutrino flavours and LHC phenomenology within the Randall–Sundrum model
G. Hou	Flavour physics and the 4th generation
M. Schmaltz	Little Higgs

Contributions to the 3rd meeting, 15–17 May 2006

	Working group 1
N. Castro	Status report on the top FCNC studies at the LHC
E. Kou	New physics effects to V_{tb} measurements in single top production at LHC
J. D'Hondt	Topological search for new flavour based physics in top quark events at the LHC
J.A. Aguilar-Saavedra	Looking for Wtb anomalous couplings in top pair decays
J. Carvalho	Studies of top quark decay asymmetries
S. Penaranda	$H \to bs$ and $b \to s\gamma$ in the MSSM with NMFV: FeynArts/FormCalc updated
S. Pukhov	Calchep for BSM physics
J.A. Aguilar-Saavedra	Discovering the Higgs boson in heavy singlet decays
S. Sultansoy	The fourth SM family: present status
G. Burdman	Signals for flavor violation in warped extra dimensions
P. Skands	The smoking gun of baryon number violation
B. Fuks	SUSY-CKM matrix determinations in SUSY electroweak processes at the LHC
C. Verzegnassi	t-Channel single top production at LHC: a realistic test of electroweak models

S. Paktinat	CMS potential for SUSY discovery in top + missing E_T final states
Z. Was	Discussion on the PHOTOS tool
P. Skands	Discussion on the SUSY Les Houches accord

Working group 2

U. Haisch	$B \to X_s \gamma$
I. Belyaev	LHCb radiative penguin
T. Iijima	Belle $B \to \tau\nu$
Y. Okada	$B \to (D)\tau\nu$ in MSSM
S. Robertson	BaBar $B \to s\nu\nu$ and $B \to l\nu$
G. Isidori	New physics benchmarks
A. Buras/C. Tarantino	Particle mixing and CPV in little Higgs models
G. Zhu	$B \to K^* ll$ in SCET
T. Huber	Electromagnetic logs in $b \to sll$
J. Berryhill	$b \to sll$: experimental status
C. Kao	Detecting Higgs bosons with muons in supergravity unified models
C.-J. Lin	Experimental results on $B \to \mu\mu$
A. Dedes	$B \to \mu\mu$ in SUSY
G. Borissov	D0 ΔM_s measurement
S. De Cecco	CDF ΔM_s measurement
A. Lenz	$\Delta \Gamma / \Gamma_{B_s}$, new results
D. Guadagnoli	ΔM_s and SUSY
M. Bona	UTfit and ΔM_s
S. T'Jampens	CKMfitter and ΔM_s
P. Ball	New physics in B_s mixing
All	Discussion on ΔM_s
C. Smith	SUSY in K decays
M. Beneke	$B \to VV$ decays can be useful
E. Conte	$\Lambda_b \to \Lambda V$
M. Vysotsky	α extraction from $B_d \to \pi\pi$ decays
O. Deschamps	$B \to \rho\pi, \rho\rho$ at LHCb
M. Bona	$B \to \rho\rho$ isospin analysis/summary of α determination
J. Rademacker	Γ extraction with 4-body D decays in $B \to DK$
S. Sultansoy	Turkish accelerator complex
D. Asner	Status of BES

Working group 3

H. Nishiguchi	Update on the status of MEG
W. Bertl	Final result of the SINDRUM II search for μ–e conversion
P. Paradisi	Probing new physics through lepton universality
M.-A. Sanchis-Lozano	Test of lepton universality in Υ decays: searching for a light Higgs boson
L. Fiorini	Testing LFV measuring $(K \to e\nu)/(K \to \mu\nu)$ in NA48: status and perspectives
O. Igonkina	Test of lepton universality in τ decay

A. van der Schaaf	Two new $\pi \to e\nu$ experiments
Y. Semertzidis	The muon $g - 2$ experiment
A. Rossi	Gauge and Yukawa mediated SUSY breaking in the triplet seesaw
S. Antusch	LFV and θ_{13} in SUSY seesaw
J. Hisano	Flavor mixing and EDMs in the SUSY models
P. Harris	The neutron EDM experiment at ILL
Y. Semertzidis	The deuteron EDM
H. Wilschut	Tests of time reversal violation (TRV) in atomic and nuclear physics—the TRImP facility nearing its completion
S. Davidson	Flavour matters in leptogenesis
L. Calibbi	LFV from SUSY-GUTs

Plenary session, Future B facilities

T. Nakada	Flavour physics at LHC
M. Yamauchi	Flavour physics at KEK super-B
M. Giorgi	Flavour physics with a linear super-B factory
K. Oide	Design of KEK super-B factory
P. Raimondi	Design of a linear super-B

Contributions to the 4th meeting, 9–11 October 2006

EDM and g-2 miniworkshop

Y. Semertzidis	Welcome/review of the report plans for EDM/$g - 2$
N. Ramsey	The history of the neutron EDM
A. Ritz	Electric dipole moments as probes of new physics
P. Harris	Review of the neutron EDMs (ILL and SNS)
K. Kirch	The neutron experiment at PSI plus the muon EDM prospects
F. Farley	How to measure $g - 2$ with 15 GeV muons
G. Onderwater	The deuteron EDM experiment
G. Venanzoni	Polarimetry for the dEDM method
W. Morse	EDM of proton and 3He
Y. Orlov	A plan of comprehensive investigation of systematic errors and spin coherence time for the deuteron resonance EDM experiment
A. Luccio	Spin and beam dynamics simulations
J.R. Guest	EDM searches on atoms with deformed nuclei: Ra-225
M. Kozlov	Theory of molecular EDM experiments
I. Masina	EDM experiments as probes of SUSY
J. Miller	Measuring the muon anomaly to 0.25 ppm
A. Hoecker	Evaluation of the hadronic vacuum polarization contribution to the muon $g - 2$
S. Redin	Hadrons at VEPP-2M
D. Leone	e^+e^- hadronic cross section measurement at DAFNE with the KLOE detector
G. Gabrielse	New measurement of the electron magnetic moment and the fine structure constant
O. Lebedev	More on EDM correlations in SUSY

	Working group 1
M. Spiropulu	CMS discovery potential for SUSY topologies
N. Krasnikov	Using the $e^{\pm}\mu^{\mp} + E_T^{miss}$ signature in the search for supersymmetry and lepton flavour violation in neutralino decays
A. Ventura	Neutralino spin measurement with ATLAS
I. Hinchliffe	Lepton flavour violation in neutralino decays
A. Gruzza	Large electroweak logarithms in heavy quark decay at LHC
C. Verzegnassi	The relevance of electroweak effects in the overall t-channel single top production at LHC
P.M. Martins Ferreira	Contributions from dimension five and six effective operators to flavour changing top physics
A. Onofre	Wtb anomalous top quark couplings
L. Benucci	FCNC top decays
F. del Aguila	Lepton mumber violation with muons at LHC
J.A. Aguilar-Saavedra	Signals of new fermions at high transverse momenta
M. Kirsanov	Detection of heavy Majorana neutrinos and right-handed bosons
G. Unel	E_6 and the Higgs boson
G. Servant	Multi-W events at the LHC
E. Ozcan	4th family physics
G. Unel	Determination of the D–d mixing angle
B. Clerbaux	CMS discovery potential for Z'/ED and spin discrimination
	Working group 2
M. Hazumi	News on superKEKB physics reach
F. Forti	Linear super-B update
F. Muheim	LHCb upgrade
A. Soni	New physics signals using exclusive radiative B-decays
R. Zwicky	$B \to K^*\gamma$ time-dependent CP asymmetry: (quasi) null test for SM
C. Eggel	Discovery potential for $B_s \to \mu^+\mu^-$ in CMS
F. Teubert	Search for the decay $B_s \to \mu^+\mu^-$ at LHCb
S. Descotes-Genon	$B_{s,d} \to KK$ using QCD Factorization and flavour symmetries
G. Bell	Higher-order QCD in exclusive B decays
D. Asner	Update on charm results
P. Spradlin	Charm physics at LHCb
C. Bobeth	New physics in $b \to sll$
F. De Fazio	$B \to K^*ll$ and extra dimensions
A. Khodjamirian	Charm resonances in $b \to sll$
U. Egede	$B \to K^*ll$ at LHCb
S. Villa	$B \to K^*ll$ and $B^+ \to \tau^+nu$ from Belle
L. Cavoto	CKM angles from BaBar
M. Hazumi	ICPV results from Belle
A. Sarti	$B \to h^+h^-$ at LHCb
A. Bevan	Rare hadronic $b \to s$ and $b \to d$ transitions (BaBar)
S. Burdin	$A_{SL}, \Delta\Gamma_s, \phi_s$

L. Wilke	Study of the decay $B_s \to J/\psi\phi$ with the CMS detector
A. Starodumov	Missing particle reconstruction using vertexing
N. Nikitine	Rare B-decay backgrounds studies—update
S. Heinemeyer	Tools (where we are)
S. Heinemeyer	Flavor benchmarks
M. Schmitt	CMS benchmark analysis
	Working group 3
S. Lavignac	Leptogenesis and LFV in type I + II seesaw mechanism
W. Rodejohann	LFV, leptogenesis and neutrino mixing in QLC scenarios
T. Yamashita	Flavour violation in "minimal" SUSY SU(5) models
C. Biggio	Unitarity in the leptonic sector
J. Miller	Prospects for a muon to electron conversion experiment at fermilab
W. Bonivento	Search for $B \to \mu e$ with LHCb
T. Shindou	How can CP phases contribute to LFV processes?
M. Aurelio Diaz	Neutrino masses and mixing in split supersymmetry
	Concluding plenary session
Y. Semertzidis	Summary of the EDM and $g-2$ miniworkshop
M. Misiak	$B_s \to X_s\gamma$ at NNLO
N. Ramsey	Contributions of magnetic resonance to other sciences

Contributions to the 5th meeting, 26–28 March 2006

	Working group 1: plenary reports
G. Burdman	Flavour and top physics, theory
N.F. Castro	Flavour and top physics, experiment
T. Lari	Sleptons
M. Klasen	Squarks
G. Unel	Exotics, I
J.A. Aguilar-Saavedra	Exotics, II
	Working group 2: plenary reports
D. Asner	Charm physics
G. Buchalla	Weak decays and QCD
J. Berryhill	Rare B decays
M. Ciuchini	$B \to s, d$ transitions, mixing, and UT angles
G. Isidori	New physics scenarios: flavour benchmarks
S. Heinemeyer	New physics scenarios: tools
	Working group 3: plenary reports
M. Raidal	The theory sections of the WG3 chapter of the report
S. Davidson	Low energy observables and phenomenological parametrization of LFV
A. Rossi	LFV in neutrino seesaw scenarios
F. Deppisch	LFV at colliders

A. van der Schaaf	LFV experiments and lepton universality
Y. Semertzidis	EDM and $g-2$ experiments
I. Bigi	The next big challenge: CP violation with charged leptons

Plenary sessions

Y. Grossman	Flavour in BSM
F. Couderc	Status and prospects for the detection of new physics at the Tevatron
T. Berry	Prospects for the observation of new physics at the LHC
U. Martyn	Prospects for flavour studies at the ILC
S. King	BSM and flavour in the lepton sector
R. Timmermans	EDMs, status and prospects
C. Gonzalez-Garcia	Probing flavour with neutrinos
A. Buras	FCNC processes in the LHC era
T. Komatsubara	Status and prospects of Kaon experiments
M. Roney	Experimental prospects for rare τ decays
J. Piedra	Tevatron heavy flavour programme: status and prospects
M. Hazumi	B factories: status and prospects
T. Nakada	The LHC heavy flavour programme
A. Stocchi	The physics potential of an e^+e^- super-B factory
K. Oide	Status of the KEK super-B project
M. Giorgi	Status of the linear super-B project
S. Stone	The physics potential of the LHCb high-lum upgrade

Contents

Flavor physics of leptons and dipole moments
M. Raidal et al. 1

Collider aspects of flavor physics at high Q
T. Lari et al. 171

B , D and K decays
G. Buchalla et al. 297

Flavor physics of leptons and dipole moments[*]

M. Raidal[1,b], A. van der Schaaf[2,b], I. Bigi[3,b], M.L. Mangano[4,a,b], Y. Semertzidis[5,b], S. Abel[6], S. Albino[7], S. Antusch[8], E. Arganda[9], B. Bajc[10], S. Banerjee[11], C. Biggio[8], M. Blanke[8,12], W. Bonivento[13], G.C. Branco[14,4], D. Bryman[15], A.J. Buras[12], L. Calibbi[16,17,18], A. Ceccucci[4], P.H. Chankowski[19], S. Davidson[20], A. Deandrea[20], D.P. DeMille[21], F. Deppisch[22], M.A. Diaz[23], B. Duling[12], M. Felcini[4], W. Fetscher[24], F. Forti[25], D.K. Ghosh[26], M. Giffels[27], M.A. Giorgi[25], G. Giudice[4], E. Goudzovskij[28], T. Han[29], P.G. Harris[30], M.J. Herrero[9], J. Hisano[31], R.J. Holt[32], K. Huitu[33], A. Ibarra[34], O. Igonkina[35,36], A. Ilakovac[37], J. Imazato[38], G. Isidori[28,39], F.R. Joaquim[9], M. Kadastik[1], Y. Kajiyama[1], S.F. King[40], K. Kirch[41], M.G. Kozlov[42], M. Krawczyk[19,4], T. Kress[27], O. Lebedev[4], A. Lusiani[25], E. Ma[43], G. Marchiori[25], A. Masiero[18], I. Masina[4], G. Moreau[44], T. Mori[45], M. Muntel[1], N. Neri[25], F. Nesti[46], C.J.G. Onderwater[47], P. Paradisi[48], S.T. Petcov[16,49], M. Picariello[50], V. Porretti[17], A. Poschenrieder[12], M. Pospelov[51], L. Rebane[1], M.N. Rebelo[14,4], A. Ritz[51], L. Roberts[52], A. Romanino[16], J.M. Roney[11], A. Rossi[18], R. Rückl[53], G. Senjanovic[54], N. Serra[13], T. Shindou[34], Y. Takanishi[16], C. Tarantino[12], A.M. Teixeira[44], E. Torrente-Lujan[55], K.J. Turzynski[56,19], T.E.J. Underwood[6], S.K. Vempati[57], O. Vives[17]

[1] National Institute for Chemical Physics and Biophysics, 10143 Tallinn, Estonia
[2] Physik-Institut der Universität Zürich, 8057 Zürich, Switzerland
[3] Physics Department, University of Notre Dame du Lac, Notre Dame, IN 46556, USA
[4] Physics Department, CERN, 1211 Geneva, Switzerland
[5] Brookhaven National Laboratory, Upton, NY 11973-5000, USA
[6] Institute for Particle Physics Phenomenology, Durham University, Durham DH1 3LE, UK
[7] II. Institute for Theoretical Physics, University of Hamburg, 22761 Hamburg, Germany
[8] Max-Planck-Institut für Physik, 80805 München, Germany
[9] Departamento de Fisica Teorica and IFT/CSIC-UAM, Universidad Autonoma de Madrid, 28049 Madrid, Spain
[10] J. Stefan Institute, 1000 Ljubljana, Slovenia
[11] Department of Physics, University of Victoria, Victoria, BC V8W 3P6, Canada
[12] Physics Department, TU Munich, 85748 Garching, Germany
[13] Università degli Studi di Cagliari and INFN Cagliari, 09042 Monserrato (CA), Italy
[14] Departamento de Física and Centro de Física Teórica de Partículas (CFTP), Instituto Superior Técnico (IST), 1049-001 Lisboa, Portugal
[15] Department of Physics and Astronomy, University of British Columbia, TRIUMF, Vancouver, BC, V6T 2A3, Canada
[16] SISSA and INFN, Sezione di Trieste, 34013 Trieste, Italy
[17] Departament de Física Teòrica, Universitat de València-CSIC, 46100 Burjassot, Spain
[18] Dipartimento di Fisica 'G. Galilei' and INFN, 35131 Padova, Italy
[19] Institute of Theoretical Physics, University of Warsaw, 00-681 Warsaw, Poland
[20] IPNL, CNRS, Université Lyon-1, 69622 Villeurbanne Cedex, France
[21] Physics Department, Yale University, New Haven, CO 06520, USA
[22] School of Physics and Astronomy, University of Manchester, Manchester M13 9PL, UK
[23] Facultad de Fisica, Pontificia Universita Catolica de Chile, Santiago 22, Chile
[24] Department of Physics, ETH Honggerberg, 8093 Zurich, Switzerland
[25] INFN and Dipartimento di Fisica, Universita di Pisa, 56127 Pisa, Italy
[26] Theoretical Physics Division, Physical Research Lab., Navrangpura, Ahmedabad 380 009, India
[27] III. Physikalisches Institut B, RWTH Aachen, 52056 Aachen, Germany
[28] Scuola Normale Superiore, 56100 Pisa, Italy
[29] Department of Physics, High Energy Physics, University of Wisconsin, Madison, WI 53706, USA
[30] Department of Physics and Astronomy, University of Sussex, Falmer, Brighton BN1 9QH, UK
[31] ICRR, University of Tokyo, Tokyo, Japan
[32] Physics Division, Argonne National Laboratory, Argonne, IL 60439-4843, USA
[33] Department of Physics, University of Helsinki, and Helsinki Institute of Physics, 00014 Helsinki, Finland
[34] Theory Group, DESY, 22603 Hamburg, Germany
[35] Physics Department, University of Oregon, Eugene, OR, USA
[36] NIKHEF, 1098 SJ Amsterdam, The Netherlands
[37] Department of Physics, University of Zagreb, 10002 Zagreb, Croatia
[38] IPNS, KEK, Ibaraki 305-0801, Japan
[39] INFN, Laboratori Nazionali di Frascati, 00044 Frascati, Italy
[40] School of Physics and Astronomy, University of Southampton, SO17 1BJ Southampton, UK
[41] Paul Scherrer Institut, 5232 Villigen, Switzerland
[42] Petersburg Nuclear Physics Institute, Gatchina 188300, Russia
[43] Department of Physics and Astronomy, University of California, Riverside, CA 92521, USA

[44]Laboratoire de Physique Théorique, UMR 8627 Université de Paris-Sud XI, 91405 Orsay Cedex, France
[45]ICEPP, University of Tokyo, Tokyo 113-0033, Japan
[46]Università dell'Aquila and INFN LNGS, 67010 L'Aquila, Italy
[47]Kernfysisch Versneller Instituut (KVI), 9747 AA Groningen, The Netherlands
[48]University of Rome "Tor Vergata" and INFN Sezione Roma II, 00133 Roma, Italy
[49]Institute of Nuclear Research and Nuclear Energy, Bulgarian Academy of Sciences, 1784 Sofia, Bulgaria
[50]INFN and Dipartimento di Fisica, Università del Salento, 73100 Lecce, Italy
[51]Department of Physics and Astronomy, University of Victoria, Victoria, BC V8P 5C2, Canada
[52]Department of Physics, Boston University, Boston, MA 02215, USA
[53]Institute for Theoretical Physics and Astrophysics, University of Würzburg, 97074 Würzburg, Germany
[54]International Centre for Theoretical Physics, Trieste, Italy
[55]Department of Physics, University of Murcia, 30100 Murcia, Spain
[56]Physics Department, University of Michigan, Ann Arbor, MI 48109, USA
[57]Centre for High Energy Physics, Indian Institute of Science, Bangalore 560012, India

Abstract This chapter of the report of the "Flavor in the era of the LHC" Workshop discusses the theoretical, phenomenological and experimental issues related to flavor phenomena in the charged lepton sector and in flavor conserving CP-violating processes. We review the current experimental limits and the main theoretical models for the flavor structure of fundamental particles. We analyze the phenomenological consequences of the available data, setting constraints on explicit models beyond the standard model, presenting benchmarks for the discovery potential of forthcoming measurements both at the LHC and at low energy, and exploring options for possible future experiments.

Contents

1 Charged leptons and fundamental dipole moments: alternative probes of the origin of flavor and CP violation 15
2 Theoretical framework and flavor symmetries . . . 18
   2.1 The flavor puzzle 18
   2.2 Flavor symmetries 19
3 Observables and their parameterization 29
   3.1 Effective operators and low scale observables . 29
   3.2 Phenomenological parameterizations of quark and lepton Yukawa couplings 35
   3.3 Leptogenesis and cosmological observables . 44
4 Organizing principles for flavor physics 46
   4.1 Grand unified theories 46
   4.2 Higher dimensional approaches 50
   4.3 Minimal flavor violation in the lepton sector . 53

5 Phenomenology of theories beyond the standard model 57
   5.1 Flavor violation in non-SUSY models directly testable at LHC 57
   5.2 Flavor and CP violation in SUSY extensions of the SM 66
   5.3 SUSY GUTs 86
   5.4 R-parity violation 97
   5.5 Higgs-mediated lepton flavor violation in supersymmetry 101
   5.6 Tests of unitarity and universality in the lepton sector 109
   5.7 EDMs from RGE effects in theories with low energy supersymmetry 116
6 Experimental tests of charged lepton universality . 122
   6.1 π decay 124
   6.2 K decay 125
   6.3 τ decay 127
7 CP violation with charged leptons 128
   7.1 μ decays 129
   7.2 CP violation in τ decays 130
   7.3 Search for T violation in $K^+ \to \pi^0 \mu^+ \nu$ decay 131
   7.4 Measurement of CP violation in ortho-positronium decay 133
8 LFV experiments 135
   8.1 Rare μ decays 135
   8.2 Searches for lepton flavor violation in τ decays 141
   8.3 $B^0_{d,s} \to e^\pm \mu^\mp$ 146
   8.4 In flight conversions 147
9 Experimental studies of electric and magnetic dipole moments 148
   9.1 Electric dipole moments 148
   9.2 Neutron EDM 151
   9.3 Deuteron EDM 153
   9.4 EDM of deformed nuclei: ^{225}Ra 156
   9.5 Electrons bound in atoms and molecules . . . 158
   9.6 Muon EDM 164

---

[*]Report of Working Group 3 of the CERN Workshop "Flavor in the era of the LHC", Geneva, Switzerland, November 2005–March 2007.

[a]e-mail: michelangelo.mangano@cern.ch

[b]Convenors

9.7 Muon $g-2$ 165
Acknowledgements 167
References 167

1 Charged leptons and fundamental dipole moments: alternative probes of the origin of flavor and CP violation

The understanding of the flavor structure and CP violation (CPV) of fundamental interactions has so far been dominated by the phenomenology of the quark sector of the standard model (SM). More recently, the observation of neutrino masses and mixing has begun extending this phenomenology to the lepton sector. While no experimental data available today link flavor and CP violation in the quark and in the neutrino sectors, theoretical prejudice strongly supports the expectation that a complete understanding should ultimately expose their common origin. Most attempts to identify the common origin, whether through grand unified (GUT) scenarios, supersymmetry (SUSY), or more exotic electroweak symmetry breaking mechanisms predict in addition testable correlations between the flavor and CP violation observables in the quark and neutrino sector on the one side, and new phenomena involving charged leptons and flavor conserving CP-odd effects on the other. This chapter of the "Flavor in the era of the LHC" report focuses precisely on the phenomenology arising from these ideas, discussing flavor phenomena in the charged lepton sector and flavor conserving CP-violating processes.

Several theoretical arguments make the studies discussed in this chapter particularly interesting.

- The charged lepton sector provides unique opportunities to test scenarios tailored to explain flavor in the quark and neutrino sectors, for example by testing correlations between neutrino mixing and the rate for $\mu \to e\gamma$ decays, as predicted by specific SUSY/GUT scenarios. Charged leptons are therefore an indispensable element of the flavor puzzle, without which its clarification could be impossible.
- The only observed source of CP violation is so far the Cabibbo–Kobayashi–Maskawa (CKM) mixing matrix. On the other hand, it is by now well established that this is not enough to explain the observed baryon asymmetry of the universe (BAU). The existence of other sources of CP violation is therefore required. CP-odd phases in neutrino mixing, directly generating the BAU through leptogenesis, are a possibility, directly affecting the charged lepton sector via, e.g., the appearance of electric dipole moments (EDMs). Likewise, EDMs could arise via CP violation in flavor conserving couplings, like phases of the gaugino fields or in extended Higgs sectors. In all cases, the observables discussed in this chapter provide essential experimental input for the understanding of the origin of CP violation.
- The excellent agreement of all flavor observables in the quark sector with the CKM picture of flavor and CP violation has recently led to the concept of minimal flavor violation (MFV). In scenarios beyond the SM (BSM) with MFV, the smallness of possible deviations from the SM is naturally built into the theory. While these schemes provide a natural setting for the observed lack of new physics (NP) signals, their consequence is often a reduced sensitivity to the underlying flavor dynamics of most observables accessible by the next generation of flavor experiments. Lepton flavor violation (LFV) and EDMs could therefore provide our only probe into this dynamics.
- Last but not least, with the exception of the magnetic dipole moments, where the SM predicts non-zero values and deviations due to new physics compete with the effect of higher order SM corrections, the observation of a non-zero value for any of the observables discussed in this chapter would be unequivocal indication of new physics. In fact, while neutrino masses and mixing can mediate lepton flavor violating transitions, as well as induce CP-odd effects, their size is such that all these effects are by many orders of magnitude smaller than anything measurable in the foreseeable future. This implies that, contrary to many of the observables considered in other chapters of this report, and although the signal interpretation may be plagued by theoretical ambiguities or systematics, there is nevertheless no theoretical systematic uncertainty to claim a discovery once a positive signal is detected.

The observables discussed here are also very interesting from the experimental point of view. They call for a very broad approach, based not only on the most visible tools of high energy physics, namely the high energy colliders, but also on a large set of smaller-scale experiments that draw from a wide variety of techniques. The emphasis of these experiments is by and large on high rates and high precision, a crucial role being played by the control of very large backgrounds and subtle systematics. A new generation of such experiments is ready to start or will start during the first part of the LHC operations. More experiments have been on the drawing board for some time, and could become reality during the LHC era if the necessary resources were made available. The synergy between the techniques and potential results provided by both the large- and small-scale experiments makes this field of research very rich and exciting and gives it a strong potential to play a key role in exploring the physics landscape in the era of the LHC.

The purpose of this document is to provide a comprehensive overview of the field, from both the theoretical and the experimental perspective. While we cover many model building aspects of neutrino physics that are directly related to the phenomenology of the quark and charged lepton sectors, for the status of the determinations of the mixing parameters and for the review of the future prospects we refer the reader to the vast existing literature, as documented for example in [1–4].

Several of the results presented are already well known, but they are nevertheless documented here to provide a self-contained review, accessible to physicists whose expertise covers only some of the many diverse aspects of this subject. Many results emerged during the workshop, including ideas on possible new experiments, further enrich this report. We present here a short outline and some highlights of the contents.

Section 2 provides the general theoretical framework that allows us to discuss flavor from a symmetry point of view. It outlines the origin of the flavor puzzles and lists the mathematical settings that have been advocated to justify or predict the hierarchies of the mixing angles in both the quark and neutrino sectors. Section 3 introduces the observables that are sensitive to flavor in the charged lepton sector and to flavor conserving CP violation, providing a unified description in terms of effective operators and effective scales for the new physics that should be responsible for them. The existing data already provide rather stringent limits on the size of these operators, as shown in several tables. We collect here in Table 1 some of the most significant benchmark results (for details, we refer to the discussion in Sect. 3.1.2). We constrain the dimensionless coefficients ϵ_i of effective operators O_i describing flavor or CP-violating interactions. Examples of these effective operators include

$$\overline{\ell}_i \sigma^{\mu\nu} \gamma_5 \ell_i F_{\mu\nu}^{\text{em}}, \qquad \overline{\ell}_i \sigma^{\mu\nu} \ell_j F_{\mu\nu}^{\text{em}}, \qquad (1.1)$$

which describe a CP-violating electric dipole moment (EDM) of lepton ℓ_i or the flavor violating decay $\ell_i \to \ell_j \gamma$, or the four-fermion operators:

$$\overline{\ell}_i \Gamma^a \ell_j \overline{q}_k \Gamma_a q_l, \qquad \overline{\ell}_i \Gamma^a \ell_j \overline{\ell}_k \Gamma_a \ell_l, \qquad (1.2)$$

where the Γ_a represent the various possible Lorentz structures. The overall normalization of the operators is chosen to reproduce the strength of transitions mediated by weak gauge bosons, assuming flavor mixing angles and CP-violating phases of order unity. The smallness of the constraints on ϵ therefore reflects either the large mass scale of flavor phenomena, or the weakness of the relative interactions.

It is clear from this table that current data are already sensitive to mass scales much larger than the electroweak scale, or to very small couplings. On the other hand, many of these constraints leave room for interesting signals coupled to the new physics at the TeV scale that can be directly discovered at the LHC. For example, a mixing of order 1 between the supersymmetric scalar partners of the charged leptons and a mass splitting among them of the order of the lepton masses is consistent with the current limits if the scalar lepton masses are just above 100 GeV, and it could lead both to their discovery at the LHC, and to observable signals at the next generation of $\ell \to \ell' \gamma$ experiments.

Most of this report will be devoted to the discussion of the phenomenological consequences of limits such as those in Table 1, setting constraints on explicit BSM models, presenting benchmarks for the discovery potential of forthcoming measurements both at the LHC and at low energy, and exploring options for future experiments aimed at increasing the reach even further.

Section 3 also introduces the phenomenological parameterizations of the quark and lepton mixing matrices that are found in the literature, emphasizing with concrete examples the correlations among the neutrino and charged lepton sectors that arise in various proposed models of neutrino masses. The section is completed by a discussion of the possible role played by leptogenesis and cosmological observables in constraining the neutrino sector.

Section 4 reviews the organizing principles for flavor physics. With a favorite dynamical theory of flavor still missing, the extended symmetries of BSM theories can provide some insight in the nature of the flavor structures of quarks and leptons, and give phenomenologically relevant constraints on low energy correlations between them. In GUT theories, for example, leptons and quarks belong to the same irreducible representations of the gauge group, and their mass matrices and mixing angles are consequently tightly related. Extra dimensional theories provide a possible dynamical origin for flavor, linking flavor to the geometry of the extra dimensions. This section also discusses the implications of models adopting for the lepton sector the same concept of MFV already explored in the case of quarks.

Section 5 discusses at length the phenomenological consequences of the many existing models, and represents the main body of this document. We cover models based on

Table 1 Bounds on CP- or flavor violating effective operators, expressed as upper limits on their dimensionless coefficients ϵ, scaled to the strength of weak interactions. For more details, in particular the overall normalization convention for the effective operators, see Sect. 3.1.2

Observable	Operator	Limit on ϵ
eEDM	$\overline{e_L}\sigma^{\mu\nu}\gamma_5 e_R F_{\mu\nu}$	$\leq 2.1 \times 10^{-12}$
$B(\mu \to e\gamma)$	$\overline{\mu}\sigma^{\mu\nu} e F_{\mu\nu}$	$\leq 3.4 \times 10^{-12}$
$B(\tau \to \mu\gamma)$	$\overline{\tau}\sigma^{\mu\nu} \mu F_{\mu\nu}$	$\leq 8.4 \times 10^{-8}$
$B(K_L^0 \to \mu^\pm e^\mp)$	$(\overline{\mu}\gamma^\mu P_L e)(\overline{s}\gamma^\mu P_L d)$	$\leq 2.9 \times 10^{-7}$

SUSY, as well as on alternative descriptions of electroweak symmetry breaking, such as little Higgs or extended Higgs sectors. In this section we discuss the predictions and the detection prospects of standard observables, such as $\ell \to \ell' \gamma$ decays or EDMs, and connect the discovery potential for these observables with the prospects for direct detection of the new massive particles at the LHC or at a future Linear Collider.

This section underlines, as is well known that the exploration of these processes has great discovery potential, since most BSM models anticipate rates that are within the reach of the forthcoming experiments. From the point of view of the synergy with collider physics, the remarkable outcome of these studies is that the sensitivities reached in the searches for rare lepton decays and dipole moments are often quite similar to those reached in direct searches at high energy. We give here some explicit examples. In $SO(10)$ SUSY GUT models, where the charged lepton mixing is induced via renormalization-group evolution of the heavy neutrinos of different generations, the observation of $B(\mu \to e\gamma)$ at the level of 10^{-13}, within the range of the just-starting MEG experiment, is suggestive of the existence of squarks and gluinos with a mass of about 1 TeV, well within the discovery reach of the LHC. Squarks and gluinos in the range of 2–2.5 TeV, at the limit of detectability for the LHC, would push $B(\mu \to e\gamma)$ down to the level of 10^{-16}. While this is well beyond the MEG sensitivity, it would well fit the ambitious goals of the next-generation $\mu \to e$ conversion experiments, strongly endorsing their plans. The decay $\mu \to e\gamma$ induced by the mixing of the scalar partners of muon and electron, and with a $B(\mu \to e\gamma)$ at the level of 10^{-13}, could give a $\chi_2^0 \to \chi_1^0 \mu^\pm e^\mp$ signal at the LHC, with up to 100 events after 300 fb^{-1}. Higher statistics and a cleaner signal would arise at a Linear Collider. Models where neutrino masses arise not from a see-saw mechanism at the GUT scale but from triplet Higgs fields at the TeV scale can be tested at the LHC, where processes like $pp \to H^{++}H^{--}$ can be detected for $m_{H^{++}}$ up to 700 GeV, using the remarkable signatures due to $B(H^{++} \to \tau^+\tau^+) \approx B(H^{++} \to \mu^+\mu^+) \approx B(H^{++} \to \mu^+\tau^+) \approx 1/3$.

Should signals of new physics be observed, alternative interpretations can be tested by exploiting different patterns of correlations that they predict among the various observables. For example, while typical SUSY scenarios predict $B(\mu \to 3e) \sim 10^{-2} B(\mu \to e\gamma)$, these branching ratios are of the same order in the case of little Higgs models with T parity. Important correlations also exist in see-saw SUSY GUT models between $B(\mu \to e\gamma)$ and $B(\tau \to \mu\gamma)$ or $B(\tau \to e\gamma)$. Furthermore, SUSY models with CP violation in the Higgs or gaugino mass matrix, be they supergravity (SUGRA) inspired or of the split-SUSY type, predict the ratio of electron and neutron EDM to be in the range of 10^{-2}–10^{-1}. Furthermore, in SUSY GUT models with see-saw mechanism correlations exist between the values of the neutron and deuteron EDMs and the heavy neutrino masses.

Section 6 discusses studies of lepton universality. The branching ratios $\Gamma(\pi \to \mu\nu)/\Gamma(\pi \to e\nu)$ and $\Gamma(K \to \mu\nu)/\Gamma(K \to e\nu)$, for example, are very well known theoretically within the SM. Ongoing experiments (at PSI and TRIUMF for the pion, and at CERN and Frascati for the kaon) test the existence of flavor-dependent charged Higgs couplings, by improving the existing accuracies by factors of order 10.

In Sect. 7 we consider CP-violating charged lepton decays, which offer interesting prospects as alternative probes of BSM phenomena. SM-allowed τ decays, such as $\tau \to \nu K\pi$, can be sensitive to new CP-violating effects. The decays being allowed by the SM, the CP-odd asymmetries are proportional to the interference of a SM amplitude with the BSM, CP-violating one. As a result, the small CP-violating amplitude contributes linearly to the rate, rather than quadratically, enhancing the sensitivity. In the specific case of $\tau \to \nu K\pi$, and for some models, a CP asymmetry at the level of 10^{-3} would correspond to $B(\tau \to \mu\gamma)$ around 10^{-8}. Another example is the CP-odd transverse polarization of the muon, P_T, in $K \to \pi\mu\nu$ decays. The current sensitivity of the KEK experiment E246, which resulted in $P_T < 5 \times 10^{-3}$ at 90% C.L., can be improved to the level of 10^{-4}, by TREK proposed at J-PARC, probing models such as multi-Higgs or R-parity-violating SUSY.

Section 8 discusses experimental searches for charged LFV processes. Transitions between e, μ, and τ might be found in the decay of almost any weakly decaying particle and searches have been performed in μ, τ, π, K, B, D, W and Z decay. Whereas the highest experimental sensitivities were reached in dedicated μ and K experiments, τ decay starts to become competitive as well. In Sect. 8 the experimental limitations to the sensitivities for the various decay modes are discussed in some detail, in particular for μ and τ decays, and some key experiments are presented. The sensitivities reached in searches for $\mu^+ \to e^+\gamma$ are limited by accidental $e^+\gamma$ coincidences and muon beam intensities have to be reduced now already. Searches for μ–e conversion, on the other hand, are limited by the available beam intensities, and large improvements in sensitivity may still be achieved. Similarly, in rare τ decays some decay modes are already background limited at the present B-factories and future sensitivities may not scale with the accumulated luminosities. Prospects of LFV decays at the LHC are limited to final states with charged leptons, such as $\tau \to 3\mu$ and $B^0_{d,s} \to e^\pm\mu^\mp$, which are discussed in detail. This section finishes with the preliminary results of a feasibility study for in-flight $\mu \to \tau$ conversions using a wide beam of high momentum muons. No working scheme emerged yet.

Section 9 covers electric and magnetic dipole moments. The muon magnetic moment has been much discussed recently, so we limit ourselves to a short review of the theoretical background and of the current and foreseeable experimental developments. In the case of EDMs, we provide an extensive description of the various theoretical approaches and experimental techniques applied to test electron and quark moments, as well as other possible sources of flavor diagonal CP-violating effects, such as the gluonic $\theta \tilde{F} F$ coupling, or CP-odd four-fermion interactions. While the experimental technique may differ considerably, the various systems provide independent and complementary information. EDMs of paramagnetic atoms such as Tl are sensitive to a combination of the fundamental electron EDM and CP-odd four-fermion interactions between nucleons and electrons. EDMs of diamagnetic atoms such as Hg are sensitive, in addition, to the intrinsic EDM of quarks, as well as to a non-zero QCD θ coupling. The neutron EDM more directly probes intrinsic quark EDMs, θ, and possible higher dimension CP-odd quark couplings. EDMs of the electron, without contamination from hadronic EDM contributions, can be tested with heavy diatomic molecules with unpaired electrons, such as YbF. In case of a positive signal the combination of measurements would help to disentangle the various contributions.

The experimental situation looks particularly promising, with several new experiments about to start or under construction. For example, new ultracold-neutron setups at ILL, PSI and Oak Ridge will increase the sensitivity to a neutron EDM by more than two orders of magnitude, to a level of about 10^{-28} e cm in 5–10 years. This sensitivity probes e.g. CP-violating SUSY phases of the order of 10^{-4} or smaller. Similar improvements are expected for the electron EDM. One of the main new ideas developed in the course of the workshop is the use of a storage ring to measure the deuteron EDM. The technical issues related to the design and construction of such an experiment, which could have a statistical sensitivity of about 10^{-29} e cm, are discussed here in some detail.

All the results presented in this document prove the great potential of this area of particle physics to shed light on one of the main puzzles of the standard model, namely the origin and properties of flavor. Low energy experiments are sensitive to scales of new physics that in several cases extend beyond several TeV. The similarity with the scales directly accessible at the LHC supports the expectation of an important synergy with the LHC collider programme, a synergy that clearly extends to future studies of the neutrino and quark sectors. The room for improvement, shown by the projections suggested by the proposed experiments, finally underscores the importance of keeping these lines of research at the forefront of the experimental high energy physics programme, providing the appropriate infrastructure, support and funding.

2 Theoretical framework and flavor symmetries

2.1 The flavor puzzle

The presence of three fermion families with identical gauge quantum numbers is a puzzle. The very origin of this replication of families constitutes the first element of the SM flavor puzzle. The second element has to do with the Yukawa interactions of those three families of fermions. While the gauge principle allows us to determine all SM gauge interactions in terms of three gauge couplings only (once the SM gauge group and the matter gauge quantum numbers have been specified), we do not have clear evidence of a guiding principle underlying the form of the 3×3 matrices describing the SM Yukawa interactions. Finally, a third element of the puzzle is represented by the peculiar pattern of fermion masses and mixing originating from those couplings.

The replication of SM fermion families can be rephrased in terms of the symmetries of the gauge part of the SM Lagrangian. The latter is in fact symmetric under a $U(3)^5$ symmetry acting on the family indexes of each of the five inequivalent SM representations forming a single SM family (q, u^c, d^c, l, e^c in Weyl notation). In other words, the gauge couplings and interactions do not depend on the (canonical) basis we choose in the flavor space of each of the five sets of fields $q_i, u_i^c, d_i^c, l, e_i^c, i = 1, 2, 3$.

This $U(3)^5$ symmetry is explicitly broken in the Yukawa sector by the fermion Yukawa matrices. It is because of this breaking that the degeneracy of the three families is broken and the fields corresponding to the physical mass eigenstates, as well as their mixing, are defined. An additional source of breaking is provided by neutrino masses. The smallness of neutrino masses is presumably due to the breaking of the accidental lepton symmetry of the SM at a scale much larger than the electroweak, in which case neutrino masses and mixing can be accounted for in the SM effective Lagrangian in terms of a dimension five operator breaking the $U(3)^5$ symmetry in the lepton doublet sector.

As mentioned, the special pattern of masses and mixing originating from the $U(3)^5$ breaking is an important element of the flavor puzzle. This pattern is quite peculiar. It suffices to mention the smallness of neutrino masses; the hierarchy of charged fermion masses relative to that of the two heavier neutrinos; the smallness of Cabibbo–Kobayashi–Maskawa mixing in the quark sector and the two large mixing angles in Pontecorvo–Maki–Nakagawa–Sakata (PMNS) matrix in the lepton sector; the mass hierarchy in the up quark sector, more pronounced than in the down quark and charged lepton sectors; the presence of a large CP-violating phase in the quark sector and the need of additional CP violation to account for baryogenesis; the approximate equality of bottom and tau masses at the scale at which the gauge couplings unify[1] and the approximate factor of 3 between the strange and muon masses, both pointing at a grand unified picture at high energy.

The origin of family replication and of the peculiar pattern of fermion masses and mixing are among the most interesting open questions in the SM, which a theory of flavor, discussed in Sect. 2, should address. As seen in Sect. 3, experiment is ahead of theory in this field. All the physical parameters describing the SM flavor structure in the quark sector have been measured with good accuracy. In the lepton sector crucial information on lepton mixing and neutrino masses is being gathered and a rich experimental program is under way to complete the picture.

Several tools are used to attack the flavor problem. Grand unified theories allow one to relate quark and lepton masses at the GUT scale and provide an appealing framework to study neutrino masses, leptogenesis, flavor models, etc. Note that in a grand unified context the $U(3)^5$ symmetry of the gauge sector is reduced (to $U(3)$ in the case in which all fermions in a family are unified in a single representation, as in $SO(10)$). Extra dimensions introduce new ways to account for the hierarchy of charged fermion masses (and in some cases for the smallness of neutrino masses) through the mechanism of localization in extra dimensions and by providing a new framework for the study of flavor symmetries. The concept of minimal flavor violation may also provide a framework for addressing flavor. The impact of those organizing principles on flavor physics is discussed in detail in Sect. 4.

From experimental point of view, however, additional handles are needed to gain more insight in the origin of flavor. Essentially this requires a discovery of new physics beyond the SM. New physics at the TeV scale may in fact be associated with an additional flavor structure, whose origin might well be related to the origin of the Yukawa couplings. Some of the present attempts to understand the pattern of fermion masses and mixing do link the flavor structure of the SM and that of the new physics sectors. In which case the search for indirect effects at low energy and for direct effects at colliders may play a primary role in clarifying our understanding of flavor. And conversely, the attempts to understand the pattern of fermion masses and mixing might lead to the prediction of new flavor physics effects. Those issues are addressed in Sect. 5.

Finally, lepton flavor physics is not just related to the lepton flavor violation or CP violation in the lepton sector but also to understanding the unitarity and universality in the lepton sector. Possible deviations from those are discussed in Sect. 5.6.

2.2 Flavor symmetries

The SM Lagrangian is $U(3)^5$ invariant in the limit in which the Yukawa couplings vanish. This might suggest that the Yukawa couplings, or at least some of them, arise from the spontaneous breaking of a subgroup of $U(3)^5$. Needless to say, the use of (spontaneously broken) symmetries as organizing principles to understand physical phenomena has been largely demonstrated in the past (chiral symmetry breaking, electroweak, etc.). In the following, we discuss the possibility of using such an approach to address the origin of the pattern of fermion masses and mixing, the constraints on the flavor structure of new physics, and to put forward expectations for flavor observables.

The spontaneously broken "flavour" or "family" symmetry can be local or global. Many (most) of the consequences of flavor symmetries are independent of this. The flavor breaking scale must be sufficiently high in such a way to suppress potentially dangerous effects associated with the new fields and interactions, in particular with the new gauge interactions (in the local case) or the unavoidable pseudo-Goldstone bosons (in the global case). In the context of an analysis in terms of effective operators of higher dimensions, a generic bound of about 10^3 TeV on the flavor scale from flavor changing neutral currents (FCNC) processes would be obtained. Nevertheless, a certain evidence for b–τ unification and the appeal of the see-saw mechanism for neutrino masses seem to suggest that these Yukawa couplings are already present near the GUT scale. This is indeed what most flavor models assume, and we shall also assume in the following.

The SM matter fields belong to specific representations of the flavor group, such that in the unbroken limit the Yukawa couplings have a particularly simple form. Typically some or all Yukawa couplings (with the possible exception of third generation ones) are not allowed. The spontaneous symmetry breaking of the flavor symmetry is provided by the vacuum expectation value (VEV) of fields often called "flavons". As the breaking presumably arises at a scale much higher than the electroweak scale, such flavons are SM singlets (or contain a SM singlet in the case of SM extensions) and typically they are only charged under the flavor symmetry. Flavor breaking is communicated dynamically to the SM fields by some interactions (possibly renormalizable, often not specified) living at a scale Λ_f not smaller than the scale of the flavor symmetry breaking. A typical example for these interactions that communicate the breaking is the exchange of heavy fermions whose mass terms respect the flavor symmetry. In that case the scale Λ_f would correspond to this fermion mass M_f. Many consequences of the flavor symmetry are actually independent of the mediation mechanism. It is therefore useful to consider an effective field theory approach below the scale Λ_f in which the

---
[1]Needless to say, precise unification requires an extension of the SM, with supersymmetry doing best from this point of view.

Table 2 Transformation of the matter superfields under the family symmetries. The ith generation SM fermion fields are grouped into the representation $\bar{5}_i = (D^c, L)_i$, $10_i = (Q, U^c, E^c)_i$, $1_i = (N^c)_i$

Field	10_3	10_2	10_1	$\bar{5}_3$	$\bar{5}_2$	$\bar{5}_1$	1_3	1_2	1_1	θ
$U(1)$	0	2	3	0	0	1	n_3^c	n_2^c	n_1^c	-1

flavor messengers have been integrated out. Once the flavon fields have acquired their VEVs, the structure of the Yukawa matrices (and other flavor parameters) can be obtained from an expansion in non-renormalizable operators involving the flavon fields and respecting the different symmetries (flavor and other symmetries) of the theory.

There are several possibilities for the flavor symmetry, local, global, accidental, continuous or discrete, Abelian or non-Abelian. Many examples are available in the literature for each of those possibilities. Some of them will be discussed in next subsections in relation to the implications considered in this study.

2.2.1 Continuous flavor symmetries

In order to provide an explicit example, we shortly discuss here one of the simplest possibilities, which goes back to the pioneering work of Froggatt–Nielsen [5]. In this model we have a $U(1)$ flavor symmetry under which the three generation of SM fields have different charges. In the simplest version we assign positive integer charges to the SM fermionic fields, the Higgs field is neutral, and we have a single flavon field θ of charge -1. The VEV of the flavon field is somewhat smaller than the mass of the heavy mediator fields M_f, so that the ratio $\epsilon = v/M_f \ll 1$. In this way the various entries in the Yukawa matrices are determined by epsilon to the power of the sum of the fermion charges with an undetermined order 1 coefficient. This mechanism explains nicely the hierarchy of fermion masses and mixing angles.

This idea is the basis for most flavor symmetries. It can be implemented in a great variety of different models. For the sake of definiteness, we show here how it works using as a concrete example a supersymmetric GUT model. Its superpotential is of the form

$$W_{\text{Yukawa}} = c_{ij}^d \epsilon^{q_i+d_j^c} Q_i D_j^c H_1 + c_{ij}^u \epsilon^{q_i+u_j^c} Q_i U_j^c H_2 \\ + c_{ij}^e \epsilon^{l_i+e_j^c} L_i E_j^c H_1 + c_{ij}^\nu \epsilon^{l_i+l_j} L_i L_j \frac{H_2 H_2}{\bar{M}}, \quad (2.1)$$

where the c's are $O(1)$ coefficients and $\bar{M}$ is the scale associated to $B - L$ breaking. The last term in this equation is an effective operator, giving Majorana masses to neutrinos, which can be generated, e.g., through a see-saw mechanism. Notice that the power of ϵ in each Yukawa coupling is proportional to the sum of the fermion charges: $Y_{ij}^u = c_{ij}^u \epsilon^{q_i+u_j^c}$,

$Y_{ij}^d = c_{ij}^d \epsilon^{q_i+d_j^c}$, etc. Hence, this mechanism explains the hierarchy of fermion masses and mixing angles through a convenient choice of charges. The value of these charges and the expansion parameter ϵ are constrained by the observed masses and angles. A convenient set of charges for example is given in Table 2. It turns out that this set of charges is the only one compatible with minimal $SU(5)$ unification. By introducing three right handed neutrinos with positive charges it is also possible to successfully realize the see-saw mechanism.

These charges give rise to the following Dirac Yukawa couplings for charged fermions at the GUT scale

$$Y_u = \begin{pmatrix} \epsilon^6 & \epsilon^5 & \epsilon^3 \\ \epsilon^5 & \epsilon^4 & \epsilon^2 \\ \epsilon^3 & \epsilon^2 & 1 \end{pmatrix}, \quad \begin{pmatrix} \epsilon^4 & \epsilon^3 & \epsilon^3 \\ \epsilon^3 & \epsilon^2 & \epsilon^2 \\ \epsilon & 1 & 1 \end{pmatrix}, \quad (2.2)$$

where $O(1)$ coefficients in each entry are understood here and in the following. With $\epsilon = O(\lambda_c)$ (the Cabibbo angle), the observed features of charged fermion masses and mixing are qualitatively well reproduced. It is known that the high energy relation $Y_e^T = Y_d$ is not satisfactory for the lighter families and should be relaxed by means of some mechanism [6–8]. The Dirac neutrino Yukawa couplings and the Majorana mass matrix of right handed neutrinos are

$$Y_\nu = \begin{pmatrix} \epsilon^{n_1^c+1} & \epsilon^{n_2^c+1} & \epsilon^{n_3^c+1} \\ \epsilon^{n_1^c} & \epsilon^{n_2^c} & \epsilon^{n_3^c} \\ \epsilon^{n_1^c} & \epsilon^{n_2^c} & \epsilon^{n_3^c} \end{pmatrix},$$

$$M_R = \begin{pmatrix} \epsilon^{2n_1^c} & \epsilon^{n_1^c+n_2^c} & \epsilon^{n_1^c+n_3^c} \\ \epsilon^{n_1^c+n_2^c} & \epsilon^{2n_2^c} & \epsilon^{n_2^c+n_3^c} \\ \epsilon^{n_1^c+n_3^c} & \epsilon^{n_2^c+n_3^c} & \epsilon^{2n_3^c} \end{pmatrix} \bar{M}. \quad (2.3)$$

Applying the see-saw mechanism to obtain the effective light neutrino mass matrix M_ν in the basis of diagonal charged lepton Yukawa couplings,[2] it is well known [9, 10] that if all right handed neutrino masses are positive the dependence on the right handed charges disappears:

$$U_{\text{PMNS}}^* m_\nu^{\text{diag}} U_{\text{PMNS}}^\dagger = m_\nu = \begin{pmatrix} \epsilon^2 & \epsilon & \epsilon \\ \epsilon & 1 & 1 \\ \epsilon & 1 & 1 \end{pmatrix} \frac{v_2^2}{\bar{M}}. \quad (2.4)$$

---

[2]Notice that going to the basis of diagonal charged leptons will only change the $O(1)$ coefficients, but not the power in ϵ of the different entries.

Experiments require $\bar{M} \sim 5 \times 10^{14}$ GeV. The features of neutrino masses and mixing are quite satisfactorily reproduced—the weak point being the tuning in the 23-determinant [9, 10] that has to be imposed. For later application, it is useful to introduce the unitary matrices which diagonalize Y_ν in the basis where both Y_e and M_R are diagonal: $V_L Y_\nu V_R = Y_\nu^{\text{diag}} \approx \text{diag}(\epsilon^{n_1^c}, \epsilon^{n_2^c}, \epsilon^{n_3^c})$. Notice that, as a consequence of the equal charges of the lepton doublets L_2 and L_3, the model predicts that V_L has a large mixing, although not necessarily maximal, in the 2–3 sector as observed in U_{PMNS}.

The literature is very rich of models based on flavor symmetries. Some references are [5, 9–40]; for more recent attempts the interested reader is referred for instance to [41–64].

2.2.2 Discrete flavor symmetries

2.2.2.1 Finite groups
Discrete flavor symmetries have gained popularity because they seem to be appropriate to address the large mixing angles observed in neutrino oscillations. To obtain a non-Abelian discrete symmetry, a simple heuristic way is to choose two specific non-commuting matrices and form all possible products. As a first example, consider the two 2×2 matrices

$$A = \begin{pmatrix} 0 & 1 \\ 1 & 0 \end{pmatrix}, \qquad B = \begin{pmatrix} \omega & 0 \\ 0 & \omega^{-1} \end{pmatrix}, \qquad (2.5)$$

where $\omega^n = 1$, i.e. $\omega = \exp(2\pi i/n)$. Since $A^2 = 1$ and $B^n = 1$, this group contains Z_2 and Z_n. For $n = 1, 2$, we obtain Z_2 and $Z_2 \times Z_2$ respectively, which are Abelian. For $n = 3$, the group generated has six elements and is in fact the smallest non-Abelian finite group S_3, the permutation group of three objects. This particular representation is not the one found in text books, but it is related to it by a unitary transformation [65], and was first used in 1990 for a model of quark mass matrices [66, 67]. For $n = 4$, the group generated has eight elements which are in fact ± 1, $\pm i\sigma_{1,2,3}$, where $\sigma_{1,2,3}$ are the usual Pauli spin matrices. This is the group of quaternions Q, which has also been used [68] for quark and lepton mass matrices. In general, the groups generated by (2.5) have $2n$ elements and may be denoted as $\Delta(2n)$.

Consider next the two 3×3 matrices:

$$A = \begin{pmatrix} 0 & 1 & 0 \\ 0 & 0 & 1 \\ 1 & 0 & 0 \end{pmatrix}, \qquad B = \begin{pmatrix} \omega & 0 & 0 \\ 0 & \omega^2 & 0 \\ 0 & 0 & \omega^{-3} \end{pmatrix}. \qquad (2.6)$$

Since $A^3 = 1$ and $B^n = 1$, this group contains Z_3 and Z_n. For $n = 1$, we obtain Z_3. For $n = 2$, the group generated has 12 elements and is A_4, the even permutation group of 4 objects, which was first used in 2001 in a model of lepton mass matrices [36, 41]. It is also the symmetry group of the tetrahedron, one of five perfect geometric solids, identified by Plato with the element "fire" [69]. In general, the groups generated by (2.6) have $3n^2$ elements and may be denoted as $\Delta(3n^2)$ [70]. They are in fact subgroups of $SU(3)$. In particular, $\Delta(27)$ has also been used [57, 71]. Generalizing to $k \times k$ matrices, we then have the series $\Delta(kn^{k-1})$. However, since there are presumably only three families, $k > 3$ is probably not of much interest.

Going back to $k = 2$, but using instead the following two matrices:

$$A = \begin{pmatrix} 0 & 1 \\ 1 & 0 \end{pmatrix}, \qquad B = \begin{pmatrix} \omega & 0 \\ 0 & 1 \end{pmatrix}. \qquad (2.7)$$

Now again $A^2 = 1$ and $B^n = 1$, but the group generated will have $2n^2$ elements. Call it $\Sigma(2n^2)$. For $n = 1$, it is just Z_2. For $n = 2$, it is D_4, i.e. the symmetry group of the square, which was first used in 2003 [47, 72]. For $k = 3$, consider

$$A = \begin{pmatrix} 0 & 1 & 0 \\ 0 & 0 & 1 \\ 1 & 0 & 0 \end{pmatrix}, \qquad B = \begin{pmatrix} \omega & 0 & 0 \\ 0 & 1 & 0 \\ 0 & 0 & 1 \end{pmatrix}, \qquad (2.8)$$

then the groups generated have $3n^3$ elements and may be denoted as $\Sigma(3n^3)$. They are in fact subgroups of $U(3)$. For $n = 1$, it is just Z_3. For $n = 2$, it is $A_4 \times Z_2$. For $n = 3$, the group $\Sigma(81)$ has been used [73] to understand the Koide formula [74] as well as lepton mass matrices [75]. In general, we have the series $\Sigma(kn^k)$.

2.2.2.2 Model recipe

1. Choose a group, e.g. S_3 or A_4, and write down its possible representations. For example S_3 has $\underline{1}, \underline{1}', \underline{2}$; A_4 has $\underline{1}, \underline{1}', \underline{1}'', \underline{3}$. Work out all product decompositions. For example $\underline{2} \times \underline{2} = \underline{1} + \underline{1}' + \underline{2}$ in S_3, and $\underline{3} \times \underline{3} = \underline{1} + \underline{1}' + \underline{1}'' + \underline{3} + \underline{3}$ in A_4.
2. Assign $(\nu, l)_{1,2,3}$ and $l^c_{1,2,3}$ to the representations of choice. To have only renormalizable interactions, it is necessary to add Higgs doublets (and perhaps also triplets and singlets) and, if so desired, neutrino singlets.
3. The Yukawa structure of the model is restricted by the choice of particle content and their representations. As the Higgs bosons acquire vacuum expectation values (which may be related by some extra or residual symmetry), the lepton mass matrices will have certain particular forms, consistent with the known values of m_e, m_μ, m_τ, etc. If the number of parameters involved is less than the number of observables, there will be one or more predictions.
4. In models with more than one Higgs doublet, flavor non-conservation will appear at some level. Its phenomenological consequences need to be worked out, to ensure

the consistency with present experimental constraints. The implications for phenomena at the TeV scale can then be explored.

5. Insisting on using only the single SM Higgs doublet requires effective non-renormalizable interactions to support the discrete flavor symmetry. In such models, there are no predictions beyond the forms of the mass matrices themselves.

6. Quarks can be considered in the same way. The two quark mass matrices m_u and m_d must be nearly aligned so that their mixing matrix involves only small angles. In contrast, the mass matrices m_ν and m_e should have different structures so that large angles can be obtained.

Some explicit examples will now be outlined.

2.2.2.3 S_3

Being the simplest, the non-Abelian discrete symmetry S_3 was used already [76] in the early days of strong interactions. There are many recent applications [55, 77–86], some of which are discussed in [87]. Typically, such models often require extra symmetries beyond S_3 to reduce the number of parameters, or assumptions of how S_3 is spontaneously and softly broken. For illustration, consider the model of Kubo et al. [77] which has recently been updated by Felix et al. [88]. The symmetry used is actually $S_3 \times Z_2$, with the assignments

$$(\nu, l), l^c, N, (\phi^+, \phi^0) \sim \underline{1} + \underline{2}, \quad (2.9)$$

and equal vacuum expectation values for the two Higgs doublets transforming as $\underline{2}$ under S_3. The Z_2 symmetry serves to eliminate four Yukawa couplings, otherwise allowed by S_3, resulting in an inverted ordering of neutrino masses with

$$\theta_{23} \simeq \pi/4, \quad \theta_{13} \simeq 0.0034, \quad m_{ee} \simeq 0.05 \text{ eV}, \quad (2.10)$$

where m_{ee} is the effective Majorana neutrino mass measured in neutrinoless double beta decay. This model relates θ_{13} to the ratio m_e/m_μ.

2.2.2.4 A_4

To understand why quarks and leptons have very different mixing matrices, A_4 turns out to be very useful. It allows the two different quark mass matrices to be diagonalized by the same unitary transformations, implying thus no mixing as a first approximation, but because of the assumed Majorana nature of the neutrinos, a large mismatch may occur in the lepton sector, thus offering the possibility of obtaining the so-called tri-bi-maximal mixing matrix [89, 90], which is a good approximation to the present data. One way of doing this is to consider the decomposition

$$U_{\text{PMNS}} = \begin{pmatrix} \sqrt{2/3} & 1/\sqrt{3} & 0 \\ -1/\sqrt{6} & 1/\sqrt{3} & -1/\sqrt{2} \\ -1/\sqrt{6} & 1/\sqrt{3} & 1/\sqrt{2} \end{pmatrix}$$

$$= \frac{1}{\sqrt{3}} \begin{pmatrix} 1 & 1 & 1 \\ 1 & \omega & \omega^2 \\ 1 & \omega^2 & \omega \end{pmatrix} \begin{pmatrix} 0 & 1 & 0 \\ 1/\sqrt{2} & 0 & -i/\sqrt{2} \\ 1/\sqrt{2} & 0 & i/\sqrt{2} \end{pmatrix}, \quad (2.11)$$

where U_{PMNS} is the observed neutrino mixing matrix and $\omega = \exp(2\pi i/3) = -1/2 + i\sqrt{3}/2$. The matrix involving ω has equal moduli for all its entries and was conjectured already in 1978 [91, 92] to be a possible candidate for the 3×3 neutrino mixing matrix.

Since $U_{\text{PMNS}} = V_e^\dagger V_\nu$, where V_e, V_ν diagonalize the matrices $m_e m_e^\dagger$, $m_\nu m_\nu^\dagger$ respectively, (2.11) may be obtained if we have

$$V_e^\dagger = \frac{1}{\sqrt{3}} \begin{pmatrix} 1 & 1 & 1 \\ 1 & \omega & \omega^2 \\ 1 & \omega^2 & \omega \end{pmatrix} \quad (2.12)$$

and

$$m_\nu = \begin{pmatrix} a+2b & 0 & 0 \\ 0 & a-b & d \\ 0 & d & a-b \end{pmatrix}$$

$$= \begin{pmatrix} 0 & 1 & 0 \\ 1/\sqrt{2} & 0 & -i/\sqrt{2} \\ 1/\sqrt{2} & 0 & i/\sqrt{2} \end{pmatrix}$$

$$\times \begin{pmatrix} a-b+d & 0 & 0 \\ 0 & a+2b & 0 \\ 0 & 0 & -a+b+d \end{pmatrix}$$

$$\times \begin{pmatrix} 0 & 1/\sqrt{2} & 1/\sqrt{2} \\ 1 & 0 & 0 \\ 0 & -i/\sqrt{2} & i/\sqrt{2} \end{pmatrix}. \quad (2.13)$$

It was discovered in Ref. [36] that (2.12) is naturally obtained with A_4 if

$$(\nu, l)_{1,2,3} \sim \underline{3}, \quad l^c_{1,2,3} \sim \underline{1} + \underline{1}' + \underline{1}'', \quad (\phi^+, \phi^0)_{1,2,3} \sim \underline{3} \quad (2.14)$$

for $\langle \phi_1^0 \rangle = \langle \phi_2^0 \rangle = \langle \phi_3^0 \rangle$. This assignment also allows m_e, m_μ, m_τ to take on arbitrary values, because there are here exactly three independent Yukawa couplings invariant under A_4. If we use this also for quarks [41], then $V_u^\dagger$ and $V_d^\dagger$ are also given by (2.12), resulting in $U_{\text{CKM}} = 1$, i.e. no mixing. This should be considered as a good first approximation because the observed mixing angles are all small. In the general case without any symmetry, we would have expected V_u and V_d to be very different.

It was later discovered in Ref. [93] that (2.13) may also be obtained with A_4, using two further assumptions. Consider the most general 3 × 3 Majorana mass matrix in the form

$$m_\nu = \begin{pmatrix} a+b+c & f & e \\ f & a+b\omega+c\omega^2 & d \\ e & d & a+b\omega^2+c\omega \end{pmatrix},$$
(2.15)

where a comes from $\underline{1}$, b from $\underline{1}'$, c from $\underline{1}''$, and (d, e, f) from $\underline{3}$ of A_4. To get (2.13), we need $e = f = 0$, i.e. the effective scalar A_4 triplet responsible for neutrino masses should have its vacuum expectation value along the $(1, 0, 0)$ direction, whereas that responsible for charged lepton masses should be $(1, 1, 1)$ as remarked earlier. This misalignment is a technical challenge to all such models [50, 94–104]. The other requirement is that $b = c$. Since they come from different representations of A_4, this is rather ad hoc. A very clever solution [50, 94] is to eliminate both, i.e. $b = c = 0$. This results in a normal ordering of neutrino masses with the prediction [96]

$$|m_{\nu_e}|^2 \simeq |m_{ee}|^2 + \Delta m_{\text{atm}}^2/9.$$
(2.16)

Other applications [60, 105–120] of A_4 have also been considered. A natural (spinorial) extension of A_4 is the binary tetrahedral group [30, 34] which is under active current discussion [64, 121–123].

Other recent applications of non-Abelian discrete flavor symmetries include those of D_4 [47, 72, 124], Q_4 [68], D_5 [125, 126], D_6 [127], Q_6 [128–130], D_7 [131], S_4 [61, 132–135], $\Delta(27)$ [57, 71], $\Delta(75)$ [15, 136], $\Sigma(81)$ [73, 75], and $B_3 \times \mathbf{Z}_2^3$ [137, 138] which has 384 elements.

2.2.3 Accidental flavor symmetries

While flavor symmetries certainly represent one of the leading approaches to understanding the pattern of fermion masses and mixing, it was recently found that the hierarchical structure of charged fermion masses and many other peculiar features of the fermion spectrum in the SM (neutrinos included) do not require a flavor symmetry to be understood, nor any other special "horizontal" dynamics involving the family indices of the SM fermions [63, 139]. Surprisingly enough, those features can in fact be recovered in a model in which the couplings of the three SM families not only are not governed by any symmetry, but are essentially anarchical (uncorrelated $\mathcal{O}(1)$ numbers) at a very high scale.

The idea is based on the hypothesis that the SM Yukawa couplings all arise from the exchange of heavy degrees of freedom (messengers) at a scale not far from the unification scale. Examples of diagrams contributing to the up and down quark Yukawa matrices are shown below, where ϕ is a SM singlet field getting a VEV. As discussed in Sects. 2.2 and 2.2.1, the same exchange mechanism is often assumed

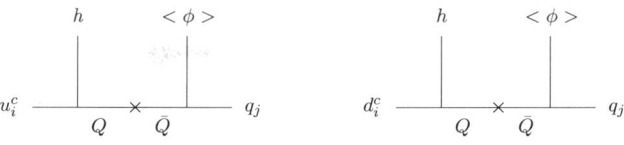

Fig. 1 Contributions to the up- and down-type quark Yukawa mass matrices, from the exchange of heavy messengers

to be at work in models with flavor symmetries. Here, however, the couplings of the heavy messengers to the SM fields are not constrained by any symmetry.[3] An hierarchy among Yukawa couplings still arises because a single set of left handed messenger fields (heavy quark doublets $Q + \bar{Q}$ in the quark sector and heavy lepton doublets $L + \bar{L}$ in the lepton sector) dominates the exchange at the heavy scale. For example, the diagrams below represents the dominant contribution to the quark Yukawa matrices. As only one field is exchanged, the Yukawa matrices have rank one. Therefore, whatever are the $\mathcal{O}(1)$ couplings in the diagram, the top and bottom Yukawa couplings are generated (at the $\mathcal{O}(1)$ level, giving large $\tan\beta$), but the first two families' ones are not, which is a good starting point to obtain a hierarchy of quark masses. This mechanism is similar to a the single right handed neutrino dominance mechanism, used in neutrino model building to obtain a hierarchical spectrum of light neutrinos [140–143]. Note that the diagonalization of the quark Yukawa matrices involves large rotations, as all the couplings are supposed to be $\mathcal{O}(1)$. However, the rotations of the up and down left handed quarks turn out to be the same (because they have same couplings to the left handed doublet messenger). Therefore, the two rotations cancel when combined in the CKM matrix, which ends up vanishing at this level.

The Yukawa couplings of the second family, and a non-vanishing V_{cb} angle, are generated by the subdominant exchange of heavier right handed messengers D^c, U^c, E^c, N^c. Altogether, the messengers form a heavy (vector-like) replica of a SM family, with the left handed fields lighter than the right handed ones. The (inter-family) hierarchy between the masses of the second and the third SM family masses arises from the (intra-family) hierarchy between left and right handed fields in the single family of messengers. In turn, in a Pati–Salam or $SO(10)$ unified model, the hierarchy between right handed and left handed fields can be easily obtained by giving mass to the messengers through a breaking of the gauge group along the T_{3R} direction. This way, the hierarchy among different families is explained in terms of the breaking of a gauge group acting on single families, with no need of flavor symmetries or other dynamics acting on the family indexes of the SM fermions.

---
[3] A discrete $\mathbf{Z}_2$ symmetry, under which *all* the three SM families (and the field ϕ) are odd, is used for the sole purpose of distinguishing the light SM fields from the heavy messengers.

It is also possible to describe the mechanism outlined above in terms of accidental flavor symmetries. In the effective theory below the scale of the right handed messengers, in fact, the Yukawa couplings of the two lighter families are "protected" by an accidental $U(2)$ symmetry. One can also consider the effective theory below the cut-off of the model, which is supposed to lie one or two orders of magnitude above the mass of the right handed messengers. In the effective theory below the cut-off, the second family gets a nonvanishing Yukawa coupling, but the Yukawa of the lightest family is still "protected" by an accidental $U(1)$ symmetry.

Surprisingly enough, a number of important features of the fermion spectrum can be obtained in this simple and economical model. The relation $|V_{cb}| \sim m_s/m_b$ is a direct consequence of the principles of this approach. The stronger mass hierarchy observed in the up quark sector is accounted for without introducing a new scale (besides the left handed and right handed messenger ones) or making the up quark sector somehow different. In spite of the absence of small coefficients, the CKM mixing angles turn out to be small. At the same time, a large atmospheric mixing can be generated in a natural way in the neutrino sector, together with normal hierarchical neutrino masses. In fact, a see-saw mechanism dominated by the single right handed (messenger) neutrino N^c is at work. The bottom and tau mass unify at the high scale, while a $B - L$ factor 3 enters the ratios of the muon and strange masses. For a detailed illustration of the model, we refer the reader to [63].

The study of FCNC and CPV effects in a supersymmetric context is still under way. Such effects might represent the distinctive signature of the model, due to the sizable radiative effects one obtains in the (23) block of the "right handed" sfermion mass matrices in both the squark and slepton sector.

2.2.4 Flavor/CP symmetries and their violation from supersymmetry breaking

While the vast literature on flavor symmetries covers a number of interesting aspects of the theory and phenomenology of flavor, we are interested here in a (non-exhaustive) review of only those aspects relevant to new physics. The relevance of flavor symmetries to new physics follows from the fact that SM extensions often contain new flavor dependent interactions. In the following we shall consider the case of supersymmetry, in which new flavor violating gaugino or higgsino interactions can be induced by possible new sources of $SU(5)^5$ breaking in the soft supersymmetry breaking terms.

While in the SM the Yukawa matrices provide the only source of flavor $(U(3)^5)$ breaking, the supersymmetric extensions of the SM are characterized by a potentially much richer flavor structure associated to the soft supersymmetry breaking Lagrangian. Unfortunately, a generic flavor structure leads to FCNC and CPV processes that can exceed the experimental bounds by up to two orders of magnitude—the so-called supersymmetric flavor and CP problem. The solution of the latter problem can lie in the supersymmetry breaking and mediation mechanism (this is the case for example of gauge mediated supersymmetry breaking) or in the constraints on the soft terms provided by flavor symmetries.

In turn, the implications of flavor symmetries on the structure of the soft terms depends on the interplay between flavor and supersymmetry breaking. Without entering the details of specific models, we can distinguish two opposite situations.

- The soft terms are flavor universal, or at least symmetric under the flavor symmetry, at the tree level, and
- flavor symmetry breaking enters the soft terms (as for the Yukawa interactions) already at the tree level, through non-renormalizable couplings to the flavon fields.

Let us consider them in greater detail.

The first possibility is that the supersymmetry breaking mechanism takes care of the FCNC and CPV problems. In the simplest case, the new sfermion masses and A-terms do not introduce new flavor structure at all. This is the case if

$$\mathbf{m}_{ij}^2 = m_0^2 \delta_{ij}, \qquad \mathbf{A}_{ij} = A_0 \delta_{ij},$$

where i, j are family indexes and the universal values m_0^2, A_0 can be different in the different sfermion sectors.[4] The breaking of the flavor symmetry is felt at the tree level only by the Yukawa matrices. Needless to say, the tree level universality of the soft terms will be spoiled by *renormalization effects* associated to interactions sensitive to Yukawa couplings [144, 145]. These effects can be enhanced by large logarithms if the scale at which the soft terms and the Yukawa interactions appear in the observable sector is sufficiently high. The radiative contributions of Yukawa couplings associated with neutrino masses (or Yukawa couplings occurring in the context of grand unification) are particularly interesting in this context, because they offer new possibilities to test flavor physics by opening a window for physics at very large scales. For example, in the minimal SUSY see-saw model only the off-diagonal elements for left-slepton soft supersymmetry breaking mass terms are generated, while in supersymmetric GUTs also the right handed slepton masses get renormalization induced flavor non-diagonal contributions. In any case, all the flavor effects induced by the soft terms can be traced back to the Yukawa couplings, which remain the only source of flavor breaking.

---

[4]This is the case for example of gauge mediation. In supergravity, supersymmetry breaking can be fully flavor blind in the case of dilaton domination. In this case, we expect the diagonal elements of the soft mass matrices to be exactly universal. However, this is not always the case. Moduli domination is often encountered, in which case fields with different modular weights receive different soft masses.

Such unavoidable effects of flavor breaking on the soft terms will be discussed in Sects. 5.2 and 5.3.

As we have just seen, the radiative contributions to soft masses represent an unavoidable but indirect effect of the physics at the origin of fermion masses and mixing. On the other hand, the mechanism generating the soft terms might not be blind to flavor symmetry breaking, in which case we might also expect flavor breaking to enter the soft terms in a more direct way. If this is the case, the soft term provide a new independent source of flavor violation. Such model-dependent *"tree level"* effects of flavor breaking on the soft terms add to the radiative effects and will be discussed in Sect. 2.2.4.1. The actual presence in the soft terms of flavor violating effects directly induced by the physics accounting for Yukawa couplings depends on the interplay of the supersymmetry breaking and the flavor generation mechanisms.

Theoretical and phenomenological [146–151] constraints on supersymmetry breaking parameters essentially force supersymmetry breaking to take place in a hidden sector with no renormalizable coupling to observable fields.[5] The soft terms are therefore often characterized by the scale Λ_{SUSY} at which supersymmetry breaking is communicated to the observable sector by some mediation mechanism. The soft terms arise in fact from non-renormalizable operators in the effective theory below Λ_{SUSY} obtained by integrating out the supersymmetry breaking messenger fields. Analogously, in the context of a theory addressing the origin of flavor, we can define a scale Λ_f at which the flavor structure arises. Let us consider for definiteness the case of flavor symmetries. The analogy with supersymmetry breaking is in this case even more pronounced. Above Λ_f, the theory is flavor symmetric. By this we mean that we can at least define conserved family numbers, perhaps part of a larger flavor symmetry. The family numbers are then spontaneously broken by the VEV of flavons that couple to observable fields through non-renormalizable interactions suppressed by the scale Λ_f.

We are now in the position to discuss the presence of "tree-level" flavor violating effects in the soft terms. A first possibility is to have $\Lambda_f \lesssim \Lambda_{\text{SUSY}}$, as for instance in the case of gravity mediation, in which we expect $\Lambda_f \lesssim M_{\text{Planck}} = \Lambda_{\text{SUSY}}$. The soft breaking terms are already present below M_{Planck}. However, the flavor symmetry is still exact at scales larger than Λ_f. Therefore, the soft terms must respect the family symmetries. At the lower scale Λ_f the effective Yukawa couplings are generated as functions of the flavon VEVs, $\langle\theta\rangle/\Lambda_f$, and analogously the soft breaking terms will also be functions of $\langle\theta\rangle/\Lambda_f$. In the $\Lambda_f \lesssim \Lambda_{\text{SUSY}}$ case, we therefore expect new "tree-level"

---

[5]The fields of the minimal supersymmetric standard model (MSSM) or its relevant extension.

sources of flavor breaking in the soft terms on top of the effects radiatively induced by the Yukawa couplings.

On the other hand, if $\Lambda_{\text{SUSY}} \ll \Lambda_f$, the soft terms are not present at the scale of flavor breaking. The prototypical example in this case is gauge mediated supersymmetry breaking (GMSB) (see [152] and references therein). At Λ_f the flavor interactions are integrated and supersymmetry is still unbroken. The only renormalizable remnant of the flavor physics below Λ_f are the Yukawa couplings. At the scale Λ_{SUSY} soft breaking terms feel flavor breaking only through the Yukawa couplings. Strictly speaking, there could also be non-renormalizable operators involving flavon fields suppressed by the heavier Λ_f. The contributions of these terms to soft masses would be proportional to $\Lambda_{\text{SUSY}}/\Lambda_f$ and therefore negligible [152]. We are then only left with the radiatively induced effects of Yukawa couplings. The qualitative arguments above show that flavor physics can provide relevant information on the interplay between the origin of supersymmetry and flavor breaking in the observable sector.

As we just saw, the family symmetry that accounts for the structure of the Yukawa couplings also constrains the structure of sfermion masses. In the limit of exact flavor symmetry, this implies family universal, or at least diagonal, sfermion mass matrices. After the breaking of the flavor symmetry giving rise to the Yukawa couplings, we can have two cases.

- The SUSY breaking mediation mechanism takes place at a scale higher or equal to the flavor symmetry breaking scale and is usually sensitive to flavor. The flavor symmetry breaking accounts for both the structure of the Yukawa couplings and the deviations of the soft breaking terms from universality. This is the general expectation in gravity mediation of the supersymmetry breaking from the hidden sector.
- The supersymmetry breaking mediation mechanism takes place at a scale much smaller than the flavor symmetry breaking scale. In this case the flavor mediation mechanism, which is flavor-blind, guarantees the universality of the soft breaking terms. The flavor symmetry breaking generates the Yukawa couplings but flavor breaking corrections in the soft mass matrices are suppressed by the ratio of the two scales. This is the case of gauge-mediation models of supersymmetry breaking [152].

We begin discussing the first case.

2.2.4.1 "Tree level" effects of flavor symmetries in supersymmetry breaking terms After the breaking of the flavor symmetry responsible for the structure of the Yukawa couplings, we can expect to have non-universal contributions to the soft breaking terms at *tree level*. Under certain conditions, mainly related to the SUSY-breaking mediation mechanism, these tree-level contributions can be sizable and have important phenomenological effects. The main example among these models where the tree level non-universality in the soft breaking terms is relevant is provided by models of supergravity mediation [153–157] (for a nice introduction see the appendix in [158]).

The structure of the scalar mass matrices when SUSY breaking is mediated by supergravity interactions is determined by the Kähler potential. We are not going to discuss here the supergravity Lagrangian; we refer the interested reader to Refs. [153–156, 158]. For our purposes, we only need to know that the Kähler potential is a non-renormalizable, real, and obviously gauge-invariant, function of the chiral superfields with dimensions of mass squared. This non-renormalizable function includes couplings with the hidden sector fields suppressed by different powers of M_{Planck}, $\phi\phi^*(1 + XX^*/M_{\text{Planck}}^2 + \cdots)$ with ϕ visible sector fields and X hidden sector fields. This Kähler potential gives rise to SUSY breaking scalar masses once a certain field of the hidden sector gets a non-vanishing F-term. The important point here is that these couplings with hidden sector fields that will eventually give rise to the soft masses are present in the theory at any scale below M_{Planck}. Below this scale, we can basically consider the hidden sector as frozen and renormalize these couplings only with visible sector interactions.

Therefore, in the following, to simplify the discussion, we concentrate only on the soft masses and treat them as couplings present at all energies below M_{Planck}. The structure of the soft mass matrices is easily understood in terms of the present symmetries. At high energies, our flavor symmetry is still an exact symmetry of the Lagrangian and therefore the soft breaking terms have to respect this symmetry [46]. At some stage, this symmetry is broken generating the Yukawa couplings in the superpotential. In the same way, the scalar masses will also receive new contributions after flavor symmetry breaking from the flavon field VEVs suppressed by mediator masses.

First we must notice that a mass term $\phi_i^\dagger \phi_i$ is clearly invariant under gauge, flavor and global symmetries and hence gives rise to a flavor diagonal contribution to the soft masses even before the family symmetry breaking.[6] Then, after flavor symmetry breaking, any invariant combination of flavon fields (VEVs) with a pair of sfermion fields, $\phi_i^\dagger \phi_j$, can also contribute to the sfermion mass matrix and will break the universality of the soft masses.

An explicit example with a continuous Abelian $U(1)$ flavor symmetry [5, 11, 13, 16, 19, 21, 44, 48, 54] was given above in Sect. 2.2.1.

---

[6] As we shall discuss in the following, these allowed contributions may be universal, the same for the different generations, as in the case of non-Abelian flavor symmetries, or they can be different for the three generations in some cases with Abelian flavor symmetries.

We turn now to the structure of the scalar mass matrices concentrating mainly on the slepton mass matrix [13, 14, 16, 43]. In this case, even before the breaking of the flavor symmetry, we have three different fields with different charges corresponding to each of the three generations. As we have seen, diagonal scalar masses are allowed by the symmetry, but being different fields, there is no reason a priori for these diagonal masses to be the same, and in general we have

$$\mathcal{L}_{m^2}^{\text{symm}} = m_1^2 \phi_1^* \phi_1 + m_2^2 \phi_2^* \phi_2 + m_3^2 \phi_3^* \phi_3. \tag{2.17}$$

Notice, however, that this situation is very dangerous, especially in the case of squarks, given that the rotation to the basis of diagonal Yukawa couplings from (2.2) will generate too large off-diagonal entries [43]. In some cases, like dilaton domination, these allowed masses can be equal avoiding this problem. In the following we assume $m_1^2 = m_2^2 = m_3^2 = m_0^2$. However, even in this case, after the breaking of the flavor symmetry we obtain new contributions proportional to the flavon VEVs that break this universality. All we have to do is to write all possible combinations of two MSSM scalar fields ϕ_i and an arbitrary number of flavon VEVs invariant under the symmetry:

$$\mathcal{L}_{m^2} = m_0^2 \bigg(\phi_1^* \phi_1 + \phi_2^* \phi_2 + \phi_3^* \phi_3 + \left(\frac{\langle\theta\rangle}{M_{\text{fl}}}\right)^{q_2-q_1} \phi_1^* \phi_2 + \left(\frac{\langle\theta\rangle}{M_{\text{fl}}}\right)^{q_3-q_1} \phi_1^* \phi_3 + \left(\frac{\langle\theta\rangle}{M_{\text{fl}}}\right)^{q_3-q_2} \phi_2^* \phi_3 + \text{h.c.} \bigg). \tag{2.18}$$

Therefore, the structure of the charged slepton mass matrix we would have in this model at the scale of flavor symmetry breaking would be (suppressing $O(1)$ coefficients):

$$m_{\tilde{L}}^2 \simeq \begin{pmatrix} 1 & \epsilon & \epsilon \\ \epsilon & 1 & 1 \\ \epsilon & 1 & 1 \end{pmatrix} m_0^2. \tag{2.19}$$

This structure has serious problems with the phenomenological bounds coming from $\mu \to e\gamma$, etc. There are other $U(1)$ examples that manage to alleviate, in part, these problems [43]. However, large LFV effects are a generic problem of these models due to the required charge assignments to reproduce the observed masses and mixing angles.

These FCNC problems in the sfermion mass matrices of Abelian symmetries were one of the main reasons for the introduction of non-Abelian flavor symmetries [18, 20]. The mechanism used in non-Abelian flavor models to generate the Yukawa couplings is again a variation of the Froggatt–Nielsen mechanism, very similar to the mechanism we have just seen for Abelian symmetries. The main difference is that in this case the left handed fermions are grouped in larger representations of the symmetry group. For instance, in a $SU(3)$ symmetry all three generations are unified in a triplet. In a $SO(3)$ flavor symmetry we can assign the three generations to a triplet or to three singlets. In a $U(2)$ flavor symmetry the third generation is a singlet and the two light generations are grouped in a doublet. Then we do not have to assign different charges to the various generations, but in exchange, we need several stages of symmetry breaking by different flavon fields with specially aligned VEVs.

We begin analyzing a non-Abelian $U(2)$ flavor symmetry. As stressed above, if the sfermions mass matrices are only constrained by a $U(1)$ flavor symmetry there is no reason why m_1^2 should be close to m_2^2 in (2.17). Unless an alignment mechanism between fermions and sfermions is available, the family symmetry should then suppress $(\tilde{m}_1^2 - \tilde{m}_2^2)/\tilde{m}^2$. At the same time, in the fermion sector, the family symmetry must suppress the Yukawa coupling of the first two families, $m_1, m_2 \ll m_3$. If the small breaking of a flavor symmetry is responsible for the smallness of $(\tilde{m}_1^2 - \tilde{m}_2^2)/\tilde{m}^2$ on one hand and of $m_1/m_3, m_2/m_3$ on the other, the symmetric limit should correspond to $\tilde{m}_1^2 = \tilde{m}_2^2$ and to $m_1 = m_2 = 0$. Interestingly enough, the largest family symmetry compatible with $SO(10)$ unification that forces $m_1 = m_2 = 0$ automatically also forces $\tilde{m}_1^2 = \tilde{m}_2^2$. This is a $U(2)$ symmetry under which the first two families transform as a doublet and the third one, as well as the Higgs, as a singlet [16, 18, 20, 24, 26].

$$\psi = \psi_a \oplus \psi_3.$$

The same conclusion can be obtained by using discrete subgroups [30, 64]. In the limit of unbroken $U(2)$, only the third generation of fermions can acquire a mass, whereas the first two generations of scalars are exactly degenerate. While the first property is not a bad approximation of the fermion spectrum, the second one is what is needed to keep FCNC and CP-violating effects under control. This observation can actually be considered as a hint that the flavor structure of the mass matrices of the fermions and of the scalars are related to each other by a symmetry principle. The same physics responsible for the peculiar pattern of fermion masses also accounts for the structure of sfermion masses.

The rank 2 of $U(2)$ allows for a two step breaking pattern:

$$U(2) \xrightarrow{\epsilon} U(1) \xrightarrow{\epsilon'} 0, \qquad (2.20)$$

controlled by two small parameters ϵ and $\epsilon' < \epsilon$, to be at the origin of the generation mass hierarchies $m_3 \gg m_2 \gg m_1$ in the fermion spectrum. Although it is natural to view $U(2)$ as a subgroup of $U(3)$, the maximal flavor group in the case of full intra-family gauge unification, $U(3)$ will be anyhow strongly broken to $U(2)$ by the large top Yukawa coupling.

A nice aspect of the $U(2)$ setting is that there is little arbitrariness in the way the symmetry breaking fields couple to the SM fermions. This is unlike what happens e.g.

with the choice of fermion charges in the cases of $U(1)$ symmetries. The Yukawa interactions transform as $(\psi_3 \psi_3)$, $(\psi_3 \psi_a)$, $(\psi_a \psi_b)$ $(a, b, c, \ldots = 1, 2)$. Hence the only relevant $U(2)$ representations for the fermion mass matrices are 1, ϕ^a, S^{ab} and A^{ab}, where S and A are symmetric and antisymmetric tensors, and the upper indices denote a $U(1)$ charge opposite to that of ψ_a. While ϕ^a and A^{ab} are both necessary, models with [20, 26] or without [24] S^{ab} are both possible.

Let us first consider the case with S^{ab}. At leading order, the flavons couple to SM fermions through $D = 5$ operators suppressed by a flavor scale Λ. Normalizing the flavons to Λ, it is convenient to choose a basis in which $\phi^2 = \mathcal{O}(\epsilon)$ and $\phi^1 = 0$, while $A^{12} = -A^{21} = \mathcal{O}(\epsilon')$. If S is present, it turns out to be automatically aligned with ϕ [27], in such a way that in the limit $\epsilon' \to 0$ a $U(1)$ subgroup is unbroken. More precisely, $S^{22} = \mathcal{O}(\epsilon)$ and all other components essentially vanish. We are then led to Yukawa matrices of the form

$$\begin{pmatrix} 0 & \epsilon' & 0 \\ -\epsilon' & \epsilon & \epsilon \\ 0 & \epsilon & 1 \end{pmatrix}. \qquad (2.21)$$

All non-vanishing entries have unknown coefficients of order unity, while still keeping $\lambda_{12} = -\lambda_{21}$. In the context of $SU(5)$ or $SO(10)$ unification, the mass relations $m_\tau \approx m_b$, $m_\mu \approx 3m_s$, $3m_e \approx m_d$ are accounted for by the choice of the transformations of A^{ab}, S^{ab} under the unified group. The stronger mass hierarchy in the up quark sector, a peculiar feature of the fermion spectrum, is then predicted, due to the interplay of the $U(2)$ and the unified gauge symmetry.

The texture in (2.21) leads to the predictions

$$\left|\frac{V_{td}}{V_{ts}}\right| = \sqrt{\frac{m_d}{m_s}}, \qquad \left|\frac{V_{ub}}{V_{cb}}\right| = \sqrt{\frac{m_u}{m_c}}. \qquad (2.22)$$

While the experimental determination of $|V_{td}/V_{ts}|$ based on one loop observables might be affected by new physics, the tree-level determination of $|V_{ub}/V_{cb}|$ is less likely to be affected and at present is significantly away from the prediction in (2.22) [29, 39]. A better agreement can be obtained by (i) relaxing the condition $\lambda_{12} = -\lambda_{21}$, (ii) allowing for small contributions to the 11, 13, 31 entries in (2.21) or by (iii) allowing for asymmetric textures [39]. The latter possibility is realized in models in which the S^{ab} flavon is not present [20].

While the model building degrees of freedom in the quark and charged lepton sector are limited, a virtue of the $U(2)$ symmetry, the neutrino sector is less constrained. This is due, in the see-saw context, to the several possible choices involved in the modelization of the singlet neutrino mass matrix. This is reflected for example in the possibility to get both small and large mixing angles [25, 28, 31, 34, 35].

In the case of an $SU(3)$ flavor symmetry, all three generations are grouped in a single triplet representation, ψ_i. In addition we have several new scalar fields (flavons) which are either triplets, $\bar{\theta}_3$, $\bar{\theta}_{23}$ and $\bar{\theta}_2$, or antitriplets, θ_3 and θ_{23}. $SU(3)_{\rm fl}$ is broken in two steps: the first step occurs when θ_3 and $\bar{\theta}_3$ get a large VEV breaking $SU(3)$ to $SU(2)$, and defining the direction of the third generation. Subsequently a smaller VEV of θ_{23} and $\bar{\theta}_{23}$ breaks the remaining symmetry and defines the second generation direction. To reproduce the Yukawa textures the large third generation Yukawa couplings require a θ_3 (and $\bar{\theta}_3$) VEV of the order of the mediator scale, $M_{\rm fl}$, while $\theta_{23}/M_{\rm fl}$ (and $\bar{\theta}_{23}/M_{\rm fl}$) have small VEVs[7] of order ε. After this breaking chain we obtain the effective Yukawa couplings at low energies through the Froggatt–Nielsen mechanism [5] integrating out heavy fields. The resulting superpotential invariant under $SU(3)$ would be

$$W_Y = H\psi_i \psi_j^c [\theta_3^i \theta_3^j + \theta_{23}^i \theta_{23}^j + \epsilon^{ikl}\bar{\theta}_{23,k}\bar{\theta}_{3,l}\theta_{23}^j(\theta_{23}\bar{\theta}_3)$$
$$+ \epsilon^{ijk}\bar{\theta}_{23,k}(\theta_{23}\bar{\theta}_3)^2 + \epsilon^{ijk}\bar{\theta}_{3,k}(\theta_{23}\bar{\theta}_3)(\theta_{23}\bar{\theta}_{23})$$
$$+ \cdots]. \qquad (2.23)$$

In this equation we can see that each of the $SU(3)$ indices of the external MSSM particles (triplets) are either saturated individually with an antitriplet flavon index (a "meson" in QCD notation) or in an antisymmetric couplings with other two triplet indices (a "baryon"). The presence of other singlets in the different term is due to the presence of additional global symmetries necessaries to ensure the correct hierarchy in the different Yukawa elements [37, 45, 46]. This structure is quite general for the different $SU(3)$ models we can build. Here we are not specially concerned with additional details and we refer to [37, 45, 46] for more complete examples. The Yukawa texture we obtain with this superpotential is the following:

$$Y^f = \begin{pmatrix} 0 & \alpha\varepsilon^3 & \beta\varepsilon^3 \\ \alpha\varepsilon^3 & \frac{\varepsilon^2}{a^2} & \gamma\frac{\varepsilon^2}{a^2} \\ \beta\varepsilon^3 & \gamma\frac{\varepsilon^2}{a^2} & 1 \end{pmatrix} a^2, \qquad (2.24)$$

with $a = \frac{\langle\theta_3\rangle}{M}$, and α, β, γ unknown coefficients of order $O(1)$.

Let us now analyze the structure of scalar soft masses. In analogy with the Abelian case, in the unbroken limit diagonal soft masses are allowed. However, the three generations belong to the same representation of the flavor symmetry and now this implies the mass is the same for the whole

---
[7]In fact, in realistic models reproducing the CKM mixing matrix, there are two different mediator scales and expansion parameters, ε in the up quark and $\bar{\varepsilon}$ in the down quark sector [37, 45, 46].

triplet. After the breaking of $SU(3)$ symmetry the scalar soft masses deviate from exact universality [46, 160–162]. Any invariant combination of flavon fields can also contribute to the sfermion masses, although flavor symmetry indices can be contracted with fermion fields. Including these corrections the leading contributions to the sfermion mass matrices are given by

$$(m_{\tilde{f}}^2)^{ij} = m_0^2 \Big(\delta^{ij} + \frac{1}{M_f^2}[\theta_3^{i\dagger}\theta_3^j + \theta_{23}^{i\dagger}\theta_{23}^j]$$
$$+ \frac{1}{M_f^4}(\epsilon^{ikl}\bar{\theta}_{3,k}\bar{\theta}_{23,l})^\dagger(\epsilon^{jmn}\bar{\theta}_{3,m}\bar{\theta}_{23,n})\Big). \quad (2.25)$$

Notice that each term inside the parentheses is trivially neutral under the symmetry because it contains always a field together with its own complex conjugate field. However, as the flavor indices of the flavon fields are contracted with the external matter fields this gives a non-trivial contribution to the sfermion mass matrices. Therefore in this model, suppressing factors of order 1 we have,

$$m_{\tilde{f}}^2 \simeq \begin{pmatrix} 1 & & \\ & 1 & \\ & & 1 \end{pmatrix} m_0^2 + \begin{pmatrix} \varepsilon^2 & 0 & 0 \\ 0 & \frac{\varepsilon^2}{a^2} & \frac{\varepsilon^2}{a^2} \\ 0 & \frac{\varepsilon^2}{a^2} & 1 \end{pmatrix} a^2 m_0^2, \quad (2.26)$$

with $a = \langle\theta_3\rangle/M_{\rm fl}$ which is still $O(1)$. In the model [37, 45, 46], the expansion parameter for right handed down quarks and charged leptons is $\bar{\varepsilon} = 0.15$. Using (2.24) and (2.26) we can obtain the slepton mass matrix in the basis of diagonal charged lepton Yukawa couplings:

$$m_{\tilde{e}_R}^2 \simeq \begin{pmatrix} 1+\bar{\varepsilon}^2 & -\bar{\varepsilon}^3 & -\bar{\varepsilon}^3 \\ -\bar{\varepsilon}^3 & 1+\bar{\varepsilon}^2 & \bar{\varepsilon}^2 \\ -\bar{\varepsilon}^3 & \bar{\varepsilon}^2 & 1 \end{pmatrix} m_0^2, \quad (2.27)$$

where we have used $a_3 \simeq \mathcal{O}(M_{\rm fl})$. Therefore that generates the order $\bar{\varepsilon}^3$ entry in the (1, 2) element. The modulo of this entry is order 3×10^{-3} at $M_{\rm GUT}$. These estimates at $M_{\rm GUT}$ are slightly reduced through renormalization group evolution to the electroweak scale and is order 1×10^{-3} at $M_{\rm W}$. This value implies that supersymmetric contribution to $\mu \to e\gamma$ is very big and can even exceed the present bounds for light slepton masses and large $\tan\beta$ if we are not in the cancellation region[163–165]. This makes this process perhaps the most promising one to find deviations from universality in flavor models. The presence of the $SU(3)$ flavor symmetry controls the structure of the sfermion mass matrices and the supersymmetric flavor problem can be nicely solved. However, interesting signals of the supersymmetric flavor structure can be found in the near future LFV experiments.

3 Observables and their parameterization

3.1 Effective operators and low scale observables

In spite of the clear success of the SM in reproducing all the known phenomenology up to energies of the order of the electroweak scale, nobody would doubt the need of a more complete theory beyond it. There remain many fundamental problems such as the experimental evidence for dark matter (DM) and neutrino masses, as well as the theoretical puzzles posed by the origin of flavor, the three generations, etc., that a complete theory should address. Therefore, we can consider the SM as the low energy effective theory of some more complete model that explains all these puzzles. Furthermore, we have strong reasons (gauge hierarchy problem, unification of couplings, dark matter candidate, etc.) to expect the appearance of new physics close to the electroweak scale. Suppose that these new particles from the more complete theory are to be found at the LHC. Experiments at lower energies $E < m_{\rm NP}$ are also sensitive [166] to this new physics (NP). Indeed the exchange of new particles can induce:

- corrections to the SM observables (such as S, T and U), and
- the appearance of *new* observables or new ($d > 4$) operators, (e.g. the flavor violating dipole operators).

Note that both effects can be parameterized by $SU(3) \times SU(2) \times U(1)$-invariant operators of mass dimension $d > 4$. We refer to these non-renormalizable operators as *effective* operators. Any NP proposed to explain new phenomena at the LHC must satisfy the experimental constraints on the effective operators it generates.

3.1.1 Effective Lagrangian approach: $\mathcal{L}_{\rm eff}$

Considering the SM as an effective theory below the scale of NP, $m_{\rm NP}$, where the heavy fields have been integrated out, we can describe the physics through an effective Lagrangian, $\mathcal{L}_{\rm eff}$. This effective Lagrangian contains all possible terms invariant under the SM gauge group and built with the SM fields. Besides the usual SM fields, we could introduce new light singlet fermions with renormalizable Yukawa couplings to the lepton doublets (and possibly small Majorana masses) to accommodate the observed neutrino masses. In this case we would have more operators allowed in the effective Lagrangian of the SM + extra light sterile states. On the assumption that the light sterile particles are weakly interacting, if present, and therefore not relevant to the LHC, we focus on the effective Lagrangian that can be constructed only from the known SM fields. Then, the effective Lagrangian at energies $E \ll m_{\rm NP}$ can be written as an expansion in $1/m_{\rm NP}$ as,

$$\mathcal{L}_{\rm eff}^{\rm SM} = \mathcal{L}_0 + \frac{1}{m_{\rm NP}}\mathcal{L}_1 + \frac{1}{m_{\rm NP}^2}\mathcal{L}_2 + \frac{1}{m_{\rm NP}^3}\mathcal{L}_3 + \cdots, \quad (3.1)$$

where $\mathcal{L}_0$ is the renormalizable SM Lagrangian containing the kinetic terms of the $U(1)$, $SU(2)$ and $SU(3)$ gauge bosons A_μ, the gauge interactions and kinetic terms of the SM fermions, $\{f\}$, and Higgs, and the Yukawa couplings of the Higgs and SM fermions. In order to fix the notation, we list the SM fermions as

$$q_i = \begin{pmatrix} u_{Li} \\ d_{Li} \end{pmatrix}, \quad \ell_i = \begin{pmatrix} \nu_{Li} \\ e_{Li} \end{pmatrix}, \quad (3.2)$$
$$u_{Ri}, \quad d_{Ri}, \quad e_{Ri},$$

where i is a flavor/family/generation index. Note that in the following we use always four-component Dirac spinors in the different Lagrangians. Explicit expressions, for $\mathcal{L}_0$ in similar notation, can be found in [167].

The different $\mathcal{L}_n$ are Lagrangians of dimension $d = 4 + n$ invariant under $SU(3) \times SU(2) \times U(1)$ and can be schematically written

$$\mathcal{L}_n = \sum_a C_a \cdot \mathcal{O}_a(H, \{f\}, \{A_\mu\}) + \text{h.c.} \quad (3.3)$$

The local operators $\mathcal{O}_a$ are gauge invariant combinations of SM fields of dimension $4 + n$. Their coefficient, which in the full Lagrangian has mass dimension $-n$, is unknown in bottom-up effective field theory, but calculable in NP models. We write this coefficient as a dimensionless C_a divided by the nth power of the mass scale of the NP mediator, $m_{\rm NP}^n$, which for new physics relevant at LHC energies would be $m_{\rm NP} \sim \sqrt{s_{\rm LHC}}$. We shall later normalize to G_F (see (3.21)).

We are mainly interested in dimension five and dimension six operators. We assume that any particles created at the LHC could generate dimension six operators, and then we can neglect higher dimension operators contributing to the same physical processes. Operators of dimension 7 include the lepton number violating operator $\epsilon_{ab}\epsilon_{cd}H^a\ell^b_{[i}\sigma^{\mu\nu}H^c\ell^d_{j]}F_{\mu\nu}$ which gives neutrino transition moments (flavor-changing dipole moments) after electroweak symmetry breaking (EWSB). At dimension 8 are two-Higgs-four-fermion operators, which can give four-fermion operators after EWSB, with a different flavor structure from the dimension six terms. We shall not analyze these operators here, but they are studied in the context of non-standard neutrino interactions [168]. Therefore, in the following, we restrict our analysis to $\mathcal{L}_1$ and $\mathcal{L}_2$.

The unique operator allowed with the standard model fields and symmetries at dimension five is $\mathcal{O}^{ij}_{\ell\ell} = \epsilon_{ab}\epsilon_{mn}H^a\overline{\ell^c}^b_i H^m \ell^n_j$ (a, b, n, m are $SU(2)$ indices). Thus we have,

$$\mathcal{L}_1 = \frac{1}{4}\kappa^{ij}_{\nu\ell\ell} \cdot \epsilon_{ab}\epsilon_{mn}H^a\overline{\ell^c}^b_i H^m \ell^n_j + \text{h.c.}, \quad (3.4)$$

where ℓ^c is the charge conjugate of the lepton doublet. After electroweak symmetry breaking, this gives rise to a Majorana mass matrix $\frac{1}{4}\kappa_{\ell\ell}^{ij}\langle H^0\rangle^2 \overline{\nu^c}_i \nu_j$ + h.c. In the neutrino mass eigenstate basis, the masses are $\kappa_{\ell\ell}^{ii}\langle H^0\rangle^2/2$. The coefficient $\kappa_{\ell\ell}^{ij} = 2Y_{ki}M_k^{-1}Y_{kj}$ is generated for instance after integrating out heavy right handed neutrinos of mass M_k in a see-saw mechanism with Yukawa coupling Y.

$\mathcal{L}_2$ is constructed with dimension-six operators built out of SM fields. An exhaustive list is given in [167], including operators with Higgs, $W^\pm$ and Z^0 external legs. Here we list operators which give interactions among leptons and photons, and leptons and quarks. We can classify the possible operators according to the external legs as follows:

- operators with a pair of leptons and an (on-shell) photon:

$$\mathcal{O}_{eB}^{ij} = \bar{\ell}_i \sigma^{\mu\nu} e_{Rj} H B_{\mu\nu},$$
$$\mathcal{O}_{eW}^{ij} = \bar{\ell}_i \sigma^{\mu\nu} \tau^I e_{Rj} H W_{\mu\nu}^I, \qquad (3.5)$$

- four-lepton operators, with Lorenz structure $\overline{L}L\overline{L}L$, $\overline{R}R\overline{R}R$ or $\overline{L}R\overline{R}L$, singlet or triplet $SU(2)$ gauge contractions (described in the operator subscript), and all possible inequivalent flavor index combinations (see Sect. 3.1.2). The $SU(2) \times U(1)$ invariant operators, with flavor indices in the superscript, are

$$\mathcal{O}_{(1)\ell\ell}^{ijkl} = (\bar{\ell}_i \gamma^\mu \ell_j)(\bar{\ell}_k \gamma_\mu \ell_l),$$
$$\mathcal{O}_{(3)\ell\ell}^{ijkl} = (\bar{\ell}_i \tau^I \gamma^\mu \ell_j)(\bar{\ell}_k \tau^I \gamma_\mu \ell_l),$$
$$\mathcal{O}_{ee}^{ijkl} = (\bar{e}_i \gamma^\mu P_R e_j)(\bar{e}_k \gamma_\mu P_R e_l), \qquad (3.6)$$
$$\mathcal{O}_{\ell e}^{ijkl} = (\bar{\ell}_i e_j)(\bar{e}_k \ell_l),$$

- two lepton two-quark operators, with Lorentz structure $\overline{L}L\overline{L}L$, $\overline{R}R\overline{R}R$ or $\overline{L}R\overline{R}L$, singlet or triplet $SU(2)$ gauge contractions (described in the operator subscript), and all possible inequivalent flavor index combinations (see Sect. 3.1.2). The $S(3) \times SU(2) \times U(1)$ invariant operators, with color indices implicit and flavor indices in the subscript, are

$$\mathcal{O}_{(1)\ell q}^{ijkl} = (\bar{\ell}_i \gamma^\mu \ell_j)(\bar{q}_k \gamma_\mu q_l),$$
$$\mathcal{O}_{(3)\ell q}^{ijkl} = (\bar{\ell}_i \tau^I \gamma^\mu \ell_j)(\bar{q}_k \tau^I \gamma_\mu q_l),$$
$$\mathcal{O}_{ed}^{ijkl} = (\bar{e}_i \gamma^\mu P_R e_j)(\bar{d}_k \gamma_\mu P_R d_l),$$
$$\mathcal{O}_{eu}^{ijkl} = (\bar{e}_i \gamma^\mu P_R e_j)(\bar{u}_k \gamma_\mu P_R u_l), \qquad (3.7)$$
$$\mathcal{O}_{\ell u}^{ijkl} = (\bar{\ell}_i u_j)(\bar{u}_k \ell_j), \quad \mathcal{O}_{\ell d}^{ijkl} = (\bar{\ell}_i d_j)(\bar{d}_k \ell_j),$$
$$\mathcal{O}_{\ell q S}^{ijkl} = (\bar{\ell}_i e_j)(\bar{q}_k u_l), \quad \mathcal{O}_{qde}^{ijkl} = (\bar{\ell}_i e_j)(\bar{d}_k q_l).$$

Therefore the Lagrangian $\mathcal{L}_2$ for leptons only is

$$\mathcal{L}_2 = C_{eB}^{ij} \cdot \mathcal{O}_{eB}^{ij} + C_{eW}^{ij} \cdot \mathcal{O}_{eW}^{ij}$$
$$+ \frac{1}{1+\delta}(C_{(1)\ell\ell}^{ijkl} \cdot \mathcal{O}_{(1)\ell\ell}^{ijkl} + C_{(3)\ell\ell}^{ijkl} \cdot \mathcal{O}_{(3)\ell\ell}^{ijkl}$$
$$+ C_{ee}^{ijkl} \cdot \mathcal{O}_{ee}^{ijkl} + 2C_{\ell e}^{ijkl} \cdot \mathcal{O}_{\ell e}^{ijkl}.) + \text{h.c.}, \qquad (3.8)$$

where we introduce the parameter δ to cancel possible factors of 2 that can arise from the +h.c.: it is 1 for $\mathcal{O}^{ij\cdots}_{\cdots} = [\mathcal{O}^{ij\cdots}_{\cdots}]^\dagger$; otherwise it is 0. The sums over i, j, k, l run over inequivalent operators, taking an operator to be inequivalent if neither it, nor its h.c., are already in the list. The factor of 2 in the definition of $\mathcal{O}_{\ell e}$ is included to compensate the 1/2 in the Fierz rearrangement below (second line of (3.13)).[8] The effective operators whose coefficients we constrain in the next section are related to those of (3.8) through an expansion in terms of the $SU(2)$ components of the fields and taking into account the electroweak symmetry breaking. For example, for the lepton operators:

$$\mathcal{O}_{eB}^{ij} = \bar{\ell}_i \sigma^{\mu\nu} e_{Rj} H B_{\mu\nu} = \cos\theta_W \langle H\rangle \bar{e}_i \sigma^{\mu\nu} P_R e_j F_{\mu\nu}^{\text{em}}, \quad (3.9)$$
$$\mathcal{O}_{eW}^{ij} = \bar{\ell}_i \sigma^{\mu\nu} \tau^I e_{Rj} H W_{\mu\nu}^I$$
$$= -\sin\theta_W \langle H\rangle \bar{e}_i \sigma^{\mu\nu} P_R e_j F_{\mu\nu}^{\text{em}}, \qquad (3.10)$$
$$\mathcal{O}_{(1)\ell\ell}^{ijkl} = (\bar{\ell}_i \gamma^\mu \ell_j)(\bar{\ell}_k \gamma_\mu \ell_l)$$
$$= (\bar{\nu}_i \gamma^\mu P_L \nu_j + \bar{e}_i \gamma^\mu P_L e_j)$$
$$\times (\bar{\nu}_k \gamma_\mu P_L \nu_l + \bar{e}_k \gamma_\mu P_L e_l), \qquad (3.11)$$
$$\mathcal{O}_{(3)\ell\ell}^{ijkl} = (\bar{\ell}_i \tau^I \gamma^\mu \ell_j)(\bar{\ell}_k \tau^I \gamma_\mu \ell_l)$$
$$= 2(\bar{\nu}_i \gamma^\mu P_L e_j)(\bar{e}_k \gamma_\mu P_L \nu_l)$$
$$+ 2(\bar{e}_i \gamma^\mu P_L \nu_j)(\bar{\nu}_k \gamma_\mu P_L e_l)$$
$$+ [(\bar{\nu}_i \gamma^\mu P_L \nu_j)(\bar{\nu}_k \gamma_\mu P_L \nu_l)$$
$$+ (\bar{e}_i \gamma^\mu P_L e_j)(\bar{e}_k \gamma_\mu P_L e_l)$$
$$- (\bar{\nu}_i \gamma^\mu P_L \nu_j)(\bar{e}_k \gamma_\mu P_L e_l)$$
$$- (\bar{e}_i \gamma^\mu P_L e_j)(\bar{\nu}_k \gamma_\mu P_L \nu_l)], \qquad (3.12)$$
$$\mathcal{O}_{\ell e}^{ijkl} = 2(\bar{\ell}_i e_j)(\bar{e}_k \ell_l)$$
$$= 2[(\bar{\nu}_i P_R e_j)(\bar{e}_k P_L \nu_l) + (\bar{e}_i P_R e_j)(\bar{e}_k P_L e_l)]$$
$$= -[(\bar{\nu}_i \gamma^\mu P_L \nu_l)(\bar{e}_k \gamma_\mu P_R e_j)$$
$$+ (\bar{e}_i \gamma^\mu P_L e_l)(\bar{e}_k \gamma_\mu P_R e_j)]. \qquad (3.13)$$

All these operators, together with $\mathcal{O}_{ee}^{ijkl}$, induce dipole moments and four-charged-lepton (4CL) vertices, as appear

---
[8]Note there will sometimes be other 2 s for identical fermions.

to the right-hand side (RHS) in the above equations. Constraints on the coefficients of the 4CL operators

$$\mathcal{O}_{PP}^{ijkl} = \frac{1}{1+\delta}(\bar{e}_i \gamma^\mu P e_j)(\bar{e}_k \gamma_\mu P e_l),$$
$$\mathcal{O}_{RL}^{ijkl} = \frac{1}{1+\delta}(\bar{e}_i \gamma^\mu P_R e_j)(\bar{e}_k \gamma_\mu P_L e_l), \quad (3.14)$$

where $P = P_R$ or P_L, are listed in Tables 4, 5, 6 and 7.

After electroweak symmetry breaking, the operators $\mathcal{O}_{eB}^{ij}$ and $\mathcal{O}_{eW}^{ij}$ become the chirality-flipping dipole moments as written in (3.9), (3.10) (where we did not include the Z–lepton–lepton operators [169]). These dipoles can be flavor conserving or transition dipole moments. The flavor diagonal operators are specially interesting because they correspond to the anomalous magnetic moments and the electric dipole moments of the different fermions. Taking $C_{e\gamma}^{ij}(q^2) = C_{eB}^{ij}(q^2)\cos\theta_W - C_{eW}^{ij}(q^2)\sin\theta_W$ as the Wilson coefficient with momentum transfer equal to q^2, we have for $q^2 = 0$,

$$\frac{C_{e\gamma}^{ii}(q^2=0)}{m_{\rm NP}^2}\langle H\rangle \bar{e}_i \sigma^{\mu\nu} P_R e_i F_{\mu\nu}^{\rm em} + \text{h.c.}$$
$$= \frac{\text{Re}\{C_{e\gamma}^{ii}(q^2=0)\}}{m_{\rm NP}^2}\langle H\rangle \bar{e}_i \sigma^{\mu\nu} e_i F_{\mu\nu}^{\rm em}$$
$$+ \frac{\text{Im}\{C_{e\gamma}^{ij}(q^2=0)\}}{m_{\rm NP}^2}\langle H\rangle i\bar{e}_i \sigma^{\mu\nu} \gamma_5 e_i F_{\mu\nu}^{\rm em}$$
$$= e\frac{a_{e_i}}{4m_{e_i}}\bar{e}_i \sigma^{\mu\nu} e_i F_{\mu\nu}^{\rm em} + \frac{i}{2}d_{e_i}\bar{e}_i \sigma^{\mu\nu}\gamma_5 e_i F_{\mu\nu}^{\rm em}, \quad (3.15)$$

with $a_{e_i} = (g_{e_i} - 2)/2$ the anomalous magnetic moment and d_{e_i} the electric dipole moment of the lepton e_i that can be found in [170].

In a given model, the coefficients of the effective operators can be obtained by matching the effective theory of (3.1) onto the model, at some matching scale (for instance, the mass scale of new particles). However, in particular models there can appear various pitfalls in constraining the generic coefficients $C^{ijkl}_{...}$. This is illustrated, for example, in the model of [171] which corresponds to adding a singlet slepton $\tilde{E}^c$ of flavor k, in R-parity violating (RPV) SUSY. In this case, after integrating out the heavy slepton we obtain the following effective operator:

$$\frac{\lambda^k_{[ij]}\lambda^{*k}_{[mn]}}{M^2}\big(\overline{(\nu_L)^c}_i e_{Lj}\big)\big(\overline{(e_L)}_n (\nu_L)^c_m\big)$$
$$= \frac{\lambda^k_{[ij]}\lambda^{*k}_{[mn]}}{2M^2}(\bar{e}_n \gamma^\mu P_L e_j)(\bar{\nu}_m \gamma_\mu P_L \nu_i), \quad (3.16)$$

where $\lambda^k_{[ij]}$ is antisymmetric in i, j because the $SU(2)$ contraction of $\ell_i \ell_j$ is antisymmetric. This is an example of operator $\mathcal{O}_{\ell\ell(1)}$, but since it is induced by singlet scalar exchange, there is no four-charged-lepton operator (compare to (3.11)). This illustrates that the bounds obtained here, by assuming that $C^{ijkl}_{...} \neq 0$ for one choice of $ijkl$ at a time, are not generic. Each process receives contributions from a sum of operators, and that sum could contain cancellations in a particular model.

Many models of new physics introduce new TeV-scale particles carrying a conserved quantum number (e.g. R-parity, T-parity...). Such particles appear in pairs at vertices, so they contribute via boxes and penguins to the four-fermion and dipole moment operators considered here. Generic formulae for the one loop contribution to a dipole moment can be found in [172], and for boxes in [173]. Extra Higgses [174, 175] would contribute to the same operators constructed from SM fields, so they are constrained by the experimental limits on the coefficients of such operators.

3.1.2 Constraints on low scale observables

In this section we present the low energy constraints on the different Wilson coefficients introduced before. Any NP found at LHC will necessarily respect the bounds presented here.

3.1.2.1 Dipole transitions After electroweak symmetry breaking, the operators of (3.9), (3.10) generate magnetic and electric dipole moments for the charged leptons. Flavor-diagonal operators give rise to anomalous magnetic moments and electric dipole moments as shown in (3.15). The anomalous magnetic moment of the electron $a_e = (g-2)_e/2$ is used to determine $\alpha_{\rm em}$. The current measurement of the muon anomalous moment $a_\mu = (g-2)_\mu/2$ deviates from the (uncertain) SM expectation by 3.2σ using e^+e^--data [176], and can be taken as a constraint, or indication on the presence of new physics. Currently there is only an upper bound on the magnetic moment of the τ from the analysis of $e^+e^- \to \tau^+\tau^-$ [170, 177]. Electric dipole moments have not yet been observed, although we have very constraining bounds specially on the electron dipole moment. In Table 3 we present the bounds of flavor diagonal dipole moments. The EDMs are discussed in detail in Sect. 5.

The bounds on off-diagonal dipole transitions are presented in Table 3. It is convenient to normalize these coefficients, $C_{e\gamma}^{ij} = C_{eB}^{ij}\cos\theta_W - C_{eW}^{ij}\sin\theta_W$, to the Fermi interactions given our ignorance on the scale of new physics $m_{\rm NP}$:

$$\frac{C_{e\gamma}^{ij}}{m_{\rm NP}^2} = \frac{4G_{\rm F}}{\sqrt{2}}\epsilon_{e\gamma}^{ij}. \quad (3.17)$$

In the literature, it is customary to use the left and right form factors for lepton flavor violating transitions defined by

Table 3 Bounds on the different dipole coefficients. Flavor diagonal dipole coefficients are given in terms of the corresponding anomalous magnetic moment, a_{e_i}, and the dipole moment, d_{e_i}. Bounds on transition moments are given in terms of the dimensionless coefficients $|\epsilon_{e\gamma}^{ij}|$ (defined in (3.17)) from the bounds on the branching ratios given in the last column. These bounds apply also both to $|\epsilon_{e\gamma}^{ij}|$ and $|\epsilon_{e\gamma}^{ji}|$. See Sect. 3.1.2 for details

(ij)	$a_i = \frac{g_i-2}{2}$	edm$_i$ (e cm)	Ref.
$\bar{e}e$	0.0011596521859(38)	$d_e \leq 1.6 \times 10^{-27}$	PDG [170, 186]
$\bar{\mu}\mu$	$11659208.0(5.4)(3.3) \times 10^{-10}$	$d_\mu \leq 2.8 \times 10^{-19}$	Muon g-2 Coll. [187, 188]
$\bar{\tau}\tau$	$-0.052 < a_\tau < 0.013$	$(-2.2 < d_\tau < 4.5) \times 10^{-17}$	LEP2 [189], Belle [190]
(ij)	$\bar{\ell}_i \sigma^{\mu\nu} e_{Rj} F^{em}_{\mu\nu}$	expt. limit	Ref.
$\bar{e}\mu$	$\leq 3.4 \times 10^{-11}$	$\leq 1.2 \times 10^{-11}$	MEGA Coll. [180]
$\bar{e}\tau$	$\leq 1.2 \times 10^{-7}$	$\leq 1.1 \times 10^{-7}$	BaBar [182]
$\bar{\mu}\tau$	$\leq 8.4 \times 10^{-8}$	$\leq 4.5 \times 10^{-8}$	Belle, BaBar [181, 191]

$$\Delta \mathcal{L}_2 = e m_{l_i} A_\mu \bar{f}_j \left[i\sigma^{\mu\nu} q_\nu \left(A_L^{ij} P_L + A_R^{ij} P_R \right) \right] f_i + \text{h.c.} \quad (3.18)$$

where f is a Dirac (4-component) fermion. The radiative decay $f_i \to f_j + \gamma$ proceeds at the rate $\Gamma = m_i^5 e^2/(16\pi) \times (|A_L^{ij}|^2 + |A_R^{ij}|^2)$ [178]. QED corrections to those decays are unusually large and may reach as much as 15% [179]. Bounds on the dimensionless coefficients $C_{e\gamma}^{ij}$ and $\epsilon_{e\gamma}^{ij}$ can be obtained by translating from A_L^{ij} and A_R^{ij}:

$$\frac{C_{e\gamma}^{ij}}{m_{NP}^2}\langle H \rangle = e \frac{m_i}{2} A_R^{ij}, \qquad \frac{C_{e\gamma}^{ji*}}{m_{NP}^2}\langle H \rangle = e \frac{m_i}{2} A_L^{ij}. \quad (3.19)$$

The experimental bounds on radiative lepton decays can be used to set bounds on these off-diagonal Wilson coefficients. The current experimental bounds are $B(\mu \to e\gamma) < 1.2 \times 10^{-11}$ [180], $B(\tau \to \mu\gamma) < 4.5 \times 10^{-8}$ [181], and $B(\tau \to e\gamma) < 1.1 \times 10^{-7}$ [182].

For the off-shell photon, $q^2 \neq 0$, there exist additional form factors,

$$\Delta \mathcal{L} = e m_{l_i} A_\mu \bar{e}_j \left[\left(g_{\mu\nu} - \frac{q_\mu q_\nu}{q^2} \right) \gamma_\nu \left(B_L^{ij} P_L + B_R^{ij} P_R \right) \right] e_i + \text{h.c.}, \quad (3.20)$$

which induce contributions to the four-fermion operators to be discussed in the next subsections. These form factors may be enhanced by a large factor compared to the on-shell photon form factors [184], $\ln(m_{NP}/m_{l_i})$, depending on the nature of new physics. Therefore, those operators become relevant for constraining new physics in R-parity violating SUSY [185] and in low-scale type-II see-saw models [184].

3.1.2.2 Four-charged-lepton operators As before, to present the bounds on the dimensionless four-charged-fermion coefficients in (3.14), we normalize them to the Fermi interactions:

$$\frac{C_{(n)\ell\ell}^{ijkl}}{m_{NP}^2} = -\frac{4G_F}{\sqrt{2}} \epsilon_{(n)\ell\ell}^{ijkl}, \qquad \frac{C_{ee}^{ijkl}}{m_{NP}^2} = -\frac{4G_F}{\sqrt{2}} \epsilon_{ee}^{ijkl},$$

$$\frac{C_{\ell e}^{ilkj}}{m_{NP}^2} = \frac{4G_F}{\sqrt{2}} \epsilon_{\ell e}^{ijkl}. \quad (3.21)$$

The current low energy constraints on the dimensionless ϵ's are shown in Tables 4, 5, 6 and 7. The rows of the tables are labeled by the flavor combination, and the column by the Lorentz structure. The numbers given in this tables correspond to the best current experimental bound on the coefficient of each operator, assuming it is the only non-zero coefficient present. The last column in the table lists the experiment setting the bound. The compositeness search limits Λ@ LEP are at 95% C.L., the decay rate bounds at 90% C.L.

Regarding the definition of the different coefficients we have to make some comments. First, note the flavor index permutation between $C_{\ell e}$ and $\epsilon_{\ell e}$:

$$C_{\ell e}^{ilkj}(\bar{\ell}_i e_l)(\bar{e}_k \ell_j) = -\frac{1}{2} \epsilon_{\ell e}^{ijkl}(\bar{\ell}_i \gamma^\mu \ell_j)(\bar{e}_k \gamma_\mu e_l). \quad (3.22)$$

There are relations between the flavor indices of the different operators. For $\mathcal{O}_{LL} = (\bar{e}\gamma^\mu P_L e)(\bar{e}\gamma_\mu P_L e)$ and $\mathcal{O}_{RR} = (\bar{e}\gamma^\mu P_R e)(\bar{e}\gamma_\mu P_R e)$ we have

$$\mathcal{O}_{PP}^{ijkl} = \mathcal{O}_{PP}^{klij}, \qquad \mathcal{O}_{PP}^{ijkl} = \mathcal{O}_{PP}^{*jilk}, \qquad \mathcal{O}_{PP}^{ijkl} = \mathcal{O}_{PP}^{ilkj}, \quad (3.23)$$

by symmetry, Hermitian conjugation and Fierz rearrangement, respectively. Therefore, the constraints on $\bar{e}e\bar{\mu}\tau$ in the first two columns of Tables 4 to 7 apply to $\epsilon_{(n)xx}^{ee\mu\tau}$, $\epsilon_{(n)xx}^{\mu\tau ee}$, $\epsilon_{(n)xx}^{*ee\tau\mu}$, $\epsilon_{(n)xx}^{*\tau\mu ee}$, $\epsilon_{(n)xx}^{e\tau\mu e}$, $\epsilon_{(n)xx}^{\mu ee\tau}$, $\epsilon_{(n)xx}^{*\tau ee\mu}$, and $\epsilon_{(n)xx}^{*e\mu\tau e}$ with $(n)xx$

Table 4 Bounds on coefficients of flavor four-lepton operators, from four-charged-lepton processes. The number is the upper bound on the dimensionless operator coefficient ϵ^{ijkl} (defined in (3.21)), arising from the measurement in the last column. The bound applies also to ϵ^{klij}. The second column is the bounds on $\epsilon^{ijkl}_{(3)\ell\ell}$, and $\epsilon^{ijkl}_{(1)\ell\ell}$ [except in the case of the bracketed limits, which are the upper bound on $\epsilon^{ijkl}_{(1)\ell\ell}$ and $2\epsilon^{ijkl}_{(1)\ell\ell}$]. The third column is the bound on $\epsilon^{ijkl}_{(1)ee}$. The bounds in these two columns apply also when the flavor indices are permuted to $jilk$ and $ilkj$. The fourth column is the bound on $\epsilon^{ijkl}_{\ell e}$ (which does not apply to the flavor permutation $ilkj$, so this is listed with a line of its own). The constraints in [brackets] apply to the two charged lepton–two neutrino operator of the same flavor structure, and arise from lepton universality in τ decays. See Sect. 3.1.2 for details

$(ijkl)$	$(\overline{e}\gamma^\mu P_L e)(\overline{e}\gamma_\mu P_L e)$	$(\overline{e}\gamma^\mu P_R e)(\overline{e}\gamma_\mu P_R e)$	$(\overline{e}\gamma_\mu P_L e)(\overline{e}\gamma^\mu P_R e)$	expt. limit	Ref.
$\overline{e}e\overline{e}e$	$(-1.8 - +2.8) \times 10^{-3}$	$(-1.8 - +2.8) \times 10^{-3}$	$(-2.4 - +4.9) \times 10^{-3}$	Λ@LEP2	[194]
$\overline{e}e\overline{\mu}\mu$	$(-7.2 - +5.2) \times 10^{-3}$	$(-7.8 - +5.8) \times 10^{-3}$	$(-9.0 - +9.6) \times 10^{-3}$	Λ@LEP2	[193, 195]
$\overline{e}\mu\overline{\mu}e$	$(-7.2 - +5,2) \times 10^{-3}$	$(-7.8 - +5.8) \times 10^{-3}$	1.3×10^{-2}	Λ, RPV@LEP2	[193, 195]
$\overline{e}e\overline{\tau}\tau$	$(-7.3 - +13) \times 10^{-3}$	$(-8.0 - +15) \times 10^{-3}$	$(-1.2 - +1.8) \times 10^{-2}$	Λ@LEP2	[193, 195]
$\overline{\tau}e\overline{e}\tau$	$(-7.3 - +13) \times 10^{-3}$	$(-8.0 - +15) \times 10^{-3}$	1.3×10^{-2}	Λ, RPV@LEP2	[193, 195]
$\overline{\mu}\mu\overline{\mu}\mu$	~ 1	~ 1	~ 1	$B(Z \to \mu\overline{\mu})$	
$\overline{\mu}\mu\overline{\tau}\tau$	~ 1 [0.0014]	~ 1	~ 1 [0.01]	$B(Z \to \mu\overline{\mu})$	
$\overline{\mu}\tau\overline{\tau}\mu$	~ 1 [0.0014]	~ 1		$B(Z \to \mu\overline{\mu})$	
$\overline{\tau}\tau\overline{\tau}\tau$	~ 1	~ 1	~ 1	$B(Z \to \tau\overline{\tau})$	

Table 5 Bounds on coefficients of four-lepton operators with $\Delta L_\alpha = -\Delta L_\beta = 1$. They apply also to flavor index permutations $klij$ and $ilkj$, except in the case of $\tau\tau e\mu$, where the bound on $\tau\mu e\tau$ in the fourth column is from μ decay and is listed separately. See the caption of Table 4 and Sect. 3.1.2 for further details

$(ijkl)$	$(\overline{e}\gamma^\mu P_L e)(\overline{e}\gamma_\mu P_L e)$	$(\overline{e}\gamma^\mu P_R e)(\overline{e}\gamma_\mu P_R e)$	$(\overline{e}\gamma_\mu P_L e)(\overline{e}\gamma^\mu P_R e)$	expt. limit
$\overline{e}e\overline{e}\mu$	7.1×10^{-7}	7.1×10^{-7}	7.1×10^{-7}	$B(\mu \to e\overline{e}e) < 10^{-12}$
$\overline{e}e\overline{e}\tau$	7.8×10^{-4}	7.8×10^{-4}	7.8×10^{-4}	$B(\tau \to e\overline{e}e) < 2 \times 10^{-7}$
$\overline{e}e\overline{\mu}\tau$	1.1×10^{-3}	1.1×10^{-3}	1.1×10^{-3}	$B(\tau \to \overline{e}e\mu) < 1.9 \times 10^{-7}$
$\overline{\mu}\mu\overline{e}\mu$	~ 1	~ 1	~ 1	$B(Z \to e\overline{\mu}) < 1.7 \times 10^{-6}$
$\overline{\mu}\mu\overline{e}\tau$	1.1×10^{-3}	1.1×10^{-3}	1.1×10^{-3}	$B(\tau \to \overline{\mu}e\mu) < 2.0 \times 10^{-7}$
$\overline{\mu}\mu\overline{\mu}\tau$	7.8×10^{-4}	7.8×10^{-4}	7.8×10^{-4}	$B(\tau \to 3\mu) < 1.9 \times 10^{-7}$
$\overline{\tau}\tau\overline{e}\mu$	~ 1 [0.05]	~ 1	~ 1 [0.05]	$B(Z \to e\overline{\mu}) < 1.7 \times 10^{-6}$
$\overline{\tau}\mu\overline{e}\tau$	~ 1 [0.05]	~ 1	[0.05]	$B(Z \to e\overline{\mu}) < 1.7 \times 10^{-6}$
$\overline{\tau}\tau\overline{e}\tau$	~ 3 [0.05]	~ 3	~ 3 [0.05]	$B(Z \to e\overline{\tau}) < 9.8 \times 10^{-6}$
$\overline{\tau}\tau\overline{\tau}\mu$	~ 3 [0.05]	~ 3	~ 3 [0.05]	$B(Z \to \tau\overline{\mu}) < 1.2 \times 10^{-5}$

Table 6 Bounds on coefficients of four-lepton operators with $\Delta L_\alpha = \Delta L_\beta = 2$. See the caption of Table 4 and Sect. 3.1.2 for details

$(ijkl)$	$(\overline{e}\gamma^\mu P_L e)(\overline{e}\gamma_\mu P_L e)$	$(\overline{e}\gamma^\mu P_R e)(\overline{e}\gamma_\mu P_R e)$	$(\overline{e}\gamma_\mu P_L e)(\overline{e}\gamma^\mu P_R e)$	expt. limit
$\overline{e}\mu\overline{e}\mu$	3.0×10^{-3}	3.0×10^{-3}	2.0×10^{-3}	$(\overline{\mu}e) \leftrightarrow (\overline{e}\mu)$
$\overline{e}\tau\overline{e}\tau$	[0.05]		[0.05]	
$\overline{\mu}\tau\overline{\mu}\tau$	[0.05]		[0.05]	

Table 7 Bounds on coefficients of four-lepton operators with $\Delta L_\alpha = \Delta L_\beta = -\frac{1}{2}\Delta L_\rho$. See the caption of Table 4 and Sect. 3.1.2 for details

$(ijkl)$	$(\overline{e}\gamma^\mu P_L e)(\overline{e}\gamma_\mu P_L e)$	$(\overline{e}\gamma^\mu P_R e)(\overline{e}\gamma_\mu P_R e)$	$(\overline{e}\gamma_\mu P_L e)(\overline{e}\gamma^\mu P_R e)$	expt. limit
$\overline{e}\mu\overline{e}\tau$	2.3×10^{-4}	2.3×10^{-4}	2.3×10^{-4}	$B(\tau \to \overline{\mu}ee) < 1.1 \times 10^{-7}$
$\overline{\mu}e\overline{\mu}\tau$	2.6×10^{-4}	2.6×10^{-4}	2.6×10^{-4}	$B(\tau \to \overline{e}\mu\mu) < 1.3 \times 10^{-7}$
$\overline{\tau}e\overline{\tau}\mu$	[0.05]		[0.05]	

equal to $(3)\ell\ell$, $(1)\ell\ell$, or $(1)ee$. Note, however that it is calculated assuming only one of these ϵ is non-zero. Similarly, the operator $\mathcal{O}_{LR}^{ijkl} = (\overline{e}_i\gamma_\mu P_L e_j)(\overline{e}_k\gamma^\mu P_R e_l)$, with coefficient $\epsilon_{\ell e}^{ijkl}$, is related by Hermitian conjugation:

$$\mathcal{O}_{LR}^{ijkl} = \mathcal{O}_{LR}^{*jilk}, \quad (3.24)$$

so again the bounds on $\epsilon_{\ell e}^{ijkl}$ apply to $\epsilon_{\ell e}^{*jilk}$. We can usually apply also these bounds to $\epsilon_{\ell e}^{klij}$ because the chirality of the fermion legs does not affect the matrix element squared, but $\epsilon_{\ell e}^{ilkj}$ is bounded separately in the tables.

The bounds from Z decays in Tables 4 and 5 are estimated from the one loop penguin diagram obtained closing two of the legs of the four-fermion operator and coupling it with the Z [192]. These bounds would be more correctly included by renormalization group mixing between the four-fermion operators and the Z–fermion–fermion operators discussed in [169]. They are listed in the tables to indicate the existence of a constraint. The bound can be applied to $\epsilon_{\ell e}^{iikl}$ and $\epsilon_{\ell e}^{ijkk}$ but it does not apply to $\epsilon_{\ell e}^{ilki}$.

Contact interaction bounds are usually quoted on the scale Λ, where

$$\epsilon_{ab}^{ijkl}\frac{4G_F}{\sqrt{2}} = \pm\frac{1}{1+\delta}\frac{4\pi}{\Lambda^2}, \quad (3.25)$$

and $\delta = 1$ for the operators $\mathcal{O}_{LL}^{eeee}$ and $\mathcal{O}_{RR}^{eeee}$ of (3.14), 0 otherwise. Since our normalization does not have this factor of 2, we have a Feynman rule $\epsilon 8G_F/\sqrt{2}$ for these operators, and correspondingly stricter bounds on the ϵ's. The bounds are the same for $\epsilon_{\ell e}^{ikki}$ and $\epsilon_{\ell e}^{kiik}$. However, contact interaction bounds are not quoted on operators of the form $(\overline{e}_i\gamma^\mu P_L e_j)(\overline{e}_j\gamma_\mu P_R e_i)$, corresponding to $\epsilon_{\ell e}^{iijj}$. Such operators are generated by sneutrino exchange in R-parity violating SUSY, so we estimate the bound $\lambda^2/m_{\tilde{\nu}}^2 < 4/(9\text{ TeV}^2)$ from the plotted constraints in [193], and impose $4|\epsilon_{ab}^{ijkl}|G_F/\sqrt{2} < \lambda^2/(2m_{\tilde{\nu}}^2)$.

Many of the 4CL operators involving two τ's are poorly constrained. In some cases, see (3.11), (3.12), new physics that generates 4CL operators also induces $(\overline{e}_i\gamma^\lambda Pe_j) \times (\overline{\nu}_k\gamma_\lambda L\nu_l)$. The coefficients of operators of the form $(\overline{\mu}\gamma^\lambda Pe)(\overline{\nu}_k\gamma_\lambda L\nu_l)$, $(\overline{\mu}\gamma^\lambda P\tau)(\overline{\nu}_k\gamma_\lambda L\nu_l)$ or $(\overline{e}\gamma^\lambda P\tau) \times (\overline{\nu}_k\gamma_\lambda L\nu_l)$, are constrained from lepton universality measurements in μ and τ decays [196]. The decay rate $\tau \to e_i\nu_k\overline{\nu}_l$ in the presence of the operators of (3.14), divided by the SM prediction for $\tau \to e_i\nu_\tau\overline{\nu}_i$, is

$$\left(1 - 2\delta_{k\tau}\delta_{il}\text{Re}\{\epsilon_{(1)\ell\ell}^{\tau\tau ii} + 2\epsilon_{(3)\ell\ell}^{\tau\tau ii}\} + \frac{4m_i}{m_\tau}\delta_{k\tau}\delta_{il}\text{Re}\{\epsilon_{\ell e}^{\tau\tau ii}\}\right.$$
$$\left. + |\epsilon_{(1)\ell\ell}^{i\tau kl}|^2 + 4|\epsilon_{(3)\ell\ell}^{i\tau kl}|^2 + |\epsilon_{\ell e}^{i\tau kl}|^2\right). \quad (3.26)$$

Within the experimental accuracy, the weak τ and μ decays verify lepton universality and agree with LEP precision measurements of m_W. Rough bounds on the ϵ's can therefore be obtained by requiring the new physics contribution to the decay rates to be less than the errors $\frac{\Delta B}{B}(\tau \to e\nu\overline{\nu}) = 0.05/17.84$, $\frac{\Delta B}{B}(\tau \to \mu\nu\overline{\nu}) = 0.05/17.36$. These are listed in the tables in [brackets]. The bracketed limit in the second column applies to $\epsilon_{(1)\ell\ell}^{ijkl}$; the bound on $\epsilon_{(3)\ell\ell}^{ijkl}$ is 1/2 the quoted number. The limit on $\epsilon_{\ell e}^{\tau e\tau\mu}$ is from its contribution to $\mu \to e\nu_\tau\overline{\nu}_\tau$.

Finally, we would like to remind the reader the various caveats to these four-fermion vertex bounds.

- The constraints are calculated "one operator at a time". This is unrealistic; new physics is likely to induce many non-renormalizable operators. In some cases, see (3.16), a symmetry in the new physics can cause cancellations such that it does not contribute to certain observables.
- The coefficients of the 4CL operators, and two ν–two charged lepton (2ν2CL) operators may differ by a factor of few, because they are induced by the exchange of different members of a multiplet, whose masses differ [197].
- The list of operators is incomplete. Perhaps some of the neglected operators give relevant constraints on new physics. For instance, bounds from lepton universality on the $(H^*\overline{\ell})\gamma^\mu\partial_\mu(H\ell)$ operator [198] are relevant to extra dimensional scenarios [199].
- Operators of dimension >6 are neglected. If the mass scale of the new physics is $\sim$TeV, then higher dimension operators with Higgs VEVs [200] such as $HH\overline{\psi}\psi\overline{\psi}\psi$ are not significantly suppressed.

3.1.2.3 Two lepton–two quark operators Once more, we normalize the coefficients of the two lepton–two quark operators in (3.6) to the Fermi interactions:

$$\frac{C_{(n)\ell q}^{ijkl}}{m_{NP}^2} = -\frac{4G_F}{\sqrt{2}}\epsilon_{(n)\ell q}^{ijkl}, \quad \frac{C_{ed}^{ijkl}}{m_{NP}^2} = -\frac{4G_F}{\sqrt{2}}\epsilon_{ed}^{ijkl},$$

$$\frac{C_{\ell d}^{ijkl}}{m_{NP}^2} = \frac{4G_F}{\sqrt{2}}\epsilon_{\ell d}^{ijkl}, \quad \frac{C_{eu}^{ijkl}}{m_{NP}^2} = -\frac{4G_F}{\sqrt{2}}\epsilon_{eu}^{ijkl},$$

$$\frac{C_{\ell u}^{ijkl}}{m_{NP}^2} = -\frac{4G_F}{\sqrt{2}}\epsilon_{\ell u}^{ijkl}, \quad \frac{C_{\ell qS}^{ijkl}}{m_{NP}^2} = -\frac{4G_F}{\sqrt{2}}\epsilon_{\ell qS}^{ijkl}, \quad (3.27)$$

$$\frac{C_{qde}^{ijkl}}{m_{NP}^2} = -\frac{4G_F}{\sqrt{2}}\epsilon_{qde}^{ijkl}.$$

The main bounds on the dimensionless ϵs are given in Tables 8 and 9. These numbers correspond to the best current experimental bound on the coefficient of each operator, assuming it is the only non-zero coefficient present. The bounds on $\epsilon_{\ell q}$ in Table 8 apply both to $\epsilon_{(1)\ell q}$ and $\epsilon_{(3)\ell q}$. These bounds have been obtained from the corresponding bounds on leptoquark couplings in Refs. [201, 202] that can be checked for further details.

Table 8 Bounds on coefficients of the left handed two quark–two lepton operators. Bound is the upper bound on the dimensionless operator coefficient ϵ^{ijkl} (defined in (3.28)), arising from the experimental determination of the observable in the next column. Bounds with a * are also valid under the exchange of the lepton indices

$(\bar{e}\gamma^\mu P_L e)(\bar{q}\gamma_\mu P_L q)$					
$(ijkl)$	Bound on $\epsilon^{ijkl}_{\ell q}$	Observable	$(ijkl)$	Bound on $\epsilon^{ijkl}_{\ell q}$	Observable
11 11	5.1×10^{-3}	R_π	22 11	5.1×10^{-3}	R_π
12 11	8.5×10^{-7}	μ–e conversion on Ti	12 12*	2.9×10^{-7}	$B(K^0_L \to \overline{\mu} e)$
ij 12	4.5×10^{-6}	$\frac{B(K^+ \to \pi^+ \bar{\nu}\nu)}{B(K^+ \to \pi^0 e^+ \nu_e)}$	ij 22	1.0	V_{cs}
ij 13	3.6×10^{-3}	V_{ub}	ij 23	4.2×10^{-2}	V_{cb}
11 23	6.6×10^{-5}	$B(B^+ \to e^+ e^- K^+)$	11 13	9.3×10^{-4}	$B(B^+ \to e^+ e^- \pi^+)$
22 23	5.4×10^{-5}	$B(B^+ \to \mu^+ \mu^- K^+)$	22 13	1.4×10^{-3}	$B(B^+ \to \mu^+ \mu^- \pi^+)$
21 23*	4.5×10^{-3}	$B(B^+ \to e^+ \mu^- K^+)$	21 13*	3.9×10^{-5}	$B(B^+ \to e^+ \mu^- \pi^+)$
12 23*	1.2×10^{-2}	$B(B^0_s \to \mu^+ e^-)$	33 12	6.6×10^{-2}	K–$\overline{K}$
22 22	6.0×10^{-2}	$\frac{B(D^+_s \to \mu^+ \nu_\mu)}{B(D^+_s \to \tau^+ \nu_\tau)}$	33 22	6.0×10^{-2}	$\frac{B(D^+_s \to \mu^+ \nu_\mu)}{B(D^+_s \to \tau^+ \nu_\tau)}$
32 23*	1.2×10^{-3}	$B(B^+ \to \mu^+ \tau^- X^+)$	33 23	9.3×10^{-3}	$B(B^+ \to \tau^+ \tau^- X^+)$

Table 9 Bounds on coefficients of the right handed vector and scalar 2 quark-2 lepton operators. Bound is the upper bound on the dimensionless operator coefficient ϵ^{ijkl} (defined in (3.28)), arising from the experimental determination of the observable in the next column. Bounds with a * are also valid under the exchange of the lepton indices

$(\bar{e}\gamma^\mu P_R e)(\bar{q}\gamma_\mu P_R q)$					
$(ijkl)$	Bound on ϵ^{ijkl}_{eu}	Observable	$(ijkl)$	Bound on ϵ^{ijkl}_{eu}	Observable
11 12	1.7×10^{-2}	$\frac{B(D^+ \to \pi^+ e^+ e^-)}{B(D^0 \to \pi^- e^+ \nu_e)}$	21 12*	1.3×10^{-2}	$\frac{B(D^+ \to \pi^+ \mu^- e^+)}{B(D^0 \to \pi^- e^+ \nu_e)}$
22 12	9.0×10^{-3}	$\frac{B(D^+ \to \pi^+ \mu^+ \mu^-)}{B(D^0 \to \pi^- e^+ \nu_e)}$	33 12	0.19	$B(D^0 - \overline{D}^0)$

$(\bar{\ell} P_R e)(\bar{d} P_L q)$					
$(ijkl)$	Bound on ϵ^{ijkl}_{qde}	Observable	$(ijkl)$	Bound on ϵ^{ijkl}_{qde}	Observable
11 11	1.5×10^{-7}	R_π	22 11	3.0×10^{-4}	R_π
12 11	5.1×10^{-3}	$B(\pi^+ \to \mu^+ \nu_e)$	12 12*	2.1×10^{-8}	$B(K^0_L \to \mu^+ e^-)$
11 12	2.7×10^{-8}	$B(K^0_L \to e^+ e^-)$	22 12	8.4×10^{-7}	$B(K^0_L \to \mu^+ \mu^-)$
22 21	1.3×10^{-2}	$B(D^+ \to \mu^+ \nu_\mu)$	22 22	1.2×10^{-2}	$\frac{B(D^+_s \to \mu^+ \nu_\mu)}{B(D^+_s \to \tau^+ \nu_\tau)}$
33 22	0.2	$\frac{B(D^+_s \to \mu^+ \nu_\mu)}{B(D^+_s \to \tau^+ \nu_\tau)}$	33 13	2.5×10^{-3}	$B(B^+ \to \tau^+ \nu_\tau)$
11 13	9.0×10^{-5}	$B(B^0 \to e^+ e^-)$	12 13*	1.2×10^{-4}	$B(B^0 \to \mu^+ e^-)$
13 13*	2.5×10^{-3}	$B(B^0 \to \tau^+ e^-)$	23 13*	3.3×10^{-3}	$B(B^0 \to \tau^+ \mu^-)$
22 13	7.5×10^{-5}	$B(B^0 \to \mu^+ \mu^-)$	11 23	6.0×10^{-4}	$B(B^0_s \to e^+ e^-)$
12 23*	2.1×10^{-4}	$B(B^0_s \to \mu^+ e^-)$	22 23	1.2×10^{-4}	$B(B^0_s \to \mu^+ \mu^-)$

3.2 Phenomenological parameterizations of quark and lepton Yukawa couplings

3.2.1 Quark sector

The quark Yukawa sector is described by the following Lagrangian:

$$\mathcal{L}_{\text{quark}} = u^c_{Ri} Y^u_{ij} Q_j \overline{H} + d^c_{Ri} Y^d_{ij} Q_j H + \text{h.c.}, \quad (3.28)$$

where $i, j = 1, 2, 3$ are generation indices, $Q_i = (d_{Li}, u_{Li})$ are the left handed quark doublets, u^c_R and d^c_R are the right handed up and down quark singlets respectively, and H is the Higgs field. On the other hand, Y^u and Y^d are complex 3×3 matrices, which can be cast by means of a singular value decomposition as

$$Y^u = V^u_R D^u_Y V^{u\dagger}_L,$$
$$Y^d = V^d_R D^d_Y V^{d\dagger}_L. \quad (3.29)$$

Here, $D_Y^u = \text{diag}(y_1^u, y_2^u, y_3^u)$ is a diagonal matrix whose entries can be chosen real and positive with $y_1^u < y_2^u < y_3^u$, and similarly for D_Y^d. $V_R^{u,d}$ and $V_L^{u,d}$ are 3×3 unitary matrices that depend on three real parameters and six phases. The unitary matrices $V_R^{u,d}$ can be absorbed in the definition of the right handed fields without any physical effect. In neutral currents the left rotations cancel out via the Glashow–Iliopoulos–Maiani (GIM) mechanism [203]. On the other hand, the redefinition of the left handed fields produces flavor mixing in the charged currents. In the physical basis where both the up and down Yukawa couplings are simultaneously diagonal, the charged current reads

$$J_{cc}^\mu = u_L^c \frac{\gamma^\mu(1-\gamma_5)}{2} (V_L^{u\dagger} V_L^d) d_L. \quad (3.30)$$

The matrix $V_L^{u\dagger} V_L^d$ can be generically written as $V_L^{u\dagger} V_L^d = \Phi_1 U_{\text{CKM}} \Phi_2$, where $\Phi_{1,2}$ are diagonal unitary matrices (thus, containing only phases) that can be absorbed by appropriate redefinitions of the left handed fields. Finally, U_{CKM} depends on three angles and one phase that cannot be removed by field redefinitions and accounts for the physical mixing between quark generations and the CP violation [204, 205]. It is usually parameterized thus:

$$U_{\text{CKM}} = \begin{pmatrix} c_{13}c_{12} & c_{13}s_{12} & s_{13}e^{-i\delta} \\ -c_{23}s_{12} - s_{23}s_{13}c_{12}e^{i\delta} & c_{23}c_{12} - s_{23}s_{13}s_{12}e^{i\delta} & s_{23}c_{13} \\ s_{23}s_{12} - c_{23}s_{13}c_{12}e^{i\delta} & -s_{23}c_{12} - c_{23}s_{13}s_{12}e^{i\delta} & c_{23}c_{13} \end{pmatrix}, \quad (3.31)$$

where $s_{ij} = \sin\theta_{ij}$, $c_{ij} = \cos\theta_{ij}$ and δ is the CP-violating phase. Experiments show a hierarchical structure in the off-diagonal entries of the CKM matrix: $|V_{ub}| \ll V_{cb} \ll V_{us}$, that can be well described by the following phenomenological parameterization of the CKM matrix, proposed by Wolfenstein [206]. It reads

$$U_{\text{CKM}} = \begin{pmatrix} 1 - \frac{\lambda^2}{2} & \lambda & A\lambda^3(\rho - i\eta) \\ -\lambda & 1 - \frac{\lambda^2}{2} & A\lambda^2 \\ A\lambda^3(1-\rho-i\eta) & -A\lambda^2 & 1 \end{pmatrix} + \mathcal{O}(\lambda^4), \quad (3.32)$$

where λ is determined with a very good precision in semileptonic K decays, giving $\lambda \simeq 0.23$, and A is measured in semileptonic B decays, giving $A \simeq 0.82$. The parameters ρ and η are more poorly measured, although a rough estimate is $\rho \simeq 0.1$, $\eta \simeq 0.3$ [207].

3.2.2 Leptonic sector with Dirac neutrinos

A Dirac mass term for the neutrinos requires the existence of three right handed neutrinos, which are singlets under the standard model gauge group. In consequence, the leptonic Lagrangian would contain in general a Majorana mass term for the right handed neutrinos, which has to be forbidden by imposing exact lepton number conservation. Then the leptonic Lagrangian reads

$$\mathcal{L}_{lep} = e_{Ri}^c Y_{ij}^e L_j \overline{H} + \nu_{Ri}^c Y_{ij}^\nu L_j H + \text{h.c.}, \quad (3.33)$$

where $L_i = (\nu_{Li}, e_{Li})$ are the left handed lepton doublets and e_R^c and ν_R^c are respectively the right handed charged lepton and neutrino singlets. Analogously to the quark sector, the Yukawa couplings can be decomposed as

$$Y^e = V_R^e D_Y^e V_L^{e\dagger}, \quad (3.34)$$

$$Y^\nu = V_R^\nu D_Y^\nu V_L^{\nu\dagger}, \quad (3.35)$$

where $V_R^{e,\nu}$ do not have any physical effect, whereas the $V_L^{e,\nu}$ have an effect in the charged current, that in the basis where the charged lepton and neutrino Yukawa couplings are simultaneously diagonal reads

$$J_{cc}^\mu = e_L^c \frac{\gamma^\mu(1-\gamma_5)}{2} (V_L^{e\dagger} V_L^\nu) \nu_L. \quad (3.36)$$

As in the case of the quark sector, the matrix $V_L^{e\dagger} V_L^\nu$ depends on three angles and six phases and can be expressed as $V_L^{e\dagger} V_L^\nu = \Phi_1 U_{\text{PMNS}} \Phi_2$. The matrices Φ_1 and Φ_2 can be absorbed by appropriate redefinitions of the left handed fields, yielding a physical mixing matrix U_{PMNS} [208, 209] that depends on three angles and one phase, and that can be parameterized by the same structure as for the quark sector, (3.31). However, the values for the angles differ substantially from the quark sector. The experimental values that result from the global fit are $\sin^2\theta_{12} = 0.26$–0.36, $\sin^2\theta_{23} = 0.38$–0.63 and $\sin^2\theta_{13} \leq 0.025$ at 2σ [210, 211]. On the other hand, the CP-violating phase δ is completely unconstrained by present experiments.

In the theory under discussion the total lepton number $L = L_e + L_\mu + L_\tau$ is conserved, but the individual lepton flavors L_l, $l = e, \mu, \tau$, are not, and LFV processes like $\mu^- \to e^- \gamma$ decay are allowed. For the neutrino masses m_{ν_j}, $j = 1, 2, 3$, satisfying the existing upper limits obtained in ^{3}H β-decay experiments, $m_j < 2.3$ eV, the $\mu^- \to e^- \gamma$ decay branching ratio is given by [212]

$$B(\mu \to e\gamma) = \frac{3\alpha}{32\pi} \left| \sum_j U_{ej}^{\text{PMNS}} U_{\mu j}^{\text{PMNS}*} \frac{m_{\nu_j}^2}{M_W^2} \right|^2, \quad (3.37)$$

where M_W is the $W^\pm$-boson mass. Thus, the $\mu^- \to e^- \gamma$ decay rate is suppressed by the factor $(m_j/M_W)^4 < 6.7 \times 10^{-43}$, which renders it unobservable. The same conclusion is valid for all other LFV decays and reactions in the minimal extension of the standard theory with light neutrino masses we are considering. The only observable manifestation of the non-conservation of the lepton charges L_l in this theory is the oscillations of neutrinos.

3.2.3 Leptonic sector with Majorana neutrinos

Neutrino masses can also be accommodated in the standard model without extending the particle content, just by adding a dimension five operator to the leptonic Lagrangian [213]:

$$\mathcal{L}_{\text{lep}} = e^c_{Ri} Y^e_{ij} L_j \overline{H} + \frac{1}{4} \kappa_{ij}(L_i H)(L_j H) + \text{h.c.} \quad (3.38)$$

with κ a 3×3 complex symmetric matrix that breaks explicitly lepton number and that has dimensions of mass^{-1}. Then, after the electroweak symmetry breaking, a Majorana mass term for neutrinos is generated:

$$m_\nu = \frac{1}{2}\kappa \langle H^0 \rangle^2. \quad (3.39)$$

This term can be diagonalized as $m_\nu = V^{\nu*}_L D_{m_\nu} V^{\nu\dagger}_L$, so that the charged current reads as in (3.36), with $V^{e\dagger}_L V^\nu_L = \Phi_1 U \Phi_2$, where the matrix U has the form of the CKM matrix, (3.31). The matrix Φ_1 containing three phases can be removed by a redefinition of the left handed charged lepton fields. However, due to the Majorana nature of the neutrinos, the matrix Φ_2 cannot be removed and is physical, yielding a leptonic mixing matrix $U_{\text{PMNS}} = U \Phi_2$ that is defined by three angles and three phases [214, 215], one associated to U, the "Dirac phase", and two associated to Φ_2, the "Majorana phases".

In the leptonic Lagrangian given by (3.38) the origin of the dimension five operator remains open. In the rest of this section, we shall review the heavy Majorana singlet (right handed) neutrino mass mechanism (type I see-saw) [216–220] and the triplet Higgs mass mechanism (type II see-saw) [215, 221–224] as the possible origins of this effective operator. The third [225] tree level realization of the operator (3.38) via triplet fermion (type III see-saw) [226] is discussed in Sect. 4.1.

3.2.3.1 Type I see-saw
In the presence of singlet right handed neutrinos, the most general Lagrangian compatible with the standard model gauge symmetry reads

$$\mathcal{L}_{\text{lep}} = e^c_{Ri} Y^e_{ij} L_j \overline{H} + \nu^c_{Ri} Y^\nu_{ij} L_j H - \frac{1}{2} \nu^{cT}_{Ri} M_{ij} \nu^c_{Rj} + \text{h.c.}, \quad (3.40)$$

where lepton number is explicitly broken by the Majorana mass term for the singlet right handed neutrinos.[9] The see-saw mechanism is implemented when $\text{eig}(M) \gg \langle H^0 \rangle$. If this is the case, at low energies the right handed neutrinos are decoupled and the theory can be well described by the effective Lagrangian for Majorana neutrinos, (3.38), with [216–220]

$$\kappa = 2 Y^{\nu T} M^{-1} Y^\nu. \quad (3.41)$$

Working in the basis where the charged lepton Yukawa matrix and the right handed mass matrix are simultaneously diagonal, it can be checked that the complete Lagrangian, (3.40), contains fifteen independent real parameters and six complex phases [229]. Of these, three correspond to the charged lepton masses, three to the right handed masses, and the remaining nine real parameters and six phases, to the neutrino Yukawa coupling. The independent parameters of the neutrino Yukawa coupling can be expressed in several ways. The most straightforward parameterization uses the singular value decomposition of the neutrino Yukawa matrix:

$$Y_\nu = V^\nu_R D^\nu_Y V^{\nu\dagger}_L, \quad (3.42)$$

where $D^\nu_Y = \text{diag}(y^\nu_1, y^\nu_2, y^\nu_3)$, with $y^\nu_i \geq 0$ and $y^\nu_1 \leq y^\nu_2 \leq y^\nu_3$. On the other hand, V^ν_L and V^ν_R are 3×3 unitary matrices, that depend in general on three real parameters and six phases. Both can be generically written as $\Phi_1 V \Phi_2$, where V has the form of the CKM matrix and $\Phi_{1,2}$ are diagonal unitary matrices (thus, containing only phases). One can check that for V^ν_R the Φ_2 matrix can be absorbed into the definition of V^ν_L, so that

$$V^\nu_R = \begin{pmatrix} e^{i\alpha^R_1} & & \\ & e^{i\alpha^R_2} & \\ & & 1 \end{pmatrix}$$
$$\times \begin{pmatrix} c^R_2 c^R_3 & c^R_2 s^R_3 & s^R_2 e^{-i\delta^R} \\ -c^R_1 s^R_3 - s^R_1 s^R_2 c^R_3 e^{i\delta^R} & c^R_1 c^R_3 - s^R_1 s^R_2 s^R_3 e^{i\delta^R} & s^R_1 c^R_2 \\ s^R_1 s^R_3 - c^R_1 s^R_2 c^R_3 e^{i\delta^R} & -s^R_1 c^R_3 - c^R_1 s^R_2 s^R_3 e^{i\delta^R} & c^R_1 c^R_2 \end{pmatrix}. \quad (3.43)$$

Similarly, for V_L the Φ_1 matrix can be absorbed into the definition of L and e_R, while keeping Y_e diagonal and real. In consequence,

$$V^\nu_L = \begin{pmatrix} c^L_2 c^L_3 & c^L_2 s^L_3 & s^L_2 e^{-i\delta^L} \\ -c^L_1 s^L_3 - s^L_1 s^L_2 c^L_3 e^{i\delta^L} & c^L_1 c^L_3 - s^L_1 s^L_2 s^L_3 e^{i\delta^L} & s^L_1 c^L_2 \\ s^L_1 s^L_3 - c^L_1 s^L_2 c^L_3 e^{i\delta^L} & -s^L_1 c^L_3 - c^L_1 s^L_2 s^L_3 e^{i\delta^L} & c^L_1 c^L_2 \end{pmatrix}$$
$$\times \begin{pmatrix} e^{i\alpha^L_1} & & \\ & e^{i\alpha^L_2} & \\ & & 1 \end{pmatrix}. \quad (3.44)$$

Therefore, in this parameterization the independent parameters in the Yukawa coupling can be identified with the three Yukawa eigenvalues, y_i, the three angles and three phases in V_L, and the three angles and three phases in

---

[9] Here we explicitly assume three generations of singlet neutrinos. For the phenomenology of a large number of singlets as predicted by string theories, see [227, 228].

V_R [229–231]. The requirement that the low energy phenomenology is successfully reproduced imposes constraints among these parameters. To be precise, the low energy leptonic Lagrangian depends just on the three charged lepton masses and the six real parameters and three complex phases of the effective neutrino mass matrix. In consequence, there are still six real parameters and three complex phases that are not determined by low energy neutrino data; this information about the high energy Lagrangian is "lost" in the decoupling of the three right handed neutrinos and cannot be recovered just from neutrino experiments.

The ambiguity in the determination of the high energy parameters can be encoded in the three right handed neutrino masses and an orthogonal complex matrix R defined as [232]

$$R = D_{\sqrt{M}}^{-1} Y_\nu U_{\text{PMNS}} D_{\sqrt{m}}^{-1} \langle H^0 \rangle, \quad (3.45)$$

so that the most general Yukawa coupling compatible with the low energy data is given by:

$$Y^\nu = D_{\sqrt{M}} R D_{\sqrt{m}} U_{\text{PMNS}}^\dagger \langle H^0 \rangle. \quad (3.46)$$

It is straightforward to check that this equation indeed satisfies the seesaw formula, (3.41). In this expression, $D_{\sqrt{m}}$ and $D_{\sqrt{M}}$ are diagonal matrices whose entries are the square roots of the light neutrino and the right handed neutrino masses, respectively, and U_{PMNS} is the leptonic mixing matrix. It is customary to parameterize R in terms of three complex angles, $\hat{\theta}_i$:

$$R = \begin{pmatrix} \hat{c}_2\hat{c}_3 & -\hat{c}_1\hat{s}_3 - \hat{s}_1\hat{s}_2\hat{c}_3 & \hat{s}_1\hat{s}_3 - \hat{c}_1\hat{s}_2\hat{c}_3 \\ \hat{c}_2\hat{s}_3 & \hat{c}_1\hat{c}_3 - \hat{s}_1\hat{s}_2\hat{s}_3 & -\hat{s}_1\hat{c}_3 - \hat{c}_1\hat{s}_2\hat{s}_3 \\ \hat{s}_2 & \hat{s}_1\hat{c}_2 & \hat{c}_1\hat{c}_2 \end{pmatrix}, \quad (3.47)$$

up to reflections, where $\hat{c}_i \equiv \cos\hat{\theta}_i$, $\hat{s}_i \equiv \sin\hat{\theta}_i$.

Whereas the physical interpretation of the right handed masses is very transparent, the meaning of R is more obscure. R can be interpreted as a dominance matrix in the sense that [233]

- R is an orthogonal transformation from the basis of the left handed leptons mass eigenstates to the one of the right handed neutrino mass eigenstates;
- if and only if an eigenvalue m_i of m_ν is dominated—in the sense already given before - by one right handed neutrino eigenstate N_j, then $|R_{ji}| \approx 1$;
- if a light pseudo-Dirac pair is dominated by a heavy pseudo-Dirac pair, then the corresponding 2×2 sector in R is a boost.

An interesting limit of this dominance behavior is the seesaw model with two right handed neutrinos (2RHN) [234, 235]. In this limit, the parameterization (3.46) still holds, with the substitutions $D_{\sqrt{M}} = \text{diag}(M_1^{-1}, M_2^{-1})$ and [236–239]

$$R = \begin{pmatrix} 0 & \cos\hat{\theta} & \xi\sin\hat{\theta} \\ 0 & -\sin\hat{\theta} & \xi\cos\hat{\theta} \end{pmatrix} \quad \text{(normal hierarchy)}, \quad (3.48)$$

$$R = \begin{pmatrix} \cos\hat{\theta} & \xi\sin\hat{\theta} & 0 \\ -\sin\hat{\theta} & \xi\cos\hat{\theta} & 0 \end{pmatrix} \quad \text{(inverted hierarchy)}, \quad (3.49)$$

with $\hat{\theta}$ a complex parameter and $\xi = \pm 1$ a discrete parameter that accounts for a discrete indeterminacy in R.

A third possible parameterization of the neutrino Yukawa coupling uses the Gram–Schmidt decomposition, in order to cast the Yukawa coupling as a product of a unitary matrix and a lower triangular matrix [240]:

$$Y^\nu = U_\triangle Y_\triangle = U_\triangle \begin{pmatrix} y_{11} & 0 & 0 \\ y_{21} & y_{22} & 0 \\ y_{31} & y_{32} & y_{33} \end{pmatrix}, \quad (3.50)$$

where the diagonal elements of $Y_\triangle$ are real. Three of the six phases in $U_\triangle$ can be absorbed into the definition of the charged leptons. Therefore, the nine real parameters and the six phases of the neutrino Yukawa coupling are identified with the three angles and three phases in $U_\triangle$ and the six real parameters and three phases in $Y_\triangle$.

In the SM extended with right handed neutrinos, the charged lepton masses and the effective neutrino mass matrix are the only source of information about the leptonic sector. However, if supersymmetry is discovered, the structure of the low energy slepton mass matrices would provide additional information about the leptonic sector, provided the mechanism of supersymmetry breaking is specified. Assuming that the slepton mass matrices are proportional to the identity at the high energy scale, quantum effects induced by the right handed neutrinos would yield at low energies a left handed slepton mass matrix with a complicated structure, whose measurement would provide additional information about the seesaw parameters [144, 145]. To be more specific, in the minimal supersymmetric seesaw model the off-diagonal elements of the low energy left handed and right handed slepton mass matrices and A-terms read, in the leading log approximation [178]

$$(m_{\tilde{L}}^2)_{ij} \simeq -\frac{1}{8\pi^2}(3m_0^2 + A_0^2) Y_{ik}^{\nu\dagger} Y_{kj}^\nu \log\frac{M_X}{M_k}, \quad (3.51)$$

$$(m_{\tilde{e}_R}^2)_{ij} \simeq 0, \quad (3.52)$$

$$(A_e)_{ij} \simeq -\frac{3}{8\pi^2} A_0 Y_e Y_{ik}^{\nu\dagger} Y_{kj}^\nu \log\frac{M_X}{M_k}, \quad (3.53)$$

where m_0 and A_0 are the universal soft supersymmetry breaking parameters at high scale M_X. Note that the diagonal elements of those mass matrices include the tree level soft mass matrix, the radiative corrections from gauge and charged lepton Yukawa interactions, and the mass contributions from F- and D-terms (which are different for charged sleptons and sneutrinos). Therefore, the measurement at low energies of rare lepton decays, electric dipole moments and slepton mass splittings would provide information about the combination

$$C_{ij} \equiv \sum_k Y_{ik}^{\nu\dagger} Y_{kj}^{\nu} \log \frac{M_X}{M_k} \equiv \left(Y_{\nu}^{\dagger} L Y_{\nu}\right)_{ij}, \quad (3.54)$$

where $L_{ij} = \log \frac{M_X}{M_i} \delta_{ij}$.

Interestingly enough, C encodes precisely the additional information needed to reconstruct the complete seesaw Lagrangian from low energy observations [241, 242] (note in particular that C is a Hermitian matrix that depends on six real parameters and three phases, which together with the nine real parameters and three phases of the neutrino mass matrix sum up to the independent fifteen real parameters and six complex phases in Y_{ν} and M).

To determine Y_{ν} and M from the low energy observables C and m_{ν}, it is convenient to define

$$\tilde{Y}^{\nu} = \text{diag}\left(\sqrt{\log \frac{M_X}{M_1}}, \sqrt{\log \frac{M_X}{M_2}}, \sqrt{\log \frac{M_X}{M_3}}\right) Y^{\nu},$$

$$\tilde{M}_k = M_k \log \frac{M_X}{M_k}, \quad (3.55)$$

so that the effective neutrino mass matrix and C now read

$$m_{\nu} = \tilde{Y}^{\nu t} \text{diag}(\tilde{M}_1^{-1}, \tilde{M}_2^{-1}, \tilde{M}_3^{-1}) \tilde{Y}^{\nu} \langle H_u^0 \rangle^2,$$

$$C = \tilde{Y}^{\nu\dagger} \tilde{Y}^{\nu}, \quad (3.56)$$

where H_u^0 is the neutral component of the up-type Higgs doublet. Using the singular value decomposition $\tilde{Y}^{\nu} = \tilde{V}_R^{\nu} \tilde{D}_Y^{\nu} \tilde{V}_L^{\nu\dagger}$, one finds that $\tilde{V}_L^{\nu\dagger}$ and $\tilde{D}_Y^{\nu}$ could be straightforwardly determined from C, since

$$C \equiv \tilde{Y}^{\nu\dagger} \tilde{Y}^{\nu} = \tilde{V}_L^{\nu\dagger} \tilde{D}_Y^2 \tilde{V}_L^{\nu}. \quad (3.57)$$

On the other hand, from $m_{\nu} = \tilde{Y}^{\nu t} \tilde{D}_M^{-1} \tilde{Y}^{\nu} \langle H_u^0 \rangle^2$ and the singular value decomposition of $\tilde{Y}^{\nu}$,

$$\tilde{D}_Y^{-1} \tilde{V}_L^{\nu*} m_{\nu} \tilde{V}_L^{\nu\dagger} \tilde{D}_Y^{-1} = \tilde{V}_R^{\nu*} \tilde{D}_M^{-1} \tilde{V}_R^{\nu\dagger}, \quad (3.58)$$

where the left hand side of this equation is known (m_{ν} is one of our inputs, and $\tilde{V}_L^{\nu}$ and $\tilde{D}_Y^{\nu}$ were obtained from (3.57)). Therefore, $\tilde{V}_R^{\nu}$ and $\tilde{D}_M$ can also be determined. This simple procedure shows that starting from the low energy observables m_{ν} and C it is possible to determine uniquely the matrices $\tilde{D}_M$ and $\tilde{Y}^{\nu} = \tilde{V}_R^{\nu} \tilde{D}_Y^{\nu} \tilde{V}_L^{\nu\dagger}$. Finally, inverting (3.56), the actual parameters of the Lagrangian M_k and Y^{ν} can be computed.

This procedure is particularly powerful in the case of the two right handed neutrino model, as the number of independent parameters involved (either at high energies or at low energies) is drastically reduced. The matrix C defined in (3.54) depends in general on six moduli and three phases. However, since the Yukawa coupling depends in the 2RHN model on only three unknown moduli and one phase, so does C, and consequently it is possible to obtain predictions on the moduli of three C-matrix elements and the phases of two C-matrix elements. Namely, from (3.46) one obtains

$$U^{\dagger} C U = U^{\dagger} \tilde{Y}^{\nu\dagger} \tilde{Y}^{\nu} U = D_{\sqrt{m}} R^{\dagger} \tilde{D}_M R D_{\sqrt{m}} / \langle H_u^0 \rangle^2, \quad (3.59)$$

where we have written $U \equiv U_{\text{PMNS}}$. Since $m_1 = 0$ in the 2RHN model,[10] it follows that $(U^{\dagger} C U)_{1i} = 0$, for $i = 1, 2, 3$, leading to three relations among the elements in C. For instance, one could derive the diagonal elements in C in terms of the off-diagonal elements:

$$C_{11} = -\frac{C_{12}^* U_{21}^* + C_{13}^* U_{31}^*}{U_{11}^*},$$

$$C_{22} = -\frac{C_{12} U_{11}^* + C_{23}^* U_{31}^*}{U_{21}^*}, \quad (3.60)$$

$$C_{33} = -\frac{C_{13} U_{11}^* + C_{23} U_{21}^*}{U_{31}^*}.$$

The observation of these correlations would be non-trivial tests of the 2RHN model.

The relations for the phases arise from the hermiticity of C, since the diagonal elements in C have to be real. Taking as the independent phase the argument of C_{12}, one can derive from (3.60) the arguments of the remaining elements:

$$e^{i \arg C_{13}} = \Big[-i \,\text{Im}(C_{12} U_{21} U_{11}^*) $$
$$\pm \sqrt{|C_{13}|^2 |U_{11}|^2 |U_{31}|^2 - \left[\text{Im}(C_{12} U_{21} U_{11}^*)\right]^2}\,\Big]$$
$$\times \left[|C_{13}| U_{31} U_{11}^*\right]^{-1},$$
$$\quad (3.61)$$
$$e^{i \arg C_{23}} = \Big[i \,\text{Im}(C_{12} U_{21} U_{11}^*) $$
$$\pm \sqrt{|C_{23}|^2 |U_{21}|^2 |U_{31}|^2 - \left[\text{Im}(C_{12} U_{21} U_{11}^*)\right]^2}\,\Big]$$
$$\times \left[|C_{23}| U_{31} U_{21}^*\right]^{-1},$$

where the ± sign has to be chosen so that the eigenvalues of C are positive. We conclude then that the C-matrix parameters C_{12}, $|C_{13}|$ and $|C_{23}|$ can be regarded as independent and can be used as an alternative parameterization of the

---

[10] Here we are assuming a neutrino spectrum with normal hierarchy. In the case with inverted hierarchy, the analysis is similar, using $m_3 = 0$.

2RHN model [243]. Together with the five moduli and the two phases of the neutrino mass matrix, we sum up to the eight moduli and the three phases necessary to reconstruct the high energy Lagrangian of the 2RHN model.

3.2.3.2 Type II seesaw The type II seesaw mechanism [215, 221–224] consists on adding to the SM particle content a Higgs triplet

$$T = \begin{pmatrix} T^0 & -\frac{1}{\sqrt{2}}T^+ \\ -\frac{1}{\sqrt{2}}T^+ & -T^{++} \end{pmatrix}. \quad (3.62)$$

Then, the leptonic potential compatible with the SM gauge symmetry reads

$$\mathcal{L}_{\text{lep}} = e^c_{Ri} Y^e_{ij} L_j \overline{H} + Y^T_{ij} L_i T L_j + \text{h.c.} \quad (3.63)$$

From this Lagrangian, it is apparent that the triplet T carries lepton number -2. If the neutral component of the triplet acquires a VEV and breaks lepton number spontaneously as happens in the Gelmini–Roncadelli model [224], the associated massless majoron rules out the model. Therefore phenomenology suggests to break lepton number explicitly via the triplet coupling to the SM Higgs boson [244]. The most general scalar potential involving one Higgs doublet and one Higgs triplet reads

$$V = m^2_H H^\dagger H + \frac{1}{2}\lambda_1 (H^\dagger H)^2 + M^2_T T^\dagger T + \frac{1}{2}\lambda_2 (T^\dagger T)^2$$
$$+ \lambda_3 (H^\dagger H)(T^\dagger T) + \mu' H^\dagger T H^\dagger, \quad (3.64)$$

where the term proportional to μ' breaks lepton number explicitly. The type II seesaw mechanism is implemented when $M_T \gg \langle H^0 \rangle$. Then the minimization of the scalar potential yields

$$\langle H^0 \rangle^2 \simeq \frac{-m^2_H}{\lambda_1 - 2\mu'^2/M^2_T}, \quad \langle T^0 \rangle \simeq \frac{-\mu' \langle H^0 \rangle^2}{M^2_T}, \quad (3.65)$$

which produce Majorana masses for the neutrinos given by

$$m_\nu = Y_T \frac{-\mu' \langle H^0 \rangle^2}{M^2_T}. \quad (3.66)$$

The Yukawa matrix Y^T has the same flavor structure as the non-renormalizable operator κ defined in (3.38) for the effective Lagrangian of Majorana neutrinos. Therefore, the parameterization of the type II seesaw model is completely identical to that case.

Supersymmetric models with low scale triplet Higgses have been extensively considered in studies of collider phenomenology [245–247]. The model [244] was first supersymmetrized in Ref. [248] as a possible scenario for leptogenesis. The requirement of a holomorphic superpotential implies introducing the triplets in a vector-like $SU(2)_W \times U(1)_Y$ representation, as $T \sim (3, 1)$ and $\bar{T} \sim (3, -1)$. The relevant superpotential terms are

$$\frac{1}{\sqrt{2}} Y^{ij}_T L_i T L_j + \frac{1}{\sqrt{2}} \lambda_1 H_1 T H_1 + \frac{1}{\sqrt{2}} \lambda_2 H_2 \bar{T} H_2$$
$$+ M_T T \bar{T} + \mu H_2 H_1, \quad (3.67)$$

where L_i are the $SU(2)_W$ lepton doublets and $H_1(H_2)$ is the Higgs doublet with hypercharge $Y = -1/2(1/2)$. Decoupling the triplet at high scale at the electroweak scale the Majorana neutrino mass matrix is given by ($v_2 = \langle H_2 \rangle$)

$$m^{ij}_\nu = Y^{ij}_T \frac{v^2_2 \lambda_2}{M_T}. \quad (3.68)$$

Note that in the supersymmetric case there is only one mass parameter, M_T, while the mass parameter μ' of the non-supersymmetric version is absent.

The couplings Y_T also induce LFV in the slepton mass matrix $m^2_{\tilde{L}}$ through renormalization group (RG) running from M_X to the decoupling scale M_T [249]. In the leading-logarithm approximation those are given by ($i \neq j$):

$$\left(m^2_{\tilde{L}}\right)_{ij} \approx \frac{-1}{8\pi^2}\left(9m^2_0 + 3A^2_0\right)\left(Y^\dagger_T Y_T\right)_{ij} \log \frac{M_X}{M_T},$$
$$\left(m^2_{\tilde{e}_R}\right)_{ij} \approx 0, \quad (3.69)$$
$$(A_e)_{ij} \approx \frac{-9}{16\pi^2} A_0 \left(Y_e Y^\dagger_T Y_T\right)_{ij} \log \frac{M_X}{M_T}.$$

Phenomenological implications of those relations will be presented in Sect. 5.

3.2.3.3 Renormalization of the neutrino mass matrix To make a connection between high scale parameters and low scale observables one needs to consider renormalization effects on neutrino masses and mixing. Below the scale where the dimension five operator is generated, the running of the neutrino mass matrix is governed by the renormalization group (RG) equation of the coupling matrix κ_ν, given by [250–253]

$$(4\pi)^2 \frac{d}{d\ln\mu}\kappa_\nu = (4\pi)^2 A_g \kappa_\nu + C_e\left(\left(Y^\dagger_e Y_e\right)^T \kappa_\nu + \kappa_\nu Y^\dagger_e Y_e\right),$$
$$(3.70)$$

where $C_e = -3/2$ for the SM and $C_e = 1$ for the MSSM. The first term does not affect the running of the neutrino mixing angles and CP violation phases; however, it affects of course the running of the neutrino mass eigenvalues. The flavor universal factor A_g is given by

$$A_g = \begin{cases} -3\alpha_2(4\pi) + \lambda + 2\,\text{tr}(3Y_u^\dagger Y_u + 3Y_d^\dagger Y_d + Y_e^\dagger Y_e) & \text{SM,} \\ -2\alpha_1(4\pi) - 6\alpha_2(4\pi) + \text{tr}(Y_u^\dagger Y_u) & \text{MSSM,} \end{cases} \quad (3.71)$$

where λ denotes the Higgs self-coupling constant and $\alpha_i = g_i^2/(4\pi)$, where g_1 and g_2 are the $U(1)_Y$ and $SU(2)$ gauge coupling constants, respectively.

Due to the smallness of the tau–Yukawa coupling in the SM, the mixing angles are not affected significantly by the renormalization group running below the generation scale of the dimension five operator. However, if the neutrino mass matrix $m_\nu = \frac{\langle\phi\rangle^2}{2}\kappa_\nu$ is realized in the seesaw scenario (type I), running effects above and between the seesaw scales can also lead to relevant running effects in the SM. Note that in the MSSM case the running of the mixing angles and CP violation phases can be large even below the seesaw scales due to the possible enhancement of the tau–Yukawa coupling by the factor $(1 + \tan\beta^2)^{1/2}$.

In order to understand generic properties of the RG evolution and to estimate the typical size of the RG effects, it is useful to consider RGEs for the leptonic mixing angles, CP phases and neutrino masses themselves, which can be derived from the RGE in (3.70). For example, below the seesaw scales, up to $\mathcal{O}(\theta_{13})$ corrections, the evolution of the mixing angles in the MSSM is given by [254] (see also [255, 256])

$$\frac{d\theta_{12}}{d\ln\mu} = \frac{-y_\tau^2}{32\pi^2}\sin 2\theta_{12} s_{23}^2 \frac{|m_1 e^{i\alpha_M} + m_2|^2}{\Delta m_{21}^2}, \quad (3.72)$$

$$\frac{d\theta_{13}}{d\ln\mu} = \frac{y_\tau^2}{32\pi^2}\sin 2\theta_{12}\sin 2\theta_{23}\frac{m_3}{\Delta m_{31}^2(1+\zeta)}$$
$$\times I(m_1, m_2, \alpha_M, \beta_M, \delta), \quad (3.73)$$

$$\frac{d\theta_{23}}{d\ln\mu} = \frac{-y_\tau^2}{32\pi^2}\frac{\sin 2\theta_{23}}{\Delta m_{31}^2}\left[c_{12}^2|m_2 e^{i\beta_M} + m_3 e^{i\alpha_M}|^2 \right.$$
$$\left. + s_{12}^2\frac{|m_1 e^{i\beta_M} + m_3|^2}{1+\zeta}\right], \quad (3.74)$$

where $I(m_1, m_2, \alpha_M, \beta_M, \delta) \equiv m_1\cos(\beta_M - \delta) - (1+\zeta) \times m_2\cos(\alpha_M - \beta_M + \delta) - \zeta m_3\cos\delta$, $s_{ij} = \sin\theta_{ij}$, $c_{ij} = \cos\theta_{ij}$, and $\zeta = \Delta m_{21}^2/\Delta m_{31}^2$. Here y_τ denotes the tau–Yukawa coupling, and one can safely neglect the contributions coming from the electron– and muon–Yukawa couplings. For the matrix P containing the Majorana phases, we use the convention $P = \text{diag}(1, e^{i\alpha_M/2}, e^{i\beta_M/2})$. In addition to the above formulae, formulae for the running of the CP phases have been derived [254]. For example, the running

of the Dirac CP-violating phase δ, observable in neutrino oscillation experiments, is given by

$$\frac{d\delta}{d\ln\mu} = \frac{Cy_\tau^2}{32\pi^2}\frac{\delta^{(-1)}}{\theta_{13}} + \frac{Cy_\tau^2}{8\pi^2}\delta^{(0)} + \mathcal{O}(\theta_{13}). \quad (3.75)$$

The coefficients $\delta^{(-1)}$ and $\delta^{(0)}$ are omitted here and can be found in [254], where also formulae for the running of the Majorana CP phases and for the neutrino mass eigenvalues (mass squared differences) can be found. From (3.75), it can be seen that the Dirac CP phase generically becomes more unstable under RG corrections for smaller θ_{13}.

In the seesaw scenario (type I), the SM or MSSM are extended by heavy right handed neutrinos and their superpartners, which are SM gauge singlets. Integrating them out below their mass scales M_R yields the dimension five operator for neutrino masses in the SM or MSSM. Above M_R, the neutrino Yukawa couplings are active, and the RGEs in the MSSM above the scales M_R are

$$(4\pi)^2\frac{d\kappa_\nu}{d\ln\mu} = \left\{-\frac{6}{5}\alpha_1(4\pi) - 6\alpha_2(4\pi) + 2\,\text{tr}(Y_\nu^\dagger Y_\nu)\right.$$
$$\left. + 6\,\text{tr}(Y_u^\dagger Y_u)\right\}\kappa_\nu + (Y_e^\dagger Y_e)^T\kappa_\nu + \kappa_\nu(Y_e^\dagger Y_e)$$
$$+ (Y_\nu^\dagger Y_\nu)^T\kappa_\nu + \kappa_\nu(Y_\nu^\dagger Y_\nu), \quad (3.76)$$

$$(4\pi)^2\frac{dM_R}{d\ln\mu} = \frac{1}{8\pi^2}\left[(Y_\nu Y_\nu^\dagger)M_R + M_R(Y_\nu Y_\nu^\dagger)^T\right], \quad (3.77)$$

$$(4\pi)^2\frac{dY_\nu}{d\ln\mu} = -Y_\nu\left[\frac{3}{5}\alpha_1(4\pi) + 3\alpha_2(4\pi)\right.$$
$$- \text{tr}(3Y_u^\dagger Y_u + Y_\nu^\dagger Y_\nu)$$
$$\left. - 3Y_\nu^\dagger Y_\nu - Y_e^\dagger Y_e\right]. \quad (3.78)$$

For non-degenerate seesaw scales, a method for dealing with the effective theories, where the heavy singlets are partly integrated out, can be found in [257]. Analytical formulae for the running of the neutrino parameters above the seesaw scales are derived in [258, 259]. The two loop beta functions can be found in Ref. [260].

The running correction to the neutrino mass matrix and its effects on the related issue have been widely analyzed (see e.g. [250–279]). We shall summarize below some of the features of RG running of the neutrino mixing parameters in the MSSM (cf. (3.72)–(3.74)).

– The RG effects are enhanced for relatively large $\tan\beta$, because the tau–Yukawa coupling becomes large.
– The mixing angles are comparatively stable with respect to the RG running in the case of *normal hierarchical* neutrino mass spectrum, $m_1 \ll m_2 \ll m_3$ even when $\tan\beta$ is large [261–267]. Nevertheless, the running effects can

have important implications facing the high precision of future neutrino oscillation experiments.

- For $m_1 \gtrsim 0.05$ eV and the case of $\tan\beta \gtrsim 10$, the RG running effects can be rather large and the leptonic mixing angles can run significantly. Particularly, the RGE effects can be very large for the solar neutrino mixing angle θ_{12} [261–267, 274, 275].
- The solar neutrino mixing angle θ_{12} at M_R depends strongly on the Majorana phase α_M [254, 267, 268, 275], which is the relative phase between m_1 and m_2, and plays very important role in the predictions of the effective Majorana mass in $(\beta\beta)_{0\nu}$-decay. The effect of RG running for θ_{12} is smallest for the CP-conserving odd case $\alpha_M = \pm\pi$, while it is significant for the CP-conserving even case $\alpha_M = 0$. For $\alpha_M = 0$ and $\tan\beta \sim 50$, for instance, we have $\tan^2\theta_{12}(M_R) \lesssim 0.5 \times \tan^2\theta_{12}(M_Z)$ for $m_1 \gtrsim 0.02$ eV.
- The RG running effect on θ_{12} due to the τ–Yukawa coupling always makes $\theta_{12}(M_Z)$ larger than $\theta_{12}(M_R)$ [267]. This constrains the models which predict the value of solar neutrino mixing angle at M_R, $\theta_{12}(M_R) > \theta_{12}(M_Z)$. For example, the bi-maximal models are strongly restricted. However, the running effects due to the neutrino Yukawa couplings are free from this feature [257]. Thus, bi-maximal models can predict the correct value of neutrino mixing angles with the neutrino Yukawa contributions [269–272].
- The RG corrections to neutrino mixing angles depend strongly on the deviation of the seesaw parameter matrix R (3.45) from identity [274]. For hierarchical light neutrinos, $m_1 \lesssim 0.01$ eV, $\tan\beta \lesssim 30$ and R non-trivial, the correction to θ_{23} and θ_{13} can be beyond their likely future experimental errors while θ_{12} is quite stable against the RG corrections [274].
- The correction to θ_{23} can be large when m_1 and/or $\tan\beta$ are/is relatively large, e.g., (i) when $m_1 \gtrsim 0.2$ eV if $\tan\beta \lesssim 10$, and (ii) for any m_1 and α_M if $\tan\beta \gtrsim 40$ [274, 275].
- The RG corrections to $\sin\theta_{13}$ can be relatively small, even for the large $\tan\beta$ if $m_1 \lesssim 0.05$ eV, and for any $m_1 \gtrsim 0.30$ eV, if $\theta_{13}(M_Z) \cong 0$ and $\alpha_M \cong 0$ (with $\beta_M = \delta = 0$). For $\alpha_M = \pi$ and $\tan\beta \sim 50$ one can have $\sin\theta_{13}(M_R) \gtrsim 0.10$ for $m_1 \gtrsim 0.08$ eV even if $\sin\theta_{13}(M_Z) = 0$ [274, 275].
- For $\tan\beta \gtrsim 30$, the value of $\Delta m_{21}^2(M_R)$ depends strongly on m_1 in the interval $m_1 \gtrsim 0.05$ eV, and on α_M, β_M, δ, and s_{13} for $m_1 \gtrsim 0.1$ eV. The dependence of $\Delta m_{31}^2(M_R)$ on m_1 and the CP phases is rather weak, unless $\tan\beta \gtrsim 40$, $m_1 \gtrsim 0.10$ eV, and $s_{13} \gg 0.05$ [275].
- Some products of the neutrino mixing parameters, such as $s_{12}c_{12}c_{23}(m_1/m_2 - e^{i\alpha_M})$ are practically stable with respect to RG running if one neglects the first and second generation charged lepton Yukawa couplings and s_{13} [268, 273, 275].

3.2.4 Quark–lepton complementarity

3.2.4.1 Golden complementarity
Quark–lepton complementarity [280–282] is based on the observation that $\theta_{12} + \theta_C$ is numerically close to $\pi/4$. Here θ_{12} is the solar neutrino mixing angle and θ_C is the Cabibbo angle. For hierarchical light neutrino masses this result is relatively stable against the renormalization effects [274]. To illustrate the idea we first review the model of exact golden complementarity.

Consider the following textures [283] for the light neutrino Majorana mass matrix m_ν and for the charged lepton Yukawa couplings Y_e:

$$m_\nu = \begin{pmatrix} 0 & m & 0 \\ m & m & 0 \\ 0 & 0 & m_{\text{atm}} \end{pmatrix},$$

$$Y_e = \begin{pmatrix} \lambda_e & 0 & 0 \\ 0 & \lambda_\mu/\sqrt{2} & \lambda_\tau/\sqrt{2} \\ 0 & -\lambda_\mu/\sqrt{2} & \lambda_\tau/\sqrt{2} \end{pmatrix}. \tag{3.79}$$

It just assumes some texture zeroes and some strict equalities among different entries. The mass eigenstates of the neutrino mass matrix are given by $m_1 = -m/\varphi$, $m_2 = m\varphi$, $m_3 = m_{\text{atm}}$, where $\varphi = (1+\sqrt{5})/2 = 1 + 1/\varphi \approx 1.62$ is known as the golden ratio [284]. Thanks to its peculiar mathematical properties this constant appears in various natural phenomena, possibly including solar neutrinos. The three neutrino mixing angles obtained from (3.79) are $\theta_{\text{atm}} = \pi/4$, $\theta_{13} = 0$ and, more importantly,

$$\tan^2\theta_{12} = 1/\varphi^2 = 0.382, \quad \text{i.e.} \quad \sin^2 2\theta_{12} = 4/5, \tag{3.80}$$

in terms of the parameter $\sin^2 2\theta_{12}$ directly measured by vacuum oscillation experiments, such as KamLAND. This prediction for θ_{12} is 1.4σ below the experimental best fit value. A positive measurement of θ_{13} might imply that the prediction for θ_{12} suffers an uncertainty up to θ_{13}.

Those properties follow from the $Z_2 \otimes Z'_2$ symmetry of the neutrino mass matrix. Explicitly $Rm_\nu R^T = m_\nu$, where

$$R = \begin{pmatrix} -1/\sqrt{5} & 2/\sqrt{5} & 0 \\ 2/\sqrt{5} & 1/\sqrt{5} & 0 \\ 0 & 0 & 1 \end{pmatrix}, \quad R' = \begin{pmatrix} 1 & 0 & 0 \\ 0 & 1 & 0 \\ 0 & 0 & -1 \end{pmatrix}, \tag{3.81}$$

and the rotations satisfy $\det R = -1$, $R \cdot R^T = 1$ and $R \cdot R = 1$. The first Z_2 is a reflection along the diagonal of the golden rectangle in the (1, 2) plane, see Fig. 2. The second Z'_2 is the $L_3 \to -L_3$ symmetry. Those symmetries allow contributions proportional to the identity matrix to be added to m_ν. This property allows one to extend this type symmetries to the quark sector.

A seesaw model with singlet neutrinos satisfying the $Z_2 \otimes Z_2'$ symmetry and giving rise to the mass matrix (3.79) is presented in [283].

Noticing that the golden prediction (3.80) satisfies with high accuracy the quark–lepton complementarity motivates one to give a golden geometric explanation also to the Cabibbo angle. $SU(5)$ unification relates the down-quark Yukawa matrix Y_d to Y_e and suggests that the up-quark Yukawa matrix Y_u is symmetric, like m_ν. One can therefore assume that Y_d is diagonal in the two first generations and that Y_u is invariant under a Z_2 reflection described by a matrix analogous to R in (3.81), but with the factors $1 \leftrightarrow 2$ exchanged. Figure 2 illustrates the geometrical meaning of two reflection axis (dashed lines): the up-quark reflection is along the diagonal of the golden rectangle tilted by $\pi/4$; note also the connection with the decomposition of the golden rectangle as an infinite sum of squares ('golden spiral'). Similarly to the neutrino case, this symmetry allows for two independent terms that can be tuned such that $m_u \ll m_c$:

$$Y_u = \lambda \begin{pmatrix} 1 & 0 & 0 \\ 0 & 1 & 0 \\ 0 & 0 & 1 \end{pmatrix} + \frac{\lambda}{\sqrt{5}} \begin{pmatrix} -2 & 1 & 0 \\ 1 & 2 & 0 \\ 0 & 0 & c \end{pmatrix}. \quad (3.82)$$

The second term fixes $\cot \theta_C = \varphi^3$, as can be geometrically seen from Fig. 2. We therefore have

$$\sin^2 2\theta_C = 1/5 \quad \text{i.e.} \quad \theta_{12} + \theta_C = \pi/4 \quad \text{i.e.}$$
$$V_{us} = \sin \theta_C = \left(1 + \varphi^6\right)^{-1/2} = 0.229. \quad (3.83)$$

This prediction is 1.9σ above the present best-fit value, $\sin \theta_C = 0.2258 \pm 0.0021$. However, as the basic elements of flavor presented here follow by construction from the 2×2 submatrices, one naturally expects that the golden prediction for V_{us} has an uncertainty at least comparable to $|V_{ub}| \sim |V_{td}| \sim$ few $\times 10^{-3}$. Thus the numerical accuracy is amazing. Should the 1.4σ discrepancy between the golden prediction (3.80) and the experimental measurement hold after final SNO and KamLAND results, analogy with the quark sector would allow one to predict the order of magnitude of neutrino mixing angle θ_{13}.

Interestingly, similar predictions on the mixing angles are obtained if some suitably chosen assumptions are made on the properties of neutral currents of quarks and leptons [285].

3.2.4.2 Correlation matrix from S_3 flavor symmetry in GUT
On more general phenomenological ground the quark–lepton complementarity [281, 282] can be described by the correlation matrix V^M between the CKM and the PMNS mixing matrices,

$$V^M = U^{\text{CKM}} \Omega U^{\text{PMNS}}, \quad (3.84)$$

where $\Omega = \text{diag}(e^{i\omega_i})$ is a diagonal matrix. In the singlet seesaw mechanism the correlation matrix V^M diagonalizes the symmetric matrix

$$\mathcal{C} = m_D^{\text{diag}} V_R^{\nu\dagger} \frac{1}{M} V_R^{\nu\star} m_D^{\text{diag}}, \quad (3.85)$$

where M is the heavy neutrino Majorana mass matrix and V_R^ν diagonalizes the neutrino Dirac matrix m_D from the right. In GUT models such as $SO(10)$ or E_6 we have intriguing relations between the Yukawa coupling of the quark sector and the one of the lepton sector. For instance, in minimal renormalizable $SO(10)$ with Higgs in the **10**, **126**, and **120**, we have $Y_e \approx Y_d^T$. In fact the flavor symmetry implies the structure of the Yukawa matrices: the equivalent entries of Y_e and Y_d are usually of the same order of magnitude. In such a case one gets

$$U^{\text{PMNS}} = \left(U^{\text{CKM}}\right)^\dagger V^M.$$

As a consequence, a S_3 flavor permutation symmetry, softly broken into S_2, gives us the prediction of $V_{13}^M = 0$ [286] and the correlations between CP-violating phases and the mixing angle θ_{12} [287].

The six generators of the S_3 flavor symmetry are the elements of the permutation group of three objects. The action of S_3 on the fields is to permute the family label of the fields. In the following we shall introduce the S_2 symmetry with respect the second and third generations. The S_2 group is an Abelian one and swap the second family $\{\mu_L, (\nu_\mu)_L, s_L, c_L, \mu_R, (\nu_\mu)_R, s_R, c_R\}$ with the third one $\{\tau_L, (\nu_\tau)_L, b_L, t_L, \tau_R, (\nu_\tau)_R, b_R, t_R\}$.

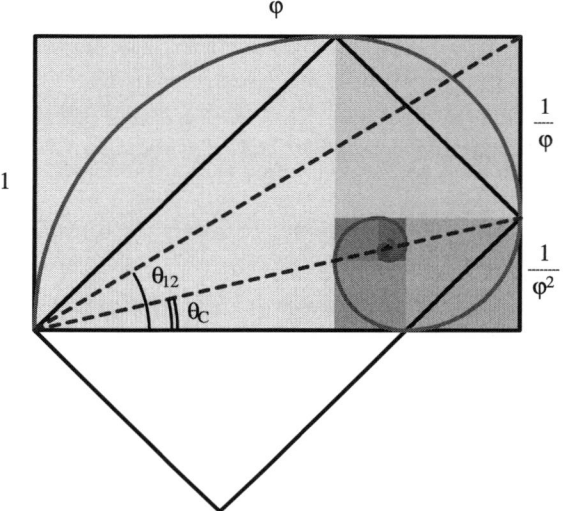

Fig. 2 Geometrical illustration of the connection between the predictions for θ_{12} and θ_C and the golden rectangle. The two *dashed lines* are the reflection axis of the Z_2 symmetry for the neutrino mass matrix and for the up quark mass matrix

Let us assume that there is an S_3 flavor symmetry at high energy, which is softly broken into S_2 [84]. In this case, before the S_3 breaking all the Yukawa matrices have the following structure:

$$Y = \begin{pmatrix} a & b & b \\ b & a & b \\ b & b & a \end{pmatrix}, \qquad (3.86)$$

where a and b independent. The S_3 symmetry implies that $(1/\sqrt{3}, 1/\sqrt{3}, 1/\sqrt{3})$ is an eigenvector of our matrix in (3.86). Moreover these kind of matrices have two equal eigenvalues. This gives us an undetermined mixing angle in the diagonalizing mixing matrices.

When S_3 is softly broken into S_2, one gets

$$Y = \begin{pmatrix} a & b & b \\ b & c & d \\ b & d & c \end{pmatrix}, \qquad (3.87)$$

with $c \approx a$ and $d \approx b$. When S_3 is broken the degeneracy is removed. In general the S_2 symmetry implies that $(0, 1/\sqrt{2}, -1/\sqrt{2})$ is an exact eigenvector of our matrix (3.87). The fact that S_3 is only softly broken into S_2 allows us to say that $(1/\sqrt{3}, 1/\sqrt{3}, 1/\sqrt{3})$ is still in a good approximation an eigenvector of Y in (3.87). Then the mixing matrix that diagonalize from the right the Yukawa mixing matrix in (3.87) is given in good approximation by the tri-bi-maximal mixing matrix (2.11).

Let us now investigate the V^M in this model. The mass matrix m_D will have the general structure in (3.87). To be more defined, let us assumed that there is an extra softly broken Z_2 symmetry under which the 1st and the 2nd families are even, while the 3rd family is odd. This extra softly broken Z_2 symmetry gives us a hierarchy between the off-diagonal and the diagonal elements of m_D, i.e. $b, d \ll a, c$. In fact if Z_2 is exact both b and d are zero. For simplicity, we assume also a quasi-degenerate spectrum for the eigenvalues of the Dirac neutrino matrix as in [288].

The right handed neutrino Majorana mass matrix is of the form

$$M = \begin{pmatrix} a & b & b' \\ b & c & d \\ b' & d & e \end{pmatrix}. \qquad (3.88)$$

Because S_3 is only softly broken into S_2 we have that $a \approx c \approx e$, and $b \approx b' \approx d$. In this approximation the M matrix is diagonalized by a U of the form in (2.11). In this case we have that m_ν is near to be S_3 and S_2 symmetric, then it is diagonalized by a mixing matrix U_ν near the tri-bi-maximal one given in (2.11). The $\mathcal{C}$ matrix is diagonalized by the mixing matrix $V_M = U_\nu U$. We obtain that V_M is a rotation in the $(1, 2)$ plane, i.e. it contains a zero in the $(1, 3)$ entry. As shown in [288], it is possible to fit the CKM and the PMNS mixing matrix within this model.

3.3 Leptogenesis and cosmological observables

3.3.1 Basic concepts and results

CP violation in the leptonic sector can have profound cosmological implications, playing a crucial role in the generation, via leptogenesis, of the observed baryon number asymmetry of the universe [289]:

$$\frac{n_B}{n_\gamma} = \left(6.1^{+0.3}_{-0.2}\right) \times 10^{-10}. \qquad (3.89)$$

In the original framework a CP asymmetry is generated through out-of-equilibrium L-violating decays of heavy Majorana neutrinos [290] leading to a lepton asymmetry $L \neq 0$. In the presence of sphaleron processes [291], which are $(B+L)$-violating and $(B-L)$-conserving, the lepton asymmetry is partially transformed to a baryon asymmetry.

The lepton number asymmetry resulting from the decay of heavy Majorana neutrinos, ε_{N_j}, was computed by several authors [292–294]. The evaluation of ε_{N_j}, involves the computation of the interference between the tree level diagram and one loop diagrams for the decay of the heavy Majorana neutrino N_j into charged leptons $l_\alpha^\pm$ ($\alpha = e, \mu, \tau$). Summing the asymmetries $\varepsilon_{N_j}^\alpha$ over charged lepton flavor, one obtains

$$\varepsilon_{N_j} = \frac{g^2}{M_W^2} \sum_{\alpha, k \neq j} \left[\mathrm{Im}\left((m_D^\dagger)_{j\alpha}(m_D)_{\alpha k}(m_D^\dagger m_D)_{jk}\right) \right. $$
$$\left. \times \frac{1}{16\pi} \left(I(x_k) + \frac{\sqrt{x_k}}{1-x_k}\right) \right] \frac{1}{(m_D^\dagger m_D)_{jj}}, \qquad (3.90)$$

where M_k denote the heavy neutrino masses, the variable x_k is defined as $x_k = \frac{M_k^2}{M_j^2}$ and $I(x_k) = \sqrt{x_k}(1 + (1 + x_k)\log(\frac{x_k}{1+x_k}))$. From (3.90) it can be seen that, when one sums over all charged leptons, the lepton number asymmetry is only sensitive to the CP-violating phases appearing in $m_D^\dagger m_D$ in the basis where M_R is diagonal. Note that this combination is insensitive to rotations of the left-hand neutrinos.

If the lepton flavors are distinguishable in the final state, it is the flavored asymmetries that are relevant [295–298]. Below $T \sim 10^{12}$ GeV, the τ Yukawa interactions are fast compared to the Hubble rate, so at least one flavor may be distinguishable. The asymmetry in family α, generated from the decay of the kth heavy Majorana neutrino depends on the combination [299] $\mathrm{Im}((m_D^\dagger m_D)_{kk'}(m_D^*)_{\alpha k}(m_D)_{\alpha k'})$ as well as on $\mathrm{Im}((m_D^\dagger m_D)_{k'k}(m_D^*)_{\alpha k}(m_D)_{\alpha k'})$. Summing over all leptonic flavors α the second term becomes real so that its imaginary part vanishes and the first term gives rise to the combination $\mathrm{Im}((m_D^\dagger m_D)_{jk}(m_D^\dagger m_D)_{jk})$ that appears in (3.90). Clearly, when one works with separate flavors the matrix U_{PMNS} does not cancel out and one is lead to the interesting possibility of having viable leptogenesis even in the case of R being a real matrix [300–303].

The simplest leptogenesis scenario corresponds to the case of heavy hierarchical neutrinos where M_1 is much smaller than M_2 and M_3. In this limit, the asymmetries generated by N_2 and N_3 are frequently ignored, because the production of N_2 and N_3 can be suppressed by kinematics (for instance, they are not produced thermally, if the re-heat temperature after inflation is $<M_2, M_3$), and the asymmetries from their decays are partially washed out [295, 304, 305]. In this hierarchical limit, the $\varepsilon_{N_1}^\alpha$ can be simplified into

$$\varepsilon_{N_1}^\alpha \simeq -\frac{3}{16\pi v^2}\left(I_{12}^\alpha \frac{M_1}{M_2} + I_{13}^\alpha \frac{M_1}{M_3}\right), \quad (3.91)$$

where

$$I_{1i}^\alpha \equiv \frac{\mathrm{Im}[(m_D^\dagger)_{1\alpha}(m_D)_{\alpha i}(m_D^\dagger m_D)_{1i}]}{(m_D^\dagger m_D)_{11}}. \quad (3.92)$$

The flavor-summed CP asymmetry ε_{N_1} can be written in terms of the parameterization equation (3.46) as

$$\varepsilon_{N_1} \approx -\frac{3}{8\pi}\frac{M_1}{v^2}\frac{\sum_i m_i^2 \mathrm{Im}(R_{1i}^2)}{\sum_i m_i |R_{1i}|^2}. \quad (3.93)$$

In this case, obviously, leptogenesis demands non-zero imaginary parts in the R matrix. It has an upper bound $|\varepsilon_{N_1}| < \varepsilon_{N_1}^{DI}$ where [306]

$$\varepsilon_{N_1}^{DI} = \frac{3}{8\pi}\frac{(m_3 - m_1)M_1}{v^2}, \quad (3.94)$$

which is proportional to M_1. So the requirement of generating a sufficient baryon asymmetry gives a lower bound on M_1 [306, 307]. Depending on the cosmological scenario, the range for minimal M_1 varies from order 10^7 GeV to 10^9 GeV [308, 309]. This bound does not move much with the inclusion of flavor effects [296, 310, 311]. In a supersymmetric world there is an upper bound $T_{\mathrm{RH}} < 10^8$ GeV on the re-heating temperature of the universe from the possible overproduction of gravitinos, the so called gravitino problem [312–315]. Together with the lower bound on M_1 the gravitino problem puts severe constraints on supersymmetric thermal leptogenesis scenarios.

However, the upper bound (3.94) is based on the (natural) assumption that higher order corrections suppressed by M_1/M_2, M_1/M_3 in (3.90) are negligible. This may not be true as explicitly demonstrated in [316] in which neutrino mass model is presented realizing $\varepsilon_{N_1} \gg \varepsilon_{N_1}^{DI}$. In such a case low scale standard thermal leptogenesis consistent with the gravitino bound is possible also for hierarchical heavy neutrinos.

Thermal leptogenesis is a rather involved thermodynamical non-equilibrium process and depends on additional parameters and on the proper treatment of thermal effects [309].

In the simplest case, the N_i are hierarchical, and N_1 decays into a combination of flavors which are indistinguishable.[11] In this case, the baryon asymmetry only depends on four parameters [306, 308, 318, 319]: the mass M_1 of the lightest heavy neutrino, together with the corresponding CP asymmetry ε_{N_1} in its decay, as well as the rescaled N_1 decay rate, or effective neutrino mass $\tilde{m}_1$ defined as

$$\tilde{m}_1 = \sum_\alpha (m_D^\dagger)_{1\alpha}(m_D)_{\alpha 1}/M_1, \quad (3.95)$$

in the weak basis where M_R is diagonal, real and positive. Finally, the baryon asymmetry depends also on the sum of all light neutrino masses squared, $\bar{m}^2 = m_1^2 + m_2^2 + m_3^2$, since it has been shown that this sum controls an important class of washout processes. If lepton flavors are distinguishable, the final baryon asymmetry depends on partial decay rates $\tilde{m}_1^\alpha$ and CP asymmetries ϵ_1^α.

The N_1 decays in the early universe at temperatures $T \sim M_1$, producing asymmetries in the distinguishable final states. A particular asymmetry will survive once washout by inverse decays go out of equilibrium. In the unflavored calculation (where lepton flavors are indistinguishable), the fraction of the asymmetry that survives is of order $\min\{1, H/\Gamma\}$, where the Hubble rate H and the N_1 total decay rate Γ are evaluated at $T = M_1$. This is usually written $H/\Gamma = m^*/\tilde{m}_1$, where [320–322]

$$m_* = \frac{16\pi^{5/2}}{3\sqrt{5}}g_*^{1/2}\frac{v^2}{M_{\mathrm{Planck}}} \simeq 10^{-3}\text{ eV}, \quad (3.96)$$

and M_{Planck} is the Planck mass ($M_{\mathrm{Planck}} = 1.2 \times 10^{19}$ GeV), $v = \langle \phi^0 \rangle/\sqrt{2} \simeq 174$ GeV is the weak scale and g_* is the effective number of relativistic degrees of freedom in the plasma and equals 106.75 in the SM case. In a flavored calculation, the fraction of a flavor asymmetry that survives can be estimated in the same way, replacing Γ by the partial decay rate.

3.3.2 Implications of flavor effects

For a long time the flavor effects in thermal leptogenesis were known [295] but their phenomenological implications were considered only in specific neutrino flavor models [235]. As discussed, in the single-flavor calculation, the most important parameters for thermal leptogenesis from N_1 decays are M_1, $\tilde{m}_1$, ϵ_{N_1} and the light neutrino mass scale. Including flavor effects gives this parameter space more dimensions (M_1, ϵ^α, $\tilde{m}_1^\alpha$), but it can still be projected onto M_1, $\tilde{m}$ space. For the readers convenience we summarize here

---

[11] This can occur above $\sim 10^{12}$ GeV, before the τ Yukawa interaction becomes fast compared to the Hubble rate, or in the case where the N_1 decay rate is faster than the charged lepton Yukawa interactions [317].

some general results on the implications of flavored leptogenesis.

In the unflavored calculation, leptogenesis does not work for degenerate light neutrinos with a mass scale above ~ 0.1 eV [323–326]. This bound does not survive in the flavored calculation, where models with a neutrino mass scale up to the cosmological bound, $\sum m_\nu < 0.68$ eV [327], can be tuned to work [296, 317].

Considering the scale of leptogenesis, flavored leptogenesis works for M_1 a factor of ~ 3 smaller in the "interesting" region of $\tilde{m} < m_{atm}$. But the lower bound on M_1, in the optimized $\tilde{m}$ region, remains $\sim 10^9$ GeV [310, 311]. A smaller M_1 could be possible for very degenerate light neutrinos [296].

An important, but disappointing, observation in single-flavor leptogenesis was the lack of a model-independent connection between CP violation for leptogenesis and PMNS phases. It was shown [328, 329] that thermal leptogenesis can work with no CP violation in U_{PMNS}, and conversely, that leptogenesis can fail in spite of phases in U_{PMNS}. In the "flavoured" leptogenesis case, it is still true that the baryon asymmetry is not sensitive to PMNS phases [330, 331] (leptogenesis can work for any value of the PMNS phases). However, interesting observations can be made in classes of models [297, 300, 302, 331].

3.3.3 Other scenarios

We have presented a brief discussion of minimal thermal leptogenesis in the context of type I seesaw with hierarchical heavy neutrinos. This scenario is the most popular one because it is generic, supported by neutrino mass mechanism and, most importantly, it has predictions for the allowed seesaw parameter space, as described above. There are many other scenarios in which leptogenesis may also be viable.

Resonant leptogenesis [293, 332] may occur when two or more heavy neutrinos are nearly degenerate in mass and in this scenario the scale of the heavy neutrino masses can be lowered whilst still being compatible with thermal leptogenesis [332–335]. Heavy neutrinos of TeV scale or below could in principle be detected at large colliders [336]. In the seesaw context low scale heavy neutrinos may follow from extra symmetry principles [334, 337–339]. Also, the SM extensions with heavy neutrinos at TeV scale or below include Kaluza–Klein modes in models with extra dimensions or extra matter content of little Higgs models.

Leptogenesis from the out-of-equilibrium decays of a Higgs triplet [244, 340, 341] is another viable scenario but requires the presence of at least two triplets for non-zero CP asymmetry. Despite the presence of gauge interactions the washout effects in this scenario are not drastically larger than those in the singlet leptogenesis scenario [341]. Hybrid leptogenesis from type I and type II seesaw can for instance occur in $SO(10)$ models [340, 342, 343]. In that case there are twelve independent CP-violating phases.

"Soft leptogenesis" [344, 345] can work in a one generational SUSY seesaw model because CP violation in this scenario comes from complex supersymmetry breaking terms. If the soft SUSY-breaking terms are of suitable size, there is enough CP violation in $\tilde{N}$–$\tilde{N}^*$ mixing to imply the observed asymmetry. Unlike non-supersymmetric triplet Higgs leptogenesis, soft leptogenesis with a triplet scalar [341, 346] can also work in the minimal supersymmetric model of type II seesaw mechanism.

A very predictive supersymmetric leptogenesis scenario is obtained if the sneutrino is playing the role of inflaton [307, 347–350]. In this scenario the universe is dominated by $\tilde{N}$. Relating $\tilde{N}$ properties to neutrino masses via the seesaw mechanism implies a lower bound $T_{RH} > 10^6$ GeV on the re-heating temperature of the universe [349]. A connection of this scenario with LFV is discussed in Sect. 5.2.

Dirac leptogenesis is another possibility considered in the literature. In this case neutrinos are of Dirac type rather than Majorana. In the original paper [351] two Higgs doublets were required and their decays create the leptonic asymmetry. Recently some authors have studied the connection between leptogenesis and low energy data with two Higgs doublets [352].

Finally, let us mention that right handed neutrinos could have been produced non-thermally in the early universe, by direct couplings to the inflation field. If this is the case, the constraints on neutrino parameters from leptogenesis depend on the details of the inflationary model [353–355].

For a recent overview of the present knowledge of neutrino masses and mixing and what can be learned about physics beyond the standard model from the various proposed neutrino experiments, see [4] and references therein.

4 Organizing principles for flavor physics

4.1 Grand unified theories

Grand unification is an attempt to unify all known interactions but gravity in a single simple gauge group. It is motivated in part by the arbitrariness of electromagnetic charge in the standard model. One has charge quantization in a purely non-Abelian theory, without an $U(1)$ factor, as in Schwinger's original idea [356] of a $SU(2)$ theory of electroweak interactions. The minimal gauge group which unifies weak and strong interactions, $SU(5)$ [357], automatically implies a quantized $U(1)$ piece too. While Dirac needed a monopole to achieve charge quantization [358], grand unification in turn predicts the existence of magnetic monopoles [359, 360]. Since it unifies quarks and leptons [361], it also predicts another remarkable phenomenon: the decay of the proton. Here we are mostly interested in GUT implications on the flavor structure of Yukawa matrices.

4.1.1 $SU(5)$: the minimal theory

The 24 gauge bosons reduce to the 12 ones of the SM plus a $SU(2)$ doublet, color triplet pair (X_μ, Y_μ) (vector leptoquarks), with $Y = 5/6$ (charges $+4/3, +1/3$) and their antiparticles. The 15 fermions of a single family in the SM fit in the $\bar{5}_F$ and 10_F anomaly-free representations of $SU(5)$, and the new super-weak interactions of leptoquarks with fermions are (α, β and γ are color indices):

$$\mathcal{L}(X,Y) = \frac{g_5}{\sqrt{2}} X_\mu^{(-4/3)\alpha}$$
$$\times \left(\bar{e}\gamma^\mu d_\alpha^c + \bar{d}_\alpha \gamma^\mu e^c - \epsilon_{\alpha\beta\gamma}\overline{u^c}^\beta \gamma^\mu u^\gamma \right)$$
$$- \frac{g_5}{\sqrt{2}} Y_\mu^{(-1/3)\alpha}$$
$$\times \left(\bar{\nu}\gamma^\mu d_\alpha^c + \bar{u}_\alpha \gamma^\mu e^c + \epsilon_{\alpha\beta\gamma}\overline{u^c}^\beta \gamma^\mu d^\gamma \right)$$
$$+ \text{h.c.}, \tag{4.1}$$

where all fermions above are explicitly left handed and $\psi^c \equiv C\bar{\psi}^T$.

The exchange of the heavy gauge bosons leads to the effective interactions suppressed by two powers of their mass m_X ($m_X \simeq m_Y$ due to $SU(2)_L$ symmetry), which preserves $B - L$, but breaks both B and L symmetries and leads to ($d = 6$) proton decay [213, 362]. From $\tau_P \gtrsim 6 \times 10^{33}$ yr [363], $m_X \gtrsim 10^{15.5}$ GeV.

The Higgs sector consists of an adjoint 24_H and a fundamental 5_H, the first breaks $SU(5) \to$ SM, the latter completes the symmetry breaking á la Weinberg–Salam. Now, $5_H = (T, D)$, where T is a color triplet and D the usual Higgs $SU(2)_L$ doublet of the SM and so the Yukawa interactions in the matrix form

$$\mathcal{L}_Y = 10_F y_u 10_F 5_H + \bar{5}_F y_d 10_F 5_H^* \tag{4.2}$$

give the quark and lepton mass matrices

$$m_u = y_u \langle D \rangle, \qquad m_d = m_e^T = y_d \langle D \rangle. \tag{4.3}$$

Note the correlation between down quarks and charged leptons [364], valid at the GUT scale, and impossible to be true for all three generations. Actually, in the SM it is wrong for all of them. It can be corrected by an extra Higgs, 45_H [6], or higher dimensional non-renormalizable interaction [7].

From (4.2), one gets also the interactions of the triplet, which lead to proton decay and thus the triplet T must be superheavy, $m_T \gtrsim 10^{12}$ GeV. The enormous split between m_T and $m_D \simeq m_W$ can be achieved through the large scale of the breaking of $SU(5)$,

$$\langle 24_H \rangle = v_X \, \text{diag}(2, 2, 2, -3, -3), \tag{4.4}$$

with $m_X^2 = m_Y^2 = \frac{25}{4} g_5^2 v_X^2$. This fine-tuning is known as the doublet–triplet problem. Whatever solution one may adopt, the huge hierarchy can be preserved in perturbation theory only by supersymmetry with low scale breaking of order TeV.

The consistency of grand unification requires that the gauge couplings of the SM unify at a single scale, in a tiny window $10^{15.5} \lesssim M_{\text{GUT}} \lesssim 10^{18}$ GeV (lower limit from proton decay, upper limit from perturbativity, i.e. to stay below M_{Pl}). Here the minimal ordinary $SU(5)$ theory described above fails badly, while the version with low energy supersymmetry does great [365–368]. Actually, one needed a heavy top quark [368], with $m_t \simeq 200$ GeV in order for the theory to work. The same is needed in order to achieve a radiative symmetry breaking of the SM gauge symmetry, where only the Higgs doublet becomes tachyonic [369, 370]. One can then define the minimal supersymmetric $SU(5)$ GUT with the three families of fermions 10_F and $\bar{5}_F$, and with 24_H and 5_H and $\bar{5}_H$ supermultiplets. It predicts $m_d = m_e^T$ at M_{GUT}, which works well for the third generation; the first two can be corrected by higher dimensional operators. Although this theory typically has a very fast $d = 5$ [150, 371–374] proton decay [375], the higher dimensional operators can easily make it in accord with experiments [376–378]. The main problem are massless neutrinos, unless one breaks R-parity (whose approximate or exact conservation must be assumed in supersymmetric $SU(5)$, contrary to some supersymmetric $SO(10)$). Other ways out include adding singlets, right handed neutrinos (type I seesaw [216–220]), or a 15_H multiplet (type II see-saw [215, 221–223]). In both cases their Yukawa are not connected to the charged sector, so it is much more appealing to go to $SO(10)$ theory, which unifies all fermions (of a single family) too, besides the interactions.

Before we move to $SO(10)$, what about ordinary non-supersymmetric $SU(5)$? In order to have $m_\nu \neq 0$ and to achieve the unification of gauge couplings one can add either (a) 15_H Higgs multiplet [379] or (b) 24_F fermionic multiplet [380]. The latter one is particularly interesting, since it leads to the mixing of the type I and type III see-saw [225, 226], with the remarkable prediction of a light $SU(2)$ fermionic triplet below TeV and $M_{\text{GUT}} \leq 10^{16}$ GeV, which offers hope both for the observable see-saw at LHC and detectable proton decay in a future generation of experiments now planned [381].

These fermionic triplets T_F would be produced in pairs through a Drell–Yan process. The production cross section for the sum of all three possible final states, $T_F^+ T_F^-$, $T_F^+ T_F^0$

and $T_F^- T_F^0$, can be read from Fig. 42 of [382]: it is approximately 20 pb for 100 GeV triplet mass, and around 40 fb for 500 GeV triplets. The triplets then decay into W or Z and a light lepton through the same Yukawa couplings that enter into the seesaw.

The clearest signature would be the three charged lepton decay of the charged triplet, but it has only a 3% branching ratio. A more promising situation is the decay into two jets with SM gauge boson invariant mass plus a charged lepton: this happens in approximately 23% of all decays. The signatures in this case is two same charge leptons plus two pairs of jets having the W or Z mass and peaks in the lepton-dijet mass. From the above estimates the cross section for such events is around 1 pb (2fb) for 100 (500) GeV triplet mass. Such signatures were suggested originally in L–R symmetric theories [383] but are quite generic of the seesaw mechanism.

4.1.2 $SO(10)$: the minimal theory of matter and gauge coupling unification

There are a number of features that make $SO(10)$ special:

– a family of fermions is unified in a 16 dimensional spinorial representation; this in turn predicts the existence of right handed neutrinos, making the implementation of the see-saw mechanism almost automatic;
– L–R symmetry [361, 384–386] is a finite gauge transformation in the form of charge conjugation. This is a consequence of both left handed fermions f_L and its charged conjugated counterparts $(f^c)_L \equiv C \overline{f}_R^T$ residing in the same representation 16_F;
– in the supersymmetric version, the matter parity $M = (-1)^{3(B-L)}$, equivalent to the R-parity $R = M(-1)^{2S}$, is a gauge transformation [387–389], a part of the centre Z_4 of $SO(10)$. In the renormalizable version of the theory it remains exact at all energies [390–392]. The lightest supersymmetric partner (LSP) is then stable and is a natural candidate for the dark matter of the universe;
– its other maximal subgroup, besides $SU(5) \times U(1)$, is $G_{PS} = SU(2)_L \times SU(2)_R \times SU(4)_C$ quark–lepton symmetry of Pati and Salam, which plays an important role in relating quark and lepton masses and mixings;
– the unification of gauge couplings can be achieved even without supersymmetry (for a recent and complete work and references therein, see [393, 394]).

Fermions belong to the spinor representation 16_F (for useful reviews on spinors and $SO(2N)$ group theory in general see [395–399]). From

$$16 \times 16 = 10 + 120 + 126, \quad (4.5)$$

the most general Yukawa sector in general contains 10_H, 120_H and $\overline{126}_H$, respectively the fundamental vector representation, the three-index antisymmetric representation and the five-index antisymmetric and anti-self-dual representation. $\overline{126}_H$ is necessarily complex, supersymmetric or not; 10_H and $\overline{126}_H$ Yukawa matrices are symmetric in generation space, while the 120_H one is antisymmetric.

The decomposition of the relevant representations under G_{PS} gives

$$\mathbf{16} = (2, 1, 4) + (1, 2, \bar{4}),$$
$$\mathbf{10} = (2, 2, 1) + (1, 1, 6),$$
$$\mathbf{120} = (2, 2, 1) + (3, 1, 6) + (1, 3, 6) + (2, 2, 15) \quad (4.6)$$
$$+ (1, 1, 10) + (1, 1, \overline{10}),$$
$$\overline{\mathbf{126}} = (3, 1, \overline{10}) + (1, 3, 10) + (2, 2, 15) + (1, 1, 6).$$

The see-saw mechanism, whether type I or II, requires $\overline{126}$: it contains both $(1, 3, 10)$ whose VEV gives a mass to ν_R (type I), and $(3, 1, \overline{10})$, which contains a color singlet, $B - L = 2$ field Δ_L, that can give directly a small mass to ν_L (type II). In $SU(5)$ language this is seen from the decomposition

$$\overline{\mathbf{126}} = 1 + 5 + 15 + \overline{45} + 50. \quad (4.7)$$

The 1 of $SU(5)$ belongs to the $(1, 3, 10)$ of G_{PS} and gives a mass for ν_R, while 15 corresponds to the $(3, 1, \overline{10})$ and gives the direct mass to ν_L.

$\overline{126}$ can be a fundamental field, or a composite of two $\overline{16}_H$ fields (for some realistic examples see for example [400–402]), or can even be induced as a two-loop effective representation built out of a 10_H and two gauge 45 dimensional representations [403–405].

Normally the light Higgs is chosen to be the smallest one, 10_H. Since $\langle 10_H \rangle = \langle (2, 2, 1) \rangle$ is a $SU(4)_C$ singlet, $m_d = m_e$ follows immediately, independently of the number of 10_H. Thus we must add either 120_H or $\overline{126}_H$ or both in order to correct the bad mass relations. Both of these fields contain $(2, 2, 15)$, which VEV alone gives the relation $m_e = -3m_d^T$.

As $\overline{126}_H$ is needed anyway for the see-saw, it is natural to take this first. The crucial point here is that in general $(2, 2, 1)$ and $(2, 2, 15)$ mix through $\langle (1, 3, 10) \rangle$ [222, 406] and thus the light Higgs is a mixture of the two. In other words, $\langle (2, 2, 15) \rangle$ in $\overline{126}_H$ is in general non-vanishing (in supersymmetry this is not automatic, but depends on the Higgs superfields needed to break $SO(10)$ at M_{GUT} or on the presence of higher dimensional operators).

If one considers all the operators allowed by $SO(10)$ for the Yukawa couplings, there are too many model parameters, and so no prediction is really possible. One option is to assume that the minimal number of parameters must be employed. It has been shown that 4 (3 of them nonrenormalizable) operators are enough in models with 10 and 45 Higgs representations only [8]. Although this is an important piece of information and it has been the starting point of a lot of model building, it is difficult to see a reason for some operators (of different dimensions) to be present and other not, without using some sort of flavor symmetry, so these type of models will not be considered in this subsection. On the other hand, a self consistent way of truncating the large number of $SO(10)$ allowed operators without relying on extra symmetries is to consider only the renormalizable ones. This is exactly what we shall assume.

In this case there are just two ways of giving mass to ν_R: by a nonzero VEV of the Higgs $\overline{126}$, or generate an effective non-renormalizable operator radiatively [403]. We shall consider in turn both of them.

4.1.2.1 Elementary $\overline{126}_H$ It is rather appealing that 10_H and $\overline{126}_H$ may be sufficient for all the fermion masses, with only two sets of symmetric Yukawa coupling matrices. The mass matrices at M_{GUT} are

$$m_d = v_{10}^d Y_{10} + v_{126}^d Y_{126}, \tag{4.8}$$

$$m_u = v_{10}^u Y_{10} + v_{126}^u Y_{126}, \tag{4.9}$$

$$m_e = v_{10}^d Y_{10} - 3v_{126}^d Y_{126}, \tag{4.10}$$

$$m_\nu = -m_D M_R^{-1} m_D + m_{\nu_L}, \tag{4.11}$$

where

$$m_D = v_{10}^u Y_{10} - 3v_{126}^u Y_{126}, \tag{4.12}$$

$$M_R = v_R Y_{126}, \tag{4.13}$$

$$m_{\nu_L} = v_L Y_{126}. \tag{4.14}$$

These relations are valid at M_{GUT}, so it is there that their validity must be tested. The analysis done so far used the results of renormalization group running from M_Z to M_{GUT} from [407, 408].

The first attempts in fitting the mass matrices assumed the domination of the type I seesaw. It was pioneered by treating CP violation perturbatively in a non-supersymmetric framework [406], and later improved with a more detailed treatment of complex parameters and supersymmetric low energy effective theory [409–411]. Nevertheless, these fits had problems to reproduce correctly the PMNS matrix parameters.

A new impetus to the whole program was given by the observation that in case type II seesaw dominates (a way to enforce it is to use a 54 dimensional Higgs representation [412]) the neutrino mass, an interesting relation in these type of models between b–τ unification and large atmospheric mixing angle can be found [413–415]. The argument is very simple and it can be traced to the relation [416]

$$m_\nu \propto m_d - m_e, \tag{4.15}$$

which follows directly from (4.8), (4.10) and (4.14), if only the second term (type II) in (4.11) is considered. Considering only the heaviest two generations as an example and taking the usually good approximation of small second generation masses and small mixing angles, one finds all the elements of the right-hand side small except the 22 element, which is proportional to the difference of two big numbers, $m_b - m_\tau$. Thus, a large neutrino atmospheric mixing angle is linked to the smallness of this 22 matrix element, and so to b–τ unification. Note that in these types of models b–τ unification is no more automatic due to the presence of the $\overline{126}$, which breaks $SU(4)_C$. It is, however, quite a good prediction of the RGE running in the case of low energy supersymmetry.

The numerical fitting was able to reproduce also a large solar mixing angle both in case of type II [417, 418] or mixed seesaw [419], predicting also a quite large $|U_{e3}| \approx 0.16$ mixing element, close to the experimental upper bound. The difficulty in fitting the CKM CP-violating phase in the first quadrant was overcome by new solutions found in [420, 421], maintaining the prediction of large $|U_{e3}| \geq 0.1$ matrix element.

All these fittings were done assuming no constraints coming from the Higgs sector. Regarding it, it was found that the minimal supersymmetric model [422–424] has only 26 model parameters [425], on top of the usual supersymmetry breaking soft terms, as in the MSSM. When one considers this minimal model, the VEVs in the mass formulae (4.8)–(4.14) are not completely arbitrary, but are connected by the restrictions of the Higgs sector. This has been first noticed in [426–428] showing a possible clash with the positive results of the unconstrained Yukawa sector studied in [420, 421]. The issue has been pursued in [429], showing that in the region of parameter space where the fermion mass fitting is successful, there are necessarily intermediate scale thresholds which spoil perturbativity of the RGE evolution of the gauge couplings.

To definitely settle the issue, two further checks should be done. (a) The χ^2 analysis used in the fitting procedure should be implemented at M_Z, not at M_{GUT}. The point is in fact that while the errors at M_Z are uncorrelated, they become strongly correlated after running to M_{GUT}, due to the large Yukawa coupling of top and possibly also of bottom, tau and neutrino. (b) Another issue is to consider also the effect of the possible increased gauge couplings on the Yukawas. Only after these two checks will be done, this minimal model could be ruled out.

A further important point is that in the case of VEVs constrained by the Higgs sector one finds from the charged fermion masses that the model predicts large $\tan\beta \simeq 40$, as confirmed by the last fits in [429]. In this regime there may be sizable corrections to the "down" fermion mass matrices from the soft SUSY breaking parameters [430]; this brings into the game also the soft SUSY breaking sector, lowering somewhat the predictivity but relaxing the difficulty in fitting the experimental data. In this scenario predictions on masses would become predictions on the soft sector.

Some topics have to be still mentioned in connection with the above: the important calculation of the mass spectrum and Clebsch–Gordan coefficients in $SO(10)$ [399, 431–439], the doublet–triplet splitting problem [440, 441], the Higgs doublet mass matrix [399, 433], the running of the gauge couplings at two loops together with threshold corrections [434], and the study of proton decay [435, 442, 443].

What if this model turns out to be wrong? There are other models on the market. The easiest idea is to add a 120 dimensional Higgs, that may also appear as a natural choice, being the last of the three allowed representations that couple with fermions. There are three different ways of doing it considered in the literature: (a) take 120 as a small, non-leading, contribution, i.e. a perturbation to the previous formulae [444–446]; (b) consider 120 on an equal footing as 10 and $\overline{126}$, but assume some extra discrete symmetry or real parameters in the superpotential, breaking CP spontaneously [447–450] (and suppressing in the first two references the dangerous $d = 5$ proton decay modes); (c) assume small $\overline{126}$ contributions to the charged fermion masses [451–454].

Another limit is to forget the 10_H altogether, as has been proposed for non-supersymmetric theories [455]. The two generation study predicts a too small ratio $m_b/m_\tau \approx 0.3$, instead of the value 0.6 that one gets by straight running. The idea is that this could get large corrections due to Dirac neutrino Yukawas [456] and the effect of finite second generation masses, as well as the inclusion of the first generation and CP-violating phases. This is worth pursuing for it provides an alternative minimal version of $SO(10)$, and after all, supersymmetry may not be there.

4.1.2.2 Radiative $\overline{126}_H$ The original idea [403] is that there is no $\overline{126}_H$ representation in the theory, but the same operator is generated by loop corrections. The representation that breaks the rank of $SO(10)$ is now 16_H, which VEV we call M_A. Generically there is a contribution to the right-handed neutrino mass at two loops:

$$M_R \approx \left(\frac{\alpha}{4\pi}\right)^2 \frac{M_A^2}{M_{\text{GUT}}} \frac{M_{\text{SUSY}}}{M_{\text{GUT}}} Y_{10}, \qquad (4.16)$$

which is too small in low energy supersymmetry (low breaking scale M_{SUSY}) as well as non-supersymmetric theories ($M_{\text{SUSY}} = M_{\text{GUT}}$, but low intermediate scale M_A required by gauge coupling unification). The only exception, proposed in [404], could be split supersymmetry [457, 458].

In the absence of $\overline{126}_H$, the charged fermion masses must be given by only 10_H and 120_H [404], together with radiative corrections. The simplest analysis of the tree order two generation case gives three interesting predictions-relations [405, 459]: (1) almost exact b–τ unification; (2) large atmospheric mixing angle related to the small quark θ_{bc} mixing angle; (3) somewhat degenerate neutrinos. For a serious numerical analysis one needs to use the RGE for the case of split supersymmetry, taking a very small $\tan\beta < 1$ to get an approximate b–τ unification [458]. One needs also some fine-tuning of the parameters to account for the small ratio $M_{\text{SUSY}}/M_{\text{GUT}} \leq 10^{-(3-4)}$ required in realistic models to have gluinos decay fast enough [460].

4.2 Higher dimensional approaches

Recently, in the context of theories with extra spatial dimensions, some new approaches toward the question of SM fermion mass hierarchy and flavor structure have arisen [461–468]. For instance, the SM fermion mass spectrum can be generated naturally by permitting the quark/lepton masses to evolve with a power-law dependence on the mass scale [465, 466]. The most studied and probably most attractive idea for generating a non-trivial flavor structure is the displacement of various SM fermions along extra dimension(s). This approach is totally different from the one discussed in Sect. 2, as it is purely geometrical and thus does not rely on the existence of any novel symmetry in the short distance theory. The displacement idea applies to the scenarios with large flat [467] or small warped [468] extra dimension(s), as we develop in the following subsections.

4.2.1 Large extra dimensions

In order to address the gauge hierarchy problem, a scenario with large flat extra dimensions has been proposed by Arkani-Hamed, Dimopoulos and Dvali (ADD) [469–471], based on a reduction of the fundamental gravity scale down to the TeV scale. In this scenario, gravity propagates in the bulk whereas SM fields live on a 3-brane. One could assume that this 3-brane has a certain thickness L along an extra dimension (as for example in [472]). Then SM fields would feel an extra dimension of size L, exactly as in a universal extra dimension (UED) model [473] (where SM fields propagate in the bulk) with one extra dimension of size L.[12]

In such a framework, the SM fermions can be localized at different positions along this extra dimension L. Then the relative displacements of quark/lepton wave function peaks produce suppression factors in the effective four-dimensional Yukawa couplings. These suppression factors

---
[12] The constraint from electroweak precision measurements is $R^{-1} \gtrsim$ 2–5 TeV, the one from direct search at LEP collider is $L^{-1} \gtrsim 5$ TeV and the expected LHC sensitivity is about $L^{-1} \sim 10$ TeV.

being determined by the overlaps of fermion wave functions (getting smaller as the distance between wave function peaks increases), they can vary with the fermion flavors and thus induce a mass hierarchy. This mechanism was first suggested in [467] and its variations have been studied in [474–484].

Let us describe this mechanism more precisely. The fermion localization can be achieved through either nonperturbative effects in string/M theory or field-theoretical methods. One field-theoretical possibility is to couple the SM fermion fields $\Psi_i(x_\mu, x_5)$ [$i = 1, \ldots, 3$ being the family index and $\mu = 1, \ldots, 4$ the usual coordinate indexes] to five dimensional scalar fields with VEV $\Phi_i(x_5)$ depending on the extra dimension (parameterized by x_5).[13] Indeed, chiral fermions are confined in solitonic backgrounds [485]. If the scalar field profile behaves as a linear function of the form $\Phi_i(x_5) = 2\mu^2 x_5 - m_i$ around its zero-crossing point $x_i^0 = m_i/2\mu^2$, the zero-mode of five dimensional fermion acquires a Gaussian wave function of typical width μ^{-1} and centered at x_i^0 along the x_5 direction: $\Psi_i^{(0)}(x_\mu, x_5) = A e^{-\mu^2(x_5 - x_i^0)^2} \psi_i(x_\mu)$, $\psi_i(x_\mu)$ being the four-dimensional fermion field and $A = (2\mu^2/\pi)^{1/4}$ a normalization factor. Then the four-dimensional Yukawa couplings between the five dimensional SM Higgs boson H and zero-mode fermions, obtained by integration on x_5 over the wall width L,[14]

$$\mathcal{S}_{\text{Yukawa}} = \int d^5x \sqrt{L} \kappa H(x_\mu, x_5) \bar{\Psi}_i^{(0)}(x_\mu, x_5) \Psi_j^{(0)}(x_\mu, x_5)$$

$$= \int d^4x \, Y_{ij} h(x_\mu) \bar{\psi}_i(x_\mu) \psi_j(x_\mu), \quad (4.17)$$

are modulated by the following effective coupling constants,

$$Y_{ij} = \int dx_5 \, \kappa A^2 e^{-\mu^2(x_5 - x_i^0)^2} e^{-\mu^2(x_5 - x_j^0)^2}$$

$$= \kappa e^{-\frac{\mu^2}{2}(x_i^0 - x_j^0)^2}. \quad (4.18)$$

It can be considered as natural to have a five dimensional Yukawa coupling constant equal to $\sqrt{L}\kappa$, where the dimensionless parameter κ is universal (in flavor and nature of fermions) and of order unity, so that the flavor structure is mainly generated by the field localization effect through the exponential suppression factor in (4.18). The remarkable feature is that, due to this exponential factor, large hierarchies can be created among the physical fermion masses, even for all fundamental parameters m_i of order of the same energy scale μ.

---

[13] Although we concentrate here on the case with only one extra dimension, for simplicity, the mechanism can be directly extended to more extra dimensions.

[14] Here, the factor $\sqrt{L}$ compensates with the Higgs component along x_5, since the Higgs boson is not localized.

This mechanism can effectively accommodate all the data on quark and charged lepton masses and mixings [486–488]. In case that right handed neutrinos are added to the SM so that neutrinos acquire Dirac masses (as those originating from Yukawa couplings (4.17)), neutrino oscillation experiment results can also be reproduced [472]. The fine-tuning, arising there on relative x_i^0 parameters, turns out to be improved when neutrinos get Majorana masses instead [489] (see also [235, 490]).

4.2.2 Small extra dimensions

Another type of higher-dimensional scenario solving the gauge hierarchy problem was suggested by Randall and Sundrum (RS) [491, 492]. There, the unique extra dimension is warped and has a size of order M_{Pl}^{-1} (M_{Pl} being the reduced Planck mass: $M_{\text{Pl}} = 2.44 \times 10^{18}$ GeV) leading to an effective gravity scale around the TeV. In the initial version, gravity propagates in the bulk and SM particles are all stuck on the TeV-brane. An extension of the original RS model was progressively proposed [493–497], motivated by its interesting features with respect to the gauge coupling unification [498–503] and dark matter problem [504, 505]. This new set-up is characterized by the presence of SM fields, except the Higgs boson (to ensure that the gauge hierarchy problem does not re-emerge), in the bulk.

In this RS scenario with bulk matter, a displacement of SM fermions along the extra dimension is also possible [468]: the effect is that the effective four-dimensional Yukawa couplings are affected by exponential suppression factors, originating from the wave function overlaps between bulk fermions and Higgs boson (confined on our TeV-brane). If the fermion localization depends on the flavor and nature of fermions, then the whole structure in flavor space can be generated by these wave function overlaps. In particular, if the top quark is located closer to the TeV-brane than the up quark, then its overlap with the Higgs boson, and thus its mass after electroweak symmetry breaking, is larger relatively to the up quark (for identical five dimensional Yukawa coupling constants).

More precisely, the fermions can acquire different localizations if each field $\Psi_i(x_\mu, x_5)$ is coupled to a distinct five dimensional mass m_i: $\int d^4x \int dx_5 \sqrt{G} \, m_i \bar{\Psi}_i \Psi_i$, G being the determinant of the RS metric. To modify the location of fermions, the masses m_i must have a non-trivial dependence on x_5, like $m_i = \text{sign}(x_5) c_i k$, where c_i are dimensionless parameters and $1/k$ is the curvature radius of anti-de Sitter space. Then the fields decompose as, $\Psi_i(x^\mu, x_5) = \sum_{n=0}^{\infty} \psi_i^{(n)}(x^\mu) f_n^i(x_5)$ [n labeling the tower of Kaluza–Klein (KK) excitations], admitting the following solution for the zero-mode wave function, $f_0^i(x_5) = e^{(2-c_i)k|x_5|}/N_0^i$, where N_0^i is a normalization factor.

The Yukawa interactions with the Higgs boson H read

$$\mathcal{S}_{\text{Yukawa}} = \int d^5x \sqrt{G} \big(Y_{ij}^{(5)} H \bar{\Psi}_{+i} \Psi_{-j} + \text{h.c.}\big)$$

$$= \int d^4x\, M_{ij} \bar{\psi}_{Li}^{(0)} \psi_{Rj}^{(0)} + \text{h.c.} + \cdots. \quad (4.19)$$

The $Y_{ij}^{(5)}$ are the five dimensional Yukawa coupling constants and the dots stand for KK mass terms. The fermion mass matrix is obtained after integrating:

$$M_{ij} = \int dx_5 \sqrt{G}\, Y_{ij}^{(5)} H f_0^i(x_5) f_0^j(x_5). \quad (4.20)$$

The $Y_{ij}^{(5)}$ can be chosen almost universal so that the quark/lepton mass hierarchies are mainly governed by the overlap mechanism. Large fermion mass hierarchies can be produced for fundamental mass parameters m_i all of order of the unique scale of the theory $k \sim M_{\text{Pl}}$.

With this mechanism, the quark masses and CKM mixing angles can be effectively accommodated [506–508], as well as the lepton masses and PMNS mixing angles in both cases where neutrinos acquire Majorana masses (via either dimension five operators [509] or the see-saw mechanism [510]) and Dirac masses (see [511], and [512, 513] for order unity Yukawa couplings leading to mass hierarchies essentially generated by the geometrical mechanism).

4.2.3 Sources of FCNC in extra dimension scenarios

GIM-violating FCNC effects in extra dimension scenarios may appear both from tree level and from loop effects.

At tree level FCNC processes can be induced by exchanges of KK excitations of neutral gauge bosons. The neutral current action of the effective four-dimensional coupling, between SM fermions $\psi_i^{(0)}(x^\mu)$ and KK excitations of any neutral gauge boson $A_\mu^{(n)}(x^\mu)$, reads in the interaction basis

$$\mathcal{S}_{\text{NC}} = g_L^{\text{SM}} \int d^4x \sum_{n=1}^\infty \bar{\psi}_{Li}^{(0)} \gamma^\mu \mathcal{C}_{Lij}^{(n)} \psi_{Lj}^{(0)} A_\mu^{(n)} + \{L \leftrightarrow R\}. \quad (4.21)$$

Therefore, FCNC interactions can be induced by the non-universality of the effective coupling constants $g_{L/R}^{\text{SM}} \times C_0^{i(n)}$ between KK modes of the gauge fields and the three SM fermion families (which have different locations along x_5).

At the loop level, KK fermion excitations may invalidate the GIM cancellation, as discussed e.g. in [511, 514] for $\ell_\alpha^\pm \to \ell_\beta^\pm \gamma$. Indeed, these excitations have KK masses which are not negligible (and thus not quasi-degenerate in family space) compared to $m_{W^\pm}$. The GIM mechanism is also invalidated by the loop contributions of the KK $W^{\pm(n)}$ modes which couple (KK level by level), e.g. to leptons in the four-dimensional theory, via an effective mixing matrix of type $V_{\text{MNS}}^{\text{eff}} = U_L^{l\dagger} \mathcal{C}_L^{(n)} U_L^\nu$ being non-unitary due to the non-universality of

$$\mathcal{C}_L^{(n)} \equiv \text{diag}\big(C_m^{1(n)}, C_m^{2(n)}, C_m^{3(n)}\big). \quad (4.22)$$

In this diagonal matrix, $C_m^{i\,(n)}$ quantifies the wave function overlap along the extra dimension between the $W^{\pm(n)}$ [$n \geq 1$] and exchanged (mth level KK) fermion $f_m^i(x_5)$ [$i = \{1, 2, 3\}$ being the generation index] (see below for more details).

The GIM mechanism for leptons can be clearly restored if the three coefficients $C_m^{i(n)}$ as well as the three KK fermion masses $m_{\text{KK}}^{i(m)}$ are equal to each other, i.e. are universal with respect to $i = \{1, 2, 3\}$ (KK level by level) [515]. Within the quark sector, on the other hand, the top quark mass cannot be totally neglected relatively to the KK up-type quark excitation scales, leading to a mass shift of the KK top quark mode from the rest of the KK up-type quark modes and removing the degeneracy among three family masses of the up quark excitations at fixed KK level (with regard to $m_{W^{\pm(n)}}$). Moreover, this means that the Yukawa interaction with the Higgs boson induces a substantial mixing of the top quark KK tower members among themselves [481, 516].

For example, the data on $b \to s\gamma$ (receiving a contribution from the exchange of a $W^{\pm(n)}$ [$n = 0, 1, \ldots$] gauge field and an up quark, or its KK excitations, at one loop-level) can be accommodated in the RS model with $m(W^{\pm(1)}) \simeq 1$ TeV, as shown in [515] using numerical methods for the diagonalization of a large dimensional mass matrix and taking into account the top quark mass effects described previously.

4.2.4 Mass bounds on Kaluza–Klein excitations

In this subsection we develop constraints on the KK gauge boson masses derived from the tree level FCNC effect described above. Our purpose is to determine whether these constraints still allow the KK gauge bosons to be sufficiently light to imply potentially visible signatures at LHC.

4.2.4.1 Large extra dimensions Let us consider the generic framework of a flat extra dimension, with a large size L, along which gravity as well as gauge bosons propagate. The SM fermions are located at different points of the fifth dimension, so that their mass hierarchy can be interpreted in term of the geometrical mechanism described in details in Sect. 4.2.1. In such a framework the exchange of the KK excitations of the gluon can bring important contributions to the K^0–$\bar{K}^0$ mixing ($\Delta F = 2$) at tree level. Indeed, the KK gluon can couple the d quark with the s quark, if these light down-quarks are displaced along the extra dimension. The obtained KK contribution to the mass splitting Δm_K in the kaon system depends on the KK gluon coupling between the s and d quarks (which is fixed by quark locations) and mainly on the mass of the first KK gluon $M_{\text{KK}}^{(1)}$. Assuming that the s, d quark locations are such that the m_s, m_d mass values are reproduced, the obtained Δm_K and also $|\varepsilon_K|$ are smaller than the associated experimental values for, respectively,

$$M_{\text{KK}}^{(1)} \gtrsim 25 \text{ TeV}, \quad \text{and} \quad M_{\text{KK}}^{(1)} \gtrsim 300 \text{ TeV}, \quad (4.23)$$

as found by the authors of [517]. The same bound coming from the D^0 meson system is weaker.

In the lepton sector the experimental upper limit on the branching ratio $B(\mu \to eee)$ imposes typically the constraint [517]

$$M_{\text{KK}}^{(1)} \gtrsim 30 \text{ TeV}, \quad (4.24)$$

since the exchange of the KK excitations of the electroweak neutral gauge bosons contributes to the decay $\mu \to eee$.

To conclude, we stress that if the extra dimensions treat families in a non-universal way (which could explain the fermion mass hierarchy), the indirect bounds from FCNC physics like the ones in (4.23)–(4.24) force the mass of the KK gauge bosons to be far from the collider reach. As a matter of fact, the LHC will be able to probe the KK excitations of gauge bosons only up to 6–7 TeV [518–521] in the present context.

4.2.4.2 Small extra dimensions In the context of the RS model with SM fields in the bulk, described in Sect. 4.2.2, the exchange of KK excitations of neutral gauge bosons (like e.g. the first Z^0 excitation: $Z^{(1)}$) also contributes to FCNC processes at tree level [468, 507, 522–526] since these KK states possess FC couplings if the different families of fermions are displaced along the warped extra dimension. There exist some configurations of fermion locations, pointed out in [513], which simultaneously reproduce all quark/lepton masses and mixing angles via the wave function effects *and* lead to amplitudes of FCNC reactions [$l_\alpha \to l_\beta l_\gamma l_\gamma$, $Z^0 \to l_\alpha l_\beta$, $P^0 - \bar{P}^0$ mixing of a generic meson P, μ–e conversion, $K^0 \to l_\alpha l_\beta$ and $K^+ \to \pi^+ \nu \nu$] compatible with the corresponding experimental constraints even for light neutral KK gauge bosons:

$$M_{\text{KK}}^{(1)} \gtrsim 1 \text{ TeV}. \quad (4.25)$$

The explanation of this result is the following. If the SM fermions with different locations are localized typically close to the Planck-brane, they have quasi-universal couplings $C_0^{i(n)}$ [cf. (4.21)] with the KK gauge bosons which have a wave function almost constant along the fifth dimension near the Planck-brane. Therefore, small FC couplings are generated in the physical basis for these fermions leading to the weak bound (4.25). The fermions from the third family, associated to heavy flavors, cannot be localized extremely close to the Planck-brane since their wave function overlap with the Higgs boson [confined on the TeV-brane] must be large in order to generate high effective Yukawa couplings. Nevertheless, this is compensated by the fact that phenomenological FCNC constraints are usually less severe in the third generation sector.

As a result, the order of lower limits on $M_{\text{KK}}^{(1)}$ coming from the considerations on both fermion mass data and FCNC processes can be as low as TeV. From the purely theoretical point of view, the favored order of magnitude for $M_{\text{KK}}^{(1)}$ is $\mathcal{O}(1)$ TeV which corresponds to a satisfactory solution for the gauge hierarchy problem. From the model building point of view one has to rely on an appropriate extension of the RS model insuring that, for light KK masses, the deviations of the electroweak precision observables do not conflict with the experimental results. The existing RS extensions, like the scenarios with brane-localized kinetic terms for fermions [527] and gauge bosons [528] (see [529, 530] for the localized gauge boson kinetic terms and [531] for the fermion ones), or the scenarios with an extended gauge symmetry (see [532–534] for different fermion charges under this broken symmetry), allow $M_{\text{KK}}^{(1)}$ to be as low as ~ 3 TeV. In such a case, one can expect a direct detection of the KK excited gauge bosons at LHC.

4.3 Minimal flavor violation in the lepton sector

4.3.1 Motivations and basic idea

Within the SM the dynamics of flavor-changing transitions is controlled by the structure of fermion mass matrices. In the quark sector, up and down quarks have mass eigenvalues which are up to 10^5 times smaller than the electroweak scale, and mass matrices which are approximately aligned. This results in the effective CKM and GIM suppressions of charged and neutral flavor violating interactions, respectively. Forcing this connection between the low energy fermion mass matrices and the flavor-changing couplings to be valid also beyond the SM, leads to new-physics scenarios with a high level of predictivity (in the flavor sector) and a natural suppression of flavor-changing transitions. The latter achievement is a key ingredient to maintain a good agreement with experiments in models where flavored degrees of freedom are expected around the TeV scale.

This is precisely the idea behind the minimal flavor violation principle [535–537]. It is a fairly general hypothesis that can be implemented in strongly-interacting theories [535], low energy supersymmetry [536, 537], multi-Higgs [537, 538] and GUT [539] models. In a model independent formulation, the MFV construction consists in identifying the flavor symmetry and symmetry breaking structure of the SM and enforce it in a more general effective theory (written in terms of SM fields and valid above the electroweak scale). In the quark sector this procedure is unambiguous: the largest group of flavor changing field transformations commuting with the gauge group is $\mathcal{G}_q = SU(3)_{Q_L} \times SU(3)_{u_R} \times SU(3)_{d_R}$, and this group is broken only by the Yukawa couplings. The invariance of the SM Lagrangian under $\mathcal{G}_q$ can be formally recovered elevating the Yukawa matrices to spurion fields with appropriate transformation properties under $\mathcal{G}_q$. The hypothesis of MFV states that these are the only spurions breaking $\mathcal{G}_q$ also beyond the SM. Within the effective theory formulation, this implies that all the higher dimensional operators constructed from SM and Yukawa fields must be (formally) invariant under $\mathcal{G}_q$. The consequences of this hypothesis in the quark sector have been extensively analyzed in the literature (see e.g. Refs. [540, 541]). Without entering into the details, we can state that the MFV hypothesis provides a plausible explanation of why no new-physics effects have been observed so far in the quark sector.

Apart from arguments based on the analogy with quarks, and despite the scarce experimental information, the definition of a minimal lepton flavor violation (MLFV) principle [542] is demanded by a severe fine-tuning problem in LFV decays of charged leptons. Within a generic effective theory approach, the radiative decays $l_i \to l_j \gamma$ proceed through the following gauge-invariant operator

$$\frac{\delta_{ij}^{RL}}{\Lambda_{\text{LFV}}^2} H^\dagger \bar{e}_R^i \sigma^{\sigma\rho} L_L^j F_{\sigma\rho}, \qquad (4.26)$$

where δ_{ij}^{RL} are the generic flavor-changing couplings and Λ_{LFV} denotes the cut-off of the effective theory. In the absence of a specific flavor structure, it is natural to expect $\delta_{ij}^{RL} = \mathcal{O}(1)$. In this case the experimental limit for $\mu \to e\gamma$ implies $\Lambda_{\text{LFV}} > 10^5$ TeV, in clear tension with the expectation of new degrees of freedom close to the TeV scale in order to stabilize the Higgs sector of the SM.

The implementation of a MFV principle in the lepton sector is not as simple as in the quark sector. The problem is that the neutrino mass matrix itself cannot be accommodated within the renormalizable part of the SM Lagrangian. The most natural way to describe neutrino masses, explaining their strong suppression, is to assume they are Majorana mass terms suppressed by the heavy scale of lepton number violation (LNV). In other words, neutrino masses are described by a non-renormalizable interaction of the type equation (3.4) suppressed by the scale $\Lambda_{\text{LNV}} \gg v = |\langle H \rangle|$. This implies that we have to face a two scale problem (presumably with the hierarchy $\Lambda_{\text{LNV}} \gg \Lambda_{\text{LFV}}$) and that we need some additional hypothesis to identify the irreducible flavor-symmetry breaking structures. As we shall illustrate in the following, we can choose whether to extend or not the field content of the SM. The construction of the effective theory based on one of these realizations of the MLFV hypothesis can be viewed as a general tool to exploit the observable consequences of a specific (minimalistic) hypothesis about the irreducible sources of lepton-flavor symmetry breaking.

4.3.2 MLFV with minimal field content

The lepton field content is the SM one: three left handed doublets L_L^i and three right handed charged lepton singlets e_R^i. The flavor symmetry group is $\mathcal{G}_l = SU(3)_{L_L} \times SU(3)_{e_R}$ and we assume the following flavor symmetry breaking Lagrangian

$$\begin{aligned}\mathcal{L}_{\text{Sym.Br.}} &= -Y_e^{ij} \bar{e}_R^i \left(H^\dagger L_L^j\right) \\ &\quad - \frac{1}{2\Lambda_{\text{LNV}}} \kappa_\nu^{ij} \left(\bar{L}_L^{ci} \tau_2 H\right)\left(H^T \tau_2 L_L^j\right) + \text{h.c.} \\ &\to -v Y_e^{ij} \bar{e}_R^i e_L^j - \frac{v^2}{2\Lambda_{\text{LNV}}} \kappa_\nu^{ij} \bar{\nu}_L^{ci} \nu_L^j + \text{h.c.} \quad (4.27)\end{aligned}$$

Here the two irreducible sources of LFV are the coefficient of dimension five LNV operator (κ_ν^{ij}) and the charged lepton Yukawa coupling (Y_e), transforming respectively as $(6,1)$ and $(\bar{3},3)$ under $\mathcal{G}_l$. An explicit realization of this scenario is provided by the so-called triplet see-saw mechanism (or see-saw of type II). This approach has the advantage of being highly predictive, but it differs in an essential way from the MFV hypothesis in the quark sector since one of the basic spurion originates from a non-renormalizable coupling.

Having identified the irreducible sources of flavor symmetry breaking and their transformation properties, we can classify the non-renormalizable operators suppressed by inverse powers of Λ_{LFV} which contribute to flavor violating processes. These operators must be invariant combinations of SM fields and the spurions Y_e and κ_ν. The complete list of the leading operators contributing to LFV decays of charged leptons is given in Refs. [542, 543]. The case of the radiative decays $l_i \to l_j \gamma$ is particularly simple since there are only two dimension six operators (operators with a structure as in (3.4), with $F_{\sigma\rho}$ replaced by the stress tensors of the $U(1)_Y$ and $SU(2)_L$ gauge groups, respectively). The MLFV hypothesis forces the flavor-changing couplings of these operators to be a spurion combination transforming as $(\bar{3},3)$ under $\mathcal{G}_l$:

$$\left(\delta_{\min}^{RL}\right)_{ij} \propto \left(Y_e \kappa_\nu^\dagger \kappa_\nu\right)_{ij} + \cdots \qquad (4.28)$$

where the dots denote terms with higher powers of Y_e or κ_ν. Up to the overall normalization, this combination can be completely determined in terms of the neutrino mass eigenvalues and the PMNS matrix. In the basis where Y_e is diagonal we can write,

$$\left(Y_e \kappa_\nu^\dagger \kappa_\nu\right)_{i \neq j} = \frac{m_{l_i}}{v} \left(\frac{\Lambda_{\text{LNV}}^2}{v^4} U_{\text{PMNS}} m_\nu^2 U_{\text{PMNS}}^\dagger \right)_{i \neq j}$$

$$\to \frac{m_{l_i}}{v} \frac{\Lambda_{\text{LNV}}^2}{v^4} \left[(U_{\text{PMNS}})_{i2} (U_{\text{PMNS}})_{j2}^* \Delta m_{\text{sol}}^2 \right.$$

$$\left. \pm (U_{\text{PMNS}})_{i3} (U_{\text{PMNS}})_{j3}^* \Delta m_{\text{atm}}^2 \right], \quad (4.29)$$

where Δm_{atm}^2 and Δm_{sol}^2 denote the squared mass differences deduced from atmospheric- and solar-neutrino data, and $+/-$ correspond to normal/inverted hierarchy, respectively. The overall factor $\Lambda_{\text{LNV}}^2/v^2$ implies that the absolute normalization of LFV rates suffers of a large uncertainty. Nonetheless, a few interesting conclusions can still be drawn [542].

- The LFV decay rates are proportional to $\Lambda_{\text{LNV}}^4/\Lambda_{\text{LFV}}^4$ and could be detected only in presence of a large hierarchy between these two scales. In particular, $\mathcal{B}(\mu \to e\gamma) > 10^{-13}$ only if $\Lambda_{\text{LNV}} > 10^9 \Lambda_{\text{LFV}}$.
- Ratios of similar LFV decay rates, such as $B(\mu \to e\gamma)/B(\tau \to \mu\gamma)$, are free from the normalization ambiguity and can be predicted in terms of neutrino masses and PMNS angles: violations of these predictions would unambiguously signal the presence of additional sources of lepton-flavor symmetry breaking. One of these predictions is the 10^{-2}–10^{-3} enhancement of $B(\tau \to \mu\gamma)$ versus $B(\mu \to e\gamma)$ shown in Fig. 3. Given the present and near-future experimental prospects on these modes, this modest enhancement implies that the $\mu \to e\gamma$ search is much more promising within this framework.
- Ratios of LFV transitions among the same two families (such as $\mu \to e\gamma$ versus $\mu \to 3e$ or $\tau \to \mu\gamma$ vs $\tau \to 3\mu$ and $\tau \to \mu e\bar{e}$) are determined by known phase space factors and ratios of various Wilson coefficients. As data will become available on different lepton flavor violating processes, if the flavor patter is consistent with the MLFV hypothesis, from these ratios it will be possible to disentangle the contributions of different operators.
- A definite prediction of the MLFV hypothesis is that the rates for decays involving light hadrons ($\pi^0 \to \mu e$, $K_L \to \mu e$, $\tau \to \mu \pi^0$, ...) are exceedingly small.

4.3.3 MLFV with extended field content

In this scenario we assume three heavy right handed Majorana neutrinos in addition to the SM fields. As a consequence, the maximal flavor group becomes $\mathcal{G}_l \times SU(3)_{\nu_R}$. In order to minimize the number of free parameters (or to maximize the predictivity of the model), we assume that the Majorana mass term for the right handed neutrinos is proportional to the identity matrix in flavor space: $(M_R)_{ij} = M_R \times \delta_{ij}$. This mass term breaks $SU(3)_{\nu_R}$ to $O(3)_{\nu_R}$ and is assumed to be the only source of LNV ($M_R \leftrightarrow \Lambda_{\text{LNV}}$).

Once the field content of model is extended, there are in principle many alternative options to define the irreducible sources of lepton flavor symmetry breaking (see e.g. Ref. [544] for an extensive discussion). However, this specific choice has two important advantages: it is predictive and closely resemble the MFV hypothesis in the quark sector. The ν_R are the counterpart of right handed up quarks and, similarly to the quark sector, the symmetry breaking sources are two Yukawa couplings of (3.40). An explicit example of MLFV with extended field content is the minimal supersymmetric standard model with degenerate right handed neutrinos.

The classification of the higher dimensional operators in the effective theory proceeds as in the minimal field content case. The only difference is that the basic spurions are now Y_ν and Y_e, transforming as $(\bar{3}, 1, 3)$ and $(\bar{3}, 3, 1)$ under $\mathcal{G}_l \times O(3)_{\nu_R}$, respectively. The determination of the spurion structures in terms of observable quantities is more involved than in the minimal field content case. In general, inverting the see-saw relation allows us to express Y_ν in terms of neutrino masses, PMNS angles and an arbitrary complex-orthogonal matrix R of (3.45) [232]. Exploiting the $O(3)_{\nu_R}$ symmetry of the MLFV Lagrangian, the real orthogonal part of R can be rotated away. We are then left with a Hermitian-orthogonal matrix H [545] which can be parameterized in terms of three real parameters (ϕ_i) which control the amount of CP violation in the right handed sector:

$$Y_\nu = \frac{M_R^{1/2}}{v} H(\phi_i) m_{\text{diag}}^{1/2} U_{\text{PMNS}}^\dagger. \quad (4.30)$$

With this parameterization for Y_ν the flavor changing coupling relevant to $l_i \to l_j \gamma$ decays reads

$$\delta_{\text{ext}}^{RL} \propto Y_e \left(Y_\nu^\dagger Y_\nu\right)$$

$$\to \frac{m_e}{v} \left(\frac{M_R}{v^2} U_{\text{PMNS}} m_{\text{diag}}^{1/2} H^2 m_{\text{diag}}^{1/2} U_{\text{PMNS}}^\dagger \right). \quad (4.31)$$

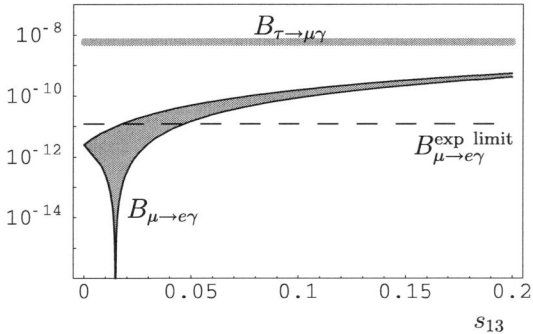

Fig. 3 $B_{l_i \to l_j \gamma} \equiv \Gamma(l_i \to l_j \gamma)/\Gamma(l_i \to l_j \nu_i \bar{\nu}_j)$ for $\mu \to e\gamma$ and $\tau \to \mu\gamma$ as a function of $\sin\theta_{13}$ in the MLFV framework with minimal field content [542]. The normalization of the vertical axis corresponds to $\Lambda_{\text{LNV}}/\Lambda_{\text{LFV}} = 10^{10}$. The *shading* is due to different values of the phase δ and the normal/inverted spectrum

In the CP-conserving limit $H \to I$ and the phenomenological predictions turns out to be quite similar to the minimal field content scenario [542]. In particular, all the general observations listed in the previous section remain valid. In the general case, i.e. for $H \neq I$, the predictivity of the model is substantially weakened. However, in principle some information about the matrix H can be extracted by studying baryogenesis through leptogenesis in the MLFV framework [546].

4.3.4 Leptogenesis

On general grounds, we expect that the tree-level degeneracy of heavy neutrinos is lifted by radiative corrections. This allows the generation of a lepton asymmetry in the interference between tree-level and one loop decays of right handed neutrinos. Following the standard leptogenesis scenario, we assume that this lepton asymmetry is later communicated to the baryon sector through sphaleron effects and that saturates the observed value of the baryon asymmetry of the universe.

The most general form of the ν_R mass splittings allowed within the MLFV framework has the following form:

$$\frac{\Delta M_R}{M_R} = c_\nu \left[Y_\nu Y_\nu^\dagger + \left(Y_\nu Y_\nu^\dagger\right)^T \right]$$
$$+ c_{\nu\nu}^{(1)} \left[Y_\nu Y_\nu^\dagger Y_\nu Y_\nu^\dagger + \left(Y_\nu Y_\nu^\dagger Y_\nu Y_\nu^\dagger\right)^T \right]$$
$$+ c_{\nu\nu}^{(2)} \left[Y_\nu Y_\nu^\dagger \left(Y_\nu Y_\nu^\dagger\right)^T \right] + c_{\nu\nu}^{(3)} \left[\left(Y_\nu Y_\nu^\dagger\right)^T Y_\nu Y_\nu^\dagger \right]$$
$$+ c_{\nu l} \left[Y_\nu Y_e^\dagger Y_e Y_\nu^\dagger + \left(Y_\nu Y_e^\dagger Y_e Y_\nu^\dagger\right)^T \right] + \cdots.$$

Even without specifying the value of the c_i, this form allows us to derive a few general conclusions [546].

- The term proportional to c_ν does not generate a CPV asymmetry, but sets the scale for the mass splittings: these are of the order of magnitude of the decay widths, realizing in a natural way the condition of resonant leptogenesis.
- The right amount of leptogenesis can be generated even with $Y_e = 0$, if all the ϕ_i are non-vanishing. However, since $Y_\nu \sim \sqrt{M_R}$, for low values of M_R ($\lesssim 10^{12}$ GeV) the asymmetry generated by the $c_{\nu l}$ term dominates. In this case η_B is typically too small to match the observed value and has a flat dependence on M_R. At $M_R \gtrsim 10^{12}$ GeV the quadratic terms $c_{\nu\nu}^{(i)}$ dominate, determining an approximate linear growth of η_B with M_R. These two regimes are illustrated in Fig. 4.

As demonstrated in Ref. [546], baryogenesis through leptogenesis is viable in MLFV models. In particular, assuming a loop hierarchy between the c_i (as expected in a perturbative scenario) and neglecting flavor-dependent effects in the Boltzmann equations (one-flavor approximation of Ref. [547]), the right size of η_B is naturally reached for $M_R \gtrsim 10^{12}$ GeV. As discussed in Ref. [301] (see also [303]), this lower bound can be weakened by the inclusion of flavor-dependent effects in the Boltzmann equations and/or by the $\tan \beta$-enhancement of Y_e occurring in two-Higgs doublet models.

From the phenomenological point of view, an important difference with respect to the CP-conserving case is the fact that non-vanishing ϕ_i change the predictions of the LFV decays, typically producing an enhancement of the $B(\mu \to e\gamma)/B(\tau \to \mu\gamma)$ ratio or the both decays separately [545]. For $M_R \gg 10^{12}$ GeV their effect is moderate and the CP-conserving predictions are recovered. The other important information following from the leptogenesis analysis is the

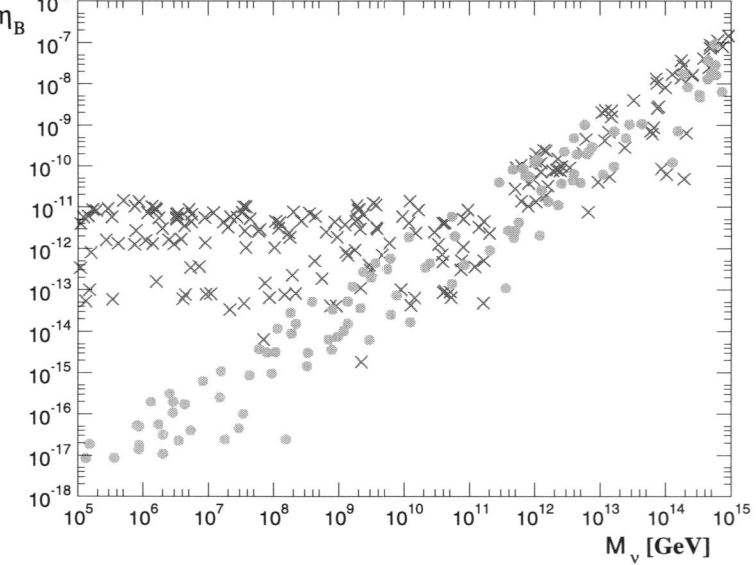

Fig. 4 Baryon asymmetry (η_B) as a function of the right handed neutrino mass scale (M_R) for $c_{\nu l} = 0$ (*dots*) and $c_{\nu l} \neq 0$ (*crosses*) in the MLFV framework with extended field content [546]

fact that the large M_R regime is favored. Assuming Λ_{LFV} to be close to the TeV scale, the M_R regime favored by leptogenesis favors a $\mu \to e\gamma$ rate within the reach of the MEG experiment [548].

4.3.5 GUT implementation

Once we accept the idea that flavor dynamics obeys a MFV principle, both in the quark and in the lepton sector, it is interesting to ask if and how this is compatible with a grand unified theory (GUT), where quarks and leptons sit in the same representations of a unified gauge group. This question has recently been addressed in [539], considering the exemplifying case of $SU(5)_{\text{gauge}}$.

Within $SU(5)_{\text{gauge}}$, the down-type singlet quarks (d^c_{iR}) and the lepton doublets (L_{iL}) belong to the $\bar{5}$ representation; the quark doublet (Q_{iL}), the up-type (u^c_{iR}) and lepton singlets (e^c_{iR}) belong to the $\mathbf{10}$ representation, and finally the right handed neutrinos (ν_{iR}) are singlet. In this framework the largest group of flavor transformation commuting with the gauge group is $\mathcal{G}_{\text{GUT}} = SU(3)_{\bar{5}} \times SU(3)_{10} \times SU(3)_1$, which is smaller than the direct product of the quark and lepton groups discussed before ($\mathcal{G}_q \times \mathcal{G}_l$). We should therefore expect some violations of the MFV+MLFV predictions either in the quark or in the lepton sector or in both.

A phenomenologically acceptable description of the low energy fermion mass matrices requires the introduction of at least four irreducible sources of $\mathcal{G}_{\text{GUT}}$ breaking. From this point of view the situation is apparently similar to the non-unified case: the four $\mathcal{G}_{\text{GUT}}$ spurions can be put in one-to-one correspondence with the low energy spurions Y_u, Y_d, Y_e, and Y_ν. However, the smaller flavor group does not allow the diagonalization of Y_d and Y_e (which transform in the same way under $\mathcal{G}_{\text{GUT}}$) in the same basis. As a result, two additional mixing matrices can appear in the expressions for flavor changing rates: $C = V^T_{e_R} V_{d_L}$ and $G = V^T_{e_L} V_{d_R}$. The hierarchical texture of the new mixing matrices is known since they reduce to the identity matrix in the limit $Y^T_e = Y_d$. Taking into account this fact, and analyzing the structure of the allowed higher-dimensional operators, a number of reasonably firm phenomenological consequences can be deduced [539]:

– There is a well defined limit in which the standard MFV scenario for the quark sector is fully recovered: $M_R \ll 10^{12}$ GeV and small $\tan\beta$ (in a two-Higgs doublet case). For $M_R \sim 10^{12}$ GeV and small $\tan\beta$, deviations from the standard MFV pattern can be expected in rare K decays but not in B physics. Ignoring fine-tuned scenarios, $M_R \gg 10^{12}$ GeV is excluded by the present constraints on quark FCNC transitions. Independently from the value of M_R, deviations from the standard MFV pattern can appear both in K and in B physics for $\tan\beta \gtrsim m_t/m_b$.

– Contrary to the non-GUT MFV framework, the rate for $\mu \to e\gamma$ (and other LFV decays) cannot be arbitrarily suppressed by lowering the average mass M_R of the heavy ν_R. This fact can easily be understood by looking at the flavor structure of the relevant effective couplings, which now assume the following form:

$$\delta^{\text{RL}}_{\text{GUT}} = c_1 Y_e Y^\dagger_\nu Y_\nu + c_2 Y_u Y^\dagger_u Y_e + c_3 Y_u Y^\dagger_u Y^T_d + \cdots. \quad (4.32)$$

In addition to the terms involving $Y_\nu \sim \sqrt{M_R}$ already present in the non-unified case, the GUT group allows also M_R-independent terms involving the quark Yukawa couplings. The latter become competitive for $M_R \lesssim 10^{12}$ GeV and their contribution is such that for $\Lambda_{\text{LFV}} \lesssim 10$ TeV the $\mu \to e\gamma$ rate is above 10^{-13} (i.e. within the reach of MEG [548]).

– Improved experimental information on $\tau \to \mu\gamma$ and $\tau \to e\gamma$ would be a powerful tool in discriminating the relative size of the standard MFV contributions versus the characteristic GUT-MFV contributions due to the different hierarchy pattern among $\tau \to \mu$, $\tau \to e$, and $\mu \to e$ transitions.

5 Phenomenology of theories beyond the standard model

5.1 Flavor violation in non-SUSY models directly testable at LHC

5.1.1 Multi-Higgs doublet models

The arbitrariness of quark masses, mixing and CP violation in the standard model stems from the fact that gauge invariance does not constrain the flavor structure of Yukawa interactions. In the SM neutrinos are strictly massless. No neutrino Dirac mass term can be introduced, due to the absence of right handed neutrinos and no Majorana mass terms can be generated, due to exact $B - L$ conservation. Since neutrinos are massless, there is no leptonic mixing in the SM, which in turn leads to separate lepton flavor conservation. Therefore, the recent observation of neutrino oscillations is evidence for physics beyond the SM. Fermion masses, mixing and CP violation are closely related to each other and also to the Higgs sector of the theory.

It has been shown that gauge theories with fermions, but without scalar fields, do not break CP symmetry [549]. A scalar (Higgs) doublet is used in the SM to break both the gauge symmetry and generate gauge boson masses as well as fermion masses through Yukawa interactions. This is known as the Higgs mechanism, which was proposed by several authors [550–553]. It predicts the existence of one neutral scalar Higgs particle—the Higgs boson. In the SM where a single Higgs doublet is introduced, it is not possible to have spontaneous CP violation since any phase in the vacuum expectation value can be eliminated by rephasing the Higgs field. Furthermore, in the SM it is also not possible to violate CP explicitly in the Higgs sector since gauge invariance together with renormalizability restrict the Higgs potential to have only quadratic and quartic terms and hermiticity constrains both of these to be real. Thus, CP violation in the SM requires the introduction of complex Yukawa couplings.

The scenario of spontaneous CP and T violation has the nice feature of putting the breakdown of discrete symmetries on the same footing as the breaking of the gauge symmetry, which is also spontaneous in order to preserve renormalizability. A simple extension of the Higgs sector that may give rise to spontaneous CP violation requires the presence of at least two Higgs doublets, and was introduced by Lee [554].

If one introduces two Higgs doublets, it is possible to have either explicit or spontaneous CP breaking. Explicit CP violation in the Higgs sector arises due to the fact that in this case there are gauge invariant terms in the Lagrangian which can have complex coefficients. Note however that the presence of complex coefficients does not always lead to explicit CP breaking.

Extensions of the SM with extra Higgs doublets are very natural since they keep the ρ parameter at tree level equal to one [555]. In multi-Higgs systems there are in general, additional sources of CP violation in the Higgs sector [556]. The most general renormalizable polynomial consistent with the $SU(2) \times U(1) \times SU(3)_c$ model with n_d Higgs doublets, ϕ_i, may be written as

$$\mathcal{L}_\phi = Y_{ab}\phi_a^\dagger \phi_b + Z_{abcd}(\phi_a^\dagger \phi_b)(\phi_c^\dagger \phi_d), \quad (5.1)$$

where repeated indices are summed. Hermiticity of $\mathcal{L}_\phi$ implies:

$$Y_{ab}^* = Y_{ba}; \qquad Z_{abcd}^* = Z_{badc}. \quad (5.2)$$

Furthermore, by construction it is obvious that:

$$Z_{abcd} = Z_{cdab}. \quad (5.3)$$

In models with more than one Higgs doublet, one has the freedom to make Higgs-basis transformations (HBT) that do not change the physical content of the model, but do change both the quadratic and the quartic coefficients. Coefficients that are complex in one Higgs basis may become real in another basis. Furthermore, a given model may have complex quartic coefficients in one Higgs basis, while they may all become real in another basis, with only the quadratic coefficients now complex, thus indicating that in that particular model CP is only softly broken. Such Higgs-basis transformations leave the Higgs kinetic energy term invariant and are of the form:

$$\phi_a \xrightarrow{\text{HBT}} \phi_a' = V_{ai}\phi_i, \qquad \phi_a^\dagger \xrightarrow{\text{HBT}} (\phi')_a^\dagger = V_{ai}^*(\phi')_i^\dagger, \quad (5.4)$$

where V is an $n_d \times n_d$ unitary matrix acting in the space of Higgs doublets. In [557] conditions for a given Higgs potential to violate CP at the Lagrangian level, expressed in terms of CP-odd Higgs-basis invariants, were derived. These conditions are expressed in terms of couplings of the unbroken Lagrangian, therefore they are relevant even at high energies, where the $SU(2) \times U(1)$ symmetry is restored. This feature renders them potentially useful for the study of baryogenesis. The derivation of these conditions follows the general method proposed in [558] and already mentioned in previous sections. The method consists of imposing invariance of the Lagrangian under the most general CP transformation of the Higgs doublets, which is a combination of a simple CP transformation for each Higgs field with a Higgs-basis transformation:

$$\phi_a \xrightarrow{\text{CP}} W_{ai}\phi_i^*; \qquad \phi_a^\dagger \xrightarrow{\text{CP}} W_{ai}^*\phi_i^T. \quad (5.5)$$

Here W is an $n_d \times n_d$ unitary matrix operating in Higgs doublets space.

A set of necessary and sufficient conditions for CP invariance in the case of two Higgs doublets have been derived [557]:

$$\begin{aligned} I_1 &\equiv \text{Tr}[Y\, Z_Y\, \widehat{Z} - \widehat{Z}\, Z_Y\, Y] = 0, \\ I_2 &\equiv \text{Tr}[Y\, Z_2\, \tilde{Z} - \tilde{Z}\, Z_2\, Y] = 0, \end{aligned} \quad (5.6)$$

where all matrices inside the parenthesis are 2×2 matrices. In the general case these are $n_d \times n_d$ matrices, and are defined by:

$$\begin{aligned} (Z_Y)_{ij} &\equiv Z_{ijmn}Y_{mn}; & \widehat{Z}_{ij} &\equiv Z_{ijmm}; \\ (Z_2)_{ij} &\equiv Z_{ipnm}Z_{mnpj}; & \tilde{Z}_{ij} &\equiv Z_{immj} \end{aligned} \quad (5.7)$$

CP-odd HBT invariants are also useful [557] to find out whether, in a given model, there is hard or soft CP breaking. One may also construct CP-odd weak basis invariants, involving $v_i \equiv \langle 0|\phi_i^0|0\rangle$, i.e., after spontaneous gauge symmetry breaking has occurred [559, 560]. Further discussions on Higgs-basis independent methods for the two-Higgs-doublet model can be found in [561–564].

So far, we have considered CP violation at the Lagrangian level in models with multi-Higgs doublets, i.e., explicit CP violation. It is also possible to derive criteria [565] to verify whether CP and T in a given model are spontaneously broken. Under T the Higgs fields ϕ_j transform as

$$T\phi_j T^{-1} = U_{jk}\phi_k, \quad (5.8)$$

where U is a unitary matrix which may mix the scalar doublets. If no extra symmetries beyond $SU(2) \times U(1)$ are present in the Lagrangian, U reduces to a diagonal matrix possibly with phases. Invariance of the vacuum under T leads to the following condition:

$$\langle 0|\phi_j^0|0\rangle = U_{jk}^* \langle 0|\phi_k^0|0\rangle^*. \quad (5.9)$$

Therefore, a set of vacua lead to spontaneous T, CP violation if there is no unitary matrix U satisfying (5.8) and (5.9) simultaneously.

Most of the previous discussion dealt with the general case of n-Higgs doublets. We analyze now the case of two Higgs doublets, where the most general gauge invariant Higgs potential can be explicitly written as

$$\begin{aligned} V_{H_2} &= m_1 \phi_1^\dagger \phi_1 + p e^{i\varphi} \phi_1^\dagger \phi_2 + p e^{-i\varphi} \phi_2^\dagger \phi_1 + m_2 \phi_2^\dagger \phi_2 \\ &+ a_1 (\phi_1^\dagger \phi_1)^2 + a_2 (\phi_2^\dagger \phi_2)^2 + b(\phi_1^\dagger \phi_1)(\phi_2^\dagger \phi_2) \\ &+ b'(\phi_1^\dagger \phi_2)(\phi_2^\dagger \phi_1) + c_1 e^{i\theta_1} (\phi_1^\dagger \phi_1)(\phi_2^\dagger \phi_1) \\ &+ c_1 e^{-i\theta_1} (\phi_1^\dagger \phi_1)(\phi_1^\dagger \phi_2) + c_2 e^{i\theta_2} (\phi_2^\dagger \phi_2)(\phi_2^\dagger \phi_1) \\ &+ c_2 e^{-i\theta_2} (\phi_2^\dagger \phi_2)(\phi_1^\dagger \phi_2) + d e^{i\delta} (\phi_1^\dagger \phi_2)^2 \\ &+ d e^{-i\delta} (\phi_2^\dagger \phi_1)^2, \quad (5.10) \end{aligned}$$

where m_i, p, a_i, b, b', c_i, and d are real and all phases are explicitly displayed. It is clear that this potential contains an excess of parameters. With the appropriate choice of Higgs basis some of these may be eliminated, without loss of generality, leaving eleven independent parameters [569–571]. The Higgs sector contains five spinless particles: three neutral and a pair of charged ones, usually denoted by h, H (CP even), A (CP odd) (or if CP is violated $h_{1,2,3}$) and $H^\pm$.

In general, models with two Higgs doublets have tree level Higgs-mediated flavor changing neutral currents (FCNC). This is a problem in view of the present stringent experimental limits on FCNC. In order to solve this problem the concept of natural flavor conservation (NFC) was introduced by imposing extra symmetries on the Lagrangian. These symmetries constrain the Yukawa couplings of the neutral scalars in such a way that the resulting neutral currents are diagonal. Glashow and Weinberg [566] and Paschos [567] have shown that the only way to achieve NFC is to ensure that only one Higgs doublet gives mass to quarks of a given charge.

In the case of two Higgs doublets the simplest solution to avoid FCNC is to require invariance of the Lagrangian under the following transformation of the Z_2 type:

$$\begin{aligned} \phi_1 &\longrightarrow \phi_1, \qquad \phi_2 \longrightarrow -\phi_2, \\ d_R &\longrightarrow d_R, \qquad u_R \longrightarrow -u_R, \end{aligned} \quad (5.11)$$

where d_R (u_R) denote the right handed down (up) quarks; all other fields remain unchanged.

It is clear from (5.10) that this symmetry eliminates explicit CP violation in the Higgs sector, since the only term of the Higgs potential with a phase that survives is the one with coefficient d, moreover a HBT of the form $\phi_1 \longrightarrow e^{i\delta/2} \phi_1$, $\phi_2 \longrightarrow \phi_2$, eliminates the phase from the Higgs potential. Furthermore, it can be shown that this symmetry also eliminates the possibility of having spontaneous CP violation.

In conclusion, models with two Higgs doublets and exact NFC cannot give rise to spontaneous CP violation. Explicit CP violation in this case requires complex Yukawa couplings leading to the Kobayashi–Maskawa mechanism with no additional source of CP violation through neutral scalar Higgs boson exchange. An interesting alternative scenario in the case of two Higgs doublets was considered in [568] with no NFC. Here CP violating Higgs FCNC are naturally suppressed through a permutation symmetry which is softly broken, still allowing for spontaneous CP violation.

Three Higgs doublet models have been considered in an attempt to introduce CP violation in an extension of the SM with NFC [566] in the Higgs sector. It was shown that indeed, in such models it is possible to violate CP in the Higgs sector either at the Lagrangian level [572] or spontaneously [573–575].

It is also possible to generate spontaneous CP violation with only one additional Higgs singlet [576], but in this case at least one isosinglet vectorial quark is required in order to generate a non-trivial phase at low energies in the Cabibbo–Kobayashi–Maskawa matrix. Such models may provide a solution to the strong CP problem of the type proposed by Nelson [577, 578] and Barr [579] as well as a common origin to all CP violations [580, 581] including the generation of the observed baryon asymmetry of the Universe. The fact that the SM cannot provide the observed baryon asymmetry [582–587], provides yet another reason to study an enlarged Higgs sector.

A lot of work has been done by many authors on possible extensions of the Higgs sector and their implications both for the hadronic and the leptonic sectors at the existing and future colliders, see e.g. [588]. Among the simplest multi-Higgs models are the two Higgs Doublet Models (2HDM) which have been analyzed in detail in many different realizations. The need to avoid potentially dangerous tree level Higgs FCNC has led to the consideration of different variants of this model with a certain discrete Z_2 symmetry imposed.

In the Type-I 2HDM the Z_2 discrete symmetry imposed on the Lagrangian is such that only one of the Higgs doublets couples to quarks and leptons. A very well known fermiophobic Higgs boson may arise in such model [589–591]. Another example is the Inert Doublet Model, with an unbroken discrete Z_2 symmetry which forbids one Higgs doublet to couple to fermions and to get a non-zero VEV [592, 593]. Physical particles related to such doublets are called "inert" particles, the lightest is stable and contributes to the Dark Matter density. In [594], the naturalness problem has been addressed in the framework of an Inert Doublet Model with a heavy (SM-like) Higgs boson. In this context Dark Matter may be composed of neutral inert Higgs bosons. Predictions are given for multilepton events with missing transverse energy at the LHC, and for the direct detection of dark matter.

The Type-II 2HDM allows one of the Higgs doublet to couple only to the right-handed up quarks while the other Higgs doublet can only couple to right handed down-type quarks and charged leptons. This is achieved by the introduction of an appropriate Z_2 symmetry, analogous to the one in (5.12). The Higgs sector of the MSSM model can be viewed as a particular realization of Type-II models but with additional constraints required by supersymmetry. Various scenarios are possible for these models—with and without decoupling of heavy Higgs particles [570, 571, 595].

Type-III 2HDM are models where, unlike in models of Type-I and II, NFC is not imposed on the Yukawa interactions. This class of models has in general scalar mediated FCNC at tree level. Various schemes have been proposed to suppress these currents, including the ad-hoc assumption that FCNC couplings are approximately given by the geometric mean of the Yukawa couplings of the two generations [596]. A very interesting alternative [597] is to have an exact symmetry of the Lagrangian which constrains FCNC couplings to be related in an exact way to the elements of the CKM matrix in such a way that FCNC are non-vanishing but naturally suppressed by the smallness of CKM mixing. Another example of Type III 2HDM is the Top Two Higgs Doublet Model which was first proposed in [598], and recently analyzed in detail in [599]. In this framework a discrete symmetry is imposed allowing only the top quark to have Yukawa couplings to one of the doublets while all other quarks and leptons have Yukawa couplings to the other doublet.

Lepton flavor violation is a feature common to many possible extensions of the SM. It can occur both through charged and neutral currents. The possibility of having lepton flavor violation in extensions of the SM, has been considered long before the discovery of neutrino masses [600, 601]. For example, in the case of multi-Higgs doublet models, it has been pointed out that even for massless neutrinos lepton flavor can be violated [602, 603]. In the context of the minimal extension of the SM, necessary to accommodate neutrino masses, where only right handed neutrinos are included LFV effects are extremely small. It is well known that the effects of LFV can be large in supersymmetry.

CLEO submitted recently a paper [604] where the ratio of the tauonic and muonic branching fractions is examined for the three $\Upsilon(1S, 2S, 3S)$ states. Agreement with expectations from lepton universality is found. The conclusion is that lepton universality is respected within the current experimental accuracy which is roughly 10%. However there is tendency for the tauonic branching fraction to turn out systematically larger than the muonic at a few per cent level.

5.1.2 Low scale singlet neutrino scenarios

In the pre-LHC era neutrino oscillations have provided some of the most robust evidence for physics beyond the SM. There are many open questions in this field; why is the absolute mass scale for the neutrinos so small with respect to the other SM particles? what is this mass scale? why is the pattern of mixing so different from the quark sector? If nature has chosen the singlet seesaw scenario [216–220] as an answer to those questions we face the prospect of never being able to produce the heavy neutrinos at a collider. Nevertheless, several extensions of this minimal see-saw scenario contain heavy neutrinos at or around the TeV scale, these include models based around the group E_6 [605, 606] and also in $SO(10)$ models [403].

Furthermore, even within the usual see-saw scenario, the observed nearly maximal mixing pattern of the light neutrinos requires further explanation. Flavor symmetries are often invoked as possible reasons for the almost tri-bi-maximal structure of the PMNS mixing matrix [607]. It is also possible that the small magnitude of the light neutrino masses is due to an approximate symmetry, allowing the right handed neutrinos to be as light as $\mathcal{O}(200\text{ GeV})$ [337].

TeV scale right handed neutrinos can also arise in radiative mechanisms of neutrino mass generation. Generically, in these models a tree-level neutrino mass is forbidden or suppressed by a symmetry but small neutrino masses may arise through loops sensitive to symmetry breaking effects [225, 608]. Indeed, several supersymmetric realizations of radiative mechanisms contain TeV scale right handed neutrinos linked to the scale of supersymmetry breaking [609, 610].

5.1.2.1 Heavy neutrinos accessible to the LHC
A low, electroweak-scale mass is not sufficient to imply that heavy neutrinos could be produced and detected at the LHC. They must have a large enough coupling (mixing) with other SM fields so that experiments will be able to distinguish their production and decay from SM background processes. In this review we concentrate on the case where heavy neutrino production and decay occurs through mixing with SM fields only. Quantitatively, we can consider a generalization of the Langacker–London parameters, $\Omega_{ll'}$, defined as

$$\Omega_{ll'} = \delta_{ll'} - \sum_{i=1}^{3} B_{li} B_{l'i}^* = \sum_{i=4}^{(3+n_R)} B_{li} B_{l'i}^*, \quad (5.12)$$

where $l, l' = e, \mu, \tau$ and B_{li} is the full $3 \times (3 + n_R)$ neutrino mixing matrix taking into account all (3 light and n_R heavy) neutrinos. The 3×3 matrix B_{li} where $i = 1, \ldots, 3$ is a good approximation to the usual PMNS matrix and $\Omega_{ll'}$ essentially measures the deviation from unitarity of the PMNS matrix.

The $\Omega_{ll'}$ are constrained by precision electroweak data [611] and the following upper limits have been set at 90% C.L.

$$\Omega_{ee} \leq 0.012, \qquad \Omega_{\mu\mu} \leq 0.0096, \qquad \Omega_{\tau\tau} \leq 0.016. \tag{5.13}$$

In addition, the off-diagonal elements of $\Omega_{ll'}$ are constrained by limits on lepton flavor violating processes such as $\tau, \mu \to e\gamma$ and $\tau, \mu \to eee$ and $\mu \to e$ conversion in nuclei [514, 612]. These limits are rather model dependent but for $M_R \gg M_W$ and $m_D \ll M_W$ (where m_D is the Dirac component of the neutrino mass matrix), the present upper bounds are [182]

$$|\Omega_{e\mu}| \leq 0.0001, \qquad |\Omega_{e\tau}| \leq 0.02, \qquad |\Omega_{\mu\tau}| \leq 0.02. \tag{5.14}$$

It has been pointed out that a heavy Majorana neutrino (N) may be produced via a DY type of mechanism at hadron colliders [608, 613–617], $pp \to W^{+*} \to \ell^+ N$, where $N \to \ell^+ W^-$, leading to lepton number violation by 2. Most of the previous studies were concentrated on the ee mode, which would result in a too week signal to be appreciable due to the recent very stringent bound $|V_{eN}|^2/m_N < 5 \times 10^{-8}$ GeV^{-1}, from the absence of the neutrinoless double beta decay. It has been recently proposed to search for the unique and clean signal, $\mu^\pm \mu^\pm + 2$ jets at the LHC [617]. It was concluded that a search at the LHC with an integrated luminosity of 100 fb^{-1} can be sensitive to a mass range of $m_N \sim$ 10–400 GeV at a 2σ level, and up to 250 GeV at a 5σ level. If this type of signal could be established, it would be even feasible to consider the search for CP violation in the heavy Majorana sector [618].

A recent analysis [619] studied more background processes including some fast detector simulations. In particular, the authors claimed a large background due to the faked leptons $b\bar{b} \to \mu^+ \mu^+$. The search sensitivity is thus reduced to 175 GeV at a 5σ level. However, the background estimate for processes such as $b\bar{b} + n$-jet has large uncertainties due to QCD perturbative calculations and kinematical acceptance. More studies remain to be done for a definitive conclusion.

5.1.2.2 Low scale model with successful baryogenesis

As a more detailed example satisfying the constraints of (5.14) we consider a model potentially accessible to colliders, where $M_R \simeq 250$ GeV which has been shown to successfully explain the baryon asymmetry of the Universe [337].

Leptogenesis has been discussed in Sect. 3.3.1. Low scale leptogenesis scenario would be possible with nearly degenerate heavy neutrinos, where self-energy effects on the leptonic asymmetries become relevant [293, 294]. In this case the CP asymmetry in the heavy neutrino decays can be resonantly enhanced [332], to the extent that the observed baryon asymmetry can be explained with heavy neutrinos as light as the electroweak scale [335, 337].

We shall consider a model with right handed neutrinos which transform under an $SO(3)$ flavor symmetry. Ignoring effects from the neutrino Yukawa couplings this symmetry is assumed to be exact at some high scale, e.g. the GUT scale M_{GUT}. This restricts the form of the heavy Majorana neutrino mass matrix at M_{GUT}

$$M_R = \mathbf{1} m_N + \delta M_S, \tag{5.15}$$

where $\delta M_S = 0$ at M_{GUT}. This form has also been considered in a class of "minimal flavor violating" models of the lepton sector [542] and naturally provides nearly degenerate heavy neutrinos compatible with resonant leptogenesis.

All other fields are singlets under this $SO(3)$ flavor symmetry and so the neutrino Yukawa couplings will break $SO(3)$ explicitly. We can still choose heavy neutrino Yukawa couplings Y^ν so that a subgroup of the $SO(3) \times U(1)_{L_e} \times U(1)_{L_\mu} \times U(1)_{L_\tau}$ flavor symmetry present without the neutrino Yukawa couplings remains unbroken. In this case a particular flavor direction can be singled out leaving $SO(2) \simeq U(1)$ unbroken. This residual $U(1)$ symmetry acts to prevent the light Majorana neutrinos from acquiring a mass. The form of the neutrino Yukawa couplings can be written

$$Y^{\nu T} = \begin{pmatrix} 0 & ae^{-i\pi/4} & ae^{i\pi/4} \\ 0 & be^{-i\pi/4} & be^{i\pi/4} \\ 0 & ce^{-i\pi/4} & ce^{i\pi/4} \end{pmatrix} + \delta Y^\nu. \tag{5.16}$$

The residual $U(1)$ symmetry is broken both by small $SO(3)$ breaking effects in the heavy Majorana mass matrix, δM_S, and by small effects parameterized by δY^ν in the Yukawa couplings. Although we shall not consider the specific origin of these effects, δM_S could arise through renormalization group running for example.

In [337], a specific model was considered where $m_N = 250$ GeV and which successfully explained the baryon asymmetry of the Universe. One of either a, b or c was constrained to be small to allow a single lepton flavor asymmetry (and subsequently a baryon asymmetry) to be generated at $T \sim 250$ GeV. The other two parameters could be as large as $\mathcal{O}(10^{-2})$. This scenario has the features necessary for a model to be visible at the LHC; heavy neutrinos with masses around $\mathcal{O}(1\text{ TeV})$ and sufficient mixing between these neutrinos and the light neutrinos to allow them to be produced from a vector boson. Specifically

$$\Omega_{ee} = \frac{|a|^2 v^2}{m_N^2}, \qquad \Omega_{\mu\mu} = \frac{|b|^2 v^2}{m_N^2}, \qquad \Omega_{\tau\tau} = \frac{|c|^2 v^2}{m_N^2}, \tag{5.17}$$

where $v = 246$ GeV is the vacuum expectation value of the Higgs field.

It should be noted that in this model the heavy neutrinos produced at the LHC would be linked indirectly with the mechanism providing light neutrinos with small masses. The light neutrinos acquire masses directly through the mechanism responsible for breaking the flavor symmetries. However, studying the properties of the heavy neutrinos accessible to the LHC would allow us to better understand the underlying symmetry protecting light neutrinos from large masses and may give us insight into the observed pattern of large mixing. In addition, further knowledge of heavy neutrinos seen at the LHC, for example small couplings with one or more lepton flavors or large, resonantly enhanced CP violation, would provide us with further information on possible explanations for the baryon asymmetry of the Universe.

5.1.3 Lepton flavor violation from the mirror leptons in little Higgs models

Little Higgs models [620–624] offer an alternative route to the solution of the little hierarchy problem. One of the most attractive models of this class is the littlest Higgs model [625] with T-parity (LHT) [626–628], where the discrete symmetry forbids tree-level corrections to electroweak observables, thus weakening the electroweak precision constraints [629]. Under this new symmetry the particles have distinct transformation properties, that is, they are either T-even or T-odd. The model is based on a two-stage spontaneous symmetry breaking occurring at the scale f and the electroweak scale v. Here the scale f is taken to be larger than about 500 GeV, which allows to expand expressions in the small parameter v/f. The additionally introduced gauge bosons, fermions and scalars are sufficiently light to be discovered at LHC and there is a dark matter candidate [630]. Moreover, the flavor structure of the LHT model is richer than the one of the SM, mainly due to the presence of three doublets of mirror quarks and three doublets of mirror leptons and their weak interactions with the ordinary quarks and leptons, as discussed in [631–633].

Now, it is well known that in the SM the FCNC processes in the lepton sector, like $\ell_i \to \ell_j \gamma$ and $\mu \to eee$, are very strongly suppressed due to tiny neutrino masses. In particular, the branching ratio for $\mu \to e\gamma$ in the SM amounts to at most 10^{-54}, to be compared with the present experimental upper bound, 1.2×10^{-11} [180], and with the one that will be available within the next two years, $\sim 10^{-13}$ [634, 635]. Results close to the SM predictions are expected within the LH model without T-parity, where the lepton sector is identical to the one of the SM and the additional $\mathcal{O}(v^2/f^2)$ corrections have only minor impact on this result. Similarly the new effects on $(g-2)_\mu$ turn out to be small [636, 637].

A very different situation is to be expected in the LHT model, where the presence of new flavor violating interactions and of mirror leptons with masses of order 1 TeV can change the SM expectations by up to 45 orders of magnitude, bringing the relevant branching ratios for lepton flavor violating (LFV) processes close to the bounds available presently or in the near future.

5.1.3.1 The model
A detailed description of the LHT model can be found in [638], where also a complete set of Feynman rules has been derived. Here we just want to state briefly the ingredients needed for the analysis of LFV decays.

The T-odd gauge boson sector consists of three heavy "partners" of the SM gauge bosons

$$W_H^\pm, \qquad Z_H, \qquad A_H, \tag{5.18}$$

with masses given to lowest order in v/f by

$$M_{W_H} = gf, \qquad M_{Z_H} = gf, \qquad M_{A_H} = \frac{g'f}{\sqrt{5}}. \tag{5.19}$$

The T-even fermion sector contains, in addition to the SM fermions, the heavy top partner T_+. On the other hand, the T-odd fermion sector [631] consists of three generations of mirror quarks and leptons with vectorial couplings under $SU(2)_L \times U(1)_Y$, that are denoted by

$$\begin{pmatrix} u_H^i \\ d_H^i \end{pmatrix}, \qquad \begin{pmatrix} \nu_H^i \\ \ell_H^i \end{pmatrix} \qquad (i = 1, 2, 3). \tag{5.20}$$

To first order in v/f the masses of up- and down-type mirror fermions are equal. Naturally, their masses are of order f. In the analysis of LFV decays, except for $K_{L,S} \to \mu e$, $K_{L,S} \to \pi^0 \mu e$, $B_{d,s} \to \ell_i \ell_j$ and $\tau \to \ell\pi, \ell\eta, \ell\eta'$, only mirror leptons are relevant.

As discussed in detail in [632], one of the important ingredients of the mirror sector is the existence of four CKM-like unitary mixing matrices, two for mirror quarks (V_{Hu}, V_{Hd}) and two for mirror leptons ($V_{H\nu}, V_{H\ell}$), that are related via

$$V_{Hu}^\dagger V_{Hd} = V_{CKM}, \qquad V_{H\nu}^\dagger V_{H\ell} = V_{PMNS}^\dagger. \tag{5.21}$$

An explicit parameterization of V_{Hd} and $V_{H\ell}$ in terms of three mixing angles and three complex (non-Majorana) phases can be found in [633].

The mirror mixing matrices parameterize flavor violating interactions between SM fermions and mirror fermions that are mediated by the heavy gauge bosons $W_H^\pm$, Z_H and A_H. The matrix notation indicates which of the light fermions of a given electric charge participates in the interaction.

In the course of the analysis of charged LFV decays it is useful to introduce the following quantities ($i = 1, 2, 3$) [639]:

$$\chi_i^{(\mu e)} = V_{H\ell}^{*ie} V_{H\ell}^{i\mu}, \qquad \chi_i^{(\tau e)} = V_{H\ell}^{*ie} V_{H\ell}^{i\tau}, \qquad (5.22)$$
$$\chi_i^{(\tau \mu)} = V_{H\ell}^{*i\mu} V_{H\ell}^{i\tau},$$

that govern $\mu \to e$, $\tau \to e$ and $\tau \to \mu$ transitions, respectively. Analogous quantities in the mirror quark sector ($i = 1, 2, 3$) [638, 641],

$$\xi_i^{(K)} = V_{Hd}^{*is} V_{Hd}^{id}, \qquad \xi_i^{(d)} = V_{Hd}^{*ib} V_{Hd}^{id}, \qquad (5.23)$$
$$\xi_i^{(s)} = V_{Hd}^{*ib} V_{Hd}^{is},$$

are needed for the analysis of the decays $K_{L,S} \to \mu e$, $K_{L,S} \to \pi^0 \mu e$ and $B_{d,s} \to \ell_i \ell_j$.

As an example, the branching ratio for the $\mu \to e\gamma$ decay contains the $\chi_i^{(\mu e)}$ factors introduced in (5.22) via the short distance function [639]

$$\bar{D}_{\text{odd}}^{\prime \mu e} = \frac{1}{4} \frac{v^2}{f^2} \sum_i \left(\chi_i^{(\mu e)} \left(D_0'(y_i) - \frac{7}{6} E_0'(y_i) \right. \right.$$
$$\left. \left. - \frac{1}{10} E_0'(y_i') \right) \right), \qquad (5.24)$$

where $y_i = (m_{Hi}^\ell / M_{W_H})^2$, $y_i' = a y_i$ with $a = 5/\tan^2\theta_W$, and explicit expressions for the functions D_0', E_0' can be found in [642].

The new parameters of the LHT model, relevant for the study of LFV decays, are

$$f, \quad m_{H1}^\ell, \quad m_{H2}^\ell, \quad m_{H3}^\ell, \quad \theta_{12}^\ell, \quad \theta_{13}^\ell, \quad \theta_{23}^\ell, \qquad (5.25)$$
$$\delta_{12}^\ell, \quad \delta_{13}^\ell, \quad \delta_{23}^\ell$$

and the ones in the mirror quark sector that can be probed by FCNC processes in K and B meson systems, as discussed in detail in [638, 641]. Once the new heavy gauge bosons and mirror fermions will be discovered and their masses measured at the LHC, the only free parameters of the LHT model will be the mixing angles θ_{ij}^ℓ and the complex phases δ_{ij}^ℓ of the matrix $V_{H\ell}$, that can be determined with the help of LFV processes. Analogous comments apply to the determination of V_{Hd} parameters in the quark sector (see [638, 641] for details on K and B physics in the LHT model).

5.1.3.2 Results LFV processes in the LHT model have for the first time been discussed in [643], where the decays $\ell_i \to \ell_j \gamma$ have been considered. Further, the new contributions to $(g - 2)_\mu$ in the LHT model have been calculated by these authors. In [639, 640] the analysis of LFV in the LHT model has been considerably extended, and includes the decays $\ell_i \to \ell_j \gamma$, $\mu \to eee$, the six three body leptonic decays $\tau^- \to \ell_i^- \ell_j^+ \ell_k^-$, the semileptonic decays $\tau \to \ell\pi, \ell\eta, \ell\eta'$ and the decays $K_{L,S} \to \mu e$, $K_{L,S} \to \pi^0 \mu e$ and $B_{d,s} \to \ell_i \ell_j$ that are flavor violating both in the quark and lepton sector. Moreover, μ–e conversion in nuclei and the flavor conserving $(g - 2)_\mu$ have been studied. Furthermore, a detailed phenomenological analysis has been performed in that paper, paying particular attention to various ratios of LFV branching ratios that will be useful for a clear distinction of the LHT model from the MSSM.

In contrast to K and B physics in the LHT model, where the SM contributions constitute a sizable and often the dominant part, the T-even contributions to LFV observables are completely negligible due to the smallness of neutrino masses and the LFV decays considered are entirely governed by mirror fermion contributions.

In order to see how large these contributions can possibly be, it is useful to consider first those decays for which the strongest constraints exist. Therefore Fig. 5 shows $B(\mu \to eee)$ as a function of $B(\mu \to e\gamma)$, obtained from a general scan over the mirror lepton parameter space, with $f = 1$ TeV. It is found that in order to fulfill the present bounds, either the mirror lepton spectrum has to be quasi-degenerate or the $V_{H\ell}$ matrix must be very hierarchical. Moreover, as shown in Fig. 6, even after imposing the constraints on $\mu \to e\gamma$ and $\mu \to eee$, the μ–e conversion rate in Ti is very likely to be found close to its current bound, and for some regions of the mirror lepton parameter space even violates this bound.

The existing constraints on LFV τ decays are still relatively weak, so that they presently do not provide a useful constraint on the LHT parameter space. However, as seen in Table 10, most branching ratios in the LHT model can

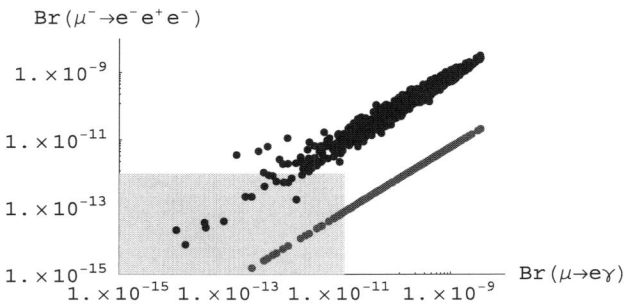

Fig. 5 Correlation between $B(\mu \to e\gamma)$ and $B(\mu \to eee)$ in the LHT model (*upper dots*) [639]. The *lower dots* represent the dipole contribution to $\mu \to eee$ separately, which, unlike in the LHT model, is the dominant contribution in the MSSM. The *grey region* is allowed by the present experimental bounds

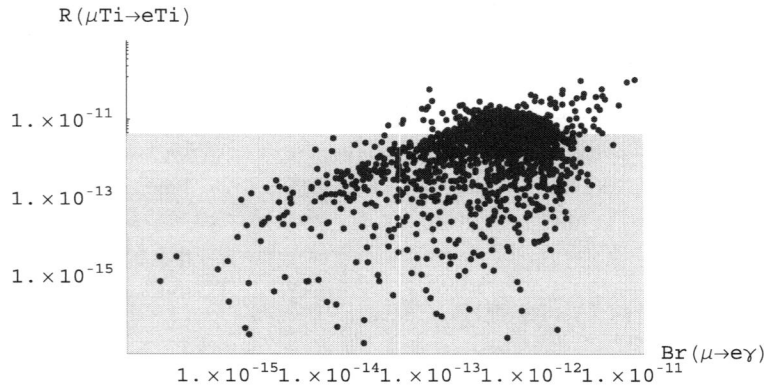

Fig. 6 $R(\mu\text{Ti} \to e\text{Ti})$ as a function of $B(\mu \to e\gamma)$, after imposing the existing constraints on $\mu \to e\gamma$ and $\mu \to eee$ [639]. The *grey region* is allowed by the present experimental bounds

Table 10 Upper bounds on LFV τ decay branching ratios in the LHT model, for two different values of the scale f, after imposing the constraints on $\mu \to e\gamma$ and $\mu \to eee$ [639]. For $f = 500$ GeV, also the bounds on $\tau \to \mu\pi, e\pi$ have been included. The current experimental upper bounds are also given. The bounds in [183] have been obtained by combining Belle [646, 647] and BaBar [182, 648] results

Decay	$f = 1000$ GeV	$f = 500$ GeV	exp. upper bound
$\tau \to e\gamma$	8×10^{-10}	1×10^{-8}	9.4×10^{-8} [183]
$\tau \to \mu\gamma$	8×10^{-10}	2×10^{-8}	1.6×10^{-8} [183]
$\tau^- \to e^- e^+ e^-$	7×10^{-10}	2×10^{-8}	2.0×10^{-7} [644]
$\tau^- \to \mu^- \mu^+ \mu^-$	7×10^{-10}	3×10^{-8}	1.9×10^{-7} [644]
$\tau^- \to e^- \mu^+ \mu^-$	5×10^{-10}	2×10^{-8}	2.0×10^{-7} [645]
$\tau^- \to \mu^- e^+ e^-$	5×10^{-10}	2×10^{-8}	1.9×10^{-7} [645]
$\tau^- \to \mu^- e^+ \mu^-$	5×10^{-14}	2×10^{-14}	1.3×10^{-7} [644]
$\tau^- \to e^- \mu^+ e^-$	5×10^{-14}	2×10^{-14}	1.1×10^{-7} [644]
$\tau \to \mu\pi$	2×10^{-9}	5.8×10^{-8}	5.8×10^{-8} [183]
$\tau \to e\pi$	2×10^{-9}	4.4×10^{-8}	4.4×10^{-8} [183]
$\tau \to \mu\eta$	6×10^{-10}	2×10^{-8}	5.1×10^{-8} [183]
$\tau \to e\eta$	6×10^{-10}	2×10^{-8}	4.5×10^{-8} [183]
$\tau \to \mu\eta'$	7×10^{-10}	3×10^{-8}	5.3×10^{-8} [183]
$\tau \to e\eta'$	7×10^{-10}	3×10^{-8}	9.0×10^{-8} [183]

reach the present experimental upper bounds, in particular for low values of f, and are very interesting in view of new experiments taking place in this and the coming decade.

The situation is different in the case of $K_L \to \mu e$, $K_L \to \pi^0 \mu e$ and $B_{d,s} \to \ell_i \ell_k$, due to the double GIM suppression in the quark and lepton sectors. E.g. $B(K_L \to \mu e)$ can reach values of at most 3×10^{-13} which is still one order of magnitude below the current bound, and $K_L \to \pi^0 \mu e$ is even by two orders of magnitude smaller. Still, measuring the rates for $K_L \to \mu e$ and $K_L \to \pi^0 \mu e$ would be desirable, as, due to their sensitivity to $\text{Re}(\xi_i^{(K)})$ and $\text{Im}(\xi_i^{(K)})$ respectively, these decays can shed light on the complex phases present in the mirror quark sector.

While the possible huge enhancements of LFV branching ratios in the LHT model are clearly interesting, such effects are common to many other NP models, such as the MSSM, and therefore cannot be used to distinguish these models. However, correlations between various branching ratios should allow a clear distinction of the LHT model from the MSSM. While in the MSSM [169, 175, 242, 649, 650] the dominant role in decays with three leptons in the final state and in μ–e conversion in nuclei is typically played by the dipole operator, in [639] it is found that this operator is basically irrelevant in the LHT model, where Z^0-penguin and box diagram contributions are much more important. As can be seen in Table 11 and also in Fig. 5 this implies a striking difference between various ratios of branching ratios in the MSSM and in the LHT model and should be very useful in distinguishing these two models. Even if for some decays this distinction is less clear when significant Higgs contributions are present [169, 175, 650], it should be easier than through high energy processes at LHC.

Another possibility to distinguish different NP models through LFV processes is given by the measurement of $\mu \to e\gamma$ with polarized muons. Measuring the angular distribution of the outgoing electrons, one can determine the

Table 11 Comparison of various ratios of branching ratios in the LHT model and in the MSSM without and with significant Higgs contributions [639]

Ratio	LHT	MSSM (dipole)	MSSM (Higgs)
$B(\mu^- \to e^-e^+e^-)/B(\mu \to e\gamma)$	0.4–2.5	$\sim 6 \times 10^{-3}$	$\sim 6 \times 10^{-3}$
$B(\tau^- \to e^-e^+e^-)/B(\tau \to e\gamma)$	0.4–2.3	$\sim 1 \times 10^{-2}$	$\sim 1 \times 10^{-2}$
$B(\tau^- \to \mu^-\mu^+\mu^-)/B(\tau \to \mu\gamma)$	0.4–2.3	$\sim 2 \times 10^{-3}$	0.06–0.1
$B(\tau^- \to e^-\mu^+\mu^-)/B(\tau \to e\gamma)$	0.3–1.6	$\sim 2 \times 10^{-3}$	0.02–0.04
$B(\tau^- \to \mu^-e^+e^-)/B(\tau \to \mu\gamma)$	0.3–1.6	$\sim 1 \times 10^{-2}$	$\sim 1 \times 10^{-2}$
$B(\tau^- \to e^-e^+e^-)/B(\tau^- \to e^-\mu^+\mu^-)$	1.3–1.7	~ 5	0.3–0.5
$B(\tau^- \to \mu^-\mu^+\mu^-)/B(\tau^- \to \mu^-e^+e^-)$	1.2–1.6	~ 0.2	5–10
$R(\mu\text{Ti} \to e\text{Ti})/B(\mu \to e\gamma)$	0.01–100	$\sim 5 \times 10^{-3}$	0.08–0.15

size of left and right handed contributions separately [651]. In addition, detecting also the electron spin would yield information on the relative phase between these two contributions [652]. We recall that the LHT model is peculiar in this respect as it does not involve any right handed contribution.

On the other hand, the contribution of mirror leptons to $(g-2)_\mu$, being a flavor conserving observable, is negligible [639, 643], so that the possible discrepancy between SM prediction and experimental data [653] cannot be cured. This should also be contrasted with the MSSM with large $\tan\beta$ and not too heavy scalars, where those corrections could be significant, thus allowing to solve the possible discrepancy between SM prediction and experimental data.

5.1.3.3 Conclusions We have seen that LFV decays open up an exciting playground for testing the LHT model. Indeed, they could offer a very clear distinction between this model and supersymmetry. Of particular interest are the ratios $B(\ell_i \to eee)/B(\ell_i \to e\gamma)$ that are $\mathcal{O}(1)$ in the LHT model but strongly suppressed in supersymmetric models even in the presence of significant Higgs contributions. Similarly, finding the μ–e conversion rate in nuclei at the same level as $B(\mu \to e\gamma)$ would point into the direction of LHT physics rather than supersymmetry.

5.1.4 Low scale triplet Higgs neutrino mass scenarios in little Higgs models

An important open issue to address in the context of little Higgs models is the origin of non-zero neutrino masses [654–658]. The neutrino mass mechanism which naturally occurs in these models is the triplet Higgs mechanism [244] which employs a scalar with the $SU(2)_L \times U(1)_Y$ quantum numbers $T \sim (3, 2)$. The existence of such a multiplet in some versions of the little Higgs models is a direct consequence of global symmetry breaking which makes the SM Higgs light. For example, in the minimal littlest Higgs model [625], the triplet Higgs with non-zero hypercharge occurs from the breaking of global $SU(5)$ down to $SO(5)$ symmetry as one of the Goldstone bosons. Its mass $M_T \sim g_s f$, where $g_s < 4\pi$ is a model dependent coupling constant in the weak coupling regime [659], is therefore predicted to be below the cut-off scale Λ, and could be within the mass reach of LHC. The present lower bound for the invariant mass of T is set by Tevatron to $M_T \geq 136$ GeV [660, 661].

Although the triplet mass scale is of order $\mathcal{O}(1)$ TeV, the observed neutrino masses can be obtained naturally. Due to the specific quantum numbers the triplet Higgs boson couples only to the left-chiral lepton doublets $L_i \sim (2, -1)$, $i = e, \mu, \tau$, via the Yukawa interactions of (3.63) and to the SM Higgs bosons via (3.64). Those interactions induce lepton flavor violating decays of charged leptons which have not been observed. The most stringent constraint on the Yukawa couplings comes from the upper limit on the tree-level decay $\mu \to eee$ and is[15] $Y_T^{ee} Y_T^{e\mu} < 3 \times 10^{-5} (M/\text{TeV})^2$ [662, 663]. Experimental bounds on the tau Yukawa couplings are much less stringent. The hierarchical light neutrino masses imply $Y_T^{ee}, Y_T^{e\mu} \ll Y_T^{\tau\tau}$ consistently with the direct experimental bounds.

Non-zero neutrino masses and mixing is presently the only experimentally verified signal of new physics beyond the SM. In the triplet neutrino mass mechanism [244] presented in Sect. 3.2.3.2 the neutrino masses are given by

$$(m_\nu)^{ij} = Y_T^{ij} v_T, \qquad (5.26)$$

where v_T is the induced triplet VEV of (3.65). It is natural that the smallness of neutrino masses is explained by the smallness of v_T. In the little Higgs models this can be achieved by requiring the Higgs mixing parameter $\mu \ll M_T$, which can be explained, for example, via shining of explicit lepton number violation from extra dimensions as shown in Refs. [664, 665], or if the triplet is related to the Dark Energy of the Universe [666, 667]. Models with additional (approximate) T-parity [626] make the smallness of v_T technically

---

[15]In little Higgs models with T-parity there exist additional sources of flavor violation from the mirror fermion sector [639, 643] discussed in the previous subsection.

natural (if the T-parity is exact, v_T must vanish). In that case $Y_T v_T \sim \mathcal{O}(0.1)$ eV while the Yukawa couplings Y can be of order charged lepton Yukawa couplings of the SM. As a result, the branching ratio of the decay $T \to WW$ is negligible. We also remind that v_T contributes to the SM oblique corrections, and the precision data fit $\hat{T} < 2 \times 10^{-4}$ [668] sets an upper bound $v_T \leq 1.2$ GeV on that parameter.

Notice the particularly simple connection between the flavor structure of light neutrinos and the Yukawa couplings of the triplet via (5.26). Therefore, independently of the overall size of the Yukawa couplings, one can predict the leptonic branching ratios of the triplet from neutrino oscillations. For the normally hierarchical light neutrino masses neutrino data implies negligible T branching fractions to electrons and $B(T^{++} \to \mu^+\mu^+) \approx B(T^{++} \to \tau^+\tau^+) \approx B(T^{++} \to \mu^+\tau^+) \approx 1/3$. Those are the final state signatures predicted by the triplet neutrino mass mechanism for collider experiments.

At LHC T^{++} can be produced singly and in pairs. The cross section of the single T^{++} production via the WW fusion process [662] $qq \to q'q'T^{++}$ scales as $\sim v_T^2$. In the context of the littlest Higgs model this process, followed by the decays $T^{++} \to W^+W^+$, was studied in Refs. [669–671]. The detailed ATLAS simulation of this channel shows [671] that in order to observe an 1 TeV T^{++}, one must have $v_T > 29$ GeV. This is in conflict with the precision physics bound $v_T \leq 1.2$ GeV as well as with the neutrino data. Therefore the WW fusion channel is not experimentally promising for the discovery of doubly charged Higgs.

On the other hand, the Drell–Yan pair production process [662, 672–678]

$$pp \to T^{++}T^{--}$$

is not suppressed by any small coupling and its cross section is known up to next to leading order [674] (possible additional contributions from new physics such as Z_H are strongly suppressed and we neglect those effects here). Followed by the lepton number violating decays $T^{\pm\pm} \to \ell^\pm\ell^\pm$, this process allows to reconstruct $T^{\pm\pm}$ invariant mass from the same charged leptons rendering the SM background to be very small in the signal region. If one also assumes that neutrino masses come from the triplet Higgs interactions, one fixes the $T^{\pm\pm}$ leptonic branching ratios. This allows to test the triplet neutrino mass model at LHC. The pure Monte Carlo study of this scenario shows [677] that T^{++} up to the mass 300 GeV is reachable in the first year of LHC ($L = 1$ fb^{-1}) and T^{++} up to the mass 800 GeV is reachable for the luminosity $L = 30$ fb^{-1}. Including the Gaussian measurement errors to the Monte Carlo the corresponding mass reaches become [677] 250 GeV and 700 GeV, respectively. The errors of those estimates of the required luminosity for discovery depend strongly on the size of statistical Monte Carlo sample of the background processes.

5.2 Flavor and CP violation in SUSY extensions of the SM

Supersymmetric models provide the richest spectrum of lepton flavor and CP-violating observables among all models. They are also among the best studied scenarios of new physics beyond the standard model. In this Section we review phenomenologically most interesting aspects of some of the supersymmetric scenarios.

5.2.1 Mass insertion approximation and phenomenology

In the low energy supersymmetric extensions of the SM the flavor and CP-violating interactions would originate from the misalignment between fermion and sfermion mass eigenstates. Understanding why all these processes are strongly suppressed is one of the major problems of low energy supersymmetry, the *supersymmetric flavor and CP problem*. The absence of deviations from the SM predictions in LFV and CPV (and other flavor changing processes in the quark sector) experiments suggests the presence of a quite small amount of fermion-sfermion misalignment. From the phenomenological point of view those effects are most easily described by the mass insertion approximation.

The relevant one loop amplitudes can be exactly written in terms of the general mass matrix of charginos and neutralinos, resulting in quite involved expressions. To obtain simple approximate expressions, it is convenient to use the so-called mass insertion method [145, 679]. This is a particularly convenient method since, in a model independent way, the tolerated deviation from alignment is quantified by the upper limits on the mass insertion δ's, defined as the small off-diagonal elements in terms of which sfermion propagators are expanded, normalized with an average sfermion mass, $\delta_{ij} = \Delta_{ij}/m_{\tilde{f}}^2$. They are of four types: δ^{LL}, δ^{RR}, δ^{RL} and δ^{LR}, according to the chiralities of the corresponding partner fermions. We shall adopt here the usual convention for the slepton mass matrix in the basis where the lepton mass matrix m_ℓ is diagonal:

$$\begin{pmatrix} \tilde{\ell}_L^\dagger & \tilde{\ell}_R^\dagger \end{pmatrix} \begin{pmatrix} m_L^2(1+\delta^{LL}) & (A^* - \mu\tan\beta)m_\ell + m_L m_R \delta^{LR} \\ (A - \mu^*\tan\beta)m_\ell + m_L m_R \delta^{LR\dagger} & m_R^2(1+\delta^{RR}) \end{pmatrix} \begin{pmatrix} \tilde{\ell}_L \\ \tilde{\ell}_R \end{pmatrix}$$

where m_L, m_R, are respectively the average real masses of the left handed and right handed sleptons and A contains only the diagonal entries the trilinear matrices at the electroweak scale. Notice that these flavor diagonal left–right mixing are always present in any MSSM and play a very important role in LFV processes. In this way, our δ^{LR} contain only the off-diagonal elements of the trilinear matrices. This definition is then slightly different from the original definition in Refs. [145, 680]. The deviations from universality are then all gathered in the different δ matrices.

Each element in these δ matrices can be tested by experiment. Searches for the decay $\ell_i \to \ell_j \gamma$ provide bounds on the absolute values of the off-diagonal (flavor violating) $|\delta_{ij}^{LL}|$, $|\delta_{ij}^{RR}|$, $|\delta_{ij}^{LR}|$ and $|\delta_{ij}^{RL}|$, while measurements of the lepton EDM (MDM), parameters and their CP-violating phases, also provide limits on the imaginary (real) part of combinations of flavor violating δ's, $\delta_{ij}^{LL}\delta_{ji}^{LR}$, $\delta_{ij}^{LR}\delta_{ji}^{RR}$, $\delta_{ij}^{LL}\delta_{ji}^{RR}$ and $\delta_{ij}^{LR}\delta_{ji}^{LR}$. Many authors have addressed the issue of the bounds on these misalignment parameters and phases in the sleptonic sector [680]. Following [163] we present the current limits on $\mu \to e\gamma$ and we analyze the impact of the planned experimental improvements on $\tau \to \mu\gamma$. In the basis where Y_ℓ is diagonal, and in the mass insertion approximation, the branching ratio of the process reads

$$B(\ell_i \to \ell_j \gamma) = 10^{-5} \times B(\ell_i \to \ell_j \bar{\nu}_j \nu_i) \frac{M_W^4}{\bar{m}_L^4}$$
$$\times \tan^2\beta \left|\delta_{ij}^{LL}\right|^2 F_{\text{SUSY}}, \quad (5.27)$$

where $F_{\text{SUSY}} = O(1)$ is a function of supersymmetric masses including both chargino and neutralino exchange (see e.g., [163], and references therein). We focus for definiteness on the mSUGRA scenario, also assuming gaugino and scalar universality at the gauge coupling unification scale and fixing μ as required by the radiative electroweak symmetry breaking.

As for LFV, Figs. 7 and 8 display the upper bounds on the $|\delta|$'s in the (M_1, m_R) plane, where M_1 and m_R are the bino

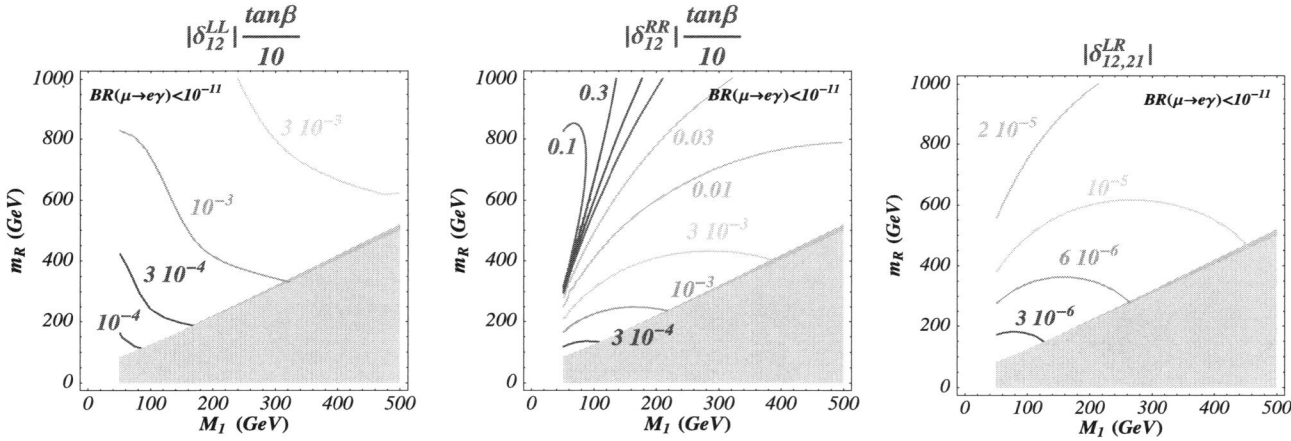

Fig. 7 Upper limits on δ_{12}'s in mSUGRA. Here M_1 and m_R are the bino and right-slepton masses, respectively

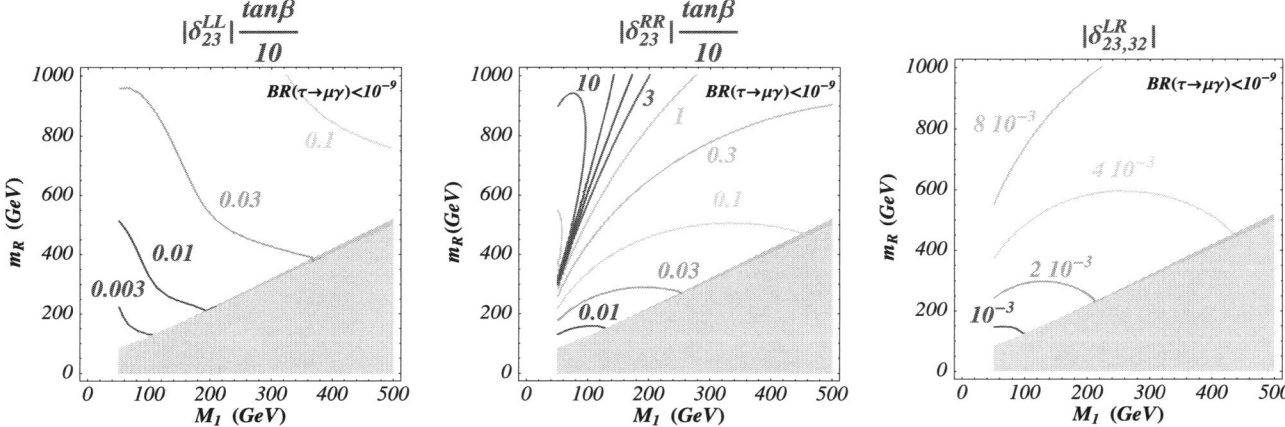

Fig. 8 Upper limits on δ_{23}'s in mSUGRA. Here M_1 and m_R are the bino and right-slepton masses, respectively

and right-slepton masses, respectively. Deviations from the mSUGRA assumptions can be estimated by means of relatively simple analytical expressions. In Figs. 7 and 8 we can see that the bounds on δ_{ji}^{RR} depend strongly and are practically absent for some values of M_1 and m_R. This fact is due to a destructive interference between the bino and bino–higgsino amplitudes [163]. On the contrary, the limits on δ_{ji}^{LL} are robust because of a constructive interference between the chargino and bino amplitudes. A weaker bound on δ_{12}^{RR} on the cancellation regions can be obtained combining the experimental information from the decays $\mu \to e\gamma$, $\mu \to eee$ and μ–e conversion in nuclei [165, 681]. The present limits on $\mu \to e\gamma$ provide interesting constraints on the related δ's. As will be discussed in the following, the present sensitivity already allow to test these δ's at the level of the radiative effects. Such a sensitivity could hopefully be reached also in future experiments on $\tau \to \mu\gamma$.

Another issue is the origin of the CP-violating phases in the leptonic EDMs. Unless the sparticle masses are increased above several TeVs, the phases in the flavor diagonal elements of the slepton left–right mass matrices (in the lepton flavor basis), in the parameters μ and A_i of supersymmetric models, have to be quite small, and this constitutes the so-called supersymmetric CP problem. For the bounds on the sources of CPV also associated to FV, like e.g. $\text{Im}(\delta_{ij}^{LL}\delta_{ji}^{RR})_{ee}$ and so on, we refer to the plots in Ref. [163].

5.2.2 Lepton flavor violation from RGE effects in SUSY seesaw model

5.2.2.1 Predictions from flavor models
Consider first the possibility that flavor and CP are exact symmetries of the soft supersymmetry breaking sector defined at the appropriate cutoff scale Λ (to be identified with the Planck scale for supergravity, the messenger mass for gauge mediation, etc). If below this scale there are flavor and CP-violating Yukawa interactions, it is well-known that in the running down to m_{SUSY} they will induce a small amount of flavor and CP violation in sparticle masses.

The Yukawa interactions associated to the fermion masses and mixing of the SM clearly violate any flavor and CP symmetries. However, with the exception of the third generation Yukawa couplings, all the entries in the Yukawa matrices are very small and the radiatively induced misalignment in the sfermion mass matrices turns out to be negligible. The Yukawa interactions of heavy states beyond the SM coupling to the SM fermions induce misalignments proportional to a proper combination of their Yukawa couplings times $\ln m_F/\Lambda$, where m_F represents the heavy state mass scale. This is the case for the seesaw interactions of the right handed neutrinos [144, 145] and/or the GUT interactions of the heavy colored triplets [682, 683] (those eventually exchanged in diagrams inducing proton decay). Notice that the observation of large mixing in light neutrino masses, may suggest the possibility that also the seesaw interactions could significantly violate flavor- and potentially also CP, in particular in view of the mechanism of leptogenesis. Remarkably, for sparticle masses not exceeding the TeV, the seesaw and colored-triplet induced radiative contributions to the LFV decays and lepton EDM might be close to or even exceed the present or planned experimental limits. Clearly, these processes constitute an important constraint on seesaw and/or GUT models.

For instance, in a type I seesaw model in the low energy basis where charged leptons are diagonal, the ij element of the left handed slepton mass matrix provides the dominant contribution in the decay $\ell_i \to \ell_j \gamma$. Assuming, for the sake of simplicity, an mSUGRA spectrum at $\Lambda = M_{\text{Pl}}$, one obtains at the leading log [178]:

$$\delta_{ij}^{LL} = \frac{(m_{ij}^2)_{LL}}{m_L^2} = -\frac{1}{8\pi^2}\frac{3m_0^2 + A_0^2}{m_L^2}C_{ij},$$

$$C_{ij} \equiv \sum_k Y_{\nu ki}^* Y_{\nu kj} \ln \frac{M_{\text{Pl}}}{M_k}, \qquad (5.28)$$

where m_0 and A_0 are respectively the universal scalar masses and trilinear couplings at M_{Pl}, m_L^2 is an average left handed slepton mass and M_k the mass of the right handed neutrino with $k = 1, 2, 3$. An experimental limit on $B(\ell_i \to \ell_j \gamma)$ corresponds to an upper bound on $|C_{ij}|$ [38, 233]. For $\mu \to e\gamma$ and $\tau \to \mu\gamma$ this bound is shown in Fig. 9 as a function of the right handed selectron mass.

The seesaw model dependence resides in C_{ij}. Notice that in the *fundamental* theory at high energy, the size of C_{ij} is determined both by the Yukawa eigenvalues and the largeness of the mixing angles of V_R, V_L, the unitary matrices which diagonalize Y_ν (in the basis where M_R and Y_e are diagonal): $V_R Y_\nu V_L = Y_\nu^{(\text{diag})}$. The left handed misalignment between neutrino and charged lepton Yukawa's is given by V_L and, due to the mild effect of the logarithm in C_{ij}, in first approximation V_L itself diagonalizes C_{ij}. If we consider hierarchical Y_ν eigenvalues, $Y_3 > Y_2 > Y_1$, the contributions from $k = 1, 2$ in (5.28) can in first approximation be neglected with respect to the contribution from the heaviest eigenvalue ($k = 3$):

$$|C_{ij}| \approx |V_{Li3}V_{Lj3}|Y_3^2 \log(M_{\text{Pl}}/M_3). \qquad (5.29)$$

Taking supersymmetric particle masses around the TeV scale, it has been shown that many seesaw models predict $|C_{\mu e}|$ and/or $|C_{\tau \mu}|$ close to the experimentally accessible range. Let us consider the predictions for the seesaw-RGE induced contribution to $\tau \to \mu\gamma$ and $\mu \to e\gamma$ in the flavor models discussed previously.

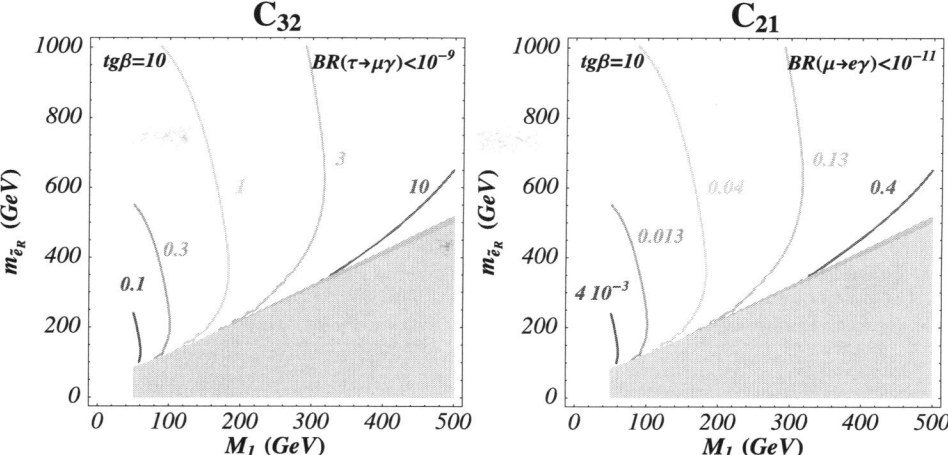

Fig. 9 Upper limit on C_{32} and C_{21} for the experimental sensitivities displayed [38]

The present experimental bound on $\tau \to \mu\gamma$ is not very strong but nevertheless promising. In models with "lopsided" Y_ν, one has $V_{L32}V_{L32} \approx 1/2$, hence $|C_{\tau\mu}| = O(4 \times Y_3^2)$, for $M_3 \simeq 4 \times 10^{15}$ GeV. This is precisely the case for the $U(1)$ flavor model discussed in Sect. 2.2, where $Y_3 \approx \epsilon^{n_3^c}$ with $\epsilon \approx 0.22$ (the Cabibbo angle). For this model, planned $\tau \to \mu\gamma$ searches could thus be successful if the heaviest right handed neutrino has null charge, $n_3^c = 0$. On the contrary, in models with small V_{L23} mixing, like in the non-Abelian models discussed previously, the seesaw-RGE induced effect is below the experimental sensitivity.

The present experimental bound on $\mu \to e\gamma$ is already very severe in constraining $|C_{\mu e}|$. For instance, if $V_L \approx V_{CKM}$, one obtains $C_{\mu e} = O(10^{-3} \times Y_3^2)$. As can be seen from Fig. 9, V_L could in future be tested at a CKM-level if $Y_3 = O(1)$ [164]. The predictions for $\mu \to e\gamma$ are however very model dependent. For the simple $U(1)$ flavor model of Sect. 2.2, the mixings of V_L are of the same order of magnitude as those of U_{PMNS} and one expects $|C_{\mu e}| = O(8 \times \epsilon^{2n_3^c + 1})$: if $n_3^c = 0$ the prediction exceeds the experimental limit, which is respected only with $n_3^c \geq 1$ [684]. On the contrary, the non-Abelian models discussed previously have $Y_3 \sim 1$, but the V_L-mixings are sufficiently small to suppress the seesaw-RGE induced effect below the present experimental level [164].

5.2.2.2 Parameter dependence for degenerate heavy neutrinos Equation (5.28) indicates that LFV in the minimal supersymmetric seesaw model depends on soft supersymmetry breaking masses as well as on the seesaw parameters. The latter can be parameterized via the heavy and light neutrino masses, the light neutrino mixing matrix and the orthogonal matrix R of (3.45). The three complex mixing angles parameterizing R can be written as $\hat\theta_j = x_j + iy_j$, $j = 1, 2, 3$. For the following numerical examples we use the mSUGRA point SPS1a [685] for SUSY breaking masses.

In the case of degenerate heavy neutrino masses, $M_i = M_R$ ($i = 1, 2, 3$), and real R, the R dependence in (5.28) and hence also in $B(l_i \to l_j\gamma)$ drops out. However, if R is complex, the LFV observables have more freedom since the dependence on y_i can be as significant as the M_R dependence, as Fig. 10 shows. For small $|y_i|$, the change in $Y_\nu^\dagger Y_\nu$ is approximately

$$\Delta_R(Y_\nu^\dagger Y_\nu) \approx U_{PMNS}\,\mathrm{diag}(\sqrt{m_i})(R^\dagger R - \mathbf{1}) \\ \times \mathrm{diag}(\sqrt{m_i})U_{PMNS}^\dagger, \quad (5.30)$$

while the renormalization effects on the soft supersymmetry breaking masses can be estimated via [273]

$$m_L^8 \simeq 0.5 m_0^2 M_{1/2}^2 (m_0^2 + 0.6 M_{1/2}^2)^2, \quad (5.31)$$

where $M_{1/2}$ is the universal gaugino mass at high scale. In certain cases, the leading logarithmic approximation fails, as pointed out in [238, 273, 275, 686].

Equation (5.30) implies three features seen in Fig. 10:

(i) Compared to the case of degenerate light neutrino masses, the y dependence in the hierarchical case is weaker.
(ii) Observables like (5.27) are larger in the case of complex R than in the case of real R. For a given M_R, even small values of y can enhance a process by orders of magnitude.
(iii) In contrast to the real R case, where $B(l_i \to l_j\gamma)$ for degenerate light neutrinos is always larger than for hierarchical light neutrinos, the relative magnitude can be reversed for complex R.

To examine the parameter dependence of rare decays at large $|y_i| > 0.1$, we extend the above analysis to the case where the y_i are independent of one another. For random values of all parameters in their full ranges, the typical be-

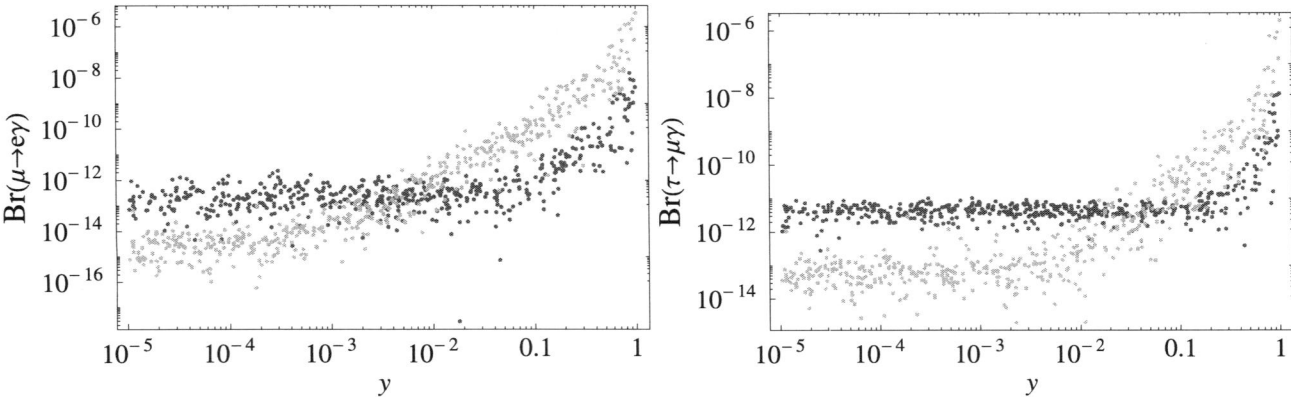

Fig. 10 (Color online) Degenerate heavy neutrinos: LFV branching ratio versus $|y_i| = y$ for fixed $M_i = M_R = 10^{12}$ GeV in mSUGRA scenario SPS1a for hierarchical (*dark red*) and degenerate (*light green*) light neutrino masses. The x_i are scattered over $0 < x_i < 2\pi$

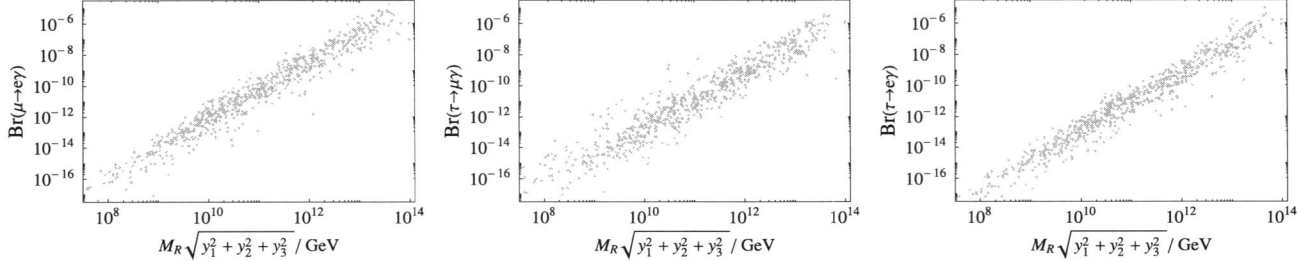

Fig. 11 Degenerate heavy neutrinos: LFV branching ratios versus $M_R\sqrt{y_1^2 + y_2^2 + y_3^2}$, for light neutrinos. The y_i are scattered logarithmically in the range $10^{-5} < |y_i| < 1$ (independently of one another) and M_R is scattered logarithmically in the range $10^{10} < M_R < M_{\text{GUT}}$. The x_i are scattered over $0 < x_i < 2\pi$

havior

$$\left|\left(Y_\nu^\dagger L Y_\nu\right)_{jk}\right|^2 \propto \begin{cases} M_R^2(C_1 y_1^2 + C_2 y_2^2 + C_3 y_3^2) & \text{deg. } \nu_L \\ M_R^2 & \text{hier. } \nu_L \end{cases}$$

$$(j \neq k), \tag{5.32}$$

is found, with $C_i = O(1)$, slightly dependent on j, k. This behavior can be seen in Fig. 11 for degenerate light neutrinos. Thus for large $|y_i|$ all rare decays may be of a similar order of magnitude. For hierarchical light neutrinos, a similar behavior is observed, but versus M_R^2 only.

5.2.2.3 Parameter dependence for hierarchical heavy neutrinos Hierarchical spectrum of heavy Majorana neutrinos, $M_1 \ll M_2 \ll M_3$, is well motivated by the arguments of light neutrino mass and mixing generation and leptogenesis. Requiring successful thermal leptogenesis puts additional constraints on the seesaw parameters and constrains the LFV observables [329]. This is the approach we take in this subsection. In particular, the relation (3.93) implies a lower bound on M_1 [306], e.g., if $\epsilon_1 > 10^{-6}$, then $M_1 > 5 \times 10^9$ GeV. Furthermore, to allow for thermal production of right handed neutrinos after inflation, one has to exclude $M_1 > 10^{11}$ GeV, at least in simple scenarios. Otherwise a too high re-heating temperature would lead to an overabundance of gravitinos, whose decays into energetic photons can spoil big bang nucleosynthesis. Details of leptogenesis have been described in Sect. 3.3.1.

Assuming hierarchical light neutrinos with $\Delta m_{\text{sol}}^2 < m_3^2 < \Delta m_{\text{atm}}^2$, the condition to reproduce the experimental baryon asymmetry, $\eta_B = (6.3 \pm 0.3) \times 10^{-10}$, puts constraints on M_1 and the R matrix [687]. This is illustrated in Fig. 12 in the M_1–x_2 plane. For $M_1 < 10^{11}$ GeV, x_2 has to approach the values $0, \pi, 2\pi$. A similar behavior is observed in the M_1–x_3 plane.

Taking $M_1 = 10^{10}$ GeV and $x_2 \approx x_3 \approx n \cdot \pi$, experimental bounds on $B(\mu \to e\gamma)$ can be used to constrain the heavy neutrino scale, here represented by the heaviest right handed neutrino mass M_3, as shown in the right plot of Fig. 12. Quantitatively, the present bound on $B(\mu \to e\gamma)$ already constrains M_3 to be smaller than $\approx 10^{13}$ GeV, while the MEG experiment at PSI is sensitive to $M_3 \leq O(10^{12})$ GeV. If no signal is observed it will be difficult to test the type I seesaw model considered here at future colliders.

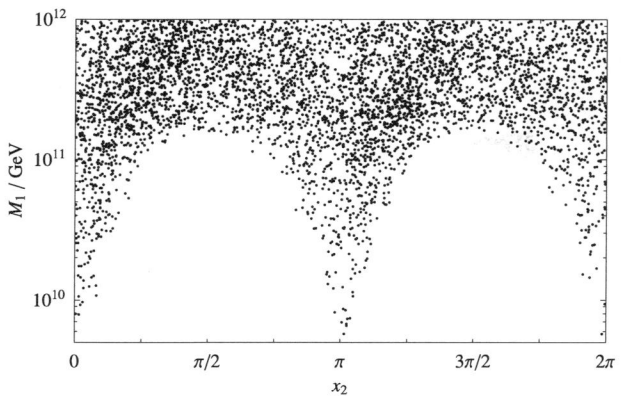
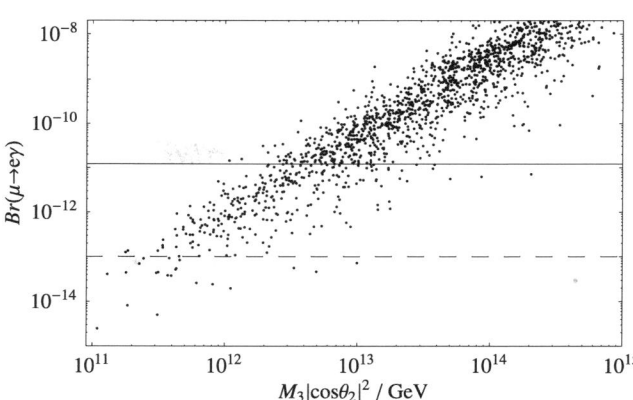

Fig. 12 Hierarchical heavy neutrinos: Region in the plane (x_2, M_1) consistent with the generation of the baryon asymmetry $\eta_B = (6.3 \pm 0.3) \times 10^{-10}$ via leptogenesis (*left*). [*Right*] $B(\mu \to e\gamma)$ versus $M_3|\cos^2\theta_2|$ in mSUGRA scenario SPS1a, for $M_1 = 10^{10}$ GeV and $x_2 \approx x_3 \approx n \cdot \pi$. All other seesaw parameters are scattered in their allowed ranges for hierarchical light and heavy neutrinos. The *solid* (*dashed*) *line* indicates the present (expected future) experimental sensitivity

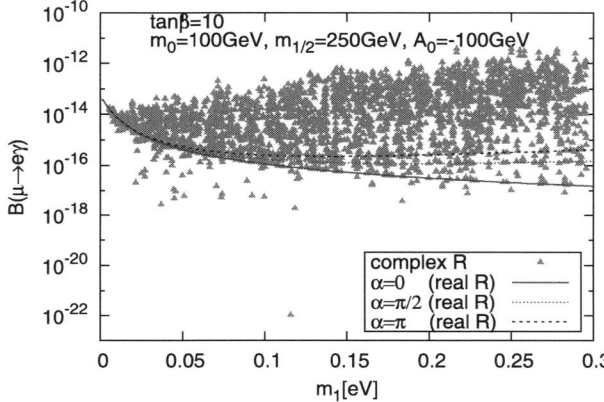
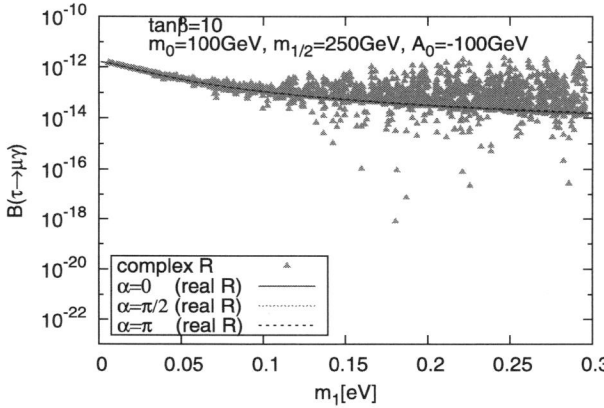

Fig. 13 The branching ratios of the LFV decays $\mu \to e + \gamma$ and $\tau \to \mu + \gamma$ versus m_1 in the cases of complex and real matrix R with $\alpha = 0; \pi/2; \pi$. The three parameters describing the matrix R [275, 545] are generated randomly. The SUSY parameters are $\tan\beta = 10$, $m_0 = 100$ GeV, $m_{1/2} = 250$ GeV, $A_0 = -100$ GeV, and the neutrino mixing parameters are $\Delta m_\odot^2 = 8.0 \times 10^{-5}$ eV^2, $\Delta m_{\text{atm}}^2 = 2.2 \times 10^{-3}$ eV^2, $\tan^2\theta_\odot = 0.4$, $\tan^2\theta_{\text{atm}} = 1$, and $\sin\theta_{13} = 0.0$. The neutrino mass spectrum at M_Z is assumed to be with normal hierarchy, $m_1(M_Z) < m_2(M_Z) < m_3(M_Z)$. The right handed neutrino mass spectrum is taken to be degenerate as $M_1 = M_2 = M_3 = 2 \times 10^{13}$ GeV [275]

5.2.2.4 Effects of renormalization of light neutrino masses on LFV

The RG running of the neutrino parameters can have an important impact on lepton flavor violating processes in MSSM extended by right handed neutrinos. In this example we assume universal soft SUSY breaking terms at GUT scale and degenerate heavy neutrinos with mass M_R. The running effects below M_R are relatively small when $\tan\beta$ is smaller than 10 and/or m_1 is much smaller than 0.05 eV. Because the combination $s_{12}c_{12}c_{23}(m_1 - m_2 e^{i\alpha_M})$, where we use the notation of Sect. 3.2.3.3, is practically stable against the RG running, and this combination is the dominant term of $(Y_\nu^\dagger Y_\nu)_{21}$ when $\alpha_M = 0$, $\theta_{13} = 0$ and $R^* = R$ are satisfied, the running effect on LFV can be neglected in this case [275]. In general, $(Y_\nu^\dagger Y_\nu)_{21}$ and $B(\mu \to e + \gamma)$ can depend strongly on θ_{13} and RG running has to be taken into account [686, 688]. Note that due to RG running, the value of θ_{13} at M_R differs from 0, even if $\theta_{13} = 0$ is assumed at low energy [275].

In many cases, the running of the neutrino parameters can significantly affect the prediction of the LFV branching ratios. In particular, for $0.05 \lesssim m_1 \lesssim 0.30$ eV, $30 \lesssim \tan\beta \lesssim 50$, the predicted $\mu \to e + \gamma$ and $\tau \to e + \gamma$ decay branching ratios, $B(\mu \to e + \gamma)$ and $B(\tau \to e + \gamma)$, can be enhanced by the effects of the RG running of θ_{ij} and m_j by 1 to 3 orders of magnitude if $\pi/4 \lesssim \alpha_M \lesssim \pi$, while $B(\tau \to \mu + \gamma)$ can be enhanced by up to a factor of 10 [275]. The effects of the running of the neutrino mixing parameters of $B(\mu \to e + \gamma)$ and $B(\tau \to e + \gamma)$ are illustrated in Fig. 13.

5.2.3 Correlations between LFV observables and collider physics

5.2.3.1 Correlations of LFV rare decays

Equations (5.27) and (5.28) imply correlations between different LFV observables. In addition to the correlations between different classes of LFV observables in the same flavor mixing channels, the assumed LFV mechanism induces also correlations among the $|(m_L)^2_{ij}|^2$ and hence among observables of different flavor mixing channels. In this framework, the ratios of the branching ratios are approximately independent of SUSY parameters:

$$\frac{B(\tau \to \mu\gamma)}{B(\mu \to e\gamma)} \propto \frac{|(Y_\nu^\dagger L Y_\nu)_{23}|^2}{|(Y_\nu^\dagger L Y_\nu)_{12}|^2}. \quad (5.33)$$

Thus the measurement of the ratio between the decay rates of the different LFV channels can provide unique information on the flavor structure of the lepton sector. The ratios of interest, such as (5.33), can exhibit, for instance, strong dependence on CP-violating parameters in neutrino Yukawa couplings [691] especially in the case of quasi-degenerate heavy RH neutrinos. As a consequence such correlations have been widely studied (see, e.g., [38, 178, 238, 242, 249, 329, 334, 649, 691, 693] and the references quoted therein).

Consequently, bounds on one LFV decay channel (process) will limit the parameter space of the LFV mechanism and thus lead to bounds on the other LFV decay channels (processes). In Fig. 14, the correlation induced by the type I seesaw mechanism between $B(\mu \to e\gamma)$ and $B(\tau \to \mu\gamma)$ is shown, and the bounds induced by the former on the latter can be easily read off. Interestingly, these bounds do not depend on whether hierarchical or quasi-degenerate heavy and light neutrinos are assumed. The present and future prospective bounds are summarized in Table 12. Note that the present upper bound on $B(\mu \to e\gamma)$ implies a stronger constraint on $B(\tau \to \mu\gamma)$ than its expected future bound.

The above results were derived in the simplifying case of a real R matrix. For complex R with $|y_i| < 1$ there is no significant change with respect to the results in Table 12 in the case of hierarchical heavy and hierarchical light neutrinos due to the weak R dependence of $B(\mu \to e\gamma)$ and $B(\tau \to \mu\gamma)$. However, for quasi-degenerate light neutrinos, $B(\tau \to \mu\gamma)$ is lowered by roughly one order of magnitude, somewhat spoiling the overlap of all scenarios observed in Fig. 14.

In Fig. 15, we display the correlation between $B(\mu \to e\gamma)$ and $B(\tau \to \mu\gamma)$ for complex R and some fixed values of M_R in the case of quasi-degenerate RH neutrino masses

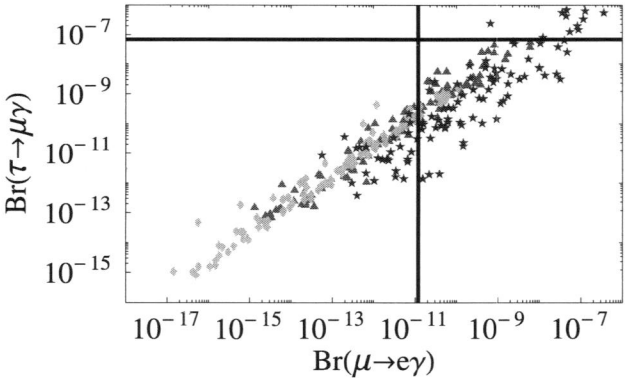

Fig. 14 $B(\tau \to \mu\gamma)$ versus $B(\mu \to e\gamma)$, in mSUGRA scenario SPS1a with neutrino parameters scattered within their experimentally allowed ranges [689]. For quasi-degenerate heavy neutrino masses, both hierarchical (*triangles*) and quasi-degenerate (*diamonds*) light neutrino masses are considered with real R and $10^{11} < M_R < 10^{14.5}$ GeV. In the case of hierarchical heavy and light neutrino masses (*stars*), the x_i are scattered over their full ranges $0 < x_i < 2\pi$ and the y_i and M_i are scattered within the bounds demanded by leptogenesis and perturbativity. Also indicated are the present experimental bounds $B(\mu \to e\gamma) < 1.2 \times 10^{-11}$ and $B(\tau \to \mu\gamma) < 6.8 \times 10^{-8}$ [191, 690]

Table 12 Present and expected future bounds on $B(\mu \to e\gamma)$ from experiment, and bounds on $B(\tau \to \mu\gamma)$ from (i) experiment (ii) the bound on $B(\mu \to e\gamma)$ together with correlations from the SUSY type I seesaw mechanism

	$B(\mu \to e\gamma)$ (exp.)	$B(\tau \to \mu\gamma)$ (exp.)	$B(\tau \to \mu\gamma)$[a]
Present	1.2×10^{-11}	6.8×10^{-8}	10^{-9}
Future	10^{-14}	10^{-9}	10^{-12}

[a] from $B(\mu \to e\gamma)$ (exp.) and SUSY seesaw

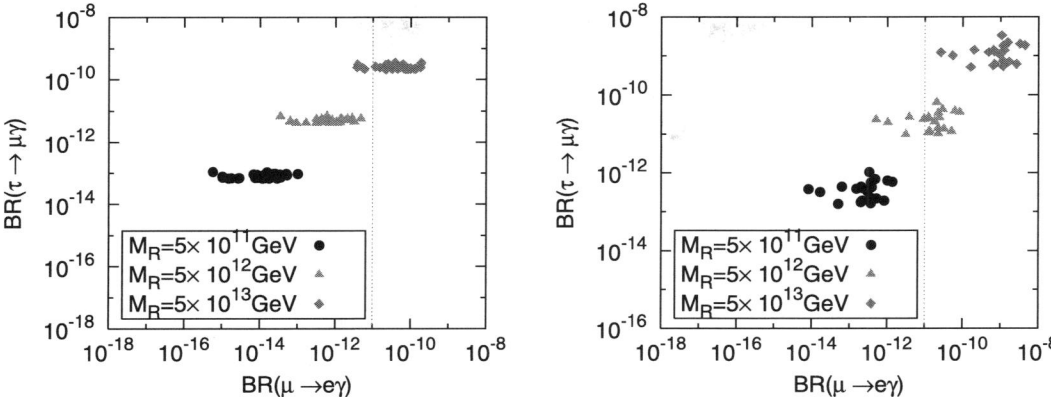

Fig. 15 The correlation between $B(\mu \to e\gamma)$ and $B(\tau \to \mu\gamma)$ for quasi-degenerate heavy neutrinos and light neutrino mass spectrum of normal hierarchical (*left panel*) and inverted hierarchical (*right panel*) type

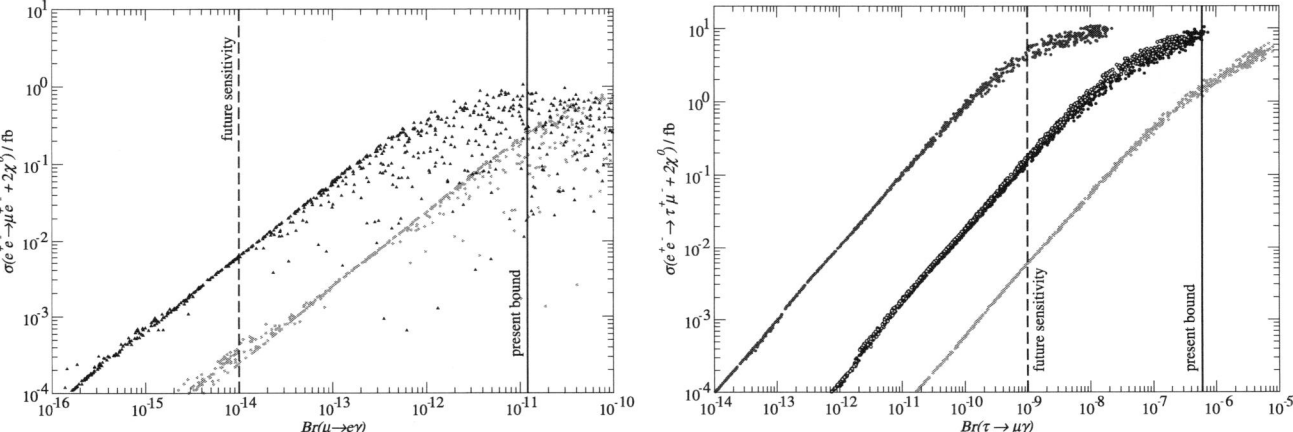

Fig. 16 Correlation of LFV LC processes and rare decays in the $e\mu$-channel (*left*) and the $\mu\tau$-channel (*right*). The seesaw parameters are scattered as in Fig. 14. The mSUGRA scenarios used are (*from left to right*): SPS1a, G' ($e\mu$) and C', B', SPS1a, I' ($\mu\tau$)

and a normal and inverted hierarchical light neutrino mass spectrum. We note that, as Fig. 15 suggests, $B(\tau \to \mu\gamma)$ is almost independent of the CP violating parameters and phases respectively in R and U, while the dependence of $B(\mu \to e\gamma)$ on the CP-violating quantities is much stronger. This is reflected, in particular, in the fact that for a fixed M_R, $B(\tau \to \mu\gamma)$ is practically constant while $B(\mu \to \mu\gamma)$ can change by 2–3 orders of magnitude.

If the $\mu \to e\gamma$ and $\tau \to \mu\gamma$ decays will be observed, the ratio of interest can give unique information on the origin of the lepton flavor violation.

5.2.3.2 LFV rare decays and linear collider processes
In high energy e^+e^- colliders [692], feasible tests of LFV are provided by the processes $e^+e^- \to \tilde{l}_a^- \tilde{l}_b^+ \to l_i^- l_j^+ + 2\tilde{\chi}_1^0$. Analogously to (5.27), one can derive the approximate expression [695]

$$\sigma\left(e^+e^- \to l_i^- l_j^+ + 2\tilde{\chi}_1^0\right)$$
$$\approx \frac{|(\delta m_L)_{ij}^2|^2}{m_{\tilde{l}}^2 \Gamma_{\tilde{l}}^2} \sigma\left(e^+e^- \to l_i^- l_i^+ + 2\tilde{\chi}_1^0\right), \quad (5.34)$$

for the production cross section in the limit of small slepton mass corrections. By comparing (5.27) with (5.34), it is immediately apparent that the linear collider (LC) processes are flavor-correlated with the rare decays considered previously. These correlations are shown in Fig. 16 for the two most important channels.

This observation implies that once the SUSY parameters are known, a measurement of, e.g., $B(\mu \to e\gamma)$ will lead to a prediction for $\sigma(e^+e^- \to \mu e + 2\tilde{\chi}_1^0)$. Quite obviously, this prediction will be independent of the specific LFV mechanism (seesaw or other). Figure 16 also demonstrates that the uncertainties in the neutrino parameters nicely drop out except at large cross sections and branching ratios.

In the previous results we have assumed a specific choice of the as yet unknown mSUGRA parameters. The results of a more systematic study of the model dependence are visualized in Fig. 17 by contour plots for $\sigma(e^+e^- \to \mu^+e^- + 2\tilde{\chi}_1^0)$ and $B(\mu \to e\gamma)$ in the m_0–$m_{1/2}$ plane with the remaining mSUGRA parameters fixed.

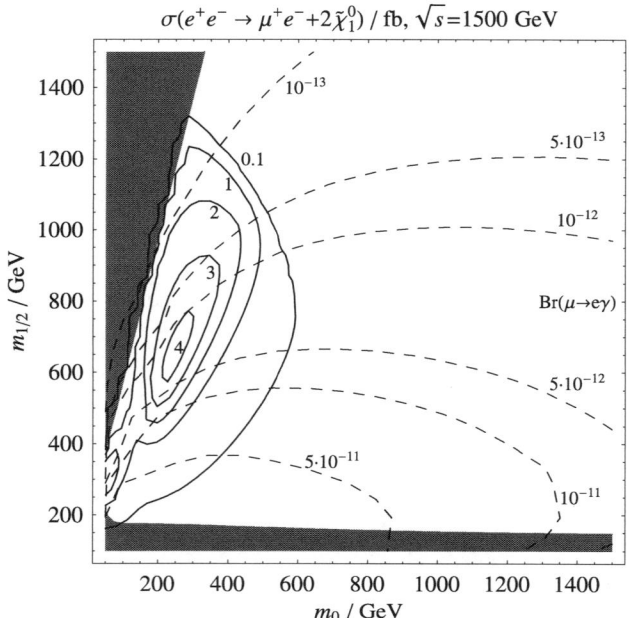

Fig. 17 Contours of the polarized cross section $\sigma(e^+e^- \to \mu^+e^- + 2\tilde{\chi}_1^0)$ (*solid*) and $B(\mu \to e\gamma)$ (*dashed*) in the m_0–$m_{1/2}$ plane. The remaining mSUGRA parameters are $A_0 = 0$ GeV, $\tan\beta = 5$, $\text{sign}(\mu) = +$. The energy and beam polarizations are $\sqrt{s_{ee}} = 1.5$ TeV, $P_{e^-} = +0.9$, $P_{e^+} = +0.7$. The neutrino oscillation parameters are fixed at their central values as given in [689], the lightest neutrino mass $m_1 = 0$ and all complex phases are set to zero, and the degenerate right handed neutrino mass scale is $M_R = 10^{14}$ GeV. The *shaded* (*red*) areas are already excluded by mass bounds from various experimental sparticle searches

5.2.3.3 LFV rare decays and LHC processes At the LHC, a feasible test of LFV is provided by squark and gluino production, followed by cascade decays of squarks and gluinos via neutralinos and sleptons [696, 697]:

$$\begin{aligned} pp &\to \tilde{q}_a\tilde{q}_b, \tilde{g}\tilde{q}_a, \tilde{g}\tilde{g}, \\ \tilde{q}_a(\tilde{g}) &\to \tilde{\chi}_2^0 q_a(g), \\ \tilde{\chi}_2^0 &\to \tilde{l}_\alpha l_\beta, \\ \tilde{l}_\alpha &\to \tilde{\chi}_1^0 l_\beta, \end{aligned} \quad (5.35)$$

where a, b run over all squark mass eigenstates, including antiparticles, and α, β are slepton (lepton) mass (flavor) eigenstates, including antiparticles. LFV can occur in the decay of the second lightest neutralino and/or the slepton, resulting in different lepton flavors, $\alpha \neq \beta$. The total cross section for the signature $l_\alpha^+ l_\beta^- + X$ can then be written as

$$\begin{aligned} &\sigma\left(pp \to l_\alpha^+ l_\beta^- + X\right) \\ &= \Bigg[\sum_{a,b} \sigma(pp \to \tilde{q}_a\tilde{q}_b) \times B(\tilde{q}_a \to \tilde{\chi}_2^0 q_a) \\ &\quad + \sum_a \sigma(pp \to \tilde{q}_a\tilde{g}) \times \big(B(\tilde{q}_a \to \tilde{\chi}_2^0 q_a) \\ &\quad + B(\tilde{g} \to \tilde{\chi}_2^0 g)\big) + \sigma(pp \to \tilde{g}\tilde{g}) \times B(\tilde{g} \to \tilde{\chi}_2^0 g)\Bigg] \\ &\quad \times B\left(\tilde{\chi}_2^0 \to l_\alpha^+ l_\beta^- \tilde{\chi}_1^0\right), \end{aligned} \quad (5.36)$$

where X can involve jets, leptons and LSPs produced by lepton flavor conserving decays of squarks and gluinos, as well as low energy proton remnants. The LFV branching ratio $B(\tilde{\chi}_2^0 \to l_\alpha^+ l_\beta^- \tilde{\chi}_1^0)$ is for example calculated in [698] in the framework of model-independent MSSM slepton mixing. In general, it involves a coherent summation over all intermediate slepton states.

Just as for the linear collider discussed in the previous section, we can correlate the expected LFV event rates at the LHC with LFV rare decays. This is shown in Fig. 18 for

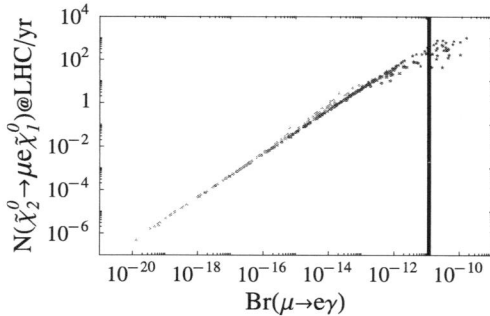
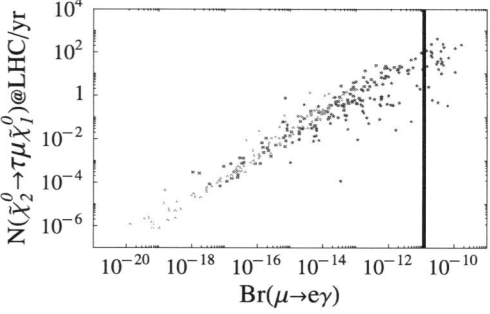

Fig. 18 (Color online) Correlation of the number of $\tilde{\chi}_2^0 \to \mu^+e^-\tilde{\chi}_1^0$ events per year at the LHC and $B(\mu \to e\gamma)$ in mSUGRA scenario C' ($m_0 = 85$ GeV, $m_{1/2} = 400$ GeV, $A_0 = 0$ GeV, $\tan\beta = 10$ GeV, $\text{sign}\,\mu = +$) for the case of hier. $\nu_{R/L}$ (*blue stars*), deg. ν_R/hier. ν_L (*red boxes*) and deg. $\nu_{R/L}$ (*green triangles*). The respective neutrino parameter scattering ranges are as in Fig. 14. An integrated LHC luminosity of 100 fb^{-1} is assumed. The current limit on $B(\mu \to e\gamma)$ is displayed by the *vertical line*

the event rates $N(\tilde{\chi}_2^0 \to \mu^+ e^- \tilde{\chi}_1^0)$ and $N(\tilde{\chi}_2^0 \to \tau^+ \mu^- \tilde{\chi}_1^0)$, respectively, originating from the cascade reactions (5.35). Both are correlated with $B(\mu \to e\gamma)$, yielding maximum rates of around 10^{2-3} per year for an integrated luminosity of (100 fb^{-1}) in the mSUGRA scenario C', consistent with the current limit on $B(\mu \to e\gamma)$.

As in the linear collider case, the correlation is approximately independent of the neutrino parameters, but highly dependent on the mSUGRA parameters. This is contemplated further in Fig. 19, comparing the sensitivity of the signature $N(\tilde{\chi}_2^0 \to \mu^+ e^- \tilde{\chi}_1^0)$ at the LHC with $B(\mu \to e\gamma)$ in the m_0–$m_{1/2}$ plane. As for the linear collider, LHC searches can be competitive with the rare decay experiments for small $m_0 \approx 200$ GeV. Tests in the large-m_0 region are again severely limited by collider kinematics.

Up to now we have considered LFV in the class of type I SUSY seesaw model described in Sect. 3.2.3.1, which is representative of models of flavor mixing in the left handed slepton sector only. However, it is instructive to analyze general mixing in the left and right handed slepton sector, independent of any underlying model for slepton flavor violation. The easiest way to achieve this is by assuming mixing between two flavors only, which can be parameterized by a mixing angle $\theta_{L/R}$ and a mass difference $(\Delta m)_{L/R}$ between the sleptons, in the case of left/right handed slepton mixing, respectively.[16] In particular, the left/right handed selectron and smuon sector is then diagonalized by

$$\begin{pmatrix} \tilde{l}_1 \\ \tilde{l}_2 \end{pmatrix} = U \cdot \begin{pmatrix} \tilde{e}_{L/R} \\ \tilde{\mu}_{L/R} \end{pmatrix}, \quad \text{with}$$

$$U = \begin{pmatrix} \cos\theta_{L/R} & \sin\theta_{L/R} \\ -\sin\theta_{L/R} & \cos\theta_{L/R} \end{pmatrix}, \quad (5.37)$$

and a mass difference $m_{\tilde{l}_2} - m_{\tilde{l}_1} = (\Delta m)_{L/R}$ between the slepton mass eigenvalues.[17] The LFV branching ratio $B(\tilde{\chi}_2^0 \to \mu^+ e^- \tilde{\chi}_1^0)$ can then be written in terms of the mixing parameters and the flavor conserving branching ratio $B(\tilde{\chi}_2^0 \to e^+ e^- \tilde{\chi}_1^0)$ as

$$B(\tilde{\chi}_2^0 \to \mu^+ e^- \tilde{\chi}_1^0)$$
$$= 2\sin^2\theta_{L/R} \cos^2\theta_{L/R} \frac{(\Delta m)_{L/R}^2}{(\Delta m)_{L/R}^2 + \Gamma_{\tilde{l}}^2}$$
$$\times B(\tilde{\chi}_2^0 \to e^+ e^- \tilde{\chi}_1^0), \quad (5.38)$$

where $\Gamma_{\tilde{l}}$ is the average width of the two sleptons involved. Maximal LFV is thus achieved by choosing $\theta_{L/R} = \pi/4$ and $(\Delta m)_{L/R} \gg \Gamma_{\tilde{l}}$. For definiteness, we use $(\Delta m)_{L/R} = 0.5$ GeV. The results of this calculation can be seen in Fig. 20, which shows contour plots of $N(\tilde{\chi}_2^0 \to \mu^+ e^- \tilde{\chi}_1^0)$ in the m_0–$m_{1/2}$ plane for maximal left and right handed slepton mixing, respectively. Also displayed are the corresponding contours of $B(\mu \to e\gamma)$. We see that the present bound $B(\mu \to e\gamma) = 10^{-11}$ still permits sizable LFV signal rates at the LHC. However, $B(\mu \to e\gamma) < 10^{-14}$ would exclude the observation of such an LFV signal at the LHC.

5.2.4 Impact of θ_{13} on LFV in SUSY seesaw

In this subsection we present the results of the LFV tau and muon decays within the SUSY singlet-seesaw context. Specifically, we consider the constrained minimal supersymmetric standard model (CMSSM) extended by three right handed neutrinos, ν_{R_i} and their corresponding SUSY partners, $\tilde{\nu}_{R_i}$ ($i = 1, 2, 3$), and use the seesaw mechanism for the neutrino mass generation. We include the predictions for the branching ratios (BRs) of two types of LFV channels, $l_j \to l_i \gamma$ and $l_j \to 3 l_i$, and compare them with the present bounds and future experimental sensitivities. We first analyze the dependence of the BRs with the most relevant SUSY-seesaw parameters, and we then focus on the particular sensitivity to θ_{13}, which we find specially interesting on the light of its potential future measurement. We

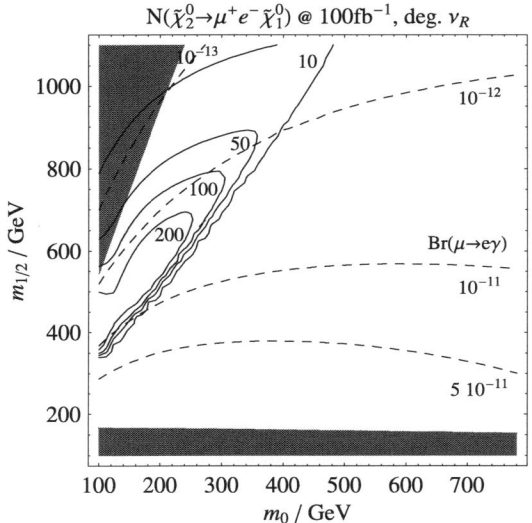

Fig. 19 Contours of the number of $\tilde{\chi}_2^0 \to \mu^+ e^- \tilde{\chi}_1^0$ events at the LHC with an integrated luminosity of 100 fb^{-1} (*solid*) and of $B(\mu \to e\gamma)$ in the m_0–$m_{1/2}$ plane. The remaining mSUGRA and neutrino oscillation parameters are as in Fig. 17. The *shaded (red) areas* are already excluded by mass bounds from various experimental sparticle searches

---

[16]Note that this is different to the approach in [698], where the slepton mass matrix elements are scattered randomly.

[17]In case of left handed mixing, the mixing angle θ_L and the mass difference $(\Delta m)_L$ are also used to describe the sneutrino sector.

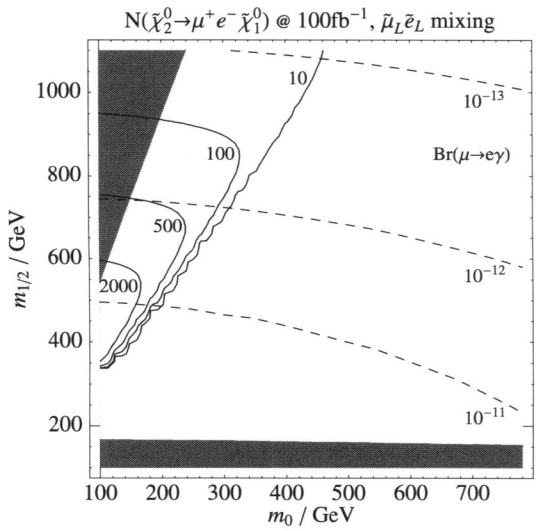
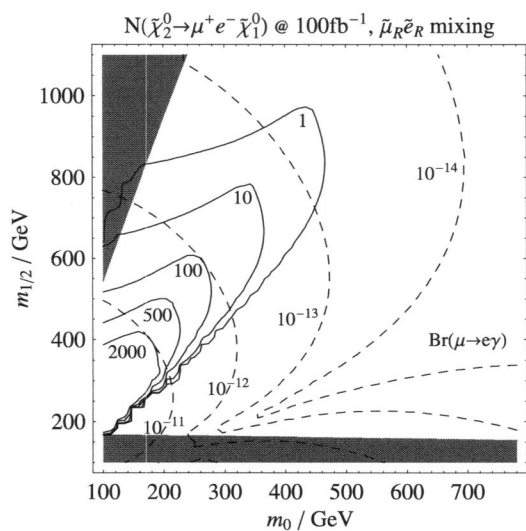

Fig. 20 Contours of the events per year $N(\tilde{\chi}_2^0 \to \mu^+ e^- \tilde{\chi}_1^0)$ at the LHC with an integrated luminosity of 100 fb^{-1} in the m_0–$m_{1/2}$ plane (*solid lines*). The remaining mSUGRA parameters are: $A_0 = -100$ GeV, $\tan\beta = 10$, $\text{sign}(\mu) = +$. The *left* and *right* panels are for maximal $\tilde{e}_L \tilde{\mu}_L$ and $\tilde{e}_R \tilde{\mu}_R$ mixing ($\theta = \pi/4$, $\Delta m = 1$ GeV), respectively. For comparison, $B(\mu \to e\gamma)$ is shown by *dashed lines*. The *shaded* (*red*) *areas* are forbidden by mass bounds from various experimental sparticle searches

further study the constraints from the requirement of successfully producing the Baryon Asymmetry of the Universe via thermal leptogenesis, which is another appealing feature of the SUSY-seesaw scenario. We conclude with the impact that a potential measurement of the leptonic mixing angle θ_{13} can have on LFV physics.

Regarding the technical aspects of the computation of the branching ratios, the most relevant points are (for details, see [649, 686]:

- It is a full one loop computation of BRs, i.e., we include all contributing one loop diagrams with the SUSY particles flowing in the loops. For the case of $l_j \to l_i \gamma$ the analytical formulas can be found in [178, 649]. For the case $l_j \to 3l_i$ the complete set of diagrams (including photon-penguin, Z-penguin, Higgs-penguin and box diagrams) and formulae are given in [649].
- The computation is performed in the physical basis for all SUSY particles entering in the loops. In other words, we do not use the mass insertion approximation (MIA).
- The running of the CMSSM-seesaw parameters from the universal scale M_X down to the electroweak scale is performed by numerically solving the full one loop renormalization group equations (RGEs) (including the extended neutrino sector) and by means of the public Fortran Code SPheno2.2.2. [699]. More concretely, we do not use the Leading Log Approximation (LLog).
- The light neutrino sector parameters that are used in 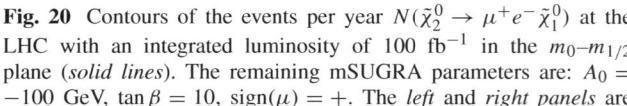 are those evaluated at the seesaw scale m_R. That is, we start with their low energy values (taken from data) and then apply the RGEs to run them up to m_R.
- We have added to the SPheno code extra subroutines that compute the LFV rates for all the $l_j \to l_i \gamma$ and $l_j \to 3l_i$ channels. We have also included additional subroutines to: Implement the requirement of successful baryogenesis (which we define as having $n_B/n_\gamma \in [10^{-10}, 10^{-9}]$) via thermal leptogenesis in the presence of upper bounds on the reheat temperature; Implement the requirement of compatibility with present bounds on lepton electric dipole moments: $\text{EDM}_{e\mu\tau} \lesssim (6.9 \times 10^{-28}, 3.7 \times 10^{-19}, 4.5 \times 10^{-17})$ e cm.

In what follows we present the main results for the case of hierarchical heavy neutrinos. We also include a comparison with present bounds on LFV rates [180, 182, 191, 644, 700] and their future sensitivities [694, 701–706]. For hierarchical heavy neutrinos, the BRs are mostly sensitive to the heaviest mass m_{N_3}, $\tan\beta$, θ_1 and θ_2 (using the R parameterization of [232]). The other input seesaw parameters m_{N_1}, m_{N_2} and θ_3 play a secondary role since the BRs do not strongly depend on them. The dependence on m_{N_1} and θ_3 appears only indirectly, once the requirement of a successful generation of baryon asymmetry of the universe (BAU) is imposed. We shall comment more on this later.

We display in Fig. 21 the predictions for $B(\mu \to e\gamma)$ and $B(\tau \to \mu\gamma)$ as a function of m_{N_3}, for a specific choice of the other input parameters. This figure clearly shows the strong sensitivity of the BRs to m_{N_3}. In fact, the BRs vary by as much as six orders of magnitude in the explored range of $5 \times 10^{11} \leq m_{N_3} \leq 5 \times 10^{14}$ GeV. Notice also that for

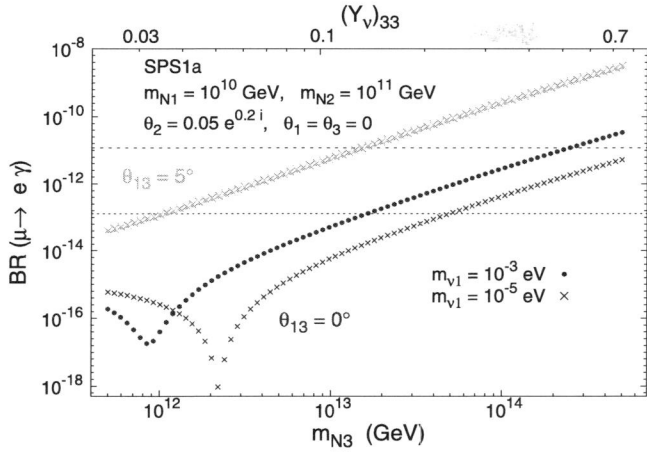
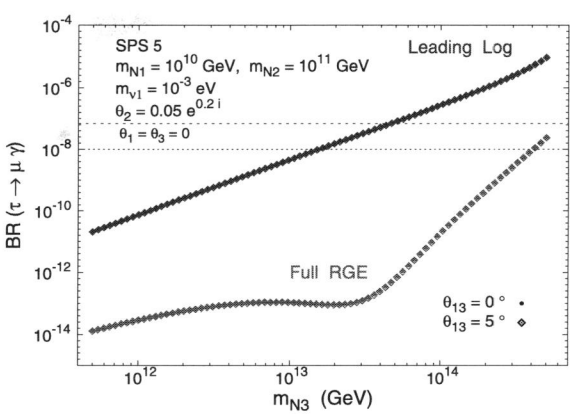

Fig. 21 On the left, $B(\mu \to e\gamma)$ as a function of m_{N_3} for SPS 1a, with $m_{\nu_1} = 10^{-5}$ eV and $m_{\nu_1} = 10^{-3}$ eV (*times, dots*, respectively), and $\theta_{13} = 0°, 5°$ (*blue/darker, green/lighter lines*). Baryogenesis is enabled by the choice $\theta_2 = 0.05e^{0.2i}$ ($\theta_1 = \theta_3 = 0$). On the upper horizontal axis we display the associated value of $(Y_\nu)_{33}$. A *dashed* (*dotted*) *horizontal line* denotes the present experimental bound (future sensi-

tivity). On the right, $B(\tau \to \mu\gamma)$ as a function of m_{N_3} for SPS5, with $m_{\nu_1} = 10^{-3}$ eV and $\theta_2 = 0.05e^{0.2i}$ ($\theta_1 = \theta_3 = 0°$). The predictions for $\theta_{13} = 0°, 5°$ are superimposed one on the top of the other. The *upper curve* is obtained using the LLog approximation and the *lower* one is the full RGE prediction. The *dashed* (*dotted*) *horizontal line* denotes the present experimental bound (future sensitivity)

the largest values of m_{N_3} considered, the predicted rates for $\mu \to e\gamma$ enter into the present experimental reach and only into the future experimental sensitivity for $\tau \to \mu\gamma$. It is also worth mentioning that by comparing our full results with the LLog predictions, we find that the LLog approximation dramatically fails in some cases. In particular, for the SPS5 point, the LLog predictions overestimate the BRs by about four orders of magnitude. For the other points SPS4, SPS1a,b and SPS2 the LLog estimate is very similar to the full result, whereas for SPS3 it underestimates the full computation by a factor of three. In general, the divergence of the LLog and the full computation occurs for low M_0 and large $M_{1/2}$ [238, 273] and/or large A_0 values [686]. The failure of the LLog is more dramatic for SUSY scenarios with large A_0. Figure 21 also shows that while in some cases (as for instance SPS1a) the behavior of the BR with m_{N_3} does follow the expected LLog approximation (BR $\sim (m_{N_3} \log m_{N_3})^2$), there are other scenarios where this is not the case. A good example is SPS5. It is also worth commenting on the deep minima of $B(\mu \to e\gamma)$ appearing in Fig. 21 for the lines associated with $\theta_{13} = 0°$. These minima are induced by the effect of the running of θ_{13}, shifting it from zero to a negative value (or equivalently $\theta_{13} > 0$ and $\delta = \pi$). In the LLog approximation, they can be understood as a cancellation occurring in the relevant quantity $Y_\nu^\dagger L Y_\nu$, with $L_{ij} = \log(M_X/m_{N_i})\delta_{ij}$. Most explicitly, the cancellation occurs between the terms proportional to $m_{N_3} L_{33}$ and $m_{N_2} L_{22}$ in the limit $\theta_{13}(m_R) \to 0^-$ (with $\theta_1 = \theta_3 = 0$). The depth of these minima is larger for smaller m_{ν_1}, as is visible in Fig. 21.

Regarding the $\tan\beta$ dependence of the BRs we obtain that, similar to what was found for the degenerate case, the BR grow as $\tan^2\beta$. The hierarchy of the BR predictions for the several SPS points is dictated by the corresponding $\tan\beta$ value, with a secondary role being played by the given SUSY spectra. We find again the following generic hierarchy: $B_{SPS4} > B_{SPS1b} \gtrsim B_{SPS1a} > B_{SPS3} \gtrsim B_{SPS2} > B_{SPS5}$.

In what concerns to the θ_i dependence of the BRs, we have found that they are mostly sensitive to θ_1 and θ_2. The BRs are nearly constant with θ_3. As has been shown in [649], the predictions for $B(\mu \to e\gamma)$, $B(\mu \to 3e)$, $B(\tau \to \mu\gamma)$ and $B(\tau \to e\gamma)$ are above their corresponding experimental bound for specific values of θ_1. Particularly, the LFV muon decay rates are well above their present experimental bounds for most of the θ_1 explored values. Notice also for SPS4 that the predicted $B(\tau \to \mu\gamma)$ rates are very close to the present experimental reach even at $\theta_1 = 0$ (that is, $R = 1$). We have also explored the dependence with θ_2 and found similar results (not shown here), with the appearance of pronounced dips at particular real values of θ_2 with the $B(\mu \to e\gamma)$, $B(\mu \to 3e)$ and $B(\tau \to \mu\gamma)$ predictions being above the experimental bounds for some θ_2 values.

We next address the sensitivity of the LFV BRs to θ_{13}. We first present the results for the simplest $R = 1$ case and then discuss how this sensitivity changes when moving from this case towards the more general case of complex R, taking into account additional constraints from the requirement of a successful BAU.

For $R = 1$, the predictions of the BRs as functions of θ_{13} in the experimentally allowed range of θ_{13}, $0° \leq \theta_{13} \leq 10°$ are illustrated in Fig. 22. In this figure we also include the present and future experimental sensitivities for all channels. We clearly see that the BRs of $\mu \to e\gamma$, $\mu \to 3e$, $\tau \to e\gamma$

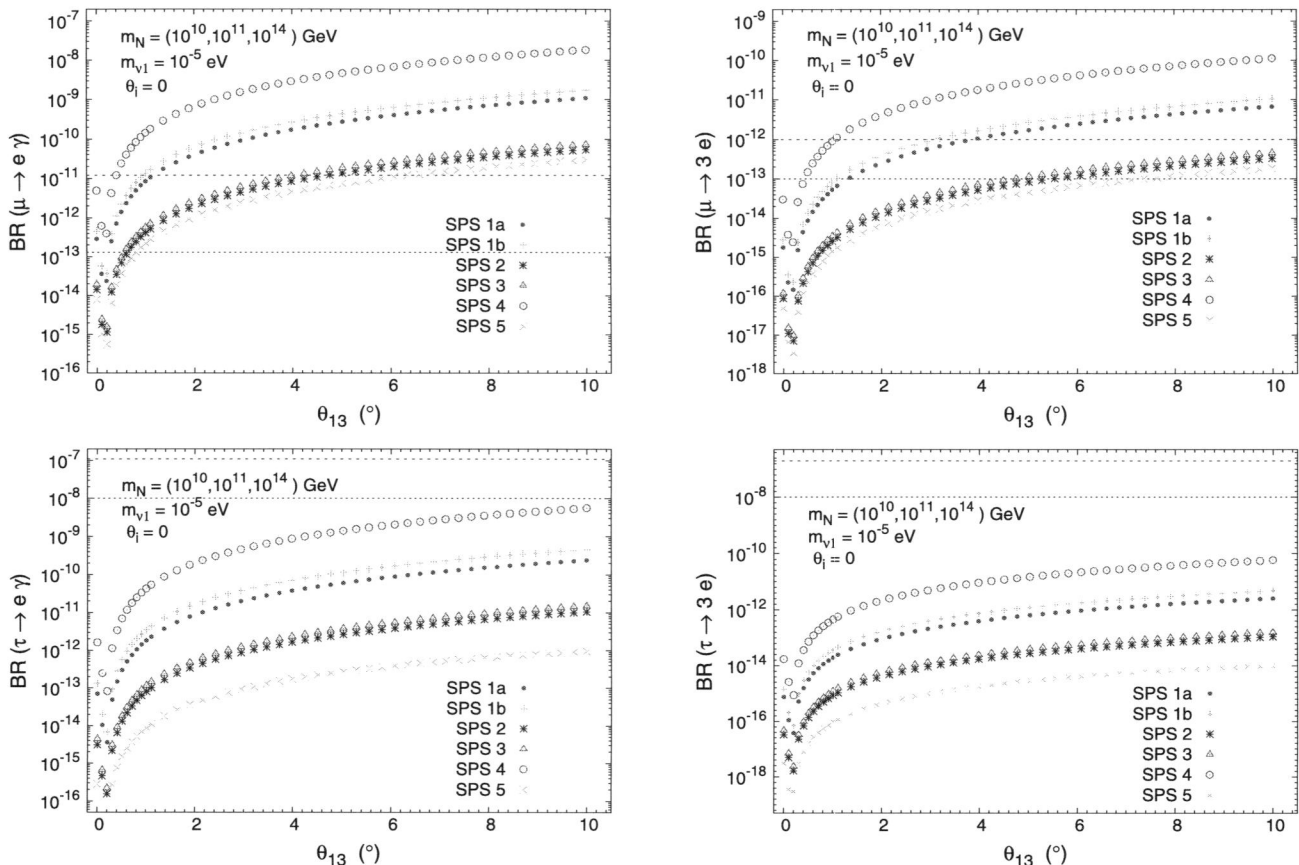

Fig. 22 $B(\mu \to e\gamma)$ and $B(\mu \to 3e)$ as a function of θ_{13} (in degrees), for SPS 1a (*dots*), 1b (*crosses*), 2 (*asterisks*), 3 (*triangles*), 4 (*circles*) and 5 (*times*). A *dashed* (*dotted*) *horizontal line* denotes the present experimental bound (future sensitivity)

and $\tau \to 3e$ are extremely sensitive to θ_{13}, with their predicted rates varying many orders of magnitude along the explored θ_{13} interval. In the case of $\mu \to e\gamma$ this strong sensitivity was previously pointed out in Ref. [707]. The other LFV channels, $\tau \to \mu\gamma$ and $\tau \to 3\mu$ (not displayed here), are nearly insensitive to this parameter. The most important conclusion from Fig. 22 is that, for this choice of parameters, the predicted BRs for both muon decay channels, $\mu \to e\gamma$ and $\mu \to 3e$, are clearly within the present experimental reach for several of the studied SPS points. The most stringent channel is manifestly $\mu \to e\gamma$ where the predicted BRs for all the SPS points are clearly above the present experimental bound for $\theta_{13} \gtrsim 5°$. With the expected improvement in the experimental sensitivity to this channel, this would happen for $\theta_{13} \gtrsim 1°$.

In addition to the small neutrino mass generation, the seesaw mechanism offers the interesting possibility of baryogenesis via leptogenesis [290]. Thermal leptogenesis is an attractive and minimal mechanism to produce a successful BAU with rates which are compatible with present data, $n_B/n_\gamma \approx (6.10 \pm 0.21) \times 10^{-10}$ [327]. In the supersymmetric version of the seesaw mechanism, it can be successfully implemented if provided that the following conditions can be satisfied. Firstly, Big Bang Nucleosynthesis gravitino problems have to be avoided, which is possible, for instance, for sufficiently heavy gravitinos. Since we consider the gravitino mass as a free parameter, this condition can be easily achieved. In any case, further bounds on the reheat temperature T_{RH} still arise from decays of gravitinos into lightest supersymmetric particles (LSPs). In the case of heavy gravitinos and neutralino LSPs masses into the range 100–150 GeV (which is the case of the present work), one obtains $T_{RH} \lesssim 2 \times 10^{10}$ GeV. In the presence of these constraints on T_{RH}, the favored region by thermal leptogenesis corresponds to small (but non-vanishing) complex R-matrix angles θ_i. For vanishing U_{MNS} CP phases the constraints on R are basically $|\theta_2|, |\theta_3| \lesssim 1$ rad(mod π). Thermal leptogenesis also constrains m_{N_1} to be roughly in the range $[10^9$ GeV, $10 \times T_{RH}]$ (see also [309, 311]). In the present work we have explicitly calculated the produced BAU in the presence of upper bounds on the reheat temperature T_{RH}. We have furthermore set as "favored BAU values" those that are within the interval $[10^{-10}, 10^{-9}]$, which contains the WMAP value, and choose the value of $m_{N_1} = 10^{10}$ GeV in some of our plots. Similar studies of the constraints from leptogenesis on LFV rates have been done in [239].

Concerning the EDMs, which are clearly non-vanishing in the presence of complex θ_i, we have checked that all the predicted values for the electron, muon and tau EDMs are well below the experimental bounds. In the following we therefore focus on complex but small θ_2 values, leading to favorable BAU, and study its effects on the sensitivity to θ_{13}. Similar results are obtained for θ_3, but for shortness are not shown here.

Figure 23 shows the dependence of the most sensitive BR to θ_{13}, $B(\mu \to e\gamma)$, on $|\theta_2|$. We consider two particular values of θ_{13}, $\theta_{13} = 0°, 5°$ and choose SPS 1a. Motivated from the thermal leptogenesis favored θ_2-regions [686], we take $0 \lesssim |\theta_2| \lesssim \pi/4$, with $\arg \theta_2 = \{\pi/8, \pi/4, 3\pi/8\}$. We display the numerical results, considering $m_{\nu_1} = 10^{-5}$ eV and $m_{\nu_1} = 10^{-3}$ eV, while for the heavy neutrino masses we take $m_N = (10^{10}, 10^{11}, 10^{14})$ GeV. There are several important conclusions to be drawn from Fig. 23. Let us first discuss the case $m_{\nu_1} = 10^{-5}$ eV. We note that one can obtain a baryon asymmetry in the range 10^{-10} to 10^{-9} for a considerable region of the analyzed $|\theta_2|$ range. Notice also that there is a clear separation between the predictions of $\theta_{13} = 0°$ and $\theta_{13} = 5°$, with the latter well above the present experimental bound. This would imply an experimental impact of θ_{13}, in the sense that the BR predictions become potentially detectable for this non-vanishing θ_{13} value. With the planned MEG sensitivity [701], both cases would be within experimental reach. However, this statement is strongly dependent on the assumed parameters, in particular m_{ν_1}. For instance, a larger value of $m_{\nu_1} = 10^{-3}$ eV, illustrated on the right panel of Fig. 23, leads to a very distinct situation regarding the sensitivity to θ_{13}. While for smaller values of $|\theta_2|$ the branching ratio displays a clear sensitivity to having θ_{13} equal or different from zero (a separation larger than

two orders of magnitude for $|\theta_2| \lesssim 0.05$), the effect of θ_{13} is diluted for increasing values of $|\theta_2|$. For $|\theta_2| \gtrsim 0.3$ the $B(\mu \to e\gamma)$ associated with $\theta_{13} = 5°$ can be even smaller than for $\theta_{13} = 0°$. This implies that in this case, a potential measurement of $B(\mu \to e\gamma)$ would not be sensitive to θ_{13}. Whether or not a SPS 1a scenario would be disfavored by current experimental data on $B(\mu \to e\gamma)$ requires a careful weighting of several aspects. Even though Fig. 23 suggests that for this particular choice of parameters only very small values of θ_2 and θ_{13} would be in agreement with current experimental data, a distinct choice of m_{N_3} (e.g. $m_{N_3} = 10^{13}$ GeV) would lead to a rescaling of the estimated BRs by a factor of approximately 10^{-2}. Although we do not display the associated plots here, in the latter case nearly the entire $|\theta_2|$ range would be in agreement with experimental data (in fact the points which are below the present MEGA bound on Fig. 23 would then lie below the projected MEG sensitivity). Regarding the other SPS points, which are not shown here, we find BRs for SPS 1b comparable to those of SPS 1a. Smaller ratios are associated with SPS 2, 3 and 5, while larger (more than one order of magnitude) BRs occur for SPS 4.

Let us now address the question of whether a joint measurement of the BRs and θ_{13} can shed some light on experimentally unreachable parameters, like m_{N_3}. The expected improvement in the experimental sensitivity to the LFV ratios supports the possibility that a BR could be measured in the future, thus providing the first experimental evidence for new physics, even before its discovery at the LHC. The prospects are especially encouraging regarding $\mu \to e\gamma$, where the experimental sensitivity will improve by at least two orders of magnitude. Moreover, and given the impressive effort on experimental neutrino physics, a measurement of θ_{13} will likely also occur in the future [708–717]. Given

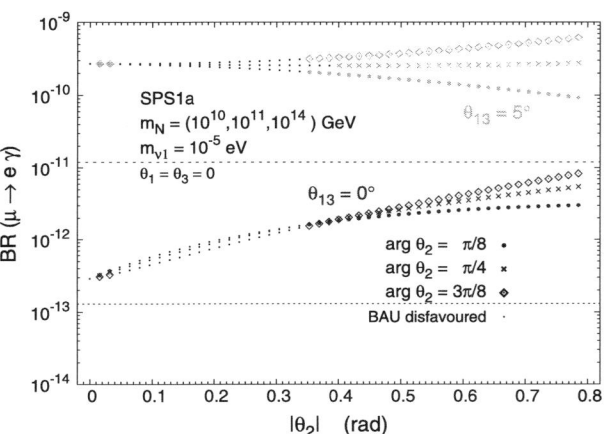
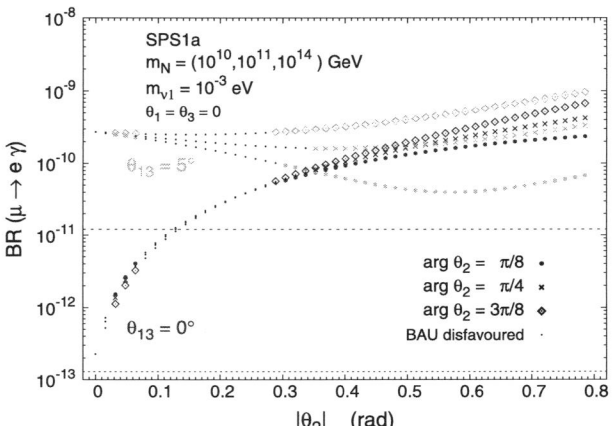

Fig. 23 $B(\mu \to e\gamma)$ as a function of $|\theta_2|$, for $\arg \theta_2 = \{\pi/8, \pi/4, 3\pi/8\}$ (*dots, times, diamonds*, respectively) and $\theta_{13} = 0°$, $5°$ (*blue/darker, green/lighter lines*). We take $m_{\nu_1} = 10^{-5}$ (10^{-3}) eV, on the *left* (*right*) panel. In all cases *black dots* represent points associated with a disfavored BAU scenario and a *dashed* (*dotted*) *horizontal line* denotes the present experimental bound (future sensitivity)

that, as previously emphasized, $\mu \to e\gamma$ is very sensitive to θ_{13}, whereas this is not the case for $B(\tau \to \mu\gamma)$, and that both BRs display the same approximate behavior with m_{N_3} and $\tan\beta$, we now propose to study the correlation between these two observables. This optimizes the impact of a θ_{13} measurement, since it allows us to minimize the uncertainty introduced from not knowing $\tan\beta$ and m_{N_3}, and at the same time offers a better illustration of the uncertainty associated with the R-matrix angles. In this case, the correlation of the BRs with respect to m_{N_3} means that, for a fixed set of parameters, varying m_{N_3} implies that the predicted point $(B(\tau \to \mu\gamma), B(\mu \to e\gamma))$ moves along a line with approximately constant slope in the $B(\tau \to \mu\gamma)$–$B(\mu \to e\gamma)$ plane. On the other hand, varying θ_{13} leads to a displacement of the point along the vertical axis.

In Fig. 24, we illustrate this correlation for SPS 1a, choosing distinct values of the heaviest neutrino mass, and we scan over the BAU-enabling R-matrix angles (setting θ_3 to zero) as

$$0 \lesssim |\theta_1| \lesssim \pi/4, \qquad -\pi/4 \lesssim \arg\theta_1 \lesssim \pi/4,$$
$$0 \lesssim |\theta_2| \lesssim \pi/4, \qquad 0 \lesssim \arg\theta_2 \lesssim \pi/4, \qquad (5.39)$$
$$m_{N_3} = 10^{12}, 10^{13}, 10^{14} \text{ GeV}.$$

We consider the following values, $\theta_{13} = 1°$, $3°$, $5°$ and $10°$, and only include in the plot the BR predictions which allow for a favorable BAU. Other SPS points have also been considered but they are not shown here for brevity (see [686]). We clearly observe in Fig. 24 that for a fixed value of m_{N_3}, and for a given value of θ_{13}, the dispersion arising from a θ_1 and θ_2 variation produces a small area rather than a point in the $B(\tau \to \mu\gamma)$–$B(\mu \to e\gamma)$ plane. The dispersion along the $B(\tau \to \mu\gamma)$ axis is of approximately one order of magnitude for all θ_{13}. In contrast, the dispersion along the $B(\mu \to e\gamma)$ axis increases with decreasing θ_{13},

ranging from an order of magnitude for $\theta_{13} = 10°$, to over three orders of magnitude for the case of small θ_{13} (1°). From Fig. 24 we can also infer that other choices of m_{N_3} (for $\theta_{13} \in [1°, 10°]$) would lead to BR predictions which would roughly lie within the diagonal lines depicted in the plot. Comparing these predictions for the shaded areas along the expected diagonal "corridor", with the allowed experimental region, allows us to conclude about the impact of a θ_{13} measurement on the allowed/excluded m_{N_3} values. The most important conclusion from Fig. 24 is that for SPS 1a, and for the parameter space defined in (5.39), an hypothetical θ_{13} measurement larger than 1°, together with the present experimental bound on the $B(\mu \to e\gamma)$, will have the impact of excluding values of $m_{N_3} \gtrsim 10^{14}$ GeV. Moreover, with the planned MEG sensitivity, the same θ_{13} measurement can further constrain $m_{N_3} \lesssim 3 \times 10^{12}$ GeV. The impact of any other θ_{13} measurement can be analogously extracted from Fig. 24.

As a final comment let us add that, remarkably, within a particular SUSY scenario and scanning over specific θ_1 and θ_2 BAU-enabling ranges for various values of θ_{13}, the comparison of the theoretical predictions for $B(\mu \to e\gamma)$ and $B(\tau \to \mu\gamma)$ with the present experimental bounds allows us to set θ_{13}-dependent upper bounds on m_{N_3}. Together with the indirect lower bound arising from leptogenesis considerations, this clearly provides interesting hints on the value of the seesaw parameter m_{N_3}. With the planned future sensitivities, these bounds would further improve by approximately one order of magnitude. Ultimately, a joint measurement of the LFV branching ratios, θ_{13} and the sparticle spectrum would be a powerful tool for shedding some light on otherwise unreachable SUSY seesaw parameters. It is clear from all this study that the interplay between LFV processes and future improvement in neutrino data is challenging for the searches of new physics.

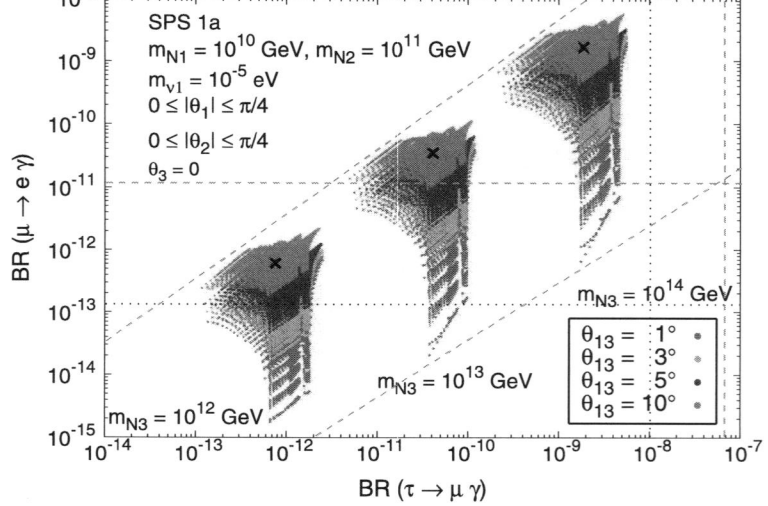

Fig. 24 (Color online) Correlation between $B(\mu \to e\gamma)$ and $B(\tau \to \mu\gamma)$ as a function of m_{N_3}, for SPS 1a. The areas displayed represent the scan over θ_i as given in (5.39). *From bottom to top*, the *colored regions* correspond to $\theta_{13} = 1°$, $3°$, $5°$ and $10°$ (*red, green, blue* and *pink*, respectively). *Horizontal* and *vertical dashed* (*dotted*) *lines* denote the experimental bounds (future sensitivities)

5.2.5 LFV in the CMSSM with constrained sequential dominance

Sequential Dominance (SD) [140, 141, 143] represents classes of neutrino models where large lepton mixing angles and small hierarchical neutrino masses can be readily explained within the seesaw mechanism. To understand how Sequential Dominance works, we begin by writing the right handed neutrino Majorana mass matrix M_{RR} in a diagonal basis as $M_{RR} = \text{diag}(M_A, M_B, M_C)$. We furthermore write the neutrino (Dirac) Yukawa matrix λ_ν in terms of (1, 3) column vectors A_i, B_i, C_i as $Y_\nu = (A, B, C)$ using left–right convention. The term for the light neutrino masses in the effective Lagrangian (after electroweak symmetry breaking), resulting from integrating out the massive right handed neutrinos, is

$$\mathcal{L}^\nu_{\text{eff}} = \frac{(\nu_i^T A_i)(A_j^T \nu_j)}{M_A} + \frac{(\nu_i^T B_i)(B_j^T \nu_j)}{M_B} + \frac{(\nu_i^T C_i)(C_j^T \nu_j)}{M_C}, \qquad (5.40)$$

where ν_i ($i = 1, 2, 3$) are the left handed neutrino fields. Sequential dominance then corresponds to the third term being negligible, the second term subdominant and the first term dominant:

$$\frac{A_i A_j}{M_A} \gg \frac{B_i B_j}{M_B} \gg \frac{C_i C_j}{M_C}. \qquad (5.41)$$

In addition, we shall shortly see that small θ_{13} and almost maximal θ_{23} require that

$$|A_1| \ll |A_2| \approx |A_2|. \qquad (5.42)$$

Without loss of generality, then, we shall label the dominant right handed neutrino and Yukawa couplings as A, the subdominant ones as B, and the almost decoupled (sub-subdominant) ones as C. Note that the mass ordering of right handed neutrinos is not yet specified. Again without loss of generality we shall order the right handed neutrino masses as $M_1 < M_2 < M_3$, and subsequently identify M_A, M_B, M_C with M_1, M_2, M_3 in all possible ways. LFV in some of these classes of SD models has been analyzed in [718]. Tri-bi-maximal *neutrino* mixing corresponds to the choice for example [719], sometimes referred to as constrained sequential dominance (CSD):

$$Y_\nu = \begin{pmatrix} 0 & be^{i\beta_2} & c_1 \\ -ae^{i\beta_3} & be^{i\beta_2} & c_2 \\ ae^{i\beta_3} & be^{i\beta_2} & c_3 \end{pmatrix}. \qquad (5.43)$$

When dealing with LFV it is convenient to work in the basis where the charged lepton mass matrix is diagonal. Let us now discuss the consequences of charged lepton corrections with a CKM-like structure, for the neutrino Yukawa matrix with CSD. By CKM-like structure we mean that V_{e_L} is dominated by a 1–2 mixing θ, i.e. that its elements $(V_{e_L})_{13}$, $(V_{e_L})_{23}$, $(V_{e_L})_{31}$ and $(V_{e_L})_{32}$ are very small compared to $(V_{e_L})_{ij}$ ($i, j = 1, 2$). After re-diagonalizing the charged lepton mass matrix, Y_ν in (5.43) becomes transformed as $Y_\nu \to V_{e_L} Y_\nu$. In the diagonal charged lepton mass basis the neutrino Yukawa matrix therefore becomes

$$Y_\nu = \begin{pmatrix} as_\theta e^{-i\lambda} e^{i\beta_3} & b(c_\theta - s_\theta e^{-i\lambda})e^{i\beta_2} & (c_1 c_\theta - c_2 s_\theta e^{-i\lambda}) \\ -ac_\theta e^{i\beta_3} & b(c_\theta + s_\theta e^{i\lambda})e^{i\beta_2} & (c_1 s_\theta e^{i\lambda} + c_2 c_\theta) \\ ae^{i\beta_3} & be^{i\beta_2} & c_3 \end{pmatrix}. \qquad (5.44)$$

After ordering M_A, M_B, M_C according to their size, there are six possible forms of Y_ν obtained from permuting the columns, with the convention always being that the dominant one is labeled by A, and so on. In particular the third column of the neutrino Yukawa matrix could be A, B or C depending on which of M_A, M_B or M_C is the heaviest. If the heaviest right handed neutrino mass is M_A then the third column of the neutrino Yukawa matrix will consist of the (re-ordered) first column of (5.44) and assuming $Y^\nu_{33} \sim 1$ we conclude that all LFV processes will be determined approximately by the first column of (5.44). Similarly if the heaviest right handed neutrino mass is M_B then we conclude that all LFV processes will be determined approximately by the second column of (5.44). Note that in both cases the ratios of branching ratios are independent of the unknown Yukawa couplings which cancel, and only depend on the charged lepton angle θ, which in the case of tri-bi-maximal neutrino mixing is related to the physical reactor angle by $\theta_{13} = \theta/\sqrt{2}$ [719, 720]. Also note that $\lambda = \delta - \pi$ where δ is the standard PDG CP-violating oscillation phase. The results for these two cases are shown in Fig. 25 [721]. The third case $M_3 = M_C$ is less predictive.

5.2.6 Decoupling of one heavy neutrino and cosmological implications

The supersymmetric seesaw model involves many free parameters. In order to correlate the model predictions for LFV processes one has to resort to some supplementary hypotheses. Here we discuss the consequences of the assumption that one of the heavy singlet neutrinos (not necessarily the heaviest one) decouples from the see-saw mechanism [238].

If the light neutrino masses are hierarchical, in which case the effects of the renormalization group (RG) running [722] of κ are negligible, at least 3 arguments support this assumption. The first one is the *naturalness of the see-saw*

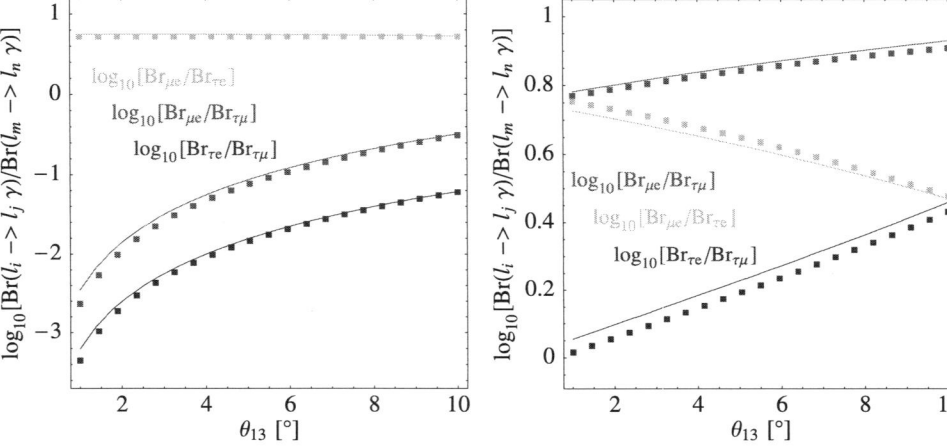

Fig. 25 Ratios of branching ratios of LFV processes $\ell_i \to \ell_j \gamma$ in CSD for $M_3 = M_A$ (*left panel*) and $M_3 = M_B$ (*right panel*) with right handed neutrino masses $M_1 = 10^8$ GeV, $M_2 = 5 \times 10^8$ GeV and $M_3 = 10^{14}$ GeV. The *solid lines* show the (naive) prediction, from the MI and LLog approximation and with RG running effects neglected, while the *dots* show the explicit numerical computation (using SPheno2.2.2. [699] extended by software packages for LFV BRs and neutrino mass matrix running [649, 686]) with universal CMSSM parameters chosen as $m_0 = 750$ GeV, $m_{1/2} = 750$ GeV, $A_0 = 0$ GeV, $\tan\beta = 10$ and $\text{sign}(\mu) = +1$. While the ratios do not significantly depend on the choice of the SUSY model, since the model-dependence has canceled out, they show a pronounced dependence on θ_{13} (and δ) in the case of $M_3 = M_A$ (and $M_3 = M_B$)

mechanism. Large mixing angles are not generic for hierarchical light neutrino masses (for a review, see [723]). They are natural only for special patterns of the matrix κ. One is a large hierarchy between one and the remaining two terms in the sum in (3.41) [38, 140, 233, 724]. This is what we call *decoupling* (one term hierarchically smaller) or *dominance* (hierarchically larger). Seesaw with only two heavy singlet neutrinos [234, 235] is the limiting case of the decoupling of N_3 with $M_3 \to \infty$ and $Y_\nu^{3A} \to 0$. The immediate consequence of decoupling is $m_{\nu_1} \ll m_{\nu_2}$ (κ has rank 2 if there are only 2 terms in the sum in (3.41)). Similarly, for dominance one has $m_{\nu_2} \ll m_{\nu_3}$.

Secondly, decoupling of the lightest singlet neutrino *alleviates the gravitino problem of leptogenesis* which in the see-saw models of neutrino masses appears to be the most natural mechanism for producing the observed baryon asymmetry of the Universe[18] (BAU). As the Universe cools down leptonic asymmetries (subsequently converted into baryon asymmetry through sphaleron transitions) $Y_\alpha \equiv (n_\alpha - \bar{n}_\alpha)/s \neq 0$ (where n_α and $\bar{n}_\alpha$ are the flavor α lepton and antilepton number densities, respectively and s is the entropy density) are produced in the decays of N_1. The final magnitudes of Y_α are proportional to the decay asymmetries $\varepsilon_{1\alpha}$, (which in turn are proportional to the heavy neutrino masses) and crucially depend on the processes which wash out the asymmetries generated by the N_1 decays. The efficiency of these processes depends on the parameters $\tilde{m}_{1\alpha} = \sum_A |R_{1A} U^*_{\alpha A}|^2 m_{\nu_A}$, where $U \equiv U_{\text{PMNS}}$ and it is the smallest (i.e. leptogenesis is most efficient) for $\tilde{m}_{1\alpha}$ in the meV range (assuming vanishing density of N_1 after re-heating and strongly hierarchical spectrum of M_A). If it is N_1 which is decoupled, there are essentially no lower bounds on $\tilde{m}_{1\alpha}$ and M_1, hence also the re-heating temperature T_{RH}, already of order 10^9 GeV are sufficient [311, 726] (see, however, e.g., [316]) to reproduce the observed BAU.[19]

Finally, one heavy singlet neutrino N_A must be decoupled if its superpartner, $\tilde{N}_A$, *plays the role of the inflaton field* [347]. In such a scenario the (s)neutrino mass M_A must be [349] 2×10^{13} GeV and the re-heating temperature following inflaton decay is given by $T_{\text{RH}} \sim \sqrt{\tilde{m}_A M_{\text{Pl}}}(M_A/\langle H \rangle)$. Requiring $T_{\text{RH}} < 10^6$ GeV (the gravitino problem) then implies $m_{\nu_A} \leq \sum_\alpha \tilde{m}_{A\alpha} < 10^{-17}$ eV. In this scenario, the leptonic asymmetries must be produced non-thermally in the inflaton decay. Decoupling of N_1 is favored because if it is $\tilde{N}_2$ or $\tilde{N}_3$ which is the inflaton the produced asymmetry may be subsequently washed out during the decays of N_1.

The assumption that N_A effectively decouples from the seesaw mechanism or that N_A effectively dominates the seesaw mechanism translates into one of the following forms

---

[18]See [290]; for a review of leptogenesis, see [324]; for a discussion of flavor effects in leptogenesis see, e.g. [296]; for recent analyses of the gravitino problem, see, e.g., [725].

[19]In contrast, for N_2 or N_3 decoupled, the washout is much stronger and M_1 has to be $\gtrsim 10^{10}$ GeV. This requires T_{RH} leading to a much larger dangerous gravitino production [727]. Lower T_{RH} is in this case possible only if N_1 and N_2 are sufficiently degenerate [728].

of R:

$$R_{\text{dec}} \simeq \Pi^{(A)} \begin{pmatrix} 1 & 0 & 0 \\ 0 & z & p \\ 0 & \mp p & \pm z \end{pmatrix} \text{ or }$$
$$R_{\text{dom}} \simeq \Pi^{(A)} \begin{pmatrix} \mp p & \pm z & 0 \\ z & p & 0 \\ 0 & 0 & 1 \end{pmatrix}, \quad (5.45)$$

where z, p are complex numbers satisfying $z^2 + p^2 = 1$ and $\Pi^{(A)}$ denotes permutation of the rows of R. Both conditions can be simultaneously satisfied for $R = \Pi^{(A)} \cdot \mathbf{1}$, known as *sequential dominance* (for a review, see, e.g. [729]).

In the framework considered, violation of the leptonic flavor is transmitted from the neutrino Yukawa couplings Y_ν to the slepton mass matrices through the RG corrections. Branching ratios of LFV decays are well described by a single mass-insertion approximation via (5.27) and (5.28). Since decoupling of N_1 is best motivated we discuss the results for LFV only in this case.[20]

The matrix R has then the first of the patterns displayed in (5.45) with $\Pi^{(1)} = \mathbf{1}$. The discussion simplifies if a technical assumption that $m_{\nu_3} M_2 < m_{\nu_2} M_3$ is made. $(\tilde{m}_L^2)_{32}$ relevant for $\tau \to \mu\gamma$ then reads

$$\left(\tilde{m}_L^2\right)_{32} \approx \frac{\kappa m_{\nu_3} M_3 U_{33} U_{23}^*}{\langle H \rangle^2} \left[\left(|z|^2 + S|p|^2\right) + \rho \frac{U_{22}^*}{U_{23}^*} x \right.$$
$$\left. + \rho \frac{U_{32}}{U_{33}} x^* + \rho^2 \frac{U_{32} U_{22}^*}{U_{33} U_{23}^*} \left(S|z|^2 + |p|^2\right) \right], \quad (5.46)$$

where $\rho = \sqrt{m_{\nu_2}/m_{\nu_3}} \sim 0.4$, $S = M_2(1 + \Delta l_2/\Delta t)/M_3 \sim M_2/M_3$ and $x = Sp^*z - z^*p$. For $(\tilde{m}_L^2)_{A1}$ relevant for $\ell_A \to e\gamma$ we get:

$$\left(\tilde{m}_L^2\right)_{A1} \approx \frac{\kappa m_{\nu_3} M_3 U_{A3} U_{12}^*}{\langle H \rangle^2} \left[\frac{U_{13}^*}{U_{12}^*} \left(|z|^2 + S|p|^2\right) + \rho x \right.$$
$$\left. + \rho^2 \frac{U_{A2}}{U_{A3}} \left(S|z|^2 + |p|^2\right) \right]. \quad (5.47)$$

Analysis of the expressions (5.46) and (5.47) leads to a number of conclusions [238]. Firstly, the branching ratios of the LFV decays depend (apart from the scales of soft supersymmetry breaking and the value of $\tan\beta$) mostly on the

---

[20] Results for N_2 decoupled are the same as for decoupled N_1 (including sub-leading effects if M_1 takes the numerical value of M_2). The same is true also for N_3 decoupled (including the case with only 2 heavy singlet neutrinos) if M_2 is numerically the same as M_3 for decoupled N_1. However, if $\tilde{N}_3$ is the inflaton the LFV decays have the rates too low to be observed. In addition, if N_3 decouples due to its very large mass its large Yukawa can, for $m_{\nu_1}/m_{\nu_3} > M_2/M_3$, still dominate the LFV effects which are then practically unconstrained by the oscillation data; some constraints can then be obtained from the limits on the electron EDM [730].

mass of the heaviest of the two un-decoupled singlet neutrinos (in this case N_3). Secondly, for fixed M_3, they depend strongly on the magnitude and phase of R_{32}, mildly on the undetermined element U_{13} of the light neutrino mixing matrix and, in addition, on the Majorana phases of U which cannot be measured in oscillation experiments [545]. The latter dependence is mild for $B(\tau \to \mu\gamma)$ but can lead to strong destructive interference *either* in $B(\mu \to e\gamma)$ or in $B(\tau \to e\gamma)$ decreasing them by several orders of magnitude. The interference effects are seen in Fig. 26(a), (b) where the predicted ranges (resulting from varying the unknown Majorana phases) of $B(\mu \to e\gamma)$ are shown as a function of $|R_{32}|$ for M_1 appropriate for the sneutrino inflation scenario, three different values of $\arg(R_{32})$ and for $m_0 = 100$ GeV, $M_{1/2} = 500$ GeV and $\tan\beta = 10$, consistent with the dark matter abundance [731]. Results for other values of these parameters can be obtained by appropriate rescalings using (5.27). For comparison, for selected values of R_{23}, we also indicate the ranges of $B(\mu \to e\gamma)$ resulting from generic form of the matrix R (constrained only by the conditions $0 < R_{12}, R_{13} < 1.5$ and $\text{Re}(Y_\nu^{AB})$, $\text{Im}(Y_\nu^{AB}) < 10$).

The bulk of the predicted values of $B(\mu \to e\gamma)$ shown in Fig. 26(a) and (b) exceed the current experimental limit. Since $M_{1/2} = 500$ GeV leads to masses of the third generation squarks above 1 TeV, suppressing $B(\mu \to e\gamma)$ by increasing the SUSY breaking scale conflicts with the stability of the electroweak scale. Moreover, as discussed in [238] in the scenario considered here generically $B(\tau \to \mu\gamma)/B(\mu \to e\gamma) \sim 0.1$. Thus the observation $\tau \to \mu\gamma$ with $B \gtrsim 10^{-9}$, accessible to future experiments would exclude this scenario.

For completeness, in Fig. 26(c) we also show predictions for $B(\mu \to e\gamma)$ in the case of N_3 dominance. $(\tilde{m}_L^2)_{21}$ is in this case controlled mainly by $|U_{13}|$. Moreover, $B(\tau \to \mu\gamma)/B(\mu \to e\gamma) \sim \max(|U_{13}|^2, \rho^4 S^2)$, while $B(\tau \to e\gamma)/B(\mu \to e\gamma) \sim 1$ allowing for experimental test of this scenario (cf. [732]). The limits $R_{32} \to 0$ in panels *a* and *b* and or $R_{21} \to 0$ in panel *c* correspond to pure sequential dominance.

In conclusion, the well motivated assumption about the decoupling/dominance of one heavy singlet neutrino significantly constrains the predictions for the LFV processes in supersymmetric model. The forthcoming experiments should be able to verify this assumption and, in consequence, to test an interesting class of neutrino mass models.

5.2.7 Triplet seesaw mechanism and lepton flavor violation

In this subsection we intend to discuss the aspect of low scale LFV in rare decays arising in the context of the triplet seesaw mechanism of Sect. 3.2.3.2. We consider both non-SUSY and SUSY versions of it. The flavor structure of the

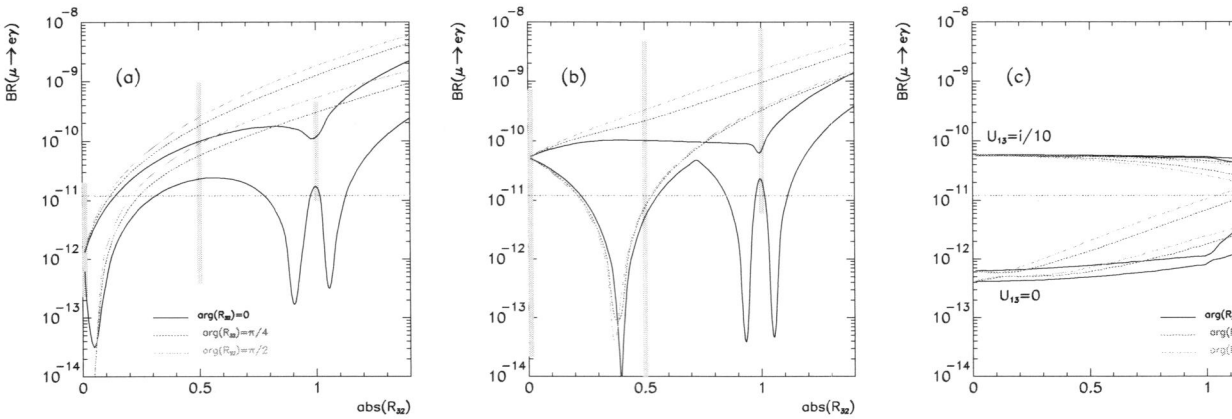

Fig. 26 (Color online) Predicted ranges of $B(\mu \to e\gamma)$ for $(M_1, M_2, M_3) = (2, 3, 50) \times 10^{13}$ GeV, $m_0 = 100$ GeV, $M_{1/2} = 500$ GeV and $\tan\beta = 10$, for the decoupling of N_1 and $U_{13} = 0$ (panel **a**) or $U_{13} = 0.1i$ (panel **b**). *Yellow ranges* show the possible variation for arbitrary form of R with $\arg(R_{32}) = 0$. *Lower* (*upper*) *pairs of lines* in the panel **c** show similar ranges for N_3 dominance for $U_{13} = 0$ ($0.1i$). The current experimental bound of 1.2×10^{-11} [180] is also shown

Fig. 27 Diagrams that contribute to the decay $\ell_j \to \ell_i \gamma$ through the exchange of the triplet scalars

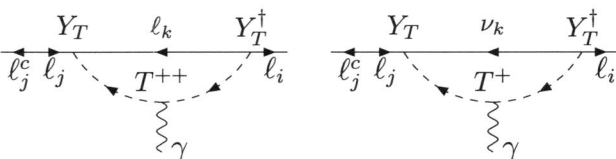

(high energy) Yukawa matrix Y_T of (3.63) is the same as that of the (low energy) neutrino mass matrix m_ν. Therefore, in the triplet seesaw scenario the neutrino mass matrix (containing nine real parameters), which can be tested in the low energy experiments, is *directly* linked to the symmetric matrix Y_T (containing also nine real parameters), modulo the ratio M_T^2/μ', see (3.66). This feature has interesting implications for LFV [249]. Collider phenomenology of the low scale triplet was discussed in Sect. 5.1.4 The triplet Lagrangian also induces LFV decays of the charged leptons through the one loop exchange of the triplet states.

The diagrams relevant for the LFV radiative decays $\ell_j \to \ell_i \gamma$ (see e.g., [733, 734]) are depicted in Fig. 27. Denoting $U_{\text{PMNS}} = V \cdot \text{diag}(1, e^{i\phi_1}, e^{i\phi_2})$, where, $\phi_{1,2}$ are the Majorana phases, those imply the following flavor structure:

$$\begin{aligned}(Y_T^\dagger Y_T)_{ij} &= \left(\frac{M_T^2}{\mu' v^2}\right)^2 (m_\nu^\dagger m_\nu)_{ij} \\ &= \left(\frac{M_T^2}{\mu' v^2}\right)^2 \left[V (m_\nu^D)^2 V^\dagger\right]_{ij},\end{aligned} \quad (5.48)$$

where $i, j = e, \mu, \tau$ are family indices. Therefore, the amount of LFV is *directly* and *univocally* expressed in terms of the low energy neutrino parameters. In particular, LFV decays depend only on 7 independent neutrino parameters (there is no dependence on the Majorana phases ϕ_i). Notice that this simple flavor structure is peculiar of the triplet seesaw case, which represents a concrete and explicit realization of the 'minimal flavor violation' hypothesis [536] in the lepton sector [542]. Indeed, according to the latter, the low energy SM Yukawa couplings are the *only source* of LFV. This is not generically the case for the seesaw mechanism realized through the exchange of the so-called 'right handed' neutrinos, where the number of independent parameters of the high energy flavor structures is twice more that of the mass matrix m_ν.

Finally, the parametric dependence of the dipole amplitude in Fig. 27 is

$$D_{ij} \approx \frac{(Y_T^\dagger Y_T)_{ij}}{16\pi^2 M_T^2} = \frac{(m_\nu^\dagger m_\nu)_{ij}}{16\pi^2 v^4} \left(\frac{M_T}{\mu'}\right)^2. \quad (5.49)$$

From the present experimental bound on $B(\mu \to e\gamma) < 1.2 \times 10^{-11}$ [180], one infers the bound $\mu' > 10^{-10} M_T$ [comparable limit is obtained from $B(\tau \to \mu\gamma)$]. We can push further our discussion considering the relative size of LFV in different family sectors:

$$\begin{aligned}\frac{(Y_T^\dagger Y_T)_{\tau\mu}}{(Y_T^\dagger Y_T)_{\mu e}} &\approx \frac{[V(m_\nu^D)^2 V^\dagger]_{\tau\mu}}{[V(m_\nu^D)^2 V^\dagger]_{\mu e}}, \\ \frac{(Y_T^\dagger Y_T)_{\tau e}}{(Y_T^\dagger Y_T)_{\mu e}} &\approx \frac{[V(m_\nu^D)^2 V^\dagger]_{\tau e}}{[V(m_\nu^D)^2 V^\dagger]_{\mu e}}.\end{aligned} \quad (5.50)$$

These ratios depend only on the neutrino parameters, while do not depend on details of the model, such as the mass scales M_T, or μ'. By taking the present best fit values of the neutrino masses and mixing angles [211, 735] provided by the analysis of the experimental data, those ratios can be explicitly expressed as

$$\frac{(Y_T^\dagger Y_T)_{\tau\mu}}{(Y_T^\dagger Y_T)_{\mu e}} \approx \left(\frac{\Delta m_A^2}{\Delta m_S^2}\right) \frac{\sin 2\theta_{23}}{\sin 2\theta_{12} \cos\theta_{23}} \sim 40,$$

$$\frac{(Y_T^\dagger Y_T)_{\tau e}}{(Y_T^\dagger Y_T)_{\mu e}} \approx -\tan\theta_{23} \sim -1, \quad (5.51)$$

where $\Delta m_A^2 (\Delta m_S^2)$ is the squared-mass difference relevant for the atmospheric (solar) neutrino oscillations. These results hold for $\theta_{13} = 0$ and for either hierarchical, quasi-degenerate or inverted hierarchical neutrino spectrum (for more details see [249, 736]). It is immediate to translate the above relations into *model-independent* predictions for ratios of LFV processes:

$$\frac{B(\tau \to \mu\gamma)}{B(\mu \to e\gamma)} \approx \left(\frac{(Y_T^\dagger Y_T)_{\tau\mu}}{(Y_T^\dagger Y_T)_{\mu e}}\right)^2 \frac{B(\tau \to \mu \nu_\tau \bar{\nu}_\mu)}{B(\mu \to e \nu_\mu \bar{\nu}_e)} \sim 300,$$

$$\frac{B(\tau \to e\gamma)}{B(\mu \to e\gamma)} \approx \left(\frac{(Y_T^\dagger Y_T)_{\tau e}}{(Y_T^\dagger Y_T)_{\mu e}}\right)^2 \frac{B(\tau \to e \nu_\tau \bar{\nu}_e)}{B(\mu \to e \nu_\mu \bar{\nu}_e)} \sim 0.2. \quad (5.52)$$

Now we focus upon the supersymmetric version of the triplet seesaw mechanism. (Just recall just that in the supersymmetric case there is only one mass parameter, M_T, while the mass parameter μ' of the non-supersymmetric version is absent from the superpotential and its role is taken by $\lambda_2 M_T$.) Regarding the aspect of LFV, in this case we have to consider besides the diagrams of Fig. 27 also the related ones with each particle in the loop replaced by its superpartner ($\ell_k \to \tilde{\ell}_k$, $T \to \tilde{T}$). Such additional contributions would cancel those in Fig. 27 in the limit of exact supersymmetry. In the presence of soft supersymmetry breaking (SSB) the cancellation is only partial and the overall result for the coefficient of the dipole amplitude behaves like

$$D_{ij} \approx \frac{(Y_T^\dagger Y_T)_{ij}}{16\pi^2} \frac{\tilde{m}^2}{M_T^4} \sim \frac{(m_\nu^\dagger m_\nu)_{ij}}{16\pi^2 (\lambda_2 v_2^2)^2} \frac{\tilde{m}^2}{M_T^2}, \quad (5.53)$$

which is suppressed with respect to the non-supersymmetric result (5.49) for $M_T > \tilde{m} \sim \mathcal{O}(10^2 \text{ GeV})$ ($\tilde{m}$ denotes an average soft breaking mass parameter). In the supersymmetric version of the triplet seesaw mechanism flavor violation can also be induced by renormalization effects via (3.69) (the complete set of RGEs of the MSSM with the triplet states have been computed in [249]). Thus in SUSY model the LFV processes can occur also in the case of very heavy triplet. In that case the relevant flavor structure responsible for LFV is again $Y_T^\dagger Y_T$ for which we have already noticed its *unambiguous* dependence on the neutrino parameters in (5.48). Clearly, we find that analogous ratios as in (5.50) hold also for the LFV entries of the soft breaking parameters, e.g.,

$$\frac{(m_{\tilde{L}}^2)_{\tau\mu}}{(m_{\tilde{L}}^2)_{\mu e}} \approx \frac{[V(m_\nu^D)^2 V^\dagger]_{\tau\mu}}{[V(m_\nu^D)^2 V^\dagger]_{\mu e}},$$

$$\frac{(m_{\tilde{L}}^2)_{\tau e}}{(m_{\tilde{L}}^2)_{\mu e}} \approx \frac{[V(m_\nu^D)^2 V^\dagger]_{\tau e}}{[V(m_\nu^D)^2 V^\dagger]_{\mu e}}. \quad (5.54)$$

Such SSB flavor violating mass parameters induce extra contributions to the LFV processes. For example, the radiative decays $\ell_j \to \ell_i \gamma$ receive also one loop contributions with the exchange of the charged-sleptons/neutralinos and sneutrinos/charginos, where the slepton masses $(m_{\tilde{L}}^2)_{ij}$ are the source of LFV. The relevant dipole terms have a parametric dependence of the form

$$D_{ij} \approx \frac{g^2}{16\pi^2} \frac{(m_{\tilde{L}}^2)_{ij}}{\tilde{m}^4} \tan\beta$$

$$\approx \frac{g^2}{16\pi^2} \frac{(m_\nu^\dagger m_\nu)_{ij}}{(\lambda_2 v_2^2)^2} \frac{M_T^2}{\tilde{m}^2} \log\left(\frac{M_G}{M_T}\right) \tan\beta. \quad (5.55)$$

Notice the inverted dependence on the ratio $\tilde{m}/M_T$ with respect to the triplet exchange contribution. Due to this feature, the MSSM sparticle induced contributions (5.55) tends to dominate over the one induced by the triplet exchange. In this case, analogous ratios as in (5.52) can be derived, i.e.,

$$\frac{B(\tau \to \mu\gamma)}{B(\mu \to e\gamma)} \approx \left(\frac{(m_{\tilde{L}}^2)_{\tau\mu}}{(m_{\tilde{L}}^2)_{\mu e}}\right)^2 \frac{B(\tau \to \mu \nu_\tau \bar{\nu}_\mu)}{B(\mu \to e \nu_\mu \bar{\nu}_e)} \sim 300,$$

$$\frac{B(\tau \to e\gamma)}{B(\mu \to e\gamma)} \approx \left(\frac{(m_{\tilde{L}}^2)_{\tau e}}{(m_{\tilde{L}}^2)_{\mu e}}\right)^2 \frac{B(\tau \to e \nu_\tau \bar{\nu}_e)}{B(\mu \to e \nu_\mu \bar{\nu}_e)} \sim 0.2. \quad (5.56)$$

(For more details see [249].)

The presence of extra $SU(2)_W$ triplet states at intermediate energy spoils the successful gauge coupling unification of the MSSM. A simple way to recover gauge coupling unification is to introduce more states X, to complete a certain representation R—such that $R = T + X$—of some unifying gauge group G, $G \supset SU(3) \times SU(2)_W \times U(1)_Y$. In general the Yukawa couplings of the states X are related to those of the triplet partners T. Indeed, this is generally the case in minimal GUT models. In this case RG effects generates not only lepton-flavor violation but also closely correlated flavor violation in the quark sector (due to the X-couplings). An explicit scenario with $G = SU(5)$ where both lepton and quark flavor violation arise from RG effects was discussed in Ref. [249]. In Sect. 5.3.2 we review a supersymmetric $SU(5)$ model for the triplet seesaw scenario.

5.3 SUSY GUTs

5.3.1 Flavor violation in the minimal supersymmetric SU(5) seesaw model

In this section we review flavor- and/or CP-violating phenomena in the minimal SUSY $SU(5)$ GUT, in which the right handed neutrinos are introduced to generate neutrino masses by the type-I seesaw mechanism [216–220]. Here, it is assumed that the Higgs doublets in this MSSM are embedded in **5**- and $\bar{\mathbf{5}}$-dimensional $SU(5)$ multiplets. Rich flavor structure is induced even in those minimal particle contents. The flavor violating SUSY breaking terms for the right handed squarks and sleptons are generated by the GUT interaction, while those are suppressed in the MSSM ($+\nu_R$) under the universal scalar mass hypothesis for the SUSY breaking terms.

The Yukawa interactions for quarks and leptons and the Majorana mass terms for the right handed neutrinos in this model are given by the following superpotential,

$$W = \frac{1}{4} Y_{ij}^u \Psi_i \Psi_j H + \sqrt{2} Y_{ij}^d \Psi_i \Phi_j \overline{H}$$
$$+ Y_{ij}^\nu \Phi_i \overline{N}_j H + \frac{1}{2} M_{N_{ij}} \overline{N}_i \overline{N}_j, \quad (5.57)$$

where Ψ and Φ are for **10**- and $\bar{\mathbf{5}}$-dimensional multiplets, respectively, and $\overline{N}$ is for the right handed neutrinos. H ($\overline{H}$) is **5**- ($\bar{\mathbf{5}}$-) dimensional Higgs multiplets. After removing the unphysical degrees of freedom, the Yukawa coupling constants in (5.57) are given as follows,

$$Y_{ij}^u = V_{ki} Y_{u_k} e^{i\varphi_{u_k}} V_{kj},$$
$$Y_{ij}^d = Y_{d_i} \delta_{ij}, \quad (5.58)$$
$$Y_{ij}^\nu = e^{i\varphi_{d_i}} U_{ij}^\star Y_{\nu_j}.$$

Here, Y_u, Y_d, Y_ν denote diagonal Yukawa couplings, φ_{u_i} and φ_{d_i} ($i = 1$–3) are CP-violating phases inherent in the SUSY $SU(5)$ GUT ($\sum_i \varphi_{u_i} = \sum_i \varphi_{d_i} = 0$). The unitary matrix V is the CKM matrix in the extension of the SM to the SUSY $SU(5)$ GUT, and each unitary matrices U and V have only a phase. When the Majorana mass matrix for the right handed neutrinos is real and diagonal in the basis of (5.59), U is the PMNS matrix measured in the neutrino oscillation experiments and the light neutrino mass eigenvalues are given as $m_{\nu_i} = Y_{\nu_i}^2 \langle H_2 \rangle^2 / M_{N_i}$, in which M_{N_i} are the diagonal components.

The colored Higgs multiplets H_c and $\overline{H}_c$ are introduced in H and $\overline{H}$ as $SU(5)$ partners of the Higgs doublets H_f and $\overline{H}_f$ in the MSSM, respectively, and they have new flavor violating interactions. Equation (5.57) is represented by the fields in the MSSM as follows,

$$W = W_{\text{MSSM}+\overline{N}}$$

$$+ \frac{1}{2} V_{ki} Y_{u_k} e^{i\varphi_{u_k}} V_{kj} Q_i Q_j H_c + Y_{u_i} V_{ij} e^{i\varphi_{d_j}} \overline{U}_i \overline{E}_j H_c$$
$$+ Y_{d_i} e^{-i\varphi_{d_i}} Q_i L_i \overline{H}_c + e^{-i\varphi_{u_i}} V_{ij}^\star Y_{d_j} \overline{U}_i \overline{D}_j \overline{H}_c$$
$$+ e^{i\varphi_{d_i}} U_{ij}^\star Y_{\nu_j} \overline{D}_i \overline{N}_j H_c. \quad (5.59)$$

Here, the superpotential in the MSSM with the right handed neutrinos is

$$W_{\text{MSSM}+\overline{N}} = V_{ji} Y_{u_j} Q_i \overline{U}_j H_f + Y_{d_i} Q_i \overline{D}_i \overline{H}_f + Y_{d_i} L_i \overline{E}_i \overline{H}_f$$
$$+ U_{ij}^\star Y_{\nu_j} L_i \overline{N}_j H_f + M_{ij} \overline{N}_i \overline{N}_j. \quad (5.60)$$

The flavor violating interactions, which are absent in the MSSM, emerge in the SUSY $SU(5)$ GUT due to existence of the colored Higgs multiplets. The colored Higgs interactions are also baryon-number violating [150, 371], and then proton decay induced by the colored Higgs exchange is a serious problem, especially in the minimal SUSY $SU(5)$ GUT [374]. However, the constraint from the proton decay depends on the detailed structure in the Higgs sector, and it is also loosened by global symmetries, such as the Peccei–Quinn symmetry and the $U(1)_R$ symmetry. Thus, we may ignore the constraint from the proton decay while we adopt the minimal Yukawa structure in (5.57).

The sfermion mass terms get sizable corrections by the colored Higgs interactions, when the SUSY-breaking terms in the MSSM are generated by dynamics above the colored Higgs masses. In the minimal supergravity scenario the SUSY breaking terms are supposed to be given at the reduced Planck mass scale (M_G). In this case, the flavor violating SUSY breaking mass terms at low energy are induced by the radiative correction, and they are qualitatively given in a flavor basis as

$$(m_{\tilde{u}_L}^2)_{ij} \simeq -V_{i3} V_{j3}^\star \frac{Y_b^2}{(4\pi)^2} (3m_0^2 + A_0^2)$$
$$\times \left(2 \log \frac{M_G^2}{M_{H_c}^2} + \log \frac{M_{H_c}^2}{M_{\text{SUSY}}^2}\right),$$

$$(m_{\tilde{u}_R}^2)_{ij} \simeq -e^{-i\varphi_{u_{ij}}} V_{i3}^\star V_{j3} \frac{2Y_b^2}{(4\pi)^2} (3m_0^2 + A_0^2) \log \frac{M_G^2}{M_{H_c}^2},$$

$$(m_{\tilde{d}_L}^2)_{ij} \simeq -V_{3i}^\star V_{3j} \frac{Y_t^2}{(4\pi)^2} (3m_0^2 + A_0^2)$$
$$\times \left(3 \log \frac{M_G^2}{M_{H_c}^2} + \log \frac{M_{H_c}^2}{M_{\text{SUSY}}^2}\right), \quad (5.61)$$

$$(m_{\tilde{d}_R}^2)_{ij} \simeq -e^{i\varphi_{d_{ij}}} U_{ik}^\star U_{jk} \frac{Y_{\nu_k}^2}{(4\pi)^2} (3m_0^2 + A_0^2) \log \frac{M_G^2}{M_{H_c}^2},$$

$$(m_{\tilde{l}_L}^2)_{ij} \simeq -U_{ik} U_{jk}^\star \frac{f_{\nu_k}^2}{(4\pi)^2} (3m_0^2 + A_0^2) \log \frac{M_G^2}{M_{N_k}^2},$$

$$(m_{\tilde{e}_R}^2)_{ij} \simeq -e^{i\varphi_{d_{ij}}} V_{3i} V_{3j}^\star \frac{3Y_t^2}{(4\pi)^2}(3m_0^2 + A_0^2)\log\frac{M_G^2}{M_{H_c}^2},$$

with $i \neq j$, where $\varphi_{u_{ij}} \equiv \varphi_{u_i} - \varphi_{u_j}$ and $\varphi_{d_{ij}} \equiv \varphi_{d_i} - \varphi_{d_j}$ and M_{H_c} is the colored Higgs mass. Here, M_{SUSY} is the SUSY-breaking scale in the MSSM, m_0 and A_0 are the universal scalar mass and trilinear coupling, respectively, in the minimal supergravity scenario. Y_t is the top quark Yukawa coupling constant while Y_b is for the bottom quark. The off-diagonal components in the right handed squarks and slepton mass matrices are induced by the colored Higgs interactions, and they depend on the CP-violating phases in the SUSY $SU(5)$ GUT with the right handed neutrinos [737].

One of the important features of the SUSY GUTs is the correlation between the leptonic and hadronic flavor violations [738, 739]. From (5.62), we get a relation

$$(m_{\tilde{d}_R}^2)_{23} \simeq e^{i\varphi_{d_{23}}}(m_{\tilde{l}_L}^2)_{23}^\star \times \left(\log\frac{M_G^2}{M_{H_c}^2} \Big/ \log\frac{M_G^2}{M_{N_3}^2}\right). \quad (5.62)$$

The right handed bottom-strange squark mixing may be tested in the B factory experiments since it affects B_s–$\bar{B}_s$ mixing, CP asymmetries in the b–s penguin processes such as $B_d \to \phi K_s$, and the mixing induced CP asymmetry in $B_d \to M_s \gamma$. (See Chap. 2.) The relation in (5.62) implies that the deviations from the standard model predictions in the b–s transition processes are correlated with $B(\tau \to \mu\gamma)$ in the SUSY $SU(5)$ GUT. We may test the model in the B factories.

In Fig. 28 we show the CP asymmetry in $B_d \to \phi K_s$ ($S_{\phi K_S}$) and $B(\tau \to \mu\gamma)$ as an example of the correlation. Here, we assume the minimal supergravity hypothesis for the SUSY breaking terms. See the caption and Ref. [738] for the input parameters and the details of the figure. It is found that $S_{\phi K_S}$ and $B(\tau \to \mu\gamma)$ are correlated and a large deviation from the standard model prediction for $S_{\phi K_S}$ is not possible due to the current bound on $B(\tau \to \mu\gamma)$ in the SUSY $SU(5)$ GUT.

In (5.62), we take the $SU(5)$-symmetric Yukawa interactions given in (5.57), while they fails to explain the fermion mass relations in the first and second generations. We have to extend the minimal model by introducing non-trivial Higgs or matter contents or the higher-dimensional operators including $SU(5)$-breaking Higgs field. These extensions may affect the prediction for the sfermion mass matrices. However, the relation in (5.62) is rather robust when the neutrino Yukawa coupling constant of the third generation is as large as those for the top and bottom quarks and the large mixing in the atmospheric neutrino oscillation comes from the lopside structure of the neutrino Yukawa coupling.

Another important feature of the SUSY GUTs is that both the left and right handed squarks and sleptons have flavor mixing terms. In this case, the hadronic and leptonic electric dipole moments (EDMs) are generated due to the flavor violation, and they may be large enough to be observed in the future EDM measurements [740]. A diagram in Fig. 29(a) generates the electron EDM even at one loop level, when the relative phase between the left and right handed slepton mixing terms is non-vanishing. While this contribution is suppressed by the flavor violation, it is compensated by a heavier fermion mass, that is, m_τ. Similar diagrams in Fig. 29(b) contribute to quark EDMs and chromo-electric dipole moments (CEDM), which induce the hadronic EDMs.

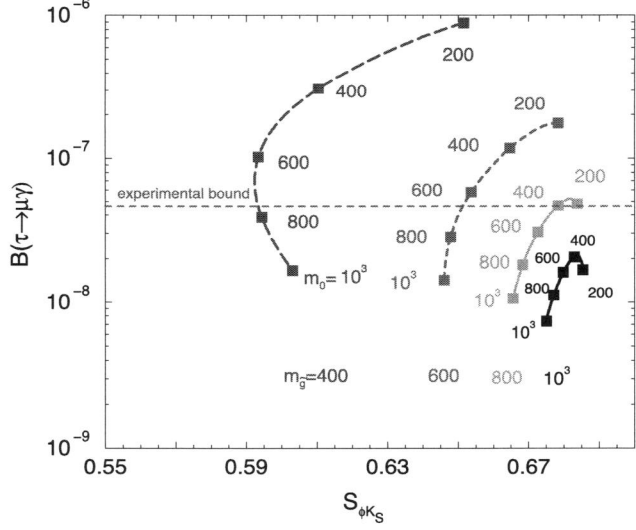

Fig. 28 $B(\tau \to \mu\gamma)$ as a function of the CP asymmetry in $B_d \to \phi K_s$ ($S_{\phi K_S}$) for fixed gluino masses $m_{\tilde{g}} = 400, 600, 800$, and 1000 GeV. Here, $\tan\beta = 10$, 200 GeV $< m_0 <$ 1 TeV, $A_0 = 0$, $m_{\nu_\tau} = 5 \times 10^{-2}$ eV, $M_{N_3} = 5 \times 10^{14}$ GeV, and $U_{32} = 1/\sqrt{2}$. $\varphi_{d_{23}}$ is taken for the deviation of $S_{\phi K_S}$ from the SM prediction to be maximum. The current experimental bound on $B(\tau \to \mu\gamma)$ [647] is also shown in the figure

Fig. 29 (a) Diagrams that generate electron EDM and (b) those that generate EDMs and CEDMs of the ith quark due to flavor violation in sfermion mass terms

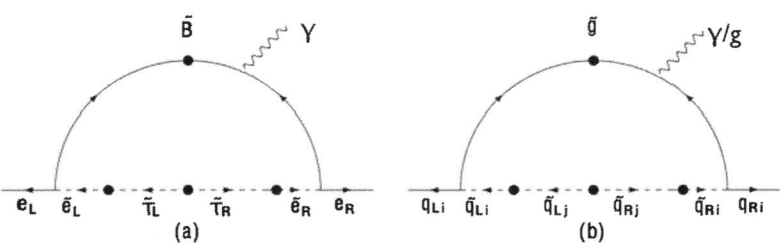

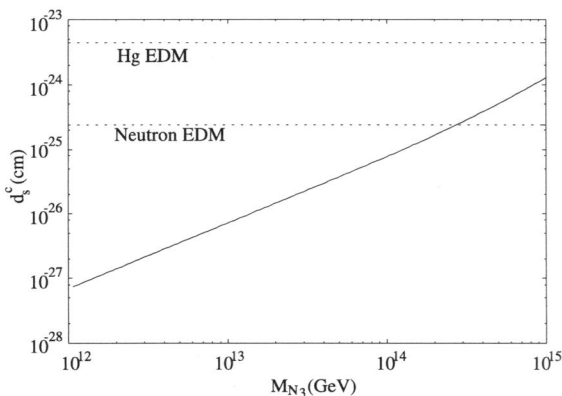
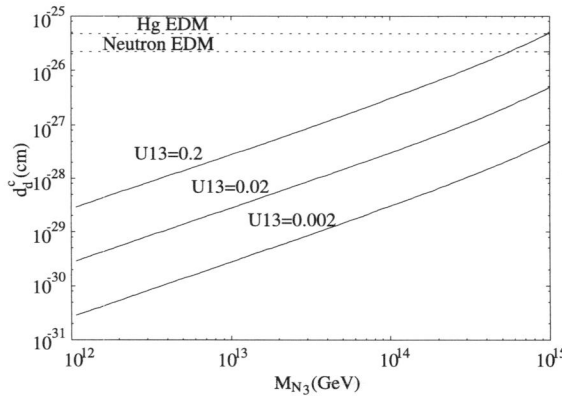

Fig. 30 CEDMs for the strange (d_s^c) and down quarks (d_d^c) as functions of the right handed tau neutrino mass, M_{N_3}. Here, $M_{H_c} = 2 \times 10^{16}$ GeV, $m_{\nu_\tau} = 0.05$ eV, $U_{23} = 1/\sqrt{2}$, and $U_{13} = 0.2$, 0.02, and 0.002. For the MSSM parameters, we take $m_0 = 500$ GeV, $A_0 = 0$, $m_{\tilde{g}} = 500$ GeV and $\tan\beta = 10$. The CP phases φ_{d_i} are taken for the CEDMs to be maximum. The upper bounds on the strange and down quark CEDMs from the mercury atom and neutron EDMs are shown in the figures

The EDM measurements are important to probe the interaction of the SUSY $SU(5)$ GUT.

In Fig. 30 the CEDMs for strange (d_s^c) and down quarks (d_d^c) are shown as functions of the right handed tau neutrino mass in the SUSY $SU(5)$ GUT with the right handed neutrinos. See the caption and Ref. [741] for the input parameters. The mercury atom EDM, which is a diamagnetic atom, is sensitive to quark CEDMs via the nuclear force, while the neutron EDM depends on them in addition to the quark EDMs. (The evaluation of the hadronic EDMs from the effective operators at the parton level is reviewed in Sect. 9.1 and also Ref. [742].) The strange quark contribution to the mercury atom EDM is suppressed by the strange quark mass. On the other hand, it is argued in Refs. [743, 744] that the strange quark component in nucleon is not negligible and the strange quark CEDM may give a sizable contribution to the neutron EDM. It implies that we may probe the different flavor mixings by measurements of the various hadronic EDMs, though the evaluation of the hadronic EDMs still has large uncertainties.

It is argued that the future measurements of neutron and deuteron EDMs may reach to levels of $\sim 10^{-28}$ e cm and $\sim 10^{-29}$ e cm, respectively. When the sensitivity of deuteron EDM is established, we may probe the new physics to the level of $d_s^c \sim 10^{-28}$ e cm and $d_d^c \sim d_u^c \sim 10^{-30}$ e cm [849]. The future measurements for the EDMs will give great impacts on the SUSY $SU(5)$ GUT with the right handed neutrinos.

5.3.2 LFV in the minimal $SU(5)$ GUT with triplet seesaw

In this section we discuss phenomenology of the minimal $SU(5)$ GUT which incorporates the triplet seesaw mechanism, previously presented in Sects. 3.2.3.2 and 5.2.7. Review of more general class of GUT models also including triplet Higgs has been given in Sect. 4.1. In GUTs based on $SU(5)$ there is no natural place for incorporating singlet neutrinos. From this point of view $SU(5)$ presents some advantage for implementing the triplet seesaw mechanism. In particular, a very predictive scenarios can be obtained in the supersymmetric case [249, 736, 745]. The triplet states T ($\overline{T}$) fit into the 15 ($\overline{15}$) representation, $15 = S + T + Z$ with S, T and Z transforming as $S \sim (6, 1, -\frac{2}{3})$, $T \sim (1, 3, 1)$, $Z \sim (3, 2, \frac{1}{6})$ under $SU(3) \times SU(2)_L \times U(1)_Y$ (the $\overline{15}$ decomposition is obvious). We shall briefly show that it is also possible to relate not only neutrino mass parameters and LFV (as shown in Sect. 5.2.7) but also sparticle and Higgs spectra and electroweak symmetry breakdown [736, 745]. For this purpose, consider that the $SU(5)$ model conserves $B - L$, so that the relevant superpotential reads

$$W_{SU(5)} = \frac{1}{\sqrt{2}}(Y_{15}\bar{5}\,15\,\bar{5} + \lambda 5_H\,\overline{15}\,5_H) + Y_5\,\bar{5}\,5_H\,10$$
$$+ Y_{10}\,10\,10\,5_H + M_5\,5_H\,\bar{5}_H + \xi X\,15\,\overline{15}, \quad (5.63)$$

where the multiplets are understood as $\bar{5} = (d^c, L)$, $10 = (u^c, e^c, Q)$ and the Higgs doublets fit with their colored partners t and $\bar{t}$, like $5_H = (t, H_2)$, $\bar{5}_H = (\bar{t}, H_1)$ and X is a singlet superfield. The $B - L$ quantum numbers are the combination $Q + \frac{4}{5}Y$ where Y are the hypercharges and $Q_{10} = \frac{1}{5}$, $Q_{\bar{5}} = -\frac{3}{5}$, $Q_{5_H(\bar{5}_H)} = -\frac{2}{5}(\frac{2}{5})$, $Q_{15} = \frac{6}{5}$, $Q_{\overline{15}} = \frac{4}{5}$, $Q_X = -2$. Both the scalar S_X and auxiliary F_X components of the superfield X are assumed to acquire a VEV through some unspecified dynamics in the hidden sector. Namely, while $\langle S_X \rangle$ only breaks $B - L$, $\langle F_X \rangle$ breaks both SUSY and $B - L$. These effects are parameterized by the superpotential mass term $M_{15}\,15\,\overline{15}$, where $M_{15} = \xi\langle S_X \rangle$, and the bilinear SSB term $-B_{15}M_{15}\,15\,\overline{15}$, with $B_{15}M_{15} = -\xi\langle F_X \rangle$. The 15 and $\overline{15}$ states act, therefore, as *messengers* of both $B - L$ and SUSY breaking to the MSSM observable sector. Once

$SU(5)$ is broken to the SM group we find, below the GUT scale M_G,

$$W = W_0 + W_T + W_{S,Z},$$

$$W_0 = Y_e e^c H_1 L + Y_d d^c H_1 Q + Y_u u^c Q H_2 + \mu H_2 H_1,$$

$$W_T = \frac{1}{\sqrt{2}}(Y_T L T L + \lambda H_2 \bar{T} H_2) + M_T T \bar{T},$$

$$W_{S,Z} = \frac{1}{\sqrt{2}} Y_S d^c S d^c + Y_Z d^c Z L + M_Z Z \bar{Z} + M_S S \bar{S}.$$
(5.64)

Here, W_0 denotes the MSSM superpotential,[21] the term W_T is responsible for neutrino mass generation [cf. (3.67)], while the couplings and masses of the colored fragments S and Z are included in $W_{S,Z}$. It is also understood that $M_T = M_S = M_Z \equiv M_{15}$. At the decoupling of the heavy states S, T, Z we obtain at tree-level the neutrino masses, given by (3.68) and at the quantum level all SSB mass parameters of the MSSM via gauge and Yukawa interactions. At one loop level, only the trilinear scalar couplings, the gaugino masses and the Higgs bilinear mass term B_H are generated:

$$A_e = \frac{3B_T}{16\pi^2} Y_e (Y_T^\dagger Y_T + Y_Z^\dagger Y_Z), \quad A_u = \frac{3B_T}{16\pi^2} |\lambda|^2 Y_u,$$

$$A_d = \frac{2B_T}{16\pi^2}(Y_Z Y_Z^\dagger + 2 Y_S Y_S^\dagger) Y_d, \quad (5.65)$$

$$M_a = \frac{7B_T}{16\pi^2} g_a^2, \quad B_H = \frac{3B_T}{16\pi^2} |\lambda|^2.$$

The scalar mass matrices instead are generated at the two-loop level and receive both gauge-mediated contributions proportional to $C_a^f g_a^4$ (C_a^f is the quadratic Casimir of the f-particle) and Yukawa-mediated ones of the form $Y_p^\dagger Y_p$ ($p = S, T, Z$). The former piece is the flavor blind contribution, which is proper of the gauge-mediated scenarios [151, 152, 746–751], while the latter ones constitute the flavor violating contributions transmitted to the SSB terms by the Yukawa's $Y_{S,T,Z}$. These contributions are mostly relevant for the mass matrices $m_{\tilde{L}}^2$ and $m_{\tilde{d}^c}^2$. For example,

$$m_{\tilde{L}}^2 = \left(\frac{B_T}{16\pi^2}\right)^2 \left[\frac{21}{10} g_1^4 + \frac{21}{2} g_2^4 - \left(\frac{27}{5} g_1^2 + 21 g_2^2\right) Y_T^\dagger Y_T \right.$$

$$- \left(\frac{21}{15} g_1^2 + 9 g_2^2 + 16 g_3^2\right) Y_Z^\dagger Y_Z$$

$$+ 18 (Y_T^\dagger Y_T)^2 + 15 (Y_Z^\dagger Y_Z)^2 + 3 \text{Tr}(Y_T^\dagger Y_T) Y_T^\dagger Y_T$$

$$+ 12 Y_Z^\dagger Y_S Y_S^\dagger Y_Z + 3 \text{Tr}(Y_Z^\dagger Y_Z) Y_Z^\dagger Y_Z$$

$$+ 9 Y_T^\dagger Y_Z^T Y_Z^* Y_T + 9 (Y_T^\dagger Y_T Y_Z^\dagger Y_Z + \text{h.c.})$$

$$\left. + 3 Y_T^\dagger Y_e^T Y_e^* Y_T + 6 Y_Z^\dagger Y_d Y_d^\dagger Y_Z \right]. \quad (5.66)$$

Since the flavor structure of $m_{\tilde{L}}^2$ is proportional to Y_T (and to Y_Z which is $SU(5)$-related to Y_T), it can be expressed in terms of the neutrino parameters [cf. (3.68)] and so the relative size of LFV in different leptonic families is predicted according to the results of (5.54).

All the soft masses have the same scaling behavior $\tilde{m} \sim B_T/(16\pi^2)$ which demands $B > 10$ TeV to fulfill the naturalness principle. This scenario appears very predictive since it contains only three free parameters: the triplet mass M_T, the effective SUSY breaking scale B_T and the coupling constant λ. The parameter space is then constrained by the experimental bounds on the Higgs boson mass, the $B(\mu \to e\gamma)$, the sfermion masses, and the requirement of radiative electroweak symmetry breaking. The phenomenological predictions more important and relevant for LHC, the B-factories [694] the incoming MEG experiment [548], the Super Flavor factory [752] or the PRISM/PRIME experiment at J-PARC [753], concern the sparticle and Higgs boson spectra and the LFV decays. Regarding the spectrum, the gluino is the heaviest sparticle while, in most of the parameter space, $\tilde{\ell}_1$ is the lightest. In the example shown in Fig. 31 the squark and slepton masses lie in the ranges 700–950 GeV and 100–300 GeV, respectively. The gluino mass is about 1.3 TeV. The chargino masses are $m_{\tilde{\chi}_1^\pm} \sim 320$ GeV and $m_{\tilde{\chi}_2^\pm} \sim 450$–550 GeV. Moreover, $m_{\tilde{\chi}_1^0} \sim 190$ GeV, $m_{\tilde{\chi}_2^0} \approx m_{\tilde{\chi}_1^\pm}$ and $m_{\tilde{\chi}_{3,4}^0} \approx m_{\tilde{\chi}_2^\pm}$. These mass ranges are within the discovery reach of the LHC.

The Higgs sector is characterized by a decoupling regime with a light SM-like Higgs boson (h) with mass in the range 110–120 GeV which is testable in the near future at LHC (mainly through the decay into 2 photons). The remaining three heavy states (H, A and $H^\pm$) have mass $m_{H,A,H^\pm} \approx$ 450–550 GeV (again, for $B_T = 20$ TeV). All the spectra increase almost linearly with B_T.

Figure 32(b) shows instead several LFV processes: $\mu \to eX$, $\mu \to e$ conversion in nuclei, $\tau \to eY$ and $\tau \to \mu Y$ ($X = \gamma, ee$, $Y = \gamma, ee, \mu\mu$). One observe that e.g., the behavior of the radiative-decay branching ratios is in agreement with the estimates given in (5.56) for $\theta_{13} = 0$. For $\theta_{13} = 0.2$ one obtains instead that $B(\tau \to \mu\gamma)/B(\mu \to e\gamma) \sim 2$ and $B(\tau \to e\gamma)/B(\mu \to e\gamma) \sim 0.1$ (the full analysis can be found in Ref. [736]). The other LFV processes shown are also correlated to the radiative ones in a model-independent way [736, 745]. The analysis shows that the future experimental sensitivity will allow to measure at most $B(\mu \to \gamma)$, $B(\mu \to 3e)$, $B(\tau \to \mu\gamma)$ and CR($\mu \to e$ Ti) for tiny θ_{13}. In particular, being $B(\tau \to \mu\gamma)/B(\mu \to e\gamma) \sim 300$, $B(\tau \to \mu\gamma)$ is expected not to exceed 3×10^{-9}, irrespective of the type of neutrino spectrum. Therefore $\tau \to \mu\gamma$ falls into

---

[21] This should be regarded as an effective approach where the Yukawa matrices Y_d, Y_e, Y_u include $SU(5)$-breaking effects needed to reproduce a realistic fermion spectrum.

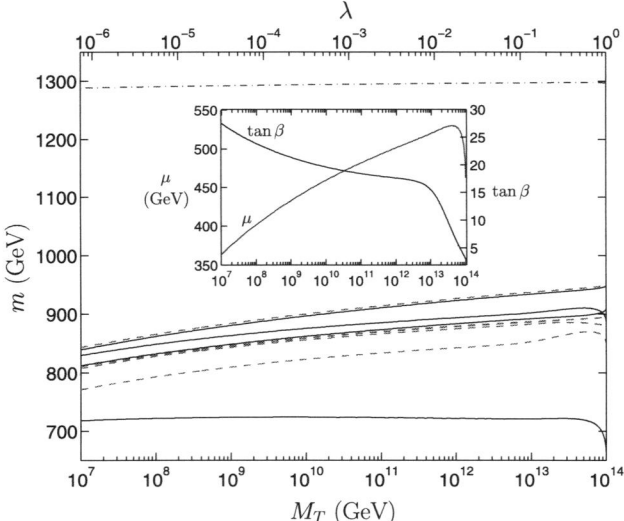
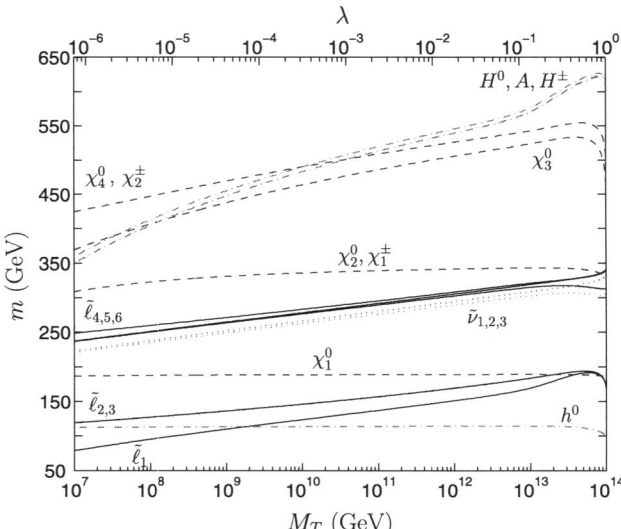

Fig. 31 (Color online) Sparticle and Higgs spectrum for $B_T = 20$ TeV. *Left panel*: squark masses, $m_{\tilde{u}}$ (*black solid line*), $m_{\tilde{d}}$ (*red dashed*) and the gluino mass (*blue dash-dotted*). In the *inner plot* $\tan\beta$ and μ are shown as obtained by the electroweak symmetry breaking conditions. *Right panel*: the masses of the charged sleptons, the sneutrinos, the charginos, the neutralinos and the Higgs bosons as the labels indicate

the LHC capability. All the decays $\tau \to \ell_i \ell_k \ell_k$ would have $B < \mathcal{O}(10^{-11})$. The predictions for the LFV branching ratios in the present scenario are summarized in Table 13.

Finally, such supersymmetric $SU(5)$ framework with a **15**, $\overline{\mathbf{15}}$ pair may be realized in contexts based on string inspired constructions [754, 755].

5.3.3 LFV from a generic SO(10) framework

The spinorial representation of the $SO(10)$, given by a 16-dimensional spinor, can accommodate all the SM model particles as well as the right handed neutrino. As discussed in Sect. 4.1, the product of two **16** matter representations can only couple to **10**, **120** or **126** representations, which can be formed by either a single Higgs field or a non-renormalizable product of representations of several Higgs fields. In either case, the Yukawa matrices resulting from the couplings to **10** and **126** are complex symmetric, whereas they are antisymmetric when the couplings are to the **120**. Thus, the most general $SO(10)$ superpotential relevant to fermion masses can be written as

$$W_{SO(10)} = Y^{10}_{ij} \mathbf{16_i}\,\mathbf{16_j}\,\mathbf{10} + Y^{126}_{ij} \mathbf{16_i}\,\mathbf{16_j}\,\mathbf{126} + Y^{120}_{ij} \mathbf{16_i}\,\mathbf{16_j}\,\mathbf{120}, \quad (5.67)$$

where i, j refer to the generation indices. In terms of the SM fields, the Yukawa couplings relevant for fermion masses are given by [756, 757]:

$$\begin{aligned}
\mathbf{16\,16\,10} &\supset \mathbf{5}(uu^c + \nu\nu^c) + \bar{\mathbf{5}}(dd^c + ee^c), \\
\mathbf{16\,16\,126} &\supset \mathbf{1}\,\nu^c\nu^c + \mathbf{15}\,\nu\nu + \mathbf{5}(uu^c - 3\nu\nu^c) \\
&\quad + \bar{\mathbf{45}}(dd^c - 3ee^c), \\
\mathbf{16\,16\,120} &\supset \mathbf{5}\,\nu\nu^c + \mathbf{45}\,uu^c + \bar{\mathbf{5}}(dd^c + ee^c) \\
&\quad + \bar{\mathbf{45}}(dd^c - 3ee^c),
\end{aligned} \quad (5.68)$$

where we have specified the corresponding $SU(5)$ Higgs representations for each of the couplings and all the fermions are left handed fields. The resulting up-type quarks and neutrinos' Dirac mass matrices can be written as

$$m^u = M^5_{10} + M^5_{126} + M^{45}_{120}, \quad (5.69)$$

$$m^\nu_D = M^5_{10} - 3M^5_{126} + M^5_{120}. \quad (5.70)$$

A simple analysis of the fermion mass matrices in the $SO(10)$ model, as detailed in (5.70) leads us to the following result: *At least one of the Yukawa couplings in $Y^\nu = v_u^{-1} m^\nu_D$ has to be as large as the top Yukawa coupling* [164]. This result holds true in general, independently of the choice of the Higgses responsible for the masses in (5.69), (5.70), provided that no accidental fine-tuned cancellations of the different contributions in (5.70) are present. If contributions from the **10**'s solely dominate, Y^ν and Y^u would be equal. If this occurs for the **126**'s, then $Y^\nu = -3Y^u$ [395]. In case both of them have dominant entries, barring a rather precisely fine-tuned cancellation between M^5_{10} and M^5_{126} in (5.70), we expect at least one large entry to be present in Y^ν. A dominant antisymmetric contribution to top quark mass

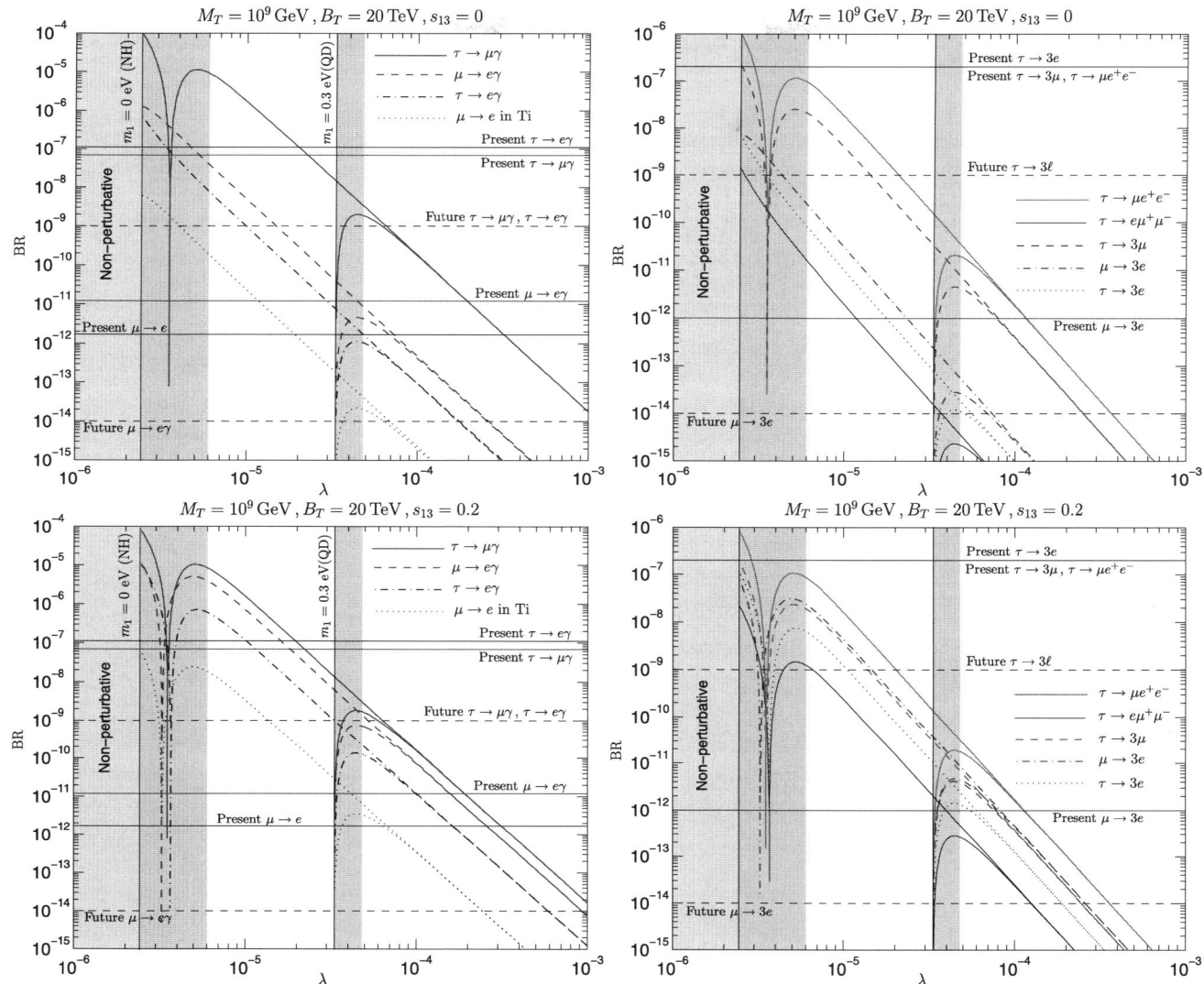

Fig. 32 Branching ratios of several LFV processes as a function of λ. The *left* (*right*) *vertical line* indicates the lower bound on λ imposed by requiring perturbativity of the Yukawa couplings $Y_{T,S,Z}$ when $m_1 = 0$ (0.3) eV [normal-hierarchical (quasi-degenerate) neutrino mass spectrum]. The *regions in green* (*grey*) are excluded by the $m_{\tilde{\ell}_1} > 100$ GeV constraint (perturbativity requirement when $m_1 = 0$)

Table 13 Expectations for the various LFV processes assuming $B(\mu \to e\gamma) = 1.2 \times 10^{-11}$. The results in parenthesis apply to the case of the inverted-hierarchical neutrino spectrum, whenever these are different from those obtained for the normal-hierarchical and quasi-degenerate ones

Decay mode	Prediction for branching ratio	
	$s_{13} = 0$	$s_{13} = 0.2$
$\tau^- \to \mu^- \gamma$	3×10^{-9}	$2(3) \times 10^{-11}$
$\tau^- \to e^- \gamma$	2×10^{-12}	$1(3) \times 10^{-12}$
$\mu^- \to e^- e^+ e^-$	6×10^{-14}	6×10^{-14}
$\tau^- \to \mu^- \mu^+ \mu^-$	7×10^{-12}	$4(6) \times 10^{-14}$
$\tau^- \to \mu^- e^+ e^-$	3×10^{-11}	$2(3) \times 10^{-13}$
$\tau^- \to e^- e^+ e^-$	2×10^{-14}	$1(3) \times 10^{-14}$
$\tau^- \to e^- \mu^+ \mu^-$	3×10^{-15}	$2(4) \times 10^{-15}$
$\mu \to e$; Ti	6×10^{-14}	6×10^{-14}

due to the **120** Higgs is phenomenologically excluded, since it would lead to at least a pair of heavy degenerate up quarks. Apart from sharing the property that at least one eigenvalue of both m^u and m_D^ν has to be large, for the rest it is clear from (5.69) and (5.70) that these two matrices are not aligned in general, and hence we may expect different mixing angles appearing from their diagonalization. This freedom is removed if one sticks to particularly simple choices of the Higgses responsible for up quark and neutrino masses.

Therefore, we see that the $SO(10)$ model with only two ten-plets would inevitably lead to small mixing in Y^ν. In fact, with two Higgs fields in symmetric representations, giving masses to the up sector and the down sector separately, it would be difficult to avoid the small CKM-like mixing in Y^ν. We shall call this case the CKM case. From here, the following mass relations hold between the quark and leptonic mass matrices at the GUT scale:[22]

$$Y^u = Y^\nu; \qquad Y^d = Y^e. \tag{5.71}$$

In the basis where charged lepton masses are diagonal, we have

$$Y^\nu = V_{\text{CKM}}^T Y_{\text{Diag}}^u V_{\text{CKM}}. \tag{5.72}$$

The large couplings in $Y^\nu \sim \mathcal{O}(Y_t)$ induce significant off-diagonal entries in $m_{\tilde{L}}^2$ through the RG evolution between M_{GUT} and the scale of the right handed Majorana neutrinos, M_{R_i}. The induced off-diagonal entries relevant to $l_j \to l_i, \gamma$ are of the order of:

$$\left(m_{\tilde{L}}^2\right)_{i \neq j} \approx -\frac{3m_0^2 + A_0^2}{8\pi^2} Y_t^2 V_{ti} V_{tj} \ln \frac{M_{\text{GUT}}}{M_{R_3}} + \mathcal{O}(Y_c^2), \tag{5.73}$$

where V_{ij} are elements of V_{CKM}, and i, j flavor indices. In this expression, the CKM angles are small but one would expect the presence of the large top Yukawa coupling to compensate such a suppression. The required right handed neutrino Majorana mass matrix, consistent with both the observed low energy neutrino masses and mixing as well as with CKM-like mixing in Y^ν is easily determined from the seesaw formula defined at the scale of right handed neutrinos.

The $B(l_i \to l_j \gamma)$ are now predictable in this case. Considering mSUGRA boundary conditions and taking $\tan\beta = 40$, we obtain that reaching a sensitivity of $\mathcal{O}(10^{-13}-10^{-14})$, as planned by the MEG experiment at PSI, for $B(\mu \to e\gamma)$ would allow us to probe the SUSY spectrum

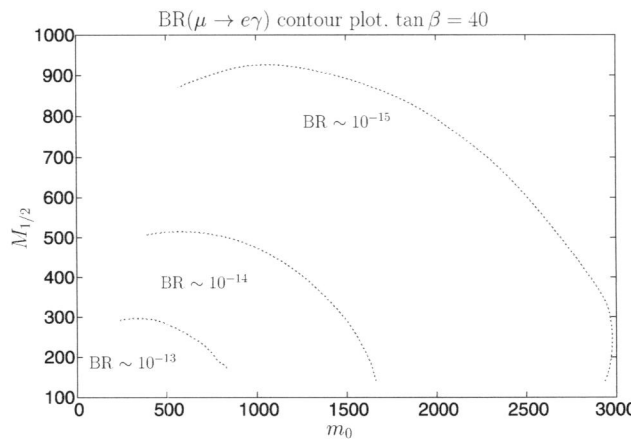

Fig. 33 Contour plot of $B(\mu \to e, \gamma)$ in the plane of the GUT-scale universal scalar and gaugino masses, $(m_0, M_{1/2})$, at $A_0 = 0$ in the CKM high $\tan\beta$ case. Note that while the plane is presently unconstrained, the planned MEG experiment sensitivity of $\mathcal{O}(10^{-13}-10^{-14})$ will be able to probe it in the $(m_0, m_{\tilde{g}}) \lesssim 1$ TeV region

completely up to $m_0 = 1200$ GeV, $M_{1/2} = 400$ GeV (notice that this corresponds to gluino and squark masses of order 1 TeV). This clearly appears from Fig. 33, which shows the $B(\mu \to e\gamma)$ contour plot in the $(m_0, M_{1/2})$ plane. Thus, in summary, though the present limits on $B(\mu \to e, \gamma)$ would not induce any significant constraints on the supersymmetry breaking parameter space, an improvement in the limit to $\sim \mathcal{O}(10^{-13}-10^{-14})$, as foreseen, would start imposing non-trivial constraints especially for the large $\tan\beta$ region.

To obtain mixing angles larger than CKM angles, asymmetric mass matrices have to be considered. In general, it is sufficient to introduce asymmetric textures either in the up sector or in the down sector. In the present case, we assume that the down sector couples to a combination of Higgs representations (symmetric and antisymmetric)[23] Φ, leading to an asymmetric mass matrix in the basis where the up sector is diagonal. As we shall see below, this would also require that the right handed Majorana mass matrix be diagonal in this basis. We have:

$$W_{SO(10)} = \frac{1}{2} Y_{ii}^{u,\nu} \mathbf{16_i} \, \mathbf{16_i} \, \mathbf{10^u} + \frac{1}{2} Y_{ij}^{d,e} \mathbf{16_i} \, \mathbf{16_j} \, \Phi$$
$$+ \frac{1}{2} Y_{ii}^R \mathbf{16_i} \, \mathbf{16_i} \, \mathbf{126},$$

where the **126**, as before, generates only the right handed neutrino mass matrix. To study the consequences of these assumptions, we see that at the level of $SU(5)$, we have

---

[22] Clearly this relation cannot hold for the first two generations of down quarks and charged leptons. One expects, small corrections due to non-renormalizable operators or suppressed renormalizable operators [6] to be invoked.

[23] The couplings of the Higgs fields in the superpotential can be either renormalizable or non-renormalizable. See [758] for a non-renormalizable example.

$$W_{SU(5)} = \frac{1}{2} Y^u_{ii} \mathbf{10_i} \, \mathbf{10_i} \, \mathbf{5_u} + Y^\nu_{ii} \bar{\mathbf{5}}_i \, \mathbf{1_i} \, \mathbf{5_u}$$
$$+ Y^d_{ij} \mathbf{10_i} \, \bar{\mathbf{5}}_j \, \bar{\mathbf{5}}_d + \frac{1}{2} M^R_{ii} \mathbf{1_i} \, \mathbf{1_i},$$

where we have decomposed the **16** into $\mathbf{10} + \bar{\mathbf{5}} + \mathbf{1}$ and $\mathbf{5_u}$ and $\bar{\mathbf{5}}_d$ are components of $\mathbf{10_u}$ and Φ respectively. To have large mixing $\sim U_{\text{PMNS}}$ in Y^ν we see that the asymmetric matrix Y^d should now give rise to both the CKM mixing as well as PMNS mixing. This is possible if

$$V_{\text{CKM}}^T Y^d U_{\text{PMNS}}^T = Y^d_{\text{Diag}}. \tag{5.74}$$

Therefore the **10** that contains the left handed downquarks would be rotated by the CKM matrix whereas the $\bar{\mathbf{5}}$ that contains the left handed charged leptons would be rotated by the U_{PMNS} matrix to go into their respective mass bases [737, 758–760]. Thus we have the following relations in the basis where charged leptons and down quarks are diagonal:

$$Y^u = V_{\text{CKM}} Y^u_{\text{Diag}} V_{\text{CKM}}^T, \tag{5.75}$$
$$Y^\nu = U_{\text{PMNS}} Y^u_{\text{Diag}}. \tag{5.76}$$

Using the seesaw formula of (3.41) and (5.76), we have

$$M_R = \text{Diag}\left\{\frac{m_u^2}{m_{\nu_1}}, \frac{m_c^2}{m_{\nu_2}}, \frac{m_t^2}{m_{\nu_3}}\right\}. \tag{5.77}$$

We now turn our attention to lepton flavor violation in this case. The branching ratio, $B(\mu \to e, \gamma)$ would now depend on

$$[Y^\nu Y^{\nu T}]_{21} = Y_t^2 U_{\mu 3} U_{e3} + Y_c^2 U_{\mu 2} U_{e2} + \mathcal{O}(h_u^2). \tag{5.78}$$

It is clear from the above that in contrast to the CKM case, the dominant contribution to the off-diagonal entries depends on the unknown magnitude of the element U_{e3} [761]. If U_{e3} is close to its present limit ~ 0.14 [762] (or at least larger than $(Y_c^2/Y_t^2)U_{e2} \sim 4 \times 10^{-5}$), the first term on the RHS of the (5.78) would dominate. Moreover, this would lead to large contributions to the off-diagonal entries in the slepton masses with $U_{\mu 3}$ of $\mathcal{O}(1)$. Thus, we have

$$\left(m_{\tilde{L}}^2\right)_{21} \approx -\frac{3m_0^2 + A_0^2}{8\pi^2} Y_t^2 U_{e3} U_{\mu 3} \ln \frac{M_{\text{GUT}}}{M_{R_3}} + \mathcal{O}(Y_c^2). \tag{5.79}$$

This contribution is larger than the CKM case by a factor of $(U_{\mu 3}U_{e3})/(V_{td}V_{ts}) \sim \mathcal{O}(10^2)$. This would mean about a factor 10^4 times larger than the CKM case in $B(\mu \to e, \gamma)$. Such enhancement with respect to the CKM case is clearly shown in the scatter plots of Fig. 34, where the CKM case is compared with the PMNS case with $U_{e3} = 0.07$. The aim of the figure is to show the capability of MEG to probe the region of mSUGRA parameter space accessible to the LHC. In fact, the plots show the value of $B(\mu \to e, \gamma)$ obtained by scanning the parameter space in the large region ($0 < m_0 < 5$ TeV, $0 < M_{1/2} < 1.5$ TeV, $-3m_0 < A_0 < +3m_0$, sign(μ)), and then keeping the points which give at least one squark lighter than 2.5 TeV (so roughly accessible to the LHC). We see that in this "LHC accessible" region the maximal case (with $U_{e3} = 0.07$) is already excluded by the MEGA limit ($B(\mu \to e, \gamma) < 1.2 \times 10^{-11}$), and therefore MEG will constrain the parameter space far beyond the LHC sensitivity for this case. If U_{e3} is very small, i.e. either zero or $\lesssim (Y_c^2/Y_t^2)U_{e2} \sim 4 \times 10^{-5}$, the second term $\propto Y_c^2$ in (5.78) would dominate, thus giving a strong suppression to the branching ratio. This could be not true, once RG effects

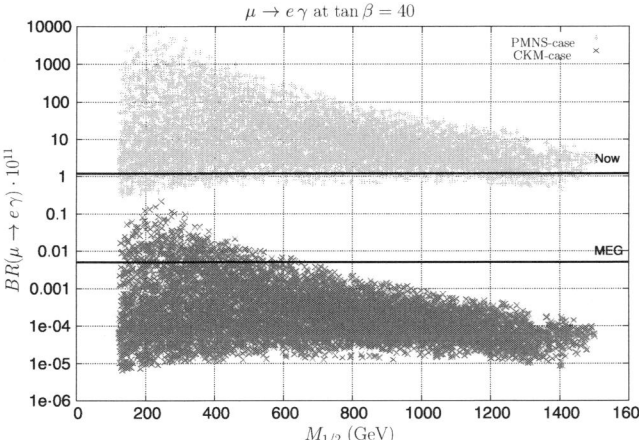
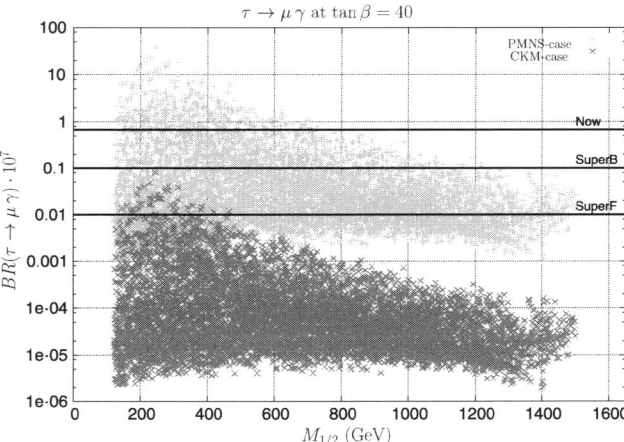

Fig. 34 Scatter plots of $B(\mu \to e, \gamma)$ (*left*) and $B(\tau \to \mu \gamma)$ (*right*) versus $M_{1/2}$ for $\tan \beta = 40$, both for the (maximal) PMNS case with $|U_{e3}| = 0.07$ and the (minimal) CKM case. The plots were obtained by scanning the SUSY parameter space in the LHC accessible region (see the text for details)

on U_{e3} itself [258, 722] are taken into account. The point is that the PMNS boundary condition (5.76) is valid at high scale. Thus, it is necessary to evolve the neutrino masses and mixing from the low energy scale, where measurements are performed, up to high energy. Such effect turns out to be not negligible in case of low energy $U_{e3} \lesssim 10^{-3}$, giving a high energy constant enhancement of order $\mathcal{O}(10^{-3})$ [688]. The consequence is that the term in (5.78) $\propto Y_t^2$ always dominates, giving a contribution to the branching ratio larger than the CKM case (which turns out to be really a "minimal" case) and bringing the most of the parameter space in the realm of MEG even for very small low energy values of U_{e3} [688, 763].

The $\tau \to \mu$ transitions are instead U_{e3}-independent probes of SUSY, whose importance was first pointed out in Ref. [732]. The off-diagonal entry in this case is given by:

$$\left(m_{\tilde{L}}^2\right)_{32} \approx -\frac{3m_0^2 + A_0^2}{8\pi^2} Y_t^2 U_{\mu 3} U_{\tau 3} \ln \frac{M_{\text{GUT}}}{M_{R_3}} + \mathcal{O}(Y_c^2). \quad (5.80)$$

In the $\tau \to \mu\gamma$ decay the situation is at the moment similarly constrained with respect to $\mu \to e\gamma$, if U_{e3} happens to be very small. The main difference is that $B(\tau \to \mu\gamma)$ does not depend on the value of U_{e3}, so that $\tau \to \mu\gamma$ will be a promising complementary channel with respect to $\mu \to e\gamma$. As far as Beauty factories [191, 644, 764] are concerned, we see from Fig. 34, that even with the present bound it is possible to rule out part of LHC accessible region in the PMNS high $\tan\beta$ regime; the planned accuracy of the SuperKEKB [694] machine $\sim \mathcal{O}(10^{-8})$ will allow to test much of high $\tan\beta$ region and will start probing the low $\tan\beta$ PMNS case, with a sensitivity to soft masses as high as $(m_0, m_{\tilde{g}}) \lesssim 900$ GeV. The situation changes dramatically if one takes into account the possibility of a Super Flavor factory: taking the sensitivity of the most promising $\tau \to \mu\gamma$ process to $\sim \mathcal{O}(10^{-9})$, the PMNS case will be nearly ruled out in the high $\tan\beta$ regime and severely constrained in the low $\tan\beta$ one; as for the CKM case we would enter the region of interest.

Let's finish with some remarks. Suppose that the LHC does find signals of low energy supersymmetry, then grand unification becomes a very appealing scenario, because of the successful unification of gauge couplings driven by the SUSY partners. Among SUSY-GUT models, an $SO(10)$ framework is much favored as it is the 'minimal' GUT to host all the fermions in a single representation and it accounts for the smallness of the observed neutrino masses by naturally including the see-saw mechanism. In the above we have addressed the issue by a generic benchmark analysis, within the ansatz that there is no fine-tuning in the neutrino Yukawa sector. We can state that LFV experiments should be able to tell us much about the structure of such a SUSY-GUT scenario. If they detect LFV processes, by their rate and exploiting the interplay between different experiments, we would be able to get hints of the structure of the unknown neutrinos' Yukawa's. On the contrary, in the case that both MEG and a future Super Flavor factory happen not to see any LFV process, only two possibilities should be left: (i) the minimal mixing, low $\tan\beta$ scenario; (ii) mSUGRA $SO(10)$ see-saw without fine-tuned Y_ν couplings is not a viable framework of physics beyond the standard model.

Actually one should remark that LFV experiments will be able to falsify some of above scenarios even in regions of the mSUGRA parameter space that are beyond the reach of LHC experiments. In this sense, the power of LFV experiments of testing/discriminating among different SUSY-GUTs models results very interesting and highly complementary to the direct searches at the LHC.

5.3.4 LFV, QFV and CPV observables in GUTs and their correlations

In a SUSY grand unified theory (GUT), quarks and leptons sit in same multiplets and are transformed ones into the others through GUT symmetry transformations. If the energy scale where the SUSY breaking terms are transmitted to the visible sector is larger then the GUT scale, as in the case of gravity mediation, such breaking terms, and in particular the sfermion mass matrices, will have to respect the underlying GUT symmetry. Hence, as already discussed in Sect. 5.3.1, the quark–lepton unification seeps also into the SUSY breaking soft sector. If the soft SUSY breaking terms respect boundary conditions which are subject to the GUT symmetry to start with, we generally expect the presence of relations among the (bilinear and trilinear) scalar terms in the hadronic and leptonic sectors [165, 739]. Such relations hold true at the (superlarge) energy scale where the correct symmetry of the theory is the GUT symmetry. After its breaking, the mentioned relations will undergo corrections which are computable through the appropriate RGE's which are related to the specific structure of the theory between the GUT and the electroweak scale (for instance, new Yukawa couplings due to the presence of RH neutrinos acting down to the RH neutrino mass scale, presence of a symmetry breaking chain with the appearance of new symmetries at intermediate scales, etc.). As a result of such a computable running, we can infer the correlations between the softly SUSY breaking hadronic and leptonic MIs at the low scale where FCNC tests are performed. Moreover, given that a common SUSY soft breaking scalar term of $\mathcal{L}_{\text{soft}}$ at scales close to M_{Planck} can give rise to RG-induced δ^q's and δ^l's at the weak scale, one may envisage the possibility to make use of the FCNC constraints on such low energy δ's to infer bounds on the soft breaking parameters of the original supergravity Lagrangian ($\mathcal{L}_{\text{sugra}}$). Indeed, for each scalar soft

Table 14 Links between various transitions between up-type, down-type quarks and charged leptons for $SU(5)$. $m_{\tilde{f}}^2$ refers to the average mass for the sfermion f, $m_{\tilde{Q}_{\text{avg}}}^2 = \sqrt{m_{\tilde{Q}}^2 m_{\tilde{d}^c}^2}$ and $m_{\tilde{L}_{\text{avg}}}^2 = \sqrt{m_{\tilde{L}}^2 m_{\tilde{e}^c}^2}$

Relations at weak-scale	Boundary conditions at M_{GUT}
$(\delta_{ij}^u)_{RR} \approx (m_{\tilde{e}^c}^2/m_{\tilde{u}^c}^2)(\delta_{ij}^l)_{RR}$	$m_{\tilde{u}^c}^2(0) = m_{\tilde{e}^c}^2(0)$
$(\delta_{ij}^q)_{LL} \approx (m_{\tilde{e}^c}^2/m_{\tilde{Q}}^2)(\delta_{ij}^l)_{RR}$	$m_{\tilde{Q}}^2(0) = m_{\tilde{e}^c}^2(0)$
$(\delta_{ij}^d)_{RR} \approx (m_{\tilde{L}}^2/m_{\tilde{d}^c}^2)(\delta_{ij}^l)_{LL}$	$m_{\tilde{d}^c}^2(0) = m_{\tilde{L}}^2(0)$
$(\delta_{ij}^d)_{LR} \approx (m_{\tilde{L}_{\text{avg}}}^2/m_{\tilde{Q}_{\text{avg}}}^2)(m_b/m_\tau)(\delta_{ij}^l)_{LR}^\star$	$A_{ij}^e = A_{ji}^d$

parameter of $\mathcal{L}_{\text{sugra}}$ one can ascertain whether the hadronic or the leptonic corresponding bound at the weak scale yields the strongest constraint at the large scale [165].

Let us consider the scalar soft breaking sector of the MSSM:

$$-\mathcal{L}_{\text{soft}} = m_{Q_{ii}}^2 \tilde{Q}_i^\dagger \tilde{Q}_i + m_{u_{ii}^c}^2 \tilde{u}_i^{c\star} \tilde{u}_i^c + m_{e_{ii}^c}^2 \tilde{e}_i^{c\star} \tilde{e}_i^c$$
$$+ m_{d_{ii}^c}^2 \tilde{d}_i^{c\star} \tilde{d}_i^c + m_{L_{ii}}^2 \tilde{L}_i^\dagger \tilde{L}_i + m_{H_1}^2 H_1^\dagger H_1$$
$$+ m_{H_2}^2 H_2^\dagger H_2 + A_{ij}^u \tilde{Q}_i \tilde{u}_j^c H_2 + A_{ij}^d \tilde{Q}_i \tilde{d}_j^c H_1$$
$$+ A_{ij}^e \tilde{L}_i \tilde{e}_j^c H_1 + (\Delta_{ij}^l)_{LL} \tilde{L}_i^\dagger \tilde{L}_j + (\Delta_{ij}^e)_{RR} \tilde{e}_i^{c\star} \tilde{e}_j^c$$
$$+ (\Delta_{ij}^q)_{LL} \tilde{Q}_i^\dagger \tilde{Q}_j + (\Delta_{ij}^u)_{RR} \tilde{u}_i^{c\star} \tilde{u}_j^c$$
$$+ (\Delta_{ij}^d)_{RR} \tilde{d}_i^{c\star} \tilde{d}_j^c + (\Delta_{ij}^e)_{LR} \tilde{e}_{Li}^\star \tilde{e}_j^c$$
$$+ (\Delta_{ij}^u)_{LR} \tilde{u}_{Li}^\star \tilde{u}_j^c + (\Delta_{ij}^d)_{LR} \tilde{d}_{Li}^\star \tilde{d}_j^c + \cdots, \quad (5.81)$$

where we have explicitly written down the various off-diagonal entries of the soft SUSY breaking matrices. Consider now that $SU(5)$ is the relevant symmetry at the scale where the above soft terms firstly show up. Then, taking into account that matter is organized into the $SU(5)$ representations $\mathbf{10} = (q, u^c, e^c)$ and $\bar{\mathbf{5}} = (l, d^c)$, one obtains the following relations

$$m_Q^2 = m_{\tilde{e}^c}^2 = m_{\tilde{u}^c}^2 = m_{\mathbf{10}}^2,$$
$$m_{\tilde{d}^c}^2 = m_L^2 = m_{\bar{\mathbf{5}}}^2, \quad (5.82)$$
$$A_{ij}^e = A_{ji}^d.$$

These equations for matrices in flavor space lead to relations between the slepton and squark flavor violating off-diagonal entries Δ_{ij}. These are:

$$\left(\Delta_{ij}^u\right)_{LL} = \left(\Delta_{ij}^u\right)_{RR} = \left(\Delta_{ij}^d\right)_{LL} = \left(\Delta_{ij}^l\right)_{RR}, \quad (5.83)$$

$$\left(\Delta_{ij}^d\right)_{RR} = \left(\Delta_{ij}^l\right)_{LL}, \quad (5.84)$$

$$\left(\Delta_{ij}^d\right)_{LR} = \left(\Delta_{ji}^l\right)_{LR} = \left(\Delta_{ij}^l\right)_{RL}^\star. \quad (5.85)$$

These GUT correlations among hadronic and leptonic scalar soft terms are summarized in the second column of Table 14.

Table 15 Links between various transitions between up-type, down-type quarks and charged leptons for PS/$SO(10)$ type models

Relations at weak-scale	Boundary conditions at M_{GUT}
$(\delta_{ij}^u)_{RR} \approx (m_{\tilde{e}^c}^2/m_{\tilde{u}^c}^2)(\delta_{ij}^l)_{RR}$	$m_{\tilde{u}^c}^2(0) = m_{\tilde{e}^c}^2(0)$
$(\delta_{ij}^q)_{LL} \approx (m_{\tilde{L}}^2/m_{\tilde{Q}}^2)(\delta_{ij}^l)_{LL}$	$m_{\tilde{Q}}^2(0) = m_{\tilde{L}}^2(0)$

Assuming that no new sources of flavor structure are present from the $SU(5)$ scale down to the electroweak scale, apart from the usual SM CKM one, one infers the relations in the first column of Table 14 at low scale. Here we have taken into account that due to their different gauge couplings "average" (diagonal) squark and slepton masses acquire different values at the electroweak scale.

Two comments are in order when looking at Table 14. First, the boundary conditions on the sfermion masses at the GUT scale (last column in Table 14) imply that the squark masses are *always* going to be larger at the weak scale compared to the slepton masses due to the participation of the QCD coupling in the RGEs. As a second remark, notice that the relations between hadronic and leptonic δ MI in Table 14 always exhibit opposite "chiralities", i.e. LL insertions are related to RR ones and vice-versa. This stems from the arrangement of the different fermion chiralities in $SU(5)$ five- and ten-plets (as it clearly appears from the final column in Table 14). This restriction can easily be overcome if we move from $SU(5)$ to left-right symmetric unified models like $SO(10)$ or the Pati–Salam (PS) case (we exhibit the corresponding GUT boundary conditions and δ MI at the electroweak scale in Table 15).

So far we have confined the discussion within the simple $SU(5)$ model, without the presence of any extra particles like right handed (RH) neutrinos. In the presence of RH neutrinos, one can envisage of two scenarios [164]: (a) with either very small neutrino Dirac Yukawa couplings and/or very small mixing present in the neutrino Dirac Yukawa matrix, (b) Large Yukawa and large mixing in the neutrino sector. In the latter case, (5.83)–(5.85) are not valid at all scales in general, as large RGE effects can significantly modify the sleptonic flavor structure while keeping the squark sector essentially unmodified; thus essentially breaking the GUT symmetric relations. In the former case where the neutrino Dirac Yukawa couplings are tiny and do not significantly

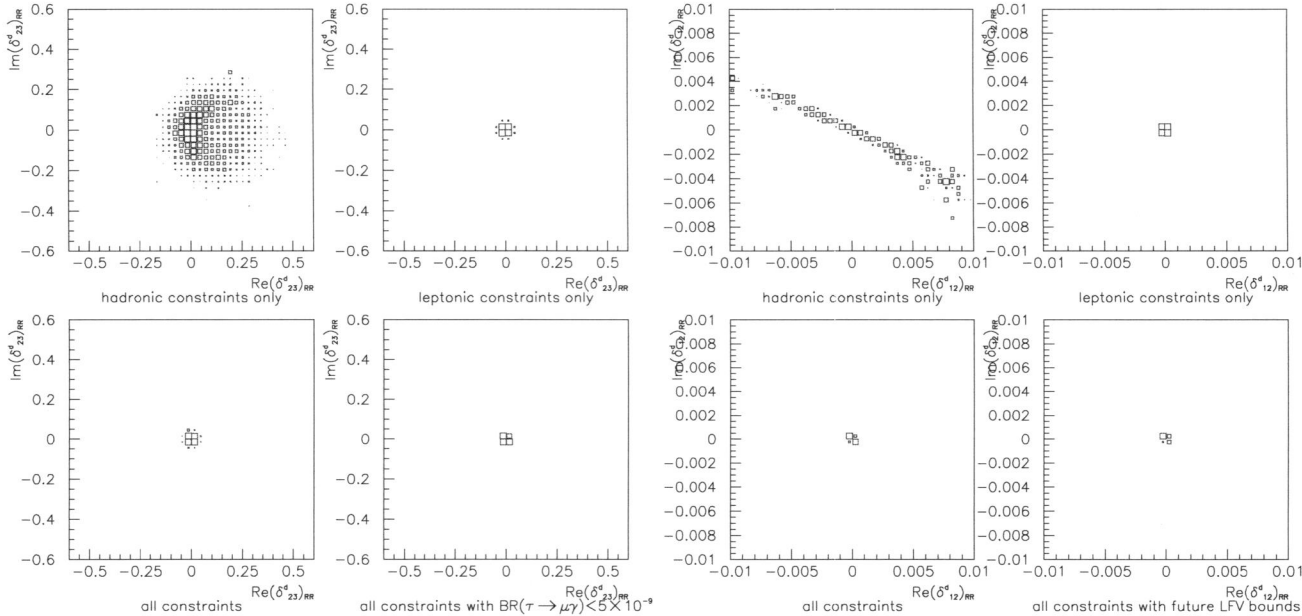

Fig. 35 *Left four panels*: allowed region for $(\delta^d_{23})_{RR}$ using constraints as indicated. *Right four panels*: the same for $(\delta^d_{12})_{RR}$. For the parameter space considered, please see the text

modify the sleptonic flavor structure, the GUT symmetric relations are expected to be valid at the weak scale. However, in both cases it is possible to say that there exists a bound on the hadronic δ parameters of the form [739]:

$$|(\delta^d_{ij})_{RR}| \geq \frac{m^2_{\tilde{L}}}{m^2_{\tilde{d}^c}} |(\delta^l_{ij})_{LL}|. \quad (5.86)$$

The situation is different if we try to translate the bound from quark to lepton MIs. An hadronic MI bound at low energy leads, after RGE evolution, to a bound on the corresponding grand unified MI at M_{GUT}, applying both to slepton and squark mass matrices. However, if the neutrino Yukawa couplings have sizable off-diagonal entries, the RGE running from M_{GUT} to M_W could still generate a new contribution to the slepton MI that exceeds this GUT bound. Therefore hadronic bounds cannot be translated to leptons unless we make some additional assumptions on the neutrino Yukawa matrices. On general grounds, given that SM contributions in the lepton sector are absent and that the branching ratios of leptonic processes constrain only the modulus of the MIs, it turns out that all the MI bounds arising from the lepton sector are circles in the $\text{Re}(\delta^d_{ij})_{AB}$–$\text{Im}(\delta^d_{ij})_{AB}$ plane and are centered at the origin.

In the following the effect of leptonic bounds on the quark mass insertions are reviewed, following the results presented in [165], where constraints on δs were studied scanning the mSUGRA parameter space in the ranges: $M_{1/2} \leq 160$ GeV, $m_0 \leq 380$ GeV, $|A_0| \leq 3m_0$ and $5 < \tan\beta < 15$. For instance, in presence of a $(\Delta^d_{23})_{LR}$ at the

GUT scale, this would have effects both in the $\tau \to \mu\gamma$ and $b \to s\gamma$ decays. Using $(\delta^d_{23})_{LR} \lesssim (m_b/m_\tau)(m^2_{\tilde{l}}/m^2_{\tilde{q}}) \times (\delta^l_{23})_{RL}$, a bound on $(\delta^l_{23})_{RL}$ from the $\tau \to \mu\gamma$ decay translates into a bound on $(\Delta^d_{23})_{LR}$ (neglecting the effects of neutrino Yukawa's the inequality transforms into equality). Thus, leptonic processes set a bound on the SUSY contributions to $B(B \to X_s\gamma)$. However, it turns out that the present leptonic bounds have no effect on the $(\delta^d_{23})_{LR}$ couplings. This is due both to the existence of strong hadronic bounds from $b \to s\gamma$ and CP asymmetries and to the relatively weak leptonic bounds here.

Similarly, in presence of a $(\Delta^d_{23})_{RR}$ at the GUT scale, the corresponding MIs at the electroweak scale are $(\delta^d_{23})_{RR}$ and $(\delta^l_{23})_{LL}$ that contribute to ΔM_{B_s} and $\tau \to \mu\gamma$ respectively (the impact of $(\Delta^d_{23})_{RR}$ on $b \to s\gamma$ and $b \to s\ell^+\ell^-$ is not relevant because of the absence of interference between SUSY and SM contributions). In Fig. 35 the allowed values of $\text{Re}(\delta^d_{23})_{RR}$ and $\text{Im}(\delta^d_{23})_{RR}$ with the different constraints are shown. The leptonic constraints are quite effective as the bound on the $B(\tau \to \mu\gamma)$ from B-factories is already very stringent, while the recent measurement of ΔM_{B_s} is less constraining. The plots correspond to $5 < \tan\beta < 15$, thus, the absolute bound on $(\delta^l_{23})_{LL}$ is set by $\tan\beta = 5$ and it scales with $\tan\beta$ as $(\delta^l_{23})_{LL} \sim (5/\tan\beta)$.[24]

---

[24]Sizable SUSY contributions to ΔM_{B_s} are still possible from the Higgs sector in the large $\tan\beta$ regime both within [765–767] and also beyond [768] the Minimal Flavor Violating framework. However, for the considered parameter space, the above effects are completely negligible.

As in the LR sector, in the LL one, there is no appreciable improvement from the inclusion of leptonic constraints. In fact, $\tau \to \mu\gamma$ is not effective to constrain $(\delta_{23}^l)_{RR}$, i.e. the leptonic MI related to $(\delta_{23}^d)_{LL}$ in this SUSY-GUTs scheme, in large portions of the parameter space because of strong cancellations among amplitudes. The analysis of the constraints on the different (δ_{13}^d) MIs gives similar results to that of the (δ_{23}^d) MIs. In this case, the hadronic constraints come mainly from ΔM_{B_d} and the different CP asymmetries measured at B-factories, while the leptonic bounds are due to the decay $\tau \to e\gamma$.

Coming to the 1–2 sector, let's see, as an example, the allowed values of $\mathrm{Re}(\delta_{12}^d)_{RR}$ and $\mathrm{Im}(\delta_{12}^d)_{RR}$. In this case, as it appears from Fig. 35, leptonic constraints, already using the present limit on $B(\mu \to e\gamma)$, are competitive and constrain the direction in which the constraint from ε_K is not effective (upper left plot). Similarly in the LR sector, even if the hadronic bounds coming from ε'/ε are quite stringent, the bounds from $\mu \to e\gamma$ are even more effective, while the LL sector results less constrained by leptonic processes, as an effect of the cancellations that $\mu \to e\gamma$ decay suffers in the RR leptonic sector.

5.4 R-parity violation

5.4.1 Introduction

In supersymmetric extensions of the standard model (SM), baryon and lepton numbers are no longer automatically protected. This is the main reason for introducing R-parity. R-parity is associated with a Z_2 subgroup of the group of continuous $U(1)$ transformations acting on the gauge superfields and the two chiral doublet Higgs superfields H_d and H_u, with their definition extended to quark and lepton superfields so that quarks and leptons carry $R = 0$ and squarks and sleptons $R = \pm 1$. One can express R-parity in terms of spin S, baryon B and lepton L number [769]:

$$\text{R-parity} = (-1)^{2S}(-1)^{3(B-L)}. \quad (5.87)$$

Taking into account the important phenomenological differences between models with and without R-parity, it is worth studying if and how R-parity can be broken. One of the main reasons to introduce R-parity is avoiding proton decay. However there are in principle other discrete or continuous symmetries that can protect proton decay while allowing for some R-parity violating couplings. In the absence of R-parity, R-parity odd terms allowed by renormalizability and gauge invariance [150] must be included in the superpotential of the minimal supersymmetric standard model,

$$W_{Rp} = \mu_i H_u L_i + \frac{1}{2}\lambda_{ijk} L_i L_j E_k^c + \lambda'_{ijk} L_i Q_j D_k^c$$
$$+ \frac{1}{2}\lambda''_{ijk} U_i^c D_j^c D_k^c, \quad (5.88)$$

where there is summation over the generation indices $i, j, k = 1, 2, 3$, and summation over gauge indices is understood. One has for example $L_i L_j E_k^c \equiv (\epsilon_{ab} L_i^a L_j^b) E_k^c = (N_i E_j - E_i N_j) E_k^c$ and $U_i^c D_j^c D_k^c \equiv \epsilon_{\alpha\beta\gamma} U_i^{\alpha c} D_j^{\beta c} D_k^{\gamma c}$, where $a, b = 1, 2$ are $SU(2)_L$ indices, $\alpha, \beta, \gamma = 1, 2, 3$ are $SU(3)_C$ indices, and ϵ_{ab} and $\epsilon_{\alpha\beta\gamma}$ are totally antisymmetric tensors (with $\epsilon_{12} = \epsilon_{123} = +1$). Gauge invariance enforces antisymmetry of the λ_{ijk} couplings in their first two indices and antisymmetry of the λ''_{ijk} couplings in their last two indices,

$$\lambda_{ijk} = -\lambda_{jik}, \quad (5.89)$$
$$\lambda''_{ijk} = -\lambda''_{ikj}. \quad (5.90)$$

The bilinear terms $\mu_i H_u L_i$ in (5.88) can be rotated away from the superpotential upon suitably redefining the lepton and Higgs superfields. However, in the presence of generic soft supersymmetry breaking terms of dimension two, bilinear R-parity violation will reappear. The fact that one can make $\mu_i = 0$ in (5.88) does not mean that the Higgs–lepton mixing associated with bilinear R-parity breaking is unphysical, but rather that there is not a unique way of parameterizing it. If R-parity is violated in the leptonic sector, no quantum numbers differentiate between lepton and Higgs superfields, and they consequently mix with each other [770]. The R-parity violation in the baryonic sector does not imply lepton flavor violation, and we do not consider such option here.

A general consequence of R-parity violation is that unless the relevant couplings are negligibly small, the supersymmetric model does not have a dark matter candidate. Thus experimental studies on dark matter will also shed light on R-parity violation.

5.4.2 Limits on couplings

Limits on R-parity violating couplings can be obtained by direct searches at colliders or requiring that the R-parity violating contribution to a given observable does not exceed the limit imposed by the precision of the experimental measurement.

On the collider side R-parity violation implies the possibility of the creation, decay or exchange of single sparticles, thus allowing new decay channels. For example, even for relatively small R-parity violating interactions, the decay of the lightest supersymmetric particle will lead to collider events departing considerably from the characteristic missing momentum signal of R-parity conserving theories. In absence of definite theoretical predictions for the values of the

45 independent trilinear Yukawa couplings Λ (λ_{ijk}, λ'_{ijk} and λ''_{ijk}), it is necessary in practice to assume a strong hierarchy among the couplings. A simplifying assumption widely used for the search at colliders is to postulate the existence of a single dominant R-parity violating coupling. When discussing specific bounds, it is necessary to choose a definite basis for quark and lepton superfields. Often it is understood that the single coupling dominance hypothesis applies in the mass eigenstate basis. It can be more natural to apply this hypothesis in the weak eigenstate basis when dealing with models in which the hierarchy among couplings originates from a flavor theory. In this case, a single process allows one to constrain several couplings, provided one has some knowledge of the rotations linking the weak eigenstate and mass eigenstate bases. Indirect bounds from loop processes typically lead to bounds on the products of two most important R-parity violating couplings, or on the sum of products of two couplings. The limits on single dominant couplings, and on products of couplings, as well as a more complete list of references, are collected in [771].

5.4.3 Spontaneous R-parity breaking

The spontaneous breaking of R-parity is characterized by an R-parity invariant Lagrangian leading to non-vanishing VEVs for some R-parity odd scalar field, which in turn generates R-parity violating terms. Such a spontaneous breakdown of R-parity generally also entails the breaking of the global $U(1)$ lepton number symmetry L which implies the existence of a massless Nambu–Goldstone real pseudoscalar boson J, the majoron. Another light scalar particle, denoted ρ, generally accompanies the majoron in the supersymmetric models. If the $U(1)$ symmetry is also explicitly broken by interaction terms in the Lagrangian, both of these particles acquire finite masses. The most severe constraints on the models with a spontaneous R-parity breaking, arise in the cases where the majoron carries electroweak gauge charges and hence is coupled to the Z bosons and to quarks and leptons. The non-singlet components contribute to the Z boson invisible width by an amount of one-half that a single light neutrino, $\delta \Gamma_{\text{inv}}^Z/6 \simeq 83$ MeV. To suppress the non-singlet components one must allow either for sufficiently small sneutrino VEVs, $v_L/M_Z \ll 1$, or for some large hierarchy of scales between v_L and the VEV parameters associated with additional electroweak singlet scalar fields [772].

However, it is not necessary that models with spontaneous R-parity violation have a majoron. Models without a majoron include a class of models with triplet Higgses, where $B - L$ is a gauge symmetry, which is necessarily spontaneously broken unless effects of non-renormalizable terms or some additional new fields are included [773]. An interesting experimental signal in these models may be a relatively light doubly charged scalar, which decays dominantly to same charge leptons (not necessarily of the same generation) [774]. Another possibility for a model without a majoron is a model where the lepton number is broken by two units explicitly, in which case the spontaneous breaking by one unit (which breaks the R-parity) does not lead to a majoron [775]. The interactions in spontaneously R-parity breaking models through the lepton number violation closely resemble explicitly R-parity breaking models with only bilinear R-parity violation. In the case of spontaneous breaking, the parameters which are free in the model with only bilinear couplings are related to each other via the sneutrino vacuum expectation value (VEV). Thus a constraint from one process affects availability of the other processes. Example bounds for such a model can be found in [776].

It is worth emphasizing that choosing single coupling dominance in the case of spontaneous breaking is not possible and in this sense, the models with spontaneous breaking are more predictive than those without.

5.4.4 Neutrino sector

The presence of non-zero couplings λ_{ijk}, λ'_{ijk} or bilinear R-parity violating parameters implies the generation of neutrino masses and mixing [777]. This is an interesting feature of R-parity violating models, but it can also be a problem, since the contribution of R-parity violating couplings may exceed by orders of magnitude the experimental bounds. Two types of contributions can be distinguished: tree-level or loop contributions.

The tree-level contributions are due to bilinear R-parity violation terms which induce a mixing between neutrinos and neutralinos [778]. This gives a massive neutrino state at tree level. When quantum corrections are included, all three neutrinos acquire a mass. The tree-level contribution arising from the neutrino–neutralino mixing can be understood, in the limit of small neutrino–neutralino mixing, as a sort of seesaw mechanism, in which the neutral gauginos and higgsinos play the role of the right handed neutrinos.

The loop contributions are induced by the trilinear R-parity violating couplings λ_{ijk} and λ'_{ijk} and by bilinear R-parity violating parameters [779]. If bilinear R-parity violation is strongly suppressed one can concentrate on the diagrams involving trilinear R-parity violating couplings only. The trilinear couplings λ_{ijk} and λ'_{ijk} contribute to each entry of the neutrino mass matrix through the lepton–slepton and quark–squark loops. The neutrino mass matrix depends therefore on a large number of trilinear R-parity violating couplings. In order to obtain a predictive model, one has to make assumptions on the structure of the trilinear couplings. In general, however, the bilinear R-parity violation contribution cannot be neglected. The presence of bilinear terms drastically modifies the calculation of one loop neutrino masses. The neutrino mass matrix receives contributions already at tree level, as discussed above, and moreover in addition to the lepton–slepton and quark–squark loops, one loop diagrams involving insertions of bilinear R-parity violating masses or slepton VEVs must be considered. One should note that the bilinear R-parity violating terms, if not suppressed, give too large loop contributions to neutrino masses.

The scenario known as bilinear R-parity violation (BRpV) corresponds to the explicit introduction of the three mass parameters μ_i in the first term in (5.88), without referring to their origin, and assuming that all the trilinear parameters are zero. The μ_i terms introduce tree-level mixing between the Higgs and lepton superfields. Therefore, they violate R-parity and lepton number, and contribute to the breaking of the $SU(2)$ symmetry by the induction of sneutrino vacuum expectation values v_i. As it was mentioned before, the mixing between neutralinos and neutrinos leads to an effective tree-level neutrino mass matrix of the form,

$$m_\nu^{0ij} = \frac{M_1 g^2 + M_2 g'^2}{4 \det M_{\chi^0}} \Lambda_i \Lambda_j, \quad (5.91)$$

where the parameters $\Lambda_i = \mu v_i + \mu_i v_d$ are proportional to the sneutrino vacuum expectation values in the basis where the μ_i terms are removed from the superpotential. Due to the symmetry of this mass matrix, only one neutrino acquires a mass. Once quantum corrections are included, this symmetry is broken, and the effective neutrino mass matrix takes the form [780],

$$m_\nu^{ij} = A \Lambda_i \Lambda_j + B(\Lambda_i \epsilon_j + \Lambda_j \epsilon_i) + C \epsilon_i \epsilon_j. \quad (5.92)$$

If the tree-level contribution dominates, as for example in SUGRA models with low values of $\tan \beta$, the atmospheric mass scale is given at tree level, and the solar mass scale is generated at one loop, explaining the hierarchy between them. Most of the time, the dominant loop in SUGRA is the one formed with bottom quarks and squarks, followed in importance by loops with charginos and neutralinos. In the tree-level dominance case the atmospheric mixing angle is well approximated by $\tan^2 \theta_{atm} = \Lambda_2^2 / \Lambda_3^2$, and the reactor angle by $\tan^2 \theta_{13} = \Lambda_1^2 / (\Lambda_2^2 + \Lambda_3^2)$. In this case, the smallness of the reactor angle is achieved with a small value of Λ_1, and the maximal mixing in the atmospheric sector with a similar value for Λ_2 and Λ_3. Supergravity scenarios where tree-level contribution does not dominate can also be found [781], in which case the previous approximations for the angles are not valid.

5.4.5 Lepton flavor violating processes at low energies

Many processes, which are either rare or forbidden in the R-parity conserving model, become possible when interactions following from the superpotential W_{R_p} in (5.88) are available. These interactions include tree-level couplings between different lepton or quark generations, as well as tree-level couplings between leptons and quarks, or leptons and Higgses.

In addition to the trilinear couplings λ and λ', bilinear couplings or spontaneous R-parity breaking contribute to the lepton flavor violating processes mentioned below through mixing.

For references about this section, see Ref. [771].

– $l_i \to l_j \gamma$, $l_i \to l_j l_k l_m$, and μ–e-conversion, and semileptonic decays of τ-leptons. The rare decays of leptons to lighter leptons are excellent probes of new physics, because they do not involve any hadronic uncertainties. Both the lepton flavor violating trilinear λ- and λ'-type couplings give rise to LFV decays $l_i \to l_j \gamma$ (loop level process with $\tilde{\nu}$–l, ν–$\tilde{l}$, or $\tilde{q}$–q' in the loop), $l_i \to l_j l_k l_m$ (tree-level process via $\tilde{\nu}$ or $\tilde{l}$), as well as for μ–e-conversion. In these processes, two non-vanishing Λ couplings are needed and usual approach is to assume a dominant product of two couplings, when determining bounds on couplings. In the μ–e-conversion, certain pairs of couplings can be probed only in the loop-level process, mediated by virtual γ or Z, which are logarithmically enhanced compared to $\mu \to e\gamma$ [185]. The hadronic contributions to the μ–e-conversion in nuclei make the theoretical error larger than in the decays without hadrons. The relatively large mass of τ allows for new semileptonic decay modes for τ. The bounds from these processes vary between $\Lambda \sim \mathcal{O}(10^{-4}$–$10^{-1})$ for 100 GeV fermion masses, and they scale as *mass*^{2}.

The experimental accuracies of the processes mentioned above are expected to increase considerably in the coming years.

– Leptonic and semileptonic decays of hadrons and top quarks. R-parity violating couplings λ'_{ijk} allow for lepton flavor violating decays of hadrons, e.g. $K_L \to e^\pm \mu^\mp$, $B_d \to \mu^+ \tau^-$, $K^+ \to \pi^+ \nu_i \nu_j$ [782], as well as semileptonic LFV top decays, e.g. $t \to \tilde{\tau}^+ b$, if kinematically allowed. The sensitivity on the couplings is restricted by the theoretical uncertainties in hadronic contributions. For 100 GeV sfermions, the bounds are $\Lambda \sim \mathcal{O}(10^{-4}$–$10^{-1})$.

5.4.6 Anomalous muon magnetic moment a_μ and electron electric dipole moment

Λ couplings affect leptons also through contributions to dipole moments. The experimental measurement of a_μ is quite precise. The theoretical calculation of the standard model contribution to a_μ contains still uncertainty, which prevents exact comparison with measurement. The contribution of R-parity violation on a_μ is small, and constrained by tiny neutrino masses.

Contribution from complex Λ to electron EDM could be large for large phases. The one loop contribution involving both bilinear and trilinear couplings is sizable for electron EDM, while one loop terms with only trilinear terms are suppressed by neutrino masses.

5.4.7 Collider signatures

The main advantage of collider studies compared to the low energy probes is that the particles can be directly produced, and thus their masses and couplings can be experimentally measured.

A major difference between R-parity conserving and breaking models from the detection point of view is the amount of missing energy. If R-parity is violated, the supersymmetric particles decay to the SM particles leaving little or no missing energy. Decays of sparticles through λ- and λ'-type couplings lead to multi-lepton final states, and λ' and λ'' to multi-jet final states. Sparticles can decay first via the R-parity conserving couplings to the lightest supersymmetric particle (LSP), which then decays via R-parity violating couplings. If e.g. a neutralino is the LSP, it may be a cascade decay product of a sfermion, chargino, or a heavier neutralino. Thus typically one gets a larger number of jets or leptons in the final state in R-parity violating than in the R-parity conserving decay. The sparticles can also decay directly to the standard model fermions via λ, λ', or λ'' couplings. Assuming all the supersymmetric particles decay inside the detector, a consequence of the decay of the LSP is that the amount of missing energy when R-parity is violated is considerably lower than in the R-parity conserving case, and only neutrinos carry the missing energy. When R-parity is violated, the LSP is not stable and need not be neutral. If then the coupling through which the LSP decays is suppressed, a long lived possibly charged particle appears, leaving a heavily ionizing, easily detectable charged track in the detector.

A simplifying assumption for the search strategy at colliders is to postulate the existence of a single dominant R-parity violating coupling. In case a non-vanishing coupling does exist with a magnitude leading to distinct phenomenology at colliders, a direct sensitivity to a long-lived LSP might be provided by the observation of displaced vertices in an intermediate range of coupling values up to $\mathcal{O}(10^{-5}\text{--}10^{-4})$. For larger Λ values the presence of R-parity violating supersymmetry will become manifest through the decay of short-lived sparticles produced by pair via gauge couplings. A possible search strategy in such cases consists of neglecting R-parity violating contributions at production in non-resonant processes. This is valid provided that the interaction strength remains sufficiently small compared to electromagnetic or weak interaction strengths, for Λ values typically below $\mathcal{O}(10^{-2}\text{--}10^{-1})$. In a similar or larger range of couplings values, R-parity violation could show up at colliders via single resonant or non-resonant production of supersymmetric particles.

For bilinear or spontaneous breaking, the lightest supersymmetric particle decays through mixing with the corresponding $R_p = +1$ particle. If the LSP is neutralino or chargino, it decays through mixing with neutrino or charged lepton, and if the LSP is a slepton it decays through mixing with the Higgs bosons, e.g. stau mixes with charged Higgs. Assuming that neutralino is the LSP, the dominant decay mode of stau is to tau and neutralino. Through mixing the charged Higgs has then a branching ratio to tau and neutralino. Thus the detection of R-parity violation includes precise measurement of the branching ratios of particles.

The main signature of BRpV is the decay of the neutralino, which decays 100% of the time into R-parity and lepton number violating modes. If squarks and sleptons are heavy and the neutralino is heavier than the gauge bosons, the neutralino decays into on-shell gauge bosons and leptons: $\chi_1^0 \to W^{\mp}\ell_i^{\pm}, Z\nu_i$. If the gauge bosons are produced off-shell, then the decay modes are $\chi_1^0 \to q\bar{q}'\ell_i^{\pm}, \ell_j^{\mp}\nu_j\ell_i^{\pm}, q\bar{q}\nu_i, \ell_j^{\pm}\ell_j^{\mp}\nu_i, \nu_j\nu_j\nu_i$. When sfermions cannot be neglected, the decay channels are the same, but squarks and sleptons contribute as intermediate particles [783]. In this model, very useful quantities are formed with ratios of branching ratios, since they can be directly linked to R-parity violating parameters and neutrino observables. We have for example,

$$\frac{B(\chi_1^0 \to q\bar{q}'\mu)}{B(\chi_1^0 \to q\bar{q}'\tau)} \approx \frac{\Lambda_2^2}{\Lambda_3^2} \approx \tan^2\theta_{\text{atm}}, \quad (5.93)$$

where the last approximation is valid in the tree-level dominance scenario. In this way, collider and neutrino measurements, coming from very different experiments, can be contrasted.

Detection possibilities and extraction of limits depend a lot on the specific model and on the collider type and energy. On general grounds a lepton–hadron collider provides both leptonic and baryonic quantum numbers in the initial state and is therefore suited for searches involving λ'. In e^+p collisions, the production of $\tilde{u}_L^j$ squarks of the jth generation via λ'_{1j1} is especially interesting as it involves a valence d quark of the incident proton. In contrast, for e^-p collisions where charge conjugate processes are accessible, the λ'_{11k} couplings become of special interest as they allow for the production, involving a valence u quark, of $\tilde{d}_R^k$ squarks of the kth generation.

The excluded regions of the parameter space for R-parity violating scenarios have been worked out from the data at LEP, HERA and Tevatron, see e.g. [784–789]. In the following we shall concentrate on the search possibilities at the LHC.

5.4.8 Hadron colliders

In hadron colliders the λ or λ' couplings can provide a viable signal. In many SUSY scenarios neutralinos and charginos are among the lightest supersymmetric particles. Their pair production or associated production of $\tilde{\chi}_1^\pm \tilde{\chi}^0$ via gauge couplings and decay via λ or λ' couplings may lead to a tri-lepton signal from each particle, providing a clean signature. One should notice that if the couplings are small, the vertex may be displaced which makes the analysis more complicated. With small enough couplings the lightest neutralino, if LSP, decays outside the detector.

If kinematically possible, gluinos and squarks are copiously produced at hadron colliders. The NLO cross section has been calculated in [790]. For $m_{\tilde{q}} > m_{\tilde{g}} > m_{\tilde{c}_L}$, the production with decay via $\lambda'_{121} \neq 0$ was studied at CDF. Also coupling λ'_{13k} from $\tilde{t}$ pair production at CDF and λ' couplings from χ_1^0 decay at D0 have been investigated.

When R-parity is violated, the supersymmetric particles can be produced singly, and thus they can be produced as resonances through R-parity violating interactions. In a hadron–hadron collider this allows one to probe for resonances in a wide mass range because of the continuous energy distribution of the colliding partons. This production mode requires non-negligible R-parity violating coupling. If a single R-parity violating coupling is dominant, the exchanged supersymmetric particle may decay through the same coupling involved in its production, giving a two fermion final state. It is also possible that the decay of the resonant SUSY particle goes through gauge interactions, giving rise to a cascade decay.

The resonant production of sneutrinos and charged sleptons (via λ' couplings) has been investigated at hadron colliders [791–795]. The production of a charged lepton with neutralino leads to a like-sign dilepton signature via λ' couplings. The production of a charged lepton with a chargino in the resonant sneutrino case decay leads to a tri-lepton final state via λ' couplings. The $\tilde{\chi}_1^0$, $\tilde{\chi}_1^\pm$, $\tilde{\nu}$ masses can be reconstructed using the tri-lepton signal.

Single production is possible also in two-body processes without resonance [796]. Sfermion production with a gauge boson has been studied in either via λ or λ'-coupling. (The process $\bar{q}_i q_j \to W^- \tilde{\nu}_k$ or $\bar{q}_i q_j \to W^+ \tilde{l}_{kL}$ can get contribution also from resonant production, but e.g. in SUGRA $m_{\tilde{l}} - m_{\tilde{\nu}} = \cos 2\beta m_W^2$ and resonance production is not kinematically viable.) Similarly via λ' or λ'' gluino can be produced with a lepton or quark, respectively. Sneutrino production with two associated jets may also provide a detectable signal [797].

Resonant production of squarks can occur via λ''-type couplings, leading eventually to jets in the final states. Although the cross sections can be considerable for these processes, the backgrounds in hadronic colliders are large, and the processes seem difficult to study [798]. In special circumstances the backgrounds can be small, e.g. for stop production in $\bar{d}_i \bar{d}_j \to \tilde{t}_1 \to b \tilde{\chi}_1^+$, with $\tilde{\chi}_i^+ \to \bar{l}_i \nu_i \tilde{\chi}_i^0$ (here it is assumed $m_{\tilde{t}_1} > m_{\chi_1^+} > m_{\chi_1^0}$, $m_{\text{top}} > m_{\chi_1^0}$). Then for λ''_{3jk}, $m_{\chi_1^0}$ is stable [799, 800]. Also single gluino production, $d_i d_j \to \tilde{g} \tilde{t}$ via resonant stop production has a good signal to background ratio for $\lambda''_{3jk} = \mathcal{O}(0.1)$ [801].

With the $t\bar{t}$ production cross section of the order of 800 pb, the LHC can be considered a top quark factory, with $\sim 10^8$ top quarks being produced per year, assuming an integrated luminosity of 100 fb^{-1}. This statistics allows for precise studies of top quark physics, in particular, for measurements of rare RpV decays. A simulation of the signal and background using ATLFAST [802], to take into account the experimental conditions prevailing at the ATLAS detector [803], was made for a top quark decaying through a $\lambda'_{\ell 31}$ coupling to $t \to \tilde{\chi}^0 \ell d$, assuming only one slepton gives the leading contribution as an intermediate state [804].

The importance of treating the top quark production and decay simultaneously $gg \to t \tilde{\chi}^0 \ell d$, rather than $\Gamma(gg \to t\bar{t}) B(t \to \tilde{\chi}^0 \ell d)$, was shown. The latest approach can underestimate the cross section by a factor of a few units, depending on the slepton mass. The reason is that the slepton forces the top quark to be off-shell, becoming the resonance itself, as can be appreciated from $\tilde{\chi}^0 \ell$ mass invariant distributions.

Two scenarios were chosen for the neutralino decay, $\tilde{\chi}^0 \to bd\nu_e$ and $\tilde{\chi}^0 \to cde$, the last one assuming a large stop–scharm mixing. The sensitivity of the LHC is presented as the significance $S/\sqrt{B}$ as a function of λ'_{131}, for slepton masses 150 and 200 GeV. The channel $t \to \tilde{\chi}^0 ed \to cdeed$ is more promising with exclusion limits at 2σ c.l. for $\lambda' > 0.03$ and observation at 5σ c.l. for $\lambda' > 0.05$, with these values slightly increasing for heavier sleptons. The $t \to \tilde{\chi}^0 ed \to bd\nu_e ed$ channel is observable only for a 150 GeV slepton mass. The significance is reduced to $\lambda' > 0.08$ at 2σ and $\lambda' > 0.15$ at 5σ level.

Since a $\lambda'_{\ell 33} \sim h_b \epsilon_\ell / \mu$ trilinear term is generated in BRpV when the ϵ_ℓ term is removed from the superpotential, we can see that the above exclusion limits for λ' are not significant in BRpV, probing only values of ϵ_ℓ parameters much larger than what is needed for neutrino oscillations.

5.5 Higgs-mediated lepton flavor violation in supersymmetry

If neutrinos are massive, one would expect LFV transitions in the Higgs sector through the decay modes $H^0 \to l_i l_j$ mediated at one loop level by the exchange of the W bosons and neutrinos. However, as for the $\mu \to e\gamma$ and the $\tau \to \mu\gamma$ case, also the $H^0 \to l_i l_j$ rates are GIM suppressed. In a supersymmetric (SUSY) framework the situation is completely different. Besides the previous contributions, supersymmetry provides new direct sources of flavor violation, namely the possible presence of off-diagonal soft terms in the slepton mass matrices and in the trilinear couplings [144]. In practice, flavor violation would originate from any misalignment between fermion and sfermion mass eigenstates. LFV processes arise at one loop level through the exchange of neutralinos (charginos) and charged sleptons (sneutrinos). The amount of the LFV is regulated by a Super-GIM mechanism that can be much less severe than in the non-supersymmetric case [144]. Another potential source of LFV in models such as the minimal supersymmetric standard model (MSSM) could be the Higgs sector, in fact, extensions of the SM containing more than one Higgs doublet generally allow flavor violating couplings of the neutral Higgs bosons. Such couplings, if unsuppressed, will lead to large flavor-changing neutral currents in direct opposition to experiments. The MSSM avoid these dangerous couplings at the tree level segregating the quark and Higgs fields so that one Higgs (H_u) can couple only to up-type quarks while the other (H_d) couples only to d-type. Within unbroken supersymmetry this division is completely natural, in fact, it is required by the holomorphy of the superpotential. However, after supersymmetry is broken, couplings of the form QU_cH_d and QD_cH_u are generated at one loop [430]. In particular, the presence of a non-zero μ term, coupled with SUSY breaking, is enough to induce non-holomorphic Yukawa interactions for quarks and leptons. For large $\tan\beta$ values the contributions to d-quark masses coming from non-holomorphic operator QD_cH_u can be equal in size to those coming from the usual holomorphic operator QD_cH_d despite the loop suppression suffered by the former. This is because the operator itself gets an additional enhancement of $\tan\beta$.

As shown in Ref. [805] the presence of these loop induced non-holomorphic couplings also leads to the appearance of flavor-changing couplings of the neutral Higgs bosons. These new couplings generate a variety of flavor-changing processes such as $B^0 \to \mu^+\mu^-$, $\bar{B}^0$–B^0 etc. [537]. Higgs-mediated FCNC can have sizable effects also in the lepton sector [806, 807]: given a source of non-holomorphic couplings, and LFV among the sleptons, Higgs-mediated LFV is unavoidable. These effects have been widely discussed in the recent literature both in a generic 2HDM [808, 809] and in supersymmetry [807, 810] frameworks. Through the study of many LFV processes as $\ell_i \to \ell_j\ell_k\ell_k$ [806, 807], $\tau \to \ell_j\eta$ [169, 810], $\ell_i \to \ell_j\gamma$ [175, 650], $\mu N \to eN$ [811], $\Phi^0 \to \ell_j\ell_k$ [174] (with $\ell_i = \tau, \mu, \ell_{j,k} = \mu, e, \Phi = h^0, H^0, A^0$) or the cross section of the $\mu N \to \tau X$ reaction [812].

5.5.1 LFV in the Higgs sector

SM extensions containing more than one Higgs doublet generally allow flavor violating couplings of the neutral Higgs bosons with fermions. Such couplings, if unsuppressed, will lead to large flavor-changing neutral currents in direct opposition to experiments. The possible solution to this problem involves an assumption about the Yukawa structure of the model. A discrete symmetry can be invoked to allow a given fermion type to couple to a single Higgs doublet, and in such case FCNC's are absent at tree level. In particular, when a single Higgs field gives masses to both types of fermions the resulting model is referred as 2HDM-I. On the other hand, when each type of fermion couples to a different Higgs doublet the model is said 2HDM-II.

In the following, we shall assume a scenario where the type-II 2HDM structure is not protected by any symmetry and is broken by loop effects (this occurs, for instance, in the MSSM).

Let us consider the Yukawa interactions for charged leptons, including the radiatively induced LFV terms [806]:

$$-\mathcal{L} \simeq \bar{l}_{Ri} Y_{l_i} H_1 \overline{L_i} + \bar{l}_{Ri} \left(Y_{l_i} \Delta_L^{ij} + Y_{l_j} \Delta_R^{ij} \right) H_2 \overline{L_j} + \text{h.c.}, \quad (5.94)$$

where H_1 and H_2 are the scalar doublets, l_{Ri} are lepton singlet for right handed fermions, L_k denote the lepton doublets and Y_{l_k} are the Yukawa couplings.

In the mass eigenstate basis for both leptons and Higgs bosons, the effective flavor violating interactions are described by the four dimension operators [806]:

$$-\mathcal{L} \simeq (2G_F^2)^{\frac{1}{4}} \frac{m_{l_i}}{c_\beta^2} \left(\Delta_L^{ij} \bar{l}_R^i l_L^j + \Delta_R^{ij} \bar{l}_L^i l_R^j \right)$$
$$\times \left(c_{\beta-\alpha} h^0 - s_{\beta-\alpha} H^0 - i A^0 \right)$$
$$+ (8G_F^2)^{\frac{1}{4}} \frac{m_{l_i}}{c_\beta^2} \left(\Delta_L^{ij} \bar{l}_R^i \nu_L^j + \Delta_R^{ij} \nu_L^i \bar{l}_R^j \right) H^\pm + \text{h.c.}, \quad (5.95)$$

where α is the mixing angle between the CP-even Higgs bosons h_0 and H_0, A_0 is the physical CP-odd boson, $H^\pm$ are the physical charged Higgs-bosons and t_β is the ratio of the vacuum expectation value for the two Higgs (where we adopt the notation, $c_x, s_x = \cos x, \sin x$ and $t_x = \tan x$). Irrespective to the mechanism of the high energy theories generating the LFV, we treat the $\Delta_{L,R}^{ij}$ terms in a model independent way. In order to constrain the $\Delta_{L,R}^{ij}$ parameters, we impose that their contributions to LFV processes do not exceed the experimental bounds [175, 650].

On the other hand, there are several models with a specific ansatz about the flavor-changing couplings. For instance, the famous multi-Higgs-doublet models proposed by Cheng and Sher [596] predict that the LFV couplings of all the neutral Higgs bosons with the fermions have the form $Hf_if_j \sim \sqrt{m_im_j}$.

In supersymmetry, the Δ^{ij} terms are induced at one loop level by the exchange of gauginos and sleptons, provided a source of slepton mixing. In the so mass insertion (MI) approximation, the expressions of $\Delta_{L,R}^{ij}$ are given by

$$\Delta_L^{ij} = -\frac{\alpha_1}{4\pi}\mu M_1 \delta_{LL}^{ij} m_L^2$$
$$\times \left[I'(M_1^2, m_R^2, m_L^2) + \frac{1}{2} I'(M_1^2, \mu^2, m_L^2) \right]$$
$$+ \frac{3}{2}\frac{\alpha_2}{4\pi}\mu M_2 \delta_{LL}^{ij} m_L^2 I'(M_2^2, \mu^2, m_L^2), \quad (5.96)$$

$$\Delta_R^{ij} = \frac{\alpha_1}{4\pi}\mu M_1 m_R^2 \delta_{RR}^{ij}\left[I'(M_1^2, \mu^2, m_R^2) - (\mu \leftrightarrow m_L) \right], \quad (5.97)$$

respectively, where μ is the Higgs mixing parameter, $M_{1,2}$ are the gaugino masses and $m_{L(R)}^2$ stands for the left–left (right–right) slepton mass matrix entry. The LFV mass insertions (MIs), i.e. $\delta_{XX}^{3\ell} = (\tilde{m}_\ell^2)_{XX}^{3\ell}/m_X^2$ ($X = L, R$), are the off-diagonal flavor changing entries of the slepton mass matrix. The loop function $I'(x, y, z)$ is such that $I'(x, y, z) = dI(x, y, z)/dz$, where $I(x, y, z)$ refers to the standard three point one loop integral which has mass dimension-2

$$I_3(x, y, z) = \frac{xy \log(x/y) + yz \log(y/z) + zx \log(z/x)}{(x-y)(z-y)(z-x)}. \quad (5.98)$$

The above expressions, i.e. (5.96), (5.97), depend only on the ratio of the SUSY mass scales and they do not decouple for large m_{SUSY}. As first shown in Ref. [174], both Δ_R^{ij} and Δ_L^{ij} couplings suffer from strong cancellations in certain regions of the parameter space due to destructive interferences among various contributions. For instance, from (5.97) it is clear that, in the Δ_R^{ij} case, such cancellations happen if $\mu = m_L$.

In the SUSY see-saw model, in the mass insertion approximation, one obtains specific values for δ_{LL}^{ij} depending on the assumptions on the flavor mixing in Y_ν [164, 707]. If the latter is of CKM size, $\delta_{LL}^{21(31)} \simeq 3 \times 10^{-5}$ and $\delta_{LL}^{32} \simeq 10^{-2}$, while in the case of the observed neutrino mixing, taking $U_{e3} = 0.07$ at about half of the current CHOOZ bound, we get $\delta_{LL}^{21(31)} \simeq 10^{-2}$ and $\delta_{LL}^{32} \simeq 10^{-1}$.

5.5.2 Phenomenology

In order to constrain the $\Delta_{L,R}^{ij}$ parameters, we impose that their contributions to LFV processes as $l_i \to l_j l_k l_k$ and $l_i \to l_j \gamma$ do not exceed the experimental bounds. At tree level, Higgs exchange contribute only to $\ell_i \to \ell_j \ell_k \ell_k$, $\tau \to \ell_j \eta$ and $\mu N \to eN$. On the other hand, a one loop Higgs exchange leads to the LFV radiative decays $\ell_i \to \ell_j \gamma$. In the following, we report the expression for the branching ratios of the above processes.

5.5.2.1 $\ell_i \to \ell_j \gamma$
The $\ell_i \to \ell_j \gamma$ process can be generated by the one loop exchange of Higgs and leptons. How-

ever, the dipole transition implies three chirality flips: two in the Yukawa vertices and one in the lepton propagator. This strong suppression can be overcome at higher order level. Going to two loop level, one has to pay the typical price of $g^2/16\pi^2$ but one can replace the light fermion masses from Yukawa vertices with the heavy fermion (boson) masses circulating in the second loop. In this case, the virtual Higgs boson couples only once to the lepton line, inducing the needed chirality flip. As a result, the two loop amplitude can provide the major effects. Naively, the ratio between the two loop fermionic amplitude and the one loop amplitude is

$$\frac{A_{l_i \to l_j \gamma}^{(2\text{-loop})f}}{A_{l_i \to l_j \gamma}^{1\text{-loop}}} \sim \frac{\alpha_{\text{em}}}{4\pi} \frac{m_f^2}{m_{l_i}^2} \log\left(\frac{m_f^2}{m_H^2}\right),$$

where $m_f = m_b, m_\tau$ is the mass of the heavy fermion circulating in the loop. We remind that in a Model II 2HDM (as SUSY) the Yukawa couplings between neutral Higgs bosons and quarks are $H\bar{t}t \sim m_t/\tan\beta$ and $H\bar{b}b \sim m_b \tan\beta$. Since the Higgs mediated LFV is relevant only at large $\tan\beta \geq 30$, it is clear that the main contributions arise from the τ and b fermions and not from the top quark. So, in this framework, $\tau \to l_j \gamma$ does not receive sizable two loop effects by heavy fermionic loops, contrary to the $\mu \to e\gamma$ case.

However, the situation can drastically change when a W boson circulates in the two loop Barr–Zee diagrams. Bearing in mind that $HW^+W^- \sim m_W$ and that pseudoscalar bosons do not couple to a W pair, it turns out that $A_{l_i \to l_j \gamma}^{(2\text{-loop})W}/A_{l_i \to l_j \gamma}^{(2\text{-loop})f} \sim m_W^2/(m_f^2 \tan\beta)$; thus, two loop W effects are expected to dominate, as it is confirmed numerically [650, 808].

As final result, the following approximate expression holds [175, 650]:

$$\frac{B(\ell_i \to \ell_j \gamma)}{B(\ell_i \to \ell_j \bar{\nu}_j \nu_\tau)}$$
$$\simeq \frac{3}{2}\frac{\alpha_{el}}{\pi}\left(\frac{m_{\ell_i}^2}{m_A^2}\right)^2 t_\beta^6 \Delta_{ij}^2 \left\{ \frac{\delta m}{m_A} \log \frac{m_{\ell_i}^2}{m_A^2} + \frac{1}{6} \right.$$
$$+ \frac{\alpha_{el}}{\pi}\left[\frac{m_W^2}{m_{\ell_i}^2}\frac{F(a_W)}{t_\beta} - \sum_{f=b,\tau} N_f q_f^2 \frac{m_f^2}{m_{\ell_i}^2}\left(\log\frac{m_f^2}{m_{\ell_i}^2} + 2\right)\right.$$
$$- \frac{N_c}{4}\left(q_{\tilde{t}}^2 \frac{m_t \mu}{t_\beta m_{\ell_i}^2} s_{2\theta_{\tilde{t}}} h(x_{\tilde{t}H})\right.$$
$$\left.\left.\left.- q_{\tilde{b}}^2 \frac{m_b A_b}{m_{\ell_i}^2} s_{2\theta_{\tilde{b}}} h(x_{\tilde{b}H})\right)\right]\right\}^2$$
$$\simeq \frac{3}{2}\frac{\alpha_{el}^3}{\pi^3} \Delta_{21}^2 t_\beta^4 \left(\frac{m_W^4}{M_H^4}\right)(F(a_W))^2, \quad (5.99)$$

where $\delta m = (m_H - m_A) \sim \mathcal{O}(m_Z^2/m_{A^0})$. The terms of the first row of (5.99) refer to one loop effects and their role is non-negligible only in τ decays. It turns out that pseudoscalar and scalar one loop amplitudes have opposite signs, so, as we have $m_A \simeq m_H$, they cancel each other to a very large extent. Since these cancellations occur, two loop effects can become important or even dominant. The two terms of the second row of (5.99) refer to two loop Barr–Zee effects induced by W and fermionic loops, respectively, while the last row of (5.99) is relative two loop Barr–Zee effects with a squark loop in the second loop. As regards the squark loop effects, it is very easy to realize that they are negligible compared to W effects. In fact, it is well known that Higgs mediated LFV can play a relevant or even a dominant role compared to gaugino mediated LFV provided that slepton masses are not below the TeV scale while maintaining the Higgs masses at the electroweak scale (and assuming large t_β values). In this context, it is natural to assume squark masses at least of the same order as the slepton masses (at the TeV scale). So, in the limit where $x_{\tilde{f}H} = m_{\tilde{f}}^2/m_H^2 \gg 1$, the loop function $h(x_{\tilde{f}H})$ is such that $(\log x_{\tilde{f}H} + 5/3)/6x_{\tilde{f}H}$ thus, even for maximum squark mixing angles $\theta_{\tilde{t},\tilde{b}}$, namely for $s_{2\theta_{\tilde{t},\tilde{b}}} = \sin 2\theta_{\tilde{t},\tilde{b}} \simeq 1$, and large A_b and μ terms, two loop squark effects remain much below the W effects, as it is straightforward to check by (5.99).

As a final result the main two loop effects are provided by the exchange of a W boson, with the loop function $F(a_W) \sim \frac{35}{16}(\log a_W)^2$ for $a_W = m_W^2/m_H^2 \ll 1$. It is noteworthy that one and two loop amplitudes have the same signs. In addition, two loops effects dominate in large portions of the parameter space, specially for large m_H values, where the mass splitting $\delta m = m_H - m_A$ decreases to zero.

5.5.2.2 $\ell_i \to \ell_j \ell_k \ell_k$

The $l_i \to l_j l_k l_k$ process can be mediated by a tree level Higgs exchange [806, 807]. However, up to one loop level, $l_i \to l_j l_k l_k$ gets additional contributions induced by $l_i \to l_j \gamma^*$ amplitudes [175, 650]. It is worth noting that the Higgs mediated monopole (chirality conserving) and dipole (chirality violating) amplitudes have the same $\tan^3 \beta$ dependence. This has to be contrasted to the non-Higgs contributions. For instance, within SUSY, the gaugino mediated dipole amplitude is proportional to $\tan \beta$ while the monopole amplitude is $\tan \beta$ independent. The expression for the Higgs mediated $l_i \to l_j l_k l_k$ can be approximated in the following way [175, 650]:

$$\frac{B(\tau \to l_j l_k l_k)}{B(\tau \to l_j \bar{\nu}_j \nu_\tau)}$$
$$\simeq \frac{m_\tau^2 m_{l_k}^2}{32 m_A^4} \Delta_{\tau j}^2 \tan^6 \beta [3 + 5\delta_{jk}]$$
$$+ \frac{\alpha_{el}}{3\pi}\left(\log \frac{m_\tau^2}{m_{l_k}^2} - 3\right) \frac{B(\tau \to l_j \gamma)}{B(\tau \to l_j \bar{\nu}_j \nu_\tau)}, \quad (5.100)$$

where we have disregarded subleading monopole effects.

5.5.2.3 $\mu N \to eN$

The $\mu \to e$ conversion in Nuclei process can be generated by a scalar operator through the tree level Higgs exchange [811]. Moreover, at one loop level, additional contributions induced by $l_i \to l_j \gamma^*$ amplitudes arise [175]; however they are subleading [175]. Finally, the following expression for $B(\mu Al \to eAl)$ is derived [811]:

$$B(\mu Al \to eAl) \simeq 1.8 \times 10^{-4} \frac{m_\mu^7 m_p^2}{v^4 m_h^4 \omega_{\text{capt}}^{Al}} \Delta_{21}^2 t_\beta^6, \quad (5.101)$$

where $\omega_{\text{capt}}^{Al} \simeq 0.7054 \times 10^6 \text{ s}^{-1}$. We observe that $B(\mu \to 3e)$ is completely dominated by the photonic $\mu \to e\gamma^*$ dipole amplitude so that $B(\mu \to eee) \simeq \alpha_{\text{em}} B(\mu \to e\gamma)$. On the other hand, tree level Higgs mediated contributions are negligible because suppressed by the electron mass through the $H(A)\bar{e}e \sim m_e$ coupling. On the contrary, $\mu N \to eN$ is not suppressed by the light constituent quark m_u and m_d but only by the nucleon masses, because the Higgs-boson coupling to the nucleon is shown to be characterized by the nucleon mass using the conformal anomaly relation [811]. In particular, the most important contribution turns out to come from the exchange of the scalar Higgs boson H which couples to the strange quark [811].

In fact, the coherent μ–e conversion process, where the initial and final nuclei are in the ground state, is expected to be enhanced by a factor of $O(Z)$ (where Z is the atomic number) compared to incoherent transition processes. Since the initial and final states are the same, the elements $\langle N|\bar{p}p|N\rangle$ and $\langle N|\bar{n}n|N\rangle$ are nothing but the proton and the neutron densities in a nucleus in the non-relativistic limit of nucleons. In this limit, the other matrix elements $\langle N|\bar{p}\gamma_5 p|N\rangle$ and $\langle N|\bar{n}\gamma_5 n|N\rangle$ vanish. Therefore, in the coherent μ–e conversion process, the dominant contributions come from the exchange of H, not A [811].

Moreover, we know that $\mu \to e\gamma^*$ (chirality conserving) monopole amplitudes are generally subdominant compared to (chirality flipping) dipole effects [175]. Note also that, the enhancement mechanism induced by Barr–Zee type diagrams is effective only for chirality flipping operators so, in the following, we shall disregard chirality conserving one loop effects.

5.5.2.4 $\tau \to \mu P$ ($P = \pi, \eta, \eta'$)

Now we consider the implications of virtual Higgs exchange for the decays $\tau \to \mu P$, where P is a neutral pseudoscalar meson ($P = \pi, \eta, \eta'$) [169, 810]. Since we assume CP conservation in the Higgs sector, only the exchange of the A Higgs boson is relevant. Moreover, in the large $\tan \beta$ limit, only the A couplings to down-type quarks are important. These can be written as

$$-i(\sqrt{2}G_{\text{F}})^{1/2} \tan \beta A(\xi_d m_d \bar{d}_R d_L + \xi_s m_s \bar{s}_R s_L$$
$$+ \xi_b m_b \bar{b}_R b_L) + \text{h.c.} \quad (5.102)$$

The parameters ξ_d, ξ_s, ξ_b are equal to one at tree level, but they can significantly deviate from this value because of higher order corrections proportional to $\tan\beta$ [537, 805], generated by integrating out superpartners. In the limit of quark flavor conservation, each ξ_q ($q = d, s, b$) has the form $\xi_q = (1 + \Delta_q \tan\beta)^{-1}$, where Δ_q appears in the loop-generated term $-h_q \Delta_q H_2^{0*} q^c q$ + h.c. [537, 805]. At energies below the bottom mass, the b-quark can be integrated out so the bilinear $-im_b b^c b$ + h.c. is effectively replaced by the gluon operator $\Omega = \frac{g_s^2}{64\pi^2}\epsilon^{\mu\nu\rho\sigma} G^a_{\mu\nu} G^a_{\rho\sigma}$, where g_s and $G^a_{\mu\nu}$ are the $SU(3)_C$ coupling constant and field strength, respectively [169]. In the limit in which the processes $\tau \to 3\mu$ and $\tau \to \mu\eta$ are both dominated by Higgs exchange, these decays are related as [169]:

$$\frac{B(\tau \to l_j \eta)}{B(\tau \to l_j \bar{\nu}_j \nu_\tau)} \simeq 9\pi^2 \left(\frac{f_\eta^8 m_\eta^2}{m_A^2 m_\tau}\right)^2 \left(1 - \frac{m_\eta^2}{m_\tau^2}\right)^2$$

$$\times \left[\xi_s + \frac{\xi_b}{3}\left(1 + \sqrt{2}\frac{f_\eta^0}{f_\eta^8}\right)\right]^2 \Delta_{3j}^2 \tan^6\beta,$$

where $m_\eta^2/m_\tau^2 \simeq 9.5 \times 10^{-2}$ and the relevant decay constants are $f_\eta^0 \sim 0.2 f_\pi$, $f_\eta^8 \sim 1.2 f_\pi$ and $f_\pi \sim 92$ MeV. In the above expression, both the contribution of the (bottom-loop induced) gluon operator Ω and the factors ξ_q were included.

For $\xi_s \sim \xi_b \sim 1$, it turns out that $B(\tau^- \to \mu^- \eta)/B(\tau^- \to \mu^- \mu^+ \mu^-) \simeq 5$, but it could also be a few times larger or smaller than that, depending on the actual values of ξ_s, ξ_b. Finally, let us compare $\tau \to \mu\eta'$ and $\tau \to \mu\pi$ with $\tau \to \mu\eta$ in the limit of Higgs exchange domination. Both ratios are suppressed, although for different reasons. The ratio $B(\tau \to \mu\pi)/B(\tau \to \mu\eta)$ is small because it is parametrically suppressed by $m_\pi^4/m_\eta^4 \sim 10^{-2}$. The ratio $B(\tau \to \mu\eta')/B(\tau \to \mu\eta)$, which seems to be $\mathcal{O}(1)$, is much smaller because the singlet and octet contributions to $\tau \to \mu\eta'$ tend to cancel against each other [169].

These results, combined with the present bound on $\tau \to \mu\eta$, imply that the Higgs mediated contribution to $B(\tau \to \mu\eta')$ and $B(\tau \to \mu\pi)$ can reach $\mathcal{O}(10^{-9})$ [169].

5.5.2.5 *Higgs* $\to \mu\tau$ The LFV Higgs $\to \mu\tau$ decays and the related phenomenology have been extensively investigated in [174]. Concerning the Higgs boson decays, we have [174]

$$B(A \to \mu^+ \tau^-) = \tan^2\beta(|\Delta_L|^2 + |\Delta_R|^2) B(A \to \tau^+ \tau^-),$$
(5.103)

where we have approximated $1/c_\beta^2 \simeq \tan^2\beta$ since non-negligible effects can only arise in the large $\tan\beta$ limit. If A is replaced with H [or h] in (5.103), the r.h.s. should also be multiplied by a factor $(c_{\beta-\alpha}/s_\alpha)^2$ [or $(s_{\beta-\alpha}/c_\alpha)^2$]. We recall that $B(A \to \mu\tau)$ can reach values of order 10^{-4}. The same holds for the 'non-standard' CP-even Higgs boson (either H or h, depending on m_A).

We now make contact with the physical observable, i.e. the $B(\Phi^0 \to \mu^+ \tau^-)$, and discuss the phenomenological implications. We outline some general features of $B(\Phi^0 \to \mu^+ \tau^-)$ at large $\tan\beta$ and the prospects for these decay channels at the Large Hadron Collider (LHC) and other colliders. Let us discuss the different Higgs bosons, as reported in [174], assuming for definiteness $\tan\beta \sim 50$, $|50\Delta|^2 \sim 10^{-3}$ ($\Delta = \Delta_L$ or Δ_R) and an integrated luminosity of 100 fb^{-1} at LHC.

If Φ^0 denotes one of the 'non-standard' Higgs bosons, we have $C_\Phi \simeq 1$ and $B(\Phi^0 \to \tau^+ \tau^-) \sim 10^{-1}$, so $B(\Phi^0 \to \mu^+ \tau^-) \sim 10^{-4}$. The main production mechanisms at LHC are bottom-loop mediated gluon fusion and associated production with $b\bar{b}$, which yield cross sections $\sigma \sim (10^3, 10^2, 20)$ pb for $m_A \sim (100, 200, 300)$ GeV, respectively. The corresponding numbers of $\Phi^0 \to \mu^+ \tau^-$ events are about $(10^4, 10^3, 2 \times 10^2)$. These estimates do not change much if the bottom Yukawa coupling Y_b is enhanced (suppressed) by radiative corrections, since in this case the enhancement (suppression) of σ would be roughly compensated by the suppression (enhancement) of $B(\Phi^0 \to \mu^+ \tau^-)$.

If Φ^0 denotes the other (more 'standard model-like') Higgs boson, the factor $C_\Phi \cdot B(\Phi^0 \to \tau^+ \tau^-)$ strongly depends on m_A, while the production cross section at LHC, which is dominated by top-loop mediated gluon fusion, is $\sigma \sim 30$ pb. For $m_A \sim 100$ GeV we may have $C_\Phi \cdot B(\Phi^0 \to \tau^+ \tau^-) \sim 10^{-1}$ and $B(\Phi^0 \to \mu^+ \tau^-) \sim 10^{-4}$, which would imply ~ 300 $\mu^+ \tau^-$ events. The number of events is generically smaller for large m_A since C_Φ scales as $1/m_A^4$, consistently with the expected decoupling of LFV effects for such a Higgs boson.

The above discussion suggests that LHC may offer good chances to detect the decays $\Phi^0 \to \mu\tau$, especially in the case of non-standard Higgs bosons. This indication should be supported by a detailed study of the background. At Tevatron the sensitivity is lower than at LHC because both the expected luminosity and the Higgs production cross sections are smaller. The number of events would be smaller by a factor 10^2–10^3. A few events may be expected also at e^+e^- or $\mu^+\mu^-$ future colliders, assuming integrated luminosities of 500 and 1 fb^{-1}, respectively. At a $\mu^+\mu^-$ collider an enhancement may occur for the non-standard Higgs bosons if radiative corrections strongly suppress Y_b, since in this case both the resonant production cross section [$\sigma \sim (4\pi/m_A^2) B(\Phi^0 \to \mu^+\mu^-)$] and the LFV branching ratios $B(\Phi^0 \to \mu^+ \tau^-)$ would be enhanced. As a result, for light m_A, hundreds of $\mu^+ \tau^-$ events could occur.

5.5.2.6 $\mu N \to \tau X$ Higgs mediated LFV effects can have also relevant impact on the cross section of the $\mu N \to \tau X$

reaction [812]. The contribution of the Higgs boson mediation to the differential cross section $\mu^- N \to \tau^- X$ is given by [812]

$$\frac{d^2\sigma}{dx\,dy} = \sum_q x f_q(x) \left\{ |\mathcal{C}_L|_q^2 \left(\frac{1-\mathcal{P}_\mu}{2}\right) + |\mathcal{C}_R|_q^2 \left(\frac{1+\mathcal{P}_\mu}{2}\right) \right\}$$
$$\times \frac{s}{8\pi} y^2, \quad (5.104)$$

where the function $f_q(x)$ is the PDF for q-quarks, $\mathcal{P}_\mu$ is the incident muon polarization such that $\mathcal{P}_\mu = +1$ and -1 correspond to the right and left handed polarization, respectively, and s is the centre-of-mass (CM) energy. The parameters x and y are defined as $x \equiv Q^2/2P \cdot q$, $y \equiv 2P \cdot q/s$, in the limit of massless tau leptons, where P is the four momentum of the target, q is the momentum transfer, and Q is defined as $Q^2 \equiv -q^2$. As seen in (5.104), experimentally, the form factors of $\mathcal{C}_L^{hH}$ and $\mathcal{C}_L^A$ ($\mathcal{C}_R^{hH}$ and $\mathcal{C}_R^A$) can be selectively studied by using purely left handed (right handed) incident muons. In SUSY models such as the MSSM with heavy right handed neutrinos, LFV is radiatively induced due to the left handed slepton mixing, which only affects $\mathcal{C}_L^{hH}$ and $\mathcal{C}_L^A$. Therefore, in the following, we focus only on those $\mathcal{C}_L^{hH}$ and $\mathcal{C}_L^A$ couplings.

The magnitudes of the effective couplings are constrained by the current experimental results of searches for LFV processes of tau decays. Therefore, both couplings are determined by the one that is more constrained, namely the pseudo-scalar coupling. It is constrained by the $\tau \to \mu \eta$ decay ($B(\tau \to \mu \eta) < 3.4 \times 10^{-7}$). Then the constraint is given on the s-associated scalar and pseudo-scalar couplings by

$$\left(|\mathcal{C}_L^A|^2 \right)_s \leq 10^{-9} \left[\text{GeV}^{-4} \right] \times B(\tau \to \mu \eta). \quad (5.105)$$

The largest values of $\mathcal{C}_L^{hH}$ and $\mathcal{C}_L^A$ can be realized with $m_{\text{SUSY}} \sim \mathcal{O}(1)$ TeV and the higgsino mass $\mu \sim \mathcal{O}(10)$ TeV [169, 810].

The cross sections of the $\mu N \to \tau X$ reaction in the DIS region is evaluated for the maximally allowed values of the effective couplings as a reference. They are plotted in Fig. 36 for different quark contributions as a function of the muon beam energy in the laboratory frame. For the PDF, CTEQ6L has been used. The target N is assumed to be a proton. For a nucleus target, the cross section would be higher, approximately by the number of nucleons in the target. The cross section sharply increases above $E_\mu \sim 50$ GeV in Fig. 36. This enhancement comes from the b-quark contribution in addition to the d- and s-quark contributions which is enhanced by a factor of m_b/m_s over the s-quark contribution. The cross section is enhanced by one order of magnitude when the muon energy changes from 50 to 100 GeV. Typically, for $E_\mu = 100$ GeV and $E_\mu = 300$ GeV, the cross section is 10^{-4} and 10^{-3} fb, respectively. With an intensity of

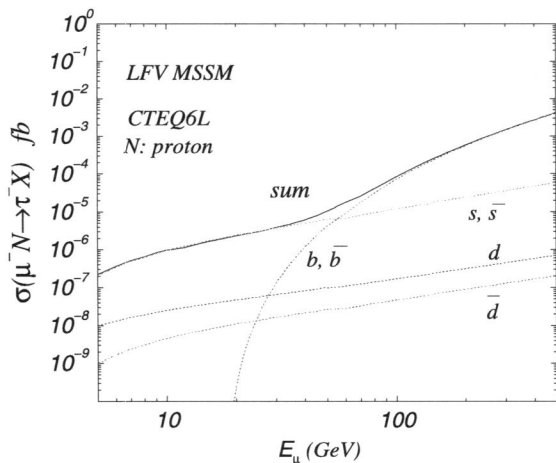

Fig. 36 Cross section of the $\mu^- N \to \tau^- X$ DIS process as a function of the muon energy for the Higgs mediated interaction [812]. It is assumed that the initial muons are purely left handed. CTEQ6L is used for the PDF

10^{20} muons per year and a target mass of 100 g/cm^2, about 10^4 (10^2) events could be expected for $\sigma(\mu N \to \tau X) = 10^{-3}$ (10^{-5}) fb, which corresponds to $E_\mu = 300$ (50) GeV from Fig. 36. This would provide good potential to improve the sensitivity by four (two) orders of magnitude from the present limit from $\tau \to \mu \eta$ decay, respectively. Such a muon intensity could be available at a future muon collider and a neutrino factory.

5.5.3 Correlations

The numerical results shown in Figs. 37 and 38 allow us to draw several observations [175, 650].

- $\tau \to l_j \gamma$ has the largest branching ratios except for a region around $m_H \sim 700$ GeV where strong cancellations among two loop effects reduce their size.[25] The following approximate relations are found:

$$\frac{B(\tau \to l_j \gamma)}{B(\tau \to l_j \eta)}$$
$$\simeq \left(\frac{\delta m}{m_A} \log \frac{m_\tau^2}{m_A^2} + \frac{1}{6} + \frac{\alpha_{el}}{\pi} \left(\frac{m_W^2}{m_\tau^2} \right) \frac{F(a_W)}{\tan \beta} \right)^2$$
$$\geq 1,$$

where the last relation is easily obtained by using the approximation for $F(z)$. If two loop effects were disregarded, then we would obtain $B(\tau \to l_j \gamma)/B(\tau \to l_j \eta) \in (1/36, 1)$ for $\delta m/m_A \in (0, 10\%)$. Two loop con-

---

[25]For a detailed discussion about the origin of these cancellations and their connection with non-decoupling properties of two loop W amplitude, see Ref. [808].

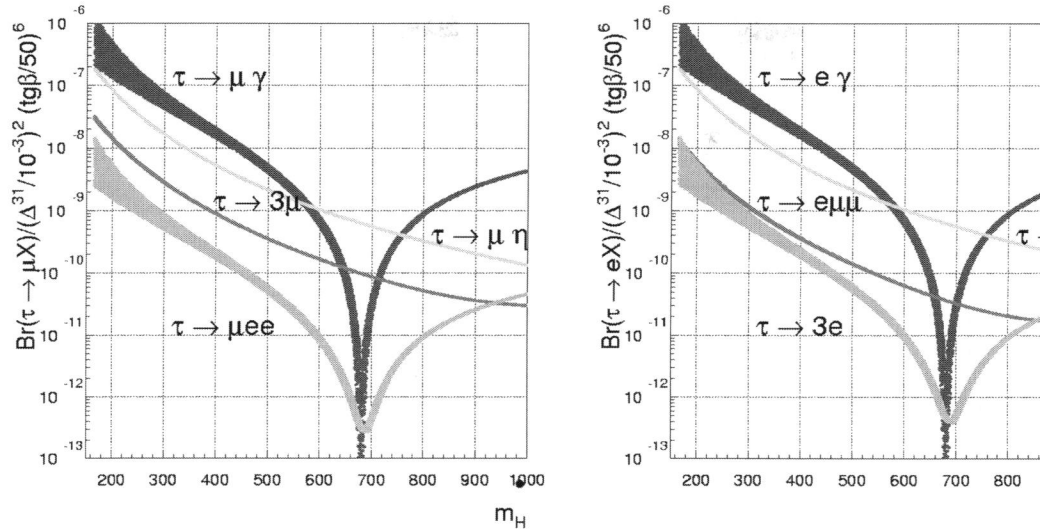

Fig. 37 Branching ratios of various $\tau \to \mu$ and $\tau \to e$ LFV processes versus the Higgs boson mass m_H in the decoupling limit as reported in [650]. $X = \gamma, \mu\mu, ee, \eta$

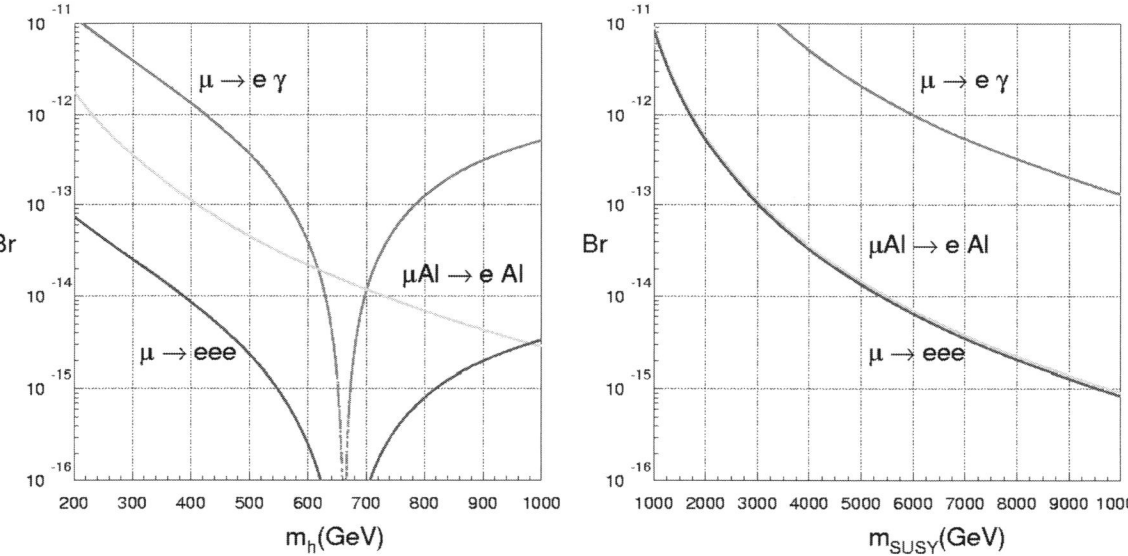

Fig. 38 *Left*: branching ratios of $\mu \to e\gamma$, $\mu \to eee$ and $\mu Al \to eAl$ in the Higgs mediated LFV case versus the Higgs boson mass m_h [175]. *Right*: branching ratios of $\mu \to e\gamma$, $\mu \to eee$ and $\mu Al \to eAl$ in the gaugino mediated LFV case versus a common SUSY mass m_{SUSY} [175]. In the figure we set $t_\beta = 50$ and $\delta^{21}_{LL} = 10^{-2}$

tributions significantly enhance $B(\tau \to l_j\gamma)$ specially for $\delta m/m_A \to 0$.

- In Fig. 37 non-negligible mass splitting $\delta m/m_A$ effects can be visible at low m_H regime through the bands of the $\tau \to l_j\gamma$ and $\tau \to l_j ee$ processes. These effects tend to vanish with increasing m_H as is correctly reproduced in Fig. 37. $\tau \to l_j\mu\mu$ does not receive visible effects by $\delta m/m_A$ terms being dominated by the tree level Higgs exchange.
- As is shown in Fig. 37 $B(\tau \to l_j\gamma)$ is generally larger than $B(\tau \to l_j\mu\mu)$; their ratio is regulated by the following approximate relation:

$$\frac{B(\tau \to l_j\gamma)}{B(\tau \to l_j\mu\mu)} \simeq \frac{36}{3+5\delta_{j\mu}} \frac{B(\tau \to l_j\gamma)}{B(\tau \to l_j\eta)} \geq \frac{36}{3+5\delta_{j\mu}},$$

where the last relation is valid only out of the cancellation region. Moreover, from the above relation it turns out that

$$\frac{B(\tau \to l_j\eta)}{B(\tau \to l_j\mu\mu)} \simeq \frac{36}{3+5\delta_{j\mu}}.$$

If we relax the condition $\xi_{s,b} = 1$, $B(\tau \to l_j \eta)$ can get values few times smaller or bigger than those in Fig. 37.

- It is noteworthy that a tree level Higgs exchange predicts that $B(\tau \to l_j ee)/B(\tau \to l_j \mu\mu) \sim m_e^2/m_\mu^2$ while, at two loop level, we obtain (out of the cancellation region):

$$\frac{B(\tau \to l_j ee)}{B(\tau \to l_j \mu\mu)} \simeq \frac{0.4}{3 + 5\delta_{j\mu}} \frac{B(\tau \to l_j \gamma)}{B(\tau \to l_j \eta)} \geq \frac{0.4}{3 + 5\delta_{j\mu}}.$$

Let us underline that, in the cancellation region, the lower bound of $B(\tau \to l_j ee)$ is given by the monopole contributions. So, in this region, $B(\tau \to l_j ee)$ is much less suppressed than $B(\tau \to l_j \gamma)$.

- The approximate relations among $\mu Al \to eAl$, $\mu \to e\gamma$ and $\mu \to eee$ branching ratios are

$$\frac{B(\mu \to e\gamma)}{B(\mu Al \to eAl)} \simeq 10^2 \left(\frac{F(a_W)}{\tan\beta}\right)^2,$$

$$\frac{B(\mu \to eee)}{B(\mu \to e\gamma)} \simeq \alpha_{el}. \quad (5.106)$$

In the above equations we retained only dominant two loop effects arising from W exchange. The exact behavior for the examined processes is reported in Fig. 38 where we can see that $\mu \to e\gamma$ gets the largest branching ratio except for a region around $m_H \sim 700$ GeV where strong cancellations among two loop effects sink its size.

The correlations among the rates of the above processes are an important signature of the Higgs-mediated LFV and allow us to discriminate between different SUSY scenarios. In fact, it is well known that, in a supersymmetric framework, besides the Higgs mediated LFV transitions, we have also LFV effects mediated by the gauginos through loops of neutralinos (charginos)–charged sleptons (sneutrinos). On the other hand, the above contributions have different decoupling properties regulated by the mass of the heaviest scalar mass (m_H) or by the heaviest mass in the slepton gaugino loops (m_{SUSY}). In principle, the m_{SUSY} and m_H masses may be unrelated, so we can always proceed by considering only the Higgs mediated effects (assuming a relatively light m_H and an heavy m_{SUSY}) or only the gaugino mediated contributions (if m_H is heavy). In the following, we are interested to make a comparison between Higgs and gaugino mediated LFV effects. In order to make the comparison as simple as possible, let us consider the simple case where all the SUSY particles are degenerate. In this case, it turns out that

$$\Delta_L^{21} \sim \frac{\alpha_2}{24\pi} \delta_{LL}^{21},$$

$$\left.\frac{B(\ell_i \to \ell_j \gamma)}{B(\ell_i \to \ell_j \bar{\nu}_j \nu_i)}\right|_{\text{Gauge}}$$

$$= \frac{2\alpha_{el}}{75\pi}\left(1 + \frac{5}{4}\tan^2\theta_W\right)^2 \left(\frac{m_W^4}{m_{SUSY}^4}\right) |\delta_{LL}^{ij}|^2 t_\beta^2,$$

$$\left.\frac{B(\ell_i \to \ell_j \gamma)}{B(\ell_i \to \ell_j \bar{\nu}_j \nu_i)}\right|_{\text{Higgs}}$$

$$\simeq 10 \frac{\alpha_{el}^3}{\pi^3}\left(\frac{\alpha_2}{24\pi}\right)^2 \left(\frac{m_W^4}{M_H^4}\right)\left(\log\frac{m_W^2}{M_H^2}\right)^4 |\delta_{LL}^{ij}|^2 t_\beta^4. \quad (5.107)$$

In Fig. 38 we report the branching ratios of the examined processes as a function of the heaviest Higgs boson mass m_H (in the Higgs LFV mediated case) or of the common SUSY mass m_{SUSY} (in the gaugino LFV mediated case). We set $t_\beta = 50$ and we consider the PMNS scenario as discussed above so that $(\delta_{LL}^{21})_{\text{PMNS}} \simeq 10^{-2}$. Sub-leading contributions proportional to $(\delta_{LL(RR)}^{23}\delta_{RR(LL)}^{31})_{\text{PMNS}}$ were neglected since, in the PMNS scenario, it turns out that $(\delta_{LL(RR)}^{23}\delta_{RR(LL)}^{31})_{\text{PMNS}}/(\delta_{LL}^{21})_{\text{PMNS}} \simeq 10^{-3}$ [707]. As we can see from Fig. 38, Higgs mediated effects start being competitive with the gaugino mediated ones when m_{SUSY} is roughly one order of magnitude larger than the Higgs mass m_H. Moreover, we stress that, both in the gaugino and in the Higgs mediated cases, $\mu \to e\gamma$ gets the largest effects. In particular, within the PMNS scenario, it turns out that Higgs mediated $B(\mu \to e\gamma) \sim 10^{-11}$ when $m_H \sim 200$ GeV and $t_\beta = 50$, that is just closed to the present experimental resolution.

The correlations among different processes predicted in the gaugino mediated case are different from those predicted in the Higgs mediated case. For instance, in the gaugino mediated scenario, $B(\tau \to l_j l_k l_k)$ gets the largest contributions by the dipole amplitudes that are $\tan\beta$ enhanced with respect to all other amplitudes resulting in a precise ratio with $B(\tau \to l_j \gamma)$, namely

$$\left.\frac{B(\ell_i \to \ell_j \ell_k \ell_k)}{B(\ell_i \to \ell_j \gamma)}\right|_{\text{Gauge}} \simeq \frac{\alpha_{el}}{3\pi}\left(\log\frac{m_\tau^2}{m_{l_k}^2} - 3\right) \simeq \alpha_{el}, \quad (5.108)$$

$$\left.\frac{B(\tau \to \ell_j ee)}{B(\tau \to \ell_j \mu\mu)}\right|_{\text{Gauge}} \simeq \frac{\log\frac{m_\tau^2}{m_e^2} - 3}{\log\frac{m_\tau^2}{m_\mu^2} - 3} \simeq 5. \quad (5.109)$$

Moreover, in the large $\tan\beta$ regime, one can find the simple theoretical relations

$$\left.\frac{B(\mu - e \text{ in Ti})}{B(\mu \to e\gamma)}\right|_{\text{Gauge}} \simeq \alpha_{el}. \quad (5.110)$$

If some ratios different from the above were discovered, then this would be clear evidence that some new process is generating the $\ell_i \to l_j$ transition, with Higgs mediation being a leading candidate.

5.5.4 Conclusions

We have reviewed the allowed rates for Higgs-mediated LFV decays in a SUSY framework. In particular, we have analyzed the decay modes of the τ, μ lepton, namely $\ell_i \to \ell_j \ell_k \ell_k$, $\ell_i \to \ell_j \gamma$, $\tau \to l_j \eta$ and $\mu N \to e N$. We have also discussed the LFV decay modes of the Higgs bosons $\Phi \to \ell_i \ell_j$ ($\Phi = h^0, H^0, A^0$) so as the impact of Higgs mediated LFV effects on the cross section of the $\mu N \to \tau X$ reaction. Analytical relations and correlations among the rates of the above processes have been established at the two loop level in the Higgs boson exchange. The correlations among the processes are a precise signature of the theory. In this respect experimental improvements in all the decay channels of the τ lepton would be very welcome. In conclusion, the Higgs-mediated contributions to LFV processes can be within the present or upcoming attained experimental resolutions and provide an important opportunity to detect new physics beyond the standard model.

5.6 Tests of unitarity and universality in the lepton sector

5.6.1 Deviations from unitarity in the leptonic mixing matrix

The presence of physics beyond the SM in the leptonic sector can generate deviations from unitarity in the mixing matrix. This is analogous to what happens in the quark sector, where the search for deviations from unitarity of the CKM matrix is considered a sensitive way to look for new physics.

In the leptonic sector a clear example of non-unitarity is given by the see-saw mechanism [216–220]. To generate naturally small neutrino masses, new heavy particles—right handed neutrinos—are added, singlet under the SM gauge group. Thus a Yukawa coupling for neutrinos can be written, as well as Majorana masses for the new heavy fields. The mass matrix of the complete theory is now an enlarged mass matrix (5×5 at least), whose diagonalization leads to small Majorana neutrino masses. The non-unitarity of the 3×3 leptonic mixing matrix can now be understood simply by observing that it is a sub-matrix of a bigger one which is unitary, since the complete theory must conserve probabilities.

Another way to see this is looking at the effective theory we obtain once the heavy fields are integrated out. The unique dimension five operator is the well-known Weinberg operator [213] which generates neutrino masses when the electroweak symmetry is broken. Masses are naturally small since they are suppressed by the mass M of the heavy particles which have been integrated out: $m_\nu \sim v^2/M$, where v is the Higgs VEV. If we go on in the expansion in effective operators, we obtain only one dimension six operator which renormalizes the kinetic energy of neutrinos. Once we perform a field redefinition to go into a mass basis with canonical kinetic terms, a non-unitary mixing matrix is obtained [813]. In minimal models deviations from unitarity generated in this way are very suppressed, since the dimension six operator is proportional to v^2/M^2. However, in more sophisticated versions of this mechanism like double (or inverse) see-saw [814] the suppression can be reduced without affecting the smallness of neutrinos masses and avoiding any fine-tuning of Yukawa couplings. In terms of effective operators, this means that it is possible to "decouple" the dimension five operator from the dimension six, permitting small neutrino masses and not so small unitarity deviations.

Usually the elements of the leptonic mixing matrix are measured using neutrino oscillation experiments assuming unitarity. No information can be extracted from electroweak decays on the individual matrix elements, due to the impossibility of detecting neutrino mass eigenstates. This is quite different from the way of measuring the CKM matrix elements. Here oscillations are important too, but since quark mass eigenstates can be tagged, direct measurements of the matrix elements can be made using electroweak decays.

The situation changes if we relax the hypothesis of unitarity of the leptonic mixing matrix. Electroweak decays acquire now an important meaning, since they can be used to constrain deviations from unitarity. Consider as an example the decay $W \to l \bar{\nu}_l$. The decay rate is modified as follows: $\Gamma = \Gamma_{\text{SM}}(NN^\dagger)_{ll}$, where N is the non-unitary leptonic mixing matrix and Γ_{SM} is the SM decay rate. This, and other electroweak processes, can therefore be used to obtain information on $(NN^\dagger)_{ll}$. Moreover, lepton flavor violating processes like $\mu \to 3e$ or μ–e conversion in nuclei can occur, while rare lepton decays like $l_i \to l_j \gamma$ can be enhanced, permitting to constrain the off-diagonal elements of $(NN^\dagger)$. Finally, universality violation effects are produced, even if the couplings are universal: for example the branching ratio of π decay (see Sect. 6) is now proportional to $(NN^\dagger)_{ee}/(NN^\dagger)_{\mu\mu}$.

In Ref. [815] all these processes have been considered, a global fit has been performed and the matrix $|(NN^\dagger)|$ has been determined (90% C.L.):

$$|NN^\dagger| \approx \begin{pmatrix} 0.994 \pm 0.005 & <7.0 \times 10^{-5} & <1.6 \times 10^{-2} \\ <7.0 \times 10^{-5} & 0.995 \pm 0.005 & <1.0 \times 10^{-2} \\ <1.6 \times 10^{-2} & <1.0 \times 10^{-2} & 0.995 \pm 0.005 \end{pmatrix}.$$
(5.111)

Similar bounds can be inferred for $|N^\dagger N|$, leading to the conclusion that deviations from unitarity in the leptonic mixing matrix are experimentally constrained to be smaller than few percent. Notice however that these bounds apply to a 3×3 mixing matrix, i.e. they constrain deviations from unitarity induced by higher energy physics which has been integrated out.[26]

However, since on the contrary the quark sector decays can only constrain the elements of $|(NN^\dagger)|$, to determine the individual elements of the leptonic mixing matrix, oscillation experiments are needed. In Ref. [815] neutrino oscillation physics is reconsidered in the case in which the mixing matrix is not unitary. The main consequence of this is that the flavor basis is no longer orthogonal, which gives rise to two physical effects:

- "zero distance" effect, i.e. flavor conversion in neutrino oscillations at $L = 0$: $P_{\nu_\alpha \nu_\beta}(E, L = 0)\hat{A} \cdot \propto |(NN^\dagger)_{\beta\alpha}|^2$;
- non-diagonal matter effects.

With the resulting formulas for neutrino oscillations, a fit to present oscillation experiments is performed, in order to determine the individual matrix elements. As in the standard case, no information at all is available on phases (four or six, depending on the nature—Dirac or Majorana—of the neutrinos), since appearance experiments would be needed. However the moduli of matrix elements can be determined, but now they are all independent, so that the free parameters are nine instead of three. The elements of the e-row can be constrained using the data from CHOOZ [816], KamLAND [817] and SNO [818], together with the information on Δm^2_{23} resulting from an analysis of K2K [819]. In contrast, less data are available for the μ-row: only those coming from K2K and Super-Kamiokande [820] on atmospheric neutrinos, and only $|N_{\mu 3}|$ and the combination $|N_{\mu 1}|^2 + |N_{\mu 2}|^2$ can be determined. No information at all is available on the τ-row. The final result is the following (3σ ranges):

$$|N| = \begin{pmatrix} 0.75\text{–}0.89 & 0.45\text{–}0.66 & <0.34 \\ [(|N_{\mu 1}|^2 + |N_{\mu 2}|^2)^{1/2} = & 0.57\text{–}0.86] & 0.57\text{–}0.86 \\ ? & ? & ? \end{pmatrix}.$$
(5.112)

Notice that, without assuming unitarity, only half of the elements can be determined from oscillation experiments alone. Adding the information from near detectors at NOMAD [821], KARMEN [822], BUGEY [823] and MINOS [824], which put bounds on $|(NN^\dagger)_{\alpha\beta}|^2$ by measuring the "zero distance" effect, the degeneracy in the μ-row can be solved, but the τ-row is still unknown.

In order to determine/constrain all the elements of the leptonic mixing matrix without assuming unitarity, data on oscillations must be combined with data from decays. The final result is

$$|N| = \begin{pmatrix} 0.75\text{–}0.89 & 0.45\text{–}0.65 & <0.20 \\ 0.19\text{–}0.55 & 0.42\text{–}0.74 & 0.57\text{–}0.82 \\ 0.13\text{–}0.56 & 0.36\text{–}0.75 & 0.54\text{–}0.82 \end{pmatrix}, \quad (5.113)$$

which can be compared to the one obtained with standard analysis [825] where similar bounds are found.

It would be good to be able to determine the elements of the mixing matrix with oscillation experiments alone, permitting thus a "direct" test of unitarity. This would be for instance a way to detect light sterile neutrinos [826]. This could be possible exploring the appearance channels for instance at future facilities under discussion, such as Super-Beams [713, 827–829], β-Beams [830] and Neutrino Factories [831, 832], where the τ-row and phases could be measured. Moreover, near detectors at neutrino factories could also improve the bounds on $(NN^\dagger)_{e\tau}$ and $(NN^\dagger)_{\mu\tau}$ by about one order of magnitude. All this information, coming from both decays and oscillation experiments, will be important not only to detect new physics, but even to discriminate among different scenarios.

5.6.2 Lepton universality

High precision electroweak tests (HPET) represent a powerful tool to probe the SM and, hence, to constrain or obtain indirect hints of new physics beyond it. A typical and relevant example of HPET is represented by the Lepton Universality (LU) breaking. Kaon and pion physics are obvious grounds where to perform such tests, for instance in the well studied $\pi_{\ell 2}$ ($\pi \to \ell \nu_\ell$) and $K_{\ell 2}$ ($K \to \ell \nu_\ell$) decays, where $l = e$ or μ.

Unfortunately, the relevance of these single decay channels in probing the SM is severely hindered by our theoretical uncertainties, which still remain at the percent level (in particular due to the uncertainties on non-perturbative quantities like f_π and f_K). This is what prevents us from fully exploiting such decay modes in constraining new physics, in spite of the fact that it is possible to obtain non-SM contributions which exceed the high experimental precision which has been achieved on those modes.

On the other hand, in the ratios R_π and R_K of the electronic and muonic decay modes $R_\pi = \Gamma(\pi \to e\nu)/\Gamma(\pi \to \mu\nu)$ and $R_K = \Gamma(K \to e\nu)/\Gamma(K \to \mu\nu)$, the hadronic uncertainties cancel to a very large extent. As a result, the SM predictions of R_π and R_K are known with excellent accuracy [833] and this makes it possible to fully exploit the great experimental resolutions on R_π [834] and R_K [834, 835] to constrain new physics effects. Given our limited predictive power on f_π and f_K, deviations from the μ–e universality represent the best hope we have at the moment to detect new physics effects in $\pi_{\ell 2}$ and $K_{\ell 2}$.

---
[26]They do not apply for instance to the case of light sterile neutrinos, where the low energy mixing matrix is larger. Indeed in this case they would be included in the sum over all light mass eigenstates contained inside $(NN^\dagger)_{ll}$ and unitarity would be restored.

The most recent NA48/2 result on R_K:

$$R_K^{\text{exp.}} = (2.416 \pm 0.043_{\text{stat.}} \pm 0.024_{\text{syst.}}) \times 10^{-5} \quad \text{NA48/2},$$

which will further improve with current analysis, significantly improves on the previous PDG value:

$$R_K^{\text{exp.}} = (2.44 \pm 0.11) \times 10^{-5}.$$

This is to be compared with the SM prediction, which reads

$$R_K^{\text{SM}} = (2.472 \pm 0.001) \times 10^{-5}.$$

The details of the experimental measurement of R_K are presented in Sect. 6.2 of this report. Denoting by $\Delta r_{\text{NP}}^{e-\mu}$ the deviation from μ–e universality in R_K due to new physics, i.e.,

$$R_K = R_K^{\text{SM}}(1 + \Delta r_{\text{NP}}^{e-\mu}), \tag{5.114}$$

the NA48/2 result requires (at the 2σ level):

$$-0.063 \leq \Delta r_{\text{NP}}^{e-\mu} \leq 0.017 \quad \text{NA48/2}.$$

In the following, we consider low energy minimal SUSY extensions of the SM (MSSM) with R parity as the source of new physics to be tested by R_K [836]. The question we intend to address is whether SUSY can cause deviations from μ–e universality in K_{l2} at a level which can be probed with the present attained experimental sensitivity, namely at the percent level. We shall show that (i) it is indeed possible for regions of the MSSM to obtain $\Delta r_{\text{NP}}^{e-\mu}$ of $\mathcal{O}(10^{-2})$ and (ii) such large contributions to $K_{\ell 2}$ do not arise from SUSY lepton flavor conserving (LFC) effects, but, rather, from LFV ones.

At first sight, this latter statement may seem rather puzzling. The $K \to e\nu_e$ and $K \to \mu\nu_\mu$ decays are LFC and one could expect that it is through LFC SUSY contributions affecting differently the two decays that one obtains the dominant source of lepton flavor non-universality in SUSY. On the other hand, one can easily guess that, whenever new physics intervenes in $K \to e\nu_e$ and $K \to \mu\nu_\mu$ to create a departure from strict SM μ–e universality, these new contributions will be proportional to the lepton masses; hence, it may happen (and, indeed, this is what occurs in the SUSY case) that LFC contributions are suppressed with respect to the LFV ones by higher powers of the first two generations lepton masses (it turns out that the first contributions to $\Delta r_{\text{NP}}^{e-\mu}$ from LFC terms arise at the cubic order in m_ℓ, with $\ell = e, \mu$). A second, important reason for such result is that among the LFV contributions to R_K one can select those which involve flavor changes from the first two lepton generations to the third one with the possibility of picking up terms proportional to the tau–Yukawa coupling which can be large in the large $\tan\beta$ regime (the parameter $\tan\beta$ denotes the ratio of Higgs vacuum expectation values responsible for the up- and down-quark masses, respectively). Moreover, the relevant one loop induced LFV Yukawa interactions are known [806] to acquire an additional $\tan\beta$ factor with respect to the tree level LFC Yukawa terms. Thus, the loop suppression factor can be (partially) compensated in the large $\tan\beta$ regime.

Finally, given the NA48/2 R_K central value below the SM prediction, one may wonder whether SUSY contributions could have the correct sign to account for such an effect. Although the above mentioned LFV terms can only add positive contributions to R_K (since their amplitudes cannot interfere with the SM one), it turns out that there exist LFC contributions arising from double LFV mass insertions (MI) in the scalar lepton propagators which can destructively interfere with the SM contribution. We shall show that there exist regions of the SUSY parameter space where the total R_K arising from all such SM and SUSY terms is indeed lower than R_K^{SM}.

Finally, we also discuss the potentiality of τ–$\mu(e)$ universality breaking in τ decays to probe new physics effects.

5.6.2.1 μ–e universality in $\pi \to \ell\nu$ and $K \to \ell\nu$ decays
Due to the V–A structure of the weak interactions, the SM contributions to $\pi_{\ell 2}$ and $K_{\ell 2}$ are helicity suppressed; hence, these processes are very sensitive to non-SM effects (such as multi-Higgs effects) which might induce an effective pseudoscalar hadronic weak current.

In particular, charged Higgs bosons ($H^\pm$) appearing in any model with two Higgs doublets (including the SUSY case) can contribute at tree level to the above processes. The relevant four-Fermi interaction for the decay of charged mesons induced by $W^\pm$ and $H^\pm$ has the following form:

$$\frac{4G_F}{\sqrt{2}} V_{ud} \left[(\bar{u}\gamma_\mu P_L d)(\bar{l}\gamma^\mu P_L \nu) \right.$$
$$\left. - \tan^2\beta \left(\frac{m_d m_l}{m_{H^\pm}^2} \right)(\bar{u} P_R d)(\bar{l} P_L \nu) \right], \tag{5.115}$$

where $P_{R,L} = (1 \pm \gamma_5)/2$. Here we keep only the $\tan\beta$ enhanced part of the $H^\pm ud$ coupling, namely the $m_d \tan\beta$ term. The decays $M \to l\nu$ (being M the generic meson) proceed via the axial-vector part of the $W^\pm$ coupling and via the pseudoscalar part of the $H^\pm$ coupling. Then, once we implement the PCACs

$$\langle 0|\bar{u}\gamma_\mu\gamma_5 d|M^-\rangle = if_M p_M^\mu,$$
$$\langle 0|\bar{u}\gamma_5 d|M^-\rangle = -if_M \frac{m_M^2}{m_d + m_u}, \tag{5.116}$$

we easily arrive at the amplitude

$$\mathcal{M}_{M \to l\nu} = \frac{G_F}{\sqrt{2}} V_{u(d,s)} f_M \bigg[m_l - m_l \tan^2 \beta \\ \times \left(\frac{m_d}{m_d + m_u}\right) \frac{m_M^2}{m_{H^\pm}^2} \bigg] \bar{l}(1-\gamma_5)\nu. \quad (5.117)$$

We observe that the SM term is proportional to m_l because of the helicity suppression while the charged Higgs term is proportional to m_l because of the Yukawa coupling. The tree level partial width is given by [806]

$$\Gamma(M^- \to l^- \bar{\nu}) = \frac{G_F^2}{8\pi} |V_{u(d,s)}|^2 f_M^2 m_M m_l^2 \left(1 - \frac{m_l^2}{m_M^2}\right) \times r_M, \quad (5.118)$$

where

$$r_M = \left[1 - \tan^2\beta \left(\frac{m_{d,s}}{m_u + m_{d,s}}\right) \frac{m_M^2}{m_{H^\pm}^2}\right]^2, \quad (5.119)$$

and where m_u is the mass of the up quark while $m_{s,d}$ stands for the down-type quark mass of the M meson ($M = K, \pi$). From (5.119) it is evident that such tree level contributions do not introduce any lepton flavor dependent correction. The first SUSY contributions violating the μ–e universality in $\pi \to \ell\nu$ and $K \to \ell\nu$ decays arise at the one loop level with various diagrams involving exchanges of (charged and neutral) Higgs scalars, charginos, neutralinos and sleptons. For our purpose, it is relevant to divide all such contributions into two classes: (i) LFC contributions where the charged meson M decays without FCNC in the leptonic sector, i.e. $M \to \ell\nu_\ell$; (ii) LFV contributions $M \to \ell_i \nu_k$, with i and k referring to different generations (in particular, the interesting case will be for $i = e, \mu$, and $k = \tau$).

5.6.2.2 *The lepton flavor conserving case* One-loop corrections to R_π and R_K include box, wave function renormalization and vertex contributions from SUSY particle exchange. The complete calculation of the μ decay in the MSSM [837] can be easily applied to meson decays.

The dominant diagrams containing one loop corrections to the lWv_l vertex have the following suppression factors (compared to the tree level graph):

- $\frac{g_2^2}{16\pi^2} \frac{m_l^2}{m_W^2} \tan\beta \frac{m_W^2}{m_h^2}$ for loops with $hW^\pm l$ exchange (with $h = H^0, h^0$),
- $\frac{g_2^2}{16\pi^2} \frac{m_l^2}{m_W^2} \tan^2\beta \frac{m_W^2}{m_h^2}$ for loops with $hH^\pm l$ exchange (with $h = H^0, h^0$ and A^0),
- $\frac{g_2^2}{16\pi^2} \frac{m_W^2}{M_{\text{SUSY}}^2}$ for loops generated by charginos/neutralinos and sleptons.

For dominant box contributions we have the following estimates:

- $\frac{g_2^2}{16\pi^2} \frac{m_d m_l}{M^2} \tan^2\beta$ for boxes with $hW^\pm l$ or $Z^0 H^\pm l$ exchange (where M is the heavier mass circulating in the loop),
- $\frac{g_2^2}{16\pi^2} \left(\frac{m_d m_l}{m_W M_{H^\pm}}\right)^2 \tan^4\beta$ for boxes with $hH^\pm l$,
- $\frac{g_2^2}{16\pi^2} \frac{m_W^2}{M_{\text{SUSY}}^2}$ for loops generated by charginos/neutralinos and sleptons (where M_{SUSY} is the heavier mass circulating in the loop).

To get a feeling of the order of magnitude of the above contributions let us show the explicit expression of the dominant Higgs contributions to the lWv_l vertex [837]:

$$\Delta r^{e-\mu}_{\text{SUSY}} = \frac{\alpha_2}{32\pi} \frac{m_\mu^2}{M_W^2} \tan^2\beta \big(-2 + I(A^0, H^\pm) \\ + c_\alpha^2 I(H^0, H^\pm) + s_\alpha^2 I(h^0, H^\pm)\big),$$

where

$$I(1,2) = \frac{1}{2} \frac{m_1^2 + m_2^2}{m_1^2 - m_2^2} \log \frac{m_1^2}{m_2^2},$$

and α is the mixing angle in the CP-even Higgs sector. Even if we assume $\tan\beta = 50$ and arbitrary relations among the Higgs boson masses we get a value for $\Delta r^{e-\mu}_{\text{SUSY}} \leq 10^{-6}$ much below the actual experimental resolution. In addition, in the large $\tan\beta$ limit, $\alpha \to 0$ and $m_{A^0} \sim m_{H^0} \sim m_{H^\pm}$ and $\Delta r^{e-\mu}_{\text{SUSY}}$ tends to vanish. The charginos/neutralinos sleptons ($\tilde{l}_{e,\mu}$) contributions to $\Delta r^{e-\mu}_{\text{SUSY}}$ are of the form

$$\Delta r^{e-\mu}_{\text{SUSY}} \sim \frac{\alpha_2}{4\pi} \left(\frac{\tilde{m}_\mu^2 - \tilde{m}_e^2}{\tilde{m}_\mu^2 + \tilde{m}_e^2}\right) \frac{m_W^2}{M_{\text{SUSY}}^2}.$$

The degeneracy of slepton masses (in particular those of the first two generations) severely suppresses these contributions. Even if we assume a quite large mass splitting among slepton masses (at the 10% level for instance) we end up with $\Delta r^{e-\mu}_{\text{SUSY}} \leq 10^{-4}$. For the box-type non-universal contributions, we find similar or even more suppressed effects compared to those we have studied. So, finally, it turns out that all these LFC contributions yield values of $\Delta r^{e-\mu}_{K\,\text{SUSY}}$ which are much smaller than the percent level required by the achieved experimental sensitivity.

On the other hand, one could wonder whether the quantity $\Delta r^{e-\mu}_{\text{SUSY}}$ can be constrained by the pion physics. In principle, the sensitivity could be even higher: from

$$R^{\text{exp.}}_\pi = (1.230 \pm 0.004) \times 10^{-4} \quad \text{PDG},$$

and by making a comparison with the SM prediction

$$R^{\text{SM}}_\pi = (1.2354 \pm 0.0002) \times 10^{-4},$$

one obtains (at the 2σ level)

$$-0.0107 \leq \Delta r_{\text{NP}}^{e-\mu} \leq 0.0022.$$

Unfortunately, even in the most favorable cases, $\Delta r_{\text{SUSY}}^{e-\mu}$ remains much below its actual experimental upper bound.

In conclusion, SUSY effects with flavor conservation in the leptonic sector can differently contribute to the $K \to e \nu_e$ and $K \to \mu \nu_\mu$ decays, hence inducing μ–e non-universality in R_K, however such effects are still orders of magnitude below the level of the present experimental sensitivity on R_K. The same conclusions hold for R_π.

5.6.2.3 The lepton flavor violating case It is well known that models containing at least two Higgs doublets generally allow for flavor violating couplings of the Higgs bosons with the fermions [838]. In the MSSM such LFV couplings are absent at tree level. However, once non-holomorphic terms are generated by loop effects (so called HRS corrections [430]) and given a source of LFV among the sleptons, Higgs-mediated (radiatively induced) $H\ell_i\ell_j$ LFV couplings are unavoidable [806, 807]. These effects have been widely discussed through the study of several processes, namely $\tau \to \ell_j \ell_k \ell_k$ [806, 807], $\tau \to \mu\eta$ [810], μ–e conversion in nuclei [811], $B \to \ell_j \tau$ [807], $H \to \ell_j \ell_k$ [174] and $\ell_i \to \ell_j \gamma$ [650].

Moreover, it has been shown [839] that Higgs-mediated LFV couplings generate a breaking of μ–e universality in purely leptonic $\pi^\pm$ and $K^\pm$ decays.

One could naively think that SUSY effects in the LFV channels $M \to \ell_i \nu_k$ are further suppressed with respect to the LFC ones. On the contrary, charged Higgs mediated SUSY LFV contributions, in particular in the kaon decays into an electron or a muon and a tau neutrino, can be strongly enhanced. The quantity which now accounts for the deviation from the μ–e universality reads

$$R_{\pi,K}^{\text{LFV}} = \frac{\sum_i \Gamma(\pi(K) \to e \nu_i)}{\sum_i \Gamma(\pi(K) \to \mu \nu_i)}, \quad i = e, \mu, \tau,$$

with the sum extended over all (anti)neutrino flavors (experimentally one determines only the charged lepton flavor in the decay products).

The dominant SUSY contributions to $R_{\pi,K}^{\text{LFV}}$ arise from the charged Higgs exchange. The effective LFV Yukawa couplings we consider are (see Fig. 39)

$$\ell H^\pm \nu_\tau \to \frac{g_2}{\sqrt{2}} \frac{m_\tau}{M_W} \Delta_R^{3l} \tan^2\beta, \quad \ell = e, \mu. \quad (5.120)$$

A crucial ingredient for the effects we are going to discuss is the quadratic dependence on $\tan\beta$ in the above coupling: one power of $\tan\beta$ comes from the trilinear scalar coupling in Fig. 39, while the second one is a specific feature of the above HRS mechanism.

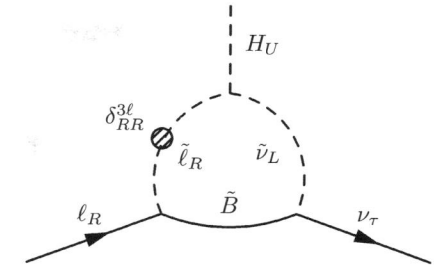

Fig. 39 Contribution to the effective $\bar{\nu}_\tau \ell_R H^+$ coupling

The $\Delta_R^{3\ell}$ terms are induced at one loop level by the exchange of bino (see Fig. 39) or bino–higgsino and sleptons. Since the Yukawa operator is of dimension four, the quantities $\Delta_R^{3\ell}$ depend only on ratios of SUSY masses, hence avoiding SUSY decoupling. In the so called MI approximation the expression of $\Delta_R^{3\ell}$ is given by:

$$\Delta_R^{3\ell} \simeq \frac{\alpha_1}{4\pi} \mu M_1 m_R^2 \delta_{RR}^{3\ell} \left[I'\left(M_1^2, \mu^2, m_R^2\right) - (\mu \leftrightarrow m_L)\right], \quad (5.121)$$

where μ is the Higgs mixing parameter, M_1 is the bino ($\tilde{B}$) mass and $m_{L(R)}^2$ stands for the left–left (right–right) slepton mass matrix entry. The LFV MIs, i.e. $\delta_{XX}^{3\ell} = (\tilde{m}_\ell^2)_{XX}^{3\ell}/m_X^2$ ($X = L, R$), are the off-diagonal flavor changing entries of the slepton mass matrix. The loop function $I'(x, y, z)$ is such that $I'(x, y, z) = dI(x, y, z)/dz$, where $I(x, y, z)$ refers to the standard three point one loop integral which has mass dimension-2.

Making use of the LFV Yukawa coupling in (5.120), it turns out that the dominant contribution to $\Delta r_{\text{NP}}^{e-\mu}$ reads [839]

$$R_K^{\text{LFV}} \simeq R_K^{\text{SM}} \left[1 + \left(\frac{m_K^4}{M_H^4}\right)\left(\frac{m_\tau^2}{m_e^2}\right) |\Delta_R^{31}|^2 \tan^6\beta \right]. \quad (5.122)$$

In (5.122) terms proportional to Δ_R^{32} are neglected given that they are suppressed by a factor m_e^2/m_μ^2 with respect to the term proportional to Δ_R^{31}.

Taking $\Delta_R^{31} \simeq 5 \times 10^{-4}$ (by means of a numerical analysis, it turns out that $\Delta_R^{3\ell} \leq 10^{-3}$ [174]), $\tan\beta = 40$ and $M_H = 500$ GeV we end up with $R_K^{\text{LFV}} \simeq R_K^{\text{SM}}(1 + 0.013)$. We see that in the large (but not extreme) $\tan\beta$ regime and with a relatively heavy $H^\pm$, it is possible to reach contributions to $\Delta r_{K\,\text{SUSY}}^{e-\mu}$ at the percent level thanks to the possible LFV enhancements arising in SUSY models.

Turning to pion physics, one could wonder whether the analogous quantity $\Delta r_{\pi\,\text{SUSY}}^{e-\mu}$ is able to constrain SUSY LFV. However, the correlation between $\Delta r_{\pi\,\text{SUSY}}^{e-\mu}$ and $\Delta r_{K\,\text{SUSY}}^{e-\mu}$:

$$\Delta r_{\pi\,\text{SUSY}}^{e-\mu} \simeq \left(\frac{m_d}{m_u + m_d}\right)^2 \left(\frac{m_\pi^4}{m_k^4}\right) \Delta r_{K\,\text{SUSY}}^{e-\mu}, \quad (5.123)$$

clearly shows that the constraints on $\Delta r^{e-\mu}_{K\,\text{SUSY}}$ force $\Delta r^{e-\mu}_{\pi\,\text{SUSY}}$ to be much below its actual experimental upper bound.

5.6.2.4 On the sign of $\Delta r^{e-\mu}_{\text{SUSY}}$ The above SUSY dominant contribution to $\Delta r^{e-\mu}_{\text{NP}}$ arises from LFV channels in the $K \to e\nu$ mode, hence without any interference effect with the SM contribution. Thus, it can only increase the value of R_K with respect to the SM expectation. On the other hand, the recent NA48/2 result exhibits a central value lower than R_K^{SM} (and, indeed, also lower than the previous PDG central value). One may wonder whether SUSY could account for such a lower R_K. Obviously, the only way it can is through terms which, contributing to the LFC $K \to l\nu_l$ channels, can interfere (destructively) with the SM contribution. We already commented that SUSY LFC contributions are subdominant. However, one can envisage the possibility of making use of the large LFV contributions to give rise to LFC ones through double LFV MI that, as a final effect, preserves the flavor.

To see this point explicitly, we report the corrections to the LFC $H^\pm \ell\nu_\ell$ vertices induced by LFV effects

$$\ell H^\pm \nu_\ell \to \frac{g_2}{\sqrt{2}} \frac{m_\ell}{M_W} \tan\beta \left(1 + \frac{m_\tau}{m_\ell} \Delta^{\ell\ell}_{RL} \tan\beta\right), \qquad (5.124)$$

where $\Delta^{\ell\ell}_{RL}$ is generated by the same diagram as in Fig. 39 but with an additional $\delta^{3\ell}_{LL}$ MI in the sneutrino propagator. In the MI approximation, $\Delta^{\ell\ell}_{RL}$ is given by

$$\Delta^{\ell\ell}_{RL} \simeq -\frac{\alpha_1}{4\pi} \mu M_1 m_L^2 m_R^2 \delta^{\ell 3}_{RR} \delta^{3\ell}_{LL} I''(M_1^2, m_L^2, m_R^2), \qquad (5.125)$$

where $I''(x,y,z) = d^2 I(x,y,z)/dy\,dz$. In the large slepton mixing case, $\Delta^{\ell\ell}_{RL}$ terms are of the same order of $\Delta^{3\ell}_R$.[27] These new effects modify the previous R_K^{LFV} expression in the following way [839]:

$$R_K^{\text{LFV}} \simeq R_K^{\text{SM}} \left[\left|1 - \frac{m_K^2}{M_H^2} \frac{m_\tau}{m_e} \Delta^{11}_{RL} t_\beta^3\right|^2 + \left(\frac{m_K^4}{M_H^4}\right)\left(\frac{m_\tau^2}{m_e^2}\right) |\Delta^{31}_R|^2 \tan^6\beta\right]. \qquad (5.126)$$

In the above expression, besides the contributions reported in (5.122), we also included the interference between SM and SUSY LFC terms (arising from a double LFV source). Setting the parameters as in the example of the above section and if $\Delta^{11}_{RL} = 10^{-4}$ we get $R_K^{\text{LFV}} \simeq R_K^{\text{SM}}(1 - 0.032)$, that is

---

[27] $\text{Im}(\delta^{13}_{RR}\delta^{31}_{LL})$ is strongly constrained by the electron electric dipole moment [163]. However, sizable contributions to R_K^{LFV} can still be induced by $\text{Re}(\delta^{13}_{RR}\delta^{31}_{LL})$.

just within the expected experimental resolution reachable by NA48/2 once all the available data will be analyzed. Finally, we remark that the above effects do not spoil the pion physics constraints.

The extension of the above results to $B \to \ell\nu$ [766] is obtained with the replacement $m_K \to m_B$, while for the $D \to \ell\nu$ case $m_K^2 \to (m_s/m_c)m_D^2$. In the most favorable scenarios, taking into account the constraints from LFV τ decays [650], spectacular order-of-magnitude enhancements for $R_B^{e/\tau}$ and $\mathcal{O}(50\%)$ deviations from the SM in $R_B^{\mu/\tau}$ are allowed [766]. There exists a stringent correlation between $R_B^{e/\tau}$ and $R_K^{e/\mu}$ so that

$$R_B^{e/\tau} \simeq \left[r_H + \frac{m_B^4}{m_K^4} \Delta r^{e-\mu}_{K\,\text{SUSY}}\right] \leq 2 \times 10^2. \qquad (5.127)$$

In particular, it turns out that $\Delta r^{e-\mu}_{K\,\text{SUSY}}$ is much more effective to constrain $R_B^{e/\tau}$ $\Gamma(B \to e\nu_\tau)$ than LFV tau decay processes.

5.6.2.5 Lepton universality in $M \to \ell\nu$ versus LFV τ decays Obviously, a legitimate worry when witnessing such a huge SUSY contribution through LFV terms is whether the bounds on LFV tau decays, like $\tau \to eX$ (with $X = \gamma, \eta, \mu\mu$), are respected [650]. Higgs mediated $B(\tau \to \ell_j X)$ and $\Delta r^{e-\mu}_{K\,\text{SUSY}}$ have exactly the same SUSY dependence; hence, we can compute the upper bounds of the relevant LFV tau decays which are obtained for those values of the SUSY parameters yielding $\Delta r^{e-\mu}_{K\,\text{SUSY}}$ at the percent level.

The most sensitive processes to Higgs mediated LFV in the τ lepton decay channels are $\tau \to \mu(e)\eta$, $\tau \to \mu(e)\mu\mu$ and $\tau \to \mu(e)\gamma$. The related branching ratios are [650]

$$\frac{B(\tau \to l_j \eta)}{B(\tau \to l_j \bar{\nu}_j \nu_\tau)} \simeq 18\pi^2 \left(\frac{f_\eta^8 m_\eta^2}{m_\tau}\right)^2 \left(1 - \frac{m_\eta^2}{m_\tau^2}\right)^2$$
$$\times \left(\frac{|\Delta^{3j}|^2 \tan^6\beta}{m_A^4}\right), \qquad (5.128)$$

where $m_\eta^2/m_\tau^2 \simeq 9.5 \times 10^{-2}$ and the relevant decay constant is $f_\eta^8 \sim 110$ MeV,

$$\frac{B(\tau \to l_j \gamma)}{B(\tau \to l_j \bar{\nu}_j \nu_\tau)} \simeq 10 \left(\frac{\alpha_{el}}{\pi}\right)^3 \tan^4\beta |\Delta_{\tau j}|^2$$
$$\times \left[\frac{m_W}{m_A} \log\left(\frac{m_W^2}{m_A^2}\right)\right]^4, \qquad (5.129)$$

$$\frac{B(\tau \to l_j \mu\mu)}{B(\tau \to l_j \bar{\nu}_j \nu_\tau)} \simeq \frac{m_\tau^2 m_\mu^2}{32} \left(\frac{|\Delta^{3j}|^2 \tan^6\beta}{m_A^4}\right)$$
$$\times (3 + 5\delta_{j\mu}), \qquad (5.130)$$

where $|\Delta^{3j}|^2 = |\Delta_L^{3j}|^2 + |\Delta_R^{3j}|^2$. It is straightforward to check that, in the large $\tan\beta$ regime, $B(\tau \to l_j \eta)$ and $B(\tau \to l_j \gamma)$ are of the same order of magnitude [650] and they are dominant compared to $B(\tau \to l_j \mu\mu)$.[28]

Given that $\Delta r_{K\,SUSY}^{e-\mu}$ and $B(\tau \to l_j X)$ have the same SUSY dependence, once we saturate the $\Delta r_{K\,SUSY}^{e-\mu}$ value (at the % level), the upper bounds on $B(\tau \to l_j X)$ (allowed by $|\Delta_R^{31}|^2$) are automatically predicted. We find that

$$B(\tau \to l_j \gamma) \sim B(\tau \to l_j \eta) \simeq 10^{-2} \left(\frac{|\Delta^{3j}|^2 \tan^6\beta}{m_A^4} \right)$$
$$\simeq 10^{-8} \Delta r_{K\,SUSY}^{e-\mu}. \quad (5.131)$$

So, employing the constraints for $\Delta r_{K\,SUSY}^{e-\mu}$ at the % level, we obtain the desired upper bounds: $B(\tau \to e\eta)$, $B(\tau \to e\gamma) \leq 10^{-10}$. Given the experimental upper bounds on the LFV τ lepton decays [764], we conclude that it is possible to saturate the upper bound on $\Delta r_{K\,SUSY}^{e-\mu}$ while remaining much below the present and expected future upper bounds on such LFV decays. There exist other SUSY contributions to LFV τ decays, like the one loop neutralino-charged slepton exchanges, for instance, where there is a direct dependence on the quantities δ_{RR}^{3j}. Given that the existing bounds on the leptonic δ_{RR} involving transitions to the third generation are rather loose [681], it turns out that also these contributions fail to reach the level of experimental sensitivity for LFV τ decays.

5.6.2.6 e–μ universality in τ decays Studying the τ–μ–e universality in the leptonic τ decays is an interesting laboratory for search for physics beyond the SM. In the SM the τ decay partial width for the leptonic modes is

$$\Gamma(\tau \to l\bar{\nu}_l\nu_\tau(\gamma))$$
$$= \frac{G_F^2 m_\tau^5}{192\pi^3} f(m_l^2/m_\tau^2)$$
$$\times \left[1 + \frac{3}{5}\frac{m_\tau^2}{M_W^2} \right] \left[1 + \frac{\alpha(m_\tau)}{\pi}\left(\frac{25}{4} - \pi^2\right) \right], \quad (5.132)$$

where $f(x) = 1 - 8x + 8x^3 - x^4 - 12x^2 \log x$ is the lepton mass correction and the last two factors are corrections from the nonlocal structure of the intermediate $W^\pm$ boson propagator and QED radiative corrections respectively. The Fermi constant G_F is determined by the muon life-time

$$G_F \equiv G_\mu = (1.16637 \pm 0.00002) \times 10^{-5} \text{ GeV}^{-2}, \quad (5.133)$$

and absorbs all the remaining electroweak radiative (loop) corrections.

The main source of non-universal contributions would be the tree level contribution from the charged Higgs boson (mass dependent couplings) and different slepton masses of the $\tilde{\mu}$, $\tilde{\tau}$ and $\tilde{e}$ sleptons exchanged in the one loop induced ℓ–W–ν_ℓ vertex. On the other hand, as discussed in previous sections, the last contribution can provide a correction that can be at most as large as 10^{-4} (in the limiting case of very light sleptons and gauginos $\sim M_W$), very far for the actual and forthcoming experimental resolutions. However, differently from the $M \to \ell\nu$ case, a tree level charged Higgs exchange breaks the lepton universality and it provides a contribution that we are going to discuss.

The deviations from $\tau - \mu - e$ universality can be conveniently discussed by studying the ratios $G_{\tau,e}/G_{\mu,e}$, $G_{\tau,\mu}/G_{\mu,e}$ and $G_{\tau,\mu}/G_{\tau,e}$, given by the ratios of the corresponding branching fractions. With the highly accurate experimental result for the $G_{\mu,e}$, the first two ratios are essentially a direct measure of non-universality in the corresponding tau decays. When the statistical error of future experiments will become negligible, the main problem for achieving maximum precision will be to reduce the systematic errors. One may expect that certain systematic errors will be canceled in the ratio $G_{\tau,\mu}/G_{\tau,e}$.

The 2004 world averaged data for the leptonic τ decay modes and τ life-time are [834, 840]

$$B^e|_{\exp} = (17.84 \pm 0.06)\%,$$
$$B^\mu|_{\exp} = (17.37 \pm 0.06)\%, \quad (5.134)$$
$$\tau_\tau = (290.6 \pm 1.1) \times 10^{-15} \text{ s}.$$

Note that the relative errors of the above measured quantities are of the 0.34–0.38%, the biggest being for the life-time. One can parameterize a possible beyond the SM contribution by a quantity Δ^l ($l = e, \mu$), defined as

$$B^l = B^l|_{SM}(1 + \Delta^l). \quad (5.135)$$

Including the W-propagator effect and QED radiative corrections, the following results for the branching ratios in the SM are obtained [840]:

$$B^e|_{SM} = (17.80 \pm 0.07)\%,$$
$$B^\mu|_{SM} = (17.32 \pm 0.07)\%. \quad (5.136)$$

Together with the experimental data this leads to the following 95% C.L. bounds on Δ^l, for the electron and muon decay mode, respectively [840]:

$$(-0.80 \leq \Delta^e \leq 1.21)\%, \quad (-0.76 \leq \Delta^\mu \leq 1.27)\%. \quad (5.137)$$

---

[28] It is remarkable that $\Delta r_{K\,SUSY}^{e-\mu} \sim |\Delta_R^{31}|^2$ while $B(\tau \to eX) \sim |\Delta_L^{31}|^2 + |\Delta_R^{31}|^2$ (with $X = \eta, \gamma$ or $\mu\mu$). In practice, $\Delta r_{K\,SUSY}^{e-\mu}$ is sensitive only to RR-type LFV terms in the slepton mass matrix while $B(\tau \to eX)$ does not distinguish between left and right sectors.

One can see that the negative contributions are constrained more strongly that the positive ones. A tree level charged Higgs exchange leads to the following contribution [841]:

$$\Gamma^{W^\pm + H^\pm}$$
$$= \Gamma^{W^\pm}\left[1 - 2\frac{m_l m_\tau \tan^2\beta}{M_{H^\pm}^2}\frac{m_l}{m_\tau}\kappa\left(\frac{m_l^2}{m_\tau^2}\right) + \frac{m_\tau^2 m_l^2 \tan^4\beta}{4M_{H^\pm}^4}\right]$$
$$\simeq \Gamma^{W^\pm}\left[1 - 1.15 \times 10^{-3}\left(\frac{200\,\text{GeV}}{M_{H^\pm}}\right)^2 \left(\frac{\tan\beta}{50}\right)^2\right],$$
(5.138)

where $\kappa(x) = \frac{g(x)}{f(x)} \simeq 0.94$ with $g(x) = 1 + 9x - 9x^2 - x^3 + 6x(1+x)\ln(x)$. In the above expression, the second term comes from the interference with the SM amplitude and it is much more important than the last one, which is suppressed by a factor $m_\tau^2 \tan^2\beta / 8 M_{H^\pm}^2$.

For the future precision of $G_{\tau,\mu}$ and $G_{\tau,e}$ measurements of order 0.1% ($G_{\mu,e}$ is known with 0.002% precision) the only effect that eventually can be observed is the slightly smaller value of $G_{\tau,\mu}$ as compared to $G_{\tau,e}$ and $G_{\mu,e}$. If measured, such effect would mean a rather precise information about MSSM: large $\tan\beta \geq 40$ and small $M_{H^\pm} \sim 200$–300 GeV.

5.6.2.7 Conclusions High precision electroweak tests, such as deviations from the SM expectations of the lepton universality breaking, represent a powerful tool to probe the SM and, hence, to constrain or obtain indirect hints of new physics beyond it. Kaon and pion physics are obvious grounds where to perform such tests, for instance in the well studied $\pi \to \ell\nu_\ell$ and $K \to \ell\nu_\ell$ decays, where $l = e$ or μ. In particular, a precise measurement of the flavor conserving $K \to \ell\nu_\ell$ decays may shed light on the size of LFV in new physics. μ–e non-universality in $K_{\ell2}$ is quite effective in constraining relevant regions of SUSY models with LFV. A comparison with analogous bounds coming from τ LFV decays shows the relevance of the measurement of R_K to probe LFV in SUSY. Moreover, τ–μ–e universality in the leptonic τ decays is an additional interesting laboratory for searching for physics beyond the SM.

5.7 EDMs from RGE effects in theories with low energy supersymmetry

EDMs probe new physics in general and in particular low energy supersymmetry. For definiteness and for simplicity, we focus here on lepton EDMs, as they are free from the theoretical uncertainties associated to the calculation of hadronic matrix elements. After a brief review of the constraints on slepton masses we discuss here a specific kind of sources of CPV, those induced radiatively by the Yukawa interactions of the heavy particles present in see-saw and/or grand unified models. It has been emphasized that these interactions could lead to LFV decays, in particular $\mu \to e\gamma$, at an observable rate; it is then natural to wonder whether this is also the case for EDMs.

As shown in Sect. 3, LFV decays, EDMs and additional contribution to MDMs all have a common origin, the dimension five dipole operator possibly induced by some new physics beyond the SM:

$$\mathcal{L}_{d=5} = \frac{1}{2}\bar{\psi}_{Ri} A_{ij} \psi_{Lj} \sigma^{\mu\nu} F_{\mu\nu} + \text{h.c.,} \quad (5.139)$$

$$B(\ell_i \to \ell_j \gamma) \propto |A_{ij}|^2, \qquad \delta a_{\ell_i} = \frac{2m_{\ell_i}}{e}\text{Re}\,A_{ii},$$
$$d_{\ell_i} = \text{Im}\,A_{ii}.$$
(5.140)

If induced at one loop, this amplitude displays a quadratic suppression with respect to the new physics mass scale, M_{NP}, and a linear dependence on the non-dimensional coupling Γ^{NP} encoding the pattern of F and CP violations (in the basis where the charged lepton mass matrix is diagonal):

$$A_{ij} \approx \frac{em_{\ell_i}}{(4\pi)^2}\frac{\Gamma^{\text{NP}}_{ij}}{M_{\text{NP}}^2}. \quad (5.141)$$

For low energy supersymmetry, the loops involve exchange of gauginos and sleptons, so that Γ^{NP} is proportional to the misalignment between leptons and sleptons, conveniently described by the flavor violating (FV) δ's of the mass insertion approximation. It is well known that the flavor conserving (FC) μ and a terms are potentially a very important source of CPV. In the expansion in powers of the FV δ's, they indeed contribute to d_{ℓ_i} at zero order:

$$\text{Im}(A_{ii}) = f_\mu m_{\ell_i} \text{Arg}(\mu) + f_a m_{\ell_i} \text{Im}(a_i)$$
$$+ f_{LLRR} \text{Im}(\delta^{LL} m_\ell \delta^{RR})_{ii} + \cdots, \quad (5.142)$$

where the various f represent supersymmetric loop functions and can be found for instance in [163]. Notice that the contribution arising at second order in the FV δ's could be even more important than the FC one, as happens for instance if CPV is always associated to FV.

Assuming no cancellations between the amplitudes, we first review briefly some limits considering for definiteness the mSUGRA scenario with $\tan\beta = 10$ and slepton masses in the range suggested by g_μ. The strong impact of $\mu \to e\gamma$ on δ^{LL} has been emphasized previously, where it was stressed that the impact of d_e is also remarkable in constraining the FC sources of CPV: $\arg\mu \leq 2 \times 10^{-3}$, $\text{Im}\,a_e/m_R \leq 0.2$. As for the other FV source in (5.142), one obtains $\text{Im}(\delta^{LL} m_\ell \delta^{RR})_{ee}/m_\tau \leq 10^{-5}$. The planned sensitivity $d_\mu \leq 10^{-23}$ e cm would also give interesting bounds: $\arg\mu \leq 10^{-1}$, $\text{Im}(\delta^{LL} m_\ell \delta^{RR})_{\mu\mu}/m_\tau \leq 10^{-1}$. Notice that, due to the lepton mass scaling law of the μ-term contribution, the present bound on $\arg\mu$ from d_e implies that the μ-term contribution to d_μ cannot exceed 2×10^{-25} e cm, below the planned projects. A positive measure of d_μ would thus signal a different source of CPV, i.e. the a_μ-term or the FV contribution. In the following we take real μ.

The a_i-terms and the FV δ's at low energy can be thought as the sum of two contributions. The first is already present at the Planck scale where soft masses are defined; we assume that this contribution is absent because of some inhibition mechanism, as could happen in supergravity. The second contribution is induced radiatively running from high to low energies by the Yukawa couplings of heavy particles[29] that potentially violate F and CP. Since LFV experiments are testing this radiatively induced misalignment, in the following we shall consider what happens for EDMs, beginning with the pure see-saw model and then adding a grand unification scenario, where heavy colored Higgs triplets are present to complete the Higgs doublets representations (in $SU(5)$ for instance they complete the 5 and $\bar{5}$). Notice that these triplets are important as in supersymmetric theories proton decays mainly through their exchange.

Consider first the case of degenerate right handed neutrinos with mass M. One can solve approximately the RGE by expanding in powers of $\ln(\Lambda/M)/(4\pi)^2$, i.e. the log of the ratio of the two scales between which the neutrino Yukawa couplings Y_ν are present times the corresponding loop factor suppression. For LFV decays δ_{ij}^{LL} is induced at 1st order and is proportional to the combination $(Y_\nu^\dagger Y_\nu)_{ij}$. In particular, $\mu \to e\gamma$ constrains $(Y_\nu^\dagger Y_\nu)_{21}$ to be small and this has a strong impact on see-saw models. To obtain an imaginary part for EDMs, one needs a non-hermitian combination of Yukawa couplings, which can be found only at 4th order: $\mathrm{Im}(Y_\nu^\dagger Y_\nu [Y_\nu^\dagger Y_\nu, Y_\ell^\dagger Y_\ell] Y_\nu^\dagger Y_\nu)_{ii}$. Such a contribution is negligibly small with respect to the present and planned experimental sensitivities.

Allowing for a non-degenerate spectrum of right handed neutrinos, EDMs get strongly enhanced while LFV decays not. The latter are simply modified by taking into account the different mass thresholds:

$$\delta_{ij}^{LL} \propto \sum_k C^k, \quad C^k \equiv Y_{\nu\,ik}^\dagger \ln\frac{\Lambda}{M_k} Y_{\nu kj}. \tag{5.143}$$

On the contrary for EDMs the see-saw induced FC and FV contributions—coming respectively from $\mathrm{Im}\,a_i$ [229, 242, 842, 843] and $\mathrm{Im}(\delta^{RR} m_\ell \delta^{LL})_{ii}$ [843, 844]—arise at 2nd and 3rd order and are proportional to the combinations [684]:

$$\mathrm{Im}(a_i) \propto \sum_{k>k'} \frac{\ln(M_k/M_{k'})}{\ln(\Lambda/M_{k'})} \mathrm{Im}(C^k C^{k'})_{ii},$$

$$\mathrm{Im}(\delta^{RR} m_\ell \delta^{LL})_{ii} \propto \sum_{k>k'} \widetilde{\ln}_{k'}^k \mathrm{Im}(C^k m_\ell^2 C^{k'})_{ii},$$

where $\widetilde{\ln}_{k'}^k$ is a logarithmic function. The FV contribution generically dominates for $\tan\beta \gtrsim 10$. Without going in the details of this formulae, we just display some representative upper estimates for the see-saw induced EDMs, considering for definiteness the g_μ region with $\tan\beta = 20$. The see-saw induced d_μ is below the planned sensitivity, $d_\mu^{SS} \lesssim 10^{-25}$ e cm. On the contrary for d_e it could be at hand, $d_e^{SS} \lesssim 0.5 \times 10^{-27}$ e cm; the expectation is however strongly model dependent and usually see-saw models that satisfy the bound from $\mu \to e\gamma$ predict a much smaller value [684]. The possibility of large d_e and its correlation with leptogenesis is discussed in [329].

Perspectives are much more interesting if there is also a stage of grand unification. In minimal supersymmetric $SU(5)$, the Higgs triplet Yukawa couplings contribute to the RGE-running for energies larger than their mass scale $M_T \sim M_{\mathrm{GUT}}$. For LFV, δ^{RR} is generated at 1st order and is proportional to a combination of the up quark Yukawa couplings [683]:

$$\delta_{ij}^{RR} \propto (Y_u^T Y_u^*)_{ij} \ln\frac{\Lambda}{M_T}. \tag{5.144}$$

Due to the weaker experimental bounds on δ^{RR}, this contribution is not very significant. On the other hand δ^{LL} is not changed, as also happens to the FC contribution to EDMs [843]. The FV contribution to EDMs is on the contrary strongly enhanced: it arises at 2nd order (also for degenerate right handed neutrinos) and is proportional to:

$$\mathrm{Im}(\delta^{RR} m_\ell \delta^{LL})_{ii} \propto \mathrm{Im}(C m_\ell Y_u^T Y_u^*)_{ii} \ln\frac{\Lambda}{M_T}. \tag{5.145}$$

As a result, considering for definiteness the g_μ region with $\tan\beta = 20$ and the representative values for triplet and right handed neutrino masses $M_T = 2 \times 10^{16}$ GeV and $M_3 = 10^{15}$ GeV, the induced d_μ is still below planned, $d_\mu^{SS5} \lesssim 5 \times 10^{-25}$ e cm, but the induced d_e could exceed by much the present limit: $d_e^{SS5} \lesssim 10^{-25}$ e cm. In turn this means that $\mathrm{Im}(e^{-i\beta} C_{13}) \lesssim 0.1$ (β being the angle of the unitarity triangle), which has of course an impact on see-saw models. Further details can be found in [843, 845]. Notice however that, in addition to the problems with light fermion masses, minimal supersymmetric $SU(5)$ is generically considered to be ruled out by proton decay induced by Higgs triplet exchange.

More realistic GUTs like $SO(10)$ succeed in suppressing the proton decay rate by introducing more Higgs triplets and enforcing a peculiar structure for their mass matrix. What are the expectation for d_e in this case? Consider what happens in a semirealistic $SO(10)$ model [845], where in

---
[29]The SM fermion Yukawa couplings induce negligible effects.

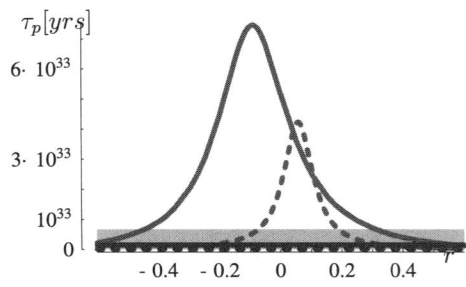

Fig. 40 (Color online) The predictions for $\tau_{p \to K\bar{\nu}}$ and d_e are shown as a function of r for the degenerate (flat *blue*) and close to pseudo Dirac (*red*) cases by taking $m_T = 10^{17}$ GeV, $\Lambda = 2 \times 10^{18}$ GeV and maximal CPV phase for d_e. The supersymmetric parameters $\tan\beta = 3$, $\tilde{M}_1 = 200$ GeV, $\bar{m}_R = 400$ GeV, have been selected. The *shaded* (*grey*) *regions* are excluded experimentally. See [845] for more details

addition to the three 16 fermion representations we introduce a couple of 10_H's containing the Higgs doublets and triplets, $10_H^u = (H_D^u, H_T^u) + (\bar{H}_D^u, \bar{H}_T^u)$, $10_H^d = (H_D^d, H_T^d) + (\bar{H}_D^d, \bar{H}_T^d)$. Up and down quark fermion masses arise when the doublets H_D^u and $\bar{H}_D^d$ acquire a VEV; in particular $Y_\nu = Y_u$, $Y_\ell = Y_d$, and also the triplet Yukawa couplings are fully determined in terms of Y_u and Y_d. As for the mass matrices of the Higgses, the doublets are diagonal in this basis, while the triplets are a priori undetermined:

$$\begin{pmatrix} \bar{H}_D^d & \bar{H}_D^u \end{pmatrix} \begin{pmatrix} \text{e.w.} & 0 \\ 0 & M_{\text{GUT}} \end{pmatrix} \begin{pmatrix} H_D^u \\ H_D^d \end{pmatrix},$$

$$\begin{pmatrix} \bar{H}_T^d & \bar{H}_T^u \end{pmatrix} M_T \begin{pmatrix} H_T^u \\ H_T^d \end{pmatrix}. \quad (5.146)$$

Let consider two limiting cases for the pattern of the triplet mass matrix, diagonal degenerate and close to pseudo-Dirac:

$$M_T^{\text{deg}} = \begin{pmatrix} 1 & 0 \\ 0 & 1 \end{pmatrix} m_T, \qquad M_T^{cpD} = \begin{pmatrix} 0 & 1 \\ 1 & r \end{pmatrix} m_T, \quad (5.147)$$

where r is a small real parameter, $r < 1$, and the exact pseudo-Dirac form corresponds to the limit $r \to 0$. Notice that the close to pseudo-Dirac form is naturally obtained in the Dimopoulos–Wilczek mechanism to solve the doublet-triplet splitting problem. The prediction for proton life-time displays a strong dependence on the structure of M_T, and only the pseudo-Dirac form is allowed, as can be seen in Fig. 40 (there is an intrinsic ambiguity due to GUT phases, so that the prediction is in between the dotted and solid curves). For EDMs on the contrary the Higgs triplets contribution to RGE is cumulative and, due to the log, mildly sensitive to the triplet mass matrix structure. In the case of $O(1)$ CPV phase (a small phase would be unnatural in this context), d_e would exceed the present bound for the values of supersymmetric parameters selected in Fig. 40. Planned searches will be a fortiori more constraining. The impact of these results go beyond the essential model described above. Indeed, the week points of the model, like the fermion mass spectra, could be addressed without changing by much the expectations for d_e. It is remarkable that *EDMs turns out to be complementary to proton decay in constraining supersymmetric GUTs*.

In the above model one obtains the relation $d_\mu/d_e \sim |V_{ts}/V_{td}|^2 \approx 25$, so that the prediction for d_μ is below the planned sensitivity. However, there are GUT models where this is not the case. For instance a significant d_μ is obtained in L-R symmetric guts [846].

5.7.1 Electron–neutron EDM correlations in SUSY

One of the questions we would like to address is whether non-zero EDM signals can constitute indirect evidence for supersymmetry. Supersymmetric models contain additional sources of CP violation compared to the SM, which induce considerable and usually too large EDMs (Fig. 41). In typical (but not all) SUSY models, the same CP-violating source induces both hadronic and leptonic EDMs such that these are correlated. The most important source is usually the CP-phase of the μ-term and, in certain non-universal scenarios, the gaugino phases. The CP-phases of the A-terms generally lead to smaller contributions.

Typical SUSY models lead to $|d_n|/|d_e| \sim \mathcal{O}(10)$–$\mathcal{O}(100)$. Thus, if both the neutron and the electron EDMs are observed, and this relation is found, it can be viewed as a clue pointing towards supersymmetry.

Since generic SUSY models suffer from the "SUSY CP problem", EDMs should be analyzed in classes of models which allow for their suppression. These include models with either small CP phases or heavy spectra. d_n–d_e EDM correlations have been analyzed in mSUGRA, the decoupling scenario with 2 heavy sfermion generations, and split

Fig. 41 One loop EDM contributions

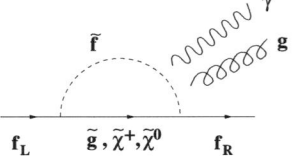

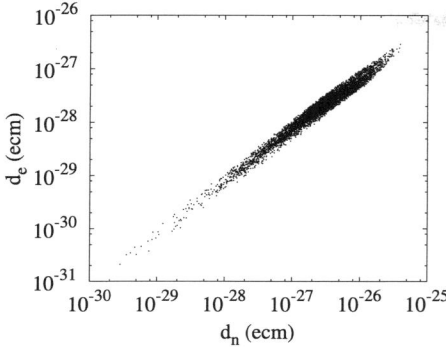

Fig. 42 EDM correlation in mSUGRA

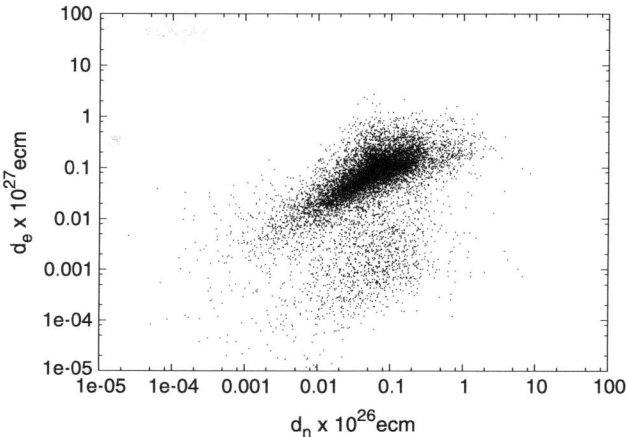

Fig. 43 EDM correlation in non-universal SUSY models

SUSY [847]. Assuming that the SUSY CP phases are all of the same order of magnitude at the GUT scale, one finds

mSUGRA: $d_e \sim 10^{-1} d_n$,
split SUSY: $d_e \sim 10^{-1} d_n$,
decoupling: $d_e \sim (10^{-1}\text{--}10^{-2}) d_n$.

These results are insensitive to $\tan\beta$ and order one variations in the mass parameters. The d_e/d_n ratio is dominated by the factor $m_e/m_q \sim 10^{-1}$, although different diagrams contribute to d_e and d_n.

An example of the d_n–d_e correlation in mSUGRA is presented in Fig. 42. There $m_0, m_{1/2}, |A|$ are varied randomly in the range [200 GeV, 1 TeV], $\tan\beta = 5$ and the phase of the μ-term ϕ_μ is taken to be in the range $[-\pi/500, \pi/500]$. The effect of the phase of the A-terms, ϕ_A, is negligible as long is it is of the same order of magnitude as ϕ_μ at the GUT scale. Clearly, the relation $d_n/d_e \sim 10$ holds for essentially all parameter values.

As the next step, we would like to see how stable these correlations are. One might expect that breaking universality at the GUT scale would completely invalidate the above results. To answer this question, we study a non-universal MSSM parameterized by

$$m_{\text{squark}}, \quad m_{\text{slepton}}, \quad M_3, \quad M_1 = M_2, \quad |A|, \\ \phi_\mu, \quad \phi_A, \quad \phi_{M_3} \quad (5.148)$$

at the GUT scale. The mass parameters are varied randomly in the range [200 GeV, 2 TeV] and the phases in the range $[-\pi/300, \pi/300]$, $\tan\beta = 5$. We find that although the correlation is not as precise as in the mSUGRA case, about 90% of the points satisfy the relation $d_n/d_e \sim 10\text{--}100$ (Fig. 43). In most of the remaining 10%, $10^4 > d_n/d_e > 10^2$, which arise when the gluino phase dominates. The reason for the correlation is that in most cases ϕ_μ is significant and induces both d_n and d_e. Apart from the factor m_q/m_e, the SUSY EDM diagrams are comparable as long as there are no large mass hierarchies in the SUSY spectrum. This means that the EDM correlation survives to a large extent, although it is possible to violate it in certain cases.

It is instructive to compare the SUSY EDM 'prediction' to those of other models. Start with the standard model. The SM background due to the CKM phase is very small, probably beyond the experimental reach. The neutron EDM can also be induced by the QCD θ-term,

$$d_n \sim 3 \times 10^{-16} \theta \; e \, \text{cm}, \quad (5.149)$$

which does not affect the electron EDM. Thus, one has $d_n \gg d_e$.

In extra dimensional models, usually there are no extra sources of CP violation and the EDM predictions are similar to the SM values. Two Higgs doublet models have additional sources of CP violation, however, the leading EDM contributions appear at 2 or 3 loops such that the typical EDM values are significantly smaller than those in SUSY models.

To conclude, we find that typical SUSY models predict $|d_n|/|d_e| \sim \mathcal{O}(10)\text{--}\mathcal{O}(100)$. Thus, if

$$d_e > d_n \quad (5.150)$$

or

$$d_e \ll d_n \quad (5.151)$$

is found, common SUSY scenarios would be disfavored, although such relations could still be obtained in baroque SUSY models.

It is interesting to consider SUSY GUT model, where CP phases in the neutrino Yukawa couplings contributes to hadronic EDMs. For instance, in $SU(5)$ SUSY GUT with right handed neutrinos, not only large mixing but also CP-violating phases in neutrino sector give significant contribution to the mixing and CP phases in the right handed scalar down sector. Though 1–2 mixing in the neutrino Yukawa coupling is strongly restricted by the $B(\mu \to e\gamma)$, 2–3 mixing in the neutrino Yukawa couplings can be significantly large and this case is interesting in B physics. Large 2–3 mixing with CP violation in neutrino sector may give a significant contribution not only to the $B(\tau \to \mu\gamma)$ but also to color EDM of s quark which may affect [848, 849] neutron and Hg EDM.

5.7.2 EDMs in split supersymmetry

Supersymmetry breaking terms involve many new sources of CP violation. Particularly worrisome are the phases associated with the invariants $\arg(A^* M_{\tilde{g}})$ and $\arg(A^* B)$, where A and B represent the usual trilinear and bilinear soft terms and $M_{\tilde{g}}$ the gaugino masses. Such phases survive in the universal limit in which all the flavor structure originates from the SM Yukawa's. If these phases are of order one, the electron and neutron EDMs induced at one loop by gaugino-sfermion exchange are typically (barring accidental cancellations [848, 850, 851]) a couple of orders of magnitude above the limits [852–855], a difficulty which is known as the supersymmetric CP problem.

Different remedies are available to this problem making the one loop sfermion contribution to the EDMs small enough, each with its pros and cons. One remedy is to have heavy enough sfermions (say heavier than 50–100 TeV to be on the safe side). Gauginos and higgsinos are not required to be heavy, and can be closer to the electroweak scale, thus preserving the supersymmetric solution to the dark matter problem and gauge coupling unification. This is the "Split" limit of the MSSM [457, 458, 856]. In this limit, the heavy sfermions suppress the dangerous one loop contributions to a negligible level. Nevertheless, some phases survive below the sfermion mass scale and, if they do not vanish for an accidental or a symmetry reason, they give rise to EDMs that are safely below the experimental limits, but sizable enough to be well within the sensitivity of the next generation of experiments [847, 856–858]. Such contributions only arise at the two-loop level, since the new phases appear in the gaugino–higgsino sector, which is not directly coupled to the SM fermions.

Besides the large EDMs, a number of additional unsatisfactory issues, all related to the presence of TeV scalars, plague the MSSM. The number of parameters exceeds 100; flavor changing neutral current processes are also one or two orders of magnitudes above the experimental limits in most of the wide parameter space; in the context of a grand unified theory, the proton decay rate associated to sfermion-mediated dimension five operators is ruled out by the Super-Kamiokande limit, at least in the minimal version of the supersymmetric $SU(5)$ model; in the supergravity context, another potential problem comes from the gravitino decay, whose rate is slow enough to interfere with primordial nucleosynthesis. While none of those issues is of course deadly—remedies are well known for each of them—it should be noted that the split solution of the supersymmetric CP problem also solves all of those issues at once. At the same time, it gives rise to a predictive framework, characterized by a rich, new phenomenology, mostly determined in terms of only 4 relevant parameters. Of course the price to be paid to make the sfermions heavy is the large fine-tuning (FT) necessary to reproduce the Higgs mass, which exacerbates the FT problem already present in the MSSM. This could be hard to accept, or not, depending on the interpretation of the FT problem, the two extreme attitudes being (i) ignoring the problem, as long as the tuning is not much worse than permille and (ii) accepting a tuning in the Higgs mass as we accept the tuning of the cosmological constant, as in split supersymmetry. The second possibility can in turn be considered as a manifestation of an anthropic selection principle [859–863].

Before moving the quantitative discussion of the effect, let us note that the pure gaugino–higgsino contribution to the EDMs, dominant in split supersymmetry and possibly near the experimental limit, might also be important in the non-split case, depending on the mechanism invoked to push the one loop sfermion contribution below the experimental limit.

5.7.2.1 Sources of CP violation in the split limit

Below the heavy sfermion mass scale, denoted generically by $\tilde{m}$, the MSSM gauginos and higgsinos, together with the SM fields constitute the field content of the model. The only interactions of gauginos and higgsinos besides the gauge ones are

$$-\mathcal{L} = \sqrt{2}(\tilde{g}_u H^\dagger \tilde{W}^a T_a \tilde{H}_u + \tilde{g}'_u Y_{H_u} H^\dagger \tilde{B} \tilde{H}_u + \tilde{g}_d H_c^\dagger \tilde{W}^a T_a \tilde{H}_d + \tilde{g}'_d Y_{H_d} H_c^\dagger \tilde{B} \tilde{H}_d) + \text{h.c.}, \quad (5.152)$$

where the Higgs–higgsino–gaugino couplings $\tilde{g}_u$, $\tilde{g}_d$, $\tilde{g}'_u$, $\tilde{g}'_d$ can be expressed in terms of the gauge couplings and $\tan\beta$ through the matching with the supersymmetric gauge interactions at the scale $\tilde{m}$, $H_c = i\sigma_2 H^*$, T_a are the $SU(2)$ generators, and $Y_{H_u} = -Y_{H_d} = 1/2$. CP violating phases can enter the effective Lagrangian below the sfermion mass scale $\tilde{m}$ through the μ-parameter, the gaugino masses M_i, $i = 1, 2, 3$, or the couplings $\tilde{g}_u$, $\tilde{g}_d$, $\tilde{g}'_u$, $\tilde{g}'_d$ (besides of course the Yukawa couplings, not relevant here). Only three combinations of the phases of the above parameters are physical, in a basis in which the Higgs VEV is in its usual form, $\langle H \rangle = (0, v)^T$, with v positive. The three combinations are $\phi_1 = \arg(\tilde{g}_u'^* \tilde{g}_d'^* M_1 \mu)$, $\phi_2 = \arg(\tilde{g}_u^* \tilde{g}_d^* M_2 \mu)$, $\xi = \arg(\tilde{g}_u \tilde{g}_d \tilde{g}_u'^* \tilde{g}_d'^*)$. Actually, the parameters above are not independent themselves. The tree-level matching with the full theory above $\tilde{m}$ gives in fact $\arg(\tilde{g}_u) = \arg(\tilde{g}'_u)$, $\arg(\tilde{g}_d) = \arg(\tilde{g}'_d)$. As a consequence, the phase ξ vanishes, thus leaving only two independent phases. Moreover, if the phases of M_1 and M_2 are equal, as in most models of supersymmetry breaking, there is actually only one CP invariant: $\phi_2 = \arg(\tilde{g}_u^* \tilde{g}_d^* M_2 \mu)$.

Fig. 44 Two loop contributions to the light SM fermion EDMs. The third diagram is for a down-type fermion f

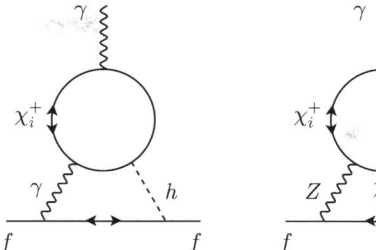
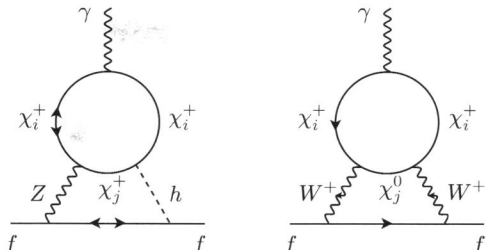

In terms of mass eigenstates, the relevant interactions are

$$-\mathcal{L} = \frac{g}{c_W} \overline{\chi_i^+} \gamma^\mu (G_{ij}^R P_R + G_{ij}^L P_L) \chi_j^+ Z_\mu$$
$$+ \left[g \overline{\chi_i^+} \gamma^\mu (C_{ij}^R P_R + C_{ij}^L P_L) \chi_j^0 W_\mu^+ \right.$$
$$\left. + \frac{g}{\sqrt{2}} \overline{\chi_i^+} (D_{ij}^R P_R + D_{ij}^L P_L) \chi_j^+ h + \text{h.c.} \right], \quad (5.153)$$

where

$$G_{ij}^L = V_{iW^+} c_W + V_{W^+j}^\dagger + V_{ih_u^+} c_{h_u^+} V_{h_u^+j}^\dagger, \quad (5.154)$$

$$-G_{ij}^{R*} = U_{iW^-} c_W - U_{W^-j}^\dagger + U_{ih_d^-} c_{h_d^-} U_{h_d^-j}^\dagger,$$

$$C_{ij}^L = -V_{iW^+} N_{jW_3}^* + \frac{1}{\sqrt{2}} V_{ih_u^+} N_{jh_u^0}^*, \quad (5.155)$$

$$C_{ij}^R = -U_{iW^-}^* N_{jW_3} - \frac{1}{\sqrt{2}} U_{ih_d^-}^* N_{jh_d^0},$$

$$g D_{ij}^R = \tilde{g}_u^* V_{ih_u^+} U_{jW^-} + \tilde{g}_d^* V_{iW^+} U_{jh_d^-}, \quad (5.156)$$

$$D^L = (D^R)^\dagger.$$

In (5.154), $c_f = T_{3f} - s_W^2 Q_f$ ($s_W^2 \equiv \sin^2\theta_W$) is the neutral current coupling coefficient of the fermion $\tilde{f}$ and, accordingly, $c_{W^\pm} = \pm\cos^2\theta_W$, $c_{h_u^+,h_d^-} = \pm(1/2 - s_W^2)$. The matrices U, V, N diagonalize the complex chargino and neutralino mass matrices, $M_+ = U^T M_+^D V$, $M_0 = N^T N_0^D N$, where $M_+^D = \text{Diag}(M_1^+, M_2^+) \geq 0$, $M_0^D = \text{Diag}(M_1^0, \ldots, M_4^0) \geq 0$ and

$$M_+ = \begin{pmatrix} M_2 & \tilde{g}_u v \\ \tilde{g}_d v & \mu \end{pmatrix},$$

$$M_0 = \begin{pmatrix} M_1 & 0 & -\tilde{g}_d' v/\sqrt{2} & \tilde{g}_u' v/\sqrt{2} \\ 0 & M_2 & \tilde{g}_d v/\sqrt{2} & -\tilde{g}_u v/\sqrt{2} \\ -\tilde{g}_d' v/\sqrt{2} & \tilde{g}_d v/\sqrt{2} & 0 & -\mu \\ \tilde{g}_u' v/\sqrt{2} & -\tilde{g}_u v/\sqrt{2} & -\mu & 0 \end{pmatrix}.$$

(5.157)

5.7.2.2 Two loop contributions to EDMs Fermion EDMs are generated only at two loops, since charginos and neutralinos, which carry the information on CP violation, are only coupled to gauge and Higgs bosons. Three diagrams contribute to the EDM of the light SM fermion f at the two-loop level. They are induced by the effective $\gamma\gamma h$, $\gamma Z h$, and γWW effective couplings and are shown in Fig. 44. The EDM d_f of the fermion f is then given by [857], where

$$d_f = d_f^{\gamma H} + d_f^{ZH} + d_f^{WW}, \quad (5.158)$$

$$d_f^{\gamma H} = \frac{e Q_f \alpha^2}{4\sqrt{2}\pi^2 s_W^2} \text{Im}(D_{ii}^R) \frac{m_f M_i^+}{M_W m_H^2} f_{\gamma H}(r_{iH}^+) \quad (5.159)$$

$$d_f^{ZH} = \frac{e(T_{3fL} - 2s_W^2 Q_f)\alpha^2}{16\sqrt{2}\pi^2 c_W^2 s_W^4} \text{Im}(D_{ij}^R G_{ji}^R - D_{ij}^L G_{ji}^L)$$
$$\times \frac{m_f M_i^+}{M_W m_H^2} f_{ZH}(r_{ZH}, r_{iH}^+, r_{jH}^+), \quad (5.160)$$

$$d_f^{WW} = \frac{e T_{3fL} \alpha^2}{8\pi^2 s_W^4} \text{Im}(C_{ij}^L C_{ij}^{R*}) \frac{m_f M_i^+ M_j^0}{M_W^4}$$
$$\times f_{WW}(r_{iW}^+, r_{jW}^0). \quad (5.161)$$

In (5.159) a sum over indices i, j is understood, Q_f is the charge of the fermion f, T_{3fL} is the third component of the weak isospin of the fermion's left handed component. Also, $r_{ZH} = (M_Z/m_H)^2$, $r_{iH}^+ = (M_i^+/m_H)^2$, $r_{iW}^+ = (M_i^+/M_W)^2$, $r_{iW}^0 = (M_i^0/M_W)^2$, where m_H is the Higgs mass, and the loop functions are given by

$$f_{\gamma H}(r) = \int_0^1 \frac{dx}{1-x} j\left(0, \frac{r}{x(1-x)}\right), \quad (5.162)$$

$$f_{ZH}(r, r_1, r_2) = \frac{1}{2} \int_0^1 \frac{dx}{x(1-x)} j\left(r, \frac{xr_1 + (1-x)r_2}{x(1-x)}\right), \quad (5.163)$$

$$f_{WW}(r_1, r_2) = \int_0^1 \frac{dx}{1-x} j\left(0, \frac{xr_1 + (1-x)r_2}{x(1-x)}\right). \quad (5.164)$$

Their analytic expressions can be found in Ref. [857]. The symmetric loop function $j(r, s)$ is defined recursively by

$$j(r) = \frac{r \log r}{r-1}, \qquad j(r,s) = \frac{j(r) - j(s)}{r-s}. \quad (5.165)$$

Equation (5.159) hold at the chargino mass scale M^+. The neutron EDM is determined as a function of the down and up quark dipoles at a much lower scale μ, at which

$$d_q(\mu) = \eta_{\mathrm{QCD}} d_q(M^+), \quad \eta_{\mathrm{QCD}} = \left[\frac{\alpha_s(M^+)}{\alpha_s(\mu)}\right]^{\gamma/2b}, \tag{5.166}$$

where the β-function coefficient is $b = 11 - 2n_q/3$ and n_q is the number of effective light quarks. The anomalous-dimension coefficient is $\gamma = 8/3$. For $\alpha_s(M_Z) = 0.118 \pm 0.004$ and $\mu = 1$ GeV (the scale of the neutron mass), the value of η_{QCD} is 0.75 for $M^+ = 1$ TeV and 0.77 for $M^+ = 200$ GeV. We expect an uncertainty of about 5% from next-to-leading order effects. This result [857] gives a QCD renormalization coefficient about a factor of 2 smaller than usually considered [864], and it agrees with the recent findings of Ref. [865].

The neutron EDM can be expressed in terms of the quark EDMs using QCD sum rules [866, 867]:

$$d_n = (1 \pm 0.5) \frac{f_\pi^2 m_\pi^2}{(m_u + m_d)(225\,\mathrm{MeV})^3}\left(\frac{4}{3}d_d - \frac{1}{3}d_u\right), \tag{5.167}$$

where $f_\pi \approx 92$ MeV and we have neglected the contribution of the quark chromo-electric dipoles, which does not arise at the two-loop level in the heavy-squark mass limit. Since d_d and d_u are proportional to the corresponding quark masses, d_n depends on the light quark masses only through the ratio m_u/m_d, for which we take the value $m_u/m_d = 0.553 \pm 0.043$.

It is instructive to consider the limit $M_i, \mu \gg M_Z, m_H$ which simplifies the EDM dependence on the CP-violating invariants $|\tilde{g}_u \tilde{g}_d / M_2 \mu| \sin\phi_2$ and $|\tilde{g}'_u \tilde{g}'_d / M_1 \mu| \sin\phi_1$. The terms depending on the second invariant are actually suppressed, so that both the electron and neutron EDM are mostly characterized by a single invariant even in the case in which the phases of M_1 and M_2 are different. The relative importance of the three contributions to d_f in (5.158) can be estimated to leading order in $\log(M_2\mu/m_H^2)$ from

$$\frac{d_f^{ZH}}{d_f^{\gamma H}} \approx \frac{(T_{3f_L} - 2s_W^2 Q_f)(3 - 4s_W^2)}{8 c_W^2 Q_f};$$

$$\frac{d_f^{WW}}{d_f^{\gamma H}} \approx -\frac{T_{3f_L}}{8 s_W^2 Q_f} \quad (M_2 = \mu). \tag{5.168}$$

Numerically, (5.168) gives $d_e^{ZH} \approx 0.05 d_e^{\gamma H}$, $d_e^{WW} \approx -0.3 d_e^{\gamma H}$ and $d_n^{ZH} \approx d_n^{\gamma H}$, $d_n^{WW} \approx -0.7 d_n^{\gamma H}$. These simple estimates show the importance of the ZH contribution to the neutron EDM.

5.7.2.3 Numerical results Let us consider a standard unified framework for the gaugino masses at the GUT scale. Using the RGEs given in Refs. [458, 868], the parameters in (5.159) can be expressed in terms of the (single) phase $\phi \equiv \phi_2$ and four positive parameters M_2, μ (evaluated at the low energy scale), $\tan\beta$, and the sfermion mass scale $\tilde{m}$. In first approximation, the dipoles depend on β and ϕ through an overall factor $\sin 2\beta \sin\phi$. The overall sfermion scale $\tilde{m}$ enters only logarithmically through the RGE equations for $\tilde{g}_{u,d}$, $\tilde{g}'_{u,d}$. The numerical results for the electron and neutrino EDMs can then conveniently be presented in the M_2–μ plane by setting $\sin 2\beta \sin\phi = 1$ (it is then sufficient to multiply the results by $\sin 2\beta \sin\phi$) and, for example, $\tilde{m} = 10^9$ GeV. Figure 45 shows the prediction for the electron EDM, the neutron EDM, and their ratio d_n/d_e. The red thick line corresponds to the present experimental limits $d_e < 1.6 \times 10^{-27}$ e cm [186], while the limit $d_n < 3 \times 10^{-26}$ e cm [869] does not impose a constraint on the parameters shown in Fig. 45.

An interesting test of split supersymmetry can be provided by a measurement of both the electron and the neutron EDMs. Indeed, in the ratio d_n/d_e the dependence on $\sin\phi$, $\tan\beta$ and $\tilde{m}$ approximately cancels out. Nevertheless, because of the different loop functions associated with the different contributions, the ratio d_n/d_e varies by $\mathcal{O}(100\%)$ when M_2 and μ are varied in the range spanned in the figures. Still, the variation of d_n/d_e is comparable to the theoretical uncertainty in (5.167), and is significantly smaller than the variation in the ordinary MSSM prediction, even in the case of universal phases [847].[30] On the other hand, the usual tight correlation between the electron and muon EDMs, $d_\mu/d_e = m_\mu/m_e$ persists.

6 Experimental tests of charged lepton universality

Lepton universality postulates that lepton interactions do not depend explicitly on lepton family number other than through their different masses and mixings. Whereas there is little doubt about the universality of electric charge there are scenarios outside the standard model in which lepton universality is violated in the interactions with W and Z bosons. Violations may also have their origin in non-SM contributions to the transition amplitudes. Such apparent violations of lepton universality can be expected in various particle decays:

– in W, Z and π decay resulting from R-parity violating extensions to the MSSM [870, 871],

---

[30] Note that the ZH contribution is missing in the analysis of the split supersymmetry case in Ref. [847], which leads to a somewhat stronger correlation between d_e and d_n.

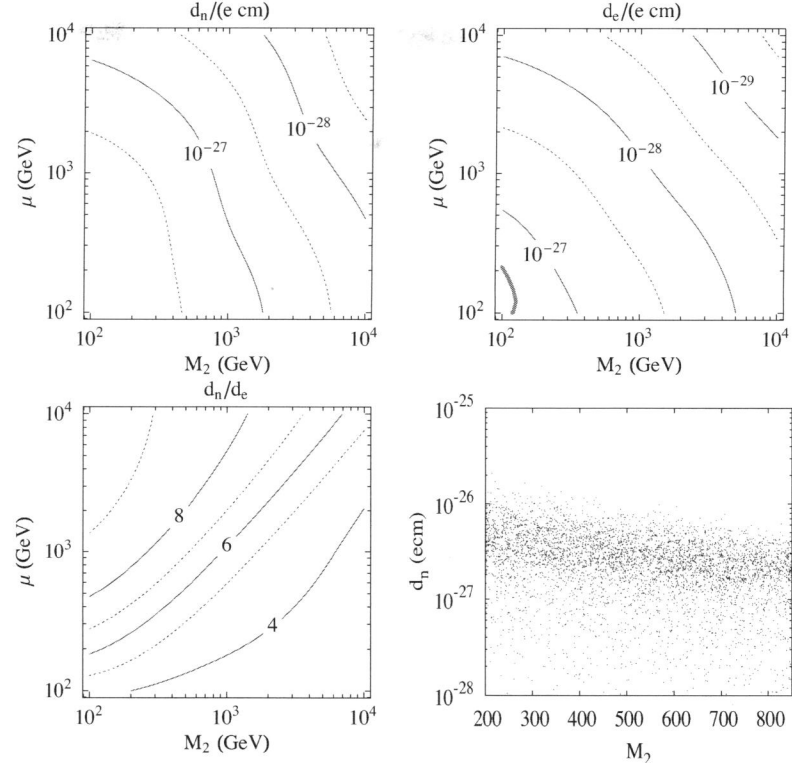

Fig. 45 Prediction for d_n, d_e, and their ratio d_n/d_e. In the contour plots we have chosen $\tan\beta = 1$, $\sin\phi = 1$, and $\tilde{m} = 10^9$ GeV. The results for d_n and d_e scale approximately linearly with $\sin 2\beta \sin\phi$, while the ratio is fairly independent of $\tan\beta$, $\sin\phi$ and $\tilde{m}$. The *red thick line* corresponds to the present experimental limit $d_e < 1.6 \times 10^{-27}$ e cm [186]. Note that the uncertainty in d_n is a factor of a few. The scatter plot shows d_n values when $M_{1,3}$ and μ are varied in the range [200 GeV, 1 TeV], m_h in [100 GeV, 300 GeV] and the CP phase in the range $[-\pi, \pi]$

- in W decay resulting from charged Higgs bosons [872, 873],
- in π decay resulting from box diagrams involving non-degenerate sleptons [874],
- in K decay resulting from LFV contributions in SUSY [839] (see Sect. 5.6),
- in Υ decay resulting from a light Higgs boson [875],
- in π and K decay from scalar interactions [876], enhanced by the strong chiral suppression of the SM amplitude for decays into $e\bar{\nu}_e$. Since these contributions result in interference terms with the SM amplitude the deviations scale with the mass M of the exchange particle like $1/M^2$ rather than $1/M^4$ as may be expected naively.

We assume the V–A Lorentz structure of the charged weak current, and parameterize universality violations by allowing for different strengths of the couplings of the individual lepton flavors:[31]

$$\mathcal{L} = \sum_{l=e,\mu,\tau} \frac{g_l}{\sqrt{2}} W_\mu \bar{\nu}_l \gamma^\mu \left(\frac{1-\gamma_5}{2}\right) l + \text{h.c.} \quad (6.1)$$

Experimental limits have recently been compiled by Loinaz et al. [877]. Results are shown in Table 16.

Following the notation of Ref. [877] one may parameterize the violations by $g_l \equiv g(1 - \epsilon_l/2)$. After introducing

[31] Still more general violations lead to deviations from the $1 - \gamma_5$ structure of the weak interaction.

Table 16 Limits on lepton universality from various processes. One should keep in mind that violations may affect the various tests differently so which constraint is best depends on the mechanism. Hypothetical non-V–A contributions, for example, would lead to larger effects in decay modes with stronger helicity suppression such as $\pi \to e\nu$ and $K \to e\nu$. Adapted from Ref. [877]. The ratios estimated from tau decays are re-calculated using PDG averages, as described in the text

Decay mode	Constraint
$W \to e\bar{\nu}_e$	$(g_\mu/g_e)_W = 0.999 \pm 0.011$
$W \to \mu\bar{\nu}_\mu$	$(g_\tau/g_e)_W = 1.029 \pm 0.014$
$W \to \tau\bar{\nu}_\tau$	
$\mu \to e\bar{\nu}_e\nu_\mu$	$(g_\mu/g_e)_\tau = 1.0002 \pm 0.0020$
$\tau \to e\bar{\nu}_e\nu_\tau$	$(g_\tau/g_e)_{\tau\mu} = 1.0012 \pm 0.0023$
$\tau \to \mu\bar{\nu}_\mu\nu_\tau$	
$\pi \to e\bar{\nu}_e$	$(g_\mu/g_e)_\pi = 1.0021 \pm 0.0016$
$\pi \to \mu\bar{\nu}_\mu$	$(g_\tau/g_e)_{\tau\pi} = 1.0030 \pm 0.0034$
$\tau \to \pi\bar{\nu}_\tau$	
$K \to e\bar{\nu}_e$	$(g_\mu/g_e)_K = 1.024 \pm 0.020$
$K \to \mu\bar{\nu}_\mu$	$(g_\tau/g_\mu)_{K\tau} = 0.979 \pm 0.017$
$\tau \to K\bar{\nu}_\tau$	

$\Delta_{ll'} \equiv \epsilon_l - \epsilon_{l'}$ the various experimental limits on deviations from lepton universality can be compared (see Fig. 46).

It is very fortunate that for most decay modes new dedicated experiments are being prepared. In the following sub-

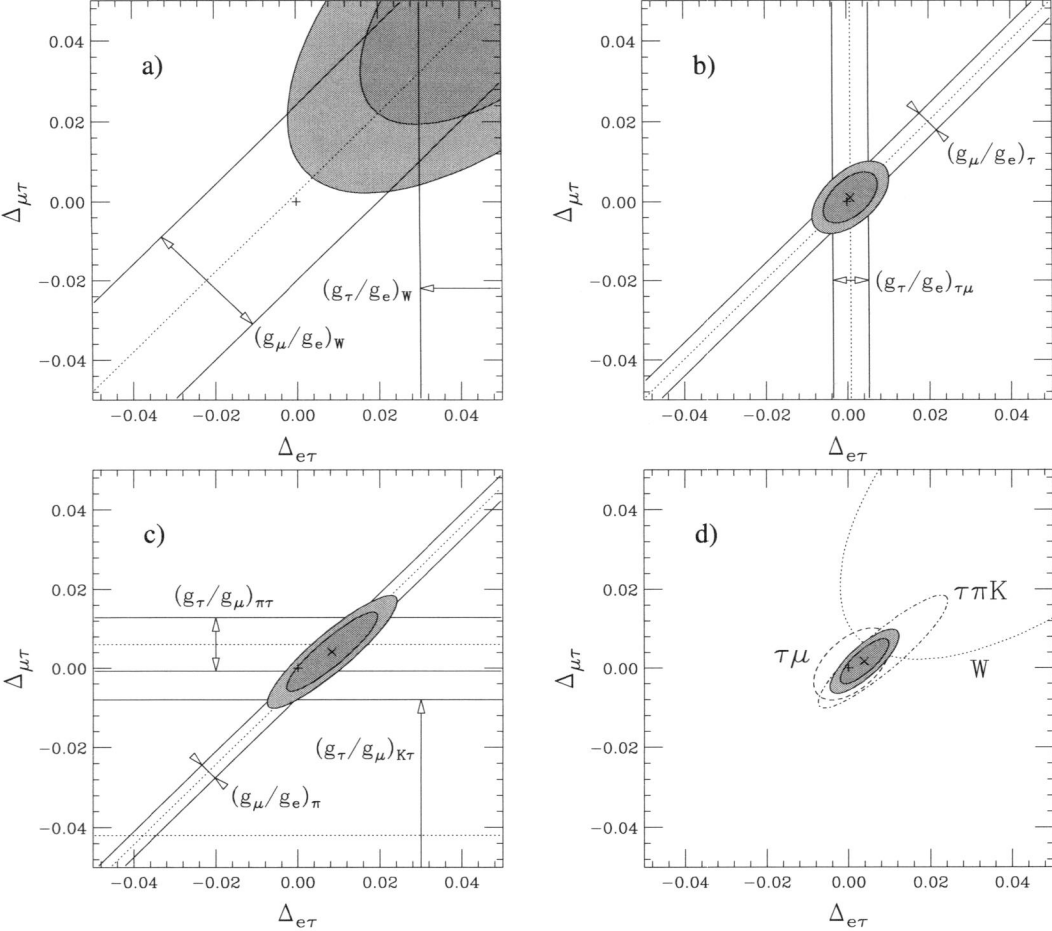

Fig. 46 Experimental constraints on violations of lepton universality from (**a**) W decay, (**b**) τ decay, (**c**) π and K decay and (**d**) the combination of (**a**)–(**c**). Parameters are defined in the text. The $\pm 1\sigma$ bands are indicated. The *shaded areas* correspond to 68% and 90% confidence levels. Results from the analysis in Ref. [877]

sections the status and prospects of these experimental tests of lepton universality are presented.

6.1 π decay

In lowest order the decay width of $\pi \to l\bar{\nu}_l$ ($l = e, \mu$) is given by:

$$\Gamma^{\text{tree}}_{\pi \to l\bar{\nu}_l} = \frac{g_l^2 g_{ud}^2 V_{ud}^2}{256\pi} \frac{f_\pi^2}{M_W^4} m_l^2 m_\pi \left(1 - \frac{m_l^2}{m_\pi^2}\right)^2. \quad (6.2)$$

By taking the branching ratio the factors affected by hadronic uncertainties cancel:

$$R^{\text{tree}}_{e/\mu} \equiv \frac{\Gamma^{\text{tree}}_{\pi \to e\bar{\nu}}}{\Gamma^{\text{tree}}_{\pi \to \mu\bar{\nu}}} = \left(\frac{g_e}{g_\mu} \times \frac{m_e}{m_\mu} \times \frac{1 - m_e^2/m_\pi^2}{1 - m_\mu^2/m_\pi^2}\right)^2.$$

Radiative corrections lower this result by 3.74(3)% [878] when assuming that final states with additional photons are included. Within the SM (i.e. $g_e = g_\mu$) this leads to:

$$R^{\text{SM}}_{e/\mu} = 1.2354(2) \times 10^{-4}. \quad (6.3)$$

Two experiments [879, 880] contribute to the present world average for the measured value:

$$R^{\text{exp}}_{e/\mu} = 1.231(4) \times 10^{-4}. \quad (6.4)$$

As a result μe universality has been tested at the level: $(g_\mu/g_e)_\pi = 1.0021(16)$.

Measurements of $R_{e/\mu}$ are based on the analysis of e^+ energy and time delay with respect to the stopping π^+. The decay $\pi \to e\nu$ is characterized by $E_{e^+} = 0.5 m_\pi c^2 = 69.3$ MeV and an exponential time distribution following the pion life-time $\tau_\pi = 26$ ns. In the case of the $\pi \to \mu\nu$ decay the 4 MeV muons, which have a range of about 1.4 mm in plastic scintillator, can be kept inside the target and are monitored by the observation of the subsequent decay $\mu \to e\nu\bar{\nu}$, which is characterized by $E_{e^+} < 0.5 m_\mu c^2 = 52.3$ MeV, and a time distribution which first grows according to the pion life-time and then falls with the muon life-time. A major systematic error is introduced by uncertainties in the low energy tail of the $\pi \to e\nu(\gamma)$ energy spectrum in the region below $0.5 m_\mu c^2$. This tail fraction typically amounts to $\approx 1\%$. The low energy tail can be studied by suppressing the $\pi \to \mu \to e$ chain by the selection of early decays and by vetoing events in which the muon is observed in the target signal. Suppression factors of typically 10^{-5} have been obtained. A study of this region is also interesting, since it might reveal the signal from a heavy sterile neutrino [881].

Although the two experiments contributing to the present world average of $R_{e/\mu}$ reached very similar statistical and systematic errors there were some significant differences. The TRIUMF experiment [880] made use of a single large NaI(Tl) crystal as main positron detector, with an energy resolution of 5% (fwhm) and a solid angle acceptance of 2.9% of 4π sr. The PSI experiment [879] used a setup of 132 identical BGO crystals with 99.8% of 4π sr acceptance and an energy resolution of 4.4% (fwhm). A large solid angle reduces the low energy tail of $\pi \to e\nu(\gamma)$ events but may also introduce a high energy tail for $\mu \to e\nu\bar{\nu}\gamma$.

Two new experiments have been approved recently aiming at a reduction of the experimental uncertainty by an order of magnitude. First results may be expected in the year 2009.

- At PSI [882] the 3π sr CsI calorimeter built for a determination of the $\pi^+ \to \pi^0 e^+ \nu$ branching ratio will be used. Large samples of $\pi \to e\nu$ decays have been recorded parasitically in the past which were used as normalization for $\pi^+ \to \pi^0 e^+ \nu$ with an accuracy of <0.3%, i.e. the level of the present experimental uncertainty of $R_{e/\mu}$. The setup was also used for the most complete studies of the radiative decays $\pi \to e\nu\gamma$ [883] and $\mu \to e\nu\bar{\nu}\gamma$ [884] done so far. Based on this experience an improvement in precision for $R_{e/\mu}$ by almost an order of magnitude is expected.
- At TRIUMF [885] a single large NaI(Tl) detector will be used again. The detector is similar in size to the one used in the previous experiment but has significantly better energy resolution. The crystal will be surrounded by CsI detectors to reduce the low energy tail of the $\pi \to e\nu$ response function. By reducing the distance between target and positron detector the geometric acceptance will be increased by an order of magnitude.

6.2 K decay

Despite the poor theoretical control over the meson decay constants, ratios of leptonic decay widths of pseudoscalar mesons such as $R_K \equiv \Gamma(K \to e\nu)/\Gamma(K \to \mu\nu)$ can be predicted with high accuracy, and have been traditionally considered as tests of the V–A structure of weak interactions through their helicity suppression and of μ–e universality. The standard model predicts [878]:

$$R_K(\text{SM}) = (2.472 \pm 0.001) \times 10^{-5} \quad (6.5)$$

to be compared with the world average [170] of published R_K measurements:

$$R_K(\exp) = (2.44 \pm 0.11) \times 10^{-5}. \quad (6.6)$$

As mentioned above the strong helicity suppression of $\Gamma(K \to e\nu)$ makes R_K sensitive to physics beyond the SM. As discussed in detail in Sect. 5.6.2.3 lepton flavor violating contributions predicted in SUSY models may lead to a deviation of R_K from the SM value in the percent range. Such contributions, arising mainly from charged Higgs exchange with large lepton flavor violating Yukawa couplings, do not decouple if SUSY masses are large and exhibit a strong dependence on $\tan\beta$. For large (but not extreme) values of this parameter, not excluded by other measurements, the interference between the SM amplitude and a double lepton-flavor violating contribution could produce a -3% effect. Other experimental constraints such as those from R_π or lepton flavor violating τ decays were shown in [839] not to be competitive with those from R_K in this scenario.

6.2.1 Preliminary NA48 results for R_K

In the original NA48/2 proposal [886] the measurement of K leptonic decays was not considered interesting enough to be mentioned. Nevertheless, triggers for such decays were implemented during the 2003 run. Since these were not very selective they had to be highly down-scaled. The data still contain about 4000 K_{e2} decays which is more than four times the previous world sample. In the analysis of these data [887] $\sim$15% background due to misidentified $K_{\mu2}$ decays was observed (see below). The preliminary result was presented at the HEP2005 Europhysics conference in Lisbon [888]:

$$R_K(\exp) = (2.416 \pm 0.043_{\text{stat}} \pm 0.024_{\text{syst}}) \times 10^{-5}, \quad (6.7)$$

marginally consistent with the SM value. While the uncertainty in this result is dominated by the statistical error, the unoptimized K_{e2} trigger and the lack of a sufficiently large control sample resulted in a $\pm0.8\%$ uncertainty.

During 2004 a 56 hours special run with simplified trigger logic at $\sim$1/4 nominal beam intensity was performed, dedicated to the collection of semileptonic $K^\pm$ decays for a measurement of $|V_{us}|$. About 4000 K_{e2} decays were extracted from these data. The preliminary result for R_K is consistent with the 2003 value with similar uncertainty although the trigger efficiencies were better known.

The NA48 apparatus includes the following subsystems relevant for the R_K measurement

- a magnetic spectrometer, composed of four drift chambers and a dipole magnet (MNP33)
- a scintillator hodoscope consisting of two planes segmented into vertical and horizontal strips, providing a fast level-1 (L1) trigger for charged particles
- a liquid krypton electromagnetic calorimeter (LKr) with an L1 trigger system.

In the analysis of the 2003-04 data K_{e2} decays were selected using two main criteria:

- $0.95 < E/pc < 1.05$ where E is the energy deposited in LKr and p is the momentum measured with the magnetic spectrometer.
- the missing mass M_X must be zero within errors, as expected for a neutrino.

The main background resulted from misidentified $K_{\mu 2}$ decays. The E/pc distribution of muons has a tail which extends to $E/pc \sim 1$ and the observed fraction of muons with $0.95 < E/pc < 1.05$ is $\sim 5 \times 10^{-6}$. $K_{\mu 2}$ background was present for $p > 25$ GeV/c where the M_X resolution provided by the magnetic spectrometer was insufficient to separate K_{e2} from $K_{\mu 2}$ decays.

6.2.2 A new measurement of $\Gamma(K \to e\nu)/\Gamma(K \to \mu\nu)$ at the SPS

During the Summer of 2007 NA62, the evolution of the NA48 experiment, has accumulated more than 100K K_{e2} decays. For this run the spectrometer momentum resolution was improved by increasing the MNP33 momentum kick from 120 to 263 MeV/c.

K_{e2} decays are selected by requiring signals from the two hodoscope planes (denoted by Q_1) and an energy deposition of at least 10 GeV in the LKr calorimeter. This trigger has an efficiency >0.99 for electron momenta $p > 15$ GeV/c. The same down-scaled Q_1 trigger was used to collect $K_{\mu 2}$ decays. The beam intensity was adjusted to obtain a total trigger rate of 10^4 Hz, which saturates the data acquisition system.

Figure 47 shows the M_X^2 versus momentum distribution for K_{e2} and $K_{\mu 2}$ decays for the 2004 data, together with the predicted distributions for the 2004 run and for the 2007 run, as obtained from a Monte Carlo simulation (for $K_{\mu 2}$ decays the electron mass is assigned to the muon). In the 2007 run, for electron momenta up to 35 GeV/c the $K_{\mu 2}$ contamination to the K_{e2} signal is reduced to a negligible level thanks to the improved spectrometer momentum resolution (see Fig. 48(a)). Using a lower limit of 15 GeV/c for the electron momentum, and taking into account the detector acceptance, this means that $\sim 43\%$ of the K_{e2} events will be kinematically background free (see Fig. 48(b)).

The fraction of $K_{\mu 2}$ faking K_{e2} decays was measured at all momenta in parallel with data taking. For this purpose

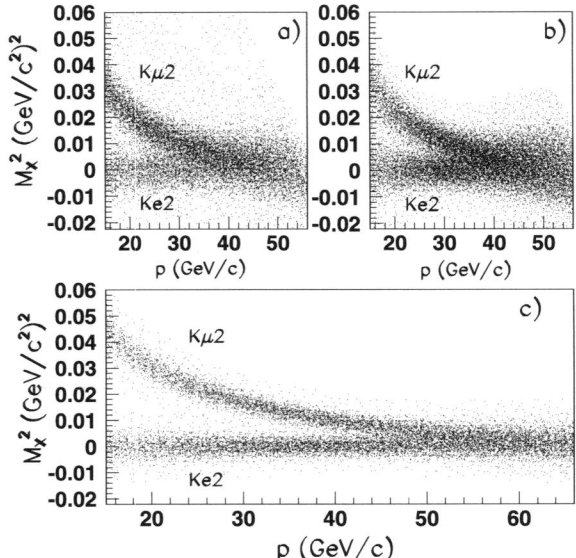

Fig. 47 Distributions of M_X^2 versus p for K_{e2} and $K_{\mu 2}$ decays. In the M_X calculation the electron mass is assumed for both processes: (**a**) measured data from the 2004 run, (**b**) Monte Carlo predictions for 2004 conditions, (**c**) Monte Carlo predictions for the conditions expected in 2007

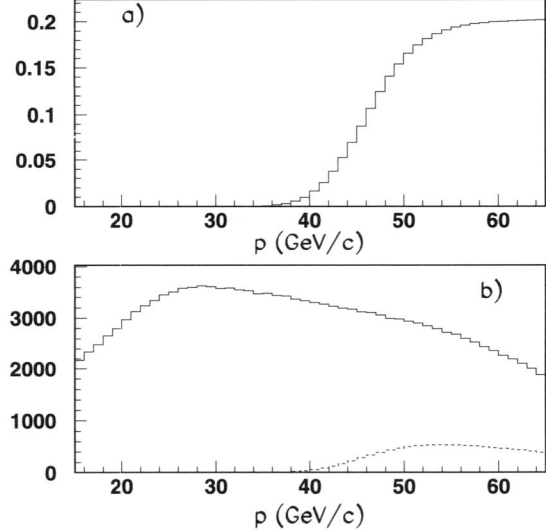

Fig. 48 (**a**) $K_{\mu 2}$ contamination in the K_{e2} sample. (**b**) Simulated momentum distributions of genuine electrons from K_{e2} decay (*full histogram*), and of fake electrons from $K_{\mu 2}$ decays (*dashed histogram*)

a ~ 5 cm thick lead plate was inserted between the two hodoscope planes covering six 6.5 cm wide vertical hodoscope counters. The requirement that charged particles traverse the lead without interacting helps to select a pure sample of $K_{\mu 2}$ decay for which the muon E/pc distribution can be directly measured for the evaluation of the $K_{\mu 2}$ contamination to the K_{e2} signal. Table 17 lists the relevant parameters describing the running conditions both for the 2004 and 2007 runs.

Table 17 Comparison of the 2004 and 2007 running conditions

	2004	2007		2004	2007
Acceptance (mr^2)	0.36×0.36	0.18×0.18	SPS duty cycle (s/s)	4.8 / 16.8	4.8 / 16.8
$\Delta\Omega$ (sr)	4×10^{-7}	1×10^{-7}	live time (days)	2.1	100
$\Delta p/p$ effective (%)	±3	±2.5	nr. of pulses	1.08×10^4	3×10^5
RMS (%)	∼3.0	∼1.8	Protons per pulse	2.5×10^{11}	1.5×10^{12}
TRIM3 x' (mr)	0	±0.3	beam momentum (GeV/c)	≈60	≈75
p_T (MeV/c)	0	±22.5	Triggers/pulse	45,000	48,000
MNP33 x' (mr)	±2.0	±3.5	Good K_{e2}/pulse	∼0.37	∼0.5
p_T (MeV/c)	±120	±263	Good K_{e2} (total)	4000	>100,000

The overall statistical error, which includes the statistical uncertainty on the background measurement, is expected to be 0.3%. The uncertainty in the trigger efficiency will be reduced to less than ±0.2%. The data collected in 2007 will provide a measurement of R_K with a total uncertainty (statistical and systematic errors combined in quadrature) of less than ±0.5%.

6.3 τ decay

There are two ways to test lepton universality in charged weak interactions using τ decays:

– the universality of all three couplings can be tested by comparing the rates of the decays $\tau \to \mu\nu\overline{\nu}$, $\tau \to e\nu\overline{\nu}$ and $\mu \to e\nu\overline{\nu}$, and
– g_τ/g_μ can be extracted by comparing $\tau \to \pi\nu$ and $\pi \to \mu\nu$.

When comparing the experimental constraints one should keep in mind the complementarity of these two tests. Whereas the purely leptonic decay modes are mediated by a transversely polarized W, the semileptonic modes involve longitudinal polarization.

6.3.1 Leptonic τ decays

The decay width of $\ell_i \to \ell_f \nu\nu$ including radiative corrections is given by [889]:

$$\Gamma(\ell_i \to \ell_f \nu\nu) = \frac{g_{\ell_i}^2 g_{\ell_f}^2}{32 m_W^2} \frac{m_{\ell_i}^5}{192\pi^3}(1 + C_{\ell_i \ell_f}), \quad (6.8)$$

where $(1 + C_{\ell_i \ell_f}) = f(x)(1 + \frac{3}{5}\frac{m_{\ell_i}^2}{M_W^2})(1 + \frac{\alpha(m_{\ell_i})}{2\pi}(\frac{25}{4} - \pi^2))$ combines weak and radiative corrections and $f(x) = 1 - 8x + 8x^3 - x^4 - 12x^2 \ln x$ with $x \equiv m_{\ell_f}^2/m_{\ell_i}^2$.

Electron–muon universality could thus be tested at the 0.2% level using:

$$\frac{g_\mu}{g_e} = \sqrt{\frac{B(\tau \to \mu\nu\nu)}{B(\tau \to e\nu\nu)} \frac{(1 + C_{\tau e})}{(1 + C_{\tau\mu})}} = 1.0002 \pm 0.0020, \quad (6.9)$$

where $C_{\tau e} = -0.004$ and $C_{\tau\mu} = -0.0313$ are the corrections from (6.8). The values of the branching ratios of leptonic τ decays are taken from [170] and are based mostly on measurements from LEP experiments. e–τ universality has been verified with similar precision:

$$\frac{g_\tau}{g_e} = \sqrt{\frac{(1 + C_{\mu e})}{(1 + C_{\tau\mu})} \frac{\tau_\mu}{\tau_\tau} \left(\frac{m_\mu}{m_\tau}\right)^5 B(\tau \to \mu\nu\nu)}$$
$$= 1.0012 \pm 0.0023, \quad (6.10)$$

where $\Gamma(\ell_i \to \ell_f \nu\nu) = B(\ell_i \to \ell_f \nu\nu)/\tau_{\ell_i}$ has been used and $C_{\mu e} = -0.0044$. The measurement of μ–τ universality can then be derived from g_τ/g_e and g_μ/g_e, giving $g_\tau/g_\mu = 1.0010 \pm 0.0023$.

The measurements used in above formulas are relatively old [170], and no input from BaBar or Belle is used. The measurements of leptonic branching fractions were done by the LEP experiments in the course of the runs at or near the Z^0 resonance [170]. The $\tau^+\tau^-$ events were selected via their topology, and the τ decay products were required to pass particle identification, using information from the calorimetry, tracking devices, time projection chambers and muon systems. The largest uncertainty on the measurement of tau branching ratios was statistical, with systematics limitations arising from the simulation and from particle identification.

The measurement of the τ life-time [170] comes from LEP experiments as well. Due to the large $\sqrt{s}$, each τ in the event has a large boost and travels 90 μm in average. However, as there is nothing but τ's produced in each event, their production vertex is unknown and has to be estimated averaging over other events or by minimizing the sum of impact parameters of both τ's decay products.

The most accurate published measurement of the τ mass [890] was done by the BES experiment, through an energy scan of the $\tau^+\tau^-$ production cross section in e^+e^- collisions around the threshold region. The collision energy scale was calibrated with J/ψ and $\psi(2S)$ resonances, with a precision of 0.25 MeV.

Therefore the major contributions to the uncertainties on the ratios g_τ/g_e and g_μ/g_e are:

- the τ leptonic branching fractions (0.3%), and
- the τ life-time (0.34%).

In the calculation above, the measurements of leptonic τ decays are taken as independent. However, there are common sources of systematic uncertainties such as uncertainties on track reconstruction, number of τ decays registered by an experiment and so on. If one measured the branching ratio $B(\tau \to e\nu\nu)/B(\tau \to \mu\nu\nu)$ directly in one experiment, as was done by ARGUS and CLEO and as is done for pion decays as well, most uncertainties would cancel. Taking the PDG average on the branching ratio [170] one obtains $g_\mu/g_e = 1.0028 \pm 0.0055$.

The following improvements can be expected in the future. The KEDR experiment is working, like BES, at the τ-pair production threshold. They plan to measure m_τ with a 0.15 MeV accuracy. A preliminary result, with accuracy comparable to BES's measurement, is available [891]. Both BaBar and Belle have accumulated large statistics of $\tau^+\tau^-$ events and should be able to perform measurements of leptonic and semileptonic τ decays, as well as to improve the measurement of the τ life-time. While the collected τ sample is much larger than at LEP, there are still significant uncertainties remaining on luminosity, tracking and particle identification. If the ratio of decay fractions is measured, then only the particle identification uncertainties will remain. Currently the electron and muon identification uncertainties for both BaBar and Belle are around 1–2%. At the B-factories the τ boost in the c.m frame is much smaller than at LEP, and in addition the energies of the e^+ and e^- beams are not the same. This leads to significant differences in the technique of the life-time measurement. In particular the 3-dimensional reconstruction of the trajectories of the decay products is poor and only the impact parameter in the plane transverse to the beams, multiplied by the polar angle of the total momentum vector of 3-prong τ decay products, can be used [892]. While the statistics allows for a very accurate measurement, the work focuses on understanding the alignment of the vertex detector and the systematics in the reconstruction of the impact parameter. The measurement of the τ mass can also be done at the B-factories. Belle has presented a mass measurement analyzing the kinematic limit of the invariant mass of 3-prong τ decays [893]. This measurement is however less precise than those of BES or KEDR.

If one takes into account recent preliminary measurements of the τ mass from the KEDR experiment [891] and of the life-time from BaBar [892], the determination of τ–e universality changes slightly to $g_\tau/g_e = 1.0021 \pm 0.0020$.

6.3.2 Hadronic τ decays

Another way to test τ–μ universality is to compare the decay rates for $\tau \to \pi\nu$ and $\pi \to \mu\nu$:

$$\frac{g_\tau^2}{g_\mu^2} = \frac{B(\tau \to \pi\nu)}{B(\pi \to \mu\nu)} \frac{\tau_\pi}{\tau_\tau} \frac{2 m_\tau m_\mu^2}{m_\pi^3} \left(\frac{m_\pi^2 - m_\mu^2}{m_\tau^2 - m_\pi^2}\right)^2 (1 + C_{\tau\pi}),$$
(6.11)

where $C_{\tau\pi} = -(1.6^{+0.9}_{-1.4})10^{-3}$ [833, 894].

Taking measurements from Ref. [170] one obtains $g_\tau/g_\mu = 0.9996 \pm 0.037$. Here the main uncertainties come from

- $\tau \to \pi\nu$ decay (1%), where the dominant contribution is due to $\tau \to \pi\pi^0\nu$ contamination and π^0 reconstruction,
- the τ life-time (0.34%), and
- the hadronic correction (0.1%).

Again, no results from the B factories are available yet, and one should expect that the large τ samples collected by BaBar and Belle will allow a significant improvement, in case the understanding of particle identification will be improved.

7 CP violation with charged leptons

There are two powerful motivations for probing CP symmetry in lepton decays:

- The discovery of CP asymmetries in B decays that are close to 100% in a sense 'de-mystifies' CP violation. For it established that complex CP phases are not intrinsically small and can be close to 90 degrees even. This de-mystification would be completed, if CP violation were found in the decays of leptons as well.
- We know that CKM dynamics, which is so successful in describing quark flavor transitions, is not relevant to baryogenesis. There are actually intriguing arguments for baryogenesis being merely a secondary effect driven by primary leptogenesis [895]. To make the latter less speculative, one has to find CP violation in dynamics of the leptonic sector.

The strength of these motivations has been well recognized in the community, as can be seen from the planned experiments to measure CP violation in neutrino oscillations and the ongoing heroic efforts to find an electron EDM. Yet there are other avenues to this goal as well that certainly are at least as difficult, namely to probe CP symmetry in muon and τ decays. Those two topics are addressed below in Sects. 7.1 and 7.2. There are also less orthodox probes, namely attempts (i) to extract an EDM for τ leptons from

$e^+e^- \to \tau^+\tau^-$, (ii) to search for a T-odd correlation in polarized ortho-positronium decays and (iii) to measure the muon transverse polarization in $K^+ \to \mu^+ \nu \pi^0$ decays. It is understood that the standard model does not produce an observable effect in any of these three cases or the other ones listed above (except for $\tau^\pm \to \nu K_S \pi^\pm$, as described below).

Concerning topic (i), one has to understand that one is searching for a CP-odd effect in an *electromagnetic* production process unlike in τ decays, which are controlled by weak forces.

In $[e^+e^-]_{\rm OP} \to 3\gamma$, topic (ii), one can construct various T-odd correlations or integrated moments between the spin vector $\vec{S}_{\rm OP}$ of polarized ortho-positronium and the momenta $\vec{k}_i$ of two of the photons that define the decay plane:

$$A_{T\rm odd} = \langle \vec{S}_{\rm OP} \cdot (\vec{k}_1 \times \vec{k}_2) \rangle,$$
$$A_{\rm CP} = \langle (\vec{S}_{\rm OP} \cdot \vec{k}_1)(\vec{S}_{\rm OP} \cdot (\vec{k}_1 \times \vec{k}_2)) \rangle. \quad (7.1)$$

– The moment $A_{T\rm odd}$ is P and CP *even*, yet T *odd*. Rather than by CP or T violation in the underlying dynamics it is generated by higher order QED processes. It has been conjectured [896] that the leading effect is formally of order α relative to the decay width due to the exchange of a photon between the two initial lepton lines. From it one has to remove the numerically leading contribution, which has to be absorbed into the bound state wave function. The remaining contribution is presumably at the sub-permille level. Alternatively $A_{T\rm odd}$ can be generated at order α^2—or at roughly the 10^{-5} level—through the interference of the lowest order decay amplitude with one where a fermion loop connects two of the photon lines.

– On the other hand the moment $A_{\rm CP}$ is odd under T as well as under P and in particular CP. Final state interactions can*not* generate a CP-odd moment with CP invariant dynamics. Observing $A_{\rm CP} \neq 0$ thus unambiguously establishes CP violation. The present experimental upper bound is around few percent; it seems feasible, see Sect. 7.4, to improve the sensitivity by more than three orders of magnitude, i.e. down to the 10^{-5} level! The caveat arises at the theoretical level: with the 'natural' scale for *weak* interference effects in positronium given by $G_F m_e^2 \sim 10^{-11}$, one needs a dramatic enhancement to obtain an observable effect.

Discussing topic (iii)—the muon transverse polarization in $K_{\mu 3}$ decays—under the heading of CP violation in the leptonic sector will seem surprising at first. Yet a general, though hand waving argument, suggests that the highly suppressed direct CP violation in nonleptonic $\Delta S = 1$—as expressed through ϵ'—rules against an observable signal even in the presence of new physics—unless the latter has a special affinity for leptons. The present status of the data and future plans are discussed in Sect. 7.3.

7.1 μ decays

The muon decay $\mu^- \to e^- \bar{\nu}_e \nu_\mu$ and its 'inverse' $\nu_\mu e^- \to \mu^- \nu_e$ are successfully described by the '$V-A$' interaction, which is a particular case of the local, derivative-free, lepton number conserving, four-fermion interaction [897]. The '$V-A$' form and the nature of the neutrinos ($\bar{\nu}_e$ and ν_e) have been determined by experiment [898–900].

The observables—energy spectra, polarizations and angular distributions—may be parameterized in terms of the dimensionless coupling constants $g^\gamma_{\varepsilon\mu}$ and the Fermi coupling constant G_F. The matrix element is

$$\mathcal{M} = \frac{4 G_F}{\sqrt{2}} \sum_{\substack{\gamma = {\rm S,V,T} \\ \varepsilon, \mu = {\rm R,L}}} g^\gamma_{\varepsilon\mu} \langle \bar{e}_\varepsilon | \Gamma^\gamma | (\nu_e)_n \rangle \langle (\bar{\nu}_\mu)_m | \Gamma_\gamma | \mu_\mu \rangle. \quad (7.2)$$

We use here the notation of Fetscher et al. [898, 901] who in turn use the sign conventions and definitions of Scheck [902]. Here $\gamma = $ S, V, T indicate a (Lorentz) scalar, vector, or tensor interaction, and the chirality of the electron or muon (right or left handed) is labeled by $\varepsilon, \mu = $ R, L. The chiralities n and m of the ν_e and the $\bar{\nu}_\mu$ are determined by given values of γ, ε and μ. The 10 complex amplitudes $g^\gamma_{\varepsilon\mu}$ and G_F constitute 19 independent parameters to be determined by experiment. The '$V-A$' interaction corresponds to $g^V_{LL} = 1$, with all other amplitudes being 0.

Experiments show the interaction to be predominantly of the vector type and left handed [$g^V_{LL}.0.96(90\%$ C.L.)] with no evidence for other couplings. The measurement of the muon life-time yields the most precise determination of the Fermi coupling constant G_F, which is presently known with a relative precision of 8×10^{-6} [903, 904]. Continued improvement of this measurement is certainly an important goal [905], since G_F is one of the fundamental parameters of the standard model.

7.1.1 T invariance in μ decays

P_{T_2}—the component of the decay positron polarization which is transverse to the positron momentum and the muon polarization—is T odd and due to the practical absence of a strong or electromagnetic final state interaction it probes T invariance. A second-generation experiment has been performed at PSI by the ETH Zürich–Cracow–PSI Collaboration [906]. They obtained, for the energy averaged transverse polarization component:

$$\langle P_{T_2} \rangle = (-3.7 \pm 7.7_{\rm stat.} \pm 3.4_{\rm syst.}) \times 10^{-3}. \quad (7.3)$$

7.1.2 Future prospects

The precision on the muon life-time can presumably be increased over the ongoing measurements by one order of magnitude [903]. Improvement in measurements of the decay parameters seems more difficult. The limits there are not given by the muon rates which usually are high enough already ($\approx 3 \times 10^8$ s^{-1} at the µE1 beam at PSI, for example), but rather by effects like positron depolarization in matter or by the small available polarization (<7%) of the electron targets used as analysers. The measurement of the transverse positron polarization might be improved with a smaller phase space (lateral beam dimension of a few millimetres or better). This experiment needs a *pulsed* beam with high polarization.

7.2 CP violation in τ decays

The betting line is that τ decays—next to the electron EDM and ν oscillations—provide the best stage to search for manifestations of CP breaking in the leptonic sector. There exists a considerable literature on the subject started by discussions on a tau-charm factory more than a decade ago [907–910] and attracting renewed interest recently [911–914] stressing the following points:

– There are many more channels than in muon decays making the constraints imposed by CPT symmetry much less restrictive.
– The τ lepton has sizable rates into multi-body final states. Due to their non-trivial kinematics asymmetries can emerge also in the final state distributions, where they are likely to be significantly larger than in the integrated widths. The channel $K_L \to \pi^+\pi^- e^+ e^-$ can illustrate this point. It commands only the tiny branching ratio of 3×10^{-7}. The forward-backward asymmetry $\langle A \rangle$ in the angle between the $\pi^+\pi^-$ and e^+e^- planes constitutes a CP *odd* observable. It has been measured by KTeV and NA48 to be truly large, namely about 13%, although it is driven by the small value of $|\epsilon_K| \sim 0.002$. I.e., one can trade branching ratio for the size an CP asymmetry.
– New physics in the form of multi-Higgs models can contribute on the tree-level like the SM W exchange.
– Some of the channels should exhibit enhanced sensitivity to new physics.
– Having polarized τ leptons provides a powerful handle on CP asymmetries and control over systematics.

These features will be explained in more detail below. It seems clear that such measurements can be performed only in e^+e^- annihilation, i.e. at the B factories running now or better still at a Super-Flavor factory, as discussed in the Working Group 2 report. There one has the added advantage that one can realistically obtain highly polarized τ leptons: This can be achieved directly by having the electron beam longitudinally polarized or more indirectly even with unpolarized beams by using the spin alignment of the produced τ pair to 'tag' the spin of the τ under study by the decay of the other τ like $\tau \to \nu\rho$.

7.2.1 $\tau \to \nu K \pi$

The most promising channels for exhibiting CP asymmetries are $\tau^- \to \nu K_S \pi^-$, $\nu K^- \pi^0$ [910]:

– Due to the heaviness of the lepton and quark flavors they are most sensitive to non-minimal Higgs dynamics while being Cabibbo suppressed in the SM.
– They can show asymmetries in the final state distributions.

The SM does generate a CP asymmetry in τ decays that should be observable. Based on known physics one can reliably predict a CP asymmetry [911]:

$$\frac{\Gamma(\tau^+ \to K_S \pi^+ \bar{\nu}) - \Gamma(\tau^- \to K_S \pi^- \nu)}{\Gamma(\tau^+ \to K_S \pi^+ \bar{\nu}) + \Gamma(\tau^- \to K_S \pi^- \nu)}$$
$$= (3.27 \pm 0.12) \times 10^{-3} \qquad (7.4)$$

due to K_S's preference for antimatter over matter. Strictly speaking, this prediction is more general than the SM: no matter what produces the CP impurity in the K_S wave function, the effect underlying (7.4) has to be present, while of course not affecting $\tau^\mp \to \nu K^\mp \pi^0$.

To generate a CP asymmetry, one needs two different amplitudes contribute coherently. This requirement is satisfied, since the $K\pi$ system can be produced from the (QCD) vacuum in a vector and scalar configuration with form factors F_V and F_S, respectively. Both are present in the data, with the vector component (mainly in the form of the K^*) dominant as expected [915]. Within the SM, there does not arise a weak phase between them on an observable level, yet it can readily be provided by a charged Higgs exchange in non-minimal Higgs models, which contributes to F_S.

A few general remarks on the phenomenology might be helpful to set the stage. For a CP violation in the underlying weak dynamics to generate an observable asymmetry in partial widths or energy distributions one needs also a relative strong phase between the two amplitudes:

$$\Gamma(\tau^- \to \nu K^- \pi^0) - \Gamma(\tau^+ \to \bar{\nu} K^+ \pi^0)$$
$$\propto \mathrm{Im}(F_H F_V^*) \, \mathrm{Im}\, g_H g_W^*, \qquad (7.5)$$

$$\frac{d}{dE_K}\Gamma(\tau^- \to \nu K^- \pi^0) - \frac{d}{dE_K}\Gamma(\tau^+ \to \bar{\nu} K^+ \pi^0)$$
$$\propto \mathrm{Im}(F_H F_V^*) \, \mathrm{Im}\, g_H g_W^*, \qquad (7.6)$$

where F_H denotes the Higgs contribution to F_S and g_H its weak coupling. This should not represent a serious restriction, since the $K\pi$ system is produced in a mass range with several resonances. If on the other hand one is searching for a T-odd correlation like

$$O_T \equiv \langle \vec{\sigma}_\tau \cdot (\vec{p}_K \times \vec{p}_\pi) \rangle, \qquad (7.7)$$

then CP violation can surface even with*out* a relative strong phase

$$O_T \propto \text{Re}(F_H F_V^*) \, \text{Im} \, g_H g_W^*. \tag{7.8}$$

Yet there is a caveat: final state interactions can generate T-odd moments even from T invariant dynamics, when one has

$$O_T \propto \text{Im}(F_H F_V^*) \, \text{Re} \, g_H g_W^*. \tag{7.9}$$

Fortunately one can differentiate between the two scenarios of (7.8), (7.9) at a B or a Super-Flavor factory, where one can compare directly the T-odd moments for the CP-conjugate pair τ^+ and τ^-:

$$O_T(\tau^+) \neq O_T(\tau^-) \implies \text{CP violation!} \tag{7.10}$$

A few numerical scenarios might illuminate the situation: a Higgs amplitude 1% or 0.1% the strength of the SM W-exchange amplitude—the former [latter] contributing [mainly] to F_S [F_V]—is safely in the 'noise' of present measurements of partial widths; yet it could conceivably create a CP asymmetry as large 1% or 0.1%, respectively. More generally a CP-odd observable in a SM allowed process is merely *linear* in a new physics amplitude, since the SM provides the other amplitude. On the other hand SM forbidden transitions—say lepton flavor violation as in $\tau \to \mu\gamma$—have to be *quadratic* in the new physics amplitude.

$$\text{CP odd} \propto |T_{\text{SM}}^* T_{\text{NP}}| \quad \text{vs.} \quad \text{LFV} \propto |T_{\text{NP}}|^2. \tag{7.11}$$

Probing CP symmetry at the 0.1% level in $\tau \to \nu K \pi$ thus has roughly the same sensitivity for a new physics amplitude as searching for $B(\tau \to \mu\gamma)$ at the 10^{-8} level.

CLEO has undertaken a pioneering search for a CP asymmetry in the angular distribution of $\tau \to \nu K_S \pi$ placing an upper bound of a few percent [916].

7.2.2 Other τ decay modes

It appears unlikely that analogous asymmetries could be observed in the Cabibbo allowed channel $\tau \to \nu \pi \pi$, yet detailed studies of $\tau \nu 3\pi/4\pi$ look promising, also because the more complex final state allows us to form T-odd correlations with unpolarized τ leptons; yet the decays of polarized τ might exhibit much larger CP asymmetries [912].

Particular attention should be paid to $\tau \to \nu K 2\pi$, which has potentially very significant additional advantages:

- One can interfere *vector* with *axial vector* $K 2\pi$ configurations.
- The larger number of kinematical variables and of specific channels should allow more internal cross checks of systematic uncertainties like detection efficiencies for positive vs. negative particles.

7.3 Search for T violation in $K^+ \to \pi^0 \mu^+ \nu$ decay

The transverse muon polarization in $K^+ \to \pi^0 \mu^+ \nu$ decay, P_T, is an excellent probe of T violation, and thus of physics beyond the standard model. Most recently the E246 experiment at the KEK proton synchrotron has set an upper bound of $|P_T| \leq 0.0050$ (90% C.L.). A next generation experiment is now being planned for the high intensity accelerator J-PARC which is aiming at more than one order of magnitude improvement in the sensitivity with $\sigma(P_T) \sim 10^{-4}$.

7.3.1 Transverse muon polarization

A non-zero value for the transverse muon polarization (P_T) in the three body decay $K \to \pi \mu \nu$ ($K_{\mu 3}$) violates T conservation with its T-odd correlation [917]. Over the last three decades dedicated experiments have been carried out in search for a non-zero P_T. Unlike other T-odd channels in e.g. nuclear beta decays, P_T in $K_{\mu 3}$ has the advantage that final state interactions (FSI), which may induce a spurious T-odd effect, are very small. With only one charged particle in the final state the FSI contribution originates only in higher loop effects and has been shown to be small. The single photon exchange contribution from two-loop diagrams was estimated more than twenty years ago as $P_T^{\text{FSI}} \leq 10^{-6}$ [918]. Quite recently two-photon exchange contributions have been studied [919]. The average value of P_T^{FSI} over the Dalitz plot was calculated to be less than 10^{-5}.

An important feature of a P_T study is the fact that the contribution from the standard model (SM) is practically zero. Since only a single element of the CKM matrix V_{us} is involved for the semileptonic $K_{\mu 3}$ decay in the SM, no CP violation appears in first order. The lowest order contribution comes from radiative corrections to the $\bar{u}\gamma_\mu(1-\gamma_5)s W^\mu$ vertex, and this was estimated to be less than 10^{-7} [920]. Therefore, non-zero P_T in the range of 10^{-3}–10^{-4} would unambiguously imply the existence of a new physics contribution [920].

Sizable P_T can be accommodated in multi-Higgs doublet models through CP violation in the Higgs sector [921–928]. P_T can be induced due to interference between charged Higgs exchange (F_S, F_P) and W exchange (F_V, F_A) as shown in Fig. 49. It is conceivable that the coupling of charged Higgs fields to leptons is strongly enhanced relative to the coupling to the up-type quarks [929] which would lead to an experimentally detectable P_T of $O(10^{-3})$. Thus, P_T could reveal a source of CP violation that escapes detection in $K \to 2\pi, 3\pi$ [920].

A number of other models also allow P_T at an observable level without conflicting with other experimental constraints, and experimental limits on P_T could thus constrain those models. Among them SUSY models with R-parity

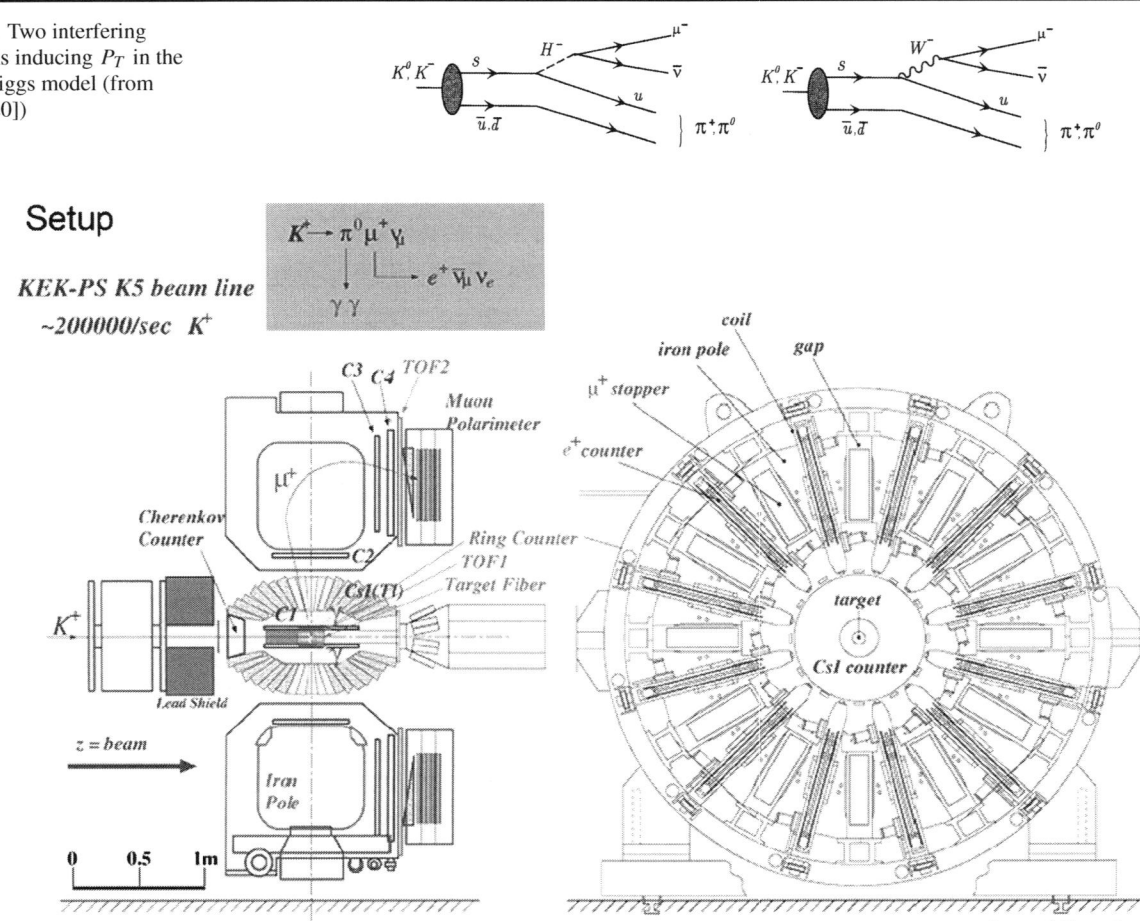

Fig. 49 Two interfering diagrams inducing P_T in the multi-Higgs model (from Ref. [920])

Fig. 50 E246 setup using the superconducting toroidal spectrometer. The elaborate detector system [933] consists of an active target (to monitor stopping K^+), a large-acceptance CsI(Tl) barrel (to detect π^0), tracking chambers (to track μ^+), and muon polarimeters (to measure P_T)

breaking [930] and a SUSY model with squark family mixing [931] should be mentioned. A recent paper [932] discusses a generic effective operator leading to a P_T expression in terms of a cut-off scale Λ and the Wilson coefficients C_S and C_T.

7.3.2 KEK E246 experiment

The most recent and highest precision P_T experiment was performed at the KEK proton synchrotron. The experiment used a stopped K^+ beam with an intensity of $\sim 10^5$/s and a setup with a superconducting toroidal spectrometer (Fig. 50). Data were taken between 1996 and 2000 for a total of 5200 hours of beam time. The determination of the muon polarization was based on a measurement of the decay positron azimuthal asymmetry in a longitudinal magnetic field using "passive polarimeters". Thanks to (i) the stopped beam method which enabled total coverage of the decay phase space and hence a forward/backward symmetric measurement with respect to the π^0 direction and (ii) the rotational-symmetric structure of the toroidal system, systematic errors could be substantially suppressed.

The T-odd asymmetry was deduced using a double ratio scheme as

$$A_T = (A_{\text{fwd}} - A_{\text{bwd}})/2, \tag{7.12}$$

where the fwd(bwd) asymmetry was calculated using the "clockwise" and "counter-clockwise" positron emission rates N_{cw} and N_{ccw} as

$$A_{\text{fwd(bwd)}} = \frac{N^{\text{cw}}_{\text{fwd(bwd)}} - N^{\text{ccw}}_{\text{fwd(bwd)}}}{N^{\text{cw}}_{\text{fwd(bwd)}} + N^{\text{ccw}}_{\text{fwd(bwd)}}}. \tag{7.13}$$

P_T was then deduced using

$$P_T = A_T / \{\alpha \langle \cos\theta_T \rangle\}' \tag{7.14}$$

with α the analyzing power and $\langle \cos\theta_T \rangle$ the average kinematic attenuation factor. The final result was [934]

$$P_T = -0.0017 \pm 0.0023(\text{stat}) \pm 0.0011(\text{syst}), \tag{7.15}$$

Table 18 Goal of the J-PARC TREK experiment compared with the E246 result

	E246 @ KEK-PS	TREK @ J-PARC
Detector	SC toroidal spectrometer	E246-upgraded
Proton beam energy	12 GeV	30 GeV
Proton intensity	1.0×10^{12}/s	6×10^{13}/s
K^+ intensity	1.0×10^{5}/s	3×10^{6}/s
Run time	$\sim 2.0 \times 10^{7}$ s	1.0×10^{7} s
$\sigma(P_T)_{\text{stat}}$	2.3×10^{-3}	$\sim 1.0 \times 10^{-4}$
$\sigma(P_T)_{\text{syst}}$	1.1×10^{-3}	$< 1.0 \times 10^{-4}$

$$\text{Im}\,\xi = -0.0053 \pm 0.0071(\text{stat}) \pm 0.0036(\text{syst}), \quad (7.16)$$

corresponding to the upper limits of $|P_T| < 0.0050$ (90% C.L.) and $|\text{Im}\,\xi| < 0.016$ (90% C.L.), respectively. Here $\text{Im}\,\xi$ is the physics parameter proportional to P_T after removal of the kinematic factor. This result constrained the three-Higgs doublet model parameter in the way of $|\text{Im}(\alpha_1 \gamma_1^*)| < 544 (M_{H_1}/\text{GeV})^2$, as the most stringent constraint on this parameter. Systematic errors were investigated thoroughly, although the total size was smaller than half of the statistical error. There were two items that could not be canceled out by any of the two cancellation mechanisms of the 12-fold azimuthal rotation and π^0-fwd/bwd: the effect from the decay plane rotation, θ_z and the misalignment of the muon magnetic field, δ_z, which should both be eliminated in the next generation J-PARC experiment.

7.3.3 The proposed J-PARC E06 (TREK) experiment

A new possible P_T experiment, E06 (TREK), at J-PARC is aiming at a sensitivity of $\sigma(P_T) \sim 10^{-4}$. J-PARC is a high intensity proton accelerator research complex now under construction in Japan with the first beam expected in 2008. In the initial phase of the machine, the main synchrotron will deliver a 9 μA proton beam at 30 GeV. A low momentum beam of 3×10^6 K^+ per second will be available for stopped K^+; this is about 30 times the beam intensity used for E246. Essentially the same detector concept will be adopted; namely the combination of a stopped K^+ beam and the toroidal spectrometer, because this system has the advantage of suppressing systematic errors by means of the double ratio measurement scheme. However, the E246 setup will be upgraded significantly. The E246 detector will be upgraded in several parts so as to accommodate the higher counting rate and to better control the systematics. The major planned upgrades are the following:

- The muon polarimeter will become an active polarimeter, providing the muon-decay vertex and the positron track, leading to an essentially background-free muon decay measurement, with an increased positron acceptance and analyzing power.
- New dipole magnets will be added, improving the field uniformity and the alignment accuracy.
- The electronics and readout of the CsI(Tl) E246 calorimeter will be replaced to maximize the counting rate, fully exploiting the intrinsic crystal speed.
- The tracking system and the active target will be improved for higher resolution and higher decay-in-flight background rejection.

As a result, 20 times higher sensitivity to P_T will be obtained after a one year run. The systematic errors will be controlled with sufficient accuracy and a final experimental error of $\sim 10^{-4}$ will be attained (see Table 18). A full description of the experiment can be found in the proposal [935].

It is now proposed to run for net 10^7s corresponding to roughly one year of J-PARC beam-time under the above mentioned beam condition. This would yield 2.4×10^9 good $K^+_{\mu 3}$ events in the π^0-fwd/bwd regions, providing an estimate of $\sigma(P_T)_{\text{stat}} = 1.35 \times 10^{-4}$. The inclusion of other π^0 regions, enabled by the adoption of the active polarimeter, would bring the statistical sensitivity further down to the 10^{-4} level. The dominant systematic errors is expected to arise from the misalignment of the polarimeter and the muon magnetic field; this will be determined from data, and Monte Carlo studies indicate a residual systematics at the 10^{-4} level.

It is proposed to run TREK in the early stage of J-PARC operation. The experimental group has already started relevant R&D for the upgrades after obtaining scientific approval, and the exact schedule will be determined after funding is granted.

7.4 Measurement of CP violation in ortho-positronium decay

CP violation in the o-Ps decay can be detected by an accurate measurement of the angular correlation between the o-Ps spin $\vec{S}_{\text{oP}}$ and the momenta of the photons from the o-Ps decay [936], as shown in (7.1). It is useful then to write the measurable quantity:

$$N(\cos\theta) = N_0(1 + C_{\text{CP}}\cos\theta), \quad (7.17)$$

with the CP violation amplitude parameter, C_{CP}, different from zero, if CP violating interactions take part in the o-Ps decay. In this equation, $N(\cos\theta)$ is the number of events with a measured value $\cos\theta \pm |\Delta(\cos(\theta))|$ (hereafter, for the sake of simplicity, it will be referred to as the $\cos\theta$ value, intending that this is measured with an uncertainty, depending on the spatial resolution of the detector). Here $\cos\theta$ is defined as the product of $\cos\theta_1$, the cosine of the angle between the $\vec{S}_{OP}$ and the unit vector in the direction of highest energy photon $\hat{k}_1$, and $\cos\theta_n$, the cosine of the angle between the $\vec{S}_{OP}$ and the unit vector in the direction perpendicular to the o-Ps decay plane, $\hat{n}$ [937].

The measured distribution $N(\cos\theta)$ should show an asymmetry given by $N(\cos\theta_+) - N(\cos\theta_-) = 2N_0 C_{CP} \times \cos\theta$, for $\cos\theta_+ = -\cos\theta_- = \cos\theta$. The quantity C_{CP} can be determined by measuring the rate of events N_+ for a given $\cos\theta_+ = \cos\theta$ and N_- for $\cos\theta_- = -\cos\theta$. In practice, N_+ is the number of events in which $\hat{k}_2$ forms an angle with $\hat{k}_1$ smaller than π, and the o-Ps spin forms an angle θ_n smaller than $\pi/2$ with the perpendicular to the o-Ps decay plane. In the N_- events, $\hat{k}_2$ forms an angle $2\pi - \theta_{12}$ with $\hat{k}_1$ and the $\vec{S}_{OP}$ forms an angle $\pi - \theta_n$ with the normal to the o-Ps decay plane. In other terms, in the N_- events the perpendicular to the o-Ps decay plane is reversed with respect to the N_+ events, by flipping the direction of $\hat{k}_2$ specularly with respect to $\hat{k}_1$. Then the measurement of the asymmetry

$$A = \frac{(N_+ - N_-)}{(N_+ + N_-)} = C_{CP} \cos\theta \qquad (7.18)$$

allows us to derive the experimental value of C_{CP}.

The measurement of the asymmetry A implies that $\cos\theta$ in (7.18) is a well defined quantity in the experiment. In turn, this implies that the o-Ps spin direction is defined. This direction can be selected using an external magnetic field $\vec{B}$, which aligns the o-Ps spin parallel ($m = 1$), perpendicular ($m = 0$) or antiparallel ($m = -1$) to the field direction. The magnetic field, in addition, perturbs and mixes the two $m = 0$ states (one for the para-Ps and the other for the o-Ps). Thus, two new states are possible for the Ps system: the perturbed singlet and the perturbed triplet states, both with $m = 0$. Their life-times depend on the $\vec{B}$ field intensity. The perturbed singlet state has a life-time shorter than 1 ns (as for the unperturbed singlet state of the para-Ps), which is not relevant in the measurement described here, because too short compared to the typical detector time resolution of 1 ns. For values of $|\vec{B}|$ of few kGauss, the perturbed o-Ps life-time can be substantially reduced [938] with respect to the unperturbed value of about 142 ns [939]. Thanks to this effect, it is possible to separate the $m = 0$ from the $m = \pm 1$ states, by measuring the o-Ps decay time. This is the time between the positron emission (by e.g., a ^{22}Na positron source) and the detection of the o-Ps decay photons. The Ps is formed in a target region, where SiO$_2$ powder is used as target material. The value of the $|\vec{B}|$ field that maximizes the decay time separation between $m = 0$ and $m = \pm 1$ states is found to be $B = 4$ kGauss, corresponding to a $m = 0$ perturbed o-Ps life-time of 30 ns.

The measurement of the asymmetry A is performed in the following way. The direction and intensity of the $\vec{B}$ field are fixed. The $\hat{k}_1$ and $\hat{k}_2$ detectors are also fixed. In this way $\cos\theta$ has a well defined value. For each event, the Ps decay time and the energies of the three photons from the o-Ps decay are measured. The off-line analysis requires the highest energy photon in the $\hat{k}_1$ detector to be within an energy range $\Delta E_1 = E_1^{max} - E_1^{min}$. The second highest energy photon must be recorded in the $\hat{k}_2$ detector within an energy range $\Delta E_2 = E_2^{max} - E_2^{min}$. Then the N_+ and N_- events are counted to determine the asymmetry in (7.18). The measurement of the asymmetry A in both the perturbed states (selected imposing short decay time, e.g., between 10 and 60 ns) and unperturbed states (selected imposing long decay time, between 60 and 170 ns) allows one to eliminate the time-independent systematics [937]. Other systematics, which are time-dependent, do not cancel out with this method and determine the final uncertainty on the C_{CP} measurement.

An improved detector with superior spatial and energy resolution, as compared to [937], is sketched in Fig. 51. It consists of a barrel of BGO crystals with the o-Ps forming region at its centre. The crystal signals are read out by avalanche photodiodes (APD), as the detector must work in the magnetic field. Improved spatial and angular resolution is obtained thanks to the smaller size of the crystal face exposed to the photons, 3×3 cm^2, and the larger barrel radius, 42 cm. Note that such a detector could also be used efficiently for PET scanning, combined with NMR diagnostic.

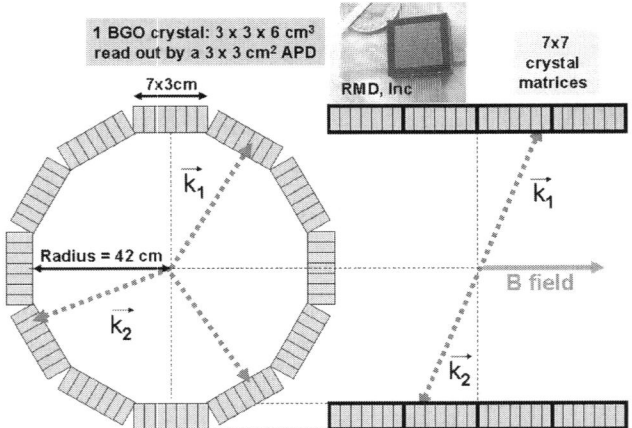

Fig. 51 Schematic view of the BGO crystal barrel calorimeter used to detect the photons from the o-Ps decay: *left*, detector front view and definition of the $\hat{k}_1$ and $\hat{k}_2$ vectors; *right*, detector side view, showing also the direction of the magnetic field $\vec{B}$

This possibility makes this detector a valuable investment also for applications in nuclear medical imaging.

With this detector configuration and a simulation of the detector response, the precision to be reached in the measurement of the $C_{\rm CP}$ parameter has been evaluated. A similar analysis was used for the event selection as described in [938], except that no veto is needed in the present configuration, thanks to the good spatial resolution of the proposed crystals. Various uncertainties affect the $C_{\rm CP}$ measurement. The time-dependent uncertainties on the asymmetry A are induced mainly by the two-photon background, which affect more strongly the events with shorter decay time, as well as by the inhomogeneity of the o-Ps formation region, which affect the measurement of the o-Ps decay time. For high event statistics (at least 10^{12} selected three photon events) the following contributions to the asymmetry measurement were found: $\Delta A_{\rm stat} \sim 10^{-6}$, $\Delta A_{\rm syst}(2\gamma\ {\rm bkgd}) \sim 10^{-6}$, $\Delta A_{\rm syst}({\rm o\text{-}Ps\ formation}) \sim 2 \times 10^{-6}$ resulting into a total uncertainty: $\Delta A_{\rm stat+syst} \sim 2.5 \times 10^{-6}$. Being $\Delta C_{\rm CP}$ related to the asymmetry total uncertainty by the relation $\Delta C_{\rm CP} = \Delta A_{\rm stat+syst}/Q$ [937] with Q, the analyzing power, evaluated to be ~ 0.5 for this detector configuration, the total uncertainty on the $C_{\rm CP}$ parameter is $\Delta C_{\rm CP} \sim 5 \times 10^{-6}$.

Although this precision is not sufficient to measure the expected standard model $C_{\rm CP}$ value of order of 10^{-9}, it is suitable to discover CP-violating terms in the order of 10^{-5}, which if detected would be signal of unexpected new physics beyond the standard model.

8 LFV experiments

Mixing of leptonic states with different family number as observed in neutrino oscillations does not necessarily imply measurable branching ratios for LFV processes involving the charged leptons. In the standard model the rates of LFV decays are suppressed relative to the dominant family-number conserving modes by a factor $(\delta m_\nu/m_W)^4$ which results in branching ratios which are out of reach experimentally. Note that a similar family changing quark decay such as $b \to s\gamma$ does obtain a very significant branching ratio of $O(10^{-4})$ due to the large top mass.

As has been discussed in great detail in this report, in almost any further extension to the standard model such as supersymmetry, grand unification or extra dimensions additional sources of LFV appear. For each scenario a large number of model calculations can be found in the literature and have been reviewed in previous sections, with predictions that may well be accessible experimentally. Improved searches for charged LFV thus may either reveal physics beyond the SM or at least lead to a significant reduction in parameter space allowed for such exotic contributions.

Charged LFV processes, i.e. transitions between e, μ, and τ, might be found in the decay of almost any weakly decaying particle. Although theoretical predictions generally depend on numerous unknown parameters these uncertainties tend to cancel in the relative strengths of these modes. Once LFV in the charged lepton sector were found, the combined information from many different experiments would allow us to discriminate between the various interpretations. Searches have been performed in μ, τ, π, K, B, D, W and Z decay. Whereas highest experimental sensitivities were reached in dedicated μ and K experiments, τ decay starts to become competitive as well.

8.1 Rare μ decays

LFV muon decays include the purely leptonic modes $\mu^+ \to e^+\gamma$ and $\mu^+ \to e^+e^+e^-$, as well as the semileptonic μ–e conversion in muonic atoms and the muonium–antimuonium oscillation. The present experimental limits are listed in Table 19.

Whereas most theoretical models favor $\mu^+ \to e^+\gamma$, this mode has a disadvantage from an experimental point of view since the sensitivity is limited by accidental $e^+\gamma$ coincidences and muon beam intensities have to be reduced now already. Searches for μ–e conversion, on the other hand, are limited by the available beam intensities and large improvements in sensitivity may still be achieved.

All recent results for μ^+ decays were obtained with "surface" muon beams containing muons originating in the decay of π^+'s that stopped very close to the surface of the pion production target, or "subsurface" beams from pion decays

Table 19 Present limits on rare μ decays

Mode	Upper limit (90% C.L.)	Year	Exp./Lab.	Ref.
$\mu^+ \to e^+\gamma$	1.2×10^{-11}	2002	MEGA / LAMPF	[180, 940]
$\mu^+ \to e^+e^+e^-$	1.0×10^{-12}	1988	SINDRUM I / PSI	[700]
$\mu^+e^- \leftrightarrow \mu^-e^+$	8.3×10^{-11}	1999	PSI	[941]
$\mu^-{\rm Ti} \to e^-{\rm Ti}$	6.1×10^{-13}	1998	SINDRUM II / PSI	[942]
$\mu^-{\rm Ti} \to e^+{\rm Ca}^*$	3.6×10^{-11}	1998	SINDRUM II / PSI	[943]
$\mu^-{\rm Pb} \to e^-{\rm Pb}$	4.6×10^{-11}	1996	SINDRUM II / PSI	[944]
$\mu^-{\rm Au} \to e^-{\rm Au}$	7×10^{-13}	2006	SINDRUM II / PSI	[945]

just below that region. Such beams are superior to conventional pion decay channels in terms of muon stop density and permit the use of relatively thin (typically 10 mg/cm^2) foils to stop the beam. Such low-mass stopping targets are required for the ultimate resolution in positron momentum and emission angle, minimal photon yield, or the efficient production of muonium in vacuum.

8.1.1 $\mu \to e\gamma$

Neglecting the positron mass the 2-body decay $\mu^+ \to e^+ \gamma$ of muons at rest is characterized by

$E_\gamma = E_e = m_\mu c^2/2 = 52.8$ MeV,

$\Theta_{e\gamma} = 180°$,

$t_\gamma = t_e$,

where t is the time of emission from the target, and Θ the opening angle between positron and photon. All $\mu \to e\gamma$ searches performed during the past three decades were limited by accidental coincidences between a positron from normal muon decay, $\mu \to e\nu\bar{\nu}$, and a photon produced in the decay of another muon, either by bremsstrahlung or by e^+e^- annihilation in flight. This background dominates by far the intrinsic background from radiative muon decay $\mu \to e\nu\bar{\nu}\gamma$. Accidental $e\gamma$ coincidences can be suppressed by testing the three conditions listed above. The vertex constraint resulting from the ability to trace back positrons and photons to an extended stopping target can further reduce background. Attempts have been made to suppress accidental coincidences by observing the low energy positron associated with the photon, but with minimal success. High muon polarization (P_μ) could help if one would limit the solid angle to accept only positrons and photons (anti)parallel to the muon spin since their rate is suppressed by the factor $1 - P_\mu$ for antiparallel emission at $E = m_\mu c^2/2$ but the reduced solid angle would have to be compensated by increased beam intensity which would raise the background again.

The most sensitive search to date was performed by the MEGA Collaboration at the Los Alamos Meson Physics Facility (LAMPF) which established an upper limit (90% C.L.) on $B(\mu \to e\gamma)$ of 1.2×10^{-11} [180, 940]. The MEG experiment [946, 947] at PSI, aims at a single-event sensitivity of $\sim 10^{-13}$–10^{-14}, and began commissioning in early 2007. A straightforward improvement factor of more than an order of magnitude in suppression of accidental background results from the DC muon beam at PSI, as opposed to the pulsed LAMPF beam which had a macro duty cycle of 7.7%. Another order of magnitude improvement is achieved by superb time resolution (≈ 0.15 ns FWHM on $t_\gamma - t_e$).

The MEG setup is shown in Fig. 52. The spectrometer magnet makes use of a novel "COBRA" (COnstant Bending RAdius) design which results in a graded magnetic field varying from 1.27 T at the centre to 0.49 T at both ends. This field distribution not only results in a constant projected bending radius for the 52.8 MeV positron, for polar emission angles θ with $|\cos\theta| < 0.35$, but also sweeps away positrons with low longitudinal momenta much faster than a constant field as used by MEGA. This design significantly reduces the instantaneous rates in the drift chambers.

The drift chambers are made of 12.5 μm thin foils supported by C-shaped carbon fibre frames which are out of the

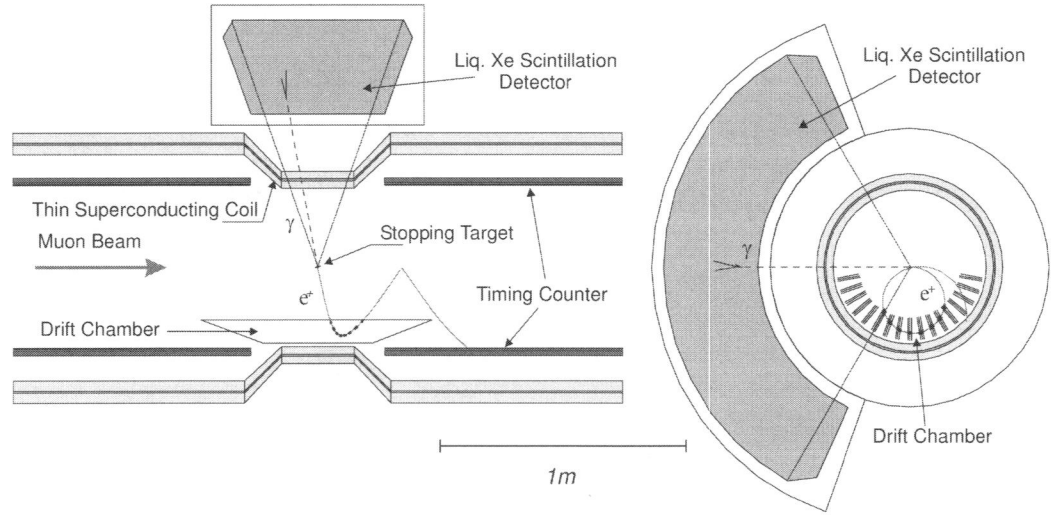

Fig. 52 Side and end views of the MEG setup. The magnetic field is shaped such that positrons are quickly swept out of the tracking region thus minimizing the load on the detectors. The cylindrical 0.8 m^3 single-cell LXe detector is viewed from all sides by 846 PMTs immersed in the LXe allowing the reconstruction of photon energy, time, conversion point and direction and the efficient rejection of pile-up signals

Fig. 53 Installing one of the timing counters into the COBRA magnet during the pilot run with the positron spectrometer at the end of 2006. The large ring is one of two Helmholtz coils used to compensate the COBRA stray field at the locations of the photomultipliers of the LXe detector

way of the positrons. The foils have "vernier" cathode pads which permit the measurement of the trajectory coordinate along the anode wires with an accuracy of about 500 µm.

There are two timing counters at both ends of the magnet (see Fig. 53), each of which consists of a layer of plastic scintillator fibers and 15 plastic scintillator bars of dimensions $4 \times 4 \times 90$ cm^3. The fibers give hit positions along the beam axis and the bars measure positron timings with a precision of $\sigma = 40$ ps. The counters are placed at large radii so only high energy positrons reach them, giving a total rate of a few 10^4/s for each bar.

High-strength Al-stabilized conductor for the magnet coil makes the magnet as thin as 0.20 X_0 radially, so that 85% of 52.8 MeV gamma rays traverse the magnet without interaction before entering the gamma detector placed outside the magnet. Whereas MEGA used rather inefficient pair spectrometers to detect the photon, MEG developed a novel liquid Xe scintillation detector, shown in Fig. 52. By viewing the scintillation light from all sides the electromagnetic shower induced by the photon can be reconstructed which allows for a precise measurement of the photon conversion point [948]. Special PMTs that work at LXe temperature ($-110°$C), persist under high pressures and are sensitive to the VUV scintillation light of LXe ($\lambda \approx 178$ nm) have been developed in collaboration with Hamamatsu Photonics. To identify and separate pile-up efficiently, fast waveform digitizing is used for all the PMT outputs.

The performance of the detector was measured with a prototype. The results are shown in Table 20. First data taking with the complete setup took place during the second half of 2007. A sensitivity of $\mathcal{O}(10^{-13})$ for the 90% C.L. upper limit in case no candidates are found should be reached after two years.

8.1.1.1 Beyond MEG Ten times larger surface muon rates than used by MEG can be achieved at PSI today already but the background suppression would have to be improved by two orders of magnitude. Accidental background N_{acc} scales with the detector resolutions as

$$N_{\text{acc}} \propto \Delta E_e \cdot \Delta t \cdot (\Delta E_\gamma \cdot \Delta \Theta_{e\gamma} \cdot \Delta x_\gamma)^2 \cdot A_T^{-1},$$

Table 20 Performance of a prototype of the MEG LXe detector at $E_\gamma = 53$ MeV

Observable	Resolution (σ)
energy	1.2%
time	65 ps
conversion point	$\approx$4 mm

with x_γ the coordinate of the photon trajectory at the target and A_T the target area. Here it is assumed that the photon can be traced back to the target with an uncertainty which is small compared to A_T. Since the angular resolution is dictated by the positron multiple scattering in the target this can be written:

$$N_{\text{acc}} \propto \Delta E_e \cdot \Delta t \cdot (\Delta E_\gamma \cdot \Delta x_\gamma)^2 \cdot \frac{d_T}{A_T},$$

with d_T the target thickness. When using a series of n target foils each of them could have a thickness of d_T/n and the beam would still be stopped. Since the area would increase like $n \cdot A_T$ the background could be reduced in proportion with $1/n^2$:

$$N_{\text{acc}} \propto \Delta E_e \cdot \Delta t \cdot (\Delta E_\gamma \cdot \Delta x_\gamma)^2 \cdot \frac{d_T/n}{n \cdot A_T},$$

so a geometry with ten targets, 1 mg/cm^2 each, would lead to the required background suppression.

8.1.2 $\mu \to 3e$

As has been discussed above the sensitivity of $\mu \to e\gamma$ searches is limited by background from accidental coincidences between a positron and a photon originating in the independent decays of two muons. Similarly, searches for the decay $\mu \to 3e$ suffer from accidental coincidences between positrons from normal muon decay and e^+e^- pairs originating from photon conversions or scattering of positrons off atomic electrons (Bhabha scattering). For this reason the muon beam should be continuous on the time scale of the muon life-time and longer. In addition to the obvious constraints on relative timing and total energy and momentum, which can be applied in $\mu \to e\gamma$ searches as well, there are powerful constraints on vertex quality and location to suppress the accidental background. Since the final state contains only charged particles the setup may consist of a magnetic spectrometer without the need for an electromagnetic calorimeter with its limited performance in terms of energy and directional resolution, rate capability, and event definition in general. On the other hand, of major concern are the high rates in the tracking system of a $\mu \to 3e$ setup which has to stand the load of the full muon decay spectrum.

8.1.2.1 The SINDRUM I experiment The present experimental limit, $B(\mu \to 3e) < 1 \times 10^{-12}$ [700], was published way back in 1988. Since no new proposals exist for this decay mode we shall analyse the prospects of an improved experiment with this SINDRUM experiment as a point of reference. A detailed description of the experiment may be found in Ref. [949].

Data were taken during six months using a 25 MeV/c subsurface beam. The beam was brought to rest with a rate of 6×10^6 μ^+/s in a hollow double-cone foam target (length 220 mm, diameter 58 mm, total mass 2.4 g). SINDRUM I is a solenoidal spectrometer with a relatively low magnetic field of 0.33 T corresponding to a transverse momentum threshold around 18 MeV/c for particles crossing the tracking system. This system consisted of five cylindrical MWPCs concentric with the beam axis. Three-dimensional space points were found by measuring the charges induced on cathode strips oriented $\pm 45°$ relative to the sense wires.

Gating times were typically 50 ns. The spectrometer acceptance for $\mu \to 3e$ was 24% of 4π sr (for a constant transition-matrix element) so the only place for a significant improvement in sensitivity would be the beam intensity.

Figure 54 shows the time distribution of the recorded $e^+e^+e^-$ triples. Apart from a prompt contribution of correlated triples one notices a dominant contribution from accidental coincidences involving low-invariant-mass e^+e^- pairs. Most of these are explained by Bhabha scattering of positrons from normal muon decay $\mu \to e\nu\bar{\nu}$. The accidental background thus scales with the target mass, but it is not obvious how to reduce this mass significantly below the 11 mg/cm^2 achieved in this search.

Figure 55 shows the vertex distribution of prompt events. One should keep in mind that most of the uncorrelated

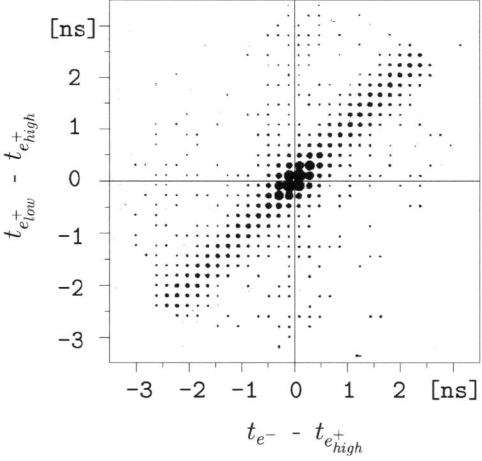

Fig. 54 Relative timing of $e^+e^+e^-$ events. The two positrons are labeled *low* and *high* according to the invariant mass when combined with the electron. One notices a contribution of correlated triples in the centre of the distribution. These events are mainly $\mu \to 3e\nu\bar{\nu}$ decays. The concentration of events along the diagonal is due to low-invariant-mass e^+e^- pairs in accidental coincidence with a positron originating in the decay of a second muon. The e^+e^- pairs are predominantly due to Bhabha scattering in the target

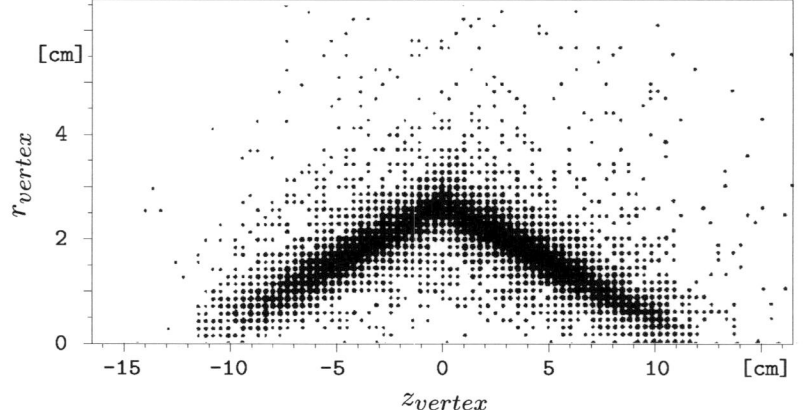

Fig. 55 Spatial distribution of the vertex fitted to prompt $e^+e^+e^-$ triples. One clearly notices the double-cone target

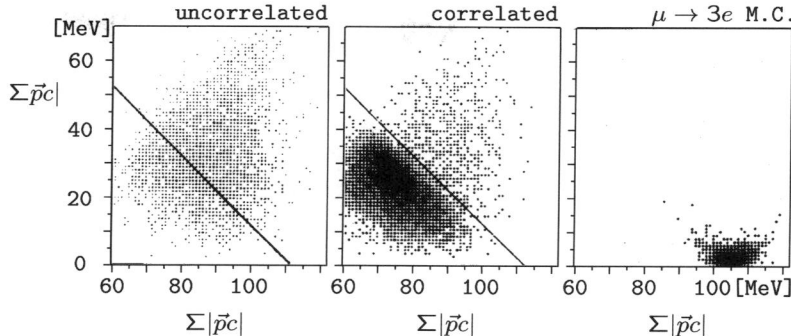

Fig. 56 Total momentum versus total energy for three event classes discussed in the text. The *line* shows the kinematic limit (within resolution) defined by $\Sigma |\vec{p}c| + |\Sigma \vec{p}c| \leq m_\mu c^2$ for any muon decay. The enhancement in the distribution of correlated triples below this limit is due to the decay $\mu \to 3e\nu\bar{\nu}$

triples contain e^+e^- pairs coming from the target and their vertex distribution will thus follow the target contour as well. This 1-fold accidental background is suppressed by the ratio of the vertex resolution (couple of mm^2) and the target area. There is no reason, other than the cost of the detection system, not to choose a much larger target. Such an increase might also help to reduce the load on the tracking detectors. Better vertex resolution would help as well. At these low energies tracking errors are dominated by multiple scattering in the first detector layer but it should be possible to gain by bringing it closer to the target.

Finally, Fig. 56 shows the distribution of total momentum versus total energy for three classes of events, (i) uncorrelated $e^+e^+e^-$ triples, (ii) correlated $e^+e^+e^-$ triples, and (iii) simulated $\mu \to 3e$ decays. The distinction between uncorrelated and correlated triples has been made on the basis of relative timing and vertex as discussed above.

8.1.2.2 How to improve? What would a $\mu \to 3e$ set-up look like that would aim at a single-event sensitivity around 10^{-16}, i.e., would make use of a beam rate around 10^{10} μ^+/s? The SINDRUM I measurement was background-free at the level of 10^{-12} with a beam of 0.6×10^7 μ^+/s. Taking into account that background would have set in at 10^{-13}, the increased stop rate would raise the background level to $\approx 10^{-10}$, so six orders of magnitude in background reduction would have to be achieved. Increasing the target size and improving the tracking resolution should bring two orders of magnitude from the vertex requirement alone. Since the dominant sources of background are accidental coincidences between two decay positrons (one of which undergoes Bhabha scattering) the background rate scales with the momentum resolution squared. Assuming an improvement by one order of magnitude, i.e., from the $\approx 10\%$ FWHM obtained by SINDRUM I to $\approx 1\%$ for a new search, one would gain two orders of magnitude from the constraint on total energy alone. The remaining factor 100 would result from the test on the collinearity of the e^+ and the e^+e^- pair.

As mentioned in Ref. [949] a dramatic suppression of background could be achieved by requiring a minimal opening angle (typically 30°) for both e^+e^- combinations. Depending on the mechanism for $\mu \to 3e$, such a cut might, however, lead to a strong loss in $\mu \to 3e$ sensitivity as well.

Whereas background levels may be under control, the question remains whether detector concepts can be developed that work at the high beam rates proposed. A large modularity will be required to solve problems of pattern recognition.

8.1.3 μ–e conversion

When negatively charged muons stop in matter they quickly form muonic atoms which reach their ground states in a time much shorter than the life-time of the atom. Muonic atoms decay mostly through *muon decay in orbit* (MIO) $\mu^-(A, Z) \to e^- \nu_\mu \bar{\nu}_e (A, Z)$ and *nuclear muon capture* (MC) $\mu^-(A, Z) \to \nu_\mu (A, Z - 1)^*$ which in lowest order may be interpreted as the incoherent sum of elementary $\mu^- p \to n \nu_\mu$ captures. The MIO rate decreases slightly for increasing values of Z (down to 85% of the free muon rate in the case of muonic gold) due to the increasing muon binding energy. The MC rate at the other hand increases roughly proportional to Z^4. The two processes have about equal rates around $Z = 12$.

When the hypothetical μ–e conversion leaves the nucleus in its ground state the nucleons act coherently, boosting the process relative to the incoherent processes with exited final states. The resulting Z dependence has been studied by several authors [950–953]. For $Z \lesssim 40$ all calculations predict a conversion probability relative to the MC rate which follows the linear rise with Z expected naively. The predictions may, however, deviate by factors 2–3 at higher Z values.

As a result of the two-body final state the electrons produced in μ–e conversion are mono-energetic and their energy is given by:

$$E_{\mu e} = m_\mu c^2 - B_\mu(Z) - R(A), \qquad (8.1)$$

where $B_\mu(Z)$ is the atomic binding energy of the muon and R is the atomic recoil energy for a muonic atom with atomic number Z and mass number A. In first approximation $B_\mu(Z) \propto Z^2$ and $R(A) \propto A^{-1}$.

8.1.3.1 Background Muon decay in orbit (MIO) constitutes an intrinsic background source which can only be suppressed with sufficient electron energy resolution. The process predominantly results in electrons with energy E_{MIO} below $m_\mu c^2/2$, the kinematic endpoint in free muon decay, with a steeply falling high energy component reaching up to $E_{\mu e}$. In the endpoint region the MIO rate varies as $(E_{\mu e} - E_{MIO})^5$ and a resolution of 1–2 MeV (FWHM) is sufficient to keep MIO background under control. Since the MIO endpoint rises at lower Z great care has to be taken to avoid low-Z contaminations in and around the target.

Another background source is due to radiative muon capture (RMC) $\mu^-(A, Z) \to \gamma(A, Z-1)^*\nu_\mu$ after which the photon creates an e^+e^- pair either internally (Dalitz pair) or through $\gamma \to e^+e^-$ pair production in the target. The RMC endpoint can be kept below $E_{\mu e}$ for selected isotopes.

Most low energy muon beams have large pion contaminations. Pions may produce background when stopping in the target through radiative pion capture (RPC) which takes place with a probability of $O(10^{-2})$. Most RPC photons have energies above $E_{\mu e}$. As in the case of RMC these photons may produce background through $\gamma \to e^+e^-$ pair production. There are various strategies to cope with RPC background:

- One option is to keep the total number of π^- stopping in the target during the live time of the experiment below 10^{4-5}. This can be achieved with the help of a moderator in the beam exploiting the range difference between pions and muons of given momentum or with a muon storage ring exploiting the difference in life-time.
- Another option is to exploit the fact that pion capture takes place at a time scale far below a nanosecond. The background can thus be suppressed with a beam counter in front of the experimental target or by using a pulsed beam selecting only delayed events.

Cosmic rays (electrons, muons, photons) are a copious source of electrons with energies around ≈ 100 MeV. With the exception of $\gamma \to e^+e^-$ pair production in the target these events can be recognized by an incoming particle. In addition, passive shielding and veto counters above the detection system help to suppress this background.

8.1.3.2 SINDRUM II The present best limits (see Table 19) have been measured with the SINDRUM II spectrometer at PSI. Most recently a search was performed on a gold target [945]. In this experiment (see Fig. 57) the pion suppression is based on the factor of two shorter range of pions as compared to muons at the selected momentum of 52 MeV/c. A simulation using the measured range distribution shows that about one in 10^6 pions cross an 8 mm thick CH_2 moderator. Since these pions are relatively slow 99.9% of them decay before reaching the gold target which is situated some 10 m further downstream. As a result pion stops in the target have been reduced to a negligible level. What remains are radiative pion capture in the degrader and $\pi^- \to e^-\bar{\nu}_e$ decay in flight shortly before entering the degrader. The resulting electrons may reach the target where they can scatter into the solid angle acceptance of the spectrometer. $\mathcal{O}(10)$ events are expected with a flat energy distribution between 80 and 100 MeV. These events are peaked in forward direction and show a time correlation with the cyclotron rf signal. To cope with this background two event classes have been introduced based on the values of polar angle and rf phase. Fig. 58 shows the corresponding momentum distributions.

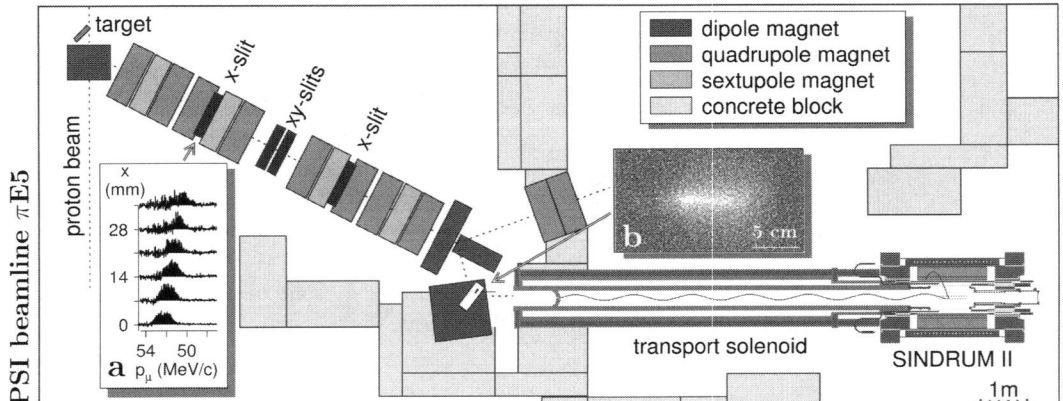

Fig. 57 Plan view of the SINDRUM II experiment. The 1 MW 590 MeV proton beam hits the 40 mm carbon production target (*top left of the figure*). The πE5 beam line transports secondary particles (π, μ, e) emitted in the backward direction to a degrader situated at the entrance of a solenoid connected axially to the SINDRUM II spectrometer. *Inset a* shows the momentum dispersion at the position of the first slit system. *Inset b* shows a cross section of the beam at the position of the beam focus

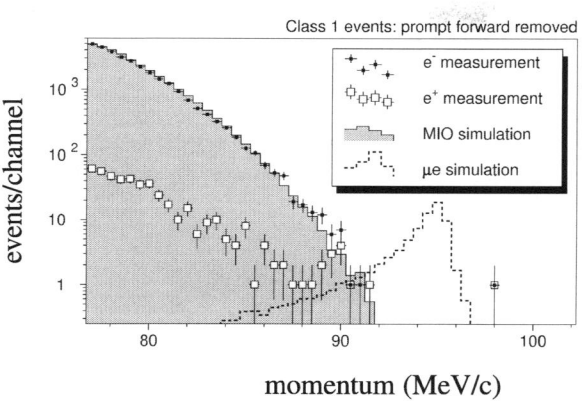
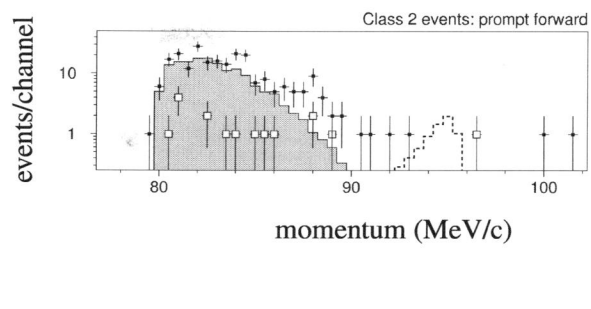

Fig. 58 Momentum distributions of electrons and positrons for two event classes described in the text. Measured distributions are compared with the results of simulations of muon decay in orbit and μ–e conversion

The spectra show no indication for μ–e conversion. The corresponding upper limit on

$$B_{\mu e} \equiv \Gamma\left(\mu^- \text{Au} \to e^- \text{Au}_{\text{g.s.}}\right) / \Gamma_{\text{capture}}\left(\mu^- \text{Au}\right)$$
$$< 7 \times 10^{-13} \quad 90\% \text{ C.L.} \quad (8.2)$$

has been obtained with the help of a likelihood analysis of the momentum distributions shown in Fig. 58 taking into account muon decay in orbit, μ–e conversion, a contribution taken from the observed positron distribution describing processes with intermediate photons such as radiative muon capture and a flat component from pion decay in flight or cosmic ray background.

8.1.3.3 New initiatives Based on a scheme originally developed during the eighties for the Moscow Meson Factory [954] μe-conversion experiments are being considered both in the USA and in Japan. The key elements are:

- A pulsed proton beam allows one to remove pion background by selecting events in a delayed time window. Proton extinction factors below 10^{-9} are needed.
- A large acceptance capture solenoid surrounding the pion production target leads to a major increase in muon flux.
- A bent solenoid transporting the muons to the experimental target results in a significant increase in momentum transmission compared to a conventional quadrupole channel. A bent solenoid not only removes neutral particles and photons but also separates electric charges.

Unfortunately, the MECO proposal at BNL [955] designed along these lines was stopped because of the high costs. Presently the possibilities are studied to perform a MECO-type of experiment at Fermilab (mu2e). There is good hope that a proton beam with the required characteristics can be produced with minor modifications to the existing accelerator complex which will become available after the Tevatron stops operation in 2009. A letter of intent is in preparation.

Further improvements are being considered for an experiment at J-PARC. To fully exploit the life-time difference to suppress pion induced background the separation has to occur in the beam line rather than after the muon has stopped since the life-time of the muonic atom may be significantly shorter than the 2.2 µs of the free muon. For this purpose a muon storage ring PRISM (Phase Rotated Intense Slow Muon source, see Fig. 59) is being considered [956] which makes use of large-acceptance fixed-field alternating-gradient (FFAG) magnets. A portion of the PRISM-FFAG ring is presently under construction as r&d project. As the name suggests the ring is also used to reduce the momentum spread of the beam (from $\approx$30% to $\approx$3%) which is achieved by accelerating late muons and decelerating early muons in RF electric fields. The scheme requires the construction of a pulsed proton beam [957] a decision about which has not been made yet. The low momentum spread of the muons allows for the use of a relatively thin target which is an essential ingredient for high resolution in the momentum measurement with the PRIME detector [704, 705].

Table 21 lists the μ^- stop rates and single-event sensitivities for the various projects discussed above.

8.2 Searches for lepton flavor violation in τ decays

Highest sensitivities to date are achieved at the B-factories and further improvements are to be expected. At the LHC the modes with three charged leptons in the final state such as $\tau \to 3\mu$ could be sufficiently clean to reach even higher sensitivity. Studies have been performed for LHCb [159] and CMS (see below).

8.2.1 B-factories

Present generation B-factories operating around the $\Upsilon(4S)$ resonance also serve as τ-factories, because the production

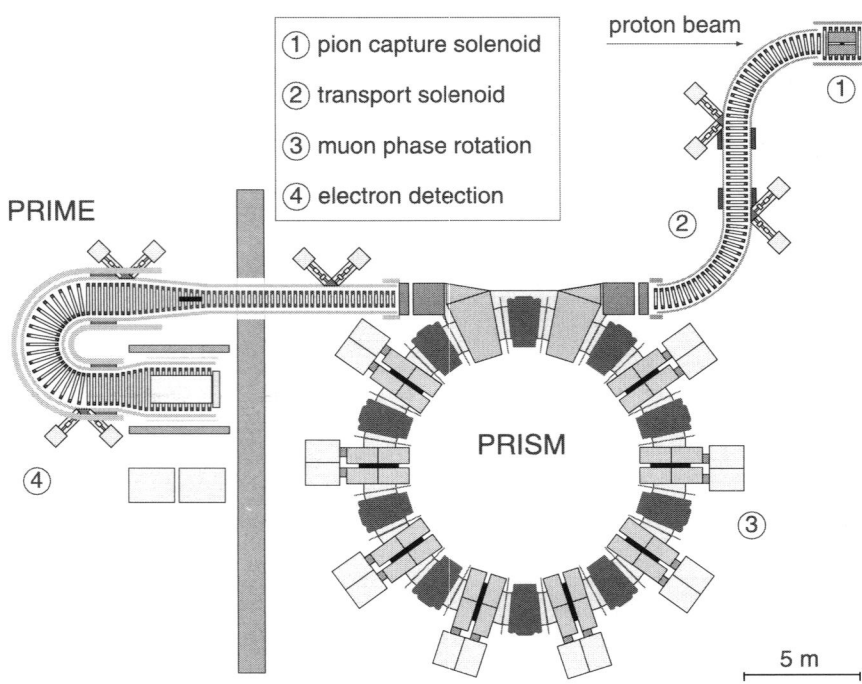

Fig. 59 Layout of PRISM/PRIME. The experimental target is situated at the entrance of the 180° bent solenoid that transports decay electrons to the detection system. See text for further explanations

Table 21 μ–e conversion searches

Project	Lab	Status	E_p [GeV]	p_μ [MeV/c]	μ^- stops [s^{-1}]	S^{a}
SINDRUM II	PSI	finished	0.6	52 ± 1	10^7	2×10^{-13}
MECO	BNL	canceled	8	45 ± 25	10^{11}	2×10^{-17}
mu2e	FNAL	under study	8	45 ± 25	0.6×10^{10}	4×10^{-17}
PRISM/PRIME	J-PARC	LOI	40	68 ± 3	10^{12}	5×10^{-19}

[a]Single-event sensitivity: value of $B_{\mu e}$ corresponding to an expectation of one observed event

cross sections $\sigma_{b\bar{b}} = 1.1$ nb and $\sigma_{\tau^+\tau^-} = 0.9$ nb are quite similar at centre-of-mass energy near 10.58 GeV. BaBar and Belle have thus been able to reach the highest sensitivity to lepton flavour violating tau decays.

Many theories beyond the standard model allow for $\tau^\pm \to \ell^\pm \gamma$ and $\tau^\pm \to \ell^\pm \ell^\mp \ell^\pm$ decays, where $\ell^- = e^-, \mu^-$, at the level of $\sim \mathcal{O}(10^{-10}$–$10^{-7})$. Examples are

- SM with additional heavy right handed Majorana neutrinos or with left handed and right handed neutral isosinglets [958];
- mSUGRA models with right handed neutrinos introduced via the see-saw mechanism [242, 959];
- supersymmetric models with Higgs exchange [174, 807] or $SO(10)$ symmetry [164, 960];
- technicolour models with non-universal Z' exchange [961].

Large neutrino mixing could induce large mixing between the supersymmetric partners of the leptons. While some scenarios predict higher rates for $\tau^\pm \to \mu^\pm \gamma$ decays, others, for example with inverted mass hierarchy for the sleptons [242], predict higher rates for $\tau^\pm \to e^\pm \gamma$ decays.

Semi-leptonic neutrino-less decays involving pseudo-scalar mesons like $\tau^\pm \to \ell^\pm P^0$, where $P^0 = \pi^0, \eta, \eta'$ may be enhanced over $\tau^\pm \to \ell^\pm \ell^\mp \ell^\pm$ decays in supersymmetric models, for example, arising out of exchange of CP-odd pseudo-scalar neutral Higgs boson, which are further enhanced by color factors associated with these decays. The large coupling of Higgs at the $s\bar{s}$ vertex enhances final states containing the η meson, giving a prediction of $B(\tau^\pm \to \mu^\pm \eta) : B(\tau^\pm \to \mu^\pm \mu^\mp \mu^\pm) : B(\tau^\pm \to \mu^\pm \gamma) = 8.4 : 1 : 1.5$ [810]. Some models with heavy Dirac neutrinos [198, 962], two Higgs doublet models, R-parity violating supersymmetric models, and flavor changing Z' models with non-universal couplings [963] allow for observable parameter space of new physics [964], while respecting the existing experimental bounds at the level of $\sim \mathcal{O}(10^{-7})$.

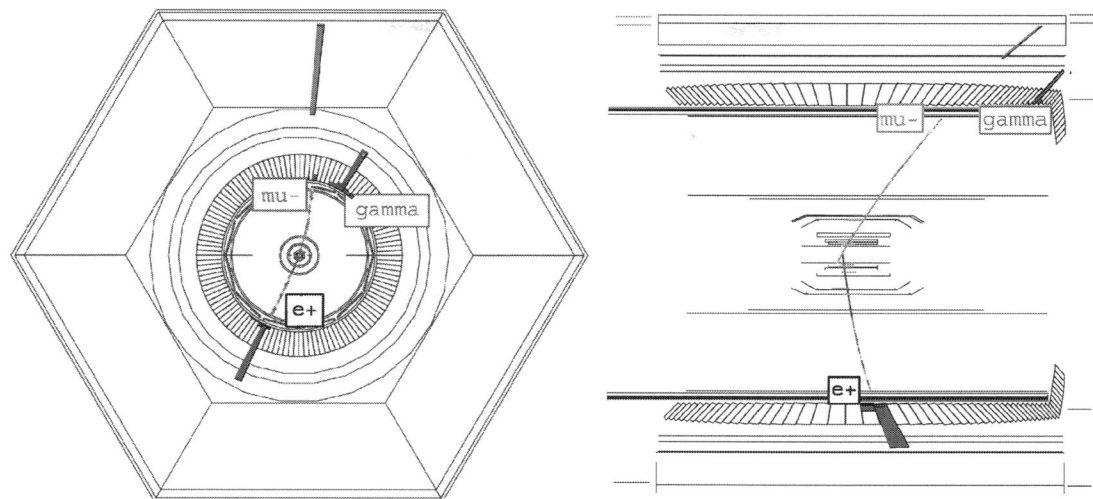

Fig. 60 Transverse and longitudinal views of a simulated $\tau \to \mu\gamma$ event in the BaBar detector. The second tau decays to $e\nu\bar{\nu}$

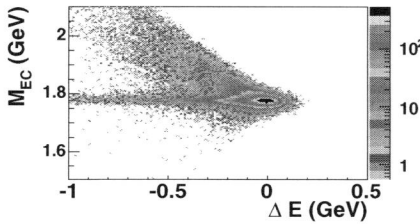

Fig. 61 m_{EC} vs. ΔE for simulated $\tau \to \mu\gamma$ events as reconstructed in the BaBar detector. The tails of distributions are due to initial state radiation and photon energy reconstruction effects. Latter causes also the shift in $\langle \Delta E \rangle$

8.2.1.1 Search strategy In the clean e^+e^- annihilation environment, the decay products of two taus produced are well separated in space as illustrated in Fig. 60.

As shown in Fig. 61 neutrino-less τ-decays have two characteristic features:

- the measured energy of τ daughters is close to half the centre-of-mass energy,
- the total invariant mass of the daughters is centered around the mass of the τ lepton.

While for $\ell\ell\ell$ modes the achieved mass resolution is excellent, the resolution (σ) of the $\ell\gamma$ final state improves from ~20 to 9 MeV by assigning the point of closest approach of the muon trajectory to the e^+e^- collision axis as the decay vertex and by using a kinematic fit with the $\mu\gamma$ CM energy constrained to $\sqrt{s}/2$ [191]. The energy resolution is typically 45 MeV with a long tail due to radiation.

The principal sources of background are radiative QED (di-muon or Bhabha) and continuum ($q\bar{q}$) events as well as $\tau^+\tau^-$ events with a mis-identified standard model decay mode. There is also some *irreducible* contribution from $\tau^+\tau^-$ events with hard initial state radiation in which one of the τ's decays into a mode with the same charged particle as the signal. For example, $\tau \to \mu\nu\bar{\nu}$ decays accompanied by a hard γ is an irreducible background in the $\tau \to \mu\gamma$ search.

The general strategy to search for the neutrino-less decays is to define a signal region, typically of size $\sim 2\sigma$, in the energy–mass plane of the τ daughters and to reduce the background expectation from well-known CM processes inside the signal region by optimizing a set of selection criteria:

- the missing momentum is consistent with the zero-mass hypothesis
- the missing momentum points inside the acceptance of the detector
- the second tau is found with the correct invariant mass
- minimal opening angle between two tau decay products
- minimal value for the highest momentum of any reconstructed track
- particle identification

The analyses are performed in a *blind* fashion by excluding events in the region of the signal box until all optimizations and systematic studies of the selection criteria have been completed. The cut values are optimized using control samples, data sidebands and Monte Carlo extrapolation to the signal region to yield the lowest expected upper limit under the no-signal hypothesis. The measured m_{EC} vs. ΔE distribution for the $\tau \to \mu\gamma$ search after applying the constraints listed above is shown in Fig. 62.

For the $\tau^\pm \to \ell^\pm P^0$ searches, the pseudo-scalar mesons (P^0) are reconstructed in the following decay modes: $\pi^0 \to \gamma\gamma$ for $\tau^\pm \to \ell^\pm \pi^0$, $\eta \to \gamma\gamma$ and $\eta \to \pi^+\pi^-\pi^0$ ($\pi^0 \to \gamma\gamma$) for $\tau^\pm \to \ell^\pm \eta$, and $\eta' \to \pi^+\pi^-\eta(\eta \to \gamma\gamma)$ and $\eta' \to \rho^0\gamma$ for $\tau^\pm \to \ell^\pm \eta'$.

8.2.1.2 Experimental results from BaBar and Belle By the beginning of 2007 BaBar and Belle had recorded integrated luminosities of $\mathcal{L} \sim 400$ and 700 fb^{-1}, respectively, which corresponds to a total of $\sim 10^9$ τ-decays. Analysis of these data samples is still ongoing and published results include only part of the data analysed. No signal has yet been observed in any of the probed channels and some limits and the corresponding integrated luminosities are summarized in Table 22. Frequentest upper limits have been calculated for the combination of the two experiments [183] using the technique of Cousins and Highland [965] following the implementation of Barlow [966].

8.2.1.3 Projection of limits to higher luminosities $B(\tau^\pm \to \mu^\pm \gamma)$ and $B(\tau^\pm \to \mu^\pm \mu^\mp \mu^\pm)$ have been lowered by five orders of magnitude over the past twenty-five years. Further significant improvements in sensitivity are expected during the next five years. Depending upon the nature of backgrounds contributing to a given search, two extreme scenarios can be envisioned in extrapolating to higher luminosities:

- If the expected background is kept below $\mathcal{O}(1)$ events, while maintaining the same efficiency $\mathcal{B}_{UL}^{90} \propto 1/\mathcal{L}$ if no signal events would be observed. In $\tau^\pm \to \mu^\pm \mu^\mp \mu^\pm$ searches, for example, the backgrounds are still quite low and the irreducible backgrounds are negligible even for projected Super B-factories.
- If there is background now already and no reduction could be achieved in the future measurements $\mathcal{B}_{UL}^{90} \propto 1/\sqrt{\mathcal{L}}$.

The $\sqrt{\mathcal{L}}$ scaling is, however, unduly pessimistic since the analyses improve steadily. Better understanding of the nature of the backgrounds will lead to a more effective separation of signal and background.

The $\tau^\pm \to \mu^\pm \gamma$ searches suffer from significant background from both $\mu^+\mu^-$ and $\tau^+\tau^-$ events and to a lesser extend from $q\bar{q}$ production. While one can expect to reduce these backgrounds with continued optimization with more luminosity at the present day B-factories, much of the background from $\tau^+\tau^-$ events is irreducible coming from $\tau \to \mu \nu \bar{\nu}$ decays accompanied by initial state radiation. This source represents about 20% of the total background in the searches performed by the BaBar experiment [191] and it is conceivable that an analysis can be developed that reduces all but this background with minimal impact on the efficiency. One could also include new selection criteria such as a cut on the polar angle of the photon which could reduce the radiative "irreducible" background by 85% with a 40% loss of signal efficiency. Table 23 summarizes the future sensitivities for various LFV decay modes.

In order to further reduce the impact of irreducible backgrounds at a future Super B-factory experiment, one can consider what is necessary to improve the mass resolution of the, e.g., μ–γ system. Currently, this resolution is limited by the γ angular resolution. Therefore improvements might be expected if the granularity of the electromagnetic calorimeter is increased.

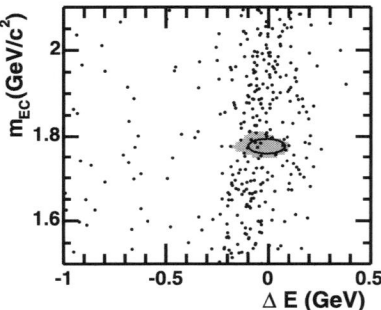

Fig. 62 Measured distribution of m_{EC} vs. ΔE for $\tau \to \mu\gamma$ reconstructed by BaBar [191]. The *shaded region* taken from Fig. 61 contains 68% of the hypothetical signal events

Table 22 Integrated luminosities and observed upper limits on the branching fractions at 90% C.L. for selected LFV tau decays by BaBar and Belle

Channel	BaBar			Belle		
	$\mathcal{L}$ (fb^{-1})	$\mathcal{B}_{UL}$ (10^{-8})	Ref.	$\mathcal{L}$ (fb^{-1})	$\mathcal{B}_{UL}$ (10^{-8})	Ref.
$\tau^\pm \to e^\pm \gamma$	232	11	[182]	535	12	[181]
$\tau^\pm \to \mu^\pm \gamma$	232	6.8	[191]	535	4.5	[181]
$\tau^\pm \to \ell^\pm \ell^\mp \ell^\pm$	92	11–33	[644]	535	2–4	[967]
$\tau^\pm \to e^\pm \pi^0$	339	13	[648]	401	8.0	[968]
$\tau^\pm \to \mu^\pm \pi^0$	339	11	[648]	401	12	[968]
$\tau^\pm \to e^\pm \eta$	339	16	[648]	401	9.2	[968]
$\tau^\pm \to \mu^\pm \eta$	339	15	[648]	401	6.5	[968]
$\tau^\pm \to e^\pm \eta'$	339	24	[648]	401	16	[968]
$\tau^\pm \to \mu^\pm \eta'$	339	14	[648]	401	13	[968]

8.2.2 CMS

So far, only $\tau \to \mu$ transitions have been studied since muons are more easily identified and the CMS trigger thresholds for muons are generally lower than for electrons. The $\tau \to \mu\gamma$ channel was studied in the past [969] both for CMS and for ATLAS but found not to be competitive with the prospects at the B-factories. The $\tau \to 3\mu$ channel looks more promising and will be discussed below.

8.2.2.1 τ production at the LHC

It is planned to operate the LHC in three different phases. After a commissioning phase the LHC will be ramped up to an initial luminosity of $\mathcal{L} = 10^{32}$ cm^{-2} s^{-1} followed by a low luminosity phase ($\mathcal{L} = 2 \times 10^{33}$ cm^{-2} s^{-1}). A high luminosity phase with $\mathcal{L} = 10^{34}$ cm^{-2} s^{-1} will start in 2010 and last for a period of several years. The integrated luminosity per year will be 10–30 and 100–300 fb^{-1} for the low and high luminosity phases, respectively [970].

The rate of τ leptons produced was estimated with the help of PYTHIA 6.227 using the parton distribution function CTEQ5L. The results are shown in Table 24. During the low luminosity phase assuming an integrated luminosity of only 10 fb^{-1} per year about 10^{12} τ leptons will be produced within the CMS detector. The dominant production sources of τ leptons at the LHC are the D_s and various B mesons. The W and the Z production sources will provide considerably less τ leptons per year, but at higher energies which is an advantage for the efficient detection of their decay products (see below).

8.2.2.2 $\tau \to 3\mu$ detection

A key feature of CMS is a 4 T magnetic field, which ensures the measurement of charged-particle momenta with a resolution of $\sigma_{p_T}/p_T = 1.5\%$ for 10 GeV muons [970] using a four-station muon system. A silicon pixel detector and tracker allow to reconstruct secondary vertices with a resolution of about 50 µm [971] and help to improve the muon reconstruction. Furthermore, CMS has an electromagnetic calorimeter (ECAL) composed of PbWO$_4$ and a copper scintillator hadronic calorimeter (HCAL). As a result of the high magnetic field and the amount of material that has to be crossed only muons with $p_T > 3$ GeV/c are accepted. The reconstruction efficiency varies between $\approx$70% at 5 GeV [972] and $\approx$98% at 100 GeV/c [970].

The two levels of the CMS trigger system are called "level 1" (L1) and "high level" (HLT). The triggers relevant for this analysis are the dedicated single and di-muon triggers. For the low luminosity phase it is planned to use as p_T thresholds for single muons 14 GeV/c at L1 and 19 GeV/c for the HLT. The thresholds for the di-muon trigger will be 3 GeV/c at L1 and 7 GeV/c for the HLT.

Most $\tau \to 3\mu$ events produced via W and Z decays will be accepted by the present triggers. Unfortunately, the low p_T of the muons from the decays of τ's originating in D_s or B decay result in a very low trigger efficiency (Fig. 63). Dedicated trigger algorithms with improved efficiency are presently being studied.

To improve the identification of low p_T muons a new method is currently under development combining the energy deposit in the ECAL, HCAL and the number of reconstructed muon track segments in the muon systems. The in-

Table 23 Expected 90% C.L. upper limits on LFV τ decays with 75 ab^{-1} assuming no signal is found and reducible backgrounds are small ($\sim \mathcal{O}(1)$ events) and the irreducible backgrounds scale as $1/\mathcal{L}$

Decay mode	Sensitivity
$\tau \to \mu\gamma$	2×10^{-9}
$\tau \to e\gamma$	2×10^{-9}
$\tau \to \mu\mu\mu$	2×10^{-10}
$\tau \to eee$	2×10^{-10}
$\tau \to \mu\eta$	4×10^{-10}
$\tau \to e\eta$	7×10^{-10}

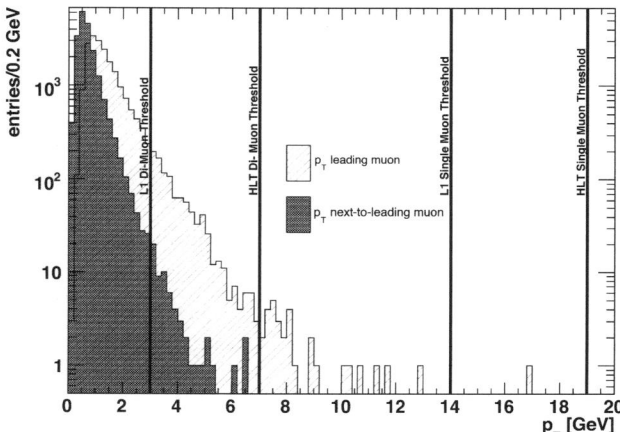

Fig. 63 p_T distributions of the leading and next-to-leading muon from the decay $\tau \to 3\mu$ at CMS. The indicated trigger thresholds for the low luminosity phase are clearly too high for the efficient detection of these events

Table 24 Number of τ leptons per year produced during the low-luminosity phase of the LHC

Production channel	$W \to \tau\nu_\tau$	$\gamma/Z \to \tau\tau$	$B^0 \to \tau X$	$B^\pm \to \tau X$	$B_s \to \tau X$	$D_s \to \tau X$
$N_\tau/10$ fb^{-1}	1.7×10^8	3.2×10^7	4.0×10^{11}	3.8×10^{11}	7.9×10^{10}	1.5×10^{12}

Fig. 64 Invariant mass distribution from the simulation of $\tau \to 3\mu$ events

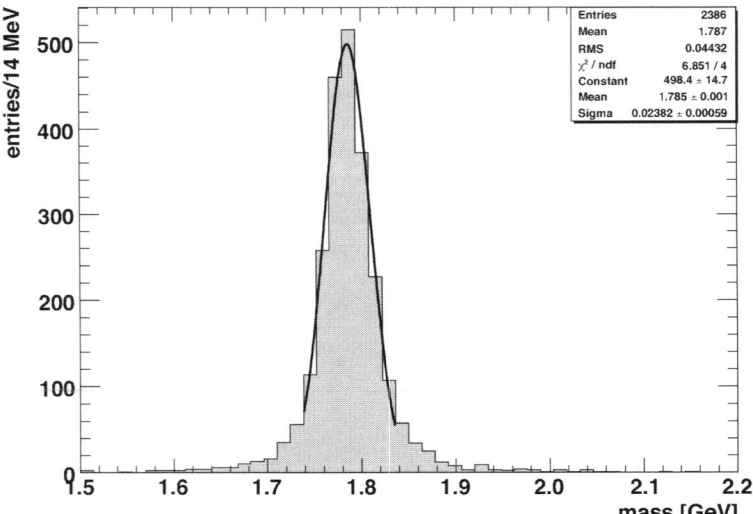

variant mass distribution of reconstructed $\tau \to 3\mu$ events is shown in Fig. 64. The resolution is about 24 MeV/c^2, which ensures a good capability to reduce background events.

8.2.2.3 Background and expected sensitivity The main sources of muons are decays of D and B mesons which are copiously produced at LHC energies. A previous study [973] suggested that these background events can be suppressed by appropriate selection criteria. The probability to misidentify an event from pile-up is small and cosmic rays can be rejected by timing. Due to the high momentum of the muons from direct W and Z decays, the contribution to the background is negligible [974].

One rare decay that can mimic the signal is $D_s \to \phi\mu\nu_\mu$ followed by a decay $\phi \to \mu\mu$. This background can be reduced by an invariant mass cut around the ϕ mass. Radiative ϕ decay $\phi \to \mu\mu\gamma$ survives this cut since the photon usually remains undetected. These radiative decays and any other heavy meson decays may be suppressed using secondary vertex properties and isolation criteria and by exploring the three-muon angular distributions. These studies are in progress.

Predictions of the achievable sensitivity are available in an older CMS Note [973]. In case no signal is observed the expected upper limit on the $\tau \to 3\mu$ branching ratio at 95% C.L. for the W source is 7.0×10^{-8} (3.8×10^{-8}) for 10 fb^{-1} (30 fb^{-1}) of collected data. For the Z source a limit of 3.4×10^{-7} and for the B meson source a limit of 2.1×10^{-7} was derived assuming an integrated luminosity of 30 fb^{-1}. The D_s source was not studied in this early paper.

Potentially including the muons from D and B meson decays may lead to significant improvements of the sensitivity. Further studies are necessary to make firm predictions.

8.3 $B^0_{d,s} \to e^\pm \mu^\mp$

The present limits $B(B^0_d \to e\mu) < 1.7 \times 10^{-7}$ [975] determined by Belle and $B(B^0_s \to e\mu) < 6.1 \times 10^{-6}$ [976] from CDF are of interest since they place bounds on the masses of two Pati–Salam leptoquarks [361] (see below). Both measurements are almost background free so significant improvements should be expected from these experiments. These decay modes have similarities with the $K^0_L \to \mu e$ decay for which an upper limit of 4.7×10^{-12} exists [977].

The prospects of a more sensitive search have been studied for the LHCb experiment [978]. Although background levels are higher, this is more than compensated by the improved single-event sensitivity. The event selection closely follows that of the $B^0_s \to \mu^+\mu^-$ decay. The dominant backgrounds come from (i) events in which two b hadrons decay into leptons combining to a fake vertex and (ii) from two-body charmless hadronic decays when the two hadrons are misidentified as leptons. Signal and background are separated on the basis of particle identification, invariant mass ($\sigma(m_B) = 50$ MeV/c^2), transverse momenta, proper distance and the isolation of the B^0 candidate from the other decay products. See Ref. [978] for details. Simulation shows that for an integrated luminosity of 2 fb^{-1} the total background can be reduced to ≈ 80 events with a selection efficiency of 1.4%. Assuming no signal would be found the 90% C.L. upper limits would be 1.6×10^{-8} and 6.5×10^{-8} for $B(B^0_d \to e^\pm \mu^\mp)$ and $B(B^0_s \to e^\pm \mu^\mp)$, respectively. These values correspond to 90% C.L. lower limits on the leptoquark mass and mixings of $90 \times F^d_{\text{mix}}$ TeV and $65 \times F^s_{\text{mix}}$ TeV, where $F^{d,s}_{\text{mix}}$ are factors taking into account generation mixing within the model. Present limits are 50 and 21 TeV, respectively (see Fig. 65).

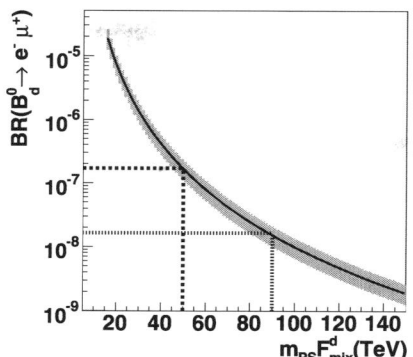 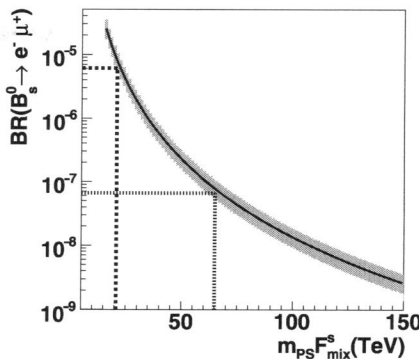

Fig. 65 90% C.L. limits on $B(B_d^0 \to e\mu)$ (*left panel*) and $B(B_s^0 \to e\mu)$ (*right panel*) and the corresponding lower limits on the products of Pati–Salam leptoquark mass and mixing. Present results are compared with results projected for LHCb for an integrated luminosity of 2 fb^{-1} in case no signal would be observed. Dashed regions indicate the theoretical uncertainties in the relation between the variables

8.4 In flight conversions

Lepton Flavor Violation could manifest itself in the conversion of high energy muons into tau leptons when scattering on nucleons in a fixed target configuration [979]. Muons can be produced much more copiously than tau leptons so $\mu \to \tau$ conversions could be more sensitive than neutrinoless $\tau \to \mu X$ decays. When considering the effective lepton flavor violating four-fermion couplings, tau decays mainly involve light quarks, so heavy quark couplings are only loosely constrained [980]. In SUSY models, muon to tau conversion could be greatly enhanced by Higgs mediation at energies where heavy quarks contribute [812].

Within the context of this workshop the experimental feasibility of such experiment has been investigated. The cross section for mu to tau lepton conversion on target has been estimated to be at most 550 ab [980] for 50 GeV muons, using an effective model independent interpretation of the tau decay LFV constraints [964] based on the 2000 data [981]. By rescaling the upper limit on $B(\tau \to \mu\pi^+\pi^-)$ to the current value [982, 983], one obtains an upper limit at 90% C.L. on the cross section of 4.7 ab. This value scales roughly linearly with the muon energy. In the context of the MSSM, the experimental data available in 2004 constrained the cross section in the range from 0.1 ab to 1 ab for muon energies from 100 to 300 GeV [812].

The following assumptions were made to assess the experimental feasibility:

- the goal is a sensitivity to the cross section corresponding to 1/10 of the present limits from tau decay, collecting at least thousand events per year;
- the active target consists of 330 planes of 300 μm thick silicon, with either strip or pixel readout;
- the target has transverse dimensions corresponding to an area of 1 m^2 and the beam is distributed homogeneously over the target.

As a consequence, 3.75×10^{19} muons/yr are needed which, assuming a 10% duty cycle and an effective data taking year of 10^7 s, corresponds to 3.75×10^{13} muons/s (peak) and 3.75×10^{12} muons/s (average).

Using the LEPTO 6.5.1 generator [984] deep inelastic muon scattering off nucleons was studied. The amount of power dissipated in the target is sustainable, and the interaction rate is 0.6 interactions per 25 ns, which is comparable to LHC experiments. Radiation levels and occupancy in the silicon active target appear to be tractable, provided pixel readout is used.

When requiring momentum transfer above 2 GeV and invariant mass of the hadronic final state above 3 GeV an effective interaction cross section of 47 nb was found. This value reduces to 15 nb when applying the level 0 trigger requirement of at least 60 GeV of hadronic energy which results in a rate of 7.7 MHz. The amount of data that needs to be extracted from the tracker for further event selection can probably be handled at such rate.

Unfortunately it appears that the required muon flux is incompatible with the operation of calorimeters as triggering and detecting devices. Assuming an LHCb-like electromagnetic calorimeter with a 2.6 cm thick lead absorber and an integration time of 25 ns, and assuming that electrons from muon decay travel unscreened for 4 m before encountering the electro-magnetic calorimeter, three high energy electrons per 25 ns integration time reach the calorimeter, preventing any effective way of triggering on electrons. Assuming an LHCb-like hadronic calorimeter structured in towers consisting of 75 layers including 13×13 cm^2 scintillating pads and 16 mm of iron each, each tower will detect 25 TeV of equivalent hadronic energy for each 25 ns of integration time just because of the muon flux energy loss. The Poisson fluctuation of the number of muons will induce a fluctuation in the detected hadronic energy per tower of about 200 GeV, preventing the use of the hadronic calorimeter as a trigger for $\mu N \to \tau X$.

In conclusion, the idea of using an intense but transversely spread muon flux to produce and detect LFV muon conversions to tau leptons does not appear feasible in this preliminary study, mainly because it does not appear possible to operate calorimeters at these rates.

9 Experimental studies of electric and magnetic dipole moments

9.1 Electric dipole moments

We review here the current status and prospects of the searches for fundamental EDMs, a flavor diagonal signal of CP violation. At the non-relativistic level, the EDM d determines a coupling of the spin to an external electric field, $\mathcal{H} \sim d\vec{E} \cdot \vec{S}$. Searches for intrinsic EDMs have a long history, stretching back to the prescient work of Purcell and Ramsey who used the neutron EDM as a test of parity in nuclear physics. At the present time, there are two primary motivations for anticipating a nonzero EDM at or near current sensitivity levels. Firstly, a viable mechanism for baryogenesis requires a new CP-odd source, which if tied to the electroweak scale necessarily has important implications for EDMs. The second is that CP-odd phases appear quite generically in models of new physics introduced for other reasons, e.g. in supersymmetric models. Indeed, it is only the limited field content of the SM which limits the appearance of CP-violation to the CKM phase and $\theta_{\rm QCD}$. The lack of any observation of a nonzero EDM has, on the flipside, provided an impressive source of constraints on new physics, and there is now a lengthy body of literature on the constraints imposed, for example, on the soft breaking sector of the MSSM. Generically, the EDMs ensure that new CP-odd phases in this sector are at most of $\mathcal{O}(10^{-1}$–$10^{-2})$, a tuning that appears rather unwarranted given the $\mathcal{O}(1)$ value of the CKM phase.

The strongest current EDM constraints are shown for three characteristic classes of observables in Table 25, and will be discussed in detail in the following.

We summarize first the details of the EDM constraints, and the induced bounds on a generic class of CP-odd operators normalized at 1 GeV, commenting on how the next generation of experiments will impact significantly on the level of sensitivity in all sectors. We then turn to a brief discussion of some of the constraints on new physics that ensue from these bounds. More detailed discussions of phenomenology of EDMs is given in the first half of this report (see e.g. Sect. 5.7).

9.1.1 CP-odd operators and electric dipole moments

We shall briefly review the relevant formulae for the observable EDMs in terms of CP-odd operators normalized at 1 GeV. Including the most significant flavor diagonal CP-odd operators (see e.g. [742]) up to dimension six, the corresponding effective Lagrangian takes the form,

$$\mathcal{L}_{\rm eff}^{1\,{\rm GeV}} = \frac{g_s^2}{32\pi^2}\bar{\theta} G^a_{\mu\nu}\tilde{G}^{\mu\nu,a} - \frac{i}{2}\sum_{i=e,u,d,s} d_i \bar{\psi}_i (F\sigma)\gamma_5 \psi_i \\ - \frac{i}{2}\sum_{i=u,d,s}\tilde{d}_i \bar{\psi}_i g_s (G\sigma)\gamma_5 \psi_i \\ + \frac{1}{3} w f^{abc} G^a_{\mu\nu}\tilde{G}^{\nu\beta,b} G^{\mu,c}_\beta \\ + \sum_{i,j=e,q} C_{ij}(\bar{\psi}_i\psi_i)(\bar{\psi}_j i\gamma_5 \psi_j) + \cdots. \quad (9.1)$$

The θ-term, as it has a dimensionless coefficient, is particularly dangerous leading to the strong CP problem and in what follows we shall invoke the axion mechanism [986] to remove this term.

The physical observables can be conveniently separated into three main categories, depending on the physical mechanisms via which an EDM can be generated: EDMs of paramagnetic atoms and molecules; EDMs of diamagnetic atoms; and the neutron EDM. The inheritance pattern for these three classes is represented schematically in Fig. 66 and, while the experimental constraints on the three classes of EDMs differ by several orders of magnitude, it is important that the actual sensitivity to the operators in (9.1) turns out to be quite comparable in all cases. This is due to various enhancements or suppression factors which are relevant in each case, primarily associated with various violations of "Schiff shielding"—the non-relativistic statement that an electric field applied to a neutral atom must necessarily be screened and thus remove any sensitivity to the EDM.

9.1.2 EDMs of paramagnetic atoms

For paramagnetic atoms, Schiff shielding is violated by relativistic effects which can in fact be very large. One has roughly [987],

$$d_{\rm para}(d_e) \sim 10\alpha^2 Z^3 d_e, \quad (9.2)$$

which for large atoms such as Thallium amounts to a huge enhancement of the field seen by the electron EDM (see e.g.

Table 25 Current constraints within three representative classes of EDMs

Class	EDM	Current bound	Ref.
Paramagnetic	^{205}Tl	$\|d_{\rm Tl}\| < 9 \times 10^{-25}\,e\,{\rm cm}$	[186]
Diamagnetic	^{199}Hg	$\|d_{\rm Hg}\| < 2 \times 10^{-28}\,e\,{\rm cm}$	[985]
Nucleon	n	$\|d_n\| < 3 \times 10^{-26}\,e\,{\rm cm}$	[869]

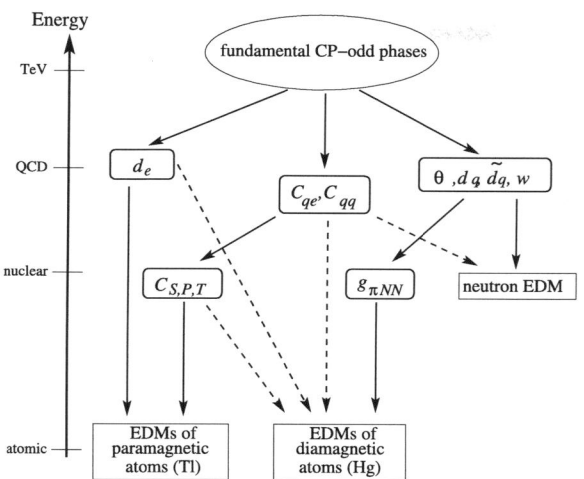
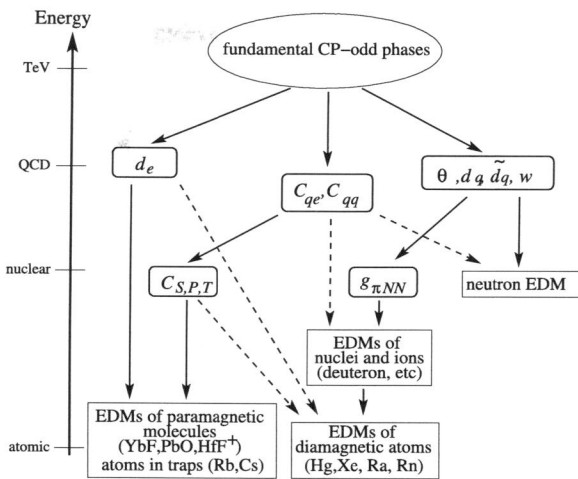

Fig. 66 A schematic plot of the hierarchy of scales between the leptonic and hadronic CP-odd sources and three generic classes of observable EDMs. The *dashed lines* indicate generically weaker dependencies in SUSY models. The current situation is given *on the left*, while *on the right* we show the dependencies of several classes of next-generation experiments

[987, 988]), which counteracts the apparently lower sensitivity of the Tl EDM bound,

$$d_{\text{Tl}} = -585 d_e - 43 \text{ GeV} \times e C_S^{\text{singlet}}. \quad (9.3)$$

We have also included here the most relevant CP-odd electron–nucleon interaction, namely $C_S \bar{e} i \gamma_5 e \bar{N} N$, which in turn is related to the semileptonic four-fermion operators in (9.1).

9.1.3 EDMs of diamagnetic atoms

For diamagnetic atoms, Schiff shielding is instead violated by the finite size of the nucleus and differences in the distribution of the charge and the EDM. However, this is a rather subtle effect,

$$d_{\text{dia}} \sim 10 Z^2 (R_N/R_A)^2 \tilde{d}_q, \quad (9.4)$$

and the suppression by the ratio of nuclear to atomic radii, R_N/R_A, generally leads to a suppression of the sensitivity to the nuclear EDM, parameterized to leading order by the Schiff moment S, by a factor of 10^3 (see e.g. [987, 988]). Thus, although the apparent sensitivity to the Hg EDM is orders of magnitude stronger than for the Tl EDM, both experiments currently have comparable sensitivity to various CP-odd operators and thus play a very complementary role. Combining the atomic $d_{\text{Hg}}(S)$, nuclear $S(\bar{g}_{\pi NN})$, and QCD $\bar{g}_{\pi NN}^{(1)}(\tilde{d}_q)$, components of the calculation [742, 988], we have

$$d_{\text{Hg}} = 7 \times 10^{-3} e(\tilde{d}_u - \tilde{d}_d) + 10^{-2} d_e + \mathcal{O}(C_S, C_{qq}), \quad (9.5)$$

where the overall uncertainty is rather large, a factor of 2–3, due to significant cancellations between various contributions. A valuable feature of d_{Hg} is its sensitivity to the triplet combination of color EDM operators $\tilde{d}_q$.

9.1.4 Neutron EDM

The neutron EDM measurement is of course not sensitive to the above atomic enhancement/suppression factors and, using the results obtained using QCD sum rule techniques [742] (see also [849, 989, 990] for alternative chiral approaches), wherein under Peccei–Quinn relaxation of the axion the contribution of sea-quarks is also suppressed at leading order:

$$d_n = (1.4 \pm 0.6)(d_d - 0.25 d_u) \\
+ (1.1 \pm 0.5) e (\tilde{d}_d + 0.5 \tilde{d}_u) \\
+ 20 \text{ MeV} \times ew + \mathcal{O}(C_{qq}). \quad (9.6)$$

Note that the proportionality to $d_q \langle \bar{q} q \rangle \sim m_q \langle \bar{q} q \rangle \sim f_\pi^2 m_\pi^2$ removes any sensitivity to the poorly known absolute value of the light quark masses.

9.1.5 Future developments

The experimental situation is currently very active, and a number of new EDM experiments, as detailed in this report, promise to improve the level of sensitivity in all three classes by one-to-two orders of magnitude in the coming years. These include: new searches for EDMs of polarizable paramagnetic molecules, which aim to exploit additional polarization effects enhancing the effective field seen by the unpaired electron by a remarkable factor of up to 10^5, and are therefore primarily sensitive to the electron EDM; new searches for the EDM of the neutron in cryogenic systems;

and also proposed searches for EDMs of charged nuclei and ions using storage rings. This latter technique clearly aims to avoid the effect of Schiff shielding and enhance sensitivity to the nuclear EDM and its hadronic constituents. A schematic summary of how a number of these new experiments will be sensitive to the set of CP-odd operators is exhibited in Fig. 66.

9.1.6 Constraints on new physics

Taking the existing bounds, and the formulae above, we obtain the following set of constraints on the CP-odd sources at 1 GeV (assuming an axion removes the dependence on $\bar{\theta}$),

$$\left| d_e + e(26 \text{ MeV})^2 \left(3 \frac{C_{ed}}{m_d} + 11 \frac{C_{es}}{m_s} + 5 \frac{C_{eb}}{m_b} \right) \right|$$
$$< 1.6 \times 10^{-27} e \text{ cm} \quad \text{from } d_{Tl}, \quad (9.7)$$

$$\left| (\tilde{d}_d - \tilde{d}_u) + \mathcal{O}(\tilde{d}_s, d_e, C_{qq}, C_{qe}) \right| < 2 \times 10^{-26} e \text{ cm}$$
$$\text{from } d_{Hg}, \quad (9.8)$$

$$\left| e(\tilde{d}_d + 0.56\tilde{d}_u) + 1.3(d_d - 0.25d_u) + \mathcal{O}(\tilde{d}_s, w, C_{qq}) \right|$$
$$< 2 \times 10^{-26} e \text{ cm} \quad \text{from } d_n, \quad (9.9)$$

where the additional $\mathcal{O}(\cdots)$ dependencies are known less precisely, but may not always be sub-leading in particular models. The precision of these results varies from 10–15% for the Tl bound, to around 50% for the neutron bound, and to a factor of a few for Hg. It is remarkable to note that, accounting for the naive mass-dependence $d_f \propto m_f$, all these constraints are of essentially the same order of magnitude and thus highly complementary. Constraints obtained in the hadronic sector using other calculational techniques differ somewhat but generally give results consistent with these within the quoted precision.

The application of these constraints to models of new physics has many facets and is discussed in several specific cases elsewhere in this report. We shall limit our attention here to just a few simple examples relevant to the motivations noted above.

9.1.7 The SUSY CP-problem

It is now rather well-known that a generic spectrum of soft SUSY-breaking parameters in the MSSM will generate EDMs via one loop diagrams [852] that violate the existing bounds by one-to-two orders of magnitude leading to the SUSY CP problem. The situation is summarized rather schematically in Fig. 67.

In many respects the situation is better described by the amount of fine tuning of the MSSM spectrum that is required to avoid these leading order contributions, and by how much the ability to avoid the EDM constraints is limited by secondary constraints from numerous, and more robust, two

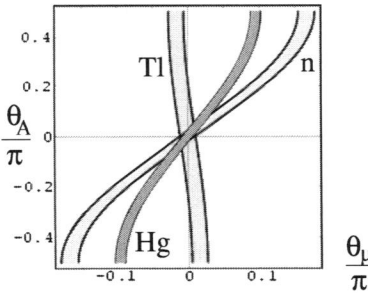

Fig. 67 Constraints on the CMSSM phases θ_A and θ_μ from a combination of the three most sensitive EDM constraints, d_n, d_{Tl} and d_{Hg}, for $M_{SUSY} = 500$ GeV, and $\tan \beta = 3$ (from [742]). The region allowed by EDM constraints is at the intersection of all three bands around $\theta_A = \theta_\mu = 0$

loop contributions [991] and four-fermion sources [992]. Indeed, if we consider two extreme cases: (i) the 2HDM, where all SUSY fermions and sfermions are very heavy; and (ii) split SUSY, where all SUSY scalars are very heavy, one finds that while one loop EDMs are suppressed, two loop contributions are already very close to the current bounds [857, 858, 992]. This bodes well for the ability of next-generation experiments to provide a comprehensive test of large SUSY phases at the electroweak scale, regardless of the detailed form of the SUSY spectrum.

9.1.8 Constraints on new SUSY thresholds

If SUSY is indeed discovered at the LHC, but with no sign of phases in the soft sector, one may instead consider the ability of EDMs to detect new supersymmetric CP-odd thresholds. At dimension five there are several R-parity conserving operators, besides those well-known examples associated with neutrino masses and baryon and lepton number violation [150, 371]. Writing the relevant dimension five superpotential as [993, 994]

$$\Delta \mathcal{W} = \frac{y_h}{\Lambda_h} H_d H_u H_d H_u + \frac{Y_{ijkl}^{qe}}{\Lambda_{qe}} (U_i Q_j) E_k L_l$$
$$+ \frac{Y_{ijkl}^{qq}}{\Lambda_{qq}} (U_i Q_j)(D_k Q_l)$$
$$+ \frac{\tilde{Y}_{ijkl}^{qq}}{\Lambda_{qq}} (U_i t^A Q_j)(D_k t^A Q_l), \quad (9.10)$$

one finds that order-one CP-odd coefficients with a generic flavor structure, particularly for the semileptonic operators, are probed by the sensitivity of d_{Tl} and d_{Hg} at the remarkable level of $\Lambda \sim 10^8$ GeV [993, 994]. This is comparable to, or better than, the corresponding sensitivity of lepton-flavor violating observables.

9.1.9 Constraints on minimal electroweak baryogenesis

As noted above, one of the primary motivations for anticipating nonzero EDMs at or near the current level of sensitivity is through the need for a viable mechanism of baryogenesis. This is clear in essentially all baryogenesis mechanisms that are tied to the electroweak scale. As a simple illustration, one can consider a *minimal* extension of the SM Higgs sector [995–997],

$$\mathcal{L}_{\dim 6} = \frac{1}{\Lambda^2}(H^\dagger H)^3 + \frac{Z_{ij}^u}{\Lambda_{\rm CP}^2}(H^\dagger H) U_i^c H Q_j$$
$$+ \frac{Z_{ij}^d}{\Lambda_{\rm CP}^2}(H^\dagger H) D_i^c H^\dagger Q_j$$
$$+ \frac{Z_{ij}^e}{\Lambda_{\rm CP}^2}(H^\dagger H) E_i^c H^\dagger L_j. \qquad (9.11)$$

The first term is required to induce a sufficiently strong first order electroweak phase transition, while the remaining operators provide the additional source (or sources) of CP-violation, where we have assumed consistency with the principle of minimal flavor violation. Modified Higgs couplings of this kind, including CP-violating effects, are currently the subject of significant research within collider physics, relevant to the LHC in particular [588], making EDM probes of models of this kind quite complementary.

As discussed in [997], such a scenario can reproduce the required baryon-to-entropy ratio, $\eta_b = 8.9 \times 10^{-11}$, while remaining consistent with the EDM bounds, provided the thresholds and the Higgs mass are quite low, e.g. 400 GeV $< \Lambda$, $\Lambda_{\rm CP} <$ 800 GeV. The EDMs in this case are generated at the two loop level, and it is clear that an improvement in EDM sensitivity by an order of magnitude would provide a conclusive test of minimal mechanisms of this form.

9.2 Neutron EDM

The neutron electric dipole moment is sensitive to many sources of CP violation. Most famously, it constrains QCD (the "strong CP problem"), but it also puts tight constraints on supersymmetry and other physics models beyond the standard model. The standard model prediction of $\sim 10^{-32}$ e cm is a factor of 10^6 below existing limits, so any convincing signal within current or foreseen sensitivity ranges will be a clear indication of physics beyond the SM.

All current nEDM experiments use NMR techniques to search for electric-field induced changes in the Larmor precession frequency of bottled ultracold neutrons. Recent results from a room-temperature apparatus at ILL yielded a new limit of $|d_n| < 2.9 \times 10^{-26}$ e cm (90% C.L.) which rules out many "natural" varieties of SUSY. Several new experiments hope to improve on this limit: two of these involve new cryogenic techniques that promise an eventual increase in sensitivity by two orders of magnitude. First results, at the level of $\sim 10^{-27}$ e cm, are to be expected within about four years.

9.2.1 ILL

A measurement of the neutron EDM was carried out at the ILL between 1996 and 2002, by a collaboration from the University of Sussex, the Rutherford Appleton Laboratory, and the ILL itself. The final published result provided a limit of $|d_n| < 2.9 \times 10^{-26}$ e cm (90% C.L.) [869]. This represents a factor of two improvement beyond the intermediate result [998] and almost a factor of four beyond the results existing prior to this experiment [999, 1000]. The collaboration, which has now expanded to include Oxford University and the University of Kure, has designed and developed "CryoEDM", a cryogenic version of the experiment that is expected to achieve two orders of magnitude improvement in sensitivity. Construction and initial testing are underway at the time of writing.

Experimental technique The room-temperature measurement was carried out using stored ultracold neutrons (i.e. having energies $\lesssim 200$ neV) from the ILL reactor. The Ramsey technique of separated oscillatory fields was used to determine the Larmor precession frequency of the neutrons within $\vec{B}$ and $\vec{E}$ fields. The signature of an EDM is a frequency shift proportional to any change in the applied electric field.

The innovative feature of this experiment was the use of a cohabiting atomic-mercury magnetometer [1001]. Spin-polarized Hg atoms shared the same volume as the neutrons, and the measurement of their precession frequency provided a continuous high-resolution monitoring of the magnetic field drift: prior to this, such drift entirely dominated the tiny $\vec{E}$-field induced frequency changes that were sought.

Systematics Analysis of the data revealed a new source of systematic error, which, as the problem of B-field drift had been virtually eliminated, became potentially the dominant error. Its origins lay in a geometric-phase (GP) effect [1002]—an unfortunate collusion between any small applied axial $\vec{B}$-field gradient and the component of $\vec{B}$ induced in the particle's rest frame by the Lorentz transformation of the electric field. This GP effect induced a frequency shift proportional to $\vec{E}$, and hence a false EDM signal. In fact, the Hg magnetometer itself was some 50 times more susceptible to this effect than were the neutrons, so the introduction of the magnetometer brought the GP systematic with it.

This effect was overcome by careful measurement of the neutron-to-Hg frequency ratios for both polarities of magnetic field, in order to determine the point nominally corresponding to zero applied axial B-field gradient, as well as by a series of auxiliary measurements to pin down small corrections due to local dipole [1003] and quadrupole fields (as well as the Earth's rotation). The final result therefore remained statistically limited.

The experiment is now complete and, as will be discussed below, the equipment will be used for further studies by another collaboration based largely at the PSI.

Still another collaboration, led by the PNPI in Russia, is developing a new room-temperature nEDM apparatus, which they plan to run at ILL. It is also intended to reach a sensitivity of $\sim 10^{-27}$ e cm, to be achieved in part by the use of multiple back-to-back measurement chambers with opposing electric fields to cancel some systematic errors.

Cryogenic experiments overview It has been known for several decades [1004] that 8.9 Å neutrons incident on superfluid ^{4}He at 0.5 K will down-scatter, transferring their energy and momentum to the helium and becoming ultracold neutrons (UCN) in the process. This so-called super-thermal UCN source provides a much higher flux than is available simply from the low energy tail of the Maxwell distribution. In addition, the immersion of the apparatus in a bath of liquid helium should allow for the provision of stronger electric fields than could be sustained *in vacuo*. The other two variables that contribute to the figure of merit for this experiment, namely the polarization and the NMR coherence time, should also be improved: the incident cold neutron beam can be very highly polarized, and the polarization remains intact during the down-scattering process; and the improved uniformity of magnetic field attainable with superconducting shields and coil will reduce depolarization during storage, while losses from up-scattering will be much reduced due to the cryogenic temperatures of the walls of the neutron storage vessels.

ILL CryoEDM experiment status The majority of the apparatus for the cryoEDM experiment has been installed at ILL, and testing is underway. UCN production via this superthermal mechanism has been demonstrated [1005], and the solid-state UCN detectors developed by the collaboration have also been shown to work well [1006]. At the time of writing, there are still some hardware problems to be resolved, in particular with components in and around the Ramsey measurement chamber. A high precision scan of the magnetic field was carried out in 2007, and measurements were made of the neutron polarization. An initial HV system will be installed in spring 2008. By the end of 2008, the system is expected to have a statistical sensitivity of $\sim 10^{-27}$ e cm.

Future plans In order to achieve optimum sensitivity, a number of improvements will need to be made:

– The superconducting magnetic shielding requires additional protective "end caps" to shield fully the ends of the superconducting solenoid.
– The current measurement chamber only has two cells: one with HV applied, and one at ground as a control. It is planned to upgrade to a four-cell chamber, with the HV applied to the central electrode, in order to be able to carry out simultaneous measurements with electric fields in opposite directions. As well as canceling several potential systematic errors, this will reduce the statistical uncertainty by doubling the number of neutrons counted.
– The ILL is preparing a new beam line with six times the currently available intensity of 8.9 Å neutrons, and wishes to transfer the experiment to that beam line in 2009. Funding for these improvements is expected to be contingent on successful running of the existing apparatus.

A sensitivity of $\sim 2 \times 10^{-28}$ e cm should be achievable within two to three years of running at the new beam line.

9.2.2 PSI

The present best limit for the neutron electric dipole moment (EDM), $|d_n| < 2.9 \times 10^{-26}$ e cm [869], was obtained by the Sussex/Rutherford/ILL Collaboration from measurements at the ILL source for ultracold neutrons [1007], as discussed in the previous section. The experiment is at this point statistically limited and also facing systematic challenges not far away [869, 1002, 1003]. In order to make further progress, both, statistical sensitivity and control of systematics, have to be improved. Gaining in statistics requires new sources for ultracold neutrons (UCN). These can be integrated into the experiment as for the new cryogenic EDM searches, delivering UCN in superfluid helium, or a multi-purpose UCN source, delivering UCN in vacuum. This high intensity UCN source is presently under construction at the Paul Scherrer Institut in Villigen, Switzerland [1008]. It is expected to become operational towards the end of 2008 and to deliver UCN densities of more than 1000 cm^{-3} to typical experiments, i.e. almost two orders of magnitude more than presently available.

The in-vacuum technique will be pushed to its limits, delivering first results in about 4 years. The following steps are planned by a sizable international collaboration [1009]:

– While the new UCN source is under construction the collaboration operates and improves the apparatus of the former Sussex/RAL/ILL Collaboration at ILL Grenoble. In order to better control the systematic issues, the magnetic field and its gradients will be monitored and stabilized using an array of laser optically pumped Cs-magnetometers

[1010, 1011]. An order of magnitude improvement compared to todays field fluctuations over the typical measurement times of 100–1000 s is certainly feasible. It is also necessary to improve the sensitivity of the Hg co-magnetometer [1012]. Other improvements of the system are with regard to UCN polarization and detection as well as upgrading the data acquisition system. The hardware efforts are accompanied by a full simulation of the system.

– It is planned to move the apparatus from ILL to PSI towards the end of 2008 in order to be ready for data taking for about two years, 2009 and 2010. In addition to the improvements of phase I, an external magnetic field stabilization system and a temperature stabilization are envisaged. Furthermore, work on developing a second co-magnetometer using a hyper-polarized noble gas species is ongoing and might further improve the systematics control. In case of a successful development, also the replacement of the Hg system together with an increase of the electric field strength may become possible. In any case, a factor of 5 gain in sensitivity is expected from the higher UCN intensity, corresponding to a limit of about 5–6×10^{-27} e cm in case the EDM is not found. In parallel to the described activities, the design of a new experimental apparatus will start in 2007. After a major design effort in 2008, set-up of the new apparatus will start in 2009.

– The new experiment will be an optimized version of the room-temperature in-vacuum approach. Another order of magnitude gain in sensitivity will be obtained by a considerable increase of the statistics due to a larger experimental volume ($\times\sqrt{5}$), a better adaption to the UCN source ($\times\sqrt{2}$), longer running time ($\times\sqrt{3}$) and by an improvement of the electric field strength ($\times 2$). Completion of the new experimental apparatus is anticipated for end of 2010, and data taking planned for 2011–2014.

The features of the experiment include

– continued use of the successful Ramsey-technique with UCN in vacuum and the apparatus at room-temperature,
– increased sensitivity due to much larger UCN statistics at the new PSI source, larger experimental volume, better polarization product and possibly larger electric field strength,
– application of a double neutron chamber system,
– improved magnetic field control and stabilization with multiple laser optically pumped Cs-magnetometers, and
– an improved co-magnetometry system.

As another very strong source for UCN is currently under construction at the FRMII in Munich, in the long run and for the optimum conditions for the experiment, the collaboration will have the opportunity to choose between PSI and FRMII.

9.2.3 SNS

A sizable US Collaboration [1013] is planning to develop a cryogenic experiment, following an early concept by Golub and Lamoreaux [1014]. It will be based at the SNS 1.4 MW spallation source at Oak Ridge. A fundamental neutron physics beam line is under construction, which will include a double monochromator to select 8.9 Å neutrons for UCN production in liquid helium.

In this experiment, spin-polarized ^{3}He will be used both as a magnetometer and as a neutron detector. The precession of the ^{3}He can, in principle, be detected with SQUID magnetometers. Meanwhile, the cross section for the absorption reaction $n + {}^3\text{He} \to p + {}^3\text{H} + 764$ keV is negligible for a total spin $J = 1$, but very large (~ 5 Mb) for $J = 0$. In consequence, a scintillation signal from this reaction will be detected with a beat frequency corresponding to the difference between the Larmor precession frequencies of the neutrons and the ^{3}He.

An application for funding to construct this experiment is currently under review. Extensive tests are underway to study, for example, the electric fields attainable in liquid helium, the ^{3}He spin relaxation time and the diffusion of ^{3}He in ^{4}He. If construction goes according to plan, commissioning will be in approximately 2013, with results following probably four or five years later. The ultimate sensitivity will be below 10^{-28} e cm.

9.3 Deuteron EDM

A new concept of investigating the EDM of bare nuclei in magnetic storage rings has been developed by the storage ring EDM Collaboration (SREC) over the past several years. The latest version of the methods analyzed turns out to be very sensitive for light (bare) nuclei and promises the best EDM experiment for θ_{QCD}, quark and quark-color EDMs.

The search for hadronic EDMs has been dominated by the search for a neutron EDM and nuclear Schiff moments in heavy diamagnetic atoms, such as ^{199}Hg. The latter depend on nuclear theory to relate the measured Schiff moment to the underlying CP violating interaction.

The sensitive 'traditional' EDM experiments are, so far, all performed on electrically neutral systems, such as the neutron, atoms, or molecules. A strong electric field is imposed, together with a weak magnetic field, and using NMR techniques, a change of the Larmor precession frequency is looked for. The application of strong electric fields precludes a straightforward use of this technique on charged particles. These particles would accelerate out of the setup, leaving little time to make an accurate measurement.

Attempt to search for an EDM on simple nuclear systems, such as the proton or deuteron, when part of an atom, are severely hindered by shielding. This so-called Schiff-screening precludes an external electric field to penetrate to the nucleus. Due to rearrangement of the atomic electrons, the net effect of the electric field on the nucleus is essentially zero. Known loop-holes include relativistic effects, non-electric components in the binding of the electrons, and an extended size of the nucleus. None of these loopholes are sufficiently strong to allow a sensitive measurement on a light atom. For hydrogen atoms, the atomic EDM resulting from a nuclear EDM is down by some seven orders of magnitude.

Nevertheless, light nuclei, and the deuteron in particular, are attractive to search for hadronic EDMs because of their relatively simple structure. Moreover, a novel experimental technique, using the motional electric field experienced by a relativistic particle when traversing a magnetic field, make it possible to directly search for EDM on charged systems, such as the (bare) deuteron.

9.3.1 Theoretical considerations

The deuteron is the simplest nucleus. It consists of a weakly bound proton and neutron in a predominantly $^3S_1$ state, with a small admixture of the D-state. From a theoretical point of view, the deuteron is especially attractive, because it is the simplest system in which the P-odd, T-odd nucleon–nucleon (NN) interaction contributes to an EDM. Moreover, the deuteron properties are well understood, so reliable and precise calculations are possible.

In [1015], a framework is presented that could serve as a starting point for the microscopic calculation of complex systems. The most general form of the interaction, based only on symmetry considerations, contains ten P- and T-odd meson-nucleon coupling constants for the lightest pseudo-scalar and vector mesons (π, ρ, η and ω).

This P-odd, T-odd interaction induces a P-wave admixture to the deuteron wave function. It is this admixture that leads to an EDM. Since the proton and neutron that make up the deuteron may also have an EDM, a disentanglement of one- and two-body contributions,

$$d_\mathcal{D} \simeq d_\mathcal{D}^{(1)} + d_\mathcal{D}^{(2)} \qquad (9.12)$$

to the EDM is necessary to uncover the underlying structure of the P-odd T-odd physics.

The two-body component is predominantly due to the polarization effect, and shows little model dependence for all leading high-quality potentials. Additional contributions arrive from meson exchange.

The one body contribution is simply the sum of the proton and neutron EDMs. The nucleon EDM has a wide variety of sources, as already discussed for the neutron. There exists no good model to describe the non-perturbative dynamics of bound quarks. A commonly used method is to evaluate hadronic loop diagrams, containing mesonic and baryonic degrees of freedom. Within the framework presented in [1015], the EDMs for the proton, neutron and deuteron are found (reproducing only the pion dependence),

$$\begin{aligned}d_p &= -0.05\bar{g}_\pi^{(0)} + 0.03\bar{g}_\pi^{(1)} + 0.14\bar{g}_\pi^{(2)} + \cdots, \\ d_n &= +0.14\bar{g}_\pi^{(0)} - 0.14\bar{g}_\pi^{(2)} + \cdots, \qquad (9.13) \\ d_\mathcal{D} &= +0.09\bar{g}_\pi^{(0)} + 0.23\bar{g}_\pi^{(1)} + \cdots.\end{aligned}$$

These dependences clearly show the complementarity of these three particles.

The magnitudes of the coupling constants can be calculated for several viable sources of CP violation. In the standard model, there is room for CP violation via the so-called $\bar{\theta}$ parameter. In the case of the nucleons, one has the relation

$$d_n \simeq -d_p \simeq 3 \times 10^{-16} \bar{\theta} \, e\,\mathrm{cm}, \qquad (9.14)$$

which yields the severe constraint $\bar{\theta} < 1 \times 10^{-10}$. For the deuteron, one finds

$$d_\mathcal{D} \simeq -10^{-16} \bar{\theta} \, e\,\mathrm{cm}. \qquad (9.15)$$

At the level of $d_\mathcal{D} \simeq 10^{-29} \, e\,\mathrm{cm}$, one probes $\bar{\theta}$ at the level of 10^{-13}. Since $\bar{\theta}$ contributes differently to the neutron and the deuteron, it is clear that both experiments are complementary. Indeed. the prediction

$$d_\mathcal{D}/d_n = -1/3 \qquad (9.16)$$

provides a beautiful check as to whether $\bar{\theta}$ is the source of the observed EDMs, should both be measured. In fact, measurement of the EDMs of the proton, deuteron and ^{3}He would allow to verify if they satisfy the relation

$$d_\mathcal{D} : d_p : d_{^3\mathrm{He}} \simeq 1 : 3 : -3. \qquad (9.17)$$

Here, it was assumed that ^{3}He has properties very similar to the neutron, which provides most of the spin.

Generic supersymmetric models contain a plethora of new particles, which may be discovered at LHC, and new CP-violating phases. Following the work by Lebedev et al. [1016] and the review by Pospelov and Ritz [742], we find that SUSY loops give rise to ordinary quark EDMs, d_q, as well as quark-color EDMs, $\tilde{d}_q$. For the neutron and deuteron one finds (with the color EDM part divided in isoscalar and isovector parts)

$$\begin{aligned}d_n &\simeq 1.4(d_d - 0.25 d_u) + 0.83 e(\tilde{d}_d + \tilde{d}_u) \\ &\quad + 0.27 e(\tilde{d}_d - \tilde{d}_u), \qquad (9.18) \\ d_\mathcal{D} &\simeq (d_d + d_u) - 0.2 e(\tilde{d}_d + \tilde{d}_u) + 6 e(\tilde{d}_d - \tilde{d}_u),\end{aligned}$$

and similar relations for e.g. the mercury EDM. The isovector part is limited to $|ec(\tilde{d}_d - \tilde{d}_u)| < 2 \times 10^{-26} \, e\,\mathrm{cm}$ by the present limit on the ^{199}Hg atom. The experimental bound on the neutron suggests that $|e(\tilde{d}_d + \tilde{d}_u)| < 4 \times 10^{-26}$ e cm, assuming the isoscalar contribution to be dominant. Also in this case, the deuteron and neutron show complementarity. This is in particular in their sensitivity to the isovector contribution, which is 20 times larger for the deuteron.

The large sensitivity to new physics (see e.g.[1016]) and the relative simplicity of calculating the nuclear wave function, make it clear that small nuclei hold great discovery potential and should therefore be vigorously pursued.

9.3.2 Experimental approach

All sensitive EDM searches are performed on neutral systems, which are (essentially) at rest. The only exception is the proposed use of molecular ions (HfF$^+$ and ThF$^+$) [1017], but also for this experiment, the motion of the molecules is not crucial.

In the recent past, several novel techniques have been proposed to use the motional electric field sensed by a particle moving through a magnetic field at relativistic velocities. The evolution of the spin orientation for a spin-1/2 particle in an electromagnetic field $(\vec{E}, \vec{B})$ is described by the so-called Thomas or BMT equation [1018]. The spin precession vector $\vec{\Omega}$, *relative* to the momentum of the particle, is given by [1019]

$$\vec{\Omega} = \frac{e}{m}\left[a\vec{B} + \left(a - \frac{1}{\gamma^2 - 1}\right)\vec{\beta} \times \vec{E} \right. $$
$$\left. + \frac{\eta}{2}\left(\vec{E} + \vec{\beta} \times \vec{B} - \frac{\gamma}{\gamma+1}\vec{\beta}(\vec{\beta}\cdot\vec{E})\right) \right] \quad (9.19)$$

with $\vec{\mu} = 2(1+a)(e/m)\vec{S}$ and $\vec{d} = \eta/2(e/m)\vec{S}$. It was assumed that $\vec{\beta}\cdot\vec{B} = 0$. The first two terms between brackets will be referred to as ω_a, whereas the last one will be referred to as ω_η.

For fast particles, the electric field in the rest frame of the particle is dominated by $\vec{\beta} \times \vec{B}$. For commonplace storage rings, this field can exceed the size of a static electric field made in the laboratory by more than an order of magnitude, thus giving the storage ring method a distinct advantage.

In a homogeneous magnetic field, $\vec{\omega}_a \propto \vec{B}$ and $\vec{\omega}_\eta \propto \vec{\beta} \times \vec{B}$ are orthogonal, leading to a small tilt in the precession plane and an second order increase in the precession frequency. Although this was used to set a limit on the muon EDM [188, 1020], it does not allow for a sensitive search.

The application of a radially oriented electric field E_r to slow down ω_a and thus to increase the tilt, was proposed in [1021]. For a field strength

$$E_r = \frac{a\beta}{1 - (1+a)\beta^2} B_z \quad (9.20)$$

the spin of an originally longitudinally polarized beam remains aligned with the momentum at all times. In this case $\vec{\beta}\cdot\vec{E} = 0$, and thus

$$\vec{\Omega} = \frac{e}{m}\frac{\eta}{2}(\vec{E} + \vec{\beta} \times \vec{B}). \quad (9.21)$$

The EDM thus manifests itself as a precession of the spin around the motional electric field $\vec{E}^* = \gamma[\vec{E} + \vec{\beta} \times \vec{B}]$, i.e. as a growing vertical polarization component parallel to $\vec{B}$. This approach can be used for all particles with a small magnetic anomaly, so that the necessary electric field strength remains feasible. Concept experiments, employing this technique, have been worked out for the muon [1022–1024] and the deuteron [1025]. Other candidate particles have been identified as well (see e.g. [1026]).

A third, most sensitive approach is reminiscent of the magnetic resonance technique introduced by Rabi [1027]. The spin is allowed to precess under the influence of a dipole magnetic field. In the original application, an oscillating magnetic field oriented perpendicular to the driving field is applied. By scanning the oscillation frequency, a resonance will be observed when the frequency of the oscillating field matches the spin precession frequency.

In this application, the oscillating magnetic field are replaced by an oscillating electric field [1028]. When at resonance, the electric field coherently interacts with the electric dipole moment. As a consequence, the polarization component along the magnetic field oscillates in the case of a sizable EDM. In practice, only the onset of the first oscillation cycle will be visible in the form of a slow growth of the vertical polarization, proportional to the EDM.

The oscillating electric field is obtained by modulating the velocity of the deuterons circulating in a magnetic field, setting up a so-called synchrotron oscillation. For a time dependent velocity $\beta(t) = \beta_0 + \delta\beta(t)$ generated by an oscillating longitudinal electric field $E_{\text{RF}}(t)$ and a constant magnetic field B, the spin evolution follows from

$$\vec{\Omega} = \frac{e}{m}\left[\left\{aB + \frac{\eta}{2}\vec{\beta}_0 \times \vec{B}\right\}\right.$$
$$\left. + \frac{\eta}{2}\left\{\delta\vec{\beta}(t) \times \vec{B} - \frac{\beta^2\gamma}{\gamma+1}\vec{E}_{\text{RF}}(t)\right\}\right]$$
$$\equiv \vec{\Omega}_0 + \delta\vec{\Omega}(t). \quad (9.22)$$

The first term yields spin precession about $\vec{\Omega}_0$, without affecting the polarization parallel to it. For $\delta\beta(t) = \delta\beta\cos(\omega t + \psi)$, and $B\delta\beta \gg \beta^2\gamma/(\gamma+1)E_{\text{RF}}$, the parallel polarization component is given by

$$\frac{dP_\parallel}{dt} \simeq \frac{e}{m}P_\circ \eta \delta\beta B \cos(\Delta\omega t + \Delta\phi), \quad (9.23)$$

with $\Delta\omega \equiv \Omega_0 - \omega$ and $\Delta\phi \equiv \phi - \psi$. The beam is assumed to have a longitudinal polarization P_0 at injection time. For $\Delta\omega = 0$ the vertical polarization will grow continuously at a rate proportional to the EDM. Maximum sensitivity is obtained for $\Delta\phi = 0$ or π, whereas for $\Delta\phi = \pi/2$ or $3\pi/2$ there is no sensitivity to the EDM. The latter will prove useful in controlling systematic errors. At the same time, the radial polarization component is given by

$$P_\perp \simeq P_0 \sin(\Omega_0 t + \phi). \quad (9.24)$$

This polarization component can be incorporated in a feedback cycle, to phase-lock the velocity modulation to the spin precession, i.e. to guarantee $\Delta\omega = 0$ and $\Delta\phi$ constant. In addition, observation of Ω_0 allows to measure or stabilize the magnetic field.

From (9.23) and (9.24), the main design criteria are easily derived, several of which are common to all other EDM experiments. They include

- high initial polarization P_0;
- large field strength $E^{\text{eff}} \propto (\delta\beta B)$;
- close control over the resonance conditions $\Delta\omega$ and phase $\Delta\phi$;
- long spin coherence time $P_\circ(t)$;
- long synchrotron coherence time $\delta\beta(t)$;
- sensitive method for independent observation of $P_\parallel$ and $P_\perp$.

The parameters of the current concept deuteron EDM ring are presented in Table 26. Coherent synchrotron oscillation can be obtained by a set of two RF cavities, one operating at a harmonic of the revolution frequency to bring the beam close to the spin-synchrotron resonance, and a second operating at the resonance frequency to create a forced oscillation.

The statistical reach of the experiment is determined by the number of particles used to determine the polarization, as well as the analyzing power of the polarimeter. The most efficient way to probe the deuteron polarization at the energy considered is by nuclear scattering. To obtain high efficiency, conventional techniques, in which a target is inserted into the beam are unsuitable. Instead, slow extraction of the beam onto a thick analyzer target is necessary. Slow extraction could be realized by exciting a weak beam resonance, or alternatively, by Coulomb scattering off a thin gas jet. The thickness of the analyzer target is optimized to yield maximum efficiency, which may reach the percent level.

The EDM will create a left-right asymmetry in the scattered particle rate, whose initial rate of growth is proportional to the EDM. False signals from, e.g., oscillating radial magnetic fields in the ring will be mitigated by varying the lattice parameters. This will change the systematic error amplitude, while leaving the EDM signal unchanged. Various features of the ring design and bunches with opposite EDM signals will be used to reduce the impact of other systematic effects.

The expected very high observability of most of the field imperfections in the experiment comes from the combination of gross amplification of the original perturbations in the control bunches, and observation and correction of the amplified parasitic growth of the vertical polarization component. This growth is many orders of magnitude more sensitive to ring imperfections than any other beam parameter. Preliminary studies shows no unmanageable sources of systematic errors at the level of the expected statistical uncertainty of 10^{-29} e cm.

There is currently great interest in EDM experiments because of their potential to find new physics complementary to and even reaching beyond that which can be found at future accelerators (LHC and beyond). The new approach described here would be the most sensitive experiment for the measurement of several possible sources of EDMs in nucleons and nuclei for the foreseeable future, if systematic uncertainties can be controlled.

9.4 EDM of deformed nuclei: ^{225}Ra

In the nuclear sector, the strongest EDM limits have been set by cell measurements which restrict the EDM of ^{199}Hg to $< 2.1 \times 10^{-28}$ e cm. A promising avenue for extending these searches is to take advantage of the large enhancements in the atomic EDM predicted for octupole deformed nuclei. One such case is ^{225}Ra, which is predicted to be two to three orders of magnitude more sensitive to T-violating interactions than ^{199}Hg. The next generation EDM search around laser-cooled and trapped ^{225}Ra is being developed by the Argon group. They have demonstrated transverse cooling, Zeeman slowing, and capturing of ^{225}Ra and ^{226}Ra atoms in a magneto-optical trap (MOT). They have measured many of the transition frequencies, life-times, hyperfine splittings

Table 26 Parameters of the concept deuteron EDM storage ring.

Parameter	Symbol	Design value
Deuteron momentum	$p_\mathcal{D}$	1500 MeV/c
Magnetic field strength	B	2 T
Bending radius	ρ	2.5 m
Length of each straight section	l	5 m
Orbit length	L	26 m
Momentum compaction	α_p	1
Cyclotron period	t_c	137 ns
Spin precession period	t_s	660 ns
Spin coherence time	τ_s	1000 s
Motional electric field	E^*/γ	375 MV/m
Synchrotron amplitude	$\delta\beta/\beta$	1%
Synchrotron harmonic	h	40
Particles per fill	N	10^{12}
Initial polarization	$P_\circ$	0.9
EDM precession rate @ $d = 10^{-26}$ e cm	ω_η	1 μrad/s

and isotope shifts of the critical transitions. This new development should enable them to launch a new generation of nuclear EDM searches. The combination of optical trapping and the use of octupole deformed nuclei should extend the reach of a new EDM search by two orders of magnitude. A non-zero EDM in diamagnetic atoms is expected to be most sensitive to a chromo-electric induced EDM effect.

Radium-225 is an especially good case for the search of the EDM because it has a relatively long life-time ($t_{1/2} = 14.9$ d), has spin 1/2 which eliminates systematic effects due to electric quadrupole coupling, is available in relatively large quantities from the decay of the long-lived ^{229}Th ($t_{1/2} = 7300$ yr), and has a well-established octupole nature. The octupole deformation enhances parity doubling of the energy levels. For example, the sensitivity to T-odd, P-odd effects in ^{225}Ra is expected to be a factor of approximately 400 larger than in ^{199}Hg, which has been used by previous searches to set the lowest limit ($<2 \times 10^{-28}$ e cm) so far on the atomic EDM. The 14.9-day half-life for ^{225}Ra is sufficiently long that measurements can be performed and systematics can be checked without resorting to an accelerator-based experiment. Nevertheless, if a ^{225}Ra beam facility were available for this experiment, approximately a hundred times more atoms could be produced which could have the impact of improving the sensitivity by yet another order of magnitude.

Laser cooling and trapping of ^{225}Ra atoms was developed in preparation of an EDM search. The laser trap allows one to collect and store the radioactive ^{225}Ra atoms that are otherwise too rare to be used for the search with conventional atomic-beam or vapor-cell type methods. Moreover, an EDM measurement on atoms in a laser trap would benefit from the advantages of high electric field, long coherence time, and a negligible so-called "$v \times E$" systematic effect.

The Argon group has demonstrated a magneto optical trap (MOT) of Ra atoms by using the $7s^2$ $^1S_0 \rightarrow 7s7p$ $^3P_1$ transition as the primary trapping transition, and $7s6d$ $^3D_1 \rightarrow 7s7p$ $^1P_1$ as the re-pump transition (see Fig. 68). They used a Ti:Sapphire ring laser system to generate the 714 nm light to excite the $7s^2$ $^1S_0 \rightarrow 7s7p$ $^3P_1$ transition.

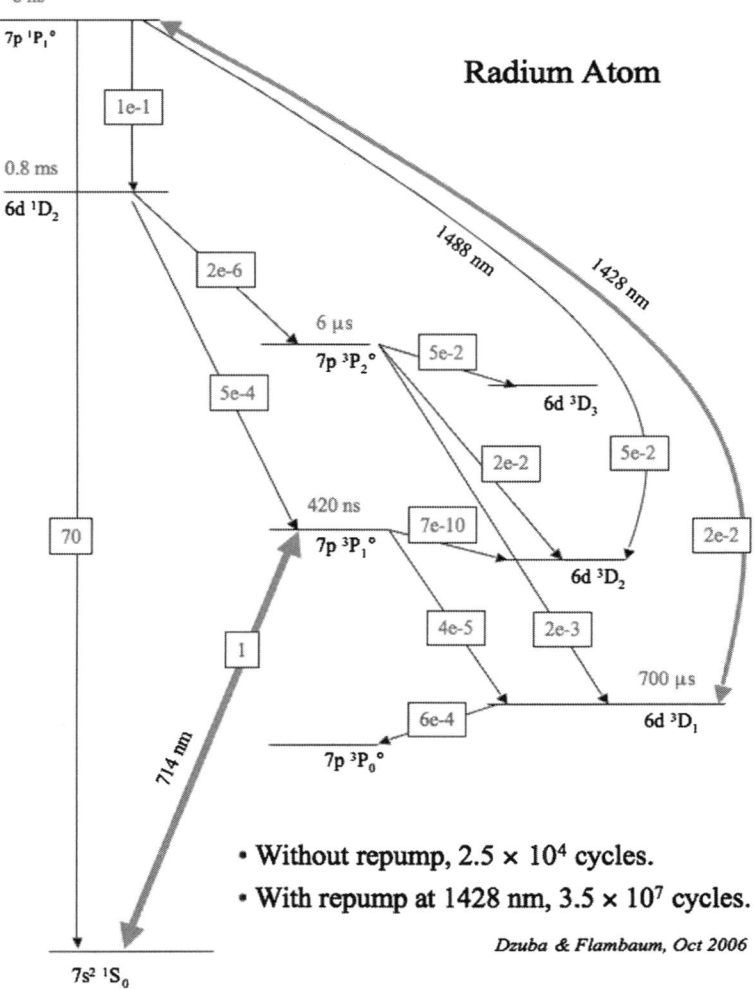

Fig. 68 Atomic level structure of radium-225 indicating the cycling transition at 714 nm and the re-pump transition at 1428 nm. The values in *boxes* indicate the relative transition probabilities

The primary leak channel from this two-level quasi-cycling system is the decay from $7s7p\,^3P_1$ to $7s6d\,^3D_1$, from which the atoms were pumped back to the ground-level via the $7s6d\,^3D_1 \rightarrow 7s7p\,^1P_1$ transition followed by a spontaneous decay from $7s7p\,^1P_1$ back to the ground-level. The re-pump was induced by laser light at 1428.6 nm generated by a diode laser. This re-pump transition can be excited for an average of 1400 times before the atom leaks to other metastable levels. Therefore, with the re-pump in place, an atom can cycle for an average of 3.5×10^7 times and stay in the MOT for at least 30 s before it leaks to dark levels. Here the MOT is used only to capture the atoms; the trapped atoms would then be transferred to an optical dipole trap for storage and measurement. They plan to achieve a life-time of 300 s in the dipole trap.

The ultimate goal of the present series of measurements is to provide a measurement that is comparable in sensitivity to the atomic EDM experiment for ^{199}Hg. Because of the enhancement from the octupole deformation of ^{225}Ra, the measurement would then be more than two orders of magnitude more sensitive to T-violating effects in the nucleus than that of the ^{199}Hg experiment. The immediate goal over the next two years is to provide an initial atomic EDM limit of $\sim 1 \times 10^{-26}\,e$ cm. Thereafter, the plan is to improve the experiment until the ultimate goal is achieved.

9.5 Electrons bound in atoms and molecules

9.5.1 Theoretical aspects

We discuss here permanent EDMs of diatomic molecules induced by the EDM of the electron and by P- and T-odd e–N neutral currents. In heavy molecules the effective electric field E_{eff} on unpaired electron(s) is many orders of magnitude higher than the external laboratory field required to polarize the molecule. As a result, the EDM of such molecules is strongly enhanced. The exact value of the enhancement factor is very sensitive to relativistic effects and to electronic correlations. In recent years several methods to calculate E_{eff} were suggested and reliable results were obtained for a number of molecules.

The study of a non-relativistic electron in a stationary state immediately leads to the zero energy shift $\delta \varepsilon$ for an atom in the external field $\boldsymbol{E}_0$ induced by the electron EDM $\boldsymbol{d}_e = d_e \boldsymbol{\sigma}$. Indeed, the average acceleration $\langle \boldsymbol{a} \rangle = 0$, so the average force $-e\langle \boldsymbol{E} \rangle = 0$. Therefore, $\delta \varepsilon = -\boldsymbol{d}_e \cdot \langle \boldsymbol{E} \rangle = 0$. This statement is known as Schiff theorem. In the relativistic case, the position-dependence of the Lorentz contraction of the electron EDM leads instead to a net overall atomic EDM [1029]. Even though $\langle \boldsymbol{E} \rangle = 0$, it still can be (and indeed is) the case that $\langle \boldsymbol{d}_e \cdot \boldsymbol{E} \rangle \neq 0$, if $\boldsymbol{d}_e$ is not spatially uniform. Taking account of the fact that the length-contracted value of $\boldsymbol{d}_e$ is NOT spatially uniform for an electron inside the Coulomb field of an atom exactly reproduces the form of the enhancement factor.

Reliable calculations of atomic energy shifts are easier with the relativistic EDM Hamiltonian for the Dirac electron, which automatically turns to zero in the non-relativistic approximation [1030]:

$$H_d = 2d_e \begin{pmatrix} 0 & 0 \\ 0 & \boldsymbol{\sigma} \end{pmatrix} \cdot \boldsymbol{E} \cong 2d_e \begin{pmatrix} 0 & 0 \\ 0 & \boldsymbol{\sigma} \end{pmatrix} \cdot \boldsymbol{E}_{\text{int}}. \quad (9.25)$$

This Hamiltonian is singular at the origin and we neglected the external field $\boldsymbol{E}_0$. Using (9.25) it is straightforward to show that the induced EDM of the heavy atom d_{at} is of the order of $10\alpha^2 Z^3 d_e$, where Z is the number of protons in the nucleus. If $Z \sim 10^2$ the atomic enhancement factor $k_{\text{at}} \equiv d_{\text{at}}/d_e \sim 10^3$. This estimate holds for atoms with an unpaired electron with $j = \frac{1}{2}$. For higher angular momentum j the centrifugal barrier strongly suppresses d_{at}.

Atomic EDM can be also induced by a scalar P, T-odd e–N neutral current [1030]:

$$H_S = i\frac{G\alpha}{2^{1/2}} Z k_S \gamma_0 \gamma_5 n(r), \quad (9.26)$$

where G is Fermi constant, γ_i are Dirac matrices, $n(r)$ is the nuclear density normalized to unity, and $Zk_S = Zk_{S,p} + Nk_{S,n}$ is the dimensionless coupling constant for a nucleus with Z protons and N neutrons. Atomic EDMs induced by the interactions (9.25), (9.26) are obviously sensitive to relativistic corrections to the wave function. Numerical calculations also show their sensitivity to correlation effects. For example, the Dirac–Fock calculation for Tl gives $d_{\text{Tl}} = -1910 d_e$ while the final answer within all order many-body perturbation theory is $d_{\text{Tl}} = -585 d_e$ (see Ref. [1030] for details). Note that the present limit on the electron EDM follows from the experiment with Tl [186].

The internal electric field in a polar molecule, $E_{\text{mol}} \sim \frac{e}{R_o^2} \sim 10^9$ V/cm, is 4–5 orders of magnitude larger than the typical laboratory field in an atomic EDM experiment. This field is directed along the molecular axis and is averaged to zero by the rotation of the molecule. The molecular axis can be polarized in the direction of the external electric field $\boldsymbol{E}_0$. One usually needs the field $E_0 \sim 10^4$ V/cm to fully polarize the heavy diatomic molecule. The corresponding molecular enhancement factor is $k_{\text{mol}} \sim k_{\text{at}} \times \frac{E_{\text{mol}}}{E_0} \sim 10^4 k_{\text{at}}$.

For closed-shell molecules all electrons are coupled and the net EDM is zero. Therefore one needs a molecule with at least one unpaired electron. Such molecules have nonzero projection Ω of electronic angular momentum on the molecular axis. Again, as in the case of atoms, for the molecules with one unpaired electron the largest enhancement corresponds to $\Omega = \frac{1}{2}$. The centrifugal barrier leads to strong suppression of the factor k_{mol} for higher values of Ω. On

Table 27 Calculated values of parameters E_{eff} and W_S from (9.27) for diatomic molecules. The question marks reflect the uncertainty in the knowledge of the ground state

Molecule	State	Ω	E_{eff} (10^9 V/cm)	W_S (kHz)	Ref.
BaF	ground	1/2	-7.5 ± 0.8	-12 ± 1	[1031, 1032]
YbF	ground	1/2	-25 ± 3	-44 ± 5	[1032, 1033]
HgF	ground	1/2	-100 ± 15	-190 ± 30	[1034]
HgH	ground	1/2	-79	-144	[1034]
PbF	ground	1/2	$+29$	$+55$	[1034]
PbO	metastable	1	-26		[1035]
HI$^+$	ground	3/2	-4		[1036]
PtH$^+$	ground (?)	3	20		[1037]
HfF$^+$	metastable (?)	1	24		[1038]

the other hand, such molecules can be polarized in a much weaker external field.

For strong external field E_0 the factor k_{mol} depends on E_0 and it is more practical to define an effective electric field on the electron E_{eff} so, that the P, T-odd energy shift for a fully polarized molecule is equal to:

$$\delta\varepsilon_{P,T} = E_{\text{eff}} d_e + \tfrac{1}{2} W_S k_S, \quad (9.27)$$

where two terms correspond to interactions (9.25) and (9.26). Calculated values of E_{eff} and W_S for a number of molecules are listed in Table 27.

An EDM experiment is currently going on with YbF molecules. This molecule has a ground state with $\Omega = \tfrac{1}{2}$. The P, T-odd parameters (9.27) were calculated with different methods by several groups, and estimates of the systematic uncertainty are available. Several other molecules and molecular ions have been suggested for the search for electron EDM including PbO, PbF, HgH, and PtH$^+$. PbF and HgH have $\Omega = \tfrac{1}{2}$ and calculations are similar to the YbF case. The ground state of PbO has closed shells and the experiment is done on the metastable state with two unpaired electrons and $\Omega = 1$. Here electronic correlations are much stronger and calculations are more difficult.

Finally, molecular ions like PtH$^+$ are less studied and even their ground states are not known exactly. It is anticipated that such ions can be trapped and a long coherence time for the EDM experiment can be achieved. Recently the first estimates of the effective field for PtH$^+$ and several other molecular ions were reported [1037]. These estimates are based on non-relativistic molecular calculations. Proper relativistic molecular calculations for these ions may be extremely challenging.

9.5.2 Experimental aspects

Over a dozen different experiments searching for the electron electric dipole moment that are under way or planned will be reviewed here. At present the experimental upper limit on d_e is [186]: $|d_e| \leq 1.6 \times 10^{-27}$ e cm, where e is the unit of electronic charge.

Most of this work is being done in small groups on university campuses. These experiments employ a wide range of technologies and conceptual approaches. Many of the latest generation of experiments promise two or more orders of magnitude improvement in statistical sensitivity, and most have means to suppress systematic errors well beyond those obtained in the previous generation.

To detect d_e, most experiments rely on the energy shift $\Delta E = -\vec{d}_e \cdot \vec{E}$ upon application of $\vec{E}$ to an electron. Until recently, most experimental searches for d_e used gas-phase paramagnetic atoms or molecules and employed the standard methods of atomic, molecular, and optical physics (laser and rf spectroscopy, optical pumping, atomic and molecular beams or vapor cells, etc.) in order to directly measure the energy shift ΔE. Recently, another class of experiments has been actively pursued, in which paramagnetic atoms bound in a solid are studied. Here the principles are rather different than for the gas-phase experiments, and techniques are more similar to those used in condensed matter physics (magnetization and electric polarization of macroscopic samples, cryogenic methods, etc.). We discuss these two classes of experiments separately.

9.5.2.1 A simple model experiment using gas-phase atoms or molecules
Experimental searches for d_e using gas-phase atoms or molecules share many broad features. Each consists of a state selector, where the initial spin state of the system is prepared; an interaction interval in which the system evolves for a time τ in an electric field $\vec{E}$ (and often a magnetic field $\vec{B} \parallel \vec{E}$ as well); and a detector to determine the final state of the spin. To understand the essential features, we consider a simple model that is readily adapted to describe most realistic experimental conditions. In this model, an "atom" of spin 1/2 with enhancement factor R, containing an unpaired electron with spin

magnetic moment μ and EDM d_e. The spin is initially prepared to lie along $\hat{x}$, i.e., is in the eigenstate $|\chi_+^x\rangle$ of spin along $\hat{x}$: $|\psi_0\rangle = |\chi_+^x\rangle \equiv \frac{1}{\sqrt{2}}\binom{1}{1}$. During the interaction interval the spin precesses about $\vec{\mathbf{E}} = \mathcal{E}\hat{z}$ and $\mathbf{B} = B\hat{z}$, in the xy plane, by angle $2\phi = -(d_e R\mathcal{E} + \mu B)\tau/\hbar$. At time τ the quantum state has then evolved to $|\psi\rangle = \frac{1}{\sqrt{2}}\binom{e^{-i\phi}}{e^{i\phi}}$. Finally, the detector measures the probability that the resulting spin state lies along $\hat{y}$. This is determined by the overlap of the wave function $|\psi\rangle$ with $|\chi_+^y\rangle \equiv \frac{1}{\sqrt{2}}\binom{1}{i}$. Hence the signal S from N detected atoms observed in time τ is $S = N|\langle\chi_+^y|\psi\rangle|^2 = N\cos^2\phi$.

The angle ϕ is the sum of a large term $\phi_1 = -\mu B\tau/(2\hbar)$ and an extremely small term $\phi_2 = -d_e R\mathcal{E}\tau/(2\hbar)$. To isolate ϕ_2 one observes S for $\vec{\mathbf{E}}$ and $\mathbf{B}$ both parallel and antiparallel. Reversing $\vec{\mathbf{E}} \cdot \mathbf{B}$ changes the relative sign of ϕ_1 and ϕ_2 and thus changes S; the largest change in S occurs by choosing B such that $\phi_1 = \pm\pi/4$. With this choice, we have $S_\pm \equiv S(\vec{\mathbf{E}} \cdot \mathbf{B} \gtrless 0) = \frac{N}{2}(1 \pm 2\phi_2)$. The minimum uncertainty in determination of the phase ϕ_2 in time τ, due to shot noise, is $\delta\phi_2 = \sqrt{1/N}$. If the experiment is repeated T/τ times for a total time of observation T, the statistical uncertainty in d_e is $\delta d_e = \sqrt{1/N_0}\sqrt{1/T\tau}|\hbar/\mathcal{E}_{\text{eff}}|$, where we used $R\mathcal{E} = \mathcal{E}_{\text{eff}}$. In practice, other "technical" noise sources can significantly increase this uncertainty, particularly fluctuations in the magnetic field. Hence, careful magnetic shielding is required in all EDM experiments.

9.5.2.2 Systematic errors The EDM is revealed by a term in the signal proportional to a P, T-odd pseudoscalar such as $\vec{\mathbf{E}} \cdot \mathbf{B}$. False terms of the same apparent form can appear even without P, T violation through a variety of experimental imperfections. The most dangerous effects appear when $\mathbf{B}$ depends on the sign of $\vec{\mathbf{E}}$, which can occur in several ways. For example, leakage currents flowing through insulators separating the electric field electrodes can generate an undesired magnetic field $\mathbf{B}_L$. Also, if the atoms or molecules have a non-zero velocity $\mathbf{v}$, a motional magnetic field $\mathbf{B}_{\text{mot}} = \frac{1}{c}\vec{\mathbf{E}} \times \mathbf{v}$ exists in addition to the applied magnetic field $\mathbf{B}$; along with various other imperfections in the system, this effect can lead to systematic errors. A related systematic effect involves geometric phases, which appear if the direction of the quantization axis (often determined by $\mathbf{B}_{\text{total}} = \mathbf{B} + \mathbf{B}_{\text{mot}}$) varies between the state selector and the analyzer [1030].

A variety of approaches are employed to deal with these and other systematics. Aside from leakage currents, most systematics depend on a combination of two or more imperfections in the experiment (i.e. misaligned or stray fields); these can be isolated by deliberately enhancing one imperfection and looking for a change in the EDM signal. Some experiments utilize, in addition to the atoms of interest, additional species as so-called "co-magnetometers". These co-magnetometer species (e.g., paramagnetic atoms with low R) are chosen to have negligible or small enhancement factors, but retain sensitivity to magnetic systematics such as those mentioned above.

In paramagnetic molecule experiments, issues with systematic effects are somewhat different. Here the ratio $\mathcal{E}_{\text{eff}}/\mathcal{E}_{\text{ext}}$ is enhanced, and relative sensitivity to magnetic systematics is correspondingly reduced. The $\vec{\mathbf{E}} \times \mathbf{v}$ effect is effectively eliminated by the large tensor Stark effect [1039] typically found in molecular states. The saturation of the molecular polarization $|P|$ (and hence $\mathcal{E}_{\text{eff}}$) leads to a well-understood non-linear dependence of the EDM signal on $\mathcal{E}_{\text{ext}}$ that can discriminate against certain systematics. Conversely, the extreme electric polarizability leads to a variety of new effects, such as a dependence of the magnetic moment μ on $\mathcal{E}_{\text{ext}}$, and geometric phase induced systematic errors related to variations in the direction of $\vec{\mathbf{E}}_{\text{ext}}$.

9.5.2.3 Experiments with gas-phase atoms and molecules
• The Berkeley thallium atomic beam experiment

This experiment gives the best current limit on d_e. In its final version [186], two pairs of vertical counter-propagating atomic beams, each consisting of Tl ($Z = 81$, $R_{\text{Tl}} = -585$ [1040]) and Na ($Z = 11$, $R_{\text{Na}} = 0.32$), were employed (see Fig. 69). Spin alignment and rotation of the $6^2P_{1/2}(F=1)$ state of Tl and the $3^2S_{1/2}(F=2, F=1)$ states of Na were accomplished, respectively, by laser optical pumping and by atomic beam magnetic resonance with separated oscillating rf fields of the Ramsey type. Detection was achieved via alignment-sensitive laser induced fluorescence. In the interaction region, with length ≈ 1 meter, the side-by-side atomic beams were exposed to nominally identical $\mathbf{B}$ fields, but opposite $\vec{\mathbf{E}}$ fields of ≈ 120 kV/cm. This provided common-mode rejection of magnetic noise and control of some systematic effects. Average thermal velocities corresponded to an interaction time $\tau \approx 2.3$ ms (1 ms) for Tl (Na) atoms. Use of counter-propagating atomic beams served to cancel all but a very small remnant of the $\vec{\mathbf{E}} \times \mathbf{v}$ effect. Various auxiliary measurements, including use of Na as a co-magnetometer, further reduced this remnant and isolated the geometric phase effect. $\mathcal{E}$ and leakage currents were measured using auxiliary measurements based on the observable quadratic Stark effect in Tl. About 5.2×10^{13} photo-electrons of signal per up/down beam pair were collected by the fluorescence detectors. The final result is $d_e = (6.9 \pm 7.4) \times 10^{-28}$ e cm, which yields the limit $|d_e| \leq 1.6 \times 10^{-27}$ e cm (90% conf.).
• Cesium vapor cell experiments

An experiment to search for d_e in a vapor cell of Cs ($Z = 55$; $R_{\text{Cs}} = 115$ [1041]) was reported by L. Hunter and co-workers [1042] at Amherst in 1989. The method is being revisited in a present-day search by led by M. Romalis at Princeton [1043]. The Amherst experiment was carried out with two glass cells, one stacked on the other in the z direction. Nominally equal and opposite $\vec{\mathbf{E}}$ fields

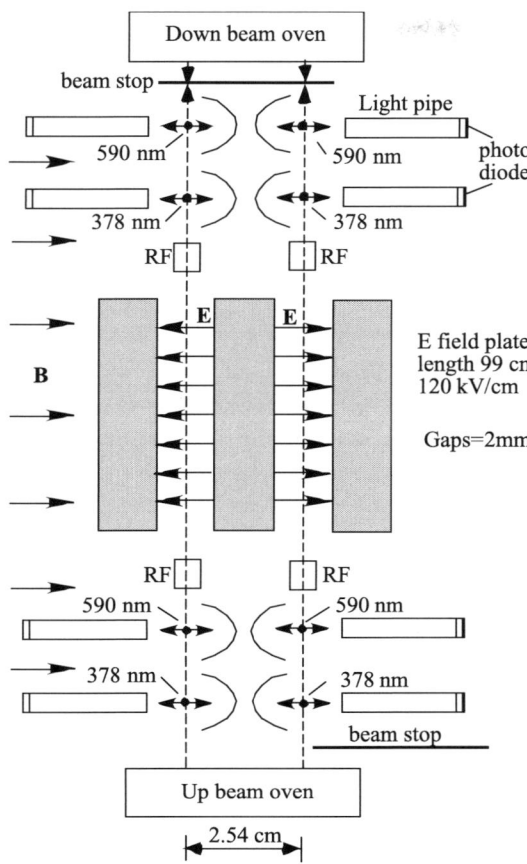

Fig. 69 Schematic diagram of the Berkeley thallium experiment [186], not to scale. Laser beams for state selection and analysis at 590 nm (for Na) and 378 nm (for Tl) are perpendicular to the page, with indicated linear polarizations. The diagram shows the up-going atomic beams active

were applied in the two cells. The cells were filled with Cs, as well as N_2 buffer gas to minimize Cs spin relaxation. Circularly polarized laser beams, directed along x, were used for spin polarization via optical pumping. Magnetic field components in all three directions were reduced to less than 10^{-7} G. Thus precession of the atomic polarization in the xy plane was nominally due to $\vec{E}$ alone. The final spin orientation was monitored by a probe laser beam directed along y. The effective interaction time was the spin relaxation time $\tau \approx 15$ ms. The signals were the intensities of the probe beams transmitted through each cell. A non-zero EDM would have been indicated by a dependence of these signals on the rotational invariant $\boldsymbol{J} \cdot (\boldsymbol{\sigma} \times \vec{\mathbf{E}})\tau$, where $\sigma, \boldsymbol{J}$ were the pump and probe circular polarizations, respectively. The most important sources of possible systematic error were leakage currents and imperfect reversal of $\vec{\mathbf{E}}$. The result was $d_e = (-1.5 \pm 5.5 \pm 1.5) \times 10^{-26} \, e$ cm.

In the new experiment at Princeton, each cell also contains ^{129}Xe at high pressure. Cs polarization is transferred to the ^{129}Xe nuclei by spin exchange collisions. Under certain conditions this coupling can also give rise to a self-compensation mechanism, where slow changes in components of magnetic field transverse to the initial polarization axis are nearly canceled by interaction between the alkali electron spin and the noble gas nuclear spin. This leaves only a signal proportional to an anomalous interaction that does not scale with the magnetic moments—for example, interaction of d_e with $\mathcal{E}_{\text{eff}}$. This mechanism (which is understood in some detail [1044, 1045]) has the potential to reduce both the effect of magnetic noise, and some systematic errors.

• Experiments with laser-cooled atoms

Laser-cooled atoms offer significant advantages for electron EDM searches. The low velocities of cold atoms yield long interaction times, and also suppress $\vec{\mathbf{E}} \times \boldsymbol{v}$ effects. However, these techniques typically yield small numbers of detectable atoms, and magnetic noise must be controlled at unprecedented levels. New systematics due to, e.g., electric forces on atoms and/or perturbations due to trapping fields (see e.g. [1046]) can appear.

Experiments based on atoms trapped in an optical lattice have been proposed by a number of investigators [1047–1049]. Two such experiments, similar in their design, are currently being developed: one led by D.S. Weiss at Pennsylvania State University and another led by D. Heinzen at the University of Texas. Both plan to use Cs atoms to detect d_e, along with Rb atoms ($Z = 37$, $R_{\text{Rb}} = 25$) as a co-magnetometer. The Texas apparatus consists of two side-by-side far-off-resonance optical dipole traps, each in a vertical 1-D lattice configuration. These traps are placed in nominally equal and opposite $\vec{\mathbf{E}}$ fields and a common $\boldsymbol{B}$ field of several mG parallel to $\vec{\mathbf{E}}$. To load the atoms into the optical lattice, cold atomic beams from 2D magneto-optical traps exterior to the shields will be captured with optical molasses between the $\vec{\mathbf{E}}$-field plates. The electric field plates will be constructed from glass coated with a transparent, conductive indium tin oxide layer.

We are aware of two other EDM experiments based on laser-cooled atoms. One employing a slow "fountain", in which Cs atoms are launched upwards and then fall back down due to gravity, has been proposed and developed by H. Gould and co-workers at the Lawrence Berkeley National Laboratory [1050]. Another, using ^{210}Fr ($\tau = 3.2$ min; $Z = 87$, $R_{\text{Fr}} = 1150$), has been proposed and is being developed by a group at the Research Center of Nuclear Physics (RCNP), Osaka University, Japan [1051].

• The YbF experiment

E.A. Hinds and co-workers [1052–1054] at Imperial College, London have developed a molecular beam experiment for investigation of d_e using YbF. Figure 70 shows the relevant energy level structure of the $X^2\Sigma_{1/2}^+$ ($v = 0$, $N = 0$) $J = 1/2$ ground state of a ^{174}YbF molecule. ^{174}Yb has nuclear spin $I_{\text{Yb}} = 0$, while $I_{\text{F}} = 1/2$; hence the $J = 1/2$ state has two hyperfine components, $F = 1$ and

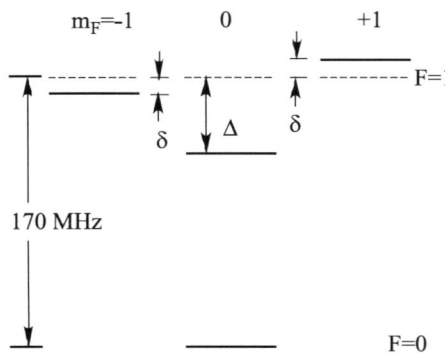

Fig. 70 Schematic diagram, not to scale, of the hyperfine structure of the $X^2\Sigma$ electronic state of ^{174}YbF in the lowest vibrational and rotational level. Δ is the tensor Stark shift. δ is the shift caused by the combination of the Zeeman effect and the effect of d_e in $\vec{E}_{\text{eff}}$

$F = 0$, separated by 170 MHz. An external electric field $\vec{E}_{\text{ext}}$ along $\hat{z}$ with magnitude $\mathcal{E}_{\text{ext}} = 8.3$ kV/cm corresponds to $\mathcal{E}_{\text{eff}} \approx 13$ GV/cm [1033, 1052–1054], which splits the $F = 1, m_F = \pm 1$ levels by $2d_e\mathcal{E}_{\text{eff}}$. In this external field, the level $F = 1, m_F = 0$ is shifted downward relative to $m_F = \pm 1$ by an amount $\Delta = 6.7$ MHz due to the large tensor Stark shift associated with the molecular electric dipole.

In the experiment, a cold beam of YbF molecules is generated by chemical reactions within a supersonic expansion of Ar or Xe carrier gas. Laser optical pumping removes all $F = 1$ state molecules, leaving only $F = 0$ remaining in the beam. Next, a 170 MHz rf magnetic field along x excites molecules from $F = 0$ to the coherent superposition $|\psi\rangle = \frac{1}{\sqrt{2}}|F = 1, m_F = 1\rangle + \frac{1}{\sqrt{2}}|1, -1\rangle$. While flying through the central interaction region of length 65 cm, the beam is exposed to parallel electric and magnetic fields $(\pm\mathcal{E}, \pm B)\hat{z}$ ($B \sim 0.1$ mGauss). Next, an rf field drives each $F = 1$ molecule back to $F = 0$. Because of the phase shift 2ϕ developed in the central region, the final population of $F = 0$ molecules is proportional to $\cos^2 \phi$. These $F = 0$ molecules are detected by laser induced fluorescence in the probe region.

The most significant systematic errors in this experiment are expected to arise from variation in the direction and magnitude of $\vec{E}$ along the beam axis. If the direction of $\vec{E}$ changes in an absolute sense, a geometric phase could be generated, and if $\vec{E}$ changes relative to B, the magnetic precession phase ϕ_1, proportional to $\vec{E}_{\text{ext}} \cdot B/|\vec{E}_{\text{ext}}|$, could be affected. A preliminary result of the YbF experiment [1052–1054], published in 2002, is $d_e = (-0.2 \pm 3.2) \times 10^{-26}$ e cm. Many significant improvements have been made since 2002, and it is likely that this experiment will yield a much more precise result in the near future.

- The PbO experiment

A search for d_e using the metastable $a(1)^3\Sigma_1$ state of PbO is being carried out at Yale [1055–1057]. The $a(1)$ state has a relatively long natural life-time: $\tau[a(1)] = 82(2)$ μs,
and can be populated in large numbers using laser excitation in a vapor cell. In this state, the level of total (rotational + electronic) angular momentum $J = 1$ contains two closely-spaced "Ω doublet" states of opposite parity, denoted as e^- and f^+. An external electric field $\vec{E}_{\text{ext}} = \mathcal{E}_{\text{ext}}\hat{z}$ mixes e^- and f^+ states with the same value of M, yielding molecular states with equal but opposite electrical polarization P. The degree of polarization $|P| \approx 1$ for $\mathcal{E}_{\text{ext}} \gtrsim 10$ V/cm. When $|P| = 1$ the effective molecular field is calculated to be $\mathcal{E}_{\text{eff}} \cong 26$ GV/cm [1035]. The opposite molecular polarization in the two Ω-doublet levels leads to a sign difference in the EDM-induced energy shift between these two levels. This difference provides an excellent opportunity for effective control of systematic errors, since comparison of the energy shifts in the upper and lower states acts as an "internal co-magnetometer" requiring only minor changes in experimental parameters to monitor.

The Yale experiment is carried out in a cell containing PbO vapor, consisting of an alumina body supporting top and bottom gold foil electrodes, and flat sapphire windows on all 4 sides. The electric field $\vec{E}_{\text{ext}} = \mathcal{E}_{\text{ext}}\hat{z}$ is quite uniform over a large cylindrical volume (diameter 5 cm, height 4 cm), and is chosen in the range 30–90 V/cm. The magnetic field B_z is chosen in the range 50–200 mG. The cell is enclosed in an oven mounted in a vacuum chamber. At the operating temperature 700 C, the PbO density is $n_{\text{PbO}} \approx 4 \times 10^{13}$ cm^{-3}.

A state with simultaneously well-defined spin and electrical polarization is populated as follows. A pulsed laser beam with z linear polarization excites the transition $X[J = 0^+] \rightarrow a(1)[J = 1^-, M = 0]$. ($X$ is the electronic ground state of PbO.) Following the laser pulse a Raman transition is driven by two microwave beams. The first, with x linear polarization, excites the upward 28.2 GHz transition $a(1)[J = 1^-, M = 0] \rightarrow a(1)[J = 2^+, M = \pm 1]$. The second, with z linear polarization and detuned to the red or blue with respect to the first by 20–60 MHz, drives the downward transition $a(1)[J = 2^+, M = \pm 1] \rightarrow a(1)[J = 1, M = \pm 1]$. The net result is that about 50% of the $J = 1^-, M = 0$ molecules are transferred to a coherent superposition of $M = \pm 1$ levels in a single desired Ω-doublet component. The subsequent spin precession (due to $\mathcal{E}$ and B) is detected by observing the frequency of quantum beats in the fluorescence that accompanies spontaneous decay to the X state. The signature of a non-zero EDM is a term in the quantum beat frequency that is proportional to $\vec{E}_{\text{ext}} \cdot B$ and that changes sign when one switches from one Ω-doublet component to the other.

The present experimental configuration is sufficient to yield statistical uncertainty comparable to the present limit on d_e in a reasonable integration time of a few weeks. However, large improvements can be made in a next generation of the experiment. In the new scheme, detection will be accomplished via absorption of a resonant microwave probe beam tuned to the 28.2 GHz transition described above. With this method, the signal-to-noise ratio is linearly proportional to the path length of the probe beam in the PbO vapor. In a second generation experiment the cell can be made ∼10 times longer than it is now, and the probe beam can pass through the cell multiple times by using suitable mirrors. Improvements in sensitivity of up to a factor of 3000 over the current generation are envisioned.

• Other molecule experiments

E. Cornell and co-workers at the Joint Institute for Laboratory Astrophysics (Boulder, Colorado) have proposed an experiment [1058] to search for d_e in the $^3\Delta_1$ electronic state of the molecular ion HfF$^+$. The premise is to take advantage of the long spin coherence times typical for trapped ion experiments with atoms, along with the large effective electric field acting on d_e in a molecule. Preliminary calculations [1036] suggest that the $^3\Delta_1$ state is a low-lying metastable state with very small Ω-doublet splittings; as in PbO, this state could thus be polarized by small external electric fields ($\lesssim$10 V/cm) to yield $\mathcal{E}_{\text{eff}} \approx 18$ GV/cm [1037]. To search for d_e, electron-spin-resonance spectroscopy, using the Ramsey method, is to be performed in the presence of rotating electric and magnetic fields. The electric field polarizes the ions and its rotation prevents them from being accelerated out of the trap. As in PbO, use of both upper and lower Ω-doublet components will yield opposite signs of the EDM signal, but nearly identical signals due to systematic effects. However, this experiment has the unique disadvantage that it is impossible to reverse the electric field: in the laboratory frame it must always point inward toward the trap center.

N. Shafer-Ray and co-workers at Oklahoma University have proposed an experiment to search for d_e in the ground $^2\Pi_{1/2}$ electronic state of PbF [1059, 1060]. The proposed scheme is similar to the YbF experiment, and the value of $\mathcal{E}_{\text{eff}}$ is also approximately the same as for YbF. The primary advantage of PbF is that its electric field-dependent magnetic moment should vanish when a suitable, large external electric field $\mathcal{E}_0 \approx 67$ kV/cm is applied [1059, 1060]. This could dramatically reduce magnetic field-related systematic errors.

9.5.2.4 Experiments with solid-state samples Recently, S. Lamoreaux [1061] revived an old idea of F. Shapiro [1062] to search for d_e by applying an electric field $\vec{\mathbf{E}}_{\text{ext}}$ to a solid sample with unpaired electron spins. If $d_e \neq 0$, at sufficiently low temperature the sample can acquire significant spin-polarization and thus a detectable magnetization along the axis of $\vec{\mathbf{E}}_{\text{ext}}$. Lamoreaux pointed out that use of modern magnetometric techniques and materials (such as Gd$_3$Ga$_5$O$_{12}$: gadolinum gallium garnet, or GGG) could yield impressive sensitivity to d_e. GGG has a number of attractive properties. Its resistivity is so high ($>10^{16}$ Ohm-cm for $T < 77$ K) that it can support large applied electric fields ($\vec{\mathbf{E}}_{\text{ext}} \approx 10$ kV/cm) with very small leakage currents. Moreover, the ion of interest in GGG, Gd^{3+} ($Z = 64$) has a non-negligible enhancement factor [1063]. A complementary experiment is being done by L. Hunter and co-workers [1064] at Amherst College. Here, a strong external magnetic field is applied to the ferrimagnetic solid Gd$_{3-x}$Y$_x$Fe$_5$O$_{12}$ (gadolinium yttrium iron garnet, or GdYIG), thus causing substantial polarization of the Gd^{3+} electron spins. If $d_e \neq 0$, this results in electric charge polarization of the sample, and thus a voltage developed across the sample that reverses with applied magnetic field.

The basic theoretical considerations that must be taken into account to estimate the expected signals [1065] in these solid-state experiments include the same types of calculations needed for free atoms. In addition, however, it is necessary to construct models for the modification of atomic electron orbitals in the solid material, as well as the response of the material to the EDM-induced perturbation of the heavy paramagnetic atom. The results of the calculations are as follows. When all Gd spins are polarized in the GdIG sample, the resulting macroscopic electric field across the sample is $\mathcal{E} = 0.7 \times 10^{-10}(d_e/10^{-27}\, e\,\text{cm})$ V/cm. A similar calculation can be used to determine the degree of spin polarization of GGG upon application of an external electric field [1061]. An externally applied electric field of 10 kV/cm yields an effective electric field $\mathcal{E}^* = -\Delta E/d_e = 3.6 \times 10^5$ V/cm acting on the EDM ([1065]; see also [1066]). The resulting magnetization M of the sample is simply related to its magnetic susceptibility χ: $M = \chi d \mathcal{E}^*/\mu_a$, where μ_a is the magnetic moment of a Gd^{3+} ion. Using the standard expression for $\chi(T)$ in a paramagnetic sample, one finds $M \approx 8 n_{\text{Gd}}(d_e \mathcal{E}^*)/(k_B T)$. Here k_B is the Boltzmann constant and T is the sample temperature. This yields a magnetic flux $\Phi = 4\pi M S$ over an area S of an infinite flat sheet. In a recent development [1066], Lamoreaux has pointed out that this type of electrically induced spin polarization can be amplified in a system that is super-paramagnetic, so that its magnetic susceptibility χ is extremely large. It appears that GdIG (GdYIG with $x = 0$) has this property at sufficiently low magnetic field. If so, the sensitivity of a magnetization measurement in GdIG at $T = 4$ K could be similar to that of GGG at much lower temperatures, greatly simplifying the required experimental techniques.

• The Indiana GGG experiment

C.Y. Liu of Indiana University has devised a prototype experiment [1067, 1068] in which two GGG disks, 4 cm in diameter and of thickness $\approx$1 cm, are sandwiched between three planar electrodes. High voltages are applied so that the electric fields in the top and bottom samples are in the same direction. If $d_e \neq 0$, a magnetic field similar to a dipole field should be generated, and this is to be detected by a flux pickup coil located in the central ground plane. The latter is designed as a planar gradiometer with 3 concentric loops, arranged to sum up the returning flux and to reject common-mode magnetic fluctuations. As the electric field polarization is modulated, the gradiometer detects the changing flux and feeds it to a SQUID sensor. The entire assembly is immersed in a liquid helium bath.

The EDM sensitivity of the prototype experiment is estimated to be $\delta d_e \approx 4 \times 10^{-26}$ e cm. Although this falls short of the ultimate desired sensitivity of 10^{-30} e cm, the prototype experiment is useful as a learning tool for solving some basic technical problems. At Indiana, a second-generation experiment is also being planned, which will operate at much lower temperatures ($\approx$10–15 mK), and will employ lower-noise SQUID magnetometers. However, questions remain as to the nature of the magnetic susceptibility χ of GGG at such low temperatures.

Some thought has gone into possible systematic effects in this system. Although crystals with inversion symmetry such as GGG and GdIG should not exhibit a linear magnetoelectric effect [1069], crystal defects and substitutional impurities can spoil this ideal. Furthermore a quadratic magnetoelectric effect does exist, and to avoid systematic errors arising from it, good control of electric field reversal is required.

• The Amherst GdYIG experiment

GdYIG is ferrimagnetic, and both Gd^{3+} ions and Fe lattices contribute to its magnetization M. Their contributions are generally of opposite sign, but at moderately low temperatures T the Fe component is roughly constant while the Gd component changes rapidly with T. There exists a "compensation" temperature T_C where the Gd and Fe magnetizations cancel each other, and the net magnetization M vanishes. For $T > T_C (< T_C)$, M is dominated by Fe (Gd). The Gd contribution to M can be reduced by replacing some Gd^{3+} ions with non-magnetic Y^{3+}. With x the average number of Y ions per unit cell, (so that $3-x$ is the average number of Gd ions per unit cell), the compensation temperature becomes $T_C = [290 - 115(3-x)]$ K. This dependence of T_C on x is exploited in the Amherst GdYIG experiment. A toroidal sample is employed, consisting of two half-toroids, each in the shape of the letter C. One "C" has $x = 1.35$ with a corresponding $T_C = 103$ K. The other "C" has $x = 1.8$ with a corresponding $T_C = 154$ K. These are joined together with copper foil electrodes at the interface. At $T = 127$ K, the magnetizations of the 2 "C's" are identical, but their Gd magnetizations are nominally opposite. When a magnetic field H is applied to the sample with a toroidal current coil, all Gd spins are nominally oriented toward the same copper electrode. Thus EDM signals from C_1 and C_2 add constructively. However below 103 K (above 154 K) the Gd magnetization is parallel (antiparallel) to M in both C's, which results in cancellation of one EDM signal by the other. Data are acquired by observing the voltage difference A (B) between the two foil electrodes for positive (negative) polarity of the applied magnetic field H. An EDM should be re-vealed by the appearance of an asymmetry $d = A - B$ that has a specific temperature dependence, as described above.

A large spurious effect has been seen that mimics an EDM signal when $T < 180$ K, but which deviates grossly from expectations for $T > 180$ K. This effect, which is associated with a component of magnetization that does not reverse with H, has so far frustrated efforts to realize the full potential of the GdYIG experiment. The best limit that has been achieved so far is [1064]: $d_e < 5 \times 10^{-24}$ e cm.

9.6 Muon EDM

The best direct upper limits for an electric dipole moment (EDM) of the muon come from the experiments measuring the muon anomalous magnetic moment ($g-2$). The CERN experiment obtained 1.1×10^{-18} e cm (95% C.L.) [1020] and the preliminary limit from Brookhaven is 2.8×10^{-19} e cm [188]. Assuming lepton universality, the electron EDM limit of $d_e < 2.2 \times 10^{-27}$ e cm [186] can be scaled by the electron to muon mass ratio, in order to obtain an indirect limit of $d_\mu < 5 \times 10^{-25}$ e cm. However, viable models exist in which the simple linear mass scaling does not apply and the value for the muon EDM could be pushed up to values in the 10^{-22} e cm region (see, e.g., [846, 1070–1072]). In order for experimental searches to become sufficiently sensitive, dedicated efforts are needed. Several years ago, a letter of intent for a dedicated experiment at JPARC [1073] was presented, proposing a new sensitive "frozen spin" method [1021, 1022]: The anomalous magnetic moment precession of the muon spin in a storage ring can be compensated by the application of a radial electric field, thus freezing the spin; a potential electric dipole moment would lead to a rotation of the spin out of the orbital plane and thus an observable up-down asymmetry which increases with time. The projected sensitivity of the proposed experiment (0.5 GeV/c muon momentum, 7 m ring radius) is 10^{-24}–10^{-25} e cm. Recently it has been pointed out that there is no immediate advantage from working at high muon momenta and a sensitive approach with a very compact setup (125 MeV/c muon momentum, 0.42 m ring radius) was outlined [1024]. Already at an existing beam line, such as the µE1 beam at PSI, a measurement with a sensitivity of better than $d_\mu \sim 5 \times 10^{-23}$ e cm within one year of data taking appears feasible. The estimates for the sensitivity assume an operation in a "one-muon-per-time" mode and the experiment would appear to be statistically limited. With an improved muon accumulation and injection scheme, the sensitivity could be further increased [1074]. Thus the compact storage ring approach at an existing facility could bring the proof of principle for the frozen spin technique and cover the next 3–4 orders of magnitude in experimental sensitivity to a possible muon EDM.

9.7 Muon $g - 2$

In his famous 1928 paper [1075–1077] Dirac pointed out that the interaction of an electron with external electric and magnetic fields may have two extra terms where "the two extra terms

$$\frac{eh}{c}(\sigma, \mathbf{H}) + i\frac{eh}{c}\rho_1(\sigma, \mathbf{E}), \tag{9.28}$$

...when divided by the factor $2m$ can be regarded as the additional potential energy of the electron due to its new degree of freedom." These terms represent the magnetic dipole (Dirac) moment and electric dipole moment interactions with the external fields.

In modern notation, for the magnetic dipole moment of the muon we have:

$$\bar{u}_\mu \left[eF_1(q^2)\gamma_\beta + \frac{ie}{2m_\mu} F_2(q^2)\sigma_{\beta\delta} q^\delta \right] u_\mu, \tag{9.29}$$

where $F_1(0) = 1$, and $F_2(0) = a_\mu$.

The magnetic dipole moment of a charged lepton can differ from its Dirac value ($g = 2$) for several reasons. Recall that the proton's g-value is 5.6 ($a_p = 1.79$), a manifestation of its quark-gluon internal structure. On the other hand, the leptons appear to have no internal structure, and the magnetic dipole moments are thought to deviate from 2 through radiative corrections, i.e. resulting from virtual particles that couple to the lepton. We should emphasize that these radiative corrections need not be limited to the standard model particles. While the current experimental uncertainty of ± 0.5 ppm on the muon anomaly is 770 times larger than that on the electron anomaly [1078], the former is far more sensitive to the effects of high mass scales. In the lowest order diagram where mass effects appear, the contribution of heavy virtual particles with mass M scales as $(m_{\text{lepton}}/M)^2$, giving the muon a factor of $(m_\mu/m_e)^2 \simeq 43000$ increase in sensitivity over the electron.

9.7.1 The standard model value of the anomalous magnetic moment

The standard model value of a lepton's anomalous magnetic moment (*the anomaly*)

$$a_\ell \equiv \frac{(g_s - 2)}{2}$$

has contributions from three different sets of radiative processes: quantum electrodynamics (QED)—with loops containing leptons (e, μ, τ) and photons; hadronic—with hadrons in vacuum polarization loops; and weak—with loops involving the bosons W, Z, and Higgs:

$$a_\ell^{\text{SM}} = a_\ell^{\text{QED}} + a_\ell^{\text{hadronic}} + a_\ell^{\text{weak}}. \tag{9.30}$$

The QED contribution has been calculated up to the leading five-loop corrections [1079]. The dominant "Schwinger term" [1080, 1081] $a^{(2)} = \alpha/2\pi$, is shown diagrammatically in Fig. 71(a). Examples of the hadronic and weak contributions are given in Fig. 71(b)–(d).

The hadronic contribution cannot be calculated directly from QCD, since the energy scale ($m_\mu c^2$) is very low, although Blum has performed a proof of principle calculation on the lattice [1082–1084]. Fortunately, dispersion theory gives a relationship between the vacuum polarization loop and the cross section for $e^+ e^- \to$ hadrons,

$$a_\mu(\text{Had}; 1) = \left(\frac{\alpha m_\mu}{3\pi}\right)^2 \int_{4m_\pi^2}^\infty \frac{ds}{s^2} K(s) R(s), \tag{9.31}$$

where

$$R \equiv \sigma_{\text{tot}}(e^+ e^- \to \text{hadrons})/\sigma_{\text{tot}}(e^+ e^- \to \mu^+ \mu^-) \tag{9.32}$$

and experimental data are used as input [1085, 1086].

The standard model value of the muon anomaly has recently been reviewed [1085], and the latest values of the contributions are given in Table 28. The sum of these contributions, adding experimental and theoretical errors in quadrature, gives

$$a_\mu^{\text{SM}(06)} = 11\,659\,1785\,(61) \times 10^{-11}, \tag{9.33}$$

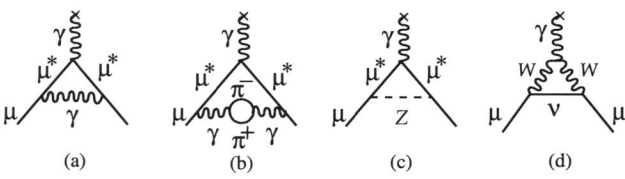

Fig. 71 The Feynman graphs for: (**a**) lowest order QED (Schwinger) term; (**b**) lowest order hadronic correction; (**c**) and (**d**) lowest order electroweak terms. The * emphasizes that in the loop the muon is off-shell. With the known limits on the Higgs mass, the contribution from the single Higgs loop is negligible

Table 28 Standard model contributions to the muon anomalous magnetic dipole moment, a_μ. All values are taken from Ref. [1085]

QED	$116\,584\,718.09 \pm 0.14_{\text{5loops}} \pm 0.08_\alpha \pm 0.04_{\text{masses}}$	$\times 10^{-11}$
Hadronic (lowest order)	$a_\mu[\text{HVP}(06)] = 6901 \pm 42_{\text{exp}} \pm 19_{\text{rad}} \pm 7_{\text{QCD}}$	$\times 10^{-11}$
Hadronic (higher order)	$a_\mu[\text{HVPh.o.}] = -97.9 \pm 0.9_{\text{exp}} \pm 0.3_{\text{rad}}$	$\times 10^{-11}$
Hadronic (light-by-light)	$a_\mu[\text{HLLS}] = 110 \pm 40$	$\times 10^{-11}$
Electroweak	$a_\mu[EW] = 154 \pm 2 \pm 1$	$\times 10^{-11}$

which should be compared with the experimental world average [187]

$$a_\mu^{\exp} = 11\,659\,2080\,(63) \times 10^{-11}. \quad (9.34)$$

One finds $\Delta a_\mu = 295(88) \times 10^{-11}$, a 3.4σ difference. It is clear that both the theoretical and the experimental uncertainty should be reduced to clarify whether there is a true discrepancy or a statistical fluctuation. We shall discuss potential improvements to the experiment below.

9.7.2 Measurement of the magnetic dipole moment

The measured value of the muon anomaly has a 0.46 ppm statistical uncertainty and a 0.28 ppm systematic uncertainty, which are combined in quadrature to obtain the total error of 0.54 ppm. To significantly improve the measured value, both errors must be reduced. We first discuss the experimental technique, and then the systematic errors.

In all but the first experiments by Garwin et al. [1087] the measurement of the magnetic anomaly made use of the spin rotation in a magnetic field relative to the momentum rotation:

$$\begin{aligned}\vec{\omega}_S &= -\frac{qg\vec{B}}{2m} - \frac{q\vec{B}}{\gamma m}(1-\gamma), \\ \vec{\omega}_C &= -\frac{q\vec{B}}{m\gamma}, \quad (9.35) \\ \vec{\omega}_a &\equiv \vec{\omega}_S - \vec{\omega}_C = -\left(\frac{g-2}{2}\right)\frac{q\vec{B}}{m} = -a_\mu\frac{q\vec{B}}{m}.\end{aligned}$$

A series of three beautiful experiments at CERN culminated in a 7.3 ppm measure of a_μ [1088]. In the third CERN experiment, a new technique was developed based on the observation that electrostatic quadrupoles could be used for vertical focusing. With the velocity transverse to the magnetic field ($\vec{\beta} \cdot \vec{B} = 0$), the spin precession formula becomes

$$\vec{\omega}_a = -\frac{q}{m}\left[a_\mu\vec{B} - \left(a_\mu - \frac{1}{\gamma^2-1}\right)\frac{\vec{\beta}\times\vec{E}}{c}\right]. \quad (9.36)$$

For $\gamma_{\text{magic}} = 29.3$ ($p = 3.09$ GeV/c), the second term vanishes so ω_a becomes independent of the electric field and the precise knowledge of the muon momentum. Also knowledge of the muon trajectories to determine the average magnetic field becomes less critical, which reduces the uncertainty in B.

This technique was used also in experiment E821 at the Brookhaven National Laboratory Alternating Gradient Synchrotron (AGS) [187, 1085]. The AGS proton beam is used to produce a beam of pions that decay to muons in an 80 m pion decay channel. Muons with p_{magic} are brought into the storage ring and stored using a fast muon kicker. Calorimeters, placed on the inner radius of the storage ring measure both the energy and arrival time of the decay electrons. Since the highest energy electrons are emitted antiparallel to the muon spin the rate of high energy electrons is modulated by the spin precession frequency:

$$N(t, E_{\text{th}}) = N_0(E_{\text{th}})e^{-t/\gamma\tau}\left[1 + A(E_{\text{th}})\cos(\omega_a t + \phi(E_{\text{th}}))\right]. \quad (9.37)$$

The time spectrum for electrons with $E > E_{\text{th}} = 1.8$ GeV is shown in [187] Fig. 72. The value of ω_a is obtained from these data using the five parameter function (9.37) as a starting point, but many additional small effects must be taken into account [187, 1085].

The magnetic field is measured with nuclear magnetic resonance (NMR) probes, and tied through calibration to the

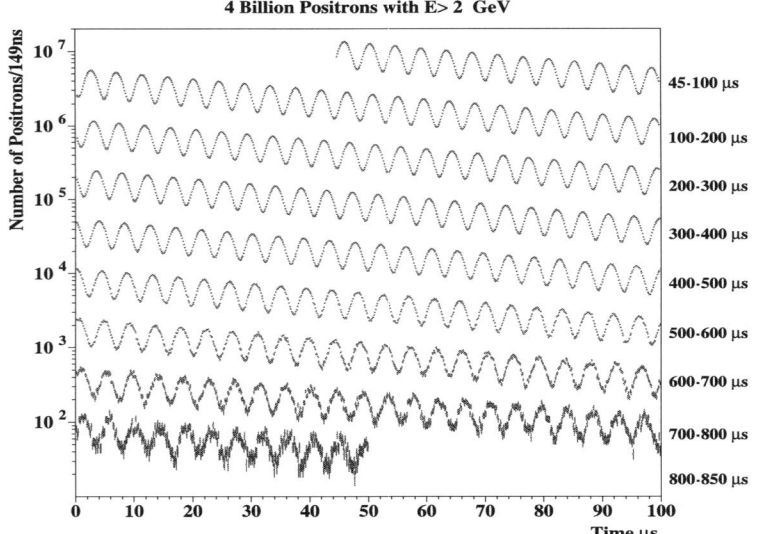

Fig. 72 The time spectrum of 10^9 positrons with energy greater than 1.8 GeV from the Y2000 run. The endpoint energy is 3.1 GeV. The time interval for each of the diagonal "wiggles" is given on the right

Larmor frequency of the free proton [187]. The anomaly is determined from

$$a_\mu = \frac{\tilde{\omega}_a/\omega_p}{\lambda - \tilde{\omega}_a/\omega_p} = \frac{\mathcal{R}}{\lambda - \mathcal{R}}, \quad (9.38)$$

where the tilde on $\tilde{\omega}_a$ indicates that the measured muon precession frequency has been adjusted for any necessary (small) corrections, such as the pitch and radial electric field corrections [1085], and $\lambda = \mu_\mu/\mu_p$ is the ratio of the muon to proton magnetic moments.

9.7.3 An improved $g - 2$ experiment

One of the major features of an upgraded experiment would be a substantially increased flux of muons into the storage ring. The BNL beam [187] took forward muons from pion decays, and selected muons 1.7% below the pion momentum. With this scheme, approximately half of the injected beam consisted of pions. An upgraded experiment would need to quadruple the quadrupoles in the pion decay channel, to increase the beam line acceptance. To decrease the hadron flash at injection one would need to go further away from the pion momentum. Alternatively one could increase the pion momentum to 5.32 GeV/c so that backward decays would produce muons at the magic momentum. Then the pion flash would be completely eliminated, which would significantly reduce the systematic error from gain instabilities.

The inflector magnet that permits the beam to enter the storage ring undeflected would need to be replaced, since the present model loses half of the beam through multiple scattering in material across the beam channel. The fast muon kicker would also need to be improved. With the significant increase in beam, the detectors would have to be segmented, new readout electronics would be needed, and a better measure of lost muons would also be needed.

To reduce the magnetic field systematic errors, significant effort will be needed to improve on the tracking of the field with time, and the calibration procedure used to tie the NMR frequency in the probes to the free proton Larmor frequency [187].

While there are technical issues to be resolved, the present technique—magic γ, electrostatic focusing, uniform magnetic field—could be pushed to below 0.1 ppm. To go further would probably require a new technique. One possibility discussed by Farley [1089] would be to use muons at much higher energy, say 15 GeV, which would increase the number of precessions that can be observed. The storage ring would consist of a small number of discrete magnets with uniform field and edge focusing and the field averaged over the orbit would be independent of orbit radius (particle momentum). The averaged field could be calibrated by injecting polarized protons and observing the proton $g - 2$ precession.

Acknowledgements The authors of this report are grateful to all the additional workshop participants who contributed with their presentations during the Working Group meetings: A. Baldini, W. Bertl, G. Colangelo, F. Farley, L. Fiorini, G. Gabrielse, J.R. Guest, A. Hoecker, J. Hosek, P. Iaydijev, Y. Kuno, S. Lavignac, D. Leone, A. Luccio, J. Miller, W. Morse, H. Nishiguchi, Y. Orlov, E. Paoloni, A. Pilaftsis, W. Porod, N. Ramsey, S. Redin, W. Rodejohann, M.-A. Sanchis-Lozano, N. Shafer-Ray, A. Soni, A. Strumia, G. Venanzoni, T. Yamashita, Z. Was, H. Wilschut.

This work was supported in part by the Marie Curie research training network "HEPTOOLS" (MRTN-CT-2006-035505).

References

1. T. Akesson et al., Eur. Phys. J. C **51**, 421 (2007). arXiv:hep-ph/0609216
2. A. Blondel et al. (eds.), ECFA/CERN studies of a European neutrino factory complex. CERN-2004-002, ECFA-04-230, April 2004
3. V. Barger et al., arXiv:0705.4396 [hep-ph]
4. R.N. Mohapatra et al., Rep. Prog. Phys. **70**, 1757 (2007). arXiv:hep-ph/0510213
5. C.D. Froggatt, H.B. Nielsen, Nucl. Phys. B **147**, 277 (1979)
6. H. Georgi, C. Jarlskog, Phys. Lett. B **86**, 297 (1979)
7. J.R. Ellis, M.K. Gaillard, Phys. Lett. B **88**, 315 (1979)
8. G. Anderson, S. Raby, S. Dimopoulos, L.J. Hall, G.D. Starkman, Phys. Rev. D **49**, 3660 (1994). arXiv:hep-ph/9308333
9. N. Irges, S. Lavignac, P. Ramond, Phys. Rev. D **58**, 035003 (1998). arXiv:hep-ph/9802334
10. J.K. Elwood, N. Irges, P. Ramond, Phys. Rev. Lett. **81**, 5064 (1998). arXiv:hep-ph/9807228
11. M. Leurer, Y. Nir, N. Seiberg, Nucl. Phys. B **398**, 319 (1993). arXiv:hep-ph/9212278
12. M. Dine, R.G. Leigh, A. Kagan, Phys. Rev. D **48**, 4269 (1993). arXiv:hep-ph/9304299
13. Y. Nir, N. Seiberg, Phys. Lett. B **309**, 337 (1993). arXiv:hep-ph/9304307
14. M. Leurer, Y. Nir, N. Seiberg, Nucl. Phys. B **420**, 468 (1994). arXiv:hep-ph/9310320
15. D.B. Kaplan, M. Schmaltz, Phys. Rev. D **49**, 3741 (1994). arXiv:hep-ph/9311281
16. A. Pomarol, D. Tommasini, Nucl. Phys. B **466**, 3 (1996). arXiv:hep-ph/9507462
17. C.D. Carone, L.J. Hall, H. Murayama, Phys. Rev. D **53**, 6282 (1996). arXiv:hep-ph/9512399
18. R. Barbieri, G.R. Dvali, L.J. Hall, Phys. Lett. B **377**, 76 (1996). arXiv:hep-ph/9512388
19. E. Dudas, C. Grojean, S. Pokorski, C.A. Savoy, Nucl. Phys. B **481**, 85 (1996). arXiv:hep-ph/9606383
20. R. Barbieri, L.J. Hall, S. Raby, A. Romanino, Nucl. Phys. B **493**, 3 (1997). arXiv:hep-ph/9610449
21. P. Binetruy, S. Lavignac, P. Ramond, Nucl. Phys. B **477**, 353 (1996). arXiv:hep-ph/9601243
22. P. Binetruy, S. Lavignac, S.T. Petcov, P. Ramond, Nucl. Phys. B **496**, 3 (1997). arXiv:hep-ph/9610481
23. Y. Nir, R. Rattazzi, Phys. Lett. B **382**, 363 (1996). arXiv:hep-ph/9603233
24. R. Barbieri, L.J. Hall, A. Romanino, Nucl. Phys. Proc. Suppl. A **52**, 141 (1997)
25. C.D. Carone, L.J. Hall, Phys. Rev. D **56**, 4198 (1997). arXiv:hep-ph/9702430
26. R. Barbieri, L.J. Hall, A. Romanino, Phys. Lett. B **401**, 47 (1997). arXiv:hep-ph/9702315
27. R. Barbieri, L. Giusti, L.J. Hall, A. Romanino, Nucl. Phys. B **550**, 32 (1999). arXiv:hep-ph/9812239

28. L.J. Hall, N. Weiner, Phys. Rev. D **60**, 033005 (1999). arXiv:hep-ph/9811299
29. R. Barbieri, L.J. Hall, A. Romanino, Nucl. Phys. B **551**, 93 (1999). arXiv:hep-ph/9812384
30. A. Aranda, C.D. Carone, R.F. Lebed, Phys. Lett. B **474**, 170 (2000). arXiv:hep-ph/9910392
31. R. Barbieri, P. Creminelli, A. Romanino, Nucl. Phys. B **559**, 17 (1999). arXiv:hep-ph/9903460
32. Z. Berezhiani, A. Rossi, Nucl. Phys. B **594**, 113 (2001). arXiv:hep-ph/0003084
33. M.C. Chen, K.T. Mahanthappa, Phys. Rev. D **62**, 113007 (2000). arXiv:hep-ph/0005292
34. A. Aranda, C.D. Carone, R.F. Lebed, Phys. Rev. D **62**, 016009 (2000). arXiv:hep-ph/0002044
35. A. Aranda, C.D. Carone, P. Meade, Phys. Rev. D **65**, 013011 (2002). arXiv:hep-ph/0109120
36. E. Ma, G. Rajasekaran, Phys. Rev. D **64**, 113012 (2001). arXiv:hep-ph/0106291
37. S.F. King, G.G. Ross, Phys. Lett. B **520**, 243 (2001). arXiv:hep-ph/0108112
38. S. Lavignac, I. Masina, C.A. Savoy, Phys. Lett. B **520**, 269 (2001). arXiv:hep-ph/0106245
39. R.G. Roberts, A. Romanino, G.G. Ross, L. Velasco-Sevilla, Nucl. Phys. B **615**, 358 (2001). arXiv:hep-ph/0104088
40. J.L. Chkareuli, C.D. Froggatt, H.B. Nielsen, Nucl. Phys. B **626**, 307 (2002). arXiv:hep-ph/0109156
41. K.S. Babu, E. Ma, J.W.F. Valle, Phys. Lett. B **552**, 207 (2003). arXiv:hep-ph/0206292
42. G.G. Ross, L. Velasco-Sevilla, Nucl. Phys. B **653**, 3 (2003). arXiv:hep-ph/0208218
43. Y. Nir, G. Raz, Phys. Rev. D **66**, 035007 (2002). arXiv:hep-ph/0206064
44. H.K. Dreiner, H. Murayama, M. Thormeier, Nucl. Phys. B **729**, 278 (2005). arXiv:hep-ph/0312012
45. S.F. King, G.G. Ross, Phys. Lett. B **574**, 239 (2003). arXiv:hep-ph/0307190
46. G.G. Ross, L. Velasco-Sevilla, O. Vives, Nucl. Phys. B **692**, 50 (2004). arXiv:hep-ph/0401064
47. W. Grimus, A.S. Joshipura, S. Kaneko, L. Lavoura, M. Tanimoto, J. High Energy Phys. **0407**, 078 (2004). arXiv:hep-ph/0407112
48. G.L. Kane, S.F. King, I.N.R. Peddie, L. Velasco-Sevilla, J. High Energy Phys. **0508**, 083 (2005). arXiv:hep-ph/0504038
49. Z. Berezhiani, F. Nesti, J. High Energy Phys. **0603**, 041 (2006). arXiv:hep-ph/0510011
50. G. Altarelli, F. Feruglio, Nucl. Phys. B **741**, 215 (2006). arXiv:hep-ph/0512103
51. O. Vives, arXiv:hep-ph/0504079
52. S. Abel, D. Bailin, S. Khalil, O. Lebedev, Phys. Lett. B **504**, 241 (2001). arXiv:hep-ph/0012145
53. J.L. Diaz-Cruz, J. Ferrandis, Phys. Rev. D **72**, 035003 (2005). arXiv:hep-ph/0504094
54. P.H. Chankowski, K. Kowalska, S. Lavignac, S. Pokorski, Phys. Rev. D **71**, 055004 (2005). arXiv:hep-ph/0501071
55. W. Grimus, L. Lavoura, J. High Energy Phys. **0508**, 013 (2005). arXiv:hep-ph/0504153
56. I. de Medeiros Varzielas, G.G. Ross, Nucl. Phys. B **733**, 31 (2006). arXiv:hep-ph/0507176
57. I. de Medeiros Varzielas, S.F. King, G.G. Ross, arXiv:hep-ph/0607045
58. I. Masina, C.A. Savoy, Nucl. Phys. B **755**, 1 (2006). arXiv:hep-ph/0603101
59. I. Masina, C.A. Savoy, Phys. Lett. B **642**, 472 (2006). arXiv:hep-ph/0606097
60. S.F. King, M. Malinsky, Phys. Lett. B **645**, 351 (2007). arXiv:hep-ph/0610250
61. C. Hagedorn, M. Lindner, R.N. Mohapatra, J. High Energy Phys. **0606**, 042 (2006). arXiv:hep-ph/0602244
62. T. Appelquist, Y. Bai, M. Piai, Phys. Lett. B **637**, 245 (2006). arXiv:hep-ph/0603104
63. L. Ferretti, S.F. King, A. Romanino, J. High Energy Phys. **0611**, 078 (2006). arXiv:hep-ph/0609047
64. F. Feruglio, C. Hagedorn, Y. Lin, L. Merlo, arXiv:hep-ph/0702194
65. E. Ma, arXiv:hep-ph/0409075
66. E. Ma, Phys. Rev. D **43**, 2761 (1991)
67. N.G. Deshpande, M. Gupta, P.B. Pal, Phys. Rev. D **45**, 953 (1992)
68. M. Frigerio, S. Kaneko, E. Ma, M. Tanimoto, Phys. Rev. D **71**, 011901 (2005). arXiv:hep-ph/0409187
69. E. Ma, Mod. Phys. Lett. A **17**, 2361 (2002). arXiv:hep-ph/0211393
70. C. Luhn, S. Nasri, P. Ramond, arXiv:hep-th/0701188
71. E. Ma, Mod. Phys. Lett. A **21**, 1917 (2006). arXiv:hep-ph/0607056
72. W. Grimus, L. Lavoura, Phys. Lett. B **572**, 189 (2003). arXiv:hep-ph/0305046
73. E. Ma, arXiv:hep-ph/0612022
74. Y. Koide, Lett. Nuovo Cimento **34**, 201 (1982)
75. E. Ma, arXiv:hep-ph/0701016
76. Y. Yamaguchi, Phys. Lett. **9**, 281 (1964)
77. J. Kubo, A. Mondragon, M. Mondragon, E. Rodriguez-Jauregui, Prog. Theor. Phys. **109**, 795 (2003) [Erratum: Prog. Theor. Phys. **114**, 287 (2005)]. arXiv:hep-ph/0302196
78. J. Kubo, H. Okada, F. Sakamaki, Phys. Rev. D **70**, 036007 (2004). arXiv:hep-ph/0402089
79. S.L. Chen, M. Frigerio, E. Ma, Phys. Rev. D **70**, 073008 (2004) [Erratum: Phys. Rev. D **70**, 079905 (2004)]. arXiv:hep-ph/0404084
80. L. Lavoura, E. Ma, Mod. Phys. Lett. A **20**, 1217 (2005). arXiv:hep-ph/0502181
81. T. Teshima, Phys. Rev. D **73**, 045019 (2006). arXiv:hep-ph/0509094
82. Y. Koide, Phys. Rev. D **73**, 057901 (2006). arXiv:hep-ph/0509214
83. R.N. Mohapatra, S. Nasri, H.B. Yu, Phys. Lett. B **639**, 318 (2006). arXiv:hep-ph/0605020
84. S. Morisi, M. Picariello, Int. J. Theor. Phys. **45**, 1267 (2006). arXiv:hep-ph/0505113
85. S. Kaneko, H. Sawanaka, T. Shingai, M. Tanimoto, K. Yoshioka, Prog. Theor. Phys. **117**, 161 (2007). arXiv:hep-ph/0609220
86. F. Feruglio, Y. Lin, arXiv:0712.1528 [hep-ph]
87. E. Ma, arXiv:hep-ph/0612013
88. O. Felix, A. Mondragon, M. Mondragon, E. Peinado, arXiv:hep-ph/0610061
89. P.F. Harrison, D.H. Perkins, W.G. Scott, Phys. Lett. B **530**, 167 (2002). arXiv:hep-ph/0202074
90. X.G. He, A. Zee, Phys. Lett. B **560**, 87 (2003). arXiv:hep-ph/0301092
91. N. Cabibbo, Phys. Lett. B **72**, 333 (1978)
92. L. Wolfenstein, Phys. Rev. D **18**, 958 (1978)
93. E. Ma, Phys. Rev. D **70**, 031901 (2004). arXiv:hep-ph/0404199
94. G. Altarelli, F. Feruglio, Nucl. Phys. B **720**, 64 (2005). arXiv:hep-ph/0504165
95. K.S. Babu, X.G. He, arXiv:hep-ph/0507217
96. E. Ma, Phys. Rev. D **72**, 037301 (2005). arXiv:hep-ph/0505209
97. A. Zee, Phys. Lett. B **630**, 58 (2005). arXiv:hep-ph/0508278
98. E. Ma, Phys. Rev. D **73**, 057304 (2006). arXiv:hep-ph/0511133
99. X.G. He, Y.Y. Keum, R.R. Volkas, J. High Energy Phys. **0604**, 039 (2006). arXiv:hep-ph/0601001
100. B. Adhikary, B. Brahmachari, A. Ghosal, E. Ma, M.K. Parida, Phys. Lett. B **638**, 345 (2006). arXiv:hep-ph/0603059

101. B. Adhikary, A. Ghosal, Phys. Rev. D **75**, 073020 (2007). arXiv:hep-ph/0609193
102. F. Yin, Phys. Rev. D **75**, 073010 (2007). arXiv:0704.3827 [hep-ph]
103. G. Altarelli, F. Feruglio, Y. Lin, Nucl. Phys. B **775**, 31 (2007). arXiv:hep-ph/0610165
104. X.G. He, Nucl. Phys. Proc. Suppl. **168**, 350 (2007). arXiv:hep-ph/0612080
105. E. Ma, Mod. Phys. Lett. A **17**, 289 (2002). arXiv:hep-ph/0201225
106. E. Ma, Mod. Phys. Lett. A **17**, 627 (2002). arXiv:hep-ph/0203238
107. K.S. Babu, T. Kobayashi, J. Kubo, Phys. Rev. D **67**, 075018 (2003). arXiv:hep-ph/0212350
108. S.L. Chen, M. Frigerio, E. Ma, Nucl. Phys. B **724**, 423 (2005). arXiv:hep-ph/0504181
109. M. Hirsch, A. Villanova del Moral, J.W.F. Valle, E. Ma, Phys. Rev. D **72**, 091301 (2005) [Erratum: Phys. Rev. D **72**, 119904 (2005)]. arXiv:hep-ph/0507148
110. E. Ma, Mod. Phys. Lett. A **20**, 1953 (2005). arXiv:hep-ph/0502024
111. E. Ma, Mod. Phys. Lett. A **20**, 2601 (2005). arXiv:hep-ph/0508099
112. E. Ma, Mod. Phys. Lett. A **20**, 2767 (2005). arXiv:hep-ph/0506036
113. E. Ma, H. Sawanaka, M. Tanimoto, Phys. Lett. B **641**, 301 (2006). arXiv:hep-ph/0606103
114. E. Ma, Mod. Phys. Lett. A **21**, 2931 (2006). arXiv:hep-ph/0607190
115. E. Ma, Mod. Phys. Lett. A **22**, 101 (2007). arXiv:hep-ph/0610342
116. L. Lavoura, H. Kuhbock, Mod. Phys. Lett. A **22**, 181 (2007). arXiv:hep-ph/0610050
117. I. de Medeiros Varzielas, S.F. King, G.G. Ross, Phys. Lett. B **644**, 153 (2007). arXiv:hep-ph/0512313
118. S. Morisi, M. Picariello, E. Torrente-Lujan, Phys. Rev. D **75**, 075015 (2007). arXiv:hep-ph/0702034
119. Y. Koide, arXiv:hep-ph/0701018
120. M. Hirsch, A.S. Joshipura, S. Kaneko, J.W.F. Valle, arXiv:hep-ph/0703046
121. P.D. Carr, P.H. Frampton, arXiv:hep-ph/0701034
122. M.C. Chen, K.T. Mahanthappa, arXiv:0705.0714 [hep-ph]
123. P.H. Frampton, T.W. Kephart, arXiv:0706.1186 [hep-ph]
124. T. Kobayashi, H.P. Nilles, F. Ploger, S. Raby, M. Ratz, Nucl. Phys. B **768**, 135 (2007). arXiv:hep-ph/0611020
125. E. Ma, Fizika B **14**, 35 (2005). arXiv:hep-ph/0409288
126. C. Hagedorn, M. Lindner, F. Plentinger, Phys. Rev. D **74**, 025007 (2006). arXiv:hep-ph/0604265
127. Y. Kajiyama, J. Kubo, H. Okada, Phys. Rev. D **75**, 033001 (2007). arXiv:hep-ph/0610072
128. P.H. Frampton, T.W. Kephart, Phys. Rev. D **51**, 1 (1995). arXiv:hep-ph/9409324
129. K.S. Babu, J. Kubo, Phys. Rev. D **71**, 056006 (2005). arXiv:hep-ph/0411226
130. J. Kubo, Phys. Lett. B **622**, 303 (2005). arXiv:hep-ph/0506043
131. S.L. Chen, E. Ma, Phys. Lett. B **620**, 151 (2005). arXiv:hep-ph/0505064
132. E. Ma, Phys. Lett. B **632**, 352 (2006). arXiv:hep-ph/0508231
133. Y. Cai, H.B. Yu, Phys. Rev. D **74**, 115005 (2006). arXiv:hep-ph/0608022
134. H. Zhang, arXiv:hep-ph/0612214
135. Y. Koide, arXiv:0705.2275 [hep-ph]
136. M. Schmaltz, Phys. Rev. D **52**, 1643 (1995). arXiv:hep-ph/9411383
137. W. Grimus, L. Lavoura, J. High Energy Phys. **0601**, 018 (2006). arXiv:hep-ph/0509239
138. W. Grimus, L. Lavoura, arXiv:hep-ph/0611149
139. S.M. Barr, Phys. Rev. D **65**, 096012 (2002). arXiv:hep-ph/0106241
140. S.F. King, Phys. Lett. B **439**, 350 (1998). arXiv:hep-ph/9806440
141. S.F. King, Nucl. Phys. B **562**, 57 (1999). arXiv:hep-ph/9904210
142. S.F. King, Nucl. Phys. B **576**, 85 (2000). arXiv:hep-ph/9912492
143. S.F. King, J. High Energy Phys. **0209**, 011 (2002). arXiv:hep-ph/0204360
144. F. Borzumati, A. Masiero, Phys. Rev. Lett. **57**, 961 (1986)
145. L.J. Hall, V.A. Kostelecky, S. Raby, Nucl. Phys. B **267**, 415 (1986)
146. S. Ferrara, L. Girardello, F. Palumbo, Phys. Rev. D **20**, 403 (1979)
147. P. Fayet, Phys. Lett. B **69**, 489 (1977)
148. E. Witten, Nucl. Phys. B **188**, 513 (1981)
149. S. Dimopoulos, H. Georgi, Nucl. Phys. B **193**, 150 (1981)
150. S. Weinberg, Phys. Rev. D **26**, 287 (1982)
151. L. Alvarez-Gaume, M. Claudson, M.B. Wise, Nucl. Phys. B **207**, 96 (1982)
152. G.F. Giudice, R. Rattazzi, Phys. Rep. **322**, 419 (1999). arXiv:hep-ph/9801271
153. E. Cremmer, S. Ferrara, L. Girardello, A. Van Proeyen, Nucl. Phys. B **212**, 413 (1983)
154. S.K. Soni, H.A. Weldon, Phys. Lett. B **126**, 215 (1983)
155. V.S. Kaplunovsky, J. Louis, Phys. Lett. B **306**, 269 (1993). arXiv:hep-th/9303040
156. A. Brignole, L.E. Ibanez, C. Munoz, Nucl. Phys. B **422**, 125 (1994) [Erratum: Nucl. Phys. B **436**, 747 (1995)]. arXiv:hep-ph/9308271
157. P.H. Chankowski, O. Lebedev, S. Pokorski, Nucl. Phys. B **717**, 190 (2005). arXiv:hep-ph/0502076
158. S.P. Martin, arXiv:hep-ph/9709356
159. M. Shapkin, Nucl. Phys. Proc. Suppl. **169**, 363 (2007)
160. G.G. Ross, O. Vives, Phys. Rev. D **67**, 095013 (2003). arXiv:hep-ph/0211279
161. S.F. King, I.N.R. Peddie, G.G. Ross, L. Velasco-Sevilla, O. Vives, J. High Energy Phys. **0507**, 049 (2005). arXiv:hep-ph/0407012
162. S. Antusch, S.F. King, M. Malinsky, arXiv:0708.1282 [hep-ph]
163. I. Masina, C.A. Savoy, Nucl. Phys. B **661**, 365 (2003). arXiv:hep-ph/0211283
164. A. Masiero, S.K. Vempati, O. Vives, Nucl. Phys. B **649**, 189 (2003). arXiv:hep-ph/0209303
165. M. Ciuchini, A. Masiero, P. Paradisi, L. Silvestrini, S.K. Vempati, O. Vives, arXiv:hep-ph/0702144
166. K. Hagiwara, S. Matsumoto, D. Haidt, C.S. Kim, Z. Phys. C **64**, 559 (1994) [Erratum: Z. Phys. C **68**, 352 (1995)]. arXiv:hep-ph/9409380
167. W. Buchmuller, D. Wyler, Nucl. Phys. B **268**, 621 (1986)
168. Z. Berezhiani, A. Rossi, Phys. Lett. B **535**, 207 (2002). arXiv:hep-ph/0111137
169. A. Brignole, A. Rossi, Nucl. Phys. B **701**, 3 (2004). arXiv:hep-ph/0404211
170. W.M. Yao et al. (Particle Data Group), J. Phys. G **33**, 1 (2006)
171. M.S. Bilenky, A. Santamaria, Nucl. Phys. B **420**, 47 (1994). arXiv:hep-ph/9310302
172. L. Lavoura, Eur. Phys. J. C **29**, 191 (2003). arXiv:hep-ph/0302221
173. T. Inami, C.S. Lim, Prog. Theor. Phys. **65**, 297 (1981) [Erratum: Prog. Theor. Phys. **65**, 1772 (1981)]
174. A. Brignole, A. Rossi, Phys. Lett. B **566**, 217 (2003). arXiv:hep-ph/0304081
175. P. Paradisi, J. High Energy Phys. **0608**, 047 (2006). arXiv:hep-ph/0601100

176. F. Jegerlehner, arXiv:hep-ph/0703125
177. G.A. Gonzalez-Sprinberg, A. Santamaria, J. Vidal, Nucl. Phys. B **582**, 3 (2000). arXiv:hep-ph/0002203
178. J. Hisano, T. Moroi, K. Tobe, M. Yamaguchi, Phys. Rev. D **53**, 2442 (1996). arXiv:hep-ph/9510309
179. A. Czarnecki, E. Jankowski, Phys. Rev. D **65**, 113004 (2002). arXiv:hep-ph/0106237
180. M.L. Brooks et al. (MEGA Collaboration), Phys. Rev. Lett. **83**, 1521 (1999). arXiv:hep-ex/9905013
181. K. Hayasaka et al. (Belle Collaboration), arXiv:0705.0650 [hep-ex]
182. B. Aubert et al. (BaBar Collaboration), Phys. Rev. Lett. **96**, 041801 (2006). arXiv:hep-ex/0508012
183. S. Banerjee, arXiv:hep-ex/0702017
184. M. Raidal, A. Santamaria, Phys. Lett. B **421**, 250 (1998). arXiv:hep-ph/9710389
185. K. Huitu, J. Maalampi, M. Raidal, A. Santamaria, Phys. Lett. B **430**, 355 (1998). arXiv:hep-ph/9712249
186. B.C. Regan, E.D. Commins, C.J. Schmidt, D. DeMille, Phys. Rev. Lett. **88**, 071805 (2002)
187. G.W. Bennett et al. (Muon g-2 Collaboration), Phys. Rev. D **73**, 072003 (2006). arXiv:hep-ex/0602035
188. R. McNabb (Muon g-2 Collaboration), arXiv:hep-ex/0407008
189. J. Abdallah et al. (DELPHI Collaboration), Eur. Phys. J. C **35**, 159 (2004). arXiv:hep-ex/0406010
190. K. Inami et al. (Belle Collaboration), Phys. Lett. B **551**, 16 (2003). arXiv:hep-ex/0210066
191. B. Aubert et al. (BaBar Collaboration), Phys. Rev. Lett. **95**, 041802 (2005). arXiv:hep-ex/0502032
192. S. Davidson, C. Pena-Garay, N. Rius, A. Santamaria, J. High Energy Phys. **0303**, 011 (2003). arXiv:hep-ph/0302093
193. M. Acciarri et al. (L3 Collaboration), Phys. Lett. B **489**, 81 (2000). arXiv:hep-ex/0005028
194. D. Bourilkov, Phys. Rev. D **64**, 071701 (2001). arXiv:hep-ph/0104165
195. G. Abbiendi et al. (OPAL Collaboration), Eur. Phys. J. C **33**, 173 (2004). arXiv:hep-ex/0309053
196. A. Pich, J.P. Silva, Phys. Rev. D **52**, 4006 (1995). arXiv:hep-ph/9505327
197. S. Bergmann, Y. Grossman, D.M. Pierce, Phys. Rev. D **61**, 053005 (2000). arXiv:hep-ph/9909390
198. M.C. Gonzalez-Garcia, J.W.F. Valle, Mod. Phys. Lett. A **7**, 477 (1992)
199. A. De Gouvea, G.F. Giudice, A. Strumia, K. Tobe, Nucl. Phys. B **623**, 395 (2002). arXiv:hep-ph/0107156
200. Z. Berezhiani, R.S. Raghavan, A. Rossi, Nucl. Phys. B **638**, 62 (2002). arXiv:hep-ph/0111138
201. S. Davidson, D.C. Bailey, B.A. Campbell, Z. Phys. C **61**, 613 (1994). arXiv:hep-ph/9309310
202. M. Herz, arXiv:hep-ph/0301079
203. S.L. Glashow, J. Iliopoulos, L. Maiani, Phys. Rev. D **2**, 1285 (1970)
204. N. Cabibbo, Phys. Rev. Lett. **10**, 531 (1963)
205. M. Kobayashi, T. Maskawa, Prog. Theor. Phys. **49**, 652 (1973)
206. L. Wolfenstein, Phys. Rev. Lett. **51**, 1945 (1983)
207. M. Battaglia et al., hep-ph/0304132
208. B. Pontecorvo, Sov. Phys. JETP **7**, 172 (1958) [Zh. Eksp. Teor. Fiz. **34**, 247 (1957)]
209. Z. Maki, M. Nakagawa, S. Sakata, Prog. Theor. Phys. **28**, 870 (1962)
210. M. Maltoni, T. Schwetz, M.A. Tortola, J.W.F. Valle, New J. Phys. **6**, 122 (2004). arXiv:hep-ph/0405172
211. G.L. Fogli, E. Lisi, A. Marrone, A. Palazzo, Prog. Part. Nucl. Phys. **57**, 742 (2006). arXiv:hep-ph/0506083
212. S.T. Petcov, Sov. J. Nucl. Phys. **25**, 340 (1977) [Yad. Fiz. **25**, 641 (1977); Errata: Yad. Fiz. **25**, 698 (1977), Yad. Fiz. **25**, 1336 (1977)]
213. S. Weinberg, Phys. Rev. Lett. **43**, 1566 (1979)
214. S.M. Bilenky, J. Hosek, S.T. Petcov, Phys. Lett. B **94**, 495 (1980)
215. J. Schechter, J.W.F. Valle, Phys. Rev. D **22**, 2227 (1980)
216. P. Minkowski, Phys. Lett. B **67**, 421 (1977)
217. T. Yanagida, in *Proceedings of the Workshop on the Baryon Number of the Universe and Unified Theories*, Tsukuba, Japan, 13–14 February 1979
218. M. Gell-Mann, P. Ramond, R. Slansky, in *Supergravity*, ed. by P. van Nieuwenhuizen, D.Z. Freedman (North Holland, Amsterdam, 1979)
219. S.L. Glashow, NATO Adv. Study Inst. Ser. B Phys. **59**, 687 (1979)
220. R.N. Mohapatra, G. Senjanovic, Phys. Rev. Lett. **44**, 912 (1980)
221. M. Magg, C. Wetterich, Phys. Lett. B **94**, 61 (1980)
222. G. Lazarides, Q. Shafi, C. Wetterich, Nucl. Phys. B **181**, 287 (1981)
223. R.N. Mohapatra, G. Senjanovic, Phys. Rev. D **23**, 165 (1981)
224. G.B. Gelmini, M. Roncadelli, Phys. Lett. B **99**, 411 (1981)
225. E. Ma, Phys. Rev. Lett. **81**, 1171 (1998). arXiv:hep-ph/9805219
226. R. Foot, H. Lew, X.G. He, G.C. Joshi, Z. Phys. C **44**, 441 (1989)
227. W. Buchmuller, K. Hamaguchi, O. Lebedev, S. Ramos-Sanchez, M. Ratz, Phys. Rev. Lett. **99**, 021601 (2007). arXiv:hep-ph/0703078
228. J.R. Ellis, O. Lebedev, arXiv:0707.3419 [hep-ph]
229. J.R. Ellis, J. Hisano, S. Lola, M. Raidal, Nucl. Phys. B **621**, 208 (2002). arXiv:hep-ph/0109125
230. S. Pascoli, S.T. Petcov, W. Rodejohann, Phys. Rev. D **68**, 093007 (2003). arXiv:hep-ph/0302054
231. J.A. Casas, A. Ibarra, F. Jimenez-Alburquerque, arXiv:hep-ph/0612289
232. J.A. Casas, A. Ibarra, Nucl. Phys. B **618**, 171 (2001). arXiv:hep-ph/0103065
233. S. Lavignac, I. Masina, C.A. Savoy, Nucl. Phys. B **633**, 139 (2002). arXiv:hep-ph/0202086
234. P.H. Frampton, S.L. Glashow, T. Yanagida, Phys. Lett. B **548**, 119 (2002). arXiv:hep-ph/0208157
235. M. Raidal, A. Strumia, Phys. Lett. B **553**, 72 (2003). arXiv:hep-ph/0210021
236. A. Ibarra, G.G. Ross, Phys. Lett. B **575**, 279 (2003). arXiv:hep-ph/0307051
237. A. Ibarra, G.G. Ross, Phys. Lett. B **591**, 285 (2004). arXiv:hep-ph/0312138
238. P.H. Chankowski, J.R. Ellis, S. Pokorski, M. Raidal, K. Turzynski, Nucl. Phys. B **690**, 279 (2004)
239. S.T. Petcov, W. Rodejohann, T. Shindou, Y. Takanishi, Nucl. Phys. B **739**, 208 (2006). arXiv:hep-ph/0510404
240. T. Endoh, T. Morozumi, T. Onogi, A. Purwanto, Phys. Rev. D **64**, 013006 (2001) [Erratum: Phys. Rev. D **64**, 059904 (2001)]. arXiv:hep-ph/0012345
241. S. Davidson, A. Ibarra, J. High Energy Phys. **0109**, 013 (2001). arXiv:hep-ph/0104076
242. J.R. Ellis, J. Hisano, M. Raidal, Y. Shimizu, Phys. Rev. D **66**, 115013 (2002). arXiv:hep-ph/0206110
243. A. Ibarra, J. High Energy Phys. **0601**, 064 (2006). arXiv:hep-ph/0511136
244. E. Ma, U. Sarkar, Phys. Rev. Lett. **80**, 5716 (1998). arXiv:hep-ph/9802445
245. K. Huitu, J. Maalampi, M. Raidal, Nucl. Phys. B **420**, 449 (1994). arXiv:hep-ph/9312235
246. K. Huitu, J. Maalampi, M. Raidal, Phys. Lett. B **328**, 60 (1994). arXiv:hep-ph/9402219
247. M. Raidal, P.M. Zerwas, Eur. Phys. J. C **8**, 479 (1999). arXiv:hep-ph/9811443

248. T. Hambye, E. Ma, U. Sarkar, Nucl. Phys. B **602**, 23 (2001). arXiv:hep-ph/0011192
249. A. Rossi, Phys. Rev. D **66**, 075003 (2002). arXiv:hep-ph/0207006
250. K.S. Babu, C.N. Leung, J.T. Pantaleone, Phys. Lett. B **319**, 191 (1993)
251. P.H. Chankowski, Z. Płuciennik, Phys. Lett. B **316**, 312 (1993)
252. S. Antusch, M. Drees, J. Kersten, M. Lindner, M. Ratz, Phys. Lett. B **519**, 238 (2001)
253. S. Antusch, M. Drees, J. Kersten, M. Lindner, M. Ratz, Phys. Lett. B **525**, 130 (2002)
254. S. Antusch, J. Kersten, M. Lindner, M. Ratz, Nucl. Phys. B **674**, 401 (2003)
255. P.H. Chankowski, W. Krolikowski, S. Pokorski, Phys. Lett. B **473**, 109 (2000)
256. J.A. Casas, J.R. Espinosa, A. Ibarra, I. Navarro, Nucl. Phys. B **573**, 652 (2000)
257. S. Antusch, J. Kersten, M. Lindner, M. Ratz, Phys. Lett. B **538**, 87 (2002)
258. S. Antusch, J. Kersten, M. Lindner, M. Ratz, M.A. Schmidt, J. High Energy Phys. **0503**, 024 (2005)
259. J.W. Mei, Phys. Rev. D **71**, 073012 (2005)
260. S. Antusch, M. Ratz, J. High Energy Phys. **0207**, 059 (2002)
261. N. Haba, N. Okamura, M. Sugiura, Prog. Theor. Phys. **103**, 367 (2000)
262. N. Haba, Y. Matsui, N. Okamura, M. Sugiura, Eur. Phys. J. C **10**, 677 (1999)
263. N. Haba, N. Okamura, Eur. Phys. J. C **14**, 347 (2000)
264. J.R. Ellis, S. Lola, Phys. Lett. B **458**, 310 (1999)
265. J.A. Casas, J.R. Espinosa, A. Ibarra, I. Navarro, Nucl. Phys. B **556**, 3 (1999)
266. J.A. Casas, J.R. Espinosa, A. Ibarra, I. Navarro, Nucl. Phys. B **569**, 82 (2000)
267. T. Miura, T. Shindou, E. Takasugi, Phys. Rev. D **66**, 093002 (2002)
268. N. Haba, Y. Matsui, N. Okamura, Eur. Phys. J. C **17**, 513 (2000)
269. S. Antusch, J. Kersten, M. Lindner, M. Ratz, Phys. Lett. B **544**, 1 (2002)
270. T. Miura, T. Shindou, E. Takasugi, Phys. Rev. D **68**, 093009 (2003)
271. S. Kanemura, K. Matsuda, T. Ota, T. Shindou, E. Takasugi, K. Tsumura, Phys. Rev. D **72**, 093004 (2005)
272. T. Shindou, E. Takasugi, Phys. Rev. D **70**, 013005 (2004)
273. S.T. Petcov, S. Profumo, Y. Takanishi, C.E. Yaguna, Nucl. Phys. B **676**, 453 (2004)
274. J.R. Ellis, A. Hektor, M. Kadastik, K. Kannike, M. Raidal, Phys. Lett. B **631**, 32 (2005). arXiv:hep-ph/0506122
275. S.T. Petcov, T. Shindou, Y. Takanishi, Nucl. Phys. B **738**, 219 (2006)
276. H.B. Nielsen, Y. Takanishi, Nucl. Phys. B **636**, 305 (2002)
277. R. Gonzalez Felipe, F.R. Joaquim, B.M. Nobre, Phys. Rev. D **70**, 085009 (2004)
278. G.C. Branco, R. Gonzalez Felipe, F.R. Joaquim, B.M. Nobre, Phys. Lett. B **633**, 336 (2006)
279. J.W. Mei, Z.Z. Xing, Phys. Rev. D **70**, 053002 (2004)
280. S.T. Petcov, A.Y. Smirnov, Phys. Lett. B **322**, 109 (1994). arXiv:hep-ph/9311204
281. M. Raidal, Phys. Rev. Lett. **93**, 161801 (2004). arXiv:hep-ph/0404046
282. H. Minakata, A.Y. Smirnov, Phys. Rev. D **70**, 073009 (2004). arXiv:hep-ph/0405088
283. Y. Kajiyama, M. Raidal, A. Strumia, arXiv:0705.4559 [hep-ph]
284. Euclid, *Elements*, Book 6, Definition 3. (A straight line is said to have been cut in extreme and mean ratio when, as the whole line is to the greater segment, so is the greater to the less)
285. Q. Duret, B. Machet, arXiv:0705.1237 [hep-ph]
286. B.C. Chauhan, M. Picariello, J. Pulido, E. Torrente-Lujan, arXiv:hep-ph/0605032
287. M. Picariello, arXiv:hep-ph/0611189
288. F. Caravaglios, S. Morisi, arXiv:hep-ph/0611078
289. C.L. Bennett et al. (WMAP Collaboration), Astrophys. J. Suppl. **148**, 1 (2003). arXiv:astro-ph/0302207
290. M. Fukugita, T. Yanagida, Phys. Lett. B **174**, 45 (1986)
291. V.A. Kuzmin, V.A. Rubakov, M.E. Shaposhnikov, Phys. Lett. B **155**, 36 (1985)
292. J. Liu, G. Segre, Phys. Rev. D **48**, 4609 (1993). arXiv:hep-ph/9304241
293. M. Flanz, E.A. Paschos, U. Sarkar, Phys. Lett. B **345**, 248 (1995) [Erratum: Phys. Lett. B **382**, 447 (1996)]. arXiv:hep-ph/9411366
294. L. Covi, E. Roulet, F. Vissani, Phys. Lett. B **384**, 169 (1996). arXiv:hep-ph/9605319
295. R. Barbieri, P. Creminelli, A. Strumia, N. Tetradis, Nucl. Phys. B **575**, 61 (2000). arXiv:hep-ph/9911315
296. A. Abada, S. Davidson, F.X. Josse-Michaux, M. Losada, A. Riotto, J. Cosmol. Astropart. Phys. **0604**, 004 (2006). arXiv:hep-ph/0601083
297. E. Nardi, Y. Nir, E. Roulet, J. Racker, J. High Energy Phys. **0601**, 164 (2006). arXiv:hep-ph/0601084
298. A. Abada, S. Davidson, A. Ibarra, F.X. Josse-Michaux, M. Losada, A. Riotto, J. High Energy Phys. **0609**, 010 (2006). arXiv:hep-ph/0605281
299. T. Fujihara, S. Kaneko, S. Kang, D. Kimura, T. Morozumi, M. Tanimoto, Phys. Rev. D **72**, 016006 (2005). arXiv:hep-ph/0505076
300. S. Pascoli, S.T. Petcov, A. Riotto, arXiv:hep-ph/0609125
301. G.C. Branco, A.J. Buras, S. Jager, S. Uhlig, A. Weiler, arXiv:hep-ph/0609067
302. G.C. Branco, R. Gonzalez Felipe, F.R. Joaquim, arXiv:hep-ph/0609297
303. S. Uhlig, arXiv:hep-ph/0612262
304. M. Plumacher, Z. Phys. C **74**, 549 (1997). arXiv:hep-ph/9604229
305. G. Engelhard, Y. Grossman, E. Nardi, Y. Nir, arXiv:hep-ph/0612187
306. S. Davidson, A. Ibarra, Phys. Lett. B **535**, 25 (2002). arXiv:hep-ph/0202239
307. K. Hamaguchi, H. Murayama, T. Yanagida, Phys. Rev. D **65**, 043512 (2002). arXiv:hep-ph/0109030
308. W. Buchmuller, P. Di Bari, M. Plumacher, Nucl. Phys. B **643**, 367 (2002). arXiv:hep-ph/0205349
309. G.F. Giudice, A. Notari, M. Raidal, A. Riotto, A. Strumia, Nucl. Phys. B **685**, 89 (2004). arXiv:hep-ph/0310123
310. S. Blanchet, P. Di Bari, arXiv:hep-ph/0607330
311. S. Antusch, A.M. Teixeira, arXiv:hep-ph/0611232
312. J.R. Ellis, J.E. Kim, D.V. Nanopoulos, Phys. Lett. B **145**, 181 (1984)
313. J.R. Ellis, D.V. Nanopoulos, S. Sarkar, Nucl. Phys. B **259**, 175 (1985)
314. J.R. Ellis, D.V. Nanopoulos, K.A. Olive, S.J. Rey, Astropart. Phys. **4**, 371 (1996). arXiv:hep-ph/9505438
315. T. Moroi, arXiv:hep-ph/9503210
316. M. Raidal, A. Strumia, K. Turzynski, Phys. Lett. B **609**, 351 (2005) [Erratum: Phys. Lett. B **632**, 752 (2006)]. arXiv:hep-ph/0408015
317. S. Blanchet, P. Di Bari, G.G. Raffelt, arXiv:hep-ph/0611337
318. W. Buchmuller, P. Di Bari, M. Plumacher, Phys. Lett. B **547**, 128 (2002). arXiv:hep-ph/0209301
319. S. Davidson, arXiv:hep-ph/0304120
320. E.W. Kolb, M.S. Turner, Front. Phys. **69**, 1 (1990)
321. W. Fischler, G.F. Giudice, R.G. Leigh, S. Paban, Phys. Lett. B **258**, 45 (1991)

322. W. Buchmuller, T. Yanagida, Phys. Lett. B **302**, 240 (1993)
323. W. Buchmuller, P. Di Bari, M. Plumacher, Nucl. Phys. B **665**, 445 (2003). arXiv:hep-ph/0302092
324. W. Buchmuller, P. Di Bari, M. Plumacher, Ann. Phys. **315**, 305 (2005). arXiv:hep-ph/0401240
325. W. Buchmuller, P. Di Bari, M. Plumacher, New J. Phys. **6**, 105 (2004). arXiv:hep-ph/0406014
326. T. Hambye, Y. Lin, A. Notari, M. Papucci, A. Strumia, Nucl. Phys. B **695**, 169 (2004). arXiv:hep-ph/0312203
327. WMAP Collaboration, D.N. Spergel et al., arXiv:astro-ph/0603449
328. G.C. Branco, T. Morozumi, B.M. Nobre, M.N. Rebelo, Nucl. Phys. B **617**, 475 (2001). arXiv:hep-ph/0107164
329. J.R. Ellis, M. Raidal, Nucl. Phys. B **643**, 229 (2002). arXiv:hep-ph/0206174
330. S. Davidson, J. Garayoa, F. Palorini, N. Rius, arXiv:0705.1503 [hep-ph]
331. S. Antusch, S.F. King, A. Riotto, J. Cosmol. Astropart. Phys. **0611**, 011 (2006). arXiv:hep-ph/0609038
332. A. Pilaftsis, Phys. Rev. D **56**, 5431 (1997). arXiv:hep-ph/9707235
333. A. Pilaftsis, Nucl. Phys. B **504**, 61 (1997). arXiv:hep-ph/9702393
334. J.R. Ellis, M. Raidal, T. Yanagida, Phys. Lett. B **546**, 228 (2002). arXiv:hep-ph/0206300
335. A. Pilaftsis, T.E.J. Underwood, Nucl. Phys. B **692**, 303 (2004). arXiv:hep-ph/0309342
336. F. del Aguila, J.A. Aguilar-Saavedra, R. Pittau, J. Phys. Conf. Ser. **53**, 506 (2006). arXiv:hep-ph/0606198
337. A. Pilaftsis, T.E.J. Underwood, Phys. Rev. D **72**, 113001 (2005). arXiv:hep-ph/0506107
338. W. Buchmuller, C. Greub, Nucl. Phys. B **381**, 109 (1992)
339. J. Gluza, Phys. Rev. D **66**, 010001 (2002). arXiv:hep-ph/0201002
340. T. Hambye, G. Senjanovic, Phys. Lett. B **582**, 73 (2004). arXiv:hep-ph/0307237
341. T. Hambye, M. Raidal, A. Strumia, Phys. Lett. B **632**, 667 (2006). arXiv:hep-ph/0510008
342. S. Antusch, S.F. King, Phys. Lett. B **597**, 199 (2004). arXiv:hep-ph/0405093
343. S. Antusch, Phys. Rev. D **76**, 023512 (2007). arXiv:0704.1591 [hep-ph]
344. Y. Grossman, T. Kashti, Y. Nir, E. Roulet, Phys. Rev. Lett. **91**, 251801 (2003). arXiv:hep-ph/0307081
345. G. D'Ambrosio, G.F. Giudice, M. Raidal, Phys. Lett. B **575**, 75 (2003). arXiv:hep-ph/0308031
346. G. D'Ambrosio, T. Hambye, A. Hektor, M. Raidal, A. Rossi, Phys. Lett. B **604**, 199 (2004). arXiv:hep-ph/0407312
347. H. Murayama, H. Suzuki, T. Yanagida, J. Yokoyama, Phys. Rev. Lett. **70**, 1912 (1993)
348. H. Murayama, T. Yanagida, Phys. Lett. B **322**, 349 (1994). arXiv:hep-ph/9310297
349. J.R. Ellis, M. Raidal, T. Yanagida, Phys. Lett. B **581**, 9 (2004). arXiv:hep-ph/0303242
350. S. Antusch, M. Bastero-Gil, S.F. King, Q. Shafi, Phys. Rev. D **71**, 083519 (2005). arXiv:hep-ph/0411298
351. K. Dick, M. Lindner, M. Ratz, D. Wright, Phys. Rev. Lett. **84**, 4039 (2000). arXiv:hep-ph/9907562
352. D. Atwood, S. Bar-Shalom, A. Soni, Phys. Lett. B **635**, 112 (2006). arXiv:hep-ph/0502234
353. T. Asaka, K. Hamaguchi, M. Kawasaki, T. Yanagida, Phys. Lett. B **464**, 12 (1999). arXiv:hep-ph/9906366
354. G.F. Giudice, A. Peloso, A. Riotto, I. Tkachev, J. High Energy Phys. **9908**, 014 (1999). arXiv:hep-ph/9905242
355. M. Endo, F. Takahashi, T.T. Yanagida, Phys. Rev. D **74**, 123523 (2006). arXiv:hep-ph/0611055
356. J.S. Schwinger, Ann. Phys. **2**, 407 (1957)
357. H. Georgi, S.L. Glashow, Phys. Rev. Lett. **32**, 438 (1974)
358. P.A.M. Dirac, Phys. Rev. **74**, 817 (1948)
359. G. 't Hooft, Nucl. Phys. B **79**, 276 (1974)
360. A.M. Polyakov, JETP Lett. **20**, 194 (1974) [Pisma Zh. Eksp. Teor. Fiz. **20**, 430 (1974)]
361. J.C. Pati, A. Salam, Phys. Rev. D **10**, 275 (1974)
362. F. Wilczek, A. Zee, Phys. Rev. Lett. **43**, 1571 (1979)
363. K. Kobayashi et al. (Super-Kamiokande Collaboration), Phys. Rev. D **72**, 052007 (2005). arXiv:hep-ex/0502026
364. A.J. Buras, J.R. Ellis, M.K. Gaillard, D.V. Nanopoulos, Nucl. Phys. B **135**, 66 (1978)
365. S. Dimopoulos, S. Raby, F. Wilczek, Phys. Rev. D **24**, 1681 (1981)
366. L.E. Ibanez, G.G. Ross, Phys. Lett. B **105**, 439 (1981)
367. M.B. Einhorn, D.R.T. Jones, Nucl. Phys. B **196**, 475 (1982)
368. W.J. Marciano, G. Senjanovic, Phys. Rev. D **25**, 3092 (1982)
369. K. Inoue, A. Kakuto, H. Komatsu, S. Takeshita, Prog. Theor. Phys. **68**, 927 (1982) [Erratum: Prog. Theor. Phys. **70**, 330 (1983)]
370. L. Alvarez-Gaume, J. Polchinski, M.B. Wise, Nucl. Phys. B **221**, 495 (1983)
371. N. Sakai, T. Yanagida, Nucl. Phys. B **197**, 533 (1982)
372. J. Hisano, H. Murayama, T. Yanagida, Nucl. Phys. B **402**, 46 (1993). arXiv:hep-ph/9207279
373. V. Lucas, S. Raby, Phys. Rev. D **55**, 6986 (1997). arXiv:hep-ph/9610293
374. T. Goto, T. Nihei, Phys. Rev. D **59**, 115009 (1999). arXiv:hep-ph/9808255
375. H. Murayama, A. Pierce, Phys. Rev. D **65**, 055009 (2002). arXiv:hep-ph/0108104
376. B. Bajc, P. Fileviez Perez, G. Senjanovic, Phys. Rev. D **66**, 075005 (2002). arXiv:hep-ph/0204311
377. B. Bajc, P. Fileviez Perez, G. Senjanovic, arXiv:hep-ph/0210374
378. D. Emmanuel-Costa, S. Wiesenfeldt, Nucl. Phys. B **661**, 62 (2003). arXiv:hep-ph/0302272
379. I. Dorsner, P.F. Perez, Nucl. Phys. B **723**, 53 (2005). arXiv:hep-ph/0504276
380. B. Bajc, G. Senjanovic, arXiv:hep-ph/0612029
381. R.J. Wilkes, arXiv:hep-ex/0507097
382. K. Cheung, C.W. Chiang, Phys. Rev. D **71**, 095003 (2005). arXiv:hep-ph/0501265
383. W.Y. Keung, G. Senjanovic, Phys. Rev. Lett. **50**, 1427 (1983)
384. R.N. Mohapatra, J.C. Pati, Phys. Rev. D **11**, 2558 (1975)
385. G. Senjanovic, R.N. Mohapatra, Phys. Rev. D **12**, 1502 (1975)
386. G. Senjanovic, Nucl. Phys. B **153**, 334 (1979)
387. R.N. Mohapatra, Phys. Rev. D **34**, 3457 (1986)
388. A. Font, L.E. Ibanez, F. Quevedo, Phys. Lett. B **228**, 79 (1989)
389. S.P. Martin, Phys. Rev. D **46**, 2769 (1992). arXiv:hep-ph/9207218
390. C.S. Aulakh, K. Benakli, G. Senjanovic, Phys. Rev. Lett. **79**, 2188 (1997). arXiv:hep-ph/9703434
391. C.S. Aulakh, A. Melfo, G. Senjanovic, Phys. Rev. D **57**, 4174 (1998). arXiv:hep-ph/9707256
392. C.S. Aulakh, A. Melfo, A. Rasin, G. Senjanovic, Phys. Lett. B **459**, 557 (1999). arXiv:hep-ph/9902409
393. N.G. Deshpande, E. Keith, P.B. Pal, Phys. Rev. D **46**, 2261 (1993)
394. N.G. Deshpande, E. Keith, P.B. Pal, Phys. Rev. D **47**, 2892 (1993). arXiv:hep-ph/9211232
395. R.N. Mohapatra, B. Sakita, Phys. Rev. D **21**, 1062 (1980)
396. F. Wilczek, A. Zee, Phys. Rev. D **25**, 553 (1982)
397. R. Slansky, Phys. Rep. **79**, 1 (1981)
398. P. Nath, R.M. Syed, Nucl. Phys. B **618**, 138 (2001). arXiv:hep-th/0109116

399. C.S. Aulakh, A. Girdhar, Int. J. Mod. Phys. A **20**, 865 (2005). arXiv:hep-ph/0204097
400. C.H. Albright, S.M. Barr, Phys. Rev. Lett. **85**, 244 (2000). arXiv:hep-ph/0002155
401. K.S. Babu, J.C. Pati, F. Wilczek, Nucl. Phys. B **566**, 33 (2000). arXiv:hep-ph/9812538
402. T. Blazek, S. Raby, K. Tobe, Phys. Rev. D **60**, 113001 (1999). arXiv:hep-ph/9903340
403. E. Witten, Phys. Lett. B **91**, 81 (1980)
404. B. Bajc, G. Senjanovic, Phys. Lett. B **610**, 80 (2005). arXiv:hep-ph/0411193
405. B. Bajc, G. Senjanovic, Phys. Rev. Lett. **95**, 261804 (2005). arXiv:hep-ph/0507169
406. K.S. Babu, R.N. Mohapatra, Phys. Rev. Lett. **70**, 2845 (1993). arXiv:hep-ph/9209215
407. H. Fusaoka, Y. Koide, Phys. Rev. D **57**, 3986 (1998). arXiv:hep-ph/9712201
408. C.R. Das, M.K. Parida, Eur. Phys. J. C **20**, 121 (2001). arXiv:hep-ph/0010004
409. K. Matsuda, Y. Koide, T. Fukuyama, Phys. Rev. D **64**, 053015 (2001). arXiv:hep-ph/0010026
410. K. Matsuda, Y. Koide, T. Fukuyama, H. Nishiura, Phys. Rev. D **65**, 033008 (2002) [Erratum: Phys. Rev. D **65**, 079904 (2002)]. arXiv:hep-ph/0108202
411. T. Fukuyama, N. Okada, J. High Energy Phys. **0211**, 011 (2002). arXiv:hep-ph/0205066
412. H.S. Goh, R.N. Mohapatra, S. Nasri, Phys. Rev. D **70**, 075022 (2004). arXiv:hep-ph/0408139
413. B. Bajc, G. Senjanović, F. Vissani, arXiv:hep-ph/0110310
414. B. Bajc, G. Senjanović, F. Vissani, Phys. Rev. Lett. **90**, 051802 (2003). arXiv:hep-ph/0210207
415. B. Bajc, G. Senjanović, F. Vissani, Phys. Rev. D **70**, 093002 (2004). arXiv:hep-ph/0402140
416. B. Brahmachari, R.N. Mohapatra, Phys. Rev. D **58**, 015001 (1998). arXiv:hep-ph/9710371
417. H.S. Goh, R.N. Mohapatra, S.P. Ng, Phys. Lett. B **570**, 215 (2003). arXiv:hep-ph/0303055
418. H.S. Goh, R.N. Mohapatra, S.P. Ng, Phys. Rev. D **68**, 115008 (2003). arXiv:hep-ph/0308197
419. B. Dutta, Y. Mimura, R.N. Mohapatra, Phys. Rev. D **69**, 115014 (2004). arXiv:hep-ph/0402113
420. S. Bertolini, M. Malinsky, Phys. Rev. D **72**, 055021 (2005). arXiv:hep-ph/0504241
421. K.S. Babu, C. Macesanu, Phys. Rev. D **72**, 115003 (2005). arXiv:hep-ph/0505200
422. T.E. Clark, T.K. Kuo, N. Nakagawa, Phys. Lett. B **115**, 26 (1982)
423. C.S. Aulakh, R.N. Mohapatra, Phys. Rev. D **28**, 217 (1983)
424. D.G. Lee, R.N. Mohapatra, Phys. Rev. D **51**, 1353 (1995). arXiv:hep-ph/9406328
425. C.S. Aulakh, B. Bajc, A. Melfo, G. Senjanović, F. Vissani, Phys. Lett. B **588**, 196 (2004). arXiv:hep-ph/0306242
426. C.S. Aulakh, arXiv:hep-ph/0506291
427. B. Bajc, A. Melfo, G. Senjanović, F. Vissani, Phys. Lett. B **634**, 272 (2006). arXiv:hep-ph/0511352
428. C.S. Aulakh, S.K. Garg, arXiv:hep-ph/0512224
429. S. Bertolini, T. Schwetz, M. Malinsky, Phys. Rev. D **73**, 115012 (2006). arXiv:hep-ph/0605006
430. L.J. Hall, R. Rattazzi, U. Sarid, Phys. Rev. D **50**, 7048 (1994). arXiv:hep-ph/9306309
431. X.G. He, S. Meljanac, Phys. Rev. D **41**, 1620 (1990)
432. D.G. Lee, Phys. Rev. D **49**, 1417 (1994)
433. B. Bajc, A. Melfo, G. Senjanović, F. Vissani, Phys. Rev. D **70**, 035007 (2004). arXiv:hep-ph/0402122
434. C.S. Aulakh, A. Girdhar, Nucl. Phys. B **711**, 275 (2005). arXiv:hep-ph/0405074
435. T. Fukuyama, A. Ilakovac, T. Kikuchi, S. Meljanac, N. Okada, Eur. Phys. J. C **42**, 191 (2005). arXiv:hep-ph/0401213
436. T. Fukuyama, A. Ilakovac, T. Kikuchi, S. Meljanac, N. Okada, J. Math. Phys. **46**, 033505 (2005). arXiv:hep-ph/0405300
437. T. Fukuyama, A. Ilakovac, T. Kikuchi, S. Meljanac, N. Okada, Phys. Rev. D **72**, 051701 (2005). arXiv:hep-ph/0412348
438. C.S. Aulakh, arXiv:hep-ph/0501025
439. C.S. Aulakh, Phys. Rev. D **72**, 051702 (2005)
440. D.G. Lee, R.N. Mohapatra, Phys. Lett. B **324**, 376 (1994). arXiv:hep-ph/9310371
441. K.S. Babu, I. Gogoladze, Z. Tavartkiladze, arXiv:hep-ph/0612315
442. H.S. Goh, R.N. Mohapatra, S. Nasri, S.P. Ng, Phys. Lett. B **587**, 105 (2004). arXiv:hep-ph/0311330
443. T. Fukuyama, A. Ilakovac, T. Kikuchi, S. Meljanac, N. Okada, J. High Energy Phys. **0409**, 052 (2004). arXiv:hep-ph/0406068
444. S. Bertolini, M. Frigerio, M. Malinsky, Phys. Rev. D **70**, 095002 (2004). arXiv:hep-ph/0406117
445. W.M. Yang, Z.G. Wang, Nucl. Phys. B **707**, 87 (2005). arXiv:hep-ph/0406221
446. B. Dutta, Y. Mimura, R.N. Mohapatra, Phys. Lett. B **603**, 35 (2004). arXiv:hep-ph/0406262
447. B. Dutta, Y. Mimura, R.N. Mohapatra, Phys. Rev. Lett. **94**, 091804 (2005). arXiv:hep-ph/0412105
448. B. Dutta, Y. Mimura, R.N. Mohapatra, Phys. Rev. D **72**, 075009 (2005). arXiv:hep-ph/0507319
449. W. Grimus, H. Kuhbock, arXiv:hep-ph/0607197
450. W. Grimus, H. Kuhbock, arXiv:hep-ph/0612132
451. C.S. Aulakh, arXiv:hep-ph/0602132
452. L. Lavoura, H. Kuhbock, W. Grimus, arXiv:hep-ph/0603259
453. C.S. Aulakh, arXiv:hep-ph/0607252
454. C.S. Aulakh, S.K. Garg, arXiv:hep-ph/0612021
455. B. Bajc, A. Melfo, G. Senjanović, F. Vissani, Phys. Rev. D **73**, 055001 (2006). arXiv:hep-ph/0510139
456. F. Vissani, A.Y. Smirnov, Phys. Lett. B **341**, 173 (1994). arXiv:hep-ph/9405399
457. N. Arkani-Hamed, S. Dimopoulos, J. High Energy Phys. **0506**, 073 (2005). arXiv:hep-th/0405159
458. G.F. Giudice, A. Romanino, Nucl. Phys. B **699**, 65 (2004) [Erratum: Nucl. Phys. B **706**, 65 (2005)]. arXiv:hep-ph/0406088
459. B. Bajc, AIP Conf. Proc. **805**, 326 (2006). arXiv:hep-ph/0602166
460. A. Arvanitaki, C. Davis, P.W. Graham, A. Pierce, J.G. Wacker, Phys. Rev. D **72**, 075011 (2005). arXiv:hep-ph/0504210
461. K. Yoshioka, Mod. Phys. Lett. A **15**, 29 (2000). arXiv:hep-ph/9904433
462. M. Bando, T. Kobayashi, T. Noguchi, K. Yoshioka, Phys. Rev. D **63**, 113017 (2001). arXiv:hep-ph/0008120
463. A. Neronov, Phys. Rev. D **65**, 044004 (2002). arXiv:gr-qc/0106092
464. N. Arkani-Hamed, L.J. Hall, D.R. Smith, N. Weiner, Phys. Rev. D **61**, 116003 (2000). arXiv:hep-ph/9909326
465. K.R. Dienes, E. Dudas, T. Gherghetta, Phys. Lett. B **436**, 55 (1998). arXiv:hep-ph/9803466
466. K.R. Dienes, E. Dudas, T. Gherghetta, Nucl. Phys. B **537**, 47 (1999). arXiv:hep-ph/9806292
467. N. Arkani-Hamed, M. Schmaltz, Phys. Rev. D **61**, 033005 (2000). arXiv:hep-ph/9903417
468. T. Gherghetta, A. Pomarol, Nucl. Phys. B **586**, 141 (2000). arXiv:hep-ph/0003129
469. N. Arkani-Hamed, S. Dimopoulos, G.R. Dvali, Phys. Lett. B **429**, 263 (1998). arXiv:hep-ph/9803315
470. I. Antoniadis, N. Arkani-Hamed, S. Dimopoulos, G.R. Dvali, Phys. Lett. B **436**, 257 (1998). arXiv:hep-ph/9804398
471. N. Arkani-Hamed, S. Dimopoulos, G.R. Dvali, Phys. Rev. D **59**, 086004 (1999). arXiv:hep-ph/9807344

472. G. Barenboim, G.C. Branco, A. de Gouvea, M.N. Rebelo, Phys. Rev. D **64**, 073005 (2001). arXiv:hep-ph/0104312
473. T. Appelquist, H.C. Cheng, B.A. Dobrescu, Phys. Rev. D **64**, 035002 (2001). arXiv:hep-ph/0012100
474. M.V. Libanov, S.V. Troitsky, Nucl. Phys. B **599**, 319 (2001). arXiv:hep-ph/0011095
475. J.M. Frere, M.V. Libanov, S.V. Troitsky, Phys. Lett. B **512**, 169 (2001). arXiv:hep-ph/0012306
476. J.M. Frere, M.V. Libanov, S.V. Troitsky, J. High Energy Phys. **0111**, 025 (2001). arXiv:hep-ph/0110045
477. M.V. Libanov, E.Y. Nougaev, J. High Energy Phys. **0204**, 055 (2002). arXiv:hep-ph/0201162
478. G.R. Dvali, M.A. Shifman, Phys. Lett. B **475**, 295 (2000). arXiv:hep-ph/0001072
479. P.Q. Hung, Phys. Rev. D **67**, 095011 (2003). arXiv:hep-ph/0210131
480. D.E. Kaplan, T.M.P. Tait, J. High Energy Phys. **0006**, 020 (2000). arXiv:hep-ph/0004200
481. D.E. Kaplan, T.M.P. Tait, J. High Energy Phys. **0111**, 051 (2001). arXiv:hep-ph/0110126
482. M. Kakizaki, M. Yamaguchi, Prog. Theor. Phys. **107**, 433 (2002). arXiv:hep-ph/0104103
483. C.H. Chang, W.F. Chang, J.N. Ng, Phys. Lett. B **558**, 92 (2003). arXiv:hep-ph/0301271
484. S. Nussinov, R. Shrock, Phys. Lett. B **526**, 137 (2002). arXiv:hep-ph/0101340
485. R. Jackiw, C. Rebbi, Phys. Rev. D **13**, 3398 (1976)
486. E.A. Mirabelli, M. Schmaltz, Phys. Rev. D **61**, 113011 (2000). arXiv:hep-ph/9912265
487. G.C. Branco, A. de Gouvea, M.N. Rebelo, Phys. Lett. B **506**, 115 (2001). arXiv:hep-ph/0012289
488. P.Q. Hung, M. Seco, Nucl. Phys. B **653**, 123 (2003). arXiv:hep-ph/0111013
489. J.M. Frere, G. Moreau, E. Nezri, Phys. Rev. D **69**, 033003 (2004). arXiv:hep-ph/0309218
490. H.V. Klapdor-Kleingrothaus, U. Sarkar, Phys. Lett. B **541**, 332 (2002). arXiv:hep-ph/0201226
491. M. Gogberashvili, Int. J. Mod. Phys. D **11**, 1635 (2002). arXiv:hep-ph/9812296
492. L. Randall, R. Sundrum, Phys. Rev. Lett. **83**, 3370 (1999). arXiv:hep-ph/9905221
493. Y. Grossman, M. Neubert, Phys. Lett. B **474**, 361 (2000). arXiv:hep-ph/9912408
494. H. Davoudiasl, J.L. Hewett, T.G. Rizzo, Phys. Lett. B **473**, 43 (2000). arXiv:hep-ph/9911262
495. A. Pomarol, Phys. Lett. B **486**, 153 (2000). arXiv:hep-ph/9911294
496. S. Chang, J. Hisano, H. Nakano, N. Okada, M. Yamaguchi, Phys. Rev. D **62**, 084025 (2000). arXiv:hep-ph/9912498
497. B. Bajc, G. Gabadadze, Phys. Lett. B **474**, 282 (2000). arXiv:hep-th/9912232
498. A. Pomarol, Phys. Rev. Lett. **85**, 4004 (2000). arXiv:hep-ph/0005293
499. L. Randall, M.D. Schwartz, J. High Energy Phys. **0111**, 003 (2001). arXiv:hep-th/0108114
500. L. Randall, M.D. Schwartz, Phys. Rev. Lett. **88**, 081801 (2002). arXiv:hep-th/0108115
501. W.D. Goldberger, I.Z. Rothstein, Phys. Rev. D **68**, 125011 (2003). arXiv:hep-th/0208060
502. K.W. Choi, I.W. Kim, Phys. Rev. D **67**, 045005 (2003). arXiv:hep-th/0208071
503. K. Agashe, A. Delgado, R. Sundrum, Ann. Phys. **304**, 145 (2003). arXiv:hep-ph/0212028
504. K. Agashe, G. Servant, Phys. Rev. Lett. **93**, 231805 (2004). arXiv:hep-ph/0403143
505. K. Agashe, G. Servant, J. Cosmol. Astropart. Phys. **0502**, 002 (2005). arXiv:hep-ph/0411254
506. S.J. Huber, Q. Shafi, Phys. Lett. B **498**, 256 (2001). arXiv:hep-ph/0010195
507. S.J. Huber, Nucl. Phys. B **666**, 269 (2003). arXiv:hep-ph/0303183
508. S. Chang, C.S. Kim, M. Yamaguchi, Phys. Rev. D **73**, 033002 (2006). arXiv:hep-ph/0511099
509. S.J. Huber, Q. Shafi, Phys. Lett. B **544**, 295 (2002). arXiv:hep-ph/0205327
510. S.J. Huber, Q. Shafi, Phys. Lett. B **583**, 293 (2004). arXiv:hep-ph/0309252
511. S.J. Huber, Q. Shafi, Phys. Lett. B **512**, 365 (2001). arXiv:hep-ph/0104293
512. G. Moreau, J.I. Silva-Marcos, J. High Energy Phys. **0601**, 048 (2006). arXiv:hep-ph/0507145
513. G. Moreau, J.I. Silva-Marcos, J. High Energy Phys. **0603**, 090 (2006). arXiv:hep-ph/0602155
514. A. Ilakovac, A. Pilaftsis, Nucl. Phys. B **437**, 491 (1995). arXiv:hep-ph/9403398
515. C.S. Kim, J.D. Kim, J.h. Song, Phys. Rev. D **67**, 015001 (2003). arXiv:hep-ph/0204002
516. F. del Aguila, M. Perez-Victoria, J. Santiago, Phys. Lett. B **492**, 98 (2000). arXiv:hep-ph/0007160
517. A. Delgado, A. Pomarol, M. Quiros, J. High Energy Phys. **0001**, 030 (2000). arXiv:hep-ph/9911252
518. T.G. Rizzo, J.D. Wells, Phys. Rev. D **61**, 016007 (2000). arXiv:hep-ph/9906234
519. I. Antoniadis, K. Benakli, M. Quiros, Phys. Lett. B **460**, 176 (1999). arXiv:hep-ph/9905311
520. P. Nath, Y. Yamada, M. Yamaguchi, Phys. Lett. B **466**, 100 (1999). arXiv:hep-ph/9905415
521. T.G. Rizzo, Phys. Rev. D **61**, 055005 (2000). arXiv:hep-ph/9909232
522. G. Burdman, Phys. Lett. B **590**, 86 (2004). arXiv:hep-ph/0310144
523. K. Agashe, G. Perez, A. Soni, Phys. Rev. Lett. **93**, 201804 (2004). arXiv:hep-ph/0406101
524. K. Agashe, G. Perez, A. Soni, Phys. Rev. D **71**, 016002 (2005). arXiv:hep-ph/0408134
525. K. Agashe, A.E. Blechman, F. Petriello, Phys. Rev. D **74**, 053011 (2006). arXiv:hep-ph/0606021
526. K. Agashe, G. Perez, A. Soni, arXiv:hep-ph/0606293
527. F. del Aguila, M. Perez-Victoria, J. Santiago, J. High Energy Phys. **0302**, 051 (2003). arXiv:hep-th/0302023
528. M. Carena, T.M.P. Tait, C.E.M. Wagner, Acta Phys. Pol. B **33**, 2355 (2002). arXiv:hep-ph/0207056
529. M. Carena, E. Ponton, T.M.P. Tait, C.E.M. Wagner, Phys. Rev. D **67**, 096006 (2003). arXiv:hep-ph/0212307
530. M. Carena, A. Delgado, E. Ponton, T.M.P. Tait, C.E.M. Wagner, Phys. Rev. D **68**, 035010 (2003). arXiv:hep-ph/0305188
531. M. Carena, A. Delgado, E. Ponton, T.M.P. Tait, C.E.M. Wagner, Phys. Rev. D **71**, 015010 (2005). arXiv:hep-ph/0410344
532. K. Agashe, A. Delgado, M.J. May, R. Sundrum, J. High Energy Phys. **0308**, 050 (2003). arXiv:hep-ph/0308036
533. K. Agashe, R. Contino, L. Da Rold, A. Pomarol, Phys. Lett. B **641**, 62 (2006). arXiv:hep-ph/0605341
534. A. Djouadi, G. Moreau, F. Richard, arXiv:hep-ph/0610173
535. R.S. Chivukula, H. Georgi, Phys. Lett. B **188**, 99 (1987)
536. L.J. Hall, L. Randall, Phys. Rev. Lett. **65**, 2939 (1990)
537. G. D'Ambrosio, G.F. Giudice, G. Isidori, A. Strumia, Nucl. Phys. B **645**, 155 (2002). hep-ph/0207036
538. A.V. Manohar, M.B. Wise, Phys. Rev. D **74**, 035009 (2006). hep-ph/0606172
539. B. Grinstein, V. Cirigliano, G. Isidori, M.B. Wise, Nucl. Phys. B **763**, 35 (2007). hep-ph/0608123
540. A.J. Buras, P. Gambino, M. Gorbahn, S. Jager, L. Silvestrini, Phys. Lett. B **500**, 161 (2001). hep-ph/0007085

541. M. Bona et al. (UTfit Collaboration), J. High Energy Phys. **0603**, 080 (2006). hep-ph/0509219
542. V. Cirigliano, B. Grinstein, G. Isidori, M.B. Wise, Nucl. Phys. B **728**, 121 (2005). hep-ph/0507001
543. V. Cirigliano, B. Grinstein, Nucl. Phys. B **752**, 18 (2006). hep-ph/0601111
544. S. Davidson, F. Palorini, Phys. Lett. B **642**, 72 (2006). hep-ph/0607329
545. S. Pascoli, S.T. Petcov, C.E. Yaguna, Phys. Lett. B **564**, 241 (2003). hep-ph/0301095
546. V. Cirigliano, G. Isidori, V. Porretti, Nucl. Phys. B **763**, 228 (2007). hep-ph/0608123
547. S. Blanchet, P. Di Bari, J. Cosmol. Astropart. Phys. **0606**, 023 (2006). hep-ph/0603107
548. M. Grassi (MEG Collaboration), Nucl. Phys. Proc. Suppl. **149**, 369 (2005)
549. W. Grimus, M.N. Rebelo, Phys. Rep. **281**, 239 (1997). arXiv:hep-ph/9506272
550. F. Englert, R. Brout, Phys. Rev. Lett. **13**, 321 (1964)
551. G.S. Guralnik, C.R. Hagen, T.W.B. Kibble, Phys. Rev. Lett. **13**, 585 (1964)
552. P.W. Higgs, Phys. Lett. **12**, 132 (1964)
553. P.W. Higgs, Phys. Rev. **145**, 1156 (1966)
554. T.D. Lee, Phys. Rev. D **8**, 1226 (1973)
555. W. Bernreuther, O. Nachtmann, Eur. Phys. J. C **9**, 319 (1999). arXiv:hep-ph/9812259
556. G.C. Branco, L. Lavoura, J.P. Silva, *CP Violation* (Clarendon, Oxford, 1999)
557. G.C. Branco, M.N. Rebelo, J.I. Silva-Marcos, Phys. Lett. B **614**, 187 (2005). arXiv:hep-ph/0502118
558. J. Bernabeu, G.C. Branco, M. Gronau, Phys. Lett. B **169**, 243 (1986)
559. L. Lavoura, J.P. Silva, Phys. Rev. D **50**, 4619 (1994). arXiv:hep-ph/9404276
560. F.J. Botella, J.P. Silva, Phys. Rev. D **51**, 3870 (1995). arXiv:hep-ph/9411288
561. S. Davidson, H.E. Haber, Phys. Rev. D **72**, 035004 (2005) [Erratum: Phys. Rev. D **72**, 099902 (2005)]. arXiv:hep-ph/0504050
562. J.F. Gunion, H.E. Haber, Phys. Rev. D **72**, 095002 (2005). arXiv:hep-ph/0506227
563. I.P. Ivanov, Phys. Lett. B **632**, 360 (2006). arXiv:hep-ph/0507132
564. H.E. Haber, D. O'Neil, Phys. Rev. D **74**, 015018 (2006). arXiv:hep-ph/0602242
565. G.C. Branco, J.M. Gerard, W. Grimus, Phys. Lett. B **136**, 383 (1984)
566. S.L. Glashow, S. Weinberg, Phys. Rev. D **15**, 1958 (1977)
567. E.A. Paschos, Phys. Rev. D **15**, 1966 (1977)
568. G.C. Branco, M.N. Rebelo, Phys. Lett. B **160**, 117 (1985)
569. L. Lavoura, Phys. Rev. D **50**, 7089 (1994). arXiv:hep-ph/9405307
570. I.F. Ginzburg, M. Krawczyk, Phys. Rev. D **72**, 115013 (2005). arXiv:hep-ph/0408011
571. J.F. Gunion, H.E. Haber, Phys. Rev. D **67**, 075019 (2003). arXiv:hep-ph/0207010
572. S. Weinberg, Phys. Rev. Lett. **37**, 657 (1976)
573. G.C. Branco, Phys. Rev. Lett. **44**, 504 (1980)
574. G.C. Branco, Phys. Rev. D **22**, 2901 (1980)
575. G.C. Branco, A.J. Buras, J.M. Gerard, Nucl. Phys. B **259**, 306 (1985)
576. L. Bento, G.C. Branco, P.A. Parada, Phys. Lett. B **267**, 95 (1991)
577. A.E. Nelson, Phys. Lett. B **136**, 387 (1984)
578. A.E. Nelson, Phys. Lett. B **143**, 165 (1984)
579. S.M. Barr, Phys. Rev. Lett. **53**, 329 (1984)
580. G.C. Branco, P.A. Parada, M.N. Rebelo, arXiv:hep-ph/0307119
581. Y. Achiman, Phys. Lett. B **599**, 75 (2004). arXiv:hep-ph/0403309
582. M.B. Gavela, P. Hernandez, J. Orloff, O. Pene, C. Quimbay, Nucl. Phys. B **430**, 382 (1994). arXiv:hep-ph/9406289
583. P. Huet, E. Sather, Phys. Rev. D **51**, 379 (1995). arXiv:hep-ph/9404302
584. G.W. Anderson, L.J. Hall, Phys. Rev. D **45**, 2685 (1992)
585. W. Buchmuller, Z. Fodor, T. Helbig, D. Walliser, Ann. Phys. **234**, 260 (1994). arXiv:hep-ph/9303251
586. K. Kajantie, M. Laine, K. Rummukainen, M.E. Shaposhnikov, Nucl. Phys. B **466**, 189 (1996). arXiv:hep-lat/9510020
587. L. Fromme, S.J. Huber, M. Seniuch, J. High Energy Phys. **0611**, 038 (2006). arXiv:hep-ph/0605242
588. E. Accomando et al., arXiv:hep-ph/0608079
589. L. Brucher, R. Santos, Eur. Phys. J. C **12**, 87 (2000). arXiv:hep-ph/9907434
590. C. Delaere (LEP Collaboration), PoS **HEP2005**, 331 (2006)
591. A.G. Akeroyd, M.A. Diaz, F.J. Pacheco, Phys. Rev. D **70**, 075002 (2004). arXiv:hep-ph/0312231
592. N.G. Deshpande, E. Ma, Phys. Rev. D **18**, 2574 (1978)
593. Q.H. Cao, E. Ma, G. Rajasekaran, arXiv:0708.2939 [hep-ph]
594. R. Barbieri, L.J. Hall, V.S. Rychkov, Phys. Rev. D **74**, 015007 (2006). arXiv:hep-ph/0603188
595. I.F. Ginzburg, M. Krawczyk, P. Osland, Nucl. Instrum. Methods A **472**, 149 (2001). arXiv:hep-ph/0101229
596. T.P. Cheng, M. Sher, Phys. Rev. D **35**, 3484 (1987)
597. G.C. Branco, W. Grimus, L. Lavoura, Phys. Lett. B **380**, 119 (1996). arXiv:hep-ph/9601383
598. A.K. Das, C. Kao, Phys. Lett. B **372**, 106 (1996). arXiv:hep-ph/9511329
599. E. Lunghi, A. Soni, arXiv:0707.0212 [hep-ph]
600. T.P. Cheng, L.F. Li, Phys. Rev. Lett. **38**, 381 (1977)
601. S.M. Bilenky, S.T. Petcov, B. Pontecorvo, Phys. Lett. B **67**, 309 (1977)
602. J.D. Bjorken, S. Weinberg, Phys. Rev. Lett. **38**, 622 (1977)
603. G.C. Branco, Phys. Lett. B **68**, 455 (1977)
604. D. Besson et al. (CLEO Collaboration), Phys. Rev. Lett. **98**, 052002 (2007). arXiv:hep-ex/0607019
605. R.N. Mohapatra, J.W.F. Valle, Phys. Rev. D **34**, 1642 (1986)
606. S. Nandi, U. Sarkar, Phys. Rev. Lett. **56**, 564 (1986)
607. P.F. Harrison, D.H. Perkins, W.G. Scott, Phys. Lett. B **458**, 79 (1999). arXiv:hep-ph/9904297
608. A. Pilaftsis, Z. Phys. C **55**, 275 (1992). arXiv:hep-ph/9901206
609. N. Arkani-Hamed, L.J. Hall, H. Murayama, D.R. Smith, N. Weiner, Phys. Rev. D **64**, 115011 (2001). arXiv:hep-ph/0006312
610. F. Borzumati, Phys. Rev. D **64**, 053005 (2001). arXiv:hep-ph/0007018
611. S. Bergmann, A. Kagan, Nucl. Phys. B **538**, 368 (1999). arXiv:hep-ph/9803305
612. A. Ioannisian, A. Pilaftsis, Phys. Rev. D **62**, 066001 (2000). arXiv:hep-ph/9907522
613. D.A. Dicus, D.D. Karatas, P. Roy, Phys. Rev. D **44**, 2033 (1991)
614. A. Datta, M. Guchait, A. Pilaftsis, Phys. Rev. D **50**, 3195 (1994). arXiv:hep-ph/9311257
615. F.M.L. Almeida, Y.D.A. Coutinho, J.A. Martins Simoes, M.A.B. do Vale, Phys. Rev. D **62**, 075004 (2000). arXiv:hep-ph/0002024
616. O. Panella, M. Cannoni, C. Carimalo, Y.N. Srivastava, Phys. Rev. D **65**, 035005 (2002). arXiv:hep-ph/0107308
617. T. Han, B. Zhang, Phys. Rev. Lett. **97**, 171804 (2006). arXiv:hep-ph/0604064
618. S. Bray, J.S. Lee, A. Pilaftsis, arXiv:hep-ph/0702294
619. F. del Aguila, J.A. Aguilar-Saavedra, R. Pittau, arXiv:hep-ph/0703261
620. N. Arkani-Hamed, A.G. Cohen, H. Georgi, Phys. Rev. Lett. **86**, 4757 (2001). arXiv:hep-th/0104005

621. H.C. Cheng, C.T. Hill, S. Pokorski, J. Wang, Phys. Rev. D **64**, 065007 (2001). arXiv:hep-th/0104179
622. N. Arkani-Hamed, A.G. Cohen, H. Georgi, Phys. Lett. B **513**, 232 (2001). arXiv:hep-ph/0105239
623. M. Schmaltz, D. Tucker-Smith, Annu. Rev. Nucl. Part. Sci. **55**, 229 (2005). arXiv:hep-ph/0502182
624. M. Perelstein, Prog. Part. Nucl. Phys. **58**, 247 (2007). arXiv:hep-ph/0512128
625. N. Arkani-Hamed, A.G. Cohen, E. Katz, A.E. Nelson, J. High Energy Phys. **0207**, 034 (2002). arXiv:hep-ph/0206021
626. S. Chang, J. High Energy Phys. **0312**, 057 (2003). arXiv:hep-ph/0306034
627. H.C. Cheng, I. Low, J. High Energy Phys. **0309**, 051 (2003). arXiv:hep-ph/0308199
628. H.C. Cheng, I. Low, J. High Energy Phys. **0408**, 061 (2004). arXiv:hep-ph/0405243
629. J. Hubisz, P. Meade, A. Noble, M. Perelstein, J. High Energy Phys. **0601**, 135 (2006). arXiv:hep-ph/0506042
630. J. Hubisz, P. Meade, Phys. Rev. D **71**, 035016 (2005). arXiv:hep-ph/0411264
631. I. Low, J. High Energy Phys. **0410**, 067 (2004). arXiv:hep-ph/0409025
632. J. Hubisz, S.J. Lee, G. Paz, J. High Energy Phys. **0606**, 041 (2006). arXiv:hep-ph/0512169
633. M. Blanke, A.J. Buras, A. Poschenrieder, S. Recksiegel, C. Tarantino, S. Uhlig, A. Weiler, Phys. Lett. B **646**, 253 (2007). arXiv:hep-ph/0609284
634. S. Yamada, Nucl. Phys. Proc. Suppl. **144**, 185 (2005)
635. http://meg.web.psi.ch/
636. S.C. Park, J.h. Song, arXiv:hep-ph/0306112
637. R. Casalbuoni, A. Deandrea, M. Oertel, J. High Energy Phys. **0402**, 032 (2004). arXiv:hep-ph/0311038
638. M. Blanke, A.J. Buras, A. Poschenrieder, S. Recksiegel, C. Tarantino, S. Uhlig, A. Weiler, J. High Energy Phys. **0701**, 066 (2007). arXiv:hep-ph/0610298
639. M. Blanke, A.J. Buras, B. Duling, A. Poschenrieder, C. Tarantino, arXiv:hep-ph/0702136
640. P.Q. Hung, Phys. Lett. B **659**, 585 (2008). arXiv:0711.0733 [hep-ph]
641. M. Blanke, A.J. Buras, A. Poschenrieder, C. Tarantino, S. Uhlig, A. Weiler, J. High Energy Phys. **0612**, 003 (2006). arXiv:hep-ph/0605214
642. A.J. Buras, arXiv:hep-ph/9806471
643. S.R. Choudhury, A.S. Cornell, A. Deandrea, N. Gaur, A. Goyal, Phys. Rev. D **75**, 055011 (2007). arXiv:hep-ph/0612327
644. B. Aubert et al. (BaBar Collaboration), Phys. Rev. Lett. **92**, 121801 (2004). arXiv:hep-ex/0312027
645. Y. Yusa et al. (Belle Collaboration), Phys. Lett. B **589**, 103 (2004). arXiv:hep-ex/0403039
646. Y. Enari et al.(Belle Collaboration), Phys. Lett. B **622**, 218 (2005). arXiv:hep-ex/0503041
647. K. Abe et al. (Belle Collaboration), arXiv:hep-ex/0609049
648. B. Aubert et al.(BaBar Collaboration), Phys. Rev. Lett. **98**, 061803 (2007). arXiv:hep-ex/0610067
649. E. Arganda, M.J. Herrero, Phys. Rev. D **73**, 055003 (2006). arXiv:hep-ph/0510405
650. P. Paradisi, J. High Energy Phys. **0602**, 050 (2006). arXiv:hep-ph/0508054
651. Y. Kuno, Y. Okada, Phys. Rev. Lett. **77**, 434 (1996). arXiv:hep-ph/9604296
652. Y. Farzan, arXiv:hep-ph/0701106
653. A. Czarnecki, W.J. Marciano, A. Vainshtein, Phys. Rev. D **67**, 073006 (2003) [Erratum: Phys. Rev. D **73**, 119901 (2006)]. arXiv:hep-ph/0212229
654. J. Lee, arXiv:hep-ph/0504136
655. T. Han, H.E. Logan, B. Mukhopadhyaya, R. Srikanth, Phys. Rev. D **72**, 053007 (2005). arXiv:hep-ph/0505260
656. F. del Aguila, M. Masip, J.L. Padilla, Phys. Lett. B **627**, 131 (2005). arXiv:hep-ph/0506063
657. S.R. Choudhury, N. Gaur, A. Goyal, Phys. Rev. D **72**, 097702 (2005). arXiv:hep-ph/0508146
658. A. Abada, G. Bhattacharyya, M. Losada, Phys. Rev. D **73**, 033006 (2006). arXiv:hep-ph/0511275
659. G.F. Giudice, C. Grojean, A. Pomarol, R. Rattazzi, arXiv:hep-ph/0703164
660. D. Acosta et al. (CDF Collaboration), Phys. Rev. Lett. **93**, 221802 (2004). arXiv:hep-ex/0406073
661. V.M. Abazov et al. (D0 Collaboration), Phys. Rev. Lett. **93**, 141801 (2004). arXiv:hep-ex/0404015
662. K. Huitu, J. Maalampi, A. Pietila, M. Raidal, Nucl. Phys. B **487**, 27 (1997). arXiv:hep-ph/9606311
663. C.X. Yue, S. Zhao, arXiv:hep-ph/0701017
664. E. Ma, M. Raidal, U. Sarkar, Phys. Rev. Lett. **85**, 3769 (2000). arXiv:hep-ph/0006046
665. E. Ma, M. Raidal, U. Sarkar, Nucl. Phys. B **615**, 313 (2001). arXiv:hep-ph/0012101
666. E. Ma, U. Sarkar, Phys. Lett. B **638**, 356 (2006). arXiv:hep-ph/0602116
667. N. Sahu, U. Sarkar, arXiv:hep-ph/0701062
668. G. Marandella, C. Schappacher, A. Strumia, Phys. Rev. D **72**, 035014 (2005). arXiv:hep-ph/0502096
669. T. Han, H.E. Logan, B. McElrath, L.T. Wang, Phys. Rev. D **67**, 095004 (2003). arXiv:hep-ph/0301040
670. T. Han, H.E. Logan, L.T. Wang, J. High Energy Phys. **0601**, 099 (2006). arXiv:hep-ph/0506313
671. G. Azuelos et al., Eur. Phys. J. C **39S2**, 13 (2005). arXiv:hep-ph/0402037
672. J.F. Gunion, C. Loomis, K.T. Pitts, in *Proceedings of 1996 DPF/DPB Summer Study on New Directions for High-Energy Physics (Snowmass 96)*, Snowmass, CO, 25 June–12 July 1996, p. LTH096. arXiv:hep-ph/9610237
673. J.F. Gunion, J. Grifols, A. Mendez, B. Kayser, F.I. Olness, Phys. Rev. D **40**, 1546 (1989)
674. M. Muhlleitner, M. Spira, Phys. Rev. D **68**, 117701 (2003). arXiv:hep-ph/0305288
675. A.G. Akeroyd, M. Aoki, Phys. Rev. D **72**, 035011 (2005). arXiv:hep-ph/0506176
676. T. Rommerskirchen, T. Hebbeker, J. Phys. G **33**, N47 (2007)
677. A. Hektor, M. Kadastik, M. Muntel, M. Raidal, L. Rebane, arXiv:0705.1495 [hep-ph]
678. T. Han, B. Mukhopadhyaya, Z. Si, K. Wang, arXiv:0706.0441 [hep-ph]
679. F. Gabbiani, A. Masiero, Nucl. Phys. B **322**, 235 (1989)
680. F. Gabbiani, E. Gabrielli, A. Masiero, L. Silvestrini, Nucl. Phys. B **477**, 321 (1996). arXiv:hep-ph/9604387
681. P. Paradisi, J. High Energy Phys. **0510**, 006 (2005). arXiv:hep-ph/0505046
682. R. Barbieri, L.J. Hall, Phys. Lett. B **338**, 212 (1994). arXiv:hep-ph/9408406
683. R. Barbieri, L.J. Hall, A. Strumia, Nucl. Phys. B **445**, 219 (1995). arXiv:hep-ph/9501334
684. I. Masina, C.A. Savoy, Phys. Rev. D **71**, 093003 (2005). arXiv:hep-ph/0501166
685. B.C. Allanach et al., in *Proceedings of the APS/DPF/DPB Summer Study on the Future of Particle Physics (Snowmass 2001)*, ed. by N. Graf, Snowmass, CO, 30 June–21 July 2001, p. P125. arXiv:hep-ph/0202233
686. S. Antusch, E. Arganda, M.J. Herrero, A.M. Teixeira, J. High Energy Phys. **0611**, 090 (2006). arXiv:hep-ph/0607263
687. F. Deppisch, H. Päs, A. Redelbach, R. Rückl, Phys. Rev. D **73**, 033004 (2006). arXiv:hep-ph/0511062
688. L. Calibbi, A. Faccia, A. Masiero, S.K. Vempati, arXiv:hep-ph/0610241

689. M. Maltoni, T. Schwetz, M.A. Tortola, J.W.F. Valle, Phys. Rev. D **68**, 113010 (2003). arXiv:hep-ph/0309130
690. S. Eidelman et al. (PDG Collaboration), Phys. Lett. B **592**, 1 (2004). arXiv:hep-ph/0310053
691. S.T. Petcov, T. Shindou, Phys. Rev. D **74**, 073006 (2006). arXiv:hep-ph/0605151
692. G. Weiglein et al. (LHC/LC Study Group), Phys. Rep. **426**, 47 (2006). arXiv:hep-ph/0410364
693. F. Deppisch, H. Päs, A. Redelbach, R. Rückl, Y. Shimizu, arXiv:hep-ph/0206122
694. A.G. Akeroyd et al. (SuperKEKB Physics Working Group), arXiv:hep-ex/0406071
695. F. Deppisch, H. Päs, A. Redelbach, R. Rückl, Y. Shimizu, Phys. Rev. D **69**, 054014 (2004). arXiv:hep-ph/0310053
696. K. Agashe, M. Graesser, Phys. Rev. D **61**, 075008 (2000). arXiv:hep-ph/9904422
697. Yu.M. Andreev, S.I. Bityukov, N.V. Krasnikov, A.N. Toropin, arXiv:hep-ph/0608176
698. A. Bartl, K. Hidaka, K. Hohenwarter-Sodek, T. Kernreiter, W. Majerotto, W. Porod, Eur. Phys. J. C **46**, 783 (2006). arXiv:hep-ph/0510074
699. W. Porod, Comput. Phys. Commun. **153**, 275 (2003). arXiv:hep-ph/0301101
700. U. Bellgardt et al. (SINDRUM Collaboration), Nucl. Phys. B **299**, 1 (1988)
701. S. Ritt (MEGA Collaboration), on the web page http://meg.web.psi.ch/docs/talks/s_ritt/mar06_novosibirsk/ritt.ppt
702. T. Iijima, talk given at the 6th Workshop on a Higher Luminosity B Factory, KEK, Tsukuba, Japan, November 2004
703. J. Aysto et al., arXiv:hep-ph/0109217
704. The PRIME Working Group, Search for the μ–e Conversion Process at an Ultimate sensitivity of the order of 10^{18} with PRISM, unpublished
705. LOI to J-PARC 50-GeV PS, LOI-25, http://psux1.kek.jp/jhf-np/LOIlist/LOIlist.html
706. Y. Kuno, Nucl. Phys. Proc. Suppl. **149**, 376 (2005)
707. A. Masiero, S.K. Vempati, O. Vives, New J. Phys. **6**, 202 (2004). arXiv:hep-ph/0407325
708. E. Ables et al. (MINOS Collaboration), Fermilab-proposal-0875
709. G.S. Tzanakos (MINOS Collaboration), AIP Conf. Proc. **721**, 179 (2004)
710. M. Komatsu, P. Migliozzi, F. Terranova, J. Phys. G **29**, 443 (2003). arXiv:hep-ph/0210043
711. P. Migliozzi, F. Terranova, Phys. Lett. B **563**, 73 (2003). arXiv:hep-ph/0302274
712. P. Huber, J. Kopp, M. Lindner, M. Rolinec, W. Winter, J. High Energy Phys. **0605**, 072 (2006). arXiv:hep-ph/0601266
713. Y. Itow et al., arXiv:hep-ex/0106019
714. A. Blondel, A. Cervera-Villanueva, A. Donini, P. Huber, M. Mezzetto, P. Strolin, arXiv:hep-ph/0606111
715. P. Huber, M. Lindner, M. Rolinec, W. Winter, arXiv:hep-ph/0606119
716. J. Burguet-Castell, D. Casper, E. Couce, J.J. Gomez-Cadenas, P. Hernandez, Nucl. Phys. B **725**, 306 (2005). arXiv:hep-ph/0503021
717. J.E. Campagne, M. Maltoni, M. Mezzetto, T. Schwetz, arXiv:hep-ph/0603172
718. T. Blazek, S.F. King, Nucl. Phys. B **662**, 359 (2003). arXiv:hep-ph/0211368
719. S.F. King, J. High Energy Phys. **0508**, 105 (2005) arXiv:hep-ph/0506297
720. S. Antusch, S.F. King, Phys. Lett. B **631**, 42 (2005). arXiv:hep-ph/0508044
721. S. Antusch, S.F. King, arXiv:0709.0666 [hep-ph]
722. P.H. Chankowski, S. Pokorski, Int. J. Mod. Phys. A **17**, 575 (2002). arXiv:hep-ph/0110249
723. G. Altarelli, F. Feruglio, Phys. Rep. **320**, 295 (1999)
724. A.Y. Smirnov, Phys. Rev. D **48**, 3264 (1993). arXiv:hep-ph/9304205
725. J. Pradler, F.D. Steffen, Phys. Rev. D **75**, 023509 (2007). arXiv:hep-ph/0608344
726. S. Davidson, J. High Energy Phys. **0303**, 037 (2003). arXiv:hep-ph/0302075
727. P.H. Chankowski, K. Turzynski, Phys. Lett. B **570**, 198 (2003). arXiv:hep-ph/0306059
728. A. Pilaftsis, Int. J. Mod. Phys. A **14**, 1811 (1999). arXiv:hep-ph/9812256
729. S.F. King, Rep. Prog. Phys. **67**, 107 (2004). arXiv:hep-ph/0310204
730. F.R. Joaquim, I. Masina, A. Riotto, arXiv:hep-ph/0701270
731. J.R. Ellis, K.A. Olive, Y. Santoso, V.C. Spanos, Phys. Lett. B **565**, 176 (2003). arXiv:hep-ph/0303043
732. T. Blazek, S.F. King, Phys. Lett. B **518**, 109 (2001). arXiv:hep-ph/0105005
733. F. Cuypers, S. Davidson, Eur. Phys. J. C **2**, 503 (1998). arXiv:hep-ph/9609487
734. E.J. Chun, K.Y. Lee, S.C. Park, Phys. Lett. B **566**, 142 (2003). arXiv:hep-ph/0304069
735. A. Strumia, F. Vissani, arXiv:hep-ph/0606054
736. F.R. Joaquim, A. Rossi, arXiv:hep-ph/0607298
737. T. Moroi, Phys. Lett. B **493**, 366 (2000)
738. J. Hisano, Y. Shimizu, Phys. Lett. B **565**, 183 (2003)
739. M. Ciuchini, A. Masiero, L. Silvestrini, S.K. Vempati, O. Vives, Phys. Rev. Lett. **92**, 071801 (2004)
740. S. Dimopoulos, L.J. Hall, Phys. Lett. B **344**, 185 (1995)
741. J. Hisano, M. Kakizaki, M. Nagai, Y. Shimizu, Phys. Lett. B **604**, 216 (2004)
742. M. Pospelov, A. Ritz, Ann. Phys. **318**, 119 (2005)
743. V.M. Khatsimovsky, I.B. Khriplovich, A.R. Zhitnitsky, Z. Phys. C **36**, 455 (1987)
744. A.R. Zhitnitsky, Phys. Rev. D **55**, 3006 (1997)
745. F.R. Joaquim, A. Rossi, arXiv:hep-ph/0604083
746. M. Dine, W. Fischler, M. Srednicki, Nucl. Phys. B **189**, 575 (1981)
747. S. Dimopoulos, S. Raby, Nucl. Phys. B **192**, 353 (1981)
748. C.R. Nappi, B.A. Ovrut, Phys. Lett. B **113**, 175 (1982)
749. S. Dimopoulos, S. Raby, Nucl. Phys. B **219**, 479 (1983)
750. M. Dine, A.E. Nelson, Phys. Rev. D **48**, 1277 (1993). arXiv:hep-ph/9303230
751. M. Dine, A.E. Nelson, Y. Shirman, Phys. Rev. D **51**, 1362 (1995). arXiv:hep-ph/9408384
752. M.A. Giorgi et al. (SuperB Group,), INFN Roadmap Report, March 2006
753. Y. Mori et al., (PRISM/PRIME Working Group), LOI at J-PARC 50-GeV PS, LOI-25. http://psux1.kek.jp/~jhf-np/LOIlist/LOIlist.html
754. M. Cvetic, I. Papadimitriou, G. Shiu, Nucl. Phys. B **659**, 193 (2003) [Erratum:Nucl. Phys. B **696**, 298 (2004)]
755. M. Cvetic, P. Langacker, arXiv:hep-th/0607238
756. R. Barbieri, D.V. Nanopoulos, G. Morchio, F. Strocchi, Phys. Lett. B **90**, 91 (1980)
757. P. Langacker, Phys. Rep. **72**, 185 (1981)
758. D. Chang, A. Masiero, H. Murayama, Phys. Rev. D **67**, 075013 (2003). arXiv:hep-ph/0205111
759. T. Moroi, J. High Energy Phys. **0003**, 019 (2000). arXiv:hep-ph/0002208
760. N. Akama, Y. Kiyo, S. Komine, T. Moroi, Phys. Rev. D **64**, 095012 (2001). arXiv:hep-ph/0104263
761. J. Sato, K. Tobe, T. Yanagida, Phys. Lett. B **498**, 189 (2001). arXiv:hep-ph/0010348
762. M. Apollonio et al.(CHOOZ Collaboration), Phys. Lett. B **466**, 415 (1999). arXiv:hep-ex/9907037

763. L. Calibbi, A. Faccia, A. Masiero, S.K. Vempati, Phys. Rev. D **74**, 116002 (2006). arXiv:hep-ph/0605139
764. K. Abe et al. (Belle Collaboration), Phys. Rev. Lett. **92**, 171802 (2004). arXiv:hep-ex/0310029
765. A.J. Buras, P.H. Chankowski, J. Rosiek, L. Slawianowska, Nucl. Phys. B **659**, 3 (2003). arXiv:hep-ph/0210145
766. G. Isidori, P. Paradisi, Phys. Lett. B **639**, 499 (2006). arXiv:hep-ph/0605012
767. E. Lunghi, W. Porod, O. Vives, Phys. Rev. D **74**, 075003 (2006). arXiv:hep-ph/0605177
768. J. Foster, K.i. Okumura, L. Roszkowski, Phys. Lett. B **641**, 452 (2006). arXiv:hep-ph/0604121
769. G.R. Farrar, P. Fayet, Phys. Lett. B **76**, 575 (1978)
770. L.J. Hall, M. Suzuki, Nucl. Phys. B **231**, 419 (1984)
771. R. Barbier et al., Phys. Rep. **420**, 1 (2005). arXiv:hep-ph/0406039
772. D. Comelli, A. Masiero, M. Pietroni, A. Riotto, Phys. Lett. B **324**, 397 (1994). arXiv:hep-ph/9310374
773. R. Kuchimanchi, R.N. Mohapatra, Phys. Rev. D **48**, 4352 (1993). arXiv:hep-ph/9306290
774. K. Huitu, J. Maalampi, Phys. Lett. B **344**, 217 (1995). arXiv:hep-ph/9410342
775. R. Kitano, K.y. Oda, Phys. Rev. D **61**, 113001 (2000). arXiv:hep-ph/9911327
776. M. Frank, K. Huitu, Phys. Rev. D **64**, 095015 (2001). arXiv:hep-ph/0106004
777. J.R. Ellis, G. Gelmini, C. Jarlskog, G.G. Ross, J.W.F. Valle, Phys. Lett. B **150**, 142 (1985)
778. S. Davidson, M. Losada, Phys. Rev. D **65**, 075025 (2002). arXiv:hep-ph/0010325
779. K.S. Babu, R.N. Mohapatra, Phys. Rev. Lett. **64**, 1705 (1990)
780. M. Hirsch, M.A. Diaz, W. Porod, J.C. Romao, J.W.F. Valle, Phys. Rev. D **62**, 113008 (2000) [Erratum: Phys. Rev. D **65**, 119901 (2002)]. arXiv:hep-ph/0004115
781. M.A. Diaz, C. Mora, A.R. Zerwekh, Eur. Phys. J. C **44**, 277 (2005). arXiv:hep-ph/0410285
782. A. Deandrea, J. Welzel, M. Oertel, J. High Energy Phys. **0410**, 038 (2004). arXiv:hep-ph/0407216
783. W. Porod, M. Hirsch, J. Romao, J.W.F. Valle, Phys. Rev. D **63**, 115004 (2001). arXiv:hep-ph/0011248
784. A. Heister et al.(ALEPH Collaboration), Eur. Phys. J. C **31**, 1 (2003). arXiv:hep-ex/0210014
785. P. Abreu et al.(DELPHI Collaboration), Phys. Lett. B **500**, 22 (2001). arXiv:hep-ex/0103015
786. P. Achard et al. (L3 Collaboration), Phys. Lett. B **524**, 65 (2002). arXiv:hep-ex/0110057
787. B. Abbott et al. (D0 Collaboration), Phys. Rev. D **62**, 071701 (2000). arXiv:hep-ex/0005034
788. S. Aid et al. (H1 Collaboration), Z. Phys. C **71**, 211 (1996). arXiv:hep-ex/9604006
789. C. Adloff et al. (H1 Collaboration), Eur. Phys. J. C **20**, 639 (2001). arXiv:hep-ex/0102050
790. W. Beenakker, R. Hopker, M. Spira, P.M. Zerwas, Phys. Rev. Lett. **74**, 2905 (1995). arXiv:hep-ph/9412272
791. J. Kalinowski, R. Ruckl, H. Spiesberger, P.M. Zerwas, Phys. Lett. B **414**, 297 (1997). arXiv:hep-ph/9708272
792. G. Moreau, E. Perez, G. Polesello, Nucl. Phys. B **604**, 3 (2001). arXiv:hep-ph/0003012
793. G. Moreau, M. Chemtob, F. Deliot, C. Royon, E. Perez, Phys. Lett. B **475**, 184 (2000). arXiv:hep-ph/9910341
794. S. Dimopoulos, L.J. Hall, Phys. Lett. B **207**, 210 (1988)
795. H.K. Dreiner, P. Richardson, M.H. Seymour, Phys. Rev. D **63**, 055008 (2001). arXiv:hep-ph/0007228
796. F. Deliot, G. Moreau, C. Royon, Eur. Phys. J. C **19**, 155 (2001). arXiv:hep-ph/0007288
797. M. Chaichian, A. Datta, K. Huitu, S. Roy, Z.h. Yu, Phys. Lett. B **594**, 355 (2004). arXiv:hep-ph/0311327
798. S. Dimopoulos, R. Esmailzadeh, L.J. Hall, G.D. Starkman, Phys. Rev. D **41**, 2099 (1990)
799. B. Allanach et al. (R Parity Working Group Collaboration), arXiv:hep-ph/9906224
800. E.L. Berger, B.W. Harris, Z. Sullivan, Phys. Rev. Lett. **83**, 4472 (1999). arXiv:hep-ph/9903549
801. M. Chaichian, K. Huitu, Z.H. Yu, Phys. Lett. B **490**, 87 (2000). arXiv:hep-ph/0007220
802. E.D. Richter-Was, L. Poggioli, ATLAS note ATLAS-PHYS-98-131 (1998)
803. W.W. Armstrong et al., ATLAS technical proposal CERN-LHCC-94-43, 1994
804. A. Belyaev, M.H. Genest, C. Leroy, R.R. Mehdiyev, J. High Energy Phys. **0409**, 012 (2004). arXiv:hep-ph/0401065
805. K.S. Babu, C.F. Kolda, Phys. Rev. Lett. **84**, 228 (2000). arXiv:hep-ph/9909476
806. K.S. Babu, C. Kolda, Phys. Rev. Lett. **89**, 241802 (2002). arXiv:hep-ph/0206310
807. A. Dedes, J.R. Ellis, M. Raidal, Phys. Lett. B **549**, 159 (2002). arXiv:hep-ph/0209207
808. D. Chang, W.S. Hou, W.Y. Keung, Phys. Rev. D **48**, 217 (1993). arXiv:hep-ph/9302267
809. M. Sher, Y. Yuan, Phys. Rev. D **44**, 1461 (1991)
810. M. Sher, Phys. Rev. D **66**, 057301 (2002). arXiv:hep-ph/0207136
811. R. Kitano, M. Koike, S. Komine, Y. Okada, Phys. Lett. B **575**, 300 (2003). arXiv:hep-ph/0308021
812. S. Kanemura, Y. Kuno, M. Kuze, T. Ota, Phys. Lett. B **607**, 165 (2005). arXiv:hep-ph/0410044
813. A. Broncano, M.B. Gavela, E. Jenkins, Phys. Lett. B **552**, 177 (2003) [Erratum: Phys. Lett. B **636**, 330 (2006)]. arXiv:hep-ph/0210271
814. M.C. Gonzalez-Garcia, J.W.F. Valle, Phys. Lett. B **216**, 360 (1989)
815. S. Antusch, C. Biggio, E. Fernandez-Martinez, M.B. Gavela, J. Lopez-Pavon, J. High Energy Phys. **0610**, 084 (2006). arXiv:hep-ph/0607020
816. M. Apollonio et al., Eur. Phys. J. C **27**, 331 (2003). hep-ex/0301017
817. T. Araki et al. (KamLAND Collaboration), Phys. Rev. Lett. **94**, 081801 (2005). hep-ex/0406035
818. B. Aharmim et al. (SNO Collaboration), Phys. Rev. C **72**, 055502 (2005). nucl-ex/0502021
819. M.H. Ahn et al. (K2K Collaboration), Phys. Rev. Lett. **90**, 041801 (2003). hep-ex/0212007
820. Y. Ashie et al. (Super-Kamiokande Collaboration), Phys. Rev. D **71**, 112005 (2005). hep-ex/0501064
821. P. Astier et al. (NOMAD Collaboration), Nucl. Phys. B **611**, 3 (2001). hep-ex/0106102
822. K. Eitel (KARMEN Collaboration), Nucl. Phys. Proc. Suppl. **91**, 191 (2001). hep-ex/0008002
823. Y. Declais et al., Nucl. Phys. B **434**, 503 (1995)
824. D.A. Petyt (MINOS Collaboration) First MINOS results from the NuMI beam. http://www-numi.fnal.gov/
825. M.C. Gonzalez-Garcia, Phys. Scr. T **121**, 72 (2005). hep-ph/0410030
826. V. Barger, S. Geer, K. Whisnant, New J. Phys. **6**, 135 (2004). arXiv:hep-ph/0407140
827. D.S. Ayres et al. (NOvA Collaboration), hep-ex/0503053
828. J.J. Gomez-Cadenas et al. (CERN Working Group on Super Beams Collaboration), hep-ph/0105297
829. J.E. Campagne, A. Cazes, Eur. Phys. J. C **45**, 643 (2006). hep-ex/0411062
830. P. Zucchelli, Phys. Lett. B **532**, 166 (2002)
831. S. Geer, Phys. Rev. D **57**, 6989 (1998) hep-ph/9712290
832. A. De Rujula, M.B. Gavela, P. Hernandez, Nucl. Phys. B **547**, 21 (1999) hep-ph/9811390

833. W.J. Marciano, A. Sirlin, Phys. Rev. Lett. **71**, 3629 (1993)
834. S. Eidelman et al. (Particle Data Group), Phys. Lett. B **592**, 1 (2004)
835. L. Fiorini (for the NA48/2 Collaboration), at ICHEP 2005
836. V.D. Barger, G.F. Giudice, T. Han, Phys. Rev. D **40**, 2987 (1989)
837. P.H. Chankowski, A. Dabelstein, W. Hollik, W.M. Mosle, S. Pokorski, J. Rosiek, Nucl. Phys. B **417**, 101 (1994)
838. J.F. Gunion, H.E. Haber, G.L. Kane, S. Dawson, *The Higgs hunter Guide* (Addison–Wesley, Reading, 1990)
839. A. Masiero, P. Paradisi, R. Petronzio, Phys. Rev. D **74**, 011701 (2006). arXiv:hep-ph/0511289
840. M. Krawczyk, D. Temes, Eur. Phys. J. C **44**, 435 (2005). arXiv:hep-ph/0410248
841. P. Krawczyk, S. Pokorski, Phys. Rev. Lett. **60**, 182 (1988)
842. J.R. Ellis, J. Hisano, M. Raidal, Y. Shimizu, Phys. Lett. B **528**, 86 (2002). arXiv:hep-ph/0111324
843. I. Masina, Nucl. Phys. B **671**, 432 (2003). arXiv:hep-ph/0304299
844. Y. Farzan, M.E. Peskin, Phys. Rev. D **70**, 095001 (2004). arXiv:hep-ph/0405214
845. I. Masina, C. Savoy, Phys. Lett. B **579**, 99 (2004). arXiv:hep-ph/0309067
846. K.S. Babu, B. Dutta, R.N. Mohapatra, Phys. Rev. Lett. **85**, 5064 (2000). arXiv:hep-ph/0006329
847. S. Abel, O. Lebedev, J. High Energy Phys. **0601**, 133 (2006)
848. T. Ibrahim, P. Nath, Phys. Rev. D **58**, 111301 (1998) [Erratum: Phys. Rev. D **60**, 099902 (1999)]. arXiv:hep-ph/9807501
849. J. Hisano, Y. Shimizu, Phys. Rev. D **70**, 093001 (2004). arXiv:hep-ph/0406091
850. M. Brhlik, L.L. Everett, G.L. Kane, J.D. Lykken, Phys. Rev. Lett. **83**, 2124 (1999). arXiv:hep-ph/9905215
851. S. Abel, S. Khalil, O. Lebedev, Phys. Rev. Lett. **86**, 5850 (2001). arXiv:hep-ph/0103031
852. J.R. Ellis, S. Ferrara, D.V. Nanopoulos, Phys. Lett. B **114**, 231 (1982)
853. S. Abel, S. Khalil, O. Lebedev, Nucl. Phys. B **606**, 151 (2001). hep-ph/0103320
854. D.A. Demir, O. Lebedev, K.A. Olive, M. Pospelov, A. Ritz, Nucl. Phys. B **680**, 339 (2004). hep-ph/0311314
855. K.A. Olive, M. Pospelov, A. Ritz, Y. Santoso, Phys. Rev. D **72**, 075001 (2005). hep-ph/0506106
856. N. Arkani-Hamed, S. Dimopoulos, G.F. Giudice, A. Romanino, Nucl. Phys. B **709**, 3 (2005). hep-ph/0409232
857. G.F. Giudice, A. Romanino, Phys. Lett. B **634**, 307 (2006). hep-ph/0510197
858. D. Chang, W.-F. Chang, W.-Y. Keung, Phys. Rev. D **71**, 076006 (2005). hep-ph/0503055
859. A.D. Linde, Phys. Lett. B **129**, 177 (1983)
860. S. Weinberg, Phys. Rev. Lett. **59**, 2607 (1987)
861. V. Agrawal, S.M. Barr, J.F. Donoghue, D. Seckel, Phys. Rev. D **57**, 5480 (1998). hep-ph/9707380
862. R. Bousso, J. Polchinski, J. High Energy Phys. **06**, 006 (2000). hep-th/0004134
863. N. Arkani-Hamed, S. Dimopoulos, S. Kachru, hep-th/0501082
864. R. Arnowitt, J.L. Lopez, D.V. Nanopoulos, Phys. Rev. D **42**, 2423 (1990)
865. G. Degrassi, E. Franco, S. Marchetti, L. Silvestrini, J. High Energy Phys. **11**, 044 (2005). hep-ph/0510137
866. M. Pospelov, A. Ritz, Phys. Rev. Lett. **83**, 2526 (1999). hep-ph/9904483
867. M. Pospelov, A. Ritz, Phys. Rev. D **63**, 073015 (2001). hep-ph/0010037
868. A. Arvanitaki, C. Davis, P.W. Graham, J.G. Wacker, Phys. Rev. D **70**, 117703 (2004). hep-ph/0406034
869. C.A. Baker et al., Phys. Rev. Lett. **97**, 131801 (2006). hep-ex/0602020
870. O. Lebedev, W. Loinaz, T. Takeuchi, Phys. Rev. D **61**, 115005 (2000)
871. M.J. Ramsey-Musolf, Phys. Rev. D **62**, 056009 (2000). arXiv:hep-ph/0004062
872. O. Lebedev, W. Loinaz, T. Takeuchi, Phys. Rev. D **62**, 055014 (2000). arXiv:hep-ph/0002106
873. J.H. Park, J. High Energy Phys. **0610**, 077 (2006)
874. M.J. Ramsey-Musolf, S. Su, arXiv:hep-ph/0612057
875. M.A. Sanchis-Lozano, Contributed paper to the Workshop on B-Factories and New Measurements, KEK, 13–14 September 2006. arXiv:hep-ph/0610046
876. B.A. Campbell, D.W. Maybury, Nucl. Phys B **709**, 419 (2005)
877. W. Loinaz, N. Okamura, S. Rayyan, T. Takeuchi, L.C.R. Wijewardhana, Phys. Rev. D **70**, 113004 (2004)
878. M. Finkemeier, Phys. Lett. B **387**, 391 (1996). arXiv:hep-ph/9505434
879. G. Czapek et al., Phys. Rev. Lett. **70**, 17 (1993)
880. D.I. Britton et al., Phys. Rev. Lett. **68**, 3000 (1992)
881. A.Y. Smirnov, R. Zukanovich Funchal, Phys. Rev. D **74**, 013001 (2006). arXiv:hep-ph/0603009
882. D. Pocanic, A. van der Schaaf (PEN Collaboration), PSI experiment R-05-01
883. M.A. Bychkov, Ph.D. thesis, University of Virginia (2005), unpublished. http://pibeta.phys.virginia.edu/docs/publications/max_thesis/thesis.pdf
884. B.A. VanDevender, Ph.D. thesis, University of Virginia, 2005, unpublished. http://pibeta.phys.virginia.edu/docs/publications/brent_diss.pdf
885. D. Bryman and T. Numao (PIENU Collaboration), TRIUMF experiment 1072
886. NA48/2 Collaboration, Report CERN/SPSC/2000-003, 16 January 2000
887. L. Fiorini, PhD thesis, Scuola Normale Superiore, Pisa, Italy, 2005, unpublished. Available from http://lfiorini.home.cern.ch/lfiorini/
888. L. Fiorini (for the NA48/2 Collaboration), in *Proc. HEP2005 Europhysics Conference*, Lisbon, 21–27 July 2005, PoS (HEP 2005) 288
889. W.J. Marciano, A. Sirlin, Phys. Rev. Lett. **61**, 1815 (1988)
890. J.Z. Bai et al.(BES Collaboration), Phys. Rev. D **53**, 20 (1996)
891. V.V. Anashin et al. (KEDR Collaboration), Nucl. Phys. Proc. Suppl. **169**, 125 (2007). arXiv:hep-ex/0611046
892. A. Lusiani (BaBar Collaboration), Nucl. Phys. Proc. Suppl. **144**, 105 (2005)
893. K. Abe et al. (Belle Collaboration), arXiv:hep-ex/0608046
894. R. Decker, M. Finkemeier, Phys. Lett. B **334**, 199 (1994)
895. W. Buchmüller, R.D. Peccei, T. Yanagida, Annu. Rev. Nucl. Part. Sci. **55**, 311 (2005)
896. W. Bernreuther, private communication
897. L. Michel, Proc. Phys. Soc. A, **63**, 514 (1950)
898. W. Fetscher, H.-J. Gerber, K.F. Johnson, Phys. Lett. B **173**, 102 (1986)
899. P. Langacker, Commun. Nucl. Part. Phys. **19** (1989)
900. A. Aktas et al. (H1 Collaboration), Phys. Lett. B **634**, 173 (2006). arXiv:hep-ex/0512060
901. W. Fetscher, H.-J. Gerber, Eur. Phys. J. C **15**, 316 (2000)
902. F. Scheck, *Electroweak and Strong Interactions* (Springer, Berlin, 1996)
903. J. Kirkby et al., PSI proposal R-99-06, 1999
904. A. Barczyk et al. (FAST Collaboration), arXiv:0707.3904 [hep-ex]
905. W. Marciano, Phys. Rev. D **60**, 093006 (1999)
906. I.C. Barnett et al., Nucl. Instrum. Methods A **455**, 329 (2000)
907. Y.S. Tsai, in *Stanford Tau Charm* (1989), pp. 0387-0393
908. Y.S. Tsai, Phys. Rev. D **51**, 3172 (1995)
909. J.H. Kuhn, E. Mirkes, Z. Phys. C **56**, 661 (1992) [Erratum: Z. Phys. C **67**, 364 (1995)]

910. J.H. Kühn, E. Mirkes, Phys. Lett. B **398**, 407 (1997)
911. I.I. Bigi, A.I. Sanda, Phys. Lett. B **625**, 47 (2005)
912. A. Datta et al., hep-ph/0610162
913. J. Bernabeu et al., Nucl. Phys. B **763**, 283 (2007)
914. D. Delepine et al., Phys. Rev. D **74**, 056004 (2006)
915. A. Pich, Int. J. Mod. Phys. A **21**, 5652 (2006)
916. S. Anderson et al. (CLEO Collaboration), Phys. Rev. Lett. **81**, 3823 (1998)
917. J.J. Sakurai, Phys. Rev. **109**, 980 (1958)
918. A.R. Zhitnitkii, Sov. J. Nucl. Phys. **31**, 529 (1980)
919. V.P. Efrosinin, I.B. Khriplovich, G.G. Kirilin, Yu.G. Kudenko, Phys. Lett. B **493**, 293 (2000). arXiv:hep-ph/0008199
920. I.I. Bigi, A.I. Sanda, *CP Violation* (Cambridge University Press, Cambridge, 2000)
921. A.I. Sanda, Phys. Rev. D **23**, 2647 (1981)
922. N.G. Deshpande, Phys. Rev. D **23**, 2654 (1981)
923. H.Y. Cheng, Phys. Rev. D **26**, 143 (1982)
924. H.Y. Cheng, Phys. Rev. D **34**, 1397 (1986)
925. I.I.Y. Bigi, A.I. Sanda, Phys. Rev. Lett. **58**, 1604 (1987)
926. M. Leurer, Phys. Rev. Lett. **62**, 1967 (1989)
927. H.Y. Cheng, Phys. Rev. D **42**, 2329 (1990)
928. G. Belanger, C.Q. Geng, Phys. Rev. D **44**, 2789 (1991)
929. R. Garisto, G.L. Kane, Phys. Rev. D **44**, 2038 (1991)
930. M. Fabbrichesi, F. Vissani, Phys. Rev. D **55**, 5334 (1997). arXiv:hep-ph/9611237
931. G.H. Wu, J.N. Ng, Phys. Lett. B **392**, 93 (1997). arXiv:hep-ph/9609314
932. W.F. Chang, J.N. Ng, arXiv:hep-ph/0512334
933. J.A. Macdonald et al., Nucl. Instrum. Methods Phys. Res. Sect. A **506**, 60 (2003)
934. M. Abe et al., Phys. Rev. D **73**, 072005 (2006)
935. J-PARC experiment proposal P06. http://j-parc.jp/NuclPart/Proposal_0606_e.html
936. W. Bernreuther, U. Low, J.P. Ma, O. Nachtmann, Z. Phys. C **41**, 143 (1988)
937. M. Skalsey, J. Van House, Phys. Rev. Lett. **67**, 1993 (1991)
938. M. Felcini, Int. J. Mod. Phys. A **19**, 3853 (2004). arXiv:hep-ex/0404041
939. O. Jinnouchi, S. Asai, T. Kobayashi, Phys. Lett. B **572**, 117 (2003). arXiv:hep-ex/0308030
940. M. Ahmed et al. (MEGA Collaboration), Phys. Rev. D **65**, 112002 (2002). arXiv:hep-ex/0111030
941. L. Willmann et al., Phys. Rev. Lett. **82**, 49 (1999). arXiv:hep-ex/9807011
942. P. Wintz, in *ICHEP 98*, Vancouver, Canada, 23–29 July 1998
943. J. Kaulard et al. (SINDRUM II Collaboration), Phys. Lett. B **422**, 334 (1998)
944. W. Honecker et al. (SINDRUM II Collaboration), Phys. Rev. Lett. **76**, 200 (1996)
945. W. Bertl et al. (SINDRUM II Collaboration), Eur. Phys. J. C **47**, 337 (2006)
946. T. Mori et al., PSI proposal R-99-5, May 1999, available as UT-ICEPP 00-02
947. A. Baldini et al., Research Proposal to INFN, Septembet 2002. http://meg.web.psi.ch/docs/
948. S. Mihara et al., Cryogenics **44**, 223 (2004)
949. W. Bertl et al. (SINDRUM Collaboration), Nucl. Phys. B **260**, 1 (1985)
950. A. Czarnecki, W.J. Marciano, K. Melnikov, AIP Conf. Proc. **549**, 938 (2002)
951. T.S. Kosmas, I.E. Lagaris, J. Phys. G **28**, 2907 (2002)
952. R. Kitano, M. Koike, Y. Okada, Phys. Rev. D **66**, 096002 (2002). arXiv:hep-ph/0203110
953. T.S. Kosmas, J.D. Vergados, O. Civitarese, A. Faessler, Nucl. Phys. A **570**, 637 (1994)
954. R.M. Dzhilkibaev, V.M. Lobashev, Sov. J. Nucl. Phys. **49**, 384 (1989) [Yad. Fiz. **49**, 622 (1989)]
955. M. Bachman et al., MECO, BNL proposal E 940, 1997
956. Letter of Intent to J-PARC, L24: The PRISM Project, a Muon Source of the World-Highest Brightness by Phase Rotation, 2003
957. Letter of Intent to J-PARC, L26: Request for a Pulsed Proton Beam Facility at J-PARC
958. G. Cvetic, C. Dib, C.S. Kim, J.D. Kim, Phys. Rev. D **66**, 034008 (2002) [Erratum: Phys. Rev. D **68** (2003) 059901]. arXiv:hep-ph/0202212
959. J.R. Ellis, M.E. Gomez, G.K. Leontaris, S. Lola, D.V. Nanopoulos, Eur. Phys. J. C **14**, 319 (2000). arXiv:hep-ph/9911459
960. T. Fukuyama, T. Kikuchi, N. Okada, Phys. Rev. D **68**, 033012 (2003). arXiv:hep-ph/0304190
961. C.X. Yuex, Y.M. Zhang, L.J. Liu, Phys. Lett. B **547**, 252 (2002). arXiv:hep-ph/0209291
962. A. Ilakovac, Phys. Rev. D **62**, 036010 (2000). arXiv:hep-ph/9910213
963. W.J. Li, Y.D. Yang, X.D. Zhang, Phys. Rev. D **73**, 073005 (2006). arXiv:hep-ph/0511273
964. D. Black, T. Han, H.J. He, M. Sher, Phys. Rev. D **66**, 053002 (2002). arXiv:hep-ph/0206056
965. R.D. Cousins, V.L. Highland, Nucl. Instrum. Methods Phys. Res. Sect. A **320**, 331 (1992)
966. R. Barlow, Comput. Phys. Commun. **149**, 97 (2002)
967. K. Abe et al. (Belle Collaboration), arXiv:0708.3272 [hep-ex]
968. Y. Miyazaki et al. (Belle Collaboration), Phys. Lett. B **648**, 341 (2007). arXiv:hep-ex/0703009
969. N.G. Unel, arXiv:hep-ex/0505030
970. M. Della Negra et al., CMS Physics Technical Design Report, vol. 1, 2006
971. T. Speer et al., CMS Note 2006/032, 2006
972. E. James et al., CMS Note 2006/010, 2006
973. R. Santinelli, M. Biasini, CMS Note 2002/037, 2002
974. R. Santinelli, Nucl. Phys. Proc. Suppl. **123**, 234 (2003)
975. M.C. Chang et al. (Belle Collaboration), Phys. Rev. D **68**, 111101 (2003). arXiv:hep-ex/0309069
976. F. Abe et al. (CDF Collaboration), Phys. Rev. Lett. **81**, 5742 (1998)
977. D. Ambrose et al. (BNL Collaboration), Phys. Rev. Lett. **81**, 5734 (1998). arXiv:hep-ex/9811038
978. W. Bonivento, N. Serra, LHCb Note 2007-028, 2007
979. S.N. Gninenko, M.M. Kirsanov, N.V. Krasnikov, V.A. Matveev, P. Nedelec, D. Sillou, M. Sher, CERN-SPSC-2004-016
980. M. Sher, I. Turan, Phys. Rev. D **69**, 017302 (2004). arXiv:hep-ph/0309183
981. D.E. Groom et al. (Particle Data Group), Eur. Phys. J. C **15**, 1 (2000)
982. B. Aubert et al. (BaBar Collaboration), Phys. Rev. Lett. **95**, 191801 (2005). arXiv:hep-ex/0506066
983. Y. Yusa et al. (Belle Collaboration), Phys. Lett. B **640**, 138 (2006). arXiv:hep-ex/0603036
984. G. Ingelman, A. Edin, J. Rathsman, Comput. Phys. Commun. **101**, 108 (1997). arXiv:hep-ph/9605286
985. M.V. Romalis, W.C. Griffith, E.N. Fortson, Phys. Rev. Lett. **86**, 2505 (2001). arXiv:hep-ex/0012001
986. R.D. Peccei, H.R. Quinn, Phys. Rev. Lett. **38**, 1440 (1977)
987. J.S.M. Ginges, V.V. Flambaum, Phys. Rep. **397**, 63 (2004). arXiv:physics/0309054
988. I.B. Khriplovich, S.K. Lamoreaux, *CP Violation Without Strangeness: Electric Dipole Moments of Particles, Atoms, and Molecules* (Springer, Berlin, 1997)
989. R.J. Crewther, P. Di Vecchia, G. Veneziano, E. Witten, Phys. Lett. B **88**, 123 (1979) [Erratum: Phys. Lett. B **91**, 487 (1980)]
990. K. Kawarabayashi, N. Ohta, Prog. Theor. Phys. **66**, 1789 (1981)
991. D. Chang, W.Y. Keung, A. Pilaftsis, Phys. Rev. Lett. **82**, 900 (1999)

992. O. Lebedev, M. Pospelov, Phys. Rev. Lett. **89**, 101801 (2002). arXiv:hep-ph/0204359
993. M. Pospelov, A. Ritz, Y. Santoso, Phys. Rev. Lett. **96**, 091801 (2006). arXiv:hep-ph/0510254
994. M. Pospelov, A. Ritz, Y. Santoso, Phys. Rev. D **74**, 075006 (2006). arXiv:hep-ph/0608269
995. C. Grojean, G. Servant, J.D. Wells, Phys. Rev. D **71**, 036001 (2005). arXiv:hep-ph/0407019
996. D. Bodeker, L. Fromme, S.J. Huber, M. Seniuch, J. High Energy Phys. **0502**, 026 (2005). arXiv:hep-ph/0412366
997. S.J. Huber, M. Pospelov, A. Ritz, Phys. Rev. D **75**, 036006 (2007). arXiv:hep-ph/0610003
998. P.G. Harris et al., Phys. Rev. Lett. **82**, 904 (1999)
999. K.F. Smith et al., Phys. Lett. B **234**, 191 (1990)
1000. I.S. Altarev et al., Phys. At. Nucl. **59**, 1152 (1996) [Yad. Fiz. **59**(N7) 1204 (1996)]
1001. K. Green et al., Nucl. Instrum. Methods A **404**, 381 (1998)
1002. J.M. Pendlebury et al., Phys. Rev. A **70**, 032102 (2004)
1003. P.G. Harris, J.M. Pendlebury, Phys. Rev. A **73**, 014101 (2006)
1004. R. Golub, J.M. Pendlebury, Phys. Lett. A **62**, 337 (1977)
1005. C.A. Baker et al., Phys. Lett. A **308-1**, 67 (2003)
1006. C.A. Baker et al., Nucl. Instrum. Methods A **501**, 517 (2003)
1007. A. Steyerl et al., Phys. Lett. A **116**, 347 (1986)
1008. http://ucn.web.psi.ch
1009. http://nedm.web.psi.ch
1010. S. Groeger et al., J. Res. NIST **110**, 179 (2005)
1011. S. Groeger et al., Appl. Phys. B **80**, 645 (2005)
1012. K. Green et al., Nucl. Instrum. Methods A **404**, 381 (1998)
1013. http://p25ext.lanl.gov/edm/edm.html
1014. R. Golub, K. Lamoreaux, Phys. Rep. **237**, 1 (1994)
1015. C.P. Liu, R.G.E. Timmermans, Phys. Rev. C **70**, 055501 (2004). arXiv:nucl-th/0408060
1016. O. Lebedev, K.A. Olive, M. Pospelov, A. Ritz, Phys. Rev. D **70**, 016003 (2004). arXiv:hep-ph/0402023
1017. R. Stutz, E. Cornell, Bull. Am. Soc. Phys. **89**, 76 (2004)
1018. J.D. Jackson, *Classical Electrodynamics*, Wiley, New York (1975)
1019. A.J. Silenko, arXiv:hep-ph/0604095
1020. J. Bailey et al. (CERN Muon Storage Ring Collaboration), J. Phys. G **4**, 345 (1978)
1021. F.J.M. Farley et al., Phys. Rev. Lett. **93**, 052001 (2004). arXiv:hep-ex/0307006
1022. Y.K. Semertzidis et al., arXiv:hep-ph/0012087
1023. J.P. Miller et al. (EDM Collaboration), AIP Conf. Proc. **698**, 196 (2004)
1024. A. Adelmann, K. Kirch, arXiv:hep-ex/0606034
1025. Y.K. Semertzidis et al. (EDM Collaboration), AIP Conf. Proc. **698** 200 (2004). arXiv:hep-ex/0308063
1026. I.B. Khriplovich, Phys. Lett. B **444**, 98 (1998). arXiv:hep-ph/9809336
1027. I.I. Rabi, J.R. Zacharias, S. Millman, P. Kusch, Phys. Rev. **53**, 318 (1938)
1028. Y.F. Orlov, W.M. Morse, Y.K. Semertzidis, Phys. Rev. Lett. **96**, 214802 (2006). arXiv:hep-ex/0605022
1029. E.D. Commins, J.D. Jackson, D.P. DeMille, Am. J. Phys. **75**, 532 (2007)
1030. E.D. Commins, Am. J. Phys. **59**, 1077 (1991)
1031. M.G. Kozlov, A.V. Titov, N.S. Mosyagin, P.V. Souchko, Phys. Rev. A **56**, R3326 (1997)
1032. M.K. Nayak, R.K. Chaudhuri, B.P. Das, Phys. Rev. A **75**, 022510 (2007)
1033. N. Mosyagin, M. Kozlov, A. Titov, J. Phys. B **31**, L763 (1998)
1034. M.G. Kozlov, L. Labzowski, J. Phys. B **28**, 1933 (1995)
1035. A.N. Petrov et al., Phys. Rev. A **72**, 022505 (2005)
1036. T.A. Isaev et al., Phys. Rev. Lett. **95**, 163004 (2005)
1037. E.R. Meyer, J.L. Bohn, M.P. Deskevitch, Phys. Rev. A **73**, 062108 (2006)
1038. A.N. Petrov, N.S. Mosyagin, T.A. Isaev, A.V. Titov, Phys. Rev. A **76**, 030501(R) (2007)
1039. M.A. Player, P.G.H. Sandars, J. Phys. B **3**, 1620 (1970)
1040. Z.W. Liu, H.P. Kelly, Phys. Rev. A **45**, R4210 (1992)
1041. W.R. Johnson, D.S. Guo, M. Idrees, J. Sapirstein, Phys. Rev. A **34**, 1043 (1986)
1042. S.A. Murthy, D. Krause, Z.L. Li, L. Hunter, Phys. Rev. Lett. **63**, 965 (1989)
1043. M.V. Romalis, private communication
1044. T.W. Kornack, M.V. Romalis, Phys. Rev. Lett. **89**, 253002 (2002)
1045. T.W. Kornack, R.K. Ghosh, M.V. Romalis, Phys. Rev. Lett. **95**, 230801 (2005)
1046. M.V. Romalis, E.N. Fortson, Phys. Rev. A **59**, 4547 (1999)
1047. C. Chin, V. Leiber, V. Vuletic, A.J. Kerman, S. Chu, Phys. Rev. A **63**, 033401 (2001)
1048. D. Heinzen, private communication
1049. D.S. Weiss, F. Fang, J. Chen, Bull. Am, Phys. Soc. APR03 J1.008 (2003)
1050. J. Amini, H. Gould, arxiv.org/abs/physics/0602011 (2006)
1051. Y. Sakemi, private communication
1052. J.J. Hudson, B.E. Sauer, M.R. Tarbutt, E.A. Hinds, Phys. Rev. Lett. **89**, 023003 (2002)
1053. M.R. Tarbutt et al., J. Phys. B **35**, 5013 (2002)
1054. B.E. Sauer, J. Wang, E. Hinds, J. Chem. Phys. **105**, 7412 (1996)
1055. D. DeMille et al., Phys. Rev. A **61**, 052507 (2000)
1056. L.R. Hunter et al., Phys. Rev. A **65**, 030501(R) (2002)
1057. D. Kawall, F. Bay, S. Bickman, Y. Jiang, D. DeMille, Phys. Rev. Lett. **92**, 133007 (2004)
1058. E. Cornell, co-workers, private communication
1059. N.E. Shafer-Ray, Phys. Rev. A **73**, 34102 (2006);
1060. N.E. Shafer-Ray, private communication
1061. S.K. Lamoreaux, Phys. Rev. A **66**, 022109 (2002)
1062. F.L. Shapiro, Sov. Phys. Usp. **11**, 345 (1968)
1063. V.A. Dzuba, O.P. Sushkov, W.R. Johnson, U.I. Safronova, Phys. Rev. A **66**, 032105 (2002)
1064. B.J. Heidenreich et al., Phys. Rev. Lett. **95**, 253004 (2005)
1065. T.N. Mukhamedjanov, V.A. Dzuba, O.P. Sushkov, Phys. Rev. A **68**, 042103 (2003)
1066. S. Lamoreaux, arXiv:physics/0701198 (2007)
1067. C.-Y. Liu, S.K. Lamoreaux, Mod. Phys. A **19**, 1235 (2004)
1068. C.-Y. Liu, private communication
1069. M. Mercier, Magnetism **6**, 77 (1974)
1070. J.L. Feng, K.T. Matchev, Y. Shadmi, Nucl. Phys. B **613**, 366 (2001). arXiv:hep-ph/0107182
1071. A. Romanino, A. Strumia, Nucl. Phys. B **622**, 73 (2002). arXiv:hep-ph/0108275
1072. A. Bartl, W. Majerotto, W. Porod, D. Wyler, Phys. Rev. D **68**, 053005 (2003). arXiv:hep-ph/0306050
1073. M. Aoki et al., J-Parc Letter of Intent, 2003
1074. A. Adelmann, K. Kirch, C.J.G. Onderwater, T. Schietinger, A. Streun, to be published
1075. P.A.M. Dirac, Proc. R. Soc. (London) A **117**, 610 (1928)
1076. P.A.M. Dirac, Proc. R. Soc. (London) A **118**, 351 (1928).
1077. P.A.M. Dirac, *The Principles of Quantum Mechanics*, 4th edn. (Oxford University Press, London, 1958). Eq. 9.28 uses Dirac's original notation
1078. B.C. Odom, D. Hanneke, B. D'Urso, G. Gabrielse, Phys. Rev. Lett. **97**, 030801 (2006) [Erratum: Phys. Rev. Lett. **99**, 039902 (2007)]
1079. T. Kinoshita, M. Nio, Phys. Rev. D **73**, 053007 (2006). arXiv:hep-ph/0512330
1080. J. Schwinger, Phys. Rev. **73** 416L (1948)
1081. J. Schwinger, Phys. Rev. **76**, 790 (1949). The former paper contains a misprint in the expression for a_e that is corrected in the longer paper

1082. T. Blum, Phys. Rev. Lett. **91**, 052001 (2003)
1083. C. Aubin, T. Blum, PoS LAT2005:089 (2005). hep-lat/0509064,
1084. M. Hayakawa et al., PoS LAT2005:353 (2005). hep-lat/0509016
1085. J.P. Miller, E. de Rafael, B.L. Roberts, Rep. Prog. Phys. **70**, 795 (2007). arXiv:hep-ph/0703049
1086. It has been proposed that the hadronic contributions could also be determined from hadronic τ decay, plus the conserved vector current hypothesis, but this prescriptions seems to have internal consistency issues which are still under study. See Ref. [1082] and references therein
1087. R.L. Garwin, L.M. Lederman, M. Weinrich, Phys. Rev. **105**, 1415 (1957)
1088. J. Bailey et al. (CERN-Mainz-Daresbury Collaboration), Nucl. Phys. B **150**, 1 (1979)
1089. F.J.M. Farley, Nucl. Instrum. Methods A **523**, 251 (2004). arXiv:hep-ex/0307024

Collider aspects of flavor physics at high Q^*

T. Lari[37,b], L. Pape[44,b], W. Porod[49,a,b], J.A. Aguilar-Saavedra[1,c], B.C. Allanach[2,c], G. Burdman[13,c], N. Castro[14,c], M. Klasen[22,c], N. Krasnikov[4,c], F. Krauss[36,c], F. Moortgat[44,c], G. Polesello[48,c], A. Tricomi[61,c], G. Ünel[54,62,c], F. del Aguila[1], J. Alwall[3], Y. Andreev[4], D. Aristizabal Sierra[5], A. Bartl[6], M. Beccaria[7,8], S. Béjar[9,10], L. Benucci[11], S. Bityukov[4], I. Borjanović[8], G. Bozzi[12], J. Carvalho[14], B. Clerbaux[15], F. de Campos[16], A. de Gouvêa[17], C. Dennis[18], A. Djouadi[19], O.J.P. Éboli[13], U. Ellwanger[19], D. Fassouliotis[20], P.M. Ferreira[21], R. Frederix[3], B. Fuks[22], J.-M. Gerard[3], A. Giammanco[3], S. Gninenko[4], S. Gopalakrishna[17], T. Goto[23], B. Grzadkowski[24], J. Guasch[25], T. Hahn[26], S. Heinemeyer[27], A. Hektor[28], M. Herquet[3], B. Herrmann[22], K. Hidaka[29], M.K. Hirsch[5], K. Hohenwarter-Sodek[6], W. Hollik[26], G.W.S. Hou[30], T. Hurth[31,32], A. Ibarra[33], J. Illana[1], M. Kadastik[28], S. Kalinin[3], C. Karafasoulis[34], M. Karagöz Ünel[18], T. Kernreiter[6], M.M. Kirsanov[35], E. Kou[3], C. Kourkoumelis[20], S. Kraml[22,31], A. Kyriakis[34], V. Lemaitre[3], G. Macorini[38,39], M.B. Magro[40], W. Majerotto[41], F. Maltoni[3], V. Matveev[4], R. Mehdiyev[42,43], M. Misiak[24,31], G. Moreau[19], M. Mühlleitner[31], M. Müntel[28], A. Onofre[14], E. Özcan[45], F. Palla[11], L. Panizzi[38,39], S. Peñaranda[46], R. Pittau[47], A. Pukhov[50], M. Raidal[28], A.R. Raklev[2], L. Rebane[28], F.M. Renard[51], D. Restrepo[52], Z. Roupas[20], R. Santos[21], S. Schumann[53], G. Servant[54,55], F. Siegert[53], P. Skands[56], P. Slavich[57,31], J. Solà[10,58], M. Spira[59], S. Sultansoy[60,43], A. Toropin[4], J. Tseng[18], J.W.F. Valle[5], F. Veloso[14], A. Ventura[7,8], G. Vermisoglou[34], C. Verzegnassi[38,39], A. Villanova del Moral[5], G. Weiglein[36], M. Yılmaz[63]

[1] Departamento de Física Teórica y del Cosmos and CAFPE, Universidad de Granada, 18071 Granada, Spain
[2] DAMTP, CMS, University of Cambridge, Cambridge CB3 0WA, UK
[3] Centre for Particle Physics and Phenomenology (CP3), Université Catholique de Louvain, 1348 Louvain-la-Neuve, Belgium
[4] Institute for Nuclear Research RAS, Moscow 117312, Russia
[5] AHEP Group, Instituto de Física Corpuscular (CSIC, Universitat de Valencia), 46071 València, Spain
[6] Institut für Theoretische Physik, Universität Wien, 1090 Vienna, Austria
[7] Dipartimento di Fisica, Università di Salento, 73100 Lecce, Italy
[8] INFN, Sezione di Lecce, Lecce, Italy
[9] Grup de Física Teòrica, Universitat Autònoma de Barcelona, 08193 Barcelona, Spain
[10] Institut de Física d'Altes Energies, Universitat Autònoma de Barcelona, 08193 Barcelona, Spain
[11] INFN and Università di Pisa, Pisa, Italy
[12] Institut für Theoretische Physik, Universität Karlsruhe, 76128 Karlsruhe, Germany
[13] Instituto de Física, Universidade de São Paulo, São Paulo, SP 05508-900, Brazil
[14] LIP, Departamento de Física, Universidade de Coimbra, 3004-516 Coimbra, Portugal
[15] Université Libre de Bruxelles, Brussels, Belgium
[16] Departamento de Física e Química, Universidade Estadual Paulista, Guaratinguetá, SP, Brazil
[17] Department of Physics and Astronomy, Northwestern University, Evanston, IL 60208, USA
[18] University of Oxford, Denys Wilkinson Building, Keble Road, Oxford OX1 3RH, UK
[19] Laboratoire de Physique Théorique, Université de Paris-Sud, 91405 Orsay, France
[20] Physics Department, University of Athens, Panepistimiopolis, Zografou, 157 84 Athens, Greece
[21] Centro de Física Teórica e Computacional, Faculdade de Ciências, Universidade de Lisboa, 1649-003 Lisbon, Portugal
[22] LPSC, Université Grenoble I/CNRS-IN2P3, 38026 Grenoble, France
[23] Theory Group, IPNS, KEK, Tsukuba 305-0801, Japan
[24] Institute of Theoretical Physics, Warsaw University, 00681 Warsaw, Poland
[25] Departament de Física Fonamental, Universitat de Barcelona, 08028 Barcelona, Spain
[26] Max-Planck-Institut für Physik, 80805 Munich, Germany
[27] Instituto de Fisica de Cantabria IFCA (CSIC-UC), 39005 Santander, Spain
[28] National Institute of Chemical Physics and Biophysics, Ravala 10, Tallinn 10143, Estonia
[29] Department of Physics, Tokyo Gakugei University, Koganei, Tokyo 184-8501, Japan
[30] Department of Physics, National Taiwan University, Taipei 10617, Taiwan
[31] Theory Division, Physics Department, CERN, 1211 Geneva, Switzerland
[32] SLAC, Stanford University, Stanford, CA 94309, USA
[33] Deutsches Elektronen-Synchrotron DESY, 22603 Hamburg, Germany
[34] Institute of Nuclear Physics, NCSR "Demokritos", Athens, Greece
[35] INR, Moscow, Russia
[36] IPPP Durham, Department of Physics, University of Durham, Durham DH1 3LE, UK
[37] Universitá degli Studi di Milano and INFN, 20133 Milano, Italy
[38] Dipartimento di Fisica Teorica, Università di Trieste, Miramare, Trieste, Italy

[39]INFN, Sezione di Trieste, 34014 Trieste, Italy
[40]Faculdade de Engenharia, Centro Universitário Fundação Santo André, Santo André, SP, Brazil
[41]Institut für Hochenergiephysik der Österreichischen Akademie der Wissenschaften, 1050 Wien, Austria
[42]Département de Physique, Université de Montréal, Montréal, Canada
[43]Institute of Physics, Academy of Sciences, Baku, Azerbaijan
[44]ETH Zurich, 8093 Zurich, Switzerland
[45]Physics and Astronomy Department, University College London, London, UK
[46]Departamento de Física Teórica, Universidad de Zaragoza, 50009 Zaragoza, Spain
[47]Dipartimento di Fisica Teorica, Università di Torino and INFN, Sezione di Torino, Italy
[48]INFN, Sezione di Pavia, 27100 Pavia, Italy
[49]Institut für Theoretische Physik und Astrophysik, Universität Würzburg, 97074 Würzburg, Germany
[50]SINP MSU, Moscow, Russia
[51]Laboratoire de Physique Théorique et Astroparticules, UMR 5207, Université Montpellier II, 34095 Montpellier Cedex 5, France
[52]Instituto de Física, Universidad de Antioquia, Antioquia, Colombia
[53]Institute for Theoretical Physics, TU Dresden, 01062 Dresden, Germany
[54]Physics Department, CERN, 1211 Geneva 23, Switzerland
[55]Service de Physique Théorique, CEA Saclay, 91191 Gif-sur-Yvette, France
[56]Theoretical Physics, Fermi National Accelerator Laboratory, Batavia, IL 60510, USA
[57]LAPTH, CNRS, UMR 5108, Chemin de Bellevue BP110, 74941 Annecy-le-Vieuy, France
[58]HEP Group, Dept. Estructura i Constituents de la Matèria, Universitat de Barcelona, 08028 Barcelona, Spain
[59]Paul Scherrer Institute, 5232 Villigen PSI, Switzerland
[60]Physics Department, TOBB University of Economics and Technology, Ankara, Turkey
[61]Diparimento di Fisica e Astronomia, Universita di Catania, 95123 Catania, Italy
[62]Physics and Astronomy Department, University of California at Irvine, Irvine, USA
[63]Physics Department, Gazi University, Ankara, Turkey

Abstract This chapter of the "Flavor in the era of LHC" workshop report discusses flavor-related issues in the production and decays of heavy states at the LHC at high momentum transfer Q, both from the experimental and the theoretical perspective. We review top quark physics, and discuss the flavor aspects of several extensions of the standard model, such as supersymmetry, little Higgs models or models with extra dimensions. This includes discovery aspects, as well as the measurement of several properties of these heavy states. We also present publicly available computational tools related to this topic.

Contents

1 Introduction 185
   1.1 The ATLAS and CMS experiments 186
2 Flavor phenomena in top quark physics 189
   2.1 Introduction 189
   2.2 Wtb vertex 190
   2.3 FCNC interactions of the top quark 195

---

*Report of Working Group 1 of the CERN Workshop "Flavor in the era of the LHC", Geneva, Switzerland, November 2005–March 2007.

[a]e-mail: porod@physik.uni-wurzburg.de

[b]Convenors

[c]Subconvenors

   2.4 New physics contributions
      to top quark production 204
3 Flavor violation in supersymmetric models 210
   3.1 Introduction 210
   3.2 Effects of lepton flavor violation
      on dilepton invariant-mass spectra 218
   3.3 Lepton flavor violation in the long lived
      stau NLSP scenario 219
   3.4 Neutralino decays in models with broken
      R-parity 220
   3.5 Reconstructing neutrino properties from
      collider experiments in a Higgs triplet
      neutrino mass model 222
   3.6 SUSY (s)lepton flavor studies with ATLAS . 223
   3.7 Using the $l^+l^- + \not{E}_T +$ jet veto signature for
      slepton detection 224
   3.8 Using the $e^\pm \mu^\mp + \not{E}_T$ signature in the
      search for supersymmetry and lepton flavor
      violation in neutralino decays 226
   3.9 Neutralino spin measurement with ATLAS . 227
   3.10 SUSY Higgs boson production and decay . . 229
   3.11 Squark/gaugino production and decay 234
   3.12 Top squark production and decay 239
   3.13 SUSY Searches at $\sqrt{s} = 14$ TeV with CMS . 246
4 Non-supersymmetric standard model extensions . 248
   4.1 Introduction 248
   4.2 New quarks 249
   4.3 New leptons: heavy neutrinos 265

4.4 New neutral gauge bosons 271
4.5 New charged gauge bosons 276
4.6 New scalars 279
5 Tools 283
5.1 Introduction 283
5.2 A summary of
    The SUSY Les Houches Accord 2 284
5.3 SuSpect, HDECAY, SDECAY
    and SUSY-HIT 292
5.4 FeynHiggs 294
5.5 FchDecay 295
5.6 MSSM NMFV in FeynArts
    and FormCalc 295
5.7 SPheno 296
5.8 SOFTSUSY 296
5.9 CalcHep for beyond
    Standard Model Physics 297
5.10 HvyN 297
5.11 PYTHIA for flavor physics at the LHC . . . 298
5.12 Sherpa for flavor physics 298
Acknowledgements 299
References 299

1 Introduction

The origin of flavor structures and CP violation remains one of the big questions in particle physics. Within the standard model (SM) the related phenomena are successfully parametrized with the help of the CKM matrix in the quark sector and the PMNS matrix in the lepton sector. In both sectors intensive studies of flavor aspects have been carried out and are still going on as discussed in the reports by the other Working Groups in this Volume. Following the unification idea, it is strongly believed that eventually both sectors can be explained by a common underlying theory of flavor. Although current SM extensions rarely include a theory of flavor, many of them tackle the flavor question with the help of some special ansatz leading to interesting predictions for future collider experiments such as the LHC.

This chapter of the "Flavor in the era of LHC" report gives a comprehensive overview of the theoretical and experimental status of: (i) How flavor physics can be explored in the production of heavy particles like the top quark or new states predicted in extensions of the SM. (ii) How flavor aspects impact the discovery and the study of the properties of these new states. Both aspects require the study of processes at high momentum transfer Q. We discuss in detail the physics of the top quark, supersymmetric models, little Higgs models, extra dimensions, grand unified models and models explaining neutrino data. From the experimental side our focus will be on ATLAS and CMS whereas LHCb will be discussed on the report by the working group 2 [1],

where B, D and K decays are discussed to obtain complementary information on flavor in the hadronic sector. Additional complementary information on flavor in the leptonic sector due to the study of leptons is discussed in the report of working group 3 [2].

Section 2 discusses flavor aspects related to the top quark, which is expected to play an important role due to its heavy mass. The LHC will be a top quark factory, allowing one to study several of its properties in great detail. The Wtb coupling is an important quantity, which in the SM is directly related to the CKM element V_{tb}. In SM extensions new couplings can be present, which can be studied with the help of the angular distribution of the top decay products and/or in single-top production. In extensions of the SM also sizable flavor-changing neutral current (FCNC) decays can be induced, such as $t \to qZ$, $t \to q\gamma$ or $t \to qg$. The SM expectations for the corresponding branching ratios are of order 10^{-14} for the electroweak decays and of order 10^{-12} for the strong one. In extensions like two-Higgs doublet models, supersymmetry or additional exotic quarks they can be up to order 10^{-4}. The anticipated sensitivity of ATLAS and CMS for these branching ratios is of order 10^{-5}. A new physics contribution will also affect single and pair production of top quarks at LHC either via loop effects or due to resonances, a discussed in the third part of this section.

In Sect. 3 we consider flavor aspects of supersymmetric models. This class of models predicts partners for the SM particles which differ in spin by $1/2$. In a supersymmetric world flavor would be described by the usual Yukawa couplings. However, we know that supersymmetry (SUSY) must be broken, in a way most commonly parameterized in terms of soft SUSY-breaking terms. After a brief overview of the additional flavor structures in the soft SUSY breaking sector we first discuss the effect of lepton flavor violation in models with conserved R-parity. In spite of the stringent constraints from low energy data such as $\mu \to e\gamma$, they can significantly modify dilepton spectra, which play an important role in the determination of the SUSY parameters. We also discuss the possibilities to discover supersymmetry using the $e^{\pm}, \mu^{\mp}$ + missing energy signature. Lepton flavor violation plays also an important role in long lived stau scenarios with the gravitino as the lightest supersymmetric particle (LSP). In models with broken R-parity, neutrino physics predicts certain ratios of branching ratios of the LSP in terms of neutrino mixing angle (in case of a gravitino LSP the prediction will be for the next to lightest SUSY particle). Here the LHC will be important to establish several consistency checks of the model. Flavor aspects affect the squark sector in several ways. Firstly one expects that the lightest squark will be the lightest stop, due to the large top Yukawa coupling. Various aspects of its properties are studied here in different scenarios. Secondly it leads to flavor-violating squark production and flavor-violating decays of squarks and gluinos despite the stringent constraints from low energy data such as $b \to s\gamma$.

Non-supersymmetric extensions of the SM, such as grand unification, little Higgs or extra dimensional models, also predict new flavor phenomena, which are presented in Sect. 4. Such SM extensions introduce new fermions (quarks and leptons), gauge bosons (charged and/or neutral) and scalars. We study the LHC capabilities to discover these new mass states, paying a special attention to how to distinguish among different theoretical models. We start with the phenomenology of additional quarks and leptons, studying in detail their production at the LHC, and the available decay channels. It turns out that particles up to a mass of 1–2 TeV can be discovered and studied. In addition to the discovery reach, we discuss the possibilities to measure their mixing with SM fermions. They are also sources of Higgs bosons (produced in their decay) and hence they can significantly enhance the Higgs discovery potential of the LHC. Extended gauge structures predict additional heavy gauge bosons and, depending on the mass hierarchy, they can either decay to new fermions or be produced in their decay. In particular, the production of heavy neutrinos can be enhanced when the SM gauge group is extended with an extra $SU(2)_R$, which predicts additional W_R bosons. We also discuss flavor aspects of the discovery of new gauge bosons. This is specially important for the case of an extra Z', which appears in any extension of the SM gauge group, and for which model discrimination is crucial. The presence or not of new W' bosons also helps identifying additional $SU(2)$ gauge structures. Finally, several SM extensions predict new scalar particles. In some cases the new scalars are involved in the neutrino mass generation mechanism, e.g. in some little Higgs models and in the Babu–Zee model, which are realizations of the type II seesaw mechanism where neutrino masses a generated via a Higgs triplet. In these two cases, high energy observables, such as the decay branching ratios of doubly charged scalars, can be related to the neutrino mixing parameters measured in neutrino oscillations.

Last but not least, computational tools play an important role in the study of flavor aspects at the LHC. In Sect. 5 we give an overview of the publicly available tools, ranging from spectrum calculators, to decay packages and Monte Carlo programs. In addition, we briefly discuss the latest version of the SUSY Les Houches Accord (2008), which serves as an interface between various programs and now includes flavor aspects.

1.1 The ATLAS and CMS experiments

The CERN Large Hadron Collider (LHC) is currently being installed in the 27-km ring previously used for the LEP e^+e^- collider. This machine will push up the high energy frontier by one order of magnitude, providing pp collisions at a center of mass energy of $\sqrt{s} = 14$ TeV.

Four main experiments will benefit from this accelerator: two general-purpose detectors, ATLAS (Fig. 1) and CMS (Fig. 2), designed to explore the physics at the TeV scale; one experiment, LHCb, dedicated to the study of B-hadrons and CP violation; and one experiment, ALICE, which will study heavy ion collisions. Here only the ATLAS and CMS experiments and their physics programs are discussed in some detail.

The main goal of these experiments is the verification of the Higgs mechanism for the electroweak symmetry breaking and the study of the "new" (i.e. non-SM) physics that is expected to manifest itself at the TeV scale to solve the hierarchy problem. The design luminosity of 10^{34} cm^{-2} s^{-1} of the new accelerator will also allow one to collect very large samples of B hadrons, W and Z gauge bosons and top quarks, allowing for stringent tests of the SM predictions.

Since this program implies the sensitivity to a very broad range of signatures and since it is not known how new physics may manifest itself, the detectors have been designed to be able to detect as many particles and signatures as possible, with the best possible precision.

In both experiments the instrumentation is placed around the interaction point over the whole solid angle, except for the LHC beam pipe. As the particles leave the interaction point, they traverse the Inner Tracker, which reconstructs the trajectories of charged particles, the Electromagnetic and Hadronic calorimeters which absorb and measure the total energy of all particles except neutrinos and muons, and the Muon Spectrometer which is used to identify and measure the momentum of muons. The presence of neutrinos (and other hypothetic weakly interacting particles) is revealed as a non-zero vector sum of the particle momenta in the plane transverse to the beam axis.

Both the Inner Tracker and the Muon Spectrometer need to be placed inside a magnetic field in order to measure the momenta of charged particles using the radius of curvature of their trajectories. The two experiments are very different in the layout they have chosen for the magnet system. In ATLAS, a solenoid provide the magnetic field for the Inner Tracker, while a system of air-core toroids outside the calorimeters provide the field for the Muon Spectrometer. In CMS, the magnetic field is provided by a single very large solenoid which contains both the Inner Tracker and the calorimeters; the muon chambers are embedded in the iron of the solenoid return yoke. The magnet layout determines the size, the weight (ATLAS is larger but lighter) and even the name of the two experiments.

The CMS Inner Detector consists of Silicon Pixel and Strip detectors, placed in a 4 T magnetic field. The ATLAS Inner Tracker is composed by a smaller number of Silicon Pixel and Strip detectors and a Transition Radiation detector

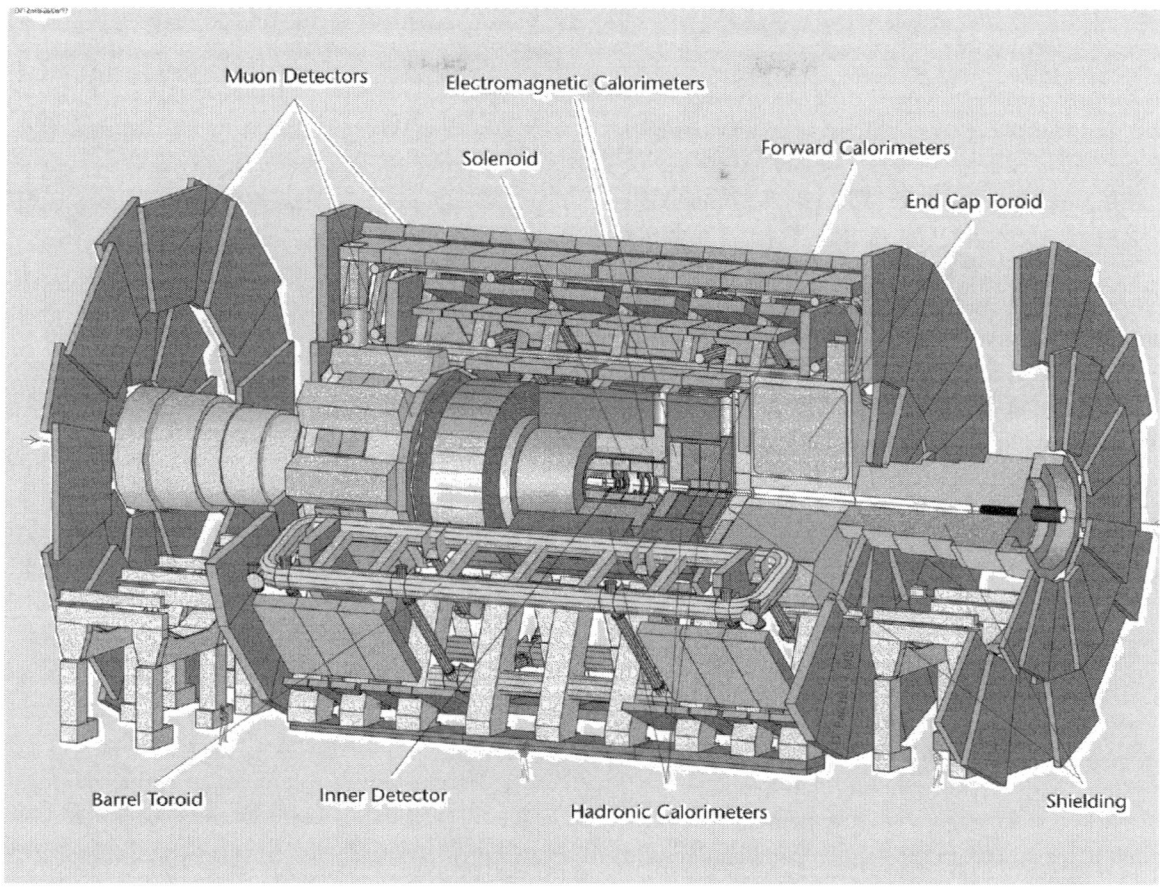

Fig. 1 An exploded view of the ATLAS detector

Fig. 2 An exploded view of the CMS detector

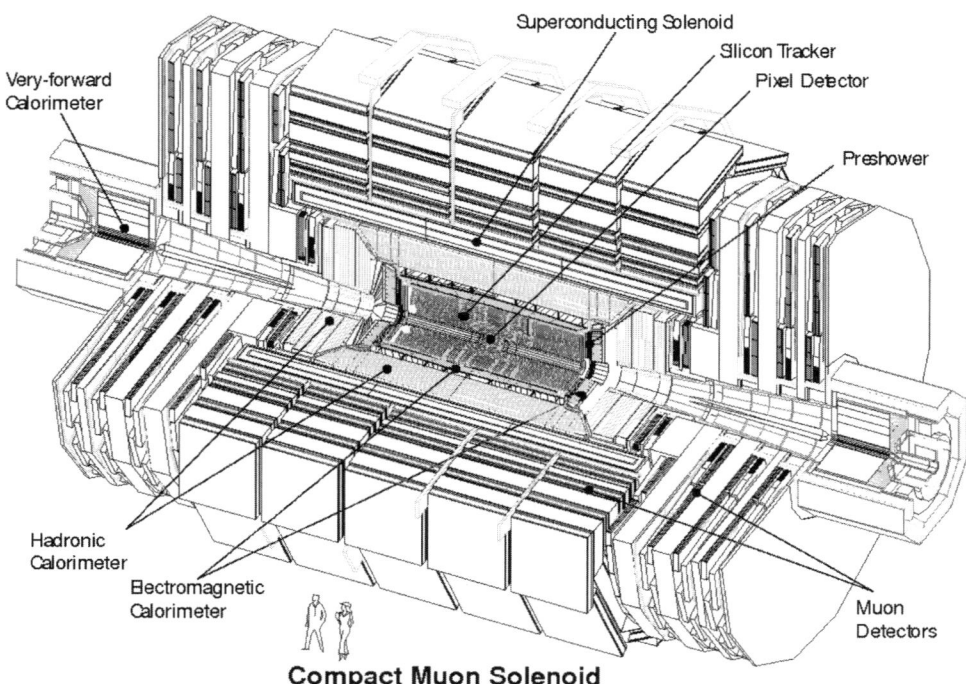

(TRT) at larger radii, inside a 2 T magnetic field. Thanks mainly to the larger magnetic field, the CMS tracker has a better momentum resolution, but the ATLAS TRT contributes to the electron/pion identification capabilities of the detector.

The CMS electromagnetic calorimeter is composed by PbWO$_4$ crystals with excellent intrinsic energy resolution ($\sigma(E)/E \sim 2\text{–}5\%/\sqrt{E(\text{GeV})}$). The ATLAS electromagnetic calorimeter is a lead/liquid argon sampling calorimeter. While the energy resolution is worse ($\sigma(E)/E \sim 10\%/\sqrt{E(\text{GeV})}$), thanks to a very fine lateral and longitudinal segmentation the ATLAS calorimeter provides more robust particle identification capabilities than the CMS calorimeter.

In both detectors the hadronic calorimetry is provided by sampling detectors with scintillator or liquid argon as the active medium. The ATLAS calorimeter has a better energy resolution for jets ($\sigma(E)/E \sim 50\%/\sqrt{E(\text{GeV})} \oplus 0.03$) than CMS ($\sigma(E)/E \sim 100\%/\sqrt{E(\text{GeV})} \oplus 0.05$) because it is thicker and has a finer sampling frequency.

The chamber stations of the CMS muon spectrometer are embedded into the iron of the solenoid return yoke, while those of ATLAS are in air. Because of multiple scattering in the spectrometer, and the larger field in the Inner Tracker the CMS muon reconstruction relies on the combination of the informations from the two systems; the ATLAS muon spectrometer can instead reconstruct the muons in stand-alone mode, though combination with the Inner Detector improves the momentum resolution at low momenta. The momentum resolution for 1 TeV muons is about 7% for ATLAS and 5% for CMS.

Muons can be unambiguously identified as they are the only particles which are capable to reach the detectors outside the calorimeters. Both detectors have also an excellent capability to identify electrons that are isolated (that is, they are outside hadronic jets). For example, ATLAS expects an electron identification efficiency of about 70% with a probability to misidentify a jet as an electron of the order of 10^{-5} [3]. The tau identification relies on the hadronic decay modes, since leptonically decaying taus cannot be separated from electrons and muons. The jets produced by hadronically decaying taus are separated from those produced by quark and gluons since they produce narrower jets with a smaller number of tracks. The capability of the ATLAS detector to separate τ jets from QCD jets is shown in Fig. 3.

The identification of the flavor of a jet produced by a quark is more difficult, and is practically limited to the identification of b jets, which are tagged by the vertex detectors using the relatively long lifetime of B mesons; the presence of a soft electron or muon inside a jet is also used to improve the b-tagging performance. In Fig. 4 the probability of mistagging a light jet as a b jet is plotted as a function of the b-tagging efficiency for the CMS detector [3]; comparable performances are expected for ATLAS.

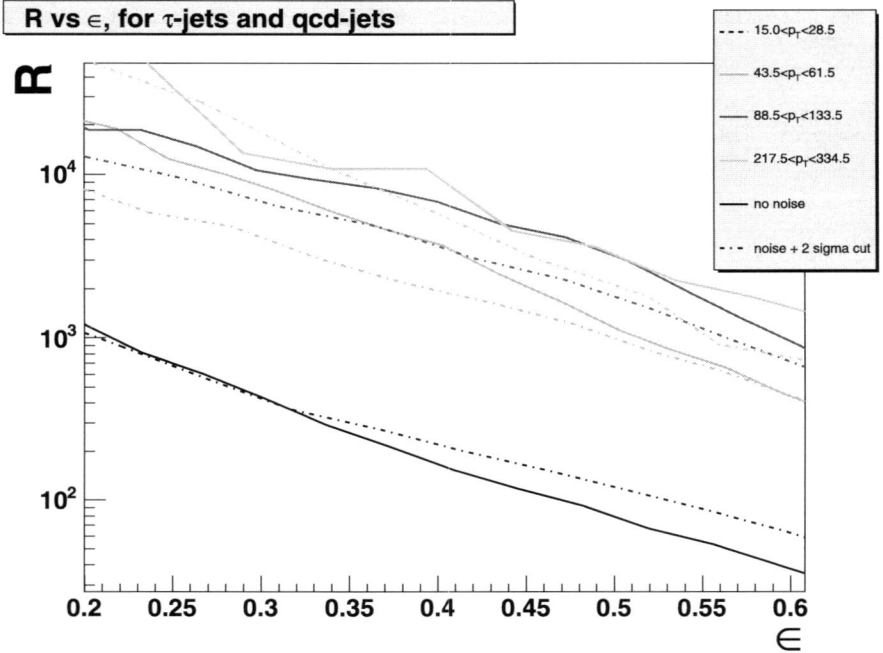

Fig. 3 The QCD jet rejection (inverse of mistagging efficiency) as a function of τ-tagging efficiency is reported for the ATLAS detector. The four *full curves* correspond to simulation without electronic noise in the calorimeters and different transverse momentum ranges, increasing from the lowest to the highest curve. The *dashed curves* correspond to simulation with electronic noise [4]

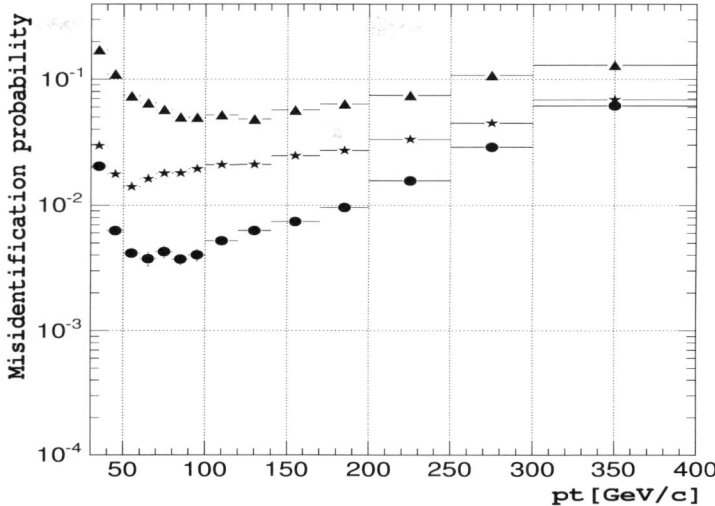

Fig. 4 The non-b jet mistagging efficiency for a fixed b-tagging efficiency of 0.5 as a function of jet transverse momentum for c jets (*triangles*), uds jets (*circles*) and gluon jets (*stars*) obtained for the CMS detector with an event sample of QCD jets and the secondary vertex tagging algorithm [5]

2 Flavor phenomena in top quark physics[1]

2.1 Introduction

The top quark is the heaviest and least studied quark of the SM. Although its properties have already been investigated at colliders [6], the available center of mass energy and the collected luminosity have not yet allowed for precise measurements, with the exception of its mass. The determination of other fundamental properties such as its couplings requires larger top samples, which will be available at the LHC. Additionally, due to its large mass, close to the electroweak scale, the top quark is believed to offer a unique window to flavor phenomena beyond the SM.

Within the SM, the Wtb vertex is due to a purely left handed current, and its size is given by the CKM matrix element V_{tb}, related to the top-bottom charged current. In a more general way, additional anomalous couplings such as right handed vectorial couplings and left and right handed tensorial couplings can also be considered. The study of the angular distribution of the top decay products at the LHC will allow for precision measurements of the structure of the Wtb vertex, providing an important probe for flavor physics beyond the SM.

In the SM there are no FCNC processes at the tree level and at one loop they can be induced by charged-current interactions, but they are suppressed by the Glashow–Iliopoulos–Maiani (GIM) mechanism [7]. These contributions limit the FCNC decay branching ratios to extremely small values in the SM. However, there are extensions of the SM which predict the presence of FCNC contributions already at the tree level and significantly enhance the top FCNC decay branching ratios [8–19]. Also loop-induced FCNCs could be greatly enhanced in some scenarios beyond the SM. In all these cases, such processes might be observed at the LHC.

In its first low luminosity phase (10 fb^{-1} per year and per experiment), the LHC will produce several million top quarks, mainly in pairs through gluon fusion $gg \to t\bar{t}$ and quark–antiquark annihilation $q\bar{q} \to t\bar{t}$, with a total cross section of $\sim$833 pb [21]. Single-top production [20, 22–26] will also occur, dominated by the t-channel process, $bq \to tq'$, with a total expected cross section of $\sim$320 pb [25, 26]. SM extensions, such as SUSY, may contribute with additional top quark production processes. The theoretical and experimental knowledge of single-top and $t\bar{t}$ production processes will result in important tests for physics beyond the SM. Moreover, besides the direct detection of new states (such as SUSY particles and Higgs bosons), new physics can also be probed via the virtual effects of the additional particles in precision observables. Finally, in addition to the potential deviations of the top couplings, it is possible that the top quark couples strongly to some sector of the new physics at the TeV scale, in such a way that the production of such states might result in new top quark signals. This possibility typically involves modifications of the top production cross sections, either for $t\bar{t}$ or single top, through the appearance of resonances or just excesses in the number of observed events. In some of these cases, the signal is directly associated with a theory of flavor, or at least of the origin of the top mass.

In this section different flavor phenomena associated to top quark physics are presented: anomalous charged and neutral top couplings, including the Wtb vertex structure and the measurement of V_{tb}; top quark FCNC processes and possible contributions of new physics to top production, including the effects of anomalous couplings in $t\bar{t}$ and single-top production; and the possible observation of resonances which strongly couple to the top quark.

---
[1] Section coordinators: G. Burdman and N. Castro.

2.2 Wtb vertex

In extensions of the SM, departures from the SM expectation $V_{tb} \simeq 0.999$ are possible [27, 28], as well as new radiative contributions to the Wtb vertex [29, 30]. These deviations might be observed in top decay processes at the LHC and can be parametrized with the effective operator formalism by considering the most general Wtb vertex (which contains terms up to dimension five) according to

$$\mathcal{L} = -\frac{g}{\sqrt{2}}\bar{b}\gamma^\mu(V_L P_L + V_R P_R)t W_\mu^-$$
$$-\frac{g}{\sqrt{2}}\bar{b}\frac{i\sigma^{\mu\nu}q_\nu}{M_W}(g_L P_L + g_R P_R)t W_\mu^- + \text{h.c.}, \quad (1)$$

with $q = p_t - p_b$ (the conventions of [31] are followed with slight simplifications in the notation). If CP is conserved in the decay, the couplings can be taken to be real.[2]

2.2.1 Wtb anomalous couplings

Within the SM, $V_L \equiv V_{tb} \simeq 1$ and V_R, g_L, g_R vanish at the tree level, while nonzero values are generated at one loop level [32]. Additional contributions to V_R, g_L, g_R are possible in SM extensions, without spoiling the agreement with low energy measurements. The measurement of $\text{Br}(b \to s\gamma)$ is an important constraint to the allowed values of the Wtb anomalous couplings.

At the LHC, the top production and decay processes will allow one to probe in detail the Wtb vertex. Top pair production takes place through the QCD interactions without involving a Wtb coupling. Additionally, it is likely that the top quark almost exclusively decays in the channel $t \to W^+b$. Therefore, its cross section for production and decay $gg, q\bar{q} \to t\bar{t} \to W^+bW^-\bar{b}$ is largely insensitive to the size and structure of the Wtb vertex. However, the angular distributions of (anti-)top decay products give information about its structure, and can then be used to trace non-standard couplings. Angular distributions relating top and antitop decay products probe not only the Wtb interactions but also the spin correlations among the two quarks produced, and thus may be influenced by new production mechanisms as well.

2.2.1.1 Constraints from B physics
Rare decays of the B-mesons as well as the $B\bar{B}$ mixing provide important constraints on the anomalous Wtb couplings because they receive large contributions from loops involving the top quark and the W boson. In fact, it is the large mass of the top quark

---

[2] A general Wtb vertex also contains terms proportional to $(p_t + p_b)^\mu$, q^μ and $\sigma^{\mu\nu}(p_t + p_b)_\nu$. Since b quarks are on shell, the W bosons decay to light particles (whose masses can be neglected) and the top quarks can be approximately assumed on-shell, these extra operators can be rewritten in terms of the ones in (1) using the Gordon identities.

that protects the corresponding FCNC amplitudes against GIM cancellation. Thus, order unity values of $V_L - V_{tb}$, V_R, g_L and g_R generically cause $\mathcal{O}(100\%)$ effects in the FCNC observables. For V_R and g_L, an additional enhancement [33, 34] by m_t/m_b occurs in the case of $\bar{B} \to X_s\gamma$, because the SM chiral suppression factor m_b/M_W gets replaced by the order unity factor m_t/M_W.

Deriving specific bounds on the anomalous Wtb couplings from loop processes requires treating them as parts of certain gauge invariant interactions. Here, we shall consider the following dimension-six operators [35]:

$$O^{V_R} = \bar{t}_R \gamma^\mu b_R (\tilde{\phi}^\dagger i D_\mu \phi) + \text{h.c.},$$
$$O^{V_L} = \bar{q}_L \tau^a \gamma^\mu q_L (\phi^\dagger \tau^a i D_\mu \phi)$$
$$\quad - \bar{q}_L \gamma^\mu q_L (\phi^\dagger i D_\mu \phi) + \text{h.c.}, \quad (2)$$
$$O^{g_R} = \bar{q}_L \sigma^{\mu\nu} \tau^a t_R \tilde{\phi} W^a_{\mu\nu} + \text{h.c.},$$
$$O^{g_L} = \bar{q}'_L \sigma^{\mu\nu} \tau^a b_R \phi W^a_{\mu\nu} + \text{h.c.},$$

where $q_L = (t_L, V_{tb}b_L + V_{ts}s_L + V_{td}d_L)$, $q'_L = (V_{tb}^* t_L + V_{cb}^* c_L + V_{ub}^* u_L, b_L)$, and ϕ denotes the Higgs doublet. Working in terms of gauge invariant operators renders the loop results meaningful, at the expense of taking into account *all* the interactions that originate from (2), not only the Wtb ones.

As an example, let us consider the $b \to s\gamma$ transition. Since it involves low momenta only, one usually treats it in the framework of an effective theory that arises from the full electroweak model (SM or its extension) after decoupling the top quark and the heavy bosons. The leading contribution to the considered decay originates from the operator

$$O_7 = \frac{e}{16\pi^2} m_b \bar{s}_L \sigma^{\mu\nu} b_R F_{\mu\nu}. \quad (3)$$

The SM value of its Wilson coefficient C_7 gets modified when the anomalous Wtb couplings are introduced. Moreover, the presence of O_7 also above the decoupling scale μ_0 becomes a necessity, because counter-terms involving O_7 renormalize the UV-divergent $b \to s\gamma$ diagrams with O^{g_L} and O^{g_R} vertices. Thus, we are led to consider the $\bar{B} \to X_s\gamma$ branching ratio as a function of not only V_L, V_R, g_L and g_R but also $C_7^{(p)}$, i.e. the "primordial" value of C_7 before decoupling. Following the approach of [36], one finds

$$\text{Br}(B \to X_s \gamma) \times 10^4$$
$$= (3.15 \pm 0.23) - 8.18(V_L - V_{tb}) + 427 V_R$$
$$\quad - 712 g_L + 1.91 g_R - 8.03 C_7^{(p)}(\mu_0)$$
$$\quad + \mathcal{O}\left[(V_L - V_{tb}, V_R, g_L, g_R, C_7^{(p)})^2\right], \quad (4)$$

for $E_\gamma > 1.6$ GeV and $\mu_0 = 160$ GeV in the $\overline{\text{MS}}$ scheme.[3] As anticipated, the coefficients at V_L and g_R are of the same order as the first (SM) term, while the coefficients at V_R and g_L get additionally enhanced. The coefficients at g_L and g_R depend on μ_0 already at the leading order, and they are well approximated by $-379 - 485 \ln \mu_0/M_W$ and $-0.87 + 4.04 \ln \mu_0/M_W$, respectively. This μ_0-dependence and the one of $C_7^{(p)}(\mu_0)$ compensate each other in (4).

Taking into account the current world average [38]:

$$\text{Br}(\bar{B} \to X_s \gamma) = \left(3.55 \pm 0.24^{+0.09}_{-0.10} \pm 0.03\right) \times 10^{-4}, \quad (5)$$

a thin layer in the five dimensional space ($V_L - V_{tb}$, V_R, g_L, g_R, $C_7^{(p)}$) is found to be allowed by $b \to s\gamma$. When one parameter at a time is varied around the origin (with the other ones turned off), quite narrow 95% CL bounds are obtained. They are listed in Table 1.

If several parameters are simultaneously turned on in a correlated manner, their magnitudes are, in principle, not bounded by $b \to s\gamma$ alone. However, the larger they are, the tighter the necessary correlation is, becoming questionable at some point.

The bounds in Table 1 have been obtained under the assumption that the non-linear terms in (4) are negligible with respect to the linear ones. If this assumption is relaxed, additional solutions to that equation arise. Such solutions are usually considered to be fine-tuned. In any case, they are expected to get excluded by a direct measurement of the Wtb anomalous couplings at the LHC (see Sect. 2.2.1.2).

Considering other processes increases the number of constraints but also brings new operators with their Wilson coefficients into the game, so long as the amplitudes undergo ultraviolet renormalization. Consequently, the analysis becomes more and more involved. Effects of V_L and V_R on $b \to sl^+l^-$ have been discussed, e.g., in [37, 39]. These analyses need to be updated in view of the recent measurements, and extended to the case of g_L and g_R. The same refers to the $B\bar{B}$ mixing, for which (to our knowledge) no dedicated calculation has been performed to date. Exclusive rare decay modes in the presence of non-vanishing V_R have been discussed in [40, 41].

2.2.1.2 ATLAS sensitivity to Wtb anomalous couplings
The polarization of the W bosons produced in the top decay is sensitive to non-standard couplings [42]. W bosons can be produced with positive, negative or zero helicity, with corresponding partial widths Γ_R, Γ_L, Γ_0 which depend on V_L, V_R, g_L and g_R. General expressions for Γ_R, Γ_L, Γ_0 in terms of these couplings can be found in [43] and were included in the program TopFit. Their absolute measurement is rather difficult, so it is convenient to consider instead the helicity fractions $F_i \equiv \Gamma_i/\Gamma$, with $\Gamma = \Gamma_R + \Gamma_L + \Gamma_0$ the total width for $t \to Wb$. Within the SM, $F_0 = 0.703$, $F_L = 0.297$, $F_R = 3.6 \times 10^{-4}$ at the tree level, for $m_t = 175$ GeV, $M_W = 80.39$ GeV, $m_b = 4.8$ GeV. We note that F_R vanishes in the $m_b = 0$ limit because the b quarks produced in top decays have left handed chirality, and for vanishing m_b the helicity and the chirality states coincide. These helicity fractions can be measured in leptonic decays $W \to \ell\nu$. Let us denote by θ_ℓ^* the angle between the charged lepton three-momentum in the W rest frame and the W momentum in the t rest frame. The normalized angular distribution of the charged lepton can be written as

$$\frac{1}{\Gamma}\frac{d\Gamma}{d\cos\theta_\ell^*} = \frac{3}{8}(1 + \cos\theta_\ell^*)^2 F_R + \frac{3}{8}(1 - \cos\theta_\ell^*)^2 F_L$$
$$+ \frac{3}{4}\sin^2\theta_\ell^* F_0, \quad (6)$$

with the three terms corresponding to the three helicity states and vanishing interference [44]. A fit to the $\cos\theta_\ell^*$ distribution allows one to extract, from experiment, the values of F_i, which are not independent but satisfy $F_R + F_L + F_0 = 1$. From these measurements one can constrain the anomalous couplings in (1). Alternatively, from this distribution one can measure the helicity ratios [43]

$$\rho_{R,L} \equiv \frac{\Gamma_{R,L}}{\Gamma_0} = \frac{F_{R,L}}{F_0}, \quad (7)$$

which are independent quantities and take the values $\rho_R = 5.1 \times 10^{-4}$, $\rho_L = 0.423$ in the SM. As for the helicity fractions, the measurement of helicity ratios sets bounds on V_R, g_L and g_R. A third and simpler method to extract information about the Wtb vertex is through angular asymmetries involving the angle θ_ℓ^*. For any fixed z in the interval $[-1, 1]$, one can define an asymmetry

$$A_z = \frac{N(\cos\theta_\ell^* > z) - N(\cos\theta_\ell^* < z)}{N(\cos\theta_\ell^* > z) + N(\cos\theta_\ell^* < z)}. \quad (8)$$

The most obvious choice is $z = 0$, giving the forward-backward (FB) asymmetry A_{FB} [31, 45].[4] The FB asymmetry is related to the W helicity fractions by

$$A_{\text{FB}} = \frac{3}{4}[F_R - F_L]. \quad (9)$$

Table 1 Agenda

	$V_L - V_{tb}$	V_R	g_L	g_R	$C_7^{(p)}(\mu_0)$
Upper bound	0.03	0.0025	0.0004	0.57	0.04
Lower bound	−0.13	−0.0007	−0.0015	−0.15	−0.14

---

[3] The negative coefficient at V_L differs from the one in Fig. 1 of [37] where an anomalous Wcb coupling was effectively included, too.

[4] Notice the difference in sign with respect to the definitions in [31, 45], where the angle $\theta_{\ell b} = \pi - \theta_\ell^*$ between the charged lepton and b quark is used.

Other convenient choices are $z = \mp(2^{2/3} - 1)$. Defining $\beta = 2^{1/3} - 1$, we have

$$z = -(2^{2/3} - 1) \rightarrow$$
$$A_z = A_+ = 3\beta[F_0 + (1 + \beta)F_R],$$
$$z = (2^{2/3} - 1) \rightarrow \quad (10)$$
$$A_z = A_- = -3\beta[F_0 + (1 + \beta)F_L].$$

Thus, A_+ (A_-) only depend on F_0 and F_R (F_L). The SM values of these asymmetries are $A_{FB} = -0.2225$, $A_+ = 0.5482$, $A_- = -0.8397$. They are very sensitive to anomalous Wtb interactions, and their measurement allows us to probe this vertex without the need of a fit to the $\cos\theta_\ell^*$ distribution. It should also be pointed out that with a measurement of two of these asymmetries the helicity fractions and ratios can be reconstructed.

In this section, the ATLAS sensitivity to Wtb anomalous couplings is reviewed. The $t\bar{t} \rightarrow W^+bW^-\bar{b}$ events in which one of the W bosons decays hadronically and the other one in the leptonic channel $W \rightarrow \ell\nu_\ell$ (with $\ell = e^\pm, \mu^\pm$), are considered as signal events.[5] Any other decay channel of the $t\bar{t}$ pair constitutes a background to this signal. Signal events have a final state topology characterized by one energetic lepton, at least four jets (including two b jets) and large transverse missing energy from the undetected neutrino. Top pair production, as well as the background from single-top production, is generated with TopReX [46]. Further backgrounds without top quarks in the final state, i.e. $b\bar{b}$, W + jets, Z/γ^* + jets, WW, ZZ and ZW production processes, are generated using PYTHIA [47]. In all cases CTEQ5L parton distribution functions (PDFs) [48] were used. Events are hadronized using PYTHIA, taking also into account both initial and final state radiation. Signal and background events are passed through the ATLAS fast simulation [49] for particle reconstruction and momentum smearing. The b jet tagging efficiency is set to 60%, which corresponds to a rejection factor of 10 (100) for c jets (light quark and gluon jets).

A two-level probabilistic analysis, based on the construction of a discriminant variable which uses the full information of some kinematical properties of the event was developed and is described elsewhere [50, 51]. After this analysis, 220024 signal events (corresponding to an efficiency of 9%) and 36271 background events (mainly from $t\bar{t} \rightarrow \tau\nu b\bar{b}q\bar{q}'$) were selected, for a luminosity of 10 fb^{-1}. The hadronic W reconstruction is done from the two non-b jets with highest transverse momentum. The mass of the hadronic top, is reconstructed as the invariant mass of the hadronic W and the b jet (among the two with highest p_T) closer to the W. The leptonic W momentum cannot be directly reconstructed due to the presence of an undetected neutrino in the final state. Nevertheless, the neutrino four-momentum can be estimated by assuming the transverse missing energy to be the transverse neutrino momentum. Its longitudinal component can then be determined, with a quadratic ambiguity, by constraining the leptonic W mass (calculated as the invariant mass of the neutrino and the charged lepton) to its known on-shell value $M_W = 80.4$ GeV. In order to solve the twofold quadratic ambiguity in the longitudinal component it is required that the hadronic and the leptonic top quarks have the minimum mass difference.

The experimentally observed $\cos\theta_\ell^*$ distribution, which includes the $t\bar{t}$ signal as well as the SM backgrounds, is affected by detector resolution, $t\bar{t}$ reconstruction and selection criteria. In order to recover the theoretical distribution, it is necessary to: (i) subtract the background; (ii) correct for the effects of the detector, reconstruction, etc. The asymmetries are measured with a simple counting of the number of events below and above a specific value of $\cos\theta_\ell^*$. This has the advantage that the asymmetry measurements are not biased by the extreme values of the angular distributions, where correction functions largely deviate from unity and special care is required.

Due to the excellent statistics achievable at the LHC, systematic errors play a crucial role in the measurement of angular distributions and asymmetries already for a luminosity of 10 fb^{-1}. A thorough discussion of the different systematic uncertainties in the determination of the correction functions is therefore compulsory. The systematic errors in the observables studied (asymmetries, helicity fractions and ratios) are estimated by simulating various reference samples and observing the differences obtained. Uncertainties originating from Monte Carlo generators, PDFs, top mass dependence, initial and final state radiation, b jet tag efficiency, jet energy scale, background cross sections, pile-up and b quark fragmentation were considered. The results of the simulation, including statistical and systematic uncertainties, are summarized in Table 2.

Table 2 Summary of the results obtained from the simulation for the observables studied, including statistical and systematic uncertainties

Observable	Result		
F_0	0.700	±0.003 (stat)	±0.019 (sys)
F_L	0.299	±0.003 (stat)	±0.018 (sys)
F_R	0.0006	±0.0012 (stat)	±0.0018 (sys)
ρ_L	0.4274	±0.0080 (stat)	±0.0356 (sys)
ρ_R	0.0004	±0.0021 (stat)	±0.0016 (sys)
A_{FB}	−0.2231	±0.0035 (stat)	±0.0130 (sys)
A_+	0.5472	±0.0032 (stat)	±0.0099 (sys)
A_-	−0.8387	±0.0018 (stat)	±0.0028 (sys)

---

[5]From now on, the W boson decaying hadronically and its parent top quark will be named "hadronic", and the W decaying leptonically and its parent top quark will be called "leptonic".

Table 3 The 1σ limits on anomalous couplings obtained from the combined measurement of $A_{\pm}, \rho_{R,L}$ are shown. In each case, the couplings which are fixed at zero are denoted by a cross

	V_R	g_L	g_R
$A_{\pm}, \rho_{R,L}$	$[-0.0195, 0.0906]$	×	×
$A_{\pm}, \rho_{R,L}$	×	$[-0.0409, 0.00926]$	×
$A_{\pm}, \rho_{R,L}$	×	×	$[-0.0112, 0.0174]$
$A_{\pm}, \rho_{R,L}$	×	$[-0.0412, 0.00944]$	$[-0.0108, 0.0175]$
$A_{\pm}, \rho_{R,L}$	$[-0.0199, 0.0903]$	×	$[-0.0126, 0.0164]$

With this results, and considering the parametric dependence of the observables on V_R, g_L and g_R [43], constraints on the anomalous couplings were set using `TopFit`. Assuming only one nonzero coupling at a time, 1σ limits from the measurement of each observable can be derived [50, 51]. These limits can be further improved by combining the measurements of the four observables $\rho_{R,L}$ and $A_{\pm}$, using the correlation matrix [51], obtained from simulation.[6] Moreover, the assumption that only one coupling is nonzero can be relaxed. However, if V_R and g_L are simultaneously allowed to be arbitrary, essentially no limits can be set on them, since for fine-tuned values of these couplings their effects on helicity fractions cancel to a large extent. In this way, values $O(1)$ of V_R and g_L are possible yielding minimal deviations on the observables studied. Therefore, in the combined limits, which are presented in Table 3, it is required that either V_R or g_L vanishes.

Finally, with the same procedure, the 68.3% confidence level (CL) regions on the anomalous couplings are obtained (Fig. 5). The boundary of the regions has been chosen as a contour of constant χ^2. In case that the probability density functions (p.d.f.) of V_R and g_L were Gaussian, the boundaries would be ellipses corresponding to $\chi^2 = 2.30$ (see for instance [52]). In our non-Gaussian case the χ^2 for which the confidence regions have 68.3% probability is determined numerically, and it is approximately 1.83 for the (g_L, g_R) plot and 1.85 for (V_R, g_R).

2.2.2 Measurement of V_{tb} in single-top production

The value of the CKM matrix element V_{tb}, is often considered to be known to a very satisfactory precision ($0.9990 < |V_{tb}| < 0.9992$ at 90% CL [53]). However, this range is determined by assuming the unitarity of the 3×3 CKM matrix, which can be violated by new physics effects. The Tevatron measurements of $R \equiv \frac{|V_{tb}|^2}{|V_{td}|^2 + |V_{ts}|^2 + |V_{tb}|^2}$ are based on the relative number of $t\bar{t}$-like events with zero, one and two tagged b jets. The resulting values for R are

---
[6]We point out that the correlations among $A_{\pm}, \rho_{R,L}$ do depend (as they must) on the method followed to extract these observables from experiment. In our case, the correlations have been derived with the same procedure used to extract $A_{\pm}, \rho_{R,L}$ from simulated experimental data.

$1.12^{+0.27}_{-0.23}$ (stat. + syst.) [54] and $1.03^{+0.19}_{-0.17}$ (stat. + syst.) [55] for CDF and DØ respectively. Note that the V_{tb} determination from R, giving $|V_{tb}| > 0.78$ at 95% CL, is obtained assuming $|V_{td}|^2 + |V_{ts}|^2 + |V_{tb}|^2 = 1$. In fact, $R \simeq 1$ only implies $|V_{tb}| \gg |V_{ts}|, |V_{td}|$. Therefore, single-top production, whose cross section is directly proportional to $|V_{tb}|$, is crucial in order to reveal the complete picture of the CKM matrix.

Recently, the DØ collaboration announced the first observation of the single-top production. The corresponding results for the t- and s-channels are [56]:

$$\sigma^{s\text{-channel}} + \sigma^{t\text{-channel}} = 4.9 \pm 1.4 \text{ pb},$$
$$\sigma^{s\text{-channel}} = 1.0 \pm 0.9 \text{ pb}, \quad (11)$$
$$\sigma^{t\text{-channel}} = 4.2^{+1.8}_{-1.4} \text{ pb}.$$

This result can be compared to the SM prediction with $|V_{tb}| = 1$ [25]: $\sigma^{s\text{-channel}}_{\text{SM}} = 0.88 \pm 0.11$ pb, $\sigma^{t\text{-channel}}_{\text{SM}} = 1.98 \pm 0.25$ pb. Taking these results into account and considering the limit $R > 0.61$ at 95% CL, excluded regions for $|V_{ti}|$ were obtained and are shown in Fig. 6a–c (see [57] for the detailed computation). From this figure, the allowed values for $|V_{ti}|$ are found to be $0 \lesssim |V_{td}| \lesssim 0.62$, $0 \lesssim |V_{ts}| \lesssim 0.62$ and $0.47 \lesssim |V_{tb}| \lesssim 1$. The new data on the single-top production provides, for the first time, the lower bound of V_{tb}. However, we have to keep in mind that the latest 95% CL upper limits on the single-top production by the CDF collaboration [58] are lower than those by DØ:

$$\sigma^{s\text{-channel}} + \sigma^{t\text{-channel}} < 2.7 \text{ pb},$$
$$\sigma^{s\text{-channel}} < 2.5 \text{ pb}, \quad (12)$$
$$\sigma^{t\text{-channel}} < 2.3 \text{ pb}.$$

Using this bound, different constraints on $|V_{ti}|$ can be found, as shown in Fig. 6d–f.

Going from Tevatron to LHC, the higher energy and luminosity will provide better possibilities for a precise determination of V_{tb}. Among all three possible production mechanisms, the t-channel ($q_W^2 < 0$) is the most promising process due to its large cross section, $\sigma \simeq 245$ pb [26, 59, 60] and V_{tb} could be determined at the 5% precision level already with 10 fb^{-1} of integrated luminosity, assuming a total error of 10% for the t-channel cross section

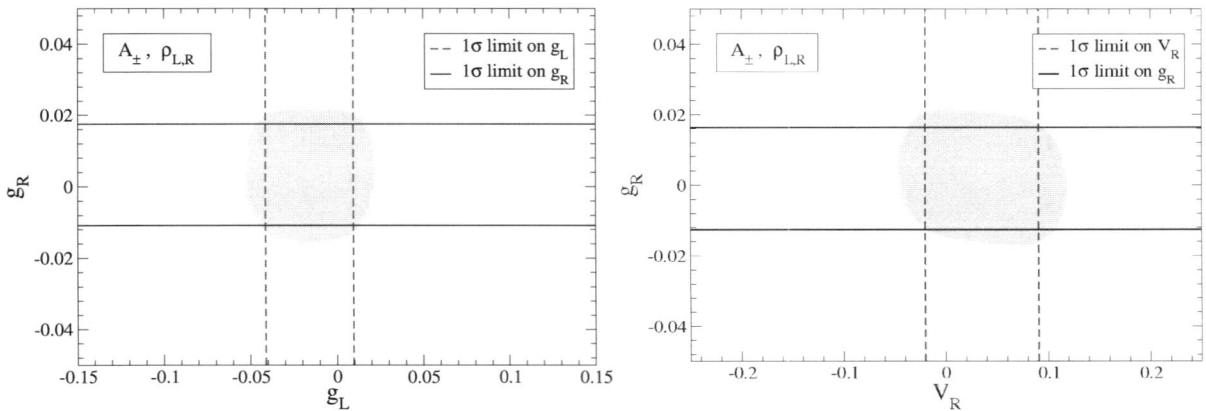

Fig. 5 68.3% CL confidence regions on anomalous couplings: g_L and g_R, for $V_R = 0$ (*left*); V_R and g_R, for $g_L = 0$ (*right*). The 1σ combined limits in Table 3 are also displayed

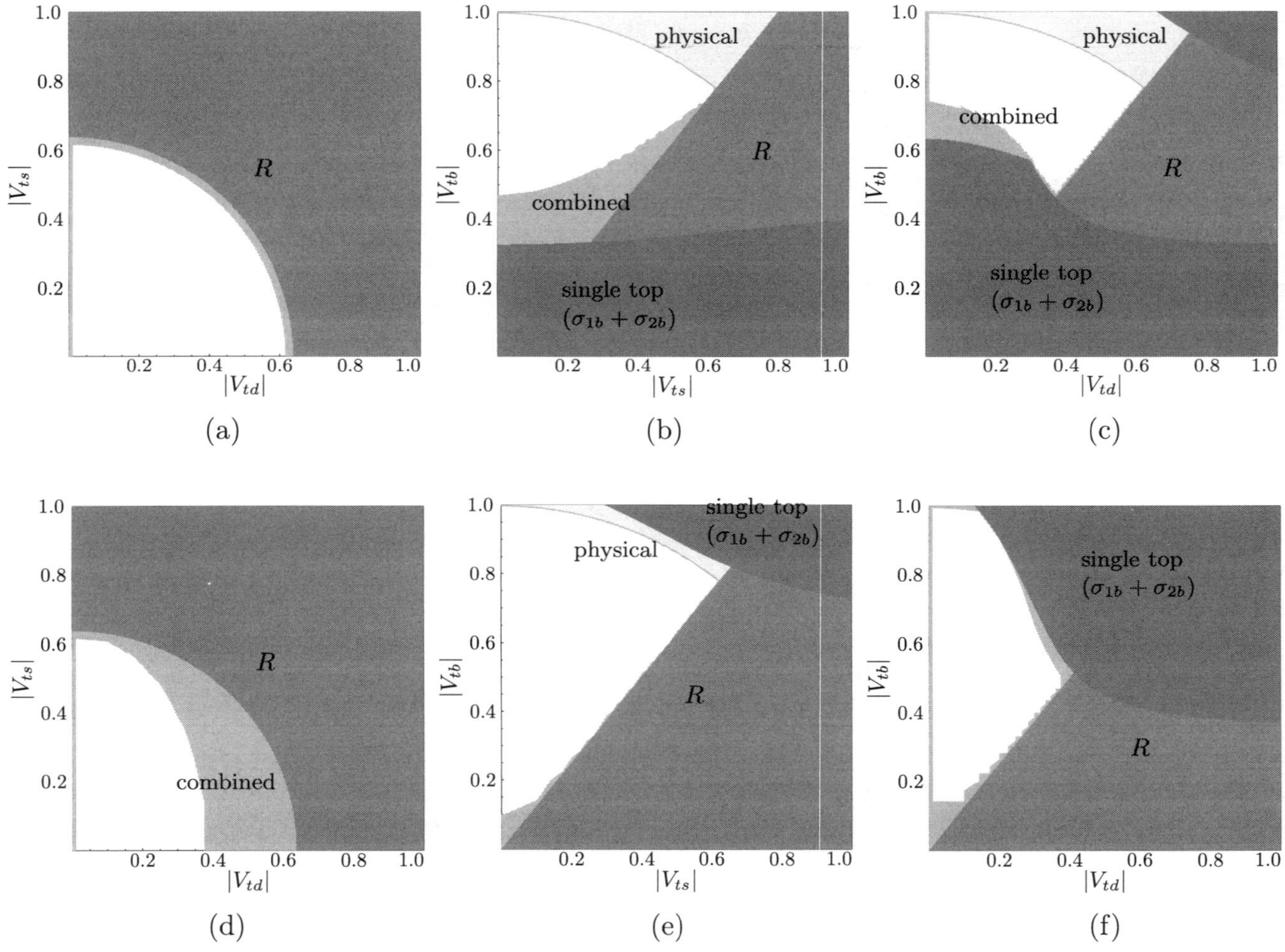

Fig. 6 Excluded regions for $|V_{td}|$, $|V_{ts}|$, and $|V_{tb}|$, obtained from the measurement of R and from the single-top production, $\sigma_{1b} + \sigma_{2b}$, at 95% CL. The figures (**a**)–(**c**) and (**d**)–(**f**) are obtained by using, respectively, the latest DØ (see (11)) and CDF (see (12)) data on the single-top production. The combination of both bounds provides an additional excluded region. The physical bound $|V_{td}|^2 + |V_{ts}|^2 + |V_{tb}|^2 < 1$ is also considered

measurement [61]. The precision of this result is limited by the systematic uncertainty and might be well improved with better understanding of the detector and background. The other channels, W-associated ($q_W^2 = M_W^2$) and s-channel ($q_W^2 > 0$), are more challenging due to a much larger systematic uncertainty. However, a measurement of these production mechanisms will also be important to further understand the nature of the top quark coupling to the weak current, especially because new physics could affect differently the different single-top production channels (see e.g. [24]).

Since V_{tb} is not known, the $V_{tb} \neq 1$ alternative should be still acceptable. If V_{tb} is considerably smaller than one, which would mean that $t(b)$ couples not only to $b(t)$ but also to the extra quarks. Thus, a measurement of $V_{tb} \neq 1$ would be an evidence for new heavy quarks. Their existence is in fact predicted by many extensions of the SM [28, 62–64] and furthermore, the current electroweak precision data allows for such possibility [65, 66]. In this class of models, the familiar 3×3 CKM matrix is a sub-matrix of a 3×4, 4×3, 4×4 or even larger matrix. Those matrices could also be constrained, e.g. by the 4×4 unitarity condition. Although the $3 \times 4/4 \times 3$ matrix, which is often induced by the vector-like quark models, breaks the GIM mechanism, the current tree-level FCNC measurements do not lead to strong constraints. However, the vector-like models with down-type quark (models with a 3×4 matrix) modify the tree-level $Zb\bar{b}$ coupling by a factor of $\cos^2\theta_{34}$, where the 3rd–4th generation mixing θ_{34} parameterizes the 3×4 matrix together with the usual CKM parameters ($\theta_{12}, \theta_{23}, \theta_{13}$). Since V_{tb} is written as $V_{tb} \simeq \cos\theta_{34}$ in the same parameterization, the measurement of R_b ratio, $R_b = \Gamma(Z \to b\bar{b})/\Gamma(Z \to \text{hadron})$, forbids V_{tb} significantly different from one in this type of models.

In the models with a singlet up-type quark (4×3 matrix case) or one complete generation (4×4 matrix case), the constraint on V_{tb} from the R_b measurement can be milder. In the SM, R_b comes from the tree diagram mentioned above and the t quark loop contribution, which is proportional to $|V_{tb}|$, is sub-dominant. If there is an extra fermion t', V_{tb} can be reduced. On the other hand, we obtain an extra loop contribution from t', proportional to $|V_{t'b}|$. In general, $V_{t'b}$ increases when V_{tb} decreases. Thus, the constraint on V_{tb} depends on the t' mass. Using the current CDF upper limit, $m_{t'} > 258$ GeV [67], it can be shown that $|V_{tb}| > 0.95$ (see Sect. 4.2.1). This result relies on the assumption that the corrections to R_b and to S, T, U parameters [68, 69] induced by loop effects are only coming from the t and t'. Therefore, more sophisticated models with an extended particle content may be less constrained. For a more precise argument in any given model, all the well measured experimental data from loop processes, such as the $B \to X_s \gamma$ branching ratio and the electroweak precision data must be comprehensively analyzed. Nevertheless it should be emphasized that the usual claim that the S parameter excludes the fourth generation is based on the assumption that $T \simeq 0$. The fourth generation model increases S and T simultaneously, and thus leaves a larger parameter space for this model than the R_b measurement alone [57, 70–72]. Further discussion on the search for extra quarks at the LHC can be found in Sect. 4 and in [57, 73].

2.3 FCNC interactions of the top quark

If the top quark has FCNC anomalous couplings to the gauge bosons, its production and decay properties will be affected. FCNC processes associated with the production [74–76] and decay [77] of top quarks have been studied at colliders and the present direct limits on the branching ratios are $\text{Br}(t \to qZ) < 7.8\%$ [74], $\text{Br}(t \to q\gamma) < 0.8\%$ [75] and $\text{Br}(t \to qg) < 13\%$ [78]. Nevertheless, the amount of data collected up to now is not comparable with the statistics expected at the LHC and thus either a discovery or an important improvement in the current limits is expected [79–82].

In the top quark sector of the SM, the small FCNC contributions limit the corresponding decay branching ratios to the gauge bosons (Z, γ and g) to below 10^{-12} [18, 83–86]. There are however extensions of the SM, like supersymmetric models including R-parity violation [8–14], multi-Higgs doublet models [15–17] and extensions with exotic (vector-like) quarks [18, 19], which predict the presence of FCNC contributions already at the tree level and significantly enhance the FCNC decay branching ratios. The theoretical predictions for the branching ratios of top FCNC decays within the SM and some of its extensions are summarized in Table 4.

In addition, theories with additional sources of FCNCs may result in flavor violation in the interactions of the scalar sector with the top quark. For example, this is the case in Topcolor-assisted Technicolor [87, 88], where tree-level FCNCs are present. In these theories the scalar sector responsible for the top quark mass can be discovered through the FCNC decay [89] $h_t \to tj$, where j is mainly a charm-quark jet. Also, and as we shall see in detail in Sect. 2.3.2, models with multi-Higgs doublets contain additional sources of flavor violation at one loop that may lead to FCNC decays of the Higgs.

Table 4 Branching ratios for FCNC top quark decays predicted by different models

Decay	SM	Two-Higgs	SUSY with R-parity violation	Exotic quarks
$t \to qZ$	$\sim 10^{-14}$	$\sim 10^{-7}$	$\sim 10^{-5}$	$\sim 10^{-4}$
$t \to q\gamma$	$\sim 10^{-14}$	$\sim 10^{-6}$	$\sim 10^{-6}$	$\sim 10^{-9}$
$t \to qg$	$\sim 10^{-12}$	$\sim 10^{-4}$	$\sim 10^{-4}$	$\sim 10^{-7}$

2.3.1 Top quark production in the effective Lagrangian approach

If a strong FCNC coupling exists associated to the top quark sector, it is expected that it influences the production of single-top events through the process $pp \to t + q, g$. This single-top production channel is thus an excellent probe for flavor phenomena beyond the SM. In this section, the phenomenology of strong flavor-changing single-top production in the effective Lagrangian approach is considered. The approach is model independent and makes use of a subset of all dimension five and six operators that preserve the gauge symmetries of the SM as written in [35]. The subset chosen contains all operators that contribute to strong FCNC including the four fermion interactions. This methodology has been used by many authors to study single-top quark production using the SM as its low energy limit but also in other models like supersymmetry, two-Higgs doublet models and others [24, 90–101].

The effective Lagrangian is a series in powers of $1/\Lambda$, Λ being the scale of new physics. We shall consider first the terms that originate from mixing with SM charged currents, that is, with diagrams with a charged boson, either as virtual particle or in the final state. These are processes of the type $pp \to (\bar{q}q) \to \bar{q}t + X$ and $pp \to (gq) \to Wt + X$ and the charge conjugate processes. Due to CKM suppression and small parton density functions, these Λ^{-2} terms are much smaller than the Λ^{-4} terms. There are several contributions of order Λ^{-4} to the cross section of single top production. These are summarized in Table 5. A more detailed discussion can be found in [102]. Cross sections for these processes were calculated in [103, 104].

The main goal of this work was to produce all cross sections and decay widths related to strong FCNC with a single top quark in a form appropriate for implementation in the TopReX generator [46]. This implies that all cross sections had to be given in differential form with the top spin taken into account. Most of the processes were already inserted in the generator (see release 4.20 of TopReX) and the remaining ones will be inserted in the near future.

In this section, a joint analysis of the results obtained in [102–104] is performed. To investigate the dependence of the cross sections on the values of the anomalous couplings, which are denoted by constants α_{ij} and β_{ij}, random values for α_{ij} and β_{ij} were generated and the resulting cross sections were plotted against the branching ratio of the top quark for the decay $t \to gu$. The motivation for doing this is simple: the top quark branching ratios for these decays may vary by as much as eight orders of magnitude, from $\sim 10^{-12}$ in the SM to $\sim 10^{-4}$ for some supersymmetric models. This quantity is therefore a good measure of whether any physics beyond that of the SM exists.

In Fig. 7 the cross sections for the processes $pp \to t + \text{jet}$ and $pp \to t + W$ via a u quark versus the branching ratio $\text{Br}(t \to gu)$ are shown. This plot was obtained by varying the constants α and β in a random way, as described before. Each combination of α and β originates a given branching ratio and a particular value for each cross section. Obviously, another set of points may generate the same value for the branching ratio but a different value for the cross section, which justifies the distribution of values of $\sigma(pp \to t + \text{jet})$ and $\sigma(pp \to t + W)$. Values of α and β for which the branching ratio varies between the SM value and the maximum value predicted by supersymmetry were chosen.[7] The cross sections for top plus jet and top plus a W boson production via a c quark are similar to these ones although smaller in value. Notice that the Wt cross section is proportional to only one of the couplings, which makes it a very attractive observable—it may allow us to impose constraints on a single anomalous coupling [102].

It should be noted that single-top production depends also on the contributions of the four fermion operators. Hence, even if the branching ratios $\text{Br}(t \to gu(c))$ are very small, there is still the possibility of having a large single-top cross section with origin in the four fermion couplings. In Fig. 7 we did not consider this possibility, setting the four-fermion couplings to zero. For a discussion on the four-fermion couplings see [103].

In Fig. 8 the cross sections for $pp \to t + Z$ and $pp \to t + \gamma$ via a u quark, versus the branching ratio $\text{Br}(t \to gu)$ are plotted. The equivalent plot with an internal c quark is similar, but the values for the cross section are much smaller. In this plot we can see that both cross sections are very small in the range of $\{\alpha\beta\}$ considered. These results imply that their contribution will hardly be seen at the LHC, unless the values for the branching ratio are peculiarly large.

Table 5 Contributions of order Λ^{-4} to the cross section of top production

Direct production	$pp \to (gq) \to t + X$
Top + jet production	$pp \to (gg) \to \bar{q}t + X$
	$pp \to (gq) \to gt + X$
	$pp \to (\bar{q}q) \to \bar{q}t + X$
	(including 4-fermion interactions)
Top + antitop production	$pp \to (gg) \to \bar{t}t + X$
	$pp \to (\bar{q}q) \to \bar{t}t + X$
top + gauge boson production	$pp \to (gq) \to \gamma t + X$
	$pp \to (gq) \to Zt + X$
	$pp \to (gq) \to Wt + X$
top + Higgs production	$pp \to (gq) \to ht + X$

---
[7]Both α/Λ^2 and β/Λ^2 were varied between 10^{-6} and 1 TeV^{-2}.

Fig. 7 Cross sections for the processes $pp \to t + \text{jet}$ (*crosses*) and $pp \to t + W$ (*stars*) via an u quark, as a function of the branching ratio $\text{Br}(t \to gu)$

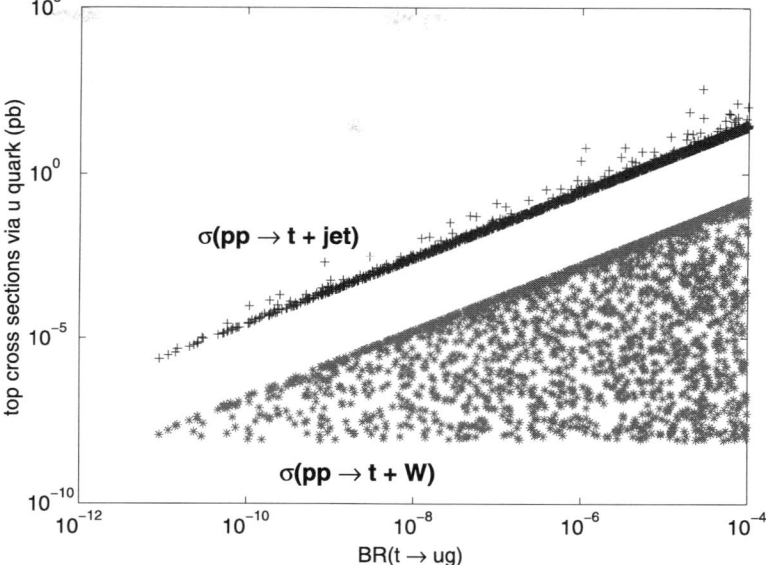

Fig. 8 Cross sections for the processes $pp \to t + Z$ (*upper line*) and $pp \to t + \gamma$ (*lower line*) via a u quark, as a function of the branching ratio $\text{Br}(t \to gu)$

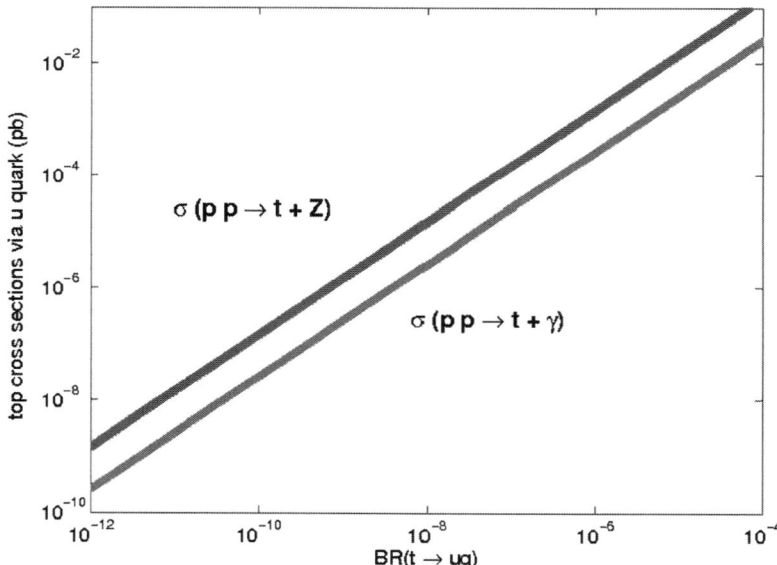

The same, in fact, could be said for $pp \to t + h$. Even for the smallest allowed SM Higgs mass, the values of the cross section for associated top and Higgs production are very small. The same holds true for the processes involving the anomalous couplings of the c quark.

The smallness of the effects of these operators in the several cross sections holds true, as well, for the top–antitop channel. In this case, even for a branching ratio $\text{Br}(t \to gu) \simeq 10^{-4}$, the contributions to the cross section $\sigma(pp \to t\bar{t})$ do not exceed, in absolute value, one picobarn.

In conclusion, the strong FCNC effective operators are constrained in their impact on several channels of top quark production. Namely, Figs. 7 and 8 illustrate that, if there are indeed strong FCNC effects on the decays of the top quark, their impact will be more significant in the single-top plus jet production channel. It is possible, according to these results, to have an excess in the cross section $\sigma(pp \to t + \text{jet})$ arising from new physics described by the operators we have considered here, at the same time obtaining results for the production of a top quark alongside a gauge and Higgs boson, or for $t\bar{t}$ production, which are entirely in agreement with the SM predictions. This reinforces the conclusion that the cross section for single-top plus jet production is an important probe for the existence of new physics beyond the SM. It is a channel extremely sensitive to the presence of that new physics, and boasts a significant excess in its cross section, whereas many other channels involving the top quark remain unchanged. Nevertheless, it may still be possible to use some of these unchanged channels, such as top plus W production, to constrain the β parameters, through the study of asymmetries such as $\sigma(pp \to tW^-) - \sigma(pp \to \bar{t}W^+)$.

2.3.2 Higgs boson FCNC decays into top quark in a general two-Higgs doublet model

The branching ratios for FCNC Higgs boson decays are at the level of 10^{-15}, for Higgs boson masses of a few hundred GeV. In this section, the FCNC decays of Higgs bosons into a top quark in a general two-Higgs doublet model (2HDM) are considered. In this model, the Higgs FCNC decay branching ratios can be substantially enhanced and perhaps can be pushed up to the visible level, particularly for h^0 which is the lightest CP-even spinless state in these models [105]. We compute the maximum branching ratios and the number of FCNC Higgs boson decay events at the LHC. The most favorable mode for production and subsequent FCNC decay is the lightest CP-even state in the type II 2HDM, followed by the other CP-even state, if it is not very heavy, whereas the CP-odd mode can never be sufficiently enhanced. The present calculation shows that the branching ratios of the CP-even states may reach 10^{-5}, and that several hundred events could be collected in the highest luminosity runs of the LHC. Some strategies to use these FCNC decays as a handle to discriminate between 2HDM and supersymmetric Higgs bosons are also pointed out.

Some work in relation with the 2HDM Higgs bosons FCNCs has already been performed in [15, 16], and, in the context of the MSSM, in [106–109]. In this work the production of any 2HDM Higgs boson ($h = h^0, H^0, A^0$) at the LHC is computed and analyzed, followed by the one-loop FCNC decay $h \to tc$. The maximum production rates of the combined cross section,

$$\sigma(pp \to h \to tc) \equiv \sigma(pp \to hX)\mathrm{Br}(h \to tc),$$

$$\mathrm{Br}(h \to tc) \equiv \frac{\Gamma(h \to t\bar{c} + \bar{t}c)}{\sum_i \Gamma(h \to X_i)}, \quad (13)$$

takes into account the restrictions from the experimental determination of the $b \to s\gamma$ branching ratio ($m_{H^\pm} \gtrsim 350$ GeV [110]), from perturbativity arguments ($0.1 \lesssim \tan\beta \lesssim 60$, where $\tan\beta$ is the ratio of the vacuum expectation values of each doublet), from the custodial symmetry ($|\delta\rho^{2\mathrm{HDM}}| \lesssim 0.1\%$) and from unitarity of the Higgs couplings. In this section a summarized explanation of the numerical analysis is given. For further details see [16, 111].

The full one-loop calculation of $\mathrm{Br}(h \to tc)$ in the type II 2HDM, as well as of the LHC production rates of these FCNC events were included. It is considered that $\mathrm{Br}(h \to tc)$ in the type I 2HDM is essentially small (for all h), and that these decays remain always invisible. The basic definitions in the general 2HDM framework can be found in [105].

The calculations were performed with the help of the numeric and algebraic programs FeynArts, FormCalc and LoopTools [112–114]. A parameter scan of the production rates over the 2HDM parameter space in the $(m_{H^\pm}, m_{h^0})$ plane was done, keeping $\tan\beta$ fixed.

In Fig. 9a–b, the $\mathrm{Br}(h^0 \to tc)$ for the lightest CP-even state (type II 2HDM) is shown. The Br is sizable, up to 10^{-5}, for the range allowed from $b \to s\gamma$. In Fig. 9c the production cross sections explicitly separated (the gluon–gluon fusion at one-loop and the $h^0 q\bar{q}$ associated production at the tree level [115, 116]) are presented. The control over $\delta\rho^{2\mathrm{HDM}}$ is displayed in Fig. 9d.

In practice, to better assess the possibility of detection at the LHC, one has to study the production rates of the FCNC events. A systematic search of the regions of parameter space with the maximum number of FCNC events for the light CP-even Higgs is presented in the form of contour lines in Fig. 9e. The dominant FCNC region for $h^0 (H^0)$ decay is where $\tan\alpha$ (α is the rotation angle which diagonalizes the matrix of the squared masses of the CP-even scalars) is large (small), $\tan\beta$ is large and $m_h \ll m_{A^0}$, with a maximum value up to few hundred events. As for the CP-odd state A^0, it plays an important indirect dynamical role on the other decays through the trilinear couplings, but its own FCNC decay rates never get a sufficient degree of enhancement due to the absence of the relevant trilinear couplings.

One should notice that in many cases one can easily distinguish whether the enhanced FCNC events stem from the dynamics of a general, unrestricted, 2HDM model, or rather from some supersymmetric mechanisms within the MSSM. In the 2HDM case the CP-odd modes $A^0 \to tc$ are completely hopeless, whereas in the MSSM they can be enhanced [106, 107, 117, 118]. Nevertheless, different ways to discriminate these rare events are discussed in [16].

The FCNC decays of the Higgs bosons into top quark final states are a potentially interesting signal, exceeding 1 fb for m_{H^+} up to 400 GeV (Fig. 9e). This however, is a small cross section once potentially important backgrounds are considered, such as Wjj and SM single-top production. A careful study of the backgrounds for this process should be carried out. If it were possible to fully reconstruct the top, then there might be hope to observe a distinctive Higgs bump in the tc channel [89].

2.3.3 single-top production by direct SUSY FCNC interactions

FCNC interactions of top quarks can provide an important indirect probe for new SUSY processes. For instance, the MSSM Higgs boson FCNC decay rates into top quark final states, e.g. $H^0, A^0 \to t\bar{c} + \bar{t}c$, can be of order 10^{-4} (see Sect. 2.3.2 and [107, 117–120]), while in the SM $\mathrm{Br}(H \to$

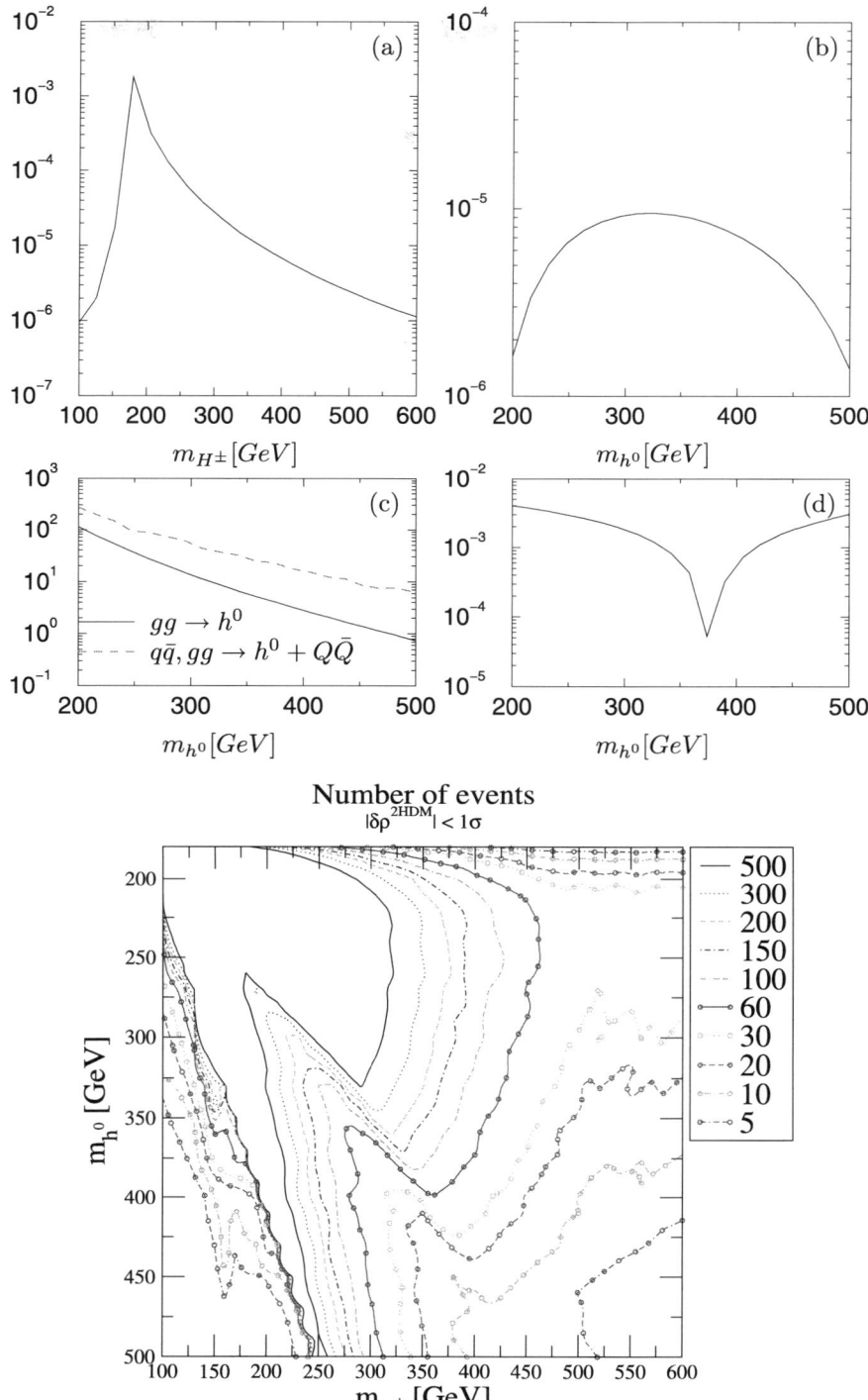

Fig. 9 (a) Br($h^0 \to tc$) versus $m_{H^\pm}$ for type II 2HDM. (b) Idem, versus m_{h^0}. (c) the production cross section (in pb) of h^0 at the LHC versus its mass. (d) $\delta\rho^{2HDM}$ versus m_{h^0}, for a fixed value of the other parameters. (e) Contour lines in the ($m_{H^\pm}, m_{h^0}$) plane for the maximum number of light CP-even Higgs FCNC events $h^0 \to tc$ produced at the LHC for 100 fb^{-1} of integrated luminosity

$t\bar{c}$) $\sim 10^{-13}$–10^{-16} (depending on the Higgs mass) [111]. There also exists the possibility to produce $t\bar{c}$ and $\bar{t}c$ final states without Higgs bosons or any other intervening particle [100, 121]. In this section it will be shown that the FCNC gluino interactions in the MSSM can actually be one efficient mechanism for direct FCNC production of top quarks [100].

In general, in the MSSM we expect terms of the form gluino–quark–squark or neutralino–fermion–sfermion, with the quark and squark having the same charge but belonging to different flavors. In the present study only the first type of terms, which are expected to be dominant, are considered. A detailed Lagrangian describing these generalized SUSY–QCD interactions mediated by gluinos can be found, e.g. in [106]. The relevant parameters are the flavor mixing coefficients δ_{ij}. In contrast to previous studies [122], in the present work, these parameters are only allowed in the LL part of the 6×6 sfermion mass matrices in flavor-chirality space. This assumption is also suggested by RG arguments [123, 124]. Thus, if M_{LL} is the LL block of a sfermion mass matrix, $\delta_{ij}(i \neq j)$ is defined as follows: $(M_{LL})_{ij} = \delta_{ij}\tilde{m}_i\tilde{m}_j$, where $\tilde{m}_i$ is the soft SUSY-breaking mass parameter corresponding to the LH squark of ith flavor [106]. The parameter δ_{23} is the one relating the 2nd and 3rd generations (therefore involving the top quark physics) and it is the less restricted one from the phenomenological point of view, being essentially a free parameter ($0 < \delta_{23} < 1$). Concretely, we have two such parameters, $\delta_{23}^{(t)LL}$ and $\delta_{23}^{(b)LL}$, for the up-type and down-type LL squark mass matrices respectively. The former enters the process under study whereas the latter enters Br($b \to s\gamma$), observable that we use to restrict our predictions on $t\bar{c} + \bar{t}c$ production. Notice that $\delta_{23}^{(b)LL}$ is related to the parameter $\delta_{23}^{(t)LL}$ because the two LL blocks of the squark mass matrices are precisely related by the CKM rotation matrix K as follows: $(\mathcal{M}_{\tilde{u}}^2)_{LL} = K(\mathcal{M}_{\tilde{d}}^2)_{LL}K^\dagger$ [125, 126].

The calculation of the full one-loop SUSY–QCD cross section $\sigma_{tc} \equiv \sigma(pp \to t\bar{c})$ using standard algebraic and numerical packages for this kind of computations [114, 127] has been performed. The typical diagrams contributing are gluon–gluon triangle loops [100]. In order to simplify the discussion it will be sufficient to quote the general form of the cross section:

$$\sigma_{tc} \sim \left(\delta_{23}^{(t)LL}\right)^2 \frac{m_t^2(A_t - \mu/\tan\beta)^2}{M_{\text{SUSY}}^4} \frac{1}{m_{\tilde{g}}^2}. \quad (14)$$

Here A_t is the trilinear top quark coupling, μ the higgsino mass parameter, $m_{\tilde{g}}$ is the gluino mass and M_{SUSY} stands for the overall scale of the squark masses [100]. The computation of σ_{tc} together with the branching ratio Br($b \to s\gamma$) in the MSSM was performed, in order to respect the experimental bounds on Br($b \to s\gamma$). Specifically, Br($b \to s\gamma$) = $(2.1$–$4.5) \times 10^{-4}$ at the 3σ level is considered [53].

In Figs. 10, 11 and 12 the main results of this analysis are presented. It can be seen that σ_{tc} is very sensitive to A_t and that it decreases with M_{SUSY} and $m_{\tilde{g}}$. As expected, it increases with $\delta_{23}^{LL} \equiv \delta_{23}^{(t)LL}$. At the maximum of σ_{tc}, it prefers $\delta_{23}^{LL} = 0.68$. The reason stems from the correlation of this maximum with the Br($b \to s\gamma$) observable. At the maximum, $2\sigma_{tc} \simeq 0.5$ pb, if we allow for relatively light gluino masses $m_{\tilde{g}} = 250$ GeV (see Fig. 12). For higher $m_{\tilde{g}}$ the cross section falls down fast; at $m_{\tilde{g}} = 500$ GeV it is already 10 times smaller. The total number of events per 100 fb^{-1} lies between 10^4–10^5 for this range of gluino masses. The fixed values of the parameters in these plots lie near the values that provide the maximum of the FCNC cross section. The dependence on μ is not shown, but it should be noticed that it decreases by $\sim 40\%$ in the allowed range $\mu = 200$–800 GeV. Values of $\mu > 800$ GeV are forbidden by Br($b \to s\gamma$). Large negative μ is also excluded by the experimental bound considered for the lightest squark mass, $m_{\tilde{q}_1} \lesssim 150$ GeV; too small $|\mu| \lesssim 200$ GeV is ruled out by the chargino mass bound $m_{\chi_1^\pm} \lesssim 90$ GeV. The approximate maximum of σ_{tc} in parameter space has been computed using an analytical procedure as described in [100].

Finally, it should be noticed that $t\bar{c}$ final states can also be produced at one-loop by the charged-current interactions within the SM. This one-loop cross section at the LHC was computed, with the result $\sigma^{\text{SM}}(pp \to t\bar{c} + \bar{t}c) = 7.2 \times 10^{-4}$ fb. It amounts to less than one event in the entire lifetime of the LHC. Consequently, evidence for such signal

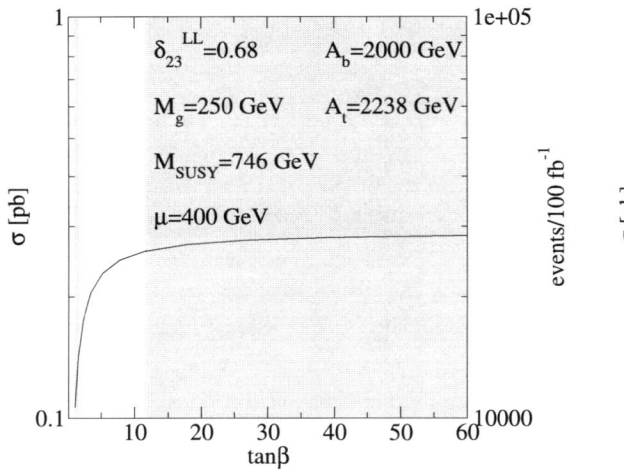
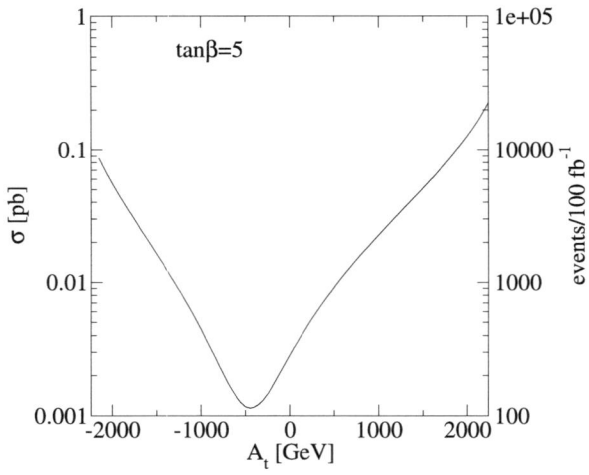

Fig. 10 σ_{tc} (in pb) and number of events per 100 fb^{-1} of integrated luminosity at the LHC, as a function of $\tan\beta$ (*left*) and A_t (*right*) for the given parameters. The shaded region is excluded by the experimental limits on Br($b \to s\gamma$)

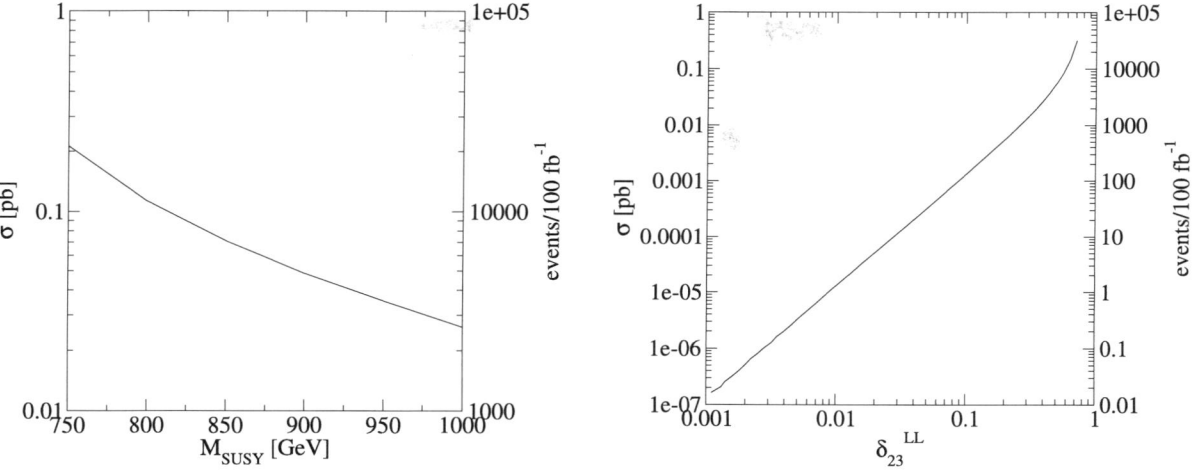

Fig. 11 σ_{tc} (in pb) and number of events per 100 fb^{-1} of integrated luminosity at the LHC, as a function of M_{SUSY} (*left*) and δ_{23}^{LL} (*right*)

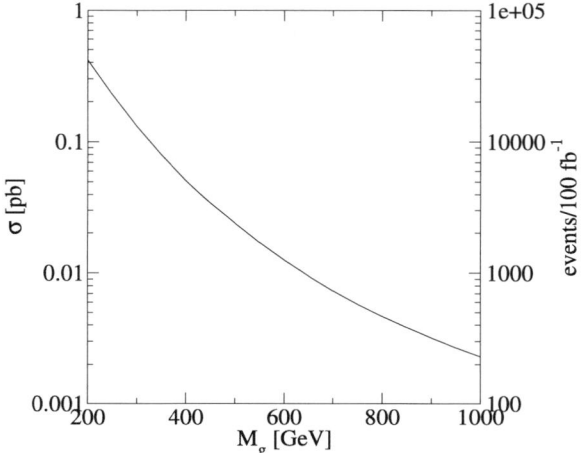

Fig. 12 σ_{tc} (in pb) and number of events per 100 fb^{-1} of integrated luminosity at the LHC, as a function of $m_{\tilde{g}}$

above the background would have to be interpreted as new physics.

The full one-loop SUSY–QCD cross section for the production of single top quark states $t\bar{c} + \bar{t}c$ at the LHC were computed. This direct production mechanism is substantially more efficient (typically a factor of 100) than the production and subsequent FCNC decay [109, 128] ($h \to t\bar{c} + \bar{t}c$) of the MSSM Higgs bosons $h = h^0, H^0, A^0$. It is important to emphasize that the detection of a significant number of $t\bar{c} + \bar{t}c$ states could be interpreted as a distinctive SUSY signature. It should be noticed however that a careful background study must be done for this channel since, unlike the Higgs decay studied in the previous section, the kinematic distributions of the signal are not likely to have a very distinctive shape compared to Wjj or SM single-top production.

2.3.4 ATLAS and CMS sensitivity to FCNC top decays

Due to the high production rate for $t\bar{t}$ pairs and single top, the LHC will allow one either to observe top FCNC decays or to establish very stringent limits on the branching ratios of such decays. In this section the study of ATLAS and CMS sensitivity to top FCNC decays is presented. A detailed description of the analysis can be found in [129, 130].

Both the CMS and ATLAS collaborations have investigated the $t \to q\gamma$ and $t \to qZ$ decay channels. Analyses have been optimized for searching FCNC decays in the $t\bar{t}$ signal, where one of the top quarks is assumed to decay through the dominant SM decay mode ($t \to bW$) and the other is assumed to decay via one of the FCNC modes. The $t\bar{t}$ final states corresponding to the different FCNC top decay modes lead to different topologies, according to the number of jets, leptons and photons. Only leptonic decay channels of Z and W bosons are considered in the analysis developed by the CMS collaboration. The ATLAS collaboration has also studied the channel corresponding to the hadronic Z decay, which is discussed elsewhere [129].

The signal is generated with TopReX [46], while PYTHIA [47] is used for background generation and modeling of quark and gluon hadronization. The generated events are passed through the fast (for ATLAS) and full (for CMS) detector simulation. Several SM processes contributing as background are studied: $t\bar{t}$ production, single-top quark production, $ZW/ZZ/WW$ + jets, $Z/W/\gamma^*$ + jets, $Zb\bar{b}$ and QCD multi-jet production.

Although ATLAS and CMS analyses differ in some details of selection procedure, they obtain the same order of magnitude for the FCNC sensitivity. In both analyses, the signal is preselected by requiring the presence of, at least, one high p_T lepton (that can be used to trigger the event) and missing energy above 20 GeV for the ATLAS analysis and above 25 GeV for the CMS analysis. Additionally, two energetic central jets from t and $\bar{t}$ decays are required. The slight differences in CMS and ATLAS thresholds reflect the differences in their sub-detectors, simulation code and reconstruction algorithms.

The CMS analysis strongly relies on b-tagging capability to distinguish the b jet from SM decay and the light jet from the anomalous one. A series of cascade selections are applied to reduce the background. For the $t \to q\gamma$ channel, the W boson is reconstructed requiring the transverse mass of the neutrino and hard lepton to be less than 120 GeV and the b jet is used to form a window mass $110 < m_{bW} < 220$ GeV. The invariant mass of the light jet and a single isolated photon with $p_T > 50$ GeV is bounded in the range [150, 200] GeV. A final selection of top back-to-back production ($\cos\phi(t\bar{t}) < -0.95$) reduces the diboson background. The $t \to qZ$ channel is extracted with the search of one Z (using same flavor-opposite charge leptons, which serve as trigger and are bound to a 10 GeV window around the Z mass) and a M_{bW} in the top mass region, with the same cuts of the previous case. One hard light jet is extracted and combined with the Z, to reveal the FCNC decay of a top recoiling against the one with SM decay ($\cos\phi(t\bar{t}) < 0$). The reconstructed FCNC top invariant-mass distributions for both channels are shown in Fig. 13.

The ATLAS collaboration has developed a probabilistic analysis for each of the considered top FCNC decay channels. In the $t \to qZ$ channel, preselected events with a reconstructed Z, large missing transverse energy and the two highest p_T jets (one b-tagged) are used to build a discriminant variable (likelihood ratio) $L_R = \ln(\prod_{i=1}^{n} P_i^S / \prod_{i=1}^{n} P_i^B)$, where $P_i^{B(S)}$ are the signal and background p.d.f., evaluated from the following physical distributions: the minimum invariant mass of the three possible combinations of two leptons (only the three highest p_T leptons were considered); the transverse momentum of the third lepton (with the leptons ordered by decreasing p_T) and the transverse momentum of the most energetic non-b jet. The discriminant variables obtained for the FCNC signal and the SM background are shown in Fig. 14 (left). For the $t \to q\gamma$ channel, preselected events are required to have one b-tag (amongst the two highest p_T jets) and at least one

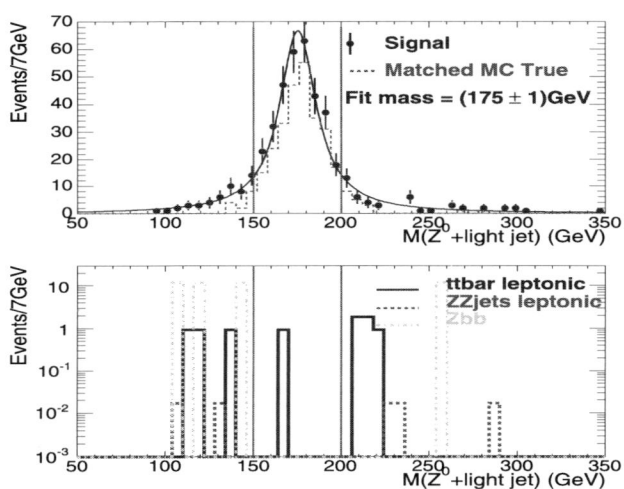
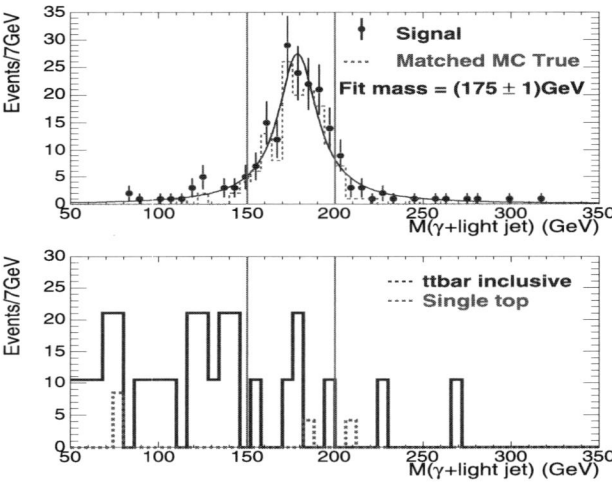
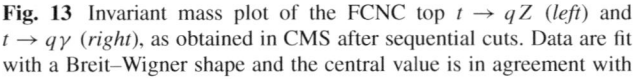

Fig. 13 Invariant mass plot of the FCNC top $t \to qZ$ (*left*) and $t \to q\gamma$ (*right*), as obtained in CMS after sequential cuts. Data are fit with a Breit–Wigner shape and the central value is in agreement with the top mass. The signal distributions obtained from reconstructed leptons and jets matched to the corresponding generated objects are also shown (*Matched MC True*)

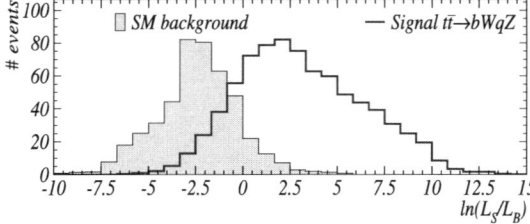
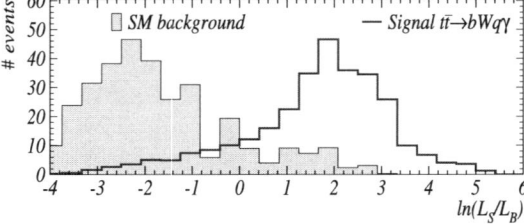

Fig. 14 Signal and background likelihood ratios, $L_R = \ln(L_S/L_B)$, obtained in ATLAS analysis for the $t \to qZ$ (*left*) and $t \to q\gamma$ (*right*) channels. The SM background (*shadow region*) is normalized to $L = 10$ fb^{-1} and the signal (*line*) is shown with arbitrary normalization

Table 6 ATLAS and CMS results for described analysis: efficiency, SM background and expected branching ratios for top FCNC decays, assuming a 5σ significance discovery ($L = 10$ fb^{-1})

	$t \to qZ$			$t \to q\gamma$		
	ϵ_S	N_B	Br (5σ)	ϵ_S	N_B	Br (5σ)
ATLAS	1.30%	0.37	13.0×10^{-4}	1.75%	3.13	1.6×10^{-4}
CMS	4.12%	1.0	11.4×10^{-4}	2.12%	54.6	5.7×10^{-4}

photon with transverse momentum above 75 GeV. For this channel, the likelihood ratio is built using the p.d.f. based on the following variables: invariant mass of the leading photon and the non-b jet; transverse momentum of the leading photon and the number of jets. The signal and background discriminant variables are shown in Fig. 14 (right). For comparison with the CMS sequential analysis, a cut on the discriminant variable (corresponding to the best $S/\sqrt{N_B}$) is applied.

Once the signal efficiency (ϵ_S) and the number of selected background events (N_B) have been obtained, branching ratio sensitivities for a signal discovery corresponding to a given significance can be evaluated. Table 6 reports the results of the two experiments, assuming an integrated luminosity of 10 fb^{-1}, 5σ discovery level and the statistical significance $\mathcal{S} = 2(\sqrt{N_B + S} - \sqrt{N_B})$ (a different definition for $\mathcal{S}$ can be found in [129]).

Having these two independent analyses, a preliminary combination of ATLAS and CMS results was performed, in order to estimate the possible LHC sensitivity to top FCNC decays. As a first attempt, the modified frequentest likelihood method (see for example [131]) is used to combine the expected sensitivity to top FCNC decays from both experiments under the hypothesis of signal absence[8] and an extrapolation to the high luminosity phase (100 fb^{-1}) is performed. These results are showed in Table 7 and indicate that a sensitivity at the level of the predictions of some new physics models (such as SUSY) can be achieved. The comparison with the current experimental limits is also shown in Fig. 15. As shown, a significant improvement on the present limits for top FCNC decays is expected at the LHC. Both collaborations have plans to assess in detail the impact of systematic uncertainties and improve the understanding of the detectors through updated simulation tools. Preliminary results indicate that the effect of theoretical systematics (as top mass, $\sigma(t\bar{t})$ and parton distribution functions) and experimental ones (such as jet/lepton energy scale and b-tagging) have an impact on the limits smaller than 30%. Thus, the order of magnitude of the results is not expected to change.

A study of the ATLAS sensitivity to FCNC $t \to qg$ decay was also presented in [129]. In this analysis, the $t\bar{t}$ production is considered, with one of the top quarks decaying into

Table 7 LHC 95% CL expected limits on $t \to qZ$ and $t \to q\gamma$ branching ratios (ATLAS and CMS preliminary combination under the hypothesis of signal absence)

Luminosity	Br($t \to qZ$)	Br($t \to q\gamma$)
10 fb^{-1}	2.0×10^{-4}	3.6×10^{-5}
100 fb^{-1}	4.2×10^{-5}	1.0×10^{-5}

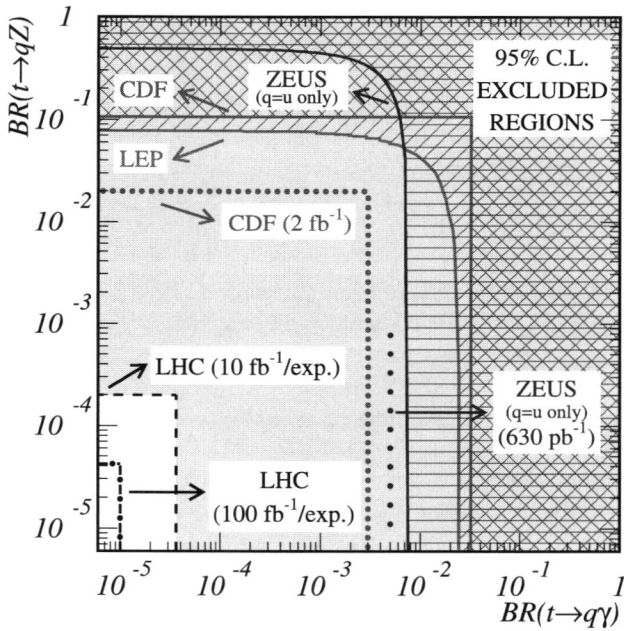

Fig. 15 The present 95% CL limits on the Br($t \to q\gamma$) versus Br($t \to qZ$) plane are shown [75, 77, 132, 133]. The expected sensitivity at HERA ($L = 630$ pb^{-1}) [75], Tevatron (run II) [134] and LHC (ATLAS and CMS preliminary combination) is also represented

qg and the other decays through the SM decay $t \to bW$. Only the leptonic decays of the W were taken into account, otherwise the final state would be fully hadronic and the signal would be overwhelmed by the QCD background. This final state is characterized by the presence of a high p_T gluon and a light jet from the FCNC decay, a b-tagged quark, one lepton and missing transverse momentum from the SM decay. As in this topology the FCNC top decay corresponds to a fully hadronic final state, a more restrictive event selection is necessary. As for the qZ and $q\gamma$ channels, a probabilistic type of analysis is adopted, using the following vari-

---
[8]For the CMS analysis a counting experiment is used, while for the ATLAS analysis the full shape of the discriminant variables was also taken into account.

Fig. 16 Signal and background likelihood ratios obtained in ATLAS analysis for the $t \to qg$ channel. The SM background (*shadow region*) is normalized to $L = 10$ fb^{-1} and the signal (*line*) is shown with arbitrary normalization

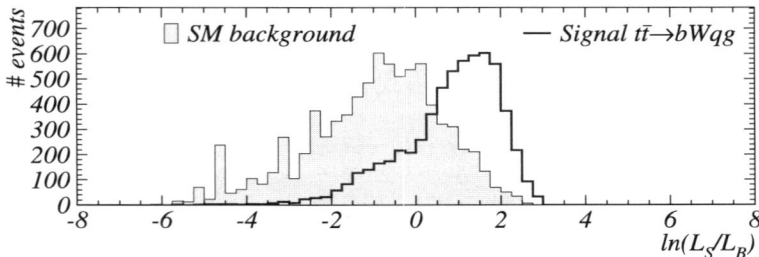

ables to build the p.d.f.: the invariant mass of the two non-b jets with highest p_T; the $b\ell\nu$ invariant mass; the transverse momenta of the b jet and of the second highest p_T non-b jet and the angle between the lepton and the leading non-b jet. The discriminant variables obtained for signal and background are shown in Fig. 16. The expected 95% CL limit on Br($t \to qg$) for $L = 10$ fb^{-1} for $L = 10$ fb^{-1} was found to be 1.3×10^{-3}. A significant improvement on this limit should be achieved by combining the results from $t\bar{t}$ production (with $t \to qg$ FCNC decay) and single-top production (see Sect. 2.3.1).

2.4 New physics contributions to top quark production

It is generally believed that the top quark, due to its large mass, can be more sensitive to new physics beyond the SM than other fermions. In particular, new processes contributing to $t\bar{t}$ and single-top production may be relevant. Single-top processes are expected to be sensitive to some SM extensions, such as SUSY. Another characteristic new process could be the production in pp collisions of an s-channel resonance decaying to $t\bar{t}$. Examples of this resonance are (i) a spin-1 leptophobic Z' boson, which would be undetectable in leptonic decay channels; (ii) Kaluza–Klein (KK) excitations of gluons or gravitons; (iii) neutral scalars. If these resonances are narrow they could be visible as a mass peak over the SM $t\bar{t}$ background. In such case, the analysis of t, $\bar{t}$ polarizations (in a suitable window around the peak) could provide essential information about the spin of the resonance. If the resonance is broad, perhaps the only way to detect it could be a deviation in $t\bar{t}$ spin correlations with respect to the SM prediction. More generally, new contributions to $t\bar{t}$ production which do not involve the exchange of a new particle in the s-channel (including, but not limited to, those mediated by anomalous couplings to the gluon) do not show up as an invariant-mass peak. In this case, the analysis of the measurement of spin correlations might provide the only way to detect new physics in $t\bar{t}$ production.

2.4.1 Potential complementary MSSM test in single-top production

At LHC, it will be possible to perform measurements of the rates of the three different single-top production processes, usually defined as t-channel, associated tW and s-channel production, with an experimental accuracy that varies with the process. From the most recent analyses one expects, qualitatively, a precision of the order of 10% for the t-channel [135], and worse accuracies for the two remaining processes. Numerically, the cross section of the t-channel is the largest one, reaching a value of approximately 250 pb [136]; for the associated production and the s-channel one expects a value of approximately 60 pb and 10 pb [137] respectively. For all the processes, the SM NLO QCD effect has been computed [26, 138], and quite recently also the SUSY QCD contribution has been evaluated [137]. Roughly, one finds for the t-channel a relative ~6% SM QCD effect and a negligible SUSY QCD component; for the associated tW production a relative ~10% SM QCD and a relative ~6% SUSY QCD effect; for the s-channel, a relative ~50% SM QCD and a negligible SUSY QCD component. As a result of the mentioned calculations, one knows the relative NLO effects of both SM and SUSY QCD. The missing part is the NLO electroweak effect. This has been computed for the two most relevant processes, i.e. the t-channel and the associated production. The NLO calculation for the s-channel is, probably, redundant given the small size of the related cross section. It is, in any case, in progress. In this section some of the results of the complete one-loop calculation of the electroweak effects in the MSSM are shown for the two processes. More precisely, eight different t-channel processes (four for single top and four for single antitop production) were considered. These processes are defined in [139]. For the associated production, the process $bg \to tW^-$ (the rate of the second process $\bar{b}g \to \bar{t}W^+$ is the same) was considered [140]. These calculations have been performed using the program LEONE, which passed three severe consistency tests described in [139, 140]. For the aim of this preliminary discussion, in this section only the obtained values of the integrated cross sections are shown, ignoring the (known) QCD effects. The integration has been performed from threshold to the effective center of mass energy ($\sqrt{\hat{s}}$), allowed to vary up to a reasonable upper limit of approximately 1 TeV. Other informations are contained in [139, 140].

Figures 17 and 18 show the obtained numerical results. In Fig. 18 (right) the discussed NLO electroweak effect was

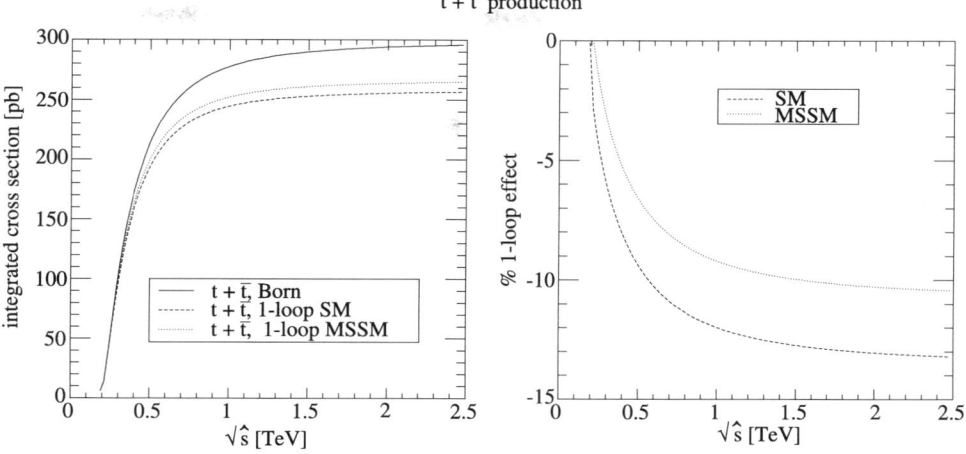

Fig. 17 Integrated cross sections for the overall t-channel production of a single top or antitop quark

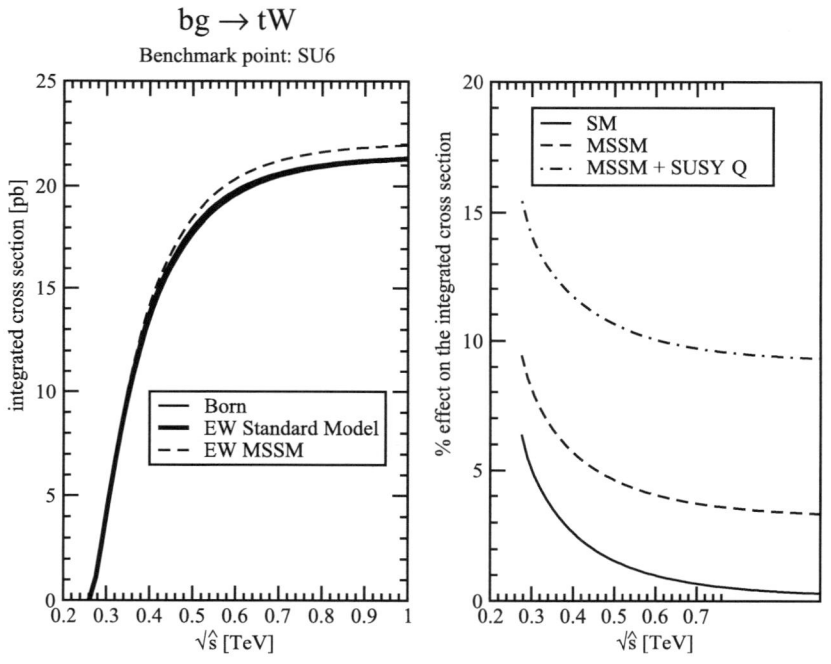

Fig. 18 Integrated cross sections for the associated production of a single top quark

added to the NLO SUSY QCD effect taken from [137]. From the figures the following main conclusions can be drawn:

1. The genuine SUSY effect in the t-channel is modest. In the most favorable case, corresponding to the ATLAS DC2 point SU6 [141], it reaches a value of approximately two percent.
2. The one-loop electroweak SM effect in the t-channel rate is large (∼13%). It is definitely larger than the NLO SM QCD effect. Its inclusion in any meaningful computational program appears to be mandatory.
3. The genuine SUSY effect in the associated production, if one limits the cross section observation to relatively low (and experimentally safe from $t\bar{t}$ background) energies (400–500 GeV), can be sizable. In the SU6 point, the combined (same sign) SUSY QCD and electroweak effects can reach a relative ten percent effect.
4. The pure electroweak SM effect in the associated production is negligible.

From the previous remarks, one can reach the final statement that, for what concerns the virtual NLO effects of the MSSM, the two processes t-channel and associated production appear to be, essentially, complementary. In this spirit, a separate experimental determination of the two rates might lead to non-trivial tests of the model.

2.4.2 Anomalous single-top production in warped extra dimensions

Randall and Sundrum have proposed the use of a non-factorizable geometry in five dimensions [142] as a solution of the hierarchy problem. The extra dimension is compactified on an orbifold S_1/Z_2 of radius r so that the bulk is a slice of anti-de Sitter space between two four dimensional boundaries. The metric depends on the five dimensional coordinate y and is given by

$$ds^2 = e^{-2\sigma(y)} \eta_{\mu\nu} dx^\mu dx^\nu - dy^2, \qquad (15)$$

where x^μ are the four dimensional coordinates, $\sigma(y) = k|y|$, with $k \sim M_P$ characterizing the curvature scale. This metric generates two effective scales: M_P and $M_P e^{-k\pi r}$. In this way, values of r not much larger than the Planck length ($kr \simeq (11\text{--}12)$) can be used in order to generate a scale $\Lambda_r \simeq M_P e^{-k\pi r} \simeq O$ (TeV) on one of the boundaries.

In the original Randall–Sundrum (RS) scenario, only gravity was allowed to propagate in the bulk, with the SM fields confined to one of the boundaries. The inclusion of matter and gauge fields in the bulk has been extensively treated in the literature [143–150]. The Higgs field must be localized on or around the TeV brane in order to generate the weak scale. As it was recognized in [147], it is possible to generate the fermion mass hierarchy from $O(1)$ flavor breaking in the bulk masses of fermions. Since bulk fermion masses result in the localization of fermion zero modes, lighter fermions should be localized toward the Planck brane, where their wave functions have an exponentially suppressed overlap with the TeV-localized Higgs, whereas fermions with order one Yukawa couplings should be localized toward the TeV brane. This constitutes a theory of fermion masses, and it has a distinct experimental signal at the LHC, as discussed below.

Since the lightest KK excitations of gauge bosons are localized toward the TeV brane, they tend to be strongly coupled to zero-mode fermions localized there. Thus, the flavor-breaking fermion localization leads to flavor-violating interactions of the KK gauge bosons, particularly with third generation quarks. For instance, the first KK excitation of the gluon, will have flavor-violating neutral couplings such as $G_\mu^{a(1)}(\bar{t}\gamma^\mu T^a q)$, where $q = u, c$.

In this section, results of a study of the flavor-violating signals of the top at the LHC are presented, following the work described in [151]. The localization of fermions in the extra dimension, and therefore their 4D masses and their couplings to the KK gauge bosons, are determined by their bulk masses. We choose a range of parameters that is consistent with the observed fermion masses and quark mixing, as well as low energy flavor and electroweak constraints. The implications for low energy flavor physics were considered in [152–154]. The bulk masses of the third generation quark doublet is fixed, as well as that of the right handed top. The following ranges were considered: $c_L^3 = [0.3, 0.4]$ and $c_R^t = [-0.4, 0.1]$, where the fermion bulk masses $c_{L,R}^f$ are expressed in units of the inverse AdS radius k. Since the latter is of the order of the Planck scale, the fermion bulk mass parameters must be naturally of order one.

The only couplings that are non-universal in practice are those of the t_R, t_L and b_L with the KK gauge bosons. All other fermions, including the right handed b quark must have localizations toward the Planck brane in order to get their small masses. The non-universality of the KK gauge boson couplings leads to tree-level flavor violation. The diagonalization of the quark mass matrix requires a change of basis for the quarks fields. In the SM, this rotation leads to the CKM matrix in the charged current, but the universality of the gauge interactions results in the GIM mechanism in the neutral currents. However, since the KK excitations of the gauge bosons are non-universal, tree-level GIM-violating couplings will appear in the physical quark basis.

The dominant non-universal effect is considered as coming from the couplings of t_R, t_L and b_L to the first KK excitation of the gluon: g_{t_R}, g_{t_L} and g_{b_L} respectively. The $SU(2)_L$ bulk symmetry implies $g_{t_L} = g_{b_L}$. For the considered range of c_L^3 and c_R^t, the following results were obtained:

$$g_{t_L} = g_{b_L} = [1.0, 2.8] g_s \qquad (16)$$

and

$$g_{t_R} = [1.5, 5] g_s, \qquad (17)$$

where g_s is the usual 4D $SU(3)_c$ coupling. The light quarks as well as the right handed b quark have

$$g_L^q = g_R^q = g_R^b \simeq -0.2 g_s, \qquad (18)$$

so they are, in practice, universally coupled, as mentioned above.

Computing the width of the intermediate KK gluon with the range of couplings obtained above, results in a range of $\Gamma_{\min.} \simeq 0.04 M_G$ and $\Gamma_{\max.} \simeq 0.35 M_G$. Then, it can be seen that the range of values for the couplings allow for rather narrow or rather broad resonances, two very different scenarios from the point of view of the phenomenology. This strong coupling of the KK gluon to the top, will also produce a $t\bar{t}$ resonance. Here we concentrate on the flavor-violating signal, since the presence of a $t\bar{t}$ resonance will not constitute proof of the flavor theory due to the difficulty in identifying resonances in the light quark channels.

In the quark mass eigenbasis the left handed up-type quarks couple to the KK gluon through the following currents: $U_L^{tt}(\bar{t}_L T^a \gamma_\mu t_L)$, $U_L^{tc}(\bar{t}_L T^a \gamma_\mu c_L)$ and $U_L^{tu}(\bar{t}_L T^a \gamma_\mu u_L)$. Similarly, the right handed up-type quarks couple through $U_R^{tt}(\bar{t}_R T^a \gamma_\mu t_R)$, $U_R^{tc}(\bar{t}_R T^a \gamma_\mu c_R)$ and $U_R^{tu}(\bar{t}_R T^a \gamma_\mu u_R)$.

Here, U_L and U_R are the left handed and right handed up-type quark rotation matrices responsible for the diagonalization of the Yukawa couplings of the up-type quarks. In what follows

$$U_L^{tu} \simeq V_{ub} \simeq 0.004, \quad (19)$$

will be conservatively assumed, and U_R^{tc} and U_R^{tu} will be taken as free parameters. Since no separation of charm from light jets is assumed, we define

$$U_R^{tq} \equiv \sqrt{\left(U_R^{tc}\right)^2 + \left(U_R^{tu}\right)^2}, \quad (20)$$

and the sensitivity of the LHC to this parameter for a given KK gluon mass is studied.

These flavor-violating interactions could be directly observed by the s-channel production of the first KK excitation of the gluon with its subsequent decay to a top and a charm or up quark. For instance, at the LHC we could have the reaction

$$pp \to G_\mu^{a(1)} \to tq, \quad (21)$$

with $q = u, c$. Thus, the Randall–Sundrum scenario with bulk matter predicts anomalous single-top production at a very high invariant mass, which is determined by the mass of the KK gluon.

In order to reduce the backgrounds, only the semileptonic decays of the top quarks were considered: $pp \to t\bar{q}(\bar{t}q) \to b\ell^+\nu_\ell\bar{q}(\bar{b}\ell^-\bar{\nu}_\ell q)$, where $\ell = e$ or μ, and $q = u, c$. Therefore, this signal exhibits one b jet, one light jet, a charged lepton and missing transverse energy. There are many SM backgrounds for this process. The dominant one is $pp \to W^\pm jj \to \ell^\pm \nu jj$ where one of the light jets is tagged as a b jet. There is also $W^\pm b\bar{b} \to \ell^\pm \nu b\bar{b}$ where one of the b jets is mistagged; single-top production via W-gluon fusion and s-channel W^*, and $t\bar{t}$ production at high invariant mass, mostly dominated by the flavor-conserving KK gluon decays.

Initially, the following jet and lepton acceptance cuts were imposed: $p_T^j > 20$ GeV, $|y_j| < 2.5$, $p_T^\ell \geq 20$ GeV, $|y_\ell| \leq 2.5$, $\Delta R_{\ell j} \geq 0.63$, $\Delta R_{\ell \ell} \geq 0.63$, where j can be either a light or a b jet. In order to further reduce the background the following additional cuts were also imposed:

1. The invariant mass of the system formed by the lepton, the b-tagged jet and the light jet was required to be within a window

$$M_{G^{(1)}} - \Delta \leq M_{bj\ell} \leq M_{G^{(1)}} + \Delta \quad (22)$$

around the first KK excitation of the gluon mass. This cut ensures that the selected events have large invariant masses, as required by the large mass of the s-channel object being exchanged. The values of Δ used in this study are presented in Table 8.

2. The transverse momentum of the light jet was required to be larger than p_{cut}, i.e.,

$$p_{j\,\text{light}} \geq p_{\text{cut}}. \quad (23)$$

Since the light jet in the signal recoils against the top forming with it a large invariant mass, it tends to be harder than the jets occurring in the background. We present in Table 8 the values for p_{cut} used in our analysis.

3. The invariant mass of the charged lepton and the b-tagged jet was also required to be *smaller* than 250 GeV:

$$M_{b\ell} \leq 250 \text{ GeV}. \quad (24)$$

This requirement is always passed by the signal, but eliminates a sizable fraction of the Wjj background. It substitutes for the full top reconstruction when the neutrino momentum is inferred, which is not used here.

In Table 9 the cross sections for signal and backgrounds for $M_G = 1$ TeV and 2 TeV are presented. The main sources

Table 8 Cuts used in the analysis (see text for details)

$M_{G^{(1)}}$ (TeV)	Δ (GeV)	p_{cut} (GeV)
1	120	350
2	250	650

Table 9 Signal and background cross sections for a KK gluon of $M_G = 1$ TeV and 2 TeV, after the successive application of the cuts defined in (22), (23) and (24). Efficiencies and b-tagging probabilities have already been included. $U_R^{tq} = 1$ was used

Process	$M_G = 1$ TeV			$M_G = 2$ TeV		
	$\sigma - (22)$	$\sigma - (23)$	$\sigma - (24)$	$\sigma - (22)$	$\sigma - (23)$	$\sigma - (24)$
$pp \to tj$	148 fb	103 fb	103 fb	5.10 fb	2.18 fb	2.18 fb
$pp \to Wjj$	243 fb	42.0 fb	21.0 fb	25.4 fb	3.79 fb	0.95 fb
$pp \to Wbb$	11.1 fb	4.07 fb	3.19 fb	0.97 fb	0.45 fb	0.06 fb
$pp \to tb$	1.53 fb	0.70 fb	0.61 fb	0.04 fb	0.02 fb	0.02 fb
$pp \to t\bar{t}$	44.4 fb	15.1 fb	14.2 fb	1.60 fb	0.29 fb	0.24 fb
Wg fusion	32.0 fb	5.23 fb	5.23 fb	1.20 fb	0.10 fb	0.10 fb

Table 10 Reach in U_R^{tq} for various integrated luminosities

M_G [TeV]	30 fb^{-1}	100 fb^{-1}	300 fb^{-1}
1	0.24	0.18	0.14
2	0.65	0.50	0.36

of backgrounds are Wjj and $t\bar{t}$ production. The signal is obtained for $U_R^{tq} = 1$ and neglecting the contributions from left handed final states, corresponding to $U_L^{tq} = 0$. Regarding the choice of bulk masses, these are fixed to obtain the minimum width which, as mentioned above, can be as small as $\Gamma_G \simeq 0.04 M_G$.

In order to evaluate the reach of the LHC, a significance of 5σ for the signal over the background is required. For a given KK gluon mass and accumulated luminosity, this can be translated into a reach in the flavor-violating parameter U_R^{tq} defined above. This is shown in Table 10. It can be seen that the LHC will be sensitive to tree-level flavor violation for KK gluon masses of up to at least 2 TeV, probing a very interesting region of values for U_R^{tq}. The reach can be somewhat better if we allow for the reconstruction of the momentum of the neutrino coming from the W decay, which typically reduces the Wjj background more drastically.

Finally, we should point out that a very similar signal exists in Topcolor-assisted Technicolor [87], where the KK gluon is replaced by the Topgluon, which has FCNC interactions with the third generation quarks [88]. The main difference between these two is that the latter is typically a broad resonance, whereas the KK gluon could be a rather narrow one, as was shown above.

2.4.3 Non-standard contributions to $t\bar{t}$ production

In $t\bar{t}$ events the top quarks are produced unpolarized at the tree level. However, the t and $\bar{t}$ spins are strongly correlated, which allows one to construct asymmetries using the angular distributions of their decay products. These spin asymmetries are dependent on the top spin. For the decay $t \to W^+ b \to \ell^+ \nu b, q\bar{q}' b$, the angular distributions of $X = \ell^+, \nu, q, \bar{q}', W^+, b$, in the top quark rest frame are given by

$$\frac{1}{\Gamma} \frac{d\Gamma}{d\cos\theta_X} = \frac{1}{2}(1 + \alpha_X \cos\theta_X) \quad (25)$$

with θ_X being the angle between the three-momentum of X in the t rest frame and the top spin direction. In the SM the spin analysing power (α_X) of the top decay products are $\alpha_{\ell^+} = \alpha_{\bar{q}'} = 1$, $\alpha_\nu = \alpha_q = -0.32$, $\alpha_{W^+} = -\alpha_b = 0.41$ at the tree level [155] (q and q' are the up- and down-type quarks, respectively, resulting from the W decay). For the decay of a top antiquark the distributions are the same, with $\alpha_{\bar{X}} = -\alpha_X$ as long as CP is conserved in the decay. One-loop corrections modify these values to $\alpha_{\ell^+} = 0.998$, $\alpha_{\bar{q}'} = 0.93$, $\alpha_\nu = -0.33$, $\alpha_q = -0.31$, $\alpha_{W^+} = -\alpha_b = 0.39$ [156–158]. We point out that in the presence of non-vanishing V_R, g_L or g_R couplings the numerical values of the constants α_X are modified, but the functional form of (25) is maintained. We have explicitly calculated them for a general CP-conserving Wtb vertex within the narrow width approximation. Explicit expressions can be found in [43]. Working in the helicity basis the double angular distribution of the decay products X (from t) and $\bar{X}'$ (from $\bar{t}$) can be written as a function of the relative number of like helicity minus opposite helicity of the $t\bar{t}$ pairs (C) [159] that measures the spin correlation between the top quark and antiquark. Its actual value depends to some extent on the PDFs used and the Q^2 scale at which they are evaluated. Using the CTEQ5L PDFs [48] and $Q^2 = \hat{s}$, (where $\hat{s}$ is the partonic center of mass energy), we find $C = 0.310$. At the one loop level, $C = 0.326 \pm 0.012$ [158].

Using the spin analysers $X, \bar{X}'$ for the respective decays of $t, \bar{t}$, one can define the asymmetries

$$A_{X\bar{X}'} \equiv \frac{N(\cos\theta_X \cos\theta_{\bar{X}'} > 0) - N(\cos\theta_X \cos\theta_{\bar{X}'} < 0)}{N(\cos\theta_X \cos\theta_{\bar{X}'} > 0) + N(\cos\theta_X \cos\theta_{\bar{X}'} < 0)}, \quad (26)$$

whose theoretical value is

$$A_{X\bar{X}'} = \frac{1}{4} C \alpha_X \alpha_{\bar{X}'}. \quad (27)$$

The angles θ_X, $\theta_{\bar{X}'}$ are measured using as spin axis the parent top (anti)quark momentum in the $t\bar{t}$ CM system. If CP is conserved in the decay, for charge conjugate decay channels we have $\alpha_{X'}\alpha_{\bar{X}} = \alpha_X \alpha_{\bar{X}'}$, so the asymmetries $A_{X'\bar{X}} = A_{X\bar{X}'}$ are equivalent. Therefore, we can sum both channels and drop the superscripts indicating the charge, denoting the asymmetries by $A_{\ell\ell'}$, $A_{\nu\ell'}$, etc. In semileptonic top decays we can select as spin analyser the charged lepton, which has the largest spin analysing power, or the neutrino, as proposed in [160]. In hadronic decays the jets corresponding to up- and down-type quarks are very difficult to distinguish, and one possibility is to use the least energetic jet in the top rest frame, which corresponds to the down-type quark 61% of the time, and has a spin analysing power $\alpha_j = 0.49$ at the tree level. An equivalent possibility is to choose the d jet by its angular distribution in the W^- rest frame [161]. In both hadronic and leptonic decays the b ($\bar{b}$) quarks can be used as well.

In the lepton + jets decay mode of the $t\bar{t}$ pair, $t\bar{t} \to \ell \nu b j j \bar{b}$ we choose the two asymmetries $A_{\ell j}$, $A_{\nu j}$, for which we obtain the SM tree-level values $A_{\ell j} = -0.0376$, $A_{\nu j} = 0.0120$. With the precision expected at LHC [50, 162], the measurements $A_{\ell j} \simeq -0.0376 \pm 0.0058$, $A_{\nu j} \simeq 0.0120 \pm$

0.0056 are feasible ($L = 10$ fb^{-1}). The dependence of these asymmetries on anomalous Wtb couplings is depicted in Fig. 19 from [43]. In the dilepton channel $t\bar{t} \to \ell\nu b \ell'\nu\bar{b}$ the asymmetries $A_{\ell\ell'}$, $A_{\nu\ell'}$, whose SM values are $A_{\ell\ell'} = -0.0775$, $A_{\nu\ell'} = 0.0247$, are selected. The uncertainty in their measurement can be estimated from [50, 162], yielding $A_{\ell\ell'} = -0.0775 \pm 0.0060$ and $A_{\nu\ell'} = 0.0247 \pm 0.0087$. Their variation when anomalous couplings are present is shown in Fig. 19. We also plot the asymmetries A_{lb}, A_{bb}, which can be measured either in the semileptonic or dilepton channel. Their SM values are $A_{lb} = 0.0314$, $A_{bb} = -0.0128$, but the experimental sensitivity has not been es-

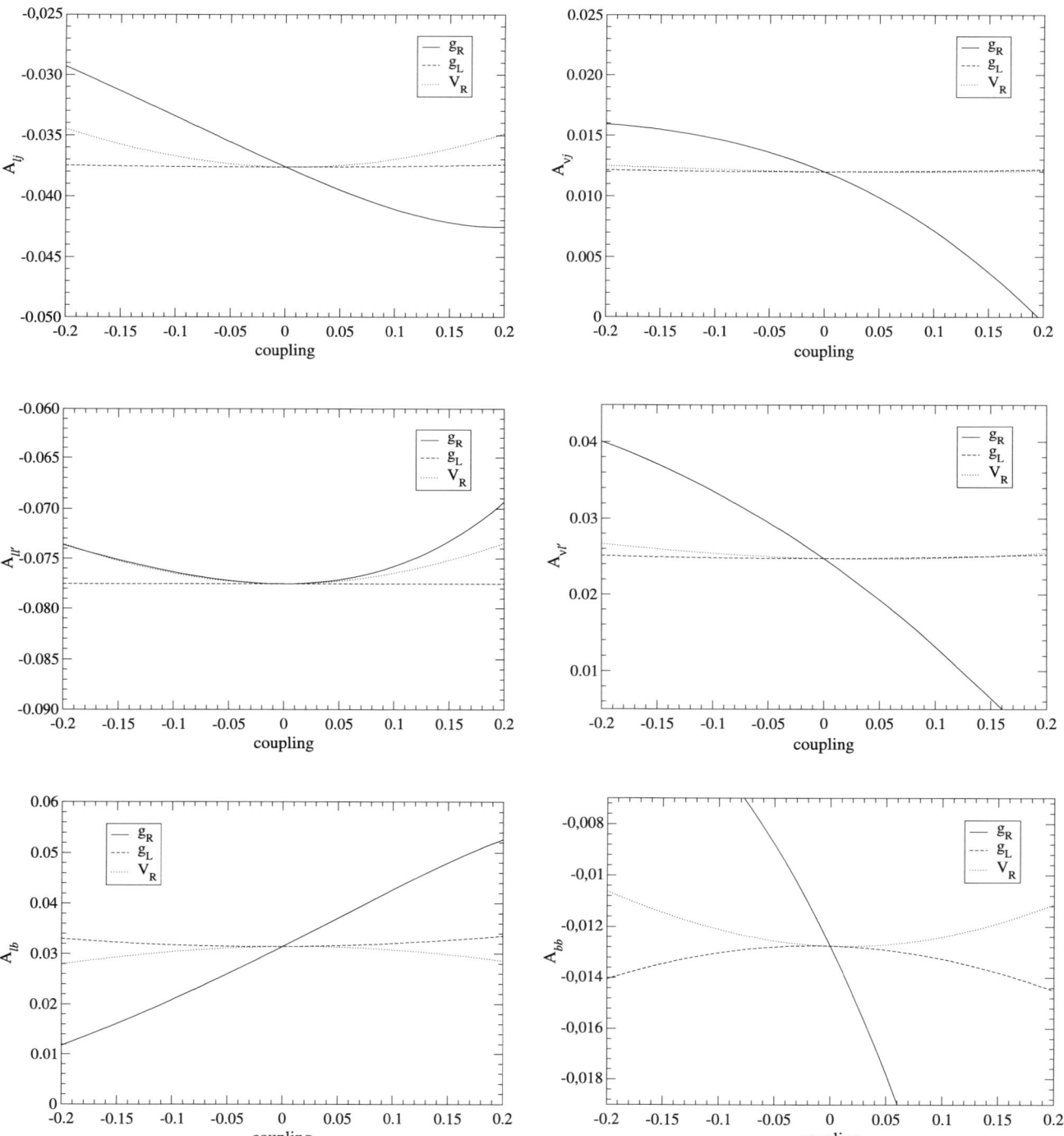

Fig. 19 Dependence of several spin correlation asymmetries on the couplings g_R, g_L and V_R, for the CP-conserving case

timated. It is expected that it may be of the order of 10% for A_{lb}, and worse for A_{bb}. The determination of the correlation factor C from these asymmetries would eventually give

$$A_{\ell\ell'} \to C = 0.310 \pm 0.024(\exp)^{+0.}_{-0.0043}$$
$$\times (\delta V_R)^{+1 \times 10^{-5}}_{-3 \times 10^{-6}} (\delta g_L)^{+7 \times 10^{-6}}_{-0.0004} (\delta g_R),$$
$$A_{\ell j} \to C = 0.310 \pm 0.045(\exp)^{+0.}_{-0.0068}$$
$$\times (\delta V_R)^{+0.0001}_{-0.0008} (\delta g_L)^{+0.0004}_{-0.0009} (\delta g_R). \quad (28)$$

The first error quoted corresponds to the experimental systematic and statistical uncertainty. The other ones are theoretical uncertainties obtained varying the anomalous couplings, one at a time. The confidence level corresponding to the intervals quoted is 68.3%. The numerical comparison of the different terms in (28) also shows that $A_{\ell j}$ and $A_{\ell\ell'}$ are much less sensitive to non-standard top couplings than observables independent of the top spin (see Sect. 2.2.1).

It is also interesting to study the relative distribution of one spin analyser from the t quark and other from the $\bar{t}$. Let $\varphi_{X\bar{X}'}$ be the angle between the three-momentum of X (in the t rest frame) and of $\bar{X}'$ (in the $\bar{t}$ rest frame). The angular distribution can be written as [158]:

$$\frac{1}{\sigma} \frac{d\sigma}{d\cos\varphi_{X\bar{X}'}} = \frac{1}{2}(1 + D\alpha_X \alpha_{\bar{X}'} \cos\varphi_{X\bar{X}'}), \quad (29)$$

with D a constant defined by this equality. From simulations, the tree-level value $D = -0.217$ is obtained, while at one loop $D = -0.238$ [158], with a theoretical uncertainty of $\sim 4\%$. Corresponding to these distributions, the following asymmetries can be built:

$$\tilde{A}_{X\bar{X}'} \equiv \frac{N(\cos\varphi_{X\bar{X}'} > 0) - N(\cos\varphi_{X\bar{X}'} < 0)}{N(\cos\varphi_{X\bar{X}'} > 0) + N(\cos\varphi_{X\bar{X}'} < 0)}$$
$$= \frac{1}{2} D\alpha_X \alpha_{\bar{X}'}. \quad (30)$$

For charge conjugate decay channels the distributions can be summed, since $\alpha_{X'}\alpha_{\bar{X}} = \alpha_X \alpha_{\bar{X}'}$ provided CP is conserved in the decay. The dependence of these asymmetries $\tilde{A}_{X\bar{X}'}$ on anomalous couplings is (within the production × decay factorization approximation) exactly the same as for the asymmetries $A_{X\bar{X}'}$ defined above. Simulations are available for $A_{\ell j}$ and $A_{\ell\ell'}$, whose theoretical SM values are $A_{\ell j} = 0.0527$, $A_{\ell\ell'} = 0.1085$. The experimental precision expected [50, 162] is $A_{\ell j} \simeq 0.0554 \pm 0.0061$, $A_{\ell\ell'} \simeq 0.1088 \pm 0.0056$. This precision is better than for $A_{\ell j}$ and $A_{\ell j}$, respectively, but still not competitive in the determination of the Wtb vertex structure.[9] Instead, we can use them

to test top spin correlations. From these asymmetries one can extract the value of D, obtaining

$$A_{\ell\ell'} \to D = -0.217 \pm 0.011(\exp)^{+0.0031}_{-0.}$$
$$\times (\delta V_R)^{+2 \times 10^{-6}}_{-8 \times 10^{-6}} (\delta g_L)^{+0.0003}_{-0.} (\delta g_R),$$
$$A_{\ell j} \to D = -0.217 \pm 0.024(\exp)^{+0.0047}_{-0.}$$
$$\times (\delta V_R)^{+0.0006}_{-9 \times 10^{-6}} (\delta g_L)^{+0.0004}_{-6 \times 10^{-5}} (\delta g_R). \quad (31)$$

The errors quoted correspond to the experimental systematic and statistical uncertainties, and the variation when one of the anomalous couplings is allowed to be nonzero. From (28) and (31) it is clear that the measurement of spin correlations is a clean probe for new $t\bar{t}$ production processes, independently of possible anomalous Wtb couplings. This is possible because the sensitivity of spin correlation asymmetries to top anomalous couplings is much weaker than for helicity fractions and related observables, discussed in Sect. 2.2.1.

3 Flavor violation in supersymmetric models[10]

3.1 Introduction

The SM explains successfully the observed flavor-violating phenomena except that for the observation in the neutrino sector one has to extend it by introducing either right handed neutrinos or additional scalars. This implies that extensions of the SM with additional flavor structures are severely constrained by the wealth of existing data in the flavor sector. Supersymmetry contains, as we shall see below, various sources of additional flavor structures. Therefore, the question arises if there can still be large flavor-violating effects in the production and decays of supersymmetric particles despite the stringent existing constraints.

Every supersymmetric model is characterized by a Kähler potential, the superpotential W and the corresponding soft SUSY breaking Lagrangian (see e.g. [163] and references therein). The first describes the gauge interactions and the other two Yukawa interactions and flavor violation. As the Kähler potential in general does not contain flavor-violating terms we shall not discuss it further. The most general superpotential containing only the SM fields and being compatible with its gauge symmetry $G_{\rm SM} = {\rm SU}(3)_c \times {\rm SU}(2)_L \times {\rm U}(1)_Y$ is given as [164, 165]:

$$W = W_{\rm MSSM} + W_{\not{R}_p}, \quad (32)$$

---

[9] Except for the case of fine-tuned cancellations, see [51].

[10] Section coordinators: M. Klasen, N. Krasnikov, T. Lari, W. Porod, and A. Tricomi.

$$W_{\text{MSSM}} = h^E_{ij} \hat{L}_i \hat{H}_d \hat{E}^c_j + h^D_{ij} \hat{Q}_i \hat{H}_d \hat{D}^c_j$$
$$+ h^U_{ij} \hat{H}_u \hat{Q}_i \hat{U}^c_j - \mu \hat{H}_d \hat{H}_u, \quad (33)$$

$$W_{\not{R}_p} = \frac{1}{2} \lambda_{ijk} \hat{L}_i \hat{L}_j \hat{E}^c_k + \lambda'_{ijk} \hat{L}_i \hat{Q}_j \hat{D}^c_k$$
$$+ \frac{1}{2} \lambda''_{ijk} \hat{U}^c_i \hat{D}^c_j \hat{D}^c_k + \epsilon_i \hat{L}_i \hat{H}_u, \quad (34)$$

where $i, j, k = 1, 2, 3$ are generation indices. $\hat{L}_i$ ($\hat{Q}_i$) are the lepton (quark) SU(2)$_L$ doublet superfields. $\hat{E}^c_j$ ($\hat{D}^c_j$, $\hat{U}^c_j$) are the electron (down- and up-quark) SU(2)$_L$ singlet superfields. h^E_{ij}, h^D_{ij}, h^U_{ij}, λ_{ijk}, λ'_{ijk}, and λ''_{ijk} are dimensionless Yukawa couplings, whereas the ϵ_i are dimensionful mass parameters. Gauge invariance implies that the first term in $W_{\not{R}_p}$ is antisymmetric in $\{i, j\}$ and the third one is antisymmetric in $\{j, k\}$. Equation (34) thus contains $9 + 27 + 9 + 3 = 48$ new terms beyond those of the minimal supersymmetric standard model (MSSM). At the level of the superpotential one can actually rotate the $(\hat{H}_d, \hat{L}_i)$ by an SU(4) transformation, so that the ϵ_i can be set to zero. However, as discussed below, this cannot be done simultaneously for the corresponding soft SUSY breaking terms and, thus, we keep them for the moment as free parameters. The soft SUSY breaking potential is given by

$$V_{\text{soft}} = V_{\text{MSSM,soft}} + V_{\not{R}_p,\text{soft}}, \quad (35)$$

$$V_{\text{MSSM}} = M^2_{L,ij} \tilde{L}_i \tilde{L}^*_j + M^2_{E,ij} \tilde{E}_i \tilde{E}^*_j$$
$$+ M^2_{Q,ij} \tilde{Q}_i \tilde{Q}^*_j + M^2_{U,ij} \tilde{U}_i \tilde{U}^*_j + M^2_{D,ij} \tilde{D}_i \tilde{D}^*_j$$
$$+ M^2_d H_d H^*_d + M^2_u H_u H^*_u - (\mu B H_d H_u + \text{h.c.})$$
$$+ \left(T^E_{ij} \tilde{L}_i H_d \tilde{E}_j + T^D_{ij} \tilde{Q}_i H_d \tilde{D}_j \right.$$
$$\left. + T^U_{ij} H_u \tilde{Q}_i \tilde{U}_j + \text{h.c.} \right), \quad (36)$$

$$V_{\not{R},\text{soft}} = \frac{1}{2} T^\lambda_{ijk} \tilde{L}_i \tilde{L}_j \tilde{E}^*_k + T^{\lambda'}_{ijk} \tilde{L}_i \tilde{Q}_j \tilde{D}^*_k$$
$$+ \frac{1}{2} T^{\lambda''}_{ijk} \tilde{U}_i \tilde{D}_j \tilde{D}_k + \epsilon_i B_i \tilde{L}_i H_u + \text{h.c.} \quad (37)$$

The mass matrices M^2_F ($F = L, E, Q, U, D$) are 3×3 hermitian matrices, whereas the T^F are general 3×3 and $3 \times 3 \times 3$ complex tensors. Obviously, the T^λ_{ijk} ($T^{\lambda''}_{ijk}$) have to be antisymmetric in the first (last) two indices due to gauge invariance. In models, where the flavor-violating terms are neglected, the T^F_{ij} terms are usually decomposed into the following products: $T^F_{ij} = A^F_{ij} h^F_{ij}$, and analogously for the trilinear terms.

The simultaneous appearance of lepton and baryon number breaking terms leads in general to a phenomenological catastrophe if all involved particles have masses of the order of the electroweak scale: rapid proton decay [164, 165]. To avoid this problem a discrete multiplicative symmetry, called R-parity (R_p), had been invented [166] which can be written as

$$R_p = (-\mathbf{1})^{3B+L+2S}, \quad (38)$$

where S is the spin of the corresponding particle. For all superfields of MSSM, the SM field has $R_p = +1$ and its superpartner has $R_p = -1$, e.g. the electron has $R_p = +1$ and the selectron has $R_p = -1$. In this way all terms in (34) are forbidden and one is left with the superpotential given in (33). To prohibit proton decay it is not necessary to forbid both type of terms but it is sufficient to forbid either the lepton or the baryon number violating terms (see e.g. [167, 168]), e.g. the baryon number terms can be forbidden by baryon triality [169]. Another possibility would be to break lepton number and thus R-parity spontaneously as discussed below. This requires, however, an enlargement of the particle content.

3.1.1 The MSSM with R-parity conservation

The existence of the soft SUSY breaking terms implies that fermions and sfermions cannot be rotated by the same rotation matrices from the electroweak basis to the mass eigenbasis. It is very convenient to work in the super-CKM basis for the squarks and to assume that h^E is diagonal and real which can be done without loss of generality. In this way the additional flavor violation in the sfermion sector is most apparent. In this way, the additional flavor violation is encoded in the mass matrices of the sfermions which reads (see also section 16 of [170]):

$$M^2_{\tilde{f}} = \begin{pmatrix} M^2_{LL} & M^{2\dagger}_{RL} \\ M^2_{RL} & M^2_{RR} \end{pmatrix}, \quad (39)$$

where the entries are 3×3 matrices. They are given by

$$M^2_{LL} = K^\dagger \hat{M}^2_Q K + m^2_u + D_{uLL}, \quad (40)$$

$$M^2_{RL} = v_d \hat{T}^U - \mu^* m_u \cot\beta, \quad (41)$$

$$M^2_{RR} = \hat{M}^2_U + m^2_u + D_{uRR} \quad (42)$$

for u-type squarks in the basis $(\tilde{u}_L, \tilde{c}_L, \tilde{t}_L, \tilde{u}_R, \tilde{c}_R, \tilde{t}_R)$. K is the CKM matrix and we have defined

$$\hat{M}^2_Q \equiv V^\dagger_d M^2_{\tilde{Q}} V_d \quad (43)$$

where V_d is the mixing matrix for the left d-quarks. $\hat{T}^U$ and $\hat{M}^2_U$ are given by a similar transformation involving the mixing matrix for left- and right handed u-quarks. The same type of notation will be kept below for d squarks and sleptons. Finally, the D-terms are given by

$$D_{fLL,RR} = \cos 2\beta M^2_Z (T^3_f - Q_f \sin^2\theta_W) \mathbb{1}. \quad (44)$$

The entries for d-type squarks in the basis $(\tilde{d}_L, \tilde{s}_L, \tilde{b}_L, \tilde{d}_R, \tilde{s}_R, \tilde{b}_R)$ read

$$M^2_{LL} = \hat{M}^2_Q + m^2_d + D_{dLL}, \tag{45}$$

$$M^2_{RL} = v_u \hat{T}^D - \mu^* m_d \tan\beta, \tag{46}$$

$$M^2_{RR} = \hat{M}^2_D + m^2_d + D_{dRR}. \tag{47}$$

For the charged sleptons one finds in the basis $(\tilde{e}_L, \tilde{\mu}_L, \tilde{\tau}_L, \tilde{e}_R, \tilde{\mu}_R, \tilde{\tau}_R)$

$$M^2_{LL} = \hat{M}^2_L + m^2_l + D_{lLL}, \tag{48}$$

$$M^2_{RL} = v_u \hat{T}^E - \mu^* m_l \tan\beta, \tag{49}$$

$$M^2_{RR} = \hat{M}^2_E + m^2_l + D_{lRR}. \tag{50}$$

Assuming that there are only left-type sneutrinos one finds for them in the basis $(\tilde{\nu}_{eL}, \tilde{\nu}_{\mu L}, \tilde{\nu}_{\tau L})$ the mass matrix

$$M^2_{LL} = \hat{M}^2_L + D_{\nu LL}. \tag{51}$$

For sleptons the relevant interaction Lagrangian, e.g. not considering the slepton Higgs or slepton gauge boson interactions, for the studies below is given in terms of mass eigenstates by

$$\mathcal{L} = \bar{l}_i \big(c^L_{ikm} P_L + c^R_{ikm} P_R \big) \tilde{\chi}^0_k \tilde{l}_m$$
$$+ \bar{l}_i \big(d^L_{ilj} P_L + d^R_{ilj} P_R \big) \tilde{\chi}^-_l \tilde{\nu}_j + \bar{\nu}_i e^R_{ikj} P_R \tilde{\chi}^0_k \tilde{\nu}_j$$
$$+ \bar{\nu}_i f^R_{ilm} P_R \tilde{\chi}^+_l \tilde{l}_m + \text{h.c.} \tag{52}$$

The specific forms of the couplings c^L_{ikm}, c^R_{ikm}, d^L_{ilj}, d^R_{ilj}, e^R_{ikj} and f^R_{ilm} can be found in [171]. The first two terms in (52) give rise to the LFV signals studied here, whereas the last one will give rise to the SUSY background because the neutrino flavor cannot be discriminated in high energy collider experiments. In particular the following decays are of primary interest:

$$\tilde{l}_j \to l_i \tilde{\chi}^0_k, \tag{53}$$

$$\tilde{\chi}^0_k \to \tilde{l}_j l_i, \tag{54}$$

$$\tilde{\chi}^0_k \to l_j l_i \tilde{\chi}^0_r. \tag{55}$$

Several studies for these decays have been performed assuming either specific high-scale models or specifying the LFV parameters at the low scale (see for instance [118, 125, 172–194]).

Performing Monte Carlo studies on the parton level it has been shown that LHC can observe SUSY LFV by studying the LFV decays of the second neutralino $\tilde{\chi}^0_2$ arising from cascade decays of gluinos and squarks, i.e. $\tilde{\chi}^0_2 \to \tilde{\ell}\ell' \to \ell'\ell''\tilde{\chi}^0_1$: signals of SUSY LFV can be extracted despite considerable backgrounds and stringent experimental bounds on flavor-violating lepton decays in case of two generation mixing in either the right or left slepton sector in the mSUGRA model [195–197]. The $\tilde{e}_R$–$\tilde{\mu}_R$ mixing case was studied in [195, 197] and the $\tilde{\mu}_L$–$\tilde{\tau}_L$ mixing case in [196].

In the (s)quark sector one has decays analogue to the ones given in (53)–(55). In addition there are decays into charginos and gluinos if kinematically allowed. Flavor effects in these decays have first been discussed in [198]. There it has been shown that one can have large effects in squark and gluino decays despite stringent constraints from B-meson physics as discussed in the "B, D and K decays" chapter. In addition, flavor mixing in the squark sector can induce flavor-violating decays of Higgs bosons as e.g. $H^0 \to bs$ [119].

In the discussion so far we have considered models where the parameters are freely given at the electroweak scale. The fact that no flavor violation in the quark sector has been found beyond SM expectations has led to the development of the concept of minimal flavor violation (MFV). The basic idea is that at a given scale the complete flavor information is encoded in the Yukawa couplings [199], e.g. that in a GUT theory the parameters at the GUT scale are given by $M^2_F = M^2_0 \mathbb{1}$ and $T_F = A_0 h_U$ with M_0 and A_0 being a real and a complex number, respectively. In such models it has been shown that the branching ratios for flavor-violating squark decays are very small and most likely not observable at LHC [200]. A similar concept has been developed for (s)leptons [201, 202]. In contrast to the squark sector one has large mixing effects in the neutrino sector which can lead to observable effects in the slepton sector at future collider experiments (see also [203] and Sect. 5.2.3 of the "Flavor physics of leptons and dipole moments" article of this volume).

3.1.2 The MSSM with broken R-parity

Recent neutrino experiments have shown that neutrinos are massive particles which mix among themselves (for a review see e.g. [204]). In contrast to leptons and quarks, neutrinos need not be Dirac particles but can be Majorana particles. In the latter case the Lagrangian contains a mass term which violates explicitly lepton number by two units. This motivates one to allow for the lepton number breaking terms in the superpotential, as they automatically imply the existence of massive neutrinos without the need of introducing right handed neutrinos and explaining their mass hierarchies [205]. The λ'' terms can still be forbidden by a discrete symmetry such as baryon triality [206].

Let us briefly comment on the number of free parameters before discussing the phenomenology in more detail. The last term in (34), $\hat{L}_i \hat{H}_u$, mixes the lepton and the Higgs superfields. In supersymmetry $\hat{L}_i$ and $\hat{H}_d$ have the same gauge and Lorentz quantum numbers and we can redefine them by a rotation in $(\hat{H}_d, \hat{L}_i)$. The terms $\epsilon_i \hat{L}_i \hat{H}_u$ can then be rotated to zero in the superpotential [205]. However, there are still the corresponding terms in the soft supersymmetry breaking Lagrangian

$$V_{\not{R},\text{soft}} = B_i \epsilon_i \tilde{L}_i H_u \tag{56}$$

which can only be rotated away if $B_i = B$ and $M^2_{H_d} = M^2_{L,i}$ [205]. Such an alignment of the superpotential terms with the soft breaking terms is not stable under the renormalization group equations [207]. Assuming an alignment at the unification scale, the resulting effects are small [207] except for neutrino masses [207–211]. Models containing only bilinear terms do not introduce trilinear terms as can easily be seen from the fact that bilinear terms have dimension of a mass whereas the trilinear ones are dimensionless. For this reason we shall keep in the following explicitly the bilinear terms in the superpotential. These couplings induce decays of the LSP violating lepton number, e.g.

$$\begin{aligned}
\tilde{\nu} &\to q\bar{q}, l^+l^-, \nu\bar{\nu}, \\
\tilde{l} &\to l^+\nu, q\bar{q}', \\
\tilde{\chi}^0_1 &\to W^\pm l^\mp, Z\nu_i, \\
\tilde{\chi}^0_1 &\to l^\pm q\bar{q}', q\bar{q}\nu_i, l^+l^-\nu_i.
\end{aligned} \tag{57}$$

How large can the branching ratio for those decay modes be? To answer this question one has to take into account existing constraints on R-parity violating parameters from low energy physics. As most of them are given in terms of trilinear couplings, we shall work for this particular considerations in the "ϵ-less" basis, e.g. rotate away the bilinear terms in the superpotential (34). Therefore, the trilinear couplings get additional contributions. Assuming, without loss of generality, that the lepton and down-type Yukawa couplings are diagonal, the trilinear couplings are given to leading order in ϵ_i/μ as [206, 212, 213]:

$$\lambda'_{ijk} \to \lambda'_{ijk} + \delta_{jk} h_{d_k} \frac{\epsilon_i}{\mu} \tag{58}$$

and

$$\begin{aligned}
&\lambda_{ijk} \to \lambda_{ijk} + \delta\lambda_{ijk}, \\
&\delta\lambda_{121} = h_e \frac{\epsilon_2}{\mu}, \quad \delta\lambda_{122} = h_\mu \frac{\epsilon_1}{\mu}, \quad \delta\lambda_{123} = 0, \\
&\delta\lambda_{131} = h_e \frac{\epsilon_3}{\mu}, \quad \delta\lambda_{132} = 0, \quad \delta\lambda_{133} = h_\tau \frac{\epsilon_1}{\mu}, \\
&\delta\lambda_{231} = 0, \quad \delta\lambda_{232} = h_\mu \frac{\epsilon_3}{\mu}, \quad \delta\lambda_{233} = h_\tau \frac{\epsilon_2}{\mu},
\end{aligned} \tag{59}$$

where we have used the fact that neutrino physics requires $|\epsilon_i/\mu| \ll 1$ [211]. An essential point to notice is that the additional contributions in (58) and (59) follow the hierarchy dictated by the down quark and charged lepton masses of the SM.

A comprehensive list of bounds on various R-parity violating parameters can be found in [214]. However, there the recent data from neutrino experiments like Super-Kamiokande [215], SNO [216] and KamLAND [217] are not taken into account. These experiments yield strong bounds on trilinear couplings involving the third generation [218, 219]. In addition also the sneutrino vacuum expectation values (VEVs) are constrained by neutrino data [211, 218]. Most of the trilinear couplings have a bound of the order $(10^{-2}–10^{-1}) \times m_{\tilde{f}}/(100 \text{ GeV})$ where $m_{\tilde{f}}$ is the mass of the sfermion in the process under considerations. The cases with stronger limits are $|\lambda'_{111}| \lesssim O(10^{-4})$ due to neutrino-less double beta decay and $|\lambda_{i33}| \simeq 5|\lambda'_{i33}| \simeq O(10^{-4})$ due to neutrino oscillation data. Moreover, neutrino oscillation data imply $|\mu^2(v_1^2 + v_2^2 + v_3^2)/\det(\mathcal{M}_{\chi^0})| \lesssim 10^{-12}$ where v_i are the sneutrino VEVs and $\det(\mathcal{M}_{\chi^0})$ is the determinant of the MSSM neutralino mass matrix.

There exists a vast literature on the effects of R-parity violation at LHC [220–226]. However, in most of these studies, in particular those considering trilinear couplings only, very often the existence of a single coupling has been assumed. However, such an assumption is only valid at a given scale as renormalization effects imply that additional couplings are present when going to a different scale via renormalization group evolution (RGE). Moreover, very often the bounds stemming from neutrino physics are not taken into account or are out-dated (e.g. assuming an MeV tau neutrino). Last but not least one should note that also in this class of models there are potential dark matter candidates, e.g. a very light gravitino [227–230].

Recently another class of models with explicitly broken R-parity has been proposed where the basic idea is that the existence of right handed neutrino superfields is the source of the μ-term of the MSSM as well as the source of neutrino masses [231]. In this case the superpotential contains only trilinear terms. Beside the usual Yukawa couplings of the MSSM the following couplings are present:

$$W_{\nu^c} = h^\nu_{ij} \hat{H}_u \hat{L}_i \hat{\nu}^c_j - \lambda_i \hat{\nu}^c_i \hat{H}_d \hat{H}_u + \frac{1}{3} \kappa^{ijk} \hat{\nu}^c_i \hat{\nu}^c_j \hat{\nu}^c_k. \tag{60}$$

Note that the second and third term break R-parity and that the sneutrino fields play the role of the gauge singlet field of the next to minimal supersymmetric standard model (NMSSM) [232–235].

3.1.3 Spontaneous R-parity violation

Up to now we have only considered explicit R-parity violation keeping the particle content of the MSSM. In the case that one enlarges the spectrum by gauge singlets one can obtain models where lepton number and, thus, R-parity, are broken spontaneously together with SU(2)⊗U(1) [236–240]. A second possibility to break R-parity spontaneously is to enlarge the gauge symmetry [241].

The most general superpotential terms involving the MSSM superfields in the presence of the SU(2)⊗U(1) singlet superfields $(\hat{v}_i^c, \hat{S}_i, \hat{\Phi})$ carrying a conserved lepton number assigned as $(-1, 1, 0)$, respectively, is given as [242]

$$\mathcal{W} = \varepsilon_{ab}\big(h_U^{ij}\hat{Q}_i^a\hat{U}_j\hat{H}_u^b + h_D^{ij}\hat{Q}_i^b\hat{D}_j\hat{H}_d^a + h_E^{ij}\hat{L}_i^b\hat{E}_j\hat{H}_d^a \\ + h_\nu^{ij}\hat{L}_i^a\hat{v}_j^c\hat{H}_u^b - \hat{\mu}\hat{H}_d^a\hat{H}_u^b - h_0\hat{H}_d^a\hat{H}_u^b\hat{\Phi}\big) \\ + h^{ij}\hat{\Phi}\hat{v}_i^c\hat{S}_j + M_R^{ij}\hat{v}_i^c\hat{S}_j + \frac{1}{2}M_\Phi\hat{\Phi}^2 + \frac{\lambda}{3!}\hat{\Phi}^3. \quad (61)$$

The first three terms together with the $\hat{\mu}$ term define the R-parity conserving MSSM, the terms in the second line only involve the SU(2)⊗U(1) singlet superfields $(\hat{v}_i^c, \hat{S}_i, \hat{\Phi})$, while the remaining terms couple the singlets to the MSSM fields. For completeness we note that lepton number is fixed via the Dirac-Yukawa h_ν connecting the right handed neutrino superfields to the lepton doublet superfields. For simplicity we assume in the discussion below that only one generation of $(\hat{v}_i^c, \hat{S}_i)$ is present.

The presence of singlets in the model is essential in order to drive the spontaneous violation of R-parity and electroweak symmetries in a phenomenologically consistent way. As in the case of explicit R-parity violation all sneutrinos obtain a VEV beside the Higgs bosons as well as the $\tilde{S}$ field and the singlet field Φ. For completeness, we want to note that in the limit where all sneutrino VEVs vanish and all singlets carrying lepton number are very heavy one obtains the NMSSM as an effective theory. The spontaneous breaking of R-parity also entails the spontaneous violation of total lepton number. This implies that one of the neutral CP-odd scalars, which we call majoron J and which is approximately given by the imaginary part of

$$\frac{\sum_i v_i^2}{Vv^2}(v_u H_u^0 - v_d H_d^0) + \sum_i \frac{v_i}{V}\tilde{v}_i + \frac{v_S}{V}S - \frac{v_R}{V}\tilde{v}^c \quad (62)$$

remains massless, as it is the Nambu–Goldstone boson associated to the breaking of lepton number. v_R and v_S are the VEVs of $\tilde{v}^c$ and $\tilde{S}$, respectively and $V = \sqrt{v_R^2 + v_S^2}$. Clearly, the presence of these additional singlets enhances further the number of neutral scalar and pseudo-scalar bosons. Explicit formulas for the mass matrices of scalar and pseudo-scalar bosons can be found e.g. in [243].

The case of an enlarged gauge symmetry can be obtained for example in left–right symmetric models, e.g. with the gauge group SU(2)$_L$ × SU(2)$_R$ × U(1)$_{B-L}$ [241]. The corresponding superpotential is given by

$$W = h_{\phi Q}\hat{Q}_L^T i\tau_2\hat{\phi}\hat{Q}_R^c + h_{\chi Q}\hat{Q}_L^T i\tau_2\hat{\chi}\hat{Q}_R^c + h_{\phi L}\hat{L}_L^T i\tau_2\hat{\phi}\hat{L}_R^c \\ + h_{\chi L}\hat{L}_L^T i\tau_2\hat{\chi}\hat{L}_R^c + h_\Delta \hat{L}_R^{cT} i\tau_2\hat{\Delta}\hat{L}_R^c \\ + \mu_1 \text{Tr}(i\tau_2\hat{\phi}^T i\tau_2\hat{\chi}) + \mu_2 \text{Tr}(\hat{\Delta}\hat{\delta}), \quad (63)$$

where the Higgs sector consists of two triplet and two bidoublet Higgs superfields with the following SU(2)$_L$ × SU(2)$_R$ × U(1)$_{B-L}$ quantum numbers:

$$\hat{\Delta} = \begin{pmatrix} \hat{\Delta}^-/\sqrt{2} & \hat{\Delta}^0 \\ \hat{\Delta}^{--} & -\hat{\Delta}^-/\sqrt{2} \end{pmatrix} \sim (\mathbf{1}, \mathbf{3}, -2),$$

$$\hat{\delta} = \begin{pmatrix} \hat{\delta}^+/\sqrt{2} & \hat{\delta}^{++} \\ \hat{\delta}^0 & -\hat{\delta}^+/\sqrt{2} \end{pmatrix} \sim (\mathbf{1}, \mathbf{3}, 2),$$

$$\hat{\phi} = \begin{pmatrix} \hat{\phi}_1^0 & \hat{\phi}_1^+ \\ \hat{\phi}_2^- & \hat{\phi}_2^0 \end{pmatrix} \sim (\mathbf{2}, \mathbf{2}, 0),$$

$$\hat{\chi} = \begin{pmatrix} \hat{\chi}_1^0 & \hat{\chi}_1^+ \\ \hat{\chi}_2^- & \hat{\chi}_2^0 \end{pmatrix} \sim (\mathbf{2}, \mathbf{2}, 0).$$

$$(64)$$

Looking at the decays of the Higgs bosons, one has to distinguish two scenarios. (i) Lepton number is gauged and thus the majoron becomes the longitudinal part of an additional neutral gauge boson. (ii) The majoron remains a physical particle in the spectrum. In the case of the enlarged gauge group there are additional doubly charged Higgs bosons H_i^{--} which have lepton number violating couplings. In e^-e^- collisions they can be produced according to

$$e^-e^- \to H_i^{--} \quad (65)$$

and have decays of the type

$$H_i^{--} \to H_j^- H_k^-, \quad (66)$$

$$H_i^{--} \to l_j^- l_k^-, \quad (67)$$

$$H_i^{--} \to \tilde{l}_j^- \tilde{l}_k^-. \quad (68)$$

In addition there exist doubly charged charginos which can have lepton flavor-violating decays:

$$\tilde{\chi}_i^{--} \to \tilde{l}_j^- l_k^-. \quad (69)$$

3.1.4 Study of supersymmetry at the LHC

If supersymmetry exists at the electroweak scale, it could hardly escape detection at the LHC. In most R-parity conserving models, the production cross section is expected to be dominated by the pair production of colored states (squarks and gluinos). These decay to lighter SUSY particles and ultimately to the LSP (lightest supersymmetric particle). If this is stable and weakly interacting, as implied by

R-parity conservation and cosmological arguments, it leaves the experimental apparatus undetected. The supersymmetric events are thus expected to show up at the LHC as an excess over SM expectations of events with several hard hadronic jets and missing energy. The LHC center of mass of 14 TeV extends the search for SUSY particles up to squark and gluino masses of 2.5 to 3 TeV [3, 244].

If squarks and gluinos are lighter than 1 TeV, as implied by naturalness arguments, this signature would be observed with high statistical significance already during the first year of running at the initial LHC luminosity of 2×10^{33} cm^{-2} s^{-1} [245]. In practice, discovery would be achieved as soon as a good understanding of the systematics on SM rates at the LHC is obtained.

A significant part of the efforts in preparation for the LHC startup is being spent in the simulations of the new physics potential. We give below a brief overview of these studies, dividing them in three categories: inclusive searches of the non-SM physics, measurement of SUSY particle masses, and measurements of other properties of SUSY particles, such as their spin or the flavor structure of their decays.

3.1.5 Inclusive searches

In these studies, the typical discovery strategy consists in searching for an excess of events with a given topology. A variety of final state signatures has been considered. Inclusive searches have mainly be carried out in the framework of mSUGRA, which has five independent parameters specified at a high energy scale: the common gaugino mass $m_{1/2}$, the common scalar mass m_0, the common trilinear coupling A_0, the ratio of the vacuum expectation values of the two Higgs doublets $\tan \beta$ and the sign of the Higgsino mixing parameter μ. The masses and decay branching ratios of the SUSY particles are then computed at the electroweak scale using the renormalization group equations, and used as input to the LHC simulation codes.

For each point of a grid covering the mSUGRA parameter space, signal events are generated at parton level and handed over to the parametrized detector simulation. The main SM background sources are simulated, where the most relevant are processes with an hard neutrino in the final state ($t\bar{t}$, W + jets, Z + jets). Multi jet QCD is also relevant because its cross section is several orders of magnitude larger than SUSY. However, it is strongly suppressed by the requirement of large transverse missing energy and it gives a significant contribution only to the final state search channels without isolated leptons. The detailed detector simulation, much more demanding in terms of computing CPU power, validates the results with parametrized detector simulations for the SM backgrounds and selected points in the mSUGRA parameter space.

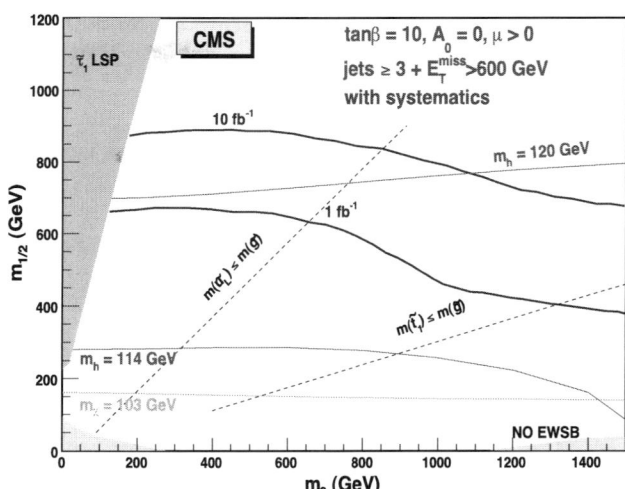

Fig. 20 CMS 5σ discovery potential using multi jets and missing transverse energy final state [5]

Cuts on missing transverse energy, the transverse momentum of jets, and other discriminating variables are optimized to give the best statistical significance for the (simulated) observed excess of events. For each integrated luminosity the regions of the parameter space for which the statistical significance exceeds the conventional discovery value of 5σ are then displayed. An example is shown in Fig. 20 for the CMS experiment [5] with similar results for ATLAS [245]. A slice of the mSUGRA parameter space is shown, for fixed $\tan \beta = 10$, $A = 0$ and $\mu > 0$. The area of parameter space favored by naturalness arguments can be explored with an integrated luminosity of only 1 fb^{-1}.

Although these results were obtained in the context of mSUGRA, the overall SUSY reach in terms of squark and gluino masses is very similar for most R-parity conserving models, provided that the LSP mass is much lower than the squark and gluino masses. This has been shown to be the case for GMSB and AMSB models [246] and even the MSSM [247].

3.1.6 Mass measurement

A first indication of the mass scale of the SUSY particles produced in the pp interaction will probably be obtained measuring the "effective mass", which is the scalar sum of transverse missing energy and the p_T of jets and leptons in the event. Such a distribution is expected to have a peak correlated with the SUSY mass scale. The correlation is strong in mSUGRA, and still usable in the more general MSSM [245].

The reconstruction of the mass spectrum of supersymmetric particles will be more challenging. Since SUSY particles would be produced in pairs, there are two undetected LSP particles in the final state, which implies that mass peaks can not be reconstructed from invariant-mass combinations, unless the mass of the LSP itself is already known.

The typical procedure consists in choosing a particular decay chain, measuring invariant-mass combinations and looking for kinematical minima and maxima. Each kinematical end point is a function of the masses of the SUSY particles in the decay chain. If enough end points can be measured, the masses of all the SUSY particles involved in the decay chain can be obtained. Once the mass of the LSP is known, mass peaks can be reconstructed.

After reducing the SM background very effectively through hard missing transverse energy cuts, the main background for this kind of measurements usually comes from supersymmetric events in which the desired decay chain is not present or was not identified correctly by the analysis. For this reason, these studies are made using data simulated for a specific point in SUSY parameter space, for which all supersymmetric production processes are simulated.

The two body decay chain $\chi_2^0 \to \tilde{l}^{\pm} l^{\pm} \to l^{\pm} l^{\pm} \chi_1^0$ is particularly promising, as it leads to a very sharp edge in the distribution of the invariant mass of the two leptons, which measures:

$$m_{\text{edge}}^2(\ell\ell) = \frac{(m_{\tilde{\chi}_2^0}^2 - m_{\tilde{\ell}_i}^2)(m_{\tilde{\ell}_i}^2 - m_{\tilde{\chi}_1^0}^2)}{m_{\tilde{\ell}_i}^2}. \tag{70}$$

The basic signature of this decay chain are two opposite-sign, same-flavor (OSSF) leptons; but two such leptons can also be produced by other processes. If the two leptons are independent of each other, one would expect equal amounts of OSSF leptons and opposite-sign, opposite-flavor (OSOF) leptons (i.e combinations $e^+\mu^-$, $e^-\mu^+$). Their distributions should also be identical, and this allows one to remove the background contribution for OSSF by subtracting the OSOF events.

Figure 21 shows the invariant mass of the two leptons obtained for SPS1A point [248] with 100 fb^{-1} of simulated ATLAS data [249]. The SM background is clearly negligible. The real background consists of other SUSY processes, which are effectively removed by the OSOF subtraction.

Several other kinematical edges can be obtained using various invariant-mass combinations involving jets and leptons. Two of such distributions are reported in Fig. 22 for the point SPS1a and 100 fb^{-1} of ATLAS simulated data [249]. Five end points, each providing a constraints on the mass of four particles, can be measured. The masses of the supersymmetric particles present in the decay chain (the left handed squark, the right handed sleptons, and the two lightest neutralinos) can thus be measured with an error between 3% (for the squark) and 12% (for the lightest neutralino) for 100 fb^{-1} of integrated luminosity.

For lepton pairs with an invariant mass near the kinematical end point, the relation

$$p_\mu(\tilde{\chi}_2^0) = \left(1 - \frac{m_{\tilde{\chi}_1^0}}{m_{ll}}\right) p_\mu(ll) \tag{71}$$

can be used to get the four-momentum of the χ_2^0, provided that the mass of the lightest neutralino has already been measured. This four-vector can then be combined with that of hadronic jets to measure the gluino and squark masses. In Fig. 23 the gluino and squark mass peaks obtained with CMS parametrized simulation are reported for another mSUGRA benchmark point, called point B [250], which is defined by $m_0 = 100$ GeV, $m_{1/2} = 250$ GeV, $A = 0$, $\mu > 0$, $\tan\beta = 10$.

Several other techniques to reconstruct the masses of supersymmetric particles have been investigated by the ATLAS and CMS collaborations. Here we shall only mention a few other possibilities:

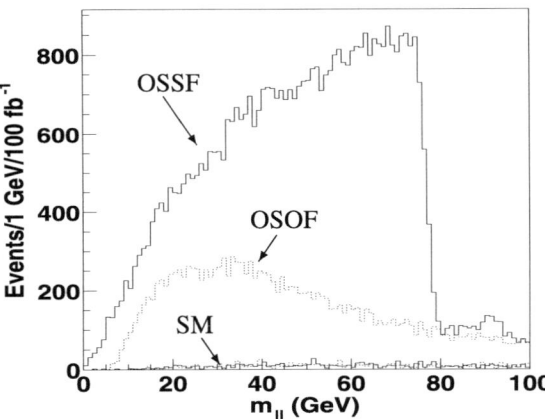
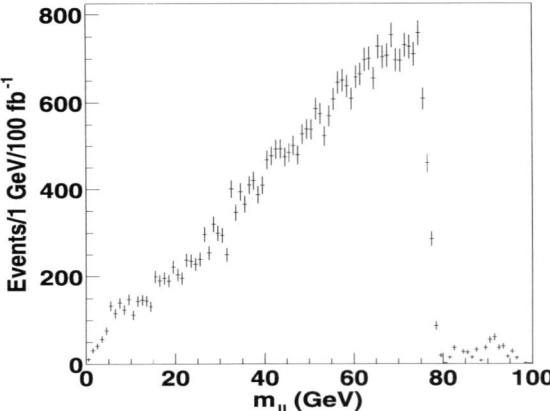

Fig. 21 Effect of subtracting background leptons, for the mSUGRA benchmark point SPS1a and an integrated luminosity of 100 fb^{-1}. In the *left plot*: the curves represent opposite-sign, same-flavor (OSSF) leptons, opposite-sign, opposite- flavor (OSOF) leptons and the SM contribution. In the *right plot*, the flavor subtraction OSSF-OSOF has been plotted: the triangular shape of the theoretical expectation is reproduced

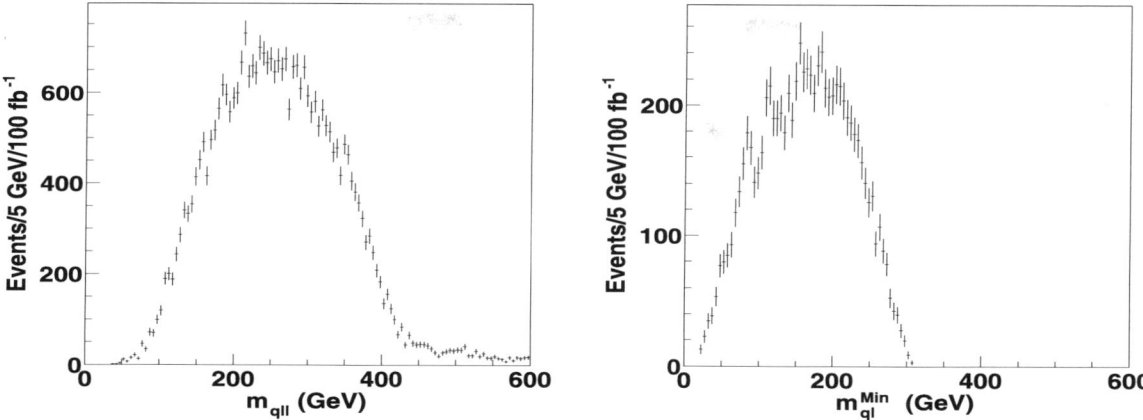

Fig. 22 Invariant mass distributions with kinematical end points, for an integrated luminosity of 100 fb^{-1}. In the *left plot* for qll combination, in the *right plot* for the maximum of ql combination

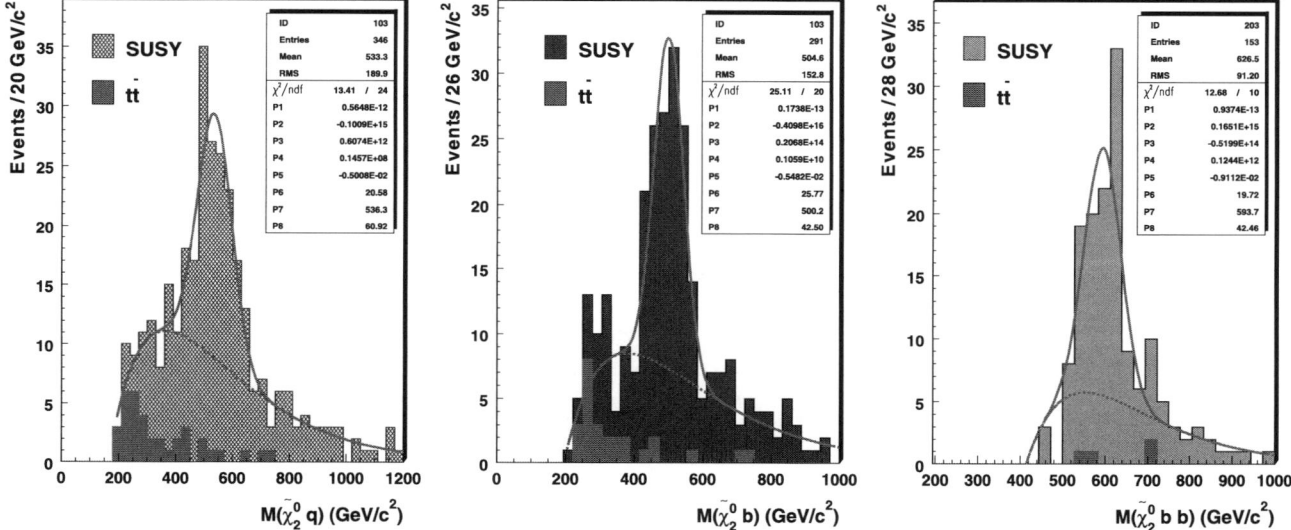

Fig. 23 Invariant mass peaks for squark (*left*), sbottom (*middle*) and gluino (*right*) at point B. The picture has been obtained using the parametrized simulation of the CMS detector. The integrated luminosity is 1 fb^{-1} for the squarks and 10 fb^{-1} for the other mass peaks

- At large $\tan\beta$ the decays into third generation leptons are dominant. The $\tau^+\tau^-$ kinematic end point is still measurable using the invariant mass of the tau visible decay products, but the expected precision is worse than that achievable with electrons and muons.
- The right handed squark often decays directly into the LSP. $\tilde{q}_R\tilde{q}_R \to q\chi_1^0 q\chi_1^0$ events can be used to reconstruct the mass of this squark. A similar technique can be used to measure the mass of left handed sleptons which decay directly into the LSP.

For the point SPS1a and an integrated luminosity of 300 fb^{-1} ATLAS expects to be able to measure at least 13 mass relations [249]. The constraints which can be put on the SUSY parameter space and on the relic density of neutralinos using these measurements are discussed in [251].

3.1.7 Flavor studies

Most studies by the LHC collaborations have focused on the discovery strategies and the measurement of the masses of SUSY particles. However, the possibility to measure other properties of the new particles, such as their spin or the branching ratios of flavor-violating decays, has also been investigated.

The measurement of the spin is interesting because it allows one to confirm the supersymmetric nature of the new particles. This measurement was investigated in [252, 253] and it is also discussed later in this article.

In the hadronic sector, the experiments are not able to discriminate the flavor of quarks of the first two generations. Hence the only possibility for flavor studies relies on

b-tagging techniques. In this report, the possibility to measure kinematical end points involving the scalar top is discussed. The scalar bottom masses may also be measured at the LHC.

The leptonic sector is more favorable from the experimental point of view, as the flavor of the three charge leptons can be identified accurately by the detectors with relatively low backgrounds. This allows for the possibility to test the presence of decays violating lepton flavor. This possibility was already discussed in early studies [196, 254, 255] and it is investigated in a few contributions to this report.

3.2 Effects of lepton flavor violation on dilepton invariant-mass spectra

In this section we discuss the effect of lepton flavor violation (LFV) on dilepton invariant-mass spectra in the decay chains

$$\tilde{\chi}_2^0 \to \tilde{\ell}_i^+ \ell_j^- \to \ell_k^+ \ell_j^- \tilde{\chi}_1^0. \tag{72}$$

In these events one studies the invariant dilepton mass spectrum $dN/dm(\ell\ell)$ with $m(\ell\ell)^2 = (p_{\ell^+} + p_{\ell^-})^2$. Its kinematical end point is used in combination with other observables to determine masses or mass differences of sparticles [256–258].

Details on the parameter dependence of flavor-violating decays can be found for example in [260]. As an example, the study point SPS1a' [259] is considered. It has a relatively light spectrum of charginos/neutralinos and sleptons, with the three lighter charged sleptons being mainly

$\tilde{\ell}_R$: $m_{\tilde{\chi}_1^0} = 97.8$ GeV, $m_{\tilde{\chi}_2^0} = 184$ GeV, $m_{\tilde{e}_1} = 125.3$ GeV, $m_{\tilde{\mu}_1} = 125.2$ GeV, $m_{\tilde{\tau}_1} = 107.4$ GeV. The underlying parameters are given in Table 11, where M_1 and M_2 are the U(1) and SU(2) gaugino mass parameters, respectively. In this example the flavor off-diagonal elements of $M_{E,\alpha\beta}^2$ ($\alpha \neq \beta$) in (50) are expected to give the most important contribution to the LFV decays of the lighter charginos, neutralinos and sleptons. We therefore discuss LFV only in the right slepton sector. To illustrate the effect of LFV on these spectra, in Fig. 24 we present invariant-mass distributions for various lepton pairs taking the following LFV parameters: $M_{E,12}^2 = 30$ GeV^2, $M_{E,13}^2 = 850$ GeV^2 and $M_{E,23}^2 = 600$ GeV^2, for which we have $(m_{\tilde{\ell}_1}, m_{\tilde{\ell}_2}, m_{\tilde{\ell}_3}) = (106.4, 125.1, 126.2)$ GeV. These parameters are chosen such that large LFV $\tilde{\chi}_2^0$ decay branching ratios are possible consistently with the experimental bounds on the rare lepton decays, for which we obtain: Br($\mu^- \to e^- \gamma$) = 9.5×10^{-12}, Br($\tau^- \to e^- \gamma$) = 1.0×10^{-7} and Br($\tau^- \to \mu^- \gamma$) = 5.2×10^{-8}. We find for the $\tilde{\chi}_2^0$ decay branching ratios: Br($e\mu$) = 1.7%, Br($e\tau$) = 3.4%, Br($\mu\tau$) = 1.8%, Br(e^+e^-) = 1%, Br($\mu^+\mu^-$) = 1.2%, Br($\tau^+\tau^-$) = 51% with Br($\ell_i\ell_j$) ≡ Br($\tilde{\chi}_2^0 \to \ell_i\ell_j\tilde{\chi}_1^0$). Note that we have summed here over all contributing sleptons.

In Fig. 24a we show the flavor-violating spectra $(100/\Gamma_{\text{tot}}) d\Gamma(\tilde{\chi}_2^0 \to \ell_i^\pm \ell_j^\mp \tilde{\chi}_1^0)/dm(\ell_i^\pm \ell_j^\mp)$ versus $m(\ell_i^\pm \ell_j^\mp)$ for the final states $\mu\tau$, $e\tau$ and $e\mu$. In cases where the final state contains a τ-lepton, one finds two sharp edges. The first one at $m \simeq 59.4$ GeV is due to an intermediate $\tilde{\ell}_1(\sim \tilde{\tau}_R)$ and the second one at $m \simeq 84.6$ GeV is due to intermediate states of the two heavier sleptons $\tilde{\ell}_2(\sim \tilde{\mu}_R)$ and $\tilde{\ell}_3(\sim \tilde{e}_R)$

Table 11 Relevant on-shell parameters for the SPS1a' [259] scenario

$\tan\beta$	10	$M_{L,11} = M_{L,22}$	184 GeV	$M_{E,33}$	111 GeV
M_1	100.1 GeV	$M_{L,33}$	182.5 GeV	A_{11}	-0.013 GeV
M_2	197.4 GeV	$M_{E,11}$	117.793 GeV	A_{22}	-2.8 GeV
μ	400 GeV	$M_{E,22}$	117.797 GeV	A_{33}	-46 GeV

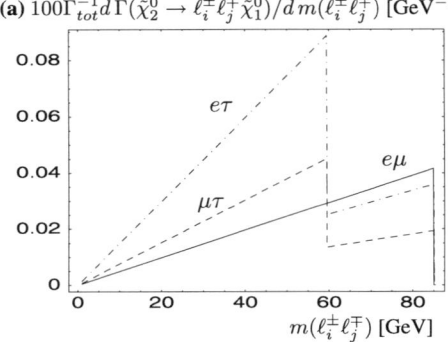
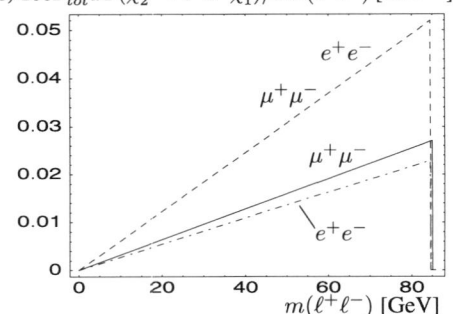

Fig. 24 Invariant mass spectra $100\Gamma_{\text{tot}}^{-1} d\Gamma(\tilde{\chi}_2^0 \to \ell_i \ell_j \tilde{\chi}_1^0)/dm(\ell_i \ell_j)$ versus $m(\ell_i \ell_j)$. In (**a**) we show the "flavor-violating" spectra summed over charges in the LFV case for the SPS1a' scenario: $e^\pm \mu^\mp$ (*full line*), $e^\pm \tau^\mp$ (*dashed dotted line*) and $\mu^\pm \tau^\mp$ (*dashed line*) and in (**b**) we show the "flavor-conserving" spectra: e^+e^- (*dashed line*) and $\mu^+\mu^-$ (*dashed line*) are for the LFC case in the SPS1a' scenario, and e^+e^- (*dashed dotted line*) and $\mu^+\mu^-$ (*full line*) are for the LFV case in the SPS1a' scenario

with $m_{\tilde{\ell}_2} \simeq m_{\tilde{\ell}_3}$. The position of the edges can be expressed in terms of the neutralino and intermediate slepton masses [256], see (70). In the case of the $e\mu$ spectrum the first edge is practically invisible because the branching ratios of $\tilde{\chi}_2^0$ into $\tilde{\ell}_1 e$ and $\tilde{\ell}_1 \mu$ are tiny for this example [260]. Note that the rate for the $e\tau$ final state is largest in this case because $|M_{E,13}^2|$ is larger than the other LFV parameters.

In Fig. 24b we show the "flavor-conserving" spectra for the final states with e^+e^- and $\mu^+\mu^-$. The dashed line corresponds to the flavor-conserving case where $M_{E,ij}^2 = 0$ for $i \neq j$. LFV causes firstly a reduction of the height of the end point peak. Secondly, it induces a difference between the $\mu^+\mu^-$ and e^+e^- spectra because the mixings among the three slepton generations are in general different from each other. The peaks at $m \simeq 59.4$ GeV in the $\mu^+\mu^-$ and e^+e^- spectra are invisible as in the $e\mu$ spectrum as the branching ratios of the corresponding flavor-violating decays are small. As for the $\tau^+\tau^-$ spectrum we remark that the height of the peak (due to the intermediate $\tilde{\ell}_1$ ($\sim \tilde{\tau}_R$)) in the $\tau^+\tau^-$ spectrum gets reduced by about 5% and that the contributions due to the intermediate $\tilde{\ell}_{2,3}$ are invisible. Moreover, the peak position gets shifted to a smaller value by about 2.7 GeV since the mass of the intermediate $\tilde{\ell}_1$ gets reduced by 1 GeV compared to the flavor-conserving case.

It is interesting to note that in the LFV case the rate of the channel $e\tau$ can be larger than those of the channels with the same flavor, e^+e^- and $\mu^+\mu^-$. Moreover, by measuring all dilepton spectra for the flavor-violating as well as flavor-conserving channels, one can make an important cross check of this LFV scenario: the first peak position of the lepton flavor-violating spectra (except the $e\mu$ spectrum) must coincide with the end point of the $\tau^+\tau^-$ spectrum and the second peak must coincide with those of the e^+e^- and $\mu^+\mu^-$ spectra.

Up to now, only the dilepton mass spectra for the SPS1a' benchmark point have been investigated in detail. Which requirements must other scenarios fulfill to obtain observable double edge structures? Obviously the kinematic condition $m_{\tilde{\chi}_s^0} > m_{\tilde{\ell}_i, \tilde{\ell}_j} > m_{\tilde{\chi}_r^0}$ must be fulfilled and sufficiently many $\tilde{\chi}_s^0$ must be produced. In addition there should be two sleptons contributing in a sizable way to the decay $\tilde{\chi}_s^0 \to \ell' \ell'' \tilde{\chi}_r^0$ and, of course, the corresponding branching ratio has to be large enough to be observed. For this the corresponding LFV entries in the slepton mass matrix have to be large enough. Moreover, also the mass difference between the two contributing sleptons has to be sufficiently large so that the difference of the positions of the two peaks is larger than the experimental resolution. In mSUGRA-like scenarios, which are characterized by a common mass m_0 for the scalars and a common gaugino mass $m_{1/2}$ at the GUT scale, the kinematic requirements (including the positions of the peaks) are fulfilled in the regions of parameter space where $m_0^2 \lesssim 0.4\, m_{1/2}^2$ and $\tan\beta \gtrsim 8$. The first condition provides for right sleptons

lighter than the $\tilde{\chi}_2^0$ and the second condition ensures that the mass difference between $\tilde{\tau}_1$ and the other two right sleptons is sufficiently large. In the region where $m_0^2 \lesssim 0.05\, m_{1/2}^2$ also the left sleptons are lighter than $\tilde{\chi}_2^0$, giving the possibility of additional structures in the dilepton mass spectra.

Details on background processes will be presented in the subsequent sections, where studies by the two experiments ATLAS and CMS are presented. Here we give a brief summary of the expected dominant background. The largest SM background is due to $t\bar{t}$ production. There is also SUSY background due to uncorrelated leptons stemming from different squark and gluino decay chains. The resulting dilepton mass distributions will, however, be smooth and decrease monotonically with increasing dilepton invariant mass as was explicitly shown in a Monte Carlo analysis in [196, 197]. It was also shown that the single edge structure can be observed over the smooth background in the $e\mu$ and $\mu\tau$ invariant-mass distributions. Therefore, the novel distributions as shown in Fig. 24, in particular the characteristic double edge structures in the $e\tau$ and $\mu\tau$ invariant-mass distributions, should be clearly visible on top of the background. Note that the usual method for background suppression, by taking the sum $N(e^+e^-) + N(\mu^+\mu^-) - N(e^\pm\mu^\mp)$, is not applicable in the case of LFV searches. Instead one has to study the individual pair mass spectra. Nevertheless, one can expect that these peaks will be well observable [261].

3.3 Lepton flavor violation in the long lived stau NLSP scenario

Supersymmetric scenarios can be roughly classified into two main classes, depending on the nature of the lightest supersymmetric particle (LSP). The most popular choice for the LSP is the neutralino, although scenarios with superweakly interacting LSP, such as the gravitino or the axino, are also compatible with all the collider experiments and cosmology. Here, we would like to concentrate on the latter class of scenarios, focusing for definiteness on the case with gravitino LSP.

Under the assumption of universality of the soft-breaking scalar, gaugino and trilinear soft terms at a high energy scale, the so-called constrained MSSM, the next-to-LSP (NLSP) can be either the lightest neutralino or the stau. If R-parity is conserved, the NLSP can only decay into the gravitino and SM particles, with a decay rate very suppressed by the gravitational interactions. As a result, the NLSP can be very long lived, with lifetimes that could be as long as seconds, minutes or even longer, mainly depending on the gravitino mass. When the NLSP is the lightest neutralino, the signatures for LFV are identical to the case with neutralino LSP, which have been extensively discussed in the literature [181–185, 188, 192, 195]. On the other hand, when the NLSP is a stau, the signatures could be very different. In this note we discuss possible signatures and propose strategies to look for LFV in future colliders in scenarios where the gravitino (or the axino) is the LSP and the stau is the NLSP [189, 262].

Motivated by the spectrum of the constrained MSSM we shall assume that the NLSP is mainly a right handed stau, although it could have some admixture of left handed stau or other leptonic flavors, and will be denoted by $\tilde{\tau}_1$. We shall also assume that next in mass in the supersymmetric spectrum are the right handed selectron and smuon, denoted by $\tilde{e}_R$ and $\tilde{\mu}_R$ respectively, also with a very small admixture of left handed states and some admixture of stau. Finally, we shall assume that next in mass are the lightest neutralino and the rest of SUSY particles. Schematically, the spectrum reads

$$m_{3/2} < m_{\tilde{\tau}_1} < m_{\tilde{e}_R,\tilde{\mu}_R} < m_{\chi_1^0}, m_{\tilde{e}_L,\tilde{\mu}_L}, m_{\tilde{\tau}_2}, \ldots . \quad (73)$$

In this class of scenarios, staus could be long lived enough to traverse several layers of the vertex detector before decaying, thus being detected as a heavily ionizing charged track. This signature is very distinctive and is not produced by any SM particle, hence the observation of heavily ionizing charged tracks would give strong support to this scenario and would allow for the search for LFV essentially without SM backgrounds.

Long lived staus could even be stopped in the detector and decay at late times, producing very energetic particles that would spring from inside the detector. Recently, prospects of collecting staus and detecting their decay products in future colliders have been discussed [263, 264]. At the LHC, cascade decays of squarks and gluinos could produce of the order of 10^6 staus per year if the sparticle masses are close to the present experimental limits [265]. Among them, $\mathcal{O}(10^3$–$10^4)$ staus could be collected by placing 1–10 kton massive material around the LHC detectors. On the other hand, at the ILC up to $\mathcal{O}(10^3$–$10^5)$ staus could be collected and studied.

If there is no LFV, the staus could only decay into taus and gravitinos, $\tilde{\tau} \to \tau \psi_{3/2}$. If on the contrary LFV exists in nature, some of the staus could decay into electrons and muons. Therefore, the detection of very energetic electrons and muons coming from inside the detector would constitute a signal of lepton flavor violation.

There are potentially two sources of background in this analysis. First, in certain regions of the SUSY parameter space selectrons or smuons could also be long lived, and the electrons and muons from their flavor-conserving decays could be mistaken for electrons and muons coming from the lepton flavor-violating decay of the stau. However, if flavor violation is large enough to be observed in these experiments, the selectron decay channel $\tilde{e} \to \tilde{\tau} ee$ is very efficient. Therefore, selectrons (and similarly, smuons) are never long lived enough to represent an important source of background. It is remarkable the interesting double role that flavor violation plays in this experiment, both as object of investigation and as crucial ingredient for the success of the experiment itself.

A second source of background for this analysis are the muons and electrons from tau decay, which could be mistaken for muons and electrons coming from the lepton flavor-violating decays $\tilde{\tau} \to \mu \psi_{3/2}$, $\tilde{\tau} \to e \psi_{3/2}$. Nevertheless, this background can be distinguished from the signal by looking at the energy spectrum: the leptons from the flavor-conserving tau decay present a continuous energy spectrum, in stark contrast with the leptons coming from the two body gravitational decay, whose energies are sharply peaked at $E_0 = (m_{\tilde{\tau},\tilde{e}}^2 + m_{\mu,e}^2 - m_{3/2}^2)/(2m_{\tilde{\tau},\tilde{e}})$. It is easy to check that only a very small fraction of the electrons and muons from the tau decay have energies close to this cut-off energy. For instance, for the typical energy resolution of an electromagnetic calorimeter, $\sigma \simeq 10\%/\sqrt{E(\text{GeV})}$, only 2×10^{-5} of the taus with energy $E_0 \sim 100$ GeV will produce electrons with energy $\simeq E_0$, within the energy resolution of the detector, which could be mistaken for electrons coming from the LFV stau decay. Therefore, for the number of NLSPs that can be typically trapped at the LHC or the ILC, the number of electrons or muons from this source of background turns out to be negligible in most instances.

Using this technique, we have estimated that at the LHC or at the future Linear Collider it would be possible to probe mixing angles in the slepton sector down to the level of $\sim 3 \times 10^{-2}$ (9×10^{-3}) at 90% confidence level if 3×10^3 (3×10^4) staus are collected [189]. A different technique, that does not require to stop the staus, was proposed in [262] for the case of an e^+e^- or e^-e^- linear collider.

3.4 Neutralino decays in models with broken R-parity

In supersymmetric models neutrino masses can be explained intrinsically supersymmetric, namely by the breaking of R-parity. The simplest way to realize this idea is to add the bilinear terms of $W_{\not{R}}$ to the MSSM superpotential W_{MSSM} (see (33) and (34)):

$$W = W_{\text{MSSM}} + \epsilon_i \hat{L}_i \hat{H}_u. \quad (74)$$

For consistency one has also to add the corresponding bilinear terms to soft SUSY breaking (see (36) and (37)) which induce small VEVs for the sneutrinos. These VEVs in turn induce a mixing between neutrinos and neutralinos, giving mass to one neutrino at tree level. The second neutrino mass is induced by loop effects (see [211, 266, 267] and references therein). The same parameters that induce neutrino masses and mixings are also responsible for the decay of the lightest supersymmetric particle (LSP). This implies that there are correlations between neutrino physics and LSP decays [213, 268–271].

In particular, the neutrino mixing angles

$$\tan^2\theta_{atm} \simeq \left(\frac{\Lambda_2}{\Lambda_3}\right)^2,$$

$$U_{e3}^2 \simeq \frac{|\Lambda_1|}{\sqrt{\Lambda_2^2 + \Lambda_3^2}}, \quad (75)$$

$$\tan^2\theta_{sol} \simeq \left(\frac{\tilde{\epsilon}_1}{\tilde{\epsilon}_2}\right)^2$$

can be related to ratios of couplings and branching ratios, for example

$$\tan^2\theta_{atm} \simeq \left|\frac{\Lambda_2}{\Lambda_3}\right|^2 \simeq \frac{\mathrm{Br}(\tilde{\chi}_1^0 \to \mu^\pm W^\mp)}{\mathrm{Br}(\tilde{\chi}_1^0 \to \tau^\pm W^\mp)}$$

$$\simeq \frac{\mathrm{Br}(\tilde{\chi}_1^0 \to \mu^\pm \bar{q} q')}{\mathrm{Br}(\tilde{\chi}_1^0 \to \tau^\pm \bar{q} q')}, \quad (76)$$

in the case of a neutralino LSP. Here $\Lambda_i = \epsilon_i v_d + \mu v_i$, v_i are the sneutrino VEVs and v_d is the VEV of H_d^0; $\tilde{\epsilon}_i = V_{ij}\epsilon_j$ where V_{ij} is the neutrino mixing matrix at tree level which is given as a function of the Λ_i. Details on the neutrino masses and mixings can be found in [211, 267].

The smallness of the R-parity violating couplings which is required by the neutrino data implies that the production and decays of the SUSY particles proceed as in the MSSM with conserved R-parity except that the LSP decays. There are several predictions for the LSP properties discussed in the literature above. Here we discuss various important examples pointing out generic features. The first observation is that the smallness of the couplings can lead to finite decay lengths of the LSP which are measurable at LHC. As an example we show in Fig. 25a the decay length of a neutralino LSP as a function of its mass. The SUSY parameters have been varied such that collider constraints as well as neutrino data are fulfilled. This is important for LHC as a secondary vertex for the neutralino decays implies that the neutralino decay products can be distinguished from the remaining leptons and jets within a cascade of decays. A first attempt to use this to establish the predicted correlation between neutralino decays and neutrino mixing angles has been presented in [272]. The finite decay length can also be used to enlarge the reach of colliders for SUSY searches as has been shown in [273] for the Tevatron and in [274] for the LHC. The fact that the decay products of the neutralino can be identified via a secondary vertex is important for the

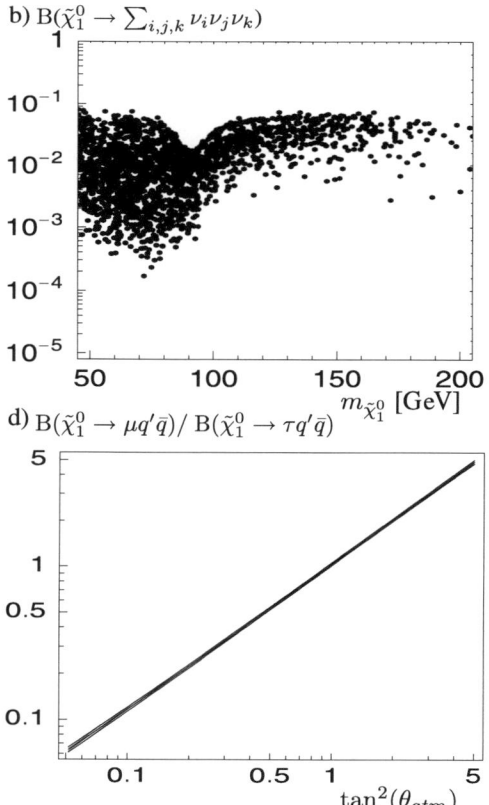

Fig. 25 Various neutralino properties: (**a**) Neutralino decay length and (**b**) invisible neutralino branching ratio summing over all neutrinos as a function of $m_{\tilde{\chi}_1^0}$; (**c**) $\mathrm{Br}(\tilde{\chi}_1^0 \to \mu q'\bar{q})/\mathrm{Br}(\tilde{\chi}_1^0 \to \tau q'\bar{q})$ scanning over the SUSY parameter and (**d**) $\mathrm{Br}(\tilde{\chi}_1^0 \to \mu q'\bar{q})/\mathrm{Br}(\tilde{\chi}_1^0 \to \tau q'\bar{q})$ for 10% variations around a fixed SUSY point as a function of $\tan^2(\theta_{atm})$

check if the predicted correlations indeed exist. As an example we show in Fig. 25c and d the ratio $\mathrm{Br}(\tilde{\chi}_1^0 \to \mu q' \bar{q})/\mathrm{Br}(\tilde{\chi}_1^0 \to \tau q' \bar{q})$ as a function of the atmospheric neutrino mixing angle $\tan^2(\theta_{\mathrm{atm}})$. In Fig. 25c a general scan is performed over the SUSY parameter space yielding a good correlation whereas in Fig. 25d the situation is shown if one assumes that the underlying SUSY parameters are known with an accuracy of 10%. The branching ratios themselves are usually of order 10%.

It is usually argued that broken R-parity implies that the missing energy signature of the MSSM is lost. This is not entirely correct if R-parity is broken via lepton number breaking as in the model discussed here. The reason is that neutrinos are not detected at LHC or ILC. This implies that the missing energy signature still is there although somewhat reduced. However, there are still cases where the LSP can decay completely invisible: $\tilde{\chi}_1^0 \to 3\nu$ or $\tilde{\nu}_i \to \nu_j \nu_k$. In Fig. 25b we see that the decay branching ratio for $\tilde{\chi}_1^0 \to 3\nu$ can go up to several per-cent. In the sneutrino case it is at most permille [213]. If one adds trilinear R-parity breaking couplings to the model, then these branching ratios will be reduced. In models with spontaneous breaking of R-parity the situation can be quite different, e.g. the invisible modes can have in total nearly 100% branching ratio [275].

As a second example, we present in Fig. 26a the decay lengths of slepton LSPs as motivated in GMSB models. Also in this case we have performed a generous scan of the SUSY parameter space. One sees that the sleptons have different decay lengths which is again useful to distinguish the various 'flavors'. However, at LHC it might be difficult to separate smuons from staus in this scenario. Provided this is possible, one could measure for example the correlation between stau decay modes and the solar-neutrino mixing angle as shown in Fig. 26b.

3.5 Reconstructing neutrino properties from collider experiments in a Higgs triplet neutrino mass model

In the previous section the neutrino masses are solely due to R-parity violation and the question arises how the situation changes if there are additional sources of neutrino masses. Therefore, a model is considered where Higgs triplets give additional contributions to the neutrino masses. It can either be obtained as a limit of spontaneous R-parity breaking models discussed in Sect. 3.1.3 or as the supersymmetric extension of the original triplet model of neutrino mass [276] with additional bilinear R-parity breaking terms [205, 237, 277]. The particle content is that of the MSSM augmented by a pair of Higgs triplet superfields, $\hat{\Delta}_u$ and $\hat{\Delta}_d$, with hypercharges $Y = +2$ and $Y = -2$, and lepton number $L = -2$ and $L = +2$, respectively. The superpotential of this model is then given by a sum of three terms,

$$W = W_{\mathrm{MSSM}} + \epsilon_i \hat{L}_i \hat{H}_u + W_\Delta, \qquad (77)$$

$$W_\Delta = \mu_\Delta \hat{\Delta}_u \hat{\Delta}_d + h_{ij} \hat{L}_i \hat{L}_j \hat{\Delta}_u. \qquad (78)$$

Additional details of the model can be found in [278]. From the analytical study of the Higgs sector, it is possible to show that the Higgs triplet VEVs are suppressed by two powers of the BRPV parameters, as already emphasized in [279].

The nonzero VEVs of this model ($v_u \equiv \langle H_u^0 \rangle$, $v_d \equiv \langle H_d^0 \rangle$, $v_i \equiv \langle \tilde{\nu}_i \rangle$, $\langle \Delta_u^0 \rangle$ and $\langle \Delta_u^0 \rangle$) produce a mixing between neutrinos, gauginos and higgsinos. For reasonable ranges of parameters, atmospheric neutrino physics is determined by the BRPV parameters, whereas the solar-neutrino mass scale depends mostly on the triplet Yukawa couplings and the triplet mass. This situation is different from the one in the model with only BRPV, where the solar-neutrino mass scale is generated by radiative corrections to neutrino masses, thus requiring $\epsilon^2/\Lambda \sim \mathcal{O}(0.1-1)$. Now, as the solar-neutrino mass scale is generated by the Higgs triplet, ϵ_i can be smaller. Using the experimentally measured values of $\tan^2 \theta_{\mathrm{atm}} \simeq 1$ and $\sin^2 2\theta_{\mathrm{CHOOZ}} \ll 1$ one can find a simple formula for the solar angle in terms of the Yukawa couplings h_{ij} of the triplet Higgs boson to the doublet leptons, which is approximately given by

$$\tan(2\theta_{\mathrm{sol}}) \simeq \frac{-2\sqrt{2}(h_{12} - h_{13})}{-2h_{11} + h_{22} + h_{33} - 2h_{23}} \equiv x. \qquad (79)$$

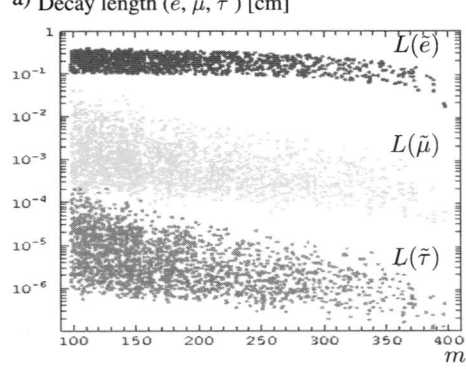
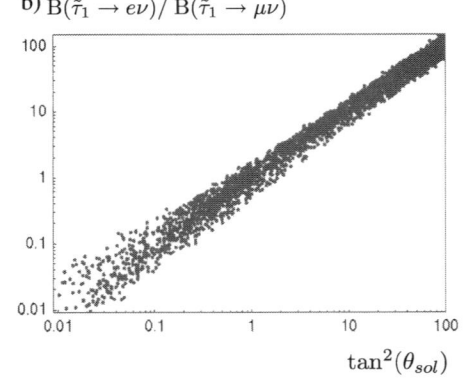

Fig. 26 Various slepton properties: (**a**) decay lengths as a function of $m_{\tilde{l}}$ and (**b**) $\mathrm{Br}(\tilde{\tau}_1 \to e\nu)/\mathrm{Br}(\tilde{\tau}_1 \to \mu\nu)$ as a function of $\tan^2(\theta_{\mathrm{sol}})$

One of the characteristic features of the triplet model of neutrino mass is the presence of doubly charged Higgs bosons Δ_u^{--}. At the LHC, the doubly charged Higgs boson can be produced in different processes, such as: (a) single production via vector boson fusion or fusion of a singly charged Higgs boson with either a vector boson or another singly charged Higgs boson; (b) pair production via a Drell–Yan process, with γ/Z exchange in the s-channel; (c) single production with a charged Higgs boson, with the exchange of W in the s-channel. In case (a) the production cross section is $\sigma(WW, WH, HH \to \Delta) = (10\text{–}1.5)$ fb for a triplet mass of $M_\Delta = (300\text{–}800)$ GeV, assuming the triplet VEV to be 9 GeV [280, 281]. However, the triplet VEV is of order eV in this model, thus suppressing this production mechanism. For (b), the production cross section is $\sigma(q\bar{q} \to \gamma/Z \to \Delta\Delta) = (5\text{–}0.05)$ fb for a triplet mass of $M_\Delta = (300\text{–}800)$ GeV [282]. The production cross section for the process (c) is $\sigma(q\bar{q}' \to W \to \Delta H) = (35\text{–}0.3)$ fb for a triplet mass of $M_\Delta = (300\text{–}800)$ GeV, where some splitting among the masses of the doubly and singly charged Higgs bosons is allowed [282, 283]. Assuming a luminosity of $\mathcal{L} = 100$ (fb year)$^{-1}$ for the LHC, the number of events for the above mentioned production processes is $\mathcal{O}(10^3\text{–}10^1)$ per year, depending on the Higgs triplet mass.

The most remarkable feature of the present model is that the decays of the doubly charged Higgs bosons can be a perfect tracer of the solar-neutrino mixing angle. Considering (79) and taking into account that the leptonic decays of the doubly charged Higgs boson are proportional to h_{ij}^2, we construct the following ratio:

$$x_{\exp} \equiv \frac{-2\sqrt{2}(\sqrt{\mathrm{Br}_{12}} - \sqrt{\mathrm{Br}_{13}})}{-2\sqrt{2\mathrm{Br}_{11}} + \sqrt{2\mathrm{Br}_{22}} + \sqrt{2\mathrm{Br}_{33}} - 2\sqrt{2\mathrm{Br}_{23}}} \quad (80)$$

with Br_{ij} denoting the measured branching ratio for the process $(\Delta_u^{--} \to l_i^- l_j^-)$. Figure 27 shows the ratio $y_{\exp}$ of the leptonic decay branching ratios of the doubly charged Higgs boson versus the solar-neutrino mixing angle. The ratio of doubly charged Higgs boson decay branching ratios is specified by the variable

$$y_{\exp} \equiv \tan^2\left(\frac{\arctan(x_{\exp})}{2}\right) \quad (81)$$

where, for the determination of $x_{\exp}$, a 10% uncertainty in the measured branching ratios has been assumed and the triplet mass has been fixed at $M_{\Delta_u} = 500$ GeV. As can be seen from the figure, there is a very strong correlation between the pattern of Higgs triplet decays and the solar-neutrino mixing angle. The 3σ range permitted by current solar and reactor neutrino data (indicated by the vertical band in Fig. 27) fixes a minimum value for $y_{\exp}$, thus requiring minimum values for the off-diagonal leptonic decay channels of the doubly charged Higgs triplet. If $\mathrm{BR}_{23} = 0$, at least either BR_{12} or BR_{13} must be larger than 0.5. On the other hand, if $\mathrm{BR}_{23} \neq 0$, then at least one of the off-diagonal branching ratios must be larger than 0.2.

As in Sect. 3.4, the decay pattern of a neutralino LSP is predicted in terms of the atmospheric neutrino mixing angle The main difference is that the ϵ_i can be smaller in this model compared to the previous one. This implies that the main decay mode $\mathrm{Br}(\tilde{\chi}_1^0 \to \nu b\bar{b})$ gets reduced [269] and the branching ratios into the final states lqq' ($l = e, \mu\tau$) increase.

3.6 SUSY (s)lepton flavor studies with ATLAS

In this section main features of Monte Carlo studies for slepton masses and spin measurements are presented as well as a study of slepton non-universality. As a reference model the SPS1a point is taken [249], which is derived from the following high-scale parameters: $m_0 = 100$ GeV, $m_{1/2} = 250$ GeV, $A_0 = -100$ GeV, $\tan\beta = 10$ and $\mathrm{sign}(\mu) = +$, where m_0 is a common scalar mass, $m_{1/2}$ a common gaugino mass, A_0 a common trilinear coupling, $\tan\beta$ the ratio of the Higgs vacuum expectation values.

Sleptons are produced either directly in pairs $\tilde{l}^+\tilde{l}^-$ or indirectly from decays of heavier charginos and neutralinos (typical mode $\tilde{\chi}_2^0 \to \tilde{l}_R l$). They can decay according to: $\tilde{l}_R \to l\tilde{\chi}_1^0$, $\tilde{l}_L \to l\tilde{\chi}_1^0$, $\tilde{l}_L \to l\tilde{\chi}_2^0$, $\tilde{l}_L \to \nu\tilde{\chi}_1^\pm$. At the end of every SUSY decay chain there is the undetectable lightest neutralino $\tilde{\chi}_1^0$, and rather than mass peaks one only measures kinematic end points in the invariant-mass distributions. Kinematic end points are a function of SUSY masses, which can then be extracted from the set of end point measurements. Fast simulation studies of left squark cascade decays $\tilde{q}_L \to \tilde{\chi}_2^0 q \to \tilde{l}_R^\pm l^\mp q \to l^+l^-q\tilde{\chi}_1^0$ ($l = e, \mu$) were performed in [249, 284]. Events with two same-flavor and opposite-sign (SFOS) leptons, at least four jets with $p_T > 150, 100, 50, 50$ GeV, effective mass $M_{\mathrm{eff}} = \sum_{i=1}^{4} p_T(\mathrm{jet}) + \slashed{E}_T > 600$ GeV and missing transverse energy $\slashed{E}_T > \max(100\ \mathrm{GeV}, 0.2 M_{\mathrm{eff}})$ were selected. Flavor

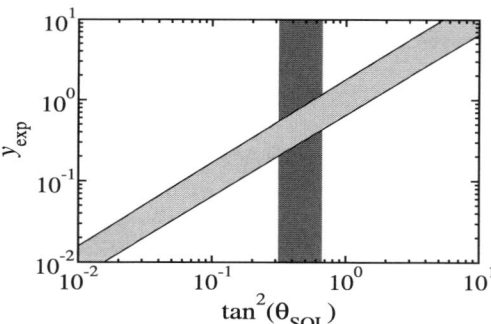

Fig. 27 Ratio of doubly charged Higgs boson leptonic decay branching ratios (assuming a 10% uncertainty) indicated by the variable $y_{\exp}$ of (81) and (80) versus the solar-neutrino mixing angle. The vertical band indicates current 3σ allowed range

subtraction $e^+e^- + \mu^+\mu^- - e^\pm\mu^\mp$ was applied. After the event selection, the SM background becomes negligible and a significant part of the SUSY background is removed. Few kinematic end points were reconstructed and fit [284]: the maximum of the distribution of the dilepton invariant mass M_{ll}^{max}, the maximum and the minimum of the distribution of the $M(llq)$ invariant mass M_{llq}^{max} and M_{llq}^{min}, the maximum of the distribution of the lower of the two l^+q, l^-q invariant masses $(M_{lq}^{low})^{max}$ and the maximum of the distribution of the higher of the two l^+q, l^-q invariant masses $(M_{lq}^{high})^{max}$. From this set of end point measurements and by taking into account the statistical fit error and the systematic error on the energy scale (1% for jets and 0.1% for leptons), SUSY masses $m_{\tilde{q}_L} = 540$ GeV, $m_{\tilde{\chi}_2^0} = 177$ GeV, $m_{\tilde{l}_R} = 143$ GeV and $m_{\tilde{\chi}_1^0} = 96$ GeV were extracted with a 6 GeV resolution for squarks and 4 GeV for non-squarks ($L = 300$ fb^{-1}).

Few experimentally challenging points in the mSUGRA parameter space constrained by the latest experimental data [285] were recently selected and studied by using a full Geant-4 simulation. Preliminary full simulation studies of left squark cascade decay for the bulk point, the coannihilation point and the focus point have been reported [286]. Events with two SFOS leptons are selected and flavor subtraction $e^+e^- + \mu^+\mu^- - e^\pm\mu^\mp$ was applied. The bulk point ($m_0 = 100$ GeV, $m_{1/2} = 300$ GeV, $A_0 = -300$ GeV, $\tan\beta = 6$, sign(μ) = +) is a typical mSUGRA point where easy SUSY discovery is expected. The end points M_{ll}^{max}, M_{llq}^{max}, M_{llq}^{min}, $(M_{lq}^{high})^{max}$ and $(M_{lq}^{low})^{max}$ were reconstructed for integrated luminosity $L = 5$ fb^{-1}. The coannihilation point ($m_0 = 70$ GeV, $m_{1/2} = 350$ GeV, $A_0 = 0$ GeV, $\tan\beta = 10$, sign(μ) = +) is challenging due to the soft leptons present in the final state. The decay of the second lightest neutralino to both left and right sleptons is open: $\tilde{\chi}_2^0 \to \tilde{l}_{L,R}l$. The end points M_{ll}^{max}, M_{llq}^{max}, $(M_{lq}^{high})^{max}$ and $(M_{lq}^{low})^{max}$ were reconstructed for integrated luminosity $L = 20$ fb^{-1}. The focus point ($m_0 = 3550$ GeV, $m_{1/2} = 300$ GeV, $A_0 = 0$ GeV, $\tan\beta = 10$, sign(μ) = +) predicts multi-TeV squark and slepton masses. Neutralinos decay directly to leptons: $\tilde{\chi}_3^0 \to l^+l^-\tilde{\chi}_1^0$, $\tilde{\chi}_2^0 \to l^+l^-\tilde{\chi}_1^0$ and dilepton end points M_{ll}^{max} were reconstructed for $L = 7$ fb^{-1}. All reconstructed end points are at the expected positions.

In the case of direct slepton production where both sleptons decay to lepton and the first lightest neutralino $\tilde{l}_L\tilde{l}_L/\tilde{l}_R\tilde{l}_R \to l^+l^-\tilde{\chi}_1^0\tilde{\chi}_1^0$, there are no end points in the invariant-mass distributions because of two missing final state particles. It is possible to estimate the slepton mass by using the so-called stransverse mass [287] variable, $M_{T2} = \min_{E_T^{miss}=E_{T1}^{miss}+E_{T2}^{miss}}\{\max\{m_T^2(p_T^{l1}, E_{T1}^{miss}), m_T^2(p_T^{l2}, E_{T2}^{miss})\}\}$. The end point of the stransverse mass distribution is a function of mass difference between slepton and the first lightest neutralino $\tilde{\chi}_1^0$. In the case of mSUGRA point SPS1a, fast simulation studies [249] show that, by using the stransverse mass, the left slepton mass $m_{\tilde{l}_L} = 202$ GeV can be estimated with the resolution of 4 GeV ($L = 100$ fb^{-1}).

Left squark cascade decays $\tilde{q}_L \to \tilde{\chi}_2^0 q \to \tilde{l}_{L,R}^\pm l^{near(\mp)} q \to l^{far(\pm)} l^{near(\mp)} q \tilde{\chi}_1^0$ are very convenient for the supersymmetric particles' spin measurement [252]. Due to slepton and squark spin-0 and neutralino $\tilde{\chi}_2^0$ spin-1/2, the invariant mass of the quark and of the first emitted ('near') lepton $M(ql^{near(\pm)})$ is charge asymmetric. The asymmetry is defined as $A = (s^+ - s^-)/(s^+ + s^-)$, $s^\pm = (d\sigma)(dM(ql^{near(\pm)}))$. Asymmetry measurements are diluted by the fact that it is usually not possible to distinguish the first emitted ('near') from the second emitted ('far') lepton. Also, squark and antisquark have opposite asymmetries and are experimentally indistinguishable, but LHC is a proton–proton collider and more squarks than antisquarks will be produced. Fast simulation studies of few points in the mSUGRA space [252, 288] show asymmetry distributions not consistent with zero, which is the direct proof of the neutralino spin-1/2 and slepton spin-0. In the case of point SPS1a, a non-zero asymmetry may be observed with 30 fb^{-1}.

For some of the points in mSUGRA space, mixing between left and right smuons is not negligible. Left–right mixing affects decay branching ratios $\tilde{\chi}_2^0 \to \tilde{l}_R l$ and charge asymmetry of invariant-mass distributions from left squark cascade decay. For the point SPS1a with modified $\tan(\beta) = 20$, fast simulation studies [288] show that different decay branching ratios for selectrons and smuons can be detected at LHC for 300 fb^{-1}.

Fast simulation studies show that SUSY masses can be extracted by using kinematic end points and stransverse mass. Preliminary full simulation analysis show that large number of mass relations can be measured for leptonic signatures with few fb^{-1} in different mSUGRA regions. What still needs to be studied more carefully is: acceptances and efficiencies for electrons and muons, calibration, trigger, optimization of cuts against SM background and fit to distributions. The asymmetry distributions are consistent with neutralino spin-1/2 and slepton spin-0. Different branching ratios for selectron and smuon, caused by smuon left–right mixing, can be detected by ATLAS.

3.7 Using the $l^+l^- + \not{E}_T +$ jet veto signature for slepton detection

The aim of this section, which is based on [289], is to study the possibility of detecting sleptons at CMS. Note the previous related papers where the sleptons detection was studied at the level of a toy detector [254, 290–293], whereas we perform a full detector simulation.

ISASUSY 7.69 [294] was used for the calculation of coupling constants and cross sections in the leading order approximation for SUSY processes. For the calculation of the next-to-leading order corrections to the SUSY cross sections the PROSPINO code [295] was used. Cross sections of the background events were calculated with PYTHIA 6.227 [47] and CompHEP 4.2pl [296]. For the considered backgrounds, the NLO corrections are known and were used. Data sets from the official production were used for the study of CMS test point LM1 and for the backgrounds $t\bar{t}$, ZZ, WW, Wt, $Zb\bar{b}$, $DY2e$, $DY2\tau$, where DY denotes Drell–Yan processes. For WZ, $DY2\mu$ and $W+$ jet backgrounds the events were generated with PYTHIA 6.227. The detector simulation and hits production were made with full CMS simulation [297], digitized and reconstructed [298]. The $DY2\mu$ and $W+$jet backgrounds were simulated with fast simulation [299].

Jets were reconstructed using an iterative cone algorithm with cone size 0.5 and their energy was corrected with the GammaJet calibration. The events are required to pass the Global Level 1 Trigger (L1), the High Level Trigger (HLT) and at least one of the following triggers: single electron, double electron, single muon, double muon. The CMS fast simulation was used for the determination of the sleptons discovery plot.

As discussed in the previous section, sleptons can be either produced at LHC directly via the Drell–Yan mechanism or in cascade decays of squarks and gluinos. The slepton production and decays described previously lead to the signature with the simplest event topology: two leptons + $\displaystyle{\not}E_T$ + jet veto. This signature arises for both direct and indirect slepton pair production. In the case of indirectly produced sleptons not only the event topology with two leptons but with single, three and four leptons is possible. Furthermore, indirect slepton production from decays of squarks and gluinos through charginos and neutralinos, can lead to an event topology with two leptons + $\displaystyle{\not}E_T$ + ($n \geq 1$) jets.

Cut-optimization led to the following selection criteria:

a. for leptons:
  - p_T—cut on leptons ($p_T^{\text{lept}} > 20$ GeV, $|\eta| < 2.4$) and lepton isolation within a $\Delta R < 0.3$ cone containing calorimeter cells and tracker;
  - effective mass of two opposite-sign and the same-flavor leptons outside the ($M_Z - 15$ GeV, $M_Z + 10$ GeV) interval;
  - $\Phi(l^+l^-) < 140°$ cut on angle between two leptons;

b. for $\displaystyle{\not}E_T$:
  - $\displaystyle{\not}E_T > 135$ GeV cut on missing E_T;
  - $\Phi(\displaystyle{\not}E_T, ll) > 170°$ cut on relative azimuthal angle between dilepton and $\displaystyle{\not}E_T$;

c. for jets:
  - jet veto cut: $N_{\text{jet}} = 0$ for a $E_T^{\text{jet}} > 30$ GeV (corrected jets) threshold in the pseudorapidity interval $|\eta| < 4.5$.

The SM backgrounds are $t\bar{t}$, WW, WZ, ZZ, Wt, $Zb\bar{b}$, DY, $W+$ jet. The main contributions come from WW and $t\bar{t}$ backgrounds. There are also internal SUSY backgrounds which arise from $\tilde{q}\tilde{q}$, $\tilde{g}\tilde{g}$ and $\tilde{q}\tilde{g}$ productions and subsequent cascade decays with jets outside the acceptance or below the threshold. Note that when we are interested in new physics discovery we have to compare the calculated number of SM background events N_{SMbg} with new physics signal events $N_{\text{new physics}} = N_{\text{slept}} + N_{\text{SUSYbg}}$, so SUSY background events increase the discovery potential of new physics.

For the point LM1 with the set of cuts for an integral luminosity $\mathcal{L} = 10$ fb^{-1} the number of signal events (direct sleptons plus sleptons from chargino/neutralino decays) is $N_S = 60$, whereas the number of SUSY background events is $N_{\text{SUSYbg}} = 4$ and the number of SM background events is $N_{\text{SMbg}} = 41$. The total signal efficiency is 1.16×10^{-4} and the background composition is 1.32×10^{-6} of the total ttbar, 1.37×10^{-5} of the total WW, 4×10^{-6} of the total WZ, 4.4×10^{-5} of the total ZZ, 8.1×10^{-6} of the total Wt, 0 of the total Zbb, DY, $W+$ jet.

The SUSY background is rather small compared to the signal, so we can assume $N_S = N_{\text{direct sleptons}} + N_{\text{chargino/neutralino}} + N_{\text{SUSYbg}} = 64$. This corresponds to significances $S_{c12} = 7.7$ and $S_{cL} = 8.3$ where the quantity S_{c12} is defined in [300] and S_{cL} in [301, 302]. Taking into account the systematic uncertainty of 23% related to the inexact knowledge of backgrounds leads to the decrease of significance S_{c12} from 7.7 to 4.3. The ratio of the numbers of background events from two different channels $N(e^+e^- + \mu^+\mu^-)/N(e^\pm\mu^\mp) = 1.37$ will be used to keep the backgrounds under control. The CMS discovery plot for two leptons + $\displaystyle{\not}E_T$ + jet veto signature is presented in Fig. 28.

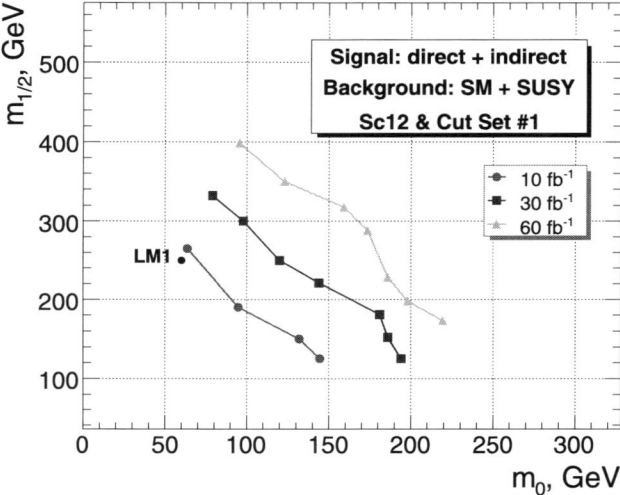

Fig. 28 Discovery plot ($\tan\beta = 10$, $\text{sign}(\mu) = +$, $A = 0$) for final states with l^+l^-, missing transverse energy and a jet veto

3.8 Using the $e^{\pm}\mu^{\mp} + \slashed{E}_T$ signature in the search for supersymmetry and lepton flavor violation in neutralino decays

The aim of this section based on [303] is the study of the possibility to detect SUSY and LFV using the $e^{\pm}\mu^{\mp} + \slashed{E}_T$ signature at CMS. The details concerning the simulations are the same as described in Sect. 3.7.

The SUSY production $pp \to \tilde{q}\tilde{q}', \tilde{g}\tilde{g}, \tilde{q}\tilde{g}$ with subsequent decays leads to the event topology $e^{\pm}\mu^{\mp} + \slashed{E}_T$. The main backgrounds contributing to the $e^{\pm}\mu^{\mp}$ events are $t\bar{t}$, ZZ, WW, WZ, Wt, $Zb\bar{b}$, $DY2\tau$, $Z+$jet. It has been found that tt background is the biggest one and it gives more than 50% contribution to the total background.

Our set of cuts is the following:

- p_T—cut on leptons ($p_T^{\text{lept}} > 20$ GeV, $|\eta| < 2.4$) and lepton isolation within $\Delta R < 0.3$ cone.
- $\slashed{E}_T > 300$ GeV cut on missing E_T.

For integrated luminosity $\mathcal{L} = 10$ fb^{-1} the number of background events with this set of cuts is $N_B = 93$. The results for various CMS study points at this luminosity are presented in Table 12.

At point LM1 the signal over background ratio is 3 and the signal efficiency is 6×10^{-4}. The background composition is 9.5×10^{-6} of the total ttbar, 3.4×10^{-6} of the total WW, 4×10^{-6} of the total WZ, 3.2×10^{-6} of the total Wt, 2.2×10^{-6} of the total $Z+$jet, 0 of the total ZZ, $Zb\bar{b}$, $DY2\tau$.

The CMS discovery plot for the $e^{\pm}\mu^{\mp} + \slashed{E}_T$ signature is presented in Fig. 29.

It has been shown in [192, 304, 305] that it is possible to look for lepton flavor violation at supercolliders through the production and decays of the sleptons. For LFV at the LHC one of the most promising processes is the LFV decay of the second neutralino [195, 197] $\tilde{\chi}_2^0 \to \tilde{l}l \to \tilde{\chi}_1^0 ll'$, where the non-zero off-diagonal component of the slepton mass matrix leads to the different flavors for the leptons in

Table 12 Number of signal events and significances S_{c12} [300] and S_{cL} [301, 302] for $\mathcal{L} = 10$ fb^{-1}

Point	N events	S_{c12}	S_{cL}
LM1	329	21.8	24.9
LM2	94	8.1	8.6
LM3	402	25.2	29.2
LM4	301	20.4	23.1
LM5	91	7.8	8.3
LM6	222	16.2	18.0
LM7	14	1.4	1.4
LM8	234	16.9	18.8
LM9	137	11.0	11.9

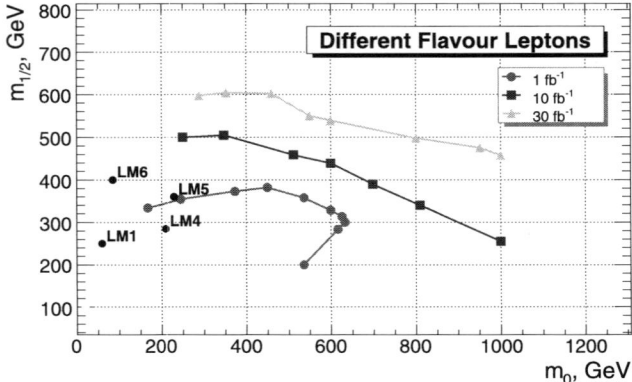

Fig. 29 Discovery plot ($\tan\beta = 10$, sign(μ) = +, $A = 0$) for the luminosities $\mathcal{L} = 1, 10, 30$ fb^{-1} for the $e^{\pm}\mu^{\mp} + \slashed{E}_T$ signature

the final state. By using the above mode, LFV in $\tilde{e}$–$\tilde{\mu}$ mixing has been investigated in [195, 197] at the parton model level and with a toy detector simulation. Here we study the prospects for LFV detection in CMS on the basis of a full simulation of both signal and background. To be specific, we study the point LM1. We assume that the LFV is due to nonzero mixing of right handed smuon and selectron. The signal of the LFV $\tilde{\chi}_2^0$ decay is two opposite-sign leptons ($e^+\mu^-$ or $e^-\mu^+$) in the final state with a characteristic edge structure. In the limit of lepton flavor conservation, the process $\tilde{\chi}_2^0 \to \tilde{l}l \to ll\tilde{\chi}_1^0$ has an edge structure for the distribution of the lepton-pair invariant mass m_{ll} and the edge mass $m_{ll}^{\max}$ is expressed by the slepton mass $m_{\tilde{l}}$ and the neutralino masses $m_{\tilde{\chi}_{1,2}^0}$ as follows:

$$\left(m_{ll}^{\max}\right)^2 = m_{\tilde{\chi}_2^0}^2 \left(1 - \frac{m_{\tilde{l}}^2}{m_{\tilde{\chi}_2^0}^2}\right) \left(1 - \frac{m_{\tilde{\chi}_1^0}^2}{m_{\tilde{l}}^2}\right). \quad (82)$$

The SUSY background for the LFV comes from uncorrelated leptons from different squark or gluino decay chains. The SM background comes mainly from

$$t\bar{t} \to bWbW \to blbl'\nu\nu'. \quad (83)$$

The Drell–Yan background from $pp \to \tau\tau \to e\mu X$ is negligible. It should be stressed that, for the signature with $e^{\pm}\mu^{\mp}$, in absence of LFV there is no edge structure for the distribution on the invariant mass $m_{\text{inv}}(e^{\pm}\mu^{\mp})$. This is present, on the other hand, with LFV, providing a clear kinematical signature. The rate for a flavor-violating decay is

$$\text{Br}(\tilde{\chi}_2^0 \to e^{\pm}\mu^{\mp}\tilde{\chi}_1^0)$$
$$= \kappa \text{Br}(\tilde{\chi}_2^0 \to e^+e^-\tilde{\chi}_1^0, \mu^+\mu^-\tilde{\chi}_1^0), \quad (84)$$

where:

$$\text{Br}(\tilde{\chi}_2^0 \to e^+e^-\tilde{\chi}_1^0, \mu^+\mu^-\tilde{\chi}_1^0)$$
$$= \text{Br}(\tilde{\chi}_2^0 \to e^+e^-\tilde{\chi}_1^0) + \text{Br}(\tilde{\chi}_2^0 \to \mu^+\mu^-\tilde{\chi}_1^0), \quad (85)$$

$$\kappa = 2x \sin^2 \theta \cos^2 \theta, \qquad (86)$$

$$x = \frac{\Delta m_{\tilde{e}\tilde{\mu}}^2}{\Delta m_{\tilde{e}\tilde{\mu}}^2 + \Gamma^2}, \qquad (87)$$

$$\mathrm{Br}(\tilde{\chi}_2^0 \to e^{\pm}\mu^{\mp})$$
$$= \mathrm{Br}(\tilde{\chi}_2^0 \to e^+\mu^-) + \mathrm{Br}(\tilde{\chi}_2^0 \to e^-\mu^+). \qquad (88)$$

Here θ is the mixing angle between $\tilde{e}_R$ and $\tilde{\mu}_R$ and Γ is the sleptons decay width. The parameter x is the measure of the quantum interference effect. There are some limits on $\tilde{e}$–$\tilde{\mu}$ mass splitting from lepton flavor-violating processes but they are not very strong.

For $\kappa = 0.25$, $\kappa = 0.1$ the distributions of the number of $e^{\pm}\mu^{\mp}$ events on the invariant mass $m_{\mathrm{inv}}(e^{\pm}\mu^{\mp})$ (see Fig. 30) clearly demonstrates the existence of the edge structure [306], i.e. the existence of the lepton flavor violation in neutralino decays. It appears that for the point LM1 the use of an additional cut

$$m_{\mathrm{inv}}(e^{\pm}\mu^{\mp}) < 85 \text{ GeV} \qquad (89)$$

reduces both the SM and SUSY backgrounds and increases the discovery potential in the LFV search. For the point LM1 we found that in the assumption of exact knowledge of the background (both the SM and SUSY backgrounds) for the integrated luminosity $\mathcal{L} = 10$ fb^{-1} it would be possible to detect LFV at 5σ level in $\tilde{\chi}_2^0$ decays for $\kappa \geq 0.04$.

3.9 Neutralino spin measurement with ATLAS

Charge asymmetries in invariant-mass distributions containing leptons can be used to prove that the neutralino spin is 1/2. This is based on a method [252] which allows one to choose between different hypotheses for spin assignment, and to discriminate SUSY from a universal extra dimensions (UED) model mimicking low energy SUSY [307, 308]. For this the decay chain

$$\tilde{q}_L \to \tilde{\chi}_2^0 q \to \tilde{l}_{L,R}^{\pm} l^{\mp} q \to l^+ l^- q \tilde{\chi}_1^0 \qquad (90)$$

will be used. In the following, the first lepton (from $\tilde{\chi}_2^0$ decay) is called *near*, and the one from slepton decay is called *far*.

Squarks and sleptons are spin-0 particles and their decays are spherically symmetric, differently from the $\tilde{\chi}_2^0$ which has spin 1/2. A charge asymmetry is expected in the invariant masses $m(ql^{\mathrm{near}(\pm)})$ formed by the quark and the near lepton. Also $m(ql^{\mathrm{far}})$ shows some small charge asymmetry [307, 308], but it is not always possible to distinguish experimentally near from far lepton, thus leading to dilution effects when measuring the $m(ql^{\mathrm{near}(\pm)})$ charge asymmetry.

In the cascade decay (90), the asymmetry in the corresponding $m(\bar{q}l)$ charge distributions is the same as the asymmetry in $m(ql)$ from $\tilde{q}_L$ decay, but with the opposite sign [309]. Though it is not possible to distinguish q from $\bar{q}$

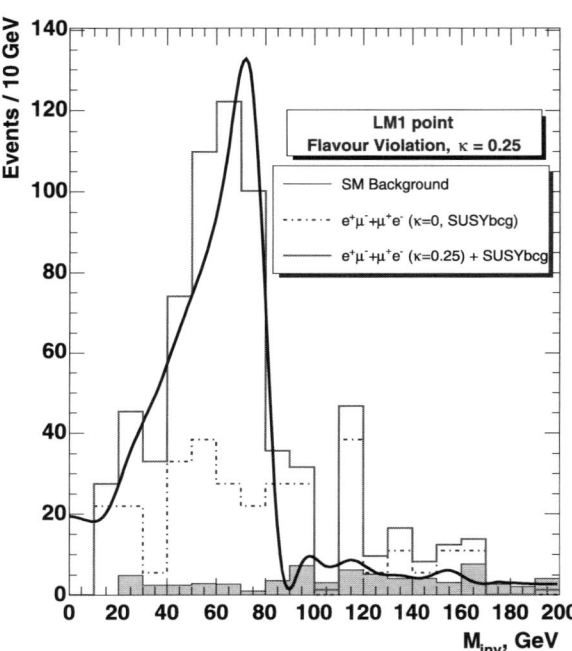
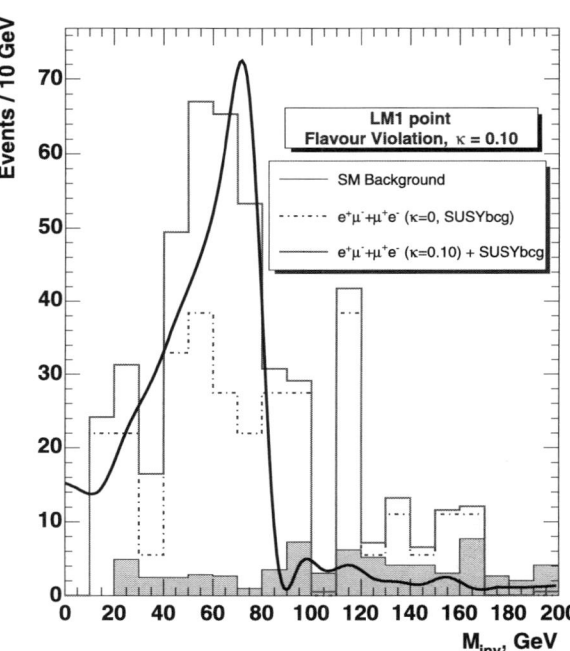

Fig. 30 The distribution of dilepton invariant mass after selection of two isolated $e^{\pm}\mu^{\mp}$ leptons with $p_T^{\mathrm{lept}} > 20$ GeV and $\not{E}_T > 300$ GeV for flavor violation parameter $k = 0.25$ (*left*) and $k = 0.1$ (*right*). The superimposed curves are fits to the invariant-mass distribution for the case of 100% LFV

at a pp collider like the LHC, more squarks than antisquarks will be produced. Here only electrons and muons are considered for analysis.

Two mSUGRA points were selected for the analysis [310]: SU1, in the stau coannihilation region ($m_0 = 70$ GeV, $m_{1/2} = 350$ GeV, $A_0 = 0$ GeV, $\tan\beta = 10$, sgn$\mu = +$) and SU3, in the bulk region ($m_0 = 100$ GeV, $m_{1/2} = 300$ GeV, $A_0 = -300$ GeV, $\tan\beta = 6$, sgn$\mu = +$). In SU1 (SU3) the LO cross section for all SUSY processes is 7.8 pb (19.3 pb), and the observability of charge asymmetry is enhanced by ~5 (~2.5) in $\tilde{q}/\tilde{\bar{q}}$ production yield.

In the SU1 point, owing to a small mass difference between $\tilde{\chi}_2^0$ and $\tilde{l}_L$ (264 and 255 GeV, respectively), the near lepton has low p_T in the $\tilde{\chi}_2^0 \to \tilde{l}_L l$ decay, while the small mass difference between $\tilde{l}_R$ and $\tilde{\chi}_1^0$ (155 GeV and 137 GeV, respectively), implies low values for far lepton's p_T in $\tilde{\chi}_2^0 \to \tilde{l}_R l$ decay. As a consequence, near and far leptons are distinguishable. Decay (90) represents ~1.6% of all SUSY production. From the three detectable particles l^+, l^-, q (where the quark hadronizes to a jet) in the final state of the $\tilde{q}_L$ decay (90) four invariant masses are formed: $m(ll)$, $m(qll)$, $m(ql^{near})$ and $m(ql^{far})$. Their kinematic maxima are given by $m(ll)^{max} = 56$ GeV ($\tilde{l}_L$), 98 GeV ($\tilde{l}_R$), $m(qll)^{max} = 614$ GeV ($\tilde{l}_L, \tilde{l}_R$), $m(ql^{near})^{max} = 181$ GeV ($\tilde{l}_L$), 583 GeV ($\tilde{l}_R$) and $m(ql^{far})^{max} = 329$ GeV ($\tilde{l}_R$), 606 GeV ($\tilde{l}_L$). In the SU3 point, only the decay $\tilde{\chi}_2^0 \to \tilde{l}_R^\pm l^\mp$ is allowed (3.8% of all SUSY production). The end points for $m(ll)$, $m(qll)$, $m(ql^{near})$ and $m(ql^{far})$ are 100, 503, 420 and 389 GeV, respectively.

Events were generated with HERWIG 6.505 [311]. SUSY samples corresponding to integrated luminosities of 100 fb^{-1} for SU1 and 30 fb^{-1} for SU3 were analyzed. Also the most relevant SM processes have been also studied, i.e. $t\bar{t}$ + jets, W + jets and Z + jets backgrounds were produced with Alpgen 2.0.5 [312]. Events were passed through a parametrized simulation of the ATLAS detector, ATLFAST [313].

In order to separate SUSY signal from SM background these *preselection* cuts were applied:

- missing transverse energy $E_T^{miss} > 100$ GeV,
- four or more jets with transverse momentum $p_T(j_1) > 100$ GeV and $p_T(j_2, j_3, j_4) > 50$ GeV,
- exactly two SFOS leptons ($p_T^{lepton} > 6$ GeV for SU1, and $p_T^{lepton} > 10$ GeV for SU3).

At this selection stage, few invariant masses are formed: the dilepton invariant mass $m(ll)$, the lepton-lepton jet invariant mass $m(jll)$, and the lepton jet invariant masses $m(jl^+)$ and $m(jl^-)$, where $l^\pm$ are the leptons and j is one of the two most energetic jets in the event. Subsequently, we require:

- $m(ll) < 100$ GeV, $m(jll) < 615$ GeV (for SU1) or $m(jll) < 500$ GeV (for SU3).

In SU1, the decays (90) with $\tilde{l}_L$ or $\tilde{l}_R$ are distinguished asking for $m(ll) < 57$ GeV or $57 < m(ll) < 100$ GeV, respectively. For SU1, in the decay (90) with $\tilde{l}_L$, the near (far) lepton is identified as the one with lower (higher) p_T, and vice versa for the decay (90) with $\tilde{l}_R$. The efficiencies and signal/background ratios after all the cuts described so far, when applied on SUSY and SM events, are shown in Table 13. Further background reduction is applied by subtracting statistically in the invariant-mass distributions events with two opposite-flavor opposite-sign (OFOS) leptons: $e^+e^- + \mu^+\mu^- - e^\pm\mu^\mp$ (SFOS-OFOS subtraction). This reduces the SUSY background by about a factor of 2 and makes SM events with uncorrelated leptons compatible with zero.

Charge asymmetries of $m(jl)$ distributions have been computed after SFOS-OFOS subtraction in the ranges [0, 220] GeV for SU1 (only for the decay (90) with $\tilde{l}_L$ and near lepton) and [0, 420] GeV for SU3. Two methods have been applied to detect the presence of a non-zero charge asymmetry:

- A non-parametric χ^2 test with respect to a constant 0 function, giving confidence level CL$_{\chi^2}$.
- A *Run Test* method [314] providing a confidence level CL$_{RT}$ for the hypothesis of a zero charge asymmetry.

The two methods are independent and are not influenced by the actual shape of the charge asymmetry. Their probabilities can be combined [314] providing a final confidence level CL$_{comb}$. In Fig. 31 charge asymmetries are reported for $m(jl^{near})_L$ in SU1 and for $m(jl)$ in SU3. With 100 fb^{-1},

Table 13 Efficiencies and S/B ratios for SUSY signal and background (SU1, SU3) and for SM background

	Efficiency (SU1)	S/B (SU1)	Efficiency (SU3)	S/B (SU3)
Signal	$(17.0 \pm 0.3)\%$		$(20.0 \pm 0.3)\%$	
SUSY background	$(0.94 \pm 0.01)\%$	0.33	$(0.75 \pm 0.01)\%$	1
$t\bar{t}$	$(2.69 \pm 0.02)\,10^{-4}$	0.18	$(3.14 \pm 0.02)\,10^{-4}$	0.9
W	$(1.4 \pm 0.9)\,10^{-5}$	~16	$(0.4 \pm 0.4)\,10^{-5}$	~300
Z	$(1.1 \pm 0.3)\,10^{-5}$	~12	$(0.9 \pm 0.2)\,10^{-5}$	~100

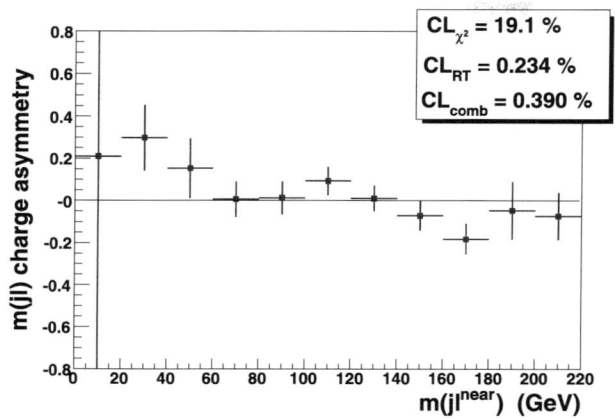
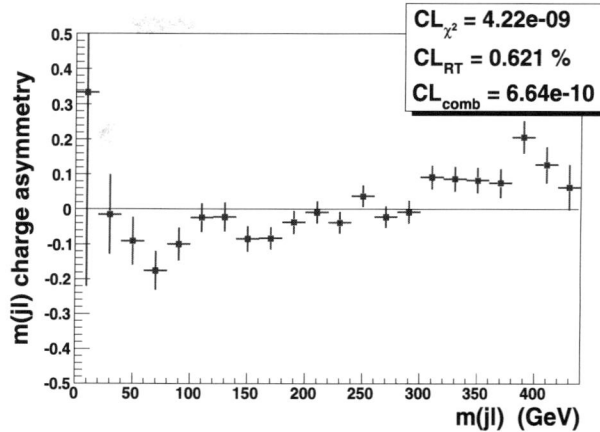

Fig. 31 Charge asymmetries for lepton jet invariant masses after same-flavor–opposite-sign—opposite-flavor–opposite-sign subtraction. *Left*: using the near lepton from the chain involving $\tilde{l}_L$ for the SU1 point. *Right*: using both near and far leptons for the SU3 point

Table 14 Confidence levels for the two methods described in the text, separately and combined, obtained on $m(jl)$ distributions for the final selected samples and for various sources of background/systematics

Analyzed sample	SU1 selection			SU3 selection		
	CL_{χ^2}	CL_{RT}	$\mathrm{CL}_{\mathrm{comb}}$	CL_{χ^2}	CL_{RT}	$\mathrm{CL}_{\mathrm{comb}}$
a. SUSY SFOS-OFOS	19.1%	0.234%	0.390%	4.22×10^{-9}	0.621%	6.64×10^{-10}
b. SUSY OFOS	57.1%	92.1%	86.4%	19.3%	93.3%	48.9%
c. SUSY SFOS bkg	30.7%	24.0%	26.6%	53.5%	30.9%	46.2%
d. SM SFOS bkg	21.4%	24.0%	20.3%	61.3%	84.1%	85.7%
e. SM OFOS bkg	73.8%	50.0%	73.7%	95.5%	30.9%	65.5%
f. SUSY wrong jet	62.8%	50.0%	67.8%	19.7%	15.9%	14.0%

in SU1 $\mathrm{CL}_{\mathrm{comb}}$ is well below 1%, while for SU3 30 fb^{-1} are enough to get a $\mathrm{CL}_{\mathrm{comb}} \sim 10^{-9}$. Different sources of background and possible systematic effects have been investigated for SU1 and SU3 samples and the obtained confidence levels are reported in Table 14 (letters **b.** to **f.**), compared to the final SUSY selected sample (letter **a.**). They refer to: selected OFOS lepton pairs (**b.**), SFOS background SUSY events (**c.**), SFOS and OFOS selected SM background events (**d.** and **e.**, respectively) and events with $m(jl)$ formed with a wrong jet (**f.**). Anyway, confidence levels are much higher than the final selected SUSY sample.

It is observed that the evidence with a 99% confidence level for a charge asymmetry needs at least 100 fb^{-1} in the case of SU1, while even less than 10 fb^{-1} would be needed for SU3 [310].

3.10 SUSY Higgs boson production and decay

FCNC interactions of neutral Higgs bosons are extremely suppressed in the SM. In the SM, one finds Br($H_{\mathrm{SM}} \to bs$) $\approx 4 \times 10^{-8}$ for $m_{H_{\mathrm{SM}}} = 114$ GeV.[11] For the neutral MSSM Higgs bosons the ratios could be of $\mathcal{O}(10^{-4}$–$10^{-3})$. Constraints from $b \to s\gamma$ data reduce these rates, though [107, 109, 119, 128]. The FCNC decays $t \to H_{\mathrm{SM}}c$ or $H_{\mathrm{SM}} \to tc$ have branching ratios of the order of 10^{-14} or less [16, 85, 107, 111], hence 10 orders of magnitude below other more conventional (and relatively well measured) FCNC processes like $b \to s\gamma$ [53]. The detection of Higgs FCNC interactions would be evidence of new physics. The MSSM introduces new sources of FCNC interactions mediated by the strongly-interacting sector.[12] They are produced by the misalignment of the quark mass matrix with the squark mass matrix, and the main parameter characterizing these interactions is the non-flavor-diagonal term in the squark-mass matrix, which we parametrize in the standard fashion [125, 126] as $(M^2)_{ij} = \delta_{ij}\tilde{m}_i\tilde{m}_j$ ($i \neq j$), $\tilde{m}_i$ being

---

[11] In the following, Br($H \to bs$) denotes the sum of the Higgs branching ratios into $b\bar{s}$ and $\bar{b}s$. The Higgs boson H stands for that of the SM, H_{SM}, or one of those of the MSSM, H^0 or A^0.

[12] For description of these interactions see e.g. [106, 125, 126] and references therein.

the flavor-diagonal mass-term of the i-flavor squark. Since there are squarks of different chiralities, there are different δ_{ij} parameters for the different chirality mixings.

3.10.1 SUSY Higgs boson flavor-changing neutral currents at the LHC

Some work in relation with the MSSM Higgs boson FCNCs has already been performed [14, 106–109, 117–120, 127, 128, 315, 316]. Here, we compute and analyze the production of any MSSM Higgs boson ($h = h^0$, H^0, A^0) at the LHC, followed by the one-loop FCNC decay $h \to bs$ or $h \to tc$, and we find the maximum production rates of the combined cross section

$$\sigma(pp \to h \to qq') \equiv \sigma(pp \to hX)\mathrm{Br}(h \to qq'),$$
$$\mathrm{Br}(h \to qq') \equiv \frac{\Gamma(h \to q\bar{q}' + \bar{q}q')}{\sum_i \Gamma(h \to X_i)}, \quad (91)$$

qq' being a pair of heavy quarks ($qq' \equiv bs$ or tc), taking into account the restrictions from the experimental determination of $\mathrm{Br}(b \to s\gamma)$ [53]. For other signals of SUSY FCNC at the LHC, without Higgs boson couplings, see Sect. 2.3.3 and [100]. For comparison of the same signal in non-SUSY models see Sect. 2.3.2 and [16, 111]. Here we assume flavor mixing only among the left squarks, since these mixing terms are expected to be the largest ones by renormalization group analysis [123].

In the following we give a summarized explanation of the computation [107, 109]. We include the full one-loop SUSY-QCD contributions to the FCNC partial decay widths $\Gamma(h \to qq')$ in the observable of (91). The Higgs sector parameters (masses and CP-even mixing angle α) have been treated using the leading m_t and $m_b \tan\beta$ approximation to the one-loop result [317–320]. The Higgs boson total decay widths $\Gamma(h \to X)$ are computed at leading order, including all the relevant channels. The MSSM Higgs boson production cross sections have been computed using the programs HIGLU 2.101 and PPHTT 1.1 [116, 321, 322]. We have used the leading order approximation for all channels. The QCD renormalization scale is set to the default value for each program. We have used the set of CTEQ4L PDF [323]. For the constraints on the FCNC parameters, we use $\mathrm{Br}(b \to s\gamma) = (2.1\text{–}4.5) \times 10^{-4}$ as the experimentally allowed range within three standard deviations [53]. We also require that the sign of the $b \to s\gamma$ amplitude is the same as in the SM [324].[13] Running quark masses $m_q(Q)$ and strong coupling constants $\alpha_s(Q)$ are used throughout, with the renormalization scale set to the decaying Higgs boson mass in the decay processes. These computations have been implemented in the computer code FchDecay [325] (see also Sect. 5.5). Given this setup, we have performed a Monte Carlo maximization [326] of the cross section in (91) over the MSSM parameter space, keeping the parameter $\tan\beta$ fixed and under the simplification that the squark and gluino soft-SUSY-breaking parameter masses are at the same scale, $m_{\tilde{q}_{L,R}} = m_{\tilde{g}} \equiv M_{\mathrm{SUSY}}$.

It is enlightening to look at the approximate leading expressions to understand the qualitative trend of the results. The SUSY–QCD contribution to the $b \to s\gamma$ amplitude can be approximated by

$$A^{\mathrm{SQCD}}(b \to s\gamma) \sim \delta_{23} m_b (\mu - A_b \tan\beta)/M_{\mathrm{SUSY}}^2, \quad (92)$$

whereas the MSSM Higgs boson FCNC effective couplings behave as

$$g_{hq\bar{q}'} \sim \delta_{23} \frac{-\mu m_{\tilde{g}}}{M_{\mathrm{SUSY}}^2} \begin{cases} \sin(\beta - \alpha_{\mathrm{eff}}) & (H^0), \\ \cos(\beta - \alpha_{\mathrm{eff}}) & (h^0), \\ 1 & (A^0). \end{cases} \quad (93)$$

The different structure of the amplitudes in (92) and (93) allows us to obtain an appreciable FCNC Higgs boson decay rate, while the prediction for $\mathrm{Br}(b \to s\gamma)$ stays inside the experimentally allowed range.

For the analysis of the bottom-strange production channel, we study first the Higgs boson branching ratio in (91). Figure 32 (left) shows the maximum value of $\mathrm{Br}(h \to bs)$ as a function of the pseudoscalar Higgs boson mass m_{A^0}. We observe that fairly large values of $\mathrm{Br}(h^0 \to bs) \sim 0.3\%$ are obtained. Table 15 (top) shows the actual values of the maximum branching ratios and the parameters that provide them for each Higgs boson. Let us discuss first the general trend, which is valid for all studied processes: the maximum is attained at large M_{SUSY} and moderate δ_{23}. The SUSY–QCD contribution to $b \to s\gamma$ in (92) decreases with M_{SUSY}, therefore to keep $\mathrm{Br}(b \to s\gamma)$ in the allowed range when M_{SUSY} is small, it has to be compensated with a low value of δ_{23}, providing a small FCNC effective coupling in (93). On the other hand, at large M_{SUSY} the second factor in (92) decreases, allowing for a larger value of δ_{23}. Thus, the first factor in (93) grows, but the second factor in (93) stays fixed (provided that $|\mu| \sim M_{\mathrm{SUSY}}$), providing overall a larger value of the effective coupling. On the other hand, a too large value of δ_{23} has to be compensated by a small value of $|\mu|/M_{\mathrm{SUSY}}$ in (92), provoking a reduction in (93). In the end, the balance of the various interactions involved produces the results of Table 15 (top).

The maximum value of the branching ratio for the lightest Higgs boson channel is obtained in the *small α_{eff} scenario* [327, 328]. In this scenario the coupling of bottom quarks to h^0 is extremely suppressed. The large value of $\mathrm{Br}(h^0 \to bs)$ is obtained because the total decay width $\Gamma(h^0 \to X)$ in the denominator of (91) tends to zero

---
[13]This constraint automatically excludes the *fine-tuned* regions of [107].

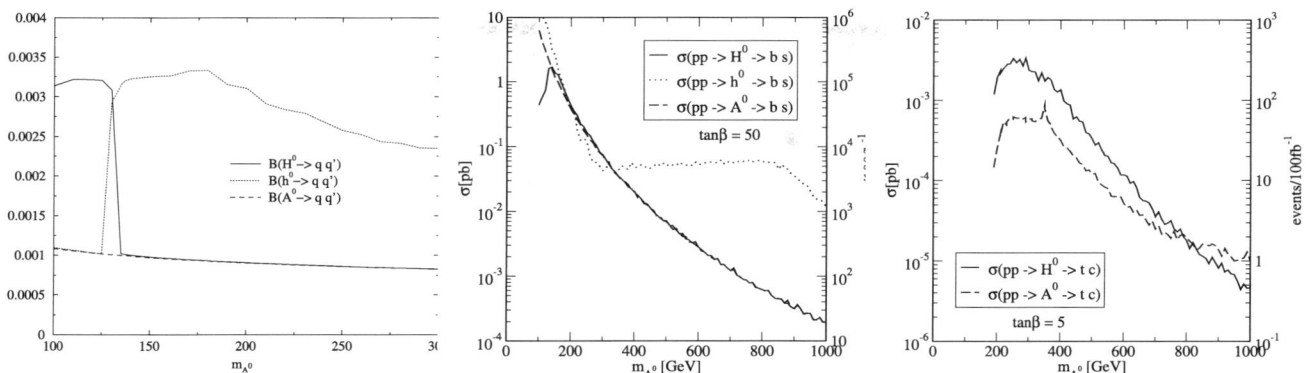

Fig. 32 *Left*: The maximum value of Br($h \to bs$) as a function of m_{A^0} for $\tan\beta = 50$. *Centre*: Maximum SUSY-QCD contributions to $\sigma(pp \to h \to bs)$ as a function of m_{A^0} for $\tan\beta = 50$. *Right*: Maximum SUSY-QCD contributions to $\sigma(pp \to h \to tc)$ as a function of m_{A^0} for $\tan\beta = 5$

Table 15 *Top*: Maximum values of Br($h \to bs$) and corresponding SUSY parameters for $m_{A^0} = 200$ GeV and $\tan\beta = 50$. *Center*: Maximum value of $\sigma(pp \to h \to bs)$ and corresponding SUSY parameters for $m_{A^0} = 200$ GeV and $\tan\beta = 50$. *Bottom*: Maximum value of $\sigma(pp \to h \to tc)$ and corresponding SUSY parameters for $m_{A^0} = 300$ GeV and $\tan\beta = 5$

h	H^0	h^0	A^0
Br($h \to bs$)	9.1×10^{-4}	3.1×10^{-3}	9.1×10^{-4}
$\Gamma(h \to X)$	11.2 GeV	1.4×10^{-3} GeV	11.3 GeV
δ_{23}	$10^{-0.43}$	$10^{-0.8}$	$10^{-0.43}$
$M_{\rm SUSY}$	1000 GeV	975 GeV	1000 GeV
A_b	-1500 GeV	-1500 GeV	-1500 GeV
μ	-460 GeV	-1000 GeV	-460 GeV
Br($b \to s\gamma$)	4.49×10^{-4}	4.48×10^{-4}	4.49×10^{-4}
$\sigma(pp \to h \to bs)$	0.45 pb	0.34 pb	0.37 pb
events/100 fb^{-1}	4.5×10^4	3.4×10^4	3.7×10^4
Br($h \to bs$)	9.3×10^{-4}	2.1×10^{-4}	8.9×10^{-4}
$\Gamma(h \to X)$	10.9 GeV	1.00 GeV	11.3 GeV
δ_{23}	$10^{-0.62}$	$10^{-1.32}$	$10^{-0.44}$
$m_{\tilde{q}}$	990 GeV	670 GeV	990 GeV
A_b	-2750 GeV	-1960 GeV	-2860 GeV
μ	-720 GeV	-990 GeV	-460 GeV
Br($b \to s\gamma$)	4.50×10^{-4}	4.47×10^{-4}	4.39×10^{-4}
$\sigma(pp \to h \to tc)$	2.4×10^{-3} pb		5.8×10^{-4} pb
events/100 fb^{-1}	240		58
Br($h \to tc$)	1.9×10^{-3}		5.7×10^{-4}
$\Gamma(h \to X)$	0.41 GeV		0.39 GeV
δ_{23}	$10^{-0.10}$		$10^{-0.13}$
$m_{\tilde{q}}$	880 GeV		850 GeV
A_t	-2590 GeV		2410 GeV
μ	-700 GeV		-930 GeV
Br($b \to s\gamma$)	4.13×10^{-4}		4.47×10^{-4}

(Fig. 32, top), and not because of a large FCNC partial decay width in its numerator [107].

The leading production channel of h^0 at the LHC at high $\tan\beta$ is the associated production with bottom quarks, and therefore the h^0 production will be suppressed when Br($h^0 \to bs$) is enhanced. We have to perform a combined analysis of the full process in (91) to obtain the maximum production rate of FCNC Higgs boson mediated events at the LHC. Figure 32 (center) shows the result of the maximization of the production cross section (91). The central column of Table 15 (center) shows that when performing the combined maximization $\Gamma(h^0 \to X)$ has a much larger value, and therefore the maximum of the combined cross section is not obtained in the *small* α_{eff} *scenario*. The number of expected events at the LHC is around 50,000 events/100 fb^{-1}. While it is a large number, the huge b-quark background at the LHC will most likely prevent its detection. Note, however, that the maximum FCNC branching ratios are around 10^{-4}–10^{-3}, which is at the same level as the already measured Br($b \to s\gamma$).

The numerical results for the tc channel are similar to the bs channel, so we focus mainly on the differences. Figure 32 (right) shows the maximum value of the production cross section $\sigma(pp \to h \to tc)$ as a function of m_{A^0}. Only the heavy neutral Higgs bosons contribute to this channel and we obtain a maximum of $\sigma^{\max}(pp \to h \to tc) \simeq 10^{-3} - 10^{-2}$ pb, which means several hundreds events per 100 fb^{-1} at the LHC. Due to the single-top quark signature they should be easier to detect than the bs channel, providing the key to a new door to study physics beyond the SM. It is now an experimental challenge to prove that these events can be effectively separated from the background.

The single-top quark FCNC signature can also be produced in other processes, like the direct production (see Sect. 2.3.3 and [100]), or other models, like the two-Higgs doublet model (see Sect. 2.3.2 and [16, 111]). In Table 16 we make a schematic comparison of these different modes. The two modes available in SUSY models probe different parts of the parameter space. While the maximum of the direct production is larger, it decreases quickly with the mass: at $M_{\text{SUSY}} = m_{\tilde{g}} \sim 800$ GeV both channels have a similar production cross section. As for the comparison with the two-Higgs doublet model, the maximum for this model is obtained in a totally different parameter set-up than the SUSY model: large $\tan \beta$, large m_{A^0}, large splitting among the Higgs boson masses, and extremal values of the CP-even Higgs mixing angle α (large/small $\tan \alpha$ for h^0/H^0). The first two conditions would produce a small value for the production in SUSY models, while the last two conditions are not possible in the SUSY parameter space. Then, the detection of a FCNC tc channel at the LHC, together with some other hint on the parameter space (large/small $\tan \beta$, m_{A^0}) would give a strong indication (or confirmation) of the underlying physics model (SUSY/non-SUSY) chosen by nature.

3.10.2 $H \to b\bar{s}$ and B-physics in the MSSM with NMFV

Here we summarize the results from a phenomenological analysis of the general constraints on flavor-changing neutral Higgs decays $H \to b\bar{s}, s\bar{b}$, set by bounds from $b \to s\gamma$ on the flavor mixing parameters in the squark mass matrices of the MSSM with non-minimal flavor violation (NMFV) and compatible with the data from $B \to X_s \mu^+ \mu^-$, assuming first one and then several types of flavor mixing contributing at a time [127]. Details of the part of the soft-SUSY-breaking Lagrangian responsible for the non-minimal squark family mixing and of the parametrization of the flavor-non-diagonal squark mass matrices are given in [114, 127] (see also Sect. 5.6 for a brief description). Previous analyses of bounds on SUSY flavor mixing parameters from $b \to s\gamma$ [329–331] have shown the importance of the interference effects between the different types of flavor violation [125, 126].

We define the dimensionless flavor-changing parameters $(\delta^u_{ab})_{23}$ ($ab = LL, LR, RL, RR$) from the flavor-off-diagonal elements of the squark mass matrices in the following way [114, 127]:

$$\begin{aligned}
\Delta^u_{LL} &\equiv (\delta^u_{LL})_{23} M_{\tilde{L},c} M_{\tilde{L},t}, \\
\Delta^u_{LR} &\equiv (\delta^u_{LR})_{23} M_{\tilde{L},c} M_{\tilde{R},t}, \\
\Delta^u_{RL} &\equiv (\delta^u_{RL})_{23} M_{\tilde{R},c} M_{\tilde{L},t}, \\
\Delta^u_{RR} &\equiv (\delta^u_{RR})_{23} M_{\tilde{R},c} M_{\tilde{R},t},
\end{aligned} \quad (94)$$

and analogously for the down sector ($\{u, c, t\} \to \{d, s, b\}$). For simplicity, we take the same values for the flavor mixing parameters in the up- and down-squark sectors: $(\delta_{ab})_{23} \equiv (\delta^u_{ab})_{23} = (\delta^d_{ab})_{23}$. The expression for the branching ratio

Table 16 Comparison of several FCNC top-charm production cross sections at the LHC, for $\sigma^{\text{SUSY}}(pp \to h \to tc)$ [this work, and [107, 109]], direct production $\sigma^{\text{SUSY}}(pp \to tc)$ (Sect. 2.3.3 and [100]), and two-Higgs doublet model $\sigma^{\text{2HDM}}(pp \to h \to tc)$ (Sect. 2.3.2 and [16, 111])

Parameter	SUSY $h \to tc$	Direct production	2HDM $h \to tc$
Maximum cross section	10^{-2}–10^{-3} pb	1 pb	5×10^{-3} pb
$\tan \beta$	Decreases fast	Insensitive	Increases fast
m_{A^0}	Decreases fast	Insensitive	Prefers large
M_{SUSY}	Prefers large	Decreases fast	–
A_t	Insensitive	Very sensitive	–
δ_{23}	Moderate	Moderate	–
Preferred channel	H^0	–	H^0/h^0
Higgs mass splitting	Given (small)	–	Prefers large

Br($B \to X_s\gamma$) to NLO is taken from [332, 333]. Besides, we assume a common value for the soft SUSY-breaking squark mass parameters, M_{SUSY}, and all the various trilinear parameters to be universal, $A \equiv A_t = A_b = A_c = A_s$ [127]. These parameters and the δs will be varied over a wide range, subject only to the requirements that all the squark masses be heavier than 100 GeV, $|\mu| > 90$ GeV and $M_2 > 46$ GeV [53]. We have chosen as a reference the following set of parameters:

$$M_{\text{SUSY}} = 800 \text{ GeV}, \quad M_2 = 300 \text{ GeV},$$
$$M_1 = \frac{5}{3}\frac{s_W^2}{c_W^2} M_2, \quad (95)$$
$$A = 500 \text{ GeV}, \quad m_A = 400 \text{ GeV},$$
$$\tan\beta = 35, \quad \mu = -700 \text{ GeV}.$$

We have modified the MSSM model file of `FeynArts` to include general flavor mixing, and added 6×6 squark mass and mixing matrices to the `FormCalc` evaluation. Both extensions are publicly available [113, 114, 334, 335]. The masses and total decay widths of the Higgs bosons were computed with `FeynHiggs` [336–339].

Next we derive the maximum values of Br($H^0 \to bs$) compatible with Br($B \to X_s\gamma$)$_{\text{exp}} = (3.3 \pm 0.4) \times 10^{-4}$ [340, 341] within three standard deviations by varying the flavor-changing parameters of the squark mass matrices. The results for the A^0 boson are very similar and we do not show them separately.

As a first step, we select one possible type of flavor violation in the squark sector, assuming that all the others vanish. The interference between different types of flavor mixing is thus ignored. We found that the flavor-off-diagonal elements are independently constrained to be at most $(\delta_{ab})_{23} \sim 10^{-3}$–$10^{-1}$. As expected [125, 126, 329–331], the bounds on $(\delta_{LR})_{23}$ are the strongest, $(\delta_{LR})_{23} \sim 10^{-3}$–$10^{-2}$. The data from $B \to X_s\mu^+\mu^-$ further constrain the parameters $(\delta_{LL})_{23}$ and $(\delta_{LR})_{23}$, the others remaining untouched. The allowed intervals for the corresponding flavor mixing parameters thus obtained are given in [127]. For our reference point (95) we find that the largest allowed value of Br($H^0 \to bs$), of $\mathcal{O}(10^{-3})$ or $\mathcal{O}(10^{-5})$, is induced by $(\delta_{RR})_{23}$ or $(\delta_{LL})_{23}$, respectively (see Fig. 33). These are the flavor-changing parameters least stringently constrained by the $b \to s\gamma$ data. Br($H^0 \to bs$) can reach $\mathcal{O}(10^{-6})$ if induced by $(\delta_{LR})_{23}$ or by $(\delta_{RL})_{23}$, the most stringently constrained flavor-changing parameter. Because of the restrictions imposed by $b \to s\gamma$, Br($H^0 \to bs$) depends very little on $(\delta_{LR})_{23}$ and $(\delta_{RL})_{23}$.

Then, we investigate the case when two off-diagonal elements of the squark mass matrix contribute simultaneously. Indeed, we performed the analysis for all possible combinations of two of the four dimensionless parameters (94). The full results are given in [127]. Figure 34 displays part of the results for our parameter set (95). Contours of constant $\Gamma(H^0 \to bs) \equiv \Gamma(H^0 \to b\bar{s}) + \Gamma(H^0 \to s\bar{b})$ are drawn for various combinations $(\delta_{ab})_{23}$–$(\delta_{cd})_{23}$ of flavor mixing parameters, which we shall refer to as "ab–cd planes" for short in the following. The colored bands represent regions experimentally allowed by $B \to X_s\gamma$. The red bands are regions disfavored by $B \to X_s\mu^+\mu^-$. The bounds on $(\delta_{LR})_{23}$, the best constrained for only one non-zero flavor-off-diagonal element, are dramatically relaxed when other

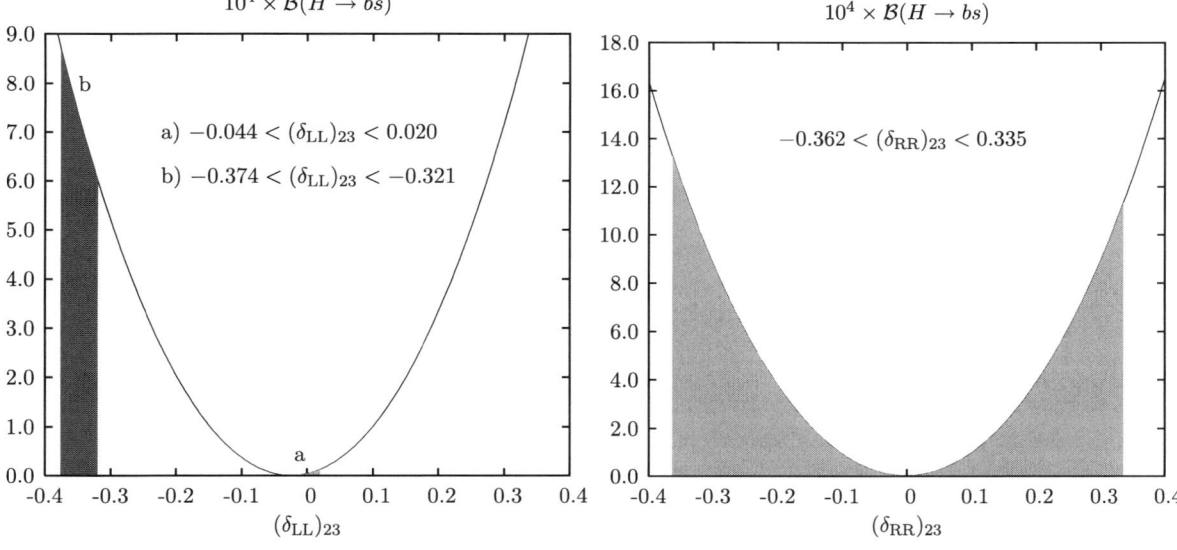

Fig. 33 Br($H^0 \to bs$) as a function of $(\delta_{LL,RR})_{23}$. The allowed intervals of these parameters determined from $b \to s\gamma$ are indicated by *shaded areas*. The left-most are, labeled a, is disfavored by $B \to X_s\mu^+\mu^-$

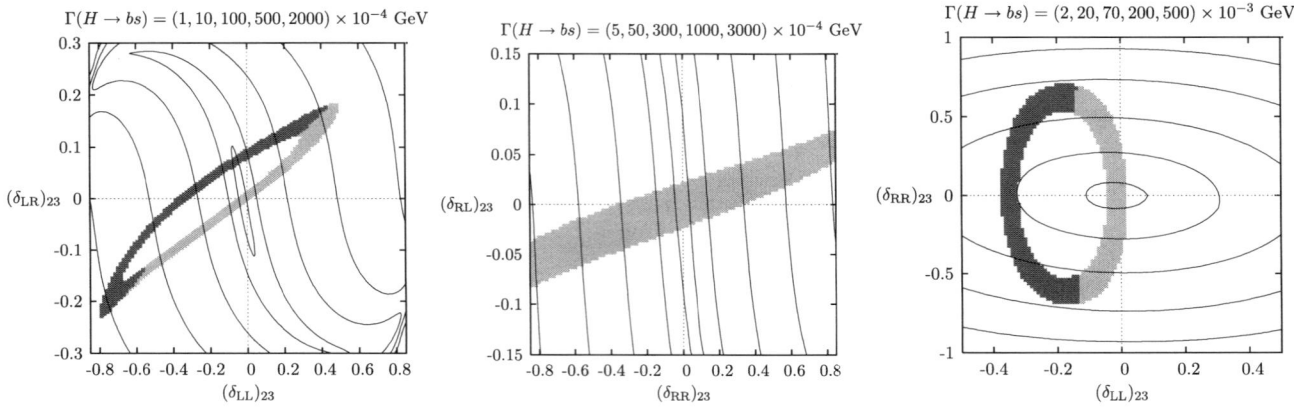

Fig. 34 Contours of constant $\Gamma(H^0 \to bs)$ in various planes of $(\delta_{ab})_{23}$. The *shaded bands* indicate regions experimentally allowed by $B \to X_s \gamma$. The *red (darker) bands* show regions disfavored by $B \to X_s \mu^+ \mu^-$

flavor-changing parameters contribute simultaneously. Values of $(\delta_{LR})_{23} \sim 10^{-1}$ are allowed. As shown in Fig. 34, large although fine-tuned values of $(\delta_{LL})_{23}$ and $(\delta_{LR})_{23}$ combined are not excluded by $b \to s\gamma$, yielding e.g. $\Gamma(H^0 \to bs)_{\max} = 0.25$ GeV for $(\delta_{LR})_{23} = -0.22$, $(\delta_{LL})_{23} = -0.8$. This translates to branching ratios compatible with experimental data of $\mathrm{Br}(H^0 \to bs)_{\max} \sim 10^{-2}$.[14] It also occurs for the *RL–RR* case. The combined effects of *RR–LL* lead to $\Gamma(H^0 \to bs)_{\max} = 0.12$ GeV for $(\delta_{RR})_{23} = 0.65$, $(\delta_{LL})_{23} = \pm 0.14$, leading to $\mathrm{Br}(H^0 \to bs)_{\max} \sim 10^{-2}$.

3.11 Squark/gaugino production and decay

Non-minimal flavor violation (NMFV) arises in the MSSM from a possible misalignment between the rotations diagonalizing the quark and squark sectors. It is conveniently parametrized in the super-CKM basis by non-diagonal entries in the squared squark mass matrices $M_{\tilde{Q}}^2$, $M_{\tilde{U}}^2$, and $M_{\tilde{D}}^2$ and the trilinear couplings A_u and A_d. Squark mixing is expected to be the largest for the second and third generations due to the large Yukawa couplings involved. In addition, stringent experimental constraints for the first generation are imposed by precise measurements of K^0–$\bar{K}^0$ and D^0–$\bar{D}^0$ mixing. Furthermore, direct searches of flavor violation depend on the possibility of flavor tagging, established experimentally only for heavy flavors. We therefore consider here only mixings of second- and third generation squarks and follow the conventions of [120].

3.11.1 Flavor-violating squark and gaugino production at the LHC

We impose mSUGRA [m_0, $m_{1/2}$, A_0, $\tan\beta$, and $\mathrm{sgn}(\mu)$] parameters at a large (grand unification) scale and use two-loop renormalization group equations and one-loop finite corrections as implemented in the computer program SPheno 2.2.2 [342] to evolve them down to the electroweak scale. At this point, we generalize the squark mass matrices by including non-diagonal terms Δ_{ij}. The scaling of these terms with the SUSY-breaking scale M_{SUSY} implies a hierarchy $\Delta_{LL} \gg \Delta_{LR,RL} \gg \Delta_{RR}$ [343]. We therefore take $\Delta_{LR,RL} = \Delta_{RR} = 0$, while $\Delta_{LL}^t = \lambda^t M_{\tilde{L}_t} M_{\tilde{L}_c}$ and $\Delta_{LL}^b = \lambda^b M_{\tilde{L}_b} M_{\tilde{L}_s}$, and assume for simplicity $\lambda = \lambda^t = \lambda^b$. The squark mass matrices are then diagonalized, and constraints from low energy and electroweak precision measurements are imposed on the corresponding theoretical observables, calculated with the computer program FeynHiggs 2.5.1 [337].

FCNC B decays and B^0–$\bar{B}^0$ mixing arise in the SM only at the one-loop level. These processes are therefore particularly sensitive to non-SM contributions entering at the same order in perturbation theory and have been intensely studied at B-factories. The most stringent constraints on SUSY-loop contributions in minimal and non-minimal flavor violation come today from the inclusive $b \to s\gamma$ decay rate as measured by BaBar, Belle, and CLEO, $\mathrm{Br}(b \to s\gamma) = (3.55 \pm 0.26) \times 10^{-4}$ [344], which affects directly the allowed squark mixing between the second and third generation [127].

Another important consequence of NMFV in the MSSM is the generation of large splittings between squark-mass eigenvalues. The splitting within isospin doublets influences the Z- and W-boson self-energies at zero-momentum $\Sigma_{Z,W}(0)$ in the electroweak ρ parameter $\Delta\rho = \Sigma_Z(0)/M_Z^2 - \Sigma_W(0)/M_W^2$ and consequently the W-boson mass M_W and the squared sine of the weak mixing angle $\sin^2\theta_W$. The latest combined fits of the Z-boson mass, width, pole asymmetry, W-boson and top quark mass constrain new physics contributions to $\Delta\rho$ to $T = -0.13 \pm 0.11$ or $\Delta\rho = -\alpha T = 0.00102 \pm 0.00086$ [344].

---

[14] Here we have used the total width of $\Gamma(H \to X) \approx 26$ GeV, $H = H^0$, A^0, for the point (95) in the MSSM with MFV.

A third observable sensitive to SUSY loop contributions is the anomalous magnetic moment $a_\mu = (g_\mu - 2)/2$ of the muon, for which recent BNL data and the SM prediction disagree by $\Delta a_\mu = (22 \pm 10) \times 10^{-10}$ [344]. In our calculation, we take into account the SM and MSSM contributions up to two loops [345, 346].

For cosmological reasons, we require the lightest SUSY particle (LSP) to be electrically neutral. We also calculate, albeit for minimal flavor violation ($\lambda = 0$) only, the cold dark matter relic density using the computer program Dark-SUSY [347] and impose a limit of $0.094 < \Omega_c h^2 < 0.136$ at 95% (2σ) confidence level. This limit has recently been obtained from the three-year data of the WMAP satellite, combined with the SDSS and SNLS survey and Baryon Acoustic Oscillation data and interpreted within a more general (11-parameter) inflationary model [348]. This range is well compatible with the older, independently obtained range of $0.094 < \Omega_c h^2 < 0.129$ [285].

Typical scans of the mSUGRA parameter space with $\tan \beta = 10$, $A_0 = 0$ and $\mu > 0$ and all experimental limits imposed at the 2σ level are shown in Fig. 35. Note that $\mu < 0$ is disfavored by $g_\mu - 2$ data, while $\Delta \rho$ only constrains the parameter space outside the mass regions shown here. In minimal flavor-violation, light SUSY scenarios such as the SPS 1a benchmark point ($m_0 = 100$ GeV, $m_{1/2} = 250$ GeV) [248] are favored $g_\mu - 2$ data. The dependence on the trilinear coupling A_0 (-100 GeV for SPS 1a, 0 GeV in our scenario) is extremely weak.

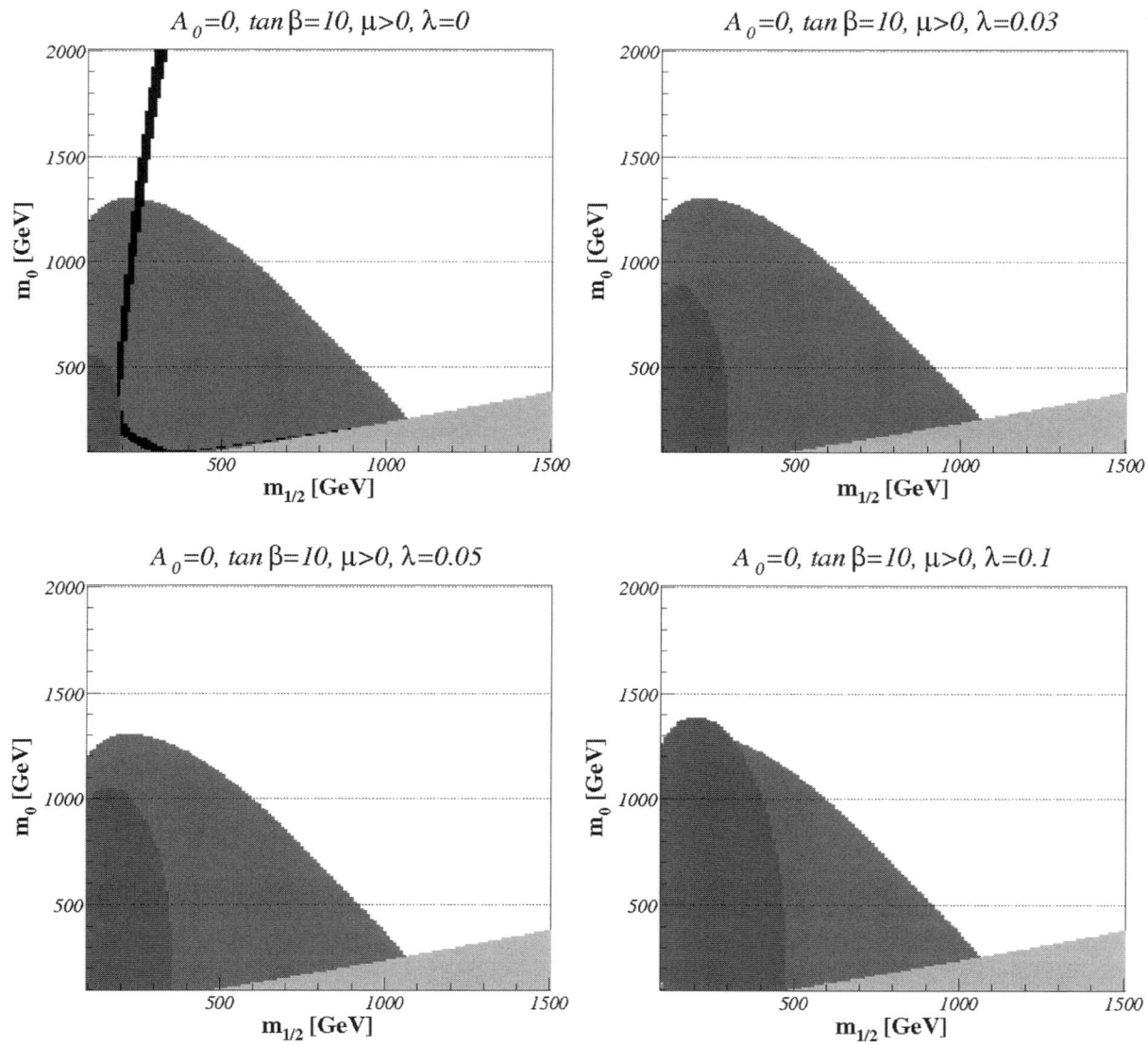

Fig. 35 Regions of mSUGRA parameter space in minimal ($\lambda = 0$) and non-minimal ($\lambda > 0$) flavor violation: favored by a_μ (*medium grey*) and WMAP (*black*); and excluded by $b \to s\gamma$ (*blue/dark grey*) and charged LSP (*orange/light grey*)

In Fig. 36 we show for our (slightly modified) SPS 1a benchmark point the dependence of the electroweak precision variables and the lightest SUSY particle masses on the NMFV parameter λ, indicating by dashed lines the ranges allowed experimentally within two standard deviations. It is interesting to see that for this benchmark point, not only the region close to minimal flavor violation ($\lambda < 0.1$) is allowed, but that there is a second allowed region at $0.4 < \lambda < 0.5$.

Next, we study in Fig. 37 the chirality and flavor decomposition of the light (1, 2) and heavy (4, 6) squarks, which changes mostly in a smooth way, but sometimes dramatically in very small intervals of λ. In particular, the second allowed region at larger λ has a quite different flavor and chirality mixture than the one at small λ.

The main result of our work is the calculation of all electroweak (and strong) squark and gaugino production channels in NMFV SUSY [349]. We show in Fig. 38 a small, but representative sample of these production cross sections: charged squark–antisquark pair production, non-diagonal squark–squark pair production, as well as chargino–squark and neutralino–squark associated production. The two $b \to s\gamma$ allowed regions ($\lambda < 0.1$ and $0.4 < \lambda < 0.5$) are indicated by vertical lines. Note that NMFV allows for a top-flavor content to be produced from non-top initial quark densities and for right handed chirality content to be produced from strong gluon or gluino exchanges. The cross sections shown here are all in the fb range and lead mostly to experimentally identifiable heavy-quark (plus missing transverse energy) final states.

In conclusion, we have performed a search in the NMFV-extended mSUGRA parameter space for regions allowed by electroweak precision data as well as cosmological constraints. In a benchmark scenario similar to SPS 1a, we find two allowed regions for second and third generation squark mixing, $\lambda < 0.1$ and $0.4 < \lambda < 0.5$, with distinct flavor and chirality content of the lightest and heaviest up- and down-type squarks. Our calculations of NMFV production cross sections at the LHC demonstrate that the corresponding squark (anti)squark pair production channels and the associated production of squarks and gauginos are very sensitive to the NMFV parameter λ. For further details see [350].

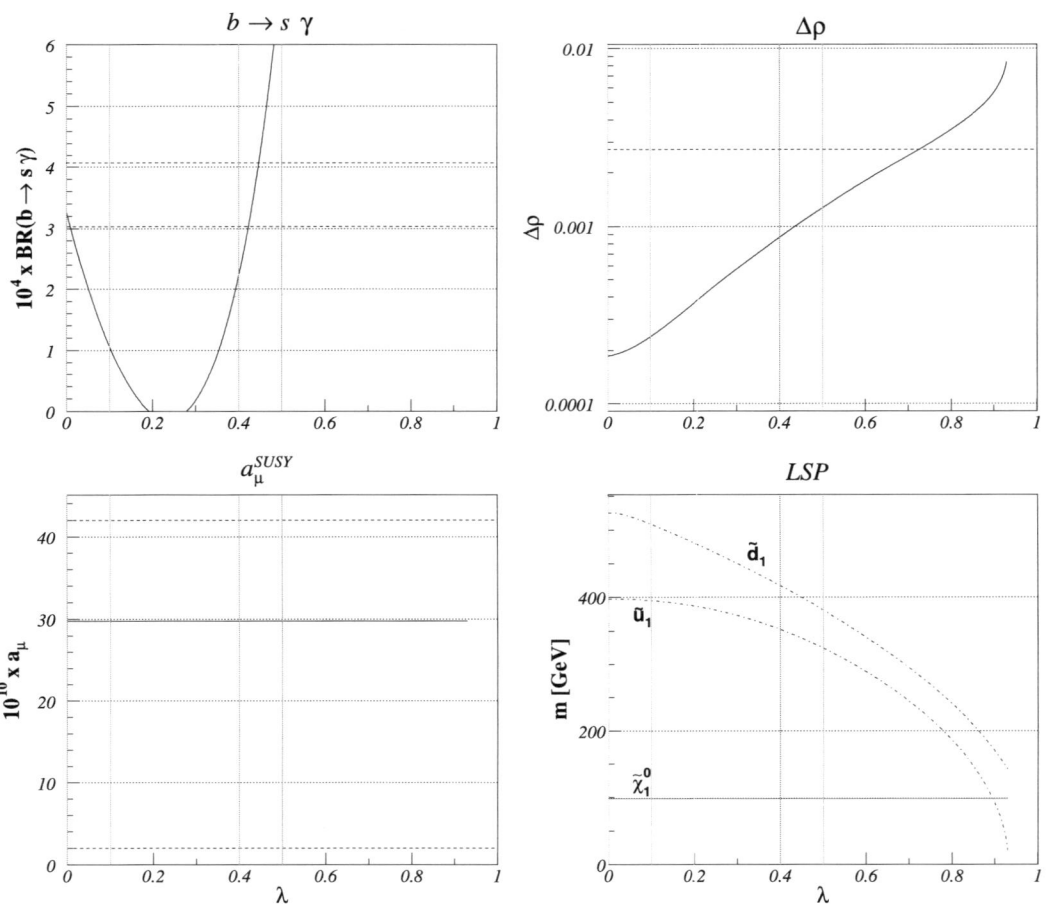

Fig. 36 Dependence of the precision variables Br($b \to s\gamma$), $\Delta\rho$, and a_μ and the lightest SUSY particle masses on the NMFV parameter λ

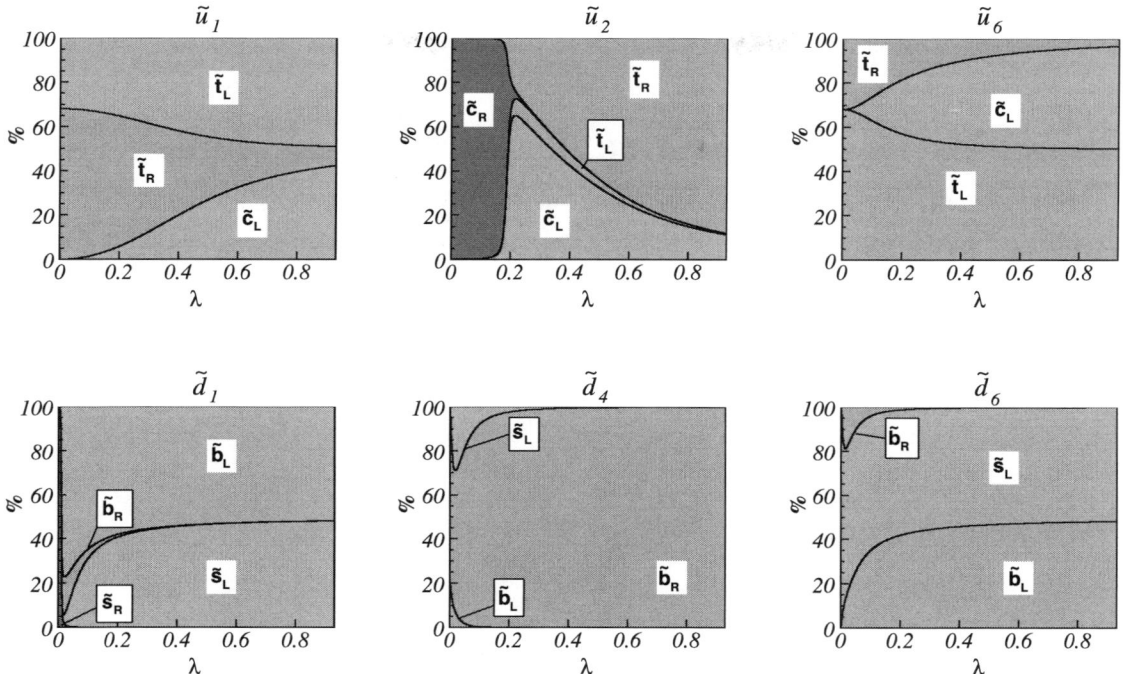

Fig. 37 Decomposition of the chirality (L, R) and flavor (c, t and s, b) content of the lightest ($\tilde{q}_1, \tilde{q}_2$) and heavier ($\tilde{q}_4, \tilde{q}_6$) up- ($q = u$) and down-type ($q = d$) squarks on the NMFV parameter λ

3.11.2 Flavor-violating squark and gluino decays

In the study of squark decays two general scenarios can be distinguished depending on the hierarchy within the SUSY spectrum:

- $m_{\tilde{g}} > m_{\tilde{q}_i}$ ($q = d, u; i = 1, \ldots, 6$): In this case the gluino will mainly decay according to

$$\tilde{g} \to d_j \tilde{d}_i, \qquad \tilde{g} \to u_j \tilde{u}_i, \tag{96}$$

with $d_j = (d, s, b)$ and $u_j = (u, c, t)$ followed by squark decays into neutralino and charginos

$$\tilde{u}_i \to u_j \tilde{\chi}_k^0, d_j \tilde{\chi}_l^+, \qquad \tilde{d}_i \to d_j \tilde{\chi}_k^0, u_j \tilde{\chi}_l^-. \tag{97}$$

In addition there can be decays into gauge- and Higgs bosons if kinematically allowed:

$$\tilde{u}_i \to Z \tilde{u}_k, H_r^0 \tilde{u}_k, W^+ \tilde{d}_j, H^+ \tilde{d}_j, \tag{98}$$

$$\tilde{d}_i \to Z \tilde{d}_k, H_r^0 \tilde{d}_k, W^- \tilde{u}_j, H^- \tilde{u}_j \tag{99}$$

where $H_r^0 = (h^0, H^0, A^0)$, $k < i, j = 1, \ldots, 6$. Due to the fact that there is left–right mixing in the sfermion mixing, one has flavor-changing neutral decays into Z-bosons at tree level.

- $m_{\tilde{g}} < m_{\tilde{q}_i}$ ($q = d, u; i = 1, \ldots, 6$): in this case the squarks decay mainly into a gluino:

$$\tilde{u}_i \to u_j \tilde{g}, \qquad \tilde{d}_i \to d_j \tilde{g} \tag{100}$$

and the gluino decays via three body decays and loop-induced two body decays into charginos and neutralinos

$$\begin{aligned}\tilde{g} &\to d_j d_i \tilde{\chi}_k^0, u_j u_i \tilde{\chi}_k^0, \\ \tilde{g} &\to u_j d_i \tilde{\chi}_l^\pm, \qquad \tilde{g} \to g \tilde{\chi}_k^0\end{aligned} \tag{101}$$

with $i, j = 1, 2, 3, l = 1, 2$ and $k = 1, 2, 3, 4$. The first two decay modes contain states with quarks of different generations.

Obviously, the flavor mixing final states of the decays listed above are constrained by the fact that all observed phenomena in rare meson decays are consistent with the SM predictions. Nevertheless, one has to check how large the branching ratios for the flavor mixing final states can be. One also has to study the impact of such final states on the discovery of SUSY as well as the determination of the underlying model parameters.

For simplicity we restrict ourselves to the mixing between second and third generation of (s)quarks. We shall take the so-called SPA point SPS1a′ [259] as a specific example which is specified by the mSUGRA parameters $m_0 = 70$ GeV, $m_{1/2} = 250$ GeV, $A_0 = -300$ GeV, $\tan \beta = 10$ and $\text{sign}(\mu) = 1$. We have checked that main features discussed below are also present in other study points, e.g. I'' and γ of [351]. At the electroweak scale (1 TeV) one gets the following data with the SPA1a′ point: $M_2 = 193$ GeV, $\mu = 403$ GeV, $m_{H^+} = 439$ GeV and $m_{\tilde{g}} = 608$ GeV. We have used the program SPheno [342] for the calculation.

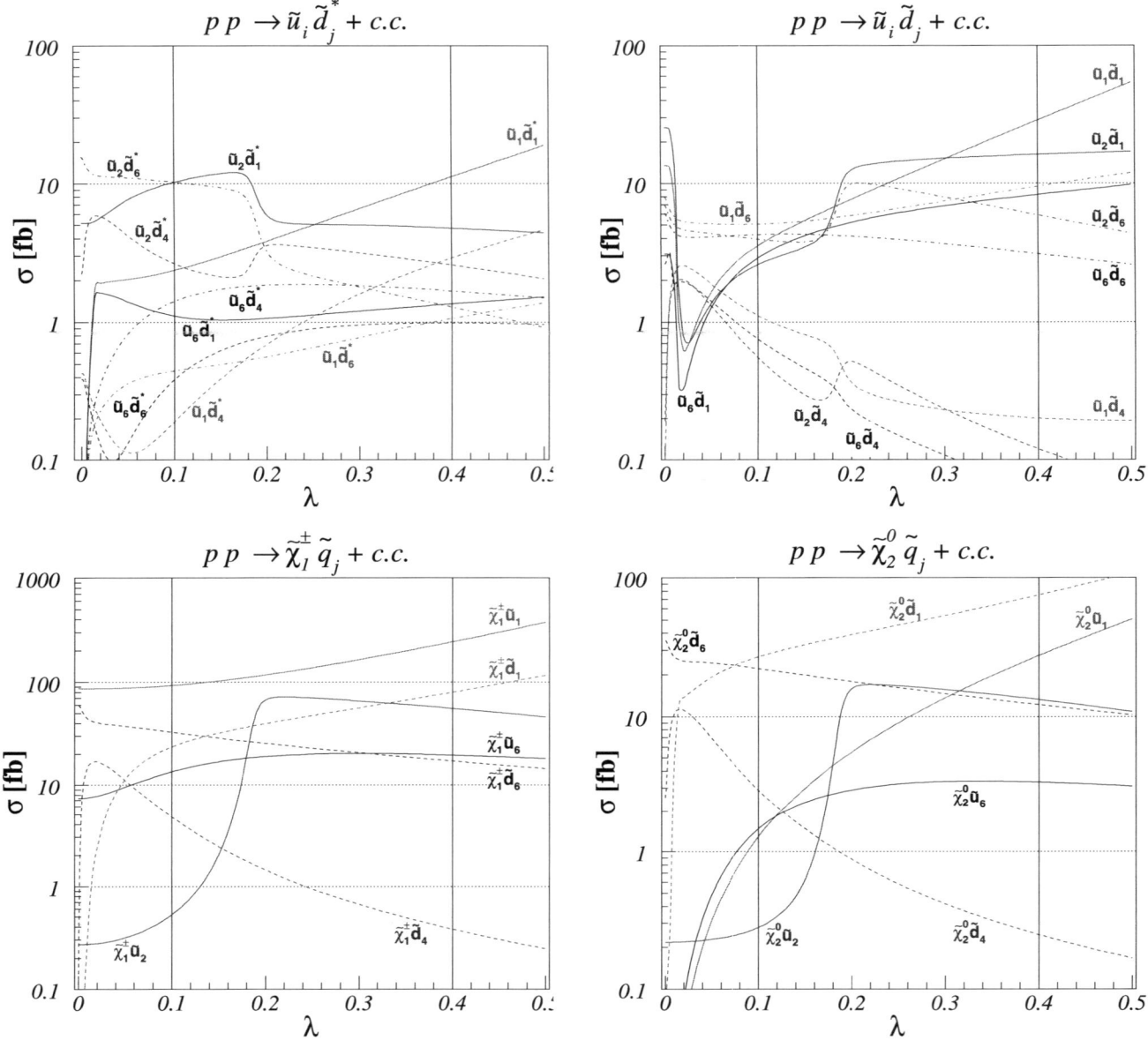

Fig. 38 Representative sample of squark and gaugino production cross sections at the LHC in NMFV

It has been shown that in minimal flavor violating scenarios the flavor-changing decay modes are quite small [200]. To get sizable flavor-changing decay branching ratios, we have added the flavor mixing parameters as given in Table 17; the resulting up-squark masses in GeV are in ascending order: 315, 488, 505, 506, 523 and 587 [GeV] whereas the resulting down-squark masses are 457, 478, 505, 518, 529, 537 [GeV]. This point is a random, but also typical one out of 20000 points fulfilling the constraints derived from the experimental measurements of the following three key observables of the $b \to s$ sector: $b \to s\gamma$, ΔM_{B_s} and $b \to sl^+l^-$. For the calculation we have used the formula

Table 17 Flavor-violating parameters in GeV^2 which are added to the SPS1a$'$ point. The corresponding values for the low energy observables are Br($b \to s\gamma$) = 3.8 × 10^{-4}, $|\Delta(M_{B_s})|$ = 19.6 ps^{-1} and Br($b \to s\mu^+\mu^-$) = 1.59 × 10^{-6}

$M^2_{Q,23}$	$M^2_{D,23}$	$M^2_{U,23}$	$v_u A^u_{23}$	$v_u A^u_{32}$	$v_d A^d_{23}$	$v_d A^d_{32}$
−18429	−37154	−32906	28104	16846	981	−853

given in [352, 353], for $b \to s\gamma$, the formula for ΔM_{B_s} given in [354] and the formula for $b \to sl^+l^-$ given in [352, 355]. Note that we have included all contributions mediated by chargino, neutralino and gluino loops as we depart here

Table 18 Branching ratios (in %) for squark and gluino decays for the point specified in Table 17. Only branching ratios larger than 1% are shown

	$\tilde{\chi}_1^0 c$	$\tilde{\chi}_1^0 t$	$\tilde{\chi}_2^0 c$	$\tilde{\chi}_2^0 t$	$\tilde{\chi}_1^+ s$	$\tilde{\chi}_1^+ b$	$\tilde{\chi}_2^+ b$	$\tilde{u}_1 Z^0$	$\tilde{u}_1 h^0$
$\tilde{u}_1$	1.4	16.8				81.1			
$\tilde{u}_2$	9.1		21.0	3.6	42.9	14.3		5.3	1.3
$\tilde{u}_3$	20.9		21.9		47.5	1.1		1.9	5.5
$\tilde{u}_6$	1.5	2.7	1.6	3.7	4.0	14.1	14.2	39.2	5.2

	$\tilde{\chi}_1^0 s$	$\tilde{\chi}_1^0 b$	$\tilde{\chi}_2^0 s$	$\tilde{\chi}_2^0 b$	$\tilde{\chi}_3^0 b$	$\tilde{\chi}_4^0 b$	$\tilde{\chi}_1^- c$	$\tilde{\chi}_1^- t$	$\tilde{u}_1 W^-$
$\tilde{d}_1$	1.4	5.7	2.7	2.8			6.5	28.1	27.3
$\tilde{d}_2$	4.2	2.9	6.3	17.8			13.4	18.8	34.8
$\tilde{d}_4$	1.8		23	3.7			41.5	5.8	20.0
$\tilde{d}_6$	77.3	15.9	4.6	3.7	2.4	2.4	7.7	5.1	40.0

	$\tilde{d}_1 s$	$\tilde{d}_1 b$	$\tilde{d}_2 s$	$\tilde{d}_2 b$	$\tilde{d}_3 d$	$\tilde{d}_4 s$	$\tilde{d}_5 d$	$\tilde{d}_6 s$	$\tilde{d}_6 b$
$\tilde{g}$	3.4	12.8	5.5	7.5	8.2	5.8	5.1	2.1	2.2
	$\tilde{u}_1 c$	$\tilde{u}_1 t$	$\tilde{u}_2 c$	$\tilde{u}_3 c$	$\tilde{u}_4 u$	$\tilde{u}_5 u$			
	1.2	14	8.8	7.9	8.2	5.5			

considerably from minimal flavor violation. The most important branching ratios for gluino and squark decays are given in Table 18. In addition the following branching ratios are larger than 1%, namely Br($\tilde{u}_6 \to \tilde{d}_1 W$) = 8.9% and Br($\tilde{u}_6 \to \tilde{d}_2 W$) = 1.8%. We have not displayed the branching ratios of the first generation nor the ones of the gluino into first generation.

It is clear from Table 18 that all listed particles have large flavor-changing decay modes. This clearly has an impact on the discovery strategy of squarks and gluinos as well as on the measurement of the underlying parameters. For example, in mSUGRA points without flavor mixing one finds usually that the left-squarks of the first two generations as well as the right squarks have similar masses. Large flavor mixing implies that there is a considerable mass splitting as can be seen by the numbers above. Therefore, the assumption of nearly equal masses should be reconsidered if sizable flavor-changing decays are discovered in squark and gluino decays.

An important part of the decay chains considered for SPS1a′ and nearby points are $\tilde{g} \to b\tilde{b}_j \to b\bar{b}\tilde{\chi}_k^0$ which are used to determine the gluino mass as well as the sbottom masses or at least their average value if these masses are close. In the analysis the existence of two b jets has been assumed, which need not to be the case as shown in the example above. Therefore, this class of analysis should be redone requiring only one b jet + one additional non-b jet to study the impact of flavor mixing on the determination of these masses.

Similar conclusions hold for the variable M_{tb}^w defined in [356]. For this variable one considers final states containing $b\tilde{\chi}_1^+$. In our example, three u-type squarks contribute with branching ratios larger than 10% in contrast to assumption that only the two stops contribute. The influence of the additional state requires for a sure a detailed Monte Carlo study which should be carried out in the future.

3.12 Top squark production and decay

Supersymmetric scenarios with a particularly light stop have been recently considered as potential candidates to provide a solid explanation of the observed baryon asymmetry of the Universe [357]. Independently of this proposal, measurements of the process of stop–chargino associated production at LHC have been considered as a rather original way of testing the usual assumptions about the supersymmetric CKM matrix [358]. In a very recent paper [359], the latter associated production process has been studied in some detail for different choices of the SUSY benchmark points, trying to find evidence for and to understand an apparently strong $\tan\beta$ dependence of the production rates. As a general feature of that study, the values of the various rates appeared, typically, smaller than one pb, to be compared with the (much) bigger rates of the stop–antistop process (see e.g. [360]).

3.12.1 Associated stop–chargino production at LHC: a light stop scenario test

Given the possible relevance of an experimental determination, it might be opportune to perform a more detailed study of the production rate size in the special light stop scenario, where one expects that the numerical value is as large as possible. Here we present the results of this study, performed at the simplest Born level given the preliminary nature of the investigation.

Fig. 39 Integrated cross sections for the process $pp \to \tilde{t}_1 \chi_1^- + X$ at the four MSSM points SU1, SU6, LS1, LS2

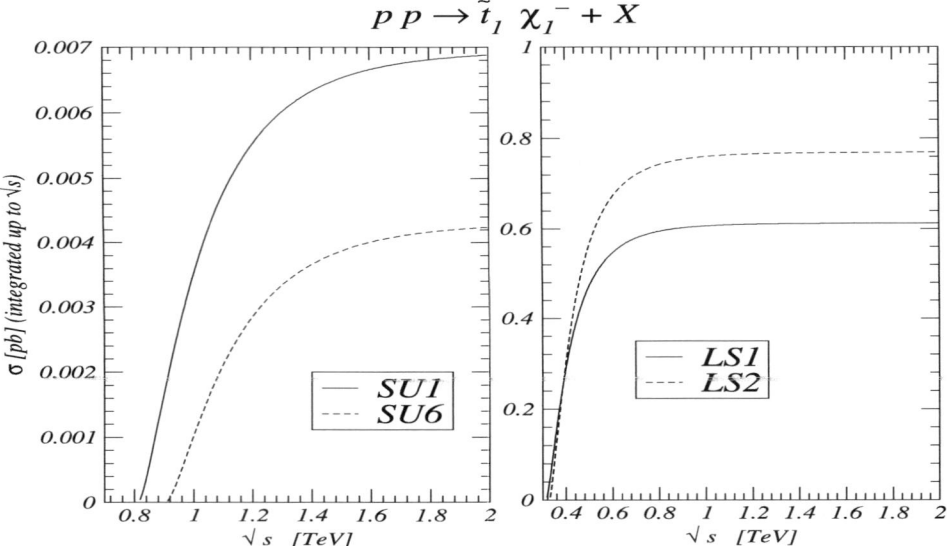

The starting point is the expression of the differential cross section, estimated at Born level in the c.m. frame of the incoming pair of the partonic process $bg \to \chi_i^- \tilde{t}_j$. Its detailed expression has been derived and discussed in [359]. The associated c.m. energy distribution (at this Born level identical to the final invariant-mass distribution) is

$$\frac{d\sigma(pp \to \tilde{t}_a \chi_i^- + X)}{d\hat{s}} = \frac{1}{S} \int_{\cos\theta_{\min}}^{\cos\theta_{\max}} d\cos\theta \, L_{bg}(\tau, \cos\theta) \frac{d\sigma_{bg \to \tilde{t}_a \chi_i^-}}{d\cos\theta}(\hat{s}), \quad (102)$$

where $\sqrt{\hat{s}}$ and $\sqrt{S}$ are the parton and total pp c.m. energies, respectively, $\tau = \hat{s}/S$, and L_{bg} is the parton process luminosity that we have evaluated using the parton distribution functions from the heavy-quark CTEQ6 set [361]. The rapidity and angular integrations are performed after imposing a cut $p_T \geq 10$ GeV.

For a preliminary analysis, we have considered the total cross section (for producing the lightest stop–chargino pair), defined as the integration of the distribution from threshold to a final energy $\sqrt{s}$ left as free variable, generally fixed by experimental considerations. To have a first feeling of the size of this quantity, we have first estimated it for two pairs of sensible MSSM benchmark points. The first pair are the ATLAS Data Challenge-2 points SU1, SU6 whose detailed description can be found in [141]. The second pair are the points LS1, LS2 introduced in [140]. These points are typical *light SUSY* scenarios and in particular share a rather small threshold energy $m_{\tilde{t}} + m_\chi$ which appears to be a critical parameter for the observability of the considered process. The main difference between SU1 and SU6 or LS1 and LS2 is the value of $\tan\beta$ (larger in SU6 and LS2). The results are shown in Fig. 39. As one sees, the various rates are essentially smaller than a pb, well below the expected stop–antistop values.

In the previous points, no special assumptions about the value of the stop mass were performed, hence keeping a conservative attitude. One sees, as expected, that the bigger rate values correspond to the lighter stop situations (LS1 and LS2). In this spirit, we have therefore considered a different MSSM point where the final stop is particularly light. More precisely, we have concentrated our analysis on the point LST2, introduced and discussed in Sect. 3.12.2 and characterized by the MSSM parameters (we list the relevant ones at Born level)

$$M_1 = \frac{5}{3} \tan^2 \theta_W M_2 = 110 \text{ GeV},$$
$$\mu = 300 \text{ GeV}, \qquad \tan\beta = 7, \qquad (103)$$
$$\tilde{t}_1 \simeq \tilde{t}_R, \qquad m_{\tilde{t}_1} = 150 \text{ GeV},$$

and consistent with the cosmological experimental bounds on the relic density. Now the threshold energy is even smaller than in the previous examples. The integrated cross section, shown in Fig. 40 reaches a maximum of about 2 pb, which might be detected by a dedicated experimental search.

3.12.2 Exploiting gluino-to-stop decays in the light stop scenario

To achieve a strong first-order electroweak phase transition in the MSSM, the lighter of the two stops, $\tilde{t}_1$, has to be lighter than the top quark [362–366]. Assuming a stable $\tilde{\chi}_1^0$ LSP, there hence exists a very interesting parameter region

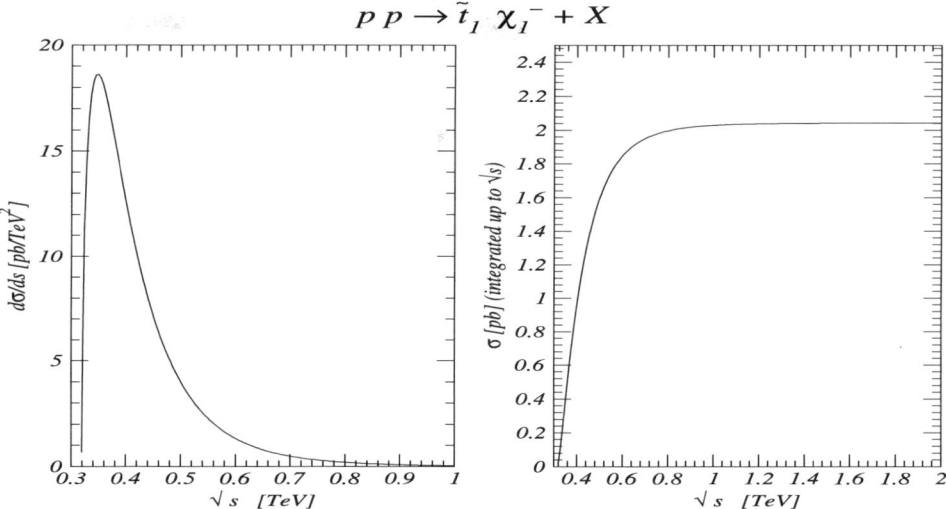

Fig. 40 Distribution $d\sigma/ds$ and integrated cross sections for the process $pp \to \tilde{t}_1 \chi_1^- + X$ at the point LST2

with a small $\tilde{\chi}_1^0$–$\tilde{t}_1$ mass difference, for which (i) coannihilation with $\tilde{t}_1$ [367, 368] leads to a viable neutralino relic density and (ii) the light stop decays dominantly into $c\tilde{\chi}_1^0$ [369].

In this case, stop pair production leads to 2 c jets + $\rlap{E}{/}_T$, a signal which is of very limited use at the LHC. One can, however, exploit [370] gluino-pair production followed by gluino decays into stops and tops: since gluinos are Majorana particles, they can decay either into $t\tilde{t}_1^*$ or $\bar{t}\tilde{t}_1$; pair produced gluinos therefore give same-sign top quarks in half of the gluino-to-stop decays. Here note that in the light stop scenario, $\tilde{g} \to t\tilde{t}_1^*$ (or $\bar{t}\tilde{t}_1$) has practically 100% branching ratio. With $\tilde{t}_1 \to c\tilde{\chi}_1^0$, $t \to bW$, and the Ws decaying leptonically, this leads to a signature of two b jets plus two same-sign leptons plus jets plus missing transverse energy:

$$pp \to \tilde{g}\tilde{g} \to bbl^+l^+ \left(\text{or } \bar{b}\bar{b}l^-l^-\right) + \text{jets} + \rlap{E}{/}_T. \quad (104)$$

In [370] we performed a case study for the 'LST1' parameter point with $m_{\tilde{\chi}_1^0} = 105$ GeV, $m_{\tilde{t}_1} = 150$ GeV, $m_{\tilde{g}} = 660$ GeV and showed that the signature (104) is easily extracted from the background. In this contribution, we focus more on the stop coannihilation region and discuss some additional issues.

We define a benchmark point 'LST2' in the stop coannihilation region by taking the parameters of LST1 and lowering the stop mass to $m_{\tilde{t}_1} = 125$ GeV. We generate signal and background events equivalent to 30 fb^{-1} of integrated luminosity and perform a fast simulation of a generic LHC detector as described in [370]. The following cuts are then applied to extract the signature of (104):

– Require two same-sign leptons (e or μ) with $p_T^{\text{lep}} > 20$ GeV.
– Require two b-tagged jets with $p_T^{\text{jet}} > 50$ GeV;
– Missing transverse energy $\rlap{E}{/}_T > 100$ GeV.

Table 19 Number of events at LST2 left after cumulative cuts for 30 fb^{-1} of integrated luminosity. "2lep, 2b" means two leptons with $p_T^{\text{lep}} > 20$ GeV plus two b jets with $p_T^{\text{jet}} > 50$ GeV. "2t" is the requirement of two tops (i.e. $m_{bl} < 160$ GeV), and "SS" that of two same-sign leptons

Cut		2lep, 2b	$\rlap{E}{/}_T$	2t	SS
Signal:	$\tilde{g}\tilde{g}$	1091	949	831	413
Background:	SM	34224	8558	8164	53
	SUSY	255	209	174	85

– Demand two combinations of the two hardest leptons and b jets that give invariant masses $m_{bl} < 160$ GeV, consistent with a top quark.

This set of cuts emphasizes the role of the same-sign top quarks in our method, and ignores the detectability of the jets initiated by the $\tilde{t}_1$ decay. Table 19 shows the effect of the cuts on both the signal and the backgrounds. Detecting in addition the (soft) c jets from the $\tilde{t}_1 \to c\tilde{\chi}_1^0$ decay, together with the excess in events with 2 c jets + $\rlap{E}{/}_T$ from stop pair production, can be used to strengthen the light stop hypothesis. A reasonable c-tagging efficiency would be very helpful in this case.

To demonstrate the robustness of the signal, we show in Fig. 41 (left) contours of 3σ, 5σ and 10σ significance[15] in the $(m_{\tilde{g}}, m_{\tilde{t}_1})$ plane. For comparison we also show as a dotted line the result of a CMS study [5], which found a reach down to 1 pb in terms of the total cross section for same-sign top production. In Fig. 41 (right), we show the decreasing significance for $m_{\tilde{g}} = 900$ GeV, as the stop–neutralino mass difference goes to zero. To be conservative,

---
[15]We define significance as $S/\sqrt{B}$, where S and B are the numbers of signal and background events.

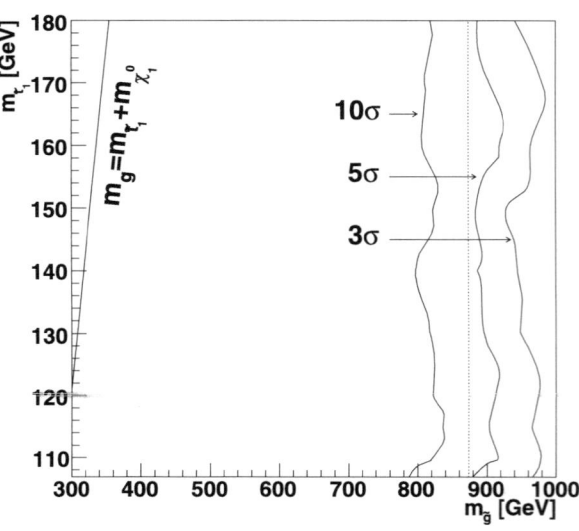 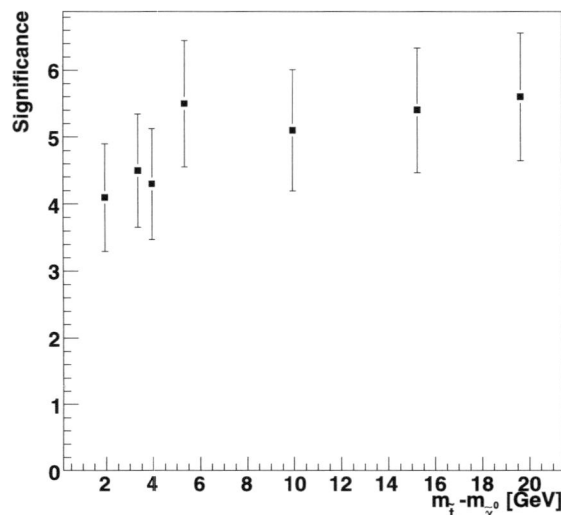

Fig. 41 Reach for the signature of (104) in the gluino–stop mass plane (*left*) and significance as a function of stop–neutralino mass difference with $m_{\tilde{g}} = 900$ GeV (*right*)

both panels in Fig. 41 assume that all squarks other than the $\tilde{t}_1$ are beyond the reach of the LHC; $\tilde{q}\tilde{q}$ and $\tilde{g}\tilde{q}$ production would increase the signal through $\tilde{q} \to \tilde{g}q$ decays (provided $m_{\tilde{q}} > m_{\tilde{g}}$) while adding only little to the background; see [370, 371] for more detail.

The usual way to determine SUSY masses in cascade decays is through kinematic end points of the invariant-mass distributions of the SM decay products, see e.g. [257, 258, 372, 373]. In our case, there are four possible end points: $m_{bl}^{\max}$, $m_{bc}^{\max}$, $m_{lc}^{\max}$ and $m_{blc}^{\max}$, of which the first simply gives a relationship between the masses of the W and the top, and the second and third are linearly dependent, so that we are left with three unknown masses and only two equations. Moreover, because of the information lost with the escaping neutrino the distributions of interest all fall very gradually to zero.

In order to nevertheless get some information on the $\tilde{\chi}_1^0$, $\tilde{t}_1$ and $\tilde{g}$ masses, we fit the whole m_{bc} and m_{lc} distributions [370, 374] and not just the end points. This requires, of course, the detection of the jets stemming from the $\tilde{t}_1$ decay. For small $m_{\tilde{t}_1} - m_{\tilde{\chi}_1^0}$ these are soft, so we demand two jets with $p_T^{\text{jet}} < 50$ GeV in addition to the cuts listed above. The results of the fits for LST2, assuming 20% c-tagging efficiency,[16] are shown in Fig. 42. The combined result of the two distributions is $m_{bc}^{\max} = 305.7 \pm 4.3$, as compared to the nominal value of $m_{bc}^{\max} \simeq 299$ GeV.

As mentioned above, the gluino-pair production leads to 50% same-sign (SS) and 50% opposite-sign (OS) top quark pairs, and hence $R = N(SS)/N(SS + OS) \simeq 0.5$ with N denoting the number of events. In contrast, in the SM one has $R \lesssim 0.01$. This offers a potential test of the Majorana nature of the gluino. The difficulty is that the number of OS leptons is completely dominated by the $t\bar{t}$ background. This can easily be seen from the last two rows of Table 19: $R \sim 0.5$ (0.02) for the signal (backgrounds) as expected; signal and backgrounds combined, however, give $R \sim 0.06$. A subtraction of the $t\bar{t}$ background as described in Sect. 3.12.3 may help to extract $R(\tilde{g}\tilde{g})$.

3.12.3 A study on the detection of a light stop squark with the ATLAS detector at the LHC

We present here an exploratory study of a benchmark model in which the stop quark has a mass of 137 GeV, and the two body decay of the stop squark into a chargino and a b quark is open. We address in detail the ability of the ATLAS experiment to separate the stop signal from the dominant SM backgrounds.

For the model under study [170] all the masses of the first two generation squarks and sleptons are set at 10 TeV, and the gaugino masses are related by the usual gaugino mass relation $M_1 : M_2 = \alpha_1 : \alpha_2$. The remaining parameters are thus defined:

$M_1 = 60.5$ GeV, $\quad \mu = 400$ GeV,

$\tan \beta = 7$, $\quad M_3 = 950$ GeV,

$m(Q_3) = 1500$ GeV, $\quad m(\tilde{t}_R) = 0$ GeV,

$m(\tilde{b}_R) = 1000$ GeV, $\quad A_t = -642.8$ GeV.

---

[16] When one or none of the remaining jets are c-tagged we pick the c jets as the hardest jets with $p_T^{\text{jet}} < 50$ GeV.

Fig. 42 Invariant-mass distributions m_{bc} (*left*) and m_{lc} (*right*) with 20% c-tagging efficiency after b-tagging (*black with error bars*) and best fit (*full black line*) for LST2. The *shaded* (*green*) *area* is the SM background and the (*blue*) *histogram* the SUSY background

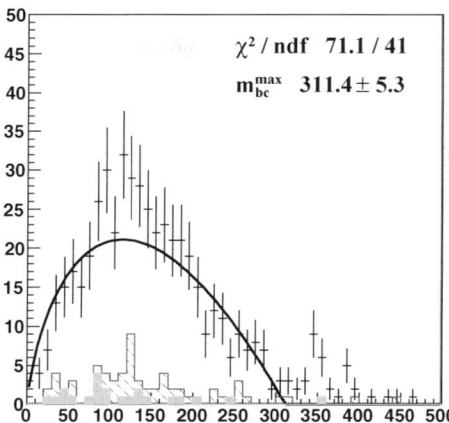
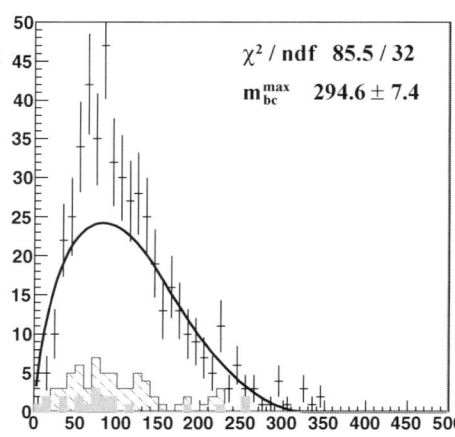

The resulting relevant masses are $m(\tilde{t}_1) = 137$ GeV, $m(\tilde{\chi}_1^\pm) = 111$ GeV, $m(\tilde{\chi}_1^0) = 58$ GeV. The $\tilde{t}_1$ decays with 100% branching ratio (BR) into $\tilde{\chi}_1^\pm b$, and $\tilde{\chi}_1^\pm$ decays with 100% BR into an off-shell W and $\tilde{\chi}_1^0$. The final state signature is therefore similar to the one for $t\bar{t}$ production: two b jets, E_T^{miss} and either 2 leptons (e, μ) (4.8% BR) or one lepton and two light jets (29% BR).

The signal cross section, calculated at NLO with the PROSPINO [360] program is 412 pb.

We analyze here the semileptonic channel, where only one of the two $\tilde{t}_1$ legs has a lepton in the final state. We apply the standard cuts for the search of the semileptonic top channel as applied in [3], but with softer requirements on the kinematics:

- One and only one isolated lepton (e, μ), $p_T^l > 20$ GeV.
- $E_T^{\text{miss}} > 20$ GeV.
- At least four jets $P_T(J_1, J_2) > 35$ GeV, $P_T(J_3, J_4) > 25$ GeV.
- Exactly two jets in the events must be tagged as b jets, they both must have $p_T > 20$ GeV. The standard ATLAS b-tagging efficiency of 60% for a rejection factor of 100 on light jets is assumed.

A total of 600k SUSY events were generated using HER-WIG 6.5 [311, 375], 1.2 M $t\bar{t}$ events using PYTHIA 6.2 [376]. The only additional background considered for this exploratory study was the associated production of a W boson with two b jets and two non-b jets, with the W decaying into e or μ. This is the dominant background for top searches at the LHC. For this process, we generated 60k events using Alpgen [312]. The number of events generated corresponds to ~ 1.8 fb^{-1}. The generated events are then passed through ATLFAST, a parametrized simulation of the AT-LAS detector [49].

After the selection cuts the efficiency for the $t\bar{t}$ background is 3.3%,[17] for $Wbbjj$ 3.1%, and for the signal 0.47%, yielding a background which is ~ 15 times higher than the signal.

An improvement of the signal/background ratio can be obtained using the minimum invariant mass of all the non-b jets with $p_T > 25$ GeV. This distribution peaks near the value of the W mass for the top background, whereas the invariant mass for the signal should be smaller than 54 GeV, which is the mass difference between the $\tilde{\chi}^\pm$ and the $\tilde{\chi}_1^0$. Requiring $m(jj) < 60$ GeV improves the signal/background ratio to 1/10, with a loss of a bit more than half the signal. We show in the left plot of Fig. 43 after this cut the distributions for the variable $m(bjj)_{\text{min}}$, i.e. the invariant mass for the combination a b-tagged jet and the two non-b jets yielding the minimum invariant mass. If the selected jets are from the decay of the stop, this invariant mass should have an end point at ~ 79 GeV, whereas the corresponding end-point should be at 175 GeV for the top background. The presence of the stop signal is therefore visible as a shoulder in the distribution as compared to the pure top contribution. A significant contribution from $Wbbjj$ is present, without a particular structure. Likewise, the variable $m(bl)_{\text{min}}$ has an end point at ~ 66 GeV for the signal and at 175 GeV for the top background, as shown in Fig. 43, and the same shoulder structure is observable. We need therefore to predict precisely the shape of the distributions for the top background in order to subtract it from the experimental distributions and extract the signal distributions.

The top background distributions can be estimated from the data themselves by exploiting the fact that we select events where one of the W from the top decays into two jets and the other decays into lepton plus neutrino. One can therefore select two pure top samples, with minimal contribution from non-top events by applying separately hard cuts on each of the two legs.

---

[17]The emission of additional hard jets at higher orders in the QCD interaction can increase the probability that the $t\bar{t}$ events satisfy the requirement of four jets. The cut efficiency is observed to increase by about 20% if MCNLO is used to generate the $t\bar{t}$ background. We do not expect such an effect to change the conclusions of the present analysis, but future studies should take it into account.

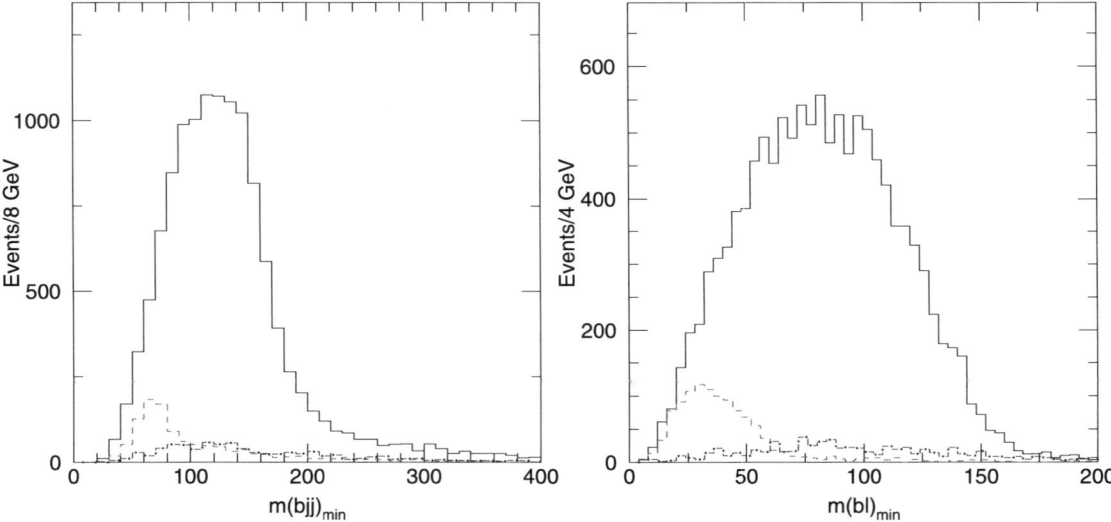

Fig. 43 Study of the stop quark signal in the minimum bjj invariant-mass distribution (*left*) and in the minimum bl invariant mass (*right*). The signal is the *dashed* (*red*) *line*, the top background the *solid* (*black*) *line*, and the Wbb background the *dot-dashed* (*blue*) *line*

- Top sample 1: the best reconstructed $bl\nu$ invariant mass is within 15 GeV of 175 GeV, and $(m_{\ell b})_{\min} > 60$ GeV in order to minimize the contribution from the stop signal. The neutrino longitudinal momentum is calculated by applying the W mass constraint.
- Top sample 2: the best reconstructed bjj mass is within 10 GeV of 175 GeV.

We assume here that we shall be able to predict the Wbb background through a combination of Monte Carlo and the study of Zbb production in the data, and we subtract this background both from the observed distributions and from the Top samples. More work is required to assess the uncertainty on this subtraction. Given the fact that this background is smaller than the signal, and it has a significantly different kinematic distribution, we expect that a 10–20% uncertainty on it will not affect the conclusions of the present analysis.

For Top sample 1, the top selection is performed by applying severe cuts on the lepton leg, it can therefore be expected that the minimum bjj invariant-mass distribution, which is built from jets from the decay of the hadronic side be essentially unaffected by the top selection cuts. This has indeed be verified to be the case [170]. The $m(bjj)$ distribution from Top sample 1 is then normalized to the observed distribution in the high mass region, where no signal is expected, and subtracted from it. A similar procedure is followed for the $m(bl)$ distribution: the top background is estimated using Top sample 2, normalized to the observed distribution in the high mass region, and subtracted from it. The results are shown in Fig. 44, with superimposed the corresponding distributions for the signal. As discussed above, we have subtracted the Wbb background from the observed distributions.

For both variables the true and measured distributions for the signal are compatible, showing the goodness of the background subtraction technique, and the expected kinematic structure is observable, even with the very small statistics generated for this analysis, corresponding to little more than one month of data taking at the initial luminosity of 10^{33} cm^{-1} s^{-1}.

Further work, outside the scope of this initial exploration, is needed on the evaluation of the masses of the involved sparticles through kinematic studies of the selected sample.

A preliminary detailed analysis of a SUSY model with a stop squark lighter than the top quark decaying into a chargino and a b jet was performed. It was shown that for this specific model after simple kinematic cuts a signal/background ratio of $\sim 1/10$ can be achieved. A new method, based on the selection of pure top samples to subtract the top background was demonstrated. Through this method it is possible to observe the kinematic structure of the stop decays, and thence to extract a measurement of the model parameters. This analysis can yield a clear signal for physics beyond the SM for just 1–2 fb^{-1}, and is therefore an excellent candidate for early discovery at the LHC.

3.12.4 Stop decay into right handed sneutrino LSP

Right handed neutrinos offer the possibility to accommodate neutrino masses. In supersymmetric models this implies the existence of right handed sneutrinos. Right handed sneutrinos are expected to be as light as other supersymmetric particles [377, 378] if the neutrinos are either Dirac fermions or if the lepton number breaking scale is at (or below) the SUSY breaking scale, assumed to be around the electroweak scale. Depending on the mechanism of SUSY breaking, the

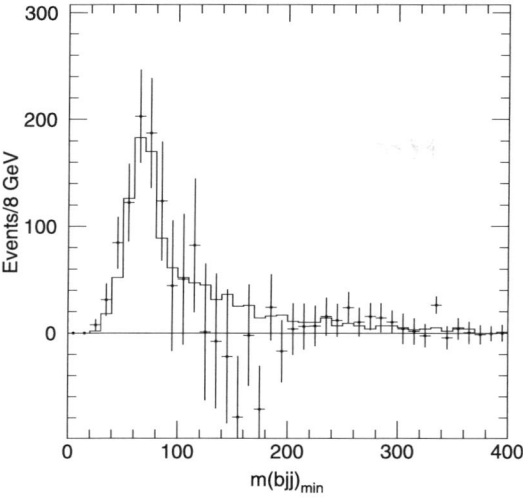
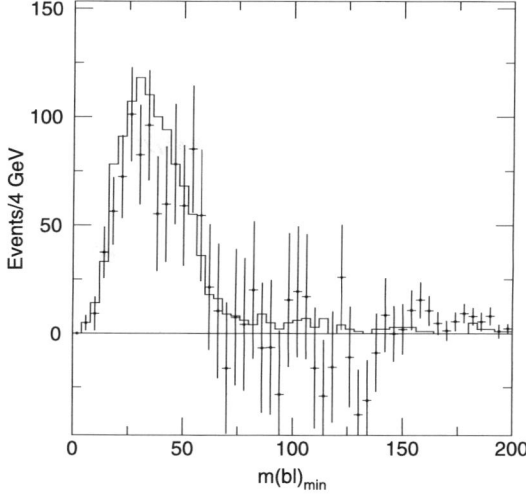

Fig. 44 *Left*: distribution of the minimum bjj invariant mass after the subtraction procedure (points with errors) superimposed to the original signal distribution (*full line*). *Right*: distribution of the minimum bl invariant mass after the subtraction procedure (points with errors) superimposed to the original signal distribution (*full line*)

lightest right handed sneutrino $\tilde{N}_R$ may be the lightest supersymmetric particle (LSP). We consider in the following such a scenario focusing on the case where the right handed stop is the next to lightest SUSY particle assuming R-parity conservation. Details on the model and other scenarios can be found in [377, 378].

As the right handed neutrino has a mass around 100 GeV, the neutrino Yukawa couplings Y_N must be very small to accommodate neutrino data: $Y_N \sim 10^{-6}$ ($Y_N \sim 10^{-12}$) in the case of Majorana neutrinos (Dirac neutrinos). This has as immediate consequence that if the SUSY breaking sneutrino trilinear "A-term" is also proportional to Y_N, the left handed and right handed sneutrinos hardly mix independent of neutrino physics because the left–right mixing term is proportional to Y_N. Decays into $\tilde{N}_R$ will give tiny decay widths as Y_N is the only coupling of $\tilde{N}_R$. For this reason, all decays of supersymmetric particles are as in the usual MSSM, but for the NLSP whose life-time can be long since it can only decay into the $\tilde{N}_R$. In the case of a stop NLSP the dominant decay mode is $\tilde{t}_1 \to b\ell^+ \tilde{N}_R$, followed by CKM suppressed ones into s and d quarks. In the limit where mixing effects for stops and charginos are neglected the corresponding matrix element squared in the rest frame of the stop reads as:

$$|T_{fi}|^2 \sim \frac{4|Y_t|^2|Y_N|^2 M_{\tilde{t}_R}^2 E_b E_l}{((p_{\tilde{t}_R} - k_b)^2 - M_{\tilde{H}}^2)^2} \frac{(1 + \cos\theta_{b\ell})}{2}, \quad (105)$$

where we have assumed that the right handed stop $\tilde{t}_R$ is the lightest stop and $\tilde{H}$ is the higgsino, E_b (E_ℓ) is the energy of the b-quark (lepton), $\theta_{b\ell}$ is the angle between the fermions. The complete formula can be found in [378]. The last factor in (105) implies that the b-quark and the lepton have a tendency to go in the same direction.

In the following we summarize the results of a Monte Carlo study at the parton level [378] using PYTHIA 6.327 [379]. We have taken $M_{\tilde{t}_R} = 225$ GeV, $M_{\tilde{N}_R} = 100$ GeV, $M_{\tilde{H}} = 250$ GeV and $Y_N = 4 \times 10^{-6}$ resulting in a mean decay length of 10 mm. Note that the stop will hadronize before decaying. However, we have neglected the related effects in this study. We have only considered direct stop pair production, and neglected stops from cascade decays, e.g. $\tilde{g} \to t\tilde{t}_R$. The signal is $pp(\bar{p}) \to \tilde{t}_R \tilde{t}_R^* \to b\ell^+ \bar{b}\ell^- + E_T^{\text{miss}}$. The dominant physics background is top quark pair production: $pp(\bar{p}) \to t\bar{t} \to bW^+ \bar{b}W^- \to b\ell^+ \bar{b}\ell^- + E_T^{\text{miss}}$, where the missing energy is due to neutrinos in the final state. We have imposed the following "Level 1" cuts: (i) fermion rapidities: $|\eta_\ell| < 2.5$, $|\eta_b| < 2.5$, (ii) p_T cuts $p_{T\ell} > 20$ GeV, $p_{Tb} > 10$ GeV and (iii) isolation cut $R_{b\ell} \equiv (\phi_b - \phi_\ell)^2 + (\eta_b - \eta_\ell)^2 > 0.4$.

Figure 45 shows various distributions for stop and top decays. Figure 45a depicts the resulting transverse displacement after including the boost of the stop. If it decays before exiting the tracking subsystem, a displaced vertex may be reconstructed through the stop decay products' 3-momenta meeting away from the primary interaction point. On each side, the b-quark itself leads to an additional displaced vertex, and its 3-momentum vector can be reconstructed from its decay products. In combination with the 3-momentum of the lepton, the stop displaced vertex can be determined. In order to reveal the displaced vertex, one must require either the b-quark or the charged lepton 3-momentum vector to miss the primary vertex. Since a pair of stops is produced, we would expect to discern two displaced vertices in the event (not counting the displaced vertices due to the b-quarks). Such an event with two displaced vertices, from each of which originates a high p_T ℓ and b-quark might

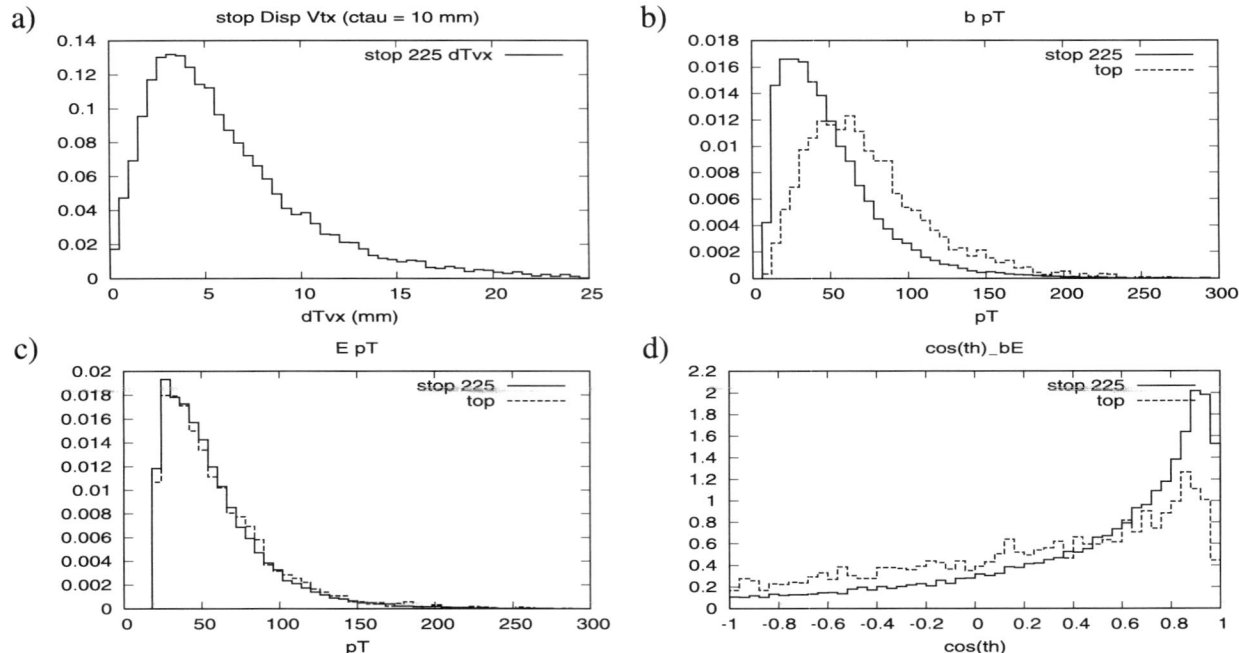

Fig. 45 Distributions of stop and top decays: (**a**) the transverse displacement of the stop (in mm), (**b**) p_T of the b-quark, (**c**) p_T of the charged lepton and (**d**) $\cos\theta_{b\ell}$, the angle between the 3-momenta $\mathbf{k}_b$ and $\mathbf{k}_\ell$

prove to be the main distinguishing characteristics of such a scenario. A cut on the displaced vertex will be very effective to separate stop events from the top background provided one can efficiently explore such cuts. We anticipate that NLSP stop searches may turn out to be physics-background free in such a case.

If the stop displaced vertex cannot be efficiently resolved, one will have to resort to more conventional analysis methods. In the remainder we explore various kinematical distributions for both the signal ($\tilde{t}_R$ pair production) and the physics background (t pair production), obtained after imposing the level 1 cuts given above. Figures 45b and c depict the p_T spectra of the produced fermions. The p_T of the b-quark from the 225 GeV stop peaks at a lower value compared to the top quark background, and therefore accepting them at high efficiency for $p_T \lesssim 40$ GeV will be very helpful in maximizing the signal acceptance. The signal and background shapes are quite similar and no simple set of p_T cuts can be made in order to significantly separate signal from background.

Figure 45d depicts the distribution of $\cos\theta_{b\ell}$, the angle between the 3-momenta $\mathbf{k}_b$ and $\mathbf{k}_\ell$, for both the signal and background. It is important to appreciate that, by default, PYTHIA generates stop decays into the three body final state according only to phase space, ignoring the angular dependence of the decay matrix element. We have reweighted PYTHIA events to include the correct angular dependence in the decay matrix element. Consistent with the expectation from (105), we see for the signal that the distribution peaks for the b-quark and charged lepton 3-momenta aligned, unlike the background. It is unfortunate that the isolation level 1 cut on the leptons removes more signal events than background events. Relaxing this constraint as much as is practical would help in this regard. Additional work will be necessary to include also the effect of spin correlations in top production and top decays so that information from the quantities $\mathbf{k}_b \cdot \mathbf{k}_{\bar{b}}$ and $\mathbf{k}_{\ell^+} \cdot \mathbf{k}_{\ell^-}$ can be exploited.

Assuming efficiencies of $\epsilon_b = 0.5$ and $\epsilon_l = 0.9$ for b-quark and lepton identification, respectively, it has been shown in [378] that stops with masses up to 500 GeV can be detected at the 5σ level for an integrated luminosity of 10 fb^{-1} if $\ell = e, \mu$ even if the displaced vertex signature is not used. Clearly, the situation will be worse in the case of $\ell = \tau$. Provided one can exploit the displaced vertex information, we expect a considerable improvement as we could not identify any physics background. Further studies are planned to investigate the questions we have touched upon here.

3.13 SUSY Searches at $\sqrt{s} = 14$ TeV with CMS

This section summarizes the recent results on SUSY searches reported in [61]. In the context of this work we refer to the generalized classification of models of new physics according to how they affect flavor physics:

- *CMFV:* Constrained minimal flavor violation [380] models: in these models the only source of quark flavor vio-

lation is the CKM matrix. Examples include minimal supergravity models with low or moderate $\tan\beta$, and models with a universal large extra dimension.
— *MFV:* Minimal flavor violation [199] models: a set of CMFV models with some new relevant operators that contribute to flavor transitions. Examples include SUSY models with large $\tan\beta$.
— *NMFV:* Next-to-minimal flavor violation [381] models: they involve third generation quarks and help to solve the flavor problems that appear in frameworks such as little Higgs, topcolor, and RS models.
— *GFV:* General Flavor Violation [382] models, which provide new sources of flavor violation. These include most of the MSSM parameter space, and almost any BSM model before flavor constraints are considered.

A useful discussion on these models can be found in [383]. The SUSY searches that are summarized here fall in the category of MFV (mSUGRA specifically) and all results are obtained with the detailed Geant-4 based CMS simulation. In the context of this workshop and in collaboration with the theory community we try to also move towards interpretation within NMVF models. Note that since the squarks and sleptons can have significant flavor-changing vertices and be complex, the connection to collider physics can be subtle indeed, the main implication being that the superpartners cannot be too heavy and that larger $\tan\beta$ is favored—with no direct signature in general. For interpretations of recent Tevatron and B-factory results the interested reader can refer for example to [384–387].

The SUSY search path has been described in the past years as a successive approximation of serial steps that move from inclusive to more exclusive measurements as follows:

— Discovery: using canonical inclusive searches.
— Characterization: putting together a picture given the channels that show excess and ratios of the observed objects (e.g. multi-leptons, photons (GMSB), ratio of same sign leptons to opposite ones, ratios of positive pairs to negative, departure from lepton universality, third generation excesses).
— Reconstruction: in canonical dark matter LSP SUSY the final state contains two neutralinos hence there is no direct mass peak due to the missing transverse energy in the event. The kinematics of the intermediate decays provide however a multitude of end points and edges that might provide mass differences and help orient towards the right mass hierarchy.
— Measurement of the underlying theory: the classical SUSY solving strategy involves more mass combinations, more decay chains, mass peaks and once the LSP mass is known the determination of the mass hierarchies, particles' spins, and eventually the model parameters. An outstanding question remains as to how many simple measurements do we need to "nail" the theory? Remember that we did not need to measure all the SM particles and their properties in order to measure the SM.

In the past three years the "inclusive" and "exclusive" *modus quaestio questio* have been approached in coincidence and in many works that range in exploitation strategy from statistical methods to fully on-shell description of unknown models and inclusion of cosmological considerations such as in [251, 388–391], to mention but a few.

It is rather safe to claim that the program of discovery and characterization will be (much) more convoluted than the one described in the serial steps above. Realistic studies of kinematic edges across even the "simple" mSUGRA parameter space show that this is a difficult job and it will take a lot of work and wisdom to do it right. Endpoint analyses by definition involve particles which are very soft in some reference frame and non-trivial issues of acceptance need to be considered.

Some of most recent SUSY searches at CMS [61], proceed in the following channels:

— Canonical inclusive
  — Multijets + $\not{E}_T$
  — μ + jets + $\not{E}_T$
  — Same-sign dimuon + $\not{E}_T$
  — Opposite-sign same-flavor dielectron and dimuon + $\not{E}_T$
  — Opposite-sign same-flavor hadronic ditau + $\not{E}_T$
  — Trileptons at high m_0
— Higher reconstructed object inclusive
  — Z^0 + $\not{E}_T$
  — Hadronic top + $\not{E}_T$
  — $h(\to b\bar{b})$ + $\not{E}_T$
— Flavor-violating
  — Opposite-sign different-flavor $e\mu$ FV neutralino decays

The attempt is to have a signature based search strategy, with educated input from theory, which as model-independent as possible. The interpretation of the search results are given in the context and parameter space of mSUGRA but re-interpreting them in different models is possible. All of the searches are including detector systematic uncertainties and a scan that provides the 5σ reach in the mSUGRA parameter space is derived for 1 and 10 fb^{-1} as shown in Fig. 46. The details of the analyses and individual search results can be found in [61].

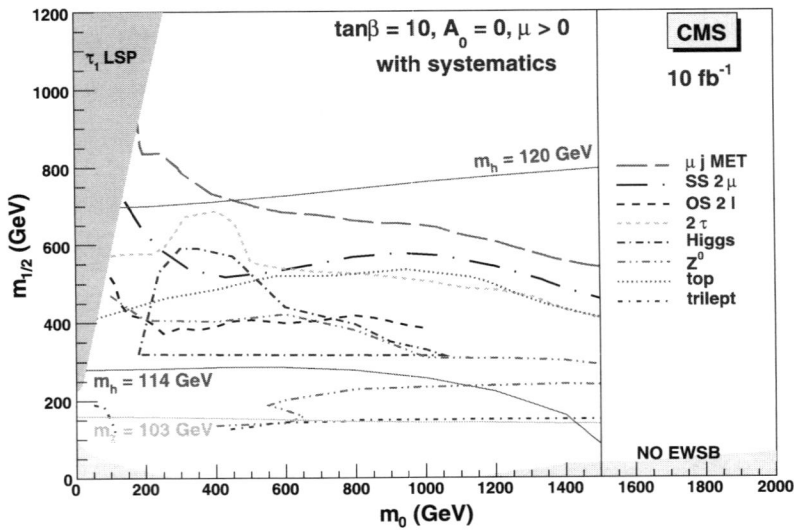
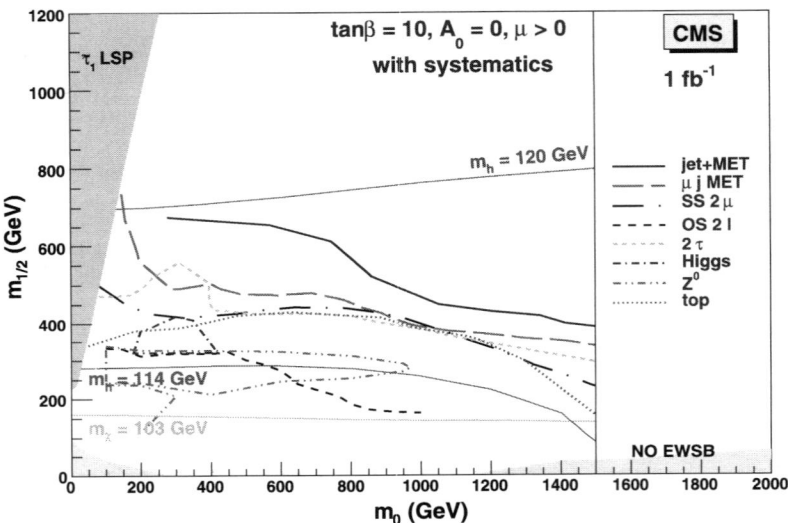

Fig. 46 5σ reach for 1 and 10 fb^{-1} at CMS in different channels

4 Non-supersymmetric standard model extensions[18]

4.1 Introduction

Although the SM has seemingly survived many stringent tests offered by both precision measurements and direct searches, it has a number of shortcomings. The most unpleasant one is the "instability" of the Higgs boson mass with respect to radiative corrections, known as the hierarchy problem. If the SM is assumed valid up to a high scale Λ of the order of the Planck mass, radiative corrections to M_h from top quark loops are of order $\delta M_h^2 \sim \Lambda^2$, i.e. much larger than M_h which is expected to be of the order of the electroweak scale. The requirement that M_h and δM_h are of the same order would imply a cut-off (and hence new physics at) $\Lambda \sim 1$–2 TeV. Some other aspects of the SM that make it unappealing as the ultimate theory of fundamental interactions (excluding gravity) are

- The lack of simplicity of the gauge structure,
- The unexpected hierarchy of fermion masses and quark mixings, and the large number of apparently free parameters necessary to describe them,
- The source of baryogenesis, which cannot be explained by the amount of CP violation present in the SM,
- The unknown mechanism behind the neutrino mass generation (neutrinos can have Dirac masses simply with the introduction of right handed fields, but present limits $m_\nu \sim 1$ eV require unnaturally small Yukawa couplings).

Therefore, the SM is believed to be the low energy limit of a more fundamental theory. Several arguments suggest

---

[18]Section coordinators: J.A. Aguilar Saavedra and G. Ünel.

that this theory may manifest itself at energies not much higher than the electroweak scale, and give support to the hope that LHC will provide signals of new physics beyond the SM.

This section deals with non-supersymmetric extensions of the SM. Among the most popular ones, we shall consider here the following ones:

1. Grand unified theories (GUTs). In these models the SM gauge group $SU(3)_c \times SU(2)_L \times U(1)_Y$ is embedded into a larger symmetry group, which is recovered at a high scale. They predict the existence of new fermions (e.g. $Q = -1/3$ singlets) and new gauge bosons (especially Z'), which may be within the reach of the LHC.
2. Little Higgs models. They address the hierarchy problem with the introduction of extra gauge symmetries and extra matter which stabilize the Higgs mass up to a higher scale $\Lambda \sim 10$ TeV. In particular, the top quark loop contribution to the Higgs mass is partially canceled with the introduction of a $Q = 2/3$ quark singlet T.
3. Theories with extra dimensions. The various extra dimensional models avoid the hierarchy problem by lowering the Planck scale in the higher dimensional theory, and some of them can explain the large hierarchies between fermion masses. The observable effect of the additional dimension is the appearance of "towers" of Kaluza–Klein (KK) excitations of fermions and bosons, with increasing masses. Depending on the model, the lightest modes can have a mass around the TeV scale and thus be produced at LHC.

It should be stressed that these SM extensions, sometimes labeled as "alternative theories" do not exclude supersymmetry (SUSY). In fact, SUSY in its minimal versions does not address some of the open questions of the SM. One example is the motivation behind the apparent gauge coupling unification. The renormalization group evolution of the coupling constants strongly suggests that they unify at a very high scale $M_{GUT} \sim 10^{15}$ GeV, and that the SM gauge group is a subgroup of a larger one, e.g. $SO(10)$, E_6 or other possibilities. Thus, SUSY can naturally coexist with GUTs. Another example of complementarity is SUSY + little Higgs models. If SUSY is broken at the TeV scale or below, it may give dangerous contributions to FCNC processes and electric dipole moments (EDMs). These contributions must be suppressed with some (well justified or not) assumption, like minimal supergravity (MSUGRA) with real parameters. These problems are alleviated if SUSY is broken at a higher scale and, up to that scale, the Higgs mass is stabilized by another mechanism, as happens in the little Higgs theories.

With the forthcoming LHC, theories beyond the SM will be tested directly via searches for new particles, and, indirectly, with measurements of the deviations from SM precision variables. Instead of studying the different SM extensions and their additional spectra separately, we follow a phenomenological/experimental approach. Thus, this section is organized according to the new particles that are expected to be produced. Section 4.2 reviews the searches for the new quarks and Sect. 4.3 for new heavy neutrinos. Studies for new gauge bosons are collected in Sects. 4.4 and 4.5, and in Sect. 4.6 some new scalar signals are presented. We do not include detailed information about the SM extensions discussed here, but refer to the original papers and dedicated reviews (see for instance [392–396]). The observation of these new particles would prove, or at least provide evidence for, the proposed theories. In this case, the identification of the underlying theory might be possible with the measurement of the couplings, production and decay modes of the new state(s). Alternatively, the non-observation of the predicted signals would disprove the models or impose lower bounds for their mass scales.

4.2 New quarks

At present, additional quarks are not required neither by experimental data nor by the consistency of the SM. But on the other hand they often appear in grand unified theories [63, 397], little Higgs models [394, 398, 399], Flavor Democracy [400] and models with extra dimensions [28, 396, 401]. Their existence is not experimentally excluded but their mixing, mainly with the lightest SM fermions, is rather constrained. They can lead to various indirect effects at low energies, and their presence could explain experimental deviations eventually found, for instance in CP asymmetries in B decays. They can also enhance flavor-changing processes, especially those involving the top quark. These issues have been dealt with in other articles of this volume. Here we are mainly concerned with their direct production and detection at LHC.

New quarks share the same electromagnetic and strong interactions of standard quarks, and thus they can be produced at LHC by $q\bar{q}$ annihilation and gluon fusion in the same way as the top quark, with a cross section which only depends on their mass, plotted in Fig. 47. Depending on their electroweak mixing with the SM fermions, they can be produced singly as well [402–404]. Their decay always takes place through electroweak interactions or interactions with scalars, and the specific decay modes available depend on the particular SM extension. Let us consider an example with N "standard" chiral generations (left handed doublets and right handed singlets under $SU(2)_L$), plus n_u up-type and n_d down-type singlets under this group.[19] While (left

---

[19]Anomaly cancellation requires that the number of lepton generations is also N. For $N > 3$ this implies additional neutrinos heavier than $M_Z/2$ to agree with the Z invisible width measurement at LEP. On the other hand, quark singlets can be introduced alone, since they do not contribute to anomalies [397].

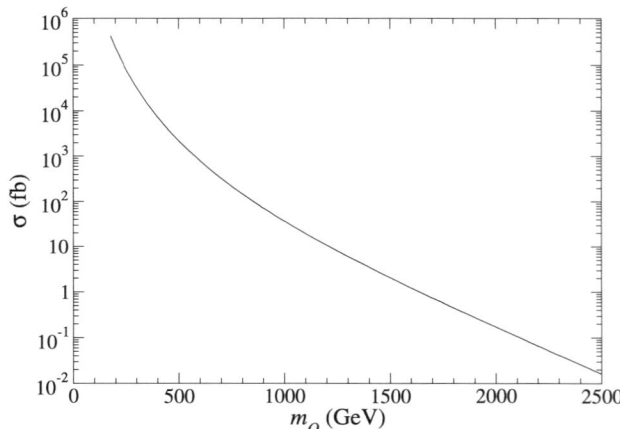

Fig. 47 Tree-level cross section for pair production of heavy quarks Q in pp collisions at $\sqrt{s} = 14$ TeV, $gg, q\bar{q} \to Q\bar{Q}$. CTEQ5L PDFs are used

handed) SU(2)$_L$ doublets couple to the $W^\pm$ and W^3 bosons, singlet fields do not. The Lagrangian in the weak eigenstate basis reads

$$\mathcal{L}_W = -\frac{g}{\sqrt{2}} [\bar{u}'_L \gamma^\mu d'_L] W^\dagger_\mu + \text{h.c.},$$
$$\mathcal{L}_Z = -\frac{g}{2c_W} [\bar{u}'_L \gamma^\mu u'_L - \bar{d}'_L \gamma^\mu d'_L - 2s_W^2 J_{\text{EM}}] Z_\mu, \quad (106)$$

where $(u', d')_L$ are the N doublets under SU(2)$_L$ and J_{EM} is the electromagnetic current which includes all (charged) quark fields. The number of mass eigenstates with charges 2/3 and −1/3 is $N_u \equiv N + n_u$, $N_d \equiv N + n_d$, respectively. The resulting weak interaction Lagrangian in the mass eigenstate basis is

$$\mathcal{L}_W = -\frac{g}{\sqrt{2}} [\bar{u}_L \gamma^\mu V d_L] W^\dagger_\mu + \text{h.c.},$$
$$\mathcal{L}_Z = -\frac{g}{2c_W} [\bar{u}_L \gamma^\mu X^u u_L - \bar{d}_L \gamma^\mu X^d d_L - 2s_W^2 J_{\text{EM}}] Z_\mu, \quad (107)$$

where $u_{L,R}$, $d_{L,R}$ are column vectors of dimensions N_u, N_d, and J_{EM} is the electromagnetic current (including all mass eigenstates). The $N_u \times N_d$ matrix V (not necessarily square) is the generalization of the 3×3 CKM matrix. The matrices $X^u = VV^\dagger$, $X^d = V^\dagger V$ have dimensions $N_u \times N_u$ and $N_d \times N_d$, respectively. In case that $n_u > 0$, the up-type mass eigenstates are mixtures of weak eigenstates with different isospin, and thus the matrix X^u is not necessarily diagonal. In other words, V is not unitary (but its 3×3 submatrix involving SM quarks is almost unitary), what prevents the GIM mechanism from fully operating. Analogous statements hold for the down sector. Therefore, models with quark singlets can have tree-level FCNC couplings to the Z boson. These couplings are suppressed by the mass of the new mass eigenstates, e.g. $X_{tc} \sim m_t m_c / m_T^2$ (with T a new

charge-2/3 quark), what forbids dangerous FCNCs in the down sector but allows for observable effects in top physics.

As within the SM, its extensions with extra quarks typically have one Higgs doublet which breaks the electroweak symmetry and originates the fermion masses. The surviving scalar field h couples to the chiral fields (through Yukawa couplings) but not to the weak eigenstate isosinglets. In the mass eigenstate basis, the scalar-quark interactions read

$$\mathcal{L}_h = -\frac{g}{2M_W} [\bar{u}_R \mathcal{M}^u X^u u_L + \bar{d}_R \mathcal{M}^d X^d d_L] h + \text{h.c.}, \quad (108)$$

with $\mathcal{M}^u$, $\mathcal{M}^d$ the diagonal mass matrices, of dimensions $N_u \times N_u$ and $N_d \times N_d$, respectively. SM extensions with extra quarks usually introduce further scalar fields, e.g. in E_6 additional scalar singlets are present, but with VEVs typically much higher than the mass scale of the new quarks and small mixing with h. Also, in supersymmetric versions of E_6 there are two Higgs doublets, in which case the generalization of (108) involves two scalar fields and the ratio of their VEVs $\tan \beta$. However, the main phenomenological features of these models can be described with the minimal scalar sector and the Lagrangian in (108). (Of course, this does not preclude that with appropriate but in principle less natural choices of parameters one can build models with a completely different behavior.) In particular, from (108) it follows that FCNC interactions with scalars have the same strength as the ones mediated by the Z boson, up to mass factors. Note also that (108) does not contradict the fact that new heavy mass eigenstates, which are mainly SU(2)$_L$ singlets, have small Yukawa couplings. For example, with an extra $Q = 2/3$ singlet the Yukawa coupling of the new mass eigenstate T is proportional to $m_T X_{TT} \simeq m_T |V_{Tb}|^2 \sim m_t / m_T^2$ (see also Sect. 4.2.1 below).

More general extensions of the SM quark sector include right handed fields transforming non-trivially under SU(2)$_L$. The simplest of such possibilities is the presence of additional isodoublets $(T, B)_{L,R}$. Their interactions are described with the right handed analogous of the terms in (107) and a generalization of (108). From the point of view of collider phenomenology, their production and decay takes place through the same channels as fourth generation or singlet quarks (with additional gauge bosons there would be additional modes). However, the constraints from low energy processes are much more stringent, since mixing with a heavy isodoublet $(T, B)_{L,R}$ can induce right handed charged currents among the known quarks, which are absent in the SM. An example of this kind is a $Wt_R b_R$ interaction, which would give a large contribution to the radiative decay $b \to s\gamma$ (see Sect. 2.2.1.1).

A heavy quark Q of either charge can decay to a lighter quark q' via charged currents, or to a lighter quark q of the same charge via FCNC couplings. The partial widths for these decays are [405]

$$\Gamma(Q \to W^+ q')$$
$$= \frac{\alpha}{16 s_W^2} |V_{Qq'}|^2 \frac{m_Q}{M_W^2} \lambda(m_Q, m_{q'}, M_W)^{1/2}$$
$$\times \left[1 + \frac{M_W^2}{m_Q^2} - 2\frac{m_{q'}^2}{m_Q^2} - 2\frac{M_W^4}{m_Q^4} + \frac{m_{q'}^4}{m_Q^4} + \frac{M_W^2 m_{q'}^2}{m_Q^4} \right],$$

$$\Gamma(Q \to Zq)$$
$$= \frac{\alpha}{32 s_W^2 c_W^2} |X_{Qq}|^2 \frac{m_Q}{M_Z^2} \lambda(m_Q, m_q, M_Z)^{1/2} \quad (109)$$
$$\times \left[1 + \frac{M_Z^2}{m_Q^2} - 2\frac{m_q^2}{m_Q^2} - 2\frac{M_Z^4}{m_Q^4} + \frac{m_q^4}{m_Q^4} + \frac{M_Z^2 m_q^2}{m_Q^4} \right],$$

$$\Gamma(Q \to hq)$$
$$= \frac{\alpha}{32 s_W^2} |X_{Qq}|^2 \frac{m_Q}{M_W^2} \lambda(m_Q, m_q, M_h)^{1/2}$$
$$\times \left[1 + 6\frac{m_q^2}{m_Q^2} - \frac{M_h^2}{m_Q^2} + \frac{m_q^4}{m_Q^4} - \frac{m_q^2 M_h^2}{m_Q^4} \right],$$

with

$$\lambda(m_Q, m, M) \equiv (m_Q^4 + m^4 + M^4 - 2m_Q^2 m^2 - 2m_Q^2 M^2 - 2m^2 M^2) \quad (110)$$

a kinematical function. (The superscripts u, d in the FCNC couplings X_{Qq} may be dropped when they are clear from the context.) Since QCD and electroweak production processes are the same for the fourth generation and exotic quarks, their decays provide the way to distinguish them. For quark singlets the neutral current decays $Q \to Zq$ are possible, and kinematically allowed (see below). Moreover, for a doublet of SM quarks (q, q') of the same generation one has $\Gamma(Q \to Zq) \simeq 1/2 \Gamma(Q \to Wq')$, for $m_Q \gg m_q, m_{q'}, M_Z, M_W$. Depending on the Higgs mass, decays $Q \to hq$ may be kinematically allowed as well, with a partial width $\Gamma(Q \to hq) \simeq 1/2 \Gamma(Q \to Wq')$ for m_Q much larger than the other masses involved. Both FCNC decays, absent for a fourth generation of heavy quarks,[20] provide clean final states in which new quark singlets could be discovered, in addition to the charged-current decays present in all cases. If the new quarks mix with the SM sector through right handed interactions with the SM gauge and Higgs bosons, the decays are the same as in (109) but replacing $V_{Qq'}$ and X_{Qq} by their right handed analogues. If the new quarks are not too heavy, the chirality of their interactions can be determined by measuring angular or energy distributions of the decay products. For instance, in a decay $T \to W^+ b \to \ell \nu b$ the charged lepton angular distribution in W rest frame (or its energy distribution in T rest frame) can be used to probe the WTb interaction (see the discussion after (112) below, and also Sect. 2.2.1.2).

Searches at Tevatron have placed the 95% CL limits $m_B \geq 128$ GeV [344] (in charged-current decays, assuming 100% branching ratio), $m_{b'} \geq 199$ GeV [406] (assuming Br$(b' \to Zb) = 1$), where b' is a charge $-1/3$ quark. If a priori assumptions on b' decays are not made, limits can be found on the branching ratios of these two channels [407–409]. In particular, it is found that for b' quarks with masses ~ 100 GeV near the LEP kinematical limit there are some windows in parameter space where b' could have escaped discovery. For a charge 2/3 quark T, the present Tevatron bound is $m_T \geq 258$ GeV [410] in charged-current decays $T \to W^+ b$, very close to the kinematical limit $m_t + M_Z$ where decays $T \to Zt$ are kinematically possible. The prospects for LHC are reviewed in the following.

4.2.1 Singlets: charge 2/3

A new up-type singlet T is expected to couple preferably to the third generation, due to the large mass of the top quark. The CKM matrix element V_{Tb} is expected to be of order m_t/m_T, although for T masses at the TeV scale or below the exact relation $V_{Tb} = m_t/m_T$ enters into conflict with latest precision electroweak data. In particular, the most stringent constraint comes from the T parameter [27]. The most recent values [344] $T = -0.13 \pm 0.11$ (for U arbitrary), $T = -0.03 \pm 0.09$ (setting $U = 0$) imply the 95% CL bounds $T \leq 0.05$, $T \leq 0.117$, respectively. The resulting limits on $|V_{Tb}|$ are plotted in Fig. 48, including for completeness the limit from R_b (plus other correlated observables like R_c, the FB asymmetries and coupling parameters) and the bound on m_T from direct searches. The mixing values obtained from the relation $V_{Tb} = \lambda m_t/m_T$ are

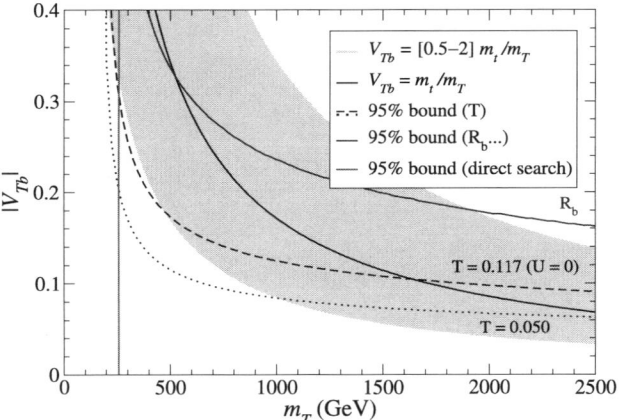

Fig. 48 95% CL bounds on $|V_{Tb}|$ from the T parameter and from R_b, and values derived from the relation $V_{Tb} = m_t/m_T$

---
[20]For 4th generation quarks neutral decays can take place radiatively, and can have sizable branching ratios if tree-level charged-current decays are very suppressed, see Sect. 4.2.4.1.

also displayed, for $\lambda = 1$ (continuous line) and $\lambda = 0.5$–2 (gray band). In this class of models the new contributions to the parameter U are very small, so it is sensible to use the less restrictive bound $T \leq 0.117$. Even in this case, mixing angles $V_{Tb} = m_t/m_T$ seem too large for T lighter than 1.7 TeV. Of course, the importance of the bound $T \leq 0.117$, and indirect bounds in general, must be neither overemphasized nor neglected. Additional new particles present in these models also contribute to T and can cancel the contribution from the new quark. But this requires fine-tuning for lower T masses and/or larger V_{Tb} mixings.

The main decays of the new quark are $T \to W^+ b$, $T \to Zt$, $T \to ht$, with partial widths given by (109). Their characteristic features are

(i) $T \to W^+ b$: The decays $W \to \ell^+ \nu$, $\ell = e, \mu$ originate very energetic charged leptons, not only due to the large T mass but also to spin effects [411]: for large m_T the charged leptons are emitted more towards the W flight direction.

(ii) $T \to Zt$: The leptonic decays $Z \to \ell^+ \ell^-$ produce a very clean final state, although with a small branching ratio.

(iii) $T \to ht$: For a light Higgs, its decay $h \to b\bar{b}$ and the decay of the top quark give a final state with three b quarks, which can be tagged to reduce backgrounds. They have an additional interest as they can produce Higgs bosons with a large cross section [405, 412].

4.2.1.1 Discovery potential In T pair production the largest m_T reach is provided by the mode $T\bar{T} \to W^+ b W^- \bar{b}$ and subsequent semileptonic decay of the $W^+ W^-$ pair, plus additional contributions from other decay modes giving the same signature plus additional jets or missing energy [73, 411]

$$T\bar{T} \to W^+ b W^- \bar{b} \to \ell^+ \nu b \bar{q} q' \bar{b},$$
$$T\bar{T} \to W^+ b h \bar{t} / h t W^- \bar{b} \to W^+ b W^- \bar{b} h$$
$$\to \ell^+ \nu b \bar{q} q' \bar{b} b\bar{b}/c\bar{c},$$
$$T\bar{T} \to W^+ b Z \bar{t} / Z t W^- \bar{b}$$
$$\to W^+ b W^- \bar{b} Z \to \ell^+ \nu b \bar{q} q' \bar{b} q'' \bar{q}''/\nu\bar{\nu}. \quad (111)$$

These signals are characterized by one energetic charged lepton, two b jets and at least two additional jets. Their main backgrounds are top pair and single-top production and $W/Zb\bar{b}$ plus jets. Charged leptons originating from $T \to Wb \to \ell\nu b$ decays are much more energetic than those from $t \to Wb \to \ell\nu b$, as has been stressed above. The charged lepton energy distribution in T (t) rest frame reads

$$\frac{1}{\Gamma}\frac{d\Gamma}{dE_\ell} = \frac{1}{(E_\ell^{\max} - E_\ell^{\min})^3}[3(E_\ell - E_\ell^{\min})^2 F_R$$
$$+ 3(E_\ell^{\max} - E_\ell)^2 F_L$$
$$+ 6(E_\ell^{\max} - E_\ell)(E_\ell - E_\ell^{\min}) F_0], \quad (112)$$

with F_i the W helicity fractions (see Sect. 2.2.1.2), which satisfy $F_L + F_R + F_0 = 1$. For the top quark they are $F_0 = 0.703$, $F_L = 0.297$, $F_R \simeq 0$, while for T with a mass of 1 TeV they are $F_0 = 0.997$, $F_L = 0.013$, $F_R \simeq 0$. It must be pointed out that for large m_T, $F_0 \simeq 1$ even when right handed WTb interactions are included; thus, the chirality of this vertex cannot be determined from these observables. The maximum and minimum energies depend on the mass of the decaying fermion, and are $E_\ell^{\min} = 18.5$ GeV, $E_\ell^{\max} = 87.4$ GeV for t, and $E_\ell^{\min} = 3.2$ GeV, $E_\ell^{\max} = 500$ GeV for T (with $m_T = 1$ TeV). The resulting energy distributions are presented in Fig. 49 (left) for the same T mass of 1 TeV. The larger mean energy in the rest frame of the parent quark is reflected in a larger transverse momentum p_T^{lep} in the laboratory frame, as can be observed in Fig. 49 (right). For the second and third decay channels in (111), denoted by (h) and (Z) respectively, the tail of the distribution is less pronounced. This is so because the charged lepton originates from $T \to Wb \to \ell\nu b$ only half of the times, and the rest comes from $t \to Wb \to \ell\nu b$ and is less energetic.

Background is suppressed by requiring large transverse momenta of the charged lepton and the jets, and with the heavy quark mass reconstruction. The reconstructed masses of the heavy quarks decaying hadronically (m_T^{had}) and semi-leptonically (m_T^{lep}) are shown in Fig. 50. For the leading decay mode $T\bar{T} \to W^+ b W^- \bar{b}$ these distributions have a peak around the true m_T value, taken here as 1 TeV, but for the additional signal contributions the events spread over a wide range. Thus, kinematical cuts on p_T^{lep}, m_T^{had} and m_T^{lep} considerably reduce the extra signal contributions.

The estimated 5σ discovery limits for 300 fb^{-1} can be summarized in Fig. 51. They also include the results from Tj (plus $\bar{T}j$) production, where the decay $T \to W^+ b$ (or $\bar{T} \to W^- \bar{b}$) also gives the highest sensitivity for large T masses [413]. The m_T reach in $T\bar{T}$ production is independent of V_{Tb}, but the Tj cross section scales with $|V_{Tb}|^2$, and thus the sensitivity of the latter process depends on V_{Tb}. T masses on the left of the vertical line can be seen with 5σ in $T\bar{T}$ production. Values of m_T and V_{Tb} over the solid curve can be seen in Tj production. The latter discovery limits have been obtained by rescaling the results for $m_T = 1$ TeV in [413, 414]. The 95% CL bounds from the T parameter (for $U = 0$ and U arbitrary) are represented by the dashed and dotted lines, respectively. Then, the yellow area (light grey in print) represents the parameter region where the new quark cannot be discovered with 5σ, and the orange triangle (dark gray) the parameters for which it can be discovered in single but not in pair production.

Several remarks are in order. The limits shown for $T\bar{T}$ and Tj only include the channel $T \to W^+ b$ (with additional

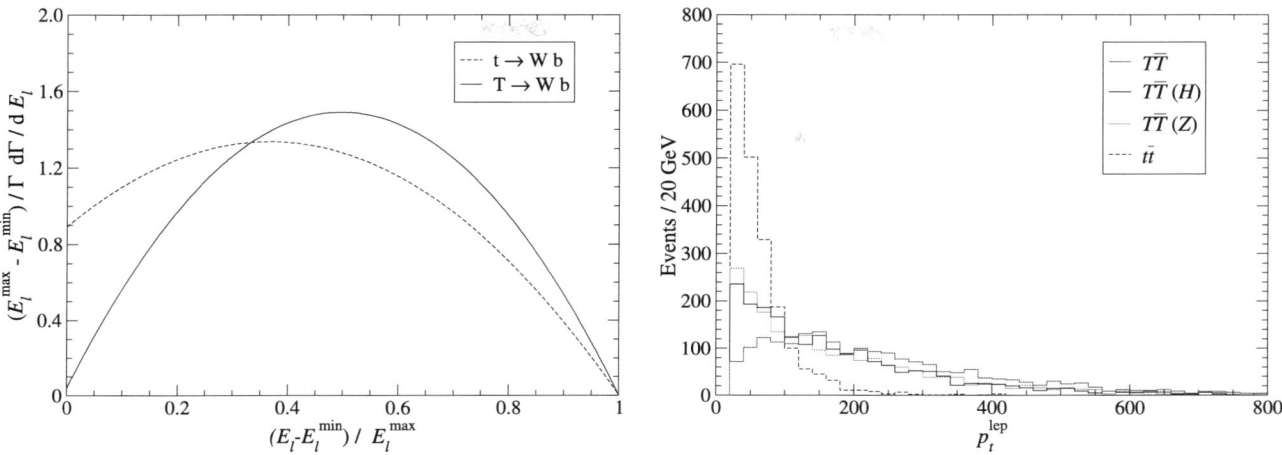

Fig. 49 *Left*: Normalized energy distributions of the charged lepton from t and T semileptonic decays, in t (T) rest frame, taking $m_T = 1$ TeV. *Right*: the resulting transverse momentum distribution in laboratory frame for the processes in (111) (*right*)

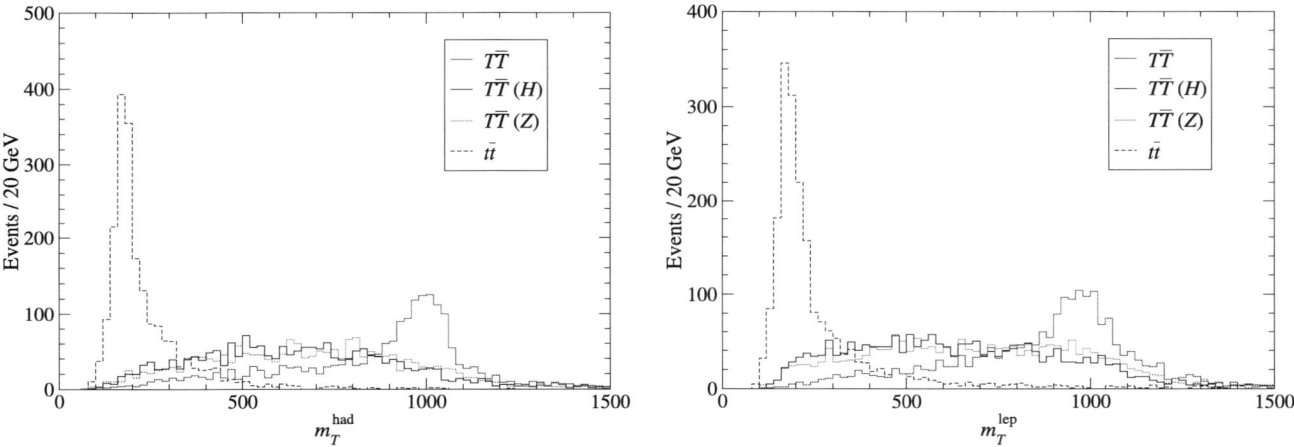

Fig. 50 Reconstructed masses of the heavy quarks decaying hadronically (*left*) and semileptonically (*right*), for the processes in (111) with $m_T = 1$ TeV, and their main background $t\bar{t}$

signal contributions giving the same final state in the former case). In both analyses the evaluation of backgrounds, e.g. $t\bar{t}$, does not include higher-order processes with extra hard jet radiation: $t\bar{t}j$, $t\bar{t}2j$, etc. These higher-order $t\bar{t}nj$ contributions may be important in the large transverse momenta region where the new quark signals are searched. Systematic uncertainties in the background are not included either, and they lower the significance with respect to the values presented here. On the other hand, additional T decay channels can be included and the event selection could be refined, e.g. by a probabilistic method, so that the limits displayed in Fig. 51 are not expected to be significantly degraded when all of these improvements are made in the analysis.

4.2.1.2 Higgs discovery from T decays Apart from the direct observation of the new quark, another exciting possibility is to discover the Higgs boson from T decays [405, 412].

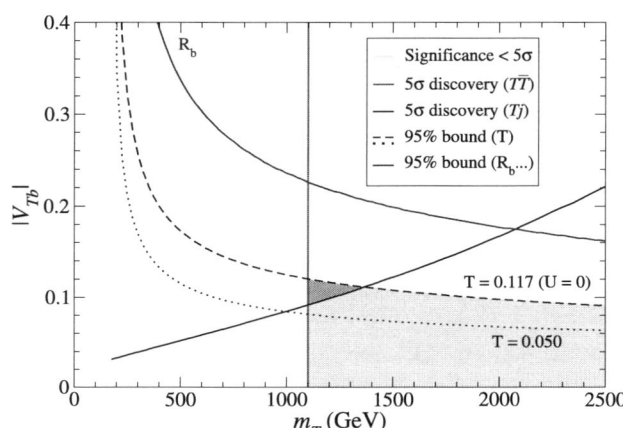

Fig. 51 Estimated 5σ discovery limits for a new charge 2/3 quark T in $T\bar{T}$ and Tj production

Very recent results from CMS have significantly lowered the expectations for the discovery of a light Higgs boson in $t\bar{t}h$ production, with $h \to b\bar{b}$. This decrease is due to a more careful calculation of the $t\bar{t}nj$ background, and to the inclusion of systematic uncertainties [61]. As a result, a light Higgs is impossible to see in this process, with a statistical significance of only $\sim 0.47\sigma$ for 30 fb^{-1} of luminosity and $M_h = 115$ GeV. But if a new quark T exists with a moderate mass, its pair production and decays,

$$\begin{aligned}
T\bar{T} &\to W^+ b h \bar{t}/h t W^- \bar{b} \to W^+ b W^- \bar{b} h \\
&\to \ell^+ \nu b \bar{q} q' \bar{b} b \bar{b}/c \bar{c}, \\
T\bar{T} &\to h t h \bar{t} \to W^+ b W^- \bar{b} h h \\
&\to \ell^+ \nu b \bar{q} q' \bar{b} b \bar{b}/c \bar{c} b \bar{b}/c \bar{c}, \\
T\bar{T} &\to Z t h \bar{t}/h t Z \bar{t} \to W^+ b W^- \bar{b} h Z \\
&\to \ell^+ \nu' b \bar{q} q' \bar{b} b \bar{b}/c \bar{c} q'' \bar{q}''/\nu \bar{\nu},
\end{aligned} \quad (113)$$

provide an additional source of Higgs bosons with a large cross section (see Fig. 47) and a total branching ratio close to 1/2. The final state is the same as in $t\bar{t}h$ production with semileptonic decay: one charged lepton, four or more b-tagged jets and two non-tagged jets. The main backgrounds are $t\bar{t}nj$ production with two b mistags and $t\bar{t}b\bar{b}$ production. The inclusion of higher-order ($n > 2$) contributions is relevant because of their increasing efficiency for larger n (the probability to have two b mistags grows with the jet multiplicity). The larger transverse momenta involved for larger n also make higher-order processes more difficult to suppress with respect to the $T\bar{T}$ signal. Lower order contributions ($n < 2$) are important as well, due to pile-up. The method followed to evaluate top pair production plus jets is to calculate $t\bar{t}nj$ for $n = 0, \ldots, 5$ with Alpgen [312] and use PYTHIA 6.4 [415] to include soft jet radiation, using the MLM matching prescription [416] to avoid double counting.

Background suppression is challenging because the higher-order $t\bar{t}nj$ backgrounds are less affected by large transverse momentum requirements. Moreover, the signal charged leptons are not so energetic, and cannot be used to discriminate signal and background as efficiently as in the previous final state. Background is suppressed with a likelihood method. Signal and background likelihood functions L_S, L_B can be built. using as variables several transverse momenta and invariant mass, as those shown in Fig. 52, as well as angles and rapidities of final state particles. (Additional details can be found in [405].) Performing cuts on these and other variables greatly improves the signal observability. For a luminosity of 30 fb^{-1}, the statistical significances obtained for the Higgs signals in final states with four, five and six b jets are [405]

$$\begin{aligned}
4\ b \text{ jets:} &\quad 6.43\sigma, \\
5\ b \text{ jets:} &\quad 6.02\sigma, \\
6\ b \text{ jets:} &\quad 5.63\sigma,
\end{aligned} \quad (114)$$

including a 20% uncertainty in the background. Additional backgrounds like electroweak $t\bar{t}b\bar{b}$ production, $t\bar{t}c\bar{c}$ (QCD and electroweak) and $W/Zb\bar{b}$ plus jets are smaller but have also been included. The combined significance is 10.45σ, a factor of 25 larger than in $t\bar{t}h$ production alone. Then, this process offers a good opportunity to quickly discover a light Higgs boson (approximately with 8 fb^{-1}) in final states containing a charged lepton and four or more b quarks. These figures are conservative, since additional signal processes $T\bar{T}nj$ have not been included in the signal evaluation. The decay channels in (113) also provide the best discovery potential for m_T relatively close to the electroweak scale. For $m_T = 500$ GeV, as assumed here, 5σ discovery of the new quark could be possible with 7 fb^{-1}.

4.2.2 Singlets: charge $-1/3$

Down-type isosinglet quark arise in the E_6 GUT models [63]. These models postulate that the group structure of the SM, $SU_C(3) \times SU_W(2) \times U_Y(1)$, originates from the breaking of the E_6 GUT scale down to the electroweak scale, and thus extend each SM family by the addition of one isosinglet down-type quark.

Following the literature, the new quarks are denoted by letters D, S, and B. The mixings between these and SM down-type quarks is responsible for the decays of the new quarks. In this study, the intra-family mixings of the new quarks are assumed to be dominant with respect to their inter-family mixings. In addition, as for the SM hierarchy, the D quark is taken to be the lightest one. For simplicity, we assume the usual CKM mixings, represented by superscript θ, to be in the up sector (an assumption that does not affect the results). The Lagrangian relevant for the down-type isosinglet quark, D, is obtained from (107), explicitly giving

$$\begin{aligned}
\mathcal{L}_\mathcal{D} =\ & \frac{\sqrt{4\pi\alpha_{em}}}{2\sqrt{2}\sin\theta_W} \big[\bar{u}^\theta \gamma_\alpha (1-\gamma_5) d \cos\phi \\
& + \bar{u}^\theta \gamma_\alpha (1-\gamma_5) D \sin\phi \big] W^\alpha \\
& - \frac{\sqrt{4\pi\alpha_{em}}}{4\sin\theta_W} \left[\frac{\sin\phi\cos\phi}{\cos\theta_W} \bar{d}\gamma_\alpha(1-\gamma_5)D \right] Z^\alpha \\
& - \frac{\sqrt{4\pi\alpha_{em}}}{12\cos\theta_W \sin\theta_W} \\
& \times \big[\bar{D}\gamma_\alpha(4\sin^2\theta_W - 3\sin^2\phi(1-\gamma_5))D\big] Z^\alpha \\
& - \frac{\sqrt{4\pi\alpha_{em}}}{12\cos\theta_W \sin\theta_W}
\end{aligned}$$

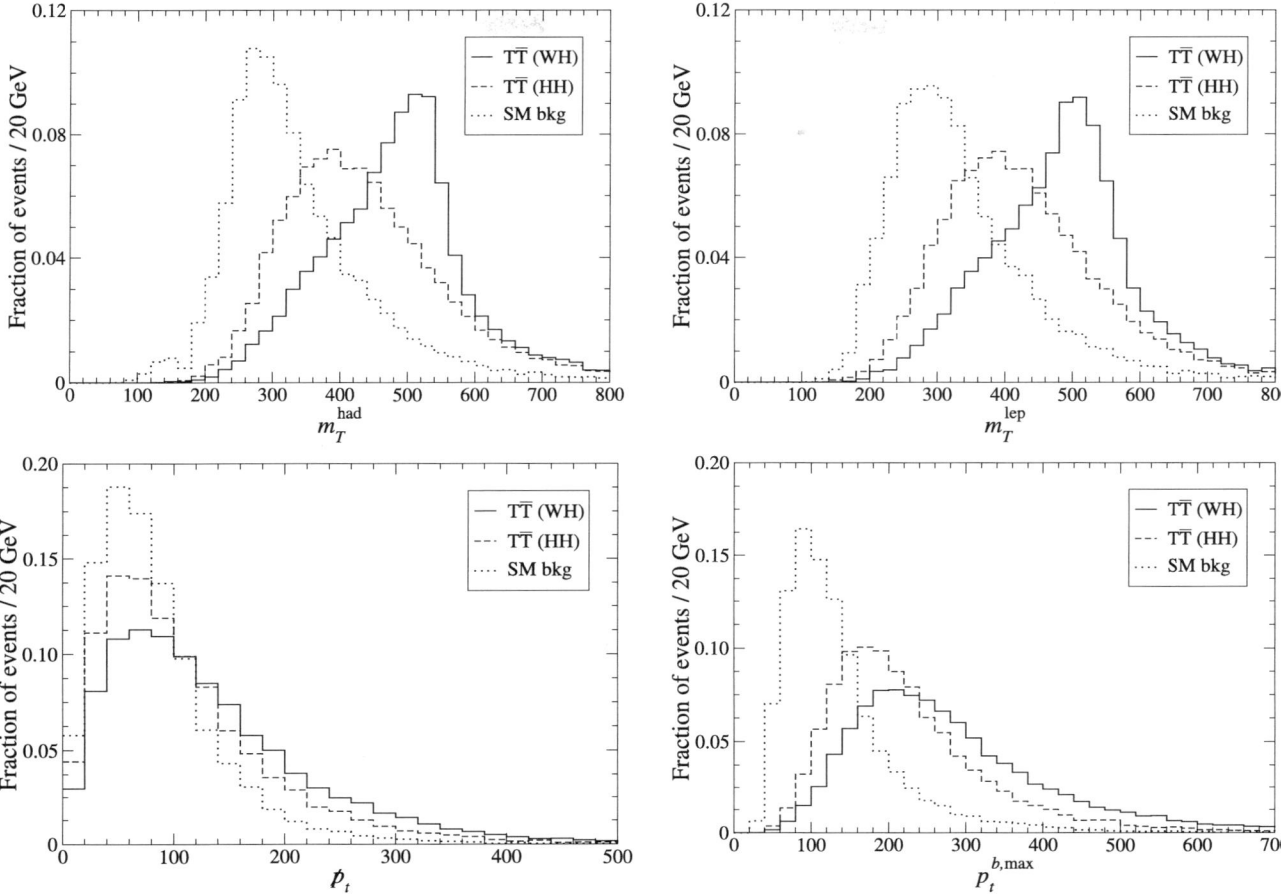

Fig. 52 Several useful variables to discriminate between heavy quark signals and background for $T\bar{T}$ production in $4b$ final states: heavy quark reconstructed masses (m_T^{had}, m_T^{lep}), missing energy ($\not{E}_T$), and maximum p_t of the b-tagged jets ($p_t^{b,\text{max}}$). The main signal processes (first two ones in (113)) are denoted by Wh, hh, respectively

$$\times \left[\bar{d}\gamma_\alpha \left(4\sin^2\theta_W - 3\cos^2\phi(1-\gamma_5)\right)d\right]Z^\alpha$$
$$+ \text{h.c.} \tag{115}$$

The measured values of V_{ud}, V_{us}, V_{ub} constrain the d and D mixing angle ϕ to $|\sin\phi| \leq 0.07$, assuming the squared sum of row elements of the new 3×4 CKM matrix equal unity (see [53] and references therein for CKM matrix related measurements). The total decay width and the contribution by neutral and charged currents were already estimated in [417]. As reported in this work, the D quark decays through a W boson with a branching ratio of 67% and through a Z boson with a branching ratio of 33%. If the Higgs boson exists, in addition to these two modes, the D quark might also decay via the $D \to hd$ channel which is available due to D–d mixing. The branching ratio of this channel for the case of $m_h = 120$ GeV and $\sin\phi = 0.05$ is calculated to be about 25%, reducing the branching ratios of the neutral and charged channels to 50% and 25%, respectively [418, 419].

4.2.2.1 The discovery potential The discovery potential of the lightest isosinglet quark has been investigated using the pair production channel which is quasi-independent of the mixing angle ϕ. The main tree-level Feynman diagrams for the pair production of D quarks at LHC are gluon fusion, and q–$\bar{q}$ annihilation. The $gD\bar{D}$ and $\gamma D\bar{D}$ vertices are the same as their SM down quark counterparts. The modification to the $Zd\bar{d}$ vertex due to d–D mixing can be neglected due to the small value of $\sin\phi$.

The Lagrangian in (115) was implemented into tree-level event generators, `CompHEP` 4.3 [420] and `MadGraph` 2.3 [421]. The impact of uncertainties in parton distribution functions (PDFs) [135] is calculated, by using different PDF sets, to be less than 10% for D quark mass values from 400 to 1400 GeV. For example at $m_D = 800$ GeV and $Q^2 = m_Z^2$, the cross section values are 450 fb (`CompHEP`, CTEQ6L1) and 468 fb (`CompHEP`, CTEQ5L) versus 449 fb (`MadGraph`, CTEQ6L1) and 459 fb (`MadGraph`, CTEQ5L) with an error of about one percent in each calculation. The largest contribution to the total cross section comes from the gluon fusion diagrams for D quark masses below 1100 GeV, while for higher D quark masses, contributions from s-channel $q\bar{q}$ annihilation subprocesses becomes dominant. For these computations, $q\bar{q}$ are assumed to be only from the first quark family since the contribution to the total cross section from $s\bar{s}$ is about 10 times smaller and the contribution from $c\bar{c}$ and $b\bar{b}$ are about 100 times smaller. The t-channel diagrams mediated by Z and W bosons, which are suppressed by the small value of $\sin\phi$ (for example 0.4 fb at $m_D = 800$ GeV) were also included in the signal generation. The very heavy isosinglet quarks are expected to immediately decay into SM particles. The cleanest signal can be obtained from both Ds decaying via a Z boson. Although it has the smallest branching ratio, the 4-lepton and 2-jet final state offers the possibility of reconstructing the invariant mass of Z bosons and thus of both D quarks. The high transverse momentum of the jets coming from the D quark decays can be used to distinguish the signal events from the background.

The D quarks in signal events were made to decay in CompHEP into SM particles. The final state particles for both signal and background events were fed into PYTHIA version 6.218 [47] for initial and final state radiation, as well as hadronization using the CompHEP to PYTHIA and MadGraph to PYTHIA interfaces provided by ATHENA 9.0.3 (the ATLAS offline software framework). To incorporate the detector effects, all event samples were processed through the ATLAS fast simulation tool, ATLFAST [422], and the final analysis has been done using physics objects that it produced. The cases of four muons, four electrons and two electrons plus two muons were separately treated to get the best reconstruction efficiency. As an example, Table 20 gives the selection efficiencies for the mixed lepton case at $m_D = 800$ GeV.

Using the convention of defining a running accelerator year as 10^7 seconds, one LHC year at the design luminosity corresponds to 100 fb^{-1}. All the signal events for this luminosity are summed and compared to the SM backgrounds, as shown in Fig. 53. It is evident that for the lowest of the considered masses, the studied channel gives an easy detection possibility, whereas for the highest mass case ($M_D = 1200$ GeV) the signal to background ratio is of the order of unity. For each D quark mass value that was considered, a Gaussian is fit to the invariant-mass distribution around the D signal peak and a polynomial to the background invariant-mass distribution. The number of accepted signal (S) and background (B) events are integrated using the fit functions in a mass window whose width is equal to 2σ around the central value of the fit Gaussian. The significance is then calculated at each mass value as $S/\sqrt{B}$, using the number of integrated events in the respective mass windows. The expected signal significance for three years of nominal LHC luminosity running is shown in Fig. 54 left hand side. The shaded band in the same plot represents the systematic errors originating from the fact that for each signal mass value, a finite number of Monte Carlo events was generated at the start of the analysis and the surviving events were selected from this event pool. For $M_D = 600$ GeV, ATLAS could observe the D quark with a significance more than 3σ before the end of the first year of low luminosity running (10 fb^{-1}/year) whereas to claim discovery with 5σ significance, it would need about 20 fb^{-1} integrated luminosity. For $M_D = 1000$ GeV, about 200 fb^{-1} integrated luminosity is necessary for a 3σ signal observation claim.

4.2.2.2 The mixing angle to SM quarks This section addresses the discovery of the isosinglet quarks via their jet associated single production at the LHC and the measurements of the mixing angle between the new and the SM quarks. The current upper limit on ϕ is $|\sin\phi| < 0.07$, allowed by the known errors on the CKM matrix elements assuming unitarity of its extended version [423]. However, in this work, a smaller thus a more conservative value, $\sin\phi = 0.045$, was considered for the calculation of the cross sections and decay widths. For other values of $\sin\phi$, both of these two quantities scale like $\sin^2\phi$. For both the signal and the background studies, the contributions from sea quarks were included. The parton distribution function CTEQ6L1 was used, and the QCD scale was set to be the mass of the D quark for both signal and background processes. The cross section for single production of the D quark for its mass up to 2 TeV and for various mixing angles is given in Fig. 55. The main tree-level signal processes are originating from the valance quarks exchanging W or Z bosons via the t-channel. The remaining processes originating from the sea quarks contribute about 20 percent to the total signal cross section.

Although the work in this section is at the generator level, various parameters of the ATLAS detector [424] such as the barrel calorimeter geometrical acceptance, minimum angular distance for jet separation and minimum transverse momentum for jets [425] were taken into account. Five mass

Table 20 The individual selection cut efficiencies ϵ for one $Z \to ee$ and one $Z \to \mu\mu$ sub-case. The subscript ℓ represents both electron and muon cases

Channel	N_ℓ	M_Z	$P_{T,\ell}$	N_{jet}	$P_{T,\text{jet}}$	$\epsilon_{\text{combined}}$
Cut	=4	=90 ± 20 GeV	$\mu(e) > 40(15)$ GeV	≥2	≥100 GeV	
ϵ signal	0.44	0.94	0.71	1	0.93	0.28
ϵ background	0.35	0.97	0.34	0.95	0.10	0.011

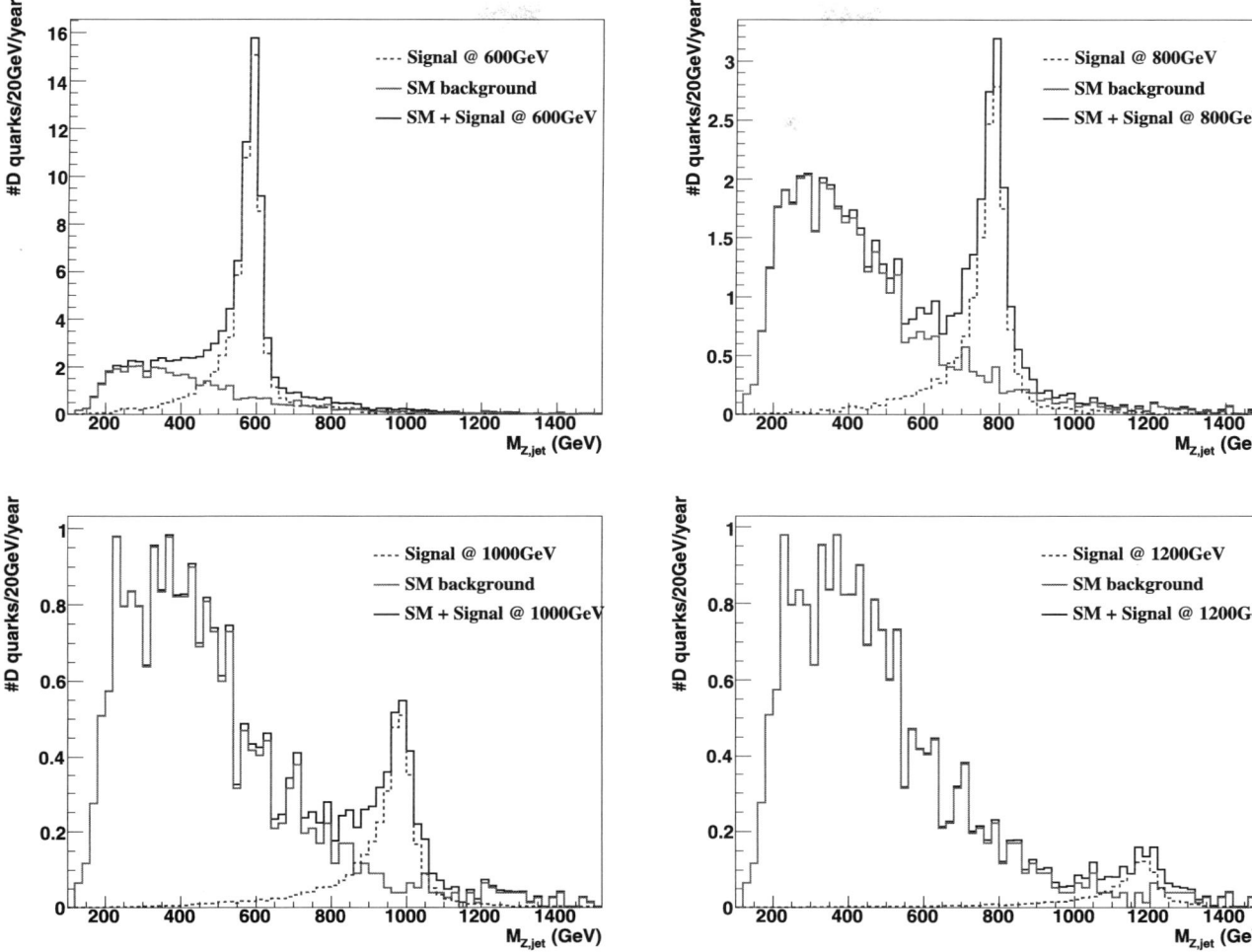

Fig. 53 Combined results for possible signal observation at $M_D = 600, 800, 1000, 1200$ GeV. The reconstructed D quark mass and the relevant SM background are plotted for a luminosity of 100 fb^{-1} which corresponds to one year of nominal LHC operation. The *dark line* shows the signal and background added, the *dashed line* is for signal only and the *light line* shows the SM background

values (400, 800, 1200, 1500 and 2000 GeV) were studied to investigate the mass dependence of the discovery potential for this channel. The cuts common to all considered mass values are

$P_{Tp} > 15$ GeV,

$|\eta_p| < 3.2$,

$|\eta_Z| < 3.2$,

$R_p > 0.4$,

$M_{Zp} = M_D \pm 20$ GeV,

where p stands for any parton; R is the cone separation angle between two partons; η_p and η_Z are pseudorapidities of a parton and Z boson respectively; and P_{Tp} is the parton transverse momentum. For each mass case, the optimal cut value is found by maximizing the significance ($S/\sqrt{B}$)

and it is used for calculating the effective cross sections presented in Table 21. To obtain the actual number of events for each mass value, the e^+e^- and $\mu^+\mu^-$ decays of the Z boson were considered. The last three rows of the same table contain the expected number of reconstructed events for both signal and background for 100 fb^{-1} of data taking. Although the lepton identification and reconstruction efficiencies are not considered, one can note that the statistical significance at $m_D = 1500$ GeV, is above 5σ after a one year at nominal luminosity.

The single production discovery results given in Table 21 can be used to investigate the mixing angle. In the event of a discovery in the single production case, the mixing angle can be obtained directly. If no discoveries are made, then the limit on the cross section can be converted to a limit curve in the D quark mass vs mixing angle plane. Therefore the angular reach for a 3σ signal is calculated by extrapolating to other $\sin\phi$ values. Figure 56 gives the mixing

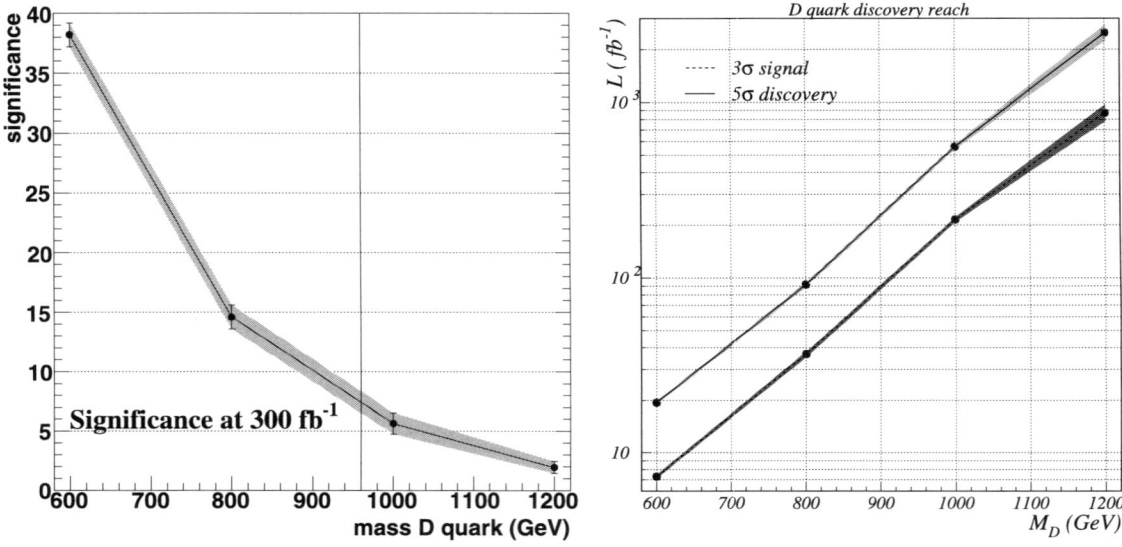

Fig. 54 On the *left*: the expected statistical significance after three years of running at nominal LHC luminosity assuming Gaussian statistics. The *vertical line* shows the limit at which the event yield drops below 10 events. On the *right*: the integrated luminosities for 3σ observation and 5 sigma discovery cases as a function of D quark mass. The bands represent uncertainties originating from finite MC sample size

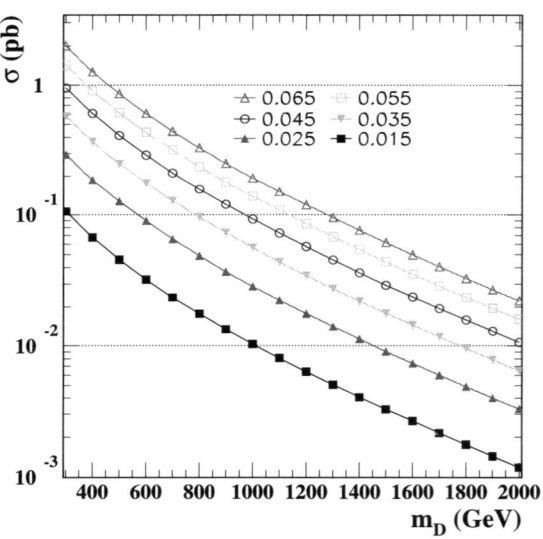

Fig. 55 Cross section in single D production as a function of D quark mass for different $\sin\phi$ values

angle versus D-quark mass plane and the 3σ reach curves for different integrated luminosities ranging from 10 fb^{-1} to 1000 fb^{-1}, which correspond to one year of low luminosity LHC operation and one year of high luminosity super-LHC operation respectively. The hashed region in the same plot is excluded using the current values of the CKM matrix elements. One should note that, this channel allows for reducing the current limit on $\sin\phi$ by half in about 100 fb^{-1} run time. The process of single production of the E$_6$ isosinglet quarks could essentially enhance the discovery potential if $\sin\phi$ exceeds 0.02. For example, with 300 fb^{-1} integrated luminosity, the 3σ discovery limit is $m_D = 2000$ GeV, if $\sin\phi = 0.03$. It should also be noted that for pair production the 3σ discovery limit was found to be about 900 GeV, independent of $\sin\phi$. If ATLAS discovers an 800 GeV D quark via pair production, single production will give the opportunity to confirm the discovery and measure the mixing angle if $\sin\phi > 0.03$. The FCNC decay channel analyzed in this paper is specific for isosinglet down-type quarks and gives the opportunity to distinguish it from other models also involving additional down-type quarks, for example the fourth SM family.

4.2.2.3 The impact on the Higgs searches The origin of the masses of SM particles is explained by using the Higgs Mechanism. The Higgs mechanism can also be preserved in E$_6$ group structure as an effective theory, although other alternatives such as dynamical symmetry breaking are also proposed [426, 427]. On the other hand, the origin of the mass of the new quarks (D, S, B) should be due to another mechanism since these are isosinglets. However, the mixing between d and D quarks will lead to decays of the latter involving h after spontaneous symmetry breaking (SSB). To find these decay channels, the interaction between the Higgs field and both down-type quarks of the first family should be considered before SSB. After SSB, the Lagrangian for the interaction between d, D quarks, and the Higgs boson becomes:

$$\mathcal{L}_h = \frac{m_D}{v} \sin^2\phi_L \bar{D}Dh$$

Table 21 The signal and background effective cross sections before the Z decay and after the optimal cuts, obtained by maximizing the $S/\sqrt{B}$, together with the D quark width in GeV for each considered mass. The number of signal and background events also the signal significance were calculated for an integrated luminosity of 100 fb^{-1}

M_D (GeV)	400	800	1200	1500	2000
Γ (GeV)	0.064	0.51	1.73	3.40	8.03
Signal (fb)	100.3	29.86	10.08	5.09	1.92
Background (fb)	2020	144	18.88	6.68	1.36
Optimal p_T cut	100	250	450	550	750
Signal events	702	209	71	36	13.5
Background events	14000	1008	132	47	9.5
Signal significance (σ)	5.9	6.6	6.1	5.2	4.37

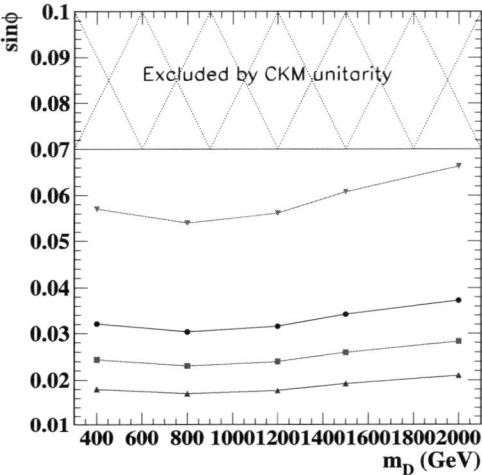

Fig. 56 3σ exclusion curves for 10, 100, 300, 1000 fb^{-1} integrated luminosities are shown from top to down

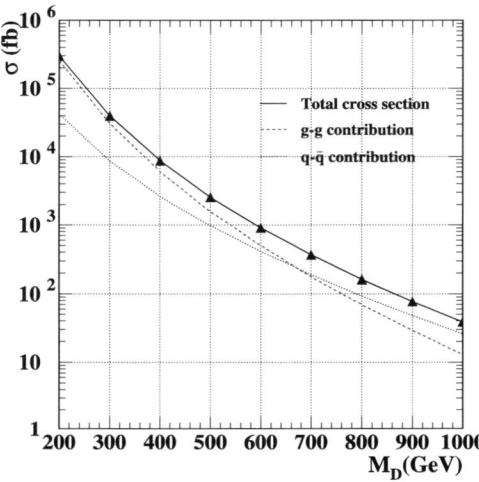

Fig. 57 Pair production of D quarks at LHC computed at tree level with CTEQ6L1 and QCD scale set at the mass of the D quark

$$-\frac{\sin\phi_L \cos\phi_L}{2v}\bar{D}[(1-\gamma^5)m_D + (1+\gamma^5)m_d]dh$$
$$-\frac{\sin\phi_L \cos\phi_L}{2v}\bar{d}[(1+\gamma^5)m_D + (1-\gamma^5)m_d]Dh$$
$$+\frac{m_d}{v}\cos^2\phi_L \bar{d}dh \qquad (116)$$

where $v = \eta/\sqrt{2}$ and $\eta = 246$ GeV is the vacuum expectation value of the Higgs field. It is seen that the D quark has a narrow width and becomes even narrower with decreasing values of ϕ since it scales through a $\sin^2\phi$-dependence. The relative branching ratios for the decay of the D quark depend on both the D quark and the Higgs mass values. For example, at the values of the D quark mass around 200 GeV and of the Higgs mass around 120 GeV one finds Br($D \rightarrow Wu$) $\sim 60\%$, Br($D \rightarrow hd$) $\sim 12\%$, Br($D \rightarrow Zd$) $\sim 28\%$, whereas as the D quark mass increases the same ratios asymptotically reach 50%, 25% and 25% respectively. As the Higgs mass increases from 120 GeV, these limit values are reached at higher D quark masses.

Depending on the masses of the D quark and the Higgs boson itself, the E_6 model could boost the overall Higgs production at the LHC. This boost is particularly interesting for the Higgs hunt, one of the main goals of the LHC experiments. For example, if the D quark mass is as low as 250 GeV, the pair production cross section at the LHC becomes as high as 10^5 fb^{-1}, which is enough to compensate for the relatively small Higgs branching ratio of 17%, as can be seen in Fig. 57. In the low mass range considered in this section (from 115 up to 135 GeV), the branching ratio $h \rightarrow b\bar{b}$ is about 70% [424]. Table 22 lists the decays involving at least one Higgs boson and the expected final state particles associated with each case. Although the case involving the Z is more suitable from the event reconstruction point of view, the focus will be on the last row, which has the highest number of expected Higgs events per year.

The full Lagrangian also involving the Higgs interaction has been implemented in a tree-level event generator, CompHEP 4.4.3 [420], to investigate the possibility of detecting the Higgs particle and reconstructing it from b jets. Assuming a light Higgs boson of mass 120 GeV, four mass values for the D quark have been taken as examples: 250 GeV, 500 GeV, 750 GeV, and 1000 GeV. 10000 signal events were produced for each mass value under consideration with the $Whjj$ final states using the CTEQ6L1 PDF set [135]. The

Table 22 For pair production of D quarks, the decay channels involving the Higgs particle. The branching ratios and the number of expected Higgs particles are calculated assuming $m_h = 120$ GeV and $m_D = 250\ (500)$ GeV

D_1	D_2	Br	#expected Higgs/100 fb^{-1}	Expected final state
$D \to hj$	$D \to hj$	0.029 (0.053)	$0.58 \times 10^6\ (2.65 \times 10^4)$	$2j 4j_b$
$D \to hj$	$D \to Zj$	0.092 (0.120)	$0.92 \times 10^6\ (3.01 \times 10^4)$	$2j 2j_b 2\ell$
$D \to hj$	$D \to Wj$	0.190 (0.235)	$1.9 \times 10^6\ (6.04 \times 10^4)$	$2j 2j_b \ell E_{T,\mathrm{miss}}$

generator level cuts on the partons, guided by the performance of the ATLAS detector, are given by

$|\eta_p| \leq 3.2,$

$p_{Tp} \geq 15 \text{ GeV},$

$R_p > 0.4,$

where η_p is the pseudorapidity for the partons giving rise to jets, $p_{T,p}$ is the transverse momentum of the partons, and R_p is the angular separation between the partons. The imposed maximum value of η requires the jets to be in the central region of the calorimeter where the jet energy resolution is optimal. The imposed lower value of p_T ensures that no jets that would eventually go undetected along the beam pipe are generated at all. The imposed lower value of R provides good separation between the two jets in the final state. Using the interface provided by CPYTH 2.3 [428], the generated particles are processed with ATHENA 11.0.41, which uses PYTHIA [47] for hadronization and ATLFAST [422] for fast detector response simulation. However, one should note that the reconstructed b jet energy and momenta were re-calibrated like in [424] to have a good match between the mean value of the reconstructed Higgs mass and its parton level value.

As for the background estimations, all the SM interactions giving the $W^{\pm} bbjj$ final state have been computed in another tree-level generator, MadGraph 2.1. [421], using the same parton level cuts and parton distribution functions. The SM background cross section is calculated to be 520 ± 11 pb. The reasons for using two separate event generators, their compatibility, and their relative merits have been discussed elsewhere [423]. The generated 40000 background events were also processed in the same way using ATLFAST for hadronization and calculation of detector effects.

The selection cuts for a D quark mass of 500 GeV, and h boson mass of 120 GeV, are given in Table 23. The invariant-mass distributions after the selection cuts are presented in Fig. 58 for 30 fb^{-1} of integrated luminosity. The signal window for D is defined as $M_D \pm 50$ GeV and for h as $M_h \pm 30$ GeV. The number of events for the signal (S) and the background (N_B) are summed in the signal windows for both signal and background to calculate the statistical significance $\sigma = S/\sqrt{S + N_B}$. For this set of parameters, it is

Table 23 Optimized event selection cuts and their efficiencies for $m_D = 500$ GeV

	Cut	ϵ signal	ϵ background
N-leptons	$=1$	0.83	0.79
N-jets	≥ 4	0.99	0.99
N-b jets	≥ 2	0.33	0.36
P_T-b jet	≥ 1 GeV	1.00	1.00
P_T-lepton	≥ 15 GeV	0.95	0.94
P_T-jet	≥ 100 GeV	0.83	0.69
$\cos\theta_{bjbj}$	≥ -0.8	0.97	0.89
M_{jj}	≥ 90 GeV	0.99	0.65
H_T	≥ 800 GeV	0.90	0.55
$|m_{D1} - m_{D2}|$	≤ 100 GeV	0.59	0.37

found that the D quark can be observed with a significance of 13.2σ and at the same time the Higgs boson with a significance of about 9.5σ. One should note that, in the SM Higgs searches, such a high statistical significance can only be reached with more than three times more data, namely with about 100 fb^{-1} of integrated luminosity.

A similar analysis was performed for the other three D quark masses: of 250, 750 and 1000 GeV. For each mass, the cut values were re-optimized to get the best statistical significance in the Higgs boson search. Figure 59 contains the 3σ and the 5σ signal significance reaches of the Higgs boson and the D quark as a function of their masses. It can be seen that, a light Higgs boson could be discovered with a 5σ statistical significance using the $D\bar{D} \to hWjj$ channel within the first year of low luminosity data taking (integrated luminosity of 10 fb^{-1}) if $m_D < 500$ GeV. Under the same conditions but with one year of design luminosity (integrated luminosity of 100 fb^{-1}), the 5σ Higgs discovery can be reached if $m_D \leq 700$ GeV. This is to be compared with the studies from the ATLAS Technical Design Report, where the most efficient channel to discover such a light Higgs is the $h \to \gamma\gamma$ decay. This search yields about 8σ signal significance with 100 fb^{-1} integrated luminosity. The presently discussed model could give the same significance (or more) with the same integrated luminosity if $m_D < 630$ GeV. Therefore, if the isosinglet quarks exist and their masses are suitable, they will provide a considerable improvement for the Higgs discovery potential.

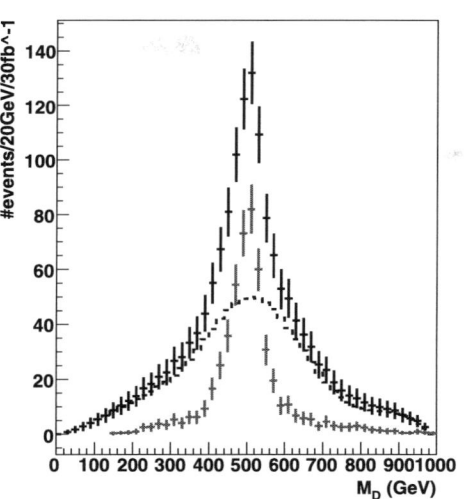
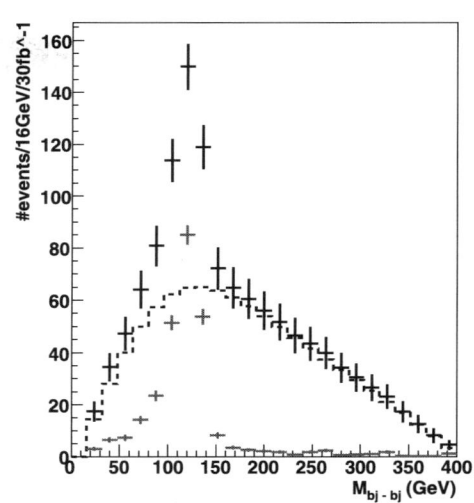

Fig. 58 Reconstructed invariant masses for the D quark (*left*) and the Higgs boson (*right*) regions, for 10 fb^{-1} of integrated luminosity. The signals are given by the *red* (*grey*) *crosses*, the full SM backgrounds by the dotted lines. The mass of the D quark was set to 500 GeV and that of the Higgs boson to 120 GeV

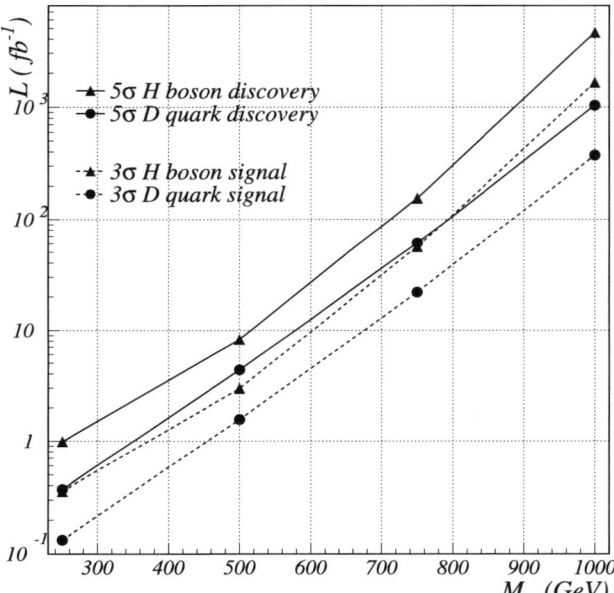

Fig. 59 The reach of ATLAS in the Higgs search for increasing D quark mass values. The *dashed lines* show the 3σ and the solid lines show the 5σ reaches of Higgs boson (*triangles*) and D quark (*circles*) searches

4.2.3 Quarks from extra dimensions: charges $-1/3$ and $5/3$

Heavy quarks of charges $(-1/3, 2/3, 5/3)$ (denoted $\tilde{q}$) are well motivated in Randall–Sundrum (RS) models with a custodial symmetry [429–433]. They are partners of the SM right handed top quark and have a mass between 500 and 1500 GeV. Their presence can be attributed to the heaviness of the top quark. This section studies the pair production of heavy $Q = -1/3$ and $Q = 5/3$ quarks, which takes place through standard QCD interactions with a cross section $\sim \mathcal{O}(10)$ pb for masses of several hundreds of GeV.

The focus is on the 4-W events, which are characteristic of the decay of new charge $-1/3$ singlets coupling to the $(t, b)_L$ doublet, in contrast with the preceding section in which the singlet D is assumed to couple to the d quark. The process under consideration is $gg, q\bar{q} \to \tilde{q}\bar{\tilde{q}} \to W^-tW^+\bar{t} \to W^-W^+bW^+W^-\bar{b}$. A straightforward trigger criterion for these events is that of a single, isolated lepton with missing E_T originating from the leptonic decay of one of the W bosons. The remaining W bosons can be reconstructed using dijet pairs. The goal in this analysis is to investigate the feasibility of multi-W reconstruction and therefore identify $\tilde{q}$ at the LHC. A simulation of this signal and its main background has been performed, and an analysis strategy outlined which distinguishes the signal from the sizable SM backgrounds [434].

There can be several $\tilde{q}$-type KK quarks in the class of composite Higgs models under consideration, leading to the same signature. Typically, in the minimal models, there is one heavy quark with electric charge $5/3$ as well as a $Q = -1/3$ quark, decaying into tW^+ and tW^- respectively, both with branching ratio essentially equal to 1. In addition, there is another bottom-type quark with tW^- branching ratio $\sim 1/2$. All these $\tilde{q}$ quarks are almost degenerate in mass. For the present model analysis, the mass of $\tilde{q}$ is taken as $m_{\tilde{q}} = 500$ GeV. The Lagrangian of the model [434] has been implemented into CalcHep 2.4.3 [435] for the simulation of $\tilde{q}$ pair production and decay through the tW channel. The actual number of 4W events coming from the pair production and decay of the other $Q = -5/3$ KK quarks, in a typical model, is taken into account by a multiplying factor.

$t\bar{t}WW$ events from $\tilde{q}$ pair production are generated with CalcHep, and they are further processed with PYTHIA 6.401 [415]. The following "trigger", applied to the generated events, is based on the lepton criteria for selecting $W \to \ell\nu$ events: at least one electron or muon with $p_T > 25$ GeV must be found within the pseudorapidity range $|\eta| < 2.4$;

then, the "missing E_T", calculated by adding all the neutrino momenta in the event and taking the component transverse to the collision axis, must exceed 20 GeV. Hadronic jets are reconstructed as they might be observed in a detector: stable charged and neutral particles within $|\eta| < 4.9$ (the range of the ATLAS hadronic calorimeter), excluding neutrinos, are first ranked in p_T order. Jets are seeded starting with the highest p_T tracks, with $p_T > 1$ GeV; softer tracks are added to the nearest existing jet, as long as they are within $\Delta R < 0.4$ of the jet centroid, where $\Delta R = \sqrt{\Delta\phi^2 + \Delta\eta^2}$. The number of jets with $p_T > 20$ GeV is shown in Fig. 60a. The signal is peaked around eight jets.

The two main backgrounds considered come from $t\bar{t}$ and $t\bar{t}h$ production. $t\bar{t}$ leads to two Ws + two bs, with four extra jets misinterpreted as coming from hadronic W decays. $t\bar{t}h$ however, can lead exactly to $4W$s and $2b$s when the Higgs mass is large enough. In this work, the Higgs mass is taken as $m_h = 115$ GeV. The background sample is dominated by $t\bar{t}$ events generated using TopReX (version 4.11) [46] and PYTHIA 6.403, with CTEQ6L parton distribution functions. The small $t\bar{t}h$ contribution to the background has been modeled with PYTHIA. As expected, the background has fewer high-p_T jets than the signal, peaking around five jets.

The number of $W/Z \to jj$ candidates (N) is counted, ensuring that jets are used only once in each event. In the heavy Higgs case with a $\tilde{q}$ mass of 500 GeV, the following sources dominate:

$N = 1$: SMW/Z processes,
$N = 2$: SM single h, WW/WZ, $t\bar{t}$,
$N = 3$: $\tilde{q}\bar{\tilde{q}} \to tWbZ \to WWZbb$,
$N = 4$: $\tilde{q}\bar{\tilde{q}} \to tWtW/tWbh/bhbh$.

In order to suppress the most common ($t\bar{t}$) SM background, the single hadronic W is eliminated by searching for a combination of two high-p_T jets whose mass falls between 70 and 90 GeV. The jets are combined in order of decreasing p_T. If a pair is found, it and the preceding pairs are removed; the dijet mass combinations of the subsequent pairs are shown in Fig. 60d. This procedure has been tested on $W + $jet simulation to ensure that it does not sculpt the combinatorial background distribution. Detailed results of the W reconstructions and consequences for $\tilde{q}$ identification are presented in [434]. The peak obtained in the dijet mass distribution suggests that it is possible to reach a signal significance beyond the 5σ level. Further investigation with more detailed simulation is required to map the discovery potential for this signal at an LHC experiment such as ATLAS, or at the ILC, and to connect the observable signal to the production cross section.

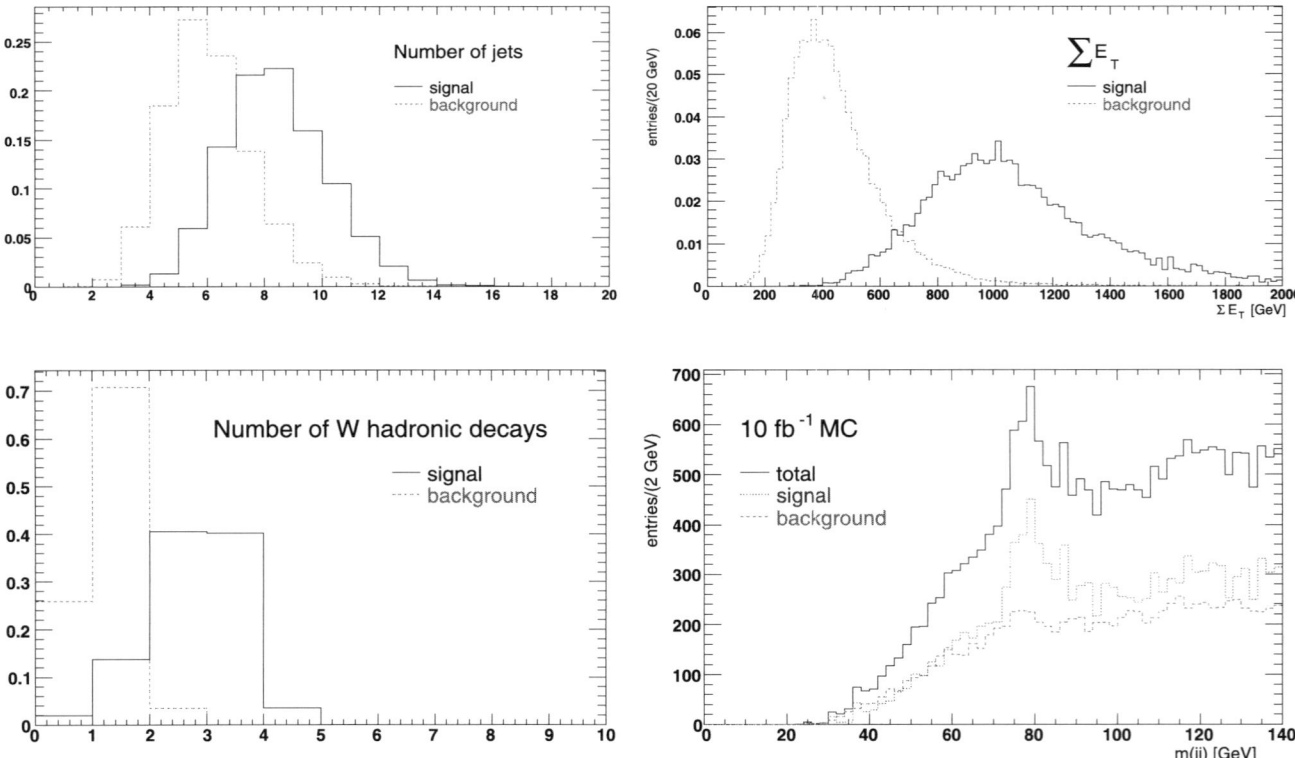

Fig. 60 *Top left* (**a**): Number of jets with $p_T > 20$ GeV. *Top right* (**b**): Scalar sum of E_T; *Bottom left* (**c**): number of Ws decaying hadronically in the event. All distributions are normalized to unit area. *Bottom right* (**d**): Dijet mass distribution after eliminating the first hadronic W candidate

4.2.4 Fourth sequential generation

The measurement of the Z invisible width implies well known constraints on the number of SM families with light neutrinos. However the discovery of neutrino masses and mixings show that the lepton sector is richer than the traditional SM. Moreover, some recent hints for new physics, mainly in CP violation effects in $b \to s$ transitions, might be accommodated with a fourth standard model family [38]. A phenomenological motivation for the existence of a fourth SM family might be attributed to the non-naturalness of the SM Yukawa couplings which vary by orders of magnitude even among the same type fermions. This consideration hints in the direction of accepting the SM as an effective theory of fundamental interactions rather than of fundamental particles. However, the electroweak theory (or SM before spontaneous symmetry breaking) itself is a theory of massless fermions where fermions with the same quantum numbers are indistinguishable. Therefore, there is no particular reason why the Yukawa couplings of a given type ($t = u, d, l, \nu$) should be different across families. If one starts with such a unique coupling coefficient per type t, for a case of n families the resulting spectrum becomes $n-1$ massless families and a single family where all particles are massive with $m = na^t \eta$ where η is the vacuum expectation value of the Higgs field. In the most simple model, where all fermions acquire mass due to a Higgs doublet, it is natural to also assume that the Yukawa couplings (therefore the masses) for different types should be comparable to each other and lie somewhere between the other couplings of EW unification:

$$a^d \approx a^u \approx a^l \approx a^\nu \approx a,$$
$$e = g_W \sin\theta_W < a/\sqrt{2} < g_Z = g_W/\cos\theta_W.$$

The measured fermion spectrum gives us a consistency check, quickly proving that the 3rd SM family can not be the singled out heavy family since $m_t \gg m_b \gg m_\tau \gg m_{\nu_\tau} \approx 0$. Therefore if the above presented naturalness assumptions are true, not only the reason behind the total number of families and the lightness of the SM neutrinos is obtained but also a set of predictions for the masses and mixings of the heavy fourth family are made through the parameterizations and fits to the extended (4×4) CKM matrix elements.

4.2.4.1 Search scenarios A recent detailed study [436] of b' and t' decay has updated old results done almost 20 years ago [437–440]. It was found that, the fourth generation while greatly enhancing FCNC top decays (see Sect. 4.2.1.1 for heavy top searches), especially $t \to cZ$ and ch, can only bring these into the borderline (10^{-6}–10^{-7}) of observability at the LHC. But the direct search for b' and t' looks far more interesting. Since $t' \to bW$ always dominates t' decay (unless the t'–b' mass difference is large), hence it can be straightforwardly discovered by a "heavy top" search, the focus will be on b'. The search scenarios are roughly separated by kinematics, i.e. whether $b' \to tW$ is allowed, and by pattern of quark mixing, i.e. whether $b' \to cW$ is suppressed with respect to the neutral decay mode.

4.2.4.2 Case $m_{b'} < m_t + M_W$ With $b' \to tW$ kinematically forbidden, it was pointed out long ago that the phenomenology is rather rich [437, 438], with the possibility of FCNC $b' \to bZ$ decay dominance, as well as the bonus that a light Higgs could be discovered via $b' \to bh$ [439, 440]. This can happen for light enough b' when $V_{cb'}$ is small enough, and has been searched for at the Tevatron. However, if the $b \to s$ CP violation indications are taken seriously, then $V_{cb'} \sim 0.12$ [441] is not small. Therefore, the $b' \to cW$ channel should be kept open. In this case, one has three scenarios:

1. $b' \to cW$ dominance—signature of $c\bar{c}W^+W^-$: for $V_{cb'}$ sizable, the lack of "charm-tagging" methods that also reject b makes this rather difficult.

2. $b' \to cW$, bZ (and bh) comparable—signature of $\bar{c}W^+bZ$ (and $\bar{c}W^+bh$, $\bar{b}bZh$): this can occur for $|V_{cb'}/V_{t'b}V_{t'b'}| \lesssim 0.005$. The measurements on the $b' \to bg$ and $b' \to b\gamma$ neutral decays [407] can motivate this choice for the CKM matrix elements ratio. The signature of $\bar{c}W^+bZ$ has never been properly studied, but shouldn't be difficult at the LHC as long as the $b' \to bZ$ branching ratio is not overly suppressed. The possible bonus of finding the Higgs makes this scenario quite attractive.

3. $b' \to tW^*$ and cW, bZ (and bh) comparable: $b' \to tW^*$ cannot be ignored above 230 GeV or so. This scenario is the most complicated, but the signature of $\bar{t}W^{*+}bZ$ is still quite tantalizing. Again, one could also expect an enhancement to Higgs searches. One should not forget that $t\bar{c}W^+W^-$ should also be considered.

Scenarios 1 and 2 form a continuum, depending on Br($b' \to bZ$).

4.2.4.3 Case $m_{b'} > m_t + M_W$ The $b' \to tW$ decay should dominate over all other modes, except when one is still somewhat restricted by kinematics while $V_{cb'}/V_{tb'}$ is very sizable. Therefore, the two available scenarios are

4. $b' \to tW$—with a signature of $t\bar{t}W^+W^-$, or $b\bar{b}W^+W^-W^+W^-$: with four W bosons plus two b jets, the signature could be striking.

5. $b' \to Wu$ or $b' \to Wc$—with a W^+W^-jj signature: the indistinguishability of the first and second family quarks in the light jets makes this signature benefit from the full b' branching ratio. Such a case is investigated in the following subsection.

It should be stressed that the standard sequential generation is considered, hence b' and t' masses should be below 800 GeV from partial wave unitarity constraints, and the mass difference between the two should be smaller or comparable to M_W. Scenario 4 and 5, together with the top-like $t' \to bW$ decay, could certainly be studied beyond 500 GeV. With such high masses, one starts to probe strong couplings. Whether there is an entirely new level of strong dynamics [72] related to the Higgs sector and what the Yukawa couplings would be, is also a rather interesting and different subject.

4.2.4.4 A case study If the fourth family is primarily mixing with the first two families, the dominant decay channels will be $t' \to W^+ s(d)$ and $b' \to W^- c(u)$. In this case, since the light quark jets are indistinguishable, the signature will be $W^+ W^- jj$ for both $t'\bar{t}'$ and $b'\bar{b}'$ pair production. According to flavor democracy, the masses of the new quarks have to be within few GeV of each other. This is also experimentally hinted by the value of the ρ parameter's value which is close to unity [53]. For such a mixing, both up- and down-type new quarks should be considered together since distinguishing between t' and b' quarks with quasi-degenerate masses in a hadron collider seems to be a difficult task. Moreover, the tree-level pair production and decay diagrams of the new b' quarks are also valid for the t' quark, provided c, u is replaced by s, d. As the model is not able to predict the masses of the new quarks, three mass values (250, 500 and 750 GeV) are considered as a mass scan. The widths of the b' and t' quarks are proportional to $|V_{b'u}|^2 + |V_{b'c}|^2$ and $|V_{t'd}|^2 + |V_{t's}|^2$ respectively. Current upper limits for corresponding CKM matrix elements are $|V_{b'u}| < 0.004$, $|V_{b'c}| < 0.044$, $|V_{t'd}| < 0.08$, and $|V_{t's}| < 0.11$. For the present case study, the common value 0.001 is used for all four elements. As the widths of the new quarks are much smaller than 1 GeV, this selection of the new CKM elements has no impact on the calculated cross sections. Table 24 gives the cross section for the $b'\bar{b}'$ or $t'\bar{t}'$ production processes which are within 1% of each other as expected. For this reason, from this point on, the b' will be considered and the results will be multiplied by two to cover both t' and b' cases. Therefore, in the final plots, the notation q_4 is used to cover both t' and b'.

To estimate the discovery possibility of the fourth family quarks, the model was implemented into a well known tree-level generator, CompHEP 4.3.3 [420]. This tool was used to simulate the pair production of the b' quarks at the LHC and their subsequent decay into SM particles. The QCD scale was set to the mass of the b' quark under study and the parton distribution function was chosen as CTEQ6L1 [135]. The generated events were fed into the ATLAS detector simulation and event reconstruction framework, ATHENA 11.0.41, using the interface program CPYTH 2.0.1 [428]. The partons were hadronized by PYTHIA 6.23 [47] and the detector response was simulated by the fast simulation software, ATLFAST [422]. The decay of the pair produced b' quarks result in two light jets (originating from the quarks and/or antiquarks of the first two SM families) and two W bosons. For the final state particles, the hadronic decay of one W boson and the leptonic (e, μ) decays of the other one are considered to ease the reconstruction.

The direct background to the signal is from SM events yielding the same final state particles. These can originate from all the SM processes which give two Ws and two non-b-tagged jets. The contributions from same sign W bosons were calculated to be substantially small. Some of the indirect backgrounds are also taken into account. These mainly included the $t\bar{t}$ pair production where the b jets from the decay of the top quark could be mistagged as a light jet. Similarly the jet associated top quark pair production $(t\bar{t}j \to W^- W^+ b\bar{b} j)$ substantially contributes to the background events as the production cross section is comparable to the pair production and only one mistagged jet would be sufficient to fake the signal events. The cross section of the next order process, namely $pp \to t\bar{t}2j$, was also calculated and has been found to be four times smaller than $t\bar{t}j$ case: therefore it was not further investigated.

The first step of the event selection was the requirement of a single isolated lepton (e or μ) of transverse momentum above 15 GeV, and at least four jets with transverse momenta above 20 GeV. The leptonically decaying W boson was reconstructed by attributing the total missing transverse momentum in the event to the lost neutrino, and using the nominal mass of the W as a constraint. The two-fold ambiguity in the longitudinal direction of the neutrino was resolved by choosing the solution with the lower neutrino energy. The four-momenta of the third and fourth most energetic jets in the event were combined to reconstruct the hadronically decaying W boson. The invariant mass of the combination of these jets was required to be less than 200 GeV. The summary of the event selection cuts and their efficiencies for both signal and background events are listed in Table 25 for a quark mass of 500 GeV.

The surviving events were used to obtain the invariant mass of the new quark. The W jet association ambiguity was resolved by selecting the combination giving the smallest mass difference between the two reconstructed quarks in the same event. The results of the reconstruction for quark masses of 500 GeV and 750 GeV are shown in Fig. 61 together with various backgrounds for integrated luminosities

Table 24 The considered quark mass values and the associated width and pair production cross sections at LHC

M_{d4}	250	500	750
Γ (GeV)	1.00×10^{-5}	8.25×10^{-5}	2.79×10^{-4}
σ (pb)	99.8	2.59	0.25

of 5 and 10 fb^{-1} respectively. The bulk of the background in both cases is due to $gg \to t\bar{t}g$ events as discussed before.

In order to extract the signal significance, an analytical function consisting of an exponential term to represent the background and a Breit–Wigner term to represent the signal resonance was fit to the total number of events in the invariant plots of Fig. 61. In both cases, the fit function is shown with the solid line, whereas the background and signal components are plotted with dashed blue and red lines, respectively. For the case of $m_{d_4} = 500$ GeV, it can be noticed that the signal function extracted from the fit slightly underestimates the true distribution. However, using the same fit functions and with 5 fb^{-1} of data, the signal significance is found to be 4.7σ. The significance is calculated after the subtraction of the estimated background: the integral area around the Breit–Wigner peak and its error are a measure of the expected number of signal events, and thus of the signal significance. A similar study with the higher mass value of 750 GeV, and with 10 fb^{-1} of data gives results with a significance of 9.4σ. This analysis has shown that the fourth family quarks with the studied mass values can be observed at the LHC with an integrated luminosity of 10 fb^{-1}. Although these results were obtained with a fast simulation, the simplistic approach in the analysis should enhance their validity.

4.2.4.5 Other possible studies The study of $\bar{c}W^+bZ$ is a relatively easy one. Due to the cleanness of the $Z \to \ell^+\ell^-$ signature, one does not need to face c jet tagging issues, and one can either have $W \to jj$ or $W \to \ell\nu$. For the latter, the offshoot is to search for $\bar{c}W^+bh$ by a $M_{b\bar{b}}$ scan with Z as standard candle. A second effort would be $\bar{t}W^{*+}bZ$, with a similar approach as above. Once experience is gained in facing c as well as W^* (relatively soft leptons or jets, or missing E_T), one could also consider $t\bar{c}WW^*$ before moving on, to the challenge of $c\bar{c}W^+W^-$. The $t\bar{t}W^+W^-$ search for heavy b' could also be pursued.

4.3 New leptons: heavy neutrinos

Models with extended matter multiplets predict additional leptons, both charged and neutral. While heavy neutral leptons (neutrinos) can be introduced to explain the smallness of the light neutrino masses in a natural way and the observed baryon asymmetry in the universe, the charged ones are not required by experiment. Here we concentrate on the neutral ones.

Heavy neutrinos with masses $m_N > M_Z$ appear in theories with extra dimensions near the TeV scale and little Higgs models, in much the same way as vector-like quarks, and in left–right models. For example, in the *simplest* little Higgs models [442], the matter content belongs to SU(3) multiplets, and the SM lepton doublets must be enlarged with one extra neutrino $N'_{\ell L}$ per family. These extra neutrinos can get a large Dirac mass of the order of the new scale $f \sim 1$ TeV if the model also includes

Table 25 Efficiencies of the selection criteria, as applied in the order listed, for the $m_q = 500$ GeV signal and the SM background

Criterion	ϵ-signal (%)	ϵ-background (%)
Single e/μ, $p_T^\ell > 15$ GeV	32	29.1
At least four jets, $p_T^j > 20$ GeV	88.3	94.2
Possible neutrino solution	71.3	73.7
$m_{jj}^W < 200$ GeV	63.5	76.0

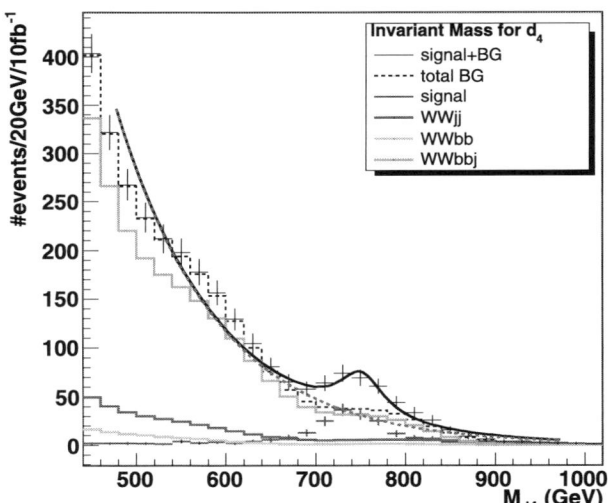

Fig. 61 Reconstructed signal and SM backgrounds for a quark of mass 500 GeV (*left*) and 750 GeV (*right*). The *colored solid lines* show SM backgrounds from various processes, the uppermost *solid black* like represents the fit to the sum of background and signal events

right handed neutrinos transforming as SU(3) singlets [443]. This mechanism provides a natural way of giving masses to the SM neutrinos, and in this framework the mixing between the light leptons and the heavy neutrinos is of order $v/\sqrt{2}f$, with $v = 246$ GeV the electroweak VEV. But besides their appearance in several specific models, heavy Majorana neutrinos are often introduced to explain light neutrino masses via the seesaw mechanism [444–447].[21] They give contributions to light neutrino masses m_ν of the order $Y^2 v^2 / 2 m_N$, where Y is a Yukawa coupling. In the minimal seesaw realization this is the only source for light neutrino masses, and the Yukawa couplings are assumed of order unity without any particular symmetry. Therefore, having $m_\nu \sim Y^2 v^2 / 2 m_N$ requires heavy masses $m_N \sim 10^{13}$ GeV to reproduce the observed light neutrino spectrum. Additionally, the light-heavy mixing is predicted to be $V_{\ell N} \sim \sqrt{m_\nu/m_N}$. These ultra-heavy particles are unobservable, and thus the seesaw mechanism is not directly testable. Nevertheless, non-minimal seesaw models can be built, with $m_N \sim 1$ TeV or smaller, if some approximate flavor symmetry suppresses the $\sim Y^2 v^2 / 2 m_N$ contribution from the seesaw [448–450]. These models can also provide a successful leptogenesis (see, for instance, [451–454]). Heavy neutrinos with masses near the electroweak scale can be produced at the next generation of colliders (see [455] for a review) if their couplings to the SM fermions and gauge bosons is not too small, or through new non-standard interactions. The most conservative point of view is to assume that heavy neutrinos are singlets under the SM gauge group and no new interactions exist, which constitutes a "minimal" scenario in this sense. On the other hand, with an extended gauge structure, for example $SU(2)_L \times SU(2)_R \times U(1)_{B-L}$ in models with left–right symmetry, additional production processes are possible, mediated by the new W' and/or Z' gauge bosons. We shall discuss these possibilities in turn.

4.3.1 Production of heavy neutrino singlets

Heavy neutrino singlets couple to the SM fields through their mixing with the SM neutrino weak eigenstates. The Lagrangian terms describing the interactions of the lightest heavy neutrino (in the mass eigenstate basis) are

$$\mathcal{L}_W = -\frac{g}{\sqrt{2}} \left(\bar{\ell} \gamma^\mu V_{\ell N} P_L N W_\mu + \bar{N} \gamma^\mu V_{\ell N}^* P_L \ell W_\mu^\dagger \right),$$

$$\mathcal{L}_Z = -\frac{g}{2 c_W} \left(\bar{\nu}_\ell \gamma^\mu V_{\ell N} P_L N + \bar{N} \gamma^\mu V_{\ell N}^* P_L \nu_\ell \right) Z_\mu, \quad (117)$$

---

[21]This mechanism, with heavy neutrino singlets under the SM gauge group, is often referred to as seesaw type I. Other possibilities to generate light neutrino masses are to introduce a scalar triplet (type II seesaw, see Sect. 4.6) or a lepton triplet (type III). In this section, heavy neutrinos are always assumed to be SM singlets.

$$\mathcal{L}_h = -\frac{g m_N}{2 M_W} \left(\bar{\nu}_\ell V_{\ell N} P_R N + \bar{N} V_{\ell N}^* P_L \nu_\ell \right) h,$$

with N the heavy neutrino mass eigenstate and V the extended MNS matrix. For Majorana N, the last terms in the Z and h interactions can be rewritten in terms of the conjugate fields. These interactions determine the N production processes, as well as its decays. The latter can happen in the channels $N \to W \ell$, $N \to Z \nu$, $N \to h \nu$. The partial widths can be straightforwardly obtained from (109) neglecting charged lepton and light neutrino masses,

$$\begin{aligned}
&\Gamma(N \to W^+ \ell^-) \\
&= \Gamma(N \to W^- \ell^+) \\
&= \frac{g^2}{64\pi} |V_{\ell N}|^2 \frac{m_N^3}{M_W^2} \left[1 - 3 \frac{M_W^4}{m_N^4} + 2 \frac{M_W^6}{m_N^6} \right], \\
&\Gamma_D(N \to Z \nu_\ell) \\
&= \frac{g^2}{128\pi c_W^2} |V_{\ell N}|^2 \frac{m_N^3}{M_Z^2} \left[1 - 3 \frac{M_Z^4}{m_N^4} + 2 \frac{M_Z^6}{m_N^6} \right], \\
&\Gamma_M(N \to Z \nu_\ell) = 2 \Gamma_D(N \to Z \nu_\ell), \\
&\Gamma_D(N \to h \nu_\ell) = \frac{g^2}{128\pi} |V_{\ell N}|^2 \frac{m_N^3}{M_W^2} \left[1 - 2 \frac{M_h^2}{m_N^2} + \frac{M_h^4}{m_N^4} \right], \\
&\Gamma_M(N \to h \nu_\ell) = 2 \Gamma_D(N \to h \nu_\ell).
\end{aligned} \quad (118)$$

The subscripts M, D refer to Majorana and Dirac heavy neutrinos, respectively, and the lepton number violating (LNV) decay $N \to W^- \ell^+$ is only possible for a Majorana N.

In the minimal seesaw the mixing angles $V_{\ell N}$ are of order $\sqrt{m_\nu/m_N}$ (and then of order 10^{-5} or smaller for $m_N > M_Z$), but in models with additional symmetries the light–heavy mixing can be decoupled from mass ratios [456]. Nevertheless, $V_{\ell N}$ are experimentally constrained to be small (this fact has already been used in order to simplify the Lagrangian above). Defining the quantities

$$\Omega_{\ell \ell'} \equiv \delta_{\ell \ell'} - \sum_{i=1}^{3} V_{\ell \nu_i} V_{\ell' \nu_i}^* = \sum_{i=1}^{3} V_{\ell N_i} V_{\ell' N_i}^* \quad (119)$$

(assuming three heavy neutrinos), limits from universality and the invisible Z width imply [457, 458]

$$\Omega_{ee} \leq 0.0054, \qquad \Omega_{\mu\mu} \leq 0.0096,$$
$$\Omega_{\tau\tau} \leq 0.016, \quad (120)$$

with a 90% confidence level (CL). In the limit of heavy neutrino masses in the TeV range, limits from lepton flavor-violating (LFV) processes require [456]

$$|\Omega_{e\mu}| \leq 0.0001, \qquad |\Omega_{e\tau}| \leq 0.01,$$
$$|\Omega_{\mu\tau}| \leq 0.01. \quad (121)$$

Additionally, for heavy Majorana neutrinos there are constraints on (V_{eN}, m_N) from the non-observation of neutrinoless double beta decay. These, however, may be evaded e.g. if two nearly degenerate Majorana neutrinos with opposite CP parities form a quasi-Dirac neutrino.

Heavy Dirac or Majorana neutrinos with a significant coupling to the electron can be best produced and seen at e^+e^- colliders in $e^+e^- \to N\nu$, which has a large cross section and whose backgrounds have moderate size [459–462]. On the other hand, a Majorana N mainly coupled to the muon or tau leptons is easier to discover at a hadronic machine like the LHC, namely in the process $q\bar{q}' \to W^* \to \ell^+ N$ (plus the charge conjugate), with subsequent decay $N \to \ell W \to \ell q\bar{q}'$. Other final states, for instance with decays $N \to Z\nu$, $N \to h\nu$, or in the production process $pp \to Z^* \to N\nu$ have much larger backgrounds. Concentrating ourselves on ℓN production with $N \to \ell W$, it is useful to classify the possible signals according to the mixing and the character of the lightest heavy neutrino.

1. For a Dirac N mixing with only one lepton flavor, the decay $N \to \ell^- W^+$ yields a $\ell^+ \ell^- W^+$ final state, with a huge SM background.
2. For a Dirac N coupled to more than one charged lepton one has also $N \to \ell'^- W^+$ with $\ell' \neq \ell$, giving the LFV signal $\ell^+ \ell'^- W^+$, which has much smaller backgrounds.
3. For a Majorana N, in addition to LNC signals, one has LNV ones arising from the decay $N \to \ell^{(\prime)+} W^-$, which also have small backgrounds.

In the following we concentrate on the case of a Majorana N coupling to the muon, which is the situation in which LHC has better discovery prospects than ILC. The most interesting signal is [463–466]

$$pp \to \mu^\pm N \to \mu^\pm \mu^\pm jj, \tag{122}$$

with two same-sign muons in the final state, and at least two jets. SM backgrounds to this LNV signal involve the production of additional leptons, either neutrinos or charged leptons (which may be missed by the detector, thus giving the final state in (122)). The main ones are $W^\pm W^\pm nj$ and $W^\pm Z nj$, where nj stands for $n = 0, \ldots$ additional jets (processes with $n < 2$ are also backgrounds due to the appearance of extra jets from pile-up). The largest reducible backgrounds are $t\bar{t}nj$, with semileptonic decay of the $t\bar{t}$ pair, and $Wb\bar{b}nj$, with leptonic W decay. In these cases, the additional same-sign muon results from the decay of a b or $\bar{b}$ quark. Only a tiny fraction of such decays produce isolated muons with sufficiently high transverse momentum but, since the $t\bar{t}nj$ and $Wb\bar{b}nj$ cross sections are so large, these backgrounds are much larger than the two previous ones. An important remark here is that the corresponding backgrounds $t\bar{t}nj$, $Wb\bar{b}nj \to e^\pm e^\pm X$ are one order of magnitude larger than the ones involving muons. The reason is that b decays produce "apparently isolated" electrons more often than muons, due to detector effects. A reliable evaluation of the $e^\pm e^\pm X$ background resulting from these processes seems to require a full simulation of the detector. Other backgrounds like Wh and Zh are negligible, with cross sections much smaller than the ones considered, $W/Zb\bar{b}$, WZ, ZZ, which give the same final states. Note also that for this heavy neutrino mass $b\bar{b}nj$, which is huge, has very different kinematics and can be eliminated. However, for $m_N < M_W$ the heavy neutrino signal and $b\bar{b}nj$ are much alike, and thus this background is the largest and most difficult to reduce. Further details can be found in [466].

Signals and backgrounds have been generated using `Alpgen` (the implementation in `Alpgen` of heavy neutrino production is discussed in Sect. 5). Events are passed through `PYTHIA` 6.4 (using the MLM prescription for jet–parton matching [416] to avoid double counting of jet radiation) and a fast simulation of the ATLAS detector. The preselection criteria used are (i) two same-sign isolated muons with pseudorapidity $|\eta| \leq 2.5$ and transverse momentum p_T larger than 10 GeV; (ii) no additional isolated charged leptons nor non-isolated muons; (iii) two jets with $|\eta| \leq 2.5$ and $p_T \geq 20$ GeV. It should be noted that requiring the absence of non-isolated muons reduces backgrounds involving Z bosons almost by a factor of two.

It must be emphasized that SM backgrounds are about two orders of magnitude larger than in previous estimations in the literature [465]. Backgrounds cannot be significantly suppressed with respect to the heavy neutrino signal using simple cuts on missing energy and muon-jet separation. Instead, a likelihood analysis has been performed [466]. Several variables are crucial in order to distinguish the signal from the backgrounds:

— The missing momentum $\not{E}_T$ (the signal does not have neutrinos in the final state).
— The separation between the second muon and the closest jet, $\Delta R_{\mu_2 j}$. For backgrounds involving b quarks this separation is rather small.
— The transverse momentum of the two muons $p_T^{\mu_1}$, $p_T^{\mu_2}$, ordered from higher (μ_1) to lower (μ_2) p_T. Backgrounds involving b quarks have one muon with small p_T.
— The b tag multiplicity (backgrounds involving b quarks often have b-tagged jets).
— The invariant mass of μ_2 and the two jets which best reconstruct a W boson, $m_{W\mu_2}$.

The distribution of these variables is presented in Fig. 62, distinguishing three likelihood classes: the signal, backgrounds with one muon from b decays, and backgrounds with both muons from W/Z decays. The $b\bar{b}$ background can be suppressed for $m_N \gtrsim 100$ GeV, and it is not shown. Additional variables like jet transverse momenta, the $\mu\mu$ invariant mass, etc. are useful, and included in the analysis.

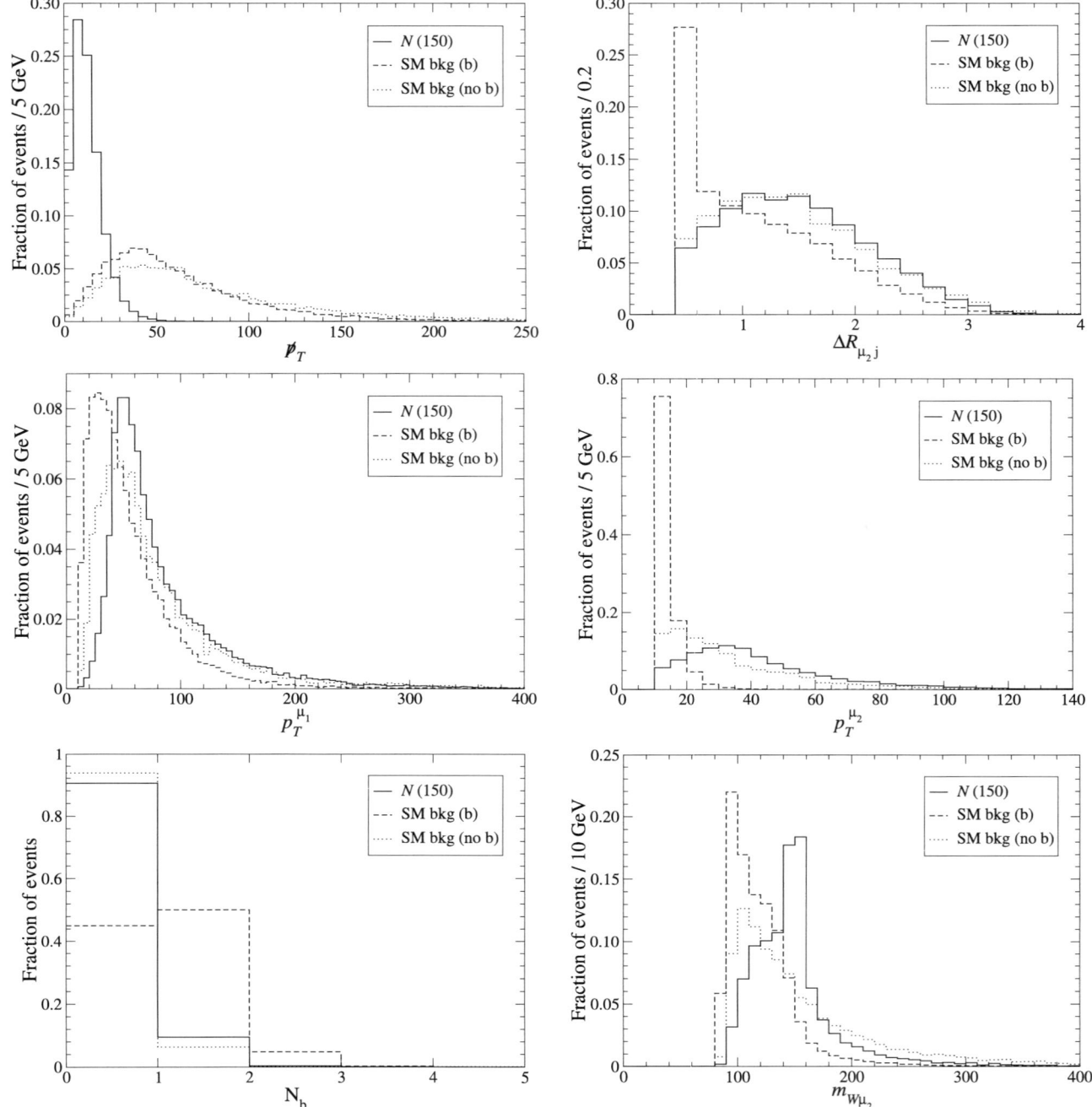

Fig. 62 Several useful variables to discriminate between the heavy neutrino signal and the backgrounds, as explained in the text

Assuming a 20% systematic uncertainty in the backgrounds (which still has to be precisely evaluated), and taking the maximum allowed mixing by low energy data, the following 5σ discovery limits are found. (i) A heavy neutrino coupling only to the muon with $|V_{\mu N}|^2 = 0.0096$ can be discovered up to masses $m_N = 200$ GeV. (ii) A heavy neutrino coupling only to the muon with $|V_{eN}|^2 = 0.0054$ can be discovered up to masses $m_N = 145$ GeV. Limits for other masses and mixing scenarios can be found in [466].

4.3.2 Heavy neutrino production from W_R decays

Models with left–right symmetry have an extended gauge structure $SU(2)_L \times SU(2)_R \times U(1)_{B-L}$ and, in addition to three new gauge bosons Z', $W_R^\pm$ (see Sects. 4.4 and 4.5) they introduce three right handed neutrinos as partners of the charged leptons in $SU(2)_R$ doublets $(N_\ell, \ell)_R$. The minimal scalar sector consists of a bidoublet and two triplets. The measurement of the T parameter and the present lower bounds on the masses of the new bosons and their mixing with the W and Z imply the hierarchy $v_L \ll (|k_1|^2 + |k_2|^2)^{1/2} \ll v_R$ among the VEVs of the bidoublet $k_{1,2}$ and the triplets $v_{L,R}$. In this situation the neutrino mass matrix exhibits a seesaw structure, heavy neutrino eigenstates N are mostly right handed and the following hierarchy is found among the couplings of the light and heavy neutrinos to the gauge bosons:

(i) $\ell \nu W$ and $\ell N W_R$ are of order unity; $\ell N W$ and $\ell \nu W_R$ are suppressed.
(ii) $\nu \nu Z$ and NNZ' are of order unity; νNZ, $\nu NZ'$, NNZ and $\nu \nu Z'$ are suppressed.

At hadron colliders the process $q\bar{q}' \to W_R \to \ell N$ [467] involves mixing angles of order unity and only one heavy particle in the final state. The best situation happens where N is lighter than W_R, so that W_R can be on its mass shell and the cross section is not suppressed by an s-channel propagator either. This is in sharp contrast with the analysis in the previous subsection, in which the process $q\bar{q}' \to W^* \to \ell N$ is suppressed by mixings and the off-shell W propagator.

Heavy neutrino production from on-shell W_R decays has been previously described in [468], and studied in detail for the ATLAS detector in [469]. Here we summarize the expectations for the CMS detector [470, 471]. Production cross sections and decay branching ratios depend on several parameters of the model. The new coupling constant g_R of $SU(2)_R$ is chosen to be equal to g_L, as happens e.g. in models with spontaneous parity breaking. Mixing between gauge bosons can be safely neglected. An additional hypothesis is that the right handed CKM matrix equals the left handed one. The heavy neutrino N is assumed to be lighter than W_R (the other two are assumed heavier) and coupling only to the electron, with a mixing angle of order unity.

For the signal event generation and calculation of cross sections, PYTHIA 6.227 is used with CTEQ5L parton distribution functions, and the model assumptions mentioned above. The analysis is focused on the W_R mass region above 1 TeV. The signal cross section, defined as the product of the total W_R production cross section times the branching ratio of W_R decay into eN, is shown in Fig. 63 as a function of m_N, for several W_R masses. For the value $M_{W_R} = 2$ TeV, the dashed line illustrates the decrease of the total cross section (due to the smaller branching ratio for $W_R \to eN$) for the case of three degenerated heavy neutrinos N_{1-3}, mixing with e, μ, τ respectively. The values $M_{W_R} = 2$ TeV, $m_N = 500$ GeV are selected as a reference point for the detailed analysis.

The detection of signal events is studied using the full CMS detector simulation and reconstruction chain. For details see [471]. The analysis proceeds through the following steps.

— Events with two isolated electrons are selected (standard isolation in the tracker is required).

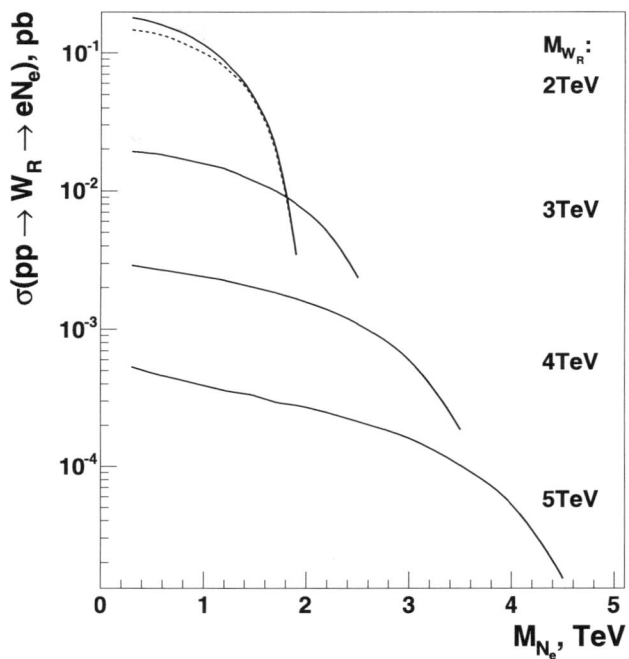

Fig. 63 Dependence of $\sigma(pp \to W_R) \times \text{Br}(W_R \to e^\pm N)$ on the heavy neutrino mass, for different values of M_{W_R}

— Events with at least two jets are selected. From these jets, the two ones with the maximum p_T are chosen.
— Using the 4-momenta of the signal jet pair and the 4-momentum of a lepton, the invariant mass $M_{ejj} = m_{N_e}^{\text{cand}}$ is calculated. Since there are two electrons, the two ejj combinations are considered. This distribution is plotted in Fig. 64. The tail above 500 GeV corresponds to a wrong choice of the electron.
— From the 4-momenta of the jet pair and the electrons, the invariant mass $M_{eejj} = M_{W_R}^{\text{cand}}$ is calculated.

Background is constituted by SM processes giving a lepton pair plus jets. The production of a Z boson plus jets has a large cross section, about 5 orders of magnitude larger than the signal. In a first approximation, this process can be simulated with PYTHIA. This background is suppressed by a cut on the lepton pair invariant mass $M_{ee} > 200$ GeV. In order to reduce the number of simulated events, it is required that the Z transverse momentum is larger than 20 GeV during the simulation, and events with sufficiently high M_{ee} are preselected at the generator level. Another background is $t\bar{t}$ production with dileptonic W^+W^- decay. It has been checked that other decay modes do not contribute significantly. Its cross section is about two orders of magnitude larger than the signal. It must be pointed out that the Majorana nature of the heavy neutrino allows one to single out the LNV final state with two like-sign leptons. This does not improve the sensitivity because, although backgrounds are smaller in this case, the signal is reduced to one half. However, in case

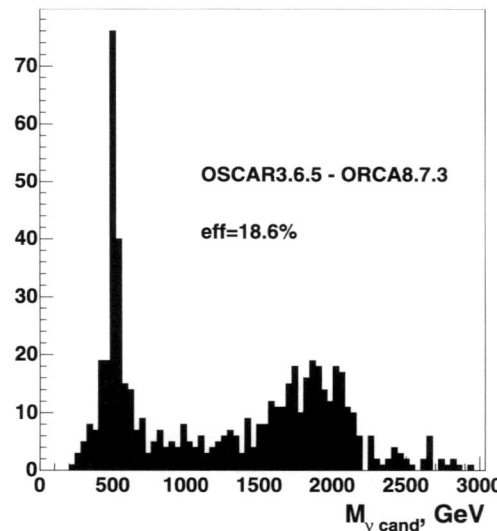

Fig. 64 Distribution of the invariant mass $m_{N_e}^{\text{cand}}$ for signal events with a heavy neutrino with $m_N = 500$ GeV. The two possible electron assignments are shown. The normalization is arbitrary

of discovery comparing events with leptons having the same and opposite charges will be an excellent cross check.

For the values $M_{W_R} = 2$ TeV, $m_N = 500$ GeV selected the reconstructed N mass peak is well visible, though the background is significant (comparable to the peak height). However, if an invariant mass $M_{eejj} > 1$ TeV is required, the background under the heavy neutrino peak drops dramatically, resulting in the mass distribution shown in Fig. 65 (left). The reconstructed W_R mass peak is shown in Fig. 65 (right).

The discovery potential is calculated using the criterion [300]

$$S = 2\left(\sqrt{N_S + N_B} - \sqrt{N_B}\right) \geq 5, \tag{123}$$

where N_S and N_B are the numbers of signal and background events respectively. The discovery limits in the (M_{W_R}, m_N) plane are shown in Fig. 66, for luminosities of 1, 10 and 30 fb^{-1}. After three years of running at low luminosity (30 fb^{-1}) this process would allow one to discover W_R and N with masses up to 3.5 and 2.3 TeV, respectively. For $M_{W_R} = 2$ TeV and $m_N = 500$ GeV discovery could be possible already after one month of running at low luminosity.

The influence of background uncertainties in these results is small since the background itself is rather small and the discovery region is usually limited by the fast drop of the signal cross section at high ratios m_N/M_{W_R} or by the fast drop of efficiency at small m_N/M_{W_R}. Signal cross section uncertainties from PDFs have been estimated by taking different PDF sets, finding changes of about 6% in the discovery region. No change of acceptance has been observed. Assuming a rather pessimistic value of 6% as the PDF uncertainty, it is easy to estimate from Fig. 63 that the uncertainty for the upper boundary of the discovery region is of 1–2%, and for the lower boundary of 2–3%.

4.3.3 Heavy neutrino pair production

New heavy neutrinos can be produced in pairs by the exchange of an s-channel neutral gauge boson. Since ZNN couplings are quadratically suppressed, NN production is only relevant when mediated by an extra Z' boson. For example, in E_6 grand unification both new Z' bosons and heavy neutrinos appear. If $M_{Z'} > 2m_N$, like-sign dilepton signals from Z' production and subsequent decay $Z' \to NN \to \ell^\pm W^\mp \ell^\pm W^\mp$ can be sizable. As it has been remarked before, like-sign dilepton signals have moderate (although not negligible) backgrounds. These are further reduced for heavier neutrino masses, when the charged leptons from the signal are more energetic and background can be suppressed demanding a high transverse momentum for both leptons.

Fig. 65 *Left*: reconstructed heavy neutrino mass peak including the SM background (textithistogram) and background only (*shaded histogram*). *Right*: the same for the W_R mass peak. In both cases an $eejj$ invariant mass above 1 TeV is required. The integrated luminosity is 30 fb^{-1}

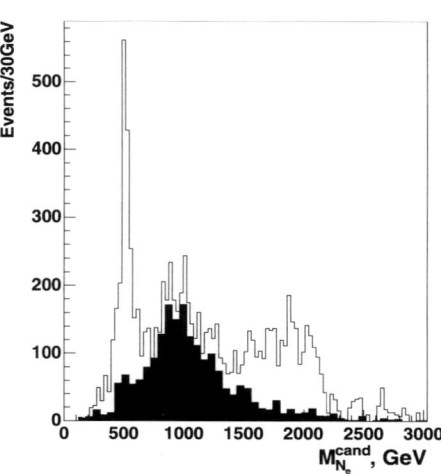
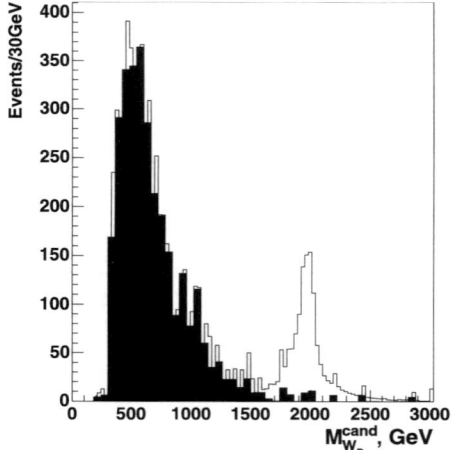

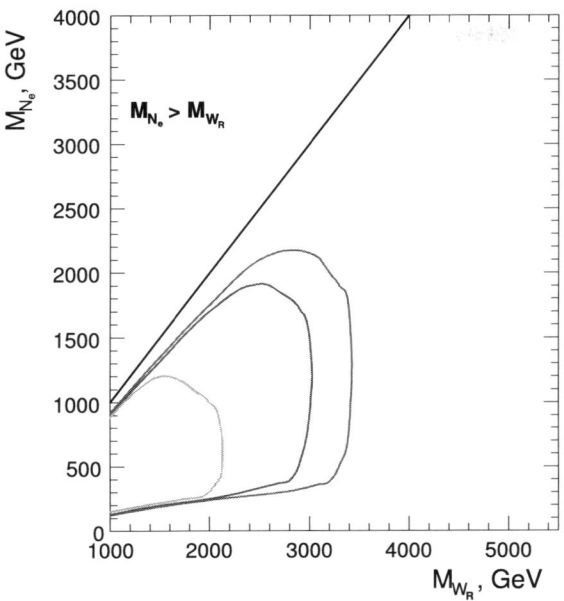

Fig. 66 CMS discovery potential for heavy Majorana neutrinos from W_R decays for integrated luminosities of 30 fb^{-1} (*red, outer contour*), 10 fb^1 (*blue, middle*) and 1 fb^{-1} (*green, inner contour*)

A striking possibility happens when the new Z' boson is leptophobic (see also the next section). If the new Z' does not couple to light charged leptons the direct limits from $p\bar{p} \to Z' \to \ell^+\ell^-$ searches at Tevatron do not apply, and the Z' could be relatively light, $M_{Z'} \gtrsim 350$ GeV. A new leptophobic Z' boson in this mass range could lead to like-sign dilepton signals observable already at Tevatron. For LHC, the 5σ sensitivity reaches $M_{Z'} = 2.5$ TeV, $m_N = 800$ GeV for a luminosity of 30 fb^{-1} [472].

To conclude this section a final comment is in order. In the three heavy neutrino production processes examined we have considered heavy Majorana neutrinos which are singlets under the SM group (seesaw type I), produced through standard or new interactions. Majorana neutrinos lead to the relatively clean LNV signature of two like-sign dileptons, but it should be pointed out that like-sign dilepton signals arise also in the other seesaw scenarios: from the single production of doubly charged scalar triplets (seesaw type II) [281], and in pair production of lepton triplets (seesaw type III) [473]. For this reason, like-sign dileptons constitute an interesting final state in which to test seesaw at LHC. Of course, additional multi-lepton signatures are characteristic of type-II (see Sect. 4.6 for a discussion on scalar triplets) and type-III seesaw, and they might help reveal the nature of seesaw at LHC.

4.4 New neutral gauge bosons

Many models beyond the SM introduce new neutral gauge bosons, generically denoted by Z'. GUTs with groups larger than SU(5) always predict the existence of at least one Z' boson. Their mass is not necessarily of the order of the unification scale $M_{\text{GUT}} \sim 10^{15}$ GeV, but on the contrary, one (or some) of these extra bosons can be "light", that is, at the TeV scale or below. Well-known examples are E_6 grand unification [63] and left–right models [474] (for reviews see also [344, 475]). Theories with extra dimensions with gauge bosons propagating in the bulk predict an infinite tower of KK excitations $Z^{(n)} = Z^{(1)}, Z^{(2)}, \ldots$, $\gamma^{(n)} = \gamma^{(1)}, \gamma^{(2)}, \ldots$. The lightest ones $Z^{(1)}, \gamma^{(1)}$, can have a mass at the TeV scale, and a phenomenology similar to Z' gauge bosons [476, 477]. Little Higgs models enlarge the SU(2)$_L \times$ U(1)$_Y$ symmetry and introduce new gauge bosons as well, e.g. in the littlest Higgs models based on [SU(2) $\times$ U(1)]^2 two new bosons Z_H, A_H appear, with masses expected in the TeV range.

The production mechanisms and decay modes of Z' bosons depend on their coupling to SM fermions.[22] These couplings are not fixed even within a class of models. For example, depending on the breaking pattern of E_6 down to the SM, the lightest Z' has different couplings to quarks and leptons or, in other words, quarks and leptons have different U(1)$'$ hypercharges. Three common breaking patterns are labeled as ψ, χ and η, and the corresponding "light" Z' as Z'_ψ, Z'_χ, Z'_η. Thus, the constraints on Z' bosons, as well as the discovery potential for future colliders refer to particular Z' models.

Present limits on Z' bosons result from precise measurements at the Z pole and above at LEP, and from the non-observation at Tevatron. Z pole measurements constrain the Z–Z' mixing, which would induce deviations in the fermion couplings to the Z. For most popular models the mixing is required to be of order of few 10^{-3} [344] (as emphasized above, limits depend on the values assumed for the Z' couplings). Measurements above the Z pole in fermion pair and W^+W^- production set constraints on the mass and mixing of the Z'. The non-observation at Tevatron in $u\bar{u}, d\bar{d} \to Z' \to \ell^+\ell^-$ sets lower bounds on $M_{Z'}$. For most common models they are of the order of 700–800 GeV [478], with an obvious dependence on the values assumed for the coupling to u, d quarks and charged leptons. LHC will explore the multi-TeV mass region and might discover a Z' with very small luminosity, for masses of the order of 1 TeV. Below we summarize the prospects for "generic" Z' bosons (for example those arising in E_6 and left–right models), which couple to quarks and leptons without any particular suppression. In this case, $u\bar{u}, d\bar{d} \to Z' \to e^+e^-, \mu^+\mu^-$ gives very clean signals and has an excellent sensitivity to

---

[22]Decays to new fermions and bosons, if any, are also possible but usually ignored in most analyses. When included they decrease the branching ratio to SM fermions, and then they lower the signal cross sections and discovery potential in the standard modes.

Fig. 67 Resonance signal (*white histograms*) and Drell–Yan background (*shaded histograms*) for KK $Z^{(1)}/\gamma^{(1)}$ boson production with $M = 4.0$ TeV (*left*), and Z'_{SM} with $M = 3.0$ TeV (*right*), with an integrated luminosity of 30 fb^{-1} (from CMS full simulation)

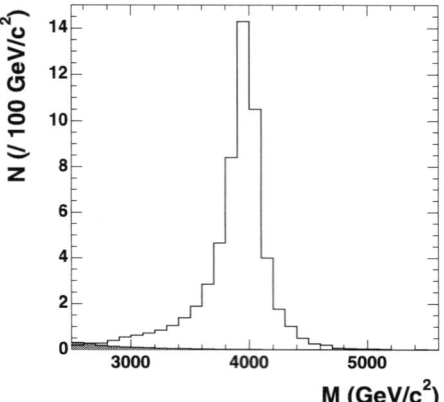
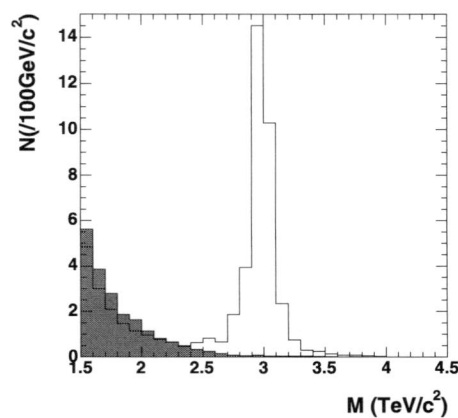

search for Z' bosons [479–482]. Then we examine the situation when lepton couplings are suppressed, in which case other Z' decay channels must be explored.

4.4.1 Z' bosons in the dilepton channel

4.4.1.1 Discovery potential The dilepton decay channel provides a clean signature of a Z' boson. The presence of this heavy particle would be detected by the observation of a resonance peak in the dilepton mass spectrum over the SM background, the largest one coming from the Drell–Yan process $q\bar{q} \to \gamma/Z \to \ell^+\ell^-$. Reducible backgrounds like QCD jets and γ jets can be suppressed mainly by applying isolation cuts and requirements on the energy deposited in the hadronic calorimeter. This is illustrated in Fig. 67 for KK excitations of the Z/γ and a "reference" Z'_{SM} (sometimes denoted as Z'_{SSM} as well) with the same couplings as the Z, in the e^+e^- decay channel. These distributions have been obtained with a full simulation of the CMS detector. More details of the analyses can be found in [483] for the e^+e^- channel and in [484, 485] for the $\mu^+\mu^-$ channel.

The discovery potential is obtained using likelihood estimators [302] suited for small event samples. The e^+e^- and $\mu^+\mu^-$ channels provide similar results, with some advantage for e^+e^- at lower Z' masses. A comparison between both is given in Fig. 68 for the E_6 Z'_ψ and the reference Z'_{SM}. For masses of 1 TeV, a luminosity of 0.1 fb^{-1} would suffice to discover the Z' bosons in most commonly used scenarios, such as Z'_ψ, Z'_χ, Z'_η mentioned above, left–right models and KK $Z^{(1)}/\gamma^{(1)}$. For a luminosity of 30 fb^{-1}, 5σ significance in the e^+e^- channel can be achieved for masses ranging up to 3.3 TeV (Z'_ψ) and 5.5 TeV ($Z^{(1)}/\gamma^{(1)}$). ATLAS studies obtain a similar sensitivity [486]. Theoretical uncertainties result from the poor knowledge of PDFs in the high x and high Q^2 domain, and from higher-order QCD and EW corrections (K factors), and they amount to 10–20%. Nevertheless, measurements of real data outside the mass peak regions will reduce this uncertainty to a large extent.

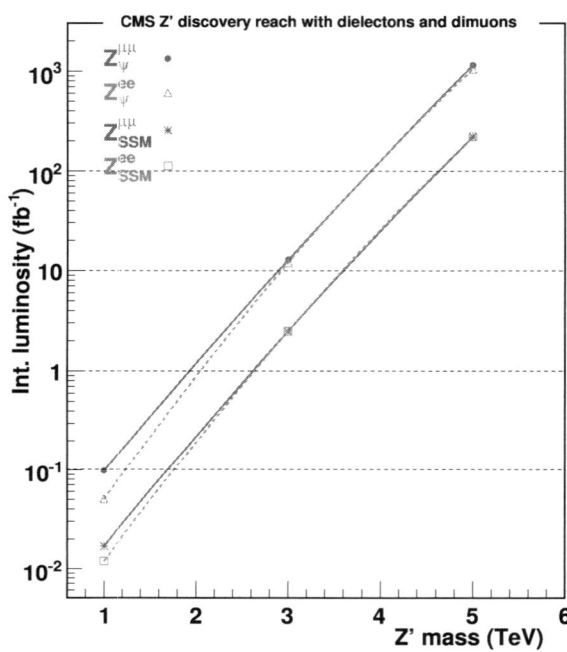

Fig. 68 5σ discovery limit as a function of the resonance mass for two examples of Z' bosons, in the e^+e^- (*red, dashed lines*) and $\mu^+\mu^-$ (*blue, solid lines*) channels (from CMS full simulation)

4.4.1.2 Z' and implications on new physics Once a new resonance decaying to $\ell^+\ell^-$ ($\ell = e, \mu$) is found, information about the underlying theory can be extracted with the study of angular distributions and asymmetries. The first step is the determination of the particle spin, what can be done with the help of the ℓ^- distribution in the $\ell^+\ell^-$ rest frame [487]. Let us denote by θ^* the angle between the final ℓ^- and the initial quark.[23] The $\cos\theta^*$ distribution is obvi-

---

[23] In pp collisions the quark direction is experimentally ambiguous because the quark can originate from either proton with equal probability. The sign ambiguity in $\cos\theta^*$ can be resolved assuming that the overall motion of the $\ell^+\ell^-$ system is in the direction of the initial quark (which gives a good estimation because the fraction of proton momentum car-

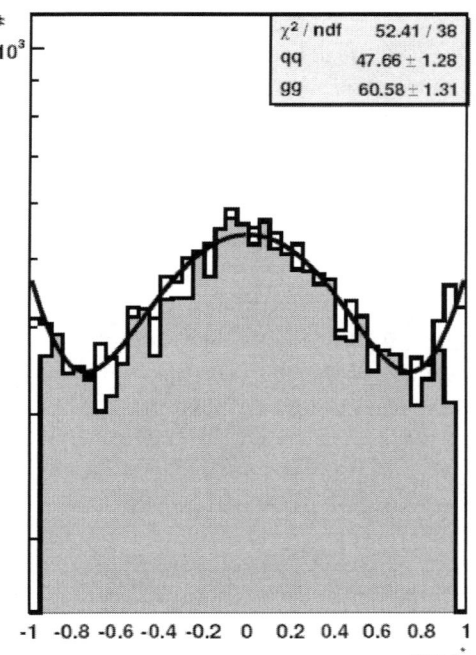
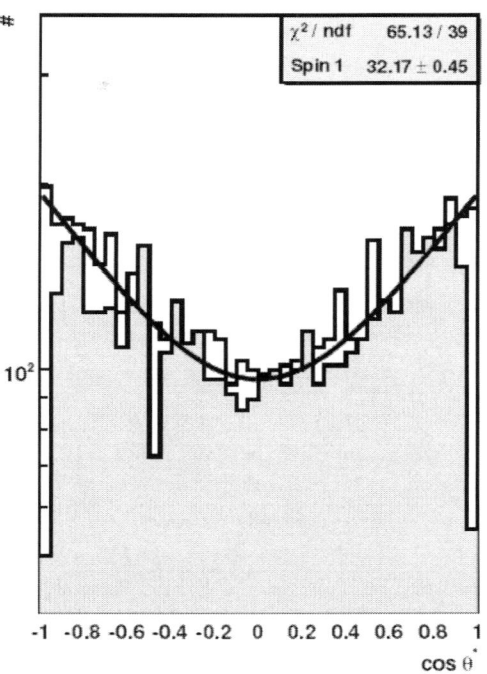

Fig. 69 Angular distributions for a 3 TeV graviton (*left*) and Z' boson (*right*) in the dimuon decay channel. Open histograms correspond to generated-level data, while colored histograms show events after full CMS detector simulation and reconstruction. Theoretical fits to Monte Carlo data are overlayed

ously flat for a scalar particle. For a spin-1 particle (γ, Z or Z') it is given by

$$\frac{d\sigma}{d\cos\theta^*} = \frac{3}{8}\left[1 + \cos^2\theta^*\right] + A_{\text{FB}}\cos\theta^*$$
$$(\gamma, Z, Z'), \qquad (124)$$

where the coefficient of the $\cos\theta^*$ term A_{FB} depends on the Z' couplings to quarks and leptons. (The $\cos\theta^*$ forward–backward asymmetry is equal to this coefficient, hence our choice of notation.) For a spin-2 graviton G the corresponding distribution is

$$\frac{d\sigma}{d\cos\theta^*} = \frac{5}{8}\left[1 - 3\epsilon_q \cos^2\theta^* + (\epsilon_g - 4\epsilon_q)\cos^4\theta^*\right]$$
$$(G). \qquad (125)$$

The constants ϵ_q and ϵ_g are the relative contributions of the two processes in which gravitons can be produced, $q\bar{q} \to G$ and $gg \to G$, which are fixed for a given mass M_G and depend on the PDFs. The method in [487] uses only the even terms in the $\cos\theta^*$ distribution (thus avoiding the dependence on the Z' model and the $\cos\theta^*$ sign ambiguity). It has been applied to the dimuon decay channel in [489]. Figure 69 shows the $\cos\theta^*$ distributions for a 3 TeV graviton and Z'. Both distributions are rather different, and the two spin hypotheses can be distinguished already with a relative small number of events. Table 26 contains, for different masses and coupling parameters c (cross sections are proportional to $|c|^2$), the integrated luminosity required to discriminate at the 2σ level between the spin-1 and spin-2 hypotheses. The cross sections for Z' bosons are assumed to be equal to the ones for gravitons with the given masses and c values. In the five cases the required signal is in the range 150–200 events, and larger for a larger number N_B of background events as one may expect. Since the production cross sections fall steeply with the mass, the integrated luminosity required for spin discrimination increases with M (and decreases for larger c). Distinguishing from the spin-0 hypothesis (a flat distribution) is harder, and requires significantly more events than discriminating spin 2 from spin 1, as discussed in [487].

It should be remarked that, apart from the direct spin determination, a Z' and a graviton can be distinguished by their decay modes. Indeed, the latter can decay to $\gamma\gamma$, and the discovery significance in this final state is equal or bet-

ried by quarks is larger in average) and taking into account the probability for a "wrong" choice. Additionally, the transverse momenta of the incoming partons is not known, and it is generally believed that optimal results are achieved by using the Collins–Soper angle θ^*_{CS} [488] as the estimation for θ^*.

Table 26 Number of signal events N_S required to discriminate at the 2σ level between the spin-1 and spin-2 hypotheses, in the presence of N_B background events (see the text). From full CMS detector simulation

M (TeV)	c	L (fb^{-1})	N_S	N_B
1.0	0.01	50	200	87
1.0	0.02	10	146	16
1.5	0.02	90	174	41
3.0	0.05	1200	154	22
3.0	0.10	290	154	22

ter than in the electron and muon channels. On the contrary, $Z' \to \gamma\gamma$ does not happen at the tree level.

The various Z' models are characterized by different parity-violating Z' couplings to quarks and leptons, reflected in different coefficients of the linear $\cos\theta^*$ term in (124). This coefficient can be measured with a technique described in [485] for the dimuon decay channel. A_{FB} is extracted using an unbinned maximum likelihood fit to events in a suitable window around the $\mu^+\mu^-$ invariant-mass peak. The fit is based on a probability density function built from several observables, including $\cos\theta^*_{CS}$ (as an estimation of the true $\cos\theta^*$). The values obtained for A_{FB} are shown in Fig. 70 for six different Z' models: the Z'_ψ, Z'_χ and Z'_η from E_6 unification, a left–right model (LRM) [474], an "alternative left–right model" (ALRM) [490] and the "benchmark" Z'_{SM}. With an integrated luminosity of 400 fb^{-1} at CMS, one can distinguish between either a Z'_χ or Z'_{ALRM} and one of the four other models with a significance level above 3σ up to a Z' mass between 2 and 2.7 TeV. One can distinguish among the four other models up to $M_{Z'} = 1 - 1.5$ TeV, whereas Z'_{ALRM} and Z'_χ are indistinguishable for $M_{Z'} > 1$ TeV.

Additional observables, like rapidity distributions [491] or the off-peak asymmetries [492] can be used to further discriminate between Z' models. We finally point out that in specific models the Z' boson may have other characteristic decay channels, which would then identify the underlying theory or provide hints towards it. One such example is the decay $Z_H, A_H \to Zh$ in little Higgs models [493], which could be observable [413]. Contrarily, in Z' models from GUTs this decay would be generically suppressed by the small Z–Z' mixing, and it is unlikely to happen.

4.4.1.3 Z' and fermion masses In models which address fermion mass generation, one can go a step further and try to relate fermion masses with other model parameters. This is the case, for instance, of extensions of the RS [142] scenario, where the SM fields (except the Higgs boson) are promoted to bulk fields. If the SM fermions acquire various localizations along the extra dimension, they provide an interpretation for the large mass hierarchies among the different flavors. Within the framework of the RS model with bulk matter, collider phenomenology and flavor physics are interestingly connected: the effective 4 dimensional couplings between KK gauge boson modes and SM fermions depend on fermion localizations along the extra dimension, which are fixed (non-uniquely) by fermion masses.[24] Here we test the observability of KK excitations of the photon and Z boson at LHC in the electron channel, $pp \to \gamma^{(n)}/Z^{(n)} \to e^+e^-$. Previous estimations for RS models are given in [149], under the simplifying assumption of a universal fermion location.

The fit of EW precision data typically imposes the bound $M_{KK} \gtrsim 10$ TeV [149, 152]. However, if the EW gauge symmetry is enlarged to $SU(2)_L \times SU(2)_R \times U(1)_X$ [429], agreement of the S, T parameters is possible for $M_{KK} \gtrsim 3$ TeV. The localization of the (t_L, b_L) doublet towards the TeV brane (necessary to generate the large top quark mass) in principle generates deviations in the Zb_Lb_L coupling (see also the next subsection), what can be avoided with a $O(3)$ custodial symmetry [432]. In the example presented here, the SM quark doublets are embedded in bidoublets $(2, 2)_{2/3}$ under the above EW symmetry, as proposed in [432] and in contrast with [429]. Motivated by having gauge representations symmetric between the quark and lepton sector, the lepton doublets are embedded into bidoublets $(2, 2)_0$. This guarantees that there are no modifications of the $Z\ell_L\ell_L$ couplings.

The simulation of $Z^{(n)}/\gamma^{(n)}$ production [494] is obtained after implementing the new processes in PYTHIA. Only $n = 0, 1, 2$ are considered, since the contributions of KK excitations with $n \geq 3$ are not significant. The cross section depends on the fermion localizations which are clearly model-dependent. In Fig. 71 we show the e^+e^- invariant-mass distribution for two different fermion localization scenarios labeled as A and B [495], both with $M_{KK} = 3$ TeV. These scenarios are in agreement with all present data on quark and lepton masses and mixings [495], in the minimal SM extension where neutrinos have Dirac masses. Furthermore, for both sets FCNC processes are below the experimental limit if $M_{KK} \gtrsim 1$ TeV. In Fig. 71 we observe that the signal can be easily extracted from the physical SM background as an excess of Drell–Yan events compared to the SM expectation.

4.4.2 Z' in hadronic channels

Z' bosons with suppressed coupling to leptons ("leptophobic" or "hadrophilic") have theoretical interest on their own. They were first introduced some time ago [496–498] on a

---

[24]Fermion masses are determined up to a global factor by the fermion localizations (which generate the large hierarchies) as well as by 3×3 matrices in flavor space with entries of order unity. Then, the relation between masses and couplings is not unique, but involves additional parameters (four 3×3 matrices).

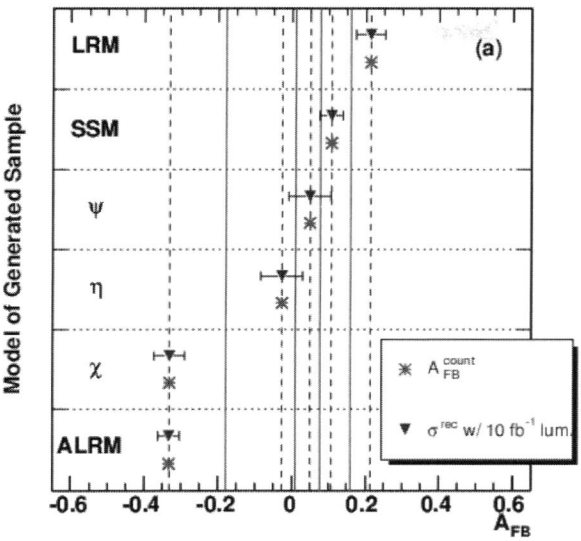
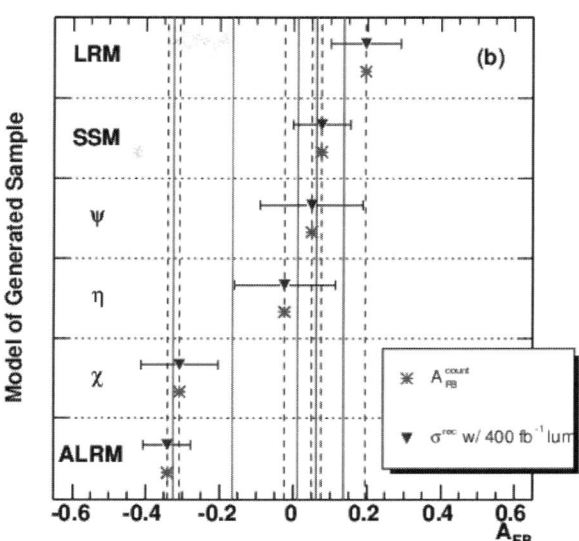

Fig. 70 Theoretical values A_{FB}^{count} (*dotted lines* and *asterisks*) and reconstructed values A_{FB}^{rec} (*triangles*) of the A_{FB} coefficient in (124), obtained for different models (see the text), with $M_{Z'} = 1$ TeV (*left*) and $M_{Z'} = 3$ TeV (*right*). The *solid vertical lines* are halfway between the adjacent values of A_{FB}^{count}. The error bars on the A_{FB}^{rec} triangles show the 1σ error scaled to 10 fb^{-1} (for $M_{Z'} = 1$ TeV) and 400 fb^{-1} (for $M_{Z'} = 3$ TeV). Obtained from CMS full detector simulation

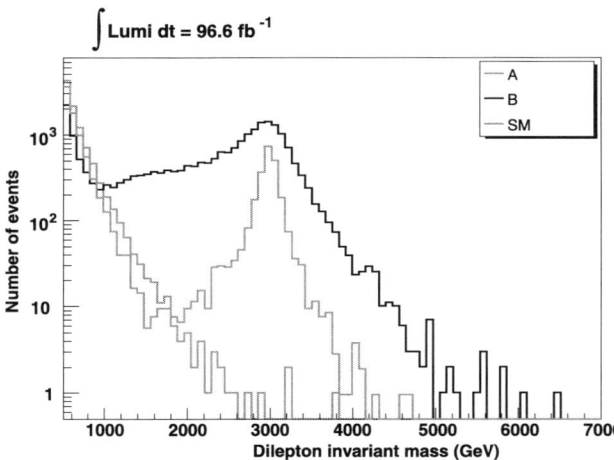

Fig. 71 Distribution of the e^+e^- invariant mass for $Z^{(n)}/\gamma^{(n)}$ production in two scenarios (A and B) for the fermion localizations and the SM background. The number of events corresponds to an integrated luminosity of 96.6 fb^{-1}

purely phenomenological basis, in an attempt to explain reported 3.5σ and 2.5σ deviations in R_b and R_c, respectively, observed by the LEP experiments at the Z pole. In order to accommodate these deviations without spoiling the good agreement for the leptonic sector, the Z' couplings to b, c were required to be much larger than those to charged leptons, so that the deviations in the Zbb, Zcc couplings induced by a small $Z - Z'$ mixing were significant for quarks but not for charged leptons. As a bonus, the introduction of leptophobic Z' bosons seemed to explain an apparent excess of jet events at large transverse momenta measured by CDF.

With more statistics available the deviations in R_b, R_c have disappeared, and SM predictions are now in good agreement with experiment. Nevertheless, a 2.7σ discrepancy in A_{FB}^b has remained until now. This deviation might well be due to some uncontrolled systematic error. But, if one accepts the A_{FB}^b measurement, explaining it with new physics contributions while keeping the good agreement for R_b is quite hard. One possibility has recently arisen in the context of RS models, where the introduction of a custodial symmetry [432] protects the Zb_Lb_L coupling from corrections due to mixing with the $Z^{(1)}$. Zb_Rb_R is allowed to receive a new contribution from mixing, which could explain the anomaly in A_{FB}^b. Alternatively, one may allow for deviations in Zb_Lb_L and Zb_Rb_R, chosen so as to fit the experimental values of R_b and A_{FB}^b [499]. The new $Z^{(1)}$ state has a mass of 2–3 TeV and suppressed couplings to charged leptons. Hence, it can be produced at LHC but mainly decays to quark-antiquark pairs. Leptophobic Z' bosons can also appear in grand unified theories as E_6 [480, 500].

Studies of the CMS sensitivity to narrow resonances in the dijet final states have been performed [501]. Experimental searches in the dijet channel are challenging because of the large QCD background and the limited dijet mass resolution. All new particles with a natural width significantly smaller than the measured dijet mass resolution should all appear as a dijet mass resonance of the same line shape in the detector. Thus, a generic analysis search has been developed to extract cross section sensitivities, which are compared to the expected cross sections from different models (excited quarks, axigluons, colorons, E_6 diquarks, color octet technirhos, W', Z', and RS gravitons), to determine the mass range for which we expect to be able to discover or exclude these models of dijet resonances. The size of the cross section is a determining factor in whether the model can be discovered, as illustrated in Fig. 72 for a sequential Z'_{SM} and other new states. For a luminosity of 10 fb^{-1} the Z'_{SM} signal is about one order of magnitude below the 5σ discovery limit for all the mass range, and a discovery is not possible. Conversely, if agreement is found with the SM expectation, Z'_{SM} masses between 2.1 and 2.5 TeV can be excluded (see Fig. 72).

For resonances decaying to $t\bar{t}$ preliminary studies have been performed in [502]. With 300 fb^{-1}, a 500 GeV resonance could be discovered for a cross section (including branching ratio to $t\bar{t}$) of 1.5 pb. For masses of 1 TeV and 3 TeV, the necessary signal cross sections are 650 and 11 fb, respectively.

4.5 New charged gauge bosons

Extensions of the SM gauge group including an additional SU(2) factor imply the existence of new bosons $W'^{\pm}$ (as well as an extra Z' boson, whose phenomenology has been described in the previous section). Two well-known examples are left–right models, in which the electroweak gauge group is SU(2)$_L \times$ SU(2)$_R \times$ U(1), and littlest Higgs models (those with group [SU(2) $\times$ U(1)]^2). As for the neutral case, the interactions of new W' bosons depend on the specific model considered. For example, in left–right models the new charged bosons (commonly denoted as W_R) have purely right handed couplings to fermions, whereas in littlest Higgs models they are purely left handed, as the ordinary W boson. Low energy limits are correspondingly different. In the former case the kaon mass difference sets a limit on the W_R mass of the order of two TeV [503]. This stringent limit is due to an enhancement of the "LR" box diagram contribution involving W and W_R exchange [504], compared to the "LL" exchange of two charged bosons with left handed couplings. On the other hand, in little Higgs models (especially in its minimal versions like the littlest Higgs model) precision electroweak data are quite constraining, and require the W' masses to be of the order of several TeV [505, 506].

4.5.1 Discovery potential

Most studies for W' discovery potential have focused on a W' boson with SM-like couplings to fermions and WZ, Wh decays suppressed. The present direct limit from Tevatron is $m_{W'} > 965$ GeV with 95% CL [507]. Previous ATLAS studies have shown that a W' boson could be observed in the leptonic decay channel $W' \to \ell \nu_\ell$, $\ell = \mu, e$, if it has a mass up to 6 TeV with 100 fb^{-1} of integrated luminosity [508]. For CMS the expectations are similar [509]. Here the possible detection of a W' signal in the muon decay channel is investigated, focusing on masses in the range 1–2.5 TeV and using the full simulation of the ATLAS detector. The signal has been generated with PYTHIA using CTEQ6L structure functions. The resulting cross sections times branching ratio, as well as the W' width for various masses, are given in Table 27 (left). The W' can be identified as a smeared Jacobian peak in the transverse mass distribution, built with the muon transverse momentum and the transverse missing energy $\not\!\!E_T$. Figure 73 shows the smearing of the edge after full simulation of the ATLAS detector.

In addition to the signal, there are contributions from the various SM backgrounds originating from the processes given in Table 27 (right). The W background is irreducible, but all the other backgrounds can be reduced applying the appropriate selections. In Table 28 the selection cuts used for the background rejection are shown.

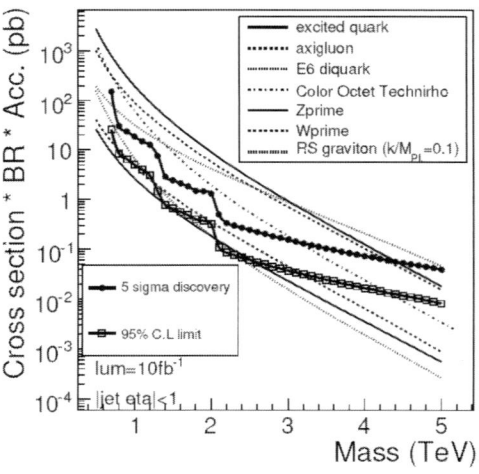

Fig. 72 5σ discovery limits (*circles*) and 95% upper bounds (*squares*) for resonances decaying to two jets, as a function of their mass. The luminosity is of 10 fb^{-1} and a full simulation of the CMS detector is used. The predictions of several models are also shown

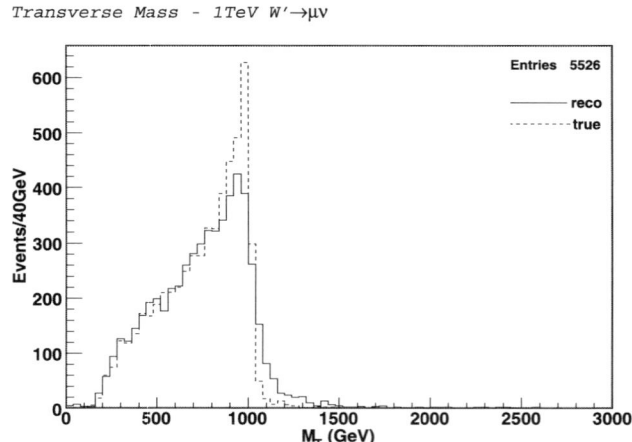

Fig. 73 Generated and reconstructed transverse mass distribution for a simulated 1 TeV W', before any detector effects and after full simulation of the ATLAS detector

Table 27 *Left*: expected cross section times branching ratio for the $W' \to \mu\nu$ signal, and total W' width. *Right*: cross section times branching ratio for the main background processes

Signal: $pp \to W' \to \mu\nu_\mu + X$		
$m_{W'}$ (TeV)	$\sigma \times$ Br(pb)	Γ_{tot} (GeV)
1.0	3.04	34.7
1.5	0.57	52.6
2.0	0.15	70.5
2.5	0.047	88.5

SM background processes	$\sigma \times$ Br(nb)
$pp \to W \to \mu\nu_\mu + X$	15
$pp \to W \to \tau\nu_\tau \to \mu\nu_\mu\nu_\tau + X$	2.6
$pp \to Z \to \mu^-\mu^+ + X$	1.5
$pp \to Z \to \tau^-\tau^+ \to \mu + X$	0.25
$pp \to t\bar{t} \to Wb W\bar{b} \to l\nu_l + X$	0.46
QCD (all dijet processes)	5×10^5

Table 28 Cross-section times branching ratio to muons and relative efficiencies after each cut. The cuts correspond to: (1) $p_T > 100$ GeV and $\not{E}_T > 50$ GeV; (2) b jet Veto; (3) JetVeto; (4) muon isolation and quality

Cut	1 TeV W'		2 TeV W'		W (off-shell)		$t\bar{t}$		Z (off-shell)	
	σ (pb)	eff (%)	σ (pb)	eff (%)	σ (pb)	eff (%)	σ (pb)	eff (%)	σ (pb)	eff (%)
1	2.52	82.8	0.126	84.0	2.04	74.4	8.878	1.93	0.251	9.89
2	2.45	80.7	0.122	81.4	1.99	72.8	1.610	0.35	0.244	9.62
3	2.23	73.3	0.104	69.4	1.95	71.1	0.966	0.21	0.237	9.34
4	2.18	71.6	0.101	67.3	1.91	69.8	0.736	0.16	0.232	9.15

The main signature of the signal is the presence of an energetic muon together with a significant missing transverse momentum in the event. When searching for a W' with mass of 1 TeV or heavier, events that contain at least one reconstructed muon with $p_T > 100$ GeV and missing transverse momentum $\not{E}_T > 50$ GeV are selected. These cuts mainly eliminate the $t\bar{t}$ background, which tends to have less energetic muons, and Z production, which in general does not have significant missing energy. Muons coming from W' decays are isolated, i.e. they do not belong to a jet. Isolated muons are identified by requiring that the calorimetric energy deposited inside the difference of a small and a bigger cone around the muon track is less than $E_{\text{cal}}^{02} - E_{\text{cal}}^{01} < 10$ GeV, where the cones '01' and '02' are determined, respectively, as the ones which have $\Delta R = 0.1, 0.2$. This double cone strategy is adopted because muons from W' decays are very energetic and therefore can have significant, almost collinear radiation. Figure 74 shows the distribution of calorimetric energy contained in the difference of the two cones for both signal and background. It is evident that the above cut reduces mainly the $t\bar{t}$ background. Events with exactly one isolated muon are selected. Z background events contain mostly two isolated muons, except for the cases where one of the muons lies in a region outside the muon spectrometer ($|\eta| > 2.7$) or is not reconstructed. These cases account for about the 30% of the high mass Z events and remain as irreducible background. QCD and $t\bar{t}$ backgrounds contain in most cases non-isolated muons coming from jets. In order to eliminate the QCD dijet background, which contains one jet misidentified as a muon, events with additional high energy jets, with $p_T > 200$ GeV, are rejected (JetVeto). The $t\bar{t}$ background is further reduced by applying a b jet veto cut (in ATLAS the jet tagging is done for jets with $p_T > 15$ GeV). Muons coming from cosmic rays and b decays are rejected with track quality criteria, what ensures that the muon track is well reconstructed. Specifically, cuts are applied on the χ^2 probability over the number of degrees of freedom and the transverse d_0 and longitudinal z_0 perigee parameters: $\text{Prob}(\chi^2)/DoF > 0.001$, $d_0/\Delta(d_0) < 10$, $z_0 < 300$ mm.

After the application of all the signal separation requirements the transverse mass distribution, shown in Fig. 75, has been statistically analyzed to determine the significance of the discovery. First, for each W' mass the transverse mass interval which gives the best discovery significance is determined. The corresponding number of signal and background events for 10 fb^{-1} are presented in Table 29. The minimum luminosity to have a 5σ significant discovery is also calculated and shown in Table 30. The significance is calculated assuming Poisson statistics. The errors in the luminosity correspond to a 5% systematic uncertainty in the signal (mainly due to the variation of PDFs) and a 20% systematic uncertainty in the background (due to several different contributions). The uncertainties in the NLO corrections (K factors) are expected to influence both the signal and the background in a similar way. The experimental systematic uncertainties are expected to be reduced only after the first data taking using the control samples of Z and W events. A control sample will also be formed in the transverse mass region between 200 and 400 GeV, which will provide the final check for the

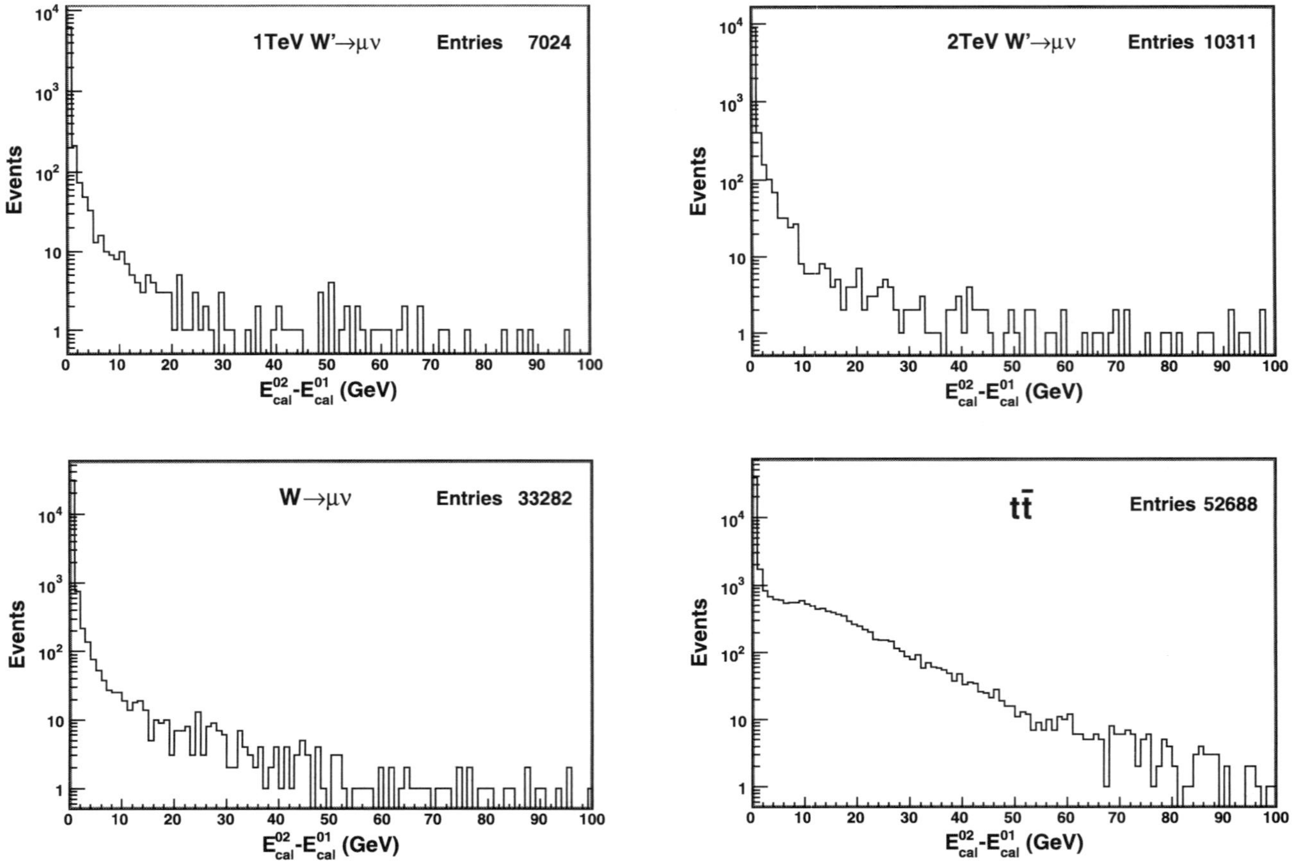

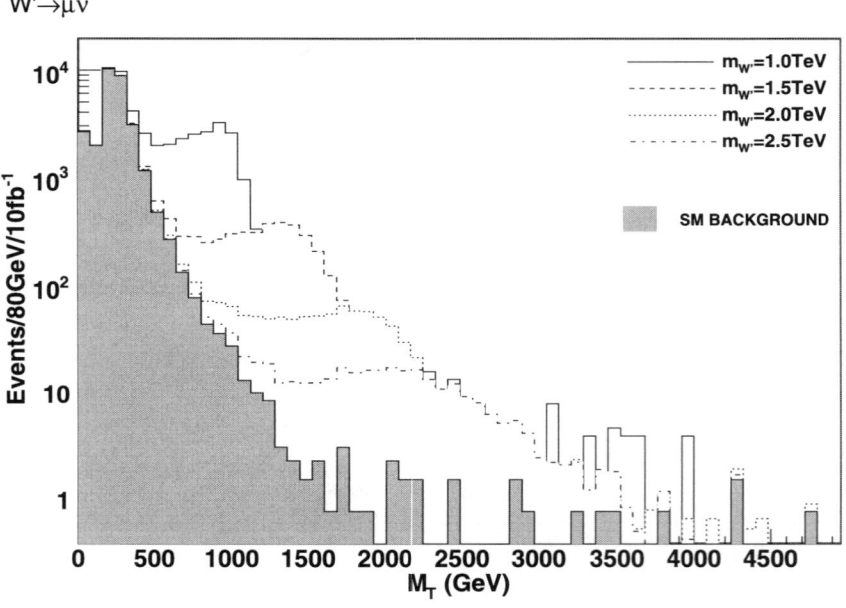

Fig. 74 Distribution of calorimetric energy contained in the difference of two cones with $\Delta R = 0.1$ and $\Delta R = 0.2$ for both signal and background events

Fig. 75 Transverse mass distribution of the SM background and W' signals corresponding to different W' masses, plotted on top of the background for an integrated luminosity of 10 fb^{-1}

systematic uncertainties collectively, concerning the scale as well as the shape of the background.

A new W' boson with SM-like couplings to fermions can be discovered with low integrated luminosity during the ini-

Table 29 Number of signal and background events expected for 10 fb^{-1} of integrated luminosity, for various W' masses. The best search windows in the transverse mass distribution (M_T) are also shown

$M_{W'}$	1.0 TeV	1.5 TeV	2.0 TeV	2.5 TeV
M_T (TeV)	0.6–1.7	0.9–2.0	1.2–2.9	1.6–3.2
Signal events	15753 ± 787	3059 ± 153	603 ± 30	225 ± 11
SM Background events	469 ± 94	76 ± 15	22 ± 5	15 ± 3

Table 30 Minimum luminosity required in order to have a 5σ discovery for various W' masses. N_S, N_B stand for the number of signal and background events, respectively, within the optimal transverse mass window

$M_{W'}$ (TeV)	Luminosity (pb^{-1})	N_S	N_B
1.0	3.0 ± 0.3	4.7	0.14
1.5	14.6 ± 1.4	4.5	0.11
2.0	84 ± 9	5.1	0.18
2.5	283 ± 31	6.4	0.42

tial LHC running. With 0.3 fb^{-1} integrated luminosity, a W' can be discovered in the ATLAS experiment with a mass up to 2.5 TeV. Imposing the additional requirement of observing at least 10 candidate signal events would rise the minimum luminosity to 0.5 fb^{-1}.

The present study so far has been performed without pileup and cavern background conditions. Both these conditions are not expected to affect much the results since the initial run will be at very low luminosity and moreover the majority of the muons of the signal concentrate in the barrel region. Nevertheless, studies for the fake reconstruction with both kinds of background are under way.

Finally, we point out that the experimental resolution for muons with p_T ranging from 0.5 to 1 TeV is about 5–10%, giving an experimental width larger than the intrinsic width, shown in Table 27 (left). Therefore, no further attempt has been made for discriminating the underlying theory based on the W' width. However, following the W' discovery, the muonic decay channel could provide valuable information concerning the FB asymmetry, which in turn could be used to discriminate between various theoretical models.

4.6 New scalars

Additional scalars appear in several theories beyond the SM to address some of the theoretical puzzles left unsolved by the SM. For example:

— 2-Higgs-doublet models, possibly relevant to the origin of the CP asymmetry,
— Little Higgs models, introduced to solve the hierarchy problem,
— Babu–Zee model, providing a source for the neutrino mass differences.

The 2-Higgs-doublet model (2HDM) contains two Higgs fields, one to give mass to SM gauge bosons and the other one remaining with CP violating terms [510]. The additional two neutral Higgs particles aim to solve the strong CP problem and explain the observed baryon asymmetry of the universe with minimum impact to the SM. Such a model can be easily investigated at LHC via either direct observation of the non-SM Higgs particles or indirectly via the enhancement to the FCNC Higgs decays involving the top quark. The details of such a discovery and of possible discrimination between the models can be found in Sect. 2.3.2.

Little Higgs models [394, 398, 399] aim to solve the hierarchy problem, without imposing a symmetry between fermions and bosons. Instead, the unwanted contributions from the loops are removed via the same spin counterparts of the involved SM particles: top quark, W and Z bosons and the Higgs itself. The coupling coefficients of these predicted particles are connected to their SM counterparts via the symmetries of the larger group embedding the SM gauge group. Depending on the selection of the embedding group, these models predict a variety of new particles. Additional charge +2/3 quarks (studied in Sect. 4.2.1), a number of spin 1 bosons and a number of scalars, with masses less than 2, 5 and 10 TeV respectively. The smallest of these symmetry groups defines the littlest Higgs model which predicts three nearly degenerate scalar particles with charges 2, 1 and 0. Experimentally, the doubly charged scalar is the most appealing one, since its manifestation would be two like-sign leptons or W bosons when produced singly [281, 413, 511], or two like-sign lepton pairs with equal invariant mass when produced in pairs [283, 413, 512]. More generally, scalar triplets appear in various type-II seesaw models. For scalar triplets in supersymmetric models see Sect. 3.5.

The Babu–Zee model, independently proposed by Zee [513] and Babu [514], proposes a particular radiative mass generation mechanism. This mechanism might help understanding the origin of neutrino masses and mixing angles which are firmly established by the neutrino oscillation experiments [515–518]. The model introduces two new charged scalars h^+ and k^{++}, both singlets under SU(2)$_L$, which couple only to leptons. Neutrino masses in this model arise at the two-loop level. Since present experimental neutrino data requires at least one neutrino to have a mass of the order of $O(0.05)$ eV [204] an estimation for the value of neutrino masses in the model indicates that for couplings f and h of order $O(1)$ (see (135)) the new scalars should have masses in the range $O(0.1–1)$ TeV [519]. The model is therefore potentially testable at the LHC.

4.6.1 Scalar triplet seesaw models

An important open issue to be addressed in the context of little Higgs models [394, 398, 399] is the origin of neutrino masses [443, 520, 521]. A neutrino mass generation mechanism which naturally occurs in these models is type II seesaw [276, 522, 523], which employs a scalar with the $SU(2)_L \times U(1)_Y$ quantum numbers $\Phi \sim (3, 2)$. The existence of such a multiplet in some little Higgs models [399, 524] is a direct consequence of the global $[SU(2) \times U(1)]^2$ symmetry breaking which makes the SM Higgs light. Although Φ is predicted to be heavier than the SM Higgs boson, the little Higgs philosophy implies that its mass could be of order $O(1)$ TeV. Due to its specific quantum numbers the triplet Higgs boson only couples to the left-chiral lepton doublets $L_i \sim (2, -1)$, $i = e, \mu, \tau$, via Yukawa interactions given by

$$\mathcal{L}_\Phi = i \bar{L}_i^c \tau_2 Y_{ij} (\tau \cdot \Phi) L_j + \text{h.c.}, \quad (126)$$

where Y_{ij} are Majorana Yukawa couplings. The interactions in (126) induce LFV decays of charged leptons which have not been observed. The most stringent constraint on the Yukawa couplings comes from the upper limit on the tree-level decay $\mu \to eee$ [281, 525]

$$Y_{ee} Y_{e\mu} < 3 \times 10^{-5} (M_{\Phi^{++}}/\text{TeV})^2, \quad (127)$$

with $M_{\Phi^{++}}$ the mass of the doubly charged scalar, constrained by direct Tevatron searches to be $M_{\Phi^{++}} \geq 136$ GeV [526, 527]. Experimental bounds on the tau Yukawa couplings are much less stringent.

According to (126), the neutral component of the triplet Higgs boson Φ^0 couples to left handed neutrinos. When it acquires a VEV v_Φ, it induces nonzero neutrino masses m_ν given by the mass matrix

$$(m_\nu)_{ij} = Y_{ij} v_\Phi. \quad (128)$$

We assume that the smallness of neutrino masses is due to the smallness of v_Φ. In this work the tau Yukawa coupling is taken to be $Y_{\tau\tau} = 0.01$, and the rest of couplings are scaled accordingly. In particular, hierarchical neutrino masses imply $Y_{ee}, Y_{e\mu} \ll Y_{\tau\tau}$, consistent with present experimental bounds.

In this framework there is a possibility to perform direct tests of the neutrino mass generation mechanism at the LHC, via pair production and subsequent decays of scalar triplets.

Here the Drell–Yan pair production of the doubly charged component

$$pp \to \Phi^{++} \Phi^{--} \quad (129)$$

is studied, followed by leptonic decays [281, 283, 512, 528–531]. Notice that in this process (i) the production cross section only depends on $M_{\Phi^{++}}$ and known SM parameters; (ii) the smallness of v_Φ in this scenario, due to the smallness of neutrino masses, implies that decays $\Phi^{++} \to W^+ W^+$ are negligible; (iii) the Φ^{++} leptonic decay branching fractions do not depend on the size of the Yukawa couplings but only on their ratios, which are known from neutrino oscillation experiments. For the normal hierarchy of neutrino masses and a very small value of the lightest neutrino mass, the triplet seesaw model predicts $\text{Br}(\Phi^{++} \to \mu^+ \mu^+) \simeq \text{Br}(\Phi^{++} \to \tau^+ \tau^+) \simeq \text{Br}(\Phi^{++} \to \mu^+ \tau^+) \simeq 1/3$. This scenario is testable at LHC experiments.

The production of the doubly charged scalar has been implemented in the PYTHIA Monte Carlo generator [47]. Final and initial state interactions and hadronization have been taken into account. Four-lepton backgrounds with high p_T leptons arise from three SM processes, namely $t\bar{t}$, $t\bar{t}Z$ and ZZ production. PYTHIA has been used to generate $t\bar{t}$ and ZZ background, while CompHEP was used to generate the $t\bar{t}Z$ background via its PYTHIA interface [420, 428]. CTEQ5L parton distribution functions have been used. Additional four-lepton backgrounds exist involving b-quarks in the final state, for example, $b\bar{b}$ production. Charged leptons from such processes are very soft, and these backgrounds can be eliminated [61]. Possible background processes from physics beyond the SM are not considered.

A clear experimental signature is obtained from the peak in the invariant mass of two like-sign muons and/or tau leptons:

$$\left(m^\pm_{\ell_1 \ell_2}\right)^2 = \left(p^\pm_{\ell_1} + p^\pm_{\ell_2}\right)^2, \quad (130)$$

where $p^\pm_{\ell_1, \ell_2}$ are the four-momenta of two like-sign leptons $\ell_1^\pm$, $\ell_2^\pm$. Since like-sign leptons originate from the decay of a doubly charged Higgs boson, their invariant-mass peaks at $m^\pm_{\ell \ell'} = M_{\Phi^{\pm\pm}}$. The four-muon final state allows one to obtain invariant masses directly from (130). In channels involving one or several tau leptons, which are seen as τ jets or secondary muons (marked as μ' below), the momenta of the latter has to be corrected according to the equation system

$$\vec{p}_{\tau_i} = k_i \vec{p}_{\text{jet}_i / \mu'_i}, \quad (131)$$

$$\vec{p}_T = \sum_i (\vec{p}_{\nu_i})_T, \quad (132)$$

$$m^+_{\ell_1 \ell_2} = m^-_{\ell_1 \ell_2}, \quad (133)$$

where i counts τ leptons, $(\vec{p}_{\nu_i})_T$ is the vector of transverse momentum of the produced neutrinos, $\vec{E}_T$ is the vector of missing transverse momentum (measured by the detector) and $k_i > 1$ are positive constants. Equation (131) describes the standard approximation that the decay products of a highly boosted τ are collinear. Equation (132) assumes missing transverse energy only to be comprised of neutrinos from τ decays. In general, it is not a high-handed simplification, because the other neutrinos in the event are much less energetic and the detector error in missing energy is order of magnitude smaller [5]. Using the first two equations it is possible to reconstruct up to two τ leptons per event. The additional requirement of (133) allows one to reconstruct a third τ, although very small measurement errors are needed.

A clear signal extraction from the SM background can be achieved using a set of selection rules imposed on a reconstructed event in the following order.

— *S1*: events with at least two positive and two negative muons or jets which have $|\eta| < 2.4$ and $p_T > 5$ GeV are selected.
— *S2*: The scalar sum of transverse momenta of the two most energetic muons or τ jets has to be larger than a certain value (depending on the Φ^{++} mass range studied).
— *S3*: If the invariant mass of a pair of opposite charge muons or τ jets is close to the Z boson mass (85–95 GeV), then the particles are eliminated from the analysis.
— *S4*: as $\Phi^{\pm\pm}$ are produced in pairs, their reconstructed invariant masses have to be equal (in each event). The condition

$$0.8 < m^{++}_{\ell_1\ell_2}/m^{--}_{\ell_1\ell_2} < 1.2 \qquad (134)$$

has been used. If the invariant masses are in this range then they are included in the histograms, otherwise it is assumed that some muon may originate from τ decay, and it is attempted to find corrections to their momenta according to the method described above.

An example of invariant-mass distribution after applying selection rules is shown in Fig. 76 for $M_{\Phi^{++}} = 500$ GeV. A tabulated example is given for $M_{\Phi^{++}} = 200$, 500 and 800 GeV in Table 31, corresponding to a luminosity $L = 30$ fb^{-1}. The strength of the S2 cut is clearly visible: almost no decrease in signal while the number of the background events descends close to its final minimum value. A peculiar behavior of S4—reducing the background, while also increasing the signal in its peak—is the effect of applying the $\tau \to \mu'$ correction method described above.

As it is seen in Table 31, the SM background can be practically eliminated. In such an unusual situation the log likelihood ratio (LLR) statistical method [131, 532] has been used to determine the 5σ discovery potential, demanding a significance larger than 5σ in 95% of "hypothetical experiments", generated using a Poisson distribution. With this criterion, Φ^{++} up to 300 GeV can be discovered in the first year of LHC ($L = 1$ fb^{-1}) and Φ^{++} up to 800 GeV can be discovered for the integrated luminosity $L = 30$ fb^{-1}. Therefore, the origin of neutrino mass can possibly be directly tested at LHC.

4.6.2 The discovery potential of the Babu–Zee model

The new charged scalars of the model introduce new gauge invariant Yukawa interactions, namely

$$\mathcal{L} = f_{\alpha\beta}\epsilon_{ij}\left(L_\alpha^{Ti} C L_\beta^j\right)h^+ + h'_{\alpha\beta}(e_\alpha^T C e_\beta)k^{++} + \text{h.c.} \qquad (135)$$

Here, L are the SM (left handed) lepton doublets, e the charged lepton singlets, α, β are generation indices and ϵ_{ij} is the completely antisymmetric tensor. Note that f is antisymmetric while h' is symmetric. Assigning $L = 2$ to h^- and k^{--}, the Lagrangian in (135) conserves lepton number. Lepton number violation in the model resides only in the following term in the scalar potential:

$$\mathcal{L} = -\mu h^+ h^+ k^{--} + \text{h.c.} \qquad (136)$$

Vacuum stability arguments can be used to derive an upper bound for the lepton number violating coupling μ [533], namely, $\mu \leq (6\pi^2)^{1/4} m_h$. The structure of (135) and (136) generates Majorana neutrino masses at the two-loop level [519, 533].

Constraints on the parameter space of the model come from neutrino physics experimental data and from the experimental upper bounds on lepton flavor violation (LFV) processes. Constraints on the antisymmetric couplings f_{xy} are entirely determined by neutrino mixing angles and depend on the hierarchy of the neutrino mass spectrum, which in this model can be normal or inverse. Analytical expressions for the ratios $\epsilon = f_{13}/f_{23}$ and $\epsilon' = f_{12}/f_{23}$, as well as numerical upper and lower bounds, were calculated in [533] and [519].

The requirement of having a large atmospheric mixing angle indicates that the symmetric Yukawa couplings h_{xy} ($x, y = \mu, \tau$) must follow the hierarchy $h_{\tau\tau} \simeq (m_\mu/m_\tau)h_{\mu\tau} \simeq (m_\mu/m_\tau)^2 h_{\mu\mu}$. The couplings h_{ee}, $h_{e\mu}$ and $h_{e\tau}$ are constrained by LFV of the type $l_a \to l_b l_c l_d$ and have to be smaller than 0.4, 4×10^{-3} and 7×10^{-2} [533]. The most relevant constraint on m_k come from the LFV processes $\tau \to 3\mu$ while for m_h is derived from $\mu \to e\gamma$. Lower bounds for both scalar masses can be found [519], the results are $m_k \gtrsim 770$ GeV, $m_h \gtrsim 200$ GeV (normal hierarchy case), and $m_h \gtrsim 900$ GeV (inverse hierarchy case). In [533] it has been estimated that at the LHC discovery of k^{++} might be possible up to masses of $m_k \leq 1$ TeV approximately. In the following it will therefore be assumed that

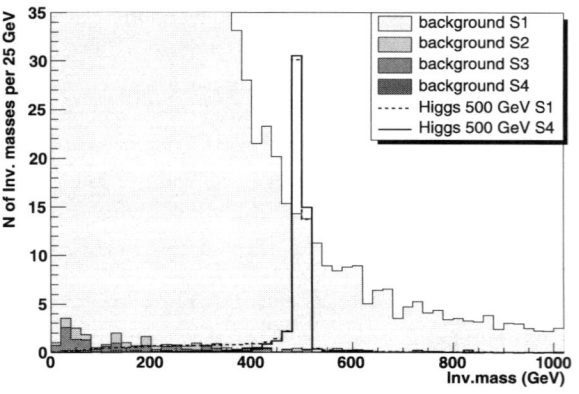
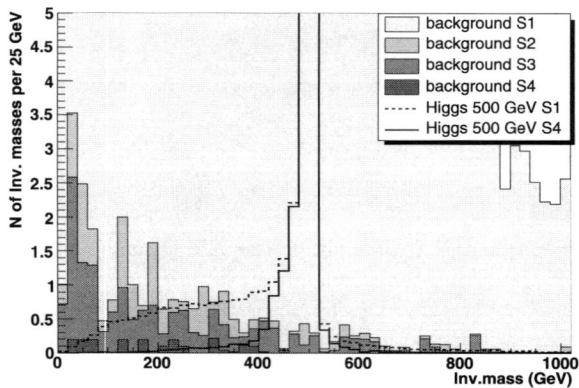

Fig. 76 Distribution of invariant masses of like-sign pairs after applying selection rules (S1–S4) for scalar mass $M_{\Phi^{++}} = 500$ GeV and the SM background ($L = 30$ fb^{-1}). The *histogram in the right panel* is a zoom of the *left histogram* to illustrate the effects of the selection rules S2–S4

Table 31 Effectiveness of the selection rules for the background and signal. All event numbers in the table are normalized for $L = 30$ fb^{-1}. The numbers in brackets mark errors at 95% confidence level for Poisson statistics. The signal increases after S4 due to the reconstructed $\tau \to \mu'$ decays

Process	N of like-sign pairs				
	N of Φ	S1	S2	S3	S4
Energy range 150–250 GeV					
$M_\Phi = 200$ GeV	4670	1534	1488	1465	1539
$t\bar{t} \to 4\ell$	–	1222 (168)	172 (8.5)	134 (6.9)	17.6 (3.7)
$t\bar{t}Z$	–	21.3 (4.0)	15.5 (1.0)	6.3 (1.2)	2.2 (1.1)
ZZ	–	95.0 (12.0)	22.5 (0.7)	9.8 (0.5)	1.7 (0.2)
Energy range 375–625 GeV					
$M_\Phi = 500$ GeV	119.2	48.4	47.5	46.8	49.5
$t\bar{t} \to 4\ell$	-	178 (28)	2.1 (0.9)	1.65 (0.87)	0.10 (0.35)
$t\bar{t}Z$	–	6.6 (1.7)	2.3 (1.0)	1.0 (1.0)	0.00 (0.1)
ZZ	–	9.4 (2.9)	1.4 (0.2)	0.68 (0.19)	0.08 (0.09)
Energy range 600–1000 GeV					
$M_\Phi = 800$ GeV	11.67	5.05	5.00	4.92	5.21
$t\bar{t} \to 4\ell$	-	77 (12)	0.00 (0.22)	0.00 (0.22)	0.00 (0.07)
$t\bar{t}Z$	–	2.6 (1.2)	0.39 (0.4)	0.39 (0.4)	0.00 (0.1)
ZZ	–	2.5 (0.8)	0.34 (0.16)	0.17 (0.09)	0.00 (0.02)

$m_k \leq 1$ TeV and, in addition, $m_h \leq 0.5$ TeV. The notation $\text{Br}(h^+ \to l_\alpha \sum_\beta \nu_\beta) = \text{Br}_h^{l_\alpha}$ and $\text{Br}(k^{++} \to l_\alpha l_\beta) = \text{Br}_k^{l_\alpha l_\beta}$ will be used. h^+ decays are governed by the parameters ϵ and ϵ'. Using the current 3σ range for neutrino mixing angles [204] it is possible to predict

$$\text{Br}_h^e = [0.13, 0.22], \qquad \text{Br}_h^\mu = [0.31, 0.50],$$
$$\text{Br}_h^\tau = [0.18, 0.35], \tag{137}$$

$$\text{Br}_h^e = [0.48, 0.50], \qquad \text{Br}_h^\mu = [0.17, 0.34],$$
$$\text{Br}_h^\tau = [0.18, 0.35]. \tag{138}$$

This is for the normal hierarchy (137) or the inverse hierarchy (138).

The doubly charged scalar decay either to two same-sign leptons or to two h^+ final states. Lepton pair final states decays are controlled by the $h_{\alpha\beta}$ Yukawa couplings while the lepton flavor-violating decay $k^{++} \to h^+ h^+$ is governed by the μ parameter (see (136)). The hierarchy among the couplings $h_{\mu\mu}, h_{\mu\tau}$ and $h_{\tau\tau}$ result in the prediction

$$\text{Br}_k^{\mu\tau}/\text{Br}_k^{\mu\mu} \simeq (m_\mu/m_\tau)^2,$$
$$\text{Br}_k^{\tau\tau}/\text{Br}_k^{\mu\mu} \simeq (m_\mu/m_\tau)^4. \tag{139}$$

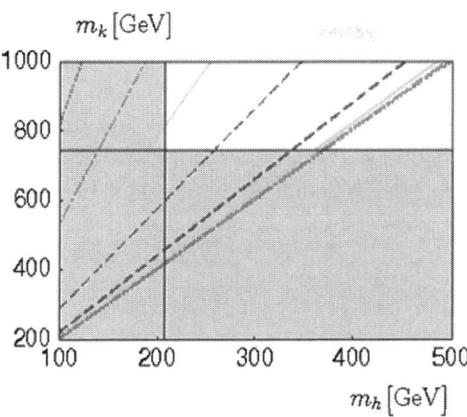
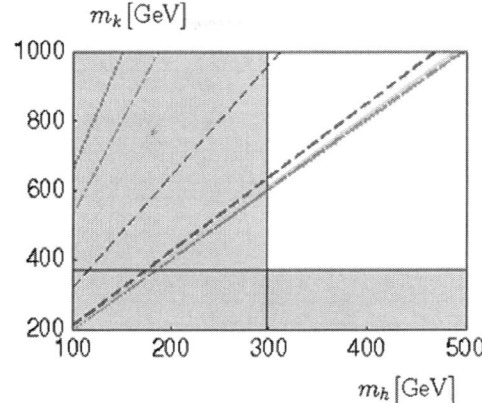

Fig. 77 Lines of constant Br($k^{++} \to h^+h^+$), assuming to the *left* $h_{\mu\mu} = 1$: Br$_k^{hh} = 0.1, 0.2, 0.3$ and 0.4 for *dotted, dash dotted, full* and *dashed line*. The *vertical line* corresponds to $m_h = 208$ GeV for which Br($\mu \to e\gamma) = 1.2 \times 10^{-11}$ and *horizontal line* to $m_k = 743$ GeV for which Br($\tau \to 3\mu) = 1.9 \times 10^{-6}$, i.e. parameter combinations to the *left/below* this line are forbidden. *Plot on the right* assumes $h_{\mu\mu} = 0.5$. Lines are for Br$_k^{hh} = 0.4, 0.5, 0.6$ and 0.7, *dotted, dash dotted, full* and *dashed line*. Again the *shaded regions* are excluded by Br($\mu \to e\gamma$) and Br($\tau \to 3\mu$)

Thus, the leptonic final states of k^{++} decays are mainly like-sign muon pairs.

Here it is important to notice that in general the decays $k^{++} \to e^+l^+$ ($l = e, \mu, \tau$) are strongly suppressed due to the LVF constraints on the h_{el} parameters. However, if the Yukawa coupling h_{ee} saturates its upper limit then electron pair final states can be possibly observed.

The branching ratio for the process $k^{++} \to h^+h^+$ reads

$$\mathrm{Br}(k^{++} \to h^+h^+) \simeq \frac{\mu^2 \beta}{m_k^2 h_{\mu\mu}^2 + \mu^2 \beta}. \tag{140}$$

Here β is the usual phase space suppression factor. From (140) it can be noted that if the process is kinematically allowed, the lepton violating coupling μ can be extracted by measuring this branching ratio. For $h_{\mu\mu} \lesssim 0.2$ the current limit on Br($\mu \to e\gamma$) rules out all $m_h \lesssim 0.5$ TeV, thus this measurement is possible only for $h_{\mu\mu} \gtrsim 0.2$. Note that smaller values of μ lead to smaller neutrino masses, thus upper bounds on the branching ratio for Br$_k^{hh}$ can be interpreted as upper limit on the neutrino mass in this model. Figure 77 shows the resulting branching ratios for two values of $h_{\mu\mu}$.

5 Tools [25]

5.1 Introduction

In this section we review the treatment of flavor aspects in publicly available calculational tools used in new physics studies at colliders. Such tools cover a wide range of applications; the wide variety of computer programs discussed here can be divided in broad classes as follows:

— Analytical precision calculations:
  the results of analytical precision calculations for specific observables, often at loop level, are coded and made available for public use. These observables are usually sets of single numbers, such as cross sections, decay widths, branching ratios etc., calculated for a specific point in the respective models parameter space. Examples for such tools covered here are HDECAY, SDECAY, FchDecay and FeynHiggs.

— Tools helping in or performing (mostly) analytical calculations:
  the best-known example of such a tool is the combination of FeynArts and FormCalc, whose treatment of flavor aspects is discussed here. FeynArts allows for a automated construction of Feynman diagrams, including higher-order effects, and the corresponding amplitude. FormCalc can then be used to evaluate the loop integrals in a semi-automated fashion.

— RGE codes:
  the RGE is solved numerically, to obtain the SUSY parameters at physical scales from the value of high energy inputs. These parameters are usually coupling constants, particle masses and widths and mixing matrices. For this purpose, a number of codes exist, here SPheno and SuSpect are presented. It should be noted that many of these RGE codes also embed a number of relevant cross sections, branching ratios etc.

— Matrix element generators/parton level generators:
  these codes calculate, in a automated fashion, cross sections for multileg tree-level processes. Usually, they are

---
[25] Section coordinators: F. Krauss, F. Moortgat, G. Polesello.

capable of generating weighted or unweighted events at the parton level, i.e. without showering or hadronization. This task is usually left for other programs, with the necessary information passed by some standard interface format [534]. Examples for this type of codes include CalcHep and HvyN.

— Full-fledged event generators:
these programs provide fully showered and hadronized events. Primary examples include PYTHIA, HERWIG, and Sherpa.

In addition interfaces are necessary to transfer data between the various programs as will be discussed in the next section.

All the programs described in this section are public and in the corresponding subsections the web pages are given from which the programs can be obtained. Updated information on the availability of the codes and the responsible persons as well as on additional packages can be found at http://www.ippp.dur.ac.uk/montecarlo/BSM/.

5.2 A summary of the SUSY Les Houches Accord 2

The states and couplings appearing in the general MSSM can be defined in a number of ways. Indeed, it is often advantageous to use different choices for different applications and hence no unique set of conventions prevails at present. In principle, this is not a problem; translations between different conventions can usually be carried out without ambiguity. From the point of view of practical application, however, such translations are, at best, tedious, and at worst they introduce an unnecessary possibility for error.

To deal with this problem, and to create a more transparent situation for non-experts, the original SUSY Les Houches Accord (SLHA1) was proposed [535]. This accord uniquely defines a set of conventions for supersymmetric models together with a common interface between codes. However, SLHA1 was designed exclusively with the MSSM with real parameters and R-parity conservation in mind. Some recent public codes [325, 342, 536–542] are either implementing extensions to this base model or are anticipating such extensions. We therefore here present extensions of the SLHA1 relevant for R-parity violation (RPV), flavor violation, and CP-violation (CPV) in the MSSM. We also consider next-to-minimal models which we shall collectively label by the acronym NMSSM. Full details of the SLHA2 agreement can be found in [543].

For simplicity, we still limit the scope of the SLHA2 in two regards: for the MSSM, we restrict our attention to *either* CPV or RPV, but not both. For the NMSSM, we define one catch-all model and extend the SLHA1 mixing only to include the new states, with CP, R-parity, and flavor still assumed conserved.

The conventions described here are a superset of those of the original SLHA1, unless explicitly stated otherwise. We use ASCII text for input and output, all dimensionful parameters are taken to be in appropriate powers of GeV, and the output formats for SLHA2 data BLOCKs follow those of SLHA1. All angles are in radians. In a few cases it has been necessary to replace the original conventions. This is clearly remarked upon in all places where it occurs, and the SLHA2 conventions then supersede the SLHA1 ones.

5.2.1 The SLHA2 conventions

5.2.1.1 Flavor violation The CKM basis is defined to be the one in which the quark mass matrix is diagonal. In the super-CKM basis [174] the squarks are rotated by exactly the same amount as their respective quark superpartners, regardless of whether this makes them (that is, the squarks) diagonal or not. Misalignment between the quark and squark sectors thus results in flavor off-diagonal terms remaining in the squark sector.

In this basis, the 6×6 squark mass matrices are defined as

$$\mathcal{L}_{\tilde{q}}^{\text{mass}} = -\Phi_u^\dagger \mathcal{M}_{\tilde{u}}^2 \Phi_u - \Phi_d^\dagger \mathcal{M}_{\tilde{d}}^2 \Phi_d, \qquad (141)$$

where $\Phi_u = (\tilde{u}_L, \tilde{c}_L, \tilde{t}_L, \tilde{u}_R, \tilde{c}_R, \tilde{t}_R)^T$ and $\Phi_d = (\tilde{d}_L, \tilde{s}_L, \tilde{b}_L, \tilde{d}_R, \tilde{s}_R, \tilde{b}_R)^T$. We diagonalize the squark mass matrices via 6×6 unitary matrices $R_{u,d}$, such that $R_{u,d} \mathcal{M}_{\tilde{u},\tilde{d}}^2 R_{u,d}^\dagger$ are diagonal matrices with increasing mass squared values. We re-define the existing PDG codes for squarks to enumerate the mass eigenstates in ascending order:

$(\tilde{d}_1, \tilde{d}_2, \tilde{d}_3, \tilde{d}_4, \tilde{d}_5, \tilde{d}_6)$
$= (1000001, 1000003, 1000005, 2000001,$
$\quad 2000003, 2000005),$
$(\tilde{u}_1, \tilde{u}_2, \tilde{u}_3, \tilde{u}_4, \tilde{u}_5, \tilde{u}_6)$
$= (1000002, 1000004, 1000006, 2000002,$
$\quad 2000004, 2000006).$

The flavor-violating parameters of the model are specified in terms of the CKM matrix together with five 3×3 matrices of soft SUSY-breaking parameters given in the super-CKM basis

$$\hat{m}_{\tilde{Q}}^2, \qquad \hat{m}_{\tilde{u}}^2, \qquad \hat{m}_{\tilde{d}}^2, \qquad \hat{T}_U, \qquad \hat{T}_D. \qquad (142)$$

Analogous rotations and definitions are used for the lepton flavor-violating parameters, in this case using the super-PMNS basis. Below, we refer to the combined basis as the super-CKM/PMNS basis.

5.2.1.2 R-parity violation We write the R-parity violating superpotential as

$$W_{\text{RPV}} = \epsilon_{ab}\left[\frac{1}{2}\hat{\lambda}_{ijk}L_i^a L_j^b \bar{E}_k + \hat{\lambda}'_{ijk}L_i^a Q_j^{bx}\bar{D}_{kx} - \hat{\kappa}_i L_i^a H_2^b\right]$$
$$+ \frac{1}{2}\hat{\lambda}''_{ijk}\epsilon^{xyz}\bar{U}_{ix}\bar{D}_{jy}\bar{D}_{kz}, \qquad (143)$$

where $x, y, z = 1,\ldots,3$ are fundamental $SU(3)_C$ indices and ϵ^{xyz} is the totally antisymmetric tensor in three dimensions with $\epsilon^{123} = +1$. In (143), $\hat{\lambda}_{ijk}$, $\hat{\lambda}'_{ijk}$ and $\hat{\kappa}_i$ break lepton number, whereas $\hat{\lambda}''_{ijk}$ violate baryon number. As in the previous section, all quantities are given in the super-CKM/super-PMNS basis. Note that in the R-parity violating case, the PMNS is an output once lepton number is violated.

The trilinear R-parity violating terms in the soft SUSY-breaking potential are

$$V_{3,\text{RPV}} = \epsilon_{ab}\left[\frac{1}{2}(\hat{T})_{ijk}\tilde{L}_{iL}^a \tilde{L}_{jL}^b \tilde{e}_{kR}^* + (\hat{T}')_{ijk}\tilde{L}_{iL}^a \tilde{Q}_{jL}^b \tilde{d}_{kR}^*\right]$$
$$+ \frac{1}{2}(\hat{T}'')_{ijk}\epsilon_{xyz}\tilde{u}_{iR}^{x*}\tilde{d}_{jR}^{y*}\tilde{d}_{kR}^{z*} + \text{h.c.} \qquad (144)$$

Note that we do not factor out the $\hat{\lambda}$ couplings (e.g. as in $\hat{T}_{ijk}/\hat{\lambda}_{ijk} \equiv A_{\lambda,ijk}$).

When lepton number is broken, additional bilinear soft SUSY-breaking potential terms can appear,

$$V_{2,\text{RPV}} = -\epsilon_{ab}\hat{D}_i\tilde{L}_{iL}^a H_2^b + \tilde{L}_{iaL}^\dagger \hat{m}^2_{\tilde{L}_i H_1} H_1^a + \text{h.c.}, \qquad (145)$$

and the sneutrinos may acquire vacuum expectation values (VEVs) $\langle\tilde{\nu}_{e,\mu,\tau}\rangle \equiv v_{e,\mu,\tau}/\sqrt{2}$. The SLHA1 defined the tree-level VEV v to be equal to $2m_Z/\sqrt{g^2 + g'^2} \sim 246$ GeV; this is now generalized to

$$v = \sqrt{v_1^2 + v_2^2 + v_e^2 + v_\mu^2 + v_\tau^2}. \qquad (146)$$

For $\tan\beta$ we maintain the SLHA1 definition, $\tan\beta = v_2/v_1$.

The Lagrangian contains the (symmetric) neutrino/neutralino mass matrix as

$$\mathcal{L}_{\tilde{\chi}^0}^{\text{mass}} = -\frac{1}{2}\tilde{\psi}^{0T}\mathcal{M}_{\tilde{\psi}^0}\tilde{\psi}^0 + \text{h.c.}, \qquad (147)$$

in the basis of 2-component spinors $\tilde{\psi}^0 = (\nu_e, \nu_\mu, \nu_\tau, -i\tilde{b}, -i\tilde{w}^3, \tilde{h}_1, \tilde{h}_2)^T$. We define the unitary 7×7 neutrino/neutralino mixing matrix N (block RVNMIX), such that:

$$-\frac{1}{2}\tilde{\psi}^{0T}\mathcal{M}_{\tilde{\psi}^0}\tilde{\psi}^0 = -\frac{1}{2}\underbrace{\tilde{\psi}^{0T}N^T}_{\tilde{\chi}^{0T}}\underbrace{N^*\mathcal{M}_{\tilde{\psi}^0}N^\dagger}_{\text{diag}(m_{\tilde{\chi}^0})}\underbrace{N\tilde{\psi}^0}_{\tilde{\chi}^0}, \qquad (148)$$

where the 7 (2-component) neutral leptons $\tilde{\chi}^0$ are defined strictly mass ordered, i.e. with the $1^{st},2^{nd},3^{rd}$ lightest corresponding to the mass entries for the PDG codes 12, 14, and 16, and the four heaviest to the PDG codes 1000022, 1000023, 1000025, 1000035.

Charginos and charged leptons may also mix in the case of L-violation. The Lagrangian contains

$$\mathcal{L}_{\tilde{\chi}^+}^{\text{mass}} = -\frac{1}{2}\tilde{\psi}^{-T}\mathcal{M}_{\tilde{\psi}^+}\tilde{\psi}^+ + \text{h.c.}, \qquad (149)$$

in the basis of 2-component spinors $\tilde{\psi}^+ = (e^+, \mu^+, \tau^+, -i\tilde{w}^+, \tilde{h}_2^+)^T$, $\tilde{\psi}^- = (e^-, \mu^-, \tau^-, -i\tilde{w}^-, \tilde{h}_1^-)^T$ where $\tilde{w}^\pm = (\tilde{w}^1 \mp \tilde{w}^2)/\sqrt{2}$. We define the unitary 5×5 charged fermion mixing matrices U, V, blocks RVUMIX, RVVMIX, such that

$$-\frac{1}{2}\tilde{\psi}^{-T}\mathcal{M}_{\tilde{\psi}^+}\tilde{\psi}^+$$
$$= -\frac{1}{2}\underbrace{\tilde{\psi}^{-T}U^T}_{\tilde{\chi}^{-T}}\underbrace{U^*\mathcal{M}_{\tilde{\psi}^+}V^\dagger}_{\text{diag}(m_{\tilde{\chi}^+})}\underbrace{V\tilde{\psi}^+}_{\tilde{\chi}^+}, \qquad (150)$$

where the generalized charged leptons $\tilde{\chi}^+$ are defined as strictly mass ordered, i.e. with the three lightest states corresponding to the PDG codes 11, 13, and 15, and the two heaviest to the codes 1000024, 1000037. For historical reasons, codes 11, 13, and 15 pertain to the negatively charged field while codes 1000024 and 1000037 pertain to the opposite charge. The components of $\tilde{\chi}^+$ in "PDG notation" would thus be $(-11,-13,-15,1000024,1000037)$. In the limit of CP conservation, U and V are chosen to be real.

R-parity violation via lepton number violation implies that the sneutrinos can mix with the Higgs bosons. In the limit of CP conservation the CP-even (-odd) Higgs bosons mix with real (imaginary) parts of the sneutrinos. We write the neutral scalars as $\phi^0 \equiv \sqrt{2}\,\text{Re}(H_1^0, H_2^0, \tilde{\nu}_e, \tilde{\nu}_\mu, \tilde{\nu}_\tau)^T$, with the mass term

$$\mathcal{L} = -\frac{1}{2}\phi^{0T}\mathcal{M}^2_{\phi^0}\phi^0, \qquad (151)$$

where $\mathcal{M}^2_{\phi^0}$ is a 5×5 symmetric mass matrix. We define the orthogonal 5×5 mixing matrix $\aleph$ (block RVHMIX) by

$$-\phi^{0T}\mathcal{M}^2_{\phi^0}\phi^0 = -\underbrace{\phi^{0T}\aleph^T}_{\Phi^{0T}}\underbrace{\aleph\mathcal{M}^2_{\phi^0}\aleph^T}_{\text{diag}(m^2_{\phi^0})}\underbrace{\aleph\phi^0}_{\Phi^0}, \qquad (152)$$

where Φ^0 are the neutral scalar mass eigenstates in strictly increasing mass order. The states are numbered sequentially by the PDG codes (25, 35, 1000012, 1000014, 1000016), regardless of flavor content.

We write the neutral pseudoscalars as $\bar{\phi}^0 \equiv \sqrt{2}\,\text{Im}(H_1^0, H_2^0, \tilde{\nu}_e, \tilde{\nu}_\mu, \tilde{\nu}_\tau)^T$, with the mass term

$$\mathcal{L} = -\frac{1}{2}\bar{\phi}^{0T}\mathcal{M}^2_{\bar{\phi}^0}\bar{\phi}^0, \qquad (153)$$

where $\mathcal{M}^2_{\bar{\phi}^0}$ is a 5×5 symmetric mass matrix. We define the 4×5 mixing matrix $\bar{\aleph}$ (block RVAMIX) by

$$-\bar{\phi}^{0T} \mathcal{M}^2_{\bar{\phi}^0} \bar{\phi}^0 = -\underbrace{\bar{\phi}^{0T} \bar{\aleph}^T}_{\bar{\Phi}^{0T}} \underbrace{\bar{\aleph} \mathcal{M}^2_{\bar{\phi}^0} \bar{\aleph}^T}_{\text{diag}(m^2_{\bar{\phi}^0})} \underbrace{\bar{\aleph} \bar{\phi}^0}_{\bar{\Phi}^0}, \qquad (154)$$

where $\bar{\Phi}^0$ are the pseudoscalar mass eigenstates in increasing mass order. The states are numbered sequentially by the PDG codes (36,1000017, 1000018,1000019), regardless of flavor composition. The Goldstone boson G^0 has been explicitly left out and the 4 rows of $\bar{\aleph}$ form a set of orthonormal vectors.

If the blocks RVHMIX, RVAMIX are present, they supersede the SLHA1 ALPHA variable/block.

The charged sleptons and charged Higgs bosons also mix in the 8×8 mass squared matrix $\mathcal{M}^2_{\phi^\pm}$, which we diagonalize by a 7×8 matrix C (block RVLMIX):

$$\mathcal{L} = -\underbrace{(H_1^{-*}, H_2^+, \tilde{e}^*_{L_i}, \tilde{e}^*_{R_j}) C^\dagger}_{\Phi^+} \underbrace{C \mathcal{M}^2_{\phi^\pm} C^\dagger}_{\text{diag}(\mathcal{M}^2_{\phi^\pm})} C \begin{pmatrix} H_1^- \\ H_2^{+*} \\ \tilde{e}_{L_k} \\ \tilde{e}_{R_l} \end{pmatrix}, \qquad (155)$$

where $i, j, k, l \in \{1, 2, 3\}$, $\alpha, \beta \in \{1, \ldots, 6\}$ and $\Phi^+ = \Phi^{-\dagger}$ are the charged scalar mass eigenstates arranged in increasing mass order. These states are numbered sequentially by the PDG codes (37,1000011,1000013,1000015, 2000011,2000013,2000015), regardless of flavor composition. The Goldstone boson G^- has been explicitly left out and the seven rows of C form a set of orthonormal vectors.

5.2.1.3 CP violation When CP symmetry is broken, quantum corrections cause mixing between the CP-even and CP-odd Higgs states. Writing the neutral scalar interaction eigenstates as $\phi^0 \equiv \sqrt{2}(\text{Re } H_1^0, \text{Re } H_2^0, \text{Im } H_1^0, \text{Im } H_2^0)^T$ we define the 3×4 mixing matrix S (block CVHMIX) by

$$-\phi^{0T} \mathcal{M}^2_{\phi^0} \phi^0 = -\underbrace{\phi^{0T} S^T}_{\Phi^{0T}} \underbrace{S^* \mathcal{M}^2_{\phi^0} S^\dagger}_{\text{diag}(m^2_{\phi^0})} \underbrace{S\phi^0}_{\Phi^0}, \qquad (156)$$

where $\Phi^0 \equiv (h_1^0, h_2^0, h_3^0)^T$ are the mass eigenstates arranged in ascending mass order; these states are numbered sequentially by the PDG codes (25,35,36), regardless of flavor composition.

For the neutralino and chargino mixing matrices, the default convention in SLHA1 is that they be real matrices. One or more mass eigenvalues may then have an apparent negative sign, which can be removed by a phase transformation on $\tilde{\chi}_i$ as explained in SLHA1 [535]. When going to CPV, the reason for introducing the negative-mass convention in the first place, namely maintaining the mixing matrices strictly real, disappears. We therefore here take all masses real and positive, with N, U, and V complex. This does lead to a nominal dissimilarity with SLHA1 in the limit of vanishing CP violation, but we note that the explicit CPV switch in MODSEL can be used to decide unambiguously which convention to follow.

For the remaining MSSM parameters we use straightforward generalizations to the complex case, see Sect. 5.2.2.4.

5.2.1.4 NMSSM We shall here define the next-to-minimal case as having exactly the field content of the MSSM with the addition of one gauge singlet chiral superfield. As to couplings and parameterizations, rather than adopting a particular choice, or treating each special case separately, below we choose instead to work at the most general level. Any particular special case can then be obtained by setting different combinations of couplings to zero. However, we do specialize to the SLHA1-like case without CP violation, R-parity violation, or flavor violation. Below, we shall use the acronym NMSSM for this class of models, but we emphasize that we understand it to relate to field content only, and not to the presence or absence of specific couplings.

In addition to the MSSM terms, the most general CP conserving NMSSM superpotential is (extending the notation of SLHA1):

$$W_{\text{NMSSM}} = W_{\text{MSSM}} - \epsilon_{ab} \lambda S H_1^a H_2^b + \frac{1}{3} \kappa S^3 + \mu' S^2 + \xi_F S, \qquad (157)$$

where W_{MSSM} is the MSSM superpotential, in the conventions of [535, (3)]. A non-zero λ in combination with a VEV $\langle S \rangle$ of the singlet generates a contribution to the effective μ term $\mu_{\text{eff}} = \lambda \langle S \rangle + \mu$, where the MSSM μ term is normally assumed to be zero, yielding $\mu_{\text{eff}} = \lambda \langle S \rangle$. The remaining terms represent a general cubic potential for the singlet; κ is dimensionless, μ' has dimension of mass, and ξ_F has dimension of mass squared. The soft SUSY-breaking terms relevant to the NMSSM are

$$V_{\text{soft}} = V_{2,\text{MSSM}} + V_{3,\text{MSSM}} + m_S^2 |S|^2 + \left(-\epsilon_{ab} \lambda A_\lambda S H_1^a H_2^b + \frac{1}{3} \kappa A_\kappa S^3 + B' \mu' S^2 + \xi_S S + \text{h.c.} \right), \qquad (158)$$

where $V_{i,\text{MSSM}}$ are the MSSM soft terms, in the conventions of [535, (5) and (7)].

At tree level, there are thus 15 parameters (in addition to m_Z which fixes the sum of the squared Higgs VEVs) that are relevant for the Higgs sector:

$$\tan\beta, \quad \mu, \quad m_{H_1}^2, \quad m_{H_2}^2, \quad m_3^2,$$
$$\lambda, \quad \kappa, \quad A_\lambda, \quad A_\kappa, \quad \mu', \quad (159)$$
$$B', \quad \xi_F, \quad \xi_S, \quad \lambda\langle S\rangle, \quad m_S^2.$$

The minimization of the effective potential imposes three conditions on these parameters, such that only 12 of them can be considered independent. For the time being, we leave it up to each spectrum calculator to decide on which combinations to accept. For the purpose of this accord, we note only that to specify a general model exactly 12 parameters from (159) should be provided in the input, including explicit zeroes for parameters desired "switched off". However, since $\mu = m_3^2 = \mu' = B' = \xi_F = \xi_S = 0$ in the majority of phenomenological constructions, for convenience we also allow for a six-parameter specification in terms of the reduced parameter list:

$$\tan\beta, \quad m_{H_1}^2, \quad m_{H_2}^2, \quad \lambda, \quad \kappa,$$
$$A_\lambda, \quad A_\kappa, \quad \lambda\langle S\rangle, \quad m_S^2. \quad (160)$$

To summarize, in addition to m_Z, the input to the accord should contain either 12 parameters from the list given in (159), including zeroes for parameters not present in the desired model, or it should contain six parameters from the list in (160), in which case the remaining six "non-standard" parameters, μ, m_3^2, μ', B', ξ_F, and ξ_F, will be assumed to be zero; in both cases the three unspecified parameters (as, e.g., $m_{H_1}^2$, $m_{H_2}^2$, and m_S^2) are assumed to be determined by the minimization of the effective potential.

The CP-even neutral scalar interaction eigenstates are $\phi^0 \equiv \sqrt{2}\,\mathrm{Re}\,(H_1^0, H_2^0, S)^T$. We define the orthogonal 3×3 mixing matrix S (block NMHMIX) by

$$-\phi^{0T}\mathcal{M}_{\phi^0}^2\phi^0 = -\underbrace{\phi^{0T}S^T}_{\Phi^{0T}}\underbrace{S\mathcal{M}_{\phi^0}^2 S^T}_{\mathrm{diag}(m_{\Phi^0}^2)}\underbrace{S\phi^0}_{\Phi^0}, \quad (161)$$

where $\Phi^0 \equiv (h_1^0, h_2^0, h_3^0)$ are the mass eigenstates ordered in mass. These states are numbered sequentially by the PDG codes (25,35,45). The format of BLOCK NMHMIX is the same as for the mixing matrices in SLHA1.

The CP-odd sector interaction eigenstates are $\bar\phi^0 \equiv \sqrt{2}\,\mathrm{Im}\,(H_1^0, H_2^0, S)^T$. We define the 2×3 mixing matrix P (block NMAMIX) by

$$-\bar\phi^{0T}\mathcal{M}_{\bar\phi^0}^2\bar\phi^0 = -\underbrace{\bar\phi^{0T}P^T}_{\bar\Phi^{0T}}\underbrace{P\mathcal{M}_{\bar\phi^0}^2 P^T}_{\mathrm{diag}(m_{\bar\Phi^0}^2)}\underbrace{P\bar\phi^0}_{\bar\Phi^0}, \quad (162)$$

where $\bar\Phi^0 \equiv (A_1^0, A_2^0)$ are the mass eigenstates ordered in mass. These states are numbered sequentially by the PDG codes (36,46). The Goldstone boson G^0 has been explicitly left out and the two rows of P form a set of orthonormal vectors.

If NMHMIX, NMAMIX blocks are present, they supersede the SLHA1 ALPHA variable/block.

The Lagrangian contains the (symmetric) 5×5 neutralino mass matrix as

$$\mathcal{L}_{\tilde\chi^0}^{\mathrm{mass}} = -\frac{1}{2}\tilde\psi^{0T}\mathcal{M}_{\tilde\psi^0}\tilde\psi^0 + \mathrm{h.c.}, \quad (163)$$

in the basis of 2-component spinors $\tilde\psi^0 = (-i\tilde b, -i\tilde w^3, \tilde h_1, \tilde h_2, \tilde s)^T$. We define the unitary 5×5 neutralino mixing matrix N (block NMNMIX), such that:

$$-\frac{1}{2}\tilde\psi^{0T}\mathcal{M}_{\tilde\psi^0}\tilde\psi^0 = -\frac{1}{2}\underbrace{\tilde\psi^{0T}N^T}_{\tilde\chi^{0T}}\underbrace{N^*\mathcal{M}_{\tilde\psi^0}N^\dagger}_{\mathrm{diag}(m_{\tilde\chi^0})}\underbrace{N\tilde\psi^0}_{\tilde\chi^0}, \quad (164)$$

where the 5 (2-component) neutralinos $\tilde\chi_i$ are defined such that the absolute value of their masses increase with i, cf. SLHA1 [535]. These states are numbered sequentially by the PDG codes (1000022, 1000023, 1000025, 1000035, 1000045).

5.2.2 Explicit proposals for SLHA2

As in the SLHA1 [535], for all running parameters in the output of the spectrum file, we propose to use definitions in the modified dimensional reduction ($\overline{\mathrm{DR}}$) scheme.

To define the general properties of the model, we propose to introduce global switches in the SLHA1 model definition block MODSEL, as follows. Note that the switches defined here are in addition to the ones in [535].

5.2.2.1 Model selection

BLOCK MODSEL

Switches and options for model selection. The entries in this block should consist of an index, identifying the particular switch in the listing below, followed by another integer or real number, specifying the option or value chosen:

3 : (Default = 0) Choice of particle content. Switches defined are
    0 : MSSM. This corresponds to SLHA1.
    1 : NMSSM. As defined here.

4 : (Default = 0) R-parity violation. Switches defined are
    0 : R-parity conserved. This corresponds to the SLHA1.
    1 : R-parity violated.

5 : (Default = 0) CP violation. Switches defined are

   0 : CP is conserved. No information even on the CKM phase is used. This corresponds to the SLHA1.

   1 : CP is violated, but only by the standard CKM phase. All other phases assumed zero.

   2 : CP is violated. Completely general CP phases allowed.

6 : (Default = 0) Flavor violation. Switches defined are

   0 : No (SUSY) flavor violation. This corresponds to the SLHA1.

   1 : Quark flavor is violated.

   2 : Lepton flavor is violated.

   3 : Lepton and quark flavor is violated.

5.2.2.2 Flavor violation

– All input SUSY parameters are given at the scale M_{input} as defined in the SLHA1 block EXTPAR, except for EXTPAR 26, which, if present, is the *pole* pseudoscalar Higgs mass, and EXTPAR 27, which, if present, is the *pole* mass of the charged Higgs boson. If no M_{input} is present, the GUT scale is used.

– For the SM input parameters, we take the Particle Data Group (PDG) definition: lepton masses are all on-shell. The light quark masses $m_{u,d,s}$ are given at 2 GeV in the $\overline{\text{MS}}$ scheme, and the heavy quark masses are given as $m_c(m_c)^{\overline{\text{MS}}}$, $m_b(m_b)^{\overline{\text{MS}}}$ and $m_t^{\text{on-shell}}$. The latter two quantities are already in the SLHA1. The others are added to SMINPUTS in the following manner:

   8 : m_{ν_3}, pole mass.

  11 : m_e, pole mass.

  12 : m_{ν_1}, pole mass.

  13 : m_μ, pole mass.

  14 : m_{ν_2}, pole mass.

  21 : $m_d(2\,\text{GeV})^{\overline{\text{MS}}}$. d quark running mass in the $\overline{\text{MS}}$ scheme.

  22 : $m_u(2\,\text{GeV})^{\overline{\text{MS}}}$. u quark running mass in the $\overline{\text{MS}}$ scheme.

  23 : $m_s(2\,\text{GeV})^{\overline{\text{MS}}}$. s quark running mass in the $\overline{\text{MS}}$ scheme.

  24 : $m_c(m_c)^{\overline{\text{MS}}}$. c quark running mass in the $\overline{\text{MS}}$ scheme.

The FORTRAN format is the same as that of SMINPUTS in SLHA1 [535].

– V_{CKM}: the input CKM matrix, in the block VCKMIN in terms of the Wolfenstein parameterization:

   1 : λ

   2 : A

   3 : $\bar{\rho}$

   4 : $\bar{\eta}$

The FORTRAN format is the same as that of SMINPUTS above.

– U_{PMNS}: the input PMNS matrix, in the block UPMNSIN. It should have the PDG parameterization in terms of rotation angles [344] (all in radians):

   1 : $\bar{\theta}_{12}$ (the solar angle)

   2 : $\bar{\theta}_{23}$ (the atmospheric mixing angle)

   3 : $\bar{\theta}_{13}$ (currently only has an upper bound)

   4 : $\bar{\delta}_{13}$ (the Dirac CP-violating phase)

   5 : α_1 (the first Majorana CP-violating phase)

   6 : α_2 (the second CP-violating Majorana phase)

The FORTRAN format is the same as that of SMINPUTS above.

– $(\hat{m}_{\tilde{Q}}^2)_{ij}^{\overline{\text{DR}}}$, $(\hat{m}_{\tilde{u}}^2)_{ij}^{\overline{\text{DR}}}$, $(\hat{m}_{\tilde{d}}^2)_{ij}^{\overline{\text{DR}}}$, $(\hat{m}_{\tilde{L}}^2)_{ij}^{\overline{\text{DR}}}$, $(\hat{m}_{\tilde{e}}^2)_{ij}^{\overline{\text{DR}}}$: the squark and slepton soft SUSY-breaking masses at the input scale in the super-CKM/PMNS basis, as defined above. They will be given in the new blocks MSQ2IN, MSU2IN, MSD2IN, MSL2IN, MSE2IN, with the same format as matrices in SLHA1. Only the "upper triangle" of these matrices should be given. If diagonal entries are present, these supersede the parameters in the SLHA1 block EXTPAR.

– $(\hat{T}_U)_{ij}^{\overline{\text{DR}}}$, $(\hat{T}_D)_{ij}^{\overline{\text{DR}}}$, and $(\hat{T}_E)_{ij}^{\overline{\text{DR}}}$: the squark and slepton soft SUSY-breaking trilinear couplings at the input scale in the super-CKM/PMNS basis. They will be given in the new blocks TUIN, TDIN, TEIN, in the same format as matrices in SLHA1. If diagonal entries are present these supersede the A parameters specified in the SLHA1 block EXTPAR [535].

For the output, the pole masses are given in block MASS as in SLHA1, and the $\overline{\text{DR}}$ and mixing parameters as follows:

– $(\hat{m}_{\tilde{Q}}^2)_{ij}^{\overline{\text{DR}}}$, $(\hat{m}_{\tilde{u}}^2)_{ij}^{\overline{\text{DR}}}$, $(\hat{m}_{\tilde{d}}^2)_{ij}^{\overline{\text{DR}}}$, $(\hat{m}_{\tilde{L}}^2)_{ij}^{\overline{\text{DR}}}$, $(\hat{m}_{\tilde{e}}^2)_{ij}^{\overline{\text{DR}}}$: the squark and slepton soft SUSY-breaking masses at scale Q in the super-CKM/PMNS basis. Will be given in the new blocks MSQ2 Q =..., MSU2 Q =..., MSD2 Q =..., MSL2 Q =..., MSE2 Q =..., with formats as the corresponding input blocks MSX2IN above.

– $(\hat{T}_U)_{ij}^{\overline{\text{DR}}}$, $(\hat{T}_D)_{ij}^{\overline{\text{DR}}}$, and $(\hat{T}_E)_{ij}^{\overline{\text{DR}}}$: The squark and slepton soft SUSY-breaking trilinear couplings in the super-CKM/PMNS basis. Given in the new blocks TU Q =...,

TDQ =..., TEQ =..., which supersede the SLHA1 blocks AD, AU, and AE, see [535].
- $(\hat{Y}_U)_{ii}^{\overline{DR}}$, $(\hat{Y}_D)_{ii}^{\overline{DR}}$, $(\hat{Y}_E)_{ii}^{\overline{DR}}$: the diagonal $\overline{DR}$ Yukawas in the super-CKM/PMNS basis, at the scale Q. Given in the SLHA1 blocks YUQ =..., YDQ =..., YEQ =..., see [535]. Note that although the SLHA1 blocks provide for off-diagonal elements, only the diagonal ones will be relevant here, due to the CKM rotation.
- The $\overline{DR}$ CKM matrix at the scale Q. It will be given in the new block(s) VCKMQ =..., with entries defined as for the input block VCKMIN above.
- The new blocks R_u = USQMIX, R_d = DSQMIX, R_e = SELMIX, and R_ν = SNUMIX connect the particle codes (= mass ordered basis) with the super-CKM/PMNS basis according to the following definitions:

cases, explicit care must be taken especially by the program writing the spectrum, but also by the one reading it, to properly arrange the rows in the order of the mass spectrum actually used.

5.2.2.3 R-parity violation The naming convention for input blocks is BLOCK RV#IN, where the '#' character represents the name of the relevant output block given below. Default inputs for all R-parity violating couplings are zero. The inputs are given at scale M_{input}, as described in SLHA1 (default is the GUT scale) and follow the output format given below (with the omission of Q =...). In addition, the known fermion masses should be given in SMINPUTS as defined above.

- The dimensionless couplings $\hat{\lambda}_{ijk}$, $\hat{\lambda}'_{ijk}$, and $\hat{\lambda}''_{ijk}$ are given in BLOCK RVLAMLLE, BLOCK RVLAMLQD,

$$\begin{pmatrix} 1000001 \\ 1000003 \\ 1000005 \\ 2000001 \\ 2000003 \\ 2000005 \end{pmatrix} = \begin{pmatrix} \tilde{d}_1 \\ \tilde{d}_2 \\ \tilde{d}_3 \\ \tilde{d}_4 \\ \tilde{d}_5 \\ \tilde{d}_6 \end{pmatrix}_{\text{massordered}} = \text{DSQMIX}_{ij} \begin{pmatrix} \tilde{d}_L \\ \tilde{s}_L \\ \tilde{b}_L \\ \tilde{d}_R \\ \tilde{s}_R \\ \tilde{b}_R \end{pmatrix}_{\text{super-CKM}}, \quad (165)$$

$$\begin{pmatrix} 1000002 \\ 1000004 \\ 1000006 \\ 2000002 \\ 2000004 \\ 2000006 \end{pmatrix} = \begin{pmatrix} \tilde{u}_1 \\ \tilde{u}_2 \\ \tilde{u}_3 \\ \tilde{u}_4 \\ \tilde{u}_5 \\ \tilde{u}_6 \end{pmatrix}_{\text{massordered}} = \text{USQMIX}_{ij} \begin{pmatrix} \tilde{u}_L \\ \tilde{c}_L \\ \tilde{t}_L \\ \tilde{u}_R \\ \tilde{c}_R \\ \tilde{t}_R \end{pmatrix}_{\text{super-CKM}}, \quad (166)$$

$$\begin{pmatrix} 1000011 \\ 1000013 \\ 1000015 \\ 2000011 \\ 2000013 \\ 2000015 \end{pmatrix} = \begin{pmatrix} \tilde{e}_1 \\ \tilde{e}_2 \\ \tilde{e}_3 \\ \tilde{e}_4 \\ \tilde{e}_5 \\ \tilde{e}_6 \end{pmatrix}_{\text{massordered}} = \text{SELMIX}_{ij} \begin{pmatrix} \tilde{e}_L \\ \tilde{\mu}_L \\ \tilde{\tau}_L \\ \tilde{e}_R \\ \tilde{\mu}_R \\ \tilde{\tau}_R \end{pmatrix}_{\text{super-PMNS}}, \quad (167)$$

$$\begin{pmatrix} 1000012 \\ 1000014 \\ 1000016 \end{pmatrix} = \begin{pmatrix} \tilde{\nu}_1 \\ \tilde{\nu}_2 \\ \tilde{\nu}_3 \end{pmatrix}_{\text{massordered}} = \text{SNUMIX}_{ij} \begin{pmatrix} \tilde{\nu}_e \\ \tilde{\nu}_\mu \\ \tilde{\nu}_\tau \end{pmatrix}_{\text{super-PMNS}}. \quad (168)$$

Note! A potential source for inconsistency arises if the masses and mixings are not calculated in the same way, e.g. if radiatively corrected masses are used with tree-level mixing matrices. In this case, it is possible that the radiative corrections to the masses shift the mass ordering relative to the tree level. This is especially relevant when near-degenerate masses occur in the spectrum and/or when the radiative corrections are large. In these

RVLAMUDD Q =... respectively. The output standard should correspond to the FORTRAN format

(1x,I2,1x,I2,1x,I2,3x,1P,E16.8,0P,3x,'#',1x,A).

where the first three integers in the format correspond to i, j, and k and the double precision number is the coupling.

- The soft SUSY-breaking couplings $\hat{T}_{ijk}$, $\hat{T}'_{ijk}$, and $\hat{T}''_{ijk}$ should be given in BLOCK RVTLLE, RVTLQD, RV-

Table 32 Summary of R-parity violating SLHA2 data blocks. Only three out of the last four blocks are independent. Which block to leave out of the input is in principle up to the user, with the caveat that a given spectrum calculator may not accept all combinations

Input block	Output block	Data
RVLAMLLEIN	RVLAMLLE	$i\ j\ k\ \hat{\lambda}_{ijk}$
RVLAMQDIN	RVLAMQD	$i\ j\ k\ \hat{\lambda}'_{ijk}$
RVLAMUDDIN	RVLAMUDD	$i\ j\ k\ \hat{\lambda}''_{ijk}$
RVTLLEIN	RVTLLE	$i\ j\ k\ \hat{T}_{ijk}$
RVTLQDIN	RVTLQD	$i\ j\ k\ \hat{T}'_{ijk}$
RVTUDDIN	RVTUDD	$i\ j\ k\ \hat{T}''_{ijk}$
NB: One of the following RV...IN blocks must be left out:		
(which one up to user and RGE code)		
RVKAPPAIN	RVKAPPA	$i\ \hat{\kappa}_i$
RVDIN	RVD	$i\ \hat{D}_i$
RVSNVEVIN	RVSNVEV	$i\ v_i$
RVM2LH1IN	RVM2LH1	$i\ \hat{m}^2_{L_iH_1}$

TUDD Q = ..., in the same format as the $\hat{\lambda}$ couplings above.

- The bilinear superpotential and soft SUSY-breaking terms $\hat{\kappa}_i$, $\hat{D}_i$, and $\hat{m}^2_{L_iH_1}$ and the sneutrino VEVs are given in BLOCK RVKAPPA, RVD, RVM2LH1, RVSNVEV Q = ... respectively, in the same format as real-valued vectors in the SLHA1.
- The input/output blocks for R-parity violating couplings are summarized in Table 32.
- The new mixing matrices that appear are described in Sect. 5.2.1.2.

As for the R-conserving MSSM, the bilinear terms (both SUSY-breaking and SUSY-respecting ones, including μ) and the VEVs are not independent parameters. They become related by the condition of electroweak symmetry breaking. This carries over to the RPV case, where not all the parameters in the input blocks RV...IN in Table 32 can be given simultaneously. Specifically, of the last 4 blocks only three are independent. One block is determined by minimizing the Higgs sneutrino potential. We do not here insist on a particular choice for which of RVKAPPAIN, RVDIN, RVSNVEVIN, and RVM2LH1IN to leave out, but leave it up to the spectrum calculators to accept one or more combinations.

5.2.2.4 CP violation When adding CP violation to the MSSM model parameters and mixing matrices, the SLHA1 blocks are understood to contain the real parts of the relevant parameters. The imaginary parts should be provided with exactly the same format, in a separate block of the same name but prefaced by IM. The defaults for all imaginary parameters will be zero.

One special case is the μ parameter. When the real part of μ is given in EXTPAR 23, the imaginary part should be given in IMEXTPAR 23, as above. However, when $|\mu|$ is determined by the conditions for electroweak symmetry breaking, only the phase φ_μ is taken as an input parameter. In this case, SLHA2 generalizes the entry MINPAR 4 to contain the cosine of the phase (as opposed to just sign(μ) in SLHA1), and we further introduce a new block IMMINPAR whose entry 4 gives the sine of the phase, that is

BLOCK MINPAR

4 : CP conserved: sign(μ).
   CP violated: $\cos\varphi_\mu = \text{Re}\,\mu/|\mu|$.

BLOCK IMMINPAR

4 : CP conserved: n/a.
   CP violated: $\sin\varphi_\mu = \text{Im}\,\mu/|\mu|$.

Note that $\cos\varphi_\mu$ coincides with sign(μ) in the CP-conserving cases.

The new 3×4 block S = CVHMIX connects the particle codes (= mass ordered basis) with the interaction basis according to the following definition:

$$\begin{pmatrix} 25 \\ 35 \\ 36 \end{pmatrix} = \begin{pmatrix} h_1^0 \\ h_2^0 \\ h_3^0 \end{pmatrix}_{\text{massordered}} = \text{CVHMIX}_{ij} \begin{pmatrix} \sqrt{2}\,\text{Re}\,H_1^0 \\ \sqrt{2}\,\text{Re}\,H_2^0 \\ \sqrt{2}\,\text{Im}\,H_1^0 \\ \sqrt{2}\,\text{Im}\,H_2^0 \end{pmatrix}.$$
(169)

In order to translate between S and other conventions, the tree-level angle α may be needed. This should be given in the SLHA1 output BLOCK ALPHA:

BLOCK ALPHA

CP conserved: α; precise definition up to spectrum calculator, see SLHA1.

CP violated: α_{tree}. Must be accompanied by the matrix S, as described above, in the block CVHMIX.

5.2.2.5 NMSSM Firstly, as described above, BLOCK MODSEL should contain the switch 3 with value 1, corresponding to the choice of the NMSSM particle content.

Secondly, for the parameters that are also present in the MSSM, we re-use the corresponding SLHA1 entries. That is, m_Z should be given in SMINPUTS entry 4 and $m_{H_1}^2, m_{H_2}^2$ can be given in the EXTPAR entries 21 and 22. $\tan\beta$ should either be given in MINPAR entry 3 (default) or EXTPAR entry 25 (user-defined input scale), as in SLHA1. If μ should be desired non-zero, it can be given in EXTPAR entry 23. The corresponding soft parameter m_3^2 can be given in EXTPAR entry 24, in the form $m_3^2/(\cos\beta\sin\beta)$, see [535].

Further, new entries in BLOCK EXTPAR have been defined for the NMSSM specific input parameters, as follows. As in the SLHA1, these parameters are all given at the common scale M_{input}, which can either be left up to the spectrum calculator or given explicitly using EXTPAR 0 [535]:

BLOCK EXTPAR

Input parameters specific to the NMSSM (i.e., in addition to the entries defined in [535])

- 61 : λ. Superpotential trilinear Higgs SH_2H_1 coupling.
- 62 : κ. Superpotential cubic S coupling.
- 63 : A_λ. Soft trilinear Higgs SH_2H_1 coupling.
- 64 : A_κ. Soft cubic S coupling.
- 65 : $\lambda\langle S\rangle$. Vacuum expectation value of the singlet (scaled by λ).
- 66 : ξ_F. Superpotential linear S coupling.
- 67 : ξ_S. Soft linear S coupling.
- 68 : μ'. Superpotential quadratic S coupling.
- 69 : $m_S'^2$. Soft quadratic S coupling (sometimes denoted $\mu'B'$).
- 70 : m_S^2. Soft singlet mass squared.

Important note: only 12 of the parameters listed in (159) should be given as input at any one time (including explicit zeroes for parameters desired "switched off"), the remaining ones being determined by the minimization of the effective potential. Which combinations to accept is left up to the individual spectrum calculator programs. Alternatively, for minimal models, six parameters of those listed in (160) should be given.

In the spectrum output, running NMSSM parameters corresponding to the EXTPAR entries above can be given in the block NMSSMRUN Q=...:

BLOCK NMSSMRUN Q=...

Output parameters specific to the NMSSM, given in the $\overline{\text{DR}}$ scheme, at the scale Q. As in the SLHA1, several of these blocks may be given simultaneously in the output, each then corresponding to a specific scale, but at least one should always be present. See the corresponding entries in EXTPAR above for definitions.

- 1 : $\lambda(Q)^{\overline{\text{DR}}}$.
- 2 : $\kappa(Q)^{\overline{\text{DR}}}$.
- 3 : $A_\lambda(Q)^{\overline{\text{DR}}}$.
- 4 : $A_\kappa(Q)^{\overline{\text{DR}}}$.
- 5 : $\lambda\langle S\rangle(Q)^{\overline{\text{DR}}}$.
- 6 : $\xi_F(Q)^{\overline{\text{DR}}}$.
- 7 : $\xi_S(Q)^{\overline{\text{DR}}}$.
- 8 : $\mu'(Q)^{\overline{\text{DR}}}$.
- 9 : $m_S'^2(Q)^{\overline{\text{DR}}}$.
- 10 : $m_S^2(Q)^{\overline{\text{DR}}}$.

The new 3×3 block $S = $ NMHMIX connects the particle codes (= mass ordered basis) for the CP-even Higgs bosons with the interaction basis according to the following definition:

$$\begin{pmatrix}25\\35\\45\end{pmatrix} = \begin{pmatrix}h_1^0\\h_2^0\\h_3^0\end{pmatrix}_{\text{massordered}} = \text{NMHMIX}_{ij}\begin{pmatrix}\sqrt{2}\,\text{Re}\,H_1^0\\\sqrt{2}\,\text{Re}\,H_2^0\\\sqrt{2}\,\text{Re}\,S\end{pmatrix}. \tag{170}$$

The new 2×3 block $S=$ NMAMIX connects the particle codes (=mass ordered basis) for the CP-odd Higgs bosons with the interaction basis according to the following definition:

$$\begin{pmatrix}36\\46\end{pmatrix} = \begin{pmatrix}A_1^0\\A_2^0\end{pmatrix}_{\text{massordered}} = \text{NMAMIX}_{ij}\begin{pmatrix}\sqrt{2}\,\text{Im}\,H_1^0\\\sqrt{2}\,\text{Im}\,H_2^0\\\sqrt{2}\,\text{Im}\,S\end{pmatrix}. \tag{171}$$

Finally, the new 5 × 5 block NMNMIX gives the neutralino mixing matrix, with the fifth mass eigenstate labeled 1000045 and the fifth interaction eigenstate being the singlino, $\tilde{s}$.

5.3 SuSpect, HDECAY, SDECAY and SUSY-HIT

5.3.1 SuSpect

The Fortran code SuSpect calculates the supersymmetric and Higgs particle spectrum in the MSSM. It deals with the "phenomenological MSSM" with 22 free parameters defined either at a low or high energy scale, with the possibility of RGE to arbitrary scales, and with constrained models with universal boundary conditions at high scales. These are the minimal supergravity (mSUGRA), the anomaly mediated SUSY breaking (AMSB) and the gauge mediated SUSY breaking (GMSB) models. The basic assumptions of the most general possible MSSM scenario are (a) minimal gauge group, (b) minimal particle content, (c) minimal Yukawa interactions and R-parity conservation, (d) minimal set of soft SUSY breaking terms. Furthermore, (i) all soft SUSY breaking parameters are real (no CP-violation); (ii) the matrices for sfermion masses and trilinear couplings are diagonal; (iii) first and second sfermion generation universality is assumed. Here and in the following we refer the reader for more details to the user's manual [544].

As for the calculation of the SUSY particle spectrum in constrained MSSMs, in addition to the choice of the input parameters, the general algorithm contains three main steps. These are (i) the RGE of parameters back and forth between the low energy scales, such as M_Z and the electroweak symmetry breaking (EWSB) scale, and the high energy scale characteristic for the various models; (ii) the consistent implementation of (radiative) EWSB; (iii) the calculation of the pole masses of the Higgs bosons and the SUSY particles, including the mixing between the current eigenstates and the radiative corrections when they are important. Here the program mainly follows the content and notation of [545], and for the leading two-loop corrections to the Higgs masses the results summarized in [546] are taken.

The necessary files for the use in SuSpect are the input file suspect2.in, the main routine suspect2.f, the routine twoloophiggs.f, which calculates the Higgs masses, as well as bsg.f for the calculation of the $b \to s\gamma$ branching ratio. The latter is needed in order to check if the results are in agreement with the experimental measurements. In the input file one can select the model to be investigated, the accuracy of the algorithm and the input data (SM fermion masses and gauge couplings). At each run SuSpect generates two output files: one easy to read, suspect2.out, and the other in the SLHA format [535].

5.3.2 HDECAY

The Fortran code HDECAY [547] calculates the decay widths and branching ratios of the SM Higgs boson, and of the neutral and charged Higgs particles of the MSSM according to the current theoretical knowledge (for reviews see [116, 548–550]). It includes:

— All kinematically allowed decay channels with branching ratios larger than 10^{-4}; apart from the two body decays also the loop-mediated, the most important three body decay modes, and in the MSSM the cascade and SUSY decay channels.
— All relevant higher-order QCD corrections to the decays into quark pairs and to the quark loop mediated decays into gluons are incorporated.
— Double off-shell decays of the CP-even Higgs bosons into massive gauge bosons, subsequently decaying into four massless fermions.
— All important 3-body decays: with off-shell heavy top quarks; with one off-shell gauge boson as well as heavy neutral Higgs decays with one off-shell Higgs boson.
— In the MSSM the complete radiative corrections in the effective potential approach with full mixing in the stop and sbottom sectors; it uses the RG improved values of the Higgs masses and couplings, the relevant NLO corrections are implemented [551, 552].
— In the MSSM, all decays into SUSY particles when kinematically allowed.
— In the MSSM, all SUSY particles are included in the loop mediated $\gamma\gamma$ and gg decay channels. In the gluonic decay modes the large QCD corrections for quark and squark loops are also included.

HDECAY has recently undergone a major upgrade. The SLHA format has been implemented, so that the program can now read in any input file in the SLHA format and also give out the Higgs decay widths and branching ratios in this format. So, the program can now be easily linked to any spectrum or decay calculator. Two remarks are in order:

(1) HDECAY calculates the higher-order corrections to the Higgs boson decays in the $\overline{MS}$ scheme whereas all scale dependent parameters read in from an SLHA input file provided by a spectrum calculator are given in the $\overline{DR}$ scheme. Therefore, HDECAY translates the input parameters from the SLHA file into the $\overline{MS}$ scheme where needed.

(2) The SLHA parameter input file only includes the MSSM Higgs boson mass values, but not the Higgs self-interactions, which are needed in HDECAY. For the time being, HDECAY calculates the missing interactions internally within the effective potential approach. This is not completely consistent with the values for the Higgs

masses, since the spectrum calculator does not necessarily do it with the same method and level of accuracy as HDECAY. The difference is of higher order, though.

5.3.3 SDECAY

The Fortran code SDECAY [553], which has implemented the MSSM in the same way as is done in SuSpect, calculates the decay widths and branching ratios of all SUSY particles in the MSSM, including the most important higher-order effects [554–556]:

- The usual two body decays for sfermions and gauginos are calculated at tree level.
- A unique feature is the possibility of calculating the SUSY-QCD corrections to the decays involving colored particles. They can amount up to several tens of per-cents in some cases. The bulk of the EW corrections has been accounted for by taking running parameters where appropriate.
- In GMSB models the two body decays into the lightest SUSY particle, the gravitino, have been implemented.
- If the two body decays are closed, multibody decays will be dominant. SDECAY calculates the three body decays of the gauginos, the gluino, the stops and sbottoms.
- Moreover, loop-induced decays of the lightest stop, the next-to-lightest neutralino and the gluino are included.
- If the three body decays are kinematically forbidden, four body decays of the lightest stop can compete with the loop-induced $\tilde{t}_1$ decay and have therefore been implemented.
- Finally, the top decays within the MSSM have been programmed.

Recently, SDECAY has been updated with some major changes being (other changes related to SUSY-HIT are listed below): (i) For reasons of shortening the output file, only non-zero branching ratios are written out in the new version. (ii) We have created common blocks for the branching ratios and total widths of the various SUSY particles.

5.3.4 SUSY-HIT

The previous three programs have been linked together in a program called SUSY-HIT [557]. Including higher-order effects in the calculations, the package allows for the consistent calculation of MSSM particle decays with the presently highest level of precision. The following files are needed to run SUSY-HIT:
Spectrum files: The spectrum can either be taken from any input file in the SLHA format or from SuSpect. In the first case, SUSY-HIT needs an SLHA input file which has to be named slhaspectrum.in. In the latter case, one needs the necessary SuSpect routines: suspect2.in, suspect2.f, twoloophiggs.f and bsg.f.

Decay files: SDECAY is the main program and now reads in susyhit.in and calls HDECAY which is now a subroutine and, in order to keep the package as small as possible, only one routine calculating the Higgs boson masses and Higgs self-couplings has been retained in HDECAY to extract the Higgs self-interaction strengths not provided by the spectrum calculators; also, HDECAY does not create any output file within the package. SDECAY passes the necessary parameters from susyhit.in to HDECAY via a newly created common block called SUSYHITIN. As before, it calls SuSpect in case the spectrum is taken from there. The SLHA parameter and spectrum input file slhaspectrum.in is read in by both HDECAY and SDECAY. The output file created by SDECAY at each run is called susyhit_slha.out if it is in the SLHA format or simply susyhit.out if it is in an output format easy to read.
Input file: The HDECAY and SDECAY input files have been merged into one input file susyhit.in. Here, first of all the user can choose among two SUSY-HIT related options.

1. The three programs SuSpect, HDECAY, SDECAY are linked and hence SuSpect provides the spectrum and the soft SUSY breaking parameters at the EWSB scale.
2. The two programs HDECAY and SDECAY are linked. The necessary input parameters are taken from a file in the SLHA format provided by any spectrum calculator.

Furthermore, various options for running the SDECAY program can be chosen, such as whether or not to include QCD corrections to two body decays, the multibody and/or loop decays, the GMSB decays and the top decays. The scale and number of loops of the running couplings can be fixed. Finally, some parameters related to HDECAY can be set, like the charm and strange quark masses, the W, Z total widths, some CKM matrix elements etc. All other necessary parameters are read in from the slhaspectrum.in input file.
Changes and how the package works: SuSpect, HDECAY and SDECAY are linked via the SLHA format. Therefore, the name of the output file provided by SuSpect has to be the same as the SLHA input file read in by HDECAY and SDECAY. We called it slhaspectrum.in. This is one of the changes made in the programs with respect to their original version. Further major changes have been made. For the complete list of changes please refer to the web page given below.
Web page: We have created a web page at the following url address: http://lappweb.in2p3.fr/~muehlleitner/SUSY-HIT/ There the user can download all files necessary for the program package as well as a makefile for compiling the programs. The latest versions of the frequently updated programs are used. Short instructions are given on how to use the programs. A file with updates and changes is provided. Finally, some examples of output files are given.

5.4 FeynHiggs

FeynHiggs is a program for computing Higgs boson masses and related observables in the (NMFV) MSSM with real or complex parameters. The observables comprise mixing angles, branching ratios, and couplings, including state-of-the-art higher-order contributions. The centerpiece is a Fortran library for use with Fortran and C/C++. Alternatively, FeynHiggs has a command-line, Mathematica, and Web interface. FeynHiggs is available from www.feynhiggs.de.

FeynHiggs [336–338, 558] is a Fortran code for the evaluation of the masses, decays and production processes of Higgs bosons in the (NMFV) MSSM with real or complex parameters. The calculation of the higher-order corrections is based on the Feynman-diagrammatic (FD) approach [338, 559–561]. At the one-loop level, it consists of a complete evaluation, including the full momentum and phase dependence, and as a further option the full 6×6 non-minimal flavor violation (NMFV) contributions [120, 127]. At the two-loop level all available corrections from the real MSSM have been included. They are supplemented by the resummation of the leading effects from the (scalar) b sector including the full complex phase dependence.

In addition to the Higgs boson masses, the program also provides results for the effective couplings and the wave function normalization factors for external Higgs bosons [562], taking into account NMFV effects from the Higgs boson self-energies. Besides the computation of the Higgs boson masses, effective couplings and wave function normalization factors, the program also evaluates an estimate for the theory uncertainties of these quantities due to unknown higher-order corrections.

Furthermore FeynHiggs contains the evaluation of all relevant Higgs boson decay widths.[26] In particular, the following quantities are calculated:

- the total width for the neutral and charged Higgs bosons,
- the branching ratios and effective couplings of the three neutral Higgs bosons to
  - SM fermions [563], $h_i \to \bar{f} f$,
  - SM gauge bosons (possibly off-shell), $h_i \to \gamma\gamma$, $ZZ^{(*)}$, $WW^{(*)}$, gg,
  - gauge and Higgs bosons, $h_i \to Z^{(*)} h_j$, $h_i \to h_j h_k$,
  - scalar fermions, $h_i \to \tilde{f}^\dagger \tilde{f}$,
  - gauginos, $h_i \to \tilde{\chi}_k^\pm \tilde{\chi}_j^\mp$, $h_i \to \tilde{\chi}_l^0 \tilde{\chi}_m^0$,
- the branching ratios and effective couplings of the charged Higgs boson to
  - SM fermions, $H^- \to \bar{f} f'$,
  - a gauge and Higgs boson, $H^- \to h_i W^-$,
  - scalar fermions, $H^- \to \tilde{f}^\dagger \tilde{f}'$,
  - gauginos, $H^- \to \tilde{\chi}_k^- \tilde{\chi}_l^0$,

---

[26]The inclusion of flavor changing decays is work in progress.

- the production cross sections of the neutral Higgs bosons at the Tevatron and the LHC in the approximation where the corresponding SM cross section is rescaled by the ratios of the corresponding partial widths in the MSSM and the SM or by the wave function normalization factors for external Higgs bosons, see [564] for further details.

For comparisons with the SM, the following quantities are also evaluated for SM Higgs bosons with the same mass as the three neutral MSSM Higgs bosons:

- the total decay width,
- the couplings and branching ratios of a SM Higgs boson to SM fermions,
- the couplings and branching ratios of a SM Higgs boson to SM gauge bosons (possibly off-shell),
- the production cross sections at the Tevatron and the LHC [564].

FeynHiggs furthermore provides results for electroweak precision observables that give rise to constraints on the SUSY parameter space (see [561] and references therein):

- the quantity $\Delta\rho$ up to the two-loop level that can be used to indicate disfavored scalar top and bottom mass combinations,
- an evaluation of M_W and $\sin^2 \theta_{\text{eff}}$, where the SUSY contributions are treated in the $\Delta\rho$ approximation (see e.g. [561]), taking into account at the one-loop level the effects of complex phases in the scalar top/bottom sector as well as NMFV effects [120],
- the anomalous magnetic moment of the muon, including a full one-loop calculation as well as leading and subleading two-loop corrections,
- the evaluation of $\text{Br}(b \to s\gamma)$ including NMFV effects [127].

Finally, FeynHiggs possesses some further features.

- Transformation of the input parameters from the $\overline{\text{DR}}$ to the on-shell scheme (for the scalar top and bottom parameters), including the full $\mathcal{O}(\alpha_s)$ and $\mathcal{O}(\alpha_{t,b})$ corrections.
- Processing of SUSY Les Houches Accord (SLHA 2) data [535, 565, 566] including the full NMFV structure. FeynHiggs reads the output of a spectrum generator file and evaluates the Higgs boson masses, branching ratios etc. The results are written in the SLHA format to a new output file.
- Predefined input files for the SPS benchmark scenarios [248] and the Les Houches benchmarks for Higgs boson searches at hadron colliders [328] are included.
- Detailed information about all the features of FeynHiggs are provided in man pages.

FeynHiggs is available from www.feynhiggs.de.

5.5 FchDecay

`FchDecay` is a computer program to compute the FCNC decay branching ratios Br($h \to bs$) and Br($h \to tc$) in the flavor violating MSSM. The input/output is performed in the SUSY Les Houches Accord II (`SLHA`) [170, 535, 567] convention (using an extension of SLHALib [565]). This program is based on the work and results of [106, 107, 109, 128, 316].

The approximations used in the computation are

- the full one-loop SUSY-QCD contributions to the FCNC partial decay widths $\Gamma(h \to bs, tc)$ is included;
- the Higgs sector parameters (masses and CP-even mixing angle α) have been treated using the leading m_t and $m_b \tan \beta$ approximation to the one-loop result;
- the Higgs bosons total decay widths $\Gamma(h \to X)$ are computed at leading order, including all the relevant channels;
- a leading order computation of Br($b \to s\gamma$) (to check the parameter space) is also included.

The code implements the flavor violating MSSM, it allows for complete intergenerational mixing in the Left-Left and Right-Right squark sector (but it does not allow for intergenerational mixing in the Left-Right sector).

The program includes a (simplified) computation of the Higgs boson masses and total decay widths, and it will write them to the output file. However:

- if the input file contains the Higgs sector parameters (masses and CP-even mixing angle α) it will use those values instead;
- if the input file contains Higgs boson decay tables, it will just add the FCNC decays to that table (instead of computing the full table).

This setup allows one to use the computations of more sophisticated programs for the Higgs boson parameters and/or total decay widths, and then run the `FchDecay` program on the resulting output file to obtain the FCNC partial decay widths.

The program is available from the web page, http://fchdecay.googlepages.com, and comes with a complete manual (detailing the included physics models, and running instructions). The authors can be reached at fchdecay@gmail.com.

5.6 MSSM NMFV in `FeynArts` and `FormCalc`

In the presence of non-minimal flavor violation (NMFV) the 2×2 mixing of the squark within each family is enlarged to a full 6×6 mixing among all three generations, such that the mixed states are

$$\tilde{u}_i = (R_u)_{ij} \begin{pmatrix} \tilde{u}_L & \tilde{c}_L & \tilde{t}_L & \tilde{u}_R & \tilde{c}_R & \tilde{t}_R \end{pmatrix}^T_j,$$
$$\tilde{d}_i = (R_d)_{ij} \begin{pmatrix} \tilde{d}_L & \tilde{s}_L & \tilde{b}_L & \tilde{d}_R & \tilde{s}_R & \tilde{b}_R \end{pmatrix}^T_j. \quad (172)$$

The matrices R_q diagonalize the mass matrices

$$M_q^2 = \begin{pmatrix} M_{LL,q}^2 & M_{LR,q}^2 \\ (M_{LR,q}^2)^* & M_{RR,q}^2 \end{pmatrix} + \Delta_q,$$

$$M^2_{AA,q} = \text{diag}(M^2_{A,q_1}, M^2_{A,q_2}, M^2_{A,q_3}), \quad (173)$$
$$\scriptstyle A=L,R$$

$$M^2_{LR,q} = \text{diag}(m_{q_1} X_{q_1}, m_{q_2} X_{q_2}, m_{q_3} X_{q_3})$$

where $q = \{u, d\}$, $\{q_1, q_2, q_3\} = u, c, t$ for the up- and d, s, b for the down-squark mass matrix and

$$M^2_{L,q_i} = M^2_{\tilde{Q}, q_i} + m^2_{q_i} + \cos 2\beta (T_3^q - Q_q s_W^2) m_Z^2,$$
$$M^2_{R,u_i} = M^2_{\tilde{U}, u_i} + m^2_{u_i} + \cos 2\beta Q_u s_W^2 m_Z^2,$$
$$M^2_{R,d_i} = M^2_{\tilde{D}, d_i} + m^2_{d_i} + \cos 2\beta Q_d s_W^2 m_Z^2, \quad (174)$$
$$X_{\{u,d\}_i} = A_{\{u,d\}_i} - \mu\{\cot\beta, \tan\beta\}.$$

The actual dimensionless input quantities δ are

$$\Delta_q = \begin{pmatrix} N^2_{LL,q} & N^2_{LR,q} \\ (N^2_{LR,q})^* & N^2_{RR,q} \end{pmatrix} \delta_q,$$

$$N^2_{AB,q} = \begin{pmatrix} M_{A,q_1} \\ M_{A,q_2} \\ M_{A,q_3} \end{pmatrix} \otimes \begin{pmatrix} M_{B,q_1} \\ M_{B,q_2} \\ M_{B,q_3} \end{pmatrix}. \quad (175)$$
$$\scriptstyle A,B=L,R$$

The new `FeynArts` model file `FVMSSM.mod` generalizes the squark couplings in `MSSM.mod` to the NMFV case. It contains the new objects

`UASf[s,s',t]` the squark mixing matrix $R_{u,d}$

`MASf[s,t]` the squark masses,

with $s, s' = 1\ldots 6$, $t = 3(u), 4(d)$.

The initialization of `MASf` and `UASf` is already built into `FormCalc`'s `model_mssm.F` but needs to be turned on by defining a preprocessor flag in `run.F`:

```
#define FLAVOUR_VIOLATION
```

The NMFV parameters $(\delta_t)_{ss'}$ are represented by the `deltaSf` array:

```
double complex deltaSf(s,s',t)
```

Since δ is a Hermitian matrix, only the entries on and above the diagonal need to be filled. For convenience, the following abbreviations can be used for individual matrix elements:

deltaLLuc = $(\delta_u)_{12}$ deltaLRuc = $(\delta_u)_{15}$
deltaRLucC = $(\delta_u)_{24}$ deltaRRuc = $(\delta_u)_{45}$
deltaLLct = $(\delta_u)_{23}$ deltaLRct = $(\delta_u)_{26}$
deltaRLctC = $(\delta_u)_{35}$ deltaRRct = $(\delta_u)_{56}$
deltaLLut = $(\delta_u)_{13}$ deltaLRut = $(\delta_u)_{16}$
deltaRLutC = $(\delta_u)_{34}$ deltaRRut = $(\delta_u)_{46}$

and analogous entries for the down sector.

Note the special treatment of the RL elements: one has to provide the complex conjugate of the element. The original lies below the diagonal and would be ignored by the eigenvalue routine.

The off-diagonal trilinear couplings A acquire non-zero entries through the relations

$$m_{q,i}(A_q)_{ij} = \left(M_q^2\right)_{i,j+3}, \quad i,j = 1\ldots 3. \quad (176)$$

In summary: NMFV effects [127] can be computed with FeynArts [112, 334, 335] and FormCalc [113]. These packages provide a high level of automation for perturbative calculations up to one loop. Compared to calculations with the MFV MSSM, only three minor changes are required:

- choosing FVMSSM.mod instead of MSSM.mod,
- setting FLAVOUR_VIOLATION in run.F,
- providing values for the deltaSf matrix.

These changes are contained in FeynArts and FormCalc, available from www.feynarts.de.

5.7 SPheno

SPheno is a program to calculate the spectrum of supersymmetric models, the decays of supersymmetric particles and Higgs bosons as well as the production cross sections of these particles in e^+e^- annihilation. Details of the algorithm used for the MSSM with real parameters and neglecting mixing between the (s)fermion generations can be found in [342]. This version can be found and downloaded from http://theorie.physik.uni-wuerzburg.de/~porod/SPheno.html. In this contribution the model extensions regarding flavor aspects are described. In the context of the MSSM the most general flavor structure as well as all CP phases are included in the RGE running and in the computation of SUSY masses at tree level as well as at the one-loop level. In the Higgs sector, the complete flavor structure is included for the calculation of the masses at the one-loop level. At the 2-loop level there is still the approximation used that the 3rd generation does not mix with the other ones. With respect to CP phases, the induced mixing between scalar and pseudoscalar Higgs bosons is not yet included. For the decays of supersymmetric particles and Higgs bosons, the complete flavor structure is taken into account at tree level

using running $\overline{DR}$ couplings to incorporate the most important loop corrections. A few examples are

$$\tilde{\chi}_i^0 \to e^{\pm}\tilde{\mu}_R^{\mp}, e^{\pm}\mu^{\mp}\tilde{\chi}_j^0, \bar{u}\tilde{c}_L, \bar{u}c\tilde{\chi}_j^0, \bar{u}b\tilde{\chi}_k^+;$$
$$\tilde{g} \to \bar{u}\tilde{c}_L, \bar{u}c\tilde{\chi}_j^0, \bar{u}b\tilde{\chi}_k^+; \quad (177)$$
$$H^+ \to \bar{b}c, \tilde{b}_1\tilde{c}_R; \quad H^0 \to \tilde{e}_R^{\pm}\tilde{\tau}_1^{\mp}. \quad (178)$$

The complete list is given in the manual. Also in the case of production in e^+e^- annihilation all flavor-off diagonal channels are available. Flavor and CP violating terms are already constrained by several experimental data. For these reason, the following observables are calculated: anomalous magnetic and electric dipole moments of leptons, the most important ones being a_μ and d_e; the leptonic rare decays $l \to l'\gamma$, $l \to 3l'$; rare decays of the Z-boson: $Z \to ll'$; the rare b decays $b \to s\gamma$, $b \to s\mu^+\mu^-$, $B_{s,d} \to \mu^+\mu^-$, $B_u^{\pm} \to \tau^{\pm}\nu$, $\delta(M_{B_{s,d}})$; and finally $\Delta\rho$.

This version of SPheno also includes extended SUSY models: (a) the NMSSM and (b) lepton number violation and thus R-parity violation. In both model classes the masses are calculated at tree level, except for the Higgs sector, where radiative corrections are included. In both cases the complete flavor structure is taken into account in the calculation of the masses, the decays of supersymmetric particles and Higgs bosons as well as in the production of these particles in e^+e^- annihilation. The low energy observables are not yet calculated in these models but the extension of the corresponding routines to include these models is foreseen for the near future.

Concerning input and output the current version of the SLHA2 accord is implemented as described in Sect. 5.2 and in [170]. The version described here is currently under testing, and the write-up of the corresponding manual is in progress. When available, the program documentation and the source code will be found on the web page given above. In the meantime a copy can be obtained be sending an email to porod@physik.uni-wuerzburg.de.

5.8 SOFTSUSY

SOFTSUSY [538] provides a SUSY spectrum in the MSSM consistent with input low energy data, and a user supplied high energy constraint. It is written in C++ with an emphasis on easy generalizability. It can produce SUSY Les Houches Accord compliant output [535], and therefore link to Monte Carlos (e.g. HERWIG [568]) or programs that calculate sparticle decays such as SDecay [553]. SOFTSUSY can be obtained from URL: http://projects.hepforge.org/softsusy. SOFTSUSY currently incorporates three-family mixing in the limit of CP conservation. The high energy constraint in SOFTSUSY upon the supersymmetry breaking terms may be completely non-universal, i.e. can have three by three-family mixing incorporated within them. All of the RGEs used to evolve the MSSM between high energy scales and the weak scale M_Z have the full three-family mixing effects incorporated at one loop in all MSSM parameters. Two-loop terms in the RGEs are included in the dominant third family approximation for speed of computation and so mixing is neglected in the two-loop terms. Currently, the smaller one-loop weak-scale threshold corrections to sparticle masses are also calculated in the dominant third family Yukawa approximation, and so family mixing is neglected within them.

The user may request that, at the weak scale, all of the quark mixing is incorporated within a symmetric up quark Yukawa matrix $(Y_U)'$, or alternatively within a symmetric down quark Yukawa matrix $(Y_D)'$. These are then related (via the SOFTSUSY conventions [538] for the Lagrangian) to the mass basis Yukawa matrices Y_U, Y_D via

$$(Y_U)' = V_{\text{CKM}}^T (Y^U) V_{\text{CKM}} \quad \text{or}$$
$$(Y_D)' = V_{\text{CKM}} (Y^D) V_{\text{CKM}}^T, \tag{179}$$

where by default V_{CKM} contains the CKM matrix in the standard parameterization with central empirical values of the input angles except for the complex phase, which is set to zero. Even if one starts at a high energy scale with a completely family-universal model (for example, mSUGRA), the off-diagonal quark Yukawa matrices induce squark mixing through RGE effects.

The second SUSY Les Houches Accord (SLHA2) has been completed recently, see Sect. 5.2. The flavor mixing aspects will be incorporated into SOFTSUSY as fast as possible, allowing for input and output of flavor mixing parameters in a common format to other programs.

5.9 CalcHep for beyond standard model physics

CalcHep [296, 435] is a package for the computation of Feynman diagrams at tree level, integration over multiparticle phase space, and partonic level event generation. The main idea of CalcHep is to make publicly available the passing on from Lagrangians to final distributions. This is done effectively with a high level of automation. CalcHep is a menu-driven system with help facilities, but it also can be used in a non-interactive batch mode.

In principle, CalcHep is restricted by tree-level calculations but there it can be applied to any model of particle interactions. CalcHep is based on the symbolic calculation of squared diagrams. To perform such a calculation it contains a built-in symbolic calculator. Calculated diagrams are transformed into a C code for further numerical evaluations. Because of the factorial increase of the number of diagrams with the number of external legs, CalcHep is restricted to $2 \to 4$ processes.

The implementation of new models for CalcHep is rather simple and can be done with help of the LanHep package. Currently, there are publicly available realizations of the SM, MSSM, NMSSM, CPVMSSM, and leptoquark model. There are also *private* realizations of models with extra dimensions and with the little Higgs model. Models with flavor violation can also be implemented in CalcHep. The code is available from the following URL: http://theory.sinp.msu.ru/~pukhov/calchep.html.

5.10 HvyN

The Monte Carlo program HvyN allows one to study heavy neutrino production processes at hadron colliders. It can be downloaded from http://www.to.infn.it/~pittau/ALPGEN_BSM.tar.gz or http://mlm.home.cern.ch/m/mlm/www/alpgen/ and it is based on the Alpgen package [312], from which inherits the main features and the interface facilities.

The code allows one to study the following three processes, where a heavy neutrino N (of Dirac or Majorana nature) is produced in association with a charged lepton:

(1) $p\overset{(-)}{p} \to \ell_1 N \to \ell_1 \ell_2 W \to \ell_1 \ell_2 f \bar{f}'$;
(2) $p\overset{(-)}{p} \to \ell_1 N \to \ell_1 \nu_{\ell_2} Z \to \ell_1 \nu_{\ell_2} f \bar{f}$;
(3) $p\overset{(-)}{p} \to \ell_1 N \to \ell_1 \nu_{\ell_2} H \to \ell_1 \nu_{\ell_2} f \bar{f}$.

The full $2 \to 4$ matrix element for the complete decay chain is implemented, so that spin correlations and finite width effects are correctly taken into account. The only relevant subprocess is

$$q \bar{q}' \to W^\star \to \ell_1 N, \tag{180}$$

followed by the full decay chain. The appropriate Lagrangian can be found in [455].

The above three processes are selected by setting an input variable (indec) to 1, 2 or 3, respectively. The flavor of the outgoing leptons, not coming from the boson decay, is controlled by two other variables il1 and il2 (the values 1, 2, 3 correspond to the first, second and third lepton family). In addition, the variable ilnv should be set to 0 (1) if a lepton number conserving (violating) process is considered. Furthermore the variable ima should be given the value 0 (1) in case of Dirac (Majorana) heavy neutrinos.

When indec = 1 and imode = 0, 1 the W decays into e and ν_e. Other decay options can be implemented at the unweighting stage according to the following options:

$1 = e \bar{\nu}_e$,
$2 = \mu \bar{\nu}_\mu$,
$3 = \tau \bar{\nu}_\tau$,
$4 = l \bar{\nu}_l (l = e, \mu, \tau)$,
$5 = q \bar{q}'$,
$6 = \text{fullyinclusive}$.

When indec = 2 the decay mode of the Z boson should be selected at the event generation level by setting the variable idf to the following values:

$0 \Rightarrow \sum_\ell \nu_\ell \bar{\nu}_\ell,$

$1 \Rightarrow \sum_\ell \ell^- \ell^+,$

$2 \Rightarrow u\bar{u}$ and $c\bar{c},$

$3 \Rightarrow d\bar{d}$ and $s\bar{s},$

$4 \Rightarrow b\bar{b},$

$11 \Rightarrow e^- e^+,$

$13 \Rightarrow \mu^- \mu^+,$

$15 \Rightarrow \tau^- \tau^+.$

When indec = 3 the following decay modes of the H boson can be selected, at the generation level, by setting the variable idf according to the following scheme:

$1 \Rightarrow \tau^- \tau^+,$

$2 \Rightarrow c\bar{c},$

$4 \Rightarrow b\bar{b}.$

5.11 PYTHIA for flavor physics at the LHC

PYTHIA [415] is a general-purpose event generator for hadronic events in e^+e^-, eh, and hh collisions (where h is any hadron or photon). The current version is always available from the PYTHIA web page, where also update notes and a number of useful sample main programs can be found. For recent brief overviews relating to SM, BSM, and Higgs physics, see [569–571], respectively. For flavor physics at the LHC, the most relevant processes in PYTHIA can be categorized as follows.

- SUSY with trilinear R-parity violation [537, 539]:
  PYTHIA includes all massive tree-level matrix elements [536] for two body sfermion decays and three body gaugino/higgsino decays. (Note: RPV production cross sections are not included.) Also, the Lund string fragmentation model has been extended to handle antisymmetric color topologies [539], allowing for a more correct treatment of baryon number flow when baryon number is violated.
- Other BSM:
  Production and decay/hadronization of (1) Charged Higgs in 2HDM and SUSY models via $\bar{q}g \to \bar{q}'H^+$, $gg/qq \to \bar{t}bH^+$, $q\bar{q} \to H^+H^-$ (including the possibility of a Z' contribution with full interference), $q\bar{q} \to H^\pm h^0/H^\pm H^0$, and $t \to bH^+$, (2) a W' (without interference with the SM W), (3) a horizontal (FCNC) gauge boson R^0 coupling between generations, e.g. $s\bar{d} \to R^0 \to \mu^- e^+$, (4) Leptoquarks L_Q via $qg \to \ell L_Q$ and $gg/q\bar{q} \to L_Q \bar{L}_Q$. (5) compositeness (e.g. u^*), (6) doubly charged Higgs bosons from L-R symmetry, (7) warped extra dimensions, and (8) a strawman technicolor model. See [415], Sects. 8.5–8.7 for details.
- Open heavy flavor production (c, b, t, b', t'): Massive matrix elements for QCD $2 \to 2$ and resonant Z/W (and Z'/W') heavy flavor production. Also includes flavor excitation and gluon splitting to massive quarks in the shower evolution [572].
- Closed heavy flavor production $(J/\psi, \Upsilon, \chi_{c,b})$: PYTHIA includes a substantial number of color singlet and (more recently) NRQCD color octet mechanisms. For details, see [415], Sect. 8.2.3.
- Hadron decays:
  A large number of c and b hadron (including -onia) decays are implemented. In both cases, most channels for which exclusive branching fractions are known are explicitly listed. For the remaining channels, either educated guesses or a fragmentation-like process determines the flavor composition of the decay products. With few exceptions, hadronic decays are then distributed according to phase space, while semileptonic ones incorporate a simple $V - A$ structure in the limit of massless decay products. See Sect. 13.3 of [415] for more details.

Additional user-defined production processes can be interfaced via the routines UPINIT and UPEVNT (see [415], Sect. 9.9), using the common Les Houches standard [534]. Flavor-violating resonance decays can also be introduced *ad hoc* via the routine PYSLHA, using SUSY Les Houches Accord decay tables [535].

5.12 Sherpa for flavor physics

Sherpa [573] is a multipurpose Monte Carlo event generator that can simulate high energetic collisions at lepton and hadron colliders. Sherpa is publicly available and the source code, potential bug-fixes, documentation material and also a Sherpa related WIKI can be found under: http://www.sherpa-mc.de. The ingredients of Sherpa especially relevant for flavor physics at the LHC are the matrix elements for corresponding hard production processes and the hadronization and decay of flavors produced.

- The matrix elements for the hard production and decay processes within Sherpa are delivered by its built-in matrix element generator AMEGIC++ [574]. At present, AMEGIC++ provides tree-level matrix elements with up to ten final state particles in the framework of the SM [575], the THDM, the MSSM [576] and the ADD model [577]. In general, the program allows all coupling constants to be complex.

The SM interactions implemented allow for the full CKM mixing of quark generations including the complex phase. The implemented set of Feynman rules for the MSSM [578, 579] also considers CKM mixing in the supersymmetrized versions of the SM weak interactions, and the interactions with charged Higgs bosons. A priori, AMEGIC++ allows for a fully general intergenerational mixing of squarks, sleptons and sneutrinos, therefore allowing for various flavor-changing interactions. However, the MSSM input parameters being obtained from the SLHA-conform files [535], only the mixing of the third generation scalar fermions is considered per default. An extension of the SLHA inputs is straightforward and should also allow one to consider complex mixing parameters. The implementation of bilinear R-parity violating supersymmetric interactions, triggering flavor violation effects as well, is currently being started. Within Sherpa the multileg matrix elements of AMEGIC++ are attached with the APACIC++ initial- and final-state parton showers [580] according to the merging algorithm of [581–584]. This procedure allows for the incorporation of parton showering and, ultimately, hadronization and hadron decay models, independent of the energy scale of the hard process.

— Hadronization within Sherpa is performed through an interface to PYTHIA's string fragmentation [376], the emerging unstable hadrons can then be treated by Sherpa's built-in hadron decay module HADRONS++. The current release, Sherpa-1.0.9, includes an early development stage, which already features complete τ-lepton decays, whereas the version currently under development includes decay tables of approximately 100 particles. Many of their decay channels, especially in the flavor relevant K, D and B decays, contain matrix elements and form factor models, while the rest are decayed isotropically according to phase space. Throughout the event chain of Sherpa spin correlations between subsequent decays are included. A proper treatment of neutral meson mixing phenomena is also being implemented.

The structure of Sherpa and its hadron decay module HADRONS++ allows for an easy incorporation of additional or customized decay matrix elements. In addition, parameters like branching ratios or form factor parametrizations can be modified by the user.

Acknowledgements We thank A. Deandrea, J. D'Hondt, A. Gruzza, I. Hinchliffe, M.M. Najafabadi, S. Paktinat, M. Spiropulu and Z. Was for their presentations in the WG1 sessions.

This work has been supported by the EU under the MRTN-CT-2004-503369 and MRTN-CT-2006-035505 network programs. A. Bartl, K. Hohenwarter-Sodek, T. Kernreiter, and W. Majerotto have been supported by the 'Fonds zur Förderung der wissenschaftlichen Forschung' (FWF) of Austria, project. No. P18959-N16, K. Hidaka has been supported by RFBR grant No. 07-02-00256. The work of N. Castro and F. Veloso has supported by Fundação para a Ciência e Tecnologia (FCT) through the grants SFRH/BD/13936/2003 and SFRH/BD/18762/2004. The work of P.M. Fereira and R. Santos is supported by FCT under contract POCI/FIS/59741/2004. P.M. Fereira is supported by FCT under contract SFRH/BPD/5575/2001. R. Santos is supported by FCT under contract SFRH/BPD/23427/2005. J. Guasch and J. Solà have been supported in part by MEC and FEDER under project 2004-04582-C02-01 and by DURSI Generalitat de Catalunya under project 2005SGR00564. The work of S. Peñaranda has been partially supported by the European Union under contract No. MEIF-CT-2003-500030, and by the I3P Contract 2005 of IFIC, CSIC. J.I. Illana acknowledges the financial support by the EU (HPRN-CT-2000-149), the Spanish MCYT (FPA2003-09298-C02-01) and Junta de Andalucía (FQM-101). The work of S. Béjar has been supported by CICYT (FPA2002-00648), by the EU (HPRN-CT-2000-00152), and by DURSI (2001SGR-00188). S. Kraml is supported by an APART (Austrian Programme of Advanced Research and Technology) grant of the Austrian Academy of Sciences. A.R. Raklev acknowledges support from the European Community through a Marie Curie Fellowship for Early Stage Researchers Training and the Norwegian Research Council. J.A. Aguilar-Saavedra acknowledges support by a MEC Ramón y Cajal contract. P. Skands has been partially supported by STFC and by Fermi Research Alliance, LLC, under Contract No. DE-AC02-07CH11359 with the United States Department of Energy. M. Misiak acknowledges support from the EU Contract MRTN-CT-2006-035482, FLAVIAnet. A. Giammanco, E. Kou and J. Alwall have been supported by the Belgian Federal Science Policy (IAP 6/11). The work of Yu. Andreev, S. Bityukov, M. Kirsanov, N. Krasnikov and A. Toropin has been supported by RFFI Grant N 07-02-00256. W. Porod has partially been supported by the German Ministry of Education and Research (BMBF) under contract 05HT6WWA.

The material in Sect. 3.6 has been presented by I. Borjanović for the ATLAS collaboration. J.A. Aguilar-Saavedra, J. Carvalho, N. Castro, A. Onofre, F. Veloso, T. Lari and G. Polesello thank members of the ATLAS collaboration for helpful discussions. They have made use of ATLAS physics analysis and simulation tools which are the result of collaboration-wide efforts.

References

1. M. Artuso et al., arXiv:0801.1833
2. M. Raidal et al., arXiv:0801.1826
3. ATLAS Collaboration, *ATLAS Detector and Physics Performance Technical Design Report*, vol. 2 (CERN, Geneva, 1999). CERN/LHCC 99-14/15
4. M. Heldmann, D. Cavalli, ATLAS Note ATL-PHYS-PUB-2006-008, 2006
5. CMS Collaboration, *CMS Physics Technical Design Report*, vol. 1. (CERN, Geneva, 2006). CMS Note CERN/LHCC 2006-001
6. A. Quadt, Eur. Phys. J. C **48**, 835 (2006)
7. S.L. Glashow, J. Iliopoulos, L. Maiani, Phys. Rev. D **2**, 1285–1292 (1970)
8. J.J. Liu, C.S. Li, L.L. Yang, L.G. Jin, Phys. Lett. B **599**, 92–101 (2004). hep-ph/0406155
9. C.S. Li, R.J. Oakes, J.M. Yang, Phys. Rev. D **49**, 4587–4594 (1994)
10. G. Eilam, A. Gemintern, T. Han, J.M. Yang, X. Zhang, Phys. Lett. B **510**, 227–235 (2001). hep-ph/0102037
11. T. Huang, J.-M. Yang, B.-L. Young, X.-M. Zhang, Phys. Rev. D **58**, 073007 (1998). hep-ph/9803334
12. J.L. Lopez, D.V. Nanopoulos, R. Rangarajan, Phys. Rev. D **56**, 3100–3106 (1997). hep-ph/9702350

13. G.M. de Divitiis, R. Petronzio, L. Silvestrini, Nucl. Phys. B **504**, 45–60 (1997). hep-ph/9704244
14. J.A. Aguilar-Saavedra, Acta Phys. Pol. B **35**, 2695–2710 (2004). hep-ph/0409342
15. B. Grzadkowski, J.F. Gunion, P. Krawczyk, Phys. Lett. B **268**, 106–111 (1991)
16. S. Béjar, J. Guasch, J. Solà, Nucl. Phys. B **600**, 21–38 (2001). hep-ph/0011091
17. D. Atwood, L. Reina, A. Soni, Phys. Rev. D **53**, 1199–1201 (1996). hep-ph/9506243
18. J.A. Aguilar-Saavedra, B.M. Nobre, Phys. Lett. B **553**, 251–260 (2003). hep-ph/0210360
19. F. del Aguila, J.A. Aguilar-Saavedra, R. Miquel, Phys. Rev. Lett. **82**, 1628–1631 (1999). hep-ph/9808400
20. *Flavor-Changing Neutral Currents: Present and Future Studies: Proceedings*, ed. by David B. Cline (World Scientific, Singapore, 1997)
21. M. Beneke et al., hep-ph/0003033
22. T. Stelzer, Z. Sullivan, S. Willenbrock, Phys. Rev. D **58**, 094021 (1998). hep-ph/9807340
23. A.S. Belyaev, E.E. Boos, L.V. Dudko, Phys. Rev. D **59**, 075001 (1999). hep-ph/9806332
24. T. Tait, C.P. Yuan, Phys. Rev. D **63**, 014018 (2001). hep-ph/0007298
25. Z. Sullivan, Phys. Rev. D **70**, 114012 (2004). hep-ph/0408049
26. J. Campbell, F. Tramontano, Nucl. Phys. B **726**, 109–130 (2005). hep-ph/0506289
27. J.A. Aguilar-Saavedra, Phys. Rev. D **67**, 035003 (2003). hep-ph/0210112
28. F. Del Aguila, J. Santiago, J. High Energy Phys. **03**, 010 (2002). hep-ph/0111047
29. J.-J. Cao, R.J. Oakes, F. Wang, J.M. Yang, Phys. Rev. D **68**, 054019 (2003). hep-ph/0306278
30. X.-L. Wang, Q.-L. Zhang, Q.-P. Qiao, Phys. Rev. D **71**, 014035 (2005). hep-ph/0501145
31. F. del Aguila, J.A. Aguilar-Saavedra, Phys. Rev. D **67**, 014009 (2003). hep-ph/0208171
32. H.S. Do, S. Groote, J.G. Korner, M.C. Mauser, Phys. Rev. D **67**, 091501 (2003). hep-ph/0209185
33. P.L. Cho, M. Misiak, Phys. Rev. D **49**, 5894–5903 (1994). hep-ph/9310332
34. K. Fujikawa, A. Yamada, Phys. Rev. D **49**, 5890–5893 (1994)
35. W. Buchmuller, D. Wyler, Nucl. Phys. B **268**, 621 (1986)
36. M. Misiak, M. Steinhauser, Nucl. Phys. B **764**, 62–82 (2007). hep-ph/0609241
37. G. Burdman, M.C. Gonzalez-Garcia, S.F. Novaes, Phys. Rev. D **61**, 114016 (2000). hep-ph/9906329
38. E. Barberio et al. (Heavy Flavor Averaging Group (HFAG) Collaboration), hep-ex/0704.3575
39. K.Y. Lee, W.Y. Song, Phys. Rev. D **66**, 057901 (2002). hep-ph/0204303
40. J.-P. Lee, K.Y. Lee, Eur. Phys. J. C **29**, 373–381 (2003). hep-ph/0209290
41. K.Y. Lee, Phys. Lett. B **632**, 99–104 (2006)
42. G.L. Kane, G.A. Ladinsky, C.P. Yuan, Phys. Rev. D **45**, 124–141 (1992)
43. J.A. Aguilar-Saavedra, J. Carvalho, N. Castro, A. Onofre, F. Veloso, Eur. Phys. J. C **50**, 519–533 (2007). hep-ph/0605190
44. R.H. Dalitz, G.R. Goldstein, Phys. Rev. D **45**, 1531–1543 (1992)
45. B. Lampe, Nucl. Phys. B **454**, 506–526 (1995)
46. S.R. Slabospitsky, L. Sonnenschein, Comput. Phys. Commun. **148**, 87–102 (2002). hep-ph/0201292
47. T. Sjostrand et al., Comput. Phys. Commun. **135**, 238–259 (2001). hep-ph/0010017
48. H.L. Lai et al. (CTEQ Collaboration), Eur. Phys. J. C **12**, 375–392 (2000). hep-ph/9903282
49. E. Richter-Was, D. Froidevaux, L. Poggioli, ATLAS Note ATL-PHYS-98-131, November, 1998
50. J.A. Aguilar-Saavedra, J. Carvalho, N. Castro, A. Onofre, F. Veloso, ATLAS Note ATL-PHYS-PUB-2006-018, 2006
51. J.A. Aguilar-Saavedra, J. Carvalho, N. Castro, A. Onofre, F. Veloso, Eur. Phys. J. C **53**, 689 (2008). arXiv:0705.3041
52. G. Cowan, *Statistical Data Analysis* (Clarendon, Oxford, 1998)
53. S. Eidelman et al. (Particle Data Group Collaboration), Phys. Lett. B **592**, 1 (2004)
54. D. Acosta et al. (CDF Collaboration), Phys. Rev. Lett. **95**, 102002 (2005). hep-ex/0505091
55. V.M. Abazov et al. (D0 Collaboration), Phys. Lett. B **639**, 616–622 (2006). hep-ex/0603002
56. V.M. Abazov (D0 Collaboration), hep-ex/0612052
57. J. Alwall et al., Eur. Phys. J. C **49**, 791–801 (2007). hep-ph/0607115
58. CDF Collaboration, CDF Note 8585, 2006
59. M.C. Smith, S. Willenbrock, Phys. Rev. D **54**, 6696–6702 (1996). hep-ph/9604223
60. T. Stelzer, Z. Sullivan, S. Willenbrock, Phys. Rev. D **56**, 5919–5927 (1997). hep-ph/9705398
61. CMS Collaboration, *CMS physics: Technical Design Report*. Physics Performance, vol. II (CERN, Geneva, 2006)
62. P. Candelas, G.T. Horowitz, A. Strominger, E. Witten, Nucl. Phys. B **258**, 46–74 (1985)
63. J.L. Hewett, T.G. Rizzo, Phys. Rep. **183**, 193 (1989)
64. N. Arkani-Hamed et al., J. High Energy Phys. **08**, 021 (2002). hep-ph/0206020
65. D. Choudhury, T.M.P. Tait, C.E.M. Wagner, Phys. Rev. D **65**, 053002 (2002). hep-ph/0109097
66. M.S. Carena, E. Ponton, J. Santiago, C.E.M. Wagner, Phys. Rev. D **76**, 035006 (2007). hep-ph/0701055
67. CDF Collaboration, CDF Note 8495, 2006
68. M.E. Peskin, T. Takeuchi, Phys. Rev. Lett. **65**, 964–967 (1990)
69. M.E. Peskin, T. Takeuchi, Phys. Rev. D **46**, 381–409 (1992)
70. V.A. Novikov, L.B. Okun, A.N. Rozanov, M.I. Vysotsky, JETP Lett. **76**, 127–130 (2002). hep-ph/0203132
71. H.-J. He, N. Polonsky, S.-F. Su, Phys. Rev. D **64**, 053004 (2001). hep-ph/0102144
72. B. Holdom, hep-ph/0606146
73. J.A. Aguilar-Saavedra, Phys. Lett. B **625**, 234–244 (2005). hep-ph/0506187
74. LEP EXOTICA Working Group Collaboration, LEP-Exotica-WG-2001-01, 2001
75. S. Chekanov et al. (ZEUS Collaboration), Phys. Lett. B **559**, 153–170 (2003). hep-ex/0302010. Results on the branching ratios: private communication by L. Bellagamba
76. A. Aktas et al. (H1 Collaboration), Eur. Phys. J. C **33**, 9–22 (2004). hep-ex/0310032
77. F. Abe et al. (CDF Collaboration), Phys. Rev. Lett. **80**, 2525–2530 (1998)
78. A.A. Ashimova, S.R. Slabospitsky, hep-ph/0604119
79. F. del Aguila, J.A. Aguilar-Saavedra, L. Ametller, Phys. Lett. B **462**, 310–318 (1999). hep-ph/9906462
80. F. del Aguila, J.A. Aguilar-Saavedra, Nucl. Phys. B **576**, 56–84 (2000). hep-ph/9909222
81. P.J. Fox, Z. Ligeti, M. Papucci, G. Perez, M.D. Schwartz, hep-ph/0704.1482
82. J.A. Aguilar-Saavedra, G.C. Branco, Phys. Lett. B **495**, 347–356 (2000). hep-ph/0004190
83. J.L. Diaz-Cruz, R. Martinez, M.A. Perez, A. Rosado, Phys. Rev. D **41**, 891–894 (1990)
84. G. Eilam, J.L. Hewett, A. Soni, Phys. Rev. D **44**, 1473–1484 (1991)
85. B. Mele, S. Petrarca, A. Soddu, Phys. Lett. B **435**, 401 (1998). hep-ph/9805498

86. C.-S. Huang, X.-H. Wu, S.-H. Zhu, Phys. Lett. B **452**, 143–149 (1999). hep-ph/9901369
87. C.T. Hill, Phys. Lett. B **345**, 483–489 (1995). hep-ph/9411426
88. G. Buchalla, G. Burdman, C.T. Hill, D. Kominis, Phys. Rev. D **53**, 5185–5200 (1996). hep-ph/9510376
89. G. Burdman, Phys. Rev. Lett. **83**, 2888–2891 (1999). hep-ph/9905347
90. E. Malkawi, T. Tait, Phys. Rev. D **54**, 5758–5762 (1996). hep-ph/9511337
91. T. Han, K. Whisnant, B.L. Young, X. Zhang, Phys. Lett. B **385**, 311–316 (1996). hep-ph/9606231
92. T. Han, K. Whisnant, B.L. Young, X. Zhang, Phys. Rev. D **55**, 7241–7248 (1997). hep-ph/9603247
93. K. Whisnant, J.-M. Yang, B.-L. Young, X. Zhang, Phys. Rev. D **56**, 467–478 (1997). hep-ph/9702305
94. M. Hosch, K. Whisnant, B.L. Young, Phys. Rev. D **56**, 5725–5730 (1997). hep-ph/9703450
95. T. Han, M. Hosch, K. Whisnant, B.-L. Young, X. Zhang, Phys. Rev. D **58**, 073008 (1998). hep-ph/9806486
96. T.G. Rizzo, Phys. Rev. D **53**, 6218–6225 (1996). hep-ph/9506351
97. T. Tait, C.P. Yuan, Phys. Rev. D **55**, 7300–7301 (1997). hep-ph/9611244
98. E. Boos, L. Dudko, T. Ohl, Eur. Phys. J. C **11**, 473–484 (1999). hep-ph/9903215
99. D. Espriu, J. Manzano, Phys. Rev. D **65**, 073005 (2002). hep-ph/0107112
100. J. Guasch, W. Hollik, S. Penaranda, J. Sola, Nucl. Phys. Proc. Suppl. **157**, 152–156 (2006). hep-ph/0601218
101. A. Arhrib, K. Cheung, C.-W. Chiang, T.-C. Yuan, Phys. Rev. D **73**, 075015 (2006). hep-ph/0602175
102. P.M. Ferreira, O. Oliveira, R. Santos, Phys. Rev. D **73**, 034011 (2006). hep-ph/0510087
103. P.M. Ferreira, R. Santos, Phys. Rev. D **73**, 054025 (2006). hep-ph/0601078
104. P.M. Ferreira, R. Santos, Phys. Rev. D **74**, 014006 (2006). hep-ph/0604144
105. J.F. Gunion, H.E. Haber, G.L. Kane, S. Dawson, *The Higgs Hunter's Guide* (Addison-Wesley, Menlo-Park, 1990)
106. J. Guasch, J. Solà, Nucl. Phys. B **562**, 3–28 (1999). hep-ph/9906268
107. S. Béjar, F. Dilmé, J. Guasch, J. Solà, J. High Energy Phys. **08**, 018 (2004). hep-ph/0402188
108. J. Guasch, hep-ph/9710267 (1997)
109. S. Béjar, J. Guasch, J. Solà, J. High Energy Phys. **10**, 113 (2005). hep-ph/0508043
110. P. Gambino, M. Misiak, Nucl. Phys. B **611**, 338–366 (2001). hep-ph/0104034
111. S. Béjar, J. Guasch, J. Solà, Nucl. Phys. B **675**, 270–288 (2003). hep-ph/0307144
112. J. Küblbeck, M. Böhm, A. Denner, Comput. Phys. Commun. **60**, 165–180 (1990)
113. T. Hahn, M. Pérez-Victoria, Comput. Phys. Commun. **118**, 153 (1999). hep-ph/9807565
114. T. Hahn, FeynArts 2.2, FormCalc and LoopTools user's guides. http://www.feynarts.de
115. M. Spira, hep-ph/9711407 (1997)
116. M. Spira, Fortschr. Phys. **46**, 203–284 (1998). hep-ph/9705337, and references therein
117. A.M. Curiel, M.J. Herrero, D. Temes, Phys. Rev. D **67**, 075008 (2003). hep-ph/0210335
118. D.A. Demir, Phys. Lett. B **571**, 193–208 (2003). hep-ph/0303249
119. A.M. Curiel, M.J. Herrero, W. Hollik, F. Merz, S. Peñaranda, Phys. Rev. D **69**, 075009 (2004). hep-ph/0312135
120. S. Heinemeyer, W. Hollik, F. Merz, S. Peñaranda, Eur. Phys. J. C **37**, 481–493 (2004). hep-ph/0403228
121. G. Eilam, M. Frank, I. Turan, hep-ph/0601253 (2006)
122. J.J. Liu, C.S. Li, L.L. Yang, L.G. Jin, Nucl. Phys. B **705**, 3–32 (2005). hep-ph/0404099
123. M.J. Duncan, Nucl. Phys. B **221**, 285 (1983)
124. M.J. Duncan, Phys. Rev. D **31**, 1139 (1985)
125. F. Gabbiani, E. Gabrielli, A. Masiero, L. Silvestrini, Nucl. Phys. B **477**, 321–352 (1996). hep-ph/9604387
126. M. Misiak, S. Pokorski, J. Rosiek, Adv. Ser. Direct. High Energy Phys. **15**, 795–828 (1998). hep-ph/9703442
127. T. Hahn, W. Hollik, J.I. Illana, S. Peñaranda, hep-ph/0512315 (2005)
128. S. Béjar, J. Guasch, J. Solà, Nucl. Phys. Proc. Suppl. **157**, 147–151 (2006). hep-ph/0601191
129. J. Carvalho, N. Castro, L.D. Chikovani, T. Djobava, J. Dodd, S. McGrath, A. Onofre, J. Parsons, F. Veloso, Eur. Phys. J. C **52**, 999–1019 (2007) [SN-ATLAS-2007-059]
130. L. Benucci, A. Kyriakis et al., CMS NOTE-2006/093, 2006
131. A.L. Read, CERN-OPEN-2000-205, 2000. Prepared for Workshop on Confidence Limits, Geneva, Switzerland, 17–18 January 2000
132. CDF Collaboration, CDF Note 8888, 2007
133. LEP Electroweak Working Group, hep-ex/0312023. http://lepewwg.web.cern.ch/LEPEWWG/Welcome.html
134. A. Juste, PoS **TOP2006**, 007 (2006). hep-ex/0603007
135. J. Pumplin et al., J. High Energy Phys. **0207**, 012 (2002). hep-ph/0201195
136. G. Mahlon, S.J. Parke, Phys. Lett. B **476**, 323–330 (2000). hep-ph/9912458
137. J.J. Zhang, C.S. Li, Z. Li, L.L. Yang, hep-ph/0610087
138. J. Campbell, R.K. Ellis, F. Tramontano, Phys. Rev. D **70**, 094012 (2004). hep-ph/0408158
139. M. Beccaria, G. Macorini, F.M. Renard, C. Verzegnassi, hep-ph/0609189
140. M. Beccaria, G. Macorini, F.M. Renard, C. Verzegnassi, Phys. Rev. D **73**, 093001 (2006). hep-ph/0601175
141. ATLAS Collaboration, ATLAS Collaboration Data Challenge 2 (DC2) points. http://paige.home.cern.ch/paige/fullsusy/romeindex.html
142. L. Randall, R. Sundrum, Phys. Rev. Lett. **83**, 3370–3373 (1999). hep-ph/9905221
143. W.D. Goldberger, M.B. Wise, Phys. Rev. D **60**, 107505 (1999). hep-ph/9907218
144. A. Pomarol, Phys. Lett. B **486**, 153–157 (2000). hep-ph/9911294
145. S. Chang, J. Hisano, H. Nakano, N. Okada, M. Yamaguchi, Phys. Rev. D **62**, 084025 (2000). hep-ph/9912498
146. Y. Grossman, M. Neubert, Phys. Lett. B **474**, 361–371 (2000). hep-ph/9912408
147. T. Gherghetta, A. Pomarol, Nucl. Phys. B **586**, 141–162 (2000). hep-ph/0003129
148. S.J. Huber, Q. Shafi, Phys. Lett. B **498**, 256–262 (2001). hep-ph/0010195
149. H. Davodiasl, J.L. Hewett, T.G. Rizzo, Phys. Rev. D **63**, 075004 (2001). hep-ph/0006041
150. J.L. Hewett, F.J. Petriello, T.G. Rizzo, J. High Energy Phys. **09**, 030 (2002). hep-ph/0203091
151. P.M. Aquino, G. Burdman, O.J.P. Eboli, Phys. Rev. Lett. **98**, 131601 (2007). hep-ph/0612055
152. G. Burdman, Phys. Rev. D **66**, 076003 (2002). hep-ph/0205329
153. G. Burdman, Phys. Lett. B **590**, 86–94 (2004). hep-ph/0310144
154. K. Agashe, G. Perez, A. Soni, Phys. Rev. Lett. **93**, 201804 (2004). hep-ph/0406101
155. M. Jezabek, Nucl. Phys. Proc. Suppl. B **37**, 197 (1994). hep-ph/9406411
156. A. Czarnecki, M. Jezabek, J.H. Kuhn, Nucl. Phys. B **351**, 70–80 (1991)
157. A. Brandenburg, Z.G. Si, P. Uwer, Phys. Lett. B **539**, 235–241 (2002). hep-ph/0205023

158. W. Bernreuther, A. Brandenburg, Z.G. Si, P. Uwer, Nucl. Phys. B **690**, 81–137 (2004). hep-ph/0403035
159. T. Stelzer, S. Willenbrock, Phys. Lett. B **374**, 169–172 (1996). hep-ph/9512292
160. M. Jezabek, J.H. Kuhn, Phys. Lett. B **329**, 317–324 (1994). hep-ph/9403366
161. G. Mahlon, S.J. Parke, Phys. Rev. D **53**, 4886–4896 (1996). hep-ph/9512264
162. F. Hubaut, E. Monnier, P. Pralavorio, K. Smolek, V. Simak, Eur. Phys. J. C **44S2**, 13–33 (2005). hep-ex/0508061
163. S.P. Martin, hep-ph/9709356
164. S. Weinberg, Phys. Rev. D **26**, 287 (1982)
165. N. Sakai, T. Yanagida, Nucl. Phys. B **197**, 533 (1982)
166. G.R. Farrar, P. Fayet, Phys. Lett. B **76**, 575–579 (1978)
167. L.E. Ibanez, G.G. Ross, Phys. Lett. B **260**, 291–295 (1991)
168. H.K. Dreiner, C. Luhn, M. Thormeier, Phys. Rev. D **73**, 075007 (2006). hep-ph/0512163
169. H.K. Dreiner, C. Luhn, H. Murayama, M. Thormeier, hep-ph/0610026
170. B.C. Allanach et al., hep-ph/0602198
171. D.J.H. Chung et al., Phys. Rep. **407**, 1–203 (2005). hep-ph/0312378
172. J. Hisano, T. Moroi, K. Tobe, M. Yamaguchi, Phys. Rev. D **53**, 2442–2459 (1996). hep-ph/9510309
173. J.F. Donoghue, H.P. Nilles, D. Wyler, Phys. Lett. B **128**, 55 (1983)
174. L.J. Hall, V.A. Kostelecky, S. Raby, Nucl. Phys. B **267**, 415 (1986)
175. F. Borzumati, A. Masiero, Phys. Rev. Lett. **57**, 961 (1986)
176. F. Gabbiani, A. Masiero, Phys. Lett. B **209**, 289–294 (1988)
177. J.S. Hagelin, S. Kelley, T. Tanaka, Nucl. Phys. B **415**, 293–331 (1994)
178. R. Barbieri, L.J. Hall, Phys. Lett. B **338**, 212–218 (1994). hep-ph/9408406
179. J. Hisano, T. Moroi, K. Tobe, M. Yamaguchi, Phys. Lett. B **391**, 341–350 (1997). hep-ph/9605296
180. J. Hisano, D. Nomura, T. Yanagida, Phys. Lett. B **437**, 351–358 (1998). hep-ph/9711348
181. N.V. Krasnikov, Phys. Lett. B **388**, 783–787 (1996). hep-ph/9511464
182. J. Hisano, M.M. Nojiri, Y. Shimizu, M. Tanaka, Phys. Rev. D **60**, 055008 (1999). hep-ph/9808410
183. D. Nomura, Phys. Rev. D **64**, 075001 (2001). hep-ph/0004256
184. M. Guchait, J. Kalinowski, P. Roy, Eur. Phys. J. C **21**, 163–169 (2001). hep-ph/0103161
185. W. Porod, W. Majerotto, Phys. Rev. D **66**, 015003 (2002). hep-ph/0201284
186. D.F. Carvalho, J.R. Ellis, M.E. Gomez, S. Lola, J.C. Romao, Phys. Lett. B **618**, 162–170 (2005). hep-ph/0206148
187. W. Porod, Czech. J. Phys. **55**, B233–B240 (2005). hep-ph/0410318
188. F. Deppisch, H. Pas, A. Redelbach, R. Ruckl, Y. Shimizu, Phys. Rev. D **69**, 054014 (2004). hep-ph/0310053
189. K. Hamaguchi, A. Ibarra, J. High Energy Phys. **02**, 028 (2005). hep-ph/0412229
190. N. Oshimo, Eur. Phys. J. C **39**, 383–388 (2005). hep-ph/0409018
191. P. Paradisi, J. High Energy Phys. **10**, 006 (2005). hep-ph/0505046
192. N. Arkani-Hamed, H.-C. Cheng, J.L. Feng, L.J. Hall, Phys. Rev. Lett. **77**, 1937–1940 (1996). hep-ph/9603431
193. N. Arkani-Hamed, J.L. Feng, L.J. Hall, H.-C. Cheng, Nucl. Phys. B **505**, 3–39 (1997). hep-ph/9704205
194. A. Bartl et al., hep-ph/0709.1157
195. K. Agashe, M. Graesser, Phys. Rev. D **61**, 075008 (2000). hep-ph/9904422
196. I. Hinchliffe, F.E. Paige, Phys. Rev. D **63**, 115006 (2001). hep-ph/0010086
197. J. Hisano, R. Kitano, M.M. Nojiri, Phys. Rev. D **65**, 116002 (2002). hep-ph/0202129
198. T. Hurth, W. Porod, Eur. Phys. J. C **33**, s764–s766 (2004). hep-ph/0311075
199. G. D'Ambrosio, G.F. Giudice, G. Isidori, A. Strumia, Nucl. Phys. B **645**, 155–187 (2002). hep-ph/0207036
200. E. Lunghi, W. Porod, O. Vives, Phys. Rev. D **74**, 075003 (2006). hep-ph/0605177
201. V. Cirigliano, B. Grinstein, G. Isidori, M.B. Wise, Nucl. Phys. B **728**, 121–134 (2005). hep-ph/0507001
202. S. Davidson, F. Palorini, Phys. Lett. B **642**, 72–80 (2006). hep-ph/0607329
203. F. Deppisch, J. Kalinowski, H. Pas, A. Redelbach, R. Ruckl, hep-ph/0401243
204. M. Maltoni, T. Schwetz, M.A. Tortola, J.W.F. Valle, New J. Phys. **6**, 122 (2004). hep-ph/0405172
205. L.J. Hall, M. Suzuki, Nucl. Phys. B **231**, 419 (1984)
206. B.C. Allanach, A. Dedes, H.K. Dreiner, Phys. Rev. D **69**, 115002 (2004). hep-ph/0309196
207. B. de Carlos, P.L. White, Phys. Rev. D **54**, 3427–3446 (1996). hep-ph/9602381
208. E. Nardi, Phys. Rev. D **55**, 5772–5779 (1997). hep-ph/9610540
209. R. Hempfling, Nucl. Phys. B **478**, 3–30 (1996). hep-ph/9511288
210. H.-P. Nilles, N. Polonsky, Nucl. Phys. B **484**, 33–62 (1997). hep-ph/9606388
211. M. Hirsch, M.A. Diaz, W. Porod, J.C. Romao, J.W.F. Valle, Phys. Rev. D **62**, 113008 (2000). hep-ph/0004115
212. H.K. Dreiner, M. Thormeier, Phys. Rev. D **69**, 053002 (2004). hep-ph/0305270
213. D. Aristizabal Sierra, M. Hirsch, W. Porod, J. High Energy Phys. **09**, 033 (2005). hep-ph/0409241
214. B.C. Allanach, A. Dedes, H.K. Dreiner, Phys. Rev. D **60**, 075014 (1999). hep-ph/9906209
215. Y. Ashie et al. (Super-Kamiokande Collaboration), Phys. Rev. D **71**, 112005 (2005). hep-ex/0501064
216. B. Aharmim et al. (SNO Collaboration), Phys. Rev. C **72**, 055502 (2005). nucl-ex/0502021
217. T. Araki et al. (KamLAND Collaboration), Phys. Rev. Lett. **94**, 081801 (2005). hep-ex/0406035
218. A. Abada, M. Losada, Phys. Lett. B **492**, 310–320 (2000). hep-ph/0007041
219. F. Borzumati, J.S. Lee, Phys. Rev. D **66**, 115012 (2002). hep-ph/0207184
220. M.C. Gonzalez-Garcia, J.C. Romao, J.W.F. Valle, Nucl. Phys. B **391**, 100–126 (1993)
221. H.K. Dreiner, G.G. Ross, Nucl. Phys. B **365**, 597–613 (1991)
222. H.K. Dreiner, M. Guchait, D.P. Roy, Phys. Rev. D **49**, 3270–3282 (1994). hep-ph/9310291
223. H. Baer, C.-h. Chen, and X. Tata, Phys. Rev. D **55**, 1466–1470 (1997). hep-ph/9608221
224. A. Bartl et al., Nucl. Phys. B **502**, 19–36 (1997). hep-ph/9612436
225. H.K. Dreiner, P. Richardson, M.H. Seymour, Phys. Rev. D **63**, 055008 (2001). hep-ph/0007228
226. X. Yin, W.-G. Ma, L.-H. Wan, Y. Jiang, Phys. Rev. D **64**, 076006 (2001). hep-ph/0107006
227. S. Borgani, A. Masiero, M. Yamaguchi, Phys. Lett. B **386**, 189–197 (1996). hep-ph/9605222
228. F. Takayama, M. Yamaguchi, Phys. Lett. B **485**, 388–392 (2000). hep-ph/0005214
229. M. Hirsch, W. Porod, D. Restrepo, J. High Energy Phys. **03**, 062 (2005). hep-ph/0503059
230. W. Buchmuller, L. Covi, K. Hamaguchi, A. Ibarra, T. Yanagida, J. High Energy Phys. **03**, 037 (2007). hep-ph/0702184
231. D.E. Lopez-Fogliani, C. Munoz, Phys. Rev. Lett. **97**, 041801 (2006). hep-ph/0508297
232. H.P. Nilles, M. Srednicki, D. Wyler, Phys. Lett. B **120**, 346 (1983)

233. J.M. Frere, D.R.T. Jones, S. Raby, Nucl. Phys. B **222**, 11 (1983)
234. J.P. Derendinger, C.A. Savoy, Nucl. Phys. B **237**, 307 (1984)
235. A.I. Veselov, M.I. Vysotsky, K.A. Ter-Martirosian, Sov. Phys. JETP **63**, 489 (1986)
236. C.S. Aulakh, R.N. Mohapatra, Phys. Lett. B **119**, 136 (1982)
237. G.G. Ross, J.W.F. Valle, Phys. Lett. B **151**, 375 (1985)
238. A. Masiero, J.W.F. Valle, Phys. Lett. B **251**, 273–278 (1990)
239. J.C. Romao, C.A. Santos, J.W.F. Valle, Phys. Lett. B **288**, 311–320 (1992)
240. M. Shiraishi, I. Umemura, K. Yamamoto, Phys. Lett. B **313**, 89–95 (1993)
241. K. Huitu, J. Maalampi, Phys. Lett. B **344**, 217–224 (1995). hep-ph/9410342
242. J.C. Romao, F. de Campos, J.W.F. Valle, Phys. Lett. B **292**, 329–336 (1992). hep-ph/9207269
243. M. Hirsch, J.C. Romao, J.W.F. Valle, A. Villanova del Moral, Phys. Rev. D **70**, 073012 (2004). hep-ph/0407269
244. S. Abdullin et al. (CMS Collaboration), J. Phys. G **28**, 469 (2002). hep-ph/9806366
245. D.R. Tovey, Eur. Phys. J. C **4**, N4 (2002)
246. A.J. Barr, C.G. Lester, M.A. Parker, B.C. Allanach, P. Richardson, J. High Energy Phys. **03**, 045 (2003). hep-ph/0208214
247. P. Aurenche, G. Belanger, F. Boudjema, J.P. Guillet, E. Pilon (eds.), Prepared for Workshop on Physics at TeV Colliders, Les Houches, France, 21 May–1 June 2001
248. B.C. Allanach et al., hep-ph/0202233
249. B.K. Gjelsten, D.J. Miller, P. Osland, G. Polesello, ATLAS Note ATL-PHYS-2004-007, 2004
250. M. Chiorboli, Czech. J. Phys. **54**, A151–A159 (2004)
251. M.M. Nojiri, G. Polesello, D.R. Tovey, J. High Energy Phys. **03**, 063 (2006). hep-ph/0512204
252. A.J. Barr, Phys. Lett. B **596**, 205–212 (2004). hep-ph/0405052
253. A.J. Barr, J. High Energy Phys. **02**, 042 (2006). hep-ph/0511115
254. S.I. Bityukov, N.V. Krasnikov, Phys. Atom. Nucl. **62**, 1213–1225 (1999). hep-ph/9712358
255. S.I. Bityukov, N.V. Krasnikov, hep-ph/9806504
256. F.E. Paige, ECONF C **960625**, SUP114 (1996). hep-ph/9609373
257. H. Bachacou, I. Hinchliffe, F.E. Paige, Phys. Rev. D **62**, 015009 (2000). hep-ph/9907518
258. B.C. Allanach, C.G. Lester, M.A. Parker, B.R. Webber, J. High Energy Phys. **09**, 004 (2000). hep-ph/0007009
259. J.A. Aguilar-Saavedra et al., Eur. Phys. J. C **46**, 43–60 (2006). hep-ph/0511344
260. A. Bartl et al., Eur. Phys. J. C **46**, 783–789 (2006). hep-ph/0510074
261. I. Hinchliffe, private communication
262. A. Ibarra, S. Roy, J. High Energy Phys. **05**, 059 (2007). hep-ph/0606116
263. K. Hamaguchi, Y. Kuno, T. Nakaya, M.M. Nojiri, Phys. Rev. D **70**, 115007 (2004). hep-ph/0409248
264. J.L. Feng, B.T. Smith, Phys. Rev. D **71**, 015004 (2005). hep-ph/0409278
265. W. Beenakker, R. Hopker, M. Spira, P.M. Zerwas, Nucl. Phys. B **492**, 51–103 (1997). hep-ph/9610490
266. J.C. Romao, M.A. Diaz, M. Hirsch, W. Porod, J.W.F. Valle, Phys. Rev. D **61**, 071703 (2000). hep-ph/9907499
267. M.A. Diaz, M. Hirsch, W. Porod, J.C. Romao, J.W.F. Valle, Phys. Rev. D **68**, 013009 (2003). hep-ph/0302021
268. B. Mukhopadhyaya, S. Roy, F. Vissani, Phys. Lett. B **443**, 191–195 (1998). hep-ph/9808265
269. W. Porod, M. Hirsch, J. Romao, J.W.F. Valle, Phys. Rev. D **63**, 115004 (2001). hep-ph/0011248
270. M. Hirsch, W. Porod, J.C. Romao, J.W.F. Valle, Phys. Rev. D **66**, 095006 (2002). hep-ph/0207334
271. M. Hirsch, W. Porod, Phys. Rev. D **68**, 115007 (2003). hep-ph/0307364
272. W. Porod, P. Skands, hep-ph/0401077
273. F. de Campos et al., Phys. Rev. D **71**, 075001 (2005). hep-ph/0501153
274. F. de Campos et al., hep-ph/0712.2156
275. M. Hirsch, W. Porod, Phys. Rev. D **74**, 055003 (2006). hep-ph/0606061
276. J. Schechter, J.W.F. Valle, Phys. Rev. D **22**, 2227 (1980)
277. J.R. Ellis, G. Gelmini, C. Jarlskog, G.G. Ross, J.W.F. Valle, Phys. Lett. B **150**, 142 (1985)
278. D. Aristizabal Sierra, M. Hirsch, J.W.F. Valle, A. Villanova del Moral, Phys. Rev. D **68**, 033006 (2003). hep-ph/0304141
279. E. Ma, Mod. Phys. Lett. A **17**, 1259–1262 (2002). hep-ph/0205025
280. G. Azuelos, K. Benslama, J. Ferland, J. Phys. G **32**, 73–92 (2006). hep-ph/0503096
281. K. Huitu, J. Maalampi, A. Pietila, M. Raidal, Nucl. Phys. B **487**, 27–42 (1997). hep-ph/9606311
282. B. Dion, T. Gregoire, D. London, L. Marleau, H. Nadeau, Phys. Rev. D **59**, 075006 (1999). hep-ph/9810534
283. A.G. Akeroyd, M. Aoki, Phys. Rev. D **72**, 035011 (2005). hep-ph/0506176
284. B.K. Gjelsten, D.J. Miller, P. Osland, J. High Energy Phys. **12**, 003 (2004). hep-ph/0410303
285. J.R. Ellis, K.A. Olive, Y. Santoso, V.C. Spanos, Phys. Lett. B **565**, 176–182 (2003). hep-ph/0303043
286. I. Borjanovic (ATLAS Collaboration), PoS **HEP2005**, 350 (2006)
287. C.G. Lester, D.J. Summers, Phys. Lett. B **463**, 99–103 (1999). hep-ph/9906349
288. T. Goto, K. Kawagoe, M.M. Nojiri, Phys. Rev. D **70**, 075016 (2004). hep-ph/0406317
289. Y. Andreev, S. Bityukov, N. Krasnikov, A. Toropin, CERN-CMS-NOTE-2006-132
290. F. del Aguila, L. Ametller, Phys. Lett. B **261**, 326–333 (1991)
291. H. Baer, C.-H. Chen, F. Paige, X. Tata, Phys. Rev. D **49**, 3283–3290 (1994). hep-ph/9311248
292. D. Denegri, L. Rurua, N. Stepanov, Detection of sleptons in cms, mass reach, CMS TN-96-059, 1996
293. Y.M. Andreev, S.I. Bityukov, N.V. Krasnikov, Phys. At. Nucl. **68**, 340–347 (2005). hep-ph/0402229
294. F.E. Paige, S.D. Protopopescu, H. Baer, X. Tata, hep-ph/0312045
295. W. Beenakker, R. Hopker, M. Spira, hep-ph/9611232
296. A. Pukhov et al., hep-ph/9908288
297. O. Oscar, Cms simulation package home page. Located at http://cmsdoc.cern.ch/oscar
298. O. Orca, Cms reconstruction package. Located at http://cmsdoc.cern.ch/orca
299. D. Acosta et al., Cms physics tdr volume 1, section 2.6: Fast simulation. CERN/LHCC-2006-001, p. 55 (2006)
300. S.I. Bityukov, N.V. Krasnikov, Mod. Phys. Lett. A **13**, 3235–3249 (1998)
301. R. Cousins, J. Mumford, V. Valuev (CMS Collaboration), Czech. J. Phys. **55**, B651–B658 (2005)
302. V. Bartsch, G. Quast, CMS Note-2005/004, 2005
303. Y.M. Andreev, S.I. Bityukov, N.V. Krasnikov, A.N. Toropin, hep-ph/0608176
304. N.V. Krasnikov, Mod. Phys. Lett. A **9**, 791–794 (1994)
305. N.V. Krasnikov, JETP Lett. **65**, 148–153 (1997). hep-ph/9611282
306. H. Baer, C.-h. Chen, F. Paige, and X. Tata. Phys. Rev. D **50**, 4508–4516 (1994). hep-ph/9404212
307. A. Datta, K. Kong, K.T. Matchev, Phys. Rev. D **72**, 096006 (2005). hep-ph/0509246
308. J.M. Smillie, B.R. Webber, J. High Energy Phys. **10**, 069 (2005). hep-ph/0507170
309. P. Richardson, J. High Energy Phys. **11**, 029 (2001). hep-ph/0110108

310. M. Biglietti et al., ATLAS Physics Note ATL-PHYS-PUB-2007-004
311. G. Corcella et al., J. High Energy Phys. **01**, 010 (2001). hep-ph/0011363
312. M.L. Mangano, M. Moretti, F. Piccinini, R. Pittau, A.D. Polosa, J. High Energy Phys. **07**, 001 (2003). hep-ph/0206293
313. E. Richter-Was, ATLAS Physics Note ATL-PHYS-98-131
314. A.G. Frodesen, O. Skjeggestad, H. Tofte, *Probability and Statistics in Particle Physics* (Universitetsforlaget, Bergen, 1979)
315. J. Guasch, J. Solà, in *Proceedings of 4th International Workshop Detectors on Linear Colliders (LCWS 99)*, Sitges, Barcelona, Spain, 28 April–5 May 1999, ed. by E. Fernández, A. Pacheco (Universitat Autònoma de Barcelona, Barcelona, 2000), pp. 196–204. hep-ph/9909503
316. S. Béjar, J. Guasch, J. Solà, in *Proceedings of the 5th International Symposium on Radiative Corrections (RADCOR 2000)*, Carmel, California, 11–15 September 2000, eConf C000911. hep-ph/0101294
317. A. Yamada, Z. Phys. C **61**, 247 (1994)
318. P. Chankowski, S. Pokorski, J. Rosiek, Nucl. Phys. B **423**, 437–496 (1994). hep-ph/9303309
319. A. Dabelstein, Z. Phys. C **67**, 495–512 (1995). hep-ph/9409375
320. A. Dabelstein, Nucl. Phys. B **456**, 25–56 (1995). hep-ph/9503443
321. M. Spira, hep-ph/9510347
322. M. Spira, A. Djouadi, D. Graudenz, P.M. Zerwas, Nucl. Phys. B **453**, 17–82 (1995). hep-ph/9504378
323. H.L. Lai et al. (CTEQ Collaboration), Phys. Rev. D **55**, 1280–1296 (1997). hep-ph/9606399
324. P. Gambino, U. Haisch, M. Misiak, Phys. Rev. Lett. **94**, 061803 (2005). hep-ph/0410155
325. http://fchdecay.googlepages.com
326. O. Brein, Comput. Phys. Commun. **170**, 42–48 (2005). hep-ph/0407340
327. M. Carena, S. Mrenna, C.E.M. Wagner, Phys. Rev. D **62**, 055008 (2000). hep-ph/9907422
328. M. Carena, S. Heinemeyer, C.E.M. Wagner, G. Weiglein, Eur. Phys. J. C **26**, 601–607 (2003). hep-ph/0202167
329. F. Borzumati, C. Greub, T. Hurth, D. Wyler, Phys. Rev. D **62**, 075005 (2000). hep-ph/9911245
330. T. Besmer, C. Greub, T. Hurth, Nucl. Phys. B **609**, 359–386 (2001). hep-ph/0105292
331. T. Besmer, C. Greub, T. Hurth, hep-ph/0111389
332. A.L. Kagan, M. Neubert, Phys. Rev. D **58**, 094012 (1998). hep-ph/9803368
333. A.L. Kagan, M. Neubert, Eur. Phys. J. C **7**, 5–27 (1999). hep-ph/9805303
334. T. Hahn, Comput. Phys. Commun. **140**, 418–431 (2001). hep-ph/0012260
335. T. Hahn, C. Schappacher, Comput. Phys. Commun. **143**, 54–68 (2002). hep-ph/0105349
336. T. Hahn, W. Hollik, S. Heinemeyer, G. Weiglein, hep-ph/0507009
337. S. Heinemeyer, W. Hollik, G. Weiglein, Comput. Phys. Commun. **124**, 76–89 (2000). hep-ph/9812320
338. S. Heinemeyer, W. Hollik, G. Weiglein, Eur. Phys. J. C **9**, 343–366 (1999). hep-ph/9812472
339. FeynHiggs webpage. http://www.feynhiggs.de
340. S. Chen et al. (CLEO Collaboration), Phys. Rev. Lett. **87**, 251807 (2001). hep-ex/0108032
341. B. Aubert et al. (BaBar Collaboration), hep-ex/0207074
342. W. Porod, Comput. Phys. Commun. **153**, 275–315 (2003). hep-ph/0301101
343. P. Brax, C.A. Savoy, Nucl. Phys. B **447**, 227–251 (1995). hep-ph/9503306
344. W.M. Yao et al. (Particle Data Group Collaboration), J. Phys. G **33**, 1–1232 (2006)
345. S. Heinemeyer, D. Stoeckinger, G. Weiglein, Nucl. Phys. B **690**, 62–80 (2004). hep-ph/0312264
346. S. Heinemeyer, D. Stoeckinger, G. Weiglein, Nucl. Phys. B **699**, 103–123 (2004). hep-ph/0405255
347. P. Gondolo et al., J. Cosmol. Astropart. Phys. **0407**, 008 (2004). astro-ph/0406204
348. J. Hamann, S. Hannestad, M.S. Sloth, Y.Y.Y. Wong, Phys. Rev. D **75**, 023522 (2007). astro-ph/0611582
349. G. Bozzi, B. Fuks, B. Herrmann, M. Klasen, Nucl. Phys. B **787**, 1 (2007)
350. G. Bozzi, B. Fuks, B. Herrmann, M. Klasen, hep-ph/0704.1826
351. A. De Roeck et al., hep-ph/0508198
352. P.L. Cho, M. Misiak, D. Wyler, Phys. Rev. D **54**, 3329–3344 (1996). hep-ph/9601360
353. T. Hurth, E. Lunghi, W. Porod, Nucl. Phys. B **704**, 56–74 (2005). hep-ph/0312260
354. A.J. Buras, P.H. Chankowski, J. Rosiek, L. Slawianowska, Nucl. Phys. B **659**, 3 (2003). hep-ph/0210145
355. T. Huber, E. Lunghi, M. Misiak, D. Wyler, Nucl. Phys. B **740**, 105–137 (2006). hep-ph/0512066
356. J. Hisano, K. Kawagoe, M.M. Nojiri, Phys. Rev. D **68**, 035007 (2003). hep-ph/0304214
357. J.M. Cline, hep-ph/0609145 (2006)
358. B. Fuks, in *Flavour of the Era of LHC Workshop*, CERN, 6–8 February 2006
359. M. Beccaria, G. Macorini, L. Panizzi, F.M. Renard, C. Verzegnassi, Phys. Rev. D **74**, 093009 (2006). hep-ph/0610075
360. W. Beenakker, M. Kramer, T. Plehn, M. Spira, P.M. Zerwas, Nucl. Phys. B **515**, 3–14 (1998). hep-ph/9710451
361. S. Kretzer, H.L. Lai, F.I. Olness, W.K. Tung, Phys. Rev. D **69**, 114005 (2004). hep-ph/0307022
362. D. Delepine, J.M. Gerard, R. Gonzalez Felipe, J. Weyers, Phys. Lett. B **386**, 183–188 (1996). hep-ph/9604440
363. M. Carena, M. Quiros, C.E.M. Wagner, Nucl. Phys. B **524**, 3–22 (1998). hep-ph/9710401
364. J.M. Cline, G.D. Moore, Phys. Rev. Lett. **81**, 3315–3318 (1998). hep-ph/9806354
365. M. Laine, K. Rummukainen, Phys. Rev. Lett. **80**, 5259–5262 (1998). hep-ph/9804255
366. C. Balazs, M. Carena, C.E.M. Wagner, Phys. Rev. D **70**, 015007 (2004). hep-ph/0403224
367. C. Boehm, A. Djouadi, M. Drees, Phys. Rev. D **62**, 035012 (2000). hep-ph/9911496
368. J.R. Ellis, K.A. Olive, Y. Santoso, Astropart. Phys. **18**, 395–432 (2003). hep-ph/0112113
369. K.-I. Hikasa, M. Kobayashi, Phys. Rev. D **36**, 724 (1987)
370. S. Kraml, A.R. Raklev, Phys. Rev. D **73**, 075002 (2006). hep-ph/0512284
371. S. Kraml, A.R. Raklev, in *Proceedings of the 14th International Conference on Supersymmetry and the Unification of Fundamental Interactions (SUSY06)*, Irvine, CA, 12–17 June 2006. hep-ph/0609293.
372. I. Hinchliffe, F.E. Paige, M.D. Shapiro, J. Soderqvist, W. Yao, Phys. Rev. D **55**, 5520–5540 (1997). hep-ph/9610544
373. C.G. Lester, CERN-THESIS-2004-003
374. D.J. Miller, P. Osland, A.R. Raklev, J. High Energy Phys. **03**, 034 (2006). hep-ph/0510356
375. S. Moretti, K. Odagiri, P. Richardson, M.H. Seymour, B.R. Webber, J. High Energy Phys. **04**, 028 (2002). hep-ph/0204123
376. T. Sjostrand, L. Lonnblad, S. Mrenna, hep-ph/0108264
377. S. Gopalakrishna, A. de Gouvea, W. Porod, J. Cosmol. Astropart. Phys. **0605**, 005 (2006). hep-ph/0602027
378. A. de Gouvea, S. Gopalakrishna, W. Porod, J. High Energy Phys. **11**, 050 (2006). hep-ph/0606296
379. T. Sjostrand, L. Lonnblad, S. Mrenna, P. Skands, hep-ph/0308153

380. A.J. Buras, Acta Phys. Pol. B **34**, 5615–5668 (2003). hep-ph/0310208
381. K. Agashe, M. Papucci, G. Perez, D. Pirjol, hep-ph/0509117
382. J. Foster, K.-i. Okumura, and L. Roszkowski. J. High Energy Phys. **03**, 044 (2006). hep-ph/0510422
383. J.D. Lykken, hep-ph/0607149
384. M. Carena, A. Menon, R. Noriega-Papaqui, A. Szynkman, C.E.M. Wagner, Phys. Rev. D **74**, 015009 (2006). hep-ph/0603106
385. G. Isidori, F. Mescia, P. Paradisi, C. Smith, S. Trine, J. High Energy Phys. **08**, 064 (2006). hep-ph/0604074
386. Z. Ligeti, M. Papucci, G. Perez, Phys. Rev. Lett **97**, 101801 (2006). hep-ph/0604112
387. T. Becher, M. Neubert, Phys. Rev. Lett. **98**, 022003 (2007). hep-ph/0610067
388. C.G. Lester, M.A. Parker, M.J. White, J. High Energy Phys. **01**, 080 (2006). hep-ph/0508143
389. B.C. Allanach, C.G. Lester, hep-ph/0705.0486
390. E.A. Baltz, P. Gondolo, J. High Energy Phys. **10**, 052 (2004). hep-ph/0407039
391. N. Arkani-Hamed et al., hep-ph/0703088
392. P. Langacker, Phys. Rep. **72**, 185 (1981)
393. M. Schmaltz, D. Tucker-Smith, Annu. Rev. Nucl. Part. Sci. **55**, 229–270 (2005). hep-ph/0502182
394. M. Perelstein, hep-ph/0512128
395. J. Hewett, M. Spiropulu, Annu. Rev. Nucl. Part. Sci. **52**, 397–424 (2002). hep-ph/0205106
396. C. Csaki, hep-ph/0404096
397. P.H. Frampton, P.Q. Hung, M. Sher, Phys. Rep. **330**, 263 (2000). hep-ph/9903387
398. N. Arkani-Hamed, A.G. Cohen, H. Georgi, Phys. Lett. B **513**, 232–240 (2001). hep-ph/0105239
399. N. Arkani-Hamed, A.G. Cohen, E. Katz, A.E. Nelson, J. High Energy Phys. **07**, 034 (2002). hep-ph/0206021
400. H. Fritzsch, J. Plankl, Phys. Lett. B **237**, 451 (1990)
401. E.A. Mirabelli, M. Schmaltz, Phys. Rev. D **61**, 113011 (2000). hep-ph/9912265
402. T. Han, H.E. Logan, B. McElrath, L.-T. Wang, Phys. Rev. D **67**, 095004 (2003). hep-ph/0301040
403. F. del Aguila, R. Pittau, Acta Phys. Pol. B **35**, 2767–2780 (2004). hep-ph/0410256
404. S. Sultansoy, G. Unel, M. Yilmaz, hep-ex/0608041
405. J.A. Aguilar-Saavedra, J. High Energy Phys. **12**, 033 (2006). hep-ph/0603200
406. A.A. Affolder et al. (CDF Collaboration), Phys. Rev. Lett. **84**, 835–840 (2000). hep-ex/9909027
407. J. Abdallah et al. (DELPHI Collaboration), Eur. Phys. J. C **50**, 507–518 (2005). hep-ex/0704.0594
408. S.M. Oliveira, R. Santos, Phys. Rev. D **68**, 093012 (2003). hep-ph/0307318
409. S.M. Oliveira, R. Santos, Acta Phys. Pol. B **34**, 5523–5530 (2003). hep-ph/0311047
410. A.A. Affolder et al. (CDF Collaboration), CDF note 8003
411. J.A. Aguilar-Saavedra, PoS **TOP2006**, 003 (2006). hep-ph/0603199
412. F. del Aguila, L. Ametller, G.L. Kane, J. Vidal, Nucl. Phys. B **334**, 1 (1990)
413. G. Azuelos et al., Eur. Phys. J. C **39S2**, 13–24 (2005). hep-ph/0402037
414. D. Costanzo, ATL-PHYS-2004-004
415. T. Sjostrand, S. Mrenna, P. Skands, J. High Energy Phys. **05**, 026 (2006). hep-ph/0603175
416. M.L. Mangano, Talk at Lund University, 2004. http://cern.ch/~mlm/talks/lund-alpgen.pdf
417. O. Cakir, M. Yilmaz, Europhys. Lett. **38**, 13–18 (1997)
418. T.C. Andre, J.L. Rosner, Phys. Rev. D **69**, 035009 (2004). hep-ph/0309254
419. S. Sultansoy, G. Unel, hep-ex/0610064
420. E. Boos et al. (CompHEP Collaboration), Nucl. Instrum. Methods A **534**, 250–259 (2004). hep-ph/0403113
421. T. Stelzer, W.F. Long, Comput. Phys. Commun. **81**, 357–371 (1994). hep-ph/9401258
422. D. Froidevaux, L. Poggioli, E. Richter-Was, ATLFAST 1.0 A package for particle-level analysis
423. R. Mehdiyev, S. Sultansoy, G. Unel, M. Yilmaz, Eur. Phys. J. C **49**, 613–622 (2007). hep-ex/0603005
424. W.W. Armstrong et al. (ATLAS Collaboration), CERN-LHCC-94-43
425. P. Jenni, M. Nordberg, M. Nessi, K. Smith, ATLAS high-level trigger, data-acquisition and control, Technical Design Report, 2003
426. Y. Hosotani, Phys. Lett. B **126**, 309 (1983)
427. B.T. McInnes, J. Math. Phys. **31**, 2094–2104 (1990)
428. A.S. Belyaev et al., hep-ph/0101232
429. K. Agashe, A. Delgado, M.J. May, R. Sundrum, J. High Energy Phys. **08**, 050 (2003). hep-ph/0308036
430. K. Agashe, G. Servant, Phys. Rev. Lett. **93**, 231805 (2004). hep-ph/0403143
431. K. Agashe, G. Servant, J. Cosmol. Astropart. Phys. **0502**, 002 (2005). hep-ph/0411254
432. K. Agashe, R. Contino, L. Da Rold, A. Pomarol, Phys. Lett. B **641**, 62–66 (2006). hep-ph/0605341
433. R. Contino, L. Da Rold, A. Pomarol, hep-ph/0612048
434. C. Dennis, M.K. Unel, G. Servant, J. Tseng, hep-ph/0701158
435. A. Pukhov, hep-ph/0412191
436. A. Arhrib, W.-S. Hou, J. High Energy Phys. **07**, 009 (2006). hep-ph/0602035
437. W.-S. Hou, R.G. Stuart, Phys. Rev. Lett. **62**, 617 (1989)
438. W.-S. Hou, R.G. Stuart, Nucl. Phys. B **320**, 277 (1989)
439. W.-S. Hou, R.G. Stuart, Phys. Lett. B **233**, 485 (1989)
440. W.-S. Hou, R.G. Stuart, Phys. Rev. D **43**, 3669–3682 (1991)
441. W.-S. Hou, M. Nagashima, A. Soddu, Phys. Rev. D **72**, 115007 (2005). hep-ph/0508237
442. M. Schmaltz, J. High Energy Phys. **08**, 056 (2004). hep-ph/0407143
443. F. del Aguila, M. Masip, J.L. Padilla, Phys. Lett. B **627**, 131–136 (2005). hep-ph/0506063
444. P. Minkowski, Phys. Lett. B **67**, 421 (1977)
445. M. Gell-Mann, P. Ramond, R. Slansky, Print-80-0576 (CERN)
446. T. Yanagida, in *Proceedings of the Workshop on the Baryon Number of the Universe and Unified Theories*, Tsukuba, Japan, 13–14 February 1979
447. R.N. Mohapatra, G. Senjanovic, Phys. Rev. Lett. **44**, 912 (1980)
448. W. Buchmuller, C. Greub, Nucl. Phys. B **363**, 345–368 (1991)
449. G. Ingelman, J. Rathsman, Z. Phys. C **60**, 243–254 (1993)
450. J. Gluza, Acta Phys. Pol. B **33**, 1735–1746 (2002). hep-ph/0201002
451. A. Pilaftsis, T.E.J. Underwood, Nucl. Phys. B **692**, 303–345 (2004). hep-ph/0309342
452. A. Pilaftsis, T.E.J. Underwood, Phys. Rev. D **72**, 113001 (2005). hep-ph/0506107
453. L. Boubekeur, T. Hambye, G. Senjanovic, Phys. Rev. Lett. **93**, 111601 (2004). hep-ph/0404038
454. A. Abada, H. Aissaoui, M. Losada, Nucl. Phys. B **728**, 55–66 (2005). hep-ph/0409343
455. F. del Aguila, J.A. Aguilar-Saavedra, R. Pittau, J. Phys. Conf. Ser. **53**, 506–527 (2006). hep-ph/0606198
456. D. Tommasini, G. Barenboim, J. Bernabeu, C. Jarlskog, Nucl. Phys. B **444**, 451–467 (1995). hep-ph/9503228
457. S. Bergmann, A. Kagan, Nucl. Phys. B **538**, 368–386 (1999). hep-ph/9803305
458. B. Bekman, J. Gluza, J. Holeczek, M. Syska, M. Zralek, Phys. Rev. D **66**, 093004 (2002). hep-ph/0207015

459. G. Azuelos, A. Djouadi, Z. Phys. C **63**, 327–338 (1994). hep-ph/9308340
460. J. Gluza, M. Zralek, Phys. Rev. D **55**, 7030–7037 (1997). hep-ph/9612227
461. F. del Aguila, J.A. Aguilar-Saavedra, A. Martinez de la Ossa, D. Meloni, Phys. Lett. B **613**, 170–180 (2005). hep-ph/0502189
462. F. del Aguila, J.A. Aguilar-Saavedra, J. High Energy Phys. **05**, 026 (2005). hep-ph/0503026
463. A. Datta, M. Guchait, A. Pilaftsis, Phys. Rev. D **50**, 3195–3203 (1994). hep-ph/9311257
464. O. Panella, M. Cannoni, C. Carimalo, Y.N. Srivastava, Phys. Rev. D **65**, 035005 (2002). hep-ph/0107308
465. T. Han, B. Zhang, Phys. Rev. Lett. **97**, 171804 (2006). hep-ph/0604064
466. F. del Aguila, J.A. Aguilar-Saavedra, R. Pittau, J. High Energy Phys. **10**, 047 (2007). hep-ph/0703261
467. W.-Y. Keung, G. Senjanovic, Phys. Rev. Lett. **50**, 1427 (1983)
468. A. Datta, M. Guchait, D.P. Roy, Phys. Rev. D **47**, 961–966 (1993). hep-ph/9208228
469. A. Ferrari et al., Phys. Rev. D **62**, 013001 (2000)
470. S.N. Gninenko, M.M. Kirsanov, N.V. Krasnikov, V.A. Matveev, hep-ph/0301140
471. S.N. Gninenko, M.M. Kirsanov, N.V. Krasnikov, V.A. Matveev, CERN-CMS-NOTE-2006-098
472. F. del Aguila, J.A. Aguilar-Saavedra, hep-ph/0705.4117
473. B. Bajc, G. Senjanovic, J. High Energy Phys. **08**, 014 (2007). hep-ph/0612029
474. P. Langacker, R.W. Robinett, J.L. Rosner, Phys. Rev. D **30**, 1470 (1984)
475. A. Leike, Phys. Rep. **317**, 143–250 (1999). hep-ph/9805494
476. I. Antoniadis, K. Benakli, M. Quiros, Phys. Lett. B **331**, 313–320 (1994). hep-ph/9403290
477. T.G. Rizzo, Phys. Rev. D **61**, 055005 (2000). hep-ph/9909232
478. A. Abulencia et al. (CDF Collaboration), Phys. Rev. Lett. **96**, 211801 (2006). hep-ex/0602045
479. V.D. Barger, N.G. Deshpande, J.L. Rosner, K. Whisnant, Phys. Rev. D **35**, 2893 (1987)
480. F. del Aguila, M. Quiros, F. Zwirner, Nucl. Phys. B **287**, 419 (1987)
481. M. Cvetic, S. Godfrey, hep-ph/9504216
482. S. Godfrey, hep-ph/0201093
483. B. Clerbaux, T. Mahmoud, C. Collard, P. Miné, CMS NOTE 2006-083, 2006
484. R. Cousins, J. Mumford, V. Valuev, CMS NOTE 2006-062, 2006
485. R. Cousins, J. Mumford, V. Valuev, CMS NOTE 2005-022, 2005
486. G. Azuelos, G. Polesello, Eur. Phys. J. C **39S2**, 1–11 (2005)
487. R. Cousins, J. Mumford, J. Tucker, V. Valuev, J. High Energy Phys. **11**, 046 (2005)
488. J.C. Collins, D.E. Soper, Phys. Rev. D **16**, 2219 (1977)
489. I. Belotelov et al., CERN-CMS-NOTE-2006-104
490. E. Ma, Phys. Rev. D **36**, 274 (1987)
491. F. del Aguila, M. Cvetic, P. Langacker, Phys. Rev. D **48**, 969–973 (1993). hep-ph/9303299
492. J.L. Rosner, Phys. Rev. D **35**, 2244 (1987)
493. G. Burdman, M. Perelstein, A. Pierce, Phys. Rev. Lett. **90**, 241802 (2003). hep-ph/0212228
494. F. Ledroit-Guillon, G. Moreau, J. Morel, in *Proceedings of the XLIrst 'Rencontres de MORIOND', Session devoted to Electroweak Interactions and Unified Theories*, La Thuile, Italy, 11–18 March 2006 (2006)
495. G. Moreau, J.I. Silva-Marcos, J. High Energy Phys. **03**, 090 (2006). hep-ph/0602155
496. P. Chiappetta, J. Layssac, F.M. Renard, C. Verzegnassi, Phys. Rev. D **54**, 789–797 (1996). hep-ph/9601306
497. G. Altarelli, N. Di Bartolomeo, F. Feruglio, R. Gatto, M.L. Mangano, Phys. Lett. B **375**, 292–300 (1996). hep-ph/9601324
498. V.D. Barger, K.-M. Cheung, P. Langacker, Phys. Lett. B **381**, 226–236 (1996). hep-ph/9604298
499. A. Djouadi, G. Moreau, F. Richard, hep-ph/0610173
500. K.S. Babu, C.F. Kolda, J. March-Russell, Phys. Rev. D **54**, 4635–4647 (1996). hep-ph/9603212
501. K. Gumus, N. Akchurin, S. Esen, R. Harris, CMS NOTE 2006-070, 2006
502. E. Cogneras, D. Pallin, ATLAS NOTE 2006-033, 2006
503. G. Barenboim, J. Bernabeu, J. Prades, M. Raidal, Phys. Rev. D **55**, 4213–4221 (1997). hep-ph/9611347
504. G. Beall, M. Bander, A. Soni, Phys. Rev. Lett. **48**, 848 (1982)
505. C. Csaki, J. Hubisz, G.D. Kribs, P. Meade, J. Terning, Phys. Rev. D **68**, 035009 (2003). hep-ph/0303236
506. J.L. Hewett, F.J. Petriello, T.G. Rizzo, J. High Energy Phys. **10**, 062 (2003). hep-ph/0211218
507. V.M. Abazov et al. (D0 Collaboration), D0note 5191-CONF
508. M.C. Cousinou, Tech. Rep. ATL-PHYS-94-059. ATL-GE-PN-59, CERN, Geneva, December 1994
509. C. Hof, T. Hebbeker, K. Höpfner, Tech. Rep. CMS-NOTE-2006-117. CERN-CMS-NOTE-2006-117, CERN, Geneva, April 2006
510. Y.L. Wu, L. Wolfenstein, Phys. Rev. Lett. **73**, 1762–1764 (1994). hep-ph/9409421
511. J. Maalampi, N. Romanenko, Phys. Lett. B **532**, 202–208 (2002). hep-ph/0201196
512. A. Hektor, M. Kadastik, M. Muntel, M. Raidal, L. Rebane, hep-ph/0705.1495
513. A. Zee, Nucl. Phys. B **264**, 99 (1986)
514. K.S. Babu, Phys. Lett. B **203**, 132 (1988)
515. Y. Fukuda et al. (Super-Kamiokande Collaboration), Phys. Rev. Lett. **81**, 1562–1567 (1998). hep-ex/9807003
516. Q.R. Ahmad et al. (SNO Collaboration), Phys. Rev. Lett. **89**, 011301 (2002). nucl-ex/0204008
517. K. Eguchi et al. (KamLAND Collaboration), Phys. Rev. Lett. **90**, 021802 (2003). hep-ex/0212021
518. M. Apollonio et al., Eur. Phys. J. C **27**, 331–374 (2003). hep-ex/0301017
519. D. Aristizabal Sierra, M. Hirsch, J. High Energy Phys. **12**, 052 (2006). hep-ph/0609307
520. T. Han, H.E. Logan, B. Mukhopadhyaya, R. Srikanth, Phys. Rev. D **72**, 053007 (2005). hep-ph/0505260
521. A. Abada, G. Bhattacharyya, M. Losada, Phys. Rev. D **73**, 033006 (2006). hep-ph/0511275
522. T.P. Cheng, L.-F. Li, Phys. Rev. D **22**, 2860 (1980)
523. G.B. Gelmini, M. Roncadelli, Phys. Lett. B **99**, 411 (1981)
524. S. Chang, J. High Energy Phys. **12**, 057 (2003). hep-ph/0306034
525. C.-X. Yue, S. Zhao, Eur. Phys. J. C **50**, 897–903 (2007). hep-ph/0701017
526. D. Acosta et al. (CDF Collaboration), Phys. Rev. Lett. **93**, 221802 (2004). hep-ex/0406073
527. V.M. Abazov et al. (D0 Collaboration), Phys. Rev. Lett. **93**, 141801 (2004). hep-ex/0404015
528. J.F. Gunion, J. Grifols, A. Mendez, B. Kayser, F.I. Olness, Phys. Rev. D **40**, 1546 (1989)
529. M. Muhlleitner, M. Spira, Phys. Rev. D **68**, 117701 (2003). hep-ph/0305288
530. T. Rommerskirchen, T. Hebbeker, J. Phys. G **33**, N47–N66 (2007)
531. T. Han, B. Mukhopadhyaya, Z. Si, K. Wang, Phys. Rev. D **76**, 075013 (2007). hep-ph/0706.0441
532. R. Barate et al. (LEP Working Group for Higgs boson searches Collaboration), Phys. Lett. B **565**, 61–75 (2003). hep-ex/0306033
533. K.S. Babu, C. Macesanu, Phys. Rev. D **67**, 073010 (2003). hep-ph/0212058
534. E. Boos et al., hep-ph/0109068
535. P. Skands et al., J. High Energy Phys. **07**, 036 (2004). hep-ph/0311123

536. H.K. Dreiner, P. Richardson, M.H. Seymour, J. High Energy Phys. **04**, 008 (2000). hep-ph/9912407
537. P.Z. Skands, Eur. Phys. J. C **23**, 173–184 (2002). hep-ph/0110137
538. B.C. Allanach, Comput. Phys. Commun. **143**, 305–331 (2002). hep-ph/0104145
539. T. Sjostrand, P.Z. Skands, Nucl. Phys. B **659**, 243 (2003). hep-ph/0212264
540. U. Ellwanger, C. Hugonie, Comput. Phys. Commun. **175**, 290–303 (2006). hep-ph/0508022
541. M. Frank, T. Hahn, S. Heinemeyer, W. Hollik, H. Rzehak, G. Weiglein, J. High Energy Phys. **02**, 047 (2007). hep-ph/0611326
542. T. Sjöstrand, S. Mrenna, P.Z. Skands, hep-ph/0710.3820
543. B.C. Allanach et al., hep-ph/0801.0045
544. A. Djouadi, J.-L. Kneur, G. Moultaka, Comput. Phys. Commun. **176**, 426–455 (2007). hep-ph/0211331
545. D.M. Pierce, J.A. Bagger, K.T. Matchev, R.-J. Zhang, Nucl. Phys. B **491**, 3–67 (1997). hep-ph/9606211
546. B.C. Allanach, A. Djouadi, J.L. Kneur, W. Porod, P. Slavich, J. High Energy Phys. **09**, 044 (2004). hep-ph/0406166
547. A. Djouadi, J. Kalinowski, M. Spira, Comput. Phys. Commun. **108**, 56–74 (1998). hep-ph/9704448
548. A. Djouadi, hep-ph/0503172
549. A. Djouadi, hep-ph/0503173
550. M. Gomez-Bock et al., J. Phys. Conf. Ser. **18**, 74–135 (2005). hep-ph/0509077
551. M.S. Carena, J.R. Espinosa, M. Quiros, C.E.M. Wagner, Phys. Lett. B **355**, 209–221 (1995). hep-ph/9504316
552. M.S. Carena, M. Quiros, C.E.M. Wagner, Nucl. Phys. B **461**, 407–436 (1996). hep-ph/9508343
553. M. Muhlleitner, A. Djouadi, Y. Mambrini, Comput. Phys. Commun. **168**, 46–70 (2005). hep-ph/0311167
554. C. Boehm, A. Djouadi, Y. Mambrini, Phys. Rev. D **61**, 095006 (2000). hep-ph/9907428
555. A. Djouadi, Y. Mambrini, Phys. Rev. D **63**, 115005 (2001). hep-ph/0011364
556. A. Djouadi, Y. Mambrini, M. Muhlleitner, Eur. Phys. J. C **20**, 563–584 (2001). hep-ph/0104115
557. A. Djouadi, M.M. Muhlleitner, M. Spira, Acta Phys. Pol. B **38**, 635–644 (2007). hep-ph/0609292
558. G. Degrassi, S. Heinemeyer, W. Hollik, P. Slavich, G. Weiglein, Eur. Phys. J. C **28**, 133–143 (2003). hep-ph/0212020
559. S. Heinemeyer, W. Hollik, G. Weiglein, Phys. Rev. D **58**, 091701 (1998). hep-ph/9803277
560. S. Heinemeyer, W. Hollik, G. Weiglein, Phys. Lett. B **440**, 296–304 (1998). hep-ph/9807423
561. S. Heinemeyer, W. Hollik, G. Weiglein, Phys. Rep. **425**, 265–368 (2006). hep-ph/0412214
562. T. Hahn, S. Heinemeyer, G. Weiglein, Nucl. Phys. B **652**, 229–258 (2003). hep-ph/0211204
563. S. Heinemeyer, W. Hollik, G. Weiglein, Eur. Phys. J. C **16**, 139–153 (2000). hep-ph/0003022
564. T. Hahn, S. Heinemeyer, F. Maltoni, G. Weiglein, S. Willenbrock, hep-ph/0607308
565. T. Hahn, hep-ph/0408283
566. T. Hahn, hep-ph/0605049
567. Susy les houches accord web page. http://home.fnal.gov/~skands/slha/
568. G. Corcella et al., hep-ph/0210213
569. M.A. Dobbs et al., hep-ph/0403045
570. P. Skands et al., in *Les Houches Workshop on Physics at TeV Colliders*, Les Houches, France, 2–20 May 2005
571. S. Heinemeyer et al., ECONF **C0508141**, ALCPG0214 (2005). hep-ph/0511332
572. E. Norrbin, T. Sjostrand, Eur. Phys. J. C **17**, 137–161 (2000). hep-ph/0005110
573. T. Gleisberg et al., J. High Energy Phys. **02**, 056 (2004). hep-ph/0311263
574. F. Krauss, R. Kuhn, G. Soff, J. High Energy Phys. **02**, 044 (2002). hep-ph/0109036
575. T. Gleisberg, F. Krauss, C.G. Papadopoulos, A. Schaelicke, S. Schumann, Eur. Phys. J. C **34**, 173–180 (2004). hep-ph/0311273
576. K. Hagiwara et al., Phys. Rev. D **73**, 055005 (2006). hep-ph/0512260
577. T. Gleisberg et al., J. High Energy Phys. **09**, 001 (2003). hep-ph/0306182
578. J. Rosiek, Phys. Rev. D **41**, 3464 (1990)
579. J. Rosiek, hep-ph/9511250
580. F. Krauss, A. Schalicke, G. Soff, Comput. Phys. Commun. **174**, 876–902 (2006). hep-ph/0503087
581. S. Catani, F. Krauss, R. Kuhn, B.R. Webber, J. High Energy Phys. **11**, 063 (2001). hep-ph/0109231
582. F. Krauss, J. High Energy Phys. **08**, 015 (2002). hep-ph/0205283
583. F. Krauss, A. Schalicke, S. Schumann, G. Soff, Phys. Rev. D **70**, 114009 (2004). hep-ph/0409106
584. A. Schalicke, F. Krauss, J. High Energy Phys. **07**, 018 (2005). hep-ph/0503281

B, *D* and *K* decays[*]

G. Buchalla[17,b], T.K. Komatsubara[43,b], F. Muheim[55,a,b], L. Silvestrini[4,b], M. Artuso[1], D.M. Asner[2], P. Ball[3], E. Baracchini[4], G. Bell[5], M. Beneke[6], J. Berryhill[7], A. Bevan[8], I.I. Bigi[9], M. Blanke[10,19], Ch. Bobeth[11], M. Bona[12], F. Borzumati[13,14], T. Browder[15], T. Buanes[16], O. Buchmüller[18], A.J. Buras[19], S. Burdin[20], D.G. Cassel[21], R. Cavanaugh[22], M. Ciuchini[23], P. Colangelo[24], G. Crosetti[25], A. Dedes[3], F. De Fazio[24], S. Descotes-Genon[26], J. Dickens[27], Z. Doležal[28], S. Dürr[29], U. Egede[30], C. Eggel[31,32], G. Eigen[16], S. Fajfer[33], Th. Feldmann[34], R. Ferrandes[24], P. Gambino[35], T. Gershon[36], V. Gibson[27], M. Giorgi[37], V.V. Gligorov[38], B. Golob[39], A. Golutvin[18,40], Y. Grossman[41], D. Guadagnoli[19], U. Haisch[42], M. Hazumi[43], S. Heinemeyer[44], G. Hiller[11], D. Hitlin[45], T. Huber[6], T. Hurth[18,46], T. Iijima[47], A. Ishikawa[48], G. Isidori[49,50], S. Jäger[18], A. Khodjamirian[34], P. Koppenburg[30], T. Lagouri[28], U. Langenegger[31], C. Lazzeroni[27], A. Lenz[51], V. Lubicz[23], W. Lucha[52], H. Mahlke[21], D. Melikhov[53], F. Mescia[50], M. Misiak[54], M. Nakao[43], J. Napolitano[56], N. Nikitin[57], U. Nierste[5], K. Oide[43], Y. Okada[43], P. Paradisi[19], F. Parodi[58], M. Patel[18], A.A. Petrov[59], T.N. Pham[60], M. Pierini[18], S. Playfer[55], G. Polesello[61], A. Policicchio[25], A. Poschenrieder[19], P. Raimondi[50], S. Recksiegel[19], P. Řezníček[28], A. Robert[62], J.L. Rosner[63], G. Ruggiero[18], A. Sarti[50], O. Schneider[64], F. Schwab[65], S. Simula[23], S. Sivoklokov[57], P. Slavich[18,66], C. Smith[67], M. Smizanska[68], A. Soni[69], T. Speer[42], P. Spradlin[38], M. Spranger[19], A. Starodumov[31], B. Stech[70], A. Stocchi[71], S. Stone[1], C. Tarantino[23], F. Teubert[18], S. T'Jampens[12], K. Toms[57], K. Trabelsi[43], S. Trine[5], S. Uhlig[19], V. Vagnoni[72], J.J. van Hunen[64], G. Weiglein[3], A. Weiler[21], G. Wilkinson[38], Y. Xie[55], M. Yamauchi[43], G. Zhu[73], J. Zupan[33], R. Zwicky[3]

[1] Syracuse University, Syracuse, NY, USA
[2] Carleton University, Ottawa, Canada
[3] Durham University, IPPP, Durham, UK
[4] Università di Roma La Sapienza and INFN, Rome, Italy
[5] Universität Karlsruhe, Karlsruhe, Germany
[6] RWTH Aachen, Aachen, Germany
[7] Fermi National Accelerator Laboratory, Batavia, IL, USA
[8] Queen Mary, University of London, London, UK
[9] University of Notre Dame, Notre Dame, IN, USA
[10] Max-Planck-Institut für Physik, München, Germany
[11] Institut für Physik, Universität Dortmund, Dortmund, Germany
[12] LAPP, Université de Savoie, IN2P3-CNRS, Annecy-le-Vieux, France
[13] ICTP, Trieste, Italy
[14] National Central University, Chung-li, Taiwan
[15] University of Hawaii at Manoa, Honolulu, HI, USA
[16] University of Bergen, Bergen, Norway
[17] Ludwig-Maximilians-Universität München, München, Germany
[18] CERN, Geneva, Switzerland
[19] Technische Universität München, Garching, Germany
[20] The University of Liverpool, Liverpool, UK
[21] Cornell University, Ithaca, NY, USA
[22] University of Florida, Gainesville, FL, USA
[23] Università di Roma Tre and INFN, Rome, Italy
[24] INFN, Bari, Italy
[25] Università di Calabria and INFN Cosenza, Cosenza, Italy
[26] LPT, CNRS/Université de Paris-Sud 11, Orsay, France
[27] University of Cambridge, Cambridge, UK
[28] IPNP, Charles University in Prague, Prague, Czech Republic
[29] NIC, FZ Jülich and DESY Zeuthen, Jülich, Germany
[30] Imperial College, London, UK
[31] ETH, Zürich, Switzerland
[32] PSI, Villigen, Switzerland
[33] Ljubljana University and Jozef Stefan Institute, Ljubljana, Slovenia
[34] Universität Siegen, Siegen, Germany
[35] Università di Torino and INFN, Torino, Italy
[36] University of Warwick, Coventry, UK

[37]Università di Pisa, SNS and INFN, Pisa, Italy
[38]University of Oxford, Oxford, UK
[39]University of Ljubljana, Ljubljana, Slovenia
[40]ITEP, Moscow, Russia
[41]Technion, Haifa, Israel
[42]Universität Zürich, Zürich, Switzerland
[43]KEK and Graduate University for Advanced Studies (Sokendai), Tsukuba, Japan
[44]IFCA, Santander, Spain
[45]CalTech, Pasadena, CA, USA
[46]SLAC, Stanford, CA, USA
[47]Nagoya University, Nagoya, Japan
[48]Saga University, Saga, Japan
[49]SNS and INFN, Pisa, Italy
[50]INFN, LNF, Frascati, Italy
[51]Universität Regensburg, Regensburg, Germany
[52]Institut für Hochenergiephysik, Österreichische Akademie der Wissenschaften, Wien, Austria
[53]Nuclear Physics Institute, Moscow State University, Moscow, Russia
[54]Warsaw University, Warsaw, Poland
[55]University of Edinburgh, Edinburgh, UK
[56]Rensselaer Polytechnic Institute, Troy, NY, USA
[57]Skobeltsin Institute of Nuclear Physics, Lomonosov Moscow State University, Moscow, Russia
[58]Università di Genova and INFN, Genova, Italy
[59]Wayne State University, Detroit, MI, USA
[60]Ecole Polytechnique, CNRS, Palaiseau, France
[61]Università di Pavia and INFN, Pavia, Italy
[62]Université de Clermont-Ferrand, Clermont-Ferrand, France
[63]Enrico Fermi Institute, University of Chicago, Chicago, IL, USA
[64]Ecole Polytechnique Fédérale de Lausanne (EPFL), Lausanne, Switzerland
[65]Universitat Autonoma de Barcelona, IFAE, Barcelona, Spain
[66]LAPTH, Annecy-le-Vieux, France
[67]Universität Bern, Bern, Switzerland
[68]Lancaster University, Lancaster, UK
[69]Brookhaven National Laboratory, Upton, NY, USA
[70]Universität Heidelberg, Heidelberg, Germany
[71]LAL, IN2P3-CNRS and Université de Paris-Sud, Orsay, France
[72]Università di Bologna and INFN, Bologna, Italy
[73]Universität Hamburg, Hamburg, Germany

Abstract The present report documents the results of Working Group 2: B, D and K decays, of the workshop on Flavor in the Era of the LHC, held at CERN from November 2005 through March 2007.

With the advent of the LHC, we will be able to probe New Physics (NP) up to energy scales almost one order of magnitude larger than it has been possible with present accelerator facilities. While direct detection of new particles will be the main avenue to establish the presence of NP at the LHC, indirect searches will provide precious complementary information, since most probably it will not be possible to measure the full spectrum of new particles and their couplings through direct production. In particular, precision measurements and computations in the realm of flavor physics are expected to play a key role in constraining the unknown parameters of the Lagrangian of any NP model emerging from direct searches at the LHC.

The aim of Working Group 2 was twofold: on the one hand, to provide a coherent up-to-date picture of the status of flavor physics before the start of the LHC; on the other hand, to initiate activities on the path towards integrating information on NP from high-p_T and flavor data.

This report is organized as follows: in Sect. 1, we give an overview of NP models, focusing on a few examples that have been discussed in some detail during the workshop, with a short description of the available computational tools for flavor observables in NP models. Section 2 contains a concise discussion of the main theoretical problem in flavor physics: the evaluation of the relevant hadronic matrix elements for weak decays. Section 3 contains a detailed discussion of NP effects in a set of flavor observables that we identified as "benchmark channels" for NP searches. The experimental prospects for flavor physics at future facilities are discussed in Sect. 4. Finally, Sect. 5 contains some assessments on the work done at the workshop and the prospects for future developments.

Contents

1 New physics scenarios 311
   1.1 Overview 311
   1.2 Model-independent approaches 312
   1.3 SUSY models 319
   1.4 Nonsupersymmetric extensions
        of the Standard Model 325
   1.5 Tools for flavor physics and beyond 328
2 Weak decays of hadrons and QCD 331
   2.1 Overview 331
   2.2 Charmless two-body B decays 331
   2.3 Light-cone QCD sum rules 337
   2.4 Lattice QCD 341
3 New physics in benchmark channels 346
   3.1 Radiative penguin decays 346
   3.2 Electroweak penguin decays 351
   3.3 Neutrino modes 361
   3.4 Very rare decays 370
   3.5 UT angles from tree decays 377
   3.6 B-meson mixing 389
   3.7 Hadronic $b \to s$ and $b \to d$ transition 398
   3.8 Kaon decays 410
   3.9 Charm physics 419
   3.10 Impact of the LHC experiments 442
4 Prospects for future facilities 443
   4.1 On the physics case of a super flavor
        factory 444
   4.2 Super$B$ proposal 450
   4.3 Accelerator design of SuperKEKB 453
   4.4 LHCb upgrade 455
5 Assessments 460
   5.1 New-physics patterns and correlations . . . 461
   5.2 Correlations between FCNC processes . . . 461
   5.3 Connection to high-energy physics 462
   5.4 Discrimination
        between new physics scenarios 476
Acknowledgements 476
References 476

---

[*]Report of Working Group 2 of the CERN Workshop "Flavor in the era of the LHC", Geneva, Switzerland, November 2005–March 2007.

[a]e-mail: f.muheim@ed.ac.uk

[b]Convenors

1 New physics scenarios[1]

1.1 Overview

The Standard Model (SM) of electroweak and strong interactions describes with an impressive accuracy all experimental data on particle physics up to energies of the order of the electroweak scale. On the other hand, we know that the SM should be viewed as an effective theory valid up to a scale $\Lambda \sim M_W$, since, among many other things, the SM does not contain a suitable candidate of dark matter and it does not account for gravitational interactions. Viewing the SM as an effective theory, however, poses a series of theoretical questions. First of all, the quadratic sensitivity of the electroweak scale on the cutoff calls for a low value of Λ, in order to avoid excessive fine tuning. Second, several of the higher-dimensional operators which appear in the SM effective Lagrangian violate the accidental symmetries of the SM. Therefore, their coefficients must be highly suppressed in order not to clash with the experimental data, in particular in the flavor sector. Unless additional suppression mechanisms are present in the fundamental theory, a cutoff around the electroweak scale is thus phenomenologically not acceptable since it generates higher-dimensional operators with large coefficients.

We are facing a formidable task: formulating a natural extension of the SM with a cutoff close to the electroweak scale and with a very strong suppression of additional sources of flavor and CP violation. While the simplest supersymmetric extensions of the SM with minimal flavor and CP violation, such as Minimal Supergravity (MSUGRA) models, seem to be the phenomenologically most viable NP options, it is fair to say that a fully consistent model of SUSY breaking has not been put forward yet. On the other hand, alternative solutions of the hierarchy problem based on extra dimensions have recently become very popular, although they have not yet been tested at the same level of accuracy as the Minimal Supersymmetric Standard Model (MSSM). Waiting for the LHC to discover new particles and shed some light on these fundamental problems, we should consider a range of NP models as wide as possible, in order to be ready to interpret the NP signals that will show up in the near future.

In the following paragraphs, we discuss how flavor and CP violation beyond the SM can be analysed on general grounds in a model-independent way. We then specialize to a few popular extensions of the SM, such as SUSY and little Higgs models, and present their most relevant aspects in view of our subsequent discussion of NP effects in flavor physics.

---

[1]Section coordinators: A.J. Buras, S. Heinemeyer, G. Isidori, Y. Okada, F. Parodi, L. Silvestrini.

1.2 Model-independent approaches

1.2.1 General considerations

In most extensions of the Standard Model (SM), the new degrees of freedom that modify the ultraviolet behavior of the theory appear only around or above the electroweak scale ($v \approx 174$ GeV). As long as we are interested in processes occurring below this scale (such as B, D and K decays), we can integrate out the new degrees of freedom and describe the new-physics effects—in full generality—by means of an Effective Field Theory (EFT) approach. The SM Lagrangian becomes the renormalizable part of a more general local Lagrangian which includes an infinite tower of higher-dimensional operators constructed in terms of SM fields and suppressed by inverse powers of a scale $\Lambda_{NP} > v$.

This general bottom-up approach allows us to analyse all realistic extensions of the SM in terms of a limited number of parameters (the coefficients of the higher-dimensional operators). The disadvantage of this strategy is that it does not allow us to establish correlations of New Physics (NP) effects at low and high energies (the scale Λ_{NP} defines the cut-off of the EFT). The number of correlations among different low-energy observables is also very limited, unless some restrictive assumptions about the structure of the EFT are employed.

The generic EFT approach is somehow the opposite of the standard top-down strategy towards NP, where a given theory—and a specific set of parameters—are employed to evaluate possible deviations from the SM. The top-down approach usually allows us to establish several correlations, both at low energies and between low- and high-energy observables. However, the price to pay is the loss of generality. This is quite a high price given our limited knowledge about the physics above the electroweak scale.

An interesting compromise between these two extreme strategies is obtained by implementing specific symmetry restrictions on the EFT. The extra constraints increase the number of correlations in low-energy observables. The experimental tests of such correlations allow us to test/establish general features of the NP model (possibly valid both at low and high energies). In particular, B, D and K decays are extremely useful in determining the flavor-symmetry breaking pattern of the NP model. The EFT approaches based on the Minimal Flavor Violation (MFV) hypothesis and its variations (MFV at large $\tan \beta$, n-MFV, ...) have exactly this goal.

In Sect. 1.2.2, we illustrate some of the main conclusions about NP effects in the flavor sector derived so far within general EFT approaches. In Sect. 1.2.3, we analyse in more detail the MFV hypothesis, discussing: (i) the general formulation and the general consequences of this hypothesis; (ii) the possible strategies to verify or falsify the MFV assumption from low-energy data; (iii) the implementation of the MFV hypothesis in more explicit beyond-the-SM frameworks, such as the Minimal Supersymmetric SM (MSSM) or Grand Unified Theories (GUTs).

1.2.2 Generic EFT approaches and the flavor problem

The NP contributions to the higher-dimensional operators of the EFT should naturally induce large effects in processes which are not mediated by tree-level SM amplitudes, such as meson–antimeson mixing ($\Delta F = 2$ amplitudes) or flavor-changing neutral-current (FCNC) rare decays. Up to now there is no evidence of deviations from the SM in these processes, and this implies severe bounds on the effective scale of various dimension-six operators. For instance, the good agreement between SM expectations and experimental determinations of K^0–$\bar{K}^0$ mixing leads to bounds above 10^4 TeV for the effective scale of $\Delta S = 2$ operators, i.e. well above the few TeV range suggested by a natural stabilization of the electroweak-symmetry breaking mechanism. Similar bounds are obtained for the scale of operators contributing to lepton-flavor violating (LFV) transitions in the lepton sector, such as $\mu \to e\gamma$.

The apparent contradiction between these two determinations of Λ is a manifestation of what in many specific frameworks (supersymmetry, technicolour, etc.) goes under the name of *flavor problem*: if we insist on the theoretical prejudice that new physics has to emerge in the TeV region, we have to conclude that the new theory possesses a highly nongeneric flavor structure. Interestingly enough, this structure has not been clearly identified yet, mainly because the SM (the low-energy limit of the new theory), does not possess an exact flavor symmetry. Within a model-independent approach, we should try to deduce this structure from data, using the experimental information on FCNC transitions to constrain its form.

1.2.2.1 Bounds on $\Delta F = 2$ operators In most realistic NP models, we can safely neglect NP effects in all cases where the corresponding effective operator is generated at the tree-level within the SM. This general assumption implies that the experimental determination of γ and $|V_{ub}|$ via tree-level processes (see Fig. 1) is free from the contamination of NP contributions. The comparison of the experimental data on meson–antimeson mixing amplitudes (both magnitudes and phases) with the theoretical SM expectations (obtained by means of the tree-level determination of the CKM matrix) allows us to derive some of the most stringent constraints on NP models.

In a wide class of beyond-the-SM scenarios, we expect sizable and uncorrelated deviations from the SM in the various $\Delta F = 2$ amplitudes.[2] As discussed by several authors [2–6], in this case, NP effects can be parameterized in

---

[2] As discussed for instance in Ref. [1], there is a rather general limit where NP effects in $\Delta F = 2$ amplitudes are expected to be the dom-

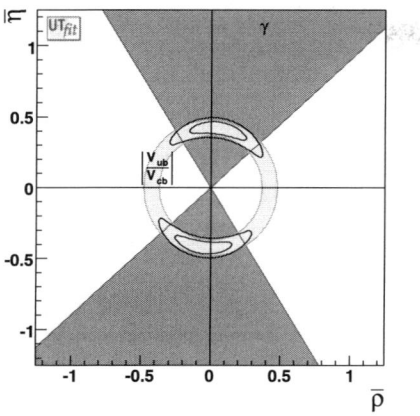

Fig. 1 Constraints on the $\bar\rho$–$\bar\eta$ plane using tree-level observables only, from Ref. [7] (see also Ref. [8])

terms of the shift induced in the B_q–$\bar{B}_q$ mixing frequencies ($q = d, s$) and in the corresponding CPV phases,

$$\frac{\langle B_q|H_{\rm eff}^{\rm full}|\bar{B}_q\rangle}{\langle B_q|H_{\rm eff}^{\rm SM}|\bar{B}_q\rangle} = C_{B_q}e^{2i\phi_{B_q}} = r_q^2 e^{2i\theta_q}, \quad (1)$$

and similarly for the neutral kaon system. The two equivalent parameterizations [(C_{B_q},ϕ_{B_q}) or (r_q,θ_q)] have been shown to facilitate the interpretation of the results of the UTfit [7] and CKMfitter [8] collaborations for the B_d case, shown in Fig. 2.

The main conclusions that can be drawn form the present analyses of new-physics effects in $\Delta F = 2$ amplitudes can be summarized as follows:

- In all the three accessible short-distance amplitudes (K^0–$\bar{K}^0$, B_d–$\bar{B}_d$ and B_s–$\bar{B}_s$), the magnitude of the new-physics amplitude cannot exceed, in size, the SM short-distance contribution. The latter is suppressed both by the GIM mechanism and by the hierarchical structure of the CKM matrix (V):

$$\mathcal{A}_{\rm SM}^{\Delta F=2} \sim \frac{G_F^2 M_W^2}{2\pi^2}(V_{ti}^* V_{tj})^2 \langle\bar{M}|(\bar{Q}_L^i\gamma^\mu Q_L^j)^2|M\rangle. \quad (2)$$

Therefore, new-physics models with TeV-scale flavored degrees of freedom and $\mathcal{O}(1)$ flavor-mixing couplings are essentially ruled out. To quantify this statement, we report here the results of the recent analysis of Ref. [9]. Writing

$$\mathcal{A}_{\rm NP}^{\Delta F=2} \sim \frac{C_{ij}^k}{\Lambda^2}\langle\bar{M}|(\bar{Q}^i\Gamma^k Q^j)^2|M\rangle, \quad (3)$$

where Γ^k is a generic Dirac and colour structure (see Ref. [9] for details), one has[3]

$$\Lambda > \begin{cases} 2\times 10^5\text{ TeV}\times |C_{12}^4|^{1/2}, \\ 2\times 10^3\text{ TeV}\times |C_{13}^4|^{1/2}, \\ 3\times 10^2\text{ TeV}\times |C_{23}^4|^{1/2}. \end{cases}$$

- As clearly shown in Fig. 3, in the B_d–$\bar{B}_d$ case, there is still room for a new-physics contribution up to the SM one. However, this is possible only if the new-physics contribution is aligned in phase with respect to the SM amplitude ($\phi_d^{\rm NP}$ close to zero). Similar, but tighter, constraints hold also for the new physics contribution to the K^0–$\bar{K}^0$ amplitude.
- Contrary to B_d–$\bar{B}_d$ and K^0–$\bar{K}^0$ amplitudes, at present there is only a very loose bound on the CPV phase of the B_s–$\bar{B}_s$ mixing amplitude. This leaves open the possibility of observing a large $\mathcal{A}_{\rm CP}(B_s \to J/\Psi\phi)$ at LHCb, which would be a clear signal of physics beyond the SM.

As we will discuss in the following, the first two items listed above find a natural explanation within the so-called hypothesis of Minimal Flavor Violation.

1.2.3 Minimal flavor violation

A very reasonable, although quite pessimistic, solution to the flavor problem is the so-called Minimal Flavor Violation (MFV) hypothesis. Under this assumption, which will be formalized in detail below, flavor-violating interactions are linked to the known structure of Yukawa couplings also beyond the SM. As a result, nonstandard contributions in FCNC transitions turn out to be suppressed to a level consistent with experiments even for $\Lambda \sim$ few TeV. One of the most interesting aspects of the MFV hypothesis is that it can naturally be implemented within the EFT approach to NP [10]. The effective theories based on this symmetry principle allow us to establish unambiguous correlations among NP effects in various rare decays. These falsifiable predictions are the key ingredients to identify in a model-independent way which are the irreducible sources of flavor symmetry breaking.

1.2.3.1 The MFV hypothesis
The pure gauge sector of the SM is invariant under a large symmetry group of flavor transformations: $\mathcal{G}_{\rm SM} = \mathcal{G}_q \otimes \mathcal{G}_\ell \otimes U(1)^5$, where

$$\begin{aligned}\mathcal{G}_q &= SU(3)_{Q_L} \otimes SU(3)_{U_R} \otimes SU(3)_{D_R}, \\ \mathcal{G}_\ell &= SU(3)_{L_L} \otimes SU(3)_{E_R},\end{aligned} \quad (4)$$

---

inant deviations from the SM in the flavor sector. This happens under the following two general assumptions: (i) the effective scale of NP is substantially higher than the electroweak scale; (ii) the dimensionless effective couplings ruling $\Delta F = 2$ transitions can be expressed as the square of the corresponding $\Delta F = 1$ coupling, without extra suppression factors.

[3] The choice $\Gamma^4 = P_L \otimes P_R$ gives the most stringent constraints. Constraints from other operators are up to one order of magnitude weaker.

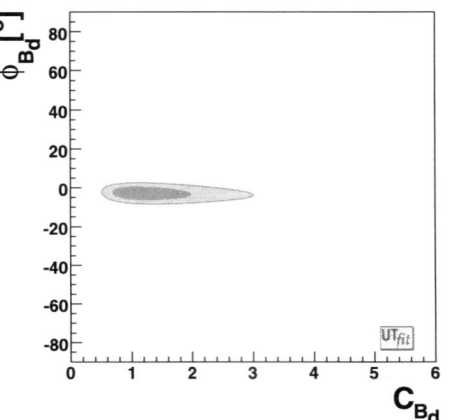
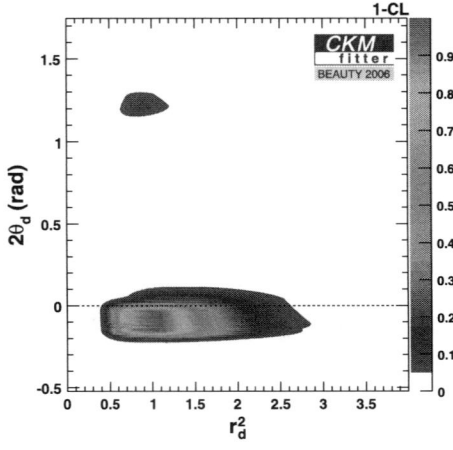

Fig. 2 Constraints on the effective parameters encoding NP effects in the B_d–$\bar{B}_d$ mixing amplitude (magnitude and phase) obtained by the UTfit [7] (*left*) and CKMfitter [8] (*right*) collaborations

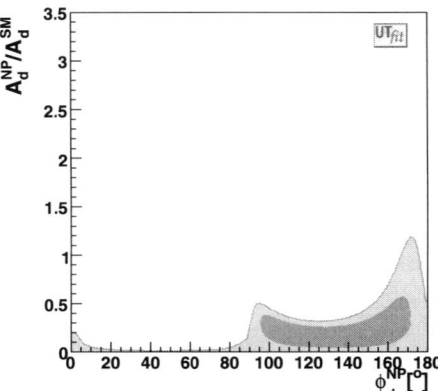
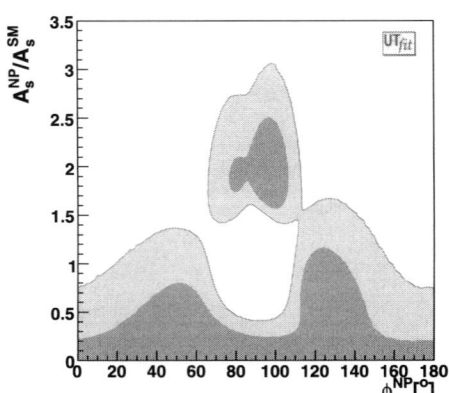

Fig. 3 Constraints on the absolute value and phase (normalized to the SM) of the new physics amplitude in B_d–$\bar{B}_d$ and B_s–$\bar{B}_s$ mixing from Ref. [9]

and three of the five $U(1)$ charges can be identified with baryon number, lepton number and hypercharge [11]. This large group and, particularly the $SU(3)$ subgroups controlling flavor-changing transitions, is explicitly broken by the Yukawa interaction

$$\mathcal{L}_Y = \bar{Q}_L Y_D D_R H + \bar{Q}_L Y_U U_R H_c + \bar{L}_L Y_E E_R H + \text{h.c.} \quad (5)$$

Since $\mathcal{G}_{\rm SM}$ is already broken within the SM, it would not be consistent to impose it as an exact symmetry beyond the SM: even if absent a the tree-level, the breaking of $\mathcal{G}_{\rm SM}$ would reappear at the quantum level because of the Yukawa interaction. The most restrictive hypothesis we can make to *protect* in a consistent way flavor mixing in the quark sector is to assume that Y_D and Y_U are the only sources of $\mathcal{G}_q$ breaking also beyond the SM. To implement and interpret this hypothesis in a consistent way, we can assume that $\mathcal{G}_q$ is indeed a good symmetry, promoting $Y_{U,D}$ to be nondynamical fields (spurions) with nontrivial transformation properties under this symmetry:

$$Y_U \sim (3, \bar{3}, 1)_{\mathcal{G}_q}, \qquad Y_D \sim (3, 1, \bar{3})_{\mathcal{G}_q}. \quad (6)$$

If the breaking of the symmetry occurs at very high energy scales—well above the TeV region where the new de-

grees of freedom necessary to stabilize the Higgs sector should appear—at low energies we would only be sensitive to the background values of the Y, i.e. to the ordinary SM Yukawa couplings. Employing the effective-theory language, we then define that an effective theory satisfies the criterion of Minimal Flavor Violation in the quark sector if all higher-dimensional operators constructed from SM and Y fields are invariant under CP and (formally) under the flavor group $\mathcal{G}_q$ [10].

According to this criterion, one should in principle consider operators with arbitrary powers of the (dimensionless) Yukawa fields. However, a strong simplification arises by the observation that all the eigenvalues of the Yukawa matrices are small, but for the top one, and that the off-diagonal elements of the CKM matrix (V_{ij}) are very suppressed. Using the $\mathcal{G}_q$ symmetry, we can rotate the background values of the auxiliary fields Y such that

$$Y_D = \lambda_d, \qquad Y_U = V^\dagger \lambda_u, \quad (7)$$

where λ are diagonal matrices and V is the CKM matrix. It is then easy to realize that, similarly to the pure SM case, the leading coupling ruling all FCNC transitions with external down-type quarks is:

$$(\lambda_{\rm FC})_{ij} = \begin{cases} (Y_U Y_U^\dagger)_{ij} \approx \lambda_t^2 V_{3i}^* V_{3j}, & i \neq j, \\ 0, & i = j. \end{cases} \quad (8)$$

The number of relevant dimension-6 effective operators is then strongly reduced (representative examples are reported in Table 1, while the complete list can be found in Ref. [10]).

1.2.3.2 Universal UT and MFV bounds on the effective operators As originally pointed out in Ref. [12], within the MFV framework several of the constraints used to determine the CKM matrix (and in particular the unitarity triangle) are not affected by NP. In this framework, NP effects are negligible not only in tree-level processes but also in a few clean observables sensitive to loop effects, such as the time-dependent CPV asymmetry in $B_d \to J/\Psi K_{L,S}$. Indeed the structure of the basic flavor-changing coupling in (8) implies that the weak CPV phase of B_d–$\bar B_d$ mixing is $\arg[(V_{td}V_{tb}^*)^2]$, exactly as in the SM. The determination of the unitarity triangle using only these clean observables (denoted Universal Unitarity Triangle) is shown in Fig. 4.[4] This construction provides a natural (a posteriori) justification of why no NP effects have been observed in the quark sector: by construction, most of the clean observables measured at B factories are insensitive to NP effects in this framework.

In Table 1, we report a few representative examples of the bounds on the higher-dimensional operators in the MFV framework. As can be noted, the built-in CKM suppression leads to bounds on the effective scale of new physics not far from the TeV region. These bounds are very similar to the bounds on flavor-conserving operators derived by precision electroweak tests. This observation reinforces the conclusion that a deeper study of rare decays is definitely needed in order to clarify the flavor problem: the experimental precision on the clean FCNC observables required to obtain bounds more stringent than those derived from precision electroweak tests (and possibly discover new physics) is typically in the 1–10% range.

Although the MFV seems to be a natural solution to the flavor problem, it should be stressed that we are still very far from having proved the validity of this hypothesis from data. A proof of the MFV hypothesis can be achieved only with a positive evidence of physics beyond the SM exhibiting the flavor pattern (link between $s \to d$, $b \to d$ and $b \to s$ transitions) predicted by the MFV assumption.

---

[4]The UUT as originally proposed in Ref. [12] includes $\Delta M_{B_d}/\Delta M_{B_s}$ and is therefore valid only in models of CMFV (see Sect. 1.2.3.3). On the other hand, removing $\Delta M_{B_d}/\Delta M_{B_s}$ from the analysis gives a UUT that is valid in any MFV scenario.

1.2.3.3 Comparison with other approaches (CMFV & n-MFV) The idea that the CKM matrix rules the strength of FCNC transitions also beyond the SM has become a very popular concept in the recent literature and has been implemented and discussed in several works (see e.g. Refs. [12–16]).

It is worth stressing that the CKM matrix represents only one part of the problem: a key role in determining the structure of FCNCs is also played by quark masses, or by the Yukawa eigenvalues. In this respect, the MFV criterion illustrated above provides the maximal protection of FCNCs (or the minimal violation of flavor symmetry), since the full structure of Yukawa matrices is preserved. At the same time, this criterion is based on a renormalization-group-invariant symmetry argument. Therefore, it can be implemented independently of any specific hypothesis about the dynamics of the new-physics framework. The only two assumptions are: (i) the flavor symmetry and the sources of its breaking; (ii) the number of light degrees of freedom of the theory (identified with the SM fields in the minimal case).

This model-independent structure does not hold in most of the alternative definitions of MFV models that can be found in the literature. For instance, the definition of Ref. [16] (denoted constrained MFV, or CMFV) contains the additional requirement that the effective FCNC operators playing a significant role within the SM are the only relevant ones also beyond the SM. This condition is realized within weakly coupled theories at the TeV scale with only one light Higgs doublet, such as the model with universal extra dimensions analysed in Ref. [17], or the MSSM with small $\tan\beta$ and small μ term. However, it does not hold in other frameworks, such as technicolour models, or the MSSM with large $\tan\beta$ and/or large μ term (see Sect. 1.2.3.6), whose low-energy phenomenology could still be described using the general MFV criterion discussed in Sect. 1.2.3.1.

Since we are still far from having proved the validity of the MFV hypothesis from data, specific less restrictive symmetry assumptions about the flavor-structure of NP can also be considered. Next-to-minimal MFV frameworks have recently been discussed in Refs. [18, 19]. As shown in Ref. [19], a convenient way to systematically analyse the possible deviations from the MFV ansatz is to introduce additional spurions of the $\mathcal{G}_{\rm SM}$ group.

1.2.3.4 MFV at large $\tan\beta$ If the Yukawa Lagrangian contains only one Higgs field, as in (5), it necessarily breaks both $\mathcal{G}_q$ and two of the $U(1)$ subgroups of $\mathcal{G}_{\rm SM}$. In models with more than one Higgs doublet, the breaking mechanisms of $\mathcal{G}_q$ and the $U(1)$ symmetries can be decoupled, allowing a different overall normalization of the $Y_{U,D}$ spurions with respect to the SM case.

A particularly interesting scenario is the two-Higgs-doublet model where the two Higgses are coupled separately

Fig. 4 Fit of the CKM unitarity triangle within the SM (*left*) and in generic extensions of the SM satisfying the MFV hypothesis (*right*) [7]

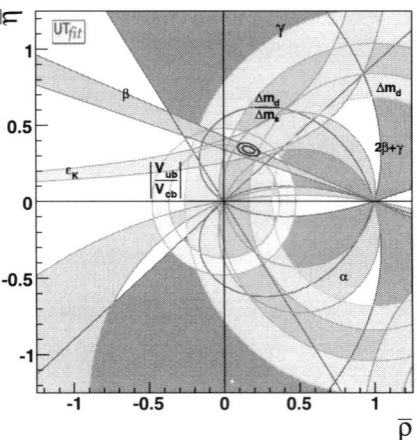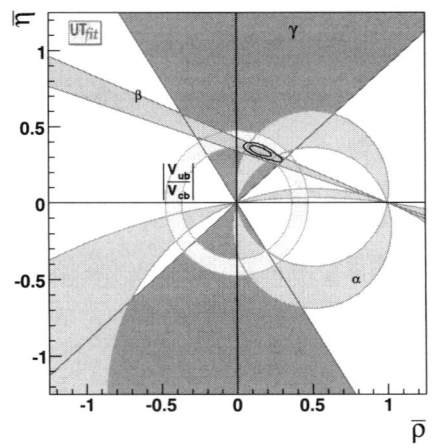

Table 1 95% C.L. bounds on the scale of representative dimension-six operators in the MFV scenario. The constraints are obtained on the single operator, with coefficient $\pm 1/\Lambda^2$ (+ or − denote constructive or destructive interference with the SM amplitude)

MFV dim-6 operator	Main observables	Λ [TeV]	
$\frac{1}{2}(\bar{Q}_L Y_U Y_U^\dagger \gamma_\mu Q_L)^2$	ϵ_K, Δm_{B_d}, Δm_{B_s}	5.9 [+]	8.8 [−]
$e H^\dagger (\bar{D}_R Y_D^\dagger Y_U Y_U^\dagger \sigma_{\mu\nu} Q_L) F_{\mu\nu}$	$B \to X_s \gamma$	5.0 [+]	9.0 [−]
$(\bar{Q}_L Y_U Y_U^\dagger \gamma_\mu Q_L)(\bar{L}_L \gamma_\mu L_L)$	$B \to (X)\ell\bar{\ell}$, $K \to \pi\nu\bar{\nu}$, $(\pi)\ell\bar{\ell}$	3.7 [+]	3.2 [−]
$(\bar{Q}_L Y_U Y_U^\dagger \gamma_\mu Q_L)(H^\dagger i D_\mu H)$	$B \to (X)\ell\bar{\ell}$, $K \to \pi\nu\bar{\nu}$, $(\pi)\ell\bar{\ell}$	2.0 [+]	2.0 [−]

to up- and down-type quarks:

$$\mathcal{L}_{Y_0} = \bar{Q}_L Y_D D_R H_D + \bar{Q}_L Y_U U_R H_U + \bar{L}_L Y_E E_R H_D + \text{h.c.} \quad (9)$$

This Lagrangian is invariant under a $U(1)$ symmetry, denoted $U(1)_{\text{PQ}}$, whose only charged fields are D_R and E_R (charge +1) and H_D (charge −1). The U_{PQ} symmetry prevents tree-level FCNCs and implies that $Y_{U,D}$ are the only sources of $\mathcal{G}_q$ breaking appearing in the Yukawa interaction (similar to the one-Higgs-doublet scenario). Coherently with the MFV hypothesis, in order to protect the good agreement between data and SM in FCNCs and $\Delta F = 2$ amplitudes, we assume that $Y_{U,D}$ are the only relevant sources of $\mathcal{G}_q$ breaking appearing in all the low-energy effective operators. This is sufficient to ensure that flavor-mixing is still governed by the CKM matrix and naturally guarantees a good agreement with present data in the $\Delta F = 2$ sector. However, the extra symmetry of the Yukawa interaction allows us to change the overall normalization of $Y_{U,D}$ with interesting phenomenological consequences in specific rare modes.

The normalization of the Yukawa couplings is controlled by $\tan\beta = \langle H_U\rangle/\langle H_D\rangle$. For $\tan\beta \gg 1$, the smallness of the b quark and τ lepton masses can be attributed to the smallness of $1/\tan\beta$ rather than to the corresponding Yukawa couplings. As a result, for $\tan\beta \gg 1$, we cannot anymore neglect the down-type Yukawa coupling. In this scenario, the determination of the effective low-energy Hamiltonian relevant to FCNC processes involves the following three steps:

- Construction of the gauge-invariant basis of dimension-six operators (suppressed by Λ^{-2}) in terms of SM fields and two Higgs doublets.
- Breaking of $SU(2) \times U(1)_Y$ and integration of the $\mathcal{O}(M_H^2)$ heavy Higgs fields.
- Integration of the $\mathcal{O}(M_W^2)$ SM degrees of freedom (top quark and electroweak gauge bosons).

These steps are well separated if we assume the scale hierarchy $\Lambda \gg M_H \gg M_W$. On the other hand, if $\Lambda \sim M_H$, the first two steps can be joined, resembling the one-Higgs-doublet scenario discussed before. The only difference is that now, at large $\tan\beta$, Y_D is not negligible, and this requires to enlarge the basis of effective dimension-six operators. From the phenomenological point of view, this implies the breaking of the strong MFV link between K- and B-physics FCNC amplitudes occurring in the one-Higgs-doublet case [10].

A more substantial modification of the one-Higgs-doublet case occurs if we allow sizable sources of $U(1)_{\text{PQ}}$ breaking. It should be pointed out that the $U(1)_{\text{PQ}}$ symmetry cannot be exact: it has to be broken at least in the scalar potential in order to avoid the presence of a massless pseudoscalar Higgs. Even if the breaking of $U(1)_{\text{PQ}}$ and $\mathcal{G}_q$ are decoupled, the presence of $U(1)_{\text{PQ}}$ breaking sources can have important implications on the structure of the Yukawa interaction. We can indeed consider new dimension-four operators such as

$$\epsilon \bar{Q}_L Y_D D_R (H_U)^c \quad \text{or} \quad \epsilon \bar{Q}_L Y_U Y_U^\dagger Y_D D_R (H_U)^c, \quad (10)$$

where ϵ denotes a generic $\mathcal{G}_q$-invariant $U(1)_{PQ}$-breaking source. Even if $\epsilon \ll 1$, the product $\epsilon \times \tan\beta$ can be $\mathcal{O}(1)$, inducing $\mathcal{O}(1)$ nondecoupling corrections to $\mathcal{L}_{Y_0}$. As discussed in specific supersymmetric scenarios, for $\epsilon \tan\beta = \mathcal{O}(1)$, the $U(1)_{PQ}$-breaking terms induce $\mathcal{O}(1)$ corrections to the down-type Yukawa couplings [20], the CKM matrix elements [21] and the charged-Higgs couplings [22–24]. Moreover, sizable FCNC couplings of the down-type quarks to the heavy neutral Higgs fields are allowed [25–30]. All these effects can be taken into account to all orders with a proper re-diagonalization of the effective Yukawa interaction [10].

Since the b-quark Yukawa coupling becomes $\mathcal{O}(1)$, the large-$\tan\beta$ regime is particularly interesting for helicity-suppressed observables in B physics. One of the clearest phenomenological consequences is a suppression (typically in the 10–50% range) of the $B \to \ell\nu$ decay rate with respect to its SM expectation [31]. Potentially measurable effects in the 10–30% range are expected also in $B \to X_s\gamma$ and ΔM_{B_s}. The most striking signature could arise from the rare decays $B_{s,d} \to \ell^+\ell^-$, whose rates could be enhanced over the SM expectations by more than one order of magnitude. An enhancement of both $B_s \to \ell^+\ell^-$ and $B_d \to \ell^+\ell^-$ respecting the MFV relation $\Gamma(B_s \to \ell^+\ell^-)/\Gamma(B_d \to \ell^+\ell^-) \approx |V_{ts}/V_{td}|^2$ would be an unambiguous signature of MFV at large $\tan\beta$.

Within the EFT approach where all the heavy degrees of freedom except the Higgs fields are integrated out, we cannot establish many other correlations among the helicity-suppressed B-physics observables. However, the scenario becomes quite predictive within a more ambitious EFT: the MSSM with MFV (see Sect. 1.2.3.6). As recently discussed in Refs. [32–34], in the MFV-MSSM with large $\tan\beta$ and heavy squarks, interesting correlations can be established among all the B-physics observables mentioned above and several flavor-conserving observables (both at low and high energies). In particular, while compatible with present B-physics constraints, this scenario can naturally resolve the long-standing $(g-2)_\mu$ anomaly and explain in a natural way why the lightest Higgs boson has not been observed yet. The predictivity, the high-sensitivity to various B-physics observables and the natural compatibility with existing data make this scenario a very interesting benchmark for correlated studies of low- and high-energy data (see Sect. 5).

1.2.3.5 MFV in grand unified theories Once we accept the idea that flavor dynamics obeys an MFV principle, at least in the quark sector, it is interesting to ask if and how this is compatible with Grand Unified Theories (GUTs), where quarks and leptons sit in the same representations of a unified gauge group. This question has recently been addressed in Ref. [35], considering the exemplifying case of $SU(5)_{\text{gauge}}$.

Within $SU(5)_{\text{gauge}}$, the down-type singlet quarks (D_R^i) and the lepton doublets (L_L^i) belong to the $\bar{\mathbf{5}}$ representation; the quark doublet (Q_L^i), the up-type (U_R^i) and lepton singlets (E_R^i) belong to the $\mathbf{10}$ representation, and finally the right-handed neutrinos (ν_R^i) are singlets. In this framework, the largest group of flavor transformation commuting with the gauge group is $\mathcal{G}_{\text{GUT}} = SU(3)_{\bar{5}} \times SU(3)_{10} \times SU(3)_1$, which is smaller than the direct product of the quark and lepton flavor groups compatible with the SM gauge sector: $\mathcal{G}_q \times \mathcal{G}_l$. We should therefore expect some violations of the MFV predictions, either in the quark sector, or in the lepton sector, or in both (a review of the MFV predictions for the lepton sector [36] can be found in the Chapter *Flavor physics of leptons and dipole moments* of this volume).

A phenomenologically acceptable description of the low-energy fermion mass matrices requires the introduction of at least four irreducible sources of $\mathcal{G}_{\text{GUT}}$ breaking. From this point of view the situation is apparently similar to the nonunified case: the four $\mathcal{G}_{\text{GUT}}$ spurions can be put in one-to-one correspondence with the low-energy spurions $Y_{U,D,E}$ plus the neutrino Yukawa coupling Y_ν (which is the only low-energy spurion in the neutrino sector assuming an approximately degenerate heavy ν_R spectrum). However, the smaller flavor group does not allow the diagonalization of Y_D and Y_E (which transform in the same way under $\mathcal{G}_{\text{GUT}}$) in the same basis. As a result, two additional mixing matrices can appear in the expressions for flavor changing rates [35]. The hierarchical texture of the new mixing matrices is known since they reduce to the identity matrix in the limit $Y_E^T = Y_D$. Taking into account this fact and analysing the structure of the allowed higher-dimensional operators, a number of reasonably firm phenomenological consequences can be deduced [35]:

— There is a well-defined limit in which the standard MFV scenario for the quark sector is fully recovered: $|Y_\nu| \ll 1$ and small $\tan\beta$. The upper bound on the neutrino Yukawa couplings implies an upper bound on the heavy neutrino masses (M_ν). In the limit of a degenerate heavy neutrino spectrum, this bound is about 10^{12} GeV. For $M_\nu \sim 10^{12}$ GeV and small $\tan\beta$, deviations from the standard MFV pattern can be expected in rare K decays but not in B physics.[5] Ignoring fine-tuned scenarios, $M_\nu \gg 10^{12}$ GeV

---

[5]The conclusion that K decays are the most sensitive probes of possible deviations from the strict MFV ansatz follows from the strong suppression of the $s \to d$ short-distance amplitude in the SM [$V_{td} V_{ts}^* = \mathcal{O}(10^{-4})$] and goes beyond the hypothesis of an underlying GUT. This is the reason why $K \to \pi\nu\bar{\nu}$ decays, which are the best probes of $s \to d$, $\Delta F = 1$ short-distance amplitudes, play a key role in any extension of the SM containing nonminimal sources of flavor symmetry breaking.

is excluded by the present constraints on quark FCNC transitions. Independently from the value of M_ν, deviations from the standard MFV pattern can appear both in K and in B physics for $\tan\beta \gtrsim m_t/m_b$.

- Contrary to the non-GUT MFV framework for the lepton sector, the rate for $\mu \to e\gamma$ and other LFV decays cannot be arbitrarily suppressed by lowering the mass of the heavy ν_R. This fact can easily be understood by noting that the GUT group allows also M_ν-independent contributions to LFV decays proportional to the quark Yukawa couplings. The latter become competitive for $M_\nu \lesssim 10^{12}$ GeV, and their contribution is such that for $\Lambda \lesssim 10$ TeV, the $\mu \to e\gamma$ rate is above 10^{-13} (i.e. within the reach of MEG [37]).
- Within this framework, improved experimental searches on $\tau \to \mu\gamma$ and $\tau \to e\gamma$ are a key tool: they are the best observables to discriminate the relative size of the non-GUT MFV contributions with respect to the GUT ones. In particular, if the quark-induced terms turn out to be dominant, the $\mathcal{B}(\tau \to \mu\gamma)/\mathcal{B}(\mu \to e\gamma)$ ratio could reach values of $\mathcal{O}(10^{-4})$, allowing $\tau \to \mu\gamma$ to be just below the present exclusion bounds.

1.2.3.6 The MFV hypothesis in the MSSM A detailed discussion of the so-called Minimal Supersymmetric extension of the SM will be presented in Sect. 1.3. Here we limit ourselves to analyse how the MFV hypothesis can be implemented in this framework and to briefly summarize its main implications.

It is first worth to recall that the adjective *minimal* in the MSSM acronyms refers to the particle content of the model and not to its flavor structure. In general, the MSSM contains a huge number of free parameters, and most of them are related to the flavor structure of the model (sfermion masses and trilinear couplings). Since the new degrees of freedom (in particular the squark fields) have well-defined transformation properties under the quark-flavor group $\mathcal{G}_q$, the MFV hypothesis can easily be implemented in this framework following the general rules outlined in Sect. 1.2.3.1: we need to consider all possible interactions compatible with (i) softly-broken supersymmetry; (ii) the breaking of $\mathcal{G}_q$ via the spurion fields $Y_{U,D}$. This allows one to express the squark mass terms and the trilinear quark–squark–Higgs couplings as follows [10, 38]:

$$\tilde{m}^2_{Q_L} = \tilde{m}^2\big(a_1 \mathbb{1} + b_1 Y_U Y_U^\dagger + b_2 Y_D Y_D^\dagger + b_3 Y_D Y_D^\dagger Y_U Y_U^\dagger$$
$$+ b_4 Y_U Y_U^\dagger Y_D Y_D^\dagger + \cdots\big), \quad (11)$$

$$\tilde{m}^2_{U_R} = \tilde{m}^2\big(a_2 \mathbb{1} + b_5 Y_U^\dagger Y_U + \cdots\big), \quad (12)$$

$$\tilde{m}^2_{D_R} = \tilde{m}^2\big(a_3 \mathbb{1} + b_6 Y_D^\dagger Y_D + \cdots\big), \quad (13)$$

$$A_U = A\big(a_4 \mathbb{1} + b_7 Y_D Y_D^\dagger + \cdots\big) Y_U, \quad (14)$$

$$A_D = A\big(a_5 \mathbb{1} + b_8 Y_U Y_U^\dagger + \cdots\big) Y_D, \quad (15)$$

where the dimensionful parameters $\tilde{m}$ and A set the overall scale of the soft-breaking terms. In (11)–(15), we have explicitly shown all independent flavor structures which cannot be absorbed into a redefinition of the leading terms (up to tiny contributions quadratic in the Yukawas of the first two families). When $\tan\beta$ is not too large and the bottom Yukawa coupling is small, the terms quadratic in Y_D can be dropped.

In a bottom-up approach, the dimensionless coefficients a_i and b_i in (11)–(15) should be considered as free parameters of the model. Note that this structure is renormalization-group invariant: the values of a_i and b_i change according to the Renormalization Group (RG) flow, but the general structure of (11)–(15) is unchanged. This is not the case if the b_i are set to zero (corresponding to the so-called hypothesis of flavor universality). If this hypothesis is set as initial condition at some high-energy scale M, then nonvanishing $b_i \sim (1/4\pi)^2 \ln M^2/\tilde{m}^2$ are generated by the RG evolution. This is for instance what happens in models with gauge-mediated supersymmetry breaking [39–41], where the scale M is identified with the mass of the hypothetical messenger particles.

Using the soft terms in (11)–(15), the physical 6×6 squark-mass matrices, after electroweak symmetry breaking, are given by

$$\tilde{M}^2_U = \begin{pmatrix} \tilde{m}^2_{Q_L} + Y_U Y_U^\dagger v_U^2 + \big(\tfrac{1}{2} - \tfrac{2}{3}s_W^2\big) M_Z^2 \cos 2\beta & (A_U - \mu Y_U \cot\beta) v_U \\ (A_U - \mu Y_U \cot\beta)^\dagger v_U & \tilde{m}^2_{U_R} + Y_U^\dagger Y_U v_U^2 + \tfrac{2}{3} s_W^2 M_Z^2 \cos 2\beta \end{pmatrix},$$

$$\tilde{M}^2_D = \begin{pmatrix} \tilde{m}^2_{Q_L} + Y_D Y_D^\dagger v_D^2 - \big(\tfrac{1}{2} - \tfrac{1}{3}s_W^2\big) M_Z^2 \cos 2\beta & (A_D - \mu Y_D \tan\beta) v_D \\ (A_D - \mu Y_D \tan\beta)^\dagger v_D & \tilde{m}^2_{D_R} + Y_D^\dagger Y_D v_D^2 - \tfrac{1}{3} s_W^2 M_Z^2 \cos 2\beta \end{pmatrix},$$

(16)

where μ is the higgsino mass parameter, and $v_{U,D} = \langle H_{U,D}\rangle$ ($\tan\beta = v_U/v_D$). The eigenvalues of these mass matrices are not degenerate; however, the mass splittings are tightly constrained by the specific (Yukawa-type) symmetry-breaking pattern.

If we are interested only in low-energy processes, we can integrate out the supersymmetric particles at one loop and project this theory into the general EFT discussed in the previous sections. In this case, the coefficients of the dimension-six effective operators written in terms of SM and Higgs fields (see Table 1) are computable in terms of the supersymmetric soft-breaking parameters. We stress that if $\tan\beta \gg 1$ (see Sect. 1.2.3.4) and/or if μ is large enough [42], the relevant operators thus obtained go beyond the restricted basis of the CMFV scenario [16]. The typical effective scale suppressing these operators (assuming an overall coefficient $1/\Lambda^2$) is

$$\Lambda \sim 4\pi\tilde{m}. \tag{17}$$

Looking at the bounds in Table 1, we then conclude that if MFV holds, the present bounds on FCNCs do not exclude squarks in the few hundred GeV mass range, i.e. well within the LHC reach.

It is finally worth recalling that the integration of the supersymmetric degrees of freedom may also lead to sizable modifications of the renormalizable operators and, in particular, of the effective Yukawa interactions. As a result, in an effective field theory with supersymmetric degrees of freedom, the relations between $Y_{U,D}$ and the physical quark masses and CKM angles are potentially modified. As already pointed out in Sect. 1.2.3.4, this effect is particularly relevant in the large $\tan\beta$ regime.

1.3 SUSY models

1.3.1 FCNC and SUSY

The generation of fermion masses and mixings ("flavor problem") gives rise to the first important distinction among theories of new physics beyond the electroweak Standard Model.

One may conceive a kind of new physics that is completely "flavor blind", i.e. new interactions that have nothing to do with the flavor structure. To provide an example of such a situation, consider a scheme where flavor arises at a very large scale (for instance the Planck mass) while new physics is represented by a supersymmetric extension of the SM with supersymmetry broken at a much lower scale and with the SUSY breaking transmitted to the observable sector by flavor-blind gauge interactions. In this case, one may think that the new physics does not cause any major change to the original flavor structure of the SM, namely that the pattern of fermion masses and mixings is compatible with the numerous and demanding tests of flavor changing neutral currents.

Alternatively, one can conceive a new physics that is entangled with the flavor problem. As an example, consider a technicolour scheme where fermion masses and mixings arise through the exchange of new gauge bosons which mix together ordinary and technifermions. Here we expect (correctly enough) new physics to have potential problems in accommodating the usual fermion spectrum with the adequate suppression of FCNC. As another example of new physics that is not flavor blind, take a more conventional SUSY model which is derived from a spontaneously broken $N = 1$ supergravity and where the SUSY breaking information is conveyed to the ordinary sector of the theory through gravitational interactions. In this case, we may expect that the scale at which flavor arises and the scale of SUSY breaking are not so different and possibly the mechanism of SUSY breaking and transmission itself is flavor-dependent. Under these circumstances, we may expect a potential flavor problem to arise, namely that SUSY contributions to FCNC processes are too large.

The potentiality of probing SUSY in FCNC phenomena was readily realized when the era of SUSY phenomenology started in the early 80's [43, 44]. In particular, the major implication that the scalar partners of quarks of the same electric charge but belonging to different generations had to share a remarkably high mass degeneracy was emphasized.

Throughout the large amount of work in the past decades, it became clearer and clearer that generically talking of the implications of low-energy SUSY on FCNC may be rather misleading. We have a minimal SUSY extension of the SM, the so-called Constrained Minimal Supersymmetric Standard Model (CMSSM), where the FCNC contributions can be computed in terms of a very limited set of unknown new SUSY parameters. Remarkably enough, this minimal model succeeds to pass all FCNC tests unscathed. To be sure, it is possible to severely constrain the SUSY parameter space, for instance using $b \to s\gamma$ in a way that is complementary to what is achieved by direct SUSY searches at colliders.

However, the CMSSM is by no means equivalent to low-energy SUSY. A first sharp distinction concerns the mechanism of SUSY breaking and transmission to the observable sector that is chosen. As we mentioned above, in models with gauge-mediated SUSY breaking (GMSB models [39, 40, 45–68]), it may be possible to avoid the FCNC threat "ab initio" (notice that this is not an automatic feature of this class of models, but it depends on the specific choice of the sector that transmits the SUSY breaking information, the so-called messenger sector). The other more "canonical" class of SUSY theories that was mentioned above has gravitational messengers and a very large scale at which SUSY breaking occurs.

In this brief discussion, we focus only on this class of gravity-mediated SUSY breaking models. Even sticking to this more limited choice, we have a variety of options with very different implications for the flavor problem: first, there exists an interesting large class of SUSY realizations where the customary R-parity (which is invoked to suppress proton decay) is replaced by other discrete symmetries which allow either baryon or lepton violating terms in the superpotential. But, even sticking to the more orthodox view of imposing R-parity, we are still left with a large variety of extensions of the MSSM at low energy. The point is that low-energy SUSY "feels" the new physics at the superlarge scale at which supergravity (i.e. local supersymmetry) broke down. In the past years, we have witnessed an increasing interest in supergravity realizations without the so-called flavor universality of the terms which break SUSY explicitly. Another class of low-energy SUSY realizations, which differ from the MSSM in the FCNC sector, is obtained from SUSY-GUT's. The interactions involving superheavy particles in the energy range between the GUT and the Planck scale bear important implications for the amount and kind of FCNC that we expect at low energy [69–71].

1.3.2 FCNC in SUSY without R-parity

It is well known that in the SM case, the imposition of gauge symmetry and the usual gauge assignment of the 15 elementary fermions of each family lead to the automatic conservation of baryon and lepton numbers (this is true at any order in perturbation theory).

On the contrary, imposing in addition to the usual $SU(3) \otimes SU(2) \otimes U(1)$ gauge symmetry an $N = 1$ global SUSY does not prevent the appearance of terms which explicitly break B or L [72, 73]. Indeed, the superpotential reads:

$$W = h^U Q H_U u^c + h^D Q H_D d^c + h^L L H_D e^c + \mu H_U H_D$$
$$+ \mu' H_U L + \lambda''_{ijk} u_i^c d_j^c d_k^c + \lambda'_{ijk} Q_i L_j d_k^c$$
$$+ \lambda_{ijk} L_i L_j e_k^c, \qquad (18)$$

where the chiral matter superfields Q, u^c, d^c, L, e^c, H_U and H_D transform under the above gauge symmetry as:

$$Q \equiv (3, 2, 1/6); \qquad u^c \equiv (\bar{3}, 1, -2/3);$$
$$d^c \equiv (\bar{3}, 1, 1/3);$$
$$L \equiv (1, 2, -1/2); \qquad e^c \equiv (1, 1, 1); \qquad (19)$$
$$H_U \equiv (1, 2, 1/2); \qquad H_D \equiv (1, 2, -1/2).$$

The couplings h^U, h^D, h^L are 3×3 matrices in the generation space; i, j and k are generation indices. Using the product of λ' and λ'' couplings, it is immediate to construct four-fermion operators leading to proton decay through the exchange of a squark. Even if one allows for the existence of λ' and λ'' couplings only involving the heaviest generation, one can show that the bound on the product $\lambda' \times \lambda''$ of these couplings is very severe (of $O(10^{-7})$) [74].

A solution is that there exists a discrete symmetry, B-parity [75–79], which forbids the B-violating terms proportional to λ'' in (18). In that case, it is still possible to produce sizable effects in FC processes. Two general features of these R-parity violating contributions are:

1. Complete loss of any correlation to the CKM elements. For instance, in the above example, the couplings λ' and λ have nothing to do with the usual angles V_{tb} and V_{ts} which appear in $b \to sl^+l^-$ in the SM.
2. Loss of correlation among different FCNC processes, which are tightly correlated in the SM. For instance, in our example, $b \to dl^+l^-$ would depend on λ' and λ parameters which are different from those appearing in B_d–$\bar{B}_d$ mixing.

In this context, it is difficult to make predictions given the arbitrariness of the large number of λ and λ' parameters. There exist bounds on each individual coupling (i.e. assuming that all the other L violating couplings are zero) [80, 81].

Obviously, the most practical way of avoiding any threat of B- and L-violating operators is to forbid <u>all</u> such terms in (18). This is achieved by imposing the usual R matter parity. This quantum number is +1 for every ordinary particle and −1 for SUSY partners. We now turn to FCNC in the framework of low-energy SUSY with R parity.

1.3.3 FCNC in SUSY with R parity—CMSSM framework

Even when R parity is imposed the FCNC challenge is not over. It is true that in this case, analogously to what happens in the SM, no tree level FCNC contributions arise. However, it is well known that this is a necessary but not sufficient condition to consider the FCNC problem overcome. The loop contributions to FCNC in the SM exhibit the presence of the GIM mechanism, and we have to make sure that in the SUSY case with R parity, some analog of the GIM mechanism is active.

To give a qualitative idea of what we mean by an effective super-GIM mechanism, let us consider the following simplified situation where the main features emerge clearly. Consider the SM box diagram responsible for K^0–$\bar{K}^0$ mixing and take only two generations, i.e. only the up and charm quarks run in the loop. In this case, the GIM mechanism yields a suppression factor of $O((m_c^2 - m_u^2)/M_W^2)$. If we replace the W boson and the up quarks in the loop with their SUSY partners and we take, for simplicity, all SUSY masses of the same order, we obtain a super-GIM factor which looks like the GIM one with the masses of the superparticles instead of those of the corresponding particles. The problem is that the up and charm squarks have masses which are much larger than those of the corresponding quarks. Hence the super-GIM factor tends to be of $O(1)$ instead of being

$O(10^{-3})$ as it is in the SM case. To obtain this small number we would need a high degeneracy between the mass of the charm and up squarks. It is difficult to think that such a degeneracy may be accidental. After all, since we invoked SUSY for a naturalness problem (the gauge hierarchy issue), we should avoid invoking a fine-tuning to solve its problems! Then one can turn to some symmetry reason. For instance, just sticking to this simple example that we are considering, one may think that the main bulk of the charm and up squark masses is the same, i.e. the mechanism of SUSY breaking should have some universality in providing the mass to these two squarks with the same electric charge. Another possibility one may envisage is that the masses of the squarks are quite high, say above few TeV's. Then even if they are not so degenerate in mass, the overall factor in front of the four-fermion operator responsible for the kaon mixing becomes smaller and smaller (it decreases quadratically with the mass of the squarks), and, consequently, one can respect the experimental result. We see from this simple example that the issue of FCNC may be closely linked to the crucial problem of how we break SUSY.

We now turn to some more quantitative considerations. We start by discussing the different degrees of concern that FCNC raise according to the specific low-energy SUSY realization one has in mind. In this section, we will consider FCNC in the CMSSM realizations. In Sect. 1.3.4, we will deal with CP-violating FCNC phenomena in the same context. After discussing these aspects in the CMSSM, we will provide bounds from FCNC and CP violation in a generic SUSY extension of the SM (Sect. 1.3.5).

Obviously the reference frame for any discussion in a specific SUSY scheme is the MSSM. Although the name seems to indicate a well-defined particle model, we can identify at least two quite different classes of low-energy SUSY models. First, we have the CMSSM, the minimal SUSY extension of the SM (i.e. with the smallest needed number of superfields) with R-parity, radiative breaking of the electroweak symmetry, universality of the soft breaking terms and simplifying relations at the GUT scale among SUSY parameters. In this *constrained* version, the MSSM exhibits only four free parameters in addition to those of the SM and is an example of a SUSY model with MFV. Moreover, some authors impose specific relations between the two parameters A and B that appear in the trilinear and bilinear scalar terms of the soft breaking sector, further reducing the number of SUSY free parameters to three. Then, all SUSY masses are just functions of these few independent parameters, and, hence, many relations among them exist.

In SUSY, there are five classes of one-loop diagrams that contribute to FCNC and CP-violating processes. They are distinguished according to the virtual particles running in the loop: W and up-quarks, charged Higgs and up-quarks, charginos and up-squarks, neutralinos and down-squarks, gluinos and down-squarks. It turns out that, in this *constrained* version of the MSSM, at low or moderate $\tan\beta$, the charged Higgs and chargino exchanges yield the dominant SUSY contributions, while at large $\tan\beta$, Higgs-mediated effects become dominant.

Obviously this very minimal version of the MSSM can be very predictive. The most powerful constraint on this minimal model in the FCNC context comes from $b \to s\gamma$ [23, 82–84]. For large values of $\tan\beta$, strong constraints are also obtained from the upper bound on $B_s \to \mu^+\mu^-$, from ΔM_s and from $B(B \to \tau\nu)$ [27–30, 32, 85]. No observable deviations from the SM predictions in other FCNC processes are expected, given the present experimental and theoretical uncertainties.

It should be kept in mind that the above stringent results strictly depend not only on the minimality of the model in terms of the superfields that are introduced but also on the "boundary" conditions that are chosen. All the low-energy SUSY masses are computed in terms of the four SUSY parameters at the Planck scale M_{Pl} through the RG evolution. If one relaxes this tight constraint on the relation of the low-energy quantities and treats the masses of the SUSY particles as independent parameters, then much more freedom is gained. This holds true even in the MSSM with MFV at small or moderate $\tan\beta$: sizable SUSY effects can be present both in meson–anti-meson mixing and in rare decays [86], in particular for light stop and charginos.

Moreover, flavor universality is by no means a prediction of low-energy SUSY. The absence of flavor universality of soft-breaking terms may result from radiative effects at the GUT scale or from effective supergravities derived from string theory. For instance, even starting with an exact universality of the soft breaking terms at the Planck scale, in a SUSY GUT scheme, one has to consider the running from this latter scale to the GUT scale. Due to the large value of the top Yukawa coupling and to the fact that quarks and lepton superfields are in common GUT multiplets, we may expect the tau slepton mass to be conspicuously different from that of the first two generation sleptons at the end of this RG running. This lack of universality at the GUT scale may lead to large violations of lepton flavor number yielding, for instance, $\mu \to e\gamma$ at a rate in the ball park of observability [87]. In the nonuniversal case, most FCNC processes receive sizable SUSY corrections, and indeed flavor physics poses strong constraints on the parameter space of SUSY models without MFV.

1.3.4 CP violation in the CMSSM

CP violation has a major potential to exhibit manifestations of physics beyond the SM. Indeed, it is quite a general feature that NP possesses new CP-violating phases in addition to the CKM phase (δ_{CKM}) or, even in those cases where this does not occur, δ_{CKM} shows up in interactions of the new particles, hence with potential departures from the SM expectations. Moreover, although the SM is able to account for the observed CP violation, the possibility of large NP contributions to CP violation in $b \to s$ transitions is still open (see Sect. 3.7 and Ref. [88] for recent reviews). The detection of CP violation in B_s mixing and the improvement of the measurements of CP asymmetries in $b \to s$ penguin decays will constitute a crucial test of the CKM picture within the SM. Again, on general grounds, we expect new physics to provide departures from the SM CKM scenario. A final remark on reasons that make us optimistic in having new physics playing a major role in CP violation concerns the matter–anti-matter asymmetry in the Universe. Starting from a baryon–anti-baryon symmetric Universe, the SM is unable to account for the observed baryon asymmetry. The presence of new CP-violating contributions when one goes beyond the SM looks crucial to produce an efficient mechanism for the generation of a satisfactory ΔB asymmetry.

The above considerations apply well to the new physics represented by low-energy supersymmetric extensions of the SM. Indeed, as we will see below, supersymmetry introduces CP-violating phases in addition to δ_{CKM}, and, even if one envisages particular situations where such extra-phases vanish, the phase δ_{CKM} itself leads to new CP-violating contributions in processes where SUSY particles are exchanged. CP violation in $b \to s$ transitions has a good potential to exhibit departures from the SM CKM picture in low-energy SUSY extensions, although, as we will discuss, the detectability of such deviations strongly depends on the regions of the SUSY parameter space under consideration.

In this section, we will deal with CP violation in the context of the CMSSM. In Sect. 1.3.5, we will discuss the CP issue in a model-independent approach.

In the CMSSM, two new "genuine" SUSY CP-violating phases are present. They originate from the SUSY parameters μ, M, A and B. The first of these parameters is the dimensionful coefficient of the $H_u H_d$ term of the superpotential. The remaining three parameters are present in the sector that softly breaks the $N = 1$ global SUSY. M denotes the common value of the gaugino masses, A is the trilinear scalar coupling, while B denotes the bilinear scalar coupling. In our notation, all these three parameters are dimensionful. The simplest way to see which combinations of the phases of these four parameters are physical [89] is to notice that for vanishing values of μ, M, A and B, the theory possesses two additional symmetries [90]. Indeed, letting B and μ vanish, a $U(1)$ Peccei–Quinn symmetry arises, which in particular rotates H_u and H_d. If M, A and B are set to zero, the Lagrangian acquires a continuous $U(1)$ R symmetry. Then we can consider μ, M, A and B as spurions which break the $U(1)_{PQ}$ and $U(1)_R$ symmetries. In this way, the question concerning the number and nature of the meaningful phases translates into the problem of finding the independent combinations of the four parameters which are invariant under $U(1)_{PQ}$ and $U(1)_R$ and determining their independent phases. There are three such independent combinations, but only two of their phases are independent. We use here the commonly adopted choice:

$$\Phi_A = \arg(A^* M), \qquad \Phi_B = \arg(B^* M). \qquad (20)$$

The main constraints on Φ_A and Φ_B come from their contribution to the electric dipole moments of the neutron and of the electron. For instance, the effect of Φ_A and Φ_B on the electric and chromoelectric dipole moments of the light quarks (u, d, s) lead to a contribution to d_N^e of order

$$d_N^e \sim 2 \left(\frac{100 \, \text{GeV}}{\tilde{m}} \right)^2 \sin \Phi_{A,B} \times 10^{-23} \, e \, \text{cm}, \qquad (21)$$

where $\tilde{m}$ here denotes a common mass for squarks and gluinos. We refer the reader to the Chapter *Flavor physics of leptons and dipole moments* of this volume for a detailed discussion of the present status of constraints on SUSY from electric dipole moments. We just remark that the present experimental bounds imply that $\Phi_{A,B}$ should be at most of $\mathcal{O}(10^{-2})$, unless one pushes SUSY masses up to $\mathcal{O}(1 \, \text{TeV})$.

In view of the previous considerations, most authors dealing with the CMSSM prefer to simply put Φ_A and Φ_B equal to zero. Actually, one may argue in favor of this choice by considering the soft breaking sector of the MSSM as resulting from SUSY breaking mechanisms which force Φ_A and Φ_B to vanish. For instance, it is conceivable that both A and M originate from the same source of $U(1)_R$ breaking. Since Φ_A "measures" the relative phase of A and M, in this case, it would "naturally" vanish. In some specific models, it has been shown [40] that through an analogous mechanism also Φ_B may vanish.

If $\Phi_A = \Phi_B = 0$, then the novelty of the CMSSM in CP-violating contributions merely arises from the presence of the CKM phase in loops with SUSY particles [89, 91–96]. The crucial point is that the usual GIM suppression, which plays a major role in evaluating ε and ε' in the SM, is replaced in the MSSM case by a super-GIM cancellation, which has the same "power" of suppression as the original GIM (see previous section). Again also in the MSSM, as it is the case in the SM, the smallness of ε and ε' is guaranteed not by the smallness of δ_{CKM} but rather by the small CKM angles and/or small Yukawa couplings. By the same token, we do not expect any significant departure of the MSSM from the SM predictions also concerning CP violation in B physics. As a matter of fact, given the large lower bounds on squark and gluino masses, one expects relatively tiny contributions of the SUSY loops in ε or ε' in comparison with the normal W loops of the SM. Let us be more detailed on this point. In the MSSM, the gluino exchange contribution to FCNC is subleading with respect to chargino ($\chi^\pm$) and charged Higgs ($H^\pm$) exchanges. Hence, when dealing with CP-violating FCNC processes in the MSSM with $\Phi_A = \Phi_B = 0$, one can confine the analysis to $\chi^\pm$ and $H^\pm$ loops. If one takes all squarks to be degenerate in mass and heavier than ~ 200 GeV, then $\chi^\pm$–$\tilde{q}$ loops are obviously severely penalized with respect to the SM W–q loops (remember that at the vertices, the same CKM angles occur in both cases). The only chance to generate sizable contributions to CP-violating phenomena is for a light stop and chargino: in this case, sizable departures from the SM predictions are possible [86].

In conclusion, the situation concerning CP violation in the MSSM case with $\Phi_A = \Phi_B = 0$ and exact universality in the soft-breaking sector can be summarized in the following way: the MSSM does not lead to any significant deviation from the SM expectation for CP-violating phenomena as d_N^e, ε, ε' and CP violation in B physics; the only exception to this statement concerns a small portion of the MSSM parameter space where a very light $\tilde{t}$ and χ^+ are present.

1.3.5 Model-independent analysis of FCNC and CP violating processes in SUSY

Given a specific SUSY model, it is in principle possible to make a full computation of all the FCNC phenomena in that context. However, given the variety of options for low-energy SUSY which was mentioned above (even confining ourselves here to models with R matter parity), it is important to have a way to extract from the whole host of FCNC processes a set of upper limits on quantities that can be readily computed in any chosen SUSY frame.

A useful model-independent parameterization of FCNC effects is the so-called mass insertion (MI) approximation [97]. It concerns the most peculiar source of FCNC SUSY contributions that do not arise from the mere supersymmetrization of the FCNC in the SM. They originate from the FC couplings of gluinos and neutralinos to fermions and sfermions [98–100]. One chooses a basis for the fermion and sfermion states where all the couplings of these particles to neutral gauginos are flavor diagonal, while the FC is exhibited by the nondiagonality of the sfermion propagators. Denoting by Δ the off-diagonal terms in the sfermion mass matrices (i.e. the mass terms relating sfermions of the same electric charge but different flavor), the sfermion propagators can be expanded as a series in terms of $\delta = \Delta/\tilde{m}^2$, where $\tilde{m}$ is the average sfermion mass. As long as Δ is significantly smaller than $\tilde{m}^2$, we can just take the first term of this expansion, and then the experimental information concerning FCNC and CP-violating phenomena translates into upper bounds on these δ's [101–104].

Obviously the above mass insertion method presents the major advantage that one does not need the full diagonalization of the sfermion mass matrices to perform a test of the SUSY model under consideration in the FCNC sector. It is enough to compute ratios of the off-diagonal over the diagonal entries of the sfermion mass matrices and compare the results with the general bounds on the δ's that we provide here from all available experimental information.

There exist four different Δ mass insertions connecting flavors i and j along a sfermion propagator: $(\Delta_{ij})_{LL}$, $(\Delta_{ij})_{RR}$, $(\Delta_{ij})_{LR}$ and $(\Delta_{ij})_{RL}$. The indices L and R refer to the helicity of the fermion partners. Instead of the dimensionful quantities Δ, it is more useful to provide bounds making use of dimensionless quantities, δ, that are obtained dividing the mass insertions by an average sfermion mass.

The comparison of several flavor-changing processes to their experimental values can be used to bound the δ's in the different sectors [104–116]. In these analyses, it is customary to consider only the dominant contributions due to gluino exchange, which give a good approximation of the full amplitude, barring accidental cancellations. In the same spirit, the bounds are usually obtained taking only one non-vanishing MI at a time, neglecting the interference among MIs. This procedure is justified a posteriori by observing that the MI bounds have typically a strong hierarchy, making the destructive interference among different MIs very unlikely.

The effective Hamiltonians for $\Delta F = 1$ and $\Delta F = 2$ transitions including gluino contributions computed in the MI approximation can be found in the literature together with the formulae of several observables [104]. Even the full NLO calculation is available for the $\Delta F = 2$ effective Hamiltonian [117, 118]. See Refs. [111–113] for the calculation of $\tan\beta$-enhanced subleading terms for several B decays in the case of general flavor violation.

In our study, we use the phenomenological constraints collected in Table 2. In particular:

Sector 1–2 The measurements of ΔM_K, ε and ε'/ε are used to constrain the $(\delta_{12}^d)_{AB}$ with $(A, B) = (L, R)$. The first two measurements, ΔM_K and ε, respectively bound the real and imaginary parts of the product $(\delta_{12}^d)(\delta_{12}^d)$. In the case of ΔM_K, given the uncertainty coming from the long-distance contribution, we use the conservative range in Table 2. The measurement of ε'/ε, on the other hand, puts a bound on Im(δ_{12}^d). This bound, however, is effective in the case of the LR MI only. Notice that, given the large hadronic uncertainties in the SM calculation of ε'/ε, we use the very loose bound on the SUSY contribution shown in Table 2. The bounds coming from the combined constraints are shown in Table 3. Notice that, here and in the other sectors, the bound on the RR MI is obtained in the presence of the radiatively-induced LL MI (see (11)). The product $(\delta_{12}^d)_{LL}(\delta_{12}^d)_{RR}$ generates left–right operators that

Table 2 Measurements and bounds used to constrain the hadronic δ^d's

Observable	Measurement/Bound	Ref.
Sector 1–2		
ΔM_K	$(0.0\text{--}5.3) \times 10^{-3}$ GeV	[119]
ε	$(2.232 \pm 0.007) \times 10^{-3}$	[119]
$|(\varepsilon'/\varepsilon)_{\text{SUSY}}|$	$< 2 \times 10^{-2}$	–
Sector 1–3		
ΔM_{B_d}	(0.507 ± 0.005) ps^{-1}	[389]
$\sin 2\beta$	0.675 ± 0.026	[389]
$\cos 2\beta$	> -0.4	[120]
Sector 2–3		
$BR(b \to (s+d)\gamma)(E_\gamma > 2.0 \text{ GeV})$	$(3.06 \pm 0.49) \times 10^{-4}$	[121]
$BR(b \to (s+d)\gamma)(E_\gamma > 1.8 \text{ GeV})$	$(3.51 \pm 0.43) \times 10^{-4}$	[122]
$BR(b \to s\gamma)(E_\gamma > 1.9 \text{ GeV})$	$(3.34 \pm 0.18 \pm 0.48) \times 10^{-4}$	[123]
$A_{\text{CP}}(b \to s\gamma)$	0.004 ± 0.036	
$BR(b \to sl^+l^-)(0.04 \text{ GeV} < q^2 < 1 \text{ GeV})$	$(11.34 \pm 5.96) \times 10^{-7}$	[124, 125]
$BR(b \to sl^+l^-)(1 \text{ GeV} < q^2 < 6 \text{ GeV})$	$(15.9 \pm 4.9) \times 10^{-7}$	[124, 125]
$BR(b \to sl^+l^-)(14.4 \text{ GeV} < q^2 < 25 \text{ GeV})$	$(4.34 \pm 1.15) \times 10^{-7}$	[124, 125]
$A_{\text{CP}}(b \to sl^+l^-)$	-0.22 ± 0.26	[119]
ΔM_{B_s}	(17.77 ± 0.12) ps^{-1}	[126]

Table 3 95% probability bounds on $|(\delta_{ij}^q)_{AB}|$ obtained for squark and gluino masses of 350 GeV. See the text for details

$|(\delta_{12}^d)_{LL,RR}|$	$|(\delta_{12}^d)_{LL=RR}|$	$|(\delta_{12}^d)_{LR}|$	$|(\delta_{12}^d)_{RL}|$
1×10^{-2}	2×10^{-4}	5×10^{-4}	5×10^{-4}
$|(\delta_{12}^u)_{LL,RR}|$	$|(\delta_{12}^u)_{LL=RR}|$	$|(\delta_{12}^u)_{LR}|$	$|(\delta_{12}^u)_{RL}|$
3×10^{-2}	2×10^{-3}	6×10^{-3}	6×10^{-3}
$|(\delta_{13}^d)_{LL,RR}|$	$|(\delta_{13}^d)_{LL=RR}|$	$|(\delta_{13}^d)_{LR}|$	$|(\delta_{13}^d)_{RL}|$
7×10^{-2}	5×10^{-3}	1×10^{-2}	1×10^{-2}
$|(\delta_{23}^d)_{LL}|$	$|(\delta_{23}^d)_{RR}|$	$|(\delta_{23}^d)_{LL=RR}|$	$|(\delta_{23}^d)_{LR,RL}|$
2×10^{-1}	7×10^{-1}	5×10^{-2}	5×10^{-3}

are enhanced both by the QCD evolution and by the matrix element (for kaons only). Therefore, the bounds on RR MIs are more stringent than the ones on LL MIs.

Sector 1–3 The measurements of ΔM_{B_d} and 2β respectively constrain the modulus and the phase of the mixing amplitude bounding the products $(\delta_{13}^d)(\delta_{13}^d)$. For the sake of simplicity, in Table 3, we show the bounds on the modulus of (δ_{13}^d) only.

Sector 2–3 This sector enjoys the largest number of constraints. The recent measurement of ΔM_{B_s} constrains the modulus of the mixing amplitude, thus bounding the products $|(\delta_{23}^d)(\delta_{23}^d)|$. Additional strong constraints come from $\Delta B = 1$ branching ratios, such as $b \to s\gamma$ and $b \to sl^+l^-$. Also for this sector, we present the bounds on the modulus of (δ_{23}^d) in Table 3.

All the bounds in Table 3 have been obtained by using the NLO expressions for the SM contributions and for SUSY where available. Hadronic matrix elements of $\Delta F = 2$ operators are taken from lattice calculations [127–130]. The values of the CKM parameters $\bar{\rho}$ and $\bar{\eta}$ are taken from the UTfit analysis in the presence of arbitrary loop-mediated NP contributions [7]. This conservative choice allows us to decouple the determination of SUSY parameters from the CKM matrix. For $b \to s\gamma$, we use NLO expressions with the value of the charm quark mass suggested by the recent NNLO calculation [376]. For the chromomagnetic contribution to ε'/ε, we have used the matrix element as estimated in Ref. [131]. The 95% probability bounds are computed using the statistical method described in Refs. [107, 132].

Concerning the dependence on the SUSY parameters, the bounds mainly depend on the gluino mass and on the "average squark mass". A mild dependence on $\tan \beta$ is introduced by the presence of double MIs $(\delta_{ij}^d)_{LL}(\delta_{jj}^d)_{LR}$ in chromomagnetic operators. This dependence however becomes sizable only for very large values of $\tan\beta$. Approximately, all bounds scale as squark and gluino masses.

1.4 Nonsupersymmetric extensions of the Standard Model

In this section, we briefly describe two most popular nonsupersymmetric extensions of the SM, paying particular attention to the flavor structure of these models. These are Little Higgs models and a model with one universal extra dimension.

1.4.1 Little Higgs models

1.4.1.1 Little hierarchy problem and Little Higgs models
The SM is in excellent agreement with the results of particle physics experiments, in particular with the electroweak (EW) precision measurements, thus suggesting that the SM cutoff scale is at least as large as 10 TeV. Having such a relatively high cutoff, however, the SM requires an unsatisfactory fine-tuning to yield a correct ($\approx 10^2$ GeV) scale for the squared Higgs mass, whose corrections are quadratic and therefore highly sensitive to the cutoff. This "little hierarchy problem" has been one of the main motivations to elaborate models of physics beyond the SM. While Supersymmetry is at present the leading candidate, different proposals have been formulated more recently. Among them, Little Higgs models play an important role, being perturbatively computable up to about 10 TeV and with a rather small number of parameters, although their predictivity can be weakened by a certain sensitivity to the unknown UV-completion of these models (see below).

In Little Higgs models [133], the Higgs is naturally light as it is identified with a Nambu–Goldstone boson (NGB) of a spontaneously broken global symmetry. An exact NGB, however, would have only derivative interactions. Gauge and Yukawa interactions of the Higgs have to be incorporated. This can be done without generating quadratically divergent one-loop contributions to the Higgs mass, through the so-called *collective symmetry breaking*.

In the following, we restrict ourselves to product-group Little Higgs models in order not to complicate the presentation. The idea of collective symmetry breaking has also been applied to simple-group models [134, 135], however the implementation is somewhat different there. (Product-group) Little Higgs models are based on a global symmetry group G, like $G = G'^N$ in the case of moose-type models [133, 136] or $G = SU(5)$ in the case of the Littlest Higgs, that is spontaneously broken to a subgroup $H \subset G$ by the vacuum condensate of a nonlinear sigma model field Σ. A subgroup of G is gauged, which contains at least two $SU(2) \times U(1)$ factors or larger groups containing such factors. The gauge group is then broken to the SM gauge group $SU(2)_L \times U(1)_Y$ by the vacuum expectation value (vev) of Σ. The potential for the Higgs field is generated radiatively, making thus the scale of the EW symmetry breaking $v \simeq 246$ GeV a loop factor smaller than the scale f, where the breaking $G \to H$ takes place.

In order to allow for a Higgs potential being generated radiatively, interaction terms explicitly breaking the global symmetry group G have to be included as well. However, these interactions have to preserve enough of the global symmetry to prevent the Higgs potential from quadratically divergent radiative contributions. Only when two or more of the corresponding coupling constants are nonvanishing, radiative corrections are allowed. In particular, only at two or higher loop level, quadratically divergent contributions appear, but these are safely small due to the loop factor in front. This mechanism is referred to as the collective symmetry breaking.

1.4.1.2 The Littlest Higgs The most economical, in matter content, Little Higgs model is the Littlest Higgs (LH) [137], where the global group $SU(5)$ is spontaneously broken into $SO(5)$ at the scale $f \approx \mathcal{O}(1$ TeV$)$ and the EW sector of the SM is embedded in an $SU(5)/SO(5)$ nonlinear sigma model. Gauge and Yukawa Higgs interactions are introduced by gauging the subgroup of $SU(5)$: $[SU(2) \times U(1)]_1 \times [SU(2) \times U(1)]_2$, with gauge couplings respectively equal to g_1, g'_1, g_2, g'_2. The key feature for the realization of collective SB is that the two gauge factors commute with a different $SU(3)$ global symmetry subgroup of $SU(5)$, which prevents the Higgs from becoming massive when the couplings of one of the two gauge factors vanish. Consequently, quadratic corrections to the squared Higgs mass involve two couplings and cannot appear at one-loop. In the LH model, the new particles appearing at the TeV scale are the heavy gauge bosons ($W_H^\pm, Z_H, A_H$), the heavy top (T) and the scalar triplet Φ.

In the LH model, significant corrections to EW observables come from tree-level heavy gauge boson contributions and the triplet vev which breaks the custodial $SU(2)$ symmetry. Consequently, EW precision tests are satisfied only for quite large values of the NP scale $f \geq (2–3)$ TeV [138, 139] that are unable to solve the little hierarchy problem. Since the LH model belongs to the class of models with Constrained Minimal Flavor Violation (CMFV) [12], the contributions of the new particles to FCNC processes turn out to be at most (10–20)% [140–146].

1.4.1.3 T-parity Motivated by reconciling the LH model with EW precision tests, Cheng and Low [147, 148] proposed to enlarge the symmetry structure of the theory by introducing a discrete symmetry called T-parity. T-parity acts as an automorphism which exchanges the $[SU(2) \times U(1)]_1$ and $[SU(2) \times U(1)]_2$ gauge factors. The invariance of the theory under this automorphism implies $g_1 = g_2$ and $g'_1 =$

g'_2. Furthermore, T-parity explicitly forbids the tree-level contributions of heavy gauge bosons and the interactions that induced the triplet vev. The custodial $SU(2)$ symmetry is restored and the compatibility with ew precision data is obtained already for smaller values of the NP scale, $f \geq 500$ GeV [149]. Another important consequence is that particle fields are T-even or T-odd under T-parity. The SM particles and the heavy top T_+ are T-even, while the heavy gauge bosons $W_H^\pm$, Z_H, A_H and the scalar triplet Φ are T-odd. Additional T-odd particles are required by T-parity: the odd heavy top T_- and the so-called mirror fermions, i.e., fermions corresponding to the SM ones but with opposite T-parity and $\mathcal{O}(1 \text{ TeV})$ mass [150].

1.4.1.4 New flavor interactions in LHT Mirror fermions are characterized by new flavor interactions with SM fermions and heavy gauge bosons, which involve two new unitary mixing matrices, in the quark sector, analogous to the CKM matrix V_{CKM} [151, 152]. They are V_{Hd} and V_{Hu}, respectively involved when the SM quark is of down- or up-type, and satisfying $V_{Hu}^\dagger V_{Hd} = V_{\text{CKM}}$ [153]. Similarly, two new mixing matrices $V_{H\ell}$ and $V_{H\nu}$ appear in the lepton sector and are respectively involved when the SM lepton is charged or a neutrino and related to the PMNS matrix [154–156] through $V_{H\nu}^\dagger V_{H\ell} = V_{\text{PMNS}}^\dagger$. Both V_{Hd} and $V_{H\ell}$ contain 3 angles, like V_{CKM} and V_{PMNS}, but 3 (non-Majorana) phases [157], i.e. two more phases than the SM matrices, that cannot be rotated away in this case.

Therefore, V_{Hd} can be parameterized as

As the LHT model, in contrast to the LH model without T-parity, does not belong to the Minimal Flavor Violation (MFV) class of models, significant effects in flavor-violating observables both in the quark and in the lepton sector are possible. This becomes evident if one looks at the contributions of mirror fermions to the short distance functions X, Y and Z that govern rare and CP-violating K and B decays. For example, the mirror fermion contribution to be added to the SM one in the X function has the following structure [158]:

$$\frac{1}{\lambda_t^{(i)}} \left[\xi_2^{(i)} F(m_{H1}, m_{H2}) + \xi_3^{(i)} F(m_{H1}, m_{H3}) \right], \quad (25)$$

where the unitarity condition $\sum_{j=1}^{3} \xi_j^{(i)} = 0$ has been used, F denotes a function of mirror fermion masses m_{Hj} ($j = 1, 2, 3$), and $\lambda_t^{(i)}$ are the well-known combinations of CKM elements with $i = K, d, s$ standing for K, B_d and B_s systems, respectively.

It is important to note that mirror fermion contributions are enhanced by a factor $1/\lambda_t^{(i)}$ and are different for K, B_d and B_s systems, thus breaking universality. As $\lambda_t^{(K)} \simeq 4 \times 10^{-4}$, whereas $\lambda_t^{(d)} \simeq 1 \times 10^{-2}$ and $\lambda_t^{(s)} \simeq 4 \times 10^{-2}$, the deviation from the SM prediction in the K system is found to be by more than an order of magnitude larger than in the B_d system and even by two orders of magnitude larger than in the B_s system. Analogous statements are valid for the Y and Z functions.

$$V_{Hd} = \begin{pmatrix} c_{12}^d c_{13}^d & s_{12}^d c_{13}^d e^{-i\delta_{12}^d} & s_{13}^d e^{-i\delta_{13}^d} \\ -s_{12}^d c_{23}^d e^{i\delta_{12}^d} - c_{12}^d s_{23}^d s_{13}^d e^{i(\delta_{13}^d - \delta_{23}^d)} & c_{12}^d c_{23}^d - s_{12}^d s_{23}^d s_{13}^d e^{i(\delta_{13}^d - \delta_{12}^d - \delta_{23}^d)} & s_{23}^d c_{13}^d e^{-i\delta_{23}^d} \\ s_{12}^d s_{23}^d e^{i(\delta_{12}^d + \delta_{23}^d)} - c_{12}^d c_{23}^d s_{13}^d e^{i\delta_{13}^d} & -c_{12}^d s_{23}^d e^{i\delta_{23}^d} - s_{12}^d c_{23}^d s_{13}^d e^{i(\delta_{13}^d - \delta_{12}^d)} & c_{23}^d c_{13}^d \end{pmatrix}, \quad (22)$$

and a similar parameterization applies to $V_{H\ell}$.

The new flavor violating interactions involving mirror fermions contain the following combinations of elements of the mixing matrices:

$$\xi_i^{(K)} = V_{Hd}^{*is} V_{Hd}^{id}, \qquad \xi_i^{(d)} = V_{Hd}^{*ib} V_{Hd}^{id},$$
$$\xi_i^{(s)} = V_{Hd}^{*ib} V_{Hd}^{is} \quad (i = 1, 2, 3) \quad (23)$$

in the quark sector, respectively for K, B_d and B_s systems, and

$$\chi_i^{(\mu e)} = V_{H\ell}^{*ie} V_{H\ell}^{i\mu}, \qquad \chi_i^{(\tau e)} = V_{H\ell}^{*ie} V_{H\ell}^{i\tau},$$
$$\chi_i^{(\tau \mu)} = V_{H\ell}^{*i\mu} V_{H\ell}^{i\tau} \quad (24)$$

that enter the leptonic transitions $\mu \to e$, $\tau \to e$ and $\tau \to \mu$, respectively.

Other LHT peculiarities are the rather small number of new particles and parameters (the SB scale f, the parameter x_L describing T_+ mass and interactions, the mirror fermion masses and V_{Hd} and $V_{H\ell}$ parameters) and the absence of new operators in addition to the SM ones. On the other hand, one has to recall that Little Higgs models are low-energy nonlinear sigma models whose unknown UV-completion introduces a theoretical uncertainty reflected by a logarithmically enhanced cut-off dependence [142, 158] in $\Delta F = 1$ processes that receive contributions from Z-penguin and box diagrams. See [142, 158] for a detailed discussion of this issue.

1.4.1.5 Phenomenological results We conclude this section with a summary of the main results found in recent LHT phenomenological studies [153, 158–161].

In the quark sector [153, 158, 159], the most evident departures from the SM predictions are found for CP-violating observables that are strongly suppressed in the SM. These are the branching ratio for $K_L \to \pi^0 \nu \bar{\nu}$ and the CP-asymmetry $S_{\psi\phi}$ that can be enhanced by an order of magnitude relative to the SM predictions. Large departures from SM expectations are also possible for $Br(K_L \to \pi^0 \ell^+ \ell^-)$ and $Br(K^+ \to \pi^+ \nu \bar{\nu})$ and the semileptonic CP-asymmetry A_{SL}^s that can be enhanced by an order of magnitude. The branching ratios for $B_{s,d} \to \mu^+\mu^-$ and $B \to X_{s,d} \nu \bar{\nu}$, instead, are modified by at most 50% and 35%, respectively, and the effects of new electroweak penguins in $B \to \pi K$ are small, in agreement with the recent data. The new physics effects in $B \to X_{s,d} \gamma$ and $B \to X_{s,d} \ell^+ \ell^-$ turn out to be below 5% and 15%, respectively, so that agreement with the data can easily be obtained. Small but still significant effects have been found in $B_{s,d}$ mass differences. In particular, a 7% suppression of ΔM_s is possible, thus improving the compatibility with the recent experimental measurement [126, 162].

The possible discrepancy between the values of $\sin 2\beta$ following directly from $A_{CP}(B_d \to \psi K_S)$ and indirectly from the usual analysis of the unitarity triangle involving $\Delta M_{d,s}$ and $|V_{ub}/V_{cb}|$ can be cured within the LHT model thanks to a new phase $\varphi_{B_d} \simeq -5°$.

The universality of NP effects, characteristic for MFV models, can be largely broken, in particular between K and $B_{s,d}$ systems. In particular, sizable departures from MFV relations between $\Delta M_{s,d}$ and $Br(B_{s,d} \to \mu^+\mu^-)$ and between $S_{\psi K_S}$ and the $K \to \pi \nu \bar{\nu}$ decay rates are possible. Similar results have been recently obtained in a model with Z'-contributions [163].

More recently, the most interesting lepton flavor violating decays have also been studied [160, 161]. These are $\ell_i \to \ell_j \gamma$ analysed in [160, 161], and $\tau \to \mu P$ (with $P = \pi, \eta, \eta'$), $\mu^- \to e^- e^+ e^-$, the six three-body decays $\tau^- \to \ell_i^- \ell_j^+ \ell_k^-$, the rate for μ–e conversion in nuclei, and the K or B decays $K_{L,S} \to \mu e$, $K_{L,S} \to \pi^0 \mu e$, $B_{d,s} \to \mu e$, $B_{d,s} \to \tau e$ and $B_{d,s} \to \tau \mu$ are studied in [161]. It was found that essentially all the rates considered can reach or approach present experimental upper bounds [164]. In particular, in order to suppress the $\mu \to e \gamma$ and $\mu^- \to e^- e^+ e^-$ decay rates and the μ–e conversion rate below the experimental upper bounds, the $V_{H\ell}$ mixing matrix has to be rather hierarchical, unless mirror leptons are quasi-degenerate. One finds [161] that the pattern of the branching ratios for LFV processes differs significantly from the one encountered in supersymmetry [165–167]. This is welcome as the distinction between supersymmetry and LHT models will be nontrivial in high-energy collider experiments. Finally, the muon anomalous magnetic moment $(g-2)_\mu$ has also been considered [160, 161], finding the result $a_\mu^{LHT} < 1.2 \times 10^{-10}$, even for the scale f as low as 500 GeV. This value is roughly a factor 5 below the current experimental uncertainty, implying that the possible discrepancy between the SM prediction and the data cannot be solved in the LHT model.

1.4.2 Universal extra dimensions

Since the work of Kaluza and Klein [168, 169] models with more than three spatial dimensions often have been used to unify the forces of nature. More recently, inspired by string theory, extra-dimensional models have been proposed to explain the origin of the TeV scale [170–179].

A simple extension of the SM including additional space dimensions is the ACD model [180] with one universal extra dimension (UED). Here all the SM fields are democratically allowed to propagate in a flat extra dimension compactified on an orbifold S^1/Z_2 of size 10^{-18} m or smaller. In general UED models, there can also be contributions from terms residing at the boundaries. Generically, these terms would violate bounds from flavor and CP violation. To be consistent with experiment, we will assume the minimal scenario where these terms vanish at the cut-off scale. The only additional free parameter then compared to the SM is the compactification scale $1/R$. Thus, all the tree level masses of the KK particles and their interactions among themselves and with the SM particles can be described in terms of $1/R$ and the parameters of the SM. In the effective four-dimensional theory, there are, in addition to the ordinary SM particles, denoted as zero ($n=0$) modes, corresponding infinite towers of KK modes ($n \geq 1$) with masses $m_{(n),KK}^2 = m_0^2 + m_n^2$, where $m_n = n/R$, and m_0 is the mass of the zero mode.

A very important property of UEDs is the conservation of KK parity that implies the absence of tree level KK contributions to low-energy processes taking place at scales $\mu \leq 1/R$. Therefore the flavor-changing neutral current (FCNC) processes like particle–anti-particle mixing, rare K and B decays and radiative decays are of particular interest. Since these processes first appear at one-loop in the SM and are strongly suppressed, the one-loop contributions from the KK modes to them could in principle be important. Also, due to conservation of KK parity, the GIM mechanism significantly improves the convergence of the sum over KK modes and thus removes the sensitivity of the calculated branching ratios to the scale $M_s \gg 1/R$ at which the higher-dimensional theory becomes nonperturbative and at which the towers of the KK particles must be cut off in an appropriate way. Since the low-energy effective Hamiltonians are governed by local operators already present in the SM and the flavor and CP violation in this model is entirely governed by the SM Yukawas, the UED model belongs to the class of models with CMFV [10, 12]. This has automatically the following important consequence for the FCNC processes considered in [17, 181–183]: the impact of the KK modes on the processes in question amounts only to the modification of the Inami–Lim one-loop functions [184], i.e. each function which in the SM depends only on m_t now also becomes a function of $1/R$:

$$F(x_t, 1/R) = F_0(x_t) + \sum_{n=1}^{\infty} F_n(x_t, x_n),$$

$$x_t = \frac{m_t^2}{m_W^2}, \ x_n = \frac{m_n^2}{m_W^2}. \tag{26}$$

1.5 Tools for flavor physics and beyond

1.5.1 Tools for flavor physics

An increasing number of calculations of flavor (related) observables is appearing, including more and more refined approaches and methods. It is desirable to have these calculations in the form of computer codes at hand. This allows us to easily use the existing knowledge for checks of the parameters/models for a phenomenological/experimental analysis or to check an independent calculation.

As a first step in this direction, we present here a collection of computer codes connected to the evaluation of flavor related observables. (A different class of codes, namely fit codes for the CKM triangle, are presented later in Sect. 1.5.3.) Some of these codes are specialized to the evaluation of a certain restricted set of observables at either low or high energies (the inclusion of codes for high-energy observables is motivated by the idea of testing a parameter space from both sides, i.e. at flavor factories and at the LHC). Others tools are devoted to the evaluation of particle spectra including NMFV effects of the MSSM or the 2HDM. Some codes allow the (essentially) arbitrary calculation of one-loop corrections including flavor effects. Finally tools are included that facilitate the hand-over of flavor parameters and observables. Following the general idea of providing the existing knowledge to the community, only codes that either are already publicly available or will become available in the near future are included. In order to be useful for the high-energy physics community, it is mandatory that the codes provide a minimum of user friendliness and support.

As a second step, it would be desirable to connect different codes (working in the same model) to each other. This could go along the lines of the SUSY Les Houches Accord [185, 186], i.e. to define a common language, a common set of input parameters. It would require the continuous effort of the various authors of the codes to comply with these definitions. Another, possibly simpler approach is to implement the tools as sub-routines, called by a master code that takes care of the correct definition of the input parameters. This is discussed in more detail in Sect. 1.5.2. It will facilitate the use of the codes also for nonexperts.

An overview of the available codes is given in Table 4. To give a better idea of the properties of each code, we also provide a list summarizing the authors, a short description, the models included, the input and output options, as well as the available literature:

1. no name

    Authors: M. Ciuchini et al. [107, 116, 187]
    Description: calculation of K–$\bar{K}$ mixing, $B_{(s)}$–$\bar{B}_{(s)}$ mixing, $b \to s\gamma$, $b \to sl^+l^-$
    Models: NMFV MSSM
    Input: electroweak-scale soft SUSY-breaking parameters
    Output: see Description, no special format
    Availability: available from the authors in the near future

Table 4 Overview about codes for the evaluation of flavor related observables; av. ≡ availability: + = available, o = planned

Name	Short description	av.
1. no name	K–$\bar{K}$ mixing, $B_{(s)}$–$\bar{B}_{(s)}$ mixing, $b \to s\gamma$, $b \to sl^+l^-$ in NMFV MSSM	o
2. no name	B physics observables in the MFV MSSM	+
3. no name	rare B and K decays in/beyond SM	o
4. SusyBSG	$B \to X_s\gamma$ in MSSM with MFV	+
5. no name	FCNC observables in MSSM	o
6. no name	FC Higgs/top decays in 2HDM I/II	o
7. no name	squark/gluino production at LO for NMFV MSSM	+
8. FeynHiggs	Higgs phenomenology in (NMFV) MSSM	+
9. FCHDECAY	FCNC Higgs decays in NMFV MSSM	+
10. FeynArts/FormCalc	(arbitrary) one-loop corrections in NMFV MSSM	+
11. SLHALib2	read/write SLHA2 data, i.e. NMFV/RPV/CPV MSSM, NMSSM	+
12. SoftSUSY	NMFV MSSM parameters from GUT scale input	+
13. SPheno	NMFV MSSM parameters from GUT scale input	+

2. no name

   Authors: G. Isidori, P. Paradisi [32]
   Description: calculation of B-physics observables
   Models: MFV MSSM
   Input: electroweak-scale soft SUSY-breaking parameters
   Output: see Description, no special format
   Availability: available from the authors upon request

3. no name

   Authors: C. Bobeth, T. Ewerth, U. Haisch [188–190]
   Description: calculation of BR's, F/B asymmetries for rare B and K decays (in/exclusive)
   Models: SM, SUSY, CMFV
   Input: SM parameters, SUSY masses, scales
   Output: see Description, no special format
   Availability: available from the authors in the near future

4. SusyBSG

   Authors: G. Degrassi, P. Gambino, P. Slavich [191]
   Description: Fortran code for $B(B \to X_s \gamma)$
   Models: SM, MSSM with MFV
   Input: see manual (SLHA(2) compatible)
   Output: see Description, no special format
   Availability: cern.ch/slavich/susybsg/home.html, manual available

5. no name

   Authors: P. Chankowski, S. Jäger, J. Rosiek [192]
   Description: calculation of various FCNC observables in the MSSM (computes 2-, 3-, 4-point Greens functions that can be used as building blocks for various amplitudes)
   Models: MSSM
   Input: MSSM Lagrangian parameters in super CKM basis (as in SLHA2)
   Output: see Description, no special format
   Availability: available from the authors in the near future

6. no name

   Authors: S. Bejar, J. Guasch [193–195]
   Description: calculation of FC decays: $\phi \to tc$, $\phi \to bs$, $t \to c\phi$ ($\phi = h, H, A$)
   Models: 2HDM type I/II (with λ_5, λ_6)
   Input: similar to SLHA2 format
   Output: similar to SLHA2 format
   Availability: available from the authors in the near future

7. no name

   Authors: G. Bozzi, B. Fuks, M. Klasen
   Description: SUSY CKM matrix determination through squark- and gaugino production at LO
   Models: NMFV MSSM
   Input: MSSM spectrum as from SUSPECT (SLHA2 compliant)
   Output: cross section (and spin asymmetry, in case) as functions of CKM parameters
   Availability: from the authors upon request

8. FeynHiggs

   Authors: S. Heinemeyer, T. Hahn, W. Hollik, H. Rzehak, G. Weiglein [199–201]
   Description: Higgs phenomenology (masses, mixings, cross sections, decay widths)
   Models: (N)MFV MSSM, CPV MSSM
   Input: electroweak-scale soft SUSY-breaking parameters (SLHA(2) compatible)
   Output: Higgs masses, mixings, cross sections, decay widths (SLHA(2) output possible)
   Availability: www.feynhiggs.de, manual available

9. FCHDECAY

   Authors: S. Bejar, J. Guasch [196–198]
   Description: $BR(\phi \to bs, tc)$ ($\phi = h, H, A$), $BR(b \to s\gamma)$, masses, mixing matrices
   Models: NMFV MSSM
   Input: via SLHA2
   Output: via SLHA2
   Availability: fchdecay.googlepages.com, manual available

10. FeynArts/FormCalc

    Authors: T. Hahn [202–204]
    Description: Compute (essentially) arbitrary one-loop corrections
    Models: NMFV MSSM, CPV MSSM
    Input: Process definition
    Output: Fortran code to compute e.g. cross-sections can be linked with SLHALib2 to obtain data from other codes
    Availability: www.feynarts.de, www.feynarts.de/formcalc, manual available

11. SLHALib2

    Authors: T. Hahn [185, 205]
    Description: read/write SLHA2 data
    Models: NMFV MSSM, RPV MSSM, CPV MSSM, NMSSM
    Input: SLHA2 input file
    Output: SLHA2 output file in the SLHA2 record
    Availability: www.feynarts.de/slha, manual available

12. SoftSUSY

    Authors: B. Allanach [206]
    Description: evaluates NMFV MSSM parameters from GUT scale input

Models: NMFV MSSM
Input: SLHA2 input file
Output: SLHA2 output file
Availability: hepforge.cedar.ac.uk/softsusy, manual available

13. Spheno

Authors: W. Porod [207]
Description: evaluates NMFV MSSM parameters from GUT scale input and some flavor obs.
Models: NMFV MSSM
Input: SLHA2 input file
Output: SLHA2 output file
Availability: ific.uv.es/~porod/SPheno.html, manual available

1.5.2 Combination of flavor physics and high-energy tools

It is desirable to connect different codes (e.g. working in the (N)MFV MSSM, as given in the previous subsection) to each other. Especially interesting is the combination of codes that provide the evaluation of (low-energy) flavor observables and others that deal with high-energy (high p_T) calculations for the same set of parameters. This combination would allow one to test the ((N)MFV MSSM) parameter space with the results from flavor experiments as well as from high-energy experiments such as ATLAS or CMS.

A relatively simple approach for the combination of different codes is their implementation as sub-routines, called by a "master code". This master code takes care of the correct definition of the input parameters for the various subroutines. This would enable e.g. experimentalists to test whether the parameter space under investigation is in agreement with various existing experimental results from both flavor and high-energy experiments.

A first attempt to develop such a "master code" has recently been started [208]. So far the flavor physics code (2) [32] and the more high-energy observable oriented code FeynHiggs [199–201] have been implemented as subroutines. The inclusion of further codes is foreseen in the near future (see [1122] for the latest developments).

The application and use of the master code would change once experimental data showing a deviation from the SM predictions is available. This can come either from the ongoing flavor experiments or latest (hopefully) from ATLAS and CMS. If such a "signal" appears at the LHC, it has to be determined to which model and to which parameters within a model it can correspond. Instead of checking parameter points (to be investigated experimentally) for their agreement with experimental data, now a scan over a chosen model could be performed. Using the master code with its subroutines, each scan point can be tested against the "signal", and preferred parameter regions can be obtained using

a χ^2 evaluation. It is obvious that the number of evaluated observables has to be as large as possible, i.e. the number of subroutines (implemented codes) should be as large as possible.

1.5.3 Fit tools

The analysis of the CKM matrix or the Unitarity Triangle (UT) requires to combine several measurements in a consistent way in order to bound the range of relevant parameters.

1.5.3.1 The UTfit package The first approach derives bounds on the parameters $\bar{\rho}$ and $\bar{\eta}$ determining the UT. The various observables, in particular ϵ_K, which parameterizes CP violation in the neutral kaon sector, the sides of the UT $|V_{ub}/V_{cb}|$, Δm_d, $\Delta m_d/\Delta m_s$ and the angles β, α and γ can be expressed as functions of $\bar{\rho}$ and $\bar{\eta}$, hence their measurements individually define probability regions in the $(\bar{\rho}, \bar{\eta})$ plane. Their combination can be achieved in a theoretically sound way in the framework of the Bayesian approach [132].

Each of the functions relates a constraint c_j (where c_j stands for ε_K, $|V_{ub}/V_{cb}|$, etc.) to $\bar{\rho}$ and $\bar{\eta}$ via a set of parameters $\mathbf{x}$, where $\mathbf{x} = \{x_1, x_2, \ldots, x_N\}$ stands for all experimentally determined or theoretically calculated quantities on which the various c_j depend,

$$c_j = c_j(\bar{\rho}, \bar{\eta}; \mathbf{x}). \tag{27}$$

The quantities c_j and $\mathbf{x}$ are affected by several uncertainties, which must be properly taken into account. The final p.d.f. obtained starting from a flat distribution of $\bar{\rho}$ and $\bar{\eta}$ is

$$f(\bar{\rho}, \bar{\eta}) \propto \int \prod_{j=1,M} f_j(\hat{c}_j|\bar{\rho}, \bar{\eta}, \mathbf{x}) \prod_{i=1,N} f_i(x_i) \, dx_i. \tag{28}$$

The integration can be done by Monte Carlo methods. There are several ways to implement a Monte Carlo integration, using different techniques to generate events.

The UTfit Collaboration has developed a software package, written in C++, that implements such a Bayesian Monte Carlo approach with the aim of performing the UT analysis. A considerable effort has been spent in order to achieve an optimal Monte Carlo generation efficiency. All the recent analyses published by the Collaboration are based on this package [7, 120, 209–211].

The UTfit code includes an interface to import job options from a set of configuration files, an interface for storing the relevant p.d.f.s inside ROOT histograms [212], tools for generating input quantities, the p.d.f.s of which cannot be expressed in simple analytical form but must be numerically defined—e.g. the current measurements of α and γ—and tools for plotting one-dimensional p.d.f.s and two-dimensional probability regions in the $(\bar{\rho}, \bar{\eta})$ plane. The

UTfit code can be easily re-adapted to solve any kind of statistical problem that can be formalized in a Bayesian inferential framework.

1.5.3.2 The CKMFitter package Another, somewhat different approach is followed by CKMFitter, an international group of experimental and theoretical particle physicists. Its goal is the phenomenology of the CKM matrix by performing a global analysis:

- within the SM, by quantifying the agreement between the data and the theory as a whole;
- within the SM, by achieving the best estimate of the theoretical parameters and the not yet measured observables;
- within an extended theoretical framework, e.g. SUSY, by searching for specific signs of new physics by quantifying the agreement between the data and the extended theory and by pinning down additional fundamental and free parameters of the extended theory.

The CKMfitter package is entirely based on the frequentest approach. The theoretical uncertainties are modeled as allowed ranges (Rfit approach) and no other a priori information is assumed where none is available. More detailed information is provided in Ref. [8] and on the CKMfitter website [213].

The source code of the CKMfitter package consists of more than 40,000 lines of Fortran code and 2000 lines of C++ code. It is publicly available on the CKMfitter website. Over the years, the fit problems became more and more complex, and the CPU time consumption increased. The global fit took about 20 hours (on one CPU). A year ago, it was decided to move to Mathematica [gain: analytical vs. numerical methods]: the global fit takes now 12 minutes. For the plots, we moved also from PAW with kumac macros to ROOT.

2 Weak decays of hadrons and QCD[6]

2.1 Overview

QCD interactions, both at short and long distances, necessarily modify the amplitudes of quark flavor processes. These interactions need to be computed sufficiently well in order to determine the parameters and mechanisms of quark flavor physics from the weak decays of hadrons observed in experiment. The standard framework is provided by the effective weak Hamiltonians

$$\mathcal{H}_{\text{eff}} \sim \sum_i C_i Q_i, \qquad (29)$$

---
[6]Section coordinator: G. Buchalla.

based on the operator product expansion and the renormalization group method. The Wilson coefficients C_i include all relevant physics from the highest scales, such as the weak scale M_W or some new physics scale, down to the appropriate scale of a given process, such as m_b for B-meson decays. This part is theoretically well under control. Theoretical uncertainties are dominated by the hadronic matrix elements of local operators Q_i. Considerable efforts are therefore devoted to calculate, estimate, eliminate or at least constrain such hadronic quantities in flavor physics applications.

This section reviews the current status of theoretical methods to treat the strong interaction dynamics in weak decays of flavored mesons, with a particular emphasis on B physics. Specific aspects of D-meson physics will be discussed in Sect. 3.9, kaons will be considered in Sect. 3.8.

The theory of charmless two-body B decays and the concept of factorization are reviewed in Sect. 2.2. The status of higher-order perturbative QCD calculations in this field is described. Universal properties of electromagnetic radiative effects in two-body B decays, which influence precision studies and isospin relations, are also discussed here. Factorization in the heavy-quark limit simplifies the matrix elements of two-body hadronic B decays considerably. In this framework, certain nonperturbative input quantities, for instance B-meson transition form factors, are in general still required. QCD sum rules on the light cone (LCSR) provide a means to compute heavy-to-light form factors at large recoil ($B \to \pi$, $B \to K^*$, etc.). The results have applications for two-body hadronic as well as rare and radiative B-meson decays. This subject is treated in Sect. 2.3. Complementary information can be obtained from lattice QCD, a general approach based on first principles, to compute nonperturbative parameters of interest to quark flavor physics. Decay constants and form factors (at small recoil) are among the most important quantities. Uncertainties arise from the limitations of the practical implementations of lattice QCD. A critical discussion of this topic and a summary of results can be found in Sect. 2.4.

2.2 Charmless two-body B decays

2.2.1 Exclusive decays and factorization

The calculation of branching fractions and CP asymmetries for charmless two-body B decays is rather involved, due to the interplay of various short- and long-distance QCD effects. Most importantly, the hadronic matrix elements of the relevant effective Hamiltonian $\mathcal{H}_{\text{eff}}^{\Delta B=1}$ [214] cannot readily be calculated from first principles. The idea of factorization is to disentangle short-distance QCD dynamics from genuinely nonperturbative hadronic effects. In order to quantify the hadronic uncertainties resulting from this procedure we have to

- establish a factorization formula in quantum field theory,
- identify and estimate the relevant hadronic input parameters.

2.2.1.1 Basic concepts of factorization We consider generic charmless B decays into a pair of mesons, $B \to M_1 M_2$, where we may think of $B \to \pi\pi$ as a typical example. The operators Q_i in the weak Hamiltonian can be written as the local product of quark currents (and electro- or chromomagnetic field strength tensors), generically denoted as $J_i^{a,b}$. In naive factorization, one assumes that also on the hadronic level the matrix element can be written as a product,

$$C_i(\mu)\langle M_1 M_2 | Q_i | B\rangle$$
$$\approx C_i(\mu)\langle M_1 | J_i^a | B\rangle \langle M_2 | J_i^b | 0\rangle + (M_1 \leftrightarrow M_2), \quad (30)$$

where $C_i(\mu)$ are Wilson coefficients, and the two matrix elements (if not zero) define the $B \to M$ form factor and the decay constant of M, respectively. The naive factorization formula (30) cannot be exact, because possible QCD interactions between M_2 and the other hadrons are neglected. On the technical level, this is reflected by an unmatched dependence on the factorization scale μ.

In order to better understand the internal dynamics in the $B \to M_1 M_2$ transition, it is useful to classify the external degrees of freedom according to their typical momentum scaling in the B-meson rest frame:

heavy b quark: $p_b \simeq m_b(1, 0_\perp, 0)$,

constituents of M_1: $p_{c1} \simeq u_i m_b/2(1, 0_\perp, +1)$,

soft spectators: $p_s \sim \mathcal{O}(\Lambda)$,

constituents of M_2: $p_{c2} \simeq v_i m_b/2(1, 0_\perp, -1)$,

where Λ is a typical hadronic scale of the order of a few 100 MeV. The index $\perp$ denotes the directions in the plane transverse to the two pion momenta, and u_i, v_i are momentum fractions satisfying $0 \leq u_i, v_i \leq 1$. Interactions of particles with momenta p_1 and p_2 imply internal virtualities of order $(p_1 \pm p_2)^2$. In Table 5, we summarize the situation for the possible interactions between the B-meson and pion constituents. We observe the emergence of two kinds of short-distance modes,

Table 5 External momentum configurations and their interactions in $B \to M_1 M_2$

	heavy	soft	coll$_1$	coll$_2$
heavy	–	heavy	hard	hard
soft	heavy	soft	hard-coll$_1$	hard-coll$_2$
coll$_1$	hard	hard-coll$_1$	coll$_1$	hard
coll$_2$	hard	hard-coll$_2$	hard	coll$_2$

- hard modes with invariant mass of order m_b,
- hard-collinear modes with energies of order $m_b/2$ and invariant mass of order $\sqrt{\Lambda m_b}$.

The systematic inclusion of these effects requires a simultaneous expansion in Λ/m_b and α_s. The leading term in the Λ/m_b expansion can be written as [215, 216]

$$\langle M_1 M_2 | Q_i | B\rangle$$
$$= F^{BM_1} f_{M_2} \int dv\, T_i^{\mathrm{I}}(v)\phi_{M_2}(v) + (M_1 \leftrightarrow M_2)$$
$$+ \hat{f}_B f_{M_1} f_{M_2} \int d\omega\, du\, dv\, T_i^{\mathrm{II}}(u,v,\omega)$$
$$\times \phi_{B_+}(\omega)\phi_{M_1}(u)\phi_{M_2}(v). \quad (31)$$

The functions ϕ_M and ϕ_{B_+} denote process-independent light-cone distribution amplitudes (LCDA) for light and heavy mesons, respectively, f_M, $\hat{f}_B$ are the corresponding decay constants, and F^{BM} is a $B \to M$ QCD form factor at $q^2 = 0$. These quantities constitute the hadronic input. The coefficient function T_i^{I} contains the effects of hard-vertex corrections as in Fig. 5(b). $T_i^{\mathrm{II}} = \mathcal{O}(\alpha_s)$ describes the hard and hard-collinear spectator interactions as in Fig. 5(c). The explicit scale dependence of the hard and hard-collinear short-distance functions T_i^{I}, T_i^{II} matches the one from the Wilson coefficients and the distribution amplitudes. The formula (31) holds for light flavor-nonsinglet pseudoscalars or longitudinally polarized vectors up to $1/m_b$ power corrections which do not, in general, factorize. Naive factorization, Fig. 5(a), is recovered in the limit $\alpha_s \to 0$ and $\Lambda/m_b \to 0$, in which T_i^{I} reduces to 1.

2.2.1.2 QCD factorization and soft-collinear effective theory (SCET) The factorization formula (31) can also be understood in the context of an effective theory for soft-collinear interactions (SCET), see for instance Refs. [217–219, 221]. Here the short-distance functions $T_i^{\mathrm{I,II}}$ arise as matching coefficients between QCD and the effective theory. The effective theory for $B \to M_1 M_2$ decays is constructed in two steps. As a consequence, the short-distance function T_i^{II} can be further factorized into a hard coefficient H_i^{II} and a hard-collinear jet function J:

$$T_i^{\mathrm{II}}(u,v,\omega) = \int dz\, H_i^{\mathrm{II}}(v,z) J(z,u,\omega). \quad (32)$$

H_i^{II} and J comprise (respectively) the contributions associated with the hard scale $\mu_b \sim m_b$ and the hard-collinear scale $\mu_{\mathrm{hc}} \sim \sqrt{m_b \Lambda}$ from Feynman diagrams that do involve the spectator and cannot be absorbed into F^{BM}. The effective theory can be used to determine the hard-collinear contributions and to resume, if desired, parametrically large logarithms $\ln \mu_b/\mu_{\mathrm{hc}}$ by renormalization group methods. We emphasize that the theoretical basis for the (diagrammatic)

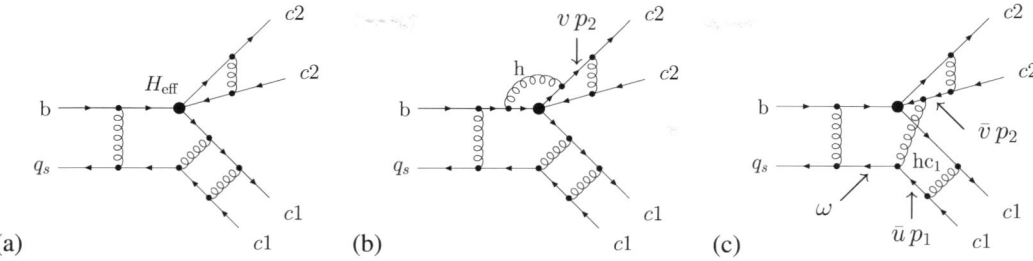

Fig. 5 Sample diagrams for QCD dynamics in $B \to M_1 M_2$ transition: (**a**) naive factorization, (**b**) vertex correction, sensitive to the momentum fraction v of collinear quarks inside the emitted pion, (**c**) spectator interactions, sensitive to the momenta of collinear quarks in both pions and of the soft spectator in the B-meson

factorization approach and SCET is *the same*. The factorization formula (31) was originally derived by a power-counting analysis of momentum regions of QCD Feynman diagrams and the resulting convolutions [215, 216]. However, in SCET the formulation of factorization proofs, the classification of power corrections of order Λ/m_b, the emergence of approximate symmetries, etc. may be more transparent [220, 221].

2.2.1.3 QCD factorization vs. "pQCD approach"
The so-called "pQCD approach" [222] follows an alternative approach to understand the strong dynamics in charmless B-decays. In contrast to QCD factorization, where the B meson form factors as well as a certain class of power corrections are identified as "nonfactorizable" quantities of order $(\alpha_s)^0$, the pQCD approach describes all contributions to the hadronic matrix elements in terms of $\mathcal{O}(\alpha_s)$ hard-scattering kernels and nonperturbative wave functions. This is achieved by introducing additional infrared prescriptions which include an exponentiation of Sudakov logarithms and a phenomenological model for transverse momentum effects. The discussion of parametric and systematic theoretical uncertainties in the pQCD approach is more difficult, because a complete NLO (i.e. $\mathcal{O}(\alpha_s^2)$) analysis of nonfactorizable effects has not yet been performed and because independent information on the hadronic input functions is not available. We will therefore not attempt a detailed review here, but instead refer to a recent phenomenological analysis [223] for details.

2.2.2 Theoretical uncertainties

2.2.2.1 Status of perturbative calculations
The calculation of the coefficient functions $T_i^{\text{I,II}}$ in SCET involves the determination of perturbative matching coefficients as well as of anomalous dimensions for effective-theory operators. The matching coefficients at order α_s have been calculated in the original BBNS papers [215, 224]. The 1-loop jet function entering T_i^{II} has been determined in [225–228]. NLO results for the spectator scattering function at order α_s^2 have been reported in [229] and will be further discussed in Sect. 2.2.3 below. One important outcome of these investigations is that the perturbative expansion at the hard-collinear scale seems to be reasonably well behaved, and the uncertainty associated with the factorization-scale dependence is under control.

2.2.2.2 Hadronic input from nonperturbative methods
Most of the theoretical information on B-meson form factors (at large recoil) and light-cone distribution amplitudes comes from the QCD sum rule approach, see Ref. [230] and references therein for a review. State-of-the-art predictions for decays into light pseudoscalars or vector mesons can be found in Refs. [231–233] and Sect. 2.3. Typically one finds (15–20)% uncertainties for form factors at $E = E_{\text{max}}$ and the $1/u$ moment of distribution amplitudes. Recently, an alternative procedure has been proposed [234] (see also Refs. [235, 236]), where sum rules are derived *within* SCET at the hard-collinear scale. In particular, this approach allows us to separate the "soft" contribution to B-meson form factors, which is found to be dominating over the spectator-scattering term.

Information on the light-cone distribution amplitude of the B-meson is encoded in the phenomenologically relevant moments

$$\lambda_B^{-1} \equiv \langle \omega^{-1} \rangle_B \equiv \int_0^\infty \frac{d\omega}{\omega} \phi_B(\omega, \mu),$$
$$\sigma_B^{(n)} \langle \omega^{-1} \rangle_B \equiv \int_0^\infty \frac{d\omega}{\omega} \ln^n \left[\frac{\mu}{\omega}\right] \phi_B(\omega, \mu). \quad (33)$$

A recent OPE analysis [237] finds $\lambda_B^{-1} = (2.09 \pm 0.24)$ GeV^{-1} and $\sigma_B^{(1)} = 1.61 \pm 0.09$ at $\mu = 1$ GeV. Similar results, with somewhat larger uncertainties, have been obtained from sum rules in Ref. [238].

2.2.2.3 BBNS approach vs. BPRS approach
So far, we have only considered the leading term in the $1/m_b$ expansion. Comparison with experimental data as well as (model-dependent) estimates show that for certain decay topologies,

power corrections may not be negligible. Different options for dealing with these (nonfactorizable) contributions lead to some ambiguity in the phenomenological analyses. The two main players are the "BBNS approach" [224, 239, 240] and the "BPRS approach" [221, 241]. A qualitative comparison of the different assumptions is given in Table 6. For more details, see Sect. 2.2.3, the original publications and the controversial discussion in [242].

The main obstacle in this context is the quantitative explanation of strong phases from final-state rescattering effects. The factorization formula predicts these phases to be either perturbative (and calculable) or power-suppressed. This qualitative picture has also been confirmed by a recent sum-rule analysis [243, 244]. However, a model-independent approach to calculate the genuinely nonperturbative rescattering effects is still lacking.

2.2.2.4 Flavor symmetries It is known for a long time (see for instance [245–247]) that approximate flavor symmetries in QCD can be used to relate branching fractions and CP asymmetries in different hadronic decay channels. In this way, the hadronic parameters can be directly extracted from experiment. For instance, in case of $B \to \pi\pi, \pi\rho, \rho\rho$ decays, the isospin analysis provides a powerful tool to constrain the CKM angle α in the SM (see Ref. [248] for a recent discussion). Isospin violation from the small quark mass difference $m_u - m_d$ and QED corrections are usually negligible. Still one has to keep in mind that long-distance radiative QED effects can be enhanced by large logarithms $\ln M_B/E_\gamma$ and compete with short-distance isospin violation from electroweak penguin operators in $\mathcal{H}_{\text{eff}}$. For instance, it has recently been shown [249] (see Sect. 2.2.4 below) that the inclusion of soft photon radiation in charged $B \to \pi\pi, \pi K$ decays can give up to 5% corrections, depending on the experimental cuts. Including hadronic states with strange quarks, one can use flavor-$SU(3)$ to get even more constraints. In general, one expects corrections to the symmetry limit to be not larger than 30% (with the possible exception of potentially large differences in nonperturbative rescattering phases), see for instance the sum-rule analysis in [250]. In the long run, one should also aim to constrain first-order $SU(3)$ corrections directly from experimental data.

2.2.3 NNLO QCD corrections

NNLO QCD corrections to the heavy-quark expansion of hadronic matrix elements for two-body charmless hadronic B-decays can be phenomenologically relevant and are important to assess the validity and perturbative stability of the factorization framework. This section gives a concise account of available results and their phenomenological impact.

2.2.3.1 Hard and hard-collinear matching coefficients
The hard coefficients T_i^{I} and H_i^{II} introduced in Sect. 2.2.1 (see (31) and (32)) are found by matching the leading momentum dependence of (respectively) QCD four- and five-point functions with a Q_i insertion to operators in SCET$_{\text{I}}$ given by products of a light (anti-)collinear quark bilinear and a heavy-light current. Schematically,

$$Q_i = \int dt\, T_i^{\text{I}}(t) \big[\bar{\chi}(tn_-)\chi(0)\big]\bigg[C_{A0}\big[\bar{\xi}(0)h_v(0)\big]$$
$$+ \frac{1}{m_b}\int ds\, C_{B1}(s)\big[\bar{\xi}(0)D_{\perp\text{hc}1}(sn_+)h_v(0)\big]\bigg]$$
$$+ \frac{1}{m_b}\int dt\, ds\, H_i^{\text{II}}(t,s)\big[\bar{\chi}(tn_-)\chi(0)\big]$$
$$\times \big[\bar{\xi}(0)D_{\perp\text{hc}1}(sn_+)h_v(0)\big], \qquad (34)$$

where certain Wilson lines and Dirac structures have been suppressed. The particular choice of heavy-light current in the first line is designed to reproduce the full QCD (not SCET) form factors; other choices of operator basis as, for instance, in the "SCET approach" [221], simply result in a reshuffling of contributions between the T_i^{I} and H_i^{II} terms. The product structure of either term together with the absence of soft-collinear interactions from the SCET$_{\text{I}}$ Lagrangian at leading power suggests factorization of both terms' hadronic matrix elements into a light-cone distribution amplitude $\langle M_2|[\bar{\chi}\chi]|0\rangle \propto \phi_{M_2}$ and (respectively) the QCD form factor F^{BM_1} and a SCET$_{\text{I}}$ nonlocal "form factor" $\Xi^{BM_1}(s)$ [251]. This expectation is indeed borne out by the finiteness of the convolutions, found in all available computations.

Table 6 Comparison of different phenomenological assumptions in BBNS and BPRS approaches

	BBNS	BPRS
Charm penguins	included in hard functions	left as complex fit parameter Δ_P
Spectator term	perturbative factorization	fit to data (two real-valued quantities ζ and ζ_J)
Ext. hadronic input	form factor and LCDA (different scenarios)	LCDA for light meson
Power corrections	model-dependent estimate (complex functions X_A and X_H)	part of systematic uncertainties

The jet function J (see (32)) arises in matching the $B1$-type current from SCET_I onto SCET_II and is known to NLO [225–228]. This matching takes the form (in position space)

$$\int d^4 x\, T\bigl(\mathcal{L}_{\text{SCET}_\text{I}}^{(1)}(x)\bigl[\bar{\xi}(0) D_{\perp\text{hc1}}(sn_+) h_v(0)\bigr]\bigr)$$
$$= \int dw\, dr\, J(s,r,w)\bigl[\bar{\xi}(rn_+)\xi(0)\bigr]\bigl[\bar{q}_s(wn_-)h_v(0)\bigr], \qquad (35)$$

where we again have suppressed Dirac structures and Wilson lines. Fourier transforming with respect to s, r, w results in $J(z, u, \omega)$ entering (31).

At leading power, all one-loop corrections to H_i^II and J and part of the two-loop contributions to T_i^I are now available. The current–current corrections to H_i^II for the $V - A \times V - A$ operators ($i = 1, 2$) have been found in Refs. [229, 252–254]. The imaginary parts of the corresponding two-loop contributions to T_i^I have been computed in Refs. [255, 256]. These are sufficient to obtain the topological tree amplitudes a_1 and a_2, involving the large Wilson coefficients $C_1 \sim 1.1$ and $C_2 \sim -0.2$ at NNLO up to an $\mathcal{O}(\alpha_s^2)$ correction to the real part of T_i^I. In particular, the imaginary part of $a_{1,2}$ is now fully known at $\mathcal{O}(\alpha_s^2)$. As it is first generated at $\mathcal{O}(\alpha_s)$, this represents a first step towards an NLO prediction of direct CP asymmetries in QCD factorization. Spectator-scattering corrections from the remaining $V - A \times V + A$ operators, as well as penguin contractions and magnetic penguin insertions, have been computed in Ref. [257]. Together they constitute the QCD penguin amplitudes a_4^p ($p = u, c$) and the colour-allowed and colour-suppressed electroweak penguin amplitudes $a_9^p \pm a_7^p$ and a_{10}^p, where the sign in front of a_7^p depends on the spins of the final-state mesons, and certain numerically enhanced power corrections ($a_{6,8}^p$, annihilation, etc.) are omitted (see, however, Sect. 2.2.3.2).

2.2.3.2 Phenomenological impact and final remarks Numerical estimates of the a_i and their uncertainties require estimating $1/m_b$ corrections, some of which are "chirally enhanced" for pseudoscalars in the final state. Of these, the scalar penguin a_6^p, and its electroweak analog a_8^p, happen to factorize at $\mathcal{O}(\alpha_s)$. NNLO corrections are not known, and their factorization is an open question. Here we use the known $\mathcal{O}(\alpha_s)$ results. Annihilation and twist-3 spectator interactions do not factorize already at LO ($\mathcal{O}(\alpha_s)$). The former are not included in any a_i but enter the physical decay amplitudes. The latter have flavor structure identical to the a_i and are by convention included as estimates. For the colour-allowed and colour-suppressed tree amplitudes a_1 and a_2, we find

$$a_1(\pi\pi) = 1.015 + [0.025 + 0.012i]_\text{V} + [? + 0.027i]_\text{VV}$$
$$- \left[\frac{r_\text{sp}}{0.485}\right]\bigl\{[0.020]_\text{LO} + [0.034 + 0.029i]_\text{HV}$$
$$+ [0.012]_\text{tw3}\bigr\}$$
$$= 0.975^{+0.034}_{-0.072} + \bigl(0.010^{+0.025}_{-0.051}\bigr)i, \qquad (36)$$

$$a_2(\pi\pi) = 0.184 - [0.153 + 0.077i]_\text{V} + [? - 0.049i]_\text{VV}$$
$$+ \left[\frac{r_\text{sp}}{0.485}\right]\bigl\{[0.122]_\text{LO} + [0.050 + 0.053i]_\text{HV}$$
$$+ [0.071]_\text{tw3}\bigr\}$$
$$= 0.275^{+0.228}_{-0.135} + \bigl(-0.073^{+0.115}_{-0.082}\bigr)i. \qquad (37)$$

In each expression, the first line gives the form-factor (vertex) contribution, the second line the spectator-scattering contribution, and the third line their sum with an estimate of the theoretical uncertainties due to hadronic input parameters (form factors, LCDAs, quark masses), power corrections, and neglected higher-order perturbative corrections as explained in detail in Ref. [257], where also the input parameter ranges employed here are given. The first two lines in (36) and (37) are decomposed into the tree (naive factorization, α_s^0), one-loop (V), and two-loop (VV) vertex correction (the question marks denote unknown real parts of order α_s^2); tree (α_s, LO), one-loop (α_s^2, HV), and twist-3 power correction (tw3) to spectator scattering. The prefactor $r_\text{sp} = (9 f_{M_1} \hat{f}_B)/(m_b F^{BM_1} \lambda_B)$ encapsulates the bulk of the hadronic uncertainties of the spectator-scattering term. Numerically, for a_1, the corrections are, both individually and in their sum, at the few-percent level such that a_1 is very close to 1 and to the naive-factorization result. On the other hand, individual corrections to a_2 are large, with a near cancellation between naive factorization and the one-loop vertex correction. a_2 is thus especially sensitive to spectator scattering and to higher-order vertex corrections. That these are all important is seen from the VV, LO, and HV numbers in (37).

Analogous expressions can be given for the remaining amplitude parameters $a_3^p \cdots a_{10}^p$ [257], except that no two-loop vertex corrections are known. Qualitatively, NNLO spectator-scattering corrections are as important for the leading-power, but small (electroweak) penguin amplitudes $a_{3,5,7,10}^p$ as they are for a_2 but are found to be small for the large electroweak penguin amplitude a_9^p. Corrections to the QCD penguin amplitude a_4^p are also small, in spite of the involvement of the large Wilson coefficient C_1. This is due to a numerical cancellation, which may be accidental. The scalar QCD and electroweak penguin amplitudes a_6^p and a_8^p are power suppressed but "chirally enhanced". NNLO corrections to them are currently unknown but might involve sizable contributions proportional to C_1, unless a similar numerical cancellation as in the case of a_4^p prevents this. This would be relevant for direct CP asymmetries in the πK system and elsewhere. For a more complete discussion, see Ref. [257].

A good fraction of NNLO corrections to the QCD factorization formula are now available. While the perturbation expansion is well behaved in all cases, some of these corrections turn out to be significant, particularly those to the colour-suppressed tree and (electroweak) penguin amplitudes. Further important corrections to the QCD and colour-suppressed EW penguin amplitudes proportional to C_1 may enter through the chirally-enhanced power corrections a_6^p and a_8^p, making their NNLO calculation an important goal.

2.2.4 QED corrections to hadronic B decays

2.2.4.1 Introduction

The large amount of data collected so far at B factories has allowed one to reach a statistical accuracy on B decays into pairs of (pseudo)scalars at a level where electromagnetic effects cannot be neglected anymore [258, 259]. On the one hand, a correct simulation of the unavoidable emission of photons from charged particles has to be included in Monte Carlo programs in order to evaluate the correct efficiency. On the other hand, a clear definition of the effective cut on (soft) photon spectra is essential for a consistent comparison both between theory and experiments and between results from different experiments.

We discuss the theoretical and experimental treatment of radiative corrections in hadronic B decays. We present analytical expressions to describe the leading effects induced by both real and virtual (soft) photons in the generic process $H \to P_1 P_2(\gamma)$, where both H and $P_{1,2}$ are scalar or pseudoscalar particles. We then discuss the procedures to be adopted in experimental analyses for a clear definition of the observables.

2.2.4.2 The scalar QED calculation

General properties of QED have been exploited in detail for most of the pure electroweak processes or in general for processes that can be fully treated in terms of perturbation theory. This is not the case of hadronic decays. However, due to the universal character of infrared QED singularities, it is possible to estimate the leading $\mathcal{O}(\alpha)$ contributions to these processes within scalar QED in the approximation of a point-like weak vertex.

The most convenient infrared-safe observable related to the process $B \to P_1 P_2$ is the photon inclusive width

$$\Gamma_{12}^{\text{incl}}(E^{\text{max}}) = \Gamma(B \to P_1 P_2 + n\gamma)|_{\sum E_\gamma < E_\gamma^{\text{max}}}$$
$$= \Gamma_{12} + \Gamma_{12+n\gamma}(E_\gamma^{\text{max}}), \quad (38)$$

namely, the width for the process $B \to P_1 P_2$ accompanied by any number of (undetected) photons with total missing energy less than or equal to E^{max} in the B meson rest frame. The infrared cut-off E_γ^{max} can be the photon energy below which the state $|P_1 P_2\rangle$ cannot be distinguished from the state $|P_1 P_2 + n\gamma\rangle$; however, in principle, it can also be chosen to be a high reference scale (up to the kinematical limit). At any order in perturbation theory, we can decompose $\Gamma_{12}^{\text{incl}}$ in terms of two theoretical quantities: the so-called nonradiative width, Γ_{12}^0, and the corresponding energy-dependent e.m. correction factor $G_{12}(E_\gamma^{\text{max}})$,

$$\Gamma_{12}^{\text{incl}}(E_\gamma^{\text{max}}) = \Gamma_{12}^0(\mu) \, G_{12}(E_\gamma^{\text{max}}, \mu). \quad (39)$$

In the limit $E_\gamma^{\text{max}} \ll M_B$, the electromagnetic correction factor can be reliably estimated within scalar QED. We define the nonradiative width $\Gamma_{12}^0(\mu)$ as

$$\Gamma_{12}^0(\mu) = \frac{\beta}{16\pi M_B} \left| \mathcal{A}_{B \to P_1 P_2}(\mu) \right|^2, \quad (40)$$

$$\beta^2 = \left[1 - (r_1 + r_2)^2\right]\left[1 - (r_1 - r_2)^2\right], \quad r_i = \frac{m_i}{M_B}, \quad (41)$$

namely the tree-level rate expressed in terms of the renormalized (scale-dependent) weak coupling. Here the m_i refer to the masses of the light mesons in the final state, and M_B is the B-meson mass. The function $G_{12}(E_\gamma^{\text{max}}, \mu)$ can be written as

$$G_{12}(E, \mu) = 1 + \frac{\alpha}{\pi} \left[b_{12} \ln\left(\frac{M_B^2}{4E^2}\right) + F_{12} \right.$$
$$\left. + \frac{1}{2} H_{12} + N_{12}(\mu) \right], \quad (42)$$

where H_{12} represents the finite term arising from virtual corrections, and F_{12} the energy-independent contribution generated by the real emission (here $E \equiv E_\gamma^{\text{max}}$):

$$\int_{E_\gamma < E} \frac{d^3 \vec{k}}{(2\pi)^3 2 E_\gamma} \sum_{\text{spins}} \left| \frac{\mathcal{A}(B \to P_1 P_2 \gamma)}{\mathcal{A}(B \to P_1 P_2)} \right|^2$$
$$= \frac{\alpha}{\pi} \left[b_{12} \ln\left(\frac{m_\gamma^2}{4E^2}\right) + F_{12} + \mathcal{O}\left(\frac{E}{M_B}\right) \right]. \quad (43)$$

As expected, after summing real and virtual corrections, the infrared logarithmic divergences cancel out in $G_{12}(E, \mu)$, giving rise to the universal $\ln(M_B/E_\gamma^{\text{max}})$ term. The scale dependence contained in $N_{12}(\mu)$ cancels out in the product $\Gamma_{12}^0 \times G_{12}$ due to the corresponding scale dependence of the weak coupling. For the explicit expressions of F_{12}, H_{12} and N_{12} and a more detailed discussion of the μ-dependence, we refer to [249]. The result thus obtained can be applied to both B and D decays.

We finally give the results for G_{+-} and G_{+0} in the limit $m_{1,2}, E \ll M_B$, which represents a convenient and very good approximation:

$$G_{+-} = 1 - \frac{\alpha}{\pi}\left\{\left[2\ln\epsilon + 1 + \ln(1-\delta^2)\right]\ln\left(\frac{4E^2}{M_B^2}\right)\right.$$
$$\left. - 4\ln\epsilon + \frac{\pi^2}{3} + 1 + \mathcal{O}(\delta)\right\}, \quad (44)$$

$$G_{+0} = 1 - \frac{\alpha}{\pi}\left\{\left[\ln\epsilon + 1 + \ln(1+\delta)\right]\ln\left(\frac{4E^2}{M_B^2}\right)\right.$$
$$\left. - 2\ln\epsilon + \frac{\pi^2}{6} - 1 + \mathcal{O}(\delta)\right\}, \quad (45)$$

where

$$\epsilon = \frac{m_1 + m_2}{2M_B}, \qquad \delta = \frac{m_1 - m_2}{m_1 + m_2}, \quad (46)$$

with $12 = +-, +0$, respectively. This approximation also serves to clarify the physical relevance of the correction factors. The logarithmic terms as well as the Coulomb correction ($\sim \pi^2$) are model-independent well-defined effects. On the other hand, the remaining constant pieces (± 1) are not meaningful in the absence of the proper UV matching, but they are subdominant and numerically rather small.

2.2.4.3 Inclusion of final-state radiation effects in an experimental analysis We will discuss in particular the inclusion of final-state radiation in the analysis of rare B decays at B factories. In this kind of environment, the efficiency is estimated through Monte Carlo simulation where QED effects are taken into account using the PHOTOS simulation package [260]. The first issue is then to check if the performances of the entire event simulation chain are the ones expected from the theory. One can thus compare the simulated $G_{12}(E_\gamma^{\max})$ function, as well as the energy and angular distribution of the generated photons (whose analytical expression can be found in [249]) and then, if needed, correct the distributions on which efficiency and parametrization of the fit variables are evaluated. Then particular care has to be taken in order to quote the results in such a way that radiation effects can be disentangled. In principle, it would be necessary to select B candidates with a specified maximum amount of $\mathcal{O}(100\text{ MeV})$ photon energy in the final states, a quantity which is difficult to reconstruct in a B factory context. Instead, one could define the data sample selecting on an observed variable which can be clearly related to the maximum allowed energy for photons $E_\gamma^{\max}$. The variable $\Delta E = E_B^* - \sqrt{s}/2$, where E_B^* is the reconstructed B candidate energy in the e^+e^- center of mass (CM) frame and $\sqrt{s}$ the total CM energy, is clearly suitable for this purpose. The ΔE window chosen for the analysis would then allow for the presence of radiated photons up to the chosen cut, providing the possibility of quoting results, e.g. on branching fractions, with a defined cut on the soft photon spectrum.

Once a result of this kind is obtained, it is easy to extract the weak couplings—which cannot be directly measured due to the intrinsic and unavoidable features of QED—employing the theoretical calculation explained in the previous section. This is very important, since the comparison between theoretical predictions and experiments can be done more efficiently in terms of weak couplings. Moreover, a meaningful comparison between different experiments can only be done in terms of weak couplings (nonradiative quantities) or in terms of inclusive widths employing the same infrared cut-off.

2.2.5 Outlook on future improvements

The improvement of our quantitative understanding of hadronic effects in charmless nonleptonic B-decays requires both experimental and theoretical efforts:

- Completion of the NNLO analyses for the factorizable vertex and hard-scattering contributions to reduce the perturbative uncertainties.
- Further improvement in hadronic input parameters (form factors, LCDA) by nonperturbative methods combined with experimental data on B- and D-meson decays.
- More systematic treatment of power-corrections.
- Better understanding of $SU(3)$-breaking effects in the analysis of B_s and $B_{u,d}$ decays.

In the future, the main limitations will probably be due to theoretical uncertainties in nonperturbative strong rescattering phases.

2.3 Light-cone QCD sum rules

2.3.1 Distribution amplitudes

Light-cone wave functions or distribution amplitudes (DA) are matrix elements defined near light-like separations connecting hadrons to their partonic constituents. They are widely used in hard exclusive processes with high momentum transfer [261], which are often dominated by light-like distances. Formally they appear in the light-cone operator product expansion (LCOPE) and can be seen as the analogue of matrix elements of local operators in the operator product expansion (OPE). The terms in the OPE are ordered according to the dimension of the operators, the terms in the LCOPE according to their twist, the dimension minus the spin. We shall discuss distribution amplitudes for light mesons, which are most relevant for the LHCb experiment.[7]

---

[7]There are of course other DA of interest. Baryon DA have recently been reviewed in [262], the photon DA is treated in [263], and a recent lecture on the B-meson DA can be found in [264].

We shall take the $K(495)$ and the $K^*(892)$ as representatives for the light pseudoscalar and vector mesons:[8]

$$\langle 0|\bar{q}(x)x_\mu \gamma^\mu \gamma_5[x,0]s(0)|K(q)\rangle$$
$$= if_K q \cdot x \int_0^1 du\, e^{-i\bar{u}q\cdot x} \phi_K(u) + O(x^2, m_K^2),$$

$$\langle 0|\bar{q}(x)x_\mu \gamma^\mu [x,0]s(0)|K^*(q,\lambda)\rangle$$
$$= (\varepsilon^{(\lambda)} \cdot x) f_{K^*} m_{K^*} \int_0^1 du\, e^{-i\bar{u}q\cdot x} \phi_K^\parallel(u) \qquad (47)$$
$$+ O(x^2, m_{K^*}^2),$$

$$\langle 0|\bar{q}(x)\sigma_{\mu\nu}[x,0]s(0)|K^*(q,\lambda)\rangle$$
$$= i\left(\varepsilon_\mu^{(\lambda)} q_\nu - \varepsilon_\nu^{(\lambda)} q_\mu\right) f_{K^*}^\perp(\mu) \int_0^1 du\, e^{-i\bar{u}q\cdot x} \phi_K^\perp(u)$$
$$+ O(x^2, m_{K^*}^2).$$

The vector x_μ is to be thought of as a vector close to the light-cone. The variable u ($\bar{u} \equiv 1 - u$) can be interpreted as the collinear momentum fraction carried by one of the constituent quarks in the meson. Corrections to the leading twist come from three sources: 1. other Dirac-structures (e.g. $\langle 0|\bar{q}(x)\gamma_5[x,0]s(0)|K(q)\rangle$), 2. higher Fock states (including an additional gluon) and 3. mass and light-cone corrections as indicated in the equations above.

The wave functions $\phi(u,\mu)$ are nonperturbative objects. Their asymptotic forms are known from perturbative QCD, $\phi(u,\mu) \xrightarrow{\mu \to \infty} 6u\bar{u}$. Use of one-loop conformal symmetry of massless QCD is made by expanding in the eigenfunctions of the evolution kernel, the Gegenbauer polynomials $C_n^{3/2}$,

$$\phi(u,\mu) = 6u\bar{u}\left(1 + \sum_{n=1}^\infty \alpha_n(\mu) C_n^{3/2}(2u-1)\right), \qquad (48)$$

where the α_n are hadronic parameters, the Gegenbauer moments. If n is odd, they vanish for particles with definite G-parity, e.g. $\alpha_{2n+1}(\pi) = 0$. For the kaon, $\alpha_{2n+1}(K) \neq 0$, which contributes to $SU(3)$ breaking. In practice the expansion is truncated after a few terms. This is motivated by the fact that the hierarchy of anomalous dimensions $\gamma_{n+1} > \gamma_n > 0$ implies $|\alpha_{n+1}| < |\alpha_n|$ at a sufficiently high scale. From concrete calculations and fits it indeed appears that the hierarchy already sets in at typical hadronic scales ~ 1 GeV. Moreover, for smooth kernels, the higher Gegenbauer moments give small contributions upon convolution much like in the familiar case of the partial wave expansion in quantum mechanics.

---

[8]In the literature, sometimes another phase convention for the vector meson states is used, where $|V\rangle_\text{other} = i|V\rangle_\text{here}$.

A different method is to model the wave-functions by using experimental and theoretical constraints. In [279], a recursive relation between the Gegenbauer moments was proposed, which involves only two additional parameters. This constitutes an alternative tool especially in cases where the conformal expansion is converging slowly.

We shall not report on higher-twist contributions here but refer to the literature [277, 278]. It should also be mentioned that higher-twist effects can be rather prominent such as in the time-dependent CP asymmetry in $B \to K^*\gamma$ via soft gluon emission [265].

2.3.1.1 Decay constants The decay constants normalize the DA. For the pseudoscalars π and K, they are well known from experiment. The decay constants of the η and η' and in general their wave functions are more complicated due to η–η' mixing and the chiral anomaly and shall not be discussed here. For the vector particles, there are two decay constants, as seen from (47). The longitudinal decay constants can be taken from experiment. For instance, for ρ^0, ω and ϕ, they are taken from $V^0 \to e^+e^-$ and, for ρ^- and K^{*-}, from $\tau^- \to V^- \nu_\tau$. It is worth noting that the difference in f_{ρ^0} and f_{ρ^-} seems consistent with the expected size of isospin breaking, whereas some time ago, there seemed to be a slight tension [269].

For the transverse decay constants $f^\perp$, one has to rely on theory. QCD sum rules provide both longitudinal and transverse decay constants [233, 266]

$$f_\rho = (206 \pm 7) \text{ MeV},$$
$$f_\rho^\perp(1 \text{ GeV}) = (165 \pm 9) \text{ MeV},$$
$$f_{K^*} = (222 \pm 8) \text{ MeV}, \qquad (49)$$
$$f_{K^*}^\perp(1 \text{ GeV}) = (185 \pm 10) \text{ MeV}.$$

In lattice QCD, there exist two quenched calculations of the ratio of decay constants [267, 268], which are consistent with the sum-rule values above. Combining all these experimental, sum-rule and lattice results we get [270]

$$f_\rho = (216 \pm 2) \text{ MeV},$$
$$f_\rho^\perp(1 \text{ GeV}) = (165 \pm 9) \text{ MeV},$$
$$f_\omega = (187 \pm 5) \text{ MeV},$$
$$f_\omega^\perp(1 \text{ GeV}) = (151 \pm 9) \text{ MeV},$$
$$f_{K^*} = (220 \pm 5) \text{ MeV}, \qquad (50)$$
$$f_{K^*}^\perp(1 \text{ GeV}) = (185 \pm 10) \text{ MeV},$$
$$f_\phi = (215 \pm 5) \text{ MeV},$$
$$f_\phi^\perp(1 \text{ GeV}) = (186 \pm 9) \text{ MeV}.$$

2.3.1.2 The first and second Gegenbauer moment As mentioned before, the first Gegenbauer moment vanishes for particles with definite G-parity. Intuitively the first Gegenbauer moment of the kaon is a measure of the average momentum fraction carried by the strange quark. Based on the constituent quark model, it is expected that $\alpha_1(K) > 0$. A negative value of this quantity [271] created some confusion and initiated reinvestigations. The sum rule used in that work is of the nondiagonal type and has a nonpositive definite spectral function, which makes the extraction of any kind of residue very unreliable. Later on, diagonal sum rules were used, and stable values were obtained [233, 272] ($\mu = 1$ GeV):

$$\alpha_1(K, \mu) = 0.06 \pm 0.03,$$
$$\alpha_1^{\parallel}(K^*, \mu) = 0.03 \pm 0.02, \quad (51)$$
$$\alpha_1^{\perp}(K^*, \mu) = 0.04 \pm 0.03,$$

although with relatively large uncertainties. An interesting alternative method was suggested in [273], where the first Gegenbauer moment was related to a quark–gluon matrix element via the equation of motion. An alternative derivation and a completion for all cases was later given in [274]. The operator equation for the kaon is

$$\frac{9}{5}\alpha_1(K) = -\frac{m_s - m_q}{m_s + m_q} + 4\frac{m_s^2 - m_q^2}{m_K^2} - 8\kappa_4(K),$$

where the twist-4 matrix element κ_4 is defined as: $\langle 0|\bar{q}(gG_{\alpha\mu})i\gamma^\mu\gamma_5 s|K(q)\rangle = iq_\alpha f_K m_K^2 \kappa_4(K)$. Similar equations exist for the longitudinal and transverse case. It is worth stressing that those operator relations are completely general and it remains to determine the twist-4 matrix elements. Attempts to determine them from QCD sum rules [273, 274] turn out to be consistent with the determinations from diagonal sum rules (51) but cannot compete in terms of the accuracy. Later lattice QCD provided the first Gegenbauer moment for the kaon DA from domain-wall fermions [275] and Wilson fermions [276], whose values agree very well with the central value of $\alpha_1(K)$ in (51) but have significantly lower uncertainty.

The second Gegenbauer moment has also been determined from diagonal sum rules for the π and K [272, 277]:

$$\alpha_2(\pi, 1 \text{ GeV}) = 0.27 \pm 0.08, \quad \frac{\alpha_2(K)}{\alpha_2(\pi)} = 1.05 \pm 0.15. \quad (52)$$

It can be seen that the $SU(3)$ breaking in the second moment is presumably moderate. Values of α_2 for the vector mesons ρ, K^* and ϕ have recently been updated in [278].

2.3.2 Heavy-to-light form factors from LCSR

Light-cone sum rules (LCSR) were developed to improve on some of the shortcomings of three-point sum rules designed to describe meson-to-meson transition form factors. The problem is that for $B \to M$ transitions, where M is a light meson, higher-order matrix elements grow with m_b rendering the OPE nonconvergent. In the case $D \to M$, three-point sum rules and LCSR yield comparable results. A review of the framework of LCSR can be found in [230].

The form factors of V and A currents for B to light pseudoscalar and vector mesons are defined as ($q = p_B - p$)

$$\langle P(p)|\bar{q}\gamma_\mu b|\bar{B}(p_B)\rangle$$
$$= f_+(q^2)\left[(p_B + p)_\mu - \frac{m_B^2 - m_P^2}{q^2}q_\mu\right]$$
$$+ f_0(q^2)\frac{m_B^2 - m_P^2}{q^2}q_\mu, \quad (53)$$

$$c_V\langle V(p,\varepsilon)|\bar{q}\gamma_\mu(1-\gamma_5)b|\bar{B}(p_B)\rangle$$
$$= \frac{2V(q^2)}{m_B + m_V}\epsilon_{\mu\nu\rho\sigma}\varepsilon^{*\nu}p_B^\rho p^\sigma$$
$$- 2im_V A_0(q^2)\frac{\varepsilon^* \cdot q}{q^2}q_\mu$$
$$- i(m_B + m_V)A_1(q^2)\left[\varepsilon_\mu^* - \frac{\varepsilon^* \cdot q}{q^2}q_\mu\right]$$
$$+ iA_2(q^2)\frac{\varepsilon^* \cdot q}{m_B + m_V}\left[(p_B + p)_\mu - \frac{m_B^2 - m_V^2}{q^2}q_\mu\right]. \quad (54)$$

The factor c_V accounts for the flavor content of particles: $c_V = \sqrt{2}$ for ρ^0, ω and $c_V = 1$ otherwise. The tensor form factors, relevant for $B \to V\gamma$ or $B \to P(V)l^+l^-$, are defined as

$$\langle P(p)|\bar{q}\sigma_{\mu\nu}q^\nu b|\bar{B}(p_B)\rangle$$
$$= \frac{if_T(q^2)}{m_B + m_P}[q^2(p + p_B)_\mu - (m_B^2 - m_P^2)q_\mu], \quad (55)$$

$$c_V\langle V(p,\varepsilon)|\bar{q}\sigma_{\mu\nu}q^\nu(1+\gamma_5)b|\bar{B}(p_B)\rangle$$
$$= 2iT_1(q^2)\epsilon_{\mu\nu\rho\sigma}\varepsilon^{*\nu}p_B^\rho p^\sigma$$
$$+ T_2(q^2)[(m_B^2 - m_V^2)\varepsilon_\mu^* - (\varepsilon^* \cdot q)(p_B + p)_\mu]$$
$$+ T_3(q^2)(\varepsilon^* \cdot q)\left[q_\mu - \frac{q^2}{m_B^2 - m_V^2}(p_B + p)_\mu\right], \quad (56)$$

with $T_1(0) = T_2(0)$. Note that the tensor form factors depend on the renormalization scale μ of the matrix element. All form factors in (53)–(56) are positive, and $\epsilon^{0123} = -1$.

LCSR allow us to obtain the form factors from a suitable correlation function for virtualities of $0 < q^2 \lesssim 14$ GeV^2. The residue in the sum rule is of the type $(f_B f_+(q^2))_{SR}$. Using the second sum rule for $(f_B)_{SR}$ to the same accuracy, the form factor is obtained as $f_+ = (f_B f_+(q^2))_{SR}/(f_B)_{SR}$, where several uncertainties cancel. The final uncertainties of the sum-rule results for the form factors are around 10% and slightly more for the $B \to K$ transitions due to the additional uncertainty in the first Gegenbauer moment. The most recent and up-to-date calculation for $B \to M$ form factors, including twist-3 radiative corrections, can be found in [231, 232]. It is not obvious how the accuracy can be significantly improved by including further corrections. One interesting option would be to calculate NNLO QCD corrections, which could first be attempted in the large-β_0 limit.

Another interesting question is whether it is possible to extend the form factor calculations to the entire physical domain $0 < q^2 < (m_B - m_{P(V)})^2$. It has been advocated by Becirevic and Kaidalov [280] to write the form factor f_+ as a dispersion relation in q^2 with a lowest-lying pole term plus a contribution from multiparticle states, which in a minimal setup can be approximated by an effective pole term at higher mass:

$$f_+(q^2) = \frac{r_1}{1 - q^2/m_1^2} + \int_{(m_B+m_P)^2}^{\infty} ds \frac{\rho(s)}{s - q^2}$$
$$\to \frac{r_1}{1 - q^2/m_1^2} + \frac{r_2}{1 - q^2/m_{\text{fit}}^2}. \quad (57)$$

In the past, it has often been popular to adopt Vector Meson Dominance (VMD), i.e. to set $r_2 = 0$. BaBar measurements of semileptonic decay spectra with five bins in the q^2-distribution now strongly disfavor simple VMD [281]. Another important point is that the fits to the parametrization (57) allow us to reproduce the results from LCSR extremely well [231, 232]. The parametrization also passes a number of consistency tests. The soft pion point $f_0(m_B^2) = f_B/f_\pi$ can be attained upon extrapolation, leading to a B-meson decay constant of $f_B \approx 205$ MeV. This is well in the ballpark of expectations and consistent with the Belle measurement of $B \to \tau\nu$. Moreover the residue $(r_1)_{f_+} = (f_{B^*} g_{BB^*\pi})/(2 m_{B^*})$, which is rather stable under the fits, agrees within ten percent with what is known from hadronic physics. Representative results are given in Table 7. More form factors can be found in (27) and Table 3 of [231] for $B \to \pi, K, \eta$ and in Table 8 of [232] for $B \to \rho, K^*, \phi, \omega$. It has to be emphasized that the $B \to K, K^*$ transitions have been evaluated before the progress in the $SU(3)$-breaking was achieved. An update would be timely and will certainly be undertaken for such important cases as $B \to K^* l^+ l^-$. In particular, for the $B \to K^* \gamma$ decay rate in the SM, it was emphasized by [282, 283] that within the framework of QCD factorization, $T_1(0)_{\text{SM-exp,QCDF}} = 0.28 \pm 0.02$. An update of $SU(3)$-breaking effects yields $T_1(0) = 0.31 \pm 0.04$ [284], which seems reasonably consistent.

In certain decay channels, such as $B \to K^* l^+ l^-$, several form factors enter at the same time. Sometimes ratios of decay rates are needed, e.g. for the extraction of $|V_{td}/V_{ts}|$ from $B \to K^* \gamma$. Simply taking the uncertainties in the individual form factors and adding them linearly could be a drastic overestimate, since parametric uncertainties, such as those from m_b, might cancel in the quantities of interest. In the former case, no efforts have been undertaken. In the latter case, a consistent evaluation [266] leads to the form factor ratio $\xi \equiv T_1^{B \to K^*}(0)/T_1^{B \to \rho}(0) = 1.17 \pm 0.09$.

2.3.3 Comparison with heavy-to-light form factors from relativistic quark models

Quark models have been frequently used in the past to estimate hadronic quantities such as form factors. They may be applied to complicated processes hardly accessible to lattice calculations and they provide connections between different processes through the wave functions of the participating hadrons. Relativistic quark models are based on a simplified picture of QCD: below the chiral symmetry breaking scale $\mu_\chi \approx 1$ GeV, quarks are treated as particles of fixed mass interacting via a relativistic potential and hadron wave functions and masses are found as solutions of three-dimensional reductions of the Bethe–Salpeter equation. The structure of the confining potential is restricted by rigorous properties of QCD, such as heavy-quark symmetry for the heavy-quark sector [285, 286] and spontaneously broken chiral symmetry for the light-quark sector [287]. The values of the constituent-quark masses and the parameters of the potential are fixed by requiring that the spectrum of observed hadron states is well reproduced [288, 289].

Various versions of the quark model were applied to the description of weak properties of heavy hadrons (see e.g. [290–292]). For instance, the weak transition form factors are given in the quark model in [293] by relativistic double spectral representations in terms of the wave functions of initial and final hadrons and the double spectral density of the corresponding Feynman diagrams with massive quarks. This approach led to very successful predictions for D decays [294, 295]. Many results for various B and B_s decays

Table 7 Form factors from light-cone sum rules

$f_+^{B \to \pi}(0)$	$T_1^{B \to \rho}(0)$	$V^{B \to \rho}(0)$	$A_0^{B \to \rho}(0)$	$A_1^{B \to \rho}(0)$	$A_2^{B \to \rho}(0)$
0.258 ± 0.031	0.267 ± 0.023	0.323 ± 0.030	0.303 ± 0.029	0.242 ± 0.023	0.221 ± 0.023

Table 8 Examples of form factors for $B \to \rho$ $[B_s \to K^*]$ from the quark model [295]

$V(0)$	$A_1(0)$	$A_2(0)$	$A_0(0)$	$T_1(0)$	$T_3(0)$
0.31 [0.44]	0.26 [0.36]	0.24 [0.32]	0.29 [0.45]	0.27 [0.39]	0.19 [0.27]

have been obtained [295–299], yielding an overall picture in agreement with other approaches, such as QCD sum rules. Table 8 gives examples of the results from [295]. A comparison between various quark models performed in [300] leads to a qualitative estimate of the overall uncertainty of some (10–15)%. The main limitation of the quark model approach is the difficulty to provide rigorous estimates of the systematic errors of the calculated hadron parameters. In this respect, quark models cannot compete with lattice gauge theory.

2.4 Lattice QCD

2.4.1 Recent results

In this section, we give a summary of recent lattice results relevant to flavor physics. The tables should be consulted with an eye on the systematics discussed in Sect. 2.4.2. For a more complete coverage, see the review talks on heavy flavor physics [301–303] and kaon physics [304–306] at the last few lattice conferences.

2.4.1.1 Decay constants
The axial-vector decay constants relevant to the $\pi \to \ell\nu$ leptonic decays

$$\langle 0|(\bar{d}\gamma_\mu\gamma_5 u)(x)|\pi(p)\rangle = if_\pi p_\mu e^{-ip\cdot x} \tag{58}$$

(and analogously for K, D, B mesons) may be evaluated on the lattice. Some recent results are collected in Table 9. The first column gives the statistical and systematic errors. The second column says whether the simulations are quenched ($N_f = 0$), or dynamical with a common m_{ud} mass only ($N_f = 2$), or with strange quark loops included ($N_f = 2 + 1$). The remaining columns indicate the light quark formulation in the sea and valence sectors and whether a continuum extrapolation has been attempted. To the quenched results, an extra 5% scale setting error should be added (see Sect. 2.4.2.1). Generally, the lattice results compare favorably to the recent experimental determinations (using the appropriate CKM element from another process) $f_D = 223(17)(03)$ MeV at CLEO [323], $f_{D_s} = 282(16)(7)$ MeV at CLEO [324], $f_{D_s} = 283(17)(16)$ MeV at BaBar [325] and $f_B = 229(^{+36}_{-31})(^{+34}_{-37})$ MeV at Belle [326]. One may also form the ratio $\sqrt{M_{D_s}}f_{D_s}/\sqrt{M_D}f_D$ to the result 1.30(12) implied by the CLEO and BaBar numbers.

2.4.1.2 Form factors
The vector form factors of semi-leptonic decays like $B \to \pi\ell\nu$ or $D \to K\ell\nu$ defined in (53) can be calculated in the range $q^2_{\min} < q^2 < q^2_{\max}$, where $q^2_{\max} = (M_B - M_\pi)^2$, $(M_D - M_K)^2$, respectively, while $q^2_{\min}$ is a soft bound (set by the cut-off effects and noise one considers tolerable). Often $f_+(0) = f_0(0)$ is used, and a parametrization is employed to extrapolate. Among the most popular are those of Bećirević–Kaidalov [280] and Ball–Zwicky [280, 281]:

$$f_+^{\mathrm{BK}}(q^2) = \frac{f}{(1-\tilde{q}^2)(1-\alpha\tilde{q}^2)},$$

$$f_0^{\mathrm{BK}}(q^2) = \frac{f}{1-\tilde{q}^2/\beta}, \tag{59}$$

$$f_+^{\mathrm{BZ}}(q^2) = \frac{f}{1-\tilde{q}^2} + \frac{r\tilde{q}^2}{(1-\tilde{q}^2)(1-\alpha\tilde{q}^2)}, \tag{60}$$

where $\tilde{q}^2 = q^2/M_{B^*}^2$ (or $q^2/M_{D^*}^2$ for D-decays), with the parameters $f = f_+(0)$, α (BK, BZ) and r (BZ). The expression in (60) is equivalent to the approximate form in (57). Some recent results, with the same meaning of the columns as before, are given in Table 10. The definition of $\mathcal{F}$ is given in [333]. Earlier work on the $B \to \pi\ell\bar{\nu}$ form factors can be found in [336–339]. For $D \to K\ell\nu$ and $D \to \pi\ell\nu$, the q^2-dependence of the form factors has been traced out by the FNAL/MILC/+ Collaboration [332] and compared to experimental results by the BES [340] and FOCUS [341] Collaborations. For $B \to \pi\ell\nu$, the q^2-dependence, as determined by the HPQCD and FNAL/MILC/+ Collaborations, is in reasonable agreement [301]. For a generic comment why the form factor at $q^2 = 0$ is not always the best thing to ask for from the lattice, see Sect. 2.4.3.

2.4.1.3 Bag parameters
On the lattice, the SM bag parameters $B_K(\mu)$ and $B_B(\mu)$ for neutral kaon and B-meson mixing

$$\langle\bar{K}^0|(\bar{s}d)_{V-A}(\bar{s}d)_{V-A}|K^0\rangle = \frac{8}{3}M_K^2 f_K^2 B_K, \tag{61}$$

$$\langle\bar{B}_q^0|(\bar{b}q)_{V-A}(\bar{b}q)_{V-A}|B_q^0\rangle = \frac{8}{3}M_{B_q}^2 f_{B_q}^2 B_{B_q}$$

$$(q = d, s) \tag{62}$$

are extracted indirectly. The measured quantities are $f_B^2 B_B$ and f_B; then the ratio is taken to obtain the quoted B_B (similar for B_K). Therefore, it makes little sense to combine B_B

Table 9 Decay constants from lattice QCD

$f_K/f_\pi = 1.24(2)$	$N_f = 2+1$	dom/dom	no	RBC/UKQCD [307]
$f_K/f_\pi = 1.218(2)(^{+11}_{-24})$	$N_f = 2+1$	stag/dom	no	NPLQCD [308]
$f_K/f_\pi = 1.208(2)(^{+07}_{-14})$	$N_f = 2+1$	stag/stag	yes	MILC [309]
$f_K/f_\pi = 1.189(7)$	$N_f = 2+1$	stag/stag	yes	HPQCD [310]
$f_{D_s} = 242(09)(10)$ MeV	$N_f = 0$	–/clov	yes	ALPHA [311]
$f_{D_s} = 240(5)(5)$ MeV	$N_f = 0$	–/clov	yes	RomeII [312]
$f_{D_s} = 249(03)(16)$ MeV	$N_f = 2+1$	stag/stag	yes	FNAL/MILC/+ [313]
$f_{D_s} = 238(11)(^{+07}_{-27})$ MeV	$N_f = 2$	clov/clov	yes	CP-PACS [301]
$f_{D_s} = 241(3)$ MeV	$N_f = 2+1$	stag/stag	yes	HPQCD [310]
$f_D = 232(7)(^{+6}_{-0})(53)$ MeV	$N_f = 0$	–/dom	no	RBC [314]
$f_D = 202(12)(^{+20}_{-25})$ MeV	$N_f = 2$	clov/clov	yes	CP-PACS [301]
$f_{D_s}/f_D = 1.05(2)(^{+0}_{-2})(6)$	$N_f = 0$	–/dom	no	RBC [314]
$f_{D_s}/f_D = 1.24(7)$	$N_f = 2+1$	stag/stag	yes	FNAL/MILC/+ [313]
$f_{D_s}/f_D = 1.162(9)$	$N_f = 2+1$	stag/stag	yes	HPQCD [310]
$f_{B_s} = 192(6)(4)$ MeV	$N_f = 0$	–/clov	yes	RomeII [312]
$f_{B_s} = 205(12)$ MeV	$N_f = 0$	–/clov	yes	ALPHA [315]
$f_{B_s} = 191(6)$ MeV	$N_f = 0$	–/clov	yes	ALPHA [316]
$f_{B_s} = 242(9)(51)$ MeV	$N_f = 2$	clov/clov	yes	CP-PACS [317]
$f_{B_s} = 217(6)(^{+37}_{-28})$ MeV	$N_f = 2$	stag/wils	yes	MILC [318]
$f_{B_s} = 260(7)(26)(8)$ MeV	$N_f = 2+1$	clov/clov	no	HPQCD [319]
$f_{B_s}/f_B = 1.179(18)(23)$	$N_f = 2$	clov/clov	yes	CP-PACS [317]
$f_{B_s}/f_B = 1.16(1)(3)(^{+4}_{-0})$	$N_f = 2$	stag/wils	yes	MILC [318]
$f_{B_s}/f_B = 1.13(3)(^{+17}_{-02})$	$N_f = 2$	clov/clov	no	JLQCD [320]
$f_{B_s}/f_B = 1.20(3)(1)$	$N_f = 2+1$	stag/stag	yes	HPQCD [321]
$f_{B_s}/f_B = 1.29(4)(6)$	$N_f = 2$	dom/dom	no	RBC [322]

from one group and f_B from another to come up with a lattice value for $f_B\sqrt{B_B}$. On the other hand,

$$\xi = \frac{f_{B_s}\sqrt{B_{B_s}}}{f_{B_d}\sqrt{B_{B_d}}} \qquad (63)$$

is benevolent from a lattice viewpoint, since it follows from the ratio of the same correlator with two different quark masses (in practice, an extrapolation $m_d \to m_d^{\text{phys}}$ is needed). Many systematic uncertainties cancel in such ratios, but the chiral extrapolation error is not reduced. It would make sense to quote the renormalization scheme and scale-independent quantity

$$\hat{B}_X = \lim_{\mu \to \infty} \alpha_s(\mu)^{2/\beta_0} \left[1 + \frac{\alpha_s}{4\pi} J_{N_f} + \cdots \right] B_X(\mu) \qquad (64)$$

with known J_{N_f}. From a perturbative viewpoint B_X and $\hat{B}_X$ are equivalent, but from a lattice perspective the latter is much better defined. Recent results for $B_K = B_K(2 \text{ GeV})$ and $B_B = B_B(m_b)$ are quoted in Table 11. Note that these values refer to bag parameters with spinor structure $VV + AA$ in the 4-fermion operator, as they appear in the SM.

2.4.1.4 BSM matrix elements There are several hadronic matrix elements for BSM operators available from the lattice. Kaon-mixing matrix elements with $VV - AA, SS + PP, SS - PP, TT$ spinor structure in the 4-fermion operator are found in [129, 130, 345, 353, 354], and $\langle \pi^0 | Q_\gamma^+ | K^0 \rangle$ is being addressed in [355]. In the literature, they go by the name of "SUSY matrix elements", but the idea is that only the (perturbatively calculated) Wilson coefficient refers to the specific BSM theory, while the (lattice-evaluated) matrix element is fully generic. Thanks to massless overlap fermions [356, 357] obeying the Ginsparg–Wilson relation [358] and hence enjoying a lattice analogue of chiral symmetry [359], it is now possible to avoid admixtures of operators with an unwanted chirality structure.

Table 10 Form factors from lattice QCD

$f_+^{K\to\pi}(0) = 0.960(5)(7)$	$N_f = 0$	–/clov	no	RomeI-Orsay [327]
$f_+^{K\to\pi}(0) = 0.952(6)$	$N_f = 2$	clov/clov	no	JLQCD [328]
$f_+^{K\to\pi}(0) = 0.968(9)(6)$	$N_f = 2$	dom/dom	no	RBC [329]
$f_+^{K\to\pi}(0) = 0.9680(16)$	$N_f = 2+1$	dom/dom	no	UKQCD/RBC [330]
$f_+^{K\to\pi}(0) = 0.962(6)(9)$	$N_f = 2+1$	stag/clov	no	FNAL/MILC/+ [331]
$f_+^{D\to\pi}(0) = 0.64(3)(6)$	$N_f = 2+1$	stag/stag	no	FNAL/MILC/+ [332]
$f_+^{D\to K}(0) = 0.73(3)(7)$	$N_f = 2+1$	stag/stag	no	FNAL/MILC/+ [332]
$f_+^{B\to\pi}(0) = 0.23(2)(3)$	$N_f = 2+1$	stag/stag	no	FNAL/MILC/+ [333]
$f_+^{B\to\pi}(0) = 0.31(5)(4)$	$N_f = 2+1$	stag/stag	yes	HPQCD [334]
$\mathcal{F}^{B\to D}(1) = 1.074(18)(16)$	$N_f = 2+1$	stag/stag	no	FNAL/MILC/+ [333]
$\mathcal{F}^{B\to D}(1) = 1.026(17)$	$N_f = 0$	–/clov	yes	RomeII [335]

Table 11 Bag parameters from lattice QCD

$B_K = 0.5746(061)(191)$	$N_f = 0$	–/dom	yes	CP-PACS [342]
$B_K = 0.55(7)$	$N_f = 0$	–/over	yes	MILC [343]
$\hat{B}_K = 0.96(10)$ [hat]	$N_f = 0$	–/wils	yes	Becirevic et al. [344]
$B_K = 0.563(21)(49)$	$N_f = 0$	–/dom	yes	RBC [345]
$B_K = 0.563(47)(07)$	$N_f = 0$	–/over	yes	BMW [129]
$\hat{B}_K = 0.789(46)$ [hat]	$N_f = 0$	–/twis	yes	ALPHA [346]
$B_K = 0.49(13)$	$N_f = 2$	clov/clov	no	UKQCD [347]
$B_K = 0.495(18)$	$N_f = 2$	dom/dom	no	RBC [348]
$B_K = 0.618(18)(135)$	$N_f = 2+1$	stag/stag	no	HPQCD/UKQCD [349]
$B_K = 0.557(12)(29)$	$N_f = 2+1$	dom/dom	no	RBC/UKQCD [350]
$B_{B_s} = 0.940(16)(22)$	$N_f = 0$	–/over	no	Orsay [351]
$B_B = 0.836(27)\binom{+56}{-62}$	$N_f = 2$	clov/clov	no	JLQCD [320]
$B_{B_s}/B_B = 1.017(16)\binom{+56}{-17}$	$N_f = 2$	clov/clov	no	JLQCD [320]
$B_{B_s}/B_B = 1.06(6)(4)$	$N_f = 2$	dom/dom	no	RBC [322]
$f_{B_s}\sqrt{\hat{B}_{B_s}} = 281(21)$ MeV	$N_f = 2+1$	stag/stag	no	HPQCD [352]
$\xi = 1.14(3)\binom{+13}{-02}$	$N_f = 2$	clov/clov	no	JLQCD [320]
$\xi = 1.33(8)(8)$	$N_f = 2$	dom/dom	no	RBC [322]

2.4.1.5 CKM matrix elements In his Lattice 2005 write-up [301], Okamoto quantifies the magnitudes of all CKM matrix elements, except $|V_{td}|$, using *exclusively lattice results* (and experimental data, of course). They are collected in Table 12. The magnitudes $|V_{ud}|, |V_{ts}|, |V_{tb}|$ may be subsequently determined if one assumes unitarity of V_{CKM}. This gives $|V_{ud}|^{\text{SM}}_{\text{Lat05}} = 0.9743(3)$, $|V_{ts}|^{\text{SM}}_{\text{Lat05}} = 3.79(53) \times 10^{-2}$ and $|V_{tb}|^{\text{SM}}_{\text{Lat05}} = 0.9992(1)$.

2.4.2 Scale setting and systematic effects

2.4.2.1 Burning $N_f + 1$ observables in N_f flavor QCD In a calculation with, say, a common ud and separate s, c quark masses, four observables must be used to set the lattice spacing and to adjust m_{ud}, m_s, m_c to their physical values (with m_{ud}, there is a practical problem, but this is immaterial to the present discussion). In general, $N_f + 1$ lattice observ-

Table 12 CKM matrix elements from lattice QCD

$|V_{us}|_{\text{Lat05}}$	$|V_{ub}|_{\text{Lat05}}$	$|V_{cd}|_{\text{Lat05}}$	$|V_{cs}|_{\text{Lat05}}$	$|V_{cb}|_{\text{Lat05}}$
0.2244(14)	$3.76(68) \times 10^{-3}$	0.245(22)	0.97(10)	$3.91(09)(34) \times 10^{-2}$

ables cannot be used to make predictions, since LQCD establishes the connection

$$\underbrace{\begin{pmatrix} M_p \\ M_\pi \\ M_K \\ M_D \\ \hline M_B \end{pmatrix}}_{\text{experiment}} \Longleftrightarrow \underbrace{\begin{pmatrix} \Lambda_{\text{QCD}} \\ m_{ud} \\ m_s \\ m_c \\ \hline m_b \end{pmatrix}}_{\text{parameters}} + \underbrace{\begin{pmatrix} f_\pi \\ f_K, B_K \\ f_D, f_{D_s} \\ \cdots \\ \hline f_B, f_{B_s} \\ B_B, B_{B_s} \end{pmatrix}}_{\text{predictions}}.$$

With infinitely precise data it would not matter which observables are sacrificed to specify the bare parameters in a given run (every observable depends a bit on each of the $N_f + 1$ parameters). In practice, the situation is different. To adjust the bare parameters in a controlled way, it is important to single out $N_f + 1$ observables that are easy to measure, do not show tremendous cut-off effects and depend strongly on one physical parameter but as weakly as possible on all other. By now it is clear that one should not use any broad resonance (e.g. the ρ), since this introduces large ambiguities [360].

Frequently, the Sommer radius r_0 [361] is used as an intermediate scale-setting quantity; e.g. the continuum limit is taken for $f_{B_s} r_0$. But the issue remains what physical distance should be identified with r_0. Typically, a quenched lattice study converts a value for $f_{B_s} r_0$ with specified statistical and systematic errors into an MeV result for f_{B_s}, assuming that r_0 is exactly 0.5 fm (the preferred value from charmonium spectroscopy), or exactly 0.47 fm (from the proton mass), or exactly 0.51 fm (from f_K). If one is interested in quenched QCD, any of these values is fine. However, if one intends to use the result for phenomenological purposes, it is more advisable to attribute a certain error to $(r_0 \text{ MeV})$ itself. For instance, one might use $r_0 = 0.49(2)$ fm. This is where the suggestion to add an extra 5% scale-setting ambiguity to most quenched results comes from. In principle, such ambiguities persist in $N_f = 2 + 1$ QCD, but they get smaller as one moves towards realistic quark masses.

2.4.2.2 Perturbative versus nonperturbative renormalization On the lattice, there are two types of renormalization. Obviously, any operator which "runs" requires renormalization. For instance, when calculating a bag parameter, the lattice result is $B_X^{\text{glue,ferm}}(a^{-1})$, where the superscript indicates the specific cut-off scheme defined by the gluon and fermion actions that have been used. In order to obtain an observable with a well-defined continuum limit, this object needs to be converted into a scheme where the pertinent scale μ is not linked to the cut-off a^{-1}. Consequently, the conversion factor in $B_X^{\overline{\text{MS}}}(\mu) = C(\mu a) B_X^{\text{glue,ferm}}(a^{-1})$ would diverge in the continuum limit, but this is immaterial, since $C(\mu a)$ is not an observable.

Besides, a finite renormalization is used for many quantities of interest. For instance, to measure f_π, one multiplies the point-like axial-vector current $A_\mu = \bar{d}\gamma_\mu \gamma_5 u$ with renormalization factor Z_A. Asymptotically (for large β), this factor behaves like $Z_A = 1 + \text{const}/\beta + O(\beta^{-2})$. Accordingly, $Z_A(\beta)$ may be calculated either in weak coupling perturbation theory or nonperturbatively. For some actions, both avenues have been pursued, and sometimes it was found that within perturbation theory it is difficult to estimate the error (there may be big shifts when going from 1-loop to 2-loop, and/or all perturbative calculations of $Z_A(\beta)$ may differ significantly from the outcome of a nonperturbative determination). The results with $N_f = 2 + 1$ staggered quarks rely on perturbation theory, and some experts fear that some of the renormalization factors may be less precisely known than what is currently believed. On the other hand, one might argue that these actions involve UV-filtering ("link-fattening") and may be less prone to such uncertainties than unfiltered ("thin-link") actions. These issues are under active investigation.

2.4.2.3 Summary of extrapolations Lattice calculations are done in a euclidean box $L^3 \times T$ with a finite lattice spacing a. From a field-theoretic viewpoint only the $T \to \infty$ limit is needed to define particle properties (to locate the pole of an Euclidean Green's function and to extract the residue, the $t \to \infty$ behavior of the correlation function $C(t)$ needs to be studied). All other limits are taken subsequently in the physical observables. A summary of all extrapolations involved is:

(1) $T \to \infty$ or removal of excited states contamination (in practice, choosing $T \gg L$ is sufficient);
(2) $a \to 0$ or removal of discretization effects (at fixed $V = L^3$ and fixed $M_{\text{had}} L$);
(3) $V \to \infty$ or removal of (spatial) finite-size effects (at fixed renormalized quark masses);
(4) $m_{ud} \to m_{ud}^{\text{phys}}$ or chiral extrapolation;
(5) $m_b \to m_b^{\text{phys}}$ or heavy-quark extrapolation/interpolation (not with Fermilab formulation).

Extrapolations 1–3 are standard in the sense that one knows how to control them. The chiral extrapolation is far from innocent, since it is not really justified to use chiral perturbation theory [362, 363] if one cannot clearly identify chiral logs in the data, and it is hard to tell such logs from lattice artifacts and finite-size effects. The entries with the smallest error bars among the $N_f = 2 + 1$ data quoted above stem from simulations with the staggered action. In such studies, the extrapolations 2 and 4 are performed by means of staggered chiral perturbation theory [364, 365], using a large number of fitting parameters. This makes it hard to judge whether the quoted error is realistic, but at least the "post processing" is done in a field-theoretic framework (no modeling). The fifth point depends on the details of the heavy-quark formulation (NRQCD, HQET, Fermilab) employed, but eventually, with $a^{-1} \simeq 10$ GeV and higher, one could use a standard relativistic action.

2.4.2.4 Conceptual issues
Besides these practical aspects, there might be conceptual issues regarding the theoretical validity of certain steps. In the past, the so-called quenched approximation has been used, where the functional determinant is neglected. While fundamentally uncontrolled, it seems to have little impact on the final result of a phenomenological study—as long as no flavor singlet quantity is measured, final-state interactions are not particularly important, and the long-distance physics involved does not exceed ~ 1 fm (i.e. for $M_\pi > 200$ MeV, which still is the case in present simulations). State-of-the-art calculations use the partially quenched framework [366–368], which, despite its name, is *not* a half-way extrapolation from quenched to unquenched. It amounts to having, besides $m_{ud}^{\text{sea}} = m_{ud}^{\text{val}}$, also data with $m_{ud}^{\text{sea}} > m_{ud}^{\text{val}}$, which typically stabilize the extrapolation to $m_{ud}^{\text{sea}} = m_{ud}^{\text{val}} = m_{ud}^{\text{phys}}$. But even with the determinant included, things remain somewhat controversial. The rooting procedure with staggered quarks (to obtain $N_f = 2 + 1$, the square-root of $\det(D_{m_{ud}}^{\text{st}})$ and the fourth-root of $\det(D_{m_s}^{\text{st}})$ is taken) has been the subject of a lively debate. Much theoretical progress on understanding its basis has been achieved—for a summary see the plenary talks on this point at the last three lattice conferences [369–372].

2.4.3 Prospects of future error bars

Future progress on the precision of lattice calculations of QCD matrix elements will hopefully come from a variety of improvements, including a growth in computer power, the development of better algorithms, the construction of better interpolating fields, and the design of better relativistic and heavy quark actions. Some of these factors are easier to forecast than others. For instance, the amount of CPU power is a rather monotonic function of time (for the lattice community as a whole, not for an individual collaboration). By contrast, progress at the algorithmic frontier comes in evolutionary steps—we have just witnessed a dramatic improvement of full QCD hybrid Monte Carlo algorithms [373]. The last two points are somewhere in between; here, every collaboration has its own preferences, which are largely driven by the kind of physics it wants to address. Below, some estimates for future error bars on quantities relevant to flavor physics will be given, but it is important to keep in mind two caveats.

The first caveat is a reminder that the anticipated percentage errors quoted below belong to a rather restricted class of observables. In the foreseeable future, lattice methods can only be competitive for processes where the following conditions hold simultaneously:

- only one hadron in initial and/or final state,
- all hadrons stable (none near thresholds),
- all valence quarks in connected graphs,
- all momenta significantly below cut-off scale $2\pi/a$.

This is the case for the quantities discussed below, but it means that quick progress on other interesting quantities, such as $f^{B \to \rho}(q^2)$, is not likely.

The second caveat concerns the role of the theoretical uncertainties, as discussed in the previous paragraph. For instance, some of the estimates given below assume that certain (finite) renormalization (i.e. matching) factors will be known at the 2-loop level. Such calculations are tedious and rely on massive computer algebra (the lattice regularization reduces the full Lorentz symmetry, resulting in a proliferation of terms). Accordingly, future progress of such calculations is difficult to predict. In the same spirit, one should mention that in the predictions discussed below, it is assumed that for $M_\pi = (250–350)$ MeV, one is in a regime where chiral perturbation theory applies and can be used to further extrapolate the lattice data to the physical pion mass. In the unlikely event that for some specific process this is not the case, the corresponding prediction would undergo substantial revision.

With these caveats in mind, it is interesting to discuss the projected error bars as they are released by some lattice groups. For instance, MILC has a detailed "road-map" of their expected percentage errors (including statistical and theoretical uncertainties) for a number of matrix elements. They are collected in the following Table 13, which they kindly provide. By far the most ambitious plans are those of HPQCD. They have just released numbers for f_{D_s} and f_{D_s}/f_D with a claimed accuracy of 1.3% and 0.8%, respectively [310]. They plan on computing f_{B_s} and f_B as well as the $B \to \pi$ form factor at $q^2 \simeq 16\,\text{GeV}^2$ to 4%. Finally, they envisage releasing the ratio f_{B_s}/f_B with 2% accuracy and ξ with 3% accuracy by the end of 2007.

In this context, it is worth pointing out that progress in other fields, in particular in experiment, has the potential to ease the task for the lattice community. For instance, quoting the vector form factor f_+ for semileptonic $B \to \pi \ell \nu$ decay at $q^2 = 0$ is not the best thing to ask for from the lattice,

Table 13 Prospects for lattice uncertainties (MILC Collab.). The $B \to \pi \ell \nu$ form factor is taken at $q^2 = 16\,\text{GeV}^2$

	Lat'06	Lat'07	2–3 yrs.	5–10 yrs.
f_{D_s}, f_{B_s}	10	7	5	3–4
f_D, f_B	11	7–8	5	4
$f_B \sqrt{B_B}$	17	8–13	4–5	3–4
ξ	–	4	3	1–2
$(B, D) \to (K, \pi)\ell\nu$	11	8	6	4
$B \to (D, D^*)\ell\nu$	4	3	2	1

since a long extrapolation is needed (see Sect. 2.4.1.2). Still, in the past, this was common practice, since there was very limited experimental information available. In the meantime, the situation has changed. Now, rather precise information on the shape of this form factor (via binned differential decay rate data $d\Gamma/dq^2$) is available, and only the absolute normalization is difficult to determine in experiment (see e.g. [374] for a detailed analysis). As a result, MILC and HPQCD give the future lattice precision attainable at $q^2 = 16$ GeV^2, i.e. at a momentum transfer which can be reached in the simulation.

3 New physics in benchmark channels

3.1 Radiative penguin decays[9]

The FCNC transitions $b \to s\gamma$ and $b \to d\gamma$ are among the most valuable probes of flavor physics. They place stringent constraints on a variety of NP models, in particular on those where the flavor-violating transition to a right-handed s- or d-quark is not suppressed, in contrast to the SM. Assuming the SM to be valid, the combination of these two processes offers a competitive way to extract the ratio of CKM matrix elements $|V_{td}/V_{ts}|$. This determination is complementary to the one from B mixing and to the one of the SM unitarity triangle based on the tree-level observables $|V_{ub}/V_{cb}|$ and the angle γ. Other interesting observables are the CP and isospin asymmetries and photon polarization. Radiative B decays are also characterized by the large impact of short-distance QCD corrections [375]. Considerable effort has gone into the calculation of these corrections, which are now approaching next-to-next-to-leading order (NNLO) accuracy [376–388]. On the experimental side, both exclusive and inclusive $b \to s\gamma$ branching ratios are known with good accuracy, $\sim 5\%$ for $B \to K^*\gamma$ and $\sim 7\%$ for $\bar{B} \to X_s\gamma$ [389], while the situation is less favorable for $b \to d\gamma$ transitions: measurements are only available for exclusive channels. Here, we shall discuss first the inclusive modes and then the exclusive ones. We shall begin with an overview of the current status of the SM calculations and later consider the situation for models of NP.

3.1.1 $\bar{B} \to X_{(s,d)}\gamma$ inclusive (theory)

The inclusive decay rate of the $\bar{B}$-meson ($\bar{B} = \bar{B}^0$ or B^-) is known to be well approximated by the perturbatively calculable partonic decay rate of the b-quark:

---

[9]Section coordinators: P. Gambino, A Golutvin.

$\Gamma(\bar{B} \to X_s\gamma)_{E_\gamma > E_0}$
$= \Gamma(b \to X_s^{\text{parton}}\gamma)_{E_\gamma > E_0} + \mathcal{O}\left(\frac{\Lambda^2}{m_b^2}, \frac{\Lambda^2}{m_c^2}, \frac{\Lambda\alpha_s}{m_b}\right),$ (65)

with $\Lambda \sim \Lambda_{\text{QCD}}$ and E_0 the photon energy cut in the $\bar{B}$-meson rest frame. The nonperturbative corrections on the r.h.s. of the above equation were analysed in Refs. [390–397]. There are also additional nonperturbative effects that become important when E_0 becomes too large ($E_0 \sim m_b/2 - \Lambda$) [398–400] or too small ($E_0 \ll m_b/2$) [401, 402].

It is convenient to consider the perturbative contribution first. At the leading order (LO), it is given by one-loop diagrams like the one in Fig. 6. Dressing this diagram with one or two virtual gluons gives examples of the next-to-leading order (NLO) and the NNLO diagrams, respectively. The gluon and light-quark bremsstrahlung must be included as well. The current experimental accuracy (see (67)) can be matched on the theoretical side only after including the NNLO QCD corrections [376].

At each order of the perturbative series in α_s, large logarithms $L = \ln M_W^2/m_b^2$ are resumed by employing a low-energy effective theory that arises after decoupling the top quark and the electroweak bosons. For example, the LO includes all $\alpha_s^n L^n$ terms, the NLO all $\alpha_s^n L^{n-1}$ terms, etc. Weak interaction vertices (operators) in this theory are either of dipole type ($\bar{s}\sigma^{\mu\nu}bF_{\mu\nu}$ and $\bar{s}\sigma^{\mu\nu}T^a bG_{\mu\nu}^a$) or contain four quarks ($[\bar{s}\Gamma b][\bar{q}\Gamma'q]$). Coupling constants at these vertices (Wilson coefficients) are first evaluated at the electroweak renormalization scale $\mu_0 \sim m_t, M_W$ by solving the so-called *matching* conditions. Next, they are evolved down to the low-energy scale $\mu_b \sim m_b$ according to the effective theory renormalization group equations (RGE). The RGE are governed by the operator *mixing* under renormalization. Finally, one computes the *matrix elements* of the operators, which in the perturbative case amounts to calculating on-shell diagrams with single insertions of the effective theory vertices.

The NNLO matching and mixing are now completely known [377–381]. The same refers to those matrix elements that involve the photonic dipole operator alone [382–386]. Matrix elements involving other operators are known at the NNLO either in the so-called large-β_0 approximation [387] or in the formal $m_c \gg m_b/2$ limit [388]. The recently published NNLO estimate [376]

$\mathcal{B}(\bar{B} \to X_s\gamma)_{E_\gamma > 1.6 \text{ GeV}} = (3.15 \pm 0.23) \times 10^{-4}$ (66)

Fig. 6 Sample LO diagram for the $b \to s\gamma$ transition

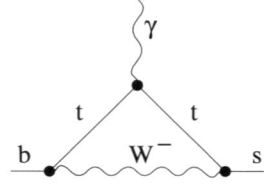

is based on this knowledge. The four types of uncertainties: nonperturbative (5%), parametric (3%), higher-order (3%) and m_c-interpolation ambiguity (3%) have been added in quadrature in (66) to obtain the total error. The main uncertainty is due to unknown $O(\alpha_s \Lambda/m_b)$ nonperturbative effects related to the matrix elements of four-quark operators (see [392]) for which no estimate exists. Similar effects related to dipole operators have been recently estimated in the vacuum insertion approximation [397].

As far as inclusive $b \to d\gamma$ decays are concerned, their measurement is quite challenging. Moreover, due to nonperturbative effects that are suppressed only by Λ_{QCD}/m_b, their theoretical accuracy is not much better than in the exclusive case. On the other hand, the experimental prospects in the exclusive case are brighter.

3.1.2 $\bar{B} \to X_{(s,d)}\gamma$ inclusive (experiment)

3.1.2.1 Present status The inclusive $b \to s\gamma$ branching fraction has been measured by BaBar, Belle and CLEO using both a sum of exclusive modes and a fully inclusive method [122, 439, 444, 445]. The inclusive measurement utilizes the continuum subtraction technique using the off-resonance data sample. In order to suppress the continuum contribution, the BaBar measurement uses lepton tags. The analyses of Belle and CLEO are untagged, and their systematic errors are dominated by continuum subtraction. The accuracy of the BaBar measurement is limited by the subtraction of backgrounds from other B decays. The Belle measurement extends the minimum photon energy down to 1.8 GeV, which covers 95% of the entire photon spectrum. All $b \to s\gamma$ branching fractions measured by BaBar, Belle and CLEO using both exclusive and inclusive methods agree well, giving a new world average of [389]

$$\mathcal{B}(\bar{B} \to X_s\gamma)_{E_\gamma > 1.6\,\text{GeV}} = (3.55 \pm 0.30) \times 10^{-4}. \quad (67)$$

This is a bit high compared to the recent NNLO calculation in (66).

The published measurements are based on only a fraction of the available statistics, but improvements with the full data set will be limited by systematic errors: from the fragmentation of the hadronic X_s in the sum of exclusive modes and from the subtraction of backgrounds in the fully inclusive method. A new method measures the spectrum of photons recoiling against a sample of fully reconstructed decays of the other B. This is currently statistics limited but should eventually have lower systematic errors. A final accuracy of 5% on the inclusive $b \to s\gamma$ branching fraction looks achievable. As for the $b \to d\gamma$ inclusive branching fraction, the measurement using a sum of exclusive modes is under study and looks to be feasible with the full datasets from the B-factories. Preliminary results have appeared in [446].

We note that the $b \to s\gamma$ spectral shape also provides valuable information on the shape functions in B meson decays. This information has been used as an input in the extraction of V_{ub} from inclusive $b \to u\ell\nu$ decays [447, 448].

Measurements of the direct CP asymmetries, published for inclusive $b \to s\gamma$ by BaBar [449] and Belle [441], show no deviation from zero. All these measurements will be statistics limited at current B-factories and will not reach the sensitivity to probe the SM prediction.

3.1.2.2 Future prospects One would expect a substantial improvement of the experimental precision for inclusive measurements at future B-factories. Studies have been performed for SuperKEKB/Belle with 50 ab^{-1} data, assuming the existing Belle detector [839]. This is probably a reasonable assumption in many cases, since the expected improvements in the detector, especially in the calorimeter, would be just sufficient to compensate for the necessity to cope with the increased background.

For the measurements that are fully statistics dominated now, it is straightforward to extrapolate to a larger integrated luminosity. The error for the direct asymmetry measurement of $b \to s\gamma$ would be $\pm 0.009(\text{stat}) \pm 0.006(\text{syst})$ for 5 ab^{-1} or $\pm 0.003(\text{stat}) \pm 0.002(\text{syst}) \pm 0.003(\text{model})$ for 50 ab^{-1}. A small systematic error implies that kaon charge asymmetries are well under control. The size of the total error is still much larger than the SM estimate, but a few percent deviation from zero due to New Physics could be identified.

One would also expect a better measurement of the branching fraction of $B \to X_s\gamma$. Although the background level is more and more severe, it would be possible to lower the E_γ bound by 0.1 GeV with roughly twice more data, and it would be possible to measure the branching fraction for $E_\gamma > 1.5$ GeV with a few ab^{-1}. Beyond that, one may need to make use of the B-tag events or $\gamma \to e^+e^-$ conversion to suppress backgrounds from continuum and neutral hadrons. Another challenging measurement would be inclusive $b \to d\gamma$ to improve our knowledge on $|V_{td}/V_{ts}|$ besides the Δm_s measurement, since the one from exclusive $B \to \rho\gamma$ will hit the theory limit soon. The first signal may be measured with 5 ab^{-1} using the sum-of-exclusive method, with a total error of $\sim 25\%$, of which the systematic error would already be dominant.

3.1.3 Exclusive $b \to (s, d)\gamma$ transitions (theory)

Whereas the inclusive modes can be essentially computed perturbatively, the treatment of exclusive channels is more complicated. QCD factorization [282, 283, 403–406] has provided a consistent framework allowing one to write the relevant hadronic matrix elements as

$$\langle V\gamma|Q_i|B\rangle = \left[T_1^{B\to V}(0) T_i^{\mathrm{I}} + \int_0^1 d\xi\, du\, T_i^{\mathrm{II}}(\xi, u) \phi_B(\xi) \phi_{2;V}^\perp(v) \right] \cdot \epsilon. \tag{68}$$

Here ϵ is the photon polarization four-vector, Q_i is one of the operators in the effective Hamiltonian for $b \to (s,d)\gamma$ transitions, $T_1^{B\to V}$ is a $B \to V$ transition form factor, and ϕ_B, $\phi_{2;V}^\perp$ are leading-twist light-cone distribution amplitudes of the B meson and the vector meson V, respectively. These quantities are universal nonperturbative objects and describe the long-distance dynamics of the matrix elements, which is factorized from the perturbative short-distance interactions included in the hard-scattering kernels T_i^{I} and T_i^{II} (see Sect. 2 for a more general discussion).

Equation (68) is sufficient to calculate observables that are of $O(1)$ in the heavy quark expansion, like $\mathcal{B}(B \to K^*\gamma)$. For $\mathcal{B}(B \to (\rho,\omega)\gamma)$, on the other hand, power-suppressed corrections play an important role, for instance weak annihilation, which is mediated by a tree-level diagram. In this case, the parametric suppression by one power of $1/m_b$ is alleviated by an enhancement factor $2\pi^2$ relative to the loop-suppressed contributions at leading order in $1/m_b$. Power-suppressed contributions also determine the time-dependent CP asymmetry in $B \to V\gamma$, see Refs. [265, 407–409], as well as isospin asymmetries [410]—all observables with a potentially large contribution from new physics. A more detailed analysis of power corrections in $B \to V\gamma$, including also B_s decays, was given in [270].

The nonperturbative quantities entering (68), i.e. $T_1^{B\to V}$ and the light-cone distribution amplitudes, at present are not provided by lattice, although this may change in the future. The most up-to-date predictions come from QCD sum rules on the light-cone, which are discussed in Sect. 2.3. In Ref. [266], the following result was obtained for the branching fraction ratio:

$$R \equiv \frac{\bar{\mathcal{B}}(B \to (\rho,\omega)\gamma)}{\bar{\mathcal{B}}(B \to K^*\gamma)}$$
$$= \frac{|V_{td}|^2}{|V_{ts}|^2} \big(0.75 \pm 0.11(\xi)$$
$$\pm 0.02(\text{UT param.}, O(1/m_b)) \big), \tag{69}$$

where $\xi \equiv T_1^{B\to K^*}(0)/T_1^{B\to \rho}(0) = 1.17 \pm 0.09$ (Sect. 2.3). The error of ξ is dominated by that of the tensor decay constants $f_{\rho,K^*}^\perp$, which currently are known to about 10% accuracy [266]; a new determination on the lattice is under way [411], which will help to reduce the error on ξ to ± 0.05. Concerning (69), two remarks are in order. First, the smallness of the $1/m_b$ correction are due to an accidental CKM suppression. Second, the $1/m_b$ corrections have a dependence on $|V_{td}/V_{ts}|$ as well, originating from a discrimination in the u- and c-loops. Equation (69) allows one to determine $|V_{td}/V_{ts}|$ from experimental data; at the time of writing (February 07), HFAG quotes $R_{\exp} = 0.028 \pm 0.005$, from which one finds $|V_{td}/V_{ts}|_{B\to V\gamma}^{\mathrm{HFAG}} = 0.192 \pm 0.014(\text{th}) \pm 0.016(\text{exp})$, which agrees very well with the results from global fits [8, 120]. The branching ratios themselves carry a larger uncertainty, because the individual $T_1^{B\to V}$ are less accurately known than their ratio. The explicit results can be found in [270]. The isospin asymmetry in $B \to K^*\gamma$ was first studied in Ref. [410] and found to be very sensitive to penguin contributions; it was updated in [270] with the result

$$A_I(K^*) = \frac{\Gamma(\bar{B}^0 \to \bar{K}^{*0}\gamma) - \Gamma(B^- \to K^{*-}\gamma)}{\Gamma(\bar{B}^0 \to \bar{K}^{*0}\gamma) + \Gamma(B^- \to K^{*-}\gamma)}$$
$$= (5.4 \pm 1.4)\%; \tag{70}$$

the present (February 07) experimental result from HFAG [389] is $(3 \pm 4)\%$. The isospin asymmetry for $B \to \rho\gamma$ depends rather crucially on the angle γ [270]. The last observable in exclusive $B \to V\gamma$ transitions to be discussed here is the time-dependent CP asymmetry, which is sensitive to the photon polarization. Photons produced from the short-distance process $b \to (s,d)\gamma$ are predominantly left-polarized, with the ratio of right- to left-polarized photons given by the helicity suppression factor $m_{s,d}/m_b$. For $B \to K^*\gamma$, where direct CP violation is doubly CKM suppressed, the CP asymmetry is given by

$$A_{\mathrm{CP}}(t) = \frac{\Gamma(\bar{B}^0(t) \to \bar{K}^{*0}\gamma) - \Gamma(B^0(t) \to K^{*0}\gamma)}{\Gamma(\bar{B}^0(t) \to \bar{K}^{*0}\gamma) + \Gamma(B^0(t) \to K^{*0}\gamma)}$$
$$= C\cos(\Delta m_B t) + S\sin(\Delta m_B t) \tag{71}$$

with $S_{K^*\gamma} = -(2 + O(\alpha_s))\sin(2\beta) m_s/m_b + \cdots \approx -3\%$ being the contribution induced by the electromagnetic dipole operator O_7. The dots denote additional contributions induced by $b \to s\gamma g$, which are not helicity suppressed but involve higher (three-particle) Fock states of the B and K^* mesons. The dominant contributions to the latter, due to c-quark loops, have been calculated in Ref. [265] from QCD sum rules on the light-cone in an expansion in inverse powers of the charm mass and updated for all other channels in [270]. A calculation of the charm-loop contribution without reference to a $1/m_c$ expansion is in preparation [412] and shows that there is a large strong phase. The u-quark loop contributions are essential for $b \to d$ transitions, since they are of the same CKM-order as the c-quark loops: a new method for their estimation was devised in [270], building on earlier ideas developed for $B \to \pi\pi$ [413]. In Table 14 we show the calculations of S for several channels.

Table 14 Standard Model predictions of the time-dependent asymmetry S (see (71)) for exclusive $b \to (s,d)\gamma$ modes

$S_{V\gamma}$	$B \to \rho$	$B \to \omega$	$B \to K^*$	$B_s \to \bar{K}^*$	$B_s \to \phi$
in %	0.2 ± 1.6	0.1 ± 1.7	$-(2.3 \pm 1.6)$	0.3 ± 1.3	$-(0.1 \pm 0.1)$

Table 15 Branching fraction measurements at BaBar, Belle and CLEO for exclusive $b \to (s,d)\gamma$ modes

Decay	$B^+ \to K^{*+}\gamma$	$B^0 \to K^{*0}\gamma$	$B^+ \to \rho^+\gamma$	$B^0 \to \rho^0\gamma$	$B^0 \to \omega\gamma$
BR/10^{-6}	40.3 ± 2.6	40.1 ± 2.0	$0.88^{+0.28}_{-0.26}$	$0.93^{+0.19}_{-0.18}$	$0.46^{+0.20}_{-0.17}$

This class of observables is interesting, because any experimental signal much larger than 2% will constitute an unambiguous signal of NP. Scenarios beyond the SM that do modify S must include the possibility of a spin-flip on the internal line which removes the helicity suppression of γ_R. Examples include left–right symmetric models and non-MFV SUSY. To date the experimental result is $S_{\text{HFAG}} = -(28 \pm 26)\%$.

3.1.4 Exclusive $b \to (s,d)\gamma$ transitions (experiment)

3.1.4.1 Present status Many exclusive $b \to (s,d)\gamma$ modes have been studied by BaBar, Belle and CLEO. Results for several important channels are collected in Table 15 [389]. The results on the $B \to \rho\gamma$, $B \to \omega\gamma$ branching fractions are still statistics limited, but by the end of the B factories it is likely that the theoretical uncertainties will be the most significant factor.

Direct CP asymmetries have been published for $B \to K^*\gamma$ and $B \to K^+\phi\gamma$ decays [440, 450, 451]. The time-dependent CP asymmetry has been measured [442, 443, 452] using the technique of projecting the K_S vertex back to the beam axis for a large sample of $B \to K^{*0}\gamma \to K_S^0 \pi^0 \gamma$ and $B \to K_S^0 \pi^0 \gamma$ decays in the high $K\pi$-mass range. In the near future, similar measurements using other exclusive radiative decay modes such as $B^0 \to K_S^0 \phi\gamma$, for which $\phi \to K^+K^-$ provides the B-decay vertex measurement, could provide similar constraints.

3.1.4.2 Future prospects at LHCb A systematic study of CP violation in radiative penguin B decays will be performed at LHCb using a dedicated high-p_T photon trigger [453]. Due to small branching ratios of order 10^{-5}–10^{-6}, their reconstruction requires a drastic suppression of backgrounds from various sources, in particular combinatorial background from $b\bar{b}$ events, containing primary and secondary vertices and characterized by high charged and neutral multiplicities.

The background suppression exploits the generic properties of beauty production in pp collisions. The large mass of beauty hadrons results in hard transverse momentum spectra of secondary particles. The large lifetime, $\langle\beta\gamma c\tau\rangle \sim 5$ mm, results in a good isolation of the B decay vertex and in the inconsistency of tracks of B-decay products with the reconstructed pp-collision vertex.

The selection procedure was optimized on the example of $B^0 \to K^{*0}\gamma \to K^+\pi^-\gamma$ decay [454], which LHCb considers as a control channel for the study of systematic errors common for radiative penguin decays. The selection cuts, based on using the two-body kinematics and various geometrical cuts on the primary and secondary vertices, were applied to 34 million fully simulated $b\bar{b}$ events. The invariant mass distribution for the selected events, shown in Fig. 7, corresponds to a data sample collected in 13 min of LHCb running at nominal luminosity of 2×10^{32} cm^{-2} s^{-1}. LHCb expects the yield for $B^0 \to K^{*0}\gamma$ decays to be 68k signal events per 2 fb^{-1} of accumulated data with background to signal ratio 0.60 ± 0.16. For $B_s \to \phi\gamma$ decays, the corresponding yield is estimated to be 11.5k with the background to signal ratio less than 0.55 at 90% C.L. The measurement of $B^0 \to K^{*0}\gamma$ decay looks also feasible at ATLAS [455].

Similar to $B^0 \to K_s^0 \pi^0 \gamma$ decays, the time-dependent CP-asymmetry sensitive to the photon polarization can also be measured in $B_s \to \phi\gamma$ decays, provided that the proper time resolution is sufficient to resolve B_s–$\bar{B}_s$ oscillations. The proper time resolution depends on the kinematics and topology of particular B_s candidates, mainly on the opening angle between kaons from ϕ decays. The sensitivity of this measurement is presently under study at LHCb.

For the future B-factory, scaling the error of the measured time-dependent CP violation asymmetry for the $B^0 \to K_s^0 \pi^0 \gamma$ channel, one would expect a statistical accuracy of about 0.1 at 5 ab^{-1} or 0.03 at 50 ab^{-1}.

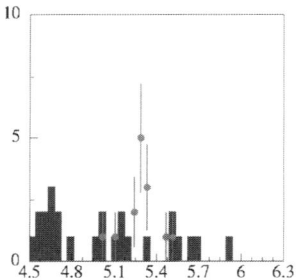

Fig. 7 The invariant mass distribution for selected $B^0 \to K^{*0}\gamma$ candidates from a $b\bar{b}$ inclusive sample. The points indicate true $B^0 \to K^{*0}\gamma$ events, and the filled histogram represents combinatorial background

LHCb also studied the possibility to measure the photon polarization in the radiative decays of polarized beauty baryons, like $\Lambda_b \to \Lambda\gamma$, using the angular asymmetry between the Λ_b spin and the photon momentum combined with the $\Lambda^0 \to p\pi$ decay polarization [456–458].

3.1.5 New Physics calculations and tools

New Physics affects the matching conditions for the Wilson coefficients of the operators in the low-energy effective theory and may even induce sizable coefficients for operators that have negligible or vanishing coefficients in the SM. The theoretical accuracy of the predictions for radiative B decays in extensions of the SM is far from the accuracy achieved in the SM. Complete NLO matching conditions are available for the MSSM with Minimal Flavor Violation (MFV) and/or large $\tan\beta$, as well as for a class of nonsupersymmetric models [425] that includes Multi-Higgs-Doublet-Models and Left-Right symmetric (LR) models. The unknown NNLO contributions to the matching conditions beyond the SM are unlikely to be numerically relevant at present.

3.1.5.1 Summary of New Physics calculations
Here is a brief summary of recent calculations and analyses in the most popular New Physics scenarios.

- **2HDMs** have been studied in full generality at NLO [83, 414, 415]. In the type-II 2HDM, $\mathcal{B}(\bar{B} \to X_s\gamma)$ places a strong bound on the mass of the charged Higgs boson, $M_{H^+} > 295$ GeV at 95% C.L., independently of the other 2HDM parameters [376]. This is much stronger than other available direct and indirect constraints on M_{H^+}.
- **MSSM.** The complete LO contributions in the MSSM have been known since the early nineties [416–423], but the NLO analysis is still incomplete to date. New sources of flavor violation generally arise in the MSSM, making a complete analysis quite complicated even at the LO [424]. While $\bar{B} \to X_s\gamma$ does place important constraints on the MSSM parameter space, they depend sensitively on the exact SUSY scenario and are hard to summarize because of the large number of parameters.
  - *MFV.* In the MFV scenario, the NLO QCD calculation of $\bar{B} \to X_s\gamma$ is now complete: the two-loop diagrams involving gluons were computed in [84, 425], and the two-loop diagrams involving gluinos were more recently computed in [426, 427]. Since weak interactions affect the squark and quark mass matrices in a different way, their simultaneous diagonalization is not RG-invariant and MFV can be imposed only at a certain renormalization scale. The results of [426, 427] therefore depend explicitly on the MFV scale, which is determined by the mechanism of SUSY breaking.
  - *Large* $\tan\beta$. In the limit of heavy superpartners, the Higgs sector of the MSSM is modified by nondecoupling effects and can differ substantially from the type-II 2HDM. Large higher-order contributions to $\bar{B} \to X_s\gamma$ in that limit originate from terms enhanced by $\tan\beta$ factors and can be taken into account to all orders in an effective Lagrangian approach [10, 23, 24, 29, 428]. In fact, large $\tan\beta$ and logs of M_{susy}/M_W have been identified in [23] as dominant NLO QCD contributions in MFV with heavy squarks. Ref. [33] recently studied the $\tan\beta$-enhanced effects when MFV is valid at the GUT scale and additional flavor violation in the squark sector is generated by the RGE of the soft SUSY-breaking parameters down to the weak scale.
  - *Beyond MFV.* In the more general case of arbitrary flavor structure in the squark sector, experimental constraints on $b \to s$ transitions have been recently studied at LO [107, 108] and including $\tan\beta$-enhanced NLO effects [111–114]: radiative decays play a central role in these analyses, and the constraints are quite strong for some of the flavor-violating parameters (see however [429] for a scenario in which radiative corrections weaken the constraints).
- **Large extra dimensions.** In these models, the contribution to $\bar{B} \to X_s\gamma$ from the Kaluza–Klein excitations of the SM particles can induce bounds on the size of the additional dimension(s). This has been studied in Refs. [17, 430] for the case of flat extra dimensions and in Refs. [431–433] for the case of warped extra dimensions.
- **Little Higgs.** In these models, the Higgs boson is regarded as the pseudo-Goldstone boson of a global symmetry that is broken spontaneously at a scale much larger than the weak scale. The most extensively studied version of the model, the Littlest Higgs, predicts the existence of heavy vector bosons, scalars and quarks. The contribution to $\bar{B} \to X_s\gamma$ from these new particles has been studied in Refs. [142, 146] for the original Littlest Higgs model and in Ref. [159] for the model in which an additional T-parity and additional particles are introduced to preserve the $SU(2)$ custodial symmetry.
- **LR models.** The contributions of left–right-symmetric models to $\bar{B} \to X_s\gamma$ are known at the NLO [425], but no recent phenomenological analysis is available.

An alternative to the analysis of $\bar{B} \to X_s\gamma$ in different models consists in constraining the Wilson coefficients of the effective theory. This *model-independent approach* has been applied combining various B-decay modes and neglecting operators that do not contribute in the SM [434, 435]. While $\mathcal{B}(\bar{B} \to X_s\gamma)$ fixes only $|C_7(m_b)|$, the sign can be learned from $B \to X_s\ell^+\ell^-$ [188].

3.1.5.2 MSSM tools for $\bar{B} \to X_s\gamma$
Several public codes (see also Sect. 1.5) that determine the MSSM mass spec-

trum and other SUSY observables contain MSSM calculations of $\mathcal{B}(\bar{B} \to X_s \gamma)$ in various approximations. In microMEGAs [436], the SM part of the calculation is performed at NLO, while the MSSM contributions are implemented following [23]. The calculation in SuSpect [437] includes also the NLO gluon corrections to the chargino contributions from [84] in the case of light squarks. In contrast, SPheno [207] and FeynHiggs [204, 438] include the SUSY contributions only at LO, but they allow for a general flavor structure in the squark sector. A computer code for the NLO QCD calculation of $\mathcal{B}(\bar{B} \to X_s \gamma)$ in the MSSM with MFV [426, 427] has recently been published [191].

3.2 Electroweak penguin decays[10]

3.2.1 Introduction

In the SM, the electroweak penguin decays $b \to s(d)\ell^+\ell^-$ are only induced at the one-loop level, leading to small branching fractions and thus a rather high sensitivity to contributions from physics beyond the SM. On the partonic level, the main contribution to the decay rates comes from the semi-leptonic operators $\mathcal{O}_9$, $\mathcal{O}_{10}$ and from the electromagnetic dipole operator $\mathcal{O}_7^\gamma$ in the effective Hamiltonian for $|\Delta B| = |\Delta S(D)| = 1$ transitions [214]. Radiative corrections induce additional sensitivity to the current–current and strong penguin operators $\mathcal{O}_{1-6}$ and $\mathcal{O}_8^g$. Part of these effects are process-independent and can be absorbed into effective Wilson coefficients. In certain regions of phase-space and for particular exclusive and inclusive observables, hadronic uncertainties are under reasonable control, and the corresponding short-distance Wilson coefficients in and beyond the SM can be tested with sufficient accuracy.

Because of their small branching fractions, these decays are experimentally challenging. Their detection requires excellent triggering and identification of leptons, with low misidentification rates for hadrons. Combinatorial backgrounds from semileptonic B and D decays must be managed, and backgrounds from long-distance contributions, such as $B \to J/\psi X_s$, must be carefully vetoed. Once identified, their interpretation (particularly the angular distributions) requires disentangling the contributing hadronic final states. Most of these experimental problems can be managed by confining studies to the simplest exclusive decay modes. Leptonic states are restricted to e^+e^- and $\mu^+\mu^-$, and hadronic states are the simplest one- or two-particle varieties, typically K, K^*, ϕ, or Λ. More inclusive studies are significantly less sensitive but have the advantage of a simpler theoretical interpretation. Fortunately, measuring fully reconstructed decays to final states with leptons (especially muons) is a strength of all future proposed B-physics experiments, hence all are capable of contributing to this topic in the LHC era.

3.2.2 Theory of electroweak penguin decays

3.2.2.1 Inclusive decays
The heavy quark expansion and the operator product expansion in the theory of inclusive $\bar{B} \to X_s \ell^+\ell^-$ decays allow one to calculate radiative QCD and QED corrections to the partonic decay rate and to parametrize and estimate power corrections to the hadronic matrix elements in a systematic way. The calculation of NNLO QCD corrections has (essentially) been completed recently [377, 379, 459–465]. These reduce the perturbative uncertainties below 10%. Also subleading $\Lambda_{\rm QCD}^2/m_c^2$ and $\Lambda_{\rm QCD}^2/m_b^2$, $\Lambda_{\rm QCD}^3/m_b^3$ corrections [390, 392, 466–470] as well as finite bremsstrahlung effects [471, 472] are available in the literature.

At this level of accuracy, QED effects become important, too. For instance, the scale ambiguity from $\alpha_{\rm em}(\mu)$ between $\mu = M_W$ and $\mu = m_b$ alone results in an uncertainty of about $\pm 4\%$. QED corrections to the Wilson coefficients have been calculated in [465], and the results for the two-loop anomalous dimension matrices have been confirmed in [473]. QED bremsstrahlung contributions where the photon is collinear with one of the outgoing leptons are enhanced by $\ln(m_b^2/m_\ell^2)$. They disappear after integration over the whole available phase space but survive and remain numerically important when q^2 is restricted to either low or high values.

A numerical analysis [473], done under the assumption of perfect separation of electrons and energetic collinear photons, results in the following branching ratios integrated in the range $1 \text{ GeV}^2 < m_{\ell\ell}^2 < 6 \text{ GeV}^2$:

$$\begin{aligned}
&\mathcal{B}(\bar{B} \to X_s \mu^+\mu^-) \\
&= \big[1.59 \pm 0.08_{\rm scale} \pm 0.06_{m_t} \pm 0.024_{C,m_c} \\
&\quad \pm 0.015_{m_b} \pm 0.02_{\alpha_s(M_Z)} \\
&\quad \pm 0.015_{\rm CKM} \pm 0.026_{\rm BR_{sl}}\big] \times 10^{-6} \\
&= (1.59 \pm 0.11) \times 10^{-6},
\end{aligned} \qquad (72)$$

$$\begin{aligned}
&\mathcal{B}(\bar{B} \to X_s e^+e^-) \\
&= \big[1.64 \pm 0.08_{\rm scale} \pm 0.06_{m_t} \pm 0.025_{C,m_c} \\
&\quad \pm 0.015_{m_b} \pm 0.02_{\alpha_s(M_Z)} \\
&\quad \pm 0.015_{\rm CKM} \pm 0.026_{\rm BR_{sl}}\big] \times 10^{-6} \\
&= (1.64 \pm 0.11) \times 10^{-6},
\end{aligned} \qquad (73)$$

where the error includes the parametric and perturbative uncertainties only. For central values and error bars of the

---
[10]Section coordinators: Th. Feldmann, J. Berryhill.

input parameters, see Table 1 of [473]. The electron and muon channels receive different contributions because of the $\ln(m_b^2/m_\ell^2)$ present in the bremsstrahlung corrections. The difference gets reduced when the BaBar and Belle angular cuts are included. One should also keep in mind that the contributions of the intermediate ψ and ψ' are assumed to be subtracted on the experimental side. Analogous results on the branching ratio in the high-$m_{\ell\ell}^2$ region and on the forward–backward asymmetry (FBA) are given in Ref. [474]. A numerical formula that gives the branching ratio for non-SM values of the relevant Wilson coefficients is given in (12) and (13) of Ref. [473].

The differential BR is sensitive to the interference of the Wilson coefficients C_7 and C_9. The FBA for the charged leptons is sensitive to the products $C_7 C_{10}$ and $C_9 C_{10}$. For instance, reversing the sign of C_7 makes the zero of the FBA disappear [434] and leads to an enhancement of the low-q^2 integrated BR:

$$\mathcal{B}(\bar{B} \to X_s \mu^+ \mu^-) = 3.11 \times 10^{-6},$$
$$\mathcal{B}(\bar{B} \to X_s e^+ e^-) = 3.19 \times 10^{-6}; \tag{74}$$

a similar value for that case has been found in [188].

3.2.2.2 Exclusive decays We focus on the theoretical description of $B \to K^* \ell^+ \ell^-$ decay as one of the phenomenologically most important examples. The double-differential spectrum may be parametrized as [475]

$$\frac{d^2 \Gamma}{dq^2 d\cos\theta_\ell} = \frac{3}{8}\left[(1+\cos^2\theta_\ell) H_T(q^2) + 2\cos\theta_\ell H_A(q^2) + 2(1-\cos^2\theta_\ell) H_L(q^2)\right]. \tag{75}$$

Here, for $\bar{B}^0$ or B^- decays, θ_ℓ is the angle between the ℓ^+ and the B-meson 3-momentum in the $\ell^+\ell^-$ c.m.s.,[11] and $q^2 = m_{\ell\ell}^2$ is the invariant mass of the lepton pair. Alternatively, the functions $H_X(q^2)$ can be expressed in terms of transversity amplitudes [476, 477]

$$H_T(q^2) = |A_{\perp,L}|^2 + |A_{\perp,R}|^2 + |A_{\parallel,L}|^2 + |A_{\parallel,R}|^2, \tag{76}$$
$$H_L(q^2) = |A_{0,L}|^2 + |A_{0,R}|^2, \tag{77}$$
$$H_A(q^2) = 2\,\mathrm{Re}\big[A_{\parallel,R} A_{\perp,R}^* - A_{\parallel,L} A_{\perp,L}^*\big]. \tag{78}$$

If the invariant mass of the lepton pair is sufficiently below the charm threshold at $q^2 = 4m_c^2$ and above the real-photon pole at $q^2 = 0$, the transversity amplitudes can be

---

[11] Different sign conventions are used in the literature.

estimated within the QCD factorization approach [283, 478, 479]:

$$A_{\perp,L/R} \simeq -A_{\parallel,L/R}$$
$$\simeq \sqrt{2} N m_B \left(1 - \frac{q^2}{m_B^2}\right)\left[\mathcal{C}_9^\perp(q^2) \mp C_{10}\right]$$
$$\times \zeta_\perp(q^2), \tag{79}$$
$$A_{0,L/R} \simeq -\frac{N m_B^2}{\sqrt{q^2}}\left(1 - \frac{q^2}{m_B^2}\right)\left[\mathcal{C}_9^\parallel(q^2) \mp C_{10}\right] \zeta_\parallel(q^2), \tag{80}$$

where the normalization factor N is defined in (3.7) in [476]. The functions $\mathcal{C}_{9,10}^\perp(q^2)$ can be calculated perturbatively in the heavy-quark limit, requiring $q^2 \lesssim \Lambda m_b \ll 4m_c^2$ [283, 478]. Large logarithms can be resumed using renormalization-group techniques in soft-collinear effective theory [479]. The form factors $\zeta_{\perp,\parallel}(q^2)$ have to be estimated from experimental data or theoretical models.[12] $1/m_b$ power corrections may be sizable and currently constitute a major source of theoretical uncertainty.

Similarly, in the region far above the charm resonances, the helicity amplitudes can be treated within heavy-quark effective theory, based on an expansion in Λ/m_b and $4m_c^2/q^2$ [480]. To first approximation one finds

$$A_{\perp,L/R} \simeq -\sqrt{2} N m_B \left(1 - \frac{q^2}{m_B^2}\right)$$
$$\times \left[\mathcal{C}_9^{\mathrm{eff}}(q^2) + \frac{2m_b m_B}{q^2} C_7^{\mathrm{eff}} \mp C_{10}\right] m_B g(q^2), \tag{81}$$
$$A_{\parallel,L/R} \simeq -\sqrt{2} N m_B \left[\mathcal{C}_9^{\mathrm{eff}}(q^2) + \frac{2m_b m_B}{q^2} C_7^{\mathrm{eff}} \mp C_{10}\right]$$
$$\times \frac{f(q^2)}{m_B}, \tag{82}$$
$$A_{0,L/R} \simeq -N m_B \frac{m_B^2 - q^2}{2 m_{K^*} \sqrt{q^2}} \left[\mathcal{C}_9^{\mathrm{eff}}(q^2) + \frac{2m_b}{m_B} C_7^{\mathrm{eff}} \mp C_{10}\right]$$
$$\times \frac{f(q^2) + (m_B^2 - q^2) a_+(q^2)}{m_B}. \tag{83}$$

Here $f(q^2), g(q^2), a_+(q^2)$ are the leading HQET form factors [480]. The effective "Wilson coefficients" C_9^{eff} are functions of the lepton invariant mass q^2 and combine short-distance dynamics encoded in Wilson coefficients and (nontrivial) long-distance dynamics at the scale m_b. In the naive

---

[12] The conventions to define the form factors $\zeta_{\perp,\parallel}$ in [479] are different from those of Ref. [478]. Therefore the explicit expressions for $\mathcal{C}_9^{\perp,\parallel}$ also differ.

factorization approximation, they are related to $\mathcal{C}_9^{\perp,\|}(q^2)$ via

$$\mathcal{C}_9^{\perp}(q^2) \approx C_9(\mu) + Y(q^2,\mu) + \frac{2m_b m_B}{q^2} C_7^{\text{eff}}(\mu) + \cdots$$

$$= C_9^{\text{eff}}(q^2) + \frac{2m_b m_B}{q^2} C_7^{\text{eff}} + \cdots, \quad (84)$$

$$\mathcal{C}_9^{\|}(q^2) \approx C_9(\mu) + Y(q^2,\mu) + \frac{2m_b}{m_B} C_7^{\text{eff}}(\mu)$$

$$= C_9^{\text{eff}}(q^2) + \frac{2m_b}{m_B} C_7^{\text{eff}} + \cdots. \quad (85)$$

(In the following, we will also use the notation $C_{9,10}(\mu = m_b) = A_{9,10}$ and $C_7^{\text{eff}}(\mu = m_b) = A_7$.)

It is to be stressed that the theoretical systematics in the kinematic regions $q^2 \ll 4m_c^2$ and $q^2 \gg 4m_c^2$ is quite different, due to the different short-distance effects to be accounted for in the calculation of $\mathcal{C}_9^{\perp,\|}(q^2)$ or $C_{7,9}^{\text{eff}}$, the independent hadronic form factors in SCET/HQET, and the different nature of (nonfactorizable) power corrections.

Experimentally, the dilepton invariant mass spectrum and the FBA are the observables of principal interest. Their theoretical expressions can be easily derived from (75). In particular, the FBA vanishes at q_0^2 if $\text{Re}[\mathcal{C}_9^{\perp}(q_0^2)] = 0$, which turns out to be very sensitive to the size and relative sign of the electroweak Wilson coefficients C_7 and C_9 [481, 482]. The theoretical predictions depend on the strategy to fix the hadronic input parameters and on the scheme to organize the perturbative expansion in QCD. The authors of [283, 478] fix the hadronic form factors from QCD sum rules [483] and calculate the short-distance coefficients in fixed-order perturbation theory. For the partially integrated branching fraction, they find

$$\int_{1\,\text{GeV}^2}^{6\,\text{GeV}^2} dq^2 \frac{dBr[B^+ \to K^{*+}\ell^+\ell^-]}{dq^2}$$

$$= \left(\frac{\zeta_\|(4\,\text{GeV}^2)}{0.66}\right)^2 \times \left(3.33^{+0.40}_{-0.31}\right) \times 10^{-7}, \quad (86)$$

where the leading dependence on one of the $B \to K^*$ form factors has been made explicit. For neutral B mesons, the result is about 10% smaller. The forward–backward asymmetry zero in this scheme comes out to be

$$q_0^2[K^{*0}] = \left(4.36^{+0.33}_{-0.31}\right) \text{GeV}^2,$$
$$q_0^2[K^{*+}] = \left(4.15^{+0.27}_{-0.27}\right) \text{GeV}^2, \quad (87)$$

with an additional uncertainty from power corrections estimated to be of order of 10%.

The authors of [479] fix the form factor $\zeta_\perp(0)$ by comparing the experimental results on $B \to K^*\gamma$ with the theoretical predictions at NLO at leading power and assuming a simple energy dependence of the form factor. Furthermore, the leading perturbative logarithms in SCET are resumed. They get a somewhat smaller value for the partially integrated branching fraction[13]

$$\int_{1\,\text{GeV}^2}^{7\,\text{GeV}^2} dq^2 \frac{dBr(B^+ \to K^{*+}\ell^+\ell^-)}{dq^2}$$

$$= \left(2.92^{+0.57}_{-0.50}|_{\zeta_\|} {}^{+0.30}_{-0.28}|_{\text{CKM}} {}^{+0.18}_{-0.20}\right) \times 10^{-7}, \quad (88)$$

which is mainly due to a smaller default value for the $B \to K^*$ form factor $\zeta_\|$ taken from [232]. The forward–backward asymmetry zero now reads

$$q_0^2 = \left(4.07^{+0.16}_{-0.13}\right) \text{GeV}^2, \quad (89)$$

where the smaller parametric uncertainties compared to (87) are traced back to the renormalization-group improvement of the perturbative series and the different strategy to fix $\zeta_\perp(q^2)$. Isospin-breaking effects between charged and neutral B decays, and potentially large hadronic uncertainties from power corrections have not been specified in [479].

As has been pointed out in [484], the K^* meson is always observed through the resonant $B \to (K\pi)\ell^+\ell^-$ decay. Depending on the considered phase-space region in the Dalitz plot, this may induce further corrections to the position of the asymmetry zero. On the other hand, it allows for an analysis of angular distributions. Following Ref. [476], one can consider the polarization fractions

$$F_L(q^2) = \frac{H_L(q^2)}{H_L(q^2) + H_T(q^2)},$$
$$F_T(q^2) = \frac{H_T(q^2)}{H_L(q^2) + H_T(q^2)} \quad (90)$$

and the K^*-polarization parameter $\alpha_{K^*}(q^2) = 2F_L/F_T - 1$. Like the FBA, these observables have smaller hadronic uncertainties (for small values of q^2), as the hadronic form-factors cancel in the ratios to first approximation [476]. Introducing the angle θ_K of the K meson relative to the B-momentum in the K^* rest frame, the triple differential decay rate reads

$$\frac{d^3\Gamma}{dq^2\, d\cos\theta_l\, d\cos\theta_K}$$

$$= \left\{\frac{9}{8} F_L \cos^2\theta_K \sin^2\theta_\ell \right.$$

$$\left. + \frac{9}{32}(1 - F_L)\sin^2\theta_K\left(1 + \cos^2\theta_\ell\right)\right\}\frac{d\Gamma}{dq^2}$$

$$+ \frac{3}{4}\sin^2\theta_K \cos\theta_\ell\left(\frac{d\Gamma_F}{dq^2} - \frac{d\Gamma_B}{dq^2}\right). \quad (91)$$

---

[13] Notice that the upper limit of integration in (88) is slightly larger than those in (86).

Finally, the remaining angle, ϕ, between the decay planes of the lepton pair and K^* meson defines the distribution [476]

$$\frac{d^2\Gamma}{dq^2 d\phi} = \frac{1}{2\pi}\left(1 + \frac{1}{2}(1-F_L)A_T^{(2)}\cos 2\phi + A_{\text{Im}}\sin 2\phi\right)\frac{d\Gamma}{dq^2}, \quad (92)$$

where the asymmetry $A_T^{(2)}(q^2)$ is sensitive to new physics from right-handed currents, and the amplitude A_{Im} is sensitive to complex phases in the hadronic matrix elements. In the SM, the asymmetry $A_T^{(2)}$ and the amplitude A_{Im} are negligible at low q^2, so the measurement of either is a precision null test.

The differential decay rate for $B \to K\ell^+\ell^-$ can be found in [478]. Within the SM, the FB asymmetry in $B \to K\ell^+\ell^-$ is highly suppressed. At hadron colliders, also the decay modes $B_s \to \phi\ell^+\ell^-$ and $B_s \to \eta'\ell^+\ell^-$ can be studied. Their theoretical description is analogous to the $B \to K^*(K)$ case, but accurate numerical studies require better knowledge of the hadronic parameters entering the B_s and $\phi(\eta, \eta')$-meson wave functions.

Baryonic decay channels, $\Lambda_b \to \Lambda^0\ell^+\ell^-$, are theoretically less well understood. So far, they have only been discussed within the (naive) factorization approximation, based on symmetry relations and model estimates for the $\Lambda_b \to \Lambda^0$ form-factors (see, e.g., [485–487]). Besides the q^2 spectrum and the FBA, the baryonic $b \to s\ell^+\ell^-$ decays offer the possibility to study various asymmetry parameters and Λ^0 polarization effects, which exhibit a particular dependence on NP effects [488–494]. Also a possible initial Λ_b polarization can be accounted for [495].

3.2.2.3 Charmonium resonances in $b \to s\ell\ell$ The calculation of inclusive and exclusive observables in $b \to s\ell^+\ell^-$ decays is complicated by the presence of long-distance contributions related to intermediate $c\bar{c}$ pairs from the 4-quark operators in the effective Hamiltonian. The effect depends on the invariant mass q^2 of the lepton pair.

For the inclusive rate, the charm quarks can be integrated out perturbatively within an OPE based on an expansion in α_s and $(1/m_c, 1/m_b)$ (with the ratio m_c/m_b kept fixed). Below the charm threshold $q^2 \ll 4m_c^2$, the expansion in $1/m_c^2$ still converges, and the inclusive decay spectrum can be described in terms of a local OPE [392, 395, 396, 466, 496, 497]. Similarly, for exclusive decays, it is possible to integrate out the intermediate charm loops perturbatively, leading to nonlocal operators whose matrix elements can be further investigated using QCDF, SCET or (light-cone) sum rules, see the discussion in Sect. 2 and [265, 394] (for the case $q^2 = 0$).

Approaching the charm threshold at $q^2 \sim 4m_c^2$, the heavy-quark expansion breaks down, both in inclusive and exclusive decays. A pragmatic solution is to ignore the $c\bar{c}$ resonance region completely by introducing "appropriate" experimental cuts on q^2. Alternatively, one may attempt to model a few resonances explicitly (in practice the J/ψ and the $\psi(2S)$), see, e.g., [482] and references therein. However, this method bears the danger of double-counting when combined with the OPE result, which can be avoided by using dispersion relations for the electromagnetic vacuum polarization [498]. Still, nonfactorizable soft interactions between the resonating charmonium system and the $B \to X_s$ transition cannot be accounted for in a systematic way at present.

For values of q^2 above the charm threshold, the invariant mass of the hadronic final state is small, and the decay rate is dominated by a few exclusive states. To trust the OPE result for the inclusive spectrum, one has to smear the experimental spectrum over a "sufficiently" large q^2 range and rely on the (semi-local) duality approximation. For the description of the exclusive channels in that region, one has to rely on an expansion in terms of $4m_c^2/q^2$ within HQET [480]. In summary, to avoid contamination from charmonium or light vector resonances, one should consider the range $1 \le q^2 \le 6$ GeV^2.

Finally, one has to mention that light-quark loops need a similar investigation in order to assess the role of light vector resonances at small values of q^2. We also should stress that while analysing the $c\bar{c}$ background in inclusive $B \to X_s l^+ l^-$ transitions, special care should be taken of the chain of $B \to J/\psi X_s$, $J/\psi \to l^+ l^- X$ decays, mimicking $b \to s l^+ l^-$ with $q^2 < m_{J/\psi}^2$.

3.2.3 Experimental studies of electroweak penguin decays

3.2.3.1 Measurements (prospects) at (super-)B factories The B-factory experiments BaBar and Belle have succeeded in measuring the $b \to s\ell^+\ell^-$ process in B decays, both exclusively [499–501] and inclusively [124, 125]. Measured observables include: total branching fractions, direct CP asymmetries, partial branching fractions vs. the dilepton q^2 and the hadronic X_s mass, and—for $B \to K^*\ell^+\ell^-$—the dilepton angular asymmetry A_{FB} vs. the dilepton q^2, the K^* longitudinal polarization vs. the dilepton q^2, and fits of the $d^2\Gamma/d\cos\theta\,dq^2$ distribution to extract experimentally A_9/A_7 and A_{10}/A_7. Upon accumulation of more data in current B factories or the proposed super B factories, it should be possible to extract most of the observables described in Sect. 3.2.2, in increasingly finer binning and precision. The expected experimental sensitivity of 50 ab^{-1} of $B \to K^*\ell^+\ell^-$ data at a super B factory is comparable to 3.3 fb^{-1} of $B^0 \to K^{*0}\mu^+\mu^-$ data at LHCb, as described below.

The optimal measurement technique is to completely reconstruct the signal B decay: selection of events with an electron or muon pair, selection of all hadrons of the appropriate X_s system (K or K^* mesons for the exclusive case and a K plus 1, 2, 3 or 4 pions for the inclusive case) and then application of the standard kinematic requirements in mass and energy for the resulting B candidate. Partial or full reconstruction requirements for the recoil B are in general suboptimal. Triggering signal events is fully efficient and particle identification is both efficient (typically (80–90)% per particle) and pure (negligible fake rates for electrons, percent level fake rates for muons and kaons) down to low particle lab momenta (0.3 GeV/c for electrons and 0.7 GeV/c for muons). Charmonium background can be efficiently vetoed by the lepton-pair mass and does not significantly contaminate the q^2 regions dominated by the short-distance physics of interest. The remaining combinatorial background, mostly from semileptonic B and D decays, is significant, but it can be reliably separated from signal by extrapolation from distributions in kinematic sidebands, typically via an unbinned maximum likelihood fit. Branching fraction results are shown in Table 16. The effective signal to background ratio for these results varies from 1:2 (inclusive) up to 2:1 (Belle $K^*\ell\ell$). Comparable sensitivity is attained for both electron and muon decay channels.

Assuming HFAG branching fractions and the efficiencies and backgrounds observed in the Belle results, the expected signal yields (and their statistical precision) per 1 ab^{-1} are 229 ± 16 (7%), 215 ± 16 (7%) and 486 ± 24 (5%) for $K\ell\ell$, $K^*\ell\ell$ and $X_s\ell\ell$, respectively. The experimental uncertainty for total branching fractions should therefore be less than or comparable to current SM theoretical uncertainties, using B-factory data alone. Direct CP violation will be bounded at the level of 5–7% with 1 ab^{-1}, and thus a Super B factory would obtain a high precision test ($\sim$1%) of the null result expected in the SM. Similar precision is expected for measuring differences in branching fractions between electron and muon channels, which is also an interesting null test of the SM [435, 503]. A possible complicating factor for the inclusive $X_s\ell\ell$ (partial) branching fractions is the necessity of an aggressive requirement on the mass M_{X_s} to be less than 1.8 GeV/c^2. Such a tight cut may introduce significant shape function effects into the interpretation of the results in the same manner as a photon energy cut does for $B \to X_s \gamma$ [504, 505]. A looser M_{X_s} requirement will have poorer precision, and thus Super B factory samples may be required to compare with the most precise predictions.

The B factories have also succeeded in accumulating large enough $B \to K^*\ell\ell$ samples to perform angular analyses as a function of dilepton mass. The angles analysed thus far include the angle, θ_ℓ, between the positive (negative) lepton and the B ($\bar{B}$) momentum in the dilepton rest frame and the angle, θ_K, of the K meson relative to the B momentum in the K^* rest frame. The integrated longitudinal K^* polarization F_L and the forward–backward asymmetry A_{FB} are related to the decay products' angular distribution via (91), which upon integration of one of the angular variables reduces to

$$\frac{d^2\Gamma}{dq^2 d\cos\theta_K} = \left\{\frac{3}{2} F_L \cos^2\theta_K + \frac{3}{4}(1-F_L)\sin^2\theta_K\right\}\frac{d\Gamma}{dq^2}, \tag{93}$$

$$\frac{d^2\Gamma}{dq^2 d\cos\theta_\ell} = \left\{\frac{3}{4} F_L \sin^2\theta_\ell + \frac{3}{8}(1-F_L)(1+\cos^2\theta_\ell) + A_{FB}\cos\theta_\ell\right\}\frac{d\Gamma}{dq^2}. \tag{94}$$

From the singly- or doubly-differential angular distributions (in a given q^2-bin) it is then possible to infer $A_{FB}(q^2)$ and $F_L(q^2)$ simultaneously. There is also the remaining angle, ϕ, between the decay planes of the lepton pair and K^* meson, which has yet to be analysed, see (92).

BaBar has measured A_{FB} and F_L in two bins of q^2 (above and below 8.4 GeV/c^2) via unbinned maximum likelihood fits to the singly-differential distributions of $\cos\theta_\ell$

Table 16 Branching fraction measurements at B factories for $b \to s\ell^+\ell^-$ decays, including integrated luminosity, signal yield, detection efficiency and the measured branching fraction over the full q^2 range. The HFAG averages are also included

Result	$\int \mathcal{L}$ (fb^{-1})	Yield	Efficiency (%)	$\mathcal{B}$ (10^{-6})
BaBar $B \to K\ell\ell$ [501]	208	46 ± 10	15 ± 1	$0.34 \pm 0.07 \pm 0.02$
Belle $B \to K\ell\ell$ [499]	253	79 ± 11	13 ± 1	$0.55 \pm 0.08 \pm 0.03$
HFAG $B \to K\ell\ell$ [502]				0.44 ± 0.05
BaBar $B \to K^*\ell\ell$ [501]	208	57 ± 14	7.9 ± 0.4	$0.78 \pm 0.19 \pm 0.11$
Belle $B \to K^*\ell\ell$ [499]	253	82 ± 11	4.6 ± 0.2	$1.65 \pm 0.23 \pm 0.11$
HFAG $B \to K^*\ell\ell$ [502]				1.17 ± 0.16
BaBar $B \to X_s\ell\ell$ [124]	82	40 ± 10	2.0 ± 0.4	$5.6 \pm 1.5 \pm 1.3$
Belle $B \to X_s\ell\ell$ [125]	140	68 ± 14	2.7 ± 0.5	$4.1 \pm 0.8 \pm 0.9$
HFAG $B \to X_s\ell\ell$ [502]				4.5 ± 1.0

Table 17 Expected statistical precision of a Super B factory, in percent, for the angular observables A_{FB} and F_L versus the integrated luminosity, integrated over various ranges of q^2

$\int \mathcal{L}$ (ab^{-1})		1	5	10	50
$K^*\ell\ell$: A_{FB}	q^2 in (1–6) GeV^2/c^4	18	8.2	5.8	2.6
	$q^2 > 10$ GeV^2/c^4	11	4.7	3.3	1.5
	All	7.9	3.5	2.5	1.1
$K^*\ell\ell$: F_L	q^2 in (1–6) GeV^2/c^4	12	5.3	3.7	1.7
	$q^2 > 10$ GeV^2/c^4	9.4	4.2	3.0	1.3
	All	7.2	3.2	2.3	1.0
$K^+\ell\ell$: A_{FB}	All	8.4	3.7	2.6	1.2

and $\cos\theta_K$, which take into account signal efficiency as a function of angle as well as background angular distributions (which are in general nonuniform and forward–backward asymmetric) [501]. Table 17 shows the expected precision for these observables extrapolated to super B luminosities, assuming HFAG branching fractions and SM predictions for $d\Gamma/dq^2$. The ultimate 50 ab^{-1} precision of the A_{FB} of $B \to K^*\ell\ell$, integrated over the theoretically preferred range of (1–6) GeV^2/c^4, is 2.6%. If this region is extended more aggressively to the original BaBar choice of (0.1–8.4) GeV^2/c^4, the signal statistics are doubled, and the precision improves to 1.8%. Similar precision is expected for F_L. Measuring integrated angular observables of these types has the advantages of model independence in their interpretation; the underlying relation between these measurements, the Wilson coefficients and the form factors can change without necessitating revision of the measurement. The averaging of multiple experimental results is also very straightforward.

Alternatively, Belle has analysed the doubly-differential distribution $d^2\Gamma/d\cos\theta_\ell\, dq^2$ and then performed a maximum likelihood fit to extract the Wilson coefficient ratios A_9/A_7 and A_{10}/A_7 from the data [500]. Using the theoretical approximation in Ref. [434] and assuming the form factor model of Ref. [482], they find

$$A_9/A_7 \simeq -15.3^{+3.4}_{-4.8} \pm 1.1,$$
$$A_{10}/A_7 \simeq 10.3^{+5.2}_{-3.5} \pm 1.8,$$
(95)

where the A_i are the leading-order Wilson coefficients. This is in agreement with the LO Standard Model predictions of -12.3 and 12.8, respectively. The dominant systematic uncertainty is from theoretical model dependence, particularly the form factor model and parametric uncertainty from m_b. This method has been studied for super B-factory luminosities, as discussed in Ref. [506]. Figure 8 shows a projection of dA_{FB}/dq^2 from a likelihood fit to the Wilson coefficients, for a simulated sample of 5 ab^{-1}, compared to A_{FB} integrated over various bins in q^2 measured from the same sample. Employing the entire range of q^2, the expected statistical precision is shown in Table 18. With (5–10) ab^{-1},

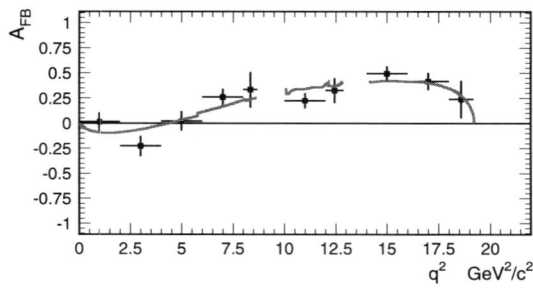

Fig. 8 Expected measurement of dA_{FB}/dq^2 for $B \to K^*\ell^+\ell^-$ (*points*) with 5 ab^{-1} of data from a super B factory; the best fit of that data for Wilson coefficients A_9 and A_{10} is superimposed (*solid line*) [506]

Table 18 Expected statistical precision for a super B factory, in percent, for Wilson coefficients A_9 and A_{10} versus the integrated luminosity, integrated over the entire range of q^2

$\int \mathcal{L}$ (ab^{-1})	1	5	10	50
A_9	25	11	7.8	3.5
A_{10}	29	13	9.2	4.1

the expected statistical uncertainty will be less than the current systematic uncertainty. The expected ultimate statistical sensitivity for 50 ab^{-1} is about 4% for each coefficient. These fits extract essentially the same information as that obtained from measuring the zero q_0^2 of dA_{FB}/dq^2 (a theoretically clean estimator of A_9/A_7), except that the distribution is analysed globally and not just in the vicinity of q_0^2; equivalent uncertainties for q_0^2 are identical to those of A_9. In order to control theoretical uncertainties, it may be necessary to restrict the fit to (1–6) GeV^2/c^4. For that measurement, the price in experimental statistics is roughly a factor of 0.6, with an even larger sacrifice in sensitivity for A_{10}, which is most relevant at high q^2.

With more data, it could also be possible to bound other Wilson coefficients which are negligible in the SM, such as those corresponding to scalar operator products or products with flipped chirality. Fitting triply- or quadruply-differential distributions with the additional decay angles

$\cos\theta_K$ and ϕ, as is currently done for large samples of $B \to VV$ decays, will also be possible.

Measuring the angular distribution of inclusive $B \to X_s\ell\ell$ decays has not yet been attempted, however with thousands of events expected at a super B factory, there will be sufficient statistics for a precise measurement of A_{FB} [507]. This is an attractive measurement, as observables such as q_0^2 are predicted more precisely than for the exclusive case (~5%). Scaling from the expected yield per ab^{-1} of 486 ± 24 and assuming the same sensitivity to A_9/A_7 per event as for the $B \to K^*\ell\ell$ Wilson coefficient fits, a 5% statistical precision for A_9/A_7 (and hence q_0^2) could be achieved with roughly 10 ab^{-1}, although again a critical issue for the precision is how wide a range of q^2 is appropriate for such fits. Understanding systematic uncertainties from a sum-of-exclusive-modes analysis will be challenging, in particular the effect of imprecise X_s fragmentation modeling on the multiply-differential efficiency.

3.2.3.2 $B_d \to K^{*0}\mu^+\mu^-$ at LHCb

The exclusive $B_d \to K^{*0}\mu^+\mu^-$ decay can be triggered and reconstructed in LHCb with high efficiency due to the clear di-muon signature and K/π separation provided by the RICH detector [508].

The selection criteria including the trigger have an efficiency of 1.1% for signal. The trigger accepts 89% of the Monte Carlo signal events, which are reconstructed offline. In 2 fb^{-1} of integrated luminosity, this selection gives an estimated signal of 7200 events with a total background of 3500 events in a ± 50 MeV/c^2 mass window around the B mass and ± 100 MeV/c^2 window around the K^{*0} mass. The branching ratio for $B_d \to K^{*0}\mu^+\mu^-$ was assumed to be 1.22×10^{-6}. The irreducible nonresonant $B_d \to K^+\pi^-\mu^+\mu^-$ background was estimated at 1730 events; the branching ratio used for this was set using a 90% upper limit estimate found from the sidebands of the K^{*0} mass in [501]. Other large components of the background are 1690 from events with two semileptonic B decays, 640 of which are from semileptonic decays of both the b and the c quarks within the same decay chain. Exclusive backgrounds from other $b \to s\mu^+\mu^-$ decays were considered and contribute at a very low level of 20 events.

The selection efficiency as a function of q^2 is flat in the region $4m_\mu^2$ to 9 GeV^2/c^4 due to the high boost of the B_d. For high q^2 values, the selection efficiency as a function of θ_l is flat, while for low q^2, the efficiency is highest around $\theta_l = \pi/2$ [509].

In addition to the well-known FB asymmetry, A_{FB}, LHCb will be able to extract information about the differential decay rate $d\Gamma/ds$ and the transversity amplitudes A_0, $A_\parallel$ and $A_\perp$ through the asymmetry $A_T^{(2)}$ and the K^{*0} longitudinal polarization F_L, see (91) and (92).

For measuring the zero point in A_{FB}, a linear fit is performed to the measured A_{FB} in the region (2–6) GeV^2/c^4 as illustrated in Fig. 9. For the resolution in the zero point of A_{FB} [509], we estimate 0.50(0.27) GeV^2/c^4 with 2(10) fb^{-1} of integrated luminosity. If the background is ignored, the resolution is 0.43(0.25) GeV^2/c^4.

The statistical errors for A_{FB}, $A_T^{(2)}$ and F_L have been estimated by performing simultaneous fits to the θ_l, θ_K and ϕ projections of the full angular distribution in 3 bins of q^2 below the ψ resonances [510]. In the theoretically favored region of $1 < q^2 < 6$ GeV^2/c^4, the resolution in $A_T^{(2)}$ is 0.42(0.16) with 2(10) fb^{-1} of integrated luminosity. See Table 19 for estimated statistical errors on all the parameters. In particular the resolution on $A_T^{(2)}$ would improve if the theoretically comfortable region could be expanded upwards from 6 GeV^2/c^4.

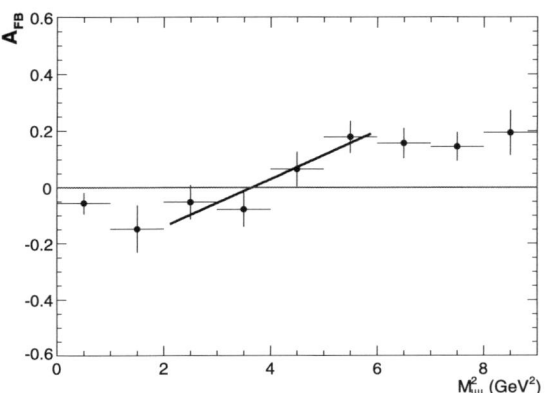
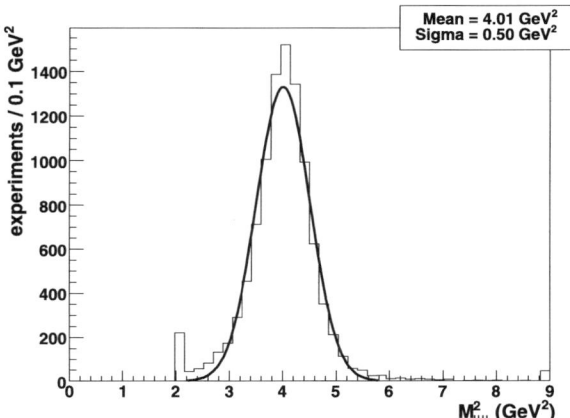

Fig. 9 The FB asymmetry in $B_d \to K^{*0}\mu^+\mu^-$ with 2 fb^{-1} of integrated luminosity at LHCb. To the *left* the FB asymmetry as a function of q^2 in a single toy Monte Carlo experiment, and to the *right* the fitted zero point location for an ensemble of Monte Carlo experiments. The peaks at 2 and 9 correspond to fits where the zero point was outside this region

Table 19 The expected resolution for measurements of the parameters A_{FB}, $A_T^{(2)}$ and F_L for the $B_d \to K^{*0}\mu^+\mu^-$ decay at LHCb in regions of the squared di-muon mass q^2 with 2 and 10 fb^{-1} of integrated luminosity

q^2 region (GeV^2/c^4)	A_{FB}		$A_T^{(2)}$		F_L	
	2 fb^{-1}	10 fb^{-1}	2 fb^{-1}	10 fb^{-1}	2 fb^{-1}	10 fb^{-1}
0.05–1.00	0.034	0.017	0.14	0.07	0.027	0.011
1.00–6.00	0.020	0.008	0.42	0.16	0.016	0.007
6.00–8.95	0.022	0.010	0.28	0.13	0.017	0.008

3.2.3.3 R_K at LHCb Reconstructing $B^+ \to K^+e^+e^-$ as well as $B^+ \to K^+\mu^+\mu^-$ allows us to extract the ratio R_K of the two branching fractions integrated over a given di-lepton mass range. The same reconstruction requirements are applied to $B^+ \to K^+\mu^+\mu^-$ and $B^+ \to K^+e^+e^-$ decay. A proper bremsstrahlung correction is essential in the latter channel. The correction for the lower reconstruction and trigger efficiency in the electron mode is extracted from $B^+ \to J/\psi K^+$ decays. The di-lepton mass range is chosen to be $4m_\mu^2 < q^2 < 6$ GeV^2/c^4 in order to avoid $c\bar{c}$ resonances (especially in the e^+e^- mode) and threshold effects due to the higher μ mass. The event yields are extracted from a fit to the $K\ell^+\ell^-$ mass distributions. Peaking backgrounds from $B^+ \to J/\psi K^+$ and $B_d \to K^*\ell^+\ell^-$ are measured using control samples and included in the fit.

The expected B candidate mass distributions are shown in Fig. 10 for five years (10 fb^{-1}) of data taking. The yields returned by the fit are given in Table 20. They are compatible with the number of true MC events. The B/S ratios are given for the full signal box within ±600 MeV around the B_u mass (shown in Fig. 10). The errors on the yields are the statistical error returned by the fit. Using these errors, one gets an error on R_K of 4.3% for 10 fb^{-1}.

3.2.3.4 Semileptonic rare B decays at ATLAS With the ATLAS experiment, NP effects in $b \to s l^+l^-$ transitions will be searched for in the branching ratio and the FB asymmetry $A_{FB}(q^2)$ between b-hadron and l^+ momenta. With baryonic decays ($\Lambda_b \to \Lambda^0 \mu^+\mu^-$), NP effects can also be extracted from Λ^0 polarization and asymmetry parameters (Figs. 2, 3, 4 from [489]), but influence of possible initial Λ_b polarization has to be accounted for [495]. Note that the measurement of the di-lepton mass spectrum is more sensitive to the ATLAS detector efficiency than to new physics.

The main part of B-physics studies will be performed in the initial LHC low-luminosity stage (3 years at $L = 10^{33}$ cm^{-2} s^{-1}). It is expected that the luminosity will vary by a factor of ∼2 during beam-coast and there will be 2–3 interactions per collision. The production rate of $b\bar{b}$ pairs at ATLAS is ∼500 kHz, which implies having 5×10^{12} $b\bar{b}$ pairs per year (10^7 seconds).

Experimental feasibility studies for rare decays of B_d^0, B_s^0, B^+ and Λ_b at ATLAS have been performed using the full detector simulation chain [511]. The decay kinematics was defined via matrix elements included into the b-physics

Table 20 Expected yields returned by the fit as described in the text

	Yield	B/S	$\sigma(m_{B_u})$
$B^+ \to K^+\mu^+\mu^-$	18774 ± 230	∼29	14 MeV/c^2
$B^+ \to K^+e^+e^-$	9240 ± 380	∼30	68 MeV/c^2

Pythia interface [512] (B_d^0, B_s^0) or using the EvtGen decay tool [513, 514] (B^+, Λ_b) with matrix elements taken from theoretical publications in [295, 434, 486, 488, 515]. The pp interactions were generated using Pythia6 [516] tuned for correct b-quark production [512]. Events were filtered at generator level to emulate the di-muonic LVL1 trigger cuts (see below), and charged tracks from the B-decays were required to fit in ATLAS tracking system capabilities ($p_T \gtrsim 0.5$ GeV, $|\eta| < 2.5$ [517]). These cuts influence the q^2 spectrum and A_{FB} shape. Study of the sample of $\Lambda_b \to \Lambda^0 \mu^+\mu^-$ events have shown that higher di-muon mass values are preferred (fraction of events with q^2 below J/ψ mass decreased from 67 to 58%) and A_{FB} is affected in the $q^2/M_b^2 < 0.1$ region (suppression by 40% of $|A_{FB}|$ was found).

The trigger system at ATLAS consists of three levels: Level 1 trigger (LVL1), Level 2 trigger (LVL2) and Event Filter (EF) [518]. LVL1 stage is based on the detection of two high-p_T muons by the fast muon trigger chambers ($p_{T\mu_1} > 6$ GeV, $p_{T\mu_2} > 4$ GeV and $|\eta_{\mu_{1,2}}| < 2.5$ driven by detector acceptance). A preliminary study of the di-muonic LVL1 performance was shown in [519]. The LVL1 rate is dominated by real di-muons giving a rate of ∼150 Hz but also by events with a single muon, doubly counted due to overlap of trigger chambers. In order to suppress the fake di-muon triggers, a system of overlap flags was introduced. The study indicated that signal rejection due to this overlap-removal algorithm is less than 0.5%. Efficiency suppression due to small di-muonic opening angles was also studied, finding the effect below 1%. Overall (75–80)% single muon and ∼60% di-muon trigger efficiency was found for the sample of $\Lambda_b \to \Lambda^0\mu^+\mu^-$ events. At the second level, the muon p_T measurement will be confirmed in the Muon Precision Chambers, Tile Calorimeter and extrapolated to the Inner Detector in order to reject muons from K/π decays. The di-muon specific detailed LVL2 and EF strategies have not yet been set up. The purpose of LVL2 is to select preliminary candidates for the B-hadrons rare decay, based

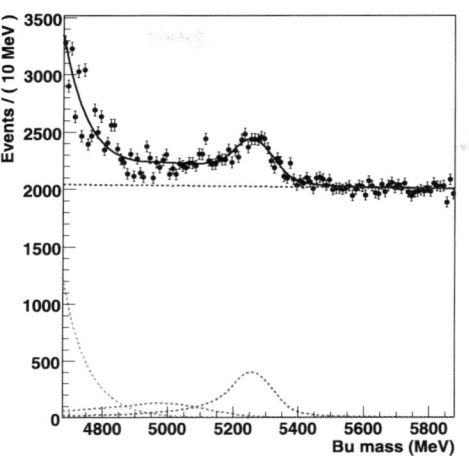 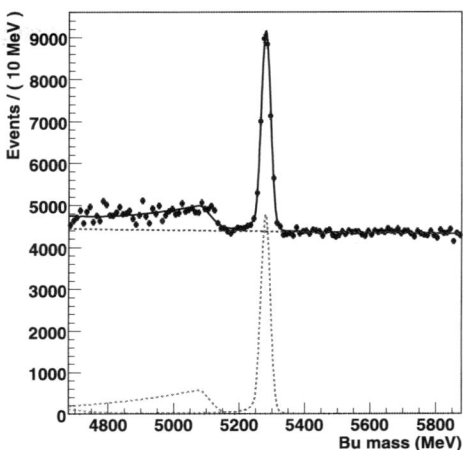

Fig. 10 Expected B^+ candidate mass distributions in the $B^+ \to K^+ e^+ e^-$ (*left*) and $B^+ \to K^+ \mu^+ \mu^-$ (*right*) modes for 10 fb^{-1} at LHCb. The *dotted lines* show the contributions from signal and specific backgrounds as extracted from the fit (see text)

on track parameters and fast calculations. A secondary fast vertex fit can optionally be used at LVL2 level to achieve a satisfactory background rejection. At the EF level, offline-like selection cuts will be applied.

The key signature of rare decays is the presence of the opposite-charge muon pair. The di-muon pair is likely to form a secondary vertex which is detached from the primary vertex. The identification of this vertex, if particularly close to the interaction point, requires well-reconstructed leptons. The event selection is done in the following order: muon and di-muon identification; secondary hadron selection; B-hadron selection. The analysis has to rely on topological variables as vertex quality, vertex separation ($c\tau_B \geq 0.5$ ps) and pointing to primary vertex constraint on the B-hadron momentum. The vertexing algorithm used is the one adopted from the CDF Collaboration [520]. Simple vertex fits are used to select secondary hadrons and di-muon candidates, while for the B-hadron, the whole cascade decay topology is fitted at once.

Due to low signal BRs, great background suppression has to be achieved. The main background source comes from beauty decays producing a muon pair in the final state. The present study based on a sample of $b\bar{b} \to X \mu_{p_T > 6(4)\,\text{GeV}} \mu_{p_T > 4\,\text{GeV}}$ events, provides upper limits for fake events as sketched in Table 21.

In Table 22, the reconstructed A_{FB} is presented for $B_d^0 \to K^{0*} \mu^+ \mu^-$ decay. We divide the q^2/M_B^2-region into three

Table 21 Expected number of events for signal and background upper limit after 30 fb^{-1} measurement

Decay	Signal	Background
$B_d^0 \to K^{0*} \mu^+ \mu^-$	2500	12000
$B_s^0 \to \phi \mu^+ \mu^-$	900	10000
$B^+ \to K^{*+} \mu^+ \mu^-$	2300	12000
$B^+ \to K^+ \mu^+ \mu^-$	4000	12000
$\Lambda_b \to \Lambda^0 \mu^+ \mu^-$	800	4000

Table 22 Averaged A_{FB} of $B_d^0 \to K^{0*} \mu^+ \mu^-$ from ATLAS simulations (not corrected for detector effects and background) at $L_{\text{int}} = 30$ fb^{-1}, its statistical precision and comparison to SM prediction

Interval of q^2/M_B^2	$-0.00 \atop -0.14$	$-0.14 \atop -0.33$	$-0.55 \atop -0.71$
Number of events	570	540	990
A_{FB}	11.8%	-6.1%	-13.7%
Statistical error	4.2%	4.3%	3.2%
SM prediction	10%	-14%	-29%

intervals: the first interval from $(2m_\mu/M_B)^2$ to the so-called "zero-point" [481], the second interval from the "zero-point" to the lower boundaries of the J/ψ and ψ' resonances and the last interval from the resonance area to $(M_B - M_{K^*})^2/M_B^2$. Data collected in 3 years of LHC operations, corresponding to 30 fb^{-1} of integrated luminosity, will be enough to confirm the Standard Model or to set strong limits on SM extensions.

An attempt to estimate the statistical errors of the branching ratio measurements has been made for $B^+ \to K^+ \mu^+ \mu^-$ and $B^+ \to K^{*+} \mu^+ \mu^-$ decays [521]. They were $\sim 3.5\%$ and $\sim 6.5\%$, respectively for $B^+ \to K^+ \mu^+ \mu^-$ and $B^+ \to K^{*+} \mu^+ \mu^-$ decays. These errors on the branching ratio measurements are much smaller than the current experimental and theoretical ones.

3.2.4 Phenomenological implications and new physics constraints

3.2.4.1 New Physics in exclusive $b \to s \ell^+ \ell^-$ induced decays The potential of SM tests and NP searches with $b \to s \ell^+ \ell^-$ transitions has been stressed and explored in several works, e.g., [507, 522] and references therein. Of particular interest for the LHC are the exclusive decays (i) $B_s \to \ell^+ \ell^-$, (ii) $B \to K^{(*)} \ell^+ \ell^-$, $B_s \to \phi \ell^+ \ell^-$, $B_s \to \eta^{(\prime)} \ell^+ \ell^-$ and (iii) $\Lambda_b \to \Lambda \ell^+ \ell^-$, where $\ell = e, \mu, (\tau)$. Decays involving additional photons, such as $B_s \to \ell^+ \ell^- \gamma$ [523] are more sensitive to the hadronic QCD dynamics than the modes (i)–(iii). They are briefly considered in Sect. 3.4. Lepton flavor violating (LFV) decays such as $b \to s e^\pm \mu^\mp$ are discussed e.g. in [524, 525] and will not be considered further here. We stress that FCNCs with final state τ-leptons are poorly constrained experimentally to date, and it would be highly desirable to fill this gap since they test third generation couplings. The latter feature is also shared by the di-neutrino final states discussed, e.g., in [526] and in Sect. 3.3.

The presence of NP can lead to modified values for the short-distance coefficients C_i, including new CP-violating phases, and the generation of new operators in the weak effective Hamiltonian. These could include chirality flipped versions of the SM operators $\mathcal{O}'_i$ (down by m_s/m_b within the SM) from right-handed currents or scalar operators from Higgs exchanges $\mathcal{O}_{S,P}$ (down by $m_\ell m_b/m_W^2$ within the SM), or tensor currents. Scenarios with *light* NP particles require additional operators, build out of the latter, see [527] for the MSSM with light sbottom and gluino. Model-independent information on $C^{(\mathrm{eff})}_{7,8,9,10}$ has been previously extracted from combined analysis of $b \to s \ell^+ \ell^-$ and radiative $b \to s\gamma, sg$ data [434, 482, 500], also including (pseudo)-scalar contributions $C_{S,P}$ [435, 528]. In this program, the study of correlations between decays and observables is an important ingredient, which enables identification of a possible SM breakdown and its sources.

The leptonic decay $\bar{B}_q^0 \to \ell^+ \ell^-$ is a smoking gun for neutral Higgs effects in SUSY models with large $\tan\beta$ and is discussed in detail in Sect. 3.4. A clean test of minimal flavor violation (MFV, see Sect. 1.2.3) is the B_d-B_s ratio $R_{\ell\ell} \equiv \mathcal{B}(\bar{B}_d^0 \to \ell^+\ell^-)/\mathcal{B}(\bar{B}_s^0 \to \ell^+\ell^-)$. In the SM and within MFV models, $0.02 \lesssim R_{\ell\ell}|_{\mathrm{SM}} \lesssim 0.05$, whereas in non-MFV scenarios, $R_{\ell\ell}$ can be $\mathcal{O}(1)$ [529]. Phases in $C_{S,P}$ are probed with time-dependent and integrated CP-asymmetries requiring lepton-polarization measurements [530–532].

Besides the measurement of branching ratios, the $\bar{B} \to K \ell^+ \ell^-$ and $\bar{B} \to K^* \ell^+ \ell^-$ decays offer a number of orthogonal observables. For instance, the latest experimental results from Belle and BaBar for these modes [500, 501, 533] already include first investigations of angular distributions. The dilepton mass (q^2) spectra of $\bar{B} \to K^{(*)} \ell^+ \ell^-$ are sensitive to the sign of $\mathrm{Re}(C_7^{\mathrm{eff}*} C_9^{\mathrm{eff}})$ and to NP contributions in $C_{9,10}$ and flipped $C'_{9,10}$ [534]—however, with rather large hadronic uncertainties from form factors and nonfactorizable long-distance effects (see Sect. 3.2.2). Using constraints on $|C_{S,P}|$ from $B_s \to \mu^+\mu^-$ [528] shows that $\bar{B} \to K^{(*)} \ell^+ \ell^-$ spectra are rather insensitive to NP effects in C_S and C_P. A dedicated study of $B \to K \ell^+ \ell^-$ angular distributions in the SM and beyond has been presented recently in [535].

The FB asymmetry for decays into light pseudoscalars, $A_{FB}(\bar{B} \to K \ell^+ \ell^-)$, vanishes in the SM. Beyond the SM, it is proportional to the lepton mass and the matrix elements of the new scalar and pseudoscalar penguin operators. The BaBar measurement of the angular distribution [501] is consistent with a zero FB asymmetry. Using model-independent constraints on $|C_{S,P}|$ from $B_s \to \mu^+\mu^-$ [528], one expects $A_{FB}(B \to K \mu^+ \mu^-) < 4\%$. Moreover, in the MSSM with large $\tan\beta$, one has $C_S \simeq -C_P$, and the FB asymmetry comes out even smaller, $A_{FB}(B \to K \ell^+ \ell^-) \lesssim 1(30)\%$ for $\ell = \mu(\tau)$ [503, 536, 537]. In contrast, for decays into light vector mesons, $A_{FB}(\bar{B} \to K^* \ell^+ \ell^-)$ is nonzero in the SM and exhibits a characteristic zero q_0^2, whose position is relatively free of hadronic uncertainties, see Sect. 3.2.2. In a general model-independent NP analysis [534, 538], the position of the zero, the magnitude and shape of $A_{FB}(\bar{B} \to K^* \ell^+ \ell^-)$ are found to depend on the modulus and phases of all Wilson coefficients. Note that also $\Lambda_b \to \Lambda \ell^+ \ell^-$ decays share the universal SM A_{FB}-zero in lowest order of the $1/m_b$ and α_s expansion [485]. In off-resonance $B \to K\pi \ell^+ \ell^-$ decays, the analogous A_{FB} zero is also sensitive to NP effects [484]. The CP-asymmetry for the FB asymmetry in $\bar{B} \to K^* \ell^+ \ell^-$ is a quasi-null test of the SM [526], with $A_{FB}^{\mathrm{CP}}|_{\mathrm{SM}} < 10^{-3}$. Sizable values can arise beyond the SM, for instance from nonstandard CP-violating Z-penguins contributing to $\arg(C_{10})$.

The (CP-averaged) isospin asymmetry in $\bar{B} \to K^* \ell^+ \ell^-$ is defined from the difference between charged and neutral B decays [539]. It vanishes in naive factorization (assuming isospin-symmetric form factors). A nonzero value arises from nonfactorizable interactions where the photon couples to the spectator quark. For small values of q^2, the isospin asymmetry can be analysed in QCDF [539]. The largest contributions are induced by the strong penguin operators $\mathcal{O}_{3-6}$, and the sign of the asymmetry depends on the sign of C_7^{eff}. Within the SM and minimal-flavor violating MSSM scenarios, the isospin asymmetry is found to be small. Sizable deviations of $A_I(\bar{B} \to K^* \ell^+ \ell^-)$ from zero would thus signal NP beyond MFV.

Following Refs. [476, 477], one can construct further observables from an angular analysis of the decay $\bar{B}^0 \to K^{*0}(\to K^-\pi^+) \ell^+ \ell^-$, see (90), (92). The SM predictions are consistent with the existing experimental data for the (integrated) value of the longitudinal K^* polarization F_L [501]. A model-independent analysis with flipped $\mathcal{O}'_7$ shows some sensitivity of the angular observables to right-handed currents [476, 477], see also [534]. The shapes of the transverse asymmetries $A_T(q^2)$ depend strongly on C_7 and C'_7, whereas NP effects in $C_{9,10}$ are rather small taking into account constraints from other B-physics data. Moreover, the zeros of $A_T^{(1,2)}(q^2)$ are sensitive to C'_7. NP can give large contributions to the polarization parameter $\alpha_{K^*}(q^2)$ and $F_{L,T}(q^2)$ in extreme scenarios, however the influence of C_9 and C_{10} is stronger, and theoretical errors are larger than in $A_T^{(1,2)}$.

The muon-to-electron ratios

$$R_H \equiv \int_{q_1}^{q_2} dq^2 \frac{d\Gamma(B \to H\mu^+\mu^-)}{dq^2}$$
$$\Big/ \int_{q_1}^{q_2} dq^2 \frac{d\Gamma(B \to He^+e^-)}{dq^2},$$
$$H = \{K, K^*\}, \qquad (96)$$

are probing for nonuniversal lepton couplings, for instance from Higgs exchange or R-parity violating interactions in SUSY models. Kinematic lepton-mass effects are tiny, $\mathcal{O}(m_\mu^2/m_b^2)$. Taking the same integration boundaries for muon and electrons, the SM predictions are rather free of hadronic uncertainties [435]

$$R_H^{SM} = 1 + \mathcal{O}(m_\mu^2/m_b^2), \quad \text{with}$$
$$R_K^{SM} = 1 \pm 0.0001, \qquad R_{K^*}^{SM} = 0.991 \pm 0.002, \qquad (97)$$

and agree with the measurements $R_K = 1.06 \pm 0.48 \pm 0.08$ and $R_{K^*} = 0.91 \pm 0.45 \pm 0.06$ [501].

Studying correlations between different observables, one may be able to discriminate between different NP models. For instance, nontrivial correlation effects appear between R_K and $\mathcal{B}(B_s \to \mu^+\mu^-)$, since $\bar{B} \to K\ell^+\ell^-$ depends on $C_{S,P} + C'_{S,P}$, whereas $\mathcal{B}(\bar{B}_q^0 \to \ell^+\ell^-)$ on $C_{S,P} - C'_{S,P}$ [435]. Also, $\mathcal{B}(B_s \to \mu^+\mu^-)$ and Δm_s are strongly correlated in the minimal flavor-violating MSSM at large $\tan\beta$ [30], whereas no such correlation occurs in models with an additional gauge singlet, like the NMSSM studied in [540]. A summary of all observables with central results is given in Table 23.

3.2.4.2 $B \to K^*\ell\ell$ and universal extra dimensions
FCNC B decays are sensitive to NP scenarios involving extra dimensions. As an example, we discuss here the possibility to constrain the model proposed in [180] (ACD model), which is an extension of the SM by a fifth (universal) extra dimension. The extra dimension is compactified to the orbifold S^1/Z_2, and all the SM fields are allowed to propagate in all dimensions. This model only requires a single additional parameter with respect to the SM, namely the radius R of the compactified extra dimension. The SM is recovered in the limit $1/R \to \infty$ where the predicted extra Kaluza–Klein particles decouple from the low-energy theory.

The effective Hamiltonian inducing $b \to s\ell^+\ell^-$, $b \to s\nu\bar{\nu}$ and $b \to s\gamma$ transitions in ACD has been computed in [17, 181]. In the case of the exclusive modes $B \to K^{(*)}\ell^+\ell^-$, $B \to K^{(*)}\nu\bar{\nu}$ and $B \to K^*\gamma$, there are several observables sensitive to $1/R$ that can be used to probe this scenario [182, 183]. At present, the most stringent experimental bound on $1/R$ comes from $B \to K^*\gamma$, leading to $1/R \geq 300$–400 GeV, depending on the assumed hadronic uncertainties.

For values of $1/R$ of the order of a few hundred GeV, one expects an enhancement of $\mathcal{B}(B \to K^{(*)}\ell^+\ell^-)$ and $\mathcal{B}(B \to K^{(*)}\nu\bar{\nu})$ with respect to the SM (of the order of 20% for $1/R = 300$ GeV) and a suppression of $\mathcal{B}(B \to K^*\gamma)$ (at the same level for $1/R = 300$ GeV). In general, the sensitivity to $1/R$ is masked by the uncertainty of the hadronic $B \to K^{(*)}$ matrix elements. A useful observable with smaller hadronic uncertainties is the position of the FB asymmetry zero in $B \to K^*\ell^+\ell^-$, which in ACD is shifted to smaller values as $1/R$ decreases, as shown in Fig. 11 (left). Another interesting quantity, which however has a more pronounced dependence on hadronic uncertainties is the position $(q^2)_{\max}$ of the maximum of the longitudinal helicity fraction of K^* in the same process; its sensitivity to $1/R$ is also shown in Fig. 11 (right).

In the case of $B \to K^{(*)}\tau^+\tau^-$ decays, τ-polarization asymmetries can be considered, in which the hadronic form factor dependence drops out for large K^* recoil energies. The transverse asymmetry decreases as $1/R$ is decreased, whereas the branching fraction increases. The combined observation of this pattern of deviations from SM results would represent a signature of the ACD scenario.

3.3 Neutrino modes[14]

Here we discuss the so-called neutrino modes. In particular, we talk about the rare SM modes $B \to X_s\nu\bar{\nu}$ and $B \to \tau\nu$. Experimentally, these modes are similar since both are associated with large missing energy. In $B \to X_s\nu\bar{\nu}$, there are the two neutrinos, in $B \to \tau\nu$, the τ decays very fast, yielding a final state with two neutrinos as well. Theoretically these two modes are different. $B \to X_s\nu\bar{\nu}$ is an FCNC process and thus occurs at one loop in the SM. $B \to \tau\nu$, on the other hand, occurs at tree level, but it is strongly suppressed for several reasons: helicity, a small CKM factor and the decay mechanism by weak annihilation $\sim 1/m_B$.

3.3.1 Neutrino modes: theory

3.3.1.1 Inclusive $b \to s\nu\bar{\nu}$ decays
Here we follow [542] with necessary updates. The FCNC decay $B \to X_s\nu\bar{\nu}$ is very sensitive to extensions of the SM and provides a unique source of constraints on some NP scenarios which predict a large enhancement of this decay mode. In particular, the $B \to X_s\nu_\tau\bar{\nu}_\tau$ mode is very sensitive to the relatively unexplored couplings of third generation fermions.

From the theoretical point of view, the decay $B \to X_s\nu\bar{\nu}$ is a very clean process. Both the perturbative α_s and the nonperturbative $1/m_b^2$ corrections are known to be small. Furthermore, in contrast to the decay $B \to X_s\ell^+\ell^-$, which suffers from (theoretical and experimental) background

---
[14] Section coordinators: Y. Grossman, T. Iijima.

Table 23 Summary of observables in $\bar{B} \to K\ell^+\ell^-$, $\bar{B} \to K^*\ell^+\ell^-$ and $\bar{B}^0_q \to \ell^+\ell^-$ decays

Observable	Comments
$d\Gamma(\bar{B} \to K^{(*)}\ell^+\ell^-)/dq^2$	Hadronic uncertainties (form factors, nonfactorizable effects, $c\bar{c}$)
	SM: depends on $|C^{\text{eff}}_{7,9,10}|$ and $\text{Re}(C^{\text{eff}*}_7 C^{\text{eff}}_9)$
	NP: sensitive to Z-penguins, $C'_{9,10}$, $\text{sgn}(C^{\text{eff}}_7)$, but not to $C^{(')}_{S,P}$
$A_{FB}(\bar{B} \to K\ell^+\ell^-)$	SM: $\simeq 0$ (quasi null test)
	NP: sensitive to $C_S + C'_S$
	using $B_s \to \mu^+\mu^-$ constraint: $<$(few % for $\mu^+\mu^-$)
$dA_{FB}(\bar{B} \to K^*\ell^+\ell^-)/dq^2$	Hadronic uncertainties
(shape and magnitude)	NP: sensitive to $\text{sgn}(C^{\text{eff}}_7)$, $\text{sgn}(C^{\text{eff}}_{10})$, Z-penguins
FB asymmetry zero	Smaller uncertainties (test of the SM)
A^{CP}_{FB}	SM: $<10^{-3}$ (quasi null test)
	NP: CP-phase in C_{10} (+ dynamic strong phase)
$dA_I(\bar{B} \to K^*\ell^+\ell^-)/dq^2$	Hadronic uncertainties
	SM: $\mathcal{O}(+10\%)$ for $q^2 \leq 2$ GeV^2; depends on $C_{5,6}$ (cf. $A_I(\bar{B} \to K^*\gamma)$)
	$\mathcal{O}(-1\%)$ for $2 \leq q^2 \leq 7$ GeV^2; depends on $C_{3,4}$
	NP: sensitive to strong penguin operators; $\text{sgn}(C^{\text{eff}}_7)$
$A^{(1,2)}_T, \alpha_{K^*}, F_{L,T}$	Smaller uncertainties (test of SM)
	NP: right-handed currents, e.g., C'_7
$R_{K^{(*)}}$	Tiny uncertainties: $<\pm 1\%$
	SM: $1 + \mathcal{O}(m^2_\mu/m^2_b)$ (common cuts)
	NP: nonuniversal lepton couplings; $C^{(')}_{S,P}$, neutral Higgs exchange
$\mathcal{B}(\bar{B}^0_q \to \ell^+\ell^-)$	Uncertainties: f_{B_q}
	SM: depends on $|C_{10}V_{tq}|$
	NP: lepton-mass effects; $C^{(')}_{S,P}$, neutral Higgs exchange
$R_{\ell\ell}$	Uncertainties: f_{B_d}/f_{B_s}
	SM: $\sim |V_{td}|^2/|V_{ts}|^2 f^2_{B_d}/f^2_{B_s}$
	NP: test of MFV

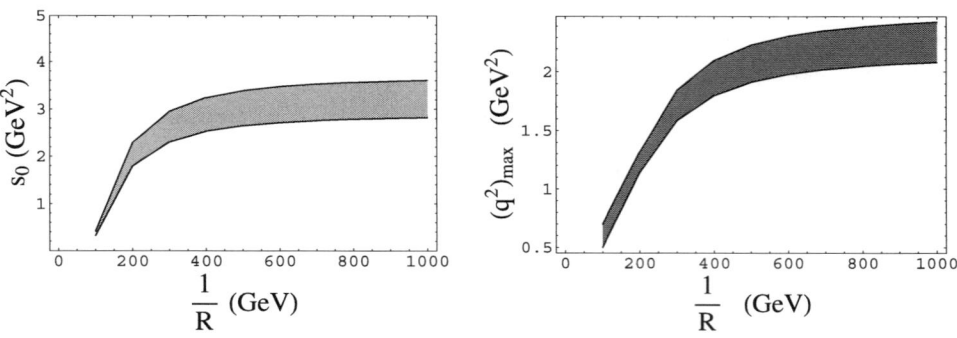

Fig. 11 Position of the zero, $s_0 \equiv q^2_0$, of A_{FB} (*left*) and of the maximum of the longitudinal K^* helicity fraction (*right*) in $B \to K^*\ell^+\ell^-$ as a function of $1/R$ in the ACD extra dimension scenario. R is the radius of the compactified extra dimension. The uncertainties only include the $B \to K^*$ form-factor dependence; nonfactorizable corrections have not been taken into account

such as $B \to X_s J/\psi \to X_s \ell^+\ell^-$, there are no important long-distance QCD contributions. Therefore, the decay $B \to X_s \nu\bar{\nu}$ is well suited to search for and constrain NP effects.

Another advantage of the $B \to X_s \nu\bar{\nu}$ mode is that the missing energy spectrum can be calculated essentially in a model-independent way. Thus, one can directly compare experimental data with the theoretical expressions as derived in specific models. Under the only assumption of two-component left-handed neutrinos, the most general form of the four-fermion interaction responsible for $B \to X_q \nu_i \bar{\nu}_j$ reads

$$\mathcal{L} = C_L O_L + C_R O_R, \tag{98}$$

where

$$O_L = [\bar{q}_L \gamma_\mu b_L][\bar{\nu}_L^i \gamma^\mu \nu_L^j],$$
$$O_R = [\bar{q}_R \gamma_\mu b_R][\bar{\nu}_L^i \gamma^\mu \nu_L^j]. \qquad (99)$$

Here L and R denote left- and right-handed components, $q = d, s$, and $i, j = e, \mu, \tau$. As the flavors of the decay products are not detected, in certain models more than one final state can contribute to the observed decay rate. Then, in principle, both C_L and C_R carry three indices q, i, j, which label the quark and neutrino flavors in the final state.

In the SM, $B \to X_s \nu \bar{\nu}$ proceeds via W-box and Z-penguin diagrams, and only O_L is present. The corresponding coefficient reads

$$C_L^{\rm SM} \simeq \frac{\sqrt{2} G_F \alpha}{\pi \sin^2 \theta_W} V_{tb}^* V_{ts} X_0(x_t),$$
$$X_0(x) = \frac{x}{8}\left[\frac{2+x}{x-1} + \frac{3x-6}{(x-1)^2} \ln x\right], \qquad (100)$$

where $x_t = m_t^2/m_W^2$. The leading $1/m_b^2$ and α_s corrections to the SM result are known. Thus, the theoretical uncertainties in the SM rate are rather small, less than $O(5\%)$. They come mainly from the uncertainties in m_t, $|V_{ts}|$ and unknown higher-order corrections. At lowest order, the missing energy spectrum in the B rest-frame is given by [541]

$$\frac{d\Gamma(B \to X_q \nu_i \bar{\nu}_j)}{dx} = \frac{m_b^5}{96\pi^3}(|C_L|^2 + |C_R|^2)\mathcal{S}(r,x). \quad (101)$$

Here we have not yet summed over the neutrino flavors. The function $\mathcal{S}(r, x)$ describes the shape of the missing energy spectrum

$$\mathcal{S}(r,x) = \sqrt{(1-x)^2 - r}\,[(1-x)(4x-1) + r(1-3x)$$
$$- 6\eta\sqrt{r}\,(1-2x-r)]. \qquad (102)$$

The dimensionless variable $x = E_{\rm miss}/m_b$ can range between $(1-r)/2 \leq x \leq 1 - \sqrt{r}$, and $r = m_s^2/m_b^2$. The parameter $\eta = -\mathrm{Re}(C_L C_R^*)/(|C_L|^2 + |C_R|^2)$ ranges between $-\frac{1}{2} \leq \eta \leq \frac{1}{2}$. Since r is very small, in practice the spectrum is independent of the relative size of C_L and C_R and therefore immune to the presence of NP.

It is convenient to define two "effective" coefficients $\tilde{C}_L$ and $\tilde{C}_R$, which can be computed in terms of the parameters of any model and are directly related to the experimental measurement. To remove the large uncertainty in the total decay rate associated with the m_b^5 factor, it is convenient to normalize $B(B \to X_s \nu \bar{\nu})$ to the semileptonic rate $B(B \to X_c e \bar{\nu})$. The contribution from $B \to X_u e \bar{\nu}$, as well as possible NP effects on the semileptonic decay rate are negligible. In constraining NP, we can also set $m_s = 0$ and neglect both order α_s and $1/m_b^2$ corrections. This is justified, since when averaged over the spectrum these effects are very small, and would affect the numerical bounds on the NP parameters only in a negligible way. For the total $B \to X_q \nu_i \bar{\nu}_j$ decay rate into all possible $q = d, s$ and $i, j = e, \mu, \tau$ final state flavors, we then obtain

$$\frac{B(B \to X \nu \bar{\nu})}{B(B \to X_c e \bar{\nu})} = \frac{\tilde{C}_L^2 + \tilde{C}_R^2}{|V_{cb}|^2 f(m_c^2/m_b^2)}, \qquad (103)$$

where $f(x) = 1 - 8x + 8x^3 - x^4 - 12x^2 \ln x$ is the usual phase-space factor, and we defined

$$\tilde{C}_L^2 = \frac{1}{8G_F^2}\sum_{q,i,j}|C_L^{qij}|^2, \qquad \tilde{C}_R^2 = \frac{1}{8G_F^2}\sum_{q,i,j}|C_R^{qij}|^2.$$
$$(104)$$

Note that channels with a different lepton flavor in the final state do not interfere. Thus, the sum among different channels is in the rate and not in the amplitude. The SM prediction, including NLO QCD corrections [214, 557, 558], is $B^{\rm SM}(B \to X_s \nu \bar{\nu}) = 4 \times 10^{-5}$.

New physics can generate new contributions to C_L and/or to C_R. Many NP models were studied in [542]. In general, there are bounds from other processes, in particular, $b \to s \ell^+ \ell^-$. In all models where these two processes are related, the NP contribution to the neutrino modes is bounded to be below the SM expectation. In that case, one needs to measure the neutrino mode at high precision in order to be able to probe these models of new physics.

The other case may be more interesting. In some models, there is an enhancement of the couplings to the third generation. Then $B \to X_s \nu \bar{\nu}$ is related only to $b \to s \tau^+ \tau^-$. This mode is very hard to measure, and thus there is no tight bound on these models. In that cases, NP could enhance the rate much above the SM rate. That is, if we find that the rate of $B \to X_s \nu \bar{\nu}$ is much above the SM rate, it will be an indication for models where the third generation is different.

3.3.1.2 Exclusive $b \to s \nu \bar{\nu}$ decays In principle, the theoretically cleanest observables are provided by inclusive decays, on the other hand, the exclusive variants will be more readily accessible in experiment. Despite the sizable theoretical uncertainties in the exclusive hadronic form factors, these processes could therefore give interesting first clues on deviations from what is expected in the SM [526]. This is particularly true if those happen to be large or if they show striking patterns. In the following, we discuss integrated observables and distributions in the invariant mass of the dilepton system, q^2, for the three-body decays $B \to M \nu \bar{\nu}$ with $M = K, K^*$. The kinematical range of q^2 is given by $0 \leq q^2 \leq (m_B - m_M)^2$. In the $B \to M \nu \bar{\nu}$ decays, q^2 is not directly measurable but it is related to the kaon energy in the B-meson rest frame, E_M, by the relation $q^2 = m_B^2 + m_M^2 - 2m_B E_M$, where $m_M \leq E_M \leq (m_B^2 + m_M^2)/(2m_B)$.

$B \to K \nu \bar{\nu}$ The dilepton spectrum of this mode is particularly simple and it is sensitive only to the combination $|C_L^\nu + C_R^\nu|^2$ [545, 546]. This is in contrast to the inclusive case where only the combination $|C_L^\nu|^2 + |C_R^\nu|^2$ entered the decay rate. In the inclusive case, all the interference terms average to zero when we sum over all the possible hadronic final states. In this way, exclusive processes are natural grounds where to perform tests of right-handed NP currents, given their interference with the purely left-handed SM current. Finally, the dilepton spectrum is [545, 546]

$$\frac{d\Gamma(B \to K \nu \bar{\nu})}{ds} = \frac{G_F^2 \alpha^2 m_B^5}{256\pi^5} |V_{ts}^* V_{tb}|^2 \lambda_K^{3/2}(s) f_+^2(s) |C_L^\nu + C_R^\nu|^2, \quad (105)$$

where we have defined the dimensionless variables $s = q^2/m_B^2$ and $r_M = m_M^2/m_B^2$ and the function

$$\lambda_M(s) = 1 + r_M^2 + s^2 - 2s - 2r_M - 2r_M s. \quad (106)$$

In the case of $M = K$, the hadronic matrix elements needed for our analysis are given by (53) with $P = K$. Up to small isospin breaking effects, which we shall neglect, the same set of form factors describes both charged ($B^- \to K^-$) and neutral ($\bar{B}^0 \to \bar{K}^0$) transitions. Thus, in the isospin limit, we get

$$\Gamma(B \to K \nu \bar{\nu}) \equiv \Gamma(B^+ \to K^+ \nu \bar{\nu}) = 2\Gamma(B^0 \to K_{L,S} \nu \bar{\nu}). \quad (107)$$

The absence of absorptive final-state interactions in this process also leads to $\Gamma(B \to K \nu \bar{\nu}) = \Gamma(\bar{B} \to \bar{K} \nu \bar{\nu})$, preventing the observation of any direct CP-violating effect. Integrating (105) over the full range of s leads to

$$\mathcal{B}(B \to K \nu \bar{\nu}) = \left(3.8^{+1.2}_{-0.6}\right) \times 10^{-6} \left|\frac{C_L^\nu + C_R^\nu}{C_L|_{\text{SM}}^\nu}\right|^2, \quad (108)$$

where the error is due to the uncertainty in the form factors.

If the experimental sensitivity on $B(B \to K \nu \bar{\nu})$ reached the 10^{-6} level, then the uncertainty due the form factors would prevent a precise extraction of $|C_L^\nu + C_R^\nu|$ from (108). This problem can be substantially reduced by relating the differential distribution of $B \to K \nu \bar{\nu}$ to the one of $B \to \pi e \nu_e$ [547, 548]:

$$\frac{d\Gamma(B \to K \nu \bar{\nu})/ds}{d\Gamma(B^0 \to \pi^- e^+ \nu_e)/ds} = \frac{3\alpha^2}{4\pi^2} \left|\frac{V_{ts}^* V_{tb}}{V_{ub}}\right|^2 \left(\frac{\lambda_K(s)}{\lambda_\pi(s)}\right)^{3/2} \left|\frac{f_+^K(s)}{f_+^\pi(s)}\right|^2 |C_L^\nu + C_R^\nu|^2. \quad (109)$$

Indeed $f_+^K(s)$ and $f_+^\pi(s)$ coincide up to $SU(3)$-breaking effects, which are expected to be small, especially far from the endpoint region. An additional uncertainty in (109) is induced by the CKM ratio $|V_{ts}^* V_{tb}|^2/|V_{ub}|^2$, which, however, can independently be determined from other processes.

$B \to K^* \nu \bar{\nu}$ A great deal of information can be obtained from the channel $B \to K^* \nu \bar{\nu}$ investigating, together with the lepton invariant mass distribution, also the FB asymmetry in the dilepton angular distribution. This may reveal effects beyond the SM that could not be observed in the analysis of the decay rate. The dilepton invariant mass spectrum of $B \to K^* \nu \bar{\nu}$ decays is sensitive to both combinations $|C_L^\nu - C_R^\nu|$ and $|C_L^\nu + C_R^\nu|$ [545, 546, 549]:

$$\frac{d\Gamma(B \to K^* \nu \bar{\nu})}{ds} = \frac{G_F^2 \alpha^2 m_B^5}{1024\pi^5} |V_{ts}^* V_{tb}|^2 \lambda_{K^*}^{1/2}(s)$$
$$\times \left\{ \frac{8s\lambda_{K^*}(s) V^2(s)}{(1+\sqrt{r_{K^*}})^2} |C_L^\nu + C_R^\nu|^2 \right.$$
$$+ \frac{1}{r_{K^*}} \left[(1+\sqrt{r_{K^*}})^2 (\lambda_{K^*}(s) + 12 r_{K^*} s) A_1^2(s) \right.$$
$$+ \frac{\lambda_{K^*}^2(s) A_2^2(s)}{(1+\sqrt{r_{K^*}})^2}$$
$$\left. \left. - 2\lambda_{K^*}(s)(1 - r_{K^*} - s) A_1(s) A_2(s) \right] |C_L^\nu - C_R^\nu|^2 \right\}, \quad (110)$$

where the form factors $A_1(s)$, $A_2(s)$ and $V(s)$ are defined in (54). Integrating (110) over the full range of s leads to

$$B(B \to K^* \nu \bar{\nu}) = \left(2.4^{+1.0}_{-0.5}\right) \times 10^{-6} \left|\frac{C_L^\nu + C_R^\nu}{C_L|_{\text{SM}}^\nu}\right|^2$$
$$+ \left(1.1^{+0.3}_{-0.2}\right) \times 10^{-5} \left|\frac{C_L^\nu - C_R^\nu}{C_L|_{\text{SM}}^\nu}\right|^2, \quad (111)$$

$$B(B \to K^* \nu \bar{\nu})|_{\text{SM}} = \left(1.3^{+0.4}_{-0.3}\right) \times 10^{-5}. \quad (112)$$

A reduction of the error induced by the poor knowledge of the form factors can be obtained by normalizing the dilepton distributions of $B \to K^* \nu \bar{\nu}$ to the one of $B \to \rho e \nu_e$ [548, 550]. This is particularly effective in the limit $s \to 0$, where the contribution proportional to $|C_L^\nu + C_R^\nu|$ (vector current) drops out.

3.3.1.3 $B \to \ell \nu$ Recently, the Belle [326] and BaBar [543] Collaborations have observed the purely leptonic decays $B^- \to \tau^- \bar{\nu}$, (120) and (121). Even if both measurements are still affected by large uncertainties, the observation of the $B^- \to \tau^- \bar{\nu}$ transition represents a fundamental step forward towards a deeper understanding of both flavor and electroweak dynamics. The precise measurement of its decay rate could provide clear evidence of NP, such as a nonstandard Higgs sector with large $\tan\beta$ [31].

Due to the $V - A$ structure of the weak interactions, the SM contributions to $B \to \ell\nu$ are helicity suppressed. Hence, these processes are very sensitive to non-SM effects (such as multi-Higgs effects) which might induce an effective pseudoscalar hadronic weak current [31]. In particular, charged Higgs bosons ($H^\pm$) appearing in any model with two Higgs doublets (including the SUSY case) can contribute at tree level to the above processes. The relevant four-Fermi interaction for the decay of charged mesons induced by $W^\pm$ and $H^\pm$ has the following form:

$$\frac{4G_F}{\sqrt{2}} V_{ub} \left[(\bar{u}\gamma_\mu P_L b)(\bar{\ell}\gamma^\mu P_L \nu) - \tan^2\beta \left(\frac{m_b m_\ell}{m_{H^\pm}^2}\right)(\bar{u} P_R b)(\bar{\ell} P_L \nu) \right], \quad (113)$$

where $P_{R,L} = (1 \pm \gamma_5)/2$. Here we keep only the $\tan\beta$-enhanced part of the $H^\pm ub$ coupling, namely the $m_b \tan\beta$ term. The decays $B \to \ell\nu$ proceed via the axial-vector part of the $W^\pm$ coupling and via the pseudoscalar part of the $H^\pm$ coupling. The amplitude then reads

$$\mathcal{A}_{B\to\ell\nu} = \frac{G_F}{\sqrt{2}} V_{ub} f_B \left[m_\ell - m_\ell \tan^2\beta \frac{m_B^2}{m_{H^\pm}^2} \right] \bar{\ell}(1-\gamma_5)\nu. \quad (114)$$

We observe that the SM term is proportional to m_ℓ because of the helicity suppression, while the charged Higgs term is proportional to m_ℓ because of the Yukawa coupling.

The SM expectation for the $B^- \to \tau^- \bar{\nu}$ branching fraction is

$$\mathcal{B}(B^- \to \tau^- \bar{\nu})^{\mathrm{SM}} = \frac{G_F^2 m_B m_\tau^2}{8\pi} \left(1 - \frac{m_\tau^2}{m_B^2}\right)^2 f_B^2 |V_{ub}|^2 \tau_B$$

$$= (1.59 \pm 0.40) \times 10^{-4}, \quad (115)$$

where we used $|V_{ub}| = (4.39 \pm 0.33) \times 10^{-3}$ from inclusive $b \to u$ semileptonic decays [389], $\tau_B = (1.643 \pm 0.010)$ ps and the recent unquenched lattice result $f_B = (0.216 \pm 0.022)$ GeV [321].

The inclusion of scalar charged currents leads to the following expression [31]:

$$R_{B\tau\nu} = \frac{\mathcal{B}(B^- \to \tau^-\bar{\nu})}{\mathcal{B}(B^- \to \tau^-\bar{\nu})^{\mathrm{SM}}} = r_H = \left[1 - \tan^2\beta \frac{m_B^2}{m_{H^\pm}^2}\right]^2. \quad (116)$$

Interestingly, in models where the two Higgs doublets are coupled separately to up- and down-type quarks, the interference between $W^\pm$ and $H^\pm$ amplitudes is necessarily *destructive*. For a natural choice of the parameters ($30 \lesssim \tan\beta \lesssim 50$, $0.5 \lesssim M_{H^\pm}/\mathrm{TeV} \lesssim 1$), (116) implies a (5–30)% suppression with respect to the SM. The corresponding expressions for the $K \to \ell\nu$ channels are obtained with the replacement $m_B \to m_K$, while for the $D \to \ell\nu$ case, $m_B^2 \to (m_s/m_c)m_D^2$. It is then easy to check that a 30% suppression of $B(B \to \tau\nu)$ should be accompanied by a 0.3% suppression (relative to the SM) in $B(D \to \ell\nu)$ and $B(K \to \ell\nu)$. At present, the theoretical uncertainty on the corresponding decay constants does not allow one to observe such effects.

Apart from the experimental error, one of the difficulties in obtaining a clear evidence of a possible deviation of $R_{B\tau\nu}$ from unity is the large parametric uncertainty induced by $|f_B|$ and $|V_{ub}|$. An interesting way to partially circumvent this problem is obtained by normalizing $B(B^- \to \tau^-\bar{\nu})$ to the B_d^0–$\bar{B}_d^0$ mass difference (ΔM_{B_d}) [32]. Neglecting the tiny isospin-breaking differences in masses, life-times and decay constants between B_d and B^- mesons, we can write [32]

$$\left.\frac{B(B^- \to \tau^-\bar{\nu})}{\tau_B \Delta M_{B_d}}\right|^{\mathrm{SM}}$$

$$= \frac{3\pi}{4\eta_B S_0(m_t^2/M_W^2)\hat{B}_{B_d}} \frac{m_\tau^2}{M_W^2}\left(1 - \frac{m_\tau^2}{m_B^2}\right)^2 \left|\frac{V_{ub}}{V_{td}}\right|^2 \quad (117)$$

$$= 1.77 \times 10^{-4} \left(\frac{|V_{ub}/V_{td}|}{0.464}\right)^2 \left(\frac{0.836}{\hat{B}_{B_d}}\right). \quad (118)$$

Following standard notation, we have denoted by $S_0(m_t^2/M_W^2)$, η_B and B_{B_d} the Wilson coefficient, the QCD correction factor and the bag parameter of the $\Delta B = 2$ operator within the SM (see e.g. Ref. [29]), using the unquenched lattice result $\hat{B}_{B_d} = 0.836 \pm 0.068$ [320] and $|V_{ub}/V_{td}| = 0.464 \pm 0.024$ from the UTfit Collaboration [210].

The ratio $R'_{B\tau\nu} = B(B^- \to \tau^-\bar{\nu})/\tau_B \Delta M_{B_d}$ could become a more stringent test of the SM in the near future, with higher statistics on the $B^- \to \tau^-\bar{\nu}$ channel. In generic extensions of the SM, the NP impact on $R_{B\tau\nu}$ and $R'_{B\tau\nu}$ is not necessarily the same. However, it should coincide if the non-SM contribution to ΔM_{B_d} is negligible, which is an excellent approximation in the class of models considered in [32].

For consistency, the $|V_{ub}/V_{td}|$ combination entering in $R'_{B\tau\nu} = B(B^- \to \tau^-\bar{\nu})/\tau_B \Delta M_{B_d}$ should be determined without using the information on ΔM_{B_d} and $B^- \to \tau^-\bar{\nu}$ (a condition that is already almost fulfilled). In the near future, one could determine this ratio with negligible hadronic uncertainties using the relation $|V_{ub}/V_{td}| = |\sin\beta_{\mathrm{CKM}}/\sin\gamma_{\mathrm{CKM}}|$.

From (116) it is evident that such tree level NP contributions, namely the r_H factor, do not introduce any lepton flavor-dependent correction, and thus departures from the SM lepton universality are not introduced. However, as pointed out in Ref. [544], this is no longer true in realistic supersymmetric frameworks if the model contains sizable sources of flavor violation in the lepton sector (a possibility that is well motivated by the large mixing angles in the neutrino sector). In the last case, we can expect observable deviations from the SM in the ratios

$$R_P^{\ell_1/\ell_2} = \frac{B(P \to \ell_1 \nu)}{B(P \to \ell_2 \nu)} \quad (119)$$

with $P = \pi, K, B$ and $\ell_{1,2} = e, \mu, \tau$. The lepton-flavor violating (LFV) effects can be quite large in e or μ modes, while in first approximation they are negligible in the τ channels. In the most favorable scenarios, taking into account the constraints from LFV τ decays [165, 166], spectacular order-of-magnitude enhancements for $R_B^{e/\tau}$ and $\mathcal{O}(100\%)$ deviations from the SM in $R_B^{\mu/\tau}$ are allowed [32]. The key ingredients that allow visible non-SM contributions in $R_P^{\mu/e}$ within the MSSM are large values of $\tan\beta$ and sizable mixing angles in the right-slepton sector such that the $P \to \ell_i \nu_j$ rate (with $i \neq j$) becomes nonnegligible.

3.3.2 Neutrino modes: experiment

Experimental prospects for neutrino modes, such as $b \to s\nu\bar{\nu}$, $B \to \tau\nu$ and $b \to c\tau\nu$, are discussed. Because of the missing multiple neutrinos in the final state, these decays lack kinematic constraints, which could be used to suppress background processes. The e^+e^- B-factories, where background is relatively low and can be reduced by reconstructing the accompanying B meson, would be the ideal place to measure these decays. We also discuss the prospect for $B \to \mu\nu$, which can be used to test the lepton universality in comparison to $B \to \tau\nu$.

Belle and BaBar have used hadronic decays to reconstruct the accompanying B (hadronic tags), for which the tagging efficiency is about 0.3(0.1)% for the charged (neutral) B meson. BaBar has used also semileptonic decays $B \to D^{(*)}\ell\nu$ (semileptonic tags) to increase the efficiency at the expense of the signal-to-noise ratio.

The present e^+e^- B-factory experiments are starting to measure some of these decays, as demonstrated by the first evidence of $B \to \tau\bar{\nu}$, which was recently reported by Belle. However, precision measurements and detection of very difficult modes, such as $b \to s\nu\bar{\nu}$, require at least a couple of tens ab^{-1} data, which can be reached only at the proposed super B-factories.

3.3.2.1 $b \to s\nu\bar{\nu}$ Presently, experimental limits on exclusive $b \to s\nu\bar{\nu}$ modes are available from Belle and BaBar. Belle has reported the result of a search for $B^- \to K^-\nu\bar{\nu}$ using a 253 fb^{-1} data sample [551]. The analysis utilizes the hadronic tags and requires that the event has neither remaining charged tracks nor neutral clusters other than the K^- candidate. Figure 12(a) shows the distribution of remaining neutral cluster energy recorded in the electromagnetic calorimeter (E_{ECL}) after all the selection cuts are applied. The signal detection efficiency is estimated to be 43% for the tagged events. In the signal region defined as $E_{ECL} < 0.3$ GeV, the expected number of signals is 0.70, assuming the SM branching fraction of $\mathcal{B}(B \to K^-\nu\bar{\nu}) = 4 \times 10^{-6}$, while the number of background estimated from the sideband data is 2.6 ± 1.6. The deduced upper limit (90% C.L.) on the branching fraction is $\mathcal{B}(B^- \to K^-\nu\bar{\nu}) < 3.6 \times 10^{-5}$. More recently, Belle has reported an upper limit of $\mathcal{B}(B^0 \to K^{*0}\nu\bar{\nu}) < 3.4 \times 10^{-4}$ from a similar analysis on a 492 fb^{-1} data sample [552].

BaBar has reported $\mathcal{B}(B^- \to K^-\nu\bar{\nu}) < 5.2 \times 10^{-5}$ by combining the hadronic and semileptonic tag events from a 82 fb^{-1} data sample [553]. The right panel of Fig. 12 shows the distribution of the remaining energy (E_{extra} in BaBar's notation) for the semileptonic tag sample. Because of the large $B^- \to D^{(*)}\ell\bar{\nu}$ branching fractions, the semileptonic tag method has a factor 2 to 3 higher efficiency than the hadronic tag method.

Based on a simple-minded extrapolation from the Belle analysis with the hadronic tags, the required integrated luminosity for observing the $B^- \to K^-\nu\bar{\nu}$ decay with $3(5)\sigma$ statistical significance is $12(33)$ ab^{-1}. The statistical precision for the branching fraction measurement will reach 18%

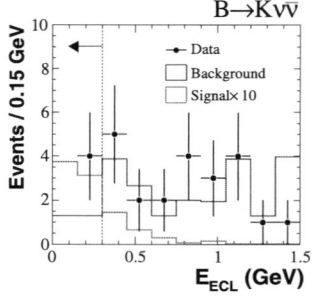 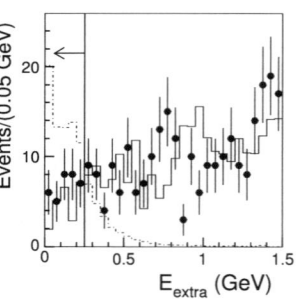

Fig. 12 Distribution of remaining energy for $B^- \to K^-\nu\bar{\nu}$ candidates: from Belle's analysis using the hadronic tag on a 253 fb^{-1} data sample (*left*) and from BaBar's analysis using the semileptonic tag on a 82 fb^{-1} data sample (*right*)

at 50 ab^{-1}. Addition of the semileptonic tag sample may improve the sensitivity (this is under investigation).

It is extremely difficult to perform an inclusive search for $b \to s\nu\bar{\nu}$. No serious studies have been made yet.

3.3.2.2 $B \to \tau\nu$

Detection of $B^- \to \tau^-\bar{\nu}$ is very similar to that of $B \to K^{(*)}\nu\bar{\nu}$, and it requires that the event has no extra charged tracks nor neutral clusters other than those from the τ decay and the accompanying B decay.

Recently Belle has reported the first evidence for $B^- \to \tau^-\bar{\nu}$ by applying the hadronic tag on a 414 fb^{-1} data sample [326]. The reconstructed τ decay modes are $\tau^- \to e^-\bar{\nu}_e\nu_\tau$, $\mu^-\bar{\nu}_\mu\nu_\tau$, $\pi^-\nu_\tau$, $\pi^-\pi^0\nu_\tau$, $\pi^-\pi^+\pi^-\nu_\tau$. The left panel of Fig. 13 presents the E_{ECL} distribution combined for all the τ decay modes, which shows an excess of events near $E_{\text{ECL}} = 0$. The number of signal (N_s) and background events (N_b) in the signal region are determined to be $N_s = 17.2^{+5.3}_{-4.7}$ and $N_b = 32.0 \pm 0.7$ by an unbinned maximum likelihood fit. The significance of the excess is 3.5σ including both statistical and systematic uncertainties. The obtained branching fraction is [326]

$$\mathcal{B}(B^- \to \tau^-\bar{\nu}) = \left[1.79^{+0.56}_{-0.49}(\text{sta})^{+0.46}_{-0.51}(\text{sys})\right] \times 10^{-4}. \quad (120)$$

BaBar has reported results of a $B^- \to \tau^-\bar{\nu}$ search using the semileptonic tag on a 288 fb^{-1} data sample [543]. The tag reconstruction efficiency is about 0.7%, depending slightly on run periods. When all the analysed τ decay modes are combined, 213 events are observed, while the background is estimated to be 191.7 ± 11.7. Since the excess is not significant, they provide an upper limit of $\mathcal{B}(B^- \to \tau^-\bar{\nu}) < 1.8 \times 10^{-4}$ (90% C.L.) and also quote the value [543]

$$\mathcal{B}(B^- \to \tau^-\bar{\nu}) = \left[0.88^{+0.68}_{-0.67}(\text{sta}) \pm 0.11(\text{sys})\right] \times 10^{-4}. \quad (121)$$

The semileptonic tag gives roughly two times higher efficiency than the hadronic tag but introduces more backgrounds.

Within the context of the SM, the product of the B meson decay constant and the magnitude of the CKM matrix element $|V_{ub}|$ is determined to be $f_B|V_{ub}| = (10.1^{+1.6}_{-1.4}(\text{sta})^{+1.3}_{-1.4}(\text{sys})) \times 10^{-4}$ GeV from the Belle result. Using the value of $|V_{ub}| = (4.39 \pm 0.33) \times 10^{-3}$ from inclusive charmless semileptonic B decay data [389], we obtain $f_B = 0.229^{+0.036}_{-0.031}(\text{sta})^{+0.034}_{-0.037}(\text{sys})$ GeV.

The charged Higgs can be constrained by comparing the measured branching fraction ($\mathcal{B}^{\text{exp}}$) to the SM value of $\mathcal{B}^{\text{SM}} = (1.59 \pm 0.40) \times 10^{-4}$, which is deduced from the above $|V_{ub}|$ value and $f_B = (0.216 \pm 0.022)$ GeV obtained from lattice QCD calculations [321]. Using the Belle result, the ratio (116) is $r_H = 1.13 \pm 0.53$, which then constrains the charged Higgs in the $(M_{H^+}, \tan\beta)$ plane, as shown in Fig. 14 (top). The hatched area indicates the region excluded at a confidence level of 95.5%.

Further accumulation of data helps to improve on both the statistical and systematic uncertainties of the branching fraction. Some of the major systematic errors, such as ambiguities in the reconstruction efficiency and the signal and background shapes, come from the limited statistics of a control sample. On the other hand, the error in the ratio r_H depends on the errors in the determination of $|V_{ub}|$ and f_B. Figure 14 (bottom) shows the expected constraint at 5 ab^{-1}, assuming the scaling of the experimental error by $1/\sqrt{L}$ (L is the luminosity) and 5% relative error for both $|V_{ub}|$ and f_B. Figure 15 presents the M_{H^+} reach at $\tan\beta = 30$ as a function of the integrated luminosity. Here the M_{H^+} reach is defined as the upper limit of the 95.5% excluded region at a given $\tan\beta$. The figure shows the expectation for three cases, $(\Delta|V_{ub}|/|V_{ub}|, \Delta f_B/f_B) = (0\%, 0\%)$, (2.5%, 2.5%) and (5%, 5%). Precise determination of $|V_{ub}|$ and f_B is desired to maximize the physics reach.

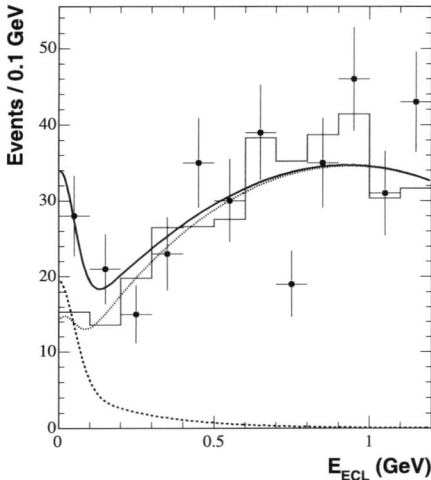

Fig. 13 Distribution of the remaining energy for $B^- \to \tau^-\bar{\nu}$ candidates: from Belle's analysis using the hadronic tag on a 414 fb^{-1} data sample (*left*) and from BaBar's analysis using the semileptonic tag on a 288 fb^{-1} data sample (*right*)

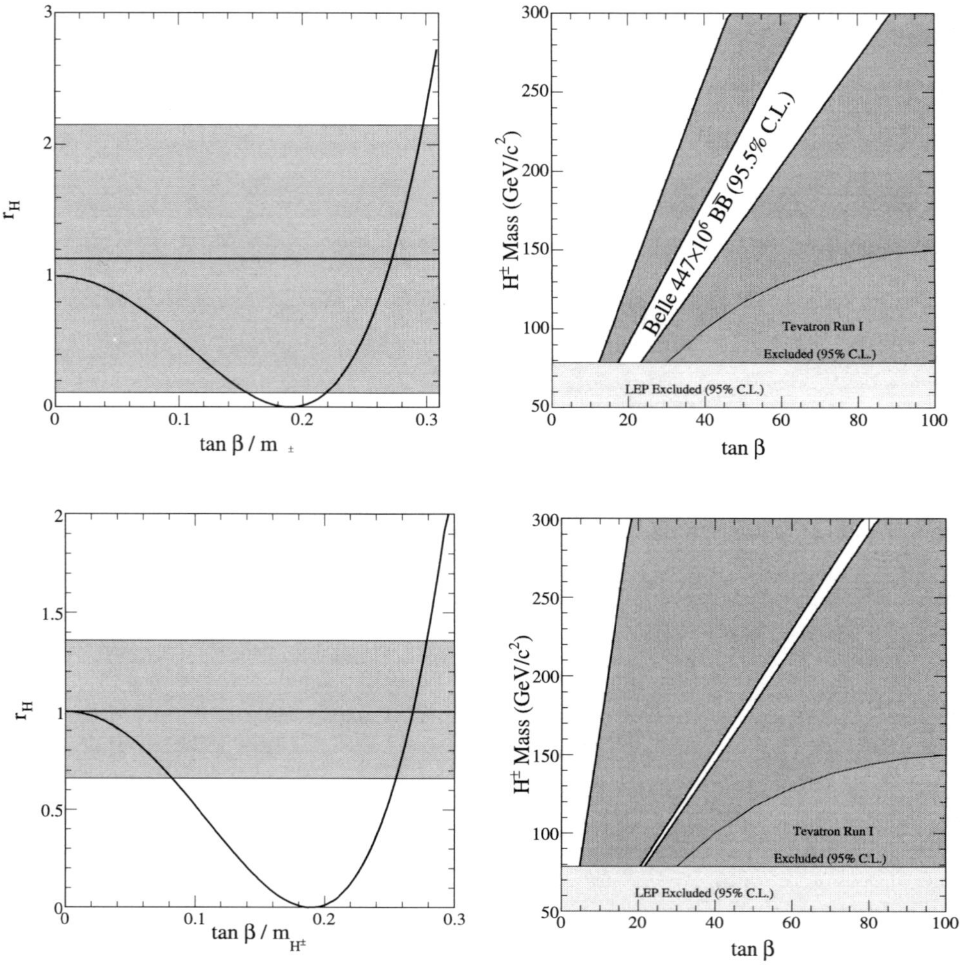

Fig. 14 The constraint on the charged Higgs; ±1σ boundary in the ratio r_H (*left*) and the 95.5% C.L. exclusion boundaries in the (M_{H^+}, tan β) plane (*right*). The *top figures* show the constraint from the present Belle result. The *bottom figures* show the expected constraints at 5 ab^{-1}

Fig. 15 Expected M_{H^+} reach at tan β = 30 as a function of the integrated luminosity. The *three curves* correspond to $(\Delta|V_{ub}|/|V_{ub}|, \Delta f_B/f_B) =$ *upper*: (0%, 0%), *middle*: (2.5%, 2.5%) and *lower*: (5%, 5%)

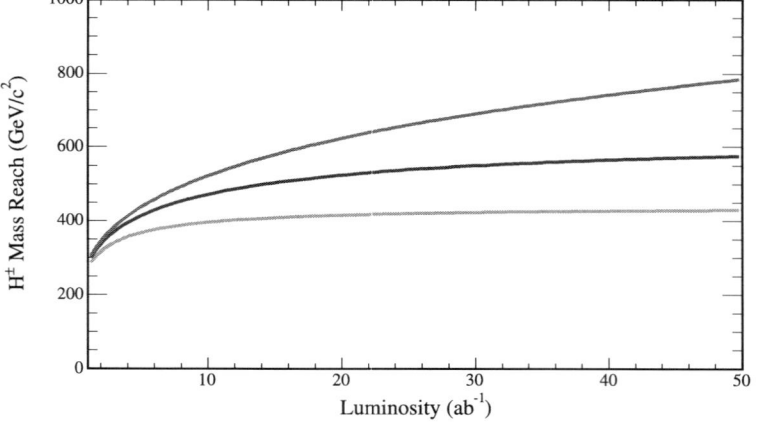

3.3.2.3 $B \to D^{(*)}\tau\nu$ The semileptonic B decay into τ final state, $B \to D^{(*)}\tau\bar{\nu}$, is also a sensitive probe for the charged Higgs. In the SM, the branching fractions are expected to be about 8×10^{-3} for $B \to D\tau\bar{\nu}$ and 1.6×10^{-2} for $B \to D^*\tau\bar{\nu}$, respectively. Because of the presence of at least two neutrinos in the final state, the reconstruction of these modes requires the reconstruction of the other B meson in the event and hence requires a larger data sample with respect to that used to measure $B \to D^{(*)}\ell\bar{\nu}$, where $\ell = \mu, e$. Figure 16 presents the expected future constraint in the (M_{H^+}, tan β) plane for a super B factory with a 5 and 50 ab^{-1} data sample.

3.3.2.4 $B \to \mu\nu$

Contrary to the $B^- \to \tau\bar{\nu}$ case, the $B^- \to \mu^-\bar{\nu}$ decay has more kinematic constraint, because it has only one neutrino in the final state and the charged lepton at a fixed energy in the B rest frame. Therefore, present analyses by Belle and BaBar take a conventional approach, where one looks for a single high momentum lepton and then inclusively reconstructs the accompanying B via a 4-vector sum of everything else in the event. The lepton momentum is smeared in the center-of-mass frame due to B momentum to give a couple of hundred MeV/c width.

The left panel of Fig. 17 shows the muon momentum distribution from the Belle analysis to search for the $B^- \to \mu^-\bar{\nu}$ decay using the conventional approach on a 253 fb^{-1} data sample. The signal detection efficiency is 2.2%. The expected number of signals based on the SM branching fraction 7.1×10^{-7} is 4.2, while the estimated background is 7.4. The reported upper limit is $\mathcal{B}(B^- \to \mu^-\bar{\nu}) \leq 1.7 \times 10^{-6}$ (90% C.L.) [554].

Recently BaBar has reported a result of the $B \to \mu\nu$ search using the hadronic tags on a 208.7 fb^{-1} data sample. In this case, as the B momentum is determined by the full reconstruction, there is no smearing in the lepton momentum. The right panel of Fig. 17 is the muon momentum distribution after all the selection cuts are applied. The signal detection efficiency is about 0.15%, an order of magnitude lower than for the conventional analysis. The reported upper limit is $\mathcal{B}(B^- \to \mu^-\bar{\nu}) \leq 7.9 \times 10^{-6}$ (90% C.L.) [555].

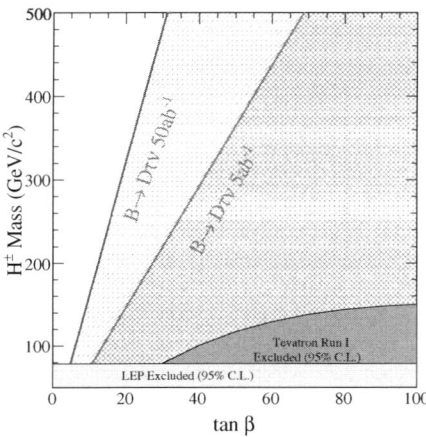

Fig. 16 Expected constraint on the charged Higgs parameters from measurements of the $B \to D\tau\bar{\nu}$ branching fraction at 5 and 50 ab^{-1}

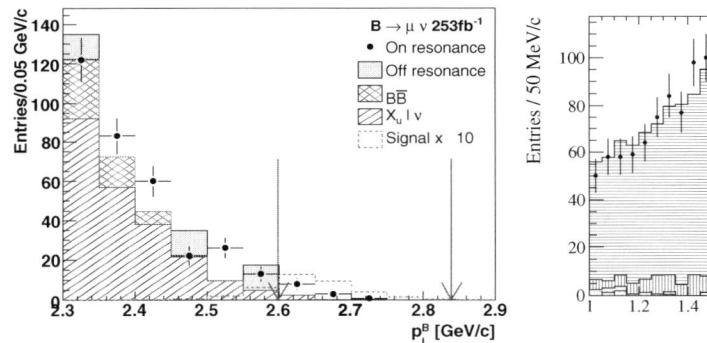
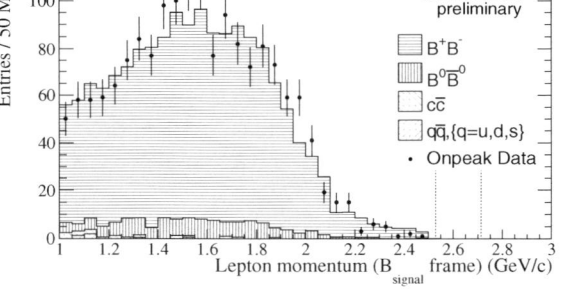

Fig. 17 Muon momentum distribution from the Belle analysis using an inclusive reconstruction of the accompanying B for a 253 fb^{-1} data sample (*left*); the same distribution from the BaBar analysis using the hadronic tags on a 208.7 fb^{-1} data sample (*right*)

Fig. 18 Expected sensitivity for $B^- \to \mu^-\bar{\nu}$ as a function of the integrated luminosity

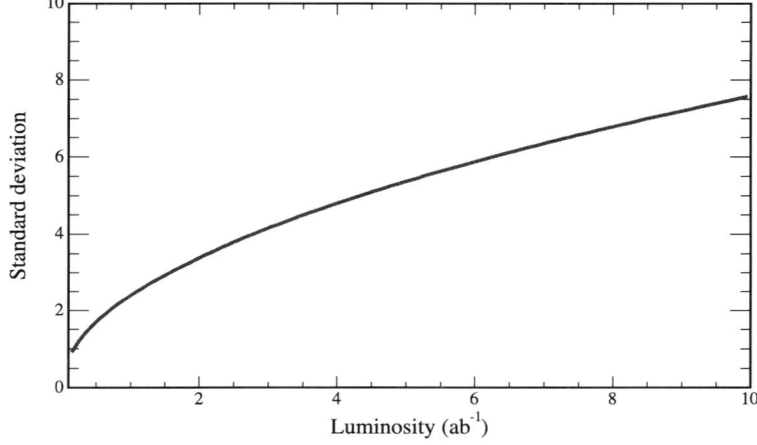

Figure 18 shows the expected statistical significance as a function of the integrated luminosity, based on a simple extrapolation from the present Belle result. Accumulation of 1.6(4.3) ab^{-1} data will allow us to detect the $B^- \to \mu^- \bar{\nu}$ signal with 3(5) statistical significance. The 50 ab^{-1} data at super B-factories will allow us to detect about 800 signal events and measure the branching fraction with about 6% statistical precision.

There are some points which need to be further studied.

- Optimization of the tagging; there may be some improvement by using the semileptonic tag in addition to the hadronic tag, especially for $B^- \to K^- \nu \bar{\nu}$, for which the impact of additional neutrinos seems to be relatively small.
- Effects of backgrounds in a high luminosity environment; future prospects are discussed so far by extrapolation from the present results, which may be too simple. In particular, the impact of higher backgrounds to the tagging efficiency and the missing energy resolution have to be more carefully examined.

3.4 Very rare decays[15]

3.4.1 Theory of $B_q \to \ell^+\ell^-$ and related decays

A particularly important class of very rare decays are the leptonic FCNC decays of a B_d or B_s meson. In addition to the electroweak loop suppression, the corresponding decay rates are helicity suppressed in the SM by a factor of m_ℓ^2/m_B^2, where m_ℓ and M_B are the masses of lepton and B meson, respectively. The effective $|\Delta B| = |\Delta S| = 1$ Hamiltonian, which describes $b \to s$ decays, already contains 17 different operators in the SM; in a generic model-independent analysis of NP, this number will exceed 100. One virtue of purely leptonic B_s decays is their dependence on a small number of operators, so that they are accessible to model-independent studies of NP. These statements, of course, equally apply to $b \to d$ transitions and leptonic B_d decays. While in the SM all six $B_q \to \ell^+\ell^-$ decays (with $q = d$ or s and $\ell = e, \mu$ or τ) are related to one another in a simple way, this is not necessarily so in models of NP. Therefore all six decay modes should be studied.

Other very rare decays, such as $B_q \to \ell^+\ell^-\ell'^+\ell'^-$, $\ell^+\ell^-\gamma$, $e^\pm\mu^\mp$, are briefly considered in Sect. 3.4.1.3 below.

3.4.1.1 $B_q \to \ell^+\ell^-$ in the Standard Model
Photonic penguins do not contribute to $B_q \to \ell^+\ell^-$ because a lepton–anti-lepton pair with zero angular momentum has charge conjugation quantum number $C = 1$, while the photon has $C = -1$. The dominant contribution stems from the Z-penguin diagram and is shown in Fig. 19.

---
[15]Section coordinators: U. Nierste, M. Smizanska.

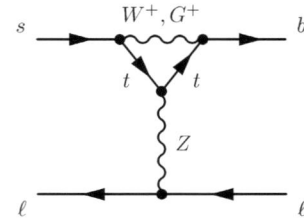

Fig. 19 Z-penguin contribution to $B_s \to \ell^+\ell^-$

There is also a box diagram with two W bosons, which is suppressed by a factor of M_W^2/m_t^2 with respect to the Z-penguin diagram. These diagrams determine the Wilson coefficient C_A of the operator

$$Q_A = \bar{b}_L \gamma^\mu q_L \bar{\ell} \gamma_\mu \gamma_5 \ell. \tag{122}$$

We will further need operators with scalar and pseudoscalar couplings to the leptons:

$$Q_S = m_b \bar{b}_R q_L \bar{\ell}\ell, \qquad Q_P = m_b \bar{b}_R q_L \bar{\ell}\gamma_5\ell. \tag{123}$$

Their coefficients C_S and C_P are determined from penguin diagrams involving the Higgs or the neutral Goldstone boson, respectively. While C_S and C_P are tiny and can be safely neglected in the SM, the situation changes dramatically in popular models of NP discussed below. The effective Hamiltonian reads

$$H_{\text{eff}} = \frac{G_F}{\sqrt{2}} \frac{\alpha}{\pi \sin^2\theta_W} V_{tb}^* V_{tq} [C_S Q_S + C_P Q_P + C_A Q_A] + \text{h.c.} \tag{124}$$

The operators Q_S', Q_P' and Q_A', where the chiralities of the quarks in the $\bar{b}q$ currents are flipped with respect to those in (122) and (123), may also become relevant in general extensions of the SM.

C_A has been determined in the next-to-leading order (NLO) of QCD [556–558]. The NLO corrections are in the percent range, and higher-order corrections play no role. C_A is commonly expressed in terms of the $\overline{\text{MS}}$ mass of the top quark, $\bar{m}_t$. A pole mass (usually quoted in the context of direct measurements and electroweak fits) of $m_t^{\text{pole}} = (171.4 \pm 2.1)$ GeV corresponds to $\bar{m}_t = (163.8 \pm 2.0)$ GeV. An excellent approximation to the NLO result for C_A, which holds with an accuracy of 5×10^{-4} for $149 < \bar{m}_t < 179$ GeV, is

$$C_A(\bar{m}_t) = 0.9636 \left[\frac{80.4 \text{ GeV}}{M_W} \frac{\bar{m}_t}{164 \text{ GeV}} \right]^{1.52}. \tag{125}$$

In the literature, $C_A(\bar{m}_t)$ is often called $Y(\bar{m}_t^2/M_W^2)$. The exact expression can be found e.g. in (16)–(18) of [558]. The branching fraction can be compactly expressed in terms of the Wilson coefficients C_A, C_S and C_P:

$$B(B_q \to \ell^+\ell^-)$$
$$= \frac{G_F^2 \alpha^2}{64\pi^3 \sin^4\theta_W} |V_{tb}^* V_{tq}|^2 \tau_{B_q} M_{B_q}^3 f_{B_q}^2 \sqrt{1 - \frac{4m_\ell^2}{M_{B_q}^2}}$$
$$\times \left[\left(1 - \frac{4m_\ell^2}{M_{B_q}^2}\right) M_{B_q}^2 C_S^2 + \left(M_{B_q} C_P - \frac{2m_\ell}{M_{B_q}} C_A\right)^2\right]. \quad (126)$$

Here f_{B_q} and τ_{B_q} are the decay constant and the lifetime of the B_q meson, respectively, and θ_W is the Weinberg angle. Since $B_q \to \ell^+\ell^-$ is a short-distance process, the appropriate value of the fine-structure constant is $\alpha = \alpha(M_Z) = 1/128$. With (125) and $C_S = C_P = 0$, (126) gives the following SM predictions:

$$B(B_s \to \tau^+\tau^-)$$
$$= (8.20 \pm 0.31) \times 10^{-7}$$
$$\times \frac{\tau_{B_s}}{1.527 \text{ ps}} \left[\frac{|V_{ts}|}{0.0408}\right]^2 \left[\frac{f_{B_s}}{240 \text{ MeV}}\right]^2, \quad (127)$$

$$B(B_s \to \mu^+\mu^-)$$
$$= (3.86 \pm 0.15) \times 10^{-9}$$
$$\times \frac{\tau_{B_s}}{1.527 \text{ ps}} \left[\frac{|V_{ts}|}{0.0408}\right]^2 \left[\frac{f_{B_s}}{240 \text{ MeV}}\right]^2, \quad (128)$$

$$B(B_s \to e^+e^-)$$
$$= (9.05 \pm 0.34) \times 10^{-14}$$
$$\times \frac{\tau_{B_s}}{1.527 \text{ ps}} \left[\frac{|V_{ts}|}{0.0408}\right]^2 \left[\frac{f_{B_s}}{240 \text{ MeV}}\right]^2, \quad (129)$$

$$B(B_d \to \tau^+\tau^-)$$
$$= (2.23 \pm 0.08) \times 10^{-8}$$
$$\times \frac{\tau_{B_d}}{1.527 \text{ ps}} \left[\frac{|V_{td}|}{0.0082}\right]^2 \left[\frac{f_{B_d}}{200 \text{ MeV}}\right]^2, \quad (130)$$

$$B(B_d \to \mu^+\mu^-)$$
$$= (1.06 \pm 0.04) \times 10^{-10}$$
$$\times \frac{\tau_{B_d}}{1.527 \text{ ps}} \left[\frac{|V_{td}|}{0.0082}\right]^2 \left[\frac{f_{B_d}}{200 \text{ MeV}}\right]^2, \quad (131)$$

$$B(B_d \to e^+e^-)$$
$$= (2.49 \pm 0.09) \times 10^{-15}$$
$$\times \frac{\tau_{B_d}}{1.527 \text{ ps}} \left[\frac{|V_{td}|}{0.0082}\right]^2 \left[\frac{f_{B_d}}{200 \text{ MeV}}\right]^2. \quad (132)$$

The dependencies on the decay constants, which have sizable theoretical uncertainties, and on the relevant CKM factors have been factored out. While $|V_{ts}|$ is well determined through the precisely measured $|V_{cb}|$, the determination of $|V_{td}|$ involves the global fit to the unitarity triangle and suffers from larger uncertainties. The residual uncertainty in (127)–(132) stems from the 2 GeV error in $\bar{m}_t$.

Alternatively, within the standard model, the CKM dependence as well as the bulk of the hadronic uncertainty may be eliminated by normalizing to the well-measured meson mass differences ΔM_{B_q}, thus trading $f_{B_q}^2$ for a (less uncertain) bag parameter $\hat{B}_q$ [559]:

$$B(B_q \to \ell^+\ell^-) = C \frac{\tau_{B_q}}{\hat{B}_q} \frac{Y^2(\bar{m}_t^2/M_W^2)}{S(\bar{m}_t^2/M_W^2)} \Delta M_q, \quad (133)$$

where S is a perturbative short-distance function, $C = 4.36 \times 10^{-10}$ includes a normalization and NLO QCD corrections, and $\ell = e, \mu$. This reduces the *total* uncertainty within the SM below the 15 percent level. (A similar formula may be written for $\ell = \tau$.)

3.4.1.2 $B_q \to \ell^+\ell^-$ and new physics
• *Additional Higgs bosons*

The helicity suppression factor of m_ℓ/M_{B_q} in front of C_A in (126) makes $B(B_q \to \ell^+\ell^-)$ sensitive to physics with new scalar or pseudoscalar interactions, which contribute to C_S and C_P. This feature renders $B_q \to \ell^+\ell^-$ highly interesting to probe models with an extended Higgs sector. Practically all weakly coupled extensions of the SM contain extra Higgs multiplets, which puts $B(B_q \to \ell^+\ell^-)$ on the center stage of indirect NP searches. Higgs bosons couple to fermions with Yukawa couplings y_f. In the SM, $y_b \propto m_b/M_W$ and $y_\ell \propto m_\ell/M_W$ are so small that Higgs penguin diagrams, in which the Z-boson of Fig. 19 is replaced by a Higgs boson, play no role. In extended Higgs sectors, the situation can be dramatically different. Models with two or more Higgs multiplets can not only accommodate Yukawa couplings of order one, they also generically contain tree-level FCNC couplings of neutral Higgs bosons. In simple two-Higgs-doublet models, these unwanted FCNC couplings are usually switched off in an ad-hoc way by imposing a discrete symmetry on the Higgs and fermion fields, which leads to the celebrated two-Higgs-doublet models of type I and type II. Here we only discuss the latter model, in which one Higgs doublet H_u only couples to up-type fermions, while the other one, H_d, solely couples to down-type fermions [560]. The parameter controlling the size of the down-type Yukawa coupling is $\tan\beta = v_u/v_d$, the ratio of the vacuum expectation values acquired by H_u and H_d. The Yukawa coupling y_f of H_d to the fermion f satisfies $y_f \sin\beta = m_f \tan\beta/v$ with $v = \sqrt{v_u^2 + v_d^2} = 174$ GeV. Hence $y_b \approx 1$ for $\tan\beta \approx 50$. The dominant contributions to C_S and C_P for large $\tan\beta$ involve charged and neutral Higgs bosons, but the final result can be solely expressed in terms of $\tan\beta$ and the charged Higgs boson mass M_{H^+} [561]:

$$C_S = C_P = \frac{m_\ell}{4M_W^2}\tan^2\beta\,\frac{\ln r}{r-1} \quad \text{with } r = \frac{M_{H^+}^2}{\bar{m}_t^2}, \qquad (134)$$

while C_A remains the same as in the SM. Although for very large values of $\tan\beta/M_{H^+}$ the branching fraction can be enhanced, the contributions in (134) typically reduce $B(B_q \to \ell^+\ell^-)$ with respect to the SM value. The decoupling for $M_{H^+} \to \infty$ is slow, e.g., for $\tan\beta = 60$ and $M_{H^+} = 500$ GeV, the new Higgs contributions reduce $B(B_q \to \ell^+\ell^-)$ by 50%!

- *Supersymmetry*

The generic Minimal Supersymmetric Standard Model (MSSM) contains many new sources of flavor violation in addition to the Yukawa couplings. These new flavor violating parameters stem from the supersymmetry-breaking terms, and their effects could easily exceed those of the CKM mechanism. In view of the success of the CKM description of flavor-changing transitions, one may supplement the MSSM with the hypothesis of MFV, which can be formulated systematically using symmetry arguments [10]. In the MFV–MSSM, the only sources of flavor violation are the Yukawa couplings, just as in the SM. In this section, the MSSM is always understood to be supplemented with the assumption of MFV. While in MFV scenarios the contributions from virtual supersymmetric particles to FCNC processes are normally smaller than the SM contribution, the situation is very different for $B_q \to \ell^+\ell^-$.

The MSSM has two Higgs doublets. At tree-level the couplings are as in the two-Higgs-doublet model of type II, because the holomorphy of the superpotential forbids the coupling of H_u to down-type fermions and that of H_d to up-type fermions. At the one-loop level, however, the situation is different, and both doublets couple to all fermions. The loop-induced couplings are proportional to the product of a supersymmetry-breaking term and the μ parameter. If $\tan\beta$ is large, the loop-induced coupling of H_u^* and the tree-level coupling of H_d give similar contributions to the masses of the down-type fermions, because the loop suppression is compensated by a factor of $\tan\beta$ [20]. In this scenario, the Higgs sector is that of a *general* two-Higgs-doublet model, which involves FCNC Yukawa couplings of the heavy neutral Higgs bosons A^0 and H^0 [25]. The Wilson coefficients C_S and C_P differ from those in (134) in two important aspects: they involve three rather than two powers of $\tan\beta$ and they depend on the mass $M_{A^0} \sim M_{H^0}$ instead of the charged Higgs boson mass. The branching ratios scale as

$$B(B_q \to \ell^+\ell^-)_{\text{SUSY}} \propto \frac{m_b^2 m_\ell^2 \tan^6\beta}{M_{A^0}^4}$$

and could, in principle, exceed the SM results in (127)–(132) by a factor of 10^3 [27]. Thus the experimental upper limit on $B(B_s \to \mu^+\mu^-)$ from the Tevatron, which is larger than $B(B_s \to \mu^+\mu^-)_{\text{SM}}$ in (128) by a factor of 25, already severely cuts into the parameter space of the MSSM. $B(B_s \to \mu^+\mu^-)$ in MSSM scenarios with large $\tan\beta$ has been studied extensively [27–30, 528, 562–564].

Very popular special cases of the MSSM are the minimal Supergravity Model (mSUGRA) [565–570] and the Constrained Minimal Supersymmetric Standard Mode (CMSSM). While the MSSM contains more than 100 parameters, mSUGRA involves only 5 additional parameters and is therefore much more predictive. In particular correlations between $B(B_s \to \mu^+\mu^-)$ and other observables emerge, for example with the anomalous magnetic moment of the muon and the mass of the lightest neutral Higgs boson [564]. Other well-motivated variants of the MSSM incorporate the parameter constraints from grand unified theories (GUTs). $B(B_s \to \mu^+\mu^-)$ is especially interesting in GUTs based on the symmetry group $SO(10)$ [564, 573, 574]. In the minimal $SO(10)$ GUT, the top and bottom Yukawa couplings y_b and y_t unify at a high scale implying that $\tan\beta$ is of order 50. While realistic $SO(10)$ models contain a nonminimal Higgs sector, any experimental information on the deviation of y_b/y_t from 1 is very desirable, as it probes the Higgs sectors of GUT theories. In conjunction with other observables like the mass difference in the B_s^0–$\bar{B}_s^0$ system [30] or $B(B^+ \to \tau^+\nu_\tau)$ [31, 32, 575], which depend in different ways on $\tan\beta$ and the masses of the non-Standard Higgs bosons and the supersymmetric particles, the measurement of $B(B_s \to \mu^+\mu^-)$ at the LHC will, within the MSSM, answer the question whether the top and bottom Yukawa couplings unify at high energies.

3.4.1.3 Other very rare decays The decays $B_q \to \ell^+\ell^-\gamma$ and $B_q \to \ell^+\ell^-\ell'^+\ell'^-$ are of little interest from a theoretical point of view. First, they are difficult to calculate, since they involve photon couplings to quarks and are thereby sensitive to soft hadron dynamics. Second, they are not helicity-suppressed, because the (real or virtual) photon can recoil against a lepton pair in a $J = 1$ state. This implies that they probe operators of the effective Hamiltonian which can more easily be studied from $B_q \to X\gamma$ and $B \to X\ell^-\ell^-$ decays. However, the absence of a helicity suppression makes $B_q \to \ell^+\ell^-\gamma$ a possible threat to $B_q \to \ell^+\ell^-$, as will be discussed in the experimental sections. A naive estimate gives $B(B_s \to \mu^+\mu^-\gamma) \sim (m_B^2/m_\mu^2)\alpha/(4\pi)B(B_s \to \mu^+\mu^-) \sim B(B_s \to \mu^+\mu^-)$, while a more detailed analysis even finds $B(B_s \to \mu^+\mu^-\gamma) > B(B_s \to \mu^+\mu^-)$ [299].

Lepton-flavor violating (LFV) decays like $B_q \to \ell^\pm\mu^\mp$ and $\ell = e, \tau$ are negligibly small in the SM. They are suppressed by two powers of m_ν/M_W, where m_ν denotes the largest neutrino mass. However, this suppression factor is absent in certain models of new physics. In supersymmetric theories with R parity (such as the MSSM), their branching ratios are smaller than those of the corresponding lepton-flavor conserving decay, e.g. $B_q \to \mu^+\mu^-$. Large effects, however, are possible in models that contain LFV tree-level couplings or leptoquarks. Here supersymmetric theories without R parity and the Pati–Salam model should be mentioned. Supersymmetry without R parity involves a plethora of new couplings, which are different for all combinations of quark and lepton flavor involved, so that no other experimental constraints prevent large effects in $B_q \to \ell^\pm \mu^\mp$. Flavor physics in the Pati–Salam model has been studied in [576].

3.4.2 Present experimental status of $B_q \to \ell^+\ell^-$ decays

The experimental searches for $B_q \to \ell^+\ell^-$ have focused on $B_s \to \mu^+\mu^-$ and $B_d \to \mu^+\mu^-$. For the e^+e^- final states, the branching fractions are suppressed with respect to $B(B \to \mu^+\mu^-)$ by $m_e^2/m_\mu^2 = 2.3 \times 10^{-5}$. The best limit that has been set is $B(B \to e^+e^-) < 61 \times 10^{-9}$ (90% C.L.) [577]. Though the branching fraction of the $\tau^+\tau^-$ mode is enhanced by a factor of 212 with respect to that of the $\mu^+\mu^-$ mode, the only experimental upper limit from BaBar is $B(B_d \to \tau^+\tau^-) < 4.1 \times 10^{-3}$ (90% C.L.) [578]. This is less sensitive than the decay $B \to \mu^+\mu^-$. Due to at least two missing neutrinos in the decays of the two τs, the reconstruction of this mode is rather difficult, since no kinematic constraint can be employed to eliminate backgrounds. At an e^+e^- super B factory, the $B_d \to \tau^+\tau^-$ mode may be observable by fully reconstructing one B meson in a hadronic mode and then searching for $B_d \to \tau^+\tau^-$ in the recoil system.

Thus, $B_{d,s} \to \mu^+\mu^-$ are the most promising modes to test the SM. Table 24 summarizes the searches for $B_s \to \mu^+\mu^-$ by different experiments in the past two decades. The 90% C.L. upper limits are shown in Fig. 20 in comparison to the SM prediction. The lowest limit of $B(B_s \to \mu^+\mu^-) < 93 \times 10^{-9}$ (95% C.L.) is obtained by the D0 experiment using about 2 fb^{-1} of $p\bar{p}$ data [579]. Using 780 pb^{-1} of $p\bar{p}$ data, CDF achieved a branching fraction upper limit of $B(B_s \to \mu^+\mu^-) < 100 \times 10^{-9}$ (95% C.L.) [580, 581]. The corresponding searches for $B_d \to \mu^+\mu^-$ are summarized in Table 25. Here, the lowest limit of $B(B_d \to \mu^+\mu^-) < 30 \times 10^{-9}$ (95% C.L.) is obtained by the CDF experiment using 780 pb^{-1} of $p\bar{p}$ data [580, 581]. The 90% C.L. upper limits are also shown in Fig. 20 in comparison with the SM prediction.

In the present CDF $B_s \to \mu^+\mu^-$ analysis, the background level is at about one event, while the branching fraction upper limit at 90% C.L. lies about a factor of 20 above

Table 24 Branching fraction upper limits (90% C.L.) for $B_s \to \mu^+\mu^-$ from different experiments

Experiment	Year	Limit [10^{-9}]	Process	Reference
D0	2007	75	$p\bar{p}$ at 1.96 TeV	[579]
CDF	2006	80	$p\bar{p}$ at 1.96 TeV	[580, 581]
CDF	2005	150	$p\bar{p}$ at 1.96 TeV	[582]
D0	2005	410	$p\bar{p}$ at 1.96 TeV	[583]
CDF	2004	580	$p\bar{p}$ at 1.96 TeV	[584]
CDF	1998	2,000	$p\bar{p}$ at 1.8 TeV	[585]
L3	1997	38,000	$e^+e^- \to Z$	[586]

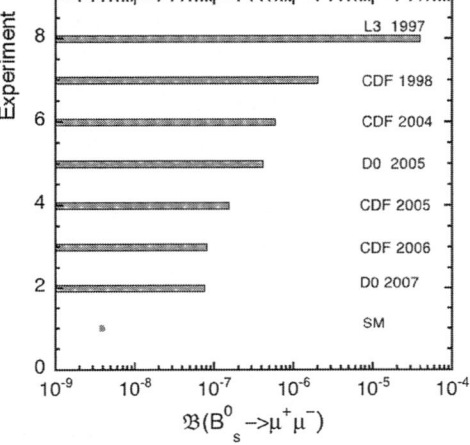
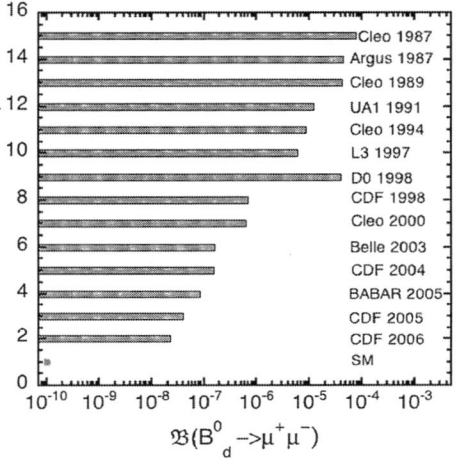

Fig. 20 Compilation of the 90% C.L. upper limits for $B(B_s \to \mu^+\mu^-)$ (*left*) and $B(B_d \to \mu^+\mu^-)$ (*right*) from different experiments in comparison to the SM prediction

Table 25 Branching fraction upper limits (90% C.L.) for $B_d \to \mu^+\mu^-$ from different experiments

Experiment	Year	Limit [10^{-9}]	Process	Reference
CDF	2006	23	$p\bar{p}$ at 1.96 TeV	[580, 581]
CDF	2005	39	$p\bar{p}$ at 1.96 TeV	[582]
BaBar	2005	83	$e^+e^- \to \Upsilon(4S)$	[577]
CDF	2004	150	$p\bar{p}$ at 1.96 TeV	[584]
Belle	2003	160	$e^+e^- \to \Upsilon(4S)$	[587]
CLEO	2000	610	$e^+e^- \to \Upsilon(4S)$	[588]
D0	1998	40,000	$p\bar{p}$ at 1.8 TeV	[589]
CDF	1998	680	$p\bar{p}$ at 1.8 TeV	[585]
L3	1997	10,000	$e^+e^- \to Z$	[586]
UA1	1991	8,300	$p\bar{p}$ at 630 GeV	[590]
ARGUS	1987	45,000	$e^+e^- \to \Upsilon(4S)$	[591]
CLEO	1987	77,000	$e^+e^- \to \Upsilon(4S)$	[592]

the SM value. Thus, any analysis attempting to reach a sensitivity at the level of the SM prediction needs a significant improvement in background rejection. Scaling the present CDF result to a luminosity of 10 fb^{-1} yields branching fraction upper limits at 90% C.L. of 6.2×10^{-9} for $B_s \to \mu^+\mu^-$ and 1.8×10^{-9} for $B_d \to \mu^+\mu^-$. A simple scaling of the BaBar result to 1 ab^{-1} yields $B(B_d \to \mu^+\mu^-) < 9 \times 10^{-9}$ (90% C.L).

3.4.3 LHC preparations for measurements of the very rare B decays

Three LHC experiments, LHCb, ATLAS and CMS, are aiming for the measurement of very rare B decays. Differences in the detector layouts lead to different strategies in data-taking, triggers and the offline selections to maximize the gain of signal events.

3.4.3.1 Luminosity conditions and triggers Whilst the nominal LHCb luminosity will be $(2-5) \times 10^{32}$ cm^{-2} s^{-1}, the forward muon stations can identify muons with low values of transverse momenta, allowing the first level trigger (L0) to collect events with one or two muons with p_T values as low as 1.1 GeV/c [593]. Because the beauty cross section grows rapidly at small transverse momenta, the lower LHCb luminosity is compensated by higher b-production. ATLAS and CMS will start to collect the exclusive di-muon B decays at a luminosity of few times 10^{33} cm^{-2} s^{-1} and will later continue at the nominal LHC luminosity of 10^{34} cm^{-2} s^{-1}. Thus rare B-decays will be recorded at all LHC luminosities. However the central detector geometries will allow muons to be recorded only above $p_T \sim (3-6)$ GeV/c at the first trigger level (L1) [594, 595].

First level triggers for the exclusive di-muon B decays in LHCb, ATLAS and CMS are summarized in Table 26. In LHCb, the strategy relies on both the single muon trigger with $p_T \geq 1.1$ GeV/c and di-muon trigger streams with

Table 26 L1(0) trigger p_T thresholds. The output trigger rates are given for a luminosity of 2×10^{32} cm^{-2} s^{-1} (LHCb) and 2×10^{33} cm^{-2} s^{-1} (ATLAS/CMS)

Experiment	L1(0) momentum cut	L1(0) rate
ATLAS 2μ	$p_T(\mu) \geq 6.0$ GeV/c	0.7 kHz
CMS 2μ	$p_T(\mu) \geq 3.0$ GeV/c	3.8 kHz
LHCb 1μ	$p_T(\mu) \geq 1.1$ GeV/c	110 kHz
LHCb 2μ	$\Sigma p_T(\mu\mu) \geq 1.3$ GeV/c	145 kHz

$\Sigma p_T(\mu\mu) \geq 1.3$ GeV/c. ATLAS and CMS will collect the majority of their signal events at 2×10^{33} cm^{-2} s^{-1} through the di-muon trigger with the muon transverse momentum thresholds 6 and 3 GeV/c, respectively. Such triggers will result in output rates of about 700 Hz and 3500 Hz for ATLAS and CMS, respectively, and about 200 kHz for LHCb.

The high level trigger (HLT) strategy is similar for all three experiments. First, one confirms the presence of trigger muon(s) by reconstructing tracks within the so called region of interest (RoI) around a muon candidate and by matching reconstructed tracks in the inner detector with tracks from the muon system. Further, cuts are applied to the muons requiring the p_T values to be above 3 GeV/c for LHCb and above 4 and 6 GeV/c for CMS and ATLAS, respectively. Then, primary and secondary vertices are reconstructed. Cuts on vertex quality $\chi^2 \leq 20$ and on the flight path of B_s candidates $L_{xy} \geq 200$ μm (ATLAS) and $L_{3D} \geq 150$ μm (CMS) are applied. LHCb (single muon stream) uses an impact parameter cut $IP(\mu) \geq 3\sigma_{IP}$ and, for the di-muon stream, the secondary vertex quality cut $\chi^2 \leq 20$. Finally, a cut on the invariant mass of the two muons is applied, $4 \leq M_{\mu\mu} \leq 6$ GeV/c^2 (ATLAS), $M_{\mu\mu} \geq 2.5$ GeV/c^2 (LHCb di-muon stream), or a mass window around the nominal B_s mass of ± 150 MeV/c^2 (CMS). The HLT rate is less than 1.7 Hz for CMS and about 660 Hz for LHCb.

A detailed description of trigger algorithms can be found in [593–595].

3.4.3.2 Offline performance and signal selection After the trigger the offline analysis faces the challenge of selecting a signal from backgrounds of similar topology. The most important offline performance parameters for the di-muon events in the kinematic ranges accepted by triggers are given in Table 27. The differences lead consequently to different selection strategies.

In ATLAS, the reconstructed di-muon invariant mass $M_{\mu\mu}$ is required to be within an interval of $(-70 \text{ MeV}/c^2, +140 \text{ MeV}/c^2)$ around the B_s mass. The isolation cut in the ATLAS experiment requires no charged tracks with $p_T \geq 0.8$ GeV/c in an angular cone $\theta \leq 15°$ around the B_s candidate. For the reconstructed vertices, the significance of the reconstructed flight path in the transverse plane defined as L_{xy}/σ_L is required to be larger than 11 and the vertex reconstruction quality parameter $\chi^2 \leq 15$. The space separation between two muon candidates is $\Delta R = \sqrt{\Delta\phi^2 + \Delta\eta^2} \leq 0.9$. Details of the study can be found in [596].

In CMS, the isolation is defined as

$$I = \frac{p_T(B_s^0)}{p_T(B_s^0) + \Sigma_{\text{trk}}|p_T|} \geq 0.85. \tag{135}$$

A value of $\Sigma_{\text{trk}}|p_T|$ is calculated for all charged tracks in a cone with $\Delta R = 1$ around the B_s candidate. For the muon separation, the value of ΔR should be in the range (0.3, 1.2). The vertex cuts are the following: $L_{xy}/\sigma_L \geq 18$ and $\chi^2 \leq 1$. The momentum of the B_s candidate should point to the primary vertex: $\cos\alpha \geq 0.995$, where α is the angle between the momentum of the B_s candidate and the vector connecting the primary and secondary vertices $\vec{V}_{\text{sec}} - \vec{V}_{\text{prim}}$. A tight mass cut is applied: $|M_{\mu\mu} - M_{B_s}| \leq 100$ MeV/c^2. Details of the study are given in [597].

In LHCb, the selection is divided into several steps [598]. First the following soft selection cuts are applied: $|M_{\mu\mu} - M_{B_s}| \leq 600$ MeV/c^2, vertex quality cut $\chi^2 \leq 14$, $IP/\sigma_{IP} \leq 6$ for the B_s candidate, secondary and primary vertex separation $|Z_{\text{sec}} - Z_{\text{prim}}|/\sigma_V \geq 0$, pointing angle $\alpha < 0.1$ rad, soft muon identification for both candidates ($\epsilon_\mu = 95\%$ and $\epsilon_\pi = 1\%$). Further on three categories of discriminant variables are introduced: Geometry (G; lifetime, B_s and μ impact parameter, distance of closest approach (DOCA) and isolation), PID (particle identification) and IM (invariant mass). These variables are used to compute the S/B ratio event by event, while no further cuts are applied. Each event is weighted with its S/B ratio in the signal sensitivity calculation. Using this method, it is expected to reconstruct about 70 signal events per 2 fb^{-1} [598]. If the previous method is combined with the requirement $G > 0.7$, with no background events left, this leads to an estimate of 20 signal events to be reconstructed in the same period as above.

In Table 28, the number of signal events is shown for each experiment for different integrated luminosities. For ATLAS/CMS, the number for 2 fb^{-1} is simply scaled from the one for 10 fb^{-1}. In the same way, the LHCb number for 10 fb^{-1} is obtained by scaling the number for 2 fb^{-1}. The CMS and ATLAS studies for 100 fb^{-1} were published in [599] and [600], respectively. In the CMS study, harder selection criteria have been applied for high luminosity, hence the reconstruction efficiency for signal events is lower with respect to lower luminosity.

3.4.3.3 Background studies The search for $B_s \to \mu^+\mu^-$ has to deal with the problem of an enormous level of background.

The largest contribution is expected to come from combinatorial background. These events consist predominantly of beauty decays, where the di-muon candidates originate either from semileptonic decays of b and $\bar{b}$ quarks or from cascade decays of one of the $b\bar{b}$ quarks. To determine the contribution of this background, LHCb simulated a sample of inclusive $b\bar{b}$ events, requiring that both b-quarks have $|\theta| < 400$ mrad, to match, on the safe side, the LHCb acceptance of 300 mrad. Nevertheless, the sample of 34 million events corresponds to only 0.16 pb^{-1}. The study of this sample, however, showed that in the sensitive region of phase space, the relevant background contains two real muons from b-decays. Hence, a specific sample of 8 million

Table 27 LHC detector performance parameters for $B \to \mu^+\mu^-$ events in the kinematic ranges of trigger acceptances. σ_{Im} is the muon track impact parameter resolution, $\sigma_{M_{\mu\mu}}$ is the $B_s \to \mu^+\mu^-$ mass resolution

Experiment	LHCb	ATLAS	CMS
p_T^μ, GeV/c	>3	>6	>4
σ_{Im}, μm	14–26	25–70	30–50
$\sigma_{M_{\mu\mu}}$, MeV/c^2	18	84	36

Table 28 Number of signal events as a function of integrated luminosity. The time after which the corresponding luminosity will be delivered is indicated in parentheses

Experiment	2 fb^{-1}	10 fb^{-1}	30 fb^{-1}	100 fb^{-1}	130 fb^{-1}
ATLAS	1.4	7.0	21.0	92	113 (4 years)
CMS	1.2	6.1	18.3	26	44 (4 years)
LHCb	20	100 (5 years)	–	–	

events was generated, corresponding to an effective luminosity of $30\,\text{pb}^{-1}$, where for both b-hadron decays a muon is required among the decay products. LHCb uses this sample to evaluate the background and extrapolates the result to a given integrated luminosity, for instance, $2\,\text{fb}^{-1}$. In the sensitive region ($G > 0.7$) [598], no background event was selected, hence an upper limit of 125 events is estimated at 90% C.L.. ATLAS simulated $b\bar{b}$ events with two muons, requiring to have transverse momenta $p_T > 6\,(4)\,\text{GeV}/c$ for the first (second) muon. In CMS, the cut for both muons was $p_T > 3\,\text{GeV}/c$. The pseudorapidity of each of the muons was required to be in the range $|\eta| < 2.4$ in agreement with the trigger acceptances. Additionally the di-muon mass was required to be in the interval $M_{\mu\mu} < 8\,\text{GeV}/c^2$ and $5 < M_{\mu\mu} < 6\,\text{GeV}/c^2$ in ATLAS and CMS, respectively. The number of background events generated with these cuts corresponds to $10\,(8)\,\text{pb}^{-1}$ for ATLAS (CMS). Both experiments evaluated the background using these samples and extrapolated the results to a given integrated luminosity. At $10\,\text{fb}^{-1}$, ATLAS expects 20 ± 12 events [601] and CMS 14^{+22}_{-14} events [597].

Due to the high sensitivity of the LHC experiments, the background composition may be changed relative to the situation at the Tevatron. In addition to combinatorial background, contributions from topologically similar rare exclusive decays as well as misidentification effects may become important. We give a classification of the different types of these potential backgrounds and several estimates of their contribution.

First, let us consider the very rare decays $B^{0\pm} \to (\pi^{0\pm}, \gamma)\mu^+\mu^-$ with branching ratios expected to be $\sim 2 \times 10^{-8}$ [299]. A background contribution may arise when the π/γ is soft and escapes detection. The di-muon invariant mass distribution has been modeled in ATLAS and CMS for cases where a $\pi^\pm$ is not reconstructed in the inner tracker, or a $\pi^0(\gamma)$ with $E_T \leq (2\text{--}4)\,\text{GeV}$ escapes detection in the electromagnetic calorimeter. Based on a full detector simulation, CMS concluded that neither of the processes $B^0 \to \gamma\mu^+\mu^-$, $B^\pm \to \pi^\pm\mu^+\mu^-$ or $B^0 \to \pi^0\mu^+\mu^-$ will contribute significantly in the signal region. ATLAS reached similar conclusions for the first two processes, while they plan to do a detailed study for the third decay. These very rare decay channels are worth studying in their own right, since some properties (for example the di-muon invariant mass spectrum) are also sensitive to NP contributions [299].

Decays into four leptons, such as $B^+_{(c)} \to \mu^+\mu^-\ell^+\nu_\ell$, are another possible background source to $B_s \to \mu^+\mu^-$. If the p_T of one of the leptons is below the detector reconstruction capabilities, then there are only two tracks observed from the B-meson vertex and the invariant mass of the di-lepton pair can be close to the $B_{d,s}$ mass. The expected branching fractions of these decays are 5×10^{-6} and 8×10^{-5} for B^+ and B_c^+, respectively [602]. Using the fast simulation tool (ATLFAST), ATLAS showed that the number of background events from $B^+ \to \mu^+\mu^-\mu^+\nu$ can be as high as 50% of the accepted signal events from $B_s \to \mu^+\mu^-$ with a SM rate. In CMS, the analysis showed that the contribution from this source is negligible. The difference is due to different mass resolutions of ATLAS and CMS. LHCb simulated a resonant mode of the four-lepton channel $B^+_{(c)} \to (J/\psi \to \mu^+\mu^-)\mu\nu$ in which two muons are coming from J/ψ. The study led to the conclusion that the background from this channel in the mass region $\pm 60\,\text{MeV}/c^2$ around the B_s mass is less than 10% of a $B_s \to \mu^+\mu^-$ signal within the SM.

The last category considered are backgrounds from B-decay channels where secondary hadrons are misidentified as muons. The simplest backgrounds come from the two-body hadronic decays $B_{d,s} \to K^\pm\pi^\mp$, $B_{d,s} \to K^\pm K^\mp$ and $B_{d,s} \to \pi^\pm\pi^\mp$. The background contribution can be estimated by assigning to each of the final-state hadrons a probability that it would be registered as a muon. This probability was obtained from full detector simulations of large samples of beauty events. Such a study has been performed at LHCb, resulting in ~ 2 events per $2\,\text{fb}^{-1}$ (in a $\pm 2\sigma$ mass window). CMS concluded that these backgrounds are negligible. ATLAS studies are in progress. Fake signal events can also be generated by semileptonic B decays such as $B^0 \to \pi^-\mu^+\nu_\mu$, which have a branching ratio $\sim 10^{-4}$. As in the previous case, background can arise from $\pi - \mu$ misidentification and a soft neutrino escaping an indirect identification. Similar channels to be accounted for are $B_s \to K^-\mu^+\nu_\mu$ and $B^+ \to K^+\mu^+\mu^-$.

3.4.3.4 LHC reach for $B_s \to \mu^+\mu^-$ The results of the signal and background studies described in the previous sections were finally used to estimate upper limits on the branching ratio of $B_s \to \mu^+\mu^-$, which are shown in Figs. 21 and 22. ATLAS and CMS used the algorithms of [164], while LHCb developed the new approach published in [598]. In all cases, the results were given at 90% confidence level as a function of integrated luminosity. The theory prediction for $B(B_s \to \mu^+\mu^-)$ shown in Figs. 21 and 22 uses the value of $f_{B_s} = (230 \pm 9)\,\text{MeV}$ extracted from the CDF measurement of $\Delta M_{B_s} = (17.8 \pm 0.1)\,\text{ps}^{-1}$. The prediction therefore assumes that NP neither affects $B_s \to \mu^+\mu^-$ nor ΔM_{B_s}. Note that the above value for f_{B_s} is also consistent with direct QCD lattice calculations (see Sect. 2.4).

After one year of data taking at the LHC the expected results from LHCb will allow us to exclude or discover NP in $B_s \to \mu^+\mu^-$. ATLAS and CMS will reach this sensitivity after three years. After LHC achieves its nominal luminosity, the ATLAS and CMS statistics will increase substantially. After five years all three experiments will be in a position to provide a measurement of the branching ratio of $B_s \to \mu^+\mu^-$.

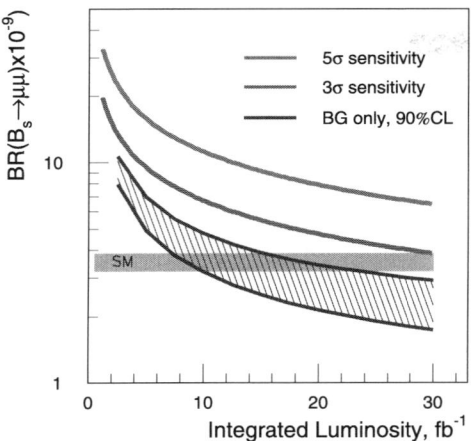

Fig. 21 Branching ratio of $B_s \to \mu^+\mu^-$ observed (3σ) or discovered (5σ) as a function of integrated luminosity for ATLAS/CMS

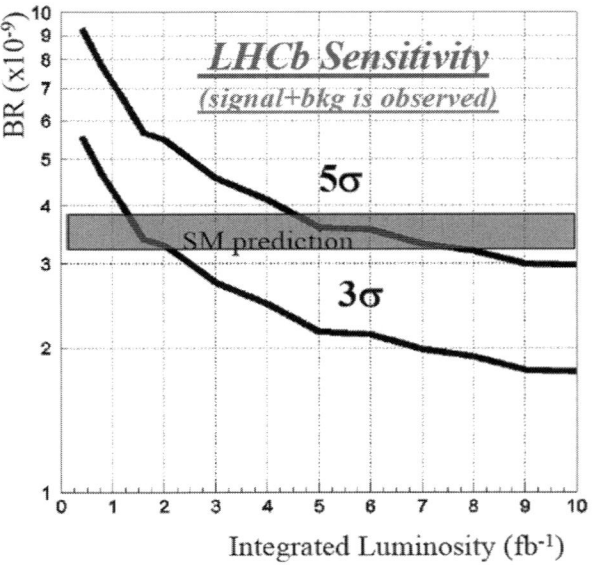

Fig. 22 Branching ratio of $B_s \to \mu^+\mu^-$ observed (3σ) or discovered (5σ) as a function of integrated luminosity for LHCb

3.4.4 Conclusions

The very rare decays $B_q \to \mu^+\mu^-$ are special in many respects. Their branching ratios are small in the SM but can be enhanced significantly in the widely studied MSSM. Leptonic meson decays belong to the physics topics that can be experimentally studied by three of the four major LHC experiments, namely LHCb, ATLAS and CMS. The LHC experiments will probe the branching fraction of $B_s \to \mu^+\mu^-$ down to the SM value and possibly reveal a smoking-gun signal of NP well ahead of the direct searches using high-p_T physics. Irrespectively of whether $B(B_s \to \mu^+\mu^-)$ is found in agreement with the SM prediction or not, the measurement will severely constrain the Higgs sector of the MSSM and will provide valuable input for LHC Higgs physics: any sizable enhancement of $B(B_s \to \mu^+\mu^-)$ implies a large value of $\tan\beta$, so that the nonstandard Higgs bosons couple strongly to b-quarks and τ-leptons. Then these Higgs bosons will be dominantly produced in association with b-jets and will decay dominantly into b-hadrons and τ-leptons.

3.5 UT angles from tree decays[16]

3.5.1 Introduction

It is very fortunate that the B system allows an almost pristine determination of all the three angles from "tree" decays. $\beta(\phi_1)$ from $J/\psi K_S$-like modes and $\gamma(\phi_3)$ from DK-type modes are genuine tree decays and are theoretically very clean. The irreducible theory error (ITE) for β is expected to be less than 1% and may be even considerably less than that [603].[17] On γ, the ITE is estimated at O(0.1%). For $\alpha(\phi_2)$, the situation with regard to theory error is a bit more complicated. Isospin analysis allows, in principle, extraction of $\alpha(\phi_2)$ from $\pi\pi$, $\rho\pi$, or $\rho\rho$, but electroweak penguin contributions (EWP) do not respect isospin. So, in each of the three channels, the EWP contributions and other isospin violations are difficult to ascertain rigorously. But given that there are three channels, it seems reasonable that the theory error even for α will be small, O(few%) (see, e.g., [605]). Given that we now have theoretical methods that will allow us to quite precisely determine all the three angles, which are fundamental parameters of the SM, it is clearly important to determine them with accuracy roughly commensurate with what the theoretical methods promise. In this section, we will summarize the current status as to our attempts to extract these three angles directly from data collected primarily through the spectacular successes of the two asymmetric B factories, followed by our guess estimates for the potential of a super B factory (SBF) with regard to this goal. Of course, LHCb will soon begin operation, and our expectations for the precisions on tree-level angle determinations from LHCb are also presented.

3.5.2 Angles from B factories of today & of tomorrow

3.5.2.1 $\beta(\phi_1)$
Measurements of CP asymmetries in the proper-time distribution of neutral B decays to CP eigenstates mediated by $b \to c\bar{c}s$ transition provide a direct measurement of $\sin 2\beta$ ($= \sin 2\phi_1$). The time-dependent decay-rate asymmetry for decays to CP eigenstates containing a charmonium and a K_S^0 meson is given by

$$A_{CP}(t) = S_{b \to c\bar{c}s} \sin(\Delta m_d t) - C_{b \to c\bar{c}s} \cos(\Delta m_d t), \quad (136)$$

---

[16] Section coordinators: M. Bona, A. Soni, K. Trabelsi, G. Wilkinson.
[17] For a more conservative (but data driven) estimate see, e.g., Ref. [604].

where Δm_d is the mass difference between the two B^0 mass eigenstates. Since these decays are dominated by a single (tree level) amplitude,[18] one expects to a very good approximation $S_{b \to c\bar{c}s} = -\eta_{CP}\sin 2\beta$ and $C_{b \to c\bar{c}s} = 0$, where η_{CP} is the CP eigenvalue of the final state.

In 2001, both BaBar and Belle Collaborations established CP violation in the B system through the $\sin 2\beta$ measurements in $b \to c\bar{c}s$ decays [606, 607].

In the latest results, the BaBar Collaboration [608], using a 348 million $B\bar{B}$ events, includes the CP-odd ($\eta_{CP} = -1$) final states $J/\psi K_S^0$, $\psi(2S)K_S^0$, $\chi_{c1}K_S^0$ and $\eta_c K_S^0$ as well as the CP-even ($\eta_{CP} = +1$) $J/\psi K_L^0$ final state. In addition, the vector–vector final state $J/\psi K^*$ with $K^* \to K_S^0 \pi^0$, which is found from an angular analysis to have η_{CP} close to $+1$ [610], is used. The Belle Collaboration [609] uses a sample of 535 million $B\bar{B}$ events where only $J/\psi K_S^0$ and $J/\psi K_L^0$ (*golden*) modes are analysed. The results for $-\eta_{CP}S_{b \to c\bar{c}s}$ and $C_{b \to c\bar{c}s}$ are given in Table 29 and in Fig. 23 and are at the 5% level for each collaboration.

The world average computed by the Heavy Flavor Averaging Group (HFAG) [502] includes also the results obtained by the ALEPH, OPAL and CDF experiments and is

$$\sin 2\beta = 0.675 \pm 0.026, \tag{137}$$

where most of the systematic uncertainties have been treated as uncorrelated. This result suggests that on the time scale of 2008, when an integrated luminosity of order of 2 fb^{-1} is expected from the B factories, the total uncertainty on $\sin 2\beta$ will be around 0.02.

The actual $\sin 2\beta$ result gives a precise constraint on the $(\bar{\rho}, \bar{\eta})$ plane, as shown in Fig. 23, and can be compared with the expected value obtained with other constraints from CP conserving quantities, and with CP violation in the kaon system, in the form of the parameter ε_K. Such comparisons have been performed by phenomenological groups: for example, the result from the global UT fit without the measurement of $\sin 2\beta$ is obtained by CKMfitter [8] to be $0.823^{+0.018}_{-0.085}$ or by UTfit [209] to be 0.759 ± 0.037. It is clear that the increased precision in the $\sin 2\beta$ measurement is now revealing some tension with the rest of the fit. This is mainly due to the actual V_{ub} value and in particular to the inclusive one, strikingly in countertendency with respect to the relatively *low* value of $\sin 2\beta$ [120].

With $\sin 2\beta$ being now a precision measurement, other analyses are being performed in order to remove the twofold ambiguity unavoidable with a sine determination.

Considering the B meson decays to the vector–vector final state $J/\psi K^{*0}$ in the case of a final state not flavor-specific ($K^{*0} \to K_S^0 \pi^0$), a time-dependent transversity analysis

---

[18]The same processes can be described by a penguin diagram which brings corrections at order $\sim \lambda^4$.

can be performed allowing sensitivity to both $\sin 2\beta$ and $\cos 2\beta$ [611]. Such analyses have been performed by both B factory experiments: from Table 30 we can remark that at present the results are dominated by large and non-Gaussian statistical errors, but nevertheless it can be said that $\cos 2\beta > 0$ is preferred by the experimental data in $J/\psi K^*$.

Finally, decays of B mesons to final states such as $D\pi^0$ are governed by $b \to c\bar{u}d$ transitions. If the final state is a CP eigenstate, i.e. $D_{CP}\pi^0$, the usual time-dependence formulae are recovered, with the sine coefficient sensitive to $\sin 2\beta$. Since there is no penguin contribution to these decays, there is even less associated theoretical uncertainty than for $b \to c\bar{c}s$ decays like $B \to J/\psi K_S^0$. When multibody D decays, such as $D \to K_S^0 \pi^+ \pi^-$, are used, a time-dependent analysis of the Dalitz plot of the neutral D decay allows a direct determination of the weak phase: 2β [612]. Such analyses have been performed by both B-factory experiments. The decays $B \to D\pi^0$, $B \to D\eta$, $B \to D\omega$, $B \to D^*\pi^0$ and $B \to D^*\eta$ are used. The daughter decays are $D^* \to D\pi^0$ and $D \to K_S^0 \pi^+ \pi^-$. The results are shown in Table 30. Again, it is clear that the data prefer $\cos 2\beta > 0$. Taken in conjunction with the $J/\psi K^*$ results, $\cos 2\beta < 0$ can be considered to be ruled out at approximately 2.3σ [209]. Time-dependent analysis of the decay $B \to D^{*+}D^{*-}K_S^0$ also prefers $\cos 2\beta > 0$.

3.5.2.2 $\alpha(\phi_2)$ The CKM unitarity angle $\alpha(=\phi_2)$, defined as $\alpha = \arg(-V_{td}V_{tb}^*/(V_{ud}V_{ub}^*))$, is a measure of the relative phase of the CKM elements V_{ub} and V_{td} in the usual parameterization of the CKM unitarity matrix. Most of the experimental information on α is extracted from measurements of the charmless decays $B \to \pi\pi$, $B \to \rho\pi$ and $B \to \rho\rho$, which can arise from the tree-level transition $b \to u(\bar{u}d)$, carrying the CKM element V_{ub} (left diagram in Fig. 24). In a simple world, where a decay mode such as $B \to \pi^+\pi^-$ is dominated by a single tree diagram, one needs only to measure the time-dependent CP asymmetry $S_{\pi\pi} = \sin 2\alpha$. However, a complication to this picture arises from the presence of loop (penguin) processes (right diagram in Fig. 24), involving different CKM matrix elements but leading to the same final states. The interference of the two diagrams then obscures the connection between the CP observables and the angle α, requiring a "tree and penguin disentanglement" strategy in the experimental program. This involves a larger set of experimental observables for the determination of the angle α that includes the time-dependent CP asymmetries S_f and C_f in B^0 decays and the branching fractions and direct CP asymmetries in both neutral and charged B decays. The net effect of the penguin amplitude is to introduce the possibility of direct CP violation ($C_f \neq 0$) and a nonzero value of $\Delta \alpha^f = \alpha_{\text{eff}}^f - \alpha$, where α_{eff}^f is determined from the relation $S_f = \sqrt{1-C_f^2}\sin 2\alpha_{\text{eff}}^f$.

Table 29 Results for the CP-violating parameters in the $b \to c\bar{c}s$ decays: $S_{b \to c\bar{c}s}$ and $C_{b \to c\bar{c}s}$. The B-factory averages are given after ICHEP 2006 as calculated by HFAG [502]. The final world averages include also the results from ALEPH, OPAL and CDF, which use only the $J/\psi K_S^0$ final state

Experiment		$-\eta_{CP} S_{b \to c\bar{c}s}$	$C_{b \to c\bar{c}s}$
BaBar	[608]	$0.710 \pm 0.034 \pm 0.019$	$0.070 \pm 0.028 \pm 0.018$
Belle	[609]	$0.642 \pm 0.031 \pm 0.017$	$-0.018 \pm 0.021 \pm 0.014$
B factory average		0.674 ± 0.026	0.012 ± 0.022
Confidence level		0.18	0.02
Average		0.675 ± 0.026	0.012 ± 0.022

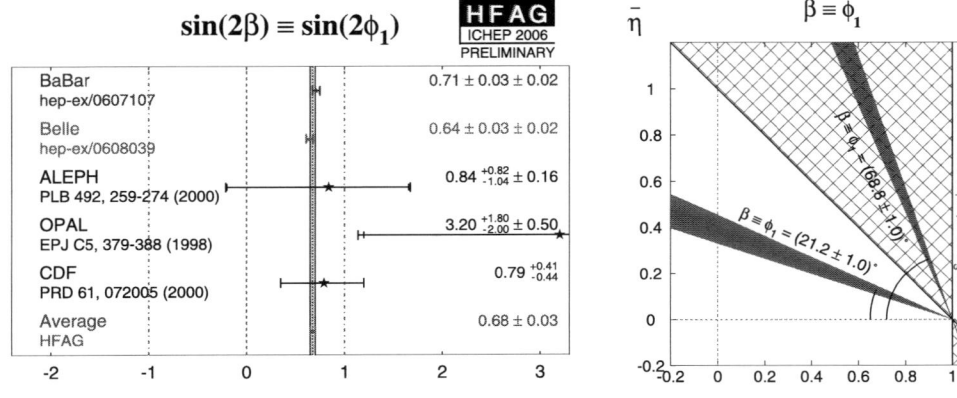

Fig. 23 World average of measurements of $S_{b \to c\bar{c}s}$ as calculated by HFAG [502] (*left*) and constraints on the $(\bar{\rho}, \bar{\eta})$ plane obtained from the average of $-\eta_{CP} S_{b \to c\bar{c}s}$ and (137) (*right*)

Table 30 Results from the B factories together with the HFAG averages [502] from the $B^0 \to J/\psi K^{*0}$ and the $B^0 \to D^{(*)}h^0$ analyses

$B^0 \to J/\psi K^{*0}$		$\sin 2\beta$	$\cos 2\beta$
BaBar	[613]	$-0.10 \pm 0.57 \pm 0.14$	$3.32^{+0.76}_{-0.96} \pm 0.27$
Belle	[614]	$-0.24 \pm 0.31 \pm 0.05$	$0.56 \pm 0.79 \pm 0.11$
Average		0.16 ± 0.28	1.64 ± 0.62
$B^0 \to D^{(*)}h^0$		$\sin 2\beta$	$\cos 2\beta$
BaBar	[615]	$0.45 \pm 0.36 \pm 0.05 \pm 0.07$	$0.54 \pm 0.54 \pm 0.08 \pm 0.18$
Belle	[616]	$0.78 \pm 0.44 \pm 0.22$	$1.87^{+0.40+0.22}_{-0.53-0.32}$
Average		0.57 ± 0.30	1.16 ± 0.42

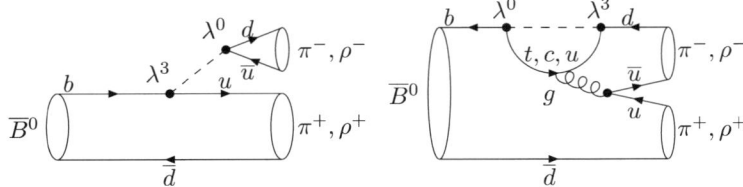

Fig. 24 The tree (*left*) and penguin (*right*) diagrams contributing to "charmless" B decays such as $B \to \pi\pi$, $B \to \rho\rho$ and $B \to \rho\pi$

For the $B \to \pi\pi$ decays, the penguin correction $\Delta\alpha^{\pi\pi}$ can be determined from an isospin analysis [246] of the decay amplitudes of the $B \to \pi\pi$ and $\bar{B} \to \pi\pi$ decays, see Fig. 25. A key element of this analysis is the branching fraction for the decay $B \to \pi^0\pi^0$, which is an indicator of the size of the penguin effects and consequently of the penguin correction $\Delta\alpha^{\pi\pi}$, which is bounded [617] by $\sin^2 \Delta\alpha^{\pi\pi} < \bar{B}(B^0 \to \pi^0\pi^0)/B(B^\pm \to \pi^\pm\pi^0)$. Ref. [211] proposes to add information on the hadronic amplitudes to the isospin analysis, for example by using the branching ratio of $B_s \to K^+K^-$ to constraint the penguin contribution (even allowing $SU(3)$ breaking effects as large as 100%).

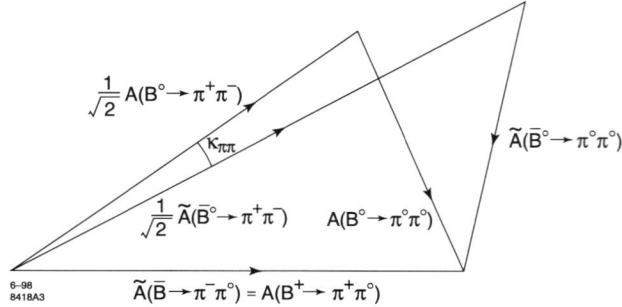

Fig. 25 Isospin triangles for the $B \to \pi\pi$ system

This would help constraining the value of α, in particular eliminating the solutions at $\alpha \sim 0$.

A system analogous to that of the $B \to \pi\pi$ decays is the family of the $B \to \rho\rho$ decays ($B^0 \to \rho^+\rho^-$, $B^+ \to \rho^+\rho^0$, $B^0 \to \rho^0\rho^0$). While in general the $B^0 \to \rho\rho$ decays can be a mixture of CP-even and CP-odd components, the angular analysis of the decay $B^0 \to \rho^+\rho^-$ (and also $B^+ \to \rho^+\rho^0$) has shown that the CP-even component (longitudinal polarization) is dominant, hence significantly simplifying the time-dependent CP analysis of the process [618, 619]. As in the case of $B \to \pi\pi$, time-dependent CP asymmetries $S^L_{\rho\rho}$ and $C^L_{\rho\rho}$ are used to determine $\alpha^{\rho\rho}_{\text{eff}}$. The branching ratio for $B^0 \to \rho^0\rho^0$ relative to $B \to \rho^+\rho^-$ and $B \to \rho^+\rho^0$ sets the scale of the penguin correction $\Delta\alpha^{\rho\rho} = \alpha^{\rho\rho}_{\text{eff}} - \alpha$, which can be determined from an isospin analysis of the decay amplitudes.

In Table 31, we present the current status of measurements used in the determination of α in the $B \to \pi\pi$ and $B \to \rho\rho$ systems [502]. Nearly all components of the isospin analysis in the $B \to \pi\pi$ system are now measured, albeit with varying degrees of precision. Also the current measurements allow for the isospin triangles to close in both systems.[19]

The fact that the branching fraction for the decay $B \to \pi^0\pi^0$ is of the same order as the branching fractions for $B^+ \to \pi^+\pi^0$ and $B^0 \to \pi^+\pi^-$ is indicative of significant contributions from penguin amplitudes in this channel. Currently the $B \to \rho^0\rho^0$ search is giving the first evidence of a signal (BaBar reporting a 3σ effect [622]) and thus a very preliminary measurement of the rate. Still, the major advantage of the $B \to \rho\rho$ system over the $\pi\pi$ one is clearly evident from the suppression of $B \to \rho^0\rho^0$ relative to $B \to \rho^+\rho^-$ and $B \to \rho^+\rho^0$ decays, implying a much smaller $\Delta\alpha$ correction and smaller related uncertainties from this source. The current $\Delta\alpha$ correction upper limits are $\Delta\alpha_{\pi\pi} < 41°$ (90% C.L.) from BaBar and $\Delta\alpha_{\rho\rho} < 21°$ (90% C.L.) from BaBar.

One other advantage of the $\rho\rho$ system is that, in contrast to $\pi^0\pi^0$, a time-dependent CP-asymmetry analysis of the $\rho^0\rho^0$ final state will be possible as soon as enough statistics are available. This feature will allow both S^{00} and C^{00} to be accessed. From a feasibility study we can foresee for the 2 ab^{-1} scenario an error of 0.3 on S^{00} and 0.25 on C^{00}. This information will greatly help in reducing the ambiguities in the α extraction from this system.

The $B \to \rho\pi$ system presents a special case with the possibility of additional handles: the final states $\rho^+\pi^-$ and $\rho^-\pi^+$, which can be reached by both B^0 and $\bar{B}^0$, have substantial overlap in the Dalitz plot; thus their amplitudes interfere and generate additional dependence on α and the strong phases of the final states. Quinn and Snyder [623] have shown that the interference effect can be exploited to extract the angle α even in the presence of penguins. This involves the amplitude analysis of the 3π Dalitz distribution.

The $\rho^\pm\pi^\mp$ final states are not CP eigenstates, and four flavor-charge configurations ($B^0(\bar{B}^0) \to \rho^\pm\pi^\mp$) must be considered. Both experiments assume that the amplitudes corresponding to these final states are dominated by the three resonances ρ^+, ρ^- and ρ^0. The ρ resonances are assumed to be the sum of the ground state $\rho(770)$ and the radial excitations $\rho(1450)$ and $\rho(1700)$. Possible contributions to the $B^0 \to \pi^+\pi^-\pi^0$ decay other than the ρ's are studied as part of the systematic uncertainties. The time-dependent analyses use a general parameterization[20] that allows one to describe the differential decay width as a linear combination of independent functions, whose coefficients are the 26 free parameters of the fit.

From the bilinear coefficients both experiments extract the quasi-two-body (Q2B) parameters. Considering only the charged bands in the Dalitz plot, the Q2B analysis involves 5 different parameters $S_{\rho\pi}$, $C_{\rho\pi}$, $\Delta S_{\rho\pi}$, $\Delta C_{\rho\pi}$ and $\mathcal{A}^{\rho\pi}_{\text{CP}}$. The first two parameterize mixing-induced CP violation related to the angle α and flavor-dependent direct CP violation, respectively. The second two are insensitive to CP violation: $\Delta S_{\rho\pi}$ is related to the strong phase difference between the amplitudes contributing to $B^0 \to \rho\pi$ decays, and $\Delta C_{\rho\pi}$ describes the asymmetry between the rates $\Gamma(B^0 \to \rho^+\pi^-) + \Gamma(\bar{B}^0 \to \rho^-\pi^+)$ and $\Gamma(B^0 \to \rho^-\pi^+) + \Gamma(\bar{B}^0 \to \rho^+\pi^-)$. Finally, $\mathcal{A}^{\rho\pi}_{\text{CP}}$ is the time-independent charge asymmetry. CP symmetry is violated if either one of the following conditions is true: $\mathcal{A}^{\rho\pi}_{\text{CP}} \neq 0$, $C_{\rho\pi} \neq 0$ or $S_{\rho\pi} \neq 0$. The first two correspond to CP violation in the decay, while the last condition is CP violation in the interference of decay amplitudes with and without B^0 mixing. In Table 32, we report the HFAG averages of the Q2B parameters provided by the experiments, which should be equivalent to determining average values directly from the averaged bilinear coefficients.

---

[19] This was not the case for the $B \to \rho\rho$ system with the pre-2006 measurements.

[20] See for details Refs. [502, 624, 625].

Table 31 Summary of measured decay properties of the $B \to \pi\pi$ and $B \to \rho\rho$ decays that are relevant to the determination of the CKM unitarity angle α. We quote here the averages updated after ICHEP 2006 as given by HFAG [502] with a total of 882 million $B\bar{B}$ pairs from BaBar (347 million events [620]) and Belle (535 million events [621]) experiments

Decay mode	BR($\times 10^6$)	S_f	C_f (or A_{CP} for B^+)
$B^0 \to \pi^+\pi^-$	5.2 ± 0.2	-0.59 ± 0.09	-0.39 ± 0.07
$B^+ \to \pi^+\pi^0$	5.7 ± 0.4	–	0.04 ± 0.05
$B^0 \to \pi^0\pi^0$	1.3 ± 0.2	–	$0.36^{+0.33}_{-0.31}$
$B^0 \to \rho^+\rho^-$	$23.1^{+3.2}_{-3.3}$		
	[$f_L = 0.968 \pm 0.023$]	-0.13 ± 0.19	-0.06 ± 0.14
$B^+ \to \rho^+\rho^0$	18.2 ± 3.0		-0.08 ± 0.10
	[$f_L = 0.912^{+0.044}_{-0.045}$]	–	
$B^0 \to \rho^0\rho^0$	1.16 ± 0.46		
	[$f_L = 0.86^{+0.12}_{-0.14}$]	–	–

Table 32 Summary of measured CP-asymmetry parameters of the $\rho\pi$ system following the convention used in [626]. We quote here the averages updated after ICHEP 2006 as given by the HFAG [502] with a total of 796 million $B\bar{B}$ pairs from BaBar (347 million events [624]) and Belle (449 million events [625]) experiments

$\rho^\pm \pi^\mp$ Q2B/Dalitz plot analysis				
$S_{\rho\pi}$	$C_{\rho\pi}$	$\Delta S_{\rho\pi}$	$\Delta C_{\rho\pi}$	$\mathcal{A}^{\rho\pi}_{\mathrm{CP}}$
0.03 ± 0.09	0.03 ± 0.07	-0.02 ± 0.10	0.36 ± 0.07	-0.13 ± 0.03
	$\mathcal{A}^{+-}_{\rho\pi}$		$\mathcal{A}^{-+}_{\rho\pi}$	
	0.11 ± 0.06		-0.19 ± 0.13	

One can transform the experimentally motivated CP parameters $\mathcal{A}^{\rho\pi}_{\mathrm{CP}}$ and $C_{\rho\pi}$ into the direct CP violation parameters $\mathcal{A}^{+-}_{\rho\pi}$ and $\mathcal{A}^{-+}_{\rho\pi}$ defined in [626]. $\mathcal{A}^{-+}_{\rho\pi}$ ($\mathcal{A}^{+-}_{\rho\pi}$) describes CP violation in B^0 decays where the ρ is emitted (not emitted) by the spectator interaction. Both experiments obtain values for $\mathcal{A}^{-+}_{\rho\pi}$ and $\mathcal{A}^{+-}_{\rho\pi}$ which are averaged in Table 32. In addition to the $B^0 \to \rho^\pm \pi^\mp$ Q2B contributions to the $\pi^+\pi^-\pi^0$ final state, there can also be a $B^0 \to \rho^0\pi^0$ component. Belle and BaBar have extracted the Q2B parameters associated with this intermediate state which average to $S_{\rho^0\pi^0} = 0.30 \pm 0.38$ and $C_{\rho^0\pi^0} = 0.12 \pm 38$ (HFAG Summer 2007).

In Fig. 26, the plots of the averages and the separate results on the various CP-violating parameters are shown: it can be seen that the two collaborations, BaBar and Belle, are still discrepant at the level of 2σ (1.5σ) in the $B \to \pi^+\pi^-$ ($B \to \rho^\pm\pi^\pm$) system. In the $\rho\rho$ system, though, some updates to the entire currently available statistics are still missing.

We can get an estimate of the current experimental value of α putting together all the analyses in all the modes. The results on the SM solution from the two fitting groups are $(92 \pm 7)°$ for the Bayesian approach [209] and $(93^{+11}_{-9})°$ for the frequentest approach [8]. From the same analyses we can also extract the SM α values using the UT fit constraints and without using the α information: $(93 \pm 6)°$ for the Bayesian approach and $(98^{+5}_{-19})°$ for the frequentest one. We can remark how the current values are in very good agreement with the expected SM values.

3.5.2.3 $\gamma(\phi_3)$

- *Measurement of γ from B decays to open charm*

The possibility of observing direct CP violation in $B \to DK$ decays was first discussed by Bigi, Carter and Sanda [627, 628]. Since then, various methods to measure the weak angle γ ($=\phi_3$) using $B \to DK$ decays have been proposed. All these methods are based on two key observations: neutral D^0 and $\bar{D}^0$ mesons can decay to a common final state, and the decay $B^+ \to DK^+$ can produce neutral D mesons of both flavors via $\bar{b} \to \bar{c}u\bar{s}$ and $\bar{b} \to \bar{u}c\bar{s}$ transitions (Fig. 27) with a relative phase θ_+ between interfering amplitudes that is the sum, $\delta_B + \gamma$, of strong and weak interaction phases. For the decay, $B^- \to DK^-$, the relative phase is $\theta_- = \delta_B - \gamma$, so both δ_B and γ can be extracted from measurements of such charge conjugate B decay modes. The feasibility of the γ measurement crucially depends on the size of r_B, the ratio of the B decay amplitudes involved ($r_B = |A(B^+ \to DK^+)|/|A(B^+ \to \bar{D}K^+)|$). The value of r_B is given by the ratio of the CKM matrix elements $|V^*_{ub}V_{cs}|/|V^*_{cb}V_{us}|$ and the colour-suppression factor and is estimated to be in the range 0.1–0.2 [629]. These methods are theoretically clean because the main contributions come from tree-level diagrams (Fig. 27).[21] Various methods have been proposed to exploit this strategy using different

---

[21] D^0–$\bar{D}^0$ mixing is neglected in the current analyses. This effect can be included though [630] and is shown to be very small within the SM [631].

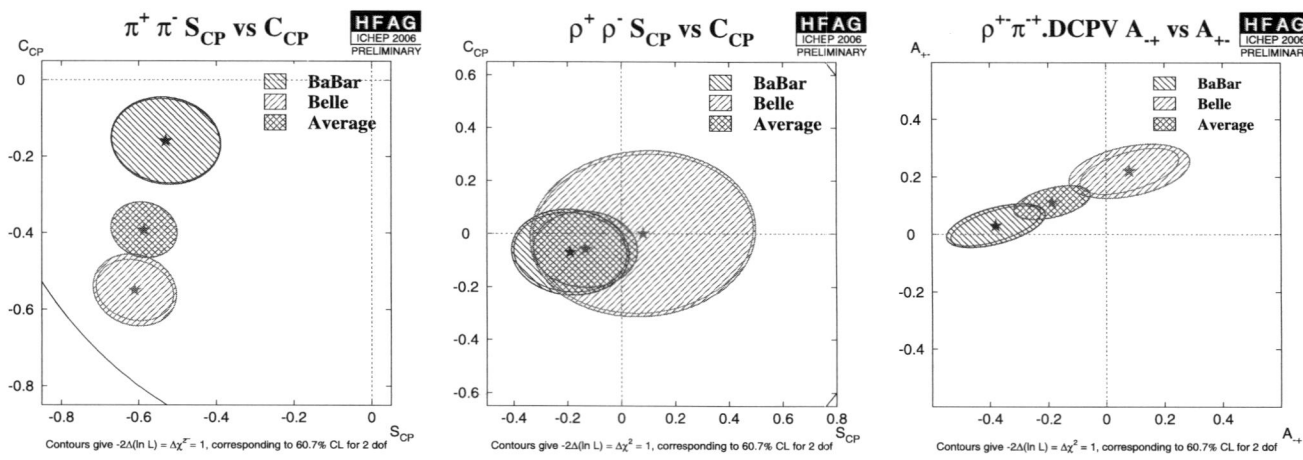

Fig. 26 The experimental results on the CP asymmetry parameters in the $\pi\pi$ (*left*), $\rho\rho$ (*center*) and $\rho\pi$ (*right*) systems, as summarized by HFAG [502]

Fig. 27 Feynman diagram of the $B^+ \to \bar{D}^0 K^+$ and $B^+ \to D^0 K^+$ decays

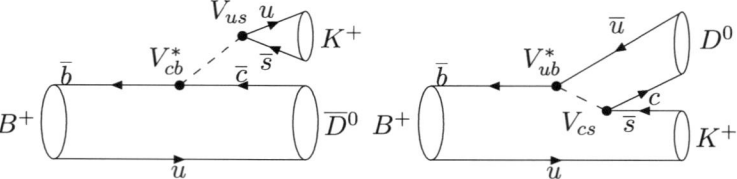

combinations of final states. These approaches include using the branching ratios of decays to CP eigenstates (GLW method [632–634]) or using doubly Cabibbo-suppressed D modes (ADS method [635]). A Dalitz plot analysis of a three-body final state of the D meson allows one to obtain all the information required for the determination of γ in a single decay mode [636–638]. Three-body final states such as $K^0_S \pi^+ \pi^-$ [637, 638] have been suggested as promising modes and give today the best estimate of the angle γ.

In the GLW method, the D is reconstructed through its decay to CP eigenstates. The experimental observables are the ratio of charge averaged partial rates, $R_{CP\pm}$, and the charge asymmetry, $A_{CP\pm}$, which are related to the model parameters through the relations $R_{CP\pm} = 1 + r_B^2 \pm 2r_B \cos\delta_B \cos\gamma$ and $A_{CP\pm} = \pm 2r_B \sin\delta_B \sin\gamma / R_{CP\pm}$. CP$_+$ refers to the CP-even final states, $\pi^+\pi^-$ and K^+K^-, and CP$_-$ refers to the CP-odd final states, $K^0_S \pi^0$, $K^0_S \phi$, $K^0_S \omega$, ... Results are available from both BaBar and Belle in the decay modes $B^\pm \to DK^\pm$, $B^\pm \to D^* K^\pm$ and $B^\pm \to DK^{*\pm}$ (Fig. 28). The errors for $R_{CP\pm}$ and $A_{CP\pm}$ are typically 10% for the most promising mode, $B^\pm \to DK^\pm$. A 3σ significance for the charge asymmetry of the $B \to DK$ mode seems to be within reach in the near future, when 1 ab^{-1} of data will be collected by each experiment. For the ADS method, using a suppressed $D \to f$ decay ($D^0 \to K^+\pi^-$, $K^+\rho^-$, $K^*\pi^-$,...), the measured quantities are the partial rate asymmetry, A_{ADS}, and the charge averaged rate, $R_{ADS} = \Gamma(B^- \to [f]_D K^-) / \Gamma(B^- \to [\bar{f}]_D K^-)$. R_{ADS}

is related to the physical parameters by the expression $r_B^2 + r_D^2 + 2r_B r_D \cos(\delta_B + \delta_D) \cos\gamma$. The overall effective branching ratio is expected to be small ($\sim 10^{-7}$), but the two interfering diagrams are of the same order of magnitude, and large asymmetries are therefore expected. The method has four unknowns: γ, r_B, $\delta_B + \delta_D$ and the amplitude ratio r_D. However, the value of r_D can be measured using decays of D mesons of known flavor. If one wants to use the ADS method alone, two modes need to be used. Of course, one can also combine one ADS mode (as an example) with one GLW CP eigenstate. No significant signal has been yet observed for the ADS modes at the B factories, so only R_{ADS} has been measured so far for the $D^{(*)} K^{(*)}$ modes (Fig. 29). These measurements will bring soon valuable constraints on r_B.

In the Dalitz method, D^0 and $\bar{D}^0$ mesons decay into the same final state $K^0_S \pi^+ \pi^-$ [637, 638] (or $K^+ \pi^- \pi^0$ [636]). Assuming no CP asymmetry in neutral D decays, the amplitude of decay as a function of Dalitz plot variables $m_+^2 = m^2_{K^0_S \pi^+}$ and $m_-^2 = m^2_{K^0_S \pi^-}$ is $M_\pm = f(m_\pm^2, m_\mp^2) + r_B e^{\pm i\gamma + i\delta_B} f(m_\mp^2, m_\pm^2)$, where $f(m_+^2, m_-^2)$ is the amplitude of the $\bar{D}^0 \to K^0_S \pi^+ \pi^-$ decay. The method has a second ambiguous solution ($\gamma + 180°$, $\delta_B + 180°$), since this transformation does not change the sum or difference of phases that are actually measured.

Results from the two B factories Belle and BaBar are available. The Belle Collaboration uses a data sample of $386 \times 10^6 B\bar{B}$ pairs [639] where the reconstructed states

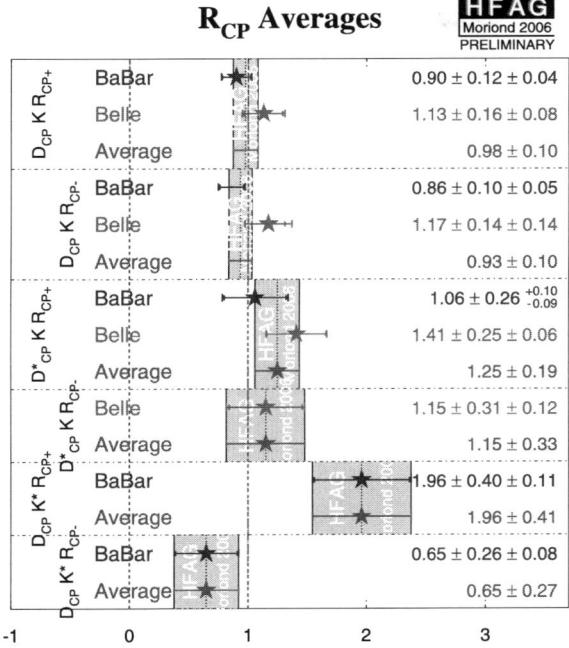
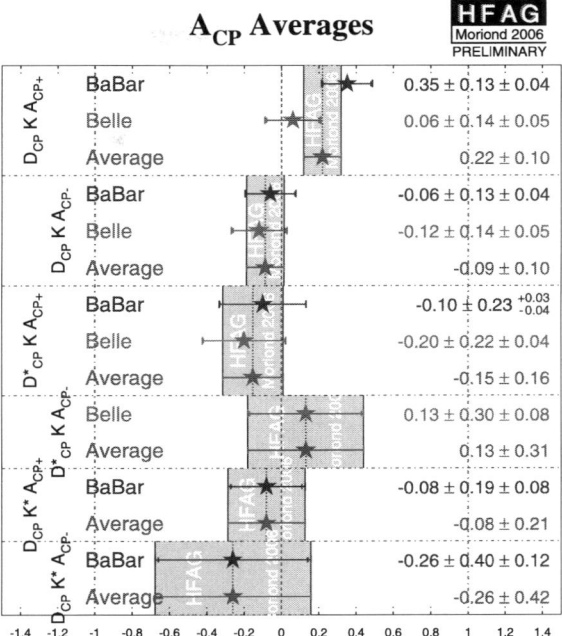

Fig. 28 $R_{CP\pm}$ and $A_{CP\pm}$ averages obtained by the B factories [389]

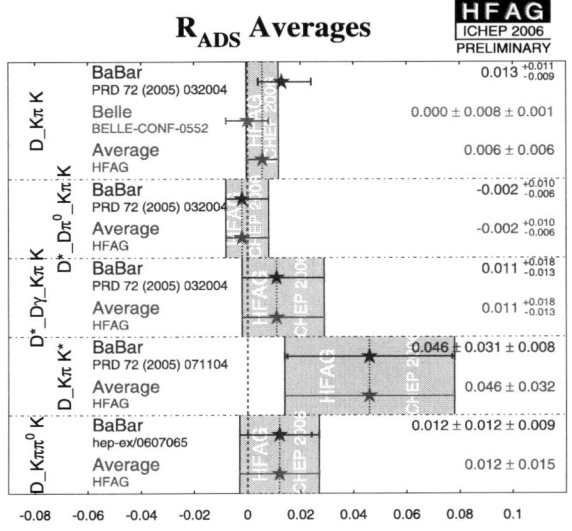

Fig. 29 R_{ADS} averages obtained by the B factories [389]

are $B^+ \to DK^+$, $B^+ \to D^*K^+$ with $D^* \to D\pi^0$ and $B^+ \to DK^{*+}$ with $K^{*+} \to K_S^0\pi^+$. Analysis by the BaBar Collaboration [640] is based on $347 \times 10^6 B\bar{B}$ pairs using $B^+ \to DK^+$ and $B^+ \to D^*K^+$ with two D^* channels: $D^* \to D\pi^0$ and $D^* \to D\gamma$ (the previous BaBar [641] publication includes also the $B^+ \to DK^{*+}$ channel, but this mode is not included in the recent update). The number of reconstructed signal events in the Belle's data are 331 ± 23, 81 ± 11 and 54 ± 8 for the $B^+ \to DK^+$, $B^+ \to D^*K^+$ and $B^+ \to DK^{*+}$ channels, respectively. BaBar finds 398 ± 23, 97 ± 13 and 93 ± 12 signal events in the $B^+ \to DK^+$, $B^+ \to D^*[D\pi^0]K^+$ and $B^+ \to D^*[D\gamma]K^+$ channels, respectively. The amplitude f is parametrized as a coherent sum of two-body decay amplitudes (16 for BaBar, 18 for Belle) plus a nonresonant decay amplitude and is determined directly in data from a large and clean sample of flavor-tagged decays produced in continuum e^+e^- annihilation. For example, Belle includes five Cabibbo-allowed amplitudes: $K^*(892)^+\pi^-$, $K^*(1410)^+\pi^-$, $K_0^*(1430)^+\pi^-$, $K_2^*(1430)^+\pi^-$ and $K^*(1680)^+\pi^-$, their doubly Cabibbo-suppressed partners, and eight channels with a K_S^0 and a $\pi\pi$ resonance: ρ, ω, $f_0(980)$, $f_2(1270)$, $f_0(1370)$, $\rho(1450)$, σ_1 and σ_2. The parameters of the σ resonances obtained in the fit are $M_{\sigma_1} = 519 \pm 6$ MeV/c^2, $\Gamma_{\sigma_1} = 454 \pm 12$ MeV/c^2, $M_{\sigma_2} = 1050 \pm 8$ MeV/c^2 and $\Gamma_{\sigma_2} = 101 \pm 7$ MeV/c^2 (the errors are statistical only), while the parameters of the other resonances are taken to be the same as in the CLEO analysis [642]. The agreement between the data and the fit result is satisfactory for the purpose of measuring γ, and the discrepancy is taken into account in the model uncertainty.

Once f is determined, a fit to $B^\pm$ data is performed to obtain the Cartesian parameters, $x_\pm = r_\pm \cos(\pm\gamma + \delta_B)$ and $y_\pm = r_\pm \sin(\pm\gamma + \delta_B)$, which have the advantage to be Gaussian-distributed, uncorrelated and unbiased (r_B is positive definite and hence exhibits a fit bias toward larger values when its central value is in a vicinity of zero) and simplify the averaging of the various measurements. Figure 30 shows the results of the separate B^+ and B^- data fits for $B \to DK$,

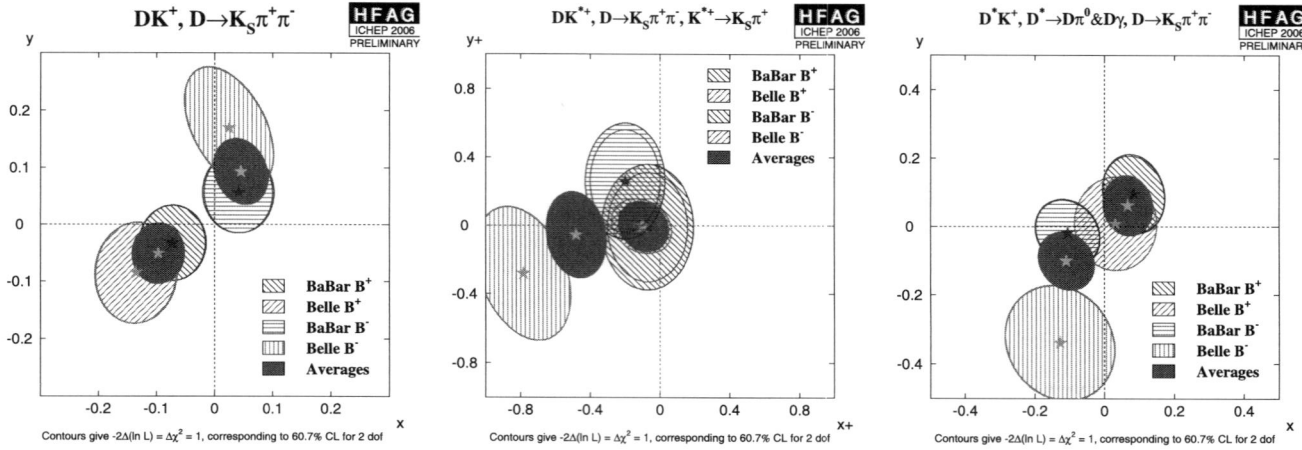

Fig. 30 Results of signal fits with free parameters $x_\pm = r\cos\theta_\pm$ and $y_\pm = r\sin\theta_\pm$ for $B^\pm \to DK^\pm$, $D^*K^\pm$ and $DK^{*\pm}$ modes from the BaBar and Belle latest publications [639, 640]. The *contours* indicate one standard deviation

D^*K and DK^* modes in the x–y plane for the BaBar and Belle Collaborations. Confidence intervals were then calculated by each experiment using a frequentest technique (the so-called Neyman ordering in the BaBar case, the Feldman and Cousins ordering [643] in the Belle case). The central values for the parameters γ, r_B and δ_B from the combined fit (using the $(x_\pm, y_\pm)$ obtained for all modes) with their one standard deviation intervals are presented in Table 33. Note that there are large correlations between the fit parameters γ and r_B. With the available data, the statistical error on γ increases with decreasing r_B and thus it depends strongly on the central value of r_B as determined by the fit. The uncertainties in the model used to parametrize the $\bar{D}^0 \to K_S^0 \pi^+ \pi^-$ decay amplitude lead to an associated systematic error in the fit result. These uncertainties arise from the fact that there is no unique choice for the set of quasi-2-body channels in the decay, as well as the various possible parameterizations of certain components, such as the nonresonant amplitude. To evaluate this uncertainty, several alternative models have been used to fit the data.

Despite similar statistical errors being obtained for $(x_\pm, y_\pm)$ in both experiments, the resulting γ error is much smaller in Belle's analysis. Since the uncertainty on γ scales roughly as $1/r_B$, the difference is explained by noticing that the BaBar $(x_\pm, y_\pm)$ measurements favor values of r_B smaller than the Belle results.

All methods (GLW, ADS and Dalitz) are sensitive to the same parameters of the B decays, and can therefore be treated in a combined fit to extract γ. Such comparisons have been performed by various phenomenological groups, such as CKMfitter [8] and UTfit [209]. The CKMfitter group using a frequentest statistical framework obtains $(77 \pm 31)°$, whereas the UTfit group with a Bayesian approach obtains $(82 \pm 19)°$. This is in agreement with the prediction from the global CKM fit (where the direct γ measurement has been excluded from the fit). As mentioned earlier, the size of the r_B parameters play a crucial role in the γ determination, and they are found to be $r_B(DK) < 0.13$, $r_B(D^*K) < 0.13$ and $r_B(DK^*) < 0.27$ at 90% C.L. by Ref. [8] and $r_B(DK) < 0.10$, $r_B(D^*K) < 0.12$ and $r_B(DK^*) < 0.26$ at 90% C.L. by Ref. [209]. All values are in agreement with the naive expectation from CKM and colour suppression.

Clearly, the precision on γ will improve with more data. However, the dependence of the sensitivity on the value

Table 33 Results of the combination of $B^+ \to DK^+$, $B^+ \to D^*K^+$, and $B^+ \to DK^{*+}$ modes for BaBar and Belle analyses. The first error is statistical, the second is systematic, and the third one is the model error. In the case of BaBar, one standard deviation constraint is given for the r_B values

Parameter	BaBar	Belle
γ	$(92 \pm 41 \pm 11 \pm 12)°$	$(53^{+15}_{-18} \pm 3 \pm 9)°$
$r_B(DK)$	< 0.140	$0.159^{+0.054}_{-0.050} \pm 0.012 \pm 0.049$
$\delta_B(DK)$	$(118 \pm 63 \pm 19 \pm 36)°$	$(146^{+19}_{-20} \pm 3 \pm 23)°$
$r_B(D^*K)$	0.017–0.203	$0.175^{+0.108}_{-0.099} \pm 0.013 \pm 0.049$
$\delta_B(D^*K)$	$(-62 \pm 59 \pm 18 \pm 10)°$	$(302^{+34}_{-35} \pm 6 \pm 23)°$
$r_B(DK^*)$		$0.564^{+0.216}_{-0.155} \pm 0.041 \pm 0.084$
$\delta_B(DK^*)$		$(243^{+20}_{-23} \pm 3 \pm 49)°$

of r_B means that we should be careful when extrapolating the present results to a higher statistics scenario. Assuming a value of r_B in the range of 0.1–0.15, the statistical error obtained by the end of the B factories (2 ab^{-1}) will be (10–15)°. The way to improve the γ sensitivity in the near future is to include more D^0 (and use of D^{*0}) modes, with combined strategies [630], use of differential spectra [644], many body modes, charm factory inputs [645, 646], along with the use of B^0 modes [644, 647]. Although at present (and until the end of B factories era) the γ accuracy in the $K_S^0 \pi^+ \pi^-$ analysis is dominated by the statistical uncertainty, the model error will eventually dominate in the context of a super B factory. Model-independent ways to extract γ have been proposed [636, 637, 648]. One way to implement this is to notice that in addition to flavor tagged $\bar{D}^0 \to K_S^0 \pi^+ \pi^-$ decays, one can use CP tagged decays to $K_S^0 \pi^+ \pi^-$ from the $\psi(3770) \to D\bar{D}$ process. Combining the two data sets, the amplitude and phase could be measured for each point on the Dalitz plot in a model-independent way. Study with MC simulations (assuming $r = 0.2$) indicates that with 50 ab^{-1} of data γ can be measured with a total accuracy of few degrees [648]. Combining all the methods with the statistics anticipated at a super B factory (50 ab^{-1}), it is expected that an error of about two degrees is obtainable (Chapter 4).

3.5.2.4 Measurement of $\sin 2\beta + \gamma$ from B decays to open charm Interference between decays with and without mixing can occur in the non-CP eigenstates $B^0 \to D^{(*)\pm} \pi^\mp (\rho^\mp)$. The Cabibbo-favored $\bar{b} \to \bar{c}$ decay amplitude interferes with the Cabibbo-suppressed $b \to u$ decay amplitude with a relative weak phase shift γ. These modes have the advantage of a relatively large branching fraction but a small ratio r of suppressed to favored amplitudes. Time-dependent asymmetries in these modes can be used to constraint $\sin(2\beta + \gamma)$ [653]: the coefficient of the $\sin(\Delta m \Delta t)$ term can be written, to a very good approximation, as $S^\pm = 2r \sin(2\beta + \gamma \pm \delta)$, where δ is the strong phase shift due to final-state interaction between the decaying mesons.

Potential competing CP-violating effects can arise from $b \to u$ transitions on the tag side if a kaon is used to tag the flavor on the other B in the event, resulting in an additional sin term $S'^\pm = 2r' \sin(2\beta + \gamma \pm \delta')$ [649]. Here, r' (δ') is the effective amplitude (phase) used to parameterize the tag side interference. To account for this term, one can rewrite $S^\pm$ as $S^\pm = (a \pm c) + b$, where $a = 2r \sin(2\beta + \gamma) \cos \delta$, $c = \cos(2\beta + \gamma)[2r \sin \delta + 2r' \sin \delta']$ and $b = 2r' \sin(2\beta + \gamma) \cos \delta'$. The results from B factories [650–652] are shown for $D\pi$ and $D^*\pi$ modes in terms of a and c in Fig. 31. CP violation would appear as $a \neq 0$. External information is however needed to determine r or δ. Naively, one can estimate $r \sim |V_{cd}^* V_{ub} / V_{ud} V_{cb}^*| \simeq 0.02$. One popular choice is the use of $SU(3)$ symmetry to obtain r by relating decay mode to B decays involving D_s mesons [653].

3.5.3 Expectations from LHCb

3.5.3.1 Introduction This section summarizes the outlook for measurements of CKM angles through tree-level processes at LHCb. All estimates are given for 2 fb^{-1} of integrated luminosity, which is a canonical year of LHCb operation. (In the summary section, extrapolations are also made to 10 fb^{-1}, which represents five years of operation.) Background estimates have been made using 34 million simulated generic $b\bar{b}$ events and, where appropriate, with specific samples of known dangerous topologies. Full details may be found in the cited LHCb notes and other references.

3.5.3.2 Measuring β with $B^0 \to J/\psi K_S^0$ The channel $B^0 \to J/\psi K_S^0$, with the J/ψ decaying to $\mu^+ \mu^-$, is relatively easy to trigger on and reconstruct at LHCb. In order to minimize systematic effects, selection cuts have been developed which impose the least possible bias on the lifetime distribution of the decaying B^0.

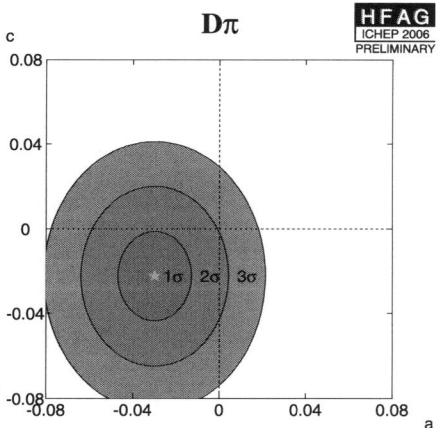
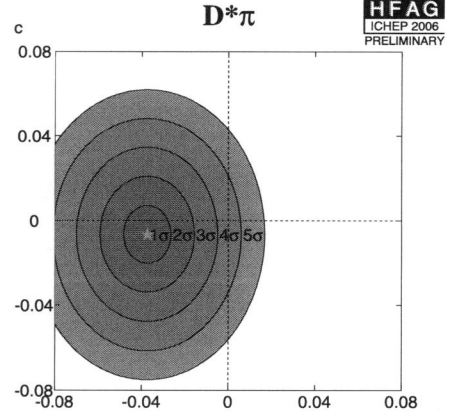

Fig. 31 Results of the a and c measurements for the $D\pi$ (*left*) and $D^*\pi$ (*right*) modes

It is estimated that 333k untagged triggered events will be collected per 2 fb^{-1} of integrated luminosity. Background studies have been performed using a large sample of generic $b\bar{b}$ events and a dedicated sample of prompt J/ψ events. The results indicate that the expected B/S ratio from the two sources is 1.1 and 7.3, respectively. The high background from prompt J/ψ's has little consequence for the $\sin 2\beta$ sensitivity, as the events are restricted to low proper times. The performance of the flavor tag is determined from the similar topology $B^0 \to J/\psi K^{*0}$ control channel. The statistical precision on $\sin 2\beta$ with 2 fb^{-1} is estimated to be 0.015. More information may be found in [666].

3.5.3.3 Measuring α with $B^0 \to \rho\pi$ and $B^0 \to \rho\rho$ at LHCb
The potential of LHCb in the decay $B^0 \to \rho\pi \to \pi^+\pi^-\pi^0$ has been studied extensively [654]. The hard spectrum of the π^0, together with the vertex constrains on the $\pi^+\pi^-$ pair, means that the decay can be well isolated from background, even in the high multiplicity environment of the LHC. A multivariate variable is built up to exploit all available discriminating variables. It is estimated that 1.4×10^4 events will be accumulated per 2 fb^{-1} of integrated luminosity. The acceptance for these events is fairly uniform over Dalitz space, apart from in the region of low ($m^2_{\pi^+\pi^0}$, $m^2_{\pi^-\pi^0}$), which is depopulated due to the minimum energy requirement on the π^0.

The background has been studied with large simulated samples of generic $b\bar{b}$ events and with specific charmless decay channels. It is concluded that the B/S ratio should not exceed one, a value which has been assumed for the subsequent sensitivity studies.

The expected precision on the angle α has been estimated using a toy Monte Carlo, taking the resolutions and acceptances from the full simulation and modeling the background as a combination of nonresonant and resonant contributions. Repeated toy experiments are performed, each of which has 10000 signal events. Various scenarios have been considered for the relative values of the penguin and tree amplitudes contributing to the final state. The results shown here assume the 'strong penguin' case [655]. An unbinned log likelihood fit is used to extract the physics parameters of interest, in particular α. The achievable precision on α varies between amplitude scenarios and fluctuates from experiment to experiment. The statistical error is below $10°$ for about 90% of experiments. The mean value is around $8°$. On about 15% of occasions, the fit converges to a pseudo-mirror solution, but these effects diminish with larger data sets. Figure 32 shows the variation in χ^2 for fits to many toy experiments as a function of α, and the average of these curves with a clear minimum seen at the input value of $\alpha = 97°$. Studies of potential systematic uncertainties indicate that it will be important to have good understanding of the ρ lineshape.

The performance of LHCb has also been investigated in the modes $B^0 \to \rho^\pm \rho^\mp$ and $B^\pm \to \rho^\pm \rho^0$. It is concluded that although significant numbers of events can be accumulated, the total event samples are similar in size to those that will come from the B factories. More promising is the decay $B^0 \to \rho^0 \rho^0$, which can be used in an isospin analysis to constrain the bias on α arising from penguin contamination in the channel $B^0 \to \rho^\pm \rho^\mp$. 1200 events will be obtained per 2 fb^{-1}, assuming a branching ratio of 1.2×10^{-6}. More details on this analysis and estimates of its impact on the α extraction within possible scenarios can be found in [654].

3.5.3.4 Measuring γ with $B \to DK$ strategies at LHCb
In principle all $B \to DK$ channels, where the D decays hadronically, carry information on the angle γ. LHCb has investigated several modes, with the emphasis on those where the decays involve charged tracks only. The presence of one or more kaons in the final state makes these decays particularly suited to LHCb, on account of its RICH system. The estimated event yields for the modes so far considered are summarized in Table 34. Background studies have been carried out using large simulation samples of generic $b\bar{b}$ events, as well as specific channels which are potential sources of contamination, for example $B \to D\pi$. In all cases, it is concluded that the background levels can be reduced to an acceptable level. More information can be found in the referenced notes. Many of the strategies that have been investigated are common to those pioneered at the B factories and discussed in Sect. 3.5.2.3.

The simplest topologies are $B \to DK$ decays where the $D^0 (\bar{D}^0)$ decays to a CP-eigenstate such as K^+K^- or

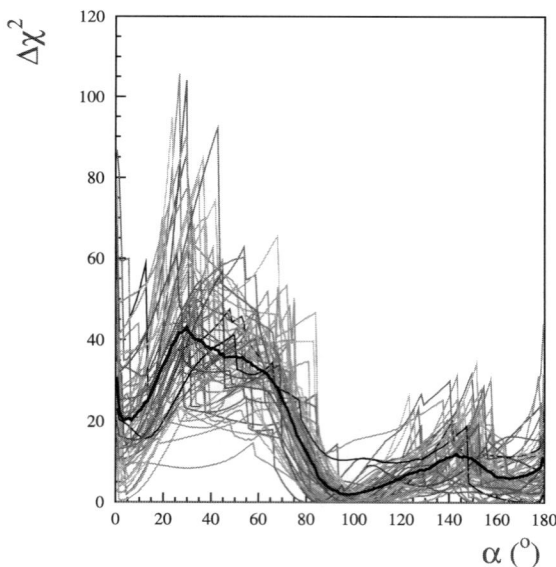

Fig. 32 Change in χ^2 with α for a fit to simulated experiments assuming the LHCb performance with 1000 signal events and a B/S ratio of 1. Each *curve* corresponds to a different experiment. Superimposed in *black* is the average of all experiments. The input value of α is $97°$

Table 34 Expected event yields and estimated background for 2 fb^{-1} in $B \to DK$ decay modes so far considered at LHCb. In the rows where two signal yields are listed, the background corresponds to that expected in either channel. All numbers come from typical scenarios presented in the references quoted in the text. The background in the $D(K_S^0 K^+ K^-)K^\pm$ final state has not yet been studied, but it is expected to be significantly smaller than that in the $D(K_S^0 \pi^+ \pi^-)K^\pm$ mode

Decay mode	Signal	Background
$B^\pm \to D(K^+K^-)K^\pm$	2600, 3200	3700 ± 1000
$B^\pm \to D(\pi^+\pi^-)K^\pm$	900, 1100	3600 ± 1500
$B^\pm \to D(K^\pm\pi^\mp)K^\pm$	28000, 28300	17500 ± 1000
$B^\pm \to D(K^\mp\pi^\pm)K^\pm$	10, 400	800 ± 500
$B^\pm \to D(K^\pm\pi^\mp\pi^+\pi^-)K^\pm$	30400, 30700	20200 ± 2500
$B^\pm \to D(K^\mp\pi^\pm\pi^+\pi^-)K^\pm$	20, 410	1200 ± 360
$B^\pm \to D(K_S^0\pi^+\pi^-)K^\pm$	5000	1000–5000 (90% C.L.)
$B^\pm \to D(K_S^0 K^+ K^-)K^\pm$	1000	/
$B^\pm \to D(K^+K^-\pi^+\pi^-)K^\pm$	1700	1500 ± 600
$B^\pm \to (D\pi^0)(K^\pm\pi^\mp)K^\pm$	16800, 16600	34300 ± 11500
$B^\pm \to (D\pi^0)(K^\mp\pi^\pm)K^\pm$	350, 100	4800 ± 3800
$B^\pm \to (D\gamma)(K^\pm\pi^\mp)K^\pm$	9400, 9300	34300 ± 11500
$B^\pm \to (D\gamma)(K^\mp\pi^\pm)K^\pm$	10, 140	4800 ± 3800
$B^0, \bar{B}^0 \to D(K^+K^-)K^{*0}, \bar{K}^{*0}$	240, 450	<1000 (90% C.L.)
$B^0, \bar{B}^0 \to D(\pi^+\pi^-)K^{*0}$	70, 140	<1000 (90% C.L.)
$B^0, \bar{B}^0 \to D(K^\pm\pi^\mp)K^{*0}, \bar{K}^{*0}$	1750, 1670	<1700 (90% C.L.)
$B^0, \bar{B}^0 \to D(K^\mp\pi^\pm)K^{*0}, \bar{K}^{*0}$	350, 260	<1700 (90% C.L.)

$\pi^+\pi^-$, or to $K^\pm\pi^\mp$. Of particular interest is the subset of highly suppressed 'ADS' decays $B^\pm \to D(K^\mp\pi^\pm)K^\pm$ where the interference effects are highest. The exact number of expected events in this mode depends on the assumption for r_B, the ratio of the interfering B decay amplitudes. Assuming a value of $r_B = 0.08$ leads to the expectation of around 400 events, integrated over B^+ and B^- channels, with a variation dependent on the value of the strong phase difference between the diagrams involved in both the B and D decays [656].

The 3-body Dalitz analysis of $K_S^0\pi^+\pi^-$ in $B \to DK$ decays has been successfully pioneered at the B factories. Here LHCb also expects to make a significant contribution with 5000 triggered and reconstructed decays per 2 fb^{-1} [658]. A technical challenge in selecting these events is presented by those K_S^0's which decay downstream of the VELO region; these decays account for around two thirds of the total sample. Although such events can be successfully reconstructed offline, this procedure is challenging to perform in the high-level trigger, where the existing track-search algorithm for K_S^0 daughters does not fit within the allocated CPU budget. It is hoped that this difficulty will be overcome. The problem is not so critical for the sister 3-body mode $D \to K_S^0 K^+ K^-$, where the two kaons offer the possibility of devising an inclusive high-level trigger selection not dependent on the finding of the K_S^0.

The 4-body modes $D \to K^\pm\pi^\mp\pi^+\pi^-$ and $D \to K^+K^-\pi^+\pi^-$ are particularly attractive to LHCb as all the decay products are prompt charged tracks. Dependent on the charge of the decaying B and the charges of the particles in the D decay, the $K\pi\pi\pi$ channel accesses four possible final states, of which the rarest two, $B^\pm \to D(K^\mp\pi^\pm\pi^+\pi^-)K^\pm$, possess large interference effects through the ADS mechanism. The expected sample size integrated over these two channels is about 400 events [659]. Provided that the sub-resonant decay structure can be fitted in a four-body amplitude analysis, these suppressed channels will provide high sensitivity to γ, either in isolation or in conjunction with the other ADS modes. An analysis of the 4-body Dalitz space of $K^+K^-\pi^+\pi^-$ accesses γ in a similar way to the 3-body self-conjugate mode $K_S^0\pi^+\pi^-$. Here 1700 events are expected [659].

Extensions of the standard $B \to DK$ strategies have also been considered at LHCb. Detailed studies have been performed of $B^0 \to DK^{*0}$, where the charge of the kaon in the $K^{*0} \to K^\pm\pi^\mp$ decay chain tags the flavor of the decaying B^0 [660]. Here both the interfering B^0 decay diagrams are colour suppressed, and hence the interference effects are higher than in the $B^\pm$ case, although the branching ratios are lower. Another method under study is $B^\pm \to D^*K^\pm$, where the D^* decays either through $D^0\pi^0$ or $D^0\gamma$. As there is a CP-conserving phase difference of π between these two paths, separation of the respective modes gives powerful additional constraints in the analysis. At LHCb, the energy of the neutral particles is too low to permit efficient selection. However, sufficient constraints exist in the decay topology to allow a full reconstruction using the charged tracks alone. Preliminary results indicate a promising performance, although there are at present insufficient Monte Carlo statistics to make a meaningful background estimate [657].

Assuming the 2 fb^{-1} event yields listed in Table 34 and the background estimates coming out of the Monte Carlo studies, full sensitivity studies have been performed for several of the analyses. The precision on γ depends on the parameters assumed. Taking $r_B = 0.08$, the statistical uncertainty is found to be $(6\text{–}10)°$ for a combined $B^\pm \to DK^\pm$ analysis involving the two-body D decay modes, and $D \to K\pi\pi$, where the resonant substructure of the latter decay is so-far neglected [656]. A similar sensitivity is found for the $B^0 \to DK^{*0}$ study involving two body modes only, where the ratio of the interfering diagrams is taken to be 0.4 [660]. Estimates have also been made of the γ sensitivity in $K_S^0\pi^+\pi^-$ [658]. Including acceptance effects and background gives a typical sensitivity of $15°$, again taking $r_B = 0.08$. At present the only available studies of $K^+K^-\pi^+\pi^-$ [661] are for signal events only. A background free analysis with the LHCb annual signal yield would have a statistical uncertainty of $14°$, also with $r_B = 0.08$. Systematic effects have not yet been considered, but it is already known from the B factories that work is needed to improve the confidence in the $D \to K_S^0\pi^+\pi^-$ decay model, an issue which is likely to be important for all the 3- and 4-body D decays.

Other decay modes remain to be investigated, for example $B^\pm \to DK^{*\pm}$, $K^{*\pm} \to K_S^0\pi^\pm$. The full power of the $B \to DK$ sensitivity will only come with a combined analysis of all accessible decay modes. The preliminary indications suggest that $B \to DK$ decays will provide LHCb's most precise value of γ, with a few degrees uncertainty being achievable with 2 fb^{-1} of data. There is no reason to expect that the experimental systematics will significantly limit this sensitivity, although more detailed studies are required. It is clear, however, that residual uncertainties associated with the understanding of the D decay in the 3- and 4-body modes could be important. A possible scenario is presented in the Summary section based on arbitrary assumptions concerning this source of uncertainty.

3.5.3.5 Measuring γ with B_s, $\bar{B}_s \to D_s^\pm K^\mp$ and B^0, $\bar{B}^0 \to D^\pm \pi^\mp$ The isolation of $B_s \to D_s^\pm K^\mp$ decays is experimentally very challenging because of the low branching ratio and the order-of-magnitude more prolific $B_s \to D_s\pi$ decay mode. The LHCb trigger system gives good performance for fully hadronic modes and selects $B_s \to D_s^\pm K^\mp$ events with an efficiency of 29%. The π–K discrimination of the RICH system reduces the $B_s \to D_s\pi$ contamination to $\sim$10%. It is estimated that the experiment will accumulate 6.2k events per 2 fb^{-1} of integrated luminosity, with a combinatoric background to signal level of <0.6 [662]. The excellent $\sim$30 fs proper time precision provided by the silicon Vertex Locator will ensure that the B_s oscillations will be well resolved and hence allow the CP asymmetries to be measured. It is estimated that the statistical precision on γ from this channel alone will be $10°$ for 2 fb^{-1}, assuming $\Delta m_s = 17.5$ ps^{-1}, $|\Delta\Gamma_s|/\Gamma_s = 0.10$ [662]. Note that this extraction requires knowledge of the weak mixing phase in the B_s system, which is imported from parallel LHCb studies performed with $B^0 \to J/\psi\phi$ decays.

A potential difficulty with the $B_s \to D_s^\pm K^\mp \gamma$ extraction arises from ambiguities. In the limit that $\Delta\Gamma_s$ is very small, the analysis returns an 8-fold ambiguity. A nonzero value of $\Delta\Gamma_s$ in principle ameliorates the problem, reducing the number of true ambiguities to four only, but even in this case, the eliminated solutions may in practice remain as false minima, on account of the limited experimental resolution. An attractive way to circumvent this difficulty is to make a combined analysis of the observables in the B_s decay and those in the U-spin symmetric $B^0 \to D^\pm\pi^\mp$ channel [663]. This approach has the added bonus of exploiting $B^0 \to D^\pm\pi^\mp$ decays in a manner which does not require knowledge of the ratio between the interfering tree diagrams, which in the B^0 system is known to be very small and hence hard to determine experimentally. LHCb will accumulate 1730k events per 2 fb^{-1} in this channel [664]. The combined analysis has the potential to reach a statistical precision of $5°$, depending on the values of the parameters involved. Any bias associated with the U-spin symmetry assumption also has a varying impact on the measurement, depending on the position in parameter space. In many scenarios, the effect is expected to be below the statistical uncertainty [665].

3.5.4 Summary

Table 35 presents a summary of the current status and the outlook for future direct measurements of the angles of the unitarity triangle from tree-dominated B decays. The last column of this table is an estimate of the ITE, which is the intrinsic error coming purely from theoretical limitations of the methods being used. It seems that for $\sin 2\beta$, at the end of the B factory era with an estimated ≈ 2 ab^{-1} of data, the experimental determination will be close to the expected theory error. In fact the theory error ($\lesssim 1\%$) is somewhat smaller, but apparently our current understanding is that experimental systematics are difficult to reduce below about $(2\text{–}3)\%$. Measurement of $\sin 2\beta$ at LHCb also looks very promising so far as the statistical error goes.

For α, although each of the three methods, $\pi\pi$, $\rho\pi$, and $\rho\rho$ will have a residual theory error due to isospin violation by EWP and/or from other sources, it is quite likely that once the experimental information with high statistics on all the three modes becomes available, the remaining intrinsic theory error will be small, $O(\text{few}\%)$. The current B factories and LHCb are expected to be able to determine α to an accuracy around $(5\text{–}8)°$, i.e. considerably worse than the ITE. A super B factory should be able to attain the level of accuracy $O(2\%) \approx$ ITE.

Unfortunately a precise determination of the angle γ is likely to remain a challenge for a long time to come. Admittedly we have been somewhat cautious in our projections for

Table 35 Unitarity Triangle from trees decays: Current status and future prospects. ITE means irreducible theory error; see text especially regarding the LHCb projections

$\int \mathcal{L} dt$	BF (Now) $\sim$1 ab^{-1}	BF(End '08) 2 ab^{-1}	LHCb 2 fb^{-1}	LHCb 10 fb^{-1}	SBF 50 ab^{-1}	ITE
$\sigma(\alpha)$	10° (11%)	7° (8%)	8.1° (9%)	4.6° (5%)	1.5° (1.6%)	O(few %)
$\sigma(\sin 2\beta)$	0.026 (4%)	0.023 (3.3%)	0.015 (2.1%)	0.007 (1%)	0.013 (2%)	$\lesssim$1%
$\sigma(\gamma)$	30° (46%)	15° (23%)	4.5° (7%)	2.4° (4%)	2° (3%)	O(0.1%)

the B factories, and there is some chance that we will gain more from combined strategies, compared to projections in this table, as additional data becomes available in the next year or two. Indeed LHCb should however be able to do at least five times better than this (i.e. an accuracy of about 2.6 degrees), with a final uncertainty dependent on the errors associated with the knowledge of the D decay structure in the modes exploited in the $B \to DK$ channels. It is interesting to note that with a SBF, and the very high statistics associated with an LHCb upgrade, the experimental error on γ could approach 1 degree, but would still be larger than that of the associated ITE.

Lastly, we must caution the reader that the LHCb numbers in Table 35 are merely illustrative values, extrapolated from present simulation studies, together with certain (in some cases) arbitrary assumptions about systematic errors. The estimated precisions for $\sin 2\beta$ contain statistical uncertainties only, as the experimental systematics are impossible to estimate properly in advance of first data. The values for α are dominated by the input from the $B^0 \to \rho\pi$ analysis, with the conservative assumption of a limiting systematic of 6°, associated with issues in the Dalitz analysis and the understanding of the ρ lineshape. The γ estimates includes inputs from the $B_s \to D_s K^\pm$, $B^\pm \to D^{(*)}(hh, hhhh)K^\pm$, $B^\pm \to D(K_S^0\pi\pi)K^\pm$ and $B^0 \to D(hh)K^*(K^+\pi^-)$ analyses. Here it is assumed that progress with the understanding of the D decay structure will result in systematics of 3° for the $D \to K_S^0\pi\pi$ mode, and twice this for the 4-body decays. An arbitrary 5° error is assigned to the B^0 channel to account for the possibility of other amplitudes contributing the $D(hh)K^+\pi^-$ final state. The $B^\pm \to DK^\pm$ inputs assume an r_B value of 0.08. Assumed quantities for other parameters are given elsewhere in Sect. 3.5.3.4.

3.6 B-meson mixing[22]

3.6.1 Introduction

During this workshop, there has been a breakthrough in the experimental study of B_s–$\bar{B}_s$ mixing with the measurement of the following quantities: the oscillation frequency Δm_s by the CDF Collaboration [126], the time-integrated untagged charge asymmetry in semileptonic B_s decays $A_{SL}^{s,\text{unt}}$ and the dimuon asymmetry A_{SL} by DØ [667, 668], the B_s lifetime from flavor-specific final states [502, 669–673], $\Delta\Gamma_s/\Gamma_s$ from the time-integrated angular analysis of $B_s \to J/\psi\phi$ decays [674] by CDF [675], supplemented by the three-dimensional constraint on Γ_s, $\Delta\Gamma_s$, and the B_s–$\bar{B}_s$ mixing phase from the time-dependent angular analysis of $B_s \to J/\psi\phi$ decays by DØ [676]. These measurements can be compared with the SM predictions and used to constrain NP contributions to the B_s–$\bar{B}_s$ mixing amplitude.

In this section, we first discuss the theoretical predictions within the SM and their uncertainties. We then present the results of a model-independent analysis of NP in B_s–$\bar{B}_s$ mixing. We discuss the implications of the experimental data for SUSY models by either allowing new sources of flavor and CP violation in the B_s sector or by considering a constrained Minimal Flavor Violation SUSY scenario. The remainder of the section is devoted to the experimental aspects of the measurements listed above and gives an outlook for the LHC.

3.6.2 Standard Model predictions

The neutral B_d and B_s mesons mix with their antiparticles leading to oscillations between the mass eigenstates. The time evolution of the neutral B–$\bar{B}$ meson pair is described, in analogy to K^0–$\bar{K}^0$ mixing, by a Schrödinger equation with the effective 2×2 Hamiltonian

$$i\frac{d}{dt}\begin{pmatrix} B_q \\ \bar{B}_q \end{pmatrix} = \left[\begin{pmatrix} M_{11}^q & M_{12}^q \\ M_{12}^{q*} & M_{11}^q \end{pmatrix} - \frac{i}{2}\begin{pmatrix} \Gamma_{11}^q & \Gamma_{12}^q \\ \Gamma_{12}^{q*} & \Gamma_{11}^q \end{pmatrix} \right] \begin{pmatrix} B_q \\ \bar{B}_q \end{pmatrix} \quad (138)$$

with $q = d, s$. The mass difference Δm_q and the width difference $\Delta\Gamma_q$ are defined as

$$\Delta m_q = m_H^q - m_L^q, \qquad \Delta\Gamma_q = \Gamma_L^q - \Gamma_H^q, \quad (139)$$

where H and L denote the Hamiltonian eigenstates with the heavier and lighter mass eigenvalue, respectively. These

---
[22]Section coordinators: V. Lubicz, J. van Hunen.

states can be written as

$$|B_q^{H,L}\rangle = \frac{1}{\sqrt{1+|(q/p)_q|^2}}(|B_q\rangle \pm (q/p)_q|\bar{B}_q\rangle). \quad (140)$$

Theoretically, the experimental observables Δm_q, $\Delta \Gamma_q$ and $|(q/p)_q|$ are related to M_{12}^q and Γ_{12}^q. In the B_d–$\bar{B}_d$ and B_s–$\bar{B}_s$ systems, the ratio Γ_{12}^q/M_{12}^q is of $\mathcal{O}(m_b^2/m_t^2) \simeq 10^{-3}$ and, neglecting terms of $\mathcal{O}(m_b^4/m_t^4)$, one has

$$\Delta m_q = 2|M_{12}^q|, \quad \frac{\Delta \Gamma_q}{\Delta m_q} = -\mathrm{Re}\left(\frac{\Gamma_{12}^q}{M_{12}^q}\right),$$
$$1 - \left|\left(\frac{q}{p}\right)_q\right| = \frac{1}{2}\mathrm{Im}\left(\frac{\Gamma_{12}^q}{M_{12}^q}\right). \quad (141)$$

The matrix elements M_{12}^q and Γ_{12}^q are related to the dispersive and the absorptive parts of the $\Delta B = 2$ transitions, respectively. Short-distance QCD corrections to these matrix elements have been computed at the NLO for both M_{12}^q [701] and Γ_{12}^q [702–704]. The long distance effects are contained in the matrix elements of four-fermion operators which have been computed with lattice QCD using various approaches to treat the b quark (HQET, NRQCD, QCD) [352, 705–710]. The corresponding bag parameters B are found to be essentially insensitive to the effect of the quenched approximation (see Sect. 2.4).

The quantity $\mathrm{Im}(\Gamma_{12}^q/M_{12}^q)$ can be measured through the CP asymmetry in B_q decays to flavor-specific final states. An important example is the semileptonic asymmetry

$$A_{SL}^s = \mathrm{Im}\left(\frac{\Gamma_{12}^q}{M_{12}^q}\right) = \frac{N(\bar{B}_s \to l^+ X) - N(B_s \to l^- X)}{N(\bar{B}_s \to l^+ X) + N(B_s \to l^- X)}. \quad (142)$$

Two updated theoretical predictions for $\Delta \Gamma_s/\Gamma_s$ and for the semileptonic asymmetry A_{SL}^s, obtained by including NLO QCD and $\mathcal{O}(1/m_b)$ [711] corrections, are

$$\Delta \Gamma_s/\Gamma_s = (7 \pm 3) \times 10^{-2},$$
$$A_{SL}^s = (2.56 \pm 0.54) \times 10^{-5} \quad [704],$$
$$\Delta \Gamma_s/\Gamma_s = (13 \pm 2) \times 10^{-2}, \quad (143)$$
$$A_{SL}^s = (2.06 \pm 0.57) \times 10^{-5} \quad [712].$$

The difference in the central values of $\Delta \Gamma_s/\Gamma_s$ is mainly due to a different choice of the operator basis [712] and is related to unknown $\mathcal{O}(\alpha_s^2)$ and $\mathcal{O}(\alpha_s/m_b)$ corrections. Although the basis chosen in [712] leads to smaller theoretical uncertainties, the shift observed in the central values may signal that the effect of higher-order corrections on $\Delta \Gamma_s/\Gamma_s$ is larger than what could have been previously estimated. We take into account this uncertainty by quoting, as final theoretical predictions in the SM, the more conservative estimate [713]

$$\Delta \Gamma_s/\Gamma_s = (11 \pm 4) \times 10^{-2},$$
$$A_{SL}^s = (2.3 \pm 0.5) \times 10^{-5}. \quad (144)$$

Concerning Δm_s, the SM predictions obtained by the UTfit and CKMfitter Collaborations are

$$\Delta m_s = (18.4 \pm 2.4)\,\mathrm{ps}^{-1} \quad [120],$$
$$\Delta m_s = (18.9^{+5.7}_{-2.8})\,\mathrm{ps}^{-1} \quad [8]. \quad (145)$$

3.6.3 B_s–$\bar{B}_s$ mixing beyond the SM

We now discuss the analysis of B_s–$\bar{B}_s$ mixing in the presence of NP contributions to the $\Delta B = 2$ effective Hamiltonian. These can be incorporated in the analysis in a model independent way, parameterizing the shift induced in the mixing frequency and phase with two parameters, C_{B_s} and $\phi_s \equiv 2\phi_{B_s}$, having in the SM expectation values of 1 and 0, respectively [2–6]:

$$C_{B_s} e^{i\phi_s} \equiv C_{B_s} e^{2i\phi_{B_s}} = \frac{(M_{12}^s)^{\mathrm{SM+NP}}}{(M_{12}^s)^{\mathrm{SM}}}. \quad (146)$$

As for the absorptive part of the B_s–$\bar{B}_s$ mixing amplitude, which is derived from the double insertion of the $\Delta B = 1$ effective Hamiltonian, it could be affected by NP effects in $\Delta B = 1$ transitions through penguin contributions. Such NP contributions were considered in [7, 210]. We shall neglect them in the present discussion. In this approximation, which is followed by most authors, NP enters B_s–$\bar{B}_s$ mixing only through the two parameters defined in (146).

Since the SM phase of Γ_{12}^s/M_{12}^s is small in comparison with the current experimental sensitivity, we shall assume in the following that CP violation in B_s mixing is dominated by the NP mixing phase ϕ_s. We then have

$$A_{SL}^s = \frac{\Delta \Gamma_s}{\Delta M_s} \tan \phi_s, \quad (147)$$

and the same NP phase ϕ_s will also govern mixing-induced CP violation in the exclusive channel $B_s \to J/\psi \phi$. Note that the phases in $A_{SL}^s = \mathrm{Im}(\Gamma_{12}^s/M_{12}^s)$ and in the $B_s \to J/\psi \phi$ asymmetry are different from each other in the SM, where $\arg(-\Gamma_{12}^s/M_{12}^s) \approx -0.004$, while the phase measured in $B_s \to J/\psi \phi$ decay is $-2\beta_s \approx -2\lambda^2 \eta \approx -0.04$ (see e.g. [674, 712]).

Making use of the experimental information described in Sect. 3.6.6, it is possible to constrain C_{B_s} and ϕ_{B_s} [7, 9, 210, 678, 712, 714, 715]. We report here the results obtained in [9].

The use of $\Delta \Gamma_s/\Gamma_s$ from the time-integrated angular analysis of $B_s \to J/\psi \phi$ decays is described for instance in

[7]. Here we use only the CDF measurement [675] as input, since the DØ analysis is now superseded by the new time-dependent study [676]. The latter provides the first direct constraint on the B_s–$\bar{B}_s$ mixing phase and also a simultaneous bound on $\Delta\Gamma_s$ and Γ_s. The time-dependent analysis determines the B_s–$\bar{B}_s$ mixing phase with a four-fold ambiguity. First of all, being untagged, it is not directly sensitive to $\sin\phi_s$, resulting in the ambiguity $(\phi_s, \cos\delta_{1,2}) \leftrightarrow (-\phi_s, -\cos\delta_{1,2})$, where $\delta_{1,2}$ represent the strong phase differences between the transverse polarization and the other ones. Second, at fixed sign of $\cos\delta_{1,2}$, there is the ambiguity $(\phi_s, \Delta\Gamma_s) \leftrightarrow (\phi_s + \pi, -\Delta\Gamma_s)$. One could be tempted to use factorization [712] or $B_d \to J/\psi K^*$ with $SU(3)$ [716] to fix the sign of $\cos\delta_{1,2}$. Unfortunately, neither factorization nor $SU(3)$ are accurate enough to draw firm conclusions on these strong phases. This is confirmed by the fact that the two approaches lead to opposite results. Waiting for future, more sophisticated experimental analyses, which could resolve this ambiguity with a technique similar to the one used by BaBar in $B_d \to J/\psi K^*$ [613], we prefer to be conservative and keep the four-fold ambiguity.

Compared to previous analyses, the additional experimental input discussed below improves considerably the determination of the phase of the B_s–$\bar{B}_s$ mixing amplitude. The fourfold ambiguity inherent in the untagged analysis of [676] is somewhat reduced by the measurements of A_{SL}^s and A_{SL} (see (150)), which slightly prefer negative values of ϕ_{B_s}. The results for C_{B_s} and ϕ_{B_s}, obtained from the general analysis allowing for NP in all sectors, are

$$C_{B_s} = 1.03 \pm 0.29,$$
$$\phi_{B_s} = (-75 \pm 14)° \cup (-19 \pm 11)° \cup (9 \pm 10)° \quad (148)$$
$$\cup (102 \pm 16)°.$$

Thus, the deviation from zero in ϕ_{B_s} is below the 1σ level, although clearly there is still ample room for values of ϕ_{B_s} very far from zero. The corresponding p.d.f. in the C_{B_s}–ϕ_{B_s} plane is shown in Fig. 33.

3.6.4 B_s–$\bar{B}_s$ mixing in SUSY with nonminimal flavor violation

The results on C_{B_s} and ϕ_{B_s} obtained above can be used to constrain any NP model. As an interesting example, we discuss here the case of SUSY with new sources of flavor and CP violation, following [118].

To fulfill our task in a model-independent way, we use the mass-insertion approximation to evaluate the gluino mediated contribution to $b \to s$ transitions. Treating off-diagonal sfermion mass terms as interactions, we perform a perturbative expansion of FCNC amplitudes in terms of mass insertions. The lowest nonvanishing order of this expansion gives an excellent approximation to the full result, given the

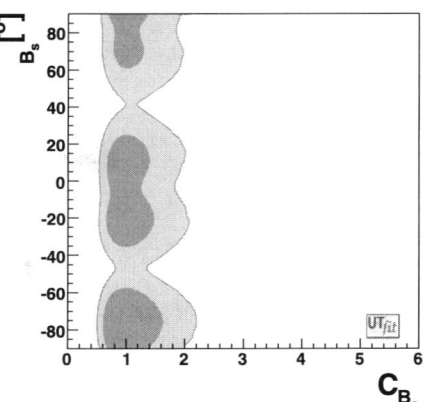

Fig. 33 Constraints on ϕ_{B_s} vs. C_{B_s} from the NP generalized analysis of Ref. [9]

tight experimental constraints on flavor-changing mass insertions. It is most convenient to work in the super-CKM basis, in which all gauge interactions carry the same flavor dependence as in the SM. In this basis, we define the mass insertions $(\delta_{ij}^d)_{AB}$ as the off-diagonal mass terms connecting down-type squarks of flavor i and j and helicity A and B, divided by the average squark mass (see Sect. 1.3).

The constraints on $(\delta_{23}^d)_{AB}$ have been studied in detail in [116] using as experimental input the branching ratios and CP asymmetries of $b \to s\gamma$ and $b \to s\ell^+\ell^-$ decays and the first measurement of B_s–$\bar{B}_s$ mixing. We perform the same analysis using the full information encoded in C_{B_s} and ϕ_{B_s} and the recently computed NLO corrections to the $\Delta B = 2$ SUSY effective Hamiltonian [118]. We refer the reader to [118] for all the details of this analysis.

For definiteness, we present here the results obtained by choosing an average squark mass of 350 GeV, a gluino mass of 350 GeV, $\mu = -350$ GeV and $\tan\beta = 3$. The dependence on μ and on $\tan\beta$ is induced by the presence of a chirality flipping, flavor conserving mass insertion proportional to $\mu \tan\beta$. In Fig. 34, we show the allowed ranges in the $\text{Re}(\delta_{23}^d)_{AB}$–$\text{Im}(\delta_{23}^d)_{AB}$ planes. The corresponding upper bounds at 95% probability are presented in Table 36.

One finds that the constraints on $(\delta_{23}^d)_{LL}$ and $(\delta_{23}^d)_{LL} = (\delta_{23}^d)_{RR}$ come from the interplay of B_s–$\bar{B}_s$ mixing with $b \to s$ decays. $(\delta_{23}^d)_{RR}$ is dominated by the information on B_s–$\bar{B}_s$ mixing, while $(\delta_{23}^d)_{LR}$ and $(\delta_{23}^d)_{RL}$ are dominated by $\Delta B = 1$ processes.

3.6.5 B_s–$\bar{B}_s$ mixing in SUSY with minimal flavor violation

As a second model-specific case for meson mixing, we mention that of SUSY with MFV. The MFV scenario is defined, in general, within the effective field theory approach of [10]. In the specific case of SUSY, the soft squark mass terms, parametrized in the previous section in terms of mass

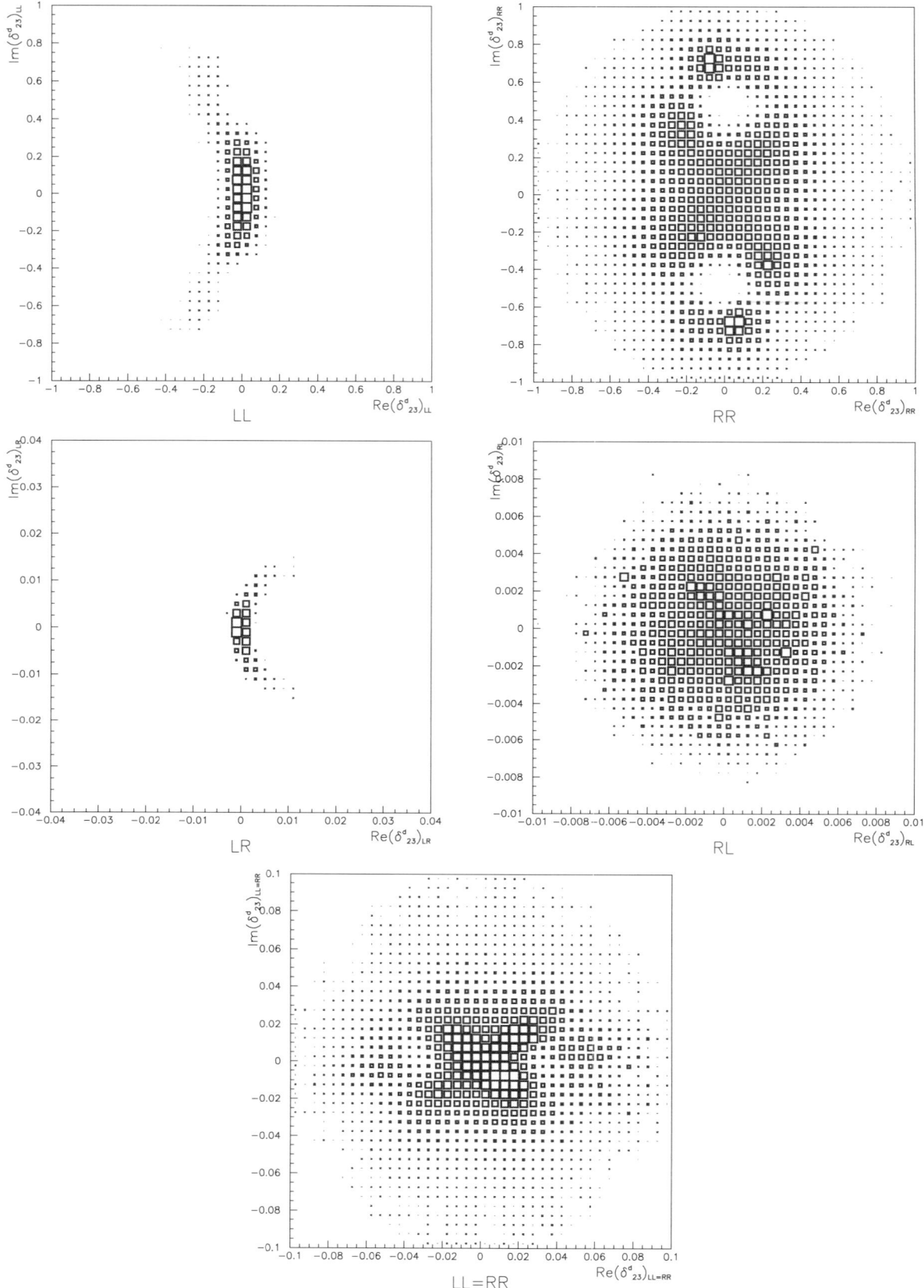

Fig. 34 Allowed range in the $\text{Re}(\delta^d_{23})_{AB}$–$\text{Im}(\delta^d_{23})_{AB}$ plane, with $AB = LL$ (*top left*), $AB = RR$ (*top right*), $AB = LR$ (*middle left*), $AB = RL$ (*middle right*) and $AB = LL$ with $(\delta^d_{23})_{LL} = (\delta^d_{23})_{RR}$ (*bottom*)

Table 36 Upper bounds (95% C.L.) on the mass insertion parameters $|(\delta^d_{23})_{AB}|$, see the text for details

$|(\delta^d_{23})_{LL}|$	$|(\delta^d_{23})_{RR}|$	$|(\delta^d_{23})_{LL=RR}|$	$|(\delta^d_{23})_{LR,RL}|$
2×10^{-1}	7×10^{-1}	5×10^{-2}	5×10^{-3}

insertions, are expanded in terms of the SM Yukawa couplings [10, 38], and the relevant parameters become the expansion coefficients. A detailed meson mixing study within this approach has been performed in [42] and for low $\tan\beta$ shows that: (i) NP contributions are *naturally* small for ΔM_s of the order of 1/ps; (ii) such contributions are always positive; (iii) if μ is not small, gluino contributions enhance (even for low $\tan\beta$) scalar operators, which then spoil the phenomenological picture of $(V-A) \times (V-A)$ dominated MFV [12]. In particular item (i) emphasizes the importance of precision determinations for lattice parameters like ξ if NP is of minimal flavor violating nature.

3.6.6 Present experimental situation

New information concerning the B_s–$\bar{B}_s$ mixing parameters became available during this workshop. The highlight was the measurement of Δm_s by DØ and CDF. The DØ experiment used the semileptonic $B_s \to D_s \mu \nu X$ decays with $D_s \to \phi \pi$, and determined a 90% confidence range for Δm_s: $17 < \Delta m_s < 21$ ps^{-1}. The initial CDF result yielded a 3σ observation of B_s–$\bar{B}_s$ mixing by making use of semileptonic and hadronic decay modes [677]. Shortly after CDF published an improved analysis [126]. In this analysis, the signal yield was increased by improving the particle identification and by using a neural network for the event selection, which allows the use of additional decay modes. Moreover the flavor tagging was improved by adding an opposite-side flavor tag based on the charge of the kaons and by the use of a neural network for the combination of the kaon, lepton and jet-charge tags. The result for Δm_s equals

$$\Delta m_s = (17.77 \pm 0.010 \pm 0.07) \text{ ps}^{-1}. \quad (149)$$

The probability that a statistical fluctuation would produce this signal is 8×10^{-8} ($> 5\sigma$ evidence). This value for Δm_s is consistent with the SM expectation, see (145). The ratio $|V_{td}/V_{ts}|$ was determined by CDF as well [126] and equals $0.2060 \pm 0.0007(\Delta m_s)^{+0.0081}_{-0.0060}(\Delta m_d + \text{theory})$.

Also information on the B_s mixing phase became available [676]. The DØ experiment performed two independent measurements of A^s_{SL}, defined in (142), using the same sign dimuon pairs [668] and time-integrated semileptonic decays $B_s \to \mu \nu D_s$ with $D_s \to \phi \pi$ [667].

The same sign dimuon asymmetry in B decays at Tevatron can be expressed as [678]

$$A_{SL} = \frac{N(b\bar{b} \to \mu^+ \mu^+ X) - N(b\bar{b} \to \mu^- \mu^- X)}{N(b\bar{b} \to \mu^+ \mu^+ X) + N(b\bar{b} \to \mu^- \mu^- X)}$$
$$= \frac{f_d Z_d A^d_{SL} + f_s Z_s A^s_{SL}}{f_d Z_d + f_s Z_s}, \quad (150)$$
$$Z_q = \frac{1}{1-y_q^2} - \frac{1}{1+x_q^2},$$
$$x_q = \Delta M_q / \Gamma_q, \qquad y_q = \Delta \Gamma_q / (2\Gamma_q).$$

Here $f_d = 0.398 \pm 0.012$ and $f_s = 0.103 \pm 0.014$ are the B_d and B_s fragmentation fractions. The measured asymmetry A_{SL} was presented by DØ in Ref. [668]:

$$A_{SL}(\text{DØ}) = A^d_{SL} + \frac{f_s Z_s}{f_d Z_d} A^s_{SL}$$
$$= -0.0092 \pm 0.0044(\text{stat.}) \pm 0.0032(\text{syst.}). \quad (151)$$

Measurements of A^d_{SL} were performed by the b factories. The average value of A^d_{SL} is [678]

$$A^d_{SL} = +0.0011 \pm 0.0055. \quad (152)$$

This leads to the value of A^s_{SL} from the same sign dimuon asymmetry:

$$A^s_{SL} = -0.0064 \pm 0.0101. \quad (153)$$

Recently DØ has also presented a time-integrated direct measurement of A^s_{SL} using semileptonic $B_s \to D^\pm \mu^\mp \nu_\mu$ decays [667]. They measure:

$$A^s_{SL} = +0.0245 \pm 0.0193(\text{stat.}) \pm 0.0035(\text{syst.}). \quad (154)$$

These two measurements of A^s_{SL} are independent, and their combination gives the charge asymmetry in semileptonic B_s decays: $A^s_{SL} = 0.0001 \pm 0.0090$ [679]. The analysis of the time-dependent angular distributions in $B_s \to J/\psi \phi$ decays [674] yields both the decay width difference $\Delta \Gamma_s$ and CP violating phase ϕ_s [676]:

$$\Delta \Gamma_s = 0.17 \pm 0.09 \pm 0.03 \text{ ps}^{-1},$$
$$\phi_s = -0.79 \pm 0.56 \pm 0.01. \quad (155)$$

Combining the results for A^s_{SL}, $\Delta \Gamma_s$, ϕ_s and using the CDF result on the mass difference Δm_s [126] gives an improved estimate for ϕ_s and $\Delta\Gamma_s$ [679]:

$$\Delta\Gamma_s = 0.13 \pm 0.09 \text{ ps}^{-1},$$
$$\phi_s = -0.70^{+0.47}_{-0.39}. \qquad (156)$$

Also new results have been released recently concerning the B_s lifetime and $\Delta\Gamma_s$. At DØ, the B_s lifetime for $B_s \to D_s \mu \nu X$ was measured to be

$$1.398 \pm 0.044(\text{stat.})^{+0.028}_{-0.025}(\text{syst.}) \text{ ps}^{-1}$$

[673]. The average B_s lifetime equals 1.466 ± 0.059 ps^{-1} [119]. CDF published the measurement of

$$\Delta\Gamma_s = \left(0.47^{+0.19}_{-0.24}(\text{stat.}) \pm 0.01(\text{syst.})\right) \text{ ps}^{-1}$$

[675].

In the near future, the LHC experiments LHCb, ATLAS and CMS will start to provide information on B_s–$\bar{B}_s$ mixing. In the following sections, the sensitivity of LHCb to the B_s mixing parameters Δm_s, $\Delta\Gamma_s$, ϕ_s and A_{SL} and the prospects for CMS will be discussed.

3.6.7 LHCb

The LHCb experiment is designed as a single-arm forward spectrometer to study b decays and CP violation. Its main characteristics are precise vertexing, efficient tracking and good particle identification. The high-precision measurements at LHCb will enable further tests of the CKM picture and probe physics beyond the SM. This is in particular true for the measurement of B_s–$\bar{B}_s$ mixing parameters such as Δm_s, $\Delta\Gamma_s$, ϕ_s and A_{SL}.

LHCb will run at a nominal luminosity of $\mathcal{L} = 2 \times 10^{32}$ cm^{-2} s^{-1}. Assuming a $b\bar{b}$ production cross-section of $\sigma_{b\bar{b}} = 500$ μb, this will correspond to an integrated luminosity of 2 fb^{-1} per nominal year of 10^7 s of data taking. All event yields quoted below are for 2 fb^{-1}. They have been obtained from a full Monte Carlo (MC) simulation of the experiment, which included the following: pileup generation, particle tracking through the detector material, detailed detector response (including timing effects such as spillover), full trigger simulation, offline reconstruction with full pattern recognition, and selection cuts. High-statistics samples of signal events have been produced for a detailed study of resolutions and efficiencies. Combinatorial background has been studied using a sample of ∼27M inclusive $b\bar{b}$ events corresponding to about 10 minutes of data taking, while identified physics background sources have been studied with large specific background samples.

3.6.7.1 Sensitivity to Δm_s from $B_s \to D_s \pi$

The mass difference Δm_s between the mass eigenstates of the B_s–$\bar{B}_s$ system is best measured as the frequency of the oscillatory behavior of the proper time distribution of flavor-tagged B_s mesons decaying to a flavor-specific final state. The best channel for this at LHCb is $B_s \to D_s \pi$, with the subsequent D_s^+ decay to $K^+ K^- \pi^+$, because of its easy topology with four charged tracks and its relatively large branching fraction of $B(B_s \to D_s \pi) \times B(D_s^+ \to K^+ K^- \pi^+) = (1.77 \pm 0.48) \times 10^{-4}$ [680]. Such decays can be detected, triggered, reconstructed and selected with a final mass resolution of ∼14 MeV/c^2 (see Fig. 35 left) and a total efficiency of about 0.4%, leading to a yield of (140k ± 40k) events in 2 fb^{-1}. After the trigger and selection, the combinatorial background is expected to be dominated by $b\bar{b}$ events and has been estimated to be less than 5% of the signal at 90% C.L., in a ±50 MeV/c^2 mass window around

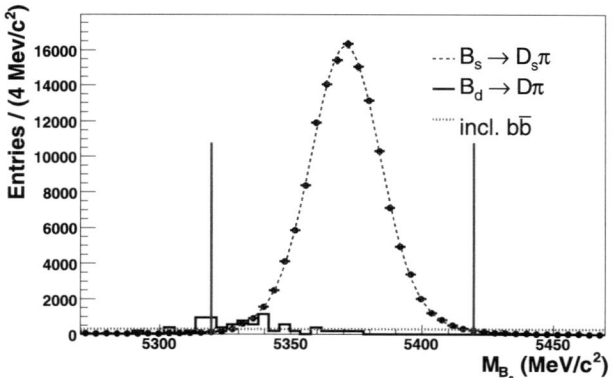
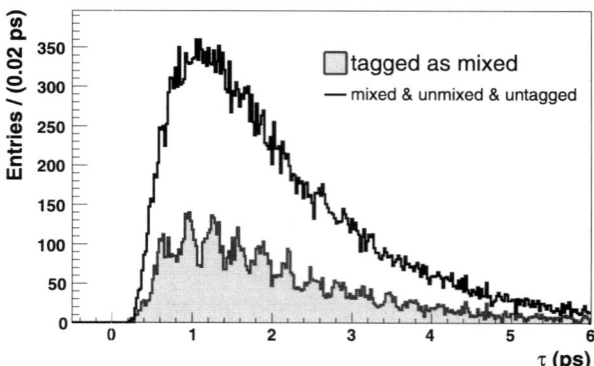

Fig. 35 *Left*: Reconstructed $B_s \to D_s \pi$ mass distribution from full MC simulation, after trigger and all selection cuts [680]. The *points with error bars* represent the signal (on an arbitrary vertical scale). The *histogram* represents the $B \to D^- \pi^+$ background and the *dotted flat line* represents the upper limit of the combinatorial background from $b\bar{b}$ events, normalized to the signal. *Right*: Reconstructed $B_s \to D_s \pi$ proper time distribution from full MC simulation of the signal, corresponding to an integrated luminosity of 0.5 fb^{-1} [680]. The *lower histogram* represents the events tagged as mixed. The background is not shown

the signal. Using the same sample of simulated $b\bar{b}$ events, the background from partially reconstructed b-hadron decays in the same mass window has been estimated to be less than 40% at 90% C.L.. This includes partially reconstructed Λ_b and B_d decays. A dedicated study showed that the background from $B \to D^-\pi^+$ decays (where one of the charged pions from the D decay could be misidentified as a kaon) is approximately 5% of the signal.

The proper time resolution, obtained on an event-by-event basis from the estimated tracking errors, typically varies between 15 and 80 fs with an average value of $\sim$40 fs (dedicated studies are being done at LHCb to model the proper time resolution [681] and to verify the estimated tracking errors [682, 683] with data). A flavor tagging power of ϵD^2 of at least 9% is achieved on the MC signal, combining several tags in a neural network: a muon or electron from the $b \to \ell$ decay of the other b-hadron, a charged kaon from the $b \to c \to s$ decay of the other b-hadron, the vertex charge of the other b-hadron, and a charged kaon accompanying the signal B_s in the fragmentation chain [684].

The statistical uncertainty on the measurement of Δm_s using an integrated luminosity of 2 fb^{-1} is expected to be ± 0.007 ps^{-1} [662]. It will be dominated by systematic uncertainties related to the determination of the proper time scale. Figure 35 (right) shows the proper time distribution from which such a measurement could be extracted.

The $B_s \to D_s\pi$ sample will play a crucial role as a control sample in all time-dependent B_s analyses; indeed it can be used to measure directly the dilution (due to flavor tagging and proper time resolution) on the $\sin(\Delta m_s t)$ and $\cos(\Delta m_s t)$ terms in time-dependent CP asymmetries. It will also be used as a normalization channel for many measurements of B_s branching fractions. More details on the selection of $B_s \to D_s\pi$ events can be found in [680].

3.6.7.2 Sensitivity to ϕ_s and $\Delta\Gamma_s$ from exclusive $\bar{b} \to \bar{c}c\bar{s}$ decays

The B_s–$\bar{B}_s$ mixing phase ϕ_s can be measured from the flavor-tagged B_s decays to CP eigenstates involving the $\bar{b} \to \bar{c}c\bar{s}$ quark-level transition. The best mode for this at LHCb is $B_s \to J/\psi\phi$. However, in this case, the vector nature of the two particles in the final state causes their relative angular momentum to take more than one value, resulting in a mixture of CP-even and CP-odd contributions. An angular analysis is therefore required to separate them on a statistical basis. This can be achieved with a simultaneous fit to the measured proper time and so-called transversity angle of the reconstructed decays. Such a fit is sensitive also to $\Delta\Gamma_s$ because of the presence of the two CP components.

The sensitivity to ϕ_s has been studied so far with the following modes:

- $B_s \to J/\psi(\mu^+\mu^-)\phi(K^+K^-)$ [685, 686];
- $B_s \to \eta_c(\pi^+\pi^-\pi^+\pi^-, \pi^+\pi^-K^+K^-, K^+K^-K^+K^-) \times \phi(K^+K^-)$ [685, 686];
- $B_s \to J/\psi(\mu^+\mu^-)\eta(\gamma\gamma, \pi^+\pi^-\pi^0)$ [685, 686];
- $B_s \to J/\psi(\mu^+\mu^-)\eta'(\eta(\gamma\gamma)\pi^+\pi^-, \rho(\pi^+\pi^-)\gamma)$ [687, 688];
- $B_s \to D_s^+(K^+K^-\pi^+)D_s^-(K^+K^-\pi^-)$ [685, 686].

The results are summarized in Table 37. For each signal event in the full simulation, the proper time and its error are estimated using a least-squares fit. The distributions of the proper time errors (scaled with the sigma of their pull distribution) are shown in Fig. 36. Most channels have a proper time resolution below 40 fs. A good proper time resolution is important for resolving the fast B_s–$\bar{B}_s$ oscillations.

The sensitivities to the B_s–$\bar{B}_s$ mixing parameters are determined by means of fast parameterized simulations, with the results of Table 37 as inputs. A large number of experiments are generated assuming the following set of parameters: $\Delta m_s = 17.5$ ps^{-1}, $\phi_s = -0.04$ rad, $\Delta\Gamma_s/\Gamma_s = 0.15$,

Table 37 Characteristics of different exclusive $\bar{b} \to \bar{c}c\bar{s}$ modes for the measurement of ϕ_s. The first 6 columns of numbers are obtained from the full MC simulation. They represent the expected number of triggered, reconstructed and selected signal events with an integrated luminosity of 2 fb^{-1} (before tagging), the background-over-signal ratio determined mainly from inclusive $b\bar{b}$ events, the B_s mass resolution, the average value of the estimated event-by-event B_s proper time error scaled by the width of its pull distribution, the flavor tagging efficiency, and the mistag probability. These parameters have been used as input to a fast MC simulation to obtain the sensitivity on ϕ_s given in the last column. The last line describes the control channel (see text)

Channel	2 fb^{-1} yield	B/S	σ_{mass} [MeV/c^2]	σ_{time} [fs]	ϵ_{tag} [%]	ω_{tag} [%]	$\sigma(\phi_s)$ [rad]
$B_s \to J/\psi\phi$	131k	0.12	14	36	57	33	0.023
$B_s \to \eta_c\phi$	3k	0.6	12	30	66	31	0.108
$B_s \to J/\psi\eta(\gamma\gamma)$	8.5k	2.0	34	37	63	35	0.109
$B_s \to J/\psi\eta(\pi^+\pi^-\pi^0)$	3k	3.0	20	34	62	30	0.142
$B_s \to J/\psi\eta'(\eta\pi^+\pi^-)$	2.2k	1.0	19	34	64	31	0.154
$B_s \to J/\psi\eta'(\rho\gamma)$	4.2k	0.4	14	29	64	31	0.080
$B_s \to D_s D_s$	4k	0.3	6	56	57	34	0.133
$B_s \to D_s\pi$	140k	0.4	14	40	63	31	–

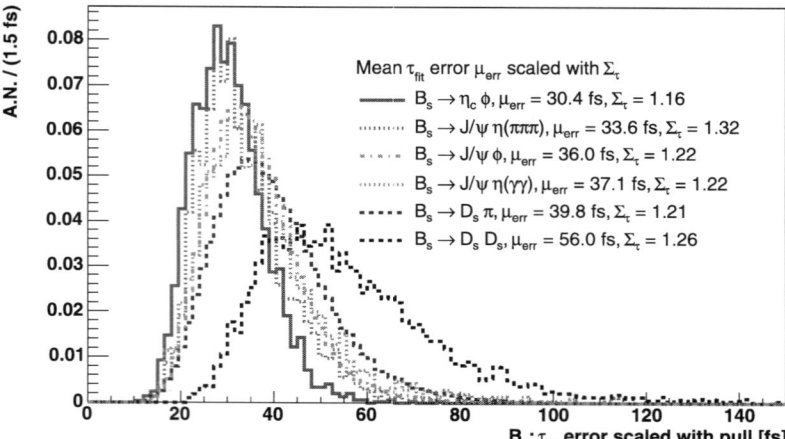

Fig. 36 Distribution of the event-by-event proper time resolution [fs] for different B_s channels, as obtained from the full MC simulation. The normalization is arbitrary

$1/\Gamma_s = 1.45$ ps, and a fraction of CP-odd component of $R_T = 0.2$ (for $B_s \to J/\psi\phi$). The different parameters are extracted by performing a likelihood fit to the mass, proper time, and transversity angle (for $B_s \to J/\psi\phi$) distributions, including a background contribution. The $\bar{b} \to \bar{c}c\bar{s}$ likelihood is simultaneously maximized with a similar likelihood for the $B_s \to D_s\pi$ control sample such as to constrain Δm_s and the mistag fraction from the data. The background properties are determined from the B_s mass sidebands. The physics parameters, extracted in the signal region with all other parameters fixed, are ϕ_s, Δm_s, $\Delta\Gamma_s/\Gamma_s$, $1/\Gamma_s$, ω_{tag}, and R_T (for $B_s \to J/\psi\phi$).

The sensitivities to ϕ_s for the different channels, obtained as the rms of the distribution of the fit results, are given in the last column of Table 37. They gently decrease with increasing $|\phi_s|$ and do not depend much on $\Delta\Gamma_s/\Gamma_s$. For instance, the statistical uncertainty on ϕ_s for $\phi_s = -0.2$ rad is ± 0.026 rad from $B_s \to J/\psi\phi$ alone, with 2 fb^{-1} [685]. The best performance is achieved with the $B_s \to J/\psi\phi$ sample, which also yields a statistical precision of ± 0.0092 on $\Delta\Gamma_s/\Gamma_s$ (2 fb^{-1}). The ϕ_s sensitivities obtained from the other modes (which are pure CP-eigenstates) are not as good but still interesting. Combining all modes, a statistical uncertainty $\sigma(\phi_s) = \pm 0.0092$ rad is expected after 10 fb^{-1}.

LHCb has the potential to perform the first significant measurement of ϕ_s, test the consistency with the SM expectations, and possibly uncover NP that may be hiding in B_s–$\bar{B}_s$ mixing.

3.6.7.3 Sensitivity to A^s_{SL} from $B_s \to D_s\mu\nu X$ and $B_s \to D_s\pi$ The CP-violating charge asymmetry A^s_{SL} is an important parameter to constrain NP contributions in B_s mixing, see Sect. 3.6.3. A^s_{SL} is accessible by measuring the charge asymmetry of the time-integrated rates of untagged B_s decays to flavor-specific final states such as $D_s^-\mu^+\nu X$ or $D_s^-\pi^+$ [689]. In LHCb, the asymmetry A^s_{SL} is measured by fitting the time-dependent decay rates. This method allows a determination of A^s_{SL} also for a nonzero production asymmetry of B_s and $\bar{B}_s$ mesons which, at the LHC, is expected to be of $\mathcal{O}(1\%)$. Based on a large sample of fully simulated inclusive $b\bar{b}$ events and a dedicated signal sample, LHCb estimates a signal yield of 1M $B_s \to D_s\mu\nu X$ events in 2 fb^{-1} of data, with a B/S ratio of about 0.36 [690]. This leads to a statistical precision of ± 0.002 on A^s_{SL} [691]. A similar analysis based on 140k $B_s \to D_s\pi$ events is expected to reach a precision of ± 0.005 with the same integrated luminosity of 2 fb^{-1} [691]. Systematic uncertainties are expected to be dominated by the detector charge asymmetry, which needs to be determined separately. A method is proposed to control the detector charge asymmetry by measuring the difference $A^s_{SL} - A^d_{SL}$ using B_s and B_d decays to the same final state, e.g. $B_s \to D_s^-\mu^+\nu X$ and $B_d \to D^-\mu^+\nu X$, where $D_s^- \to K^+K^-\pi^-$ and $D^- \to K^+K^-\pi^-$.

3.6.7.4 Correcting for trigger biases in lifetime fitting at LHCb Lifetime measurements at LHCb will help for the detector calibration and provide tests of theoretical predictions based on the heavy-quark expansion. In order to exploit the full range of decays available at LHCb, it is important to have a method for fitting lifetimes in hadronic channels, which are biased by the impact parameter cuts in the trigger. We have investigated a Monte-Carlo independent method to take into account the trigger effects. The method is based on calculating event-by-event acceptance functions from the decay geometry and does not require any external input. Current results with the method are given in [692]. The method for the case of two-body decays is described in [693].

The decay $B_d \to D^-\pi^+$ has an expected yield of 1.34M events per 2 fb^{-1}. The S/B ratio is expected to be around 5 [664]. Fitting the B_d lifetime with 60k toy Monte Carlo signal events achieves a statistical precision of 0.007 ps, while fitting to 60k signal and 15k background events achieves a precision of 0.009 ps (the current world average is 1.530 ± 0.009 ps [119]). A similar result is seen in data generated with the full LHCb detector simulation [692]. Therefore, although the systematic errors associated with this method are unknown at the moment, we can expect a very good measurement of the B_d lifetime using the decay $B_d \to D^- \pi^+$.

3.6.8 CMS

3.6.8.1 Sensitivity to $\Delta \Gamma_s$

Also at CMS the decay $B_s \to J/\psi \phi \to \mu^+ \mu^- K^+ K^-$ is being studied [694]. Several important background processes have been identified. The prompt J/ψ production is the main source of background at trigger level, since it represents a dominant contribution to the Level-1 dimuon trigger rate. For the offline selection, the main background is the inclusive decay $b \to J/\psi X$. The decay $B_d \to J/\psi K^{*0} \to \mu^+ \mu^- K^+ \pi^-$ is of particular concern, since the pion can be mistaken to be a kaon, and hence the decay be misidentified as $B_s \to J/\psi \phi$. Furthermore, the final state of this B_d decay also displays a time-dependent angular distribution similar to that of the B_s decay under study, with different physical parameters. The B_s decay chain is selected at Level-1 by the dimuon trigger. The latter demands two muons with a transverse momentum above 3 GeV/c, and the additional requirement that these muons have opposite charge can be used.

In the HLT [695], b candidates are identified by doing a partial reconstruction of the decay products in the tracker in restricted tracking regions and imposing invariant mass and vertex requirements [696].

The HLT selection of the decay $B_s \to J/\psi \phi$ has been separated in two steps. In the first, called Level 2, J/ψ candidates with a displaced vertex are identified. Tracks are then reconstructed in the tracking regions defined by the Level 1 muon candidates, and all track pairs of opposite charge for which the invariant mass is within 150 MeV/c^2 of the world-average J/ψ mass are retained. To remove the prompt J/ψ background, the two muon candidates are then fitted to a common decay vertex, and the significance of the transverse decay length is required to be above 3. With this selection, the accepted rate is reduced to approximately 15 Hz, with 80% of the J/ψ originating in the decay of b hadrons.

Next, at Level 3, a further reduction is achieved by doing a full reconstruction of the B_s decay. To reconstruct the kaons, the tracking region is chosen around the direction of each J/ψ candidate. Assigning the kaon mass to the reconstructed tracks, all oppositely charged track pairs for which the invariant mass is within 20 MeV/c^2 of the world-average mass of the ϕ meson are retained for a resolution in the invariant mass of the ϕ meson of 4.5 MeV/c^2. With the two muon candidates, the four-track invariant mass is required to be within 200 MeV/c^2 of the world-average mass of the B_s meson. The resolution in the invariant mass of the B_s meson is found to be 65 MeV/c^2. Here as well, a vertex fit of the four tracks is performed, imposing a similar requirement as above. The total rate for this selection is well below 0.1 Hz, and a yield of approximately 456000 signal events can be expected within 30 fb^{-1} of data.

In the offline selection, candidates are reconstructed by combining two muons of opposite charge with two further tracks of opposite charge. As CMS does not possess a particle identification system suitable for this measurement, all measured tracks have to be considered as possible kaon candidates, which adds a substantial combinatorial background. A kinematic fit is made, where the four tracks are constrained to come from a common vertex and the invariant mass of the two muons is constrained to be equal to the mass of the J/ψ. With this fit, a resolution on the invariant mass of the B_s meson of 14 MeV/c^2 is found. The invariant mass of the two kaons is required to be within 8 MeV/c^2 of the world-average mass of the ϕ meson.

With this selection, a yield of approximately 327000 signal events can be expected within 30 fb^{-1} of data, with a background of 39000 events. These do not include a requirement on the four-track invariant mass of the candidates, since the sidebands could be used later in the analysis. However, only a small fraction of these events are directly under the B_s peak, and even a simple cut will reduce the number of background events by a significant factor.

The measurement of the width difference $\Delta \Gamma_s$ can now be done on this sample of untagged B_s candidates. As mentioned earlier, the $J/\psi \phi$ final state is an admixture of CP-even and CP-odd states, and an angular analysis is required [674, 697]. As the CP-even and CP-odd components have different angular dependences and different time evolutions, the different parameters can be measured by performing an unbinned maximum likelihood fit on the observed time evolution of the angular distribution. In the absence of background and without distortion, the p.d.f. describing the data would be the original differential decay rate. The distortion of this distribution by the detector acceptance, trigger efficiency and the different selection criteria must be taken into account by an efficiency function modeling the effect of the decay length requirements and the distortion of the angular distribution.

A sample corresponding to an integrated luminosity of 1.3 fb^{-1} was considered, which allows us to have a realistic ratio of misidentified $B_d \to J/\psi K^*$ and signal events. With the low number of background events that remain after all selection requirements, an accurate modeling of the background is not possible, neither of its angular distribution nor of its time-dependent efficiency. Therefore the background events are simply added to the data set, and their expected distribution is not included in the p.d.f. used in the fit. The p.d.f. then simply describes the B_s distribution. With such a fit, in which the invariant mass of the candidates is not

Table 38 Results of the maximum likelihood fit for an integrated luminosity of 1.3 fb^{-1} (signal and background)

Parameter	Input value	Result	Stat. error	Sys. error	Total error	Rel. error
$|A_0(0)|^2$	0.57	0.5823	0.0061	0.0152	0.0163	2.8%
$|A_\|(0)|^2$	0.217	0.2130	0.0077	0.0063	0.0099	4.6%
$|A_\perp(0)|^2$	0.213	0.2047	0.0065	0.0099	0.0118	5.8%
$\bar{\Gamma}_s$	0.712 ps^{-1}	0.7060 ps^{-1}	0.0080 ps^{-1}	0.0227 ps^{-1}	0.0240 ps^{-1}	3.4%
$\Delta\Gamma_s$	0.142 ps^{-1}	0.1437 ps^{-1}	0.0255 ps^{-1}	0.0113 ps^{-1}	0.0279 ps^{-1}	19%
$\Delta\Gamma_s/\Gamma_s$	0.2	0.2036	0.0374	0.0173	0.0412	20%

taken into account, a restriction on the invariant mass of the candidates should obviously be made. Choosing a window of ± 36 MeV/c^2 around the world-average B_s mass reduces the number of B_d background events by another 59%, while reducing the number of signal candidates by only 2.9%. The result of the fit is given in Table 38, where both the statistical and expected systematic uncertainties are quoted. A first measurement of the width difference of the weak eigenstates could thus be made with an uncertainty of 20%. On a larger sample corresponding to an integrated luminosity of 10 fb^{-1}, it is foreseen that the statistical uncertainty would be reduced to 0.011.

3.6.8.2 Missing particles in the reconstruction The best way to study the B_s–$\bar{B}_s$ oscillations is to have a fully reconstructed final state of the B_s decay. The disadvantage of such decay channels is the limited statistics. Many more signal events can be collected in semileptonic decays as $B_s \to D_s^- \ell^+ \nu$. Due to the missing neutrino in this decay, the B_s momentum, and hence the proper-time resolution for the B_s, is less precise than in the fully reconstructed case, even if a correction (k-factor) is applied. However, recently a new method (ν-reco) has been proposed [698], which allows us to calculate the neutrino momentum with the help of vertex information.

In order to verify the ν-reco method, an MC simulation has been developed to study B_s–$\bar{B}_s$ mixing in the semileptonic decay mode. Kinematical cuts, track parameters and vertex positions (primary and secondary) have been simulated according to typical hadron collider detector conditions [670, 673, 699, 700]. The proper time resolution obtained is $\sigma = 132$ fs with the k-factor method and $\sigma = 91$ fs with the ν-reco method.

3.7 Hadronic $b \to s$ and $b \to d$ transition[23]

FCNC processes can occur only at the loop level in the SM and therefore are potentially sensitive to new virtual particles. In particular, hadronic FCNC B decays are sensitive to NP contributions to penguin operators. Among these decays, the penguin-dominated $b \to s\bar{q}q$ transitions are the most promising [717–721]. However, an accurate evaluation of the SM amplitudes is required in order to disentangle NP contributions. Unfortunately hadronic uncertainties hinder a pristine calculation of the decay amplitudes. In this chapter, various theoretical approaches to the calculation of the hadronic uncertainties are discussed. In addition, the present experimental status is presented together with prospects at B-factories and LHCb.

3.7.1 Theoretical estimates of ΔS with factorization

In the following, we quantify $\Delta S_f \equiv -\eta_f S_f - \sin(2\beta)$, where S_f is the sin-term of the time-dependent CP asymmetry, based on QCD factorization [215, 216] calculations of the $B \to f$ decay amplitudes. We may write the decay amplitude as

$$A(\bar{B} \to f) = V_{cb}V_{cs}^* a_f^c + V_{ub}V_{us}^* a_f^u \propto 1 + e^{-i\gamma} d_f, \quad (157)$$

where $d_f = \epsilon_{\text{KM}} a_f^u/a_f^c \equiv \epsilon_{\text{KM}} \hat{d}_f$ and $\epsilon_{\text{KM}} = |V_{ub}V_{us}^*/(V_{cb}V_{cs}^*)| \sim 0.025$. The expectation that ΔS_f is small derives from the CKM suppression ϵ_{KM} and the expectation that the ratio of hadronic amplitudes, $\hat{d}_f$, is not much larger than 1. Then

$$\Delta S_f = 2\epsilon_{\text{KM}} \text{Re}(\hat{d}_f) \cos(2\beta) \sin\gamma + O(d_f^2). \quad (158)$$

QCD factorization calculations of ΔS_f for various final states have been performed at leading order [722] and next-to-leading order [240, 723, 724]. Other factorization-inspired calculations can be found in [241, 725]. The results are generally in good agreement with one another. The following is primarily an update of [723]. Ref. [724] also discusses an estimate of long-distance rescattering effects. Since the significance of the model underlying this estimate is unclear, these (small) effects will not be included here.

The hadronic amplitudes a_f^p are sums of "topological" amplitudes, referring to colour-allowed tree (T), colour-suppressed tree (C), QCD penguin (P^p), singlet penguin (S^p), electroweak penguin (P_{EW}^p, $P_{\text{EW},C}^p$) and annihilation contributions. The numerical analysis below takes into account all flavor amplitudes following [240], but it suffices to focus on a few dominant terms to understand the qualitative

---
[23]Section coordinators: M. Ciuchini, F. Muheim.

features of the result. Then, for the various final states, the relevant hadronic amplitude ratio is given by

$$\pi^0 K_S: \quad \hat{d}_f \sim \frac{[-P^u] + [C]}{[-P^c]},$$

$$\rho^0 K_S: \quad \hat{d}_f \sim \frac{[P^u] - [C]}{[P^c]},$$

$$\eta' K_S: \quad \hat{d}_f \sim \frac{[-P^u] - [C]}{[-P^c]},$$

$$\phi K_S: \quad \hat{d}_f \sim \frac{[-P^u]}{[-P^c]}, \quad (159)$$

$$\eta K_S: \quad \hat{d}_f \sim \frac{[P^u] + [C]}{[P^c]},$$

$$\omega K_S: \quad \hat{d}_f \sim \frac{[P^u] + [C]}{[P^c]}.$$

The convention here is that the quantities in square brackets have positive real parts. (Recall from (158) that ΔS_f mainly requires the real part of $\hat{d}_f$.) In factorization, $\text{Re}[P^u/P^c]$ is near unity, roughly independent of the particular final state, hence ΔS_f receives a nearly universal, small and *positive* contribution of about $2\epsilon_{\text{KM}} \cos(2\beta) \sin\gamma \approx 0.03$. On the contrary, the magnitudes and signs of the penguin amplitudes' real parts can be very different. Hence the influence of the colour-suppressed tree amplitude C determines the difference in ΔS_f between the different modes. For $(\pi^0, \eta, \omega)K_S$, the effect of C is constructive, but for $(\rho, \eta')K_S$, it is destructive. However, the magnitude of $\text{Re}[P_c]$ is much larger for $\eta' K_S$ than for ρK_S, hence $\text{Re}(\hat{d}_f)$ remains small and positive for $\eta' K_S$ but becomes negative for ρK_S.

The result of the calculation of ΔS_f is shown in Table 39. The columns labeled "ΔS_f (Theory)" use the input parameters (CKM parameters, strong coupling, quark masses, form factors, decay constants, moments of light-cone distribution amplitudes) summarized in Table 1 of [240]. The uncertainty estimate is computed by adding in quadrature the individual parameter uncertainties. The result displays the anticipated pattern. The variation of the central value from the nearly universal contribution of approximately ϵ_{KM} is due to $\text{Re}[C/P^c]$, and the error comes primarily from this quantity. It is therefore dominated by the uncertainty in the hard-spectator scattering contribution to C and the penguin annihilation contribution to P^c. In general one expects the prediction of the asymmetry S_f in factorization to be more accurate than the prediction of the direct CP asymmetry C_f, since S_f is determined by $\text{Re}(a_f^u/a_f^c)$ which is large and calculated at next-to-leading order. The resultant error on ΔS_f is roughly of the size of ΔS_f itself. Quadratic addition of theoretical errors may not always lead to a conservative error estimate. Therefore we also perform a random scan of the allowed theory parameter space, taking the minimal and maximal value of an observable attained in this scan to define its predicted range. In doing so, we discard all theoretical parameter sets which give CP-averaged branching fractions not compatible within 3 sigma with the experimental data, that is we require $8.1 < 10^6 \, Br(\pi^0 K^0) < 11.8$, $2.5 < 10^6 \, Br(\rho^0 K^0) < 8.2$, $5.3 < 10^6 \, Br(\phi K^0) < 11.9$, $2.9 < 10^6 \, Br(\omega K^0) < 7.5$, $0.2 < 10^6 \, Br(\eta K^0) < 2.4$. Note that we do not require the theoretical parameters to reproduce the $\eta' K^0$ branching fraction for reasons explained in [723]. The resulting ranges for ΔS_f from a scan of 200000 theoretical parameter sets are shown in the columns labeled "ΔS_f [Range]" in Table 39. It is seen that the ranges are not much different from those obtained by adding parameter uncertainties in quadrature—except for the ηK_S final state. For ηK_S, large negative values of ΔS_f originate from small regions of the parameter space, where by cancellations the leading penguin amplitude P_c becomes very small. This leads to large amplifications of C/P^c and hence of ΔS_f. Except for the case of ηK_S, these parameter space regions are excluded by the lower limits on the branching fractions.

Factorization-based calculations of two-body final states with scalar mesons and three-body final states are on a less solid footing than the final states discussed above. The following estimates have been obtained for the three-kaon modes [726]:

$$\Delta S_{K^+K^-K_S} = 0.06^{+0.08}_{-0.02}, \qquad \Delta S_{K_S K_S K_S} = 0.06^{+0.00}_{-0.00}. \quad (160)$$

The quoted error should be regarded with due caution.

Table 39 Comparison of theoretical and experimental results for ΔS_f

Mode	ΔS_f (Theory)	ΔS_f [Range]	Mode	ΔS_f (Theory)	ΔS_f [Range]
$\pi^0 K_S$	$0.07^{+0.05}_{-0.04}$	$[+0.03, 0.13]$	$\rho^0 K_S$	$-0.08^{+0.08}_{-0.12}$	$[-0.29, 0.01]$
$\eta' K_S$	$0.01^{+0.01}_{-0.01}$	$[+0.00, 0.03]$	ϕK_S	$0.02^{+0.01}_{-0.01}$	$[+0.01, 0.05]$
ηK_S	$0.10^{+0.11}_{-0.07}$	$[-0.76, 0.27]$	ωK_S	$0.13^{+0.08}_{-0.08}$	$[+0.02, 0.21]$

In conclusion, QCD calculations of the time-dependent CP asymmetry in hadronic $b \to s$ transitions yield only small corrections to the expectation $-\eta_f S_f \approx \sin(2\beta)$. With the exception of the $\rho^0 K_S$ final state, the correction ΔS_f is positive. The effect and theoretical uncertainty is particularly small for the two final states ϕK_S and $\eta' K_S$ [240]. The final-state dependence of ΔS_f is ascribed to the colour-suppressed tree amplitude. It appears difficult to constrain ΔS_f theory-independently by other observables. In particular, the direct CP asymmetries or the charged decays corresponding to $f = M K_S$ probe hadronic quantities other than those relevant to ΔS_f if these observables take values in the expected range. Here M stands for a charged light meson. Large deviations from expectations such as large direct CP asymmetries would clearly indicate a defect in our understanding of hadronic physics, but even then the quantitative implications for S_f would be unclear. A hadronic interpretation of large ΔS_f would probably involve an unknown long-distance effect that discriminates strongly between the up- and charm-penguin amplitude resulting in an enhancement of the up-penguin amplitude. No model is known that could plausibly produce such an effect.

3.7.2 Theoretical estimates of ΔS from three-body decays

While a possibility of constraining the CKM weak phase from three-body $\Delta S = 1$ B decays has been raised a long time ago [727], a discussion of three-body final states as probes of CKM phase has gained more momentum only recently with the experimental advances. The present experimental situation that includes measurements of time-dependent CP asymmetries in $B^0 \to K_S K_S K_S$, $B^0 \to \pi^0 \pi^0 K_S$ and $B^0 \to K^+ K^- K_{S,L}$ is summarized in Table 40. The quoted CP asymmetries are phase space (dps) integrated quantities with

$$S_f^{\text{3-body}} \equiv (1 - 2f_+) \sin 2\beta^{\text{eff}}$$
$$= \frac{2 \operatorname{Im} \int dps (e^{-2i\beta} A_f \bar{A}_f^*)}{\int dps |A_f|^2 + \int dps |\bar{A}_f|^2}. \quad (161)$$

Here f_+ is the CP-even component fraction, while A_f and $\bar{A}_f$ denote the $A(B^0 \to f)$ and $A(\bar{B}^0 \to f)$ amplitudes respectively. While $B^0 \to K_S K_S K_S$ and $B^0 \to \pi^0 \pi^0 K_S$ are decays into completely CP even final states [728], the decay $B^0 \to K^+ K^- K_S$ has both components but is still mostly CP-even with $f_+ \sim 0.9$. This is obtained either from isospin analysis from $B^+ \to K_S K_S K_S$ decay assuming penguin dominance [729–733] or directly from angular analysis [734], in agreement with each other.

A $\Delta S = 1$ B decay amplitude can be in general decomposed in terms of "tree" ($\sim V_{ub}^* V_{us}$) and "penguin" ($\sim V_{cb}^* V_{cs}$) contributions as shown in (157) for the case of two-body $\bar{B}$ decays. An expression analogous to (158) holds for ΔS_f, here given by

$$\Delta S_f = \sin 2\beta^{\text{eff}} - \sin 2\beta = 2 \cos 2\beta \sin \gamma \operatorname{Re}(\xi_f), \quad (162)$$

where $\sin 2\beta^{\text{eff}}$ is defined in (161), and the ratio

$$\xi_f \equiv \frac{V_{ub}^* V_{us}}{V_{cb}^* V_{cs}} \frac{\int dps\, T_f^* P_f}{\int dps\, P_f^* P_f}, \quad (163)$$

suitably averaged over the final phase space, replaces the ratio d_f defined in the previous section for two-body decays. In addition, the direct CP asymmetries are given by

$$C_f = -2 \sin \gamma \operatorname{Im}(\xi_f). \quad (164)$$

The difference ΔS_f was analysed using $SU(3)$ flavor symmetries [729, 736, 737] and was calculated in a model-dependent way in Ref. [726]. The approach is based on flavor $SU(3)$ and exploits the fact that the related $\Delta S = 0$ final states, f', are more sensitive to the "tree" amplitudes which are CKM enhanced when compared to the $\Delta S = 1$ amplitudes (because $V_{us} < V_{ud}$). However, "penguin" amplitudes are CKM suppressed (because $V_{cs} \to V_{cd}$). This then leads to a bound on ξ_f of the form

$$\xi_f < \lambda \sum_{f'} a_{f'} \sqrt{\frac{Br(f')}{Br(f)}}, \quad (165)$$

where $\lambda = 0.22$, $a_{f'}$ are the coefficients arising from $SU(3)$ Clebsch–Gordan coefficients, and the sum is over $\Delta S = 0$ final states f'. The bounds are better if less modes enter the sum, which can be achieved through a dynamical assumption of small annihilation-like amplitudes. This then gives

$$\xi_{K^+ K^- K^0} < 1.02 \quad [736], \qquad \xi_{K_S K_S K_S} < 0.31 \quad [737], \quad (166)$$

with bounds for a number of other modes listed in [736]. These are only very conservative upper bounds not at all

Table 40 Measured CP asymmetries in $B^0 \to 3P$ decays [502]

Mode	$\sin(2\beta^{\text{eff}})$	C_f
$K_S K_S K_S$ [609, 735]	0.51 ± 0.21	-0.23 ± 0.15
$\pi^0 \pi^0 K_S$ [829]	$-0.84 \pm 0.71 \pm 0.08$	$0.27 \pm 0.52 \pm 0.13$
$K^+ K^- K_{S,L}$ [732, 734]	$0.58 \pm 0.13^{+0.12}_{-0.09}$	0.15 ± 0.09

indicative of the expected size $\xi_f \sim \lambda^2 T_f/P_f$. One also expects $\xi_{K^+K^-K^0} < \xi_{K_S K_S K_S}$, since in the latter case, all the tree operator contributions are OZI suppressed as the final state does not contain valence u-quarks. This expectation was confirmed by a model-dependent calculation that combined QCD factorization with heavy-meson chiral perturbation theory [726]. This approach is valid only in a region of phase space where one of the light mesons is slow and the other two are very energetic, while for the remaining phase space, a model for the form factors was used. Ref. [726] then obtains

$$\Delta S_{K_S K_S K_S} = 0.02, \qquad \Delta S_{K^+K^-K_S} \lesssim O(0.1). \quad (167)$$

An argument exists that the latter could be smaller [738], but one should also keep in mind the comment at the end of the previous section.

A different use of three-body final states is provided by the time-dependent Dalitz plot analysis with a fit to quasi-two body resonant modes. Interferences between resonances then fix relative strong phases giving additional experimental information. In this way, BaBar was able to resolve the $\beta \to \pi/2 - \beta$ discrete ambiguity using a $B^0 \to K^+K^-K_{S,L}$ Dalitz plot analysis [739]. The interference of CP-even and CP-odd contributions leads to a $\cos 2\beta^{\text{eff}}$ term (with $\beta^{\text{eff}} \to \beta$ in the limit of no tree pollution). Another example is measuring phases of $\Delta I = 1$ amplitudes of $B \to (K^*\pi)_{I=1/2,3/2}$, $B_s \to (K^*\bar K)_{I=1}$ and $B_s \to (\bar K^*K)_{I=1}$ from resonance interferences in $B \to K\pi\pi$ and $B_s \to K\bar K\pi$. This then gives information on CKM parameters complementary to other methods [740–742]. Using $SU(3)$ hadronic uncertainties due to electroweak penguin operators O_9 and O_{10} were shown to be very small in $B \to K\pi\pi$ and $B_s \to K\pi\pi$ and somewhat larger in $B_s \to K\bar K\pi$ [742]. The first processes imply a precise linear relation between $\bar\rho$ and $\bar\eta$, with a measurable slope and an intercept at $\bar\eta = 0$ involving a theoretical error of 0.03. The decays $B_s \to K\pi\pi$ permit a measurement of γ involving a theoretical error below a degree. Furthermore, while time-dependence is required when studying B^0 decays at the $\Upsilon(4S)$, it may not be needed when studying B_s decays at hadronic colliders.

3.7.3 Flavor symmetries and estimates of $b \to s$ transitions

Decomposing the $B \to MM$ amplitudes in terms of flavor $SU(3)$ or isospin reduced matrix elements leads to relations between different amplitudes, since the effective weak Hamiltonian usually transform only under a subset of all possible representations [743]. The group-theoretical approach based on reduced matrix elements [245, 744, 745] is equivalent to a diagrammatic approach of topological amplitudes [746–750]. In the latter, it is easier to introduce dynamical assumptions such as neglecting annihilation-like amplitudes. These were shown to be $1/m_b$ suppressed for decays into nonisosinglets [751], while not all of them are $1/m_b$ suppressed if η, η' occur in the final state (see Appendix C of [241]).

The $SU(3)$ approach has been used in global fits to the experimentally measured $B \to PP$ and $B \to PV$ decays [752–761] in which both the values of hadronic parameters as well as the value of weak phase γ are determined. However, in order to obtain a stable fit, a number of dynamical assumptions are needed. In the most recent fit to $B \to PP$ [756], t-quark dominance in penguin amplitudes and negligible annihilation-like topologies (also for isosinglets) were assumed. Both β and γ were determined with central values slightly above the CKMfitter and UTfit determinations. Allowing for a new weak phase in P_{EW} for $\Delta S = 1$ modes leads to statistically significant reduction of χ^2, while choosing this phase to be zero does give the size of $|P_{\text{EW}}|$ in excellent agreement with the Neubert–Rosner relation [762–765]. A large strong phase difference $\arg(C/T) \sim -60°$ was found, while expected to be $1/m_b$ suppressed from QCD factorization and SCET [221, 240, 766]. As stressed in Ref. [767], the direct CP asymmetries $A_{\text{CP}}(B^0 \to K^+\pi^-)$ and $A_{\text{CP}}(B^+ \to K^+\pi^0)$ would be of the same sign for $\arg(C/T)$ small, which is excluded at 4.7σ at present.

Assumption of negligible annihilation topologies used in $SU(3)$ fits can be tested by comparing $B^0 \to K^0\bar K^0$, $B^+ \to K^+\bar K^0$, where annihilation is CKM enhanced, with $B^+ \to K^0\pi^+$ [768, 769]. $SU(3)$ breaking has been addressed in [756, 770] showing a small effect on the values of extracted parameters. Further tests of $SU(3)$ breaking or searches of NP will be possible using B_s decays [771–776], with the first CDF measurement of $Br(B_s \to K^+K^-)$ leading the way [777]. Errors due to the dynamical assumptions can be reduced if fits are made to only a subset of modes, e.g. to $\pi\pi, \pi K$ [756, 760, 770, 778–781]. Furthermore, dynamical assumptions can be avoided entirely if only a set of modes related through U-spin is used [771, 782, 783]. This leads to stable fits, while giving γ with a theoretical error of a few degrees. Further studies of $SU(3)$ breaking effects are called for, though.

Because of the different CKM hierarchy of tree and penguin amplitudes in $\Delta S = 1$ and $\Delta S = 0$ decays, tree pollution in $\Delta S = 1$ decays can be bounded using $SU(3)$ related $\Delta S = 0$ modes [729]. Correlated bounds on ΔS_f and C_f for $\eta' K_S$ and $\pi^0 K_S$ final states have been presented in [784–787]. Such a model-independent bound on $\Delta S_{\phi K_S}$ is not available at present, since many more $\Delta S = 0$ modes enter, some of which have not been measured yet [788].

Very precise relations between $\Delta S = 1$ $B \to \pi K$ CP asymmetries or decay rates can be obtained using isospin decompositions. The sum rule between decay widths $\Gamma(K^0\pi^+) + \Gamma(K^+\pi^-) = 2\Gamma(K^+\pi^0) + 2\Gamma(K^0\pi^0)$ [789,

790] (equivalent to $R_n = R_c$ [791]) is violated by CKM doubly suppressed terms calculable in $1/m_b$ expansion [221, 240, 241, 766], while harder to calculate isospin-breaking corrections cancel to first order [792]. The sum rule $\Delta(K^+\pi^-) + \Delta(K^0\pi^+) - 2\Delta(K^+\pi^0) - 2\Delta(K^0\pi^0) = 0$ for the rate differences $\Delta(f) = \Gamma(\bar{B} \to \bar{f}) - \Gamma(B \to f)$ is valid in the isospin limit and is thus violated by EWP. However, these corrections vanish in the $SU(3)$, $m_b \to \infty$ limit making the sum rule very precise [793].

3.7.4 Applications of U-spin symmetry to B_d and B_s decays

The current data in B physics suggests that B_d decays agree well with SM predictions, while B_s decays remain poorly known and might be affected by NP. Within the SM, the CKM mechanism correlates the electroweak part of these transitions, but quantitative predictions are difficult due to hadronic effects. The latter can be estimated relying on the approximate $SU(3)$-flavor symmetry of QCD: information on hadronic effects, extracted from data in one channel, can be exploited in other channels related by flavor symmetry, leading to more accurate predictions within the SM.

In addition to isospin symmetry, an interesting theoretical tool is provided by U-spin symmetry, which relates d- and s-quarks [771]. Indeed, this symmetry holds for long- and short-distances and does not suffer from electroweak corrections, making it a valuable instrument to analyse processes with significant penguins and thus a potential sensitivity to NP. However, due to the significant difference $m_s - m_d$, U-spin breaking corrections of order 30% may occur, depending on the processes.

As a first application of U-spin, relations were obtained between $B_d \to \pi^+\pi^-$ and $B_s \to K^+K^-$. This led to correlations among the observables in the two decays such as branching ratios and CP asymmetries [771–774] and to a prediction for $BR(B_s \to K^+K^-) = (35^{+73}_{-20}) \times 10^{-6}$ [780]. These results helped to investigate the potential of such decays to discover NP [775, 794]. Unfortunately, the accuracy of the method is limited not only by the persistent discrepancy between BaBar and Belle on $B_d \to \pi^+\pi^-$ CP asymmetries but also by poorly known U-spin corrections. In these analyses, the ratio of tree contributions $R_c = |T^s_{K\pm}/T^d_{\pi\pm}|$ was taken from QCD sum rules as 1.76 ± 0.17 [250] (updated to $1.52^{+0.18}_{-0.14}$ [272]). In addition, the ratio of penguin-to-tree ratios $\xi = |(P^s_{K\pm}/T^s_{K\pm})/(P^d_{\pi\pm}/T^d_{\pi\pm})|$ was assumed equal to 1 [780] or 1 ± 0.2 [775, 794] in agreement with rough estimates within QCD factorization (QCDF) [795].

Indeed QCDF may complement flavor symmetries by a more accurate study of short-distance effects. However, QCDF cannot predict some significant $1/m_B$-suppressed long-distance effects, which have to be estimated through models. Recently, it was proposed to combine QCDF and U-spin in the decays mediated by penguin operators $B_d \to K^0\bar{K}^0$ and $B_s \to K^0\bar{K}^0$ [796] and in their vector–vector analogues [797].

First, tree (T^{d0}) and penguin (P^{d0}) contributions to $B_d \to K^0\bar{K}^0$ can be determined by combining the currently available data with $|T^{d0} - P^{d0}|$, which can be accurately computed in QCDF because long-distance effects, seen as infrared divergences, cancel in this difference. U-spin suggests accurate relations between these hadronic parameters in $B_d \to K^0\bar{K}^0$ and those in $B_s \to K^0\bar{K}^0$. Actually, we expect similar long-distance effects since the $K^0\bar{K}^0$ final state is invariant under the d–s exchange. Short distances are also related since the two processes are mediated by penguin operators through diagrams with the same topologies. U-spin breaking arises only in a few places: factorizable corrections encoded in $f = [M^2_{B_s} F^{B_s \to K}(0)]/[M^2_{B_d} F^{B_d \to K}(0)]$ and non-factorizable corrections from weak annihilation and spectator scattering. Because of these expected tight relations, QCDF can be relied upon to assess U-spin breaking between the two decays. Indeed, up to the factorizable factor f, penguin (as well as tree) contributions to both decays are numerically very close. Penguins in $B_d \to K^0\bar{K}^0$ and $B_s \to K^+K^-$ should have very close values as well, whereas no such relation exists for the (CKM-suppressed) tree contribution to the latter, to be estimated in QCDF.

These relations among hadronic parameters, inspired by U-spin considerations and quantified within QCD factorization, can be exploited to determine the tree and penguin contributions to $B_s \to KK$ decays and the corresponding observables. In particular, one gets $BR(B_s \to K^0\bar{K}^0) = (18 \pm 7 \pm 4 \pm 2) \times 10^{-6}$ and $BR(B_s \to K^+\bar{K}^-) = (20 \pm 8 \pm 4 \pm 2) \times 10^{-6}$, in very good agreement with the latest CDF measurement. The same method provides significantly improved determinations of the U-spin breaking ratios $\xi = 0.83 \pm 0.36$ and $R_c = 2.2 \pm 0.7$. These results have been exploited to determine the impact of supersymmetric models on these decays [798].

New results on $B \to K$ form factors and on the $B_d \to K^0\bar{K}^0$ branching ratio and direct CP-asymmetry should lead to a significant improvement of the predictions in the B_s sector. The potential of other pairs of nonleptonic B_d and B_s decays remains to be investigated.

3.7.5 Applications of the RGI parametrization to $b \to s$ transitions

Few general parameterizations of the $\Delta B = 1$ hadronic amplitudes exist in the literature. Here we use the parametrization proposed in [799] which decomposes decay amplitudes in terms of Renormalization-Group-Invariant (RGI) parameters. For our purpose, we just need to recall a few basic facts about the classification of RGI's. First of all, we have six nonpenguin parameters, containing only non-penguin contractions of the current–current operators $Q_{1,2}$: emission parameters $E_{1,2}$, annihilation parameters $A_{1,2}$ and Zweig-suppressed emission–annihilation parameters $EA_{1,2}$. Then, we have four parameters containing only penguin contractions of the current–current operators $Q_{1,2}$ in the GIM-suppressed combination $Q_{1,2}^c - Q_{1,2}^u$: P_1^{GIM} and Zweig suppressed P_{2-4}^{GIM}. Finally, we have four parameters containing penguin contractions of current–current operators $Q_{1,2}^c$ (the so-called charming penguins [800]) and all possible contractions of penguin operators Q_{3-12}: $P_{1,2}$ and the Zweig-suppressed $P_{3,4}$. In the following, Zweig-suppressed parameters are neglected. We refer the reader to the original reference for details. We can then write schematically the $b \to s$ decay amplitude as

$$\mathcal{A}(B \to F) = -V_{ub}^* V_{us} \sum (T_i + P_i^{GIM}) - V_{tb}^* V_{ts} \sum P_i, \tag{168}$$

where $T_i = \{E_i, A_i, EA_i\}$ are not present in pure-penguin decays.

The idea developed in [801] is to write down the RGI parameters as the sum of their expression in the infinite mass limit, for example using QCD factorization, plus an arbitrary contribution corresponding to subleading terms in the power expansion. These additional contributions are then determined by a fit to the experimental data. In $b \to s$ penguins, the dominant power-suppressed correction is given by charming penguins, and the corresponding parameter can be determined with high precision from data and is found to be compatible with a Λ/m_b correction to factorization [801]. However, nondominant corrections, for example GIM penguin parameters in $b \to s$ decays, can be extracted from data only in a few cases (for example in $B \to K\pi$ decays). Yet predictions for ΔS_f depend crucially on these corrections, so that one needs external input to constrain them. One interesting avenue is to extract the support of GIM penguins from $SU(3)$-related channels ($b \to d$ penguins), in which they are not Cabibbo-suppressed, and to use this support, including a possible large $SU(3)$ breaking of 100%, in the fit of $b \to s$ penguin decays. Alternatively, one can omit the calculation in factorization and fit directly the RGI parameters from the experimental data, instead of fitting the power-suppressed corrections [604, 802].

Compared to factorization approaches, general parameterizations have less predictive power but are more general. In particular, they tend to overestimate the theoretical uncertainty and are thus best suited to search for NP in a conservative way. In addition, these methods have the advantage that for several channels, the predicted ΔS decreases with the experimental uncertainty in BR's and CP asymmetries of $b \to s$ and $SU(3)$-related $b \to d$ penguins.

In the analysis reported here [88, 803], we vary the absolute values of the subdominant amplitudes in the range $[0, UL]$ (while the phases are unconstrained) and study the dependence of the predictions on the upper limit UL. For example, we show in Fig. 37 the effect of changing the up-

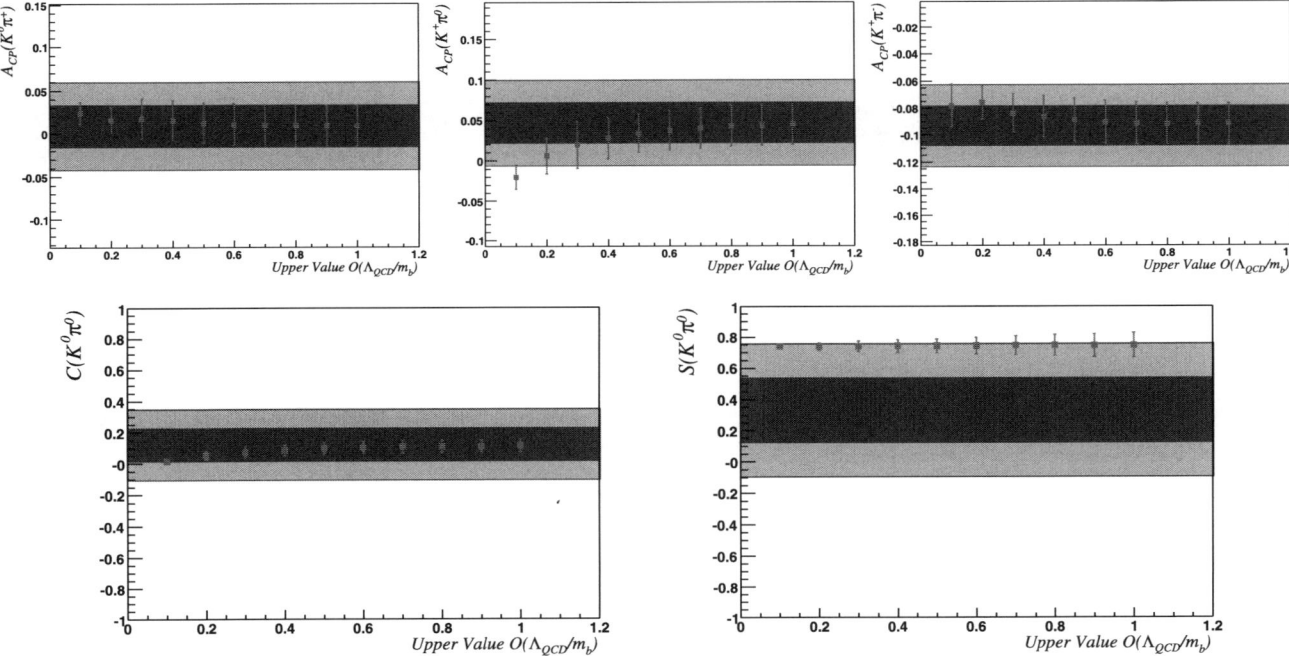

Fig. 37 CP asymmetries for $B \to K\pi$ decays, obtained varying subdominant contributions in the range [0, UV], with the upper value UV scanned between zero and one (in units of E_1). For comparison, the experimental 68% (95%) probability range is given by the *dark* (*light*) band

per limit of the range in which subdominant terms are varied on the prediction of some observables in $B \to K\pi$ decays. It can be seen that reasonable subdominant terms make any $K\pi$ puzzle disappear. Furthermore, the prediction of $S_{\pi^0 K_S}$ has small theoretical error and is quite stable against the effect of subdominant terms.

In Table 41, we collect predictions for ΔS_f obtained using the method sketched above for $UL = 0.5$ (in units of the leading amplitude), as suggested by the $SU(3)$-related modes $B \to KK$. Notice that the theoretical uncertainty is smaller for $B \to \pi^0 K_S$, because the number of observables in the $B \to K\pi$ system is sufficient to constrain efficiently the hadronic parameters. This means that the theoretical error can be kept under control by improving the experimental data in these channels. On the other hand, the information on $B \to \phi K_S$ is not sufficient to bound the subleading terms, and this results in a relatively large theoretical uncertainty that cannot be decreased without additional input on hadronic parameters. Furthermore, using $SU(3)$ to constrain $\Delta S_{\phi K_S}$ is difficult, because the number of amplitudes involved is very large [245, 736, 737, 788].

The ideal situation would be represented by a pure penguin decay for which the information on P_i^{GIM} is available with minimal theoretical input. Such a situation is realized by the pure penguin decays $B_s \to K^{0(*)} \bar{K}^{0(*)}$. An upper bound for the P_i^{GIM} entering this amplitude can be obtained from the $SU(3)$-related channels $B_d \to K^{0(*)} \bar{K}^{0(*)}$. Then, even adding a generous 100% $SU(3)$ breaking and an arbitrary strong phase, it is possible to have full control over the theoretical error in ΔS [802].

3.7.6 $b \to s$ transitions in the MSSM

In this section, we discuss phenomenological effects of the new sources of flavor and CP violation in $b \to s$ processes that arise in the squark sector [104, 108, 109, 804–823] of the MSSM. In general, in the MSSM, squark masses are neither flavor-universal nor aligned to quark masses, so that they are not flavor diagonal in the super-CKM basis, in which quark masses are diagonal and all neutral current vertices are flavor diagonal. The ratios of off-diagonal squark mass terms to the average squark mass define four new sources of flavor violation in the $b \to s$ sector: the mass insertions $(\delta_{23}^d)_{AB}$ with $A, B = L, R$ referring to the helicity of the corresponding quarks. These δ's are in general complex, so that they also violate CP. One can think of them as additional CKM-type mixings arising from the SUSY sector. Assuming that the dominant SUSY contribution comes

Table 41 Predictions for ΔS_f using the RGI parametrization

$\Delta S_{\pi^0 K_S}$	$(2.4 \pm 5.9) \times 10^{-2}$	$\Delta S_{\eta' K_S}$	$(-0.7 \pm 5.4) \times 10^{-2}$
$\Delta S_{\phi K_S}$	$(0.4 \pm 9.2) \times 10^{-2}$	$\Delta S_{\rho^0 K_S}$	$(-6.2 \pm 8.4) \times 10^{-2}$
$\Delta S_{\omega K_S}$	$(5.6 \pm 10.7) \times 10^{-2}$		

from the strong interaction sector, i.e. from gluino exchange, all FCNC processes can be computed in terms of the SM parameters plus the four δ's plus the relevant SUSY parameters: the gluino mass $m_{\tilde{g}}$, the average squark mass $m_{\tilde{q}}$, $\tan\beta$ and the μ parameter. The impact of additional SUSY contributions such as chargino exchange has been discussed in detail in Ref. [816]. We consider only the case of small or moderate $\tan\beta$, since for large $\tan\beta$, the constraints from $B_s \to \mu^+ \mu^-$ and Δm_s preclude the possibility of having large effects in $b \to s$ hadronic penguin decays [28, 29, 32, 34, 114, 115, 813].

Barring accidental cancellations, one can consider one single δ parameter, fix the SUSY masses and study the phenomenology. The constraints on δ's come at present from $B \to X_s \gamma$, $B \to X_s l^+ l^-$ and from the $B_s - \bar{B}_s$ mixing amplitude. We refer the reader to Refs. [88, 107, 116, 824] for all the details and results of this analysis.

Fixing as an example $m_{\tilde{g}} = m_{\tilde{q}} = |\mu| = 350$ GeV and $\tan\beta = 3$, one obtains the following constraints on δ's:

$$|(\delta_{23}^d)_{LL}| < 2 \times 10^{-1}, \qquad |(\delta_{23}^d)_{RR}| < 7 \times 10^{-1}, \qquad (169)$$
$$|(\delta_{23}^d)_{RL,LR}| < 5 \times 10^{-3}.$$

Notice that all constraints scale approximately linearly with the squark and gluino masses.

Having the present experimental bounds on the δ's, we can turn to the evaluation of the time-dependent CP asymmetries. The uncertainty in the calculation of SUSY effects is larger than the SM one. Following [107], we use QCDF enlarging the range for power-suppressed contributions to annihilation chosen in [240] as suggested in [825]. We warn the reader about the large theoretical uncertainties that affect this evaluation.

In Fig. 38, we present the results for $S_{\phi K_s}$, $S_{\pi^0 K_s}$, $S_{\eta' K_s}$ and $S_{\omega K_s}$. They do not show a sizable dependence on the sign of μ or on $\tan\beta$ for the chosen range of SUSY parameters. We see that:

– deviations from the SM expectations are possible in all channels, and the present experimental central values can be reproduced;
– deviations are more easily generated by LR and RL insertions, due to the enhancement mechanism discussed above;
– as noticed in [826, 827], the correlation between S_{PP} and S_{PV} depends on the chirality of the NP contributions. For example, we show in Fig. 39 the correlation between $S_{K_S\phi}$ and $S_{K_S\pi^0}$ for the four possible choices for mass insertions. We see that the $S_{K_S\phi}$ and $S_{K_S\pi^0}$ are correlated for LL and LR mass insertions and anticorrelated for RL and RR mass insertions.

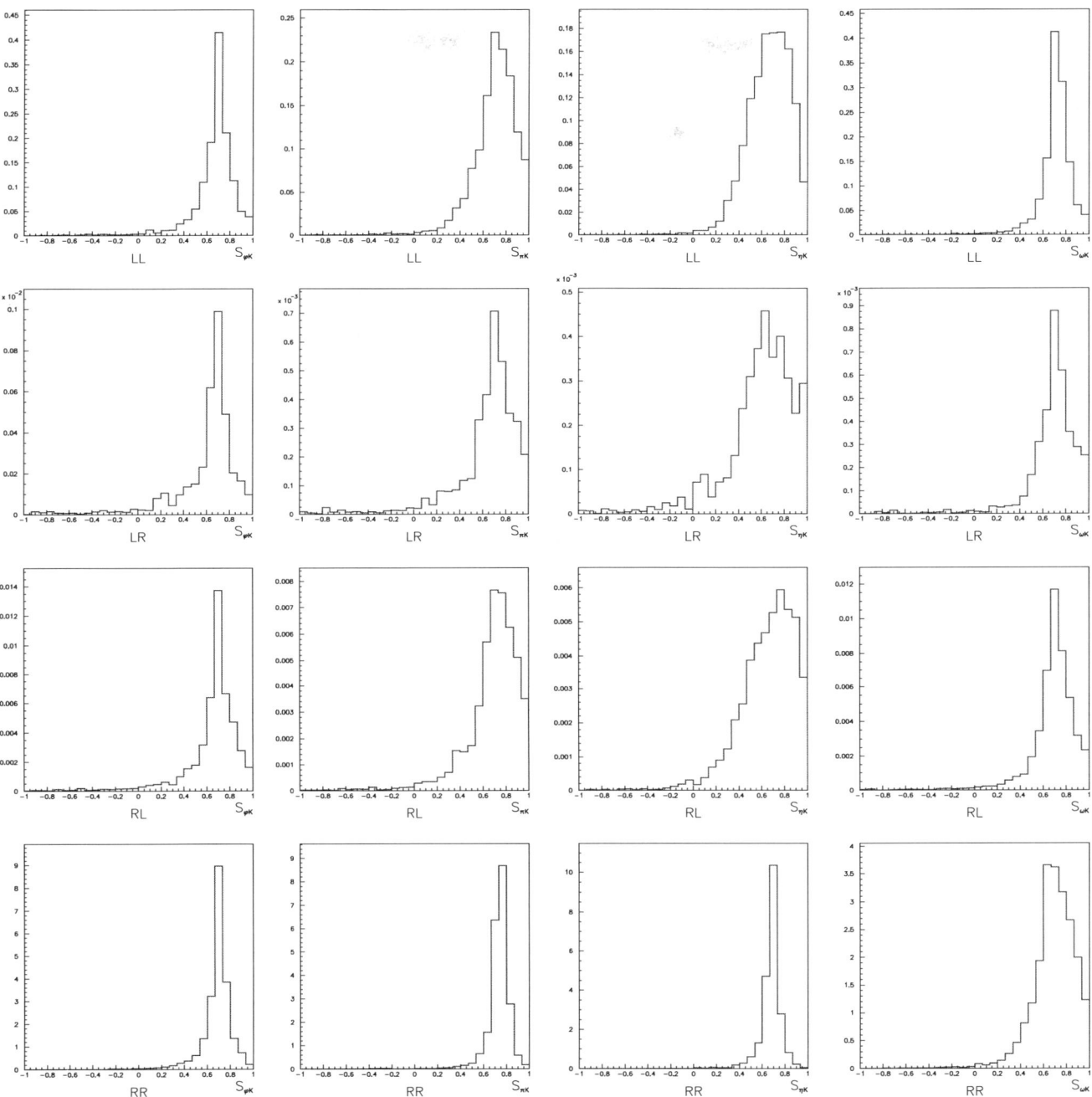

Fig. 38 Probability density functions for $S_{\phi K_s}$, $S_{\pi^0 K_s}$, $S_{\eta' K_s}$ and $S_{\omega K_s}$ induced by $(\delta^d_{23})_{AB}$ with $A, B = \{L, R\}$

An interesting issue is the scaling of SUSY effects in S_f with squark and gluino masses. Similarly to the constraints from other processes, the dominant SUSY contribution to S_f scales linearly with SUSY masses as long as $m_{\tilde{g}} \sim m_{\tilde{q}} \sim \mu$. This means that there is no decoupling of SUSY contributions to S_f as long as the constraints from other processes can be satisfied with $\delta < 1$. The bounds on LL and RR mass insertions quickly reach the physical boundary at $\delta = 1$. On the other hand, LR and RL are well below that bound. Chirality flipping LR and RL mass insertions cannot become too large in order to avoid charge and colour breaking minima and unbounded from below directions in the scalar potential [828]. Nevertheless, it is easy to check that the flavor bounds used above are stronger for SUSY masses above the TeV scale. We conclude that LR and RL mass insertions can give observable effects to S_f for SUSY masses within the reach of LHC and even above.

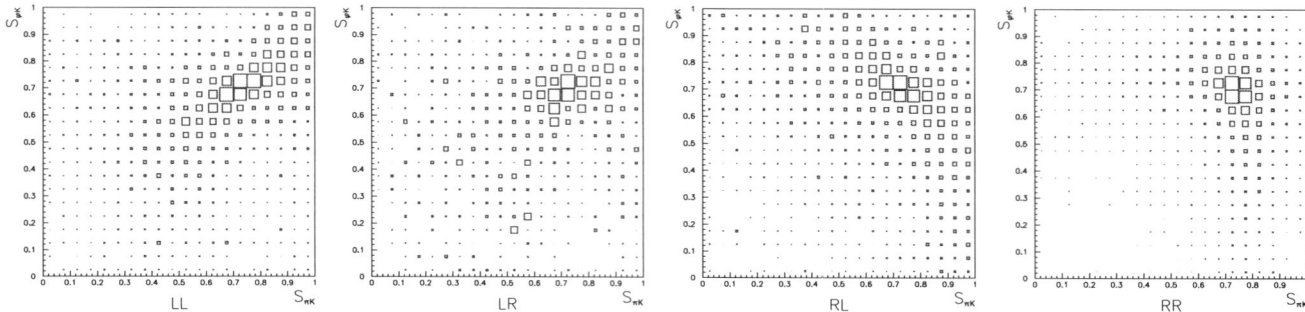

Fig. 39 Correlation between $S_{\phi K_s}$ and $S_{\pi^0 K_s}$ for LL, LR, RL and RR mass insertions

3.7.7 Experimental status and future prospects for time-dependent CP violation in hadronic $b \to s(d)$ transitions

CP asymmetries in B^0 and B_s decays that are governed by the $b \to s$ transition are very sensitive to new CP-violating phases beyond the Standard Model (SM). There are a few golden modes that are practically free from hadronic uncertainties; examples include $B^0 \to \phi K_S^0$, $\eta' K_S^0$, $K_S^0 K_S^0 K_S^0$ and $B_s^0 \to \phi\phi$, see Fig. 40. Precise measurements for these decays have been among the most important topics of quark flavor physics in the last few years and will also remain crucially important in the future.

At the B factories, the decay chain $\Upsilon(4S) \to B^0 \bar{B}^0 \to f_{CP} f_{tag}$ is used to measure time-dependent CP asymmetries, where one of the B mesons decays at time t_{CP} to a final state f_{CP} and the other decays at time t_{tag} to a final state f_{tag} that distinguishes between B^0 and $\bar{B}^0$. The rate of this decay chain has a time dependence [627, 628] given by

$$\mathcal{P}(\Delta t) = e^{-|\Delta t|/\tau_{B^0}} 4\tau_{B^0}$$
$$\times \left\{ 1 + q \cdot \left[\mathcal{S} \sin(\Delta m_d \Delta t) + \mathcal{A} \cos(\Delta m_d \Delta t) \right] \right\}. \quad (170)$$

Here $\mathcal{S}$ and $\mathcal{A}$ are CP-violation parameters, τ_{B^0} is the B^0 lifetime, Δm_d is the mass difference between the two B^0 mass eigenstates, $\Delta t = t_{CP} - t_{tag}$, and the b-flavor charge $q = +1(-1)$ when the tagging B meson is a B^0 ($\bar{B}^0$). To a good approximation, the SM predicts $\mathcal{S} = -\xi_f - -\sin 2\phi_1$ and $\mathcal{A} = 0$ for both tree transitions (e.g. $b \to c\bar{c}s$) and penguin transitions (e.g. $b \to s\bar{s}s$) unless V_{ub} or V_{td} is involved in the decay amplitude. Here $\xi_f = +1(-1)$ corresponds to CP-even (-odd) final states.

BaBar and Belle have accumulated more than 10^9 $B\bar{B}$ pairs with both experiments combined and have measured time-dependent CP asymmetries in various B^0 decays that are dominated by the $b \to s$ transition. Details of the measurements are described elsewhere [609, 732, 735, 739, 829–832]; here we briefly explain the essence of the measurements. Branching fractions for these charmless decay modes are typically around 10^{-5} ignoring daughter branching fractions. Efficient continuum suppression using sophisticated techniques such as Fisher discriminants, likelihood ratios and neural network has been performed to keep a reasonable signal-to-noise ratio. The flavor of the accompanying B meson is identified from inclusive properties of remaining particles; information from primary and secondary leptons, charged kaons, Λ baryons, slow and fast pions is combined by using a neural network (BaBar) or a lookup-table (Belle). A typical effective efficiency for flavor tagging is 30% in both cases. Good understanding of the vertex resolution function is obtained by using large-statistics control samples such as $B \to D^{(*)}\pi$, $D^*\ell\nu$ etc. Lifetime and mixing measurements with a precision of $\mathcal{O}(1)\%$ are obtained as byproducts.

The present status of the measurements is summarized in Fig. 41. Although the result for each individual mode does not significantly differ from the SM expectation (i.e. $\mathcal{S}_{J/\psi K^0}$), most of the $\mathcal{S}$ values are smaller than the SM expectation. When all the $b \to s$ modes are combined, the result differs from the SM expectation by $1.1\,\sigma$.[24] Combining the results of all the $b \to s$ modes is naive as the theoretical uncertainties vary considerably amongst the modes. Much more data are needed to firmly establish a new CP-violating phase beyond the SM for each golden mode.

Measurements of the $\mathcal{A}$ terms yield values consistent with zero, i.e. consistent with the SM at the moment. Nonzero $\mathcal{A}$ requires a strong phase difference between the SM amplitude and the NP amplitude. Therefore it is possible to observe significant deviations from the SM for $\mathcal{S}$ while $\mathcal{A}$ is consistent with zero. Also, since $\mathcal{A}$ is not calculable precisely, in general it is hard to obtain quantitative information from the measurements of $\mathcal{A}$ terms. An exception is the $B^0 \to K^0\pi^0$ decay. Thanks to a precise sum rule based on the isospin symmetry [793], the value for $\mathcal{A}_{K^0\pi^0}$ can be

---

[24]Due to the highly non-Gaussian errors of the result from $B^0 \to f_0 K_S^0$ with $f_0 \to \pi^+\pi^-$ and the fact that this result has a significant effect on the χ^2 of the naive $b \to s$ penguin average, this outlying point is excluded.

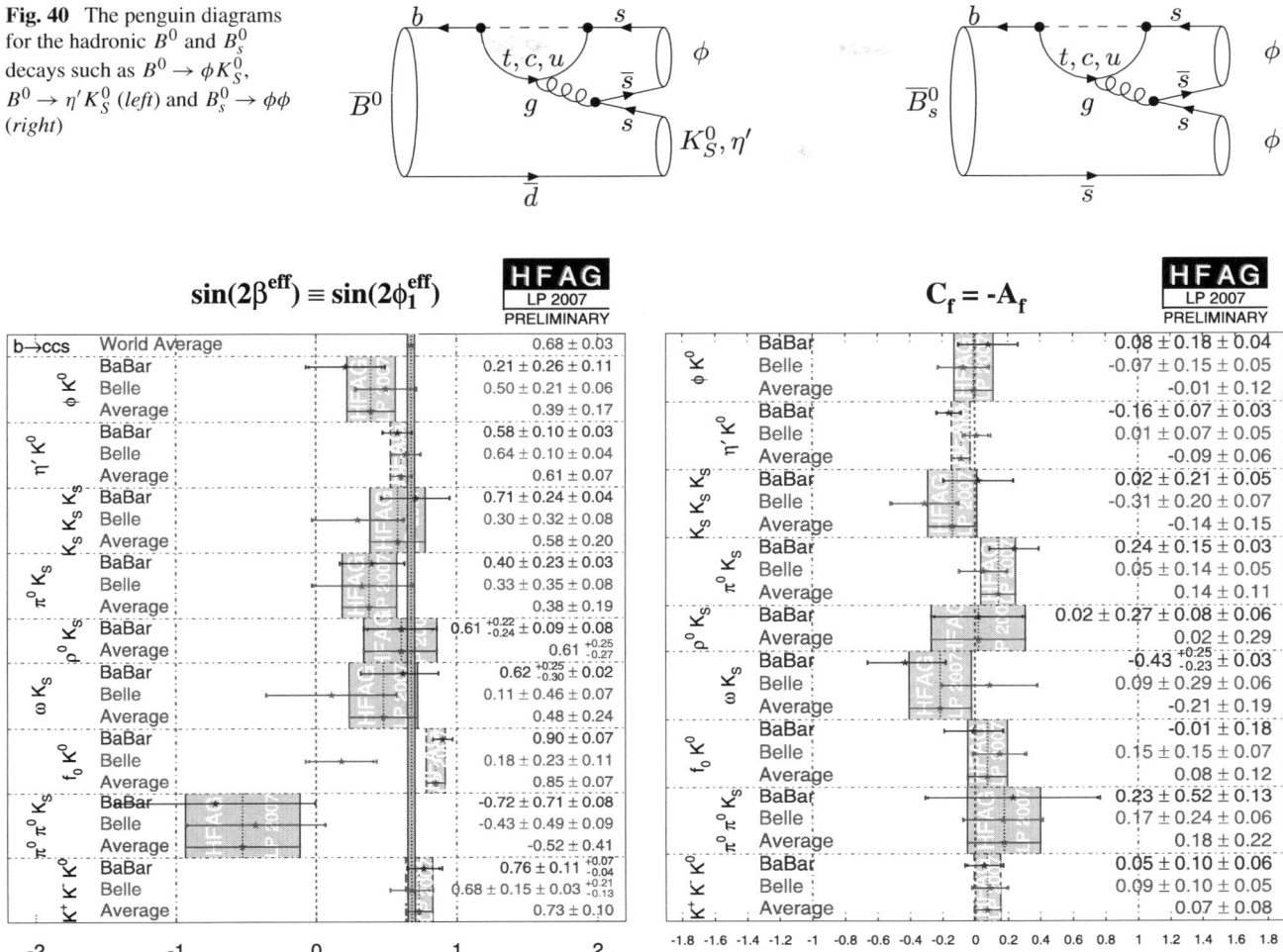

Fig. 40 The penguin diagrams for the hadronic B^0 and B_s^0 decays such as $B^0 \to \phi K_S^0$, $B^0 \to \eta' K_S^0$ (*left*) and $B_s^0 \to \phi\phi$ (*right*)

Fig. 41 Summary of the experimental results for time-dependent CP asymmetries from BaBar and Belle as of August 2007

predicted within the SM from measurements of branching fractions and CP asymmetries of the other $B \to K\pi$ decays; $\mathcal{A}_{K^0\pi^0} = -0.16 \pm 0.04$ is predicted, while measurements yield $\mathcal{A}_{K^0\pi^0} = -0.12 \pm 0.11$.

Due to further CKM-suppression, CP asymmetry measurements for modes dominated by the $b \to d$ transition require even higher statistics than those required for the studies of the $b \to s$ transition. The only measurement available at the moment is $\mathcal{S}_{B^0 \to K_S^0 K_S^0} = -1.28^{+0.80+0.11}_{-0.73-0.16}$ [833], where the first error is statistic and the second error is systematic.

In the near future, the LHCb experiment will probe new CP violating phases beyond the SM in $b \to s$ transitions. With the copious production of B_s^0 mesons, LHCb will be able to study $b \to s$ transitions using the decay $B_s^0 \to \phi\phi$, see Fig. 40. In the SM, the CP-violating phase $\mathcal{S}_{\phi\phi}$ for $B_s^0 \to \phi\phi$ is expected to be very close to zero as there is a cancellation of the B_s^0 mixing and decay phases [834].

In the LHCb experiment, the reconstruction efficiency for $B_s^0 \to \phi\phi$ is expected to be larger than for $B^0 \to \phi K_S^0$ which compensates for the four times smaller fraction of b-quarks to hadronize into a B_s^0 meson. In addition, flavor tagging is also favorable for B_s^0 decays where the same-side kaon tagging contributes significantly to the effective flavor tagging efficiency. From a full simulation LHCb expects a yield of 3100 reconstructed $B_s^0 \to \phi\phi$ events in a 2 fb^{-1} data sample with a background to signal ratio $B/S < 0.8$ at 90% C.L. [835]. The $\mathcal{S}_{\phi\phi}$ sensitivity has been studied using a toy Monte Carlo, taking the resolutions and acceptances from the full simulation. A unbinned likelihood fit is performed on 500 toy data sets. This is used to extract $\mathcal{S}_{\phi\phi}$ and all other physical parameters which cannot be determined from elsewhere. In a 2 fb^{-1} data set, $\mathcal{S}_{\phi\phi}$ can be measured with a precision of $\sigma(\mathcal{S}_{\phi\phi}) = 0.11$ (statistical error only). After about 5 years of data taking, LHCb is expected to accumulate a data sample of 10 fb^{-1}, which will give a statistical uncertainty of $\sigma(\mathcal{S}_{\phi\phi}) = 0.05$ [835].

In a similar study, LHCb investigated the decay $B^0 \to \phi K_S^0$. A yield of 920 events is expected in 2 fb^{-1} of integrated luminosity with a background to signal ratio 0.3 <

$B/S < 1.1$ at 90% C.L. The sensitivity for the CP violating asymmetry $\sin 2\beta^{\rm eff}$ is 0.23 (0.10) in a 2 (10) fb^{-1} data sample [836].

Table 42 lists the expected CP reach at LHCb and a super-B factory for the theoretically cleanest $b \to s$ decay modes. We expect that the precision will be better by an order of magnitude than now. Such measurements will thus allow us to detect effects from physics beyond the SM even if the mass scale of NP is $\mathcal{O}(1)$ TeV.

3.7.8 Two-body hadronic B decay results from the B-factories

This class of B decays manifests a wide range of interesting phenomena, from direct CP violation and broken $SU(3)$ symmetry constraints on the SM uncertainties in measurements of the unitary triangle angles to the amplitude hierarchy found in decays to final states containing two spin-one particles (vector or axial-vector mesons, V and A, respectively).

The only direct CP violation signal observed by the B-factories is in the $B_d^0 \to K^\pm \pi^\mp$ channel. In contrast to the small effect observed in kaon decay, the direct CP asymmetry in $B_d^0 \to K^\pm \pi^\mp$ is large: -0.093 ± 0.015 [620, 838]. The quest for additional signals of direct CP violation in B meson decays is ongoing in a plethora of different channels [502]. The next goals of the B-factories are to observe direct CP violation in the decay of $B_u^\pm$ mesons and other B_d^0 channels.

The B-factories have recently observed CPV in $B_d^0 \to \eta' K^0$ decays [609, 832]. These $b \to s$ penguin processes are probes of NP and have the most precisely measured time-dependent CP asymmetry parameters of all of the penguin modes. Any deviation ΔS of the measured asymmetry parameter $S_{\eta' K^0}$ from $\sin 2\beta$ is an indication of NP (For example, see [507, 839]). In addition to relying on theoretical calculations of the SM pollution to these decays [241, 723, 726], it is possible to experimentally constrain the SM pollution using $SU(3)$ symmetry [784]. This requires precision knowledge of the branching fractions of the B_d^0 meson decays to the following pseudo-scalar pseudo-scalar (PP) final states: $\pi^0 \pi^0$, $\pi^0 \eta$, $\pi^0 \eta'$, $\eta\eta$, $\eta'\eta$, $\eta'\eta'$ final states [840, 841]. The related decays $B_{u,d} \to \eta' \rho$ and $B_{u,d} \to \eta' K^*$ [842, 843] can also be used to understand the standard model contributions to $B_d^0 \to \eta' K^0$ decays and the hierarchy of ηK^0 to $\eta' K^0$ decays.

The angular analysis of $B \to VV$ decays provides eleven observables (six amplitudes and five relative phases) that can be used to test theoretical calculations [611]. The hierarchy of A_0, A_+ and A_- amplitudes obtained from a helicity (or A_0, $A_\parallel$ and $A_\perp$ in the transversity basis) analysis of such decays allows one to search for possible right-handed currents in any NP contribution to the total amplitude. For low statistics studies, a simplified angular analysis is performed, where one measures the fraction of longitudinally polarized events defined as $f_L = |A_0|^2 / \sum |A_i|^2$. Tree dominated decays such as $B_d^0 \to \rho^+ \rho^-$ have $f_L \sim 1.0$ [844, 845]. Current data for penguin dominated processes ($\phi K^*(892)$ [846, 847], $K^*(892)\rho$ [848, 849]) that are observed to have nontrivial values of f_L can be accommodated in the SM. In addition to this, one can search for T-odd CP violating asymmetries in triple products constructed from the angular distributions [850]. It has also been suggested that nonstandard model effects could be manifest in a number of other observables [851]. The measured rates of electroweak penguin dominated B decays to final states involving a ϕ meson are also probes of NP [852]. The study of $B \to AV$ decays also provides this rich set of observables to study, however current results only yield an upper limit on $B_d^0 \to a_1^\pm \rho^\mp$ decays [853]. BaBar have recently studied the angular distribution for the vector-tensor decay $B_d^0 \to \phi K^*(1430)$ [846].

3.7.9 $B \to h^+ h'^-$ decays at LHCb

The charmless decays of B mesons to two-body modes have been extensively studied at the B-factories. Even if the current knowledge in the B_d and B_u sectors starts to be quite constrained, the B_s sector still remains an open field. At present, by using a displaced vertex trigger CDF has already collected an interesting sample of $B \to h^+ h'^-$ decays

Table 42 CP reach at LHCb [1048] and at a super-B factory for the $b \to s$ decay modes that are theoretically cleanest. The estimated accuracy from the B factories (2 ab^{-1}) is given for comparison. We assume an integrated luminosity of 10 fb^{-1} for LHCb and 50 ab^{-1} for a super B factory, which are the goals of the experiments. Errors for LHCb are statistical only. Projections for the super B factory are from Ref. [837] and include both statistical and systematic uncertainties and $\Delta \sin 2\phi_1 \equiv \sin 2\phi_1^{\rm eff} - \sin 2\phi_1$

Mode	Observable	B factories 2 ab^{-1}	LHCb 10 fb^{-1}	Super B factory 50 ab^{-1}
$B^0 \to \phi K^0$	$\Delta \sin 2\phi_1$	0.13	0.10	0.029
$B^0 \to \eta' K^0$	$\Delta \sin 2\phi_1$	0.05	–	0.020
$B^0 \to K_S^0 K_S^0 K_S^0$	$\Delta \sin 2\phi_1$	0.15	–	0.037
$B_s^0 \to \phi\phi$	$\mathcal{S}_{\phi\phi}$	–	0.05	–

[854], providing a first observation of the two-body mode $B_s \to K^+K^-$. However it will most likely not be able to perform precision measurements of the time-dependent CP asymmetry of the $B_s \to K^+K^-$ decay.

The LHCb experiment, thanks to the large beauty production cross section at the LHC and to its excellent vertexing and triggering capabilities, will be able to collect huge samples of $B \to h^+h'^-$ decays [855]. Furthermore, its particle identification system, composed in particular by two RICH detectors, will allow one to disentangle various $B \to h^+h'^-$ modes with a purity exceeding 90% as well as high efficiency. The PID capabilities of LHCb are clearly visible in Fig. 42, which shows the distribution of the $\pi^+\pi^-$ invariant mass from Monte Carlo samples of $B \to h^+h'^-$ modes, before and after the employment of the PID information.

In order to calibrate the PID response, LHCb will make use of a dedicated trigger line—not making use of PID information in order not to introduce biases—intended to collect very large samples of D^* decay chains to charged kaons and pions. In order to reject combinatorial background, the event selection is based on a series of cuts, optimized by means of a multivariate technique, which include the transverse momenta and the impact parameter significances of the charged legs with respect to the primary vertex, the χ^2 of the common vertex fit, the transverse momentum, the impact parameter significance and the distance of flight significance of the candidate b-hadron and the invariant mass (the resolution for the $B \to h^+h'^-$ modes is expected to be about 18 MeV/c^2). The event yields and background-to-signal ratios estimated using a full GEANT4 based simulation are reported in Table 43.

In order to measure CP violation from the time-dependent CP asymmetries, other key ingredients are the tagging capability and the propertime resolution, the latter being particularly relevant to resolve the fast B_s oscillations. The effective tagging power for a B_d decay at LHCb, according to full simulations, is expected to be about 5%, while for a B_s decay, it is significantly larger, due to the larger efficiency of the same side kaon tagging, and is about 9%. The calibration of the tagging power for $B \to h^+h'^-$ modes will be performed by using the flavor-specific modes $B_d \to K^+\pi^-$ and $B_s \to \pi^+K^-$. As far as the propertime resolution is concerned, it is predicted by the full simulation to be about 40 fs, and it will be calibrated on data by using large samples of $J/\psi \to \mu^+\mu^-$ decays collected through a dedicated di-muon trigger line thought not to introduce biases in the J/ψ propertime.

The direct CP asymmetries of the flavor specific $B \to h^+h'^-$ modes can be measured without a time-dependent fit and without the need of tagging the B meson. The sta-

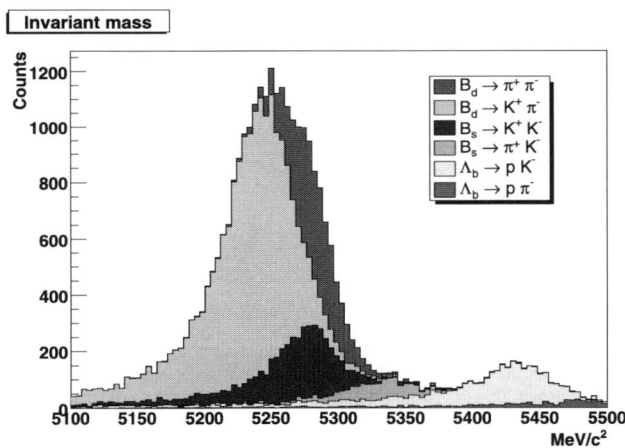 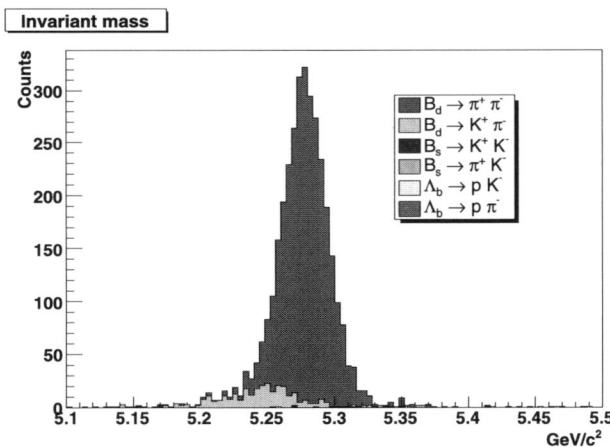

Fig. 42 *Left*: $\pi^+\pi^-$ invariant mass distribution for $B \to h^+h'^-$ decays expected at LHCb, obtained without using PID information. *Right*: same plot after PID cuts are applied

Table 43 Annual yields and background-to-signal ratios for $B \to h^+h'^-$ decays at LHCb [855]

Channel	Assumed BR	Annual yield	B/S (combinatorial)	B/S (two-body)
$B_d^0 \to \pi^+\pi^-$	4.8	36000	0.46	0.08
$B_d^0 \to K^+\pi^-$	18.5	138000	0.14	0.02
$B_s^0 \to \pi^+K^-$	4.8	10000	1.92	0.54
$B_s^0 \to K^+K^-$	18.5	36000	<0.06	0.08
$\Lambda_b \to p\pi^-$	4.8	9000	1.66	0.11
$\Lambda_b \to pK^-$	18.5	32000	<0.08	0.02

tistical sensitivity on the charge asymmetry corresponding to a running time of 10^7 s at the nominal LHCb luminosity 2×10^{32} cm^{-2} s^{-1} ("one nominal LHCb year" in the following) is 0.003 for the $B_d \to K^+\pi^-$ decay and 0.02 for the $B_s \to \pi^+ K^-$ decay. In order to extract the direct (C) and mixing-induced (S) CP violation terms from the time-dependent decay rates of the $B_d \to \pi^+\pi^-$ and $B_s \to K^+ K^-$ and estimate the statistical sensitivity, we performed unbinned maximum likelihood fits on fast Monte Carlo data sets which parametrize the decay rates according to the outcomes of the full simulation. The expected sensitivity for C and S, corresponding to one nominal LHCb year, both for the $B_d \to \pi^+\pi^-$ and $B_s \to K^+ K^-$ channels, is about 0.04.

According to the method proposed in [771], the employment of the U-spin symmetry allows us to combine the measurements of C and S for the $B_d \to \pi^+\pi^-$ and $B_s \to K^+ K^-$ modes in order to extract the γ angle. Assuming a perfect U-spin symmetry, we predict a sensitivity on γ for a nominal LHCb year around $5°$. If a 20% U-spin breaking is taken into account, the sensitivity deteriorates up to about $10°$, still not spoiling the method of its predictive capabilities on γ. Being these modes characterized by the presence of loops inside the penguins, they could reveal NP effects, pointing to a value of γ in contrast with the one determined from pure tree-level decays, such as $B \to DK$ modes.

In Table 43, LHCb also reports expected yields for Λ_b baryon decays. An additional application of the Λ_b baryon that has been considered is testing CP and T symmetries using the decay modes $\Lambda_b \to \Lambda V$, where $V = J/\psi, \rho^0, \omega$. This is discussed in [856].

3.8 Kaon decays[25]

3.8.1 Introduction

The rare decays $K^+ \to \pi^+ \nu\bar{\nu}$ and $K_L \to \pi^0 \nu\bar{\nu}$ play an important role in the search for the underlying mechanism of flavor mixing and CP violation [857–860]. As such, they are excellent probes of physics beyond the Standard Model (SM). Among the many rare K- and B-decays, the $K^+ \to \pi^+\nu\bar{\nu}$ and $K_L \to \pi^0\nu\bar{\nu}$ modes are unique, since their SM branching ratios can be computed to an exceptionally high degree of precision, not matched by any other FCNC process involving quarks.

The main reason for the exceptional theoretical cleanness of the $K^+ \to \pi^+\nu\bar{\nu}$ and $K_L \to \pi^0\nu\bar{\nu}$ decays is the fact that, within the SM, these processes are mediated by electroweak amplitudes of $\mathcal{O}(G_F^2)$, described by Z^0-penguins and box diagrams which exhibit a power-like GIM mechanism. This property implies a severe suppression of nonperturbative effects, which is generally not the case for meson decays receiving contributions of $\mathcal{O}(G_F \alpha_s)$ (gluon penguins) and/or $\mathcal{O}(G_F \alpha_{\rm em})$ (photon penguins), which therefore have only a logarithmic GIM mechanism. A related important virtue, following from this peculiar electroweak structure, is the fact that $K \to \pi\nu\bar{\nu}$ amplitudes can be described in terms of a single effective operator, namely

$$Q_{sd}^{\nu\bar{\nu}} = (\bar{s}_L \gamma^\mu d_L)(\bar{\nu}_L \gamma_\mu \nu_L). \tag{171}$$

The hadronic matrix elements of $Q_{sd}^{\nu\bar{\nu}}$ relevant for $K \to \pi\nu\bar{\nu}$ amplitudes can be extracted directly from the well-measured $K^+ \to \pi^0 e^+ \nu$ decay, including the leading isospin breaking (IB) corrections [861]. The estimation of the matrix elements is improved and extended [862] beyond the leading order analysis.

In the case of $K_L \to \pi^0 \nu\bar{\nu}$, which is CP-violating and dominated by the dimension-six top quark contribution, the SM Short-Distance (SD) dynamics is then encoded in a perturbatively calculable real function X that multiplies the CKM factor $\lambda_t = V_{ts}^* V_{td}$. In the case of $K^+ \to \pi^+\nu\bar{\nu}$, also a charm-quark contribution proportional to $\lambda_c = V_{cs}^* V_{cd}$ has to be taken into account, but the recent NNLO QCD calculation of the dimension-six charm quark corrections [863, 864] and the progress in the evaluation of dimension-eight charm and long-distance (LD) up quark effects [865] elevated the theoretical cleanness of $K^+ \to \pi^+\nu\bar{\nu}$ almost to the level of $K_L \to \pi^0 \nu\bar{\nu}$. More details will be given in Sect. 3.8.2.

The important virtue of $K \to \pi\nu\bar{\nu}$ decays is that their clean theoretical character remains valid in essentially all extensions of the SM and that $Q_{sd}^{\nu\bar{\nu}}$, due to the special properties of the neutrinos, remains the only relevant operator. Consequently, in most SM extensions, the NP contributions to $K^+ \to \pi^+\nu\bar{\nu}$ and $K_L \to \pi^0\nu\bar{\nu}$ can be parametrized by just two parameters, the magnitude and the phase of the function, in a model-independent manner [866]

$$X = |X|e^{i\theta_X} \tag{172}$$

that multiplies λ_t in the relevant effective Hamiltonian. In the SM, $|X| = X_{\rm SM}$ and $\theta_X = 0$.

The parameters $|X|$ and θ_X can be extracted from $\mathcal{B}(K_L \to \pi^0\nu\bar{\nu})$ and $\mathcal{B}(K^+ \to \pi^+\nu\bar{\nu})$ without hadronic uncertainties, while the function X can be calculated in any extension of the SM within perturbation theory. Of particular interest is the ratio

$$\frac{\mathcal{B}(K_L \to \pi^0\nu\bar{\nu})}{\mathcal{B}(K_L \to \pi^0\nu\bar{\nu})_{\rm SM}} = \left|\frac{X}{X_{\rm SM}}\right|^2 \left[\frac{\sin(\beta - \theta_X)}{\sin\beta}\right]^2. \tag{173}$$

Bearing in mind that $\beta \approx 21.4°$, (173) shows that $K_L \to \pi^0\nu\bar{\nu}$ is a very sensitive function of the new phase θ_X. The

---
[25]Section coordinators: A.J. Buras, T.K. Komatsubara.

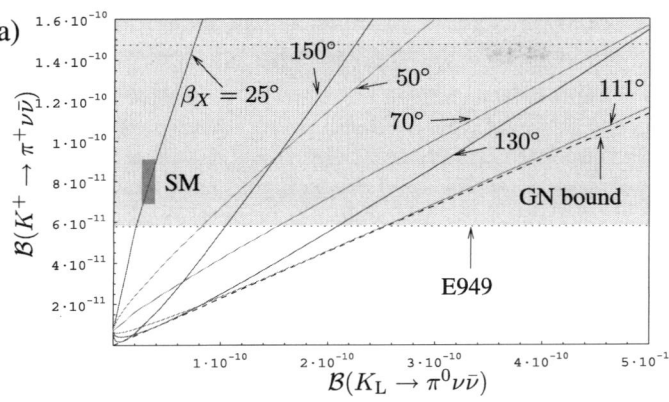
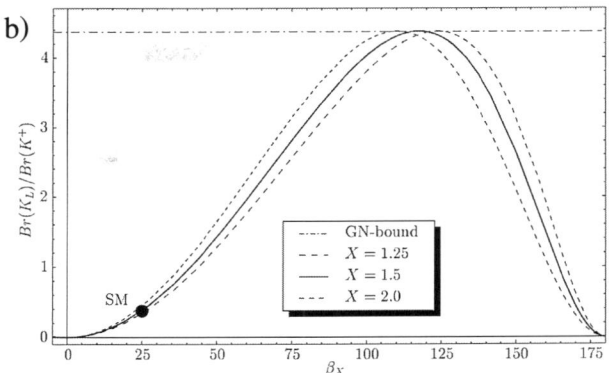

Fig. 43 (a) $\mathcal{B}(K^+ \to \pi^+ \nu \bar{\nu})$ vs. $\mathcal{B}(K_L \to \pi^0 \nu \bar{\nu})$ for various values of $\beta_X = \beta - \theta_X$ (including E949 data) [780]. The *dotted horizontal lines* indicate the lower part of the experimental range [867–869] and the *grey area* the SM prediction. We also show the Grossman–Nir (GN) bound [870]. (b) The ratio of the $K \to \pi \nu \bar{\nu}$ branching ratios as a function of β_X for $|X| = 1.25, 1.5, 2.0$. The *horizontal line* is again the GN bound

pattern of the two $K \to \pi \nu \bar{\nu}$ branching ratios as a function of θ_X is illustrated in Fig. 43a. We note that the ratio of the two modes shown in Fig. 43b depends very mildly on $|X|$ and therefore provides an excellent tool to extract the non-standard CP-violating phase θ_X.

An interesting and complementary window to $|\Delta S| = 1$ SD transitions is provided by the $K_L \to \pi^0 \ell^+ \ell^-$ system ($\ell = \mu, e$). While the latter is theoretically not as clean as the $K \to \pi \nu \bar{\nu}$ system, it is sensitive to different types of SD operators. The $K_L \to \pi^0 \ell^+ \ell^-$ decay amplitudes have three main ingredients: (i) a clean direct-CP-violating (CPV) component determined by SD dynamics; (ii) an indirect-CPV term due to K^0–$\bar{K}^0$ mixing; (iii) an LD CP-conserving (CPC) component due to two-photon intermediate states. Although generated by very different dynamics, these three components are of comparable size and can be computed (or indirectly determined) to good accuracy within the SM [871, 872]. In the presence of nonvanishing NP contributions, the combined measurements of $K \to \pi \nu \bar{\nu}$ and $K \to \pi^0 \ell^+ \ell^-$ decays provide a unique tool to distinguish among different NP models.

The following discussion concentrates on the $K \to \pi \nu \bar{\nu}$ and $K_L \to \pi^0 \ell^+ \ell^-$ decays in the SM (Sects. 3.8.2 and 3.8.3) and its most popular extensions (Sects. 3.8.4 and 3.8.5). In Sect. 3.8.6, we stress the complementarity of K- and B-physics as well as the interplay with the high-p_T physics at the LHC. Recent theoretical updates on kaon decays are found in [873–875]. Experimental programs at CERN and J-PARC are described in Sects. 3.8.7 and 3.8.8, respectively. The current experimental status is summarized in Table 44.

3.8.2 $K^+ \to \pi^+ \nu \bar{\nu}$ and $K_L \to \pi^0 \nu \bar{\nu}$ in the SM

After summation over the three lepton families the SM branching ratios for the $K \to \pi \nu \bar{\nu}$ decays can be written as

$$\mathcal{B}(K^+ \to \pi^+ \nu \bar{\nu})_{\rm SM} = \kappa_+ \left[\left(\frac{{\rm Im} \lambda_t}{\lambda^5} X_{\rm SM} \right)^2 + \left(\frac{{\rm Re} \lambda_t}{\lambda^5} X_{\rm SM} + \frac{{\rm Re} \lambda_c}{\lambda} (P_c + \delta P_{c,u}) \right)^2 \right], \quad (174)$$

$$\mathcal{B}(K_L \to \pi^0 \nu \bar{\nu})_{\rm SM} = \kappa_L \left(\frac{{\rm Im} \lambda_t}{\lambda^5} X_{\rm SM} \right)^2, \quad (175)$$

where $\lambda = |V_{us}|$, while $\kappa_+ = (5.26 \pm 0.06) \times 10^{-11} (\lambda/0.225)^8$ and $\kappa_L = (2.29 \pm 0.03) \times 10^{-10} (\lambda/0.225)^8$ [879] include the leading IB corrections in relating $K \to \pi \nu \bar{\nu}$ to $K^+ \to \pi^0 e^+ \nu$ [861]. The dimension-six top-quark contribution $X_{\rm SM} = 1.464 \pm 0.041$ [863, 864] accounts for around 63% and almost 100% of the total rates. It is known to NLO [557, 558], with a scale uncertainty of about 1%. In $K^+ \to \pi^+ \nu \bar{\nu}$, dimension-six charm quark corrections and subleading dimension-eight charm and LD up quark effects, characterized by $P_c = 0.38 \pm 0.04$ [863, 864] and

Table 44 Current experimental results or limits for rare K decay branching fractions

$B(K^+ \to \pi^+ \nu \bar{\nu})$	$B(K_L \to \pi^0 \nu \bar{\nu})$	$B(K_L \to \pi^0 e^+ e^-)$	$B(K_L \to \pi^0 \mu^+ \mu^-)$
$(1.47^{+1.30}_{-0.89}) \times 10^{-10}$ [867–869]	$<6.7 \times 10^{-8}$ [876]	$<2.8 \times 10^{-10}$ [877]	$<3.8 \times 10^{-10}$ [878]

$\delta P_{c,u} = 0.04 \pm 0.02$ [865], amount to a moderate 33% and a mere 4%. Light quark contributions are negligible in the case of the $K_L \to \pi^0 \nu \bar{\nu}$ decay [880].

Taking into account all the indirect constraints from the latest global UT fit, the SM predictions for the two $K \to \pi \nu \bar{\nu}$ rates read

$$\begin{aligned}\mathcal{B}(K^+ \to \pi^+ \nu \bar{\nu})_{\text{SM}} &= (8.4 \pm 1.0) \times 10^{-11}, \\ \mathcal{B}(K_L \to \pi^0 \nu \bar{\nu})_{\text{SM}} &= (2.7 \pm 0.4) \times 10^{-11}.\end{aligned} \quad (176)$$

The quoted central value of $K^+ \to \pi^+ \nu \bar{\nu}$ corresponds to $m_c = 1.3$ GeV, and the given error breaks down as follows: residual scale uncertainties (13%), m_c (22%), CKM, α_s, and m_t (37%), and matrix-elements from $K^+ \to \pi^0 e^+ \nu$ and light quark contributions (28%). The main source of uncertainty in $K_L \to \pi^0 \nu \bar{\nu}$ is parametric (74%), while the impact of scales (11%) and IB (15%) is subdominant. SM predictions for $K \to \pi \nu \bar{\nu}$ with total uncertainties at the level of 5% or below are thus possible through a better knowledge of m_c, of the IB in the $K \to \pi$ form factors, and/or by a lattice study [881] of higher-dimensional and LD contributions.

While the determination of $|V_{td}|$, $\sin 2\beta$, and γ from the $K \to \pi \nu \bar{\nu}$ system is without doubt still of interest, with the slow progress in measuring the relevant branching ratios and much faster progress in the extraction of the angle γ from the $B_s \to DK$ system to be expected at the LHC, the role of the $K \to \pi \nu \bar{\nu}$ system will shift towards the search for NP rather than the determination of the CKM parameters.

In fact, determining the UT from tree-level dominated K- and B-decays and thus independently of NP will allow us to find the "true" values of the CKM parameters. Inserting these, hopefully accurate, values in (174) and (175) will allow us to obtain very precise SM predictions for the rates of both rare K-decays. A comparison with future data on $K \to \pi \nu \bar{\nu}$ may then give a clear signal of potential NP contributions in a theoretically clean environment. Even deviations by 20% from the SM expectations could be considered as signals of NP, while such a conclusion cannot be drawn in most other decays, in which the theoretical errors are at least 10%.

3.8.3 $K_L \to \pi^0 \ell^+ \ell^-$ in the SM

As mentioned in the introduction, the $K_L \to \pi^0 \ell^+ \ell^-$ amplitudes have three main components. The interesting direct-CPV component, proportional to $\text{Im}\lambda_t$, is generated by Z^0-, γ-penguins and box diagrams and is SD dominated. It is encoded by local dimension-six vector $Q_{7V} = (\bar{s}d)_V(\bar{\ell}\ell)_V$ and axial-vector $Q_{7A} = (\bar{s}d)_V(\bar{\ell}\ell)_A$ operators, whose Wilson coefficients $y_{7V,7A}$ are known to NLO [882]. The former produces the $\ell^+\ell^-$ pair in a 1^{--} state, the latter both in 1^{++} and 0^{-+} states. As in the $K \to \pi \nu \bar{\nu}$ case, the corresponding hadronic matrix elements are obtained precisely from $K_{\ell 3}$ decays [861].

The other two components are of electromagnetic origin and are dominated by LD dynamics. These contributions cannot be computed from first principles. However, they can be related to measurable quantities within Chiral Perturbation Theory (CHPT). The indirect CPV amplitude, $\mathcal{A}(K_L \approx \varepsilon K_1 \to \pi^0 \gamma^* \to \pi^0 \ell^+ \ell^-)$, is determined [883]—up to a sign ambiguity—by the measurements of $\mathcal{B}(K_S \to \pi^0 \ell^+ \ell^-)$. In this case, the $\ell^+\ell^-$ pair is produced in a 1^{--} state and interferes with the SD contribution of Q_{7V}. As discussed in [871, 884], various theoretical arguments point toward a constructive interference. Finally, the CPC contribution ($K_L \to \pi^0 \gamma^* \gamma^* \to \pi^0 \ell^+ \ell^-$) produces the $\ell^+\ell^-$ pair either in a helicity-suppressed 0^{++} state or in a phase-space suppressed 2^{++} state. Within CHPT, only the 0^{++} state is produced at LO through the finite two-loop process $K_L \to \pi^0 P^+ P^- \to \pi^0 \gamma \gamma \to \pi^0 \ell^+ \ell^-$ ($P = \pi, K$). Higher-order corrections are estimated using $K_L \to \pi^0 \gamma \gamma$ experimental data for both the 0^{++} and 2^{++} contributions [871, 872].

Altogether, the branching ratios can be expressed as [871, 872]

$$\mathcal{B}(K_L \to \pi^0 \ell^+ \ell^-) = \left(C_{\text{dir}}^\ell \pm C_{\text{int}}^\ell |a_S| + C_{\text{mix}}^\ell |a_S|^2 + C_{\gamma\gamma}^\ell\right) \times 10^{-12}, \quad (177)$$

where the C_i are reported in Table 45, $w_{7A,7V} = \text{Im}(\lambda_t y_{7A,7V})/\text{Im}\lambda_t$, and $|a_S| = 1.2 \pm 0.2$ is fixed from $\mathcal{B}^{\text{exp}}(K_S \to \pi^0 \ell^+ \ell^-)$ [885, 886]. Using the SM values of $y_{7A,7V}$ [882], the predicted rates are

$$\begin{aligned}\mathcal{B}_{\text{SM}}^{e^+e^-} &= 3.54^{+0.98}_{-0.85}(1.56^{+0.62}_{-0.49}) \times 10^{-11}, \\ \mathcal{B}_{\text{SM}}^{\mu^+\mu^-} &= 1.41^{+0.28}_{-0.26}(0.95^{+0.22}_{-0.21}) \times 10^{-11}\end{aligned} \quad (178)$$

for constructive (destructive) interference. Currently, the theory error (see Fig. 46a) is dominated by the uncertainty on $|a_S|$. Better measurements of $\mathcal{B}(K_S \to \pi^0 \ell^+ \ell^-)$ would thus be very welcome. Also, better measurements of $K_L \to \pi^0 \gamma \gamma$ would help in reducing the error on the 0^{++} and 2^{++} contributions. Alternatively, they can be partially cut away through energy cuts or Dalitz plot analyses [871, 872, 887]. As shown in Fig. 46a, the irreducible theoretical errors on

Table 45 Numerical coefficients for the evaluation of $\mathcal{B}(K_L \to \pi^0 \ell^+ \ell^-)$ as given in (177)

	C_{dir}^ℓ	C_{int}^ℓ	C_{mix}^ℓ	$C_{\gamma\gamma}^\ell$
$\ell = e$	$(4.62 \pm 0.24)(w_{7V}^2 + w_{7A}^2)$	$(11.3 \pm 0.3)w_{7V}$	14.5 ± 0.5,	≈ 0
$\ell = \mu$	$(1.09 \pm 0.05)(w_{7V}^2 + 2.32 w_{7A}^2)$	$(2.63 \pm 0.06)w_{7V}$	3.36 ± 0.20	5.2 ± 1.6

these modes can be pushed below the 10% level, allowing very significant tests of flavor physics.

The integrated forward–backward (or lepton-energy) asymmetry (see references in [887]) generated by the interference between CPC and CPV amplitudes cannot be reliably estimated at present for $\ell = e$ because of the poor theoretical control on the 2^{++} contribution. In the case of A_{FB}^μ, the situation is better, since the 2^{++} part is negligible. One has $A_{FB}^\mu \approx 20\%$ (-12%) for constructive (destructive) interference. Interestingly, though the error is large, A_{FB}^μ can be used to fix the sign of a_S.

Let us close with a short comment on $K_L \to \mu^+\mu^-$. Here the SD part is CPC and has recently been evaluated at NNLO [888]. The much larger LD contribution proceeds via two photons. While its absorptive part is fixed from $K_L \to \gamma\gamma$, its dispersive part is difficult to estimate, requiring unknown counterterms in CHPT [889]. Moreover, in this case, the two-photon LD amplitude interferes with the SD one (they both produce a lepton pair in a 0^{-+} state). This interference, which depends on the sign of $\mathcal{A}(K_L \to \gamma\gamma)$, is presumably constructive [890], and better measurements of $K_S \to \pi^0 \gamma\gamma$ or $K^+ \to \pi^+ \gamma\gamma$ could settle this sign. However, even with the help of this information it is difficult to reduce the theoretical error below $\sim 50\%$ of the SD contribution.

3.8.4 $K^+ \to \pi^+ \nu\bar{\nu}$ and $K_L \to \pi^0 \nu\bar{\nu}$ beyond the SM

Minimal flavor violation In models with Minimal Flavor Violation (MFV) [10, 12], both decays are, like in the SM, governed by a single real function X that can take a different value than in the SM due to new particle exchange in the relevant Z^0-penguin and box diagrams (see Fig. 43a). Restricting first our discussion to the so-called constrained MFV (CMFV) (see [891]), in which strong correlations between K- and B-decays exist, one finds that the branching ratios for $K^+ \to \pi^+ \nu\bar{\nu}$ and $K_L \to \pi^0 \nu\bar{\nu}$ cannot be much larger than their SM values given in (176). The 95% probability bounds read [190]

$$\mathcal{B}(K^+ \to \pi^+ \nu\bar{\nu})_{\text{CMFV}} \leq 11.9 \times 10^{-11},$$
$$\mathcal{B}(K_L \to \pi^0 \nu\bar{\nu})_{\text{CMFV}} \leq 4.6 \times 10^{-11}. \quad (179)$$

Explicit calculations in a model with one Universal Extra Dimension (UED) [181] and in the Littlest Higgs model without T-parity [142] give explicit examples of this scenario with the branching ratios within 20% of the SM expectations. The latest detailed analysis of $K \to \pi \nu\bar{\nu}$ in the MSSM with MFV can be found in [879].

Probably the most interesting property of this class of models is a theoretically clean determination of the angle β of the standard UT, which utilizes both branching ratios and is independent of the value of X [892, 893]. Consequently, this determination is universal within the class of MFV models, and any departure of the resulting value of β from the corresponding one measured in B-decays would signal non-MFV interactions.

Littlest Higgs model with T-parity The structure of $K \to \pi \nu\bar{\nu}$ decays in the Littlest Higgs model with T-parity (LHT) differs notably from the one found in MFV models due to the presence of mirror quarks and leptons that interact with the light fermions through the exchange of heavy charged ($W_H^\pm$) and neutral (Z_H^0, A_H^0) gauge bosons. The mixing matrix V_{Hd} that governs these interactions can differ from V_{CKM}, which implies the presence of non-MFV interactions. Instead of a single real function X that is universal within the K-, B_d- and B_s-systems in MFV models, one now has three functions

$$X_K = |X_K|e^{i\theta_K}, \qquad X_d = |X_d|e^{i\theta_d}, \qquad X_s = |X_s|e^{i\theta_s}$$
(180)

that due to the presence of mirror fermions can have different phases and magnitudes.

Moreover, it is important to note that mirror fermion contributions are enhanced by a CKM factor $1/\lambda_t^{(i)}$ with $i = K, d, s$ for the K-, B_d- and B_s-systems, respectively. As $\lambda_t^{(K)} \simeq 4 \times 10^{-4}$, whereas $\lambda_t^{(d)} \simeq 1 \times 10^{-2}$ and $\lambda_t^{(s)} \simeq 4 \times 10^{-2}$, the deviation from the SM prediction in the K-system is found to be by more than an order of magnitude larger than in the B_d-system and even by two orders of magnitude larger than in the B_s-system. This possibility can have a major impact on the $K \to \pi \nu\bar{\nu}$ system, since the correlations between K- and B-decays are partially lost and the presence of a large phase θ_K can change the pattern of these decays from the one observed in MFV. A detailed analysis [158] shows that both branching ratios can depart significantly from their SM values and can be as high as 5.0×10^{-10}. As shown in Fig. 44a, there are two branches of allowed values with strong correlations between both branching ratios within a given branch. In the lower branch, only $\mathcal{B}(K^+ \to \pi^+ \nu\bar{\nu})$ can differ substantially from the SM expectations reaching values well above the present central experimental value. In the second branch, $\mathcal{B}(K_L \to \pi^0 \nu\bar{\nu})$ and $\mathcal{B}(K^+ \to \pi^+ \nu\bar{\nu})$ can be as high as 5.0×10^{-10} and 2.3×10^{-10}, respectively. Moreover, $\mathcal{B}(K_L \to \pi^0 \nu\bar{\nu})$ can be larger than $\mathcal{B}(K^+ \to \pi^+ \nu\bar{\nu})$, which is excluded within MFV models. Other features distinguishing this model from MFV are thoroughly discussed in [158].

Supersymmetry Within the MSSM with R-parity conservation, sizable nonstandard contributions to $K \to \pi \nu\bar{\nu}$ decays can be generated if the soft-breaking terms have a non-MFV structure. The leading amplitudes giving rise to large effects are induced by: (i) chargino/up-squark loops [131,

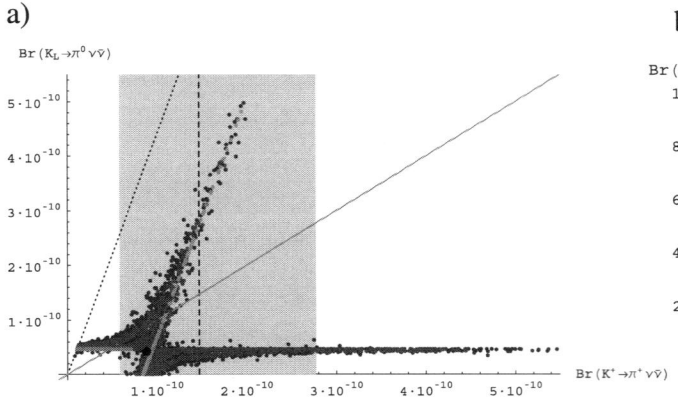

Fig. 44 (**a**) $\mathcal{B}(K_L \to \pi^0 \nu\bar{\nu})$ vs. $\mathcal{B}(K^+ \to \pi^+ \nu\bar{\nu})$ in the LHT model [158]. The *shaded area* represents the experimental 1σ-range for $\mathcal{B}(K^+ \to \pi^+ \nu\bar{\nu})$. The GN bound is displayed by the *dotted line*, while the *solid line* separates the two areas where $\mathcal{B}(K_L \to \pi^0 \nu\bar{\nu})$ is larger or smaller than $\mathcal{B}(K^+ \to \pi^+ \nu\bar{\nu})$. (**b**) $\mathcal{B}(K_L \to \pi^0 e^+ e^-)$ (*upper curve*) and $\mathcal{B}(K_L \to \pi^0 \mu^+ \mu^-)$ (*lower curve*) as functions of $\mathcal{B}(K_L \to \pi^0 \nu\bar{\nu})$ in the LHT model [158]

866, 894, 895]; (ii) charged Higgs/top quark loops [896]. In the first case, large effects are generated if the left–right mixing (A term) of the up-squarks has a non-MFV structure [10]. In the second case, deviations from the SM are induced by non-MFV terms in the right–right down sector, provided that the ratio of the two Higgs vacuum expectation values ($\tan\beta = v_u/v_d$) is large ($\tan\beta \sim 30$–50).

The effective Hamiltonian encoding SD contributions in the general MSSM has the following structure:

$$\mathcal{H}_{\text{eff}}^{(\text{SD})} \propto \sum_{l=e,\mu,\tau} V_{ts}^* V_{td} \big[X_L (\bar{s}_L \gamma^\mu d_L)(\bar{\nu}_{lL} \gamma_\mu \nu_{lL})$$
$$+ X_R (\bar{s}_R \gamma^\mu d_R)(\bar{\nu}_{lL} \gamma_\mu \nu_{lL}) \big], \quad (181)$$

where the SM case is recovered for $X_R = 0$ and $X_L = X_{\text{SM}}$. In general, both X_R and X_L are nonvanishing, and the misalignment between quark and squark flavor structures implies that they are both complex quantities. Since the $K \to \pi$ matrix elements of $(\bar{s}_L \gamma^\mu d_L)$ and $(\bar{s}_R \gamma^\mu d_R)$ are equal, the combination $X_L + X_R$ allows us to describe all the SD contributions to $K \to \pi \nu\bar{\nu}$ decays. More precisely, we can simply use the SM expressions for the branching ratios in (174) to (175) with the replacement

$$X_{\text{SM}} \to X_{\text{SM}} + X_L^{\text{SUSY}} + X_R^{\text{SUSY}}. \quad (182)$$

In the limit of almost degenerate superpartners, the leading chargino/up-squarks contribution is [895]

$$X_L^{\chi^\pm} \approx \frac{1}{96} \left[\frac{(\delta_{LR}^u)_{23}(\delta_{RL}^u)_{31}}{\lambda_t} \right] = \frac{1}{96\lambda_t} \frac{(\tilde{M}_u^2)_{2L3R}(\tilde{M}_u^2)_{3R1L}}{(\tilde{M}_u^2)_{LL}(\tilde{M}_u^2)_{RR}}. \quad (183)$$

As pointed out in [895], a remarkable feature of the above result is that no extra $\mathcal{O}(M_W/M_{\text{SUSY}})$ suppression and no explicit CKM suppression is present (as it happens in the chargino/up-squark contributions to other processes). Furthermore, the (δ_{LR}^u)-type mass insertions are not strongly constrained by other B- and K-observables. This implies that large departures from the SM expectations in $K \to \pi \nu\bar{\nu}$ decays are allowed, as confirmed by the complete analyses in [192, 879]. As illustrated in Fig. 45a, $K \to \pi \nu\bar{\nu}$ are the best observables to determine/constrain from experimental data the size of the off-diagonal (δ_{LR}^u) mass insertions or, equivalently, the up-type trilinear terms A_{i3} [$(\tilde{M}_u^2)_{iL3R} \approx m_t A_{i3}$]. Their measurement is therefore extremely interesting also in the LHC era.

In the large $\tan\beta$ limit, the charged Higgs/top quark exchange leads to [896]

$$X_R^{H^\pm} \approx \left[\left(\frac{m_s m_d t_\beta^2}{2 M_W^2} \right) \right.$$
$$\left. + \frac{(\delta_{RR}^d)_{31}(\delta_{RR}^d)_{32}}{\lambda_t} \left(\frac{m_b^2 t_\beta^2}{2 M_W^2} \right) \frac{\epsilon_{RR}^2 t_\beta^2}{(1+\epsilon_i t_\beta)^4} \right]$$
$$\times f_H(y_{tH}), \quad (184)$$

where $y_{tH} = m_t^2/M_H^2$, $f_H(x) = x/4(1-x) + x \log x/4(x-1)^2$ and $\epsilon_{i,RR} t_\beta = \mathcal{O}(1)$ for $t_\beta = \tan\beta \sim 50$. The first term of (184) arises from MFV effects, and its potential $\tan\beta$ enhancement is more than compensated by the smallness of $m_{d,s}$. The second term on the r.h.s. of (184), which would appear only at the three-loop level in a standard loop expansion, can be largely enhanced by the $\tan^4\beta$ factor and does not contain any suppression due to light quark masses. Similarly to the double mass-insertion mechanism of (183), also in this case the potentially leading effect is the one generated when two off-diagonal squark mixing terms replace the two CKM factors V_{ts} and V_{td}.

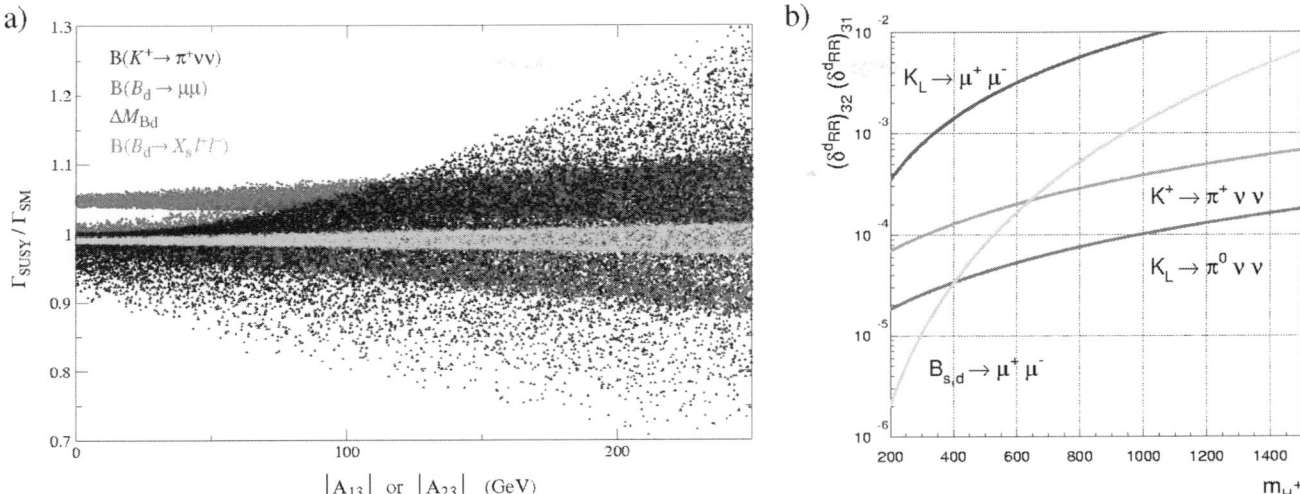

Fig. 45 Supersymmetric contributions to $K \to \pi \nu \bar{\nu}$. (**a**) Dependence of various FCNC observables (normalized to their SM value) on the up-type trilinear terms A_{13} and A_{23} for $A_{ij} \leq \lambda A_0$ and $\tan\beta = 2$–4 (other key parameters in GeV: $\mu = 500 \pm 10$, $M_2 = 300 \pm 10$, $M_{\tilde{u}_R} = 600 \pm 20$, $M_{\tilde{q}_L} = 800 \pm 20$, $A_0 = 1000$) [879]. (**b**) Sensitivity to $(\delta^d_{RR})_{23}(\delta^d_{RR})_{31}$ of various rare K- and B-decays as a function of M_{H^+}, setting $\tan\beta = 50$, $\mu < 0$ and assuming almost degenerate superpartners (the bounds from the two $K \to \pi \nu \bar{\nu}$ modes are obtained assuming a 10% measurement of their branching ratios, while the $B_{s,d} \to \mu^+\mu^-$ bounds refer to the present experimental limits [896])

The coupling of the $(\bar{s}_R \gamma^\mu d_R)(\bar{\nu}_L \gamma_\mu \nu_L)$ effective FCNC operator generated by charged-Higgs/top quark loops is phenomenologically relevant only at large $\tan\beta$ and with non-MFV right–right soft-breaking terms: a specific but well-motivated scenario within grand-unified theories (see e.g. [897, 898]). These nonstandard effects do not vanish in the limit of heavy squarks and gauginos and have a slow decoupling with respect to the charged-Higgs boson mass. As shown in [896], the B-physics constraints still allow a large room of nonstandard effects in $K \to \pi \nu \bar{\nu}$ even for flavor-mixing terms of CKM size (see Fig. 45b).

3.8.5 $K_L \to \pi^0 \ell^+ \ell^-$ beyond the SM

Within the SM, $K_L \to \pi^0 e^+ e^-$ and $K_L \to \pi^0 \mu^+ \mu^-$ decays have a very similar dynamics but for the different lepton masses. This makes them an ideal probe of NP effects when taken in combination [872, 887]. Moreover, $K_L \to \pi^0 \mu^+ \mu^-$ is sensitive to Higgs-induced helicity-suppressed operators, to which $K \to \pi \nu \bar{\nu}$ (and $K_L \to \pi^0 e^+ e^-$) are blind.

NP with SM operators In many scenarios, such as enhanced electroweak penguins (EEWP) [780], the MSSM at moderate $\tan\beta$ [899], Little Higgs models (LHT) [158], UED [181] and leptoquark models [900], NP only modifies the strength of the SM operators, without introducing new structures. In general, these models induce larger effects for $K_L \to \pi^0 \nu \bar{\nu}$ than for $K_L \to \pi^0 \ell^+ \ell^-$. Still, the latter modes should not be disregarded as they offer the possibility to disentangle effects in the vector and axial-vector currents. Indeed, Q_{7A} produces the final lepton pair also in a helicity-suppressed 0^{-+} state, hence contributes differently to $K_L \to \pi^0 e^+ e^-$ and $K_L \to \pi^0 \mu^+ \mu^-$, while the Q_{7V} contributions are identical for both modes (up to phase-space corrections, and assuming lepton flavor universality) [872].

As a consequence, the area spanned in the $\mathcal{B}(K_L \to \pi^0 e^+ e^-) - \mathcal{B}(K_L \to \pi^0 \mu^+ \mu^-)$ plane for arbitrary $w_{7A,7V}$ is nontrivial, see Fig. 46b. Taking all errors into account, this translates into the bounds $0.1 + 0.24 \mathcal{B}^{ee} \leq \mathcal{B}^{\mu\mu} \leq 0.6 + 0.58 \mathcal{B}^{ee}$ with $\mathcal{B}^{\ell\ell} = \mathcal{B}(K_L \to \pi^0 \ell^+ \ell^-) \times 10^{11}$ [887].

Usually, in specific models, there are correlations between the effects of NP on Q_{7V} and Q_{7A} operators. In the MSSM at moderate $\tan\beta$, the dominant effect is due to chargino contributions to Z^0- and γ-penguins [131, 866, 894, 895] sensitive to the double up-squark mass insertions. Since Z^0- and γ-penguins are correlated, so are Q_{7V} and Q_{7A}, and only a subregion of the red area can be reached. This is true whether or not there are new CP-phases. Interestingly, in the LHT model [158], the contributions to w_{7V} cancel each other to a large extent, leading to a quasi one-to-one correspondence, see Fig. 44b. This constitutes a powerful test of the model. In the case of MFV, the overall effect is found to be always smaller than for $K_L \to \pi^0 \nu \bar{\nu}$, with a maximum enhancement w.r.t. the SM of about 10% [879]. Finally, the contribution of the dipole operator $(\bar{s}\sigma^{\mu\nu}d)F_{\mu\nu}$ can be absorbed into w_{7V} [131], and NP contributions of this type cannot be singled out.

NP with new operators NP could of course also induce new operators. A systematic analysis of the impact of all possible dimension-six semileptonic operators on

$K_L \to \pi^0 \ell^+ \ell^-$ can be found in [887]. Here we concentrate on the most interesting case of (pseudo-)scalar operators $Q_S = (\bar{s}d)(\bar{\ell}\ell)$ and $Q_P = (\bar{s}d)(\bar{\ell}\gamma_5\ell)$ inducing a CPC (CPV) contribution. These operators are enhanced in the MSSM at large $\tan\beta$ where they originate from neutral Higgs exchanges and are sensitive to down-squark mass insertions [563]. Being helicity-suppressed, they affect only the muon mode and can lead to a clear signal outside the red region in Fig. 46b. Of course, in the MSSM, the $(\bar{s}\gamma_5 d)(\bar{\ell}\ell)$ and $(\bar{s}\gamma_5 d)(\bar{\ell}\gamma_5\ell)$ operators contributing to $K_L \to \ell^+\ell^-$ are also generated. Interestingly, the current $\mathcal{B}(K_L \to \mu^+\mu^-)^{\text{exp}}$ still leaves open the large yellow region in Fig. 46b, when combined with general $Q_{7V,7A}$ operators.

Finally, note that tree-level leptoquark exchange [900] or sneutrino exchange in SUSY without R-parity [901–904] can also induce (pseudo-)scalar operators but without helicity-suppression. However, to evade the strong constraint from $\mathcal{B}(K_L \to e^+e^-)^{\text{exp}} = (9^{+6}_{-4}) \times 10^{-12}$, one would need to invoke a large breaking of lepton-flavor universality to have a visible effect in $K_L \to \pi^0 \mu^+\mu^-$.

3.8.6 Conclusions on the theoretical prospects

Rare K-decays are excellent probes of NP. Firstly, their exceptional cleanness allows one to access very high energy scales. As stressed recently in [35, 158, 860, 879], NP could be seen in rare K-decays without significant signals in $B_{d,s}$-decays and, in specific scenarios, even without new particles within the LHC reach. Secondly, if LHC finds NP, its energy scale will be fixed. Then, the combined measurements of the four rare K-modes would help in discriminating among NP models. For instance, we have seen that specific correlations exist in MFV or LHT, which can be used as powerful tests (see Fig. 44). Further, in all cases, the information extracted from the four modes is essential to establish the NP flavor structure in the $s \to d$ sector, as illustrated in the MSSM at both moderate (see Fig. 45a) and large $\tan\beta$ (see Figs. 45b and 46b). Rare K-decays are thus an integral part, along with B-physics and collider observables, of the grand project of reconstructing the NP model from data. Experimentally, together with these very rare modes, improving bounds on forbidden decays (e.g. $K \to \pi e \mu$) can be interesting. Also, rare K-decays would benefit from experimental progress in (less rare) radiative K-decays like $K_S \to \pi^0 \ell^+\ell^-$ (see Fig. 46a). For all these reasons, it is very important to pursue ambitious K-physics programs in the era of the LHC.

3.8.7 Program at CERN

The proposed experiment NA62 (formerly NA48/3) at CERN-SPS [905] aims to collect about 80 $K^+ \to \pi^+ \nu\bar{\nu}$ events with an excellent signal over background ratio in two years of running, allowing for a 10% measurement of the branching ratio of the $K^+ \to \pi^+ \nu\bar{\nu}$ decay. The data taking should start in 2010. NA62 will replace the NA48 apparatus at CERN and will make use of the existing beam line. The layout of the experiment is sketched in Fig. 47.

The experiment proposes to exploit a kaon decay in flight technique to achieve 10% of signal acceptance. An intense 400 GeV/c proton beam, extracted from the SPS, produces a secondary charged beam by impinging on a Be target. A 100 m long beam line selects a 75 GeV/c momentum beam with a 1% RMS momentum band. This beam covers

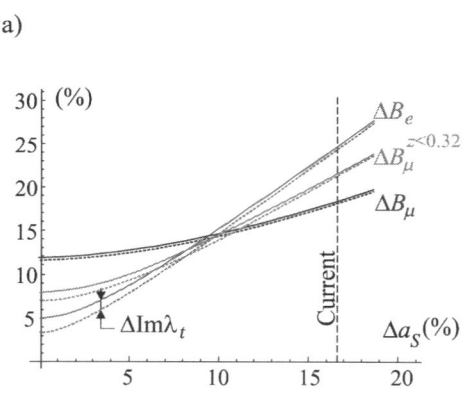

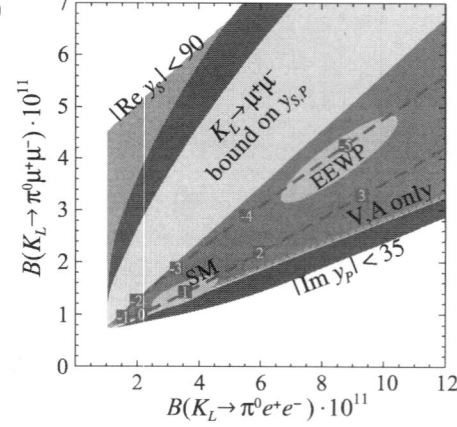

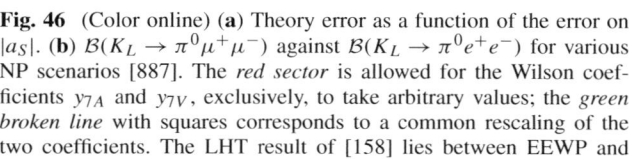

Fig. 46 (Color online) (**a**) Theory error as a function of the error on $|a_S|$. (**b**) $\mathcal{B}(K_L \to \pi^0 \mu^+\mu^-)$ against $\mathcal{B}(K_L \to \pi^0 e^+e^-)$ for various NP scenarios [887]. The *red sector* is allowed for the Wilson coefficients y_{7A} and y_{7V}, exclusively, to take arbitrary values; the *green broken line* with squares corresponds to a common rescaling of the two coefficients. The LHT result of [158] lies between EEWP and V, A only. *Light blue* (*dark blue*) corresponds to arbitrary $y_{7A,7V}$ together with $|\text{Re } y_S| < 90$ ($|\text{Im } y_P| < 35$), respectively, while the *yellow region* corresponds to $y_{7A,7V,S,P}$ arbitrary but compatible with the $\mathcal{B}(K_L \to \mu^+\mu^-)$ measurement, where y_S and y_S are the coefficients for scalar and pseudoscalar operators

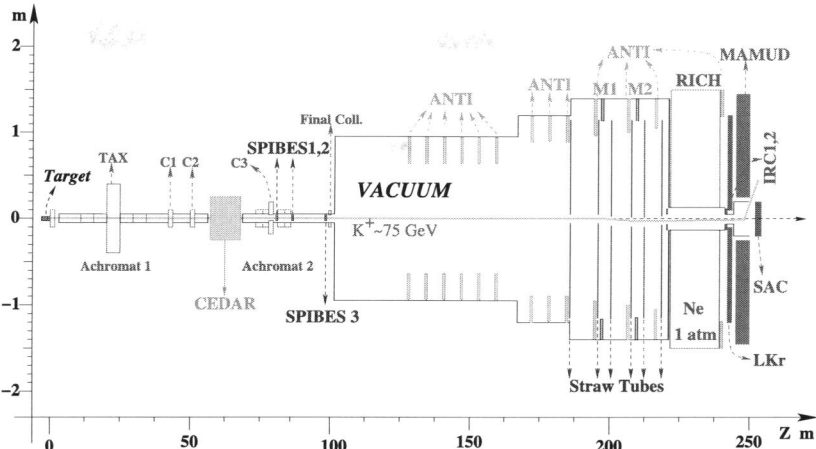

Fig. 47 Layout of the NA62 (NA48/3) experiment

a 16 cm² area, has an average rate of about 800 MHz and is composed by 6% of K^+ and 94% of π^+, e^+ and protons. A differential Cerenkov counter (CEDAR) placed along the beam line ensures a positive kaon identification. The beam enters in a 80 m long decay region evacuated at a level of 10^{-6} mbar, enough to avoid sizable background from the interaction of the particles with the residual gas. The kaon decay rate in the decay region is about 6 MHz: it provides about 10^{13} K^+ decays in two years of data taking, assuming 100 days as running time at 60% of efficiency, which is a very realistic estimate based on the decennial NA48 experience at the SPS.

The experimental signature of a $K^+ \to \pi^+ \nu \bar{\nu}$ is one reconstructed positive track in the downstream detector. The squared missing mass allows a kinematical separation between the signal and about 90% of the total background (see Fig. 48). The precise kinematical reconstruction of the event requires a performing tracking system for the beam particles and the charged decay products of the kaons.

The beam tracker consists of three Si pixels stations (SPIBES) having a surface of 36×48 mm². The charged particle rate on each station is about 60 MHz cm^{-2} on average. The stations are made up by 300×300 µm² pixels, 300 µm thick and containing the sensor and the chip bump-bonded on it. At least 200 ps time resolution per station is required to provide a suitable tag of the kaon track. A mistagging of the kaon, in fact, may be a source of background, because it spoils the resolution of the reconstructed squared missing mass.

Six straw chambers, 0.5% radiation length thick, placed in the same vacuum of the decay region form the downstream spectrometer. Two magnets provide a redundant measurement of the particle momentum, useful to keep the non-Gaussian tails of the reconstruction under control. The central hole of each station, which lets the undecayed beam pass through, must be displaced in the bending plane of the magnets according to the path of the 75 GeV/c positive beam. This configuration allows the tracker to be used as a veto for negative particles up to 60 GeV/c, needed for the rejection of backgrounds like $K^+ \to \pi^+ \pi^- e^+ \nu$. A reduced size prototype will be built and tested in 2007.

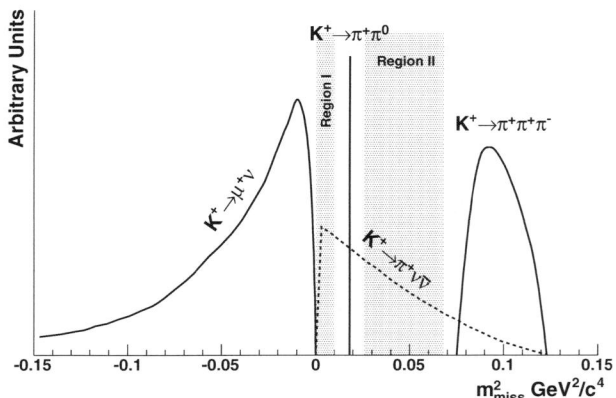
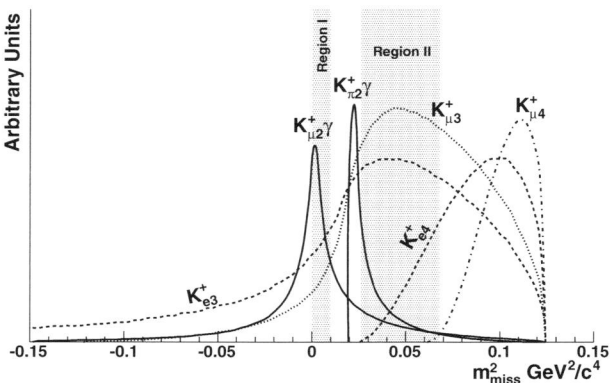

Fig. 48 Squared missing mass for kaon decays. The squared missing mass is defined as the square of the difference between the 4-momentum of the kaon and of the decayed track in the hypothesis that it is a pion

A system of γ vetoes, a μ veto and a RICH complement the tracking system to guarantee a 10^{13} level of background rejection.

A 18 m long RICH located after the spectrometer and filled with Ne at atmospheric pressure is the core of the $e^+/\pi/\mu$ separation. A 11 cm radius beam pipe crosses the RICH, and two tilted mirrors at the end reflect the Cerenkov light toward an array of about 2000 phototubes placed in the focal plane. Simulations showed that enough photoelectrons can be collected per track to achieve a better than 3σ π/μ separation between 15 and 35 GeV/c. The RICH provides also the timing of the downstream track with a 100 ps time resolution. The construction and test of a full length prototype is planned for 2007.

A combination of calorimeters covering up to 50 mrad serves to identify the photons. Ring-shaped calorimeters, most of them laying in the high vacuum of the decay region, cover the angular region between 10 and 50 mrad. Tests on prototypes built using lead scintillator tiles and scintillating fibers are scheduled for 2007 at a tagged γ facility at LNF. The existing NA48 liquid krypton calorimeter (LKr) [906] is intended to be used as a veto for γ down to 1 mrad. Data taken by NA48/2 in 2004 and a test run performed in 2006 using a tagged γ beam at CERN show that the LKr matches our requests in terms of efficiency. A program of consolidation and update of the readout electronics of the LKr is under way. Small calorimeters around the beam pipe and behind the muon veto cover the low angle region.

Six meters of alternated plates of iron and extruded scintillators form a hadronic sampling calorimeter (MAMUD) able to provide a 10^5 μ rejection. An aperture in the center lets the beam pass through, and a magnetic field inside deflects the beam out of the acceptance of the last γ veto.

Simulations of the whole apparatus based on GEANT3 and GEANT4 showed that 10% signal acceptance are safely achievable. The use of the RICH constrains the accepted pion track within the (15, 35) GeV/c momentum range. The higher cut is an important loss of signal acceptance but assures that events like $K^+ \to \pi^+\pi^0$ deposit at least 40 GeV of electromagnetic energy, making their rejection easier. The simulations indicate that a 10% background level is nearly achievable.

The overall experimental design requires a sophisticated technology for which an intense R&D program is started. Actually we propose an experiment able to reach a sensitivity of 10^{-12} per event, employing existing infrastructure and detectors at CERN.

3.8.8 Program at J-PARC

The Japan Proton Accelerator Research Complex (J-PARC) [907] is a new facility being constructed in the Tokai area of Japan as a joint project of High Energy Accelerator Research Organization (KEK) and Japan Atomic Energy Agency. Slow-extracted proton beam, which is of 30 GeV and whose intensity is 2×10^{14} protons per 0.7-s spill every 3.3 s at the Phase-1, is transported to the experimental area called NP Hall (Fig. 49). The proton beam hits the target and produces a variety of secondary particles, including low-energy K^+'s and K_L's.

The first PAC meeting for Nuclear and Particle Physics Experiments at J-PARC was held in the early summer of 2006 [908]. Concerning kaon physics, two proposals, "Measurement of T-violating Transverse Muon Polarization in $K^+ \to \pi^0\mu^+\nu$ Decays" and "Proposal for $K_L \to \pi^0\nu\bar{\nu}$ Experiment at J-Parc" received scientific approval. The latter proposal on the $K_L \to \pi^0\nu\bar{\nu}$ decay is discussed in this section; the former one is discussed in Chapter *Flavor physics of leptons and dipole moments* of this volume.

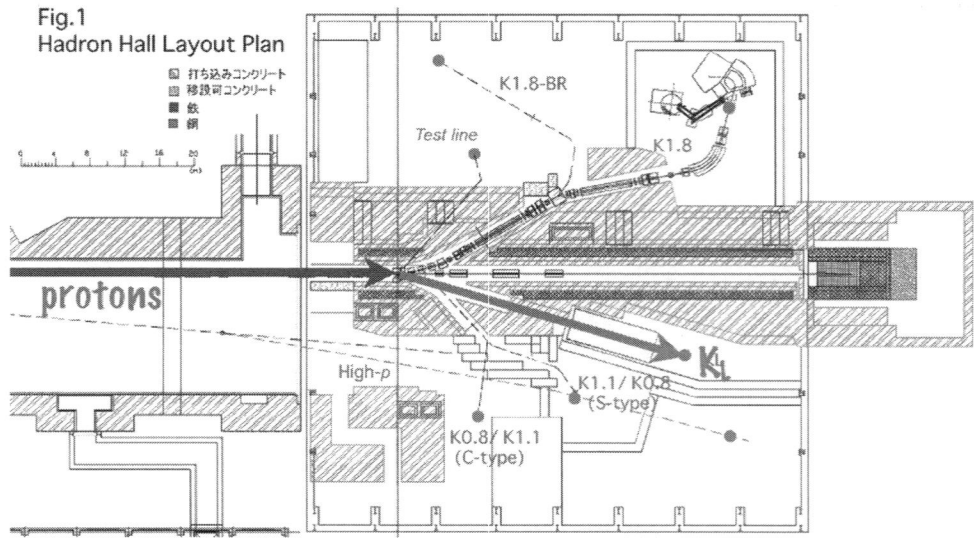

Fig. 49 A plan for the layout of NP Hall at J-PARC

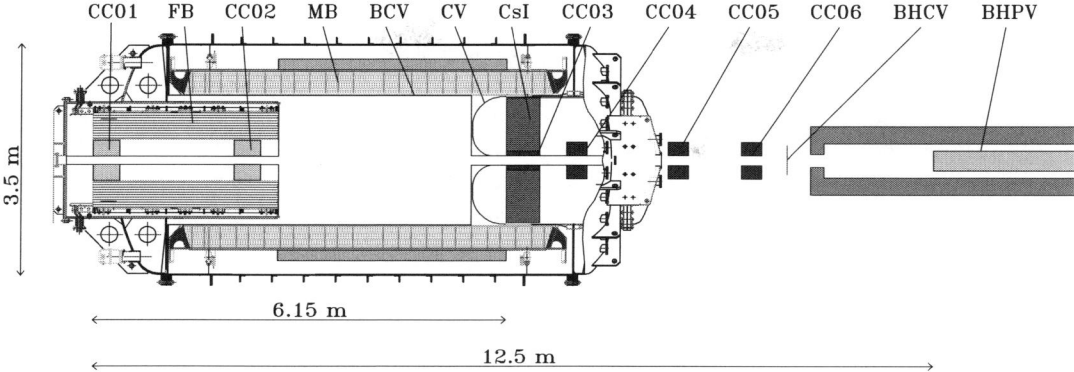

Fig. 50 Schematic view of the detector setup for the E14 experiment at J-PARC

The branching ratio for $K_L \to \pi^0 \nu \bar{\nu}$ is predicted to be $(2.7 \pm 0.4) \times 10^{-11}$ in the Standard Model, while the experimental upper limit, 6.7×10^{-8} at the 90% confidence level, is currently set by the E391a Collaboration at the KEK 12-GeV PS using the data collected during the second period of data taking [876]. E391a was the first dedicated experiment for $K_L \to \pi^0 \nu \bar{\nu}$ and aimed to be a pilot experiment. The new proposal at J-PARC [909] is to measure the branching ratio with an uncertainty less than 10% and takes a step-by-step approach to achieve this goal.

The common T1 target on the A-line and the beamline with a 16-degree extraction angle, as shown in Fig. 49, will be used in the first stage of the experiment (E14). Survey of a new neutral beamline in the first year of J-PARC commissioning and operation is essential to understand the beam-related issues at J-PARC. The E14 experiment will be performed by the date of "5 years of LHC" ($\sim$2012/2013); the goal is to make the first observation of the decay. In the current simulation, 3.5 SM events with 1.8×10^{21} protons on target in total are expected with the S/N ratio of 1.4. The beamline elements and the detector of E391a will be re-used by imposing necessary modifications. A schematic view of the detector setup is shown in Fig. 50. In particular, the undoped CsI crystals in the calorimeter for measuring the two photons from π^0 in $K_L \to \pi^0 \nu \bar{\nu}$ will be replaced with the smaller-size and longer crystals used in the Fermilab KTeV experiment (Fig. 51); discussions on the loan of the crystals are in progress. The technique of waveform digitization will be used on the outputs of the counters in the detector to distinguish pile-up signals from legitimate two-photon signals under the expected high-rate conditions. A new extra photon detection system to reduce the $K_L \to \pi^0 \pi^0$ background will cover the regions in or around the neutral beam.

After the E14 experiment establishes the experimental techniques to achieve the physics goal, the beamline and the detector will be upgraded for the next stage. More than 100 SM events (equivalent to a single event sensitivity of less

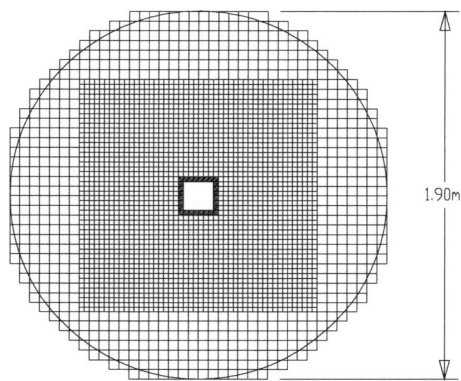

Fig. 51 Layout of the calorimeter for the J-PARC K_L experiment with the KTeV CsI crystals

than 3×10^{-13}) with an S/N ratio of 4.8 will be accumulated by the era of a "super B-factory" ($\sim$2020).

3.9 Charm physics[26]

3.9.1 *Case for continuing charm studies in a nutshell*

While nobody can doubt the seminal role that charm studies played for the evolution and acceptance of the SM, conventional wisdom is less enthused about their future. Yet on closer examination a strong case emerges in two respects, both of which are based on the weak phenomenology predicted by the SM for charm:

- to gain new insights into and make progress in establishing theoretical control over QCD's nonperturbative dynamics, which will also calibrate our theoretical tools for B studies;
- to use charm transitions as a novel window into NP.

Lessons from the first item will have an obvious impact on the tasks listed under the second one. They might actually

---
[26] Section coordinators: D.M. Asner, S. Fajfer.

be of great value even beyond QCD if the NP anticipated for the TeV scale is of strongly interacting variety.

Detailed analyses of leptonic and semileptonic decays of charm hadrons provide a challenging testbed for validating lattice QCD, which is the only known framework with the promise for a truly quantitative treatment of charm hadrons that can be improved *systematically*.

While significant 'profit' can be 'guaranteed' for the first item, the situation is less clear concerning the second one, the search for NP. While it had to be expected that no sign of NP would show up at the present level of experimental sensitivity, no clear-cut benchmark has been set at which level NP could emerge with even odds. In that sense, one is dealing with hypothesis-generating rather than probing research. It will be essential to harness the statistical power of the LHC for high-quality charm studies.

Yet the situation is much more promising than it seems at first glance. NP scenarios in general induce FCNCs that a priori have little reason to be as much suppressed as in the SM. More specifically, they could be substantially stronger for up-type than for down-type quarks; this can happen in particular in models which have to reduce strangeness changing neutral currents below phenomenologically acceptable levels by some alignment mechanism.

In such scenarios, charm plays a unique role among the up-type quarks u, c and t; for only charm allows the full range of probes for NP in general and flavor-changing neutral currents in particular: (i) Since top quarks do not hadronize [910], there can be no T^0–$\bar{T}^0$ oscillations. More generally, hadronization, while hard to bring under theoretical control, enhances the observability of CP violation. (ii) As far as u quarks are concerned, π^0, η and η' decay electromagnetically, not weakly. They are their own antiparticles and thus cannot oscillate. CP asymmetries are mostly ruled out by CPT invariance.

Our basic contention can then be formulated as follows: *Charm transitions provide a unique portal for a novel access to flavor dynamics with the experimental situation being a priori quite favorable (apart from the absence of Cabibbo suppression). Yet even that handicap can be overcome by statistics.*

The truly committed reader can find more nourishment for her/his curiosity in several recent reviews [911–913].

These points alluded to above will be addressed in somewhat more detail in the following sections.

3.9.2 Charm mixing

Prior observations of mixing in all down-type quark mixing systems puts charm physics in a unique position in the modern investigations of flavor physics as the system where the first evidence for the phenomena has emerged only recently (just before the publication of this document). Results of these studies are addressed after a short phenomenological introduction.

The SM contributions to charm mixing are suppressed to $\tan^2 \theta_c \approx 5\%$, because D^0 decays are Cabibbo favored. The GIM cancellation could further suppress mixing through off-shell intermediate states to 10^{-2}–10^{-6}. SM predictions for charm mixing rates span several orders of magnitude [913–917]. Fortunately, CP violation in mixing is $\mathcal{O}(10^{-6})$ in the SM, so CP violation involving D^0–$\bar{D}^0$ oscillations is a reliable probe of NP.

Charm physics studies are complementary to the corresponding programs in bottom or strange systems due to the fact that D^0–$\bar{D}^0$ mixing is influenced by the dynamical effects of *down-type particles*.

Effective $\Delta C = 2$ interactions generate contributions to the effective operators that change a D^0 state into a $\bar{D}^0$ state, leading to the mass eigenstates

$$|D_{\frac{1}{2}}\rangle = p|D^0\rangle \pm q|\bar{D}^0\rangle, \qquad R_m^2 = \left|\frac{q}{p}\right|^2, \qquad (185)$$

where the complex parameters p and q are obtained from diagonalizing the D^0–$\bar{D}^0$ mass matrix with $|p|^2 + |q|^2 = 1$. If CP-violation in mixing is neglected, p becomes equal to q, so $|D_{\frac{1}{2}}\rangle$ become CP eigenstates, $CP|D_\pm\rangle = \pm|D_\pm\rangle$.

The time evolution of a D^0 or $\bar{D}^0$ is conventionally described by an effective Hamiltonian which is non-Hermitian and allows the mesons to decay. We write

$$i\frac{\partial}{\partial t}\begin{bmatrix}|D^0(t)\rangle \\ |\bar{D}^0(t)\rangle\end{bmatrix} = \left(\boldsymbol{M} - \frac{i}{2}\boldsymbol{\Gamma}\right)\begin{bmatrix}|D^0(t)\rangle \\ |\bar{D}^0(t)\rangle\end{bmatrix},$$

where $\boldsymbol{M}$ and $\boldsymbol{\Gamma}$ are 2×2 matrices. We invoke CPT invariance so that $M_{11} = M_{22} \equiv M$ and $\Gamma_{11} = \Gamma_{22} \equiv \Gamma$. The eigenvalues of this Hamiltonian are

$$\lambda_{1,2} = M_{1,2} - \frac{i}{2}\Gamma_{1,2} \equiv \left(M - \frac{i}{2}\Gamma\right) \pm \frac{q}{p}\left(M_{12} - \frac{i}{2}\Gamma_{12}\right),$$

where $M_{1,2}$ are the masses of the $D_{1,2}$ and $\Gamma_{1,2}$ are their decay widths, and

$$\frac{q}{p} = \sqrt{\frac{M_{12}^* - \frac{i}{2}\Gamma_{12}^*}{M_{12} - \frac{i}{2}\Gamma_{12}}}.$$

The mass and width splittings between these eigenstates are given by

$$x \equiv \frac{m_1 - m_2}{\Gamma}, \qquad y \equiv \frac{\Gamma_1 - \Gamma_2}{2\Gamma}, \qquad R_M = \frac{x^2 + y^2}{2}. \qquad (186)$$

These parameters are experimentally observable and can be studied using a variety of methods to be discussed below.

SM and all reasonable models of NP predict $x, y \ll 1$ [913–917], which influences the available strategies for those measurements.

3.9.3 Semileptonic decays

The most natural way to search for charm mixing is to employ semileptonic decays. It is also not the most sensitive way, as it is only sensitive to R_M, a quadratic function of x and y. The use of the D^0 semileptonic decays for the mixing search involves the measurement of the time-dependent or time-integrated rate for the wrong-sign (WS) decays of D, where $c \to \bar{c} \to \bar{s}\ell^-\bar{\nu}$, relative to the right-sign (RS) decay rate, $c \to s\ell^+\nu$. Decays $D^0 \to K^{(*)-}\ell^+\bar{\nu}$ have been experimentally searched for [918–922]. Although the time-integrated rate is measured, several experiments use the time dependence of D^0 decays to increase the sensitivity. Currently the best sensitivity is reached by the Belle experiment, $R_M = (0.20 \pm 0.47 \pm 0.14) \times 10^{-3}$, using 253 fb^{-1} of data in $e^\pm$ mode only. Projecting to a possible 2 ab^{-1}, one can hope for a sensitivity of about $\pm 0.2 \times 10^{-3}$, including also systematic uncertainty.

3.9.4 Nonleptonic decays to non-CP eigenstates

A decay mode providing one of the best sensitivities to the mixing parameters is $D^0 \to K^+\pi^-$. Time-dependent studies allow separation of the direct doubly-Cabibbo suppressed (DCS) $D^0 \to K^+\pi^-$ amplitude from the mixing contribution $D^0 \to \bar{D}^0 \to K^+\pi^-$ [923, 924],

$$\Gamma[D^0 \to K^+\pi^-]$$
$$= e^{-\Gamma t}|A_{K^-\pi^+}|^2[R_D + \sqrt{R_D}R_m(y'\cos\phi - x'\sin\phi)\Gamma t$$
$$+ R_m^2 R_M^2 (\Gamma t)^2], \quad (187)$$

where R_D is the ratio of DCS and Cabibbo-favored (CF) decay rates. Since x and y are small, the best constraint comes from the linear terms in t that are also linear in x and y. A direct extraction of x and y from (187) is not possible due to the unknown relative strong phase $\delta_{K\pi}$ of DCS and CF amplitudes, as $x' = x\cos\delta_{K\pi} + y\sin\delta_{K\pi}$, $y' = y\cos\delta_{K\pi} - x\sin\delta_{K\pi}$. This phase can be measured independently (see CLEO-c result in Sect. 3.9.8). The corresponding formula can also be written [925] for $\bar{D}^0$ decay with $x' \to -x'$ and $R_m \to R_m^{-1}$.

Experimentally, this method of D^0 mixing search requires a good understanding of the detector decay time resolution to model correctly the measured distribution. Several experiments performed fits to disentangle the individual contributions in (187) [926–932]. The most recent study by BaBar Collaboration [933] finds an evidence for nonzero values of the mixing parameters. The preliminary 95% C.L. contours of the measured values are shown in Fig. 52. In terms of single parameter errors to be used for projections the most accurate is the measurement by Belle, using 400 fb^{-1} of data. Several fits to decay time distributions are performed; assuming that the CP violation is negligible, the result is $x'^2 = (0.18^{+0.21}_{-0.23}) \times 10^{-3}$, $y' = (0.6^{+4.0}_{-3.9}) \times 10^{-3}$ and $R_D = (3.64 \pm 0.17) \times 10^{-3}$, where the errors are statistical only. Projections of the 95% C.L. (x'^2, y') contour to the axes yield confidence intervals of $x'^2 < 0.72 \times 10^{-3}$ and $y' \in [-9.9, 6.8] \times 10^{-3}$. With a 2 ab^{-1} data sample, a statistical accuracy of 0.1×10^{-3} and 2×10^{-3} can be expected for x'^2 and y', respectively, similar to the current systematic uncertainties; a large contribution to the latter is due to the background modeling, the understanding of which might improve with a larger data sample as well.

CDF has demonstrated the potential of experiments at hadron colliders to make mixing-related measurements using hadronic decays through the recent study of WS $D^0 \to$

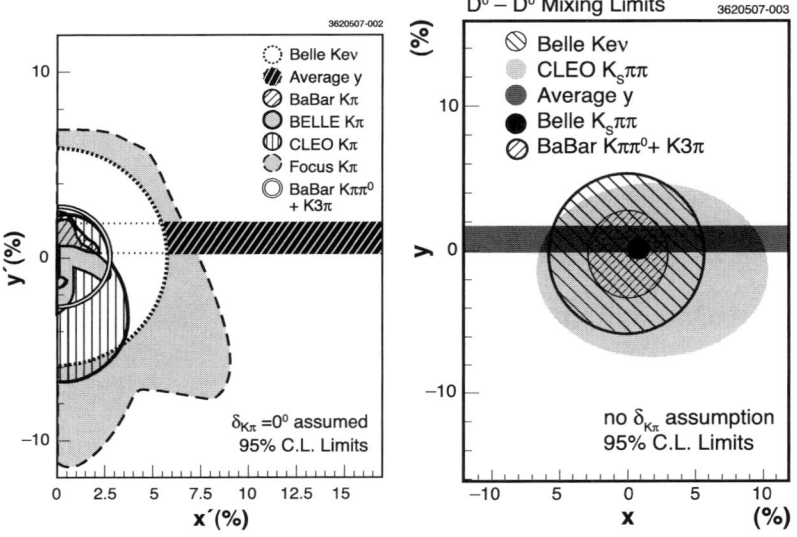

Fig. 52 Allowed regions in the x' vs y' plane (*left*) and x vs y for the measurements described in the text. We assume $\delta_{K\pi} = 0$ to place the y results in x' vs y'. A nonzero $\delta_{K\pi}$ would rotate the $D^0 \to$ CP eigenstates (y results) confidence region clockwise about the origin by δ

$K^+\pi^-$ events [934]. Using the distinctive $D^* \to D^0\pi$ signature and an integrated luminosity of $0.35\,\text{fb}^{-1}$, a sample of around 2000 WS decays have been accumulated with a background to signal level of order 1. The ratio of WS to RS decays is found to be $[4.05 \pm 0.21(\text{stat}) \pm 0.11(\text{syst})] \times 10^{-3}$. This ratio is equivalent to R_D in the limit that x' and y' are zero, and CP violation is negligible. Provided that the systematic uncertainties can continue to be kept under control, the full Tevatron dataset of several fb^{-1} will give a more precise result for R_D than the B-factories, under the stated assumption. More interesting results are to be expected should it prove possible to perform a time-dependent measurement.

LHCb expects to collect very high statistics in all charged two-body D^0 decays through the inclusion of a dedicated $D^* \to D^0(hh')\pi$ filter in the experiment's high-level trigger [935]. In one year of operation at nominal luminosity ($2\,\text{fb}^{-1}$), 0.2 million WS and 50 million RS $K\pi$ events will be written to tape, where the triggered D^* has originated from a B decay. A similar number of decays are expected where the D^* is produced in the primary event vertex.

In a mixing analysis, it is necessary to measure the proper lifetime of the decaying D^0. LHCb's good vertexing allows the decay point of the D^0 to be well determined, and also the production point in the case of D^*'s produced in the primary vertex. For that sample where the D^* arises from a B decay, it is necessary to vertex the D^0 direction with other B decay products in order to find the production point, a procedure which entails a loss in efficiency. Additional cuts are needed to enhance the purity of the WS signal and combat the most significant background source, where the wrong 'slow pion' is associated with a genuine D^0. This contamination is dangerous for the reason that is the charge of the slow pion which tags the initial flavor of the D^0 meson. After this selection, 46,500 WS decays are expected from B events per $2\,\text{fb}^{-1}$, with a background to signal ratio of around 2.5.

These performance figures have been used as input to a 'toy Monte Carlo' study to determine LHCb's sensitivity to the mixing parameters, including both the effects of background and the estimated proper time resolution and acceptance. The study was performed for event yields corresponding to $10\,\text{fb}^{-1}$ of integrated luminosity, that is 5 years of operation at nominal operation. It was found that with such a sample, LHCb will have a statistical sensitivity on x'^2 and y' of 0.6×10^{-4} and 0.9×10^{-3}, respectively. Further work is needed to identify and combat the possible sources of systematic uncertainty.

3.9.5 Multi-body hadronic D^0 decays

In multi-body hadronic D^0 decays, possible differences in the resonant structure between the CF and DCS decays must be taken into account and, as discussed below, be exploited. The time integrated relative rates $R_{WS} = \Gamma(D^0 \to K^+\pi^-(n\pi))/\Gamma(D^0 \to K^-\pi^+(n\pi))$, which assuming negligible CP violation equal to $R_D + \sqrt{R_D}y' + (x'^2 + y'^2)/2$, have been measured for $n\pi = \pi^0, \pi^+\pi^-$ [937, 942, 946, 947]. For the latter mode, Belle measures $R_{WS}(K\pi\pi\pi) = (0.320 \pm 0.018 \pm 0.013)\%$. Assuming a particular value of x' in combination with the previous equation gives an allowed band in the (R_D, y') plane; however, one should note that the value of x' is decay mode dependent. Studies with $D^0 \to K^\mp\pi^\pm\pi^-\pi^+$ events will also be possible at LHCb, where plans are under consideration to extend the $D^* \to D^0(h^+h'^-)\pi$ high-level trigger stream to include charged 4-body D^0 decays. The foreseen event yields would be similar to those anticipated for the $D^0 \to K^\mp\pi^\pm$ case.

The BaBar Collaboration studied the time-dependence of the above multi-body decay modes [948]. Since the possible mixing contribution followed by CF decay needs to be distinguished from the DCS decays, the sensitivity of the measurement is increased by selecting regions of phase space where the ratio of the two is the largest. The preliminary value of R_M, which is not affected by this selection, is found to be $R_M = (0.023 \pm {}^{0.018}_{0.014} \pm 0.004)\%$ ($R_M < 0.054\%$ at 95% C.L. using a Bayesian approach) in the $D^0 \to K^+\pi^-\pi^0$ mode, and without selecting a region of phase-space $R_{WS}(K\pi\pi^0) = (0.214 \pm 0.008 \pm 0.008)\%$ is obtained. By combining the obtained $\delta \log \mathcal{L}(R_M)$ curve with the one from the study of the $D^0 \to K^+\pi - \pi^+\pi^-$ channel $R_M = (0.020 \pm {}^{0.011}_{0.010})\%$ ($R_M < 0.042\%$ at 95% C.L. using a Bayesian approach) is obtained (stat. uncertainty only). The combined data are compatible with the no-mixing hypothesis at the 2.1% C.L.

3.9.6 Time-dependent Dalitz-plot analysis

Due to the strong variation of the interference effects over the $D^0 \to K^+\pi^-(n\pi)$ phase-space, a Dalitz analysis of these modes can give further insight into the D^0 mixing. Such an analysis has been performed for $D^0 \to K_S\pi^-\pi^+$ channel by CLEO Collaboration [949], and recently results from Belle Collaboration became available [950]. Different intermediate states contributing to $K_S\pi^-\pi^+$ (CP even or odd, like K_Sf_0 or $K_S\rho^0$, or flavor eigenstates, like $K^*(892)^+\pi^-$), that can be determined by inspection of the Dalitz plane, contribute differently to the decay time distribution of $D^0 \to K_S\pi^-\pi^+$. A simultaneous fit of the Dalitz and decay time distributions is used to determine the mixing parameters $x = (0.80 \pm 0.29 \pm 0.17)\%$ and $y = (0.33 \pm 0.24 \pm 0.15)\%$. Important systematic error arises due to the uncertainty of the model used for the description of the Dalitz structure (around $\pm 0.15\%$ and $\pm 0.10\%$ on x and y, respectively). Projecting the amount of data used in the analysis ($540\,\text{fb}^{-1}$) to the amount possibly available to the B-factories in the future ($2\,\text{ab}^{-1}$), the statistical precision on each parameter could be improved to ~0.15%. Hence the systematic error, receiving contributions from the uncertainty of the t distribution modeling (similar as for the case of $D^0 \to K^+\pi^-$ decays) as well as from the Dalitz model, will need to be studied carefully.

3.9.7 Nonleptonic decays to CP eigenstates

D^0 mixing can be measured by comparing the lifetimes extracted from the analysis of D decays into the CP-even and CP-odd final states. In practice, the lifetime measured in D decays into CP-even final state f_{CP}, such as $K^+K^-, \pi^+\pi^-, \phi K_S$, etc., is compared to the one obtained from a measurement of decays to a non-CP eigenstate, such as $K^-\pi^+$. This analysis is also sensitive to a *linear* function of y via

$$y_{CP} = \frac{\tau(D \to K^-\pi^+)}{\tau(D \to K^+K^-)} - 1 = y\cos\phi - x\sin\phi\left[\frac{R_m^2 - 1}{2}\right], \tag{188}$$

where ϕ is a CP-violating phase. In the limit of vanishing CP violation, $y_{CP} = y$. This measurement requires precise determination of lifetimes. It profits from some cancellation of the systematic uncertainties in the ratio $\tau(K^-\pi^+)/\tau(f_{CP})$. To date CP = +1 final states K^+K^- and $\pi^+\pi^-$ have been used [951–957].

In the course of preparation of this document, the Belle Collaboration obtained a new result on y_{CP} using 540 fb^{-1} of data [957]. It represents evidence for the D^0–$\bar{D}^0$ mixing with $y_{CP} = 1.31 \pm 0.32 \pm 0.25\%$ differing from zero by 3.2 standard deviations.

With the currently available statistical samples at the B-factories, the statistical uncertainty of the measurements using the D^{*+} tag is comparable to the systematic one. The latter arises mainly from an imperfect modeling of the t distribution of the background (although the overall background level is small, and the systematic uncertainty due to this source might decrease with increased data sample) and from a possible noncancellation of systematic errors on individual lifetime measurements. With the final B-factories' data set, one can hope for a total uncertainty on y_{CP} of around $\pm 0.25\%$. To this, systematic error contributes $\pm 0.10\%$ if the sources expected to scale with the luminosity are taken into account.

LHCb intends to make an important contribution to the measurements of a nonzero value of y_{CP} through the high statistics available from the D^* trigger and the excellent particle identification capabilities of its RICH system. A sample of 1.6×10^6 $D^0 \to K^+K^-$ events is expected from B decays alone after all selection cuts. The expected sensitivity to y_{CP} from this source with 5 years of data is 0.5×10^{-3}.

3.9.8 Quantum-correlated final states

The construction of tau-charm factories introduces new *time-independent* methods that are sensitive to a linear function of y. One can use the fact that heavy meson pairs produced in the decays of heavy quarkonium resonances have the useful property that the two mesons are in the CP-correlated states [958, 959]. For instance, by tagging one of the mesons as a CP eigenstate, a lifetime difference may be determined by measuring the leptonic branching ratio of the other meson. The final states reachable by neutral charmed mesons are determined by a set of selection rules according to the initial virtual photon quantum numbers $J^{PC} = 1^{--}$ [959, 960]. Currently, the decay rates of several singly-tagged (only a single meson is fully reconstructed) and doubly-tagged (both mesons reconstructed) final states of the $D^0\bar{D}^0$ pairs are measured at CLEO-c [961], where the individual fractions depend on the mixing parameters y and R_M, D^0 branching fractions and phases between DCS and CF decays. Types of decays considered include semileptonic decays and decays to flavor and CP eigenstates. The above parameters are determined from a fit to the efficiency-corrected yields using 281 pb^{-1} of data, with the preliminary results most relevant to the D^0 mixing $y = -0.058 \pm 0.066$, $R_M = (1.7 \pm 1.5) \times 10^{-3}$ and $\cos\delta_{K\pi} = 1.09 \pm 0.66$. The systematic uncertainties, expected to be of smaller size, are being evaluated. At CLEO-c, the precision of results is expected to be reduced by increasing the data sample by a factor of three, increasing the number of CP eigenstate modes and using constraints from other measurements of D^0 branching fractions. The same method will be exploited by BES III with an expected data sample of 20 fb^{-1}. Statistical uncertainty could be reduced to $\sigma(y) \sim 0.002$, $\sigma(R_M) \sim 0.2 \times 10^{-3}$ and $\sigma(\cos\delta_{K\pi}) \sim 0.02$.

3.9.9 Summary of experimental D mixing results

The constraints in x' vs. y' and x vs y are shown in Fig. 52. Approximate uncertainties of the measured quantities, as expected from the data samples assumed above, are shown in Table 46. The errors shown include scaled statistical errors from the most precise existing measurements and estimates of possible systematic uncertainties.

As a simple illustration of the projected results, a χ^2 minimization in terms of the mixing parameters x and y and of $\cos\delta_{K\pi}$ can be performed. For the unknown true values $x = 5 \times 10^{-3}$, $y = 1 \times 10^{-2}$ and $\delta_{K\pi} = 0°$, one finds the central 68% C.L. intervals of $x \in [3, 7] \times 10^{-3}$, $y \in [0.85, 1.15] \times 10^{-2}$ and $\delta_{K\pi} \in [-12°, 12°]$. In some cases, the p.d.f.'s for the estimated parameters are significantly non-Gaussian.

The HFAG charm decays subgroup [389] is preparing world averages of all the charm measurements. For charm

Table 46 Approximate expected precision (σ) on the measured quantities using methods described in the text for the integrated luminosity of 10 fb^{-1} at LHCb, 2 ab^{-1} at the B-factories at 10 GeV and 20 fb^{-1} at BESIII running at charm threshold. The LHCb numbers do not include the effect of systematic errors but neglect the contribution of events from prompt charm production. Entries marked '/' in the LHCb column are where expected performance numbers are not yet available

Mode	Observable	LHCb (10 fb^{-1})	B-factories (2 ab^{-1})	$\psi(3770)$ (20 fb^{-1})
$D^0 \to K^{(*)-}\ell^+\bar{\nu}$	R_M	/	0.2×10^{-3}	
$D^0 \to K^+\pi^-$	x'^2	0.6×10^{-4}	1.5×10^{-4}	
	y'	0.9×10^{-3}	2.5×10^{-3}	
$D^0 \to K^+K^-$	y_{CP}	0.5×10^{-3}	3×10^{-3}	
$D^0 \to K_S^0\pi^+\pi^-$	x	/	2×10^{-3}	
	y	/	2×10^{-3}	
$\psi(3770) \to D^0\bar{D}^0$	x^2			3×10^{-4}
	y			4×10^{-3}
	$\cos\delta$			0.05

mixing, the averages not only take into account correlations between measurements but combine the multidimensional likelihood functions associated with each measurement. A very preliminary average is available [389] giving $x = (8.7^{+3.0}_{-3.4}) \times 10^{-3}$ and $y = (6.6^{+2.1}_{-2.0}) \times 10^{-3}$. Allowing for CP violation, the very preliminary average is $x = (8.4^{+3.2}_{-3.4}) \times 10^{-3}$ and $y = (6.9 \pm 2.1) \times 10^{-3}$.

The constraints in the x vs y plane are shown in Fig. 53. The significance of the oscillation effect exceeds 5σ.

The interpretation of the new results in terms of NP is inconclusive. It is not yet clear whether the effect is caused by $x = 0$ or $y = 0$ or both, although the latter is favored, as shown in Table 47. Both an upgraded LHCb and a high luminosity super B-factory will be able to observe both life-

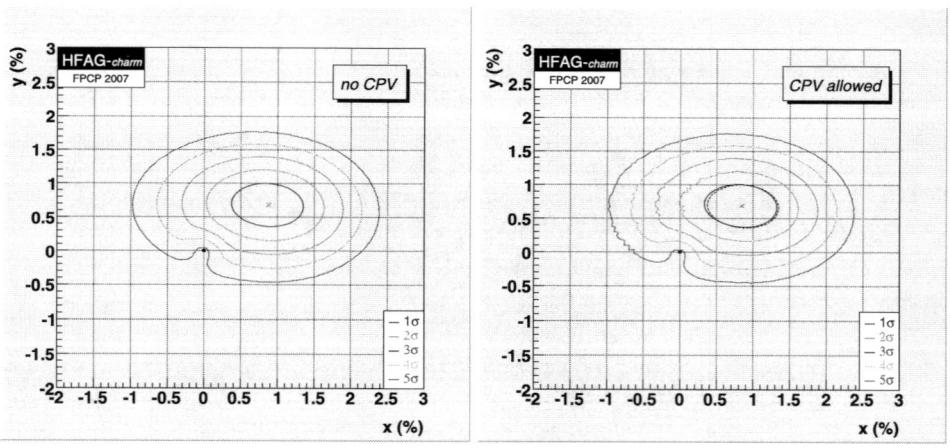

Fig. 53 All charm mixing measurements are combined by HFAG [389] to provide constraints in the x vs. y plane. Contours (1 through 5σ) of the allowed region are shown. The significance of the oscillation effect exceeds 5σ

Table 47 Approximate expected precision (σ) on the measured quantities using methods described in the text for the integrated luminosity of 100 fb^{-1} at an upgraded LHCb, 75 ab^{-1} at a super B-factory at 10 GeV and 200 fb^{-1} at a super B-factory running at charm threshold. The upgraded LHCb numbers are merely the results from Table 46 scaled to the new integrated luminosity

Mode	Observable	LHCb (100 fb^{-1})	Super B (75 ab^{-1})	$\psi(3770)$ (200 fb^{-1})
$D^0 \to K^+\pi^-$	x'^2	2.0×10^{-5}	3×10^{-5}	
	y'	2.8×10^{-4}	7×10^{-4}	
$D^0 \to K^+K^-$	y_{CP}	1.5×10^{-4}	5×10^{-4}	
$D^0 \to K_S^0\pi^+\pi^-$	x	/	5×10^{-4}	
	y	/	5×10^{-4}	
$\psi(3770) \to D^0\bar{D}^0$	x^2			$<0.2 \times 10^{-4}$
	y			$(1-2) \times 10^{-3}$
	$\cos\delta$			<0.05

time and mass differences in the D^0 system if they lie in the range of SM predictions.

A serious limitation in the interpretation of charm oscillations in terms of NP is the theoretical uncertainty on the SM prediction. Nonetheless, if oscillations occur at the level suggested by the recent results, this will open the window to searches for CP asymmetries that do provide unequivocal NP signals.

3.9.10 New Physics contributions to D mixing

As one can see from the previous discussion, mixing in the charm system is very small. As it turns out, theoretical predictions of x and y in the SM are very uncertain, from a percent to orders of magnitude smaller [914–917, 962–964]. Thus, NP contributions are difficult to distinguish in the absence of large CP violation in mixing.

In order to see how NP might affect the mixing amplitude, it is instructive to consider off-diagonal terms in the neutral D mass matrix,

$$\left(M - \frac{i}{2}\Gamma\right)_{12}$$
$$= \frac{1}{2M_D}\langle \bar{D}^0|\mathcal{H}_w^{\Delta C=-2}|D^0\rangle$$
$$+ \frac{1}{2M_D}\sum_n \frac{\langle \bar{D}^0|\mathcal{H}_w^{\Delta C=-1}|n\rangle\langle n|\mathcal{H}_w^{\Delta C=-1}|D^0\rangle}{M_D - E_n + i\epsilon}, \quad (189)$$

where $\mathcal{H}_w^{\Delta C=-1}$ is the effective $|\Delta C| = 1$ Hamiltonian. Since all new physics particles are much heavier than the SM ones, the most natural place for NP to affect mixing amplitudes is in the $|\Delta C| = 2$ contribution, which corresponds to a local interaction at the charm quark mass scale.

As can be seen from Fig. 54, predictions for x vary by orders of magnitude for different models. It is interesting to note that some models *require* large signals in the charm system if mixing and FCNCs in the strange and beauty systems are to be small (e.g. the SUSY alignment model).

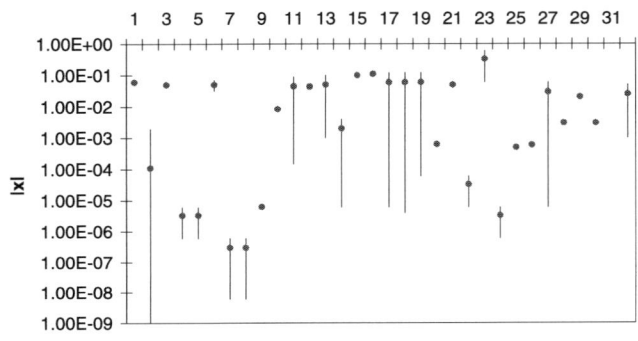

Fig. 54 NP predictions for $|x|$. Horizontal line references are tabulated in Table 5 of Refs. [914–917]

The local $|\Delta C| = 2$ interaction cannot, however, affect $\Delta \Gamma_D$, because it does not have an absorptive part. Thus, naively, NP cannot affect the lifetime difference y. This is, however, not quite correct. Consider a D^0 decay amplitude which includes a small NP contribution, $A[D^0 \to n] = A_n^{(SM)} + A_n^{(NP)}$. Here, $A_n^{(NP)}$ is assumed to be smaller than the current experimental uncertainties on those decay rates. Then it is a good approximation to write y as

$$y \simeq \sum_n \frac{\rho_n}{\Gamma_D} A_n^{(SM)} \bar{A}_n^{(SM)} + 2\sum_n \frac{\rho_n}{\Gamma_D} A_n^{(NP)} \bar{A}_n^{(SM)}. \quad (190)$$

The SM contribution to y is known to vanish in the limit of exact flavor $SU(3)$. Moreover, the first-order correction is also absent, so the SM contribution arises only as a *second*-order effect. Thus, those NP contributions which do not vanish in the flavor $SU(3)$ limit must determine the lifetime difference there, even if their contributions are tiny in the individual decay amplitudes [965]. A simple calculation reveals that NP contribution to y can be as large as several percent in R-parity-violating SUSY models or as small as $\sim 10^{-10}$ in the models with interactions mediated by charged Higgs particles [965]. Assuming that the projected precisions on x, y and $\cos(\delta_{K\pi})$ discussed below are achieved, a range of NP models can be ruled out. On the other hand, the uncertainty of SM predictions for the mixing parameters can in some scenarios (positive measurement, $y > x$) make the identification of NP contribution difficult. It is important to make a precise determination of individual parameters, using all the experimental methods mentioned (and possibly new ones) in order to pin down possible cracks in the SM.

3.9.11 D mixing impact on the CKM angle γ/ϕ_3

Beside the importance of the mixing in the charm sector per-se discussed above, the results of mentioned measurements can also have an impact on the determination of the UT angle γ/ϕ_3. Several proposed methods for measuring γ/ϕ_3 use the interference between $B^- \to D^0 K^-$ and $B^- \to \bar{D}^0 K^-$ which occurs when both D^0 and $\bar{D}^0$ decay to the same final state [627, 632, 633, 636, 645, 646].

The quantity sensitive to the angle γ/ϕ_3 is the asymmetry $A_{DK} = [Br(B^- \to f_D K^-) - Br(B^+ \to \bar{f}_D K^+)]/[Br(B^- \to f_D K^-) + Br(B^+ \to \bar{f}_D K^+)]$, where f_D denotes the common final state of D^0 and $\bar{D}^0$. A_{DK} can be expressed as

$$A_{DK} = \frac{2 r_B r_D e^{-\epsilon} \sin(\delta_B + \delta_D) \sin \gamma/\phi_3}{r_B^2 + r_D^2 + 2 r_B r_D e^{-\epsilon} \cos(\delta_B + \delta_D) \cos \gamma/\phi_3}, \quad (191)$$

where δ_B is the difference of the strong phases in decays $B^- \to D^0 K^-$ and $B^- \to \bar{D}^0 K^-$, δ_D is the difference of the strong phases for $D^0 \to f_D$ and $\bar{D}^0 \to f_D$, r_B is the ratio of amplitudes $|\mathcal{A}(B^- \to \bar{D}^0 K^-)|/|\mathcal{A}(B^- \to D^0 K^-)|$, and r_D is the ratio $|\mathcal{A}(D^0 \to f_D)|/|\mathcal{A}(\bar{D}^0 \to f_D)|$. The dilution factor $e^{-\epsilon}$ arises if $x, y \neq 0$.

In case of nonnegligible D^0 mixing, the time integrated interference term between $\mathcal{A}(D^0 \to f_D)$ and $\mathcal{A}(\bar{D}^0 \to f_D)$ depends on x and y, resulting in [631]

$$\epsilon = \frac{1}{8}(x^2 + y^2)\left(\frac{1}{r_D^2} + r_D^2\right) - \frac{1}{4}(x^2 \cos 2\delta_D + y^2 \sin 2\delta_D). \tag{192}$$

Using f_D which is a CP eigenstate [632, 633] (the case where $f_D = K_S^0 \pi^+ \pi^-$ is discussed in Sect. 3.9.27.1) and neglecting CP violation in D^0 decays, the above expressions simplify due to $r_D = 1$, $\delta_D = 0$, and thus $\epsilon = y^2/4$. For $f = K^+K^-, \pi^+\pi^-$, the asymmetry A_{DK} is measured to be $0.06 \pm 0.14 \pm 0.05$ using an integrated luminosity of 250 fb^{-1} [966]. Projecting the result to 2 ab^{-1}, the expected statistical accuracy is ± 0.05. An uncertainty on y of 2%, on the other hand, reflects in an error of $\sigma(A_{DK}) \approx 5 \times 10^{-5}$ using the above equations (conservatively assuming that $r_B = 0.25$ and $\sin \delta_B = \sin \phi_3 = 1$). It is thus save to conclude that neglecting the effect of D^0 mixing in this method of γ/ϕ_3 determination is appropriate.

Besides f_D being a CP eigenstate, the final state can be chosen to arise from DCS decays [636, 645, 646]. In this case, the strong phase δ_D enters the expressions. To illustrate the effect of δ_D on extraction of the angle γ/ϕ_3, one can envisage usage of two distinct final states, for example the above mentioned $f = K^+K^-, \pi^+\pi^-$ and $K^+\pi^-$, which can also be reached from either D^0 or $\bar{D}^0$. For the former, the same asymmetry A_{DK} can be measured, while for the latter, the ratio $R_{DK} = Br(B^- \to D_{\text{sup}} K^-)/Br(B^- \to D_{\text{fav}} K^-)$ is also sensitive to γ/ϕ_3. Here, D_{sup} denotes DCS decays $D^0 \to K^+\pi^-$, and D_{fav} stands for $D^0 \to K^-\pi^+$. R_{DK} depends on the unknown angles:

$$R_{DK} = r_B^2 + r_D^2 + 2r_B r_D \cos(\delta_B + \delta_D) \cos \gamma/\phi_3 \tag{193}$$

with $r_D = (6.2 \pm 0.1) \times 10^{-2}$ [119]. Assuming that r_B is known, measuring A_{DK} and R_{DK} constrains possible ranges for δ_B and γ/ϕ_3. Knowledge of δ_D clearly helps in limiting the $(\gamma/\phi_3, \delta_B)$ allowed region. We can use the projected result $A_{DK} = 0.06 \pm 0.05$ and the ratio $R_{DK} = (2.3 \pm 1.5 \pm 0.1) \times 10^{-2}$ as obtained using 250 fb^{-1} of data [967]. Hence one can expect $R_{DK} = (2.3 \pm 0.6) \times 10^{-2}$ with the final B-factories data set. The approximate two-dimensional 68% C.L. contour obtained by plotting the corresponding χ^2 of the two projected measurements as a function of γ/ϕ_3 and δ_B is shown in Fig. 55. The left plot shows the allowed region for the current value of $\delta_D = (0 \pm 1.15)$ rad [961]. To show the effect of an improved knowledge of the D-meson decays strong phase, the value $\delta_D = (0 \pm 0.45)$ rad (see Table 46) is used in the right plot. The allowed region of the unknown angles is significantly reduced, although it should be noted that the actual region strongly depends on the central values of δ_D as well as r_B (for the latter, the value 0.12 was used in the plots).

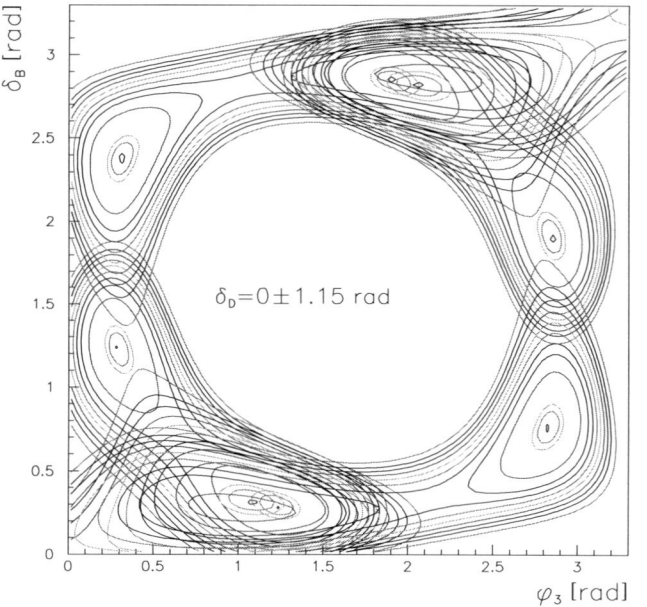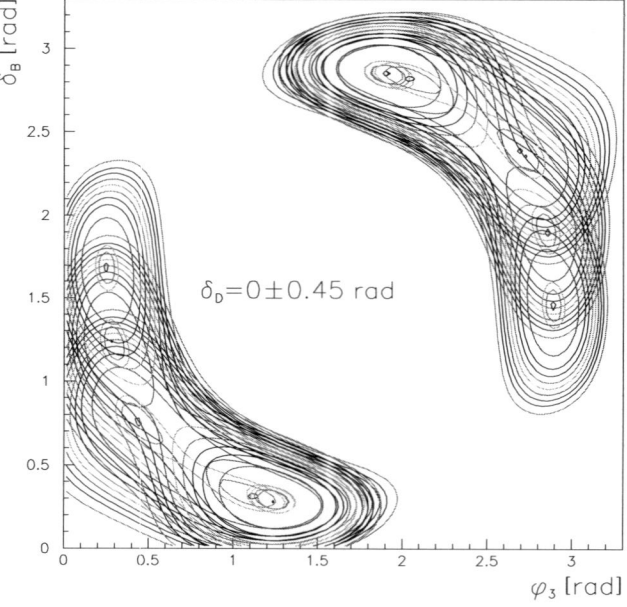

Fig. 55 The 68% C.L. contour for γ/ϕ_3 and δ_B using the projected results of measurements described in the text. The strong phase difference δ_D between $D^0 \to K^+\pi^-/K^-\pi^+$ decays is assumed to have the values marked in the plots

3.9.12 CP violation with and without oscillations

3.9.13 Theoretical overview

Most factors favor or even call for dedicated searches for CP violation in charm transitions:

⊕ Since baryogenesis implies the existence of NP in CP-violating dynamics, it would be unwise not to undertake dedicated searches for CP asymmetries in charm decays, where the 'background' from known physics is between absent and small: for within the SM, the effective weak phase is highly diluted, namely $\sim\mathcal{O}(\lambda^4)$, and it can arise only in *singly-Cabibbo-suppressed* transitions, where one expects asymmetries to reach the $\mathcal{O}(0.1\%)$ level; significantly larger values would signal NP. *Any* asymmetry in *Cabibbo-allowed or doubly-suppressed* channels requires the intervention of NP—except for $D^\pm \to K_S \pi^\pm$ [912], where the CP impurity in K_S induces an asymmetry of 3.3×10^{-3}. One should keep in mind that in going from Cabibbo-allowed to Cabibbo singly- and doubly-suppressed channels, the SM rate is *suppressed* by factors of about twenty and four hundred, respectively:

$$\Gamma_{SM}(H_c \to [S=-1]) : \Gamma_{SM}(H_c \to [S=0])$$
$$: \Gamma_{SM}(H_c \to [S=+1])$$
$$\simeq 1 : 1/20 : 1/400. \quad (194)$$

One would expect that this suppression will enhance the visibility of NP.

⊕ Strong phase shifts required for *direct* CP violation to emerge in partial widths are in general large as are the branching ratios into relevant modes; while large final-state interactions complicate the interpretation of an observed signal in terms of the microscopic parameters of the underlying dynamics, they enhance its observability.

⊕ Since the SM provides many amplitudes for charm decays, CP asymmetries can be linear in NP amplitudes, thus increasing sensitivity to the latter.

⊕ Decays to final states of *more than* two pseudoscalar or one pseudoscalar and one vector meson contain more dynamical information than given by their widths; their distributions, as described by Dalitz plots or T odd moments, can exhibit CP asymmetries that might be considerably larger than those for the width. This will be explained in a bit more detail later on.

⊕ The distinctive channel $D^{\pm*} \to D\pi^\pm$ provides a powerful tag on the flavor identity of the neutral D meson.

⊖ The 'fly in the ointment' is that D^0–$\bar{D}^0$ oscillations are on the slow side.

⊕ Nevertheless one should take on this challenge. For CP violation involving D^0–$\bar{D}^0$ oscillations is a reliable probe of NP: the asymmetry is controlled by $\sin \Delta m_D t \cdot \mathrm{Im}(q/p)\bar{\rho}(D \to f)$. Within the SM, both factors are small, namely $\sim\mathcal{O}(10^{-3})$, making such an asymmetry unobservably tiny—unless there is NP; for a recent NP model, see [433]. One should note that this observable is *linear* in x_D rather than quadratic as for CP insensitive quantities like $D^0(t) \to l^- X$. D^0–$\bar{D}^0$ oscillations, CP violation and NP might thus be discovered simultaneously in a transition. We will return to this point below.

⊖ Honesty compels us to concede there is no attractive, let alone compelling scenario of NP for charm transitions whose footprints should not be seen also in B decays.

⊕ It is all too often overlooked that CPT invariance can provide nontrivial constraints on CP asymmetries. For it imposes equality not only on the masses and total widths of particles and antiparticles but also on the widths for 'disjoint' *sub*sets of channels. 'Disjoint' subsets are the decays to final states that can*not* rescatter into each other. Examples are semileptonic vs. nonleptonic modes with the latter subdivided further into those with strangeness $S = -1, 0, +1$. Observing a CP asymmetry in one channel one can then infer in which other channels the 'compensating' asymmetries have to arise [912].

3.9.14 Direct CP violation in partial rates

CP violation in $\Delta C = 1$ dynamics can be searched for by comparing partial widths for CP-conjugate channels. For an observable effect, two conditions have to be satisfied simultaneously: a transition must receive contributions from two coherent amplitudes with (a) different weak and (b) different strong phases as well. While condition (a) is just the requirement of CP violation in the underlying dynamics, condition (b) is needed to make the relative weak phase observable. Since the decays of charm hadrons proceed in the nearby presence of many hadronic resonances inducing virulent final state interactions (FSI), requirement (b) is in general easily met; thus it provides no drawback for the *observability* of a CP asymmetry—albeit it does for its *interpretation*.

As already mentioned, CKM dynamics does not support any CP violation in Cabibbo-allowed and doubly suppressed channels due to the absence of a second weak amplitude; the only exception are modes containing a K_S (or K_L) like $D^+ \to K_S \pi^+$ vs. $D^- \to K_S \pi^-$ which have to exhibit an asymmetry of 0.0032 reflecting the CP impurity in the K_S (or K_L) wave function. In once-Cabibbo-suppressed transitions one expects CP asymmetries, albeit highly diluted ones of order $\lambda^4 \sim 10^{-3}$.

While we have good information on the size of the weak phase, we do not know how to predict the size of the relevant matrix elements and strong phases in a reliable way. Even if a direct CP asymmetry larger than about 10^{-3} were observed in a Cabibbo-suppressed mode, say even as large as 10^{-2}, at present, we could not claim such a signal to establish the intervention of NP. A judicious exercise in 'theoretical engineering' could, however, solve our conundrum.

3.9.15 Theoretical engineering

CP asymmetries in integrated partial widths depend on hadronic matrix elements and (strong) phase shifts, neither of which can be predicted accurately. However the craft of theoretical engineering can be practiced with profit here. One makes an ansatz for the general form of the matrix elements and phase shifts that are included in the description of $D \to PP$, PV, VV etc. channels, where P and V denote pseudoscalar and vector mesons, and fits them to the measured branching ratios on the Cabibbo-allowed, once and twice forbidden level. If one has sufficiently accurate and comprehensive data, one can use these fitted values of the hadronic parameters to predict CP asymmetries. Such analyses have been undertaken in the past [968], but the data base was not as broad and precise as one would like. *CLEO-c and BESIII measurements will certainly lift such studies to a new level of reliability.*

3.9.16 CP violation in final-state distributions

Once the final state in $D \to f$ is more complex than a pair of pseudoscalar mesons or a pseudoscalar plus a vector meson, it contains more dynamical information than given by the modulus of its amplitude, since its kinematics are no longer trivial. CP asymmetries in final-state distributions can be substantially larger than in integrated partial widths.

The simplest such case is given by decays into three pseudoscalar mesons, for which Dalitz plots analyses represent a very sensitive tool with the phase information they yield. They require large statistics; yet once those have been obtained, the return is very substantial. For the constraints, one has on a Dalitz plot population provide us with powerful weapons to control systematic uncertainties.

Such phenomenological advantages of having more complex final states apply also for four-body etc. final states. Measuring T odd moments with

$$O_T \stackrel{T}{\Longrightarrow} -O_T \quad (195)$$

is an efficient way to make use of data with limited statistics. A simple example for a final state with four mesons a, b, c and d is given by $O_T = \langle \vec{p}_c \cdot (\vec{p}_a \times \vec{p}_b) \rangle$.

While FSI are not necessary for the emergence of such effects—unlike the situation for partial width asymmetries—, they can fake a signal of T violation with T being an *anti*linear operator; yet that can be disentangled by comparing T odd moments for CP conjugate modes [969]:

$$O_T(D \to f) \neq -O_T(\bar{D} \to \bar{f}) \implies \text{CP violation.} \quad (196)$$

A dramatic example for CP violation manifesting itself in a final-state distribution much more dramatically than in a partial width has been found in K_L decays. Consider the rare mode $K_L \to \pi^+\pi^- e^+ e^-$ and define by ϕ the angle between the $\pi^+\pi^-$ and e^+e^- planes. The differential width has the general form

$$\frac{d\Gamma}{d\phi}(K_L \to \pi^+\pi^- e^+ e^-)$$
$$= \Gamma_1 \cos^2\phi + \Gamma_2 \sin^2\phi + \Gamma_3 \cos\phi \sin\phi. \quad (197)$$

Upon integrating over ϕ the Γ_3 term drops out from the total width, which thus is given in terms of $\Gamma_{1,2}$ with Γ_3 representing a forward–backward asymmetry:

$$\langle A \rangle \equiv \frac{\int_0^{\pi/2} \frac{d\Gamma}{d\phi} - \int_{\pi/2}^{\pi} \frac{d\Gamma}{d\phi}}{\int_0^{\pi} \frac{d\Gamma}{d\phi}} = \frac{2\Gamma_3}{\pi(\Gamma_1 + \Gamma_2)}. \quad (198)$$

Under P and T, one has $\cos\phi \sin\phi \to -\cos\phi \sin\phi$. Accordingly $\langle A \rangle$ and Γ_3 constitute a T odd correlation, while $\Gamma_{1,2}$ are T even. Γ_3 is driven by the CP impurity ϵ_K in the kaon wave function. $\langle A \rangle$ has been measured to be large in full agreement with theoretical predictions [970]:

$$\langle A \rangle = 0.138 \pm 0.022. \quad (199)$$

One should note that this observable is driven by $|\epsilon_K| \simeq 0.0023$.

D decays can be treated in an analogous way. Consider the Cabibbo-suppressed channel[27]

$$\stackrel{(-)}{D} \to K \bar{K} \pi^+ \pi^- \quad (200)$$

and define ϕ to be the angle between the $K\bar{K}$ and $\pi^+\pi^-$ planes. Then one has

$$\frac{d\Gamma}{d\phi}(D \to K\bar{K}\pi^+\pi^-)$$
$$= \Gamma_1 \cos^2\phi + \Gamma_2 \sin^2\phi + \Gamma_3 \cos\phi \sin\phi, \quad (201)$$
$$\frac{d\Gamma}{d\phi}(\bar{D} \to K\bar{K}\pi^+\pi^-)$$
$$= \bar{\Gamma}_1 \cos^2\phi + \bar{\Gamma}_2 \sin^2\phi + \bar{\Gamma}_3 \cos\phi \sin\phi. \quad (202)$$

As before, the partial width for $D[\bar{D}] \to K\bar{K}\pi^+\pi^-$ is given by $\Gamma_{1,2}[\bar{\Gamma}_{1,2}]$; $\Gamma_1 \neq \bar{\Gamma}_1$ or $\Gamma_2 \neq \bar{\Gamma}_2$ represents direct CP violation in the partial width. Γ_3 & $\bar{\Gamma}_3$ constitute T odd correlations. By themselves they do not necessarily indicate CP vi-

---

[27] This mode can exhibit direct CP violation even within the SM.

olation, since they can be induced by strong final-state interactions. However

$$\Gamma_3 \neq \bar{\Gamma}_3 \implies \text{CP violation!} \quad (203)$$

It is quite possible or even likely that a difference in Γ_3 vs. $\bar{\Gamma}_3$ is significantly larger than in Γ_1 vs. $\bar{\Gamma}_1$ or Γ_2 vs. $\bar{\Gamma}_2$. Furthermore one can expect that differences in detection efficiencies can be handled by comparing Γ_3 with $\Gamma_{1,2}$ and $\bar{\Gamma}_3$ with $\bar{\Gamma}_{1,2}$.

3.9.17 CP asymmetries involving oscillations

For final states that are common to D^0 and $\bar{D}^0$ decays, one can search for CP violation manifesting itself with the help of D^0–$\bar{D}^0$ oscillations in qualitative—though certainly not quantitative—analogy to $B_d \to J/\psi K_S$. Such common states can be CP eigenstates—like $D^0 \to K^+K^-/\pi^+\pi^-/K_S\eta^{(\prime)}$—but do not have to be: two very promising candidates are $D^0 \to K_S\pi^+\pi^-$, where one can bring the full Dalitz plot machinery to bear, and $D^0 \to K^+\pi^-$ vs. $\bar{D}^0 \to K^-\pi^+$, since its SM amplitude is doubly-Cabibbo-suppressed. Undertaking *time-dependent* Dalitz plot studies requires a higher initial overhead, yet in the long run this should pay handsome dividends exactly, since Dalitz analyses can invoke many internal correlations that in turn serve to control systematic uncertainties.

Searching for such effects with the required sensitivity (see below) will be quite challenging. Nevertheless one should take on this challenge. For CP violation involving D^0–$\bar{D}^0$ oscillations is a reliable probe of NP: the asymmetry is controlled by $\sin \Delta m_D t \cdot \text{Im}(q/p)\bar{\rho}(D \to f)$. Within the SM, both factors are small, namely $\sim \mathcal{O}(10^{-3})$, making such an asymmetry unobservably tiny—unless there is NP; for a recent NP model, see [433]. One should note that this observable is *linear* in x_D rather than quadratic as for CP-insensitive quantities like $D^0(t) \to l^- X$. D^0–$\bar{D}^0$ oscillations, CP violation and NP might thus be discovered simultaneously in a transition.

3.9.18 Experimental searches for CP violation

Let the amplitude for D^0 to decay to a final state f be written as

$$A_f \equiv \langle f | \mathcal{H}_{\text{int}} | D^0 \rangle,$$

where $\mathcal{H}_{\text{int}}$ is the interaction Hamiltonian responsible for $D^0 \to f$. If CP is conserved, that is if $[\mathcal{H}_{\text{int}}, CP] = 0$, then we can clearly write

$$\begin{aligned} A_f &= \langle f | (\text{CP})^\dagger (\text{CP}) \mathcal{H}_{\text{int}} | D^0 \rangle \\ &= \langle f | (\text{CP})^\dagger \mathcal{H}_{\text{int}} (\text{CP}) | D^0 \rangle \\ &= -\langle \bar{f} | \mathcal{H}_{\text{int}} | \bar{D}^0 \rangle \equiv -\bar{A}_{\bar{f}}, \end{aligned} \quad (204)$$

where $\bar{f}$ is the conjugate final state to f. Consequently, a measurement that shows $\Gamma(D^0 \to f) \neq \Gamma(\bar{D}^0 \to \bar{f})$ is a demonstration that CP is violated in this decay.

Most CP violation results are from the FNAL fixed target experiments E791 and FOCUS and from the CLEO experiment and search for direct CP violation. The CP violation asymmetry is defined as

$$A_{\text{CP}} \equiv \frac{\Gamma(D \to f) - \Gamma(\bar{D} \to \bar{f})}{\Gamma(D \to f) + \Gamma(\bar{D} \to \bar{f})}. \quad (205)$$

A few results from CLEO, BaBar and Belle experiments consider CP violation in mixing. Typically, precisions of a few percent are obtained [119]. No evidence for CP violation is observed consistent with SM expectations.

Certainly very large samples will be available from hadron colliders. From an existing CDF measurement [971] it is possible to anticipate yields of over 0.5–1 million $D^0 \to K^+K^-$ events being available with the likely final Tevatron integrated luminosity of 5–10 fb^{-1}. This sample will have an intrinsic statistical precision of $\leq 0.2\%$. With the higher production cross-section and its dedicated D^* trigger, LHCb will accumulate samples of up to 10 million tagged events in each year of nominal operation [935]. The RICH system will ensure a low background, and these decays will be complemented by those selected in the $D^0 \to \pi^+\pi^-$ mode. In order to exploit these enormous statistics, it will be necessary to pay great attention to systematics biases. Initial state asymmetries and detector asymmetries will be the main concerns.

3.9.18.1 Three-body decays Direct CP violation searches in analyses of charm decays to three-body final states are more complicated than two-body decays. Three methods have been used to search for CP asymmetries: (1) integrate over phase space and construct A_{CP} as in two-body decays; (2) examine CP asymmetry in the quasi-two-body resonances; (3) perform a full Dalitz-plot analysis for D and $\bar{D}$ separately. The Dalitz-plot analysis procedure [936] allows increased sensitivity to CP violation by probing decay amplitudes rather than the decay rate. E791 [937], FOCUS [938] and BaBar [939] have analysed $D^+ \to K^+K^-\pi^+$ using method (1). E791 and BaBar have also analysed $D^+ \to K^-K^+\pi^+$ using method (2). FOCUS has a Dalitz-plot analysis in progress [940]. The $D^+ \to K^+K^-\pi^+$ Dalitz plot is well described by eight quasi-two-body decay channels. A signature of CP violation in charm Dalitz-plot analyses is different amplitudes and phases for D and $\bar{D}$ samples. No evidence for CP violation is observed.

The decay $D^{*+} \to D^0\pi^+$ enables the discrimination between D^0 and $\bar{D}^0$. The CLEO Collaboration has searched for CP violation integrated across the Dalitz plot in $D^0 \to K^\mp\pi^\pm\pi^0$ [941, 942], $K_S^0\pi^+\pi^-$ [943] and $\pi^+\pi^-\pi^0$ [944] decays. No evidence of CP violation has been observed.

CLEO has considered CP violation more generally in a simultaneous fit to the $D^0 \to K_S^0 \pi^+\pi^-$ and $\bar{D}^0 \to K_S^0 \pi^+\pi^-$ Dalitz plots. The possibility of interference between CP-conserving and CP-violating amplitudes provides a more sensitive probe of CP violation. The constraints on the square of the CP-violating amplitude obtained in the resonant submodes of $D^0 \to K_S^0 \pi^+\pi^-$ range from 3.5×10^{-4} to 28.4×10^{-4} (95% C.L.) [943].

3.9.18.2 Four-body decays FOCUS has searched for T-violation using the four-body decay modes $D^0 \to K^+K^-\pi^+\pi^-$ [969]. As described in Sect. 3.9.16, a T-odd correlation can be formed with the momenta, $C_T \equiv (\vec{p}_{K^+} \cdot (\vec{p}_{\pi^+} \times \vec{p}_{\pi^-}))$. Under time-reversal, $C_T \to -C_T$, however $C_T \neq 0$ does not establish T-violation. Since time reversal is implemented by an anti-unitary operator, $C_T \neq 0$, it can be induced by FSI [945]. This ambiguity can be resolved by measuring $\bar{C}_T \equiv (\vec{p}_{K^+} \cdot (\vec{p}_{\pi^+} \times \vec{p}_{\pi^-}))$ in $\bar{D}^0 \to K^+K^-\pi^+\pi^-$; $C_T \neq \bar{C}_T$ establishes T violation. FOCUS reports a preliminary asymmetry $A_T = 0.075 \pm 0.064$ from a sample of $\sim$400 decays. More restrictive constraints are anticipated from CLEO-c, where in 281 pb^{-1} a sample of 2300 $D^\pm \to K_S^0 K^\pm \pi^+\pi^-$ have been accumulated.

3.9.19 Experiments exploiting quantum correlations

Most high-statistics measurements of D^0 decay employ "flavor tagging", through the sign of the slow pion in $D^* \to \pi_{\text{slow}} D$. That is, if combined with a slow π^+ to make a D^{*+}, the neutral D meson is a D^0. Conversely, a slow π^- implies a $\bar{D}^0$.

An entirely different way to tag flavor and CP is to exploit quantum correlations in $D^0\bar{D}^0$ production in e^+e^- annihilation [958–960].

The production process $e^+e^- \to \psi(3770) \to D^0\bar{D}^0$ produces an eigenstate of $CP+$ in the first step, since the $\psi(3770)$ has J^{PC} equal to 1^{--}. Now consider the case where both the D^0 and $\bar{D}^0$ decay into CP eigenstates. Then the decays $\psi(3770) \to f_+^i f_+^j$ or $f_-^i f_-^j$ are forbidden, where f_+ denotes a CP$^+$ eigenstate, and f_- denotes a CP$^-$ eigenstate. This is because $CP(f_\pm^i f_\pm^j) = (-1)^\ell = -1$ for $\ell = 1$ $\psi(3770)$. Hence, if a final state such as $(K^+K^-)(\pi^+\pi^-)$ is observed, one immediately has evidence of CP violation. Moreover, all CP$^+$ and CP$^-$ eigenstates can be summed over for this measurement. The expected sensitivity to direct CP violation is $\sim 1\%$. This measurement can also be performed at higher energies, where the final state $D^{*0}\bar{D}^{*0}$ is produced. When either D^* decays into a π^0 and a D^0, the situation is the same as above. When the decay is $D^{*0} \to \gamma D^0$, the CP parity is changed by a multiplicative factor of -1 and all decays $f_+^i f_-^j$ violate CP [945]. Additionally, CP asymmetries in CP even initial states depend linearly on x allowing sensitivity to CP violation in mixing of $\sim 3\%$.

For e^+e^- machines running at the $\psi(3770)$, the D mesons are produced with very little momentum in the laboratory. Hence, their flight distance is virtually impossible to determine, and we instead measure time-integrated decay rates. From Ref. [960] we have

$$\Gamma(j,k) = Q_M |A(j,k)|^2 + R_M |B(j,k)|^2, \quad (206)$$

where

$$A(j,k) \equiv A_j \bar{A}_k - \bar{A}_j A_k$$

is the "unmixed" contribution to the decay rate, and

$$B(j,k) \equiv \frac{p}{q} A_j A_k - \frac{q}{p} \bar{A}_j \bar{A}_k$$

is the contribution from D^0–$\bar{D}^0$ mixing. The integrations also yield the factors

$$Q_M = \frac{1}{2}\left[\frac{1}{1-y^2} + \frac{1}{1+x^2}\right] \approx 1 - \frac{x^2-y^2}{2},$$

$$R_M = \frac{1}{2}\left[\frac{1}{1-y^2} - \frac{1}{1+x^2}\right] \approx \frac{x^2+y^2}{2}.$$

Mixing does not occur if the eigenstates of the decay Hamiltonian have the same mass and width, i.e. $x = y = 0$. In any case, we expect $R_M \ll Q_M \approx 1$. Nevertheless, mixing would result in the second term of (206), and it is here that one obtains sensitivity to CP violation through $q \neq p$. This will be exploited at CLEO-c and eventually to a greater extent at BES III.

3.9.20 Benchmarks for future searches

Since the primary goal is to establish the intervention of NP, one 'merely' needs a sensitivity level above the reach of the SM; 'merely' does not mean that it can easily be achieved. As far as *direct* CP violation is concerned—in partial widths as well as in final-state distributions—this means asymmetries down to the 10^{-3} or even 10^{-4} level in Cabibbo-allowed channels and 1% level or better in twice Cabibbo-suppressed modes; in Cabibbo-once-suppressed decays one wants to reach the 10^{-3} range, although CKM dynamics can produce effects of that order, because future advances might sharpen the SM predictions—and one will get them along with the other channels. For *time-dependent* asymmetries in $D^0 \to K_S\pi^+\pi^-$, K^+K^-, $\pi^+\pi^-$ etc. and in $D^0 \to K^+\pi^-$, one should strive for the $\mathcal{O}(10^{-4})$ and $\mathcal{O}(10^{-3})$ levels, respectively.

Statisticswise these are not utopian goals considering the very large event samples foreseen at LHCb.

When probing asymmetries below the $\sim 1\%$ level, one has to struggle against systematic uncertainties, in particular since detectors are made from matter. There are three powerful weapons in this struggle: (i) Resolving the time evolution of asymmetries that are controlled by x_D and y_D, which requires excellent microvertex detectors; (ii) Dalitz plot consistency checks; (iii) quantum statistics constraints on distributions, T odd moments etc. [958, 960].

3.9.21 Rare decays

Searches for rare-decay processes have played an important role in the development of the SM. FCNC processes have been studied extensively for K and B mesons in both K^0–$\bar{K}^0$ and B^0–$\bar{B}^0$ mixing and in rare FCNC decays. The corresponding processes in the charm sector has received less attention, and the experimental upper limits are currently above SM predictions. Short-distance FCNC processes in charm decays are much more highly suppressed by the GIM mechanism than the corresponding down-type quark decays because of the large top quark mass.

Observation of D^+ FCNC decays $D^+, D_s^+ \to \pi^+ l^+ l^-$ and $K^+ l^+ l^-$ could therefore provide an indication of NP or of unexpectedly large rates for long-distance SM processes like $D^+ \to \pi^+ V$, $V \to l^+ l^-$, with a real or virtual vector meson V. Detailed description on rare charm decays can be found in Refs. [911, 913]. The charm meson radiative decays are also very important to understand final state interaction which may enhance the decay rates. In Refs. [911, 913], the decay rates of $D \to V\gamma$ (V can be ϕ, ω, ρ and K^*) had been estimated to be 10^{-5}–10^{-6}, which can be reached at BES-III and the B-factories.

3.9.22 Inclusive $c \to u$ transitions

The $s \to d$ and $b \to s$ transitions offer the possibility to investigate effects of NP in the down-type quark sector. The $c \to u$ transition, however, gives a chance to study effects of NP in the up-type quark sector. In the SM, the contribution coming from the penguin diagrams in $c \to u\gamma$ transition is strongly GIM suppressed giving a branching ratio of order 10^{-18} [972]. The QCD-corrected effective Lagrangian gives $BR(c \to u\gamma) \simeq 3 \times 10^{-8}$ [973, 974]. A variety of models beyond the standard model were investigated, and it was found that the gluino exchange diagrams [975] within general minimal supersymmetric SM (MSSM) might lead to the enhancement

$$\frac{BR(c \to u\gamma)_{\text{MSSM}}}{BR(c \to u\gamma)_{\text{SM}}} \simeq 10^2. \tag{207}$$

Within SM, the $c \to ul^+l^-$ amplitude is given by the γ and Z penguin diagrams and W box diagram at one-loop electroweak order in the SM. It is dominated by the light quark contributions in the loop. The leading order rate for the inclusive $c \to ul^+l^-$ calculated within SM [976] was found to be suppressed by QCD corrections in [911]. The inclusion of the renormalization group equations for the Wilson coefficients gave an additional significant suppression [977] leading to the rates $\Gamma(c \to ue^+e^-)/\Gamma_{D^0} = 2.4 \times 10^{-10}$ and $\Gamma(c \to u\mu^+\mu^-)/\Gamma_{D^0} = 0.5 \times 10^{-10}$. These transitions are largely driven by virtual photon at low dilepton mass m_{ll}.

The leading contribution to $c \to ul^+l^-$ in general MSSM with the conserved R parity comes from one-loop diagrams with gluino and squarks in the loop [911, 975, 976]. It proceeds via virtual photon and significantly enhances the $c \to ul^+l^-$ spectrum at small dilepton mass m_{ll}. The authors of [911] have investigated a SUSY extension of the SM with R parity breaking and they found that it can modify the rate. Using the most recent CLEO [978] results for the $D^+ \to \pi^+ e^+ e^-$, one can set the bound for the product of the relevant parameters entering the R-parity violating $\tilde{\lambda}'_{22k}\tilde{\lambda}'_{21k} \simeq 0.001$ (assuming that the mass of squark $M_{\tilde{D}_k} \simeq 100$ GeV). This bound give the rates $BR_R(c \to ue^+e^-) \simeq 1.6 \times 10^{-8}$ and $BR_R(c \to u\mu^+\mu^-) \simeq 1.8 \times 10^{-8}$.

Recently, the effects of Littlest Higgs models were investigated in rare D decays [145], and it was found that there is a new tree level coupling which gives a $c \to uZ$ transition. However, that effect is insignificant due to the parameters constrained by the present electroweak data (see Ref. [25] in [145]). A number of models of NP contain an extra up-type heavy quark [980] causing the appearance of the FCNCs at tree level for the up-quark sector. The Lagrangian which describes this FCNC interaction is given by

$$\mathcal{L}_{\text{NC}} = \frac{g}{\cos\theta_W} Z_\mu \big(J_{W^3}^\mu - \sin^2\theta_W J_{EM}^\mu\big), \tag{208}$$

where J_{EM}^μ is the same electromagnetic current as in the SM, while $J_{W^3}^\mu$ is given by

$$J_{W^3}^\mu = \frac{1}{2}\bar{U}_L^m \gamma^\mu \Omega U_L^m - \frac{1}{2}\bar{D}_L^m \gamma^\mu D_L^m \tag{209}$$

with $L = \frac{1}{2}(1 - \gamma_5)$ and mass eigenstates $U_L^m = (u_L, c_L, t_L, T_L)^T$, $D_L^m = (d_L, s_L, b_L)^T$. The neutral current for the down-type quarks is the same as in the SM, while the up sector has additional currents (see Ref. [145]). The unitarity conditions of the CKM matrix might constrain this coupling. However, the present bound on Δm in D^0–$\bar{D}^0$ mixing limits the parameter describing the cuZ vertex to be $\Omega_{uc} \simeq 0.004$, giving the more strict limit on that parameter. The invariant dilepton mass distribution of the $c \to ul^+l^-$ distribution is only moderately enhanced.

3.9.23 Exclusive rare D decays

The study of exclusive D meson rare decay modes is very difficult due to the dominance of the long-distance effects [145, 911–913, 972–984]. The $D \to V\gamma$ decay rates were calculated in [913, 972, 981, 983]. The long-distance contribution is induced by the effective nonleptonic $|\Delta c| = 1$ weak Lagrangian. In calculations of [983], the long-distance effects were determined using a heavy meson chiral Lagrangian. The factorization approximation has been used for the calculation of weak transition elements. The results of [972] obtained within a different framework are in very good agreement with the results of [983]. In Table 48, the branching ratios of $D \to V\gamma$ decays [983] are given. The uncertainty is due to relative unknown phases of various contributions. Although the branching ratios are dominated by the long-distance contributions, the size of the short-distance contribution can be extracted from the difference of the decay widths $\Gamma(D^0 \to \rho^0\gamma)$ and $\Gamma(D^0 \to \omega\gamma)$ [982]. Namely, the long-distance mechanism $c\bar{u} \to d\bar{d}\gamma$ screens the $c\bar{u} \to u\bar{u}\gamma$ transition in $D^0 \to \rho^0\gamma$ and $D^0 \to \omega\gamma$, the ρ^0 and ω mesons being mixtures of $u\bar{u}$ and $d\bar{d}$. Fortunately, the LD contributions are mostly canceled in the ratio

$$R = \frac{BR(D^0 \to \rho^0\gamma) - BR(D^0 \to \omega\gamma)}{BR(D^0 \to \omega\gamma)}$$
$$\propto \mathrm{Re}\, \frac{A(D^0 \to u\bar{u}\gamma)}{A(D^0 \to d\bar{d}\gamma)}, \qquad (210)$$

which is proportional to the SD amplitude $A(D^0 \to u\bar{u}\gamma)$ driven by $c \to u\gamma$. This ratio is $R_{SM} = (6 \pm 15)\%$ in Ref. [982] and can be enhanced up to $\mathcal{O}(1)$ in the MSSM. In addition to the $c \to u\gamma$ searches in the charm meson decays, in Ref. [976], it was suggested to search for this transition in the decay $B_c \to B_u^*\gamma$, where the long-distance contribution is much smaller.

The inclusive $c \to ul^+l^-$ process can be tested in the rare decays $D \to \mu^+\mu^-$, $D \to P(V)l^+l^-$ [911, 913, 976, 977]. The branching ratio for the rare decay $D \to \mu^+\mu^-$ is very small in the SM. The detailed treatment of this decay rate [911] gives $Br(D \to \mu^+\mu^-) \simeq 3 \times 10^{-13}$ [911]. This decay rate can be enhanced within a study which considers SUSY with R-parity breaking effects [911, 912]. Using the bound $\tilde{\lambda}'_{22k}\tilde{\lambda}'_{21k} \simeq 0.001$, one obtains the limit $Br(D \to \mu^+\mu^-)_R \simeq 4 \times 10^{-7}$, a value which would be accessible at LHCb [935]. The $D \to P(V)l^+l^-$ decays offer another possibility to study the $c \to ul^+l^-$ transition in charm sector. The $D^+ \to \pi^+l^+l^-$ and $D^0 \to \rho^0 e^+e^-$ decay modes are simplest to be accessed by experiment [145]. The effects

Table 48 Predicted branching ratios for $D \to V\gamma$ decays

$D \to V\gamma$	BR
$D^0 \to \bar{K}^{*0}\gamma$	$[6–36] \times 10^{-5}$
$D_s^+ \to \rho^+\gamma$	$[20–80] \times 10^{-5}$
$D^0 \to \rho^0\gamma$	$[0.1–1] \times 10^{-5}$
$D^0 \to \omega\gamma$	$[0.1–0.9] \times 10^{-5}$
$D^0 \to \phi\gamma$	$[0.4–1.9] \times 10^{-5}$
$D^+ \to \rho^+\gamma$	$[0.4–6.3] \times 10^{-5}$
$D_s^+ \to K^{*+}\gamma$	$[1.2–5.1] \times 10^{-5}$
$D^+ \to K^{*+}\gamma$	$[0.3–4.4] \times 10^{-6}$
$D^0 \to K^{*0}\gamma$	$[0.3–2.0] \times 10^{-6}$

of SUSY with R parity violation were studied in [911]. The recent experimental results of [978] restrict the R parity violating parameters found in [911] more than one order of magnitude.

The most appropriate decay modes for the experimental searches of the NP coming from the FCNC tree level current are $D^+ \to \pi^+l^+l^-$ and $D^0 \to \rho^0 e^+e^-$. The total rate for $D \to Xl^+l^-$ is dominated by the long-distance resonant contributions at dilepton mass $m_{ll} = m_\rho$, m_ω, m_ϕ, and even the largest contributions from NP are not expected to affect the total rate significantly [911, 976]. NP could only modify the SM differential spectrum at low m_{ll} below ρ or the spectrum at high m_{ll} above ϕ. In the case of $D \to \pi l^+l^-$ differential decay distribution, there is a broad region at high m_{ll} (see Fig. 56), which presents a unique possibility to study the $c \to ul^+l^-$ transition [145, 976]. In Table 49, we present the branching ratios for $D^+ \to \pi^+e^+e^-$ and $D^0 \to \rho^0 l^+l^-$, giving the SM short-distance, long-distance contributions, as well as the effects of NP arising from the existence of one extra up-type quark. The total rates in the SM and NP scenarios are completely dominated by the resonant long-distance contribution $D \to XV_0 \to Xl^+l^-$ [145, 911]. The SM short-distance contribution for $D^0 \to \rho^0 l^+l^-$ (see Fig. 56) is not shown, since it is completely negligible in comparison to the long-distance contribution. The forward–backward asymmetry for $D^0 \to \rho^0 l^+l^-$ vanishes in SM, while it is reaching 0.05 in an NP model with extra up-type quark, as given in Fig. 57. Such an asymmetry is still small and will be difficult to observe in present or planned experiments given that the rate itself is already small.

3.9.24 Experimental results

There are a large number of FCNC charm decays including radiative, fully leptonic decays, lepton flavor violating (LFV) and lepton number violating (LNV) that have been measured experimentally.

Belle has reported the observation of the decay $D^0 \to \phi\gamma$. This is the first observation of a flavor-changing radiative decay of a charmed meson. The Cabibbo- and colour-suppressed decays $D^0 \to \phi\pi^0$, $\phi\eta$ are also observed for the first time. The branching fractions are $\mathcal{B}(D^0 \to \phi\gamma) = [2.60^{+0.70}_{-0.61}{}^{+0.15}_{-0.17}] \times 10^{-5}$ (somewhat higher than predicted in Table 48), $\mathcal{B}(D^0 \to \phi\pi^0) = [8.01 \pm 0.26 \pm 0.47] \times 10^{-4}$ and $\mathcal{B}(D^0 \to \phi\eta) = [1.48 \pm 0.47 \pm 0.09] \times 10^{-4}$.

Recently, CLEO-c reported the branching fraction of the resonant decay $\mathcal{BR}(D^+ \to \pi^+\phi \to \pi^+e^+e^-) = (2.8 \pm 1.9 \pm 0.2) \times 10^{-6}$ [978]. The LNV or LFV decays $D^+ \to \pi^-l^+l^+$, $K^-l^+l^+$ and $\pi^+\mu^+e^-$ are forbidden in the SM. Past searches have set upper limits for the dielectron and dimuon decay modes [119].

The BaBar Collaboration has recently reported on FCNC decays of the form $D^+/D_s^+/\Lambda_c^+ \to \pi^+/K^+/p\,\ell^+\ell'^-$, where the two leptons, ℓ^+ and ℓ'^-, can each be either an

Table 49 Branching ratios for the decays probing the $c \to u l^+ l^-$ transition

Br	Short-distance contribution only		Total rate $\simeq$ long-distance contr.	Experiment
	SM	SM + NP		
$D^+ \to \pi^+ e^+ e^-$	6×10^{-12}	8×10^{-9}	1.9×10^{-6}	$<7.4 \times 10^{-6}$
$D^+ \to \pi^+ \mu^+ \mu^-$	6×10^{-12}	8×10^{-9}	1.9×10^{-6}	$<8.8 \times 10^{-6}$
$D^0 \to \rho^0 e^+ e^-$	negligible	5×10^{-10}	1.6×10^{-7}	$<1.0 \times 10^{-4}$
$D^0 \to \rho^0 \mu^+ \mu^-$	negligible	5×10^{-10}	1.5×10^{-7}	$<2.2 \times 10^{-5}$

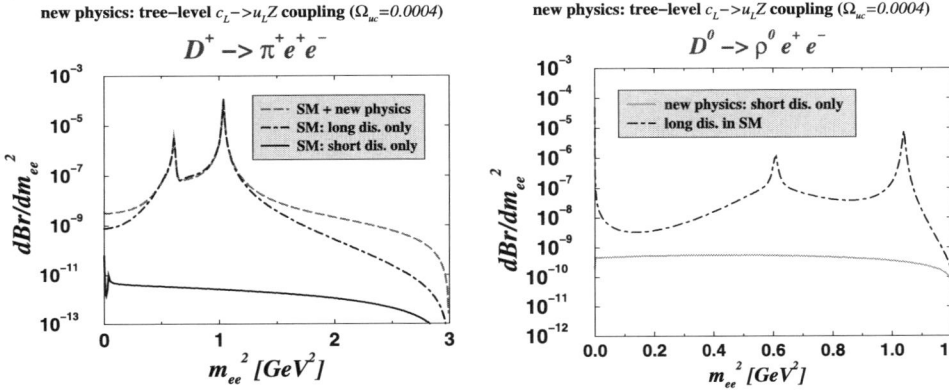

Fig. 56 The dilepton mass distribution dBr/dm_{ee}^2 for the decay $D^+ \to \pi^+ e^+ e^-$ as a function of the dilepton mass square $m_{ee}^2 = (p_+ + p_-)^2$ (*left*) and the dilepton mass distribution for $D^0 \to \rho^0 e^+ e^-$ (*right*)

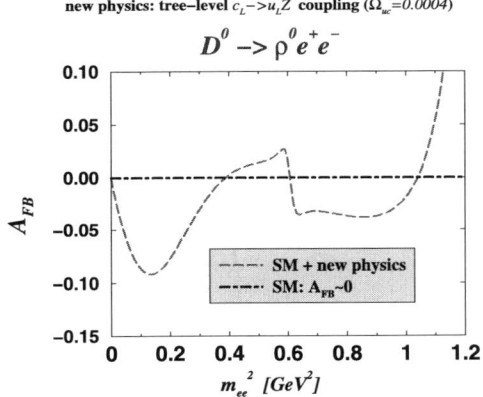

Fig. 57 The figure shows the forward–backward asymmetry for $D^0 \to \rho^0 e^+ e^-$

electron or a muon. Upper limits are set at the 90% C.L. between 4×10^{-6} and 40×10^{-6} on the SM and LFV processes [979].

In Table 50, the current limits and expected sensitivities at BES-III are summarized for D^+ and D^0, respectively.

3.9.25 Precision CKM physics

Precision measurements of the CKM matrix continue to be of great interest, despite impressive strides in determining its parameters [7–9, 120, 209–211]. We first give an overview of ways in which studies of charm can help this effort. More details on some aspects are given in subsequent subsections.

In Sect. 3.9.26, we discuss direct measurements of the CKM elements governing $c \to d$ and $c \to s$ transitions. We then turn in Sect. 3.9.27 to ways in which charm can be of help in determining the remaining elements. An elementary constraint on new physics is discussed in Sect. 3.9.28, while Sect. 3.9.29 summarizes.

3.9.26 Direct determinations

3.9.26.1 V_{ud}, V_{us}, and unitarity The parameter $V_{us} = \lambda$ is measured (with some recent contributions playing a key role) to be 0.2257 ± 0.0021 [119]. To sufficient accuracy, one then expects $V_{ud} = \sqrt{1 - |V_{us}|^2} = 0.9742 \pm 0.0005$, since $|V_{ub}| \simeq 0.004$ and hence its square can be neglected in the unitarity relation $|V_{ud}|^2 + |V_{us}|^2 + |V_{ub}|^2 = 1$. The experimental value for V_{ud}, based primarily upon comparing beta-decays of certain nuclei to muon decays, is $V_{ud} = 0.97377 \pm 0.00027$, so unitarity is adequately satisfied for the first row.

3.9.26.2 V_{cd} For the first column, one expects $|V_{ud}|^2 + |V_{cd}|^2 + |V_{td}|^2 = 1$. With the value of V_{ud} quoted above and $|V_{td}| \simeq 0.008$, one then expects $|V_{cd}| = 0.227 \pm 0.001$. This is to be compared with the value 0.230 ± 0.011 obtained from neutrino interactions [119] and $0.213 \pm 0.008 \pm 0.021$ from charm semileptonic decays [992]. The first error is experimental, and the second one is associated with uncertainty in the form factor. Measurements of the branching fractions for $D \to \pi \ell \nu$ decay are improving somewhat

Table 50 Current and projected 90% C.L. upper limits on rare D^+ and D^0 decay modes at BES-III with 20 fb^{-1} data at $\psi(3770)$ peak

Mode	Reference experiment	Best upper limits (10^{-6})	BES-III ($\times 10^{-6}$)	Mode	Reference experiment	Best upper limits (10^{-6})	BES-III ($\times 10^{-6}$)
D^+				D^0			
$\pi^+ e^+ e^-$	CLEO-c [978]	7.4	0.03	$\gamma\gamma$	CLEO [985]	28	0.05
$\pi^+ \mu^+ \mu^-$	FOCUS [986]	8.8	0.03	$\mu^+\mu^-$	D0 [987]	2.4	0.03
$\pi^+ \mu^\pm e^\mp$	BaBar [979]	5.9/10.8	0.03	$\mu^+ e^-$	E791 [988]	8.1	0.03
$\pi^- e^+ e^+$	CLEO-c [978]	3.6	0.03	$e^+ e^-$	E791 [988]	6.2	0.03
$\pi^- \mu^+ \mu^+$	FOCUS [986]	4.8	0.03	$\pi^0 \mu^+\mu^-$	E653 [989]	180	0.05
$\pi^- \mu^+ e^+$	E791 [988]	50	0.03	$\pi^0 \mu^+ e^-$	CLEO [990]	86	0.05
$K^+ e^+ e^-$	CLEO-c [978]	6.2	0.03	$\pi^0 e^+ e^-$	CLEO [990]	45	0.05
$K^+ \mu^+ \mu^-$	FOCUS [986]	9.2	0.03	$K_S \mu^+\mu^-$	E653 [989]	260	0.1
$K^+ \mu^\pm e^\mp$	BaBar [979]	5.9/5.7	0.03	$K_S \mu^+ e^-$	CLEO [990]	100	0.1
$K^- e^+ e^+$	CLEO-c [978]	4.5	0.03	$K_S e^+ e^-$	CLEO [990]	110	0.1
$K^- \mu^+ \mu^+$	FOCUS [986]	13	0.03	$\eta \mu^+\mu^-$	CLEO [990]	530	0.1
$K^- \mu^+ e^+$	E687 [991]	130	0.03	$\eta \mu^+ e^-$	CLEO [990]	100	0.1
				$\eta e^+ e^-$	CLEO [990]	110	0.1

(Sect. 3.9.29.2), so the precision of $|V_{cd}|$ from this source will improve. However, from the current uncertainties in $\mathcal{B}(D \to \pi \ell \nu)$ it is clear that one will not be able to match the precision of the unitarity test for the first row of the CKM matrix anytime soon. *Given* CKM unitarity, which says to sufficient accuracy that we should expect the value of $|V_{cd}|$ mentioned above, one can use it to constrain form factors in semileptonic charm decays and compare them with lattice QCD calculations.

3.9.26.3 V_{cs}

A similar philosophy applies to the CKM element V_{cs}. Unitarity applied to the second column of the CKM matrix implies $|V_{cs}| = \sqrt{1 - |V_{us}|^2 - |V_{ts}|^2}$. Taking the experimental value of V_{us} mentioned above and the unitarity-based estimate $V_{ts} \simeq -V_{cb}$, we estimate $|V_{cs}| = 0.9733 \pm 0.0006$. This precision will not be matched by experiment soon. The best measurements come from semileptonic charm decays and yield $|V_{cs}| = 0.957 \pm 0.017 \pm 0.093$, with the second error coming from uncertainty in the form factor. Again, assuming unitarity, one will be able to subject lattice gauge theory predictions to important tests.

3.9.27 Indirect tests

3.9.27.1 V_{ub}

The primary difficulty in measuring the matrix element V_{ub} is that it must be extracted from b semileptonic decays which proceed to charm all but 2% of the time. Inclusive methods must rely on kinematic separation techniques, the oldest of which is the study of leptons with energies beyond the endpoint for $b \to c \ell \nu$. Exclusive decays such as $B \to \pi \ell \nu$ and $B \to \rho \ell \nu$ do not share this problem, but one must understand the corresponding form factors.

Tests of form factors in *charm* decays predicted by lattice gauge theories can help validate predictions for B decays.

The phase of V_{ub}^* (γ or ϕ_3 in the standard parameterizations) can be measured in several ways with the help of information from charm decays. These help, for example, in using decays such as $B \to D_{CP} K$ decays to learn γ. For D modes such as $K_S \pi^+ \pi^-$, $\pi^+ \pi^- \pi^0$, $K^+ K^- \pi^0$, and $K_S K^\pm \pi^\mp$, Dalitz plots yield information on CP-eigenstate and flavor-eigenstate modes and their relative phases [993].

The interference of $b \to c\bar{u}s$ (real) and $b \to u\bar{c}s$ ($\sim e^{-i\gamma}$) subprocesses in $B^- \to D^0 K^-$ and $B^- \to \bar{D}^0 K^-$, respectively, is sensitive to the weak phase γ. This interference may be probed by studying common decay products of D^0 and $\bar{D}^0$ into neutral D CP eigenstates or into doubly-Cabibbo-suppressed modes [627, 632, 633, 636, 645, 646].

As one example, the decays $B^\pm \to K^\pm (K^{*+} K^-)_D$ and $B^\pm \to K^\pm (K^{*-} K^+)_D$ provide information on γ if the relative (strong) phase between $D^0 \to K^{*+} K^-$ and $D^0 \to K^{*-} K^+$ is known [994]. One can learn this relative phase from the study of $D^0 \to K^+ K^- \pi^0$, since both final states occur and interfere with one another where K^{*+} and K^{*-} bands cross on the Dalitz plot [995]. This method was used recently by the CLEO Collaboration [996] to show that this interference was predominantly destructive in the overlap region.

As another example, one can determine γ using $B^\pm \to DK^\pm$ followed by $D \to K_S \pi^+ \pi^-$, $K_S K^+ K^-$, $K_S \pi^+ \pi^- \pi^0$ [637, 997]. Recent high-statistics studies have been performed by BaBar [641] and Belle [639]. The precision of these measurements will eventually be limited by the understanding of the $D \to K_S^0 \pi^+ \pi^-$ Dalitz plot. K-matrix descriptions of the $\pi\pi$ S-wave may yield improved models of charm Dalitz plots, and these models will be tested using the CP tagged sample of charm decays at CLEO-c and later at BES-III. The model uncertainty, which is currently $\pm 10°$, may be reduced to a few degrees.

Model independent methods [648, 998] use CP tagged $K_S^0 \pi^+ \pi^-$ and $D\bar{D} \to (K_S^0 \pi^+ \pi^-)^2$ to control the Dalitz plot model uncertainty. Analyses underway at CLEO-c are expected to control this systematic uncertainty on γ/ϕ_3 to a few degrees.

3.9.27.2 V_{cb} The semileptonic decays of B mesons to D or D^* mesons are one source of information about the element V_{cb}, but one must understand form factors satisfactorily. Lattice gauge theories make predictions for such form factors; the validation of lattice form factor predictions in charm decays again is a key ingredient in establishing credibility of the $B \to D^{(*)}$ form factor predictions. Moreover, under some circumstances, it is helpful to have precise information about D branching ratios to specific final states, which detailed charm studies can provide.

3.9.27.3 V_{td} and $|V_{td}/V_{ts}|$ The mixing of B^0 and $\bar{B}^0$ is governed primarily by the CKM product $|V_{tb}^* V_{td}|$. If unitarity is assumed, $|V_{tb}|$ is very close to 1, so the dominant CKM source of uncertainty is $|V_{td}|$. However, the matrix element of the short-distance operator inducing the $b\bar{d} \to d\bar{b}$ transition contains an unknown factor $f_B^2 B_B$, where f_B is the B meson decay constant, while $B_B = \mathcal{O}(1)$ is known as the "bag constant" or "vacuum saturation factor" and expresses the degree to which the vacuum intermediate state dominates the transition. The corresponding mixing of strange B's and their antiparticles is governed by $|V_{tb}^* V_{ts}|$ and $f_{B_s}^2 B_{B_s}$.

Lattice gauge theories predict not only f_B and f_{B_s} (as well as the constants B_B and B_{B_s}) but also the decay constants f_D and f_{D_s} for charmed mesons. Thus, the study of charmed meson decay constants (Sect. 3.9.29.1) and their ratios and comparison with lattice predictions can shed indirect light on quantities of interest in determining the CKM matrix elements V_{td} and V_{ts}.

To give one example of the role charm measurements can play, it is expected on rather general grounds [999] that f_{B_s}/f_B and f_{D_s}/f_D are equal to within a few percent. Now, the ratio f_{B_s}/f_B is a key ingredient in the extraction of $|V_{td}/V_{ts}|$ from measurements of B^0–$\bar{B}^0$ and B_s^0–$\bar{B}_s^0$ mixing. The determination of Ref. [677] utilized the estimate $(f_{B_s}\sqrt{B_{B_s}}/f_B\sqrt{B_B}) = 1.21^{+0.047}_{-0.035}$ from the lattice [301]. With a sufficiently good measurement of f_{D_s}/f_D and the theoretical input (again, from the lattice) that $B_{B_s}/B_B \simeq 1$, one could check the lattice prediction or simply substitute an experimental measurement for it.

3.9.28 New physics constraint

To see how great an impact even modest improvements in testing CKM unitarity in the charm sector would have, we consider a model in which a fourth family (t', b') of quarks is added to the usual three, with neutrinos heavy enough to evade the constraint $N_\nu = 3$ due to invisible Z decays. Unitarity relations involving the first two rows and columns of the expanded 4×4 CKM matrix allow us to calculate the following 90% C.L. upper limits using the best-measured quantities mentioned above:

$$|V_{ub'}| = \sqrt{1 - |V_{ud}|^2 - |V_{us}|^2 - |V_{ub}|^2} \le 0.05, \quad (211)$$

$$|V_{cb'}| = \sqrt{1 - |V_{cd}|^2 - |V_{cs}|^2 - |V_{cb}|^2} \le 0.5, \quad (212)$$

$$|V_{t'd}| = \sqrt{1 - |V_{ud}|^2 - |V_{cd}|^2 - |V_{td}|^2} \le 0.07, \quad (213)$$

$$|V_{t's}| = \sqrt{1 - |V_{us}|^2 - |V_{cs}|^2 - |V_{ts}|^2} \le 0.5. \quad (214)$$

The poor quality of the bounds on $|V_{cb'}|$ and $|V_{t's}|$ is largely due to the 10% error on $|V_{cs}|$, which translates to errors of 0.18 on $|V_{cb}|^2$ and $|V_{td}|^2$ and 90% C.L. upper limits on them of about 1/4. Thus improved measurements of V_{cs} could have a great impact on closing a rather gaping window for new physics or even revealing it.

3.9.29 Summary of overview

The above examples show that charmed particle studies have a large role to play in precision CKM physics, affecting nearly all the elements of the CKM matrix. In turn, precision CKM physics is important as a clue to the very origin of quark masses, since the CKM matrix arises from the same physics which generates those masses.

3.9.29.1 Leptonic decays Purely leptonic decays of charm mesons are of prime importance for checks of theoretical QCD calculations and searches for NP. Extraction of precise CKM information from neutral B mixing requires precision knowledge of the ratio of decay constants for B_s and B^0 [214]. While QCD calculations provide this estimate, the uncertainties are large, and the methods need to checked by seeing if they can reproduce charm measurements. Leptonic decays proceed in the SM by annihilation of the charm quark and spectator antiquark into a virtual W^+ that transforms to a lepton-antineutrino pair, as shown for the D^+ meson in Fig. 58.

In the SM, the decay width is given by [1000]

$$\Gamma(D^+ \to \ell^+ \nu) = \frac{G_F^2}{8\pi} f_{D^+}^2 m_\ell^2 M_{D^+} \left(1 - \frac{m_\ell^2}{M_{D^+}^2}\right)^2 |V_{cd}|^2, \quad (215)$$

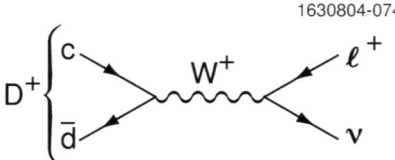

Fig. 58 The decay diagram for $D^+ \to \ell^+ \nu$

where M_{D^+} is the D^+ mass, m_ℓ is the mass of the final-state lepton, $|V_{cd}|$ is a CKM matrix element assumed to be equal to $|V_{us}|$, and G_F is the Fermi coupling constant. (The same formula applies to $D_s^+ \to \ell^+ \nu$ decays with the replacement of D_s^+ mass and $|V_{cs}|$.)

NP can affect the expected widths; any undiscovered charged bosons would interfere with the SM W^+. These effects may be difficult to ascertain, since they would simply change the values of the f_i's. The ratio $f_{D_s^+}/f_{D^+}$ is much better predicted in the SM than the values individually, so deviations seen here could point to beyond the SM charged bosons. For example, Akeroyd [1001] predicts that the presence of a charged Higgs boson would suppress this ratio significantly.

We can also measure the ratio of decay rates to different leptons, and the predictions then are fixed only by well-known masses. For example, for $\tau^+\nu$ to $\mu^+\nu$:

$$R \equiv \frac{\Gamma(D^+ \to \tau^+\nu)}{\Gamma(D^+ \to \mu^+\nu)} = \frac{m_{\tau^+}^2 (1 - \frac{m_{\tau^+}^2}{M_{D^+}^2})^2}{m_{\mu^+}^2 (1 - \frac{m_{\mu^+}^2}{M_{D^+}^2})^2}. \qquad (216)$$

Any deviation from this formula would be a manifestation of physics beyond the SM. This could occur if any other charged intermediate boson existed that coupled to leptons differently than mass-squared. Then the couplings would be different for muons and τ's. This would be a manifest violation of lepton universality, which has identical couplings of the muon, the tau, and the electron to the gauge bosons (γ, Z^0 and $W^\pm$) [1002]. (We note that in some models of supersymmetry, the charged Higgs boson couples as mass-squared to the leptons and therefore its presence would not cause a deviation from (216) [31].)

The CLEO-c Collaboration has published a result for f_{D^+} [323, 1003]. Several results have been obtained for $f_{D_s^+}$, the most precise being a preliminary result from CLEO-c. To measure f_{D^+}, CLEO-c uses a "double-tag" method, possible because at an e^+e^- centre-of-mass energy of 3770 GeV, the location of the ψ'' resonance, D^+D^- final states are produced without any extra particles. Here one D^- is fully reconstructed, and then there are enough kinematic constraints (energy and momentum) to search for $D^+ \to \mu^+\nu$ by constructing the missing mass-squared (MM^2) opposite the D^- and the muon, which should peak at the essentially zero neutrino mass-squared. Explicitly,

$$\text{MM}^2 = (E_{\text{beam}} - E_{\mu^+})^2 - (-\boldsymbol{p}_{D^-} - \boldsymbol{p}_{\mu^+})^2, \qquad (217)$$

where $\boldsymbol{p}_{D^-}$ is the three-momentum of the fully reconstructed D^-. The CLEO-c MM^2 distribution is shown in Fig. 59. The peak near zero contains 50 signal events of which 2.8 are estimated background.

The resulting rate is

$$\mathcal{B}(D^+ \to \mu^+\nu) = \left(4.40 \pm 0.66^{+0.09}_{-0.12}\right) \times 10^{-4}. \qquad (218)$$

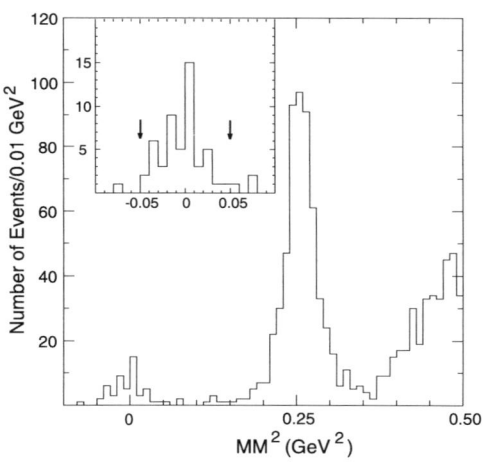
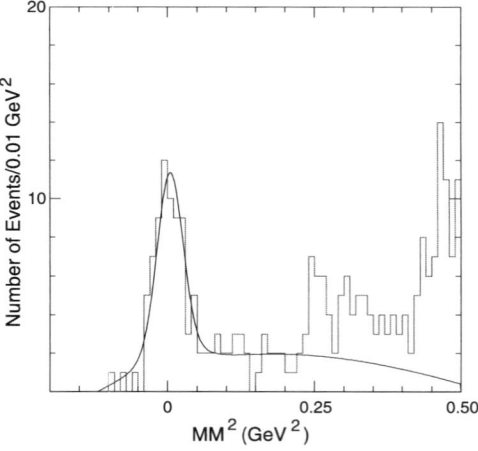

Fig. 59 CLEO-c missing mass-squared distributions. (*Left*) Using D^- tags and one additional opposite sign charged track depositing <300 MeV (consistent with a muon) in the calorimeter and no extra energetic clusters. The *inset* shows the signal region for $D^+ \to \mu^+\nu$ enlarged; the defined signal region is shown between the two arrows. (*Right*) Using D_s^- tags but allowing any energy deposit in the calorimeter (consistent with muon or pion). The *curve* is the predicted shape for the sum $D_s^+ \to \mu^+\nu + D_s^+ \to \tau^+\nu, \tau^+ \to \pi^+\nu$ normalized to the data for MM^2 < 0.2 GeV^2

The decay constant f_{D^+} is then obtained from (215) using 1.040 ± 0.007 ps as the D^+ lifetime [119] and $|V_{cd}| = 0.2238 \pm 0.0029$, giving

$$f_{D^+} = \left(222.6 \pm 16.7^{+2.8}_{-3.4}\right) \text{ MeV}. \tag{219}$$

CLEO-c also sets limits on $\mathcal{B}(D^+ \to e^+\nu_e) < 2.4 \times 10^{-5}$ [323, 1003] and $\mathcal{B}(D^+ \to \tau^+\nu)$ branching ratio to $<2.1 \times 10^{-3}$ (90% C.L.) [1004]. These limits are consistent with SM expectations.

Before turning to theoretical prediction of f_{D^+}, we discuss the current status of $D_s^+ \to \mu^+\nu$. Results here have been obtained by several experiments [119]. However, these results have been subject to sizable systematic errors, the largest of which usually is the uncertainty on $\mathcal{B}(D_s^+ \to \phi\pi^+)$ that is important because the measurements are usually normalized by taking the ratio of the observed number of $\ell^+\nu$ events to $\phi\pi^+$ events.

CLEO-c eliminates this uncertainty by making absolute measurements directly. Data are obtained near 4.170 GeV. Here the cross-section for $D_s^{*\pm}D_s^{\mp}$ is ~ 1 nb. Both $\mu^+\nu$ and $\tau^+\nu$ decays are examined with two different decay modes of the τ^+ used, $\pi^+\bar{\nu}$ and $e^+\nu\bar{\nu}$. The MM^2 distribution for the sum of $D_s^+ \to \mu^+\nu + D_s^+ \to \tau^+\nu$, $\tau^+ \to \pi^+\nu$ is shown on the right side of Fig. 59. Analysing these samples separately, they find the ratio R from (216) is consistent with the SM expectation of 9.72. Combining both gives a measurement using (215) of $f_{D_s} = 282 \pm 16 \pm 7$ MeV. CLEO-c also uses the $D_s^+ \to \tau^+\nu, \tau \to e^+\nu\bar{\nu}$ to find $f_{D_s} = 278 \pm 17 \pm 12$ MeV. Combining the two results gives

$$f_{D_s} = 280.1 \pm 11.6 \pm 6.0 \text{ MeV}. \tag{220}$$

Using only the $D_s^+ \to \tau^+\nu, \tau \to e^+\nu\bar{\nu}$ and the $D_s^+ \to \mu^+\nu$, CLEO-c finds

$$R = \frac{\Gamma(D_s^+ \to \tau^+\nu)}{\Gamma(D_s^+ \to \mu^+\nu)} = 9.9 \pm 1.7 \pm 0.7, \tag{221}$$

again consistent with the SM expectation. Furthermore CLEO-c also sets limits on $\mathcal{B}(D_s^+ \to e^+\nu_e) < 3.1 \times 10^{-4}$.

The branching fractions, modes and derived values of $f_{D_s^+}$ from all measurements are listed in Table 51. Most measurements of $D_s^+ \to \ell^+\nu$ are normalized with respect to $\mathcal{B}(D_s^+ \to \phi\pi^+)$. These measurements are difficult to average because of the uncertainty in this scale, and we do not attempt this here. We can extract a value for ratio using the CLEO-c measurements only, since the scale error is absent:

$$f_{D_s^+}/f_{D^+} = 1.26 \pm 0.11 \pm 0.03. \tag{222}$$

Theoretical calculations of $f_{D_s^+}$, f_{D^+} and the ratio $\frac{f_{D_s^+}}{f_{D^+}}$ are listed in Table 52. While the CLEO-c decay constant results are slightly higher than most theoretical expectations, the ratio is quite consistent with lattice gauge theory and most other models. Furthermore, no deviations from SM expectations are found in the ratio of decay rates for various lepton species.

3.9.29.2 Semileptonic decays The study of semileptonic charm decays has several important ramifications. Figure 60 shows the Feynman diagram describing these decays. It shows that the matrix element describing these decays can be expressed as the product of a leptonic current, unaffected by strong interactions, and a hadronic current, where the nonperturbative QCD effects are generally modeled with form factors. Theoretical predictions for these form factors have been derived in the framework of quark models, QCD sum rules and lattice QCD. Thus the study of inclusive and exclusive semileptonic decay branching fractions and form factors provides the experimental constraints needed to assess whether theoretical calculations are reliable and feature well-understood errors.

On the other hand, once computational techniques developed to predict relevant form factors demonstrate that

Table 51 Measurements of $f_{D_s^+}$. Results have been updated for new values of the D_s lifetime. ALEPH uses both measurements to derive a value for the decay constant

Exp.	Mode	$\mathcal{B}$	$\mathcal{B}_{\phi\pi}$ (%)	$f_{D_s^+}$ (MeV)
CLEO-c	$\mu^+\nu$	$(6.57 \pm 0.90 \pm 0.34) \times 10^{-3}$		$281 \pm 19 \pm 7$
CLEO-c	$\tau^+\nu, \tau \to \pi\nu$	$(7.1 \pm 1.4 \pm 0.3) \times 10^{-2}$		$296 \pm 29 \pm 7$
CLEO-c	$\tau^+\nu, \tau \to e\nu\nu$	$(6.29 \pm 0.78 \pm 0.52) \times 10^{-2}$		$278 \pm 17 \pm 12$
CLEO-c	combined	–		$280.1 \pm 11.6 \pm 6.0$
CLEO [1005]	$\mu^+\nu$	$(6.2 \pm 0.8 \pm 1.3 \pm 1.6) \times 10^{-3}$	3.6 ± 0.9	$273 \pm 19 \pm 27 \pm 33$
BEATRICE [1006]	$\mu^+\nu$	$(8.3 \pm 2.3 \pm 0.6 \pm 2.1) \times 10^{-3}$	3.6 ± 0.9	$315 \pm 43 \pm 12 \pm 39$
ALEPH [1007]	$\mu^+\nu$	$(6.8 \pm 1.1 \pm 1.8) \times 10^{-3}$	3.6 ± 0.9	$285 \pm 19 \pm 40$
ALEPH [1007]	$\tau^+\nu$	$(5.8 \pm 0.8 \pm 1.8) \times 10^{-2}$		
OPAL [1008]	$\tau^+\nu$	$(7.0 \pm 2.1 \pm 2.0) \times 10^{-3}$		$286 \pm 44 \pm 41$
L3 [1009]	$\tau^+\nu$	$(7.4 \pm 2.8 \pm 1.6 \pm 1.8) \times 10^{-3}$		$302 \pm 57 \pm 32 \pm 37$
BaBar [325]	$\mu^+\nu$	$(6.7 \pm 0.8 \pm 0.3 \pm 0.7) \times 10^{-3}$	4.7 ± 0.5	$283 \pm 17 \pm 7 \pm 14$

Table 52 Theoretical predictions of f_{D^+} and $f_{D_s^+}/f_{D^+}$. QL indicates quenched lattice calculations

Model	$f_{D_s^+}$ (MeV)	f_{D^+} (MeV)	$f_{D_s^+}/f_{D^+}$
Lattice ($n_f = 2 + 1$) [313]	$249 \pm 3 \pm 16$	$201 \pm 3 \pm 17$	$1.24 \pm 0.01 \pm 0.07$
QL (Taiwan) [1010]	$266 \pm 10 \pm 18$	$235 \pm 8 \pm 14$	$1.13 \pm 0.03 \pm 0.05$
QL (UKQCD) [709]	$236 \pm 8^{+17}_{-14}$	$210 \pm 10^{+17}_{-16}$	$1.13 \pm 0.02^{+0.04}_{-0.02}$
QL [1011]	$231 \pm 12^{+6}_{-1}$	$211 \pm 14^{+2}_{-12}$	1.10 ± 0.02
QCD Sum Rules [1012]	205 ± 22	177 ± 21	$1.16 \pm 0.01 \pm 0.03$
QCD Sum Rules [1013]	235 ± 24	203 ± 20	1.15 ± 0.04
Quark Model [1014]	268	234	1.15
Quark Model [1015]	248 ± 27	230 ± 25	1.08 ± 0.01
Potential Model [1016, 1017]	241	238	1.01
Isospin Splittings [1018]		262 ± 29	

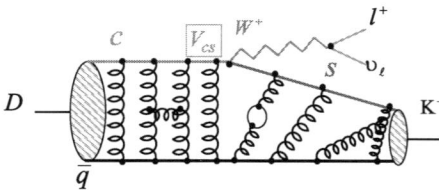

Fig. 60 Feynman diagram for the semileptonic decay of charmed mesons. The QCD nonperturbative effects are described by q^2-dependent form factors

they can achieve reliable results with well understood errors, these data allow precise determinations of the CKM matrix elements V_{cs} and V_{cd}. Moreover a combination of charm and beauty semileptonic decay studies can be used to determine V_{ub}.

3.9.30 Branching fractions

We are now progressing towards a complete precision determination of the absolute inclusive and exclusive charm semileptonic branching fractions. Inclusive semileptonic widths can provide some information on weak annihilation diagrams [912]. Finally, better knowledge of the inclusive positron spectra can be used to improved modeling of the "cascade" decays $b \to c \to se^+\nu_e$ and thus it affects the precision of several measurements of b decays.

CLEO-c uses the two tagging modes with lowest background ($\bar{D}^0 \to K^+\pi^-$ and $D^- \to K^+\pi^-\pi^-$) to measure the inclusive D^0 and D^+ semileptonic branching fractions [1019]. The kinematic constraints available through the use of D tagged samples from data taken at the $\psi(3770)$ provide a unique tool to select a pure sample of electrons/positrons coming from D semileptonic decays. They obtain $\mathcal{B}(D^0 \to X\ell\nu_e) = (6.46 \pm 0.17 \pm 0.13)\%$ and $\mathcal{B}(D^+ \to X\ell\nu_e) = (16.13 \pm 0.20 \pm 0.33)\%$. The inclusive branching fractions can be translated into inclusive semileptonic widths Γ_{D^+} and Γ_{D^0}, using the well-known D lifetimes [119]. These widths are expected to be equal, modulo isospin violations, and indeed the measured ratio $\Gamma^{sl}_{D^+}/\Gamma^{sl}_{D^0} = 0.985 \pm 0.028 \pm 0.015$: thus isospin violations are limited to be below $\sim 3\%$.

BES-II [340, 1020] and CLEO-c [1021, 1022] have recently published data on exclusive semileptonic branching fractions. BES-II results are based on 33 pb^{-1}; the CLEO-c published data are based on the first 57 pb^{-1}, preliminary results included in this report are based on 281 pb^{-1}.

The variable $U \equiv E_\text{miss} - |c\vec{p}_\text{miss}|$, where E_miss and $\vec{p}_\text{miss}$ represent the missing energy and momentum of the D meson decaying semileptonically, is used to select signal events. This variable is a non-Lorentz invariant version of MM^2. Table 53 summarizes the recent data, as well as the averages reported in the PDG 2006 [119].

A comparison between the inclusive branching fractions of the D^+ and D^0 mesons with the sum of the measured exclusive branching fractions determines whether there are unobserved semileptonic decay modes. The corresponding sums of exclusive branching fractions are: $\sum_i \mathcal{B}(D^0 \to X_i \ell\nu_e) = 6.1 \pm 0.2 \pm 0.2$ and $\sum_i \mathcal{B}(D^+ \to X_i \ell\nu_e) = 15.1 \pm 0.50 \pm 0.50$; the measured exclusive modes are consistent with saturating the inclusive widths, although there is some room left for higher multiplicity modes. In particular, CLEO-c also provides the first evidence for $D^0 \to K^-\pi^+\pi^-e^+\nu_e$ [1023]. They study the MM^2 inferred from the missing energy and momentum in the event, and they obtain the preliminary branching fractions:

$$\mathcal{B}(D^0 \to K^-\pi^+\pi^-e^+\nu_e) = \left(2.9^{+1.5}_{-1.1} \pm 0.5\right) \times 10^{-4}, \quad (223)$$

$$\mathcal{B}(D^0 \to K_1(1270)e^+\nu_e) \times \mathcal{B}(K_1(1270) \to K^-\pi^+\pi^-)$$
$$= \left(2.2^{+1.4}_{-1.0} \pm 0.2\right) \times 10^{-4}. \quad (224)$$

This branching fraction is about at the level predicted by Isgur and Scora [292] and is consistent with the expectation that charm semileptonic decays are dominated by the pseudoscalar and vector lowest mass resonances.

Table 53 CLEO-c branching fractions and new world averages

D^+ mode	Recent data $\mathcal{B}$ (%)	PDG 2006	D^0 mode	Recent data $\mathcal{B}$ (%)	PDG 2006
$\bar{K}^0 e^+ \nu_e$	$8.86 \pm 0.17 \pm 0.20$	8.7 ± 0.5	$K^- e^+ \nu_e$	$3.58 \pm 0.05 \pm 0.05$	3.47 ± 0.13
$\pi^0 e^+ \nu_e$	$0.397 \pm 0.027 \pm 0.028$	0.44 ± 0.06	$\pi^- e^+ \nu_e$	$0.309 \pm 0.012 \pm 0.006$	0.262 ± 0.026
$\eta e^+ \nu_e$	$0.129 \pm 0.019 \pm 0.07$		$K^{*-} e^+ \nu_e$	$2.16 \pm 0.15 \pm 0.08$	2.16 ± 0.16
$\bar{K}^{*0} e^+ \nu_e$	$5.56 \pm 0.27 \pm 0.23$	5.61 ± 0.31	$\rho^- e^+ \nu_e$	$0.156 \pm 0.016 \pm 0.009$	0.194 ± 0.41
$\rho^0 e^+ \nu_e$	$0.232 \pm 0.020 \pm 0.012$	0.22 ± 0.04			
$\omega e^+ \nu_e$	$0.149 \pm 0.027 \pm 0.005$	$0.16^{+0.07}_{-0.06}$			

Finally, D semileptonic decays are a tool to explore light quark spectroscopy. For example, a few years ago, the FOCUS Collaboration reported some evidence for an s-wave interference effect in the decay amplitude of $D^+ \to K^{*0} \mu^+ \nu_\mu$ [1024]. This observation can shed some light on our understanding of the elusive scalar meson κ. This observation has been recently confirmed by CLEO-c in the channel $D^+ \to K^{*0} e^+ \nu_e$ [1025]. This study will acquire soon a broader scope when CLEO-c will pursue similar analyses in the D_s system.

3.9.31 Form factors for $D \to K(\pi)\ell\nu$ and $D \to K^*(\rho)\ell\nu$

Recently, nonquenched lattice QCD calculations for $D \to K\ell\bar{\nu}$ and $D \to \pi\ell\nu$ have been reported [332]. The chiral extrapolation is performed at fixed $E = \vec{v} \cdot \vec{p}_P$, where E is the energy of the light meson in the centre-of-mass D frame, $\vec{v}$ is the unit 4-velocity of the D meson, and $\vec{p}_P$ is the 4-momentum of the light hadron P (K or π). The results are presented in terms of the parametrization originally proposed by Becirevic and Kaidalov (BK) [280]:

$$f_+(q^2) = \frac{F}{(1-\tilde{q}^2)(1-\alpha\tilde{q}^2)}, \qquad (225)$$
$$f_0(q^2) = \frac{F}{1-\tilde{q}^2/\beta},$$

where q^2 is the 4-momentum of the electron-ν pair, $\tilde{q}^2 = q^2/m_{D_x^*}^2$, $F = f_+(0)$, and α and β are fit parameters. This formalism models the effects of higher mass resonances other than the dominant spectroscopic pole (D_s^{*+} for the $K\ell\nu$ final state and D^{*+} for $\pi\ell\nu$ [1026]).

Table 54 shows the fit results obtained from FOCUS [341], CLEO III [1027], Belle [1028] and BaBar [1029] compared to the lattice QCD predictions [332]. In addition, all these experiments perform a single pole fit, traditionally used because of the conventional ansatz of several quark models [1030], and the BK parametrization discussed before. In Table 55, we include preliminary results of fits obtained with the simple pole model by CLEO-c. All of these experiments obtain very good fits also with simple pole form

Table 54 Measured shape parameter α compared to lattice QCD predictions

	$\alpha(D^0 \to K\ell\nu)$	$\alpha(D^0 \to \pi\ell\nu)$
Lattice QCD [332]	$0.5 \pm 0.04 \pm 0.07$	$0.44 \pm 0.04 \pm 0.07$
FOCUS [341]	$0.28 \pm 0.08 \pm 0.07$	
CLEOIII [1027]	$0.36 \pm 0.10^{+0.03}_{-0.07}$	$0.37^{+0.20}_{-0.31} \pm 0.15$
Belle [1028]	$0.40 \pm 0.12 \pm 0.09$	$0.03 \pm 0.27 \pm 0.13$
BaBar [1029]	$0.43 \pm 0.03 \pm 0.04$	

Table 55 Measured values of M_{pole}

	$M_{\text{pole}}(D^0 \to K\ell\nu)$ (GeV)	$M_{\text{pole}}(D^0 \to \pi\ell\nu)$ (GeV)
FOCUS [341]	$1.93 \pm 0.05 \pm 0.03$	$1.91^{+0.30}_{-0.15} \pm 0.07$
CLEOIII [1027]	$1.89 \pm 0.05^{+0.04}_{-0.03}$	$1.86^{+0.10+0.07}_{-0.06-0.03}$
Belle [1028]	$1.88 \pm 0.06 \pm 0.03$	$2.01 \pm 0.13 \pm 0.04$
BaBar [1029]	$1.854 \pm 0.016 \pm 0.020$	
CLEO-c [1023]	$1.96 \pm 0.03 \pm 0.01$	$1.95 \pm 0.04 \pm 0.02$

factors; however the simple pole fit does not yield the expected spectroscopic mass. This may hint that other higher-order resonances are contributing to the form factors [1026]. It has been argued [1031] that even the BK parametrization is too simple and that a three parameter form factor is more appropriate. This issue can be resolved by larger data samples, with better sensitivity to the curvature of the form factor near the high recoil region.

In experimental studies of $D \to K^*(\rho)\ell\nu$, usually a single pole parametrization of form factors was used. Following the Becirevic–Kaidalov approach [1032, 1033], a new parametrization of the relevant form factors was given by

$$A_1(q^2) = \frac{A_1(0)}{1 - b'x}, \qquad A_2(q^2) = \frac{A_2(0)}{(1 - b'x)(1 - b''x)},$$
$$A_0(q^2) = \frac{A_0(0)}{(1 - y)(1 - a'y)},$$
$$V(q^2) = \frac{A_1(0)}{\xi(1 - x)(1 - ax)}.$$

This parametrization takes into account all known scaling properties of the decay to light vector semileptonic transition. The study of nonparametric determination of helicity amplitudes in the semileptonic $D \to K^*(\rho)\ell\nu$ decays will shed more light on the corresponding decays in B physics.

3.9.32 Lattice QCD checks

By combining the information of the measured leptonic and semileptonic widths, the ratio

$$R_{sl} = \sqrt{\frac{\Gamma(D^+ \to \mu^+ \nu_\mu)}{\Gamma(D \to \pi e \nu_e)}}, \quad (226)$$

independent of $|V_{cd}|$, can be evaluated, which is a pure check of the theory. We assume isospin symmetry, and thus $\Gamma(D \to \pi e^+ \nu_e) = \Gamma(D^0 \to \pi^- e^+ \nu_e) = 2\Gamma(D^+ \to \pi^0 e^+ \nu_e)$. For the theoretical inputs, we use the recent unquenched lattice QCD calculations in three flavors [313], as they reflect the state of the art of the theory and have been evaluated in a consistent manner. The theory ratio is

$$R_{sl}^{th} = \sqrt{\frac{\Gamma^{th}(D^+ \to \mu^+ \nu_\mu)}{\Gamma^{th}(D \to \pi e \nu_e)}} = 0.212 \pm 0.028. \quad (227)$$

The quoted error is evaluated through a careful study of the theory statistical and systematic uncertainties, assuming Gaussian errors. The corresponding experimental R_{sl}^{exp} is calculated using the CLEO-c f_D and isospin averaged $\Gamma(D \to \pi e^+ \nu_e)$:

$$R_{sl}^{exp} = \sqrt{\frac{\Gamma^{exp}(D^+ \to \mu^+ \nu)}{\Gamma^{exp}(D \to \pi e \nu_e)}} = 0.237 \pm 0.019. \quad (228)$$

The theory and data are in good agreement, though the errors need to be reduced both in theory and experiment to validate the theory at the needed level of precision (~ 1–3%).

3.9.32.1 Hadronic decays
While the dynamical issues are considerably more complex in nonleptonic than in semileptonic decays—both a blessing and a curse—, the available theoretical tools are more limited. For inclusive rates like lifetimes, one can turn to expansions in powers of $1/m_c$ to obtain at least a semi-quantitative description. For exclusive modes, we have 'Old Faithful,' namely quark models, but also QCD sum rules and chiral dynamics with the latter two (in contrast to the first one) firmly rooted in QCD. Lattice QCD, usually perceived as panacea, faces much more daunting challenges in dealing with nonleptonic charm transitions than for semileptonic modes due to the central role played by strong final-state interactions. Yet comprehensive measurements can teach us valuable lessons that can enlighten us about light flavor spectroscopy and also serve as cross checks on B studies. Below we list some core examples for such lessons.

3.9.32.2 Lifetime ratios
Heavy quark theory (HQT) allows us to describe inclusive decays of charm hadrons through an expansion in powers of $1/m_c$ implemented by the OPE. With the charm quark mass m_c exceeding ordinary hadronic scales merely by a moderate amount, the expansion parameter is not much smaller than unity. In the description of fully integrated widths like lifetimes, the leading nonperturbative contributions arise in order $1/m_c^2$ rather than $1/m_c$, which might be their saving grace. Indeed the resulting theoretical description of the lifetime ratios for the seven weakly decaying $C = 1$ charm hadrons has been remarkably successful [912]. Note that these seven charm lifetimes vary by a factor of 15, while the four singly-beautiful hadrons differ by less than 30%. The B_c meson is shorter lived by a factor of three than the other four beauty hadrons—not surprisingly, since it represents a glorified charm decay.

The same framework allows us to predict also the lifetimes of the $C = 2$ double-heavy baryons Ξ_{cc}, Ω_{cc} and even the $C = 3$ Ω_{ccc} [912]:

$$\begin{aligned}\tau(\Xi_{cc}^{++}) &\sim 0.35 \text{ ps}, & \tau(\Xi_{cc}^+) &\sim 0.07 \text{ ps}, \\ \tau(\Omega_{cc}^+) &\sim 0.1 \text{ ps}, & \tau(\Omega_{ccc}^{++}) &\sim 0.14 \text{ ps}.\end{aligned} \quad (229)$$

The SELEX Collaboration has found tantalizing evidence for $\Xi_{cc}^{+,++}$ baryons all decaying with ultrashort lifetimes below 0.03 ps. This feature cannot be accommodated in HQT. *If* confirmed, one would have to view the apparent successes of the HQT description of the $C = 1$ lifetimes as mere coincidences.

3.9.32.3 Absolute branching ratios
Precision absolute branching fraction measurements are difficult due to normalization and systematic effects. Only one *golden mode* is needed to anchor the rest for each state. A desire to use all-charged final states necessitates use of some three-body modes where proper modeling of the Dalitz structure is needed to ensure an accurate efficiency simulation. These results serve not only to normalize charm physics but also much B physics due to dominance of $b \to c$ decays. For example, charm branching fractions affect $B \to D^*\ell\nu$, used to extract V_{cb}.

Near-threshold $D\bar{D}$ pairs from $\psi(3770)$ decays and $D_s^{*\pm}D_s^{\mp}$ produced at 4170 MeV from CLEO-c now provide the best precision. Systematics are controlled, and normalization provided with tagging: studying one D vs. a fully-reconstructed *tag* $\bar{D}$. Precision on the golden modes $D^0 \to K^-\pi^+$ and $D^+ \to K^-\pi^+\pi^+$ results are limited by uncertainties of about 1% per track [1034] from tracking-finding and particle-identification efficiencies. Further studies [1035] are reducing these to less than 0.5% per track. Current statistical precision for $D_s^+ \to K^+K^-\pi^+$ decays [1035] is 5%; final CLEO-c accuracy should be about 3%, limited by statistics. Producing a useful new result for the popular $D_s^+ \to \phi\pi^+$ mode is complicated by several factors: a nonresonant contribution under the ϕ, Breit–Wigner tails of the ϕ, treatment of nearby resonances like the $f(980)$ and lack of detail in existing publications. The merit of such studies goes beyond determining the branching ratio for $D_s^+ \to \phi\pi^+$ and learning about hadronic resonances (see below). Their greatest impact might come in precision analyses of $B_d \to \phi K_S$ and its CP asymmetries.

3.9.32.4 Dalitz plot studies & light flavor spectroscopy

Dalitz plot studies represent powerful analysis tools that are deservedly experiencing a renaissance in heavy flavor decays. Constructing a satisfactory description of the Dalitz plot populations allows one to extract the maximal amount of information from the data in a self-consistent way. One has to keep in mind, though, that a priori different parameterizations can be chosen; one has to make a judicious choice based on theoretical considerations. Along with better theoretical descriptions of the decay rate, improved treatments of background and efficiency may also be needed.

One important application concerns the spectroscopy of light flavor hadrons, i.e. those made up from u, d and s quarks. Modes like $D_{(s)} \to 3\pi, 3K, K\pi\pi, K\bar{K}\pi$ offer more than a treasure trove of additional data: since the final state evolves from a well-defined initial one, we know some quantum numbers of the overall system. Finding evidence for, say, a $\pi\pi$ resonance like the σ in Cabibbo-favored D and Cabibbo-suppressed D_s modes with parameters consistent with what is inferred from low-energy $\pi\pi$ scattering would constitute a powerful validation for the σ being a bona fide resonance.

Such lessons possess considerable intrinsic value. The latter is greatly amplified, since these insights will turn out to be of great help in understanding B decays into the analogous final states when searching for CP asymmetries there.

3.9.32.5 QCD sum rules

More than twenty years ago a pioneering analysis of D and D_s decays into two-body final states of the PP and PV type was performed by Blok and Shifman through a novel application of QCD sum rules. Those are—unlike quark models—genuinely based on the QCD. Their drawback, as for most applications of QCD sum rules, is that one has to allow for an irreducible theoretical uncertainty of about 20%; furthermore they are very labor intensive. The authors of Ref. [1036] assumed $SU(3)_{\text{fl}}$ symmetry to make their analysis manageable—clearly a source of significant theoretical uncertainty. It would be marvelous if some courageous minds would take up the challenge of updating and extending this study.

3.9.32.6 On theoretical engineering

Even without reliable predictions for exclusive nonleptonic widths, it makes a lot of sense to measure as many as precisely as possible on the Cabibbo-allowed, once and twice suppressed levels. It can provide vital input into searches for direct CP violation in charm decays.

CP asymmetries in integrated partial widths depend on hadronic matrix elements and (strong) phase shifts, neither of which can be predicted accurately. However the craft of theoretical engineering can be practiced with profit here. One makes an ansatz for the general form of the matrix elements and phase shifts that are included in the description of $D \to PP, PV, VV$ etc. channels, where P and V denote pseudoscalar and vector mesons, and fits them to the measured branching ratios on the Cabibbo-allowed, once and twice forbidden level. If one has sufficiently accurate and comprehensive data, one can use these fitted values of the hadronic parameters to predict CP asymmetries. Such analyses have been undertaken in the past [968] and more recently by [1037–1042], but the data base was not as broad and precise as one would like. CLEO-c and BESIII measurements will certainly lift such studies to a new level of reliability.

Similar information can be obtained in a more subtle and model independent way using quantum entanglement in [958]

$$e^+e^- \to \psi(3770) \to D^0 \bar{D}^0 \qquad (230)$$

and observing the subsequent decay of the neutral D mesons into final states like $f(D) = K^-\pi^+, K^+\pi^-, K^+K^-, \pi^+\pi^-$. Since the $D^0\bar{D}^0$ pair forms a coherent system, one can extract the strong phases reliably. This procedure is described in detail in Sect. 3.9.2.

3.9.32.7 Time dependent Dalitz studies

Tracking three-body channels like $D^0 \to K\bar{K}\pi, K_S^0\pi\pi$ through time-dependent Dalitz plot studies is a very powerful way to look for NP through CP asymmetries involving D^0–$\bar{D}^0$ oscillations, as described in more detail in Sects. 3.9.2 and 3.9.12.

3.9.33 Summary on ongoing and future charm studies

Even accepting for the moment that the SM can provide a complete description of all charm transitions, detailed and comprehensive measurements of the latter will continue to teach us important and quite possible even novel lessons on QCD. Those lessons are of considerable intellectual value and would also prepare us if the anticipated NP driving the electroweak phase transition were of the strongly interacting variety.

Yet most definitely those lessons will sharpen both our experimental and theoretical tools for studying B decays and thus will be essential in saturating the discovery potential for NP there. Analyses of (semi)leptonic charm decays will yield powerful validation challenges to LQCD that, if passed successfully, will be of great benefit to extractions of $|V_{ub}|$ in particular. Careful studies of three-body final states in charm decays will yield useful constraints in analyses of the corresponding B modes and their CP asymmetries. The relevant measurements can be made at the tau-charm, the B and super-flavor factories. Yet there is one area in *this* context where hadronic experiments and in particular LHCb can make important contributions, namely in the search for and observation of doubly-heavy charm baryons of the $[ccq]$ type and their lifetimes.

The study of charm dynamics was crucial in establishing the SM paradigm. Even so it is conceivable that another revolution might originate there in particular by observing non-SM type CP violation with and without oscillations. For on one hand the SM predicts practically zero results (except for direct CP violation in Cabibbo-suppressed channels), and on the other hand FCNCs might well be considerably less suppressed for up- than for down-type quarks. Charm is the only up-type quark that allows the full range of searches for CP violation. Modes like $D^0 \to K^+K^-$, $K^+\pi^-$ have the potential to exhibit (time-dependent) CP asymmetries that—if observed—would establish the presence of NP. Likewise for asymmetries in final-state distributions like Dalitz plots or for T odd moments. Again especially LHCb appears well positioned to bring the statistical muscle of the LHC to bear on analysing these transitions.

3.10 Impact of the LHC experiments[28]

3.10.1 Overview

The LHC will start operating in 2008. This will allow the LHC experiments—ATLAS, CMS and LHCb—to make substantial progress in heavy flavor physics and possibly to open a window to new physics beyond the Standard Model. LHCb is a heavy flavor physics experiment designed to make precision measurements of CP violation and of rare decays of B hadrons. The general purpose experiments ATLAS and CMS also have a B physics programme, which will be carried out mainly during the first years of LHC operation with lower luminosity. The large cross section of 500 μb for $b\bar{b}$-quark production in pp collisions at 14 TeV centre-of-mass energy will allow the LHC experiments to collect much larger data samples of B hadrons than previously available.

Many of the expected LHC results have been reported in Sect. 3 of this report. Here we summarize a few of the anticipated highlights and provide the interested reader with a guide to more detailed discussions in Sects. 3.1 to 3.9. We also present the LHCb detector and illustrate how the different sub-detectors are crucial to achieve the expected performance on selected decays.

---

[28] Section coordinator: F. Muheim.

3.10.2 The LHCb experiment

LHCb is a dedicated heavy flavor experiment at the LHC. The LHCb detector is a single arm forward spectrometer which exploits the fact that a large fraction of the $b\bar{b}$ cross section is in the forward region. The LHCb experiment will operate at a luminosity of 2 to 5×10^{32} cm^{-2}s^{-1} and expects to accumulate a data sample of ~ 10 fb^{-1} over the next five years.

The LHCb detector layout is shown in Fig. 61. A silicon vertex detector (VELO) will be used to determine very precisely the decay length of B mesons. A typical proper time resolution of about 40 fs will be achieved for fully reconstructed decays. Charged tracks are momentum analysed by a dipole magnet, and their trajectories are recorded in four tracking stations. LHCb also features excellent particle identification: two Ring Imaging Cherenkov (RICH) detectors are employed to distinguish pions from kaons over a large momentum range; an electromagnetic calorimeter (ECAL) will measure electrons and photons; and muons are cleanly identified in the muon stations. A challenging task is to discriminate the interesting events from the much more copious minimum bias events at a hadron collider. A first-level hardware trigger (L0) operating at the interaction rate of 40 MHz is triggering on collisions containing muons with large transverse momenta, $p_T > 1$ GeV/c, hadrons which deposit a transverse energy $E_T > 3.5$ GeV in the hadron calorimeter (HCAL) and electrons or photons above an ECAL threshold of $E_T > 2.5$ GeV. The L0 trigger reduces the rate of interactions to below 1.1 MHz, at which all LHCb data will be read out. Events will then be examined by the High Level Trigger (HLT) running on a large computer farm. The HLT will reduce the output rate to 2 kHz which will be written to storage.

3.10.3 Expected highlights from LHC results

In the SM, flavor changing neutral current (FCNC) transitions are suppressed as these only occur through loop diagrams. These processes are thus sensitive to contributions from new heavy particles to the virtual loops and pose excellent probes to new physics (NP) beyond the SM. The LHC experiments will collect very large data samples of B and charm hadrons. These will allow them to probe NP at much increased sensitivities and to make precision tests of CP violation which, in the SM, arises solely through the CKM mechanism.

FCNC $b \to s$ transitions are an exciting NP probe and have been studied extensively. The very rare decay $B_s \to \mu^+\mu^-$ has a well-predicted SM rate which could be enhanced considerably in many NP models. As shown for LHCb in Fig. 61, the three LHC experiments all have excellent muon identification systems to identify the final-state

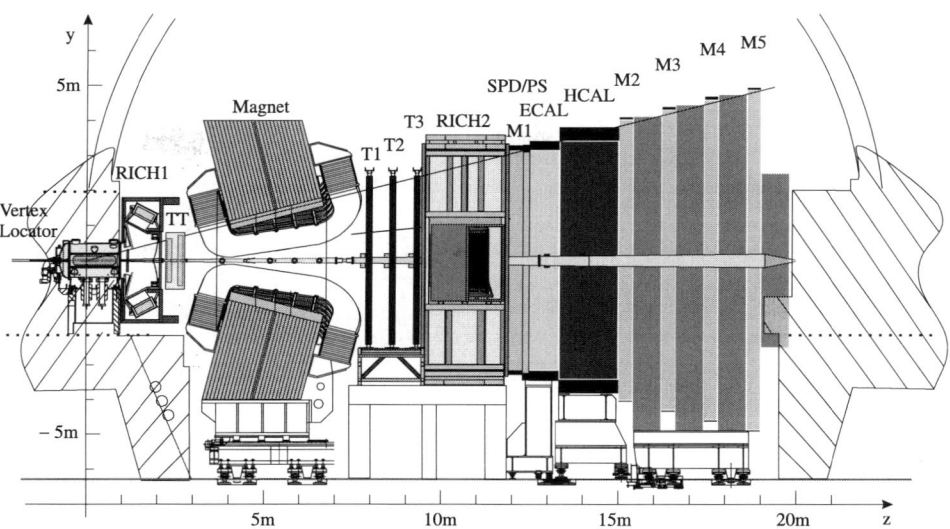

Fig. 61 Schematic of the LHCb detector layout, showing the Vertex Locator (VELO), the dipole magnet, the two RICH detectors, the four tracking stations TT and T1–T3, the Scintillating Pad Detector (SPD), Preshower (PS), Electromagnetic (ECAL) and Hadronic (HCAL) calorimeters and the five muon stations M1–M5

particles of this powerful NP probe. The ATLAS, CMS and LHCb trigger and selection efficiencies as well as sensitivities for $B_s \to \mu^+\mu^-$ are described in detail in Sect. 3.4.3. The muon detectors will also enable the LHC experiments to investigate electroweak penguin transitions in the decays $B_d \to K^{*0}\mu^+\mu^-$ and $\Lambda_b \to \Lambda^0\mu^+\mu^-$. The sensitivity to the forward–backward asymmetry A_{FB}, the transversity asymmetry $A_T^{(2)}$ and the K^{*0} longitudinal polarization F_L as well as the Λ^0 polarization from LHCb and ATLAS are presented in Sect. 3.2.3. LHCb is equipped with an electromagnetic calorimeter, which will be employed to trigger and reconstruct the photon of the radiative penguin decays $B^0 \to K^{*0}\gamma$ and $B_s \to \phi\gamma$. This is discussed in detail in Sect. 3.1.4.2.

The LHCb experiment will be able to measure the CP violating weak phase ϕ_s in the interference of B_s mixing and $B_s \to J/\psi\phi$ decays down to the SM prediction. This will require the measurement of a time-dependent CP asymmetry, and the fast oscillations of the B_s meson need to be resolved. This ϕ_s determination will be made possible with the excellent vertex resolution of the VELO detector. In addition, the flavor of the B_s at production must be determined. Good flavor tagging is also required. This will be achieved with the RICH detectors which will cleanly identify charged kaons from the $b \to c \to s$ decays of the other B hadron or accompanying the signal B_s in the fragmentation chain. The sensitivity of measurements of ϕ_s, the B_s oscillation frequency Δm_s and the lifetime difference $\Delta\Gamma_s$ expected from LHCb and CMS are discussed in detail in Sects. 3.6.7 and 3.6.8.

LHCb will also perform precise measurements of the angles of the CKM unitarity triangle (α, β and γ). The expected sensitivities are presented in Sect. 3.5.3. The CKM angle γ can be measured using two interfering diagrams in neutral and charged $B \to DK$ decays as well as in $B_s \to D_s^\mp K^\pm$ decays. The excellent kaon-pion separation from the RICH will greatly facilitate these analyses. A precision of a few degrees is expected, which is significantly better than current results. CP asymmetries in hadronic $b \to s$ transitions are sensitive to new physics in penguin loops. The best modes for LHCb are the decays $B_s \to \phi\phi$ and $B^0 \to \phi K_S^0$, as these are experimentally accessible and have small theoretical uncertainties. The expected sensitivities are discussed in Sect. 3.7.7. LHCb will also measure γ in loops with two-body hadronic decays $B \to hh'$, which is presented in Sect. 3.7.9.

In addition, the LHCb experiment will reconstruct large samples of charm mesons. This will substantially improve the precision of the D^0 mixing parameters, and a detailed discussion is given in Sects. 3.9.4 to 3.9.10.

4 Prospects for future facilities[29]

There are several new facilities for flavor physics discussed in the community among which the Super Flavor Factories (SFF) and the upgrade of the LHCb experiment are the most important ones for B physics. These are analysed in this chapter (for future kaon and charm physics facilities, see also Sects. 3.8 and 3.9).

The physics case of an SFF is worked out in Sect. 4.1. All opportunities of such a facility in B, charm and τ lepton physics are discussed. Then the two existing proposals for such a machine, namely Super$B$ and SuperKEKB, are presented in Sects. 4.2 and 4.3, respectively. Finally, the physics, detector and accelerator issues of a possible future upgrade of the LHCb experiment are discussed in Sect. 4.4.

---

[29] Section coordinator: T. Hurth.

4.1 On the physics case of a super flavor factory

We summarize the physics case of a high-luminosity e^+e^- flavor factory collecting an integrated luminosity of (50–75) ab^{-1}. Many NP sensitive measurements involving B and D mesons and τ leptons, unique to a SFF, can be performed with excellent sensitivity to new particles with masses up to ∼100 (or even ∼1000) TeV. Flavor- and CP-violating couplings of new particles that may be discovered at the LHC can be measured in most scenarios, even in unfavorable cases assuming MFV. Together with the LHC, an SFF, following either the SuperKEKB or the Super$B$ proposal, could be soon starting the project of reconstructing the NP Lagrangian.

4.1.1 Introduction

In spite of the tremendous success of the SM, it is fair to say that the flavor sector of the SM is much less understood than its gauge sector. Masses and mixing of the quarks and leptons, which have a significant but unexplained hierarchy pattern, enter as free parameters to be determined experimentally. In fact, while symmetries shape the gauge sector, no principle governs the flavor structure of the SM Lagrangian. Yukawa interactions provide a phenomenological description of the flavor processes which, while successful so far, leaves most fundamental questions unanswered. Hence the need to go beyond the SM.

Indeed the search for evidence of physics beyond the SM is the main goal of particle physics in the next decades. The LHC at CERN will start soon looking for the Higgs boson, the last missing building block of the SM. At the same time, it will intensively search for NP, for which there are solid theoretical motivations related to the quantum stabilization of the Fermi scale to expect an appearance at energies around 1 TeV. However, pushing the high-energy frontier, i.e. increasing the available centre-of-mass energy in order to produce and observe new particles, is only one way to look for NP, the other being high-precision studies of rare processes. New particles could reveal themselves through their virtual effects in processes involving only standard particles as has been the case several times in the history of particle physics. Flavor physics is the best candidate as a tool for NP searches through quantum effects for several reasons. FCNCs, neutral meson–anti-meson mixing and CP violation occur at the loop level in the SM and therefore are potentially subject to $\mathcal{O}(1)$ NP virtual corrections. In addition, quark flavor violation in the SM is governed by the weak interaction and suppressed by the small quark mixing angles. Both these features are not necessarily shared by NP which, in such cases, could produce very large effects. Indeed, the inclusion in the SM of generic NP flavor-violating terms with natural $\mathcal{O}(1)$ couplings is known to violate present experimental constraints unless the NP scale is pushed up to (10–100) TeV depending on the flavor sector. This difference between the NP scale emerging from flavor physics and the one suggested by Higgs physics could be a problem for model builders (the so-called flavor problem), but it clearly indicates that flavor physics has the potential to push the explored NP scale in the 100 TeV region. On the other hand, if the NP scale is indeed close to 1 TeV, the flavor structure of NP must be highly nontrivial, and the experimental determination of the flavor-violating couplings is particularly interesting. Any NP model established at the TeV scale to solve the gauge hierarchy problem includes new flavored particles and new flavor- and CP-violating parameters. Therefore, such a model must provide a solution also to the flavor and CP problems. This may be related to other interesting questions. For instance, in supersymmetry, the flavor problem is directly linked to the crucial issue of supersymmetry breaking. Similar problems also occur in models of extra-dimensions (flavor properties of Kaluza–Klein states), Technicolour models (flavor couplings of Techni-fermions), little-Higgs models (flavor couplings of new gauge bosons and fermions) and multi-Higgs models (CP-violating Higgs couplings). Once NP is found at the TeV scale, precision measurements of flavor- and CP-violating observables would shed light on the detailed structure of the underlying model.

In the light of the above considerations, an SFF, following the recent proposals for SuperKEKB (see Sect. 4.3 and Ref. [839]) and Super$B$ (see Sect. 4.2 and Ref. [1045]), has one mission: to search for NP in the flavor sector exploiting a huge leap in integrated luminosity and the wide range of observables that it can measure. However, this goal can be pursued in different ways depending on whether evidence of NP has been found at the time an SFF starts taking data, or not.

In both cases, an SFF can search for evidence of NP irrespective of the values of the new particle masses and of the unknown flavor-violating couplings. A first set is given by measurements of observables which are predicted by the SM with small uncertainty, including those which are vanishingly small (null tests). Among them, there are the flavor-violating τ decays, direct CP asymmetries in $B \to X_{s+d}\gamma$, in τ decays and in some nonleptonic D decays, CP violation in neutral charm meson mixing, the dilepton invariant mass at which the forward–backward asymmetry of $B \to X_s \ell^+ \ell^-$ vanishes, and lepton universality violating B and τ decays. Any deviation, as small as an SFF could measure, from its SM value of any observable in this set could be ascribed to NP with essentially no uncertainty. A second set of NP-sensitive observables, including very interesting decays such as $b \to s$ penguin-dominated nonleptonic B decays, $B \to \tau\nu$, $B \to D^{(*)}\tau\nu$, $B \to K^*\gamma$, $B \to \rho\gamma$, and many others, require more accurate determinations of SM contributions and improved control of the hadronic uncertainties with respect to what we can do today in order to match the experimental precision achievable at an SFF and to allow for an unambiguous identification of an NP signal. The error on the SM can be reduced using the improved determination of the CKM matrix provided by an SFF itself. This can be achieved using generalized CKM fits which allow for a 1% determination of the CKM parameters using tree-level and $\Delta F = 2$ processes even in the presence of generic NP contributions. As far as hadronic uncertainties are concerned, the extrapolation of our present knowledge and techniques shows that it is possible to reach the required accuracy by the time an SFF will be running using improved lattice QCD results obtained with next-generation computers [1045] and/or bounding the theoretical uncertainties with data-driven methods exploiting the huge SFF data sample.

Finally, it must be emphasized that while an SFF will perform detailed studies of beauty, charm and tau lepton physics, the results will be highly complementary to those on several important observables related to B_s meson oscillations, kaon and muon decays that will be measured elsewhere. Most benchmark charm measurements, in particular interesting NP-related measurements such as CP violation in charm mixing, will still be statistics-limited after the CLEOc, BESIII and B factory projects are completed and can only be pursued to their ultimate precision at an SFF. Operation at the $\Upsilon(5S)$ resonance provides the possibility of exploiting the clean e^+e^- environment to measure B_s^0 decays with neutral particles in the final state, thus of complementing the measurements at LHCb. An SFF has sensitivity for τ physics that is far superior to any other existing or proposed experiment, and the physics reach can be extended even further by the possibility to operate with polarized beams. It is particularly noteworthy that the combined information on μ and τ flavor violating decays that will be provided by MEG [1043] together with an SFF can shed light on the mechanism responsible for lepton flavor violation.

4.1.2 Experimental sensitivities

An SFF with integrated luminosity of (50–75) ab^{-1} can perform a wide range of important measurements and dramatically improve upon the results from the current generation of B factories. Many of these measurements cannot be made in a hadronic environment and are unique to an SFF. In Table 56, we give indicative estimates of the precision on some of the most important observables that can be achieved by an SFF with integrated luminosity of (50–75) ab^{-1}. Here we have not attempted to comment on the whole range of measurements that can be performed by such a machine but instead focus on channels with the greatest phenomenological impact. For more details, including a wide range of additional measurements, we guide the reader to the existing reports [1044–1047, 1059], where also all original references are given.

The most important measurements within the CKM metrology are the UT angles, the angle β (also known as ϕ_1) measured using mixing-induced CP violation in $B^0 \to J/\psi K^0$, the angle α (ϕ_2) measured using rates and asymmetries in $B \to \pi\pi$, $\rho\pi$ and $\rho\rho$, and the angle γ (ϕ_3) measured using rates and asymmetries in $B \to D^{(*)} K^{(*)}$ decays, using final states accessible to both D^0 and $\bar{D}^0$. Moreover, an SFF will improve our knowledge of the lengths of the UT sides. In particular, the CKM matrix element $|V_{ub}|$ will be precisely measured through both inclusive and exclusive semileptonic $b \to u$ decays.

Among the measurements sensitive for NP, there are the mixing-induced CP violation parameters in charmless hadronic B decays dominated by the $b \to s$ penguin transition, $S(\phi K^0)$, $S(\eta' K^0)$ and $S(K_S^0 K_S^0 K_S^0)$. Within the Standard Model, these give the same value of $\sin(2\beta)$ that is determined in $B^0 \to J/\psi K^0$ decays, up to a level of theoret-

Table 56 Expected sensitivity that can be achieved on some of the most important observables by an SFF with integrated luminosity of (50–75) ab^{-1}. The range of values given allow for possible variation in the total integrated luminosity, in the accelerator and detector design and in limiting systematic effects. For further details, refer to [1045, 1047]

Observable	SFF sensitivity
$\sin(2\beta)$ $(J/\psi K^0)$	0.005–0.012
γ $(B \to D^{(*)} K^{(*)})$	$(1–2)°$
α $(B \to \pi\pi, \rho\rho, \rho\pi)$	$(1–2)°$
$|V_{ub}|$ (exclusive)	$(3–5)\%$
$|V_{ub}|$ (inclusive)	$(2–6)\%$
$\bar{\rho}$	$(1.7–3.4)\%$
$\bar{\eta}$	$(0.7–1.7)\%$
$S(\phi K^0)$	0.02–0.03
$S(\eta' K^0)$	0.01–0.02
$S(K_S^0 K_S^0 K_S^0)$	0.02–0.04
ϕ_D	$(1–3)°$
$\mathcal{B}(B \to \tau\nu)$	$(3–4)\%$
$\mathcal{B}(B \to \mu\nu)$	$(5–6)\%$
$\mathcal{B}(B \to D\tau\nu)$	$(2–2.5)\%$
$\mathcal{B}(B \to \rho\gamma)/\mathcal{B}(B \to K^*\gamma)$	$(3–4)\%$
$A_{CP}(b \to s\gamma)$	0.004–0.005
$A_{CP}(b \to (s+d)\gamma)$	0.01
$S(K_S^0 \pi^0 \gamma)$	0.02–0.03
$S(\rho^0 \gamma)$	0.08–0.12
$A^{FB}(B \to X_s \ell^+\ell^-)\, s_0$	$(4–6)\%$
$\mathcal{B}(B \to K\nu\bar{\nu})$	$(16–20)\%$
$\mathcal{B}(\tau \to \mu\gamma)$	$(2–8) \times 10^{-9}$
$\mathcal{B}(\tau \to \mu\mu\mu)$	$(0.2–1) \times 10^{-9}$
$\mathcal{B}(\tau \to \mu\eta)$	$(0.4–4) \times 10^{-9}$

ical uncertainty that is estimated to be ∼(2–5)% within factorization. (The theoretical error in these and other modes, such as $B \to K_S \pi^0$, can be also bounded with data-driven methods [88, 778–781]. Presently these give larger uncertainties but will become more precise as more data is available.) Many extensions of the SM result in deviations from this prediction. Another distinctive probe of new sources of CP violation is ϕ_D, the CP violating phase in neutral D meson mixing, which is negligible in the SM and can be precisely measured using, for example, $D \to K_S^0 \pi^+ \pi^-$ decays. Furthermore, branching fractions for leptonic and semileptonic B decays are sensitive to charged Higgs exchange. In particular these modes are sensitive to new physics, even in the unfavorable MFV scenario, with a large ratio of the Higgs vacuum expectation values, $\tan \beta$. Measurements of rare radiative and electroweak penguin processes are well known to be particularly sensitive to NP: the ratio of branching fractions $\mathcal{B}(B \to \rho \gamma)/\mathcal{B}(B \to K^* \gamma)$ depends on the ratio of CKM matrix parameters $|V_{td}/V_{ts}|$, with additional input from lattice QCD. Within the SM, this result must be consistent with constraints from the UT fits. The inclusive CP asymmetries $A_{CP}(b \to s\gamma)$ or $A_{CP}(b \to (s+d)\gamma)$ are predicted in the SM to be small or exactly zero respectively with well-understood theoretical uncertainties. The mixing-induced CP asymmetry in radiative $b \to s$ transitions measured for example through $S(K_S^0 \pi^0 \gamma)$ is sensitive to the emitted photon polarization. Within the SM, the photon is strongly polarized, and the mixing-induced asymmetry small but new right-handed currents can break this prediction even without the introduction of any new CP violating phase. Similarly, $S(\rho^0 \gamma)$ probes radiative $b \to d$ transitions. The dilepton invariant mass squared s at which the forward–backward asymmetry in the distribution of $B \to X_s \ell^+ \ell^-$ decays is zero (denoted $A^{FB}(B \to X_s \ell^+ \ell^-) s_0$), for which the theoretical uncertainty of the SM prediction is small, is sensitive to NP in electroweak penguin operators; finally, the branching fraction for the rare electroweak penguin decay $B \to K \nu \bar{\nu}$ is an important probe for NP even if this appears only well above the electroweak scale. An SFF also allows for the measurement of branching ratios of lepton flavor violating τ decays, such as $\tau \to \mu\gamma$, $\tau \to \mu\mu\mu$ and $\tau \to \mu\eta$. Within the SM, these are negligibly small, but many models of new physics create observable lepton flavor violation signatures.

The sensitivities of these measurements to NP effects may be shown by a few examples: In Fig. 62, we show a simulation of the time-dependent asymmetry in $B^0 \to \phi K^0$, compared to that for $B^0 \to J/\psi K^0$. The events are generated using the current central values of the measurements. With the precision of an SFF and the present central values, the difference between the two data sets is larger than the theoretical expectation, showing evidence of NP contributions.

In Fig. 63, we show how lepton flavor violation in the decay $\tau \to \mu\gamma$ may be discovered at an SFF. The simulation corresponds to a branching fraction of $\mathcal{B}(\tau \to \mu\gamma) = 10^{-8}$, which is within the range predicted by many new physics models. The signal is clearly observable, and well within the reach of an SFF. The simulation includes the effects of irreducible background from initial state radiation photons, though improvements in the detector and in the analysis may lead to better control of this limitation. Other lepton flavor violating decay modes, such as $\tau \to \mu\mu\mu$, do not suffer from this background and have correspondingly cleaner experimental signatures.

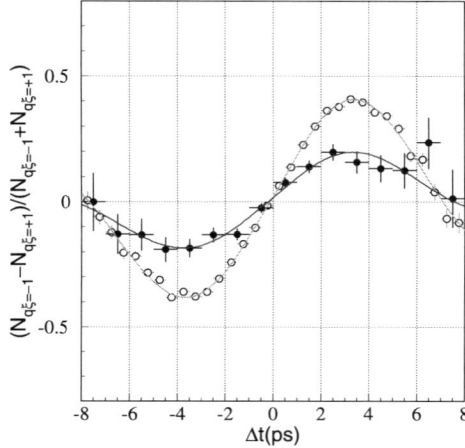

Fig. 62 Simulation of NP effects in $B^0 \to \phi K^0$, as could be observed by an SFF. The open circles show simulated $B^0 \to J/\psi K^0$ events, the *filled circles* show simulated $B^0 \to \phi K^0$ events. Both have *curves* showing fit results superimposed (from [1047])

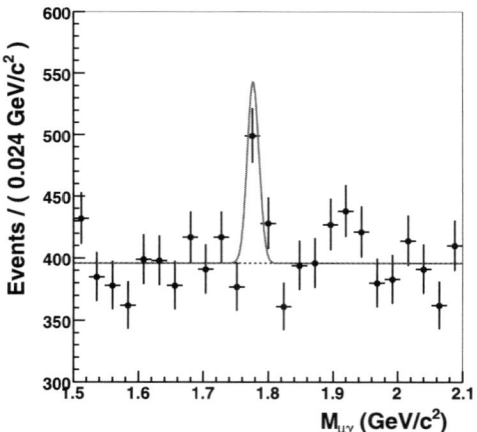

Fig. 63 Monte Carlo simulation of the appearance of $\tau \to \mu\gamma$ at an SFF. A clear peak in the $\mu\gamma$ invariant mass distribution is visible above the background. The branching fraction used in the simulation is $\mathcal{B}(\tau \to \mu\gamma) = 10^{-8}$, an order of magnitude below the current upper limit. With 75 ab^{-1} of data, the significance of such a decay is expected to exceed 5σ

The differences between the SFF physics programme and those of the current B factories are striking. At an SFF, measurements of known rare processes such as $b \to s\gamma$ or CP violation in hadronic $b \to s$ penguin transitions such as $B^0 \to \phi K_S^0$ will be advanced to unprecedented precision. Channels which are just being observed in the existing data, such as $B^0 \to \rho^0 \gamma$, $B^+ \to \tau^+ \nu_\tau$ and $B \to D^{(*)} \tau \nu$ will become precision measurements at an SFF. Furthermore, detailed studies of decay distributions and asymmetries that cannot be performed with the present statistics will enable the sensitivity to NP to be significantly improved. Another salient example lies in D^0–$\bar{D}^0$ oscillations: the current evidence for charm mixing, which cannot be interpreted in terms of NP, opens the door for precise measurements of the CP violating phase in charm mixing, which is known to be zero in the SM with negligible uncertainty.

In addition, these measurements will be accompanied by dramatic discoveries of new modes and processes. These will include decays such as $B \to K \nu \bar{\nu}$, which is the signature of the theoretically clean quark level process $b \to s \nu \bar{\nu}$. The high statistics and clean environment of an SFF allow for the accompanying B meson to be fully reconstructed in a hadronic decay mode, which then in turn allows a one-charged prong rare decay to be isolated. Another example is $B^+ \to \pi^+ \ell^+ \ell^-$, the most accessible $b \to d \ell^+ \ell^-$ process. These decays are the next level beyond $b \to s \ell^+ \ell^-$ decays, which were first observed in the B-factory era. Such significant advances will result in a strong phenomenological impact of the SFF physics programme.

Since an SFF will take data in the LHC era, it is reasonable to ask how the physics reach compares with the B physics potential of the LHC experiments, most notably LHCb. By 2014, the LHCb experiment is expected to have accumulated 10 fb^{-1} of data from pp collisions at a luminosity of $\sim 2 \times 10^{32}$ cm^{-2} s^{-1}. In the following, we assume the most recent estimates of LHCb sensitivity with that data set [1048]. Note that LHCb is planning an upgrade where they would run at 10 times the initial design luminosity and record a data sample of about 100 fb^{-1}, see Sect. 4.4 and [1049].

The most striking outcome of any comparison between SFF and LHCb is that the strengths of the two experiments are largely complementary. For example, the large boost of the B hadrons produced at LHCb allows studies of the oscillations and mixing-induced CP violation of B_s mesons, while many of the measurements that constitute the primary physics motivation for an SFF cannot be performed in the hadronic environment, including rare decay modes with missing energy such as $B^+ \to \ell^+ \nu_\ell$ and $B^+ \to K^+ \nu \bar{\nu}$. Measurements of the CKM matrix elements $|V_{ub}|$ and $|V_{cb}|$ and inclusive analyses of processes such as $b \to s\gamma$ also benefit greatly from the SFF environment. At LHCb, the reconstruction efficiencies are reduced for channels containing several neutral particles and for studies where the B decay vertex must be determined from a K_S^0 meson. Consequently, an SFF has unique potential to measure the photon polarization via mixing-induced CP violation in $B^0 \to K_S^0 \pi^0 \gamma$. Similarly, an SFF is well placed to study possible NP effects in hadronic $b \to s$ penguin decays as it can measure precisely the CP asymmetries in many B_d^0 decay modes including ϕK^0, $\eta' K^0$, $K_S^0 K_S^0 K_S^0$ or $K_S^0 \pi^0$. While LHCb will have limited capability for these channels, it can achieve complementary measurements using decay modes such as $B_s^0 \to \phi \gamma$ and $B_s^0 \to \phi \phi$ for radiative and hadronic $b \to s$ transitions, respectively. Where there is overlap, the strength of the SFF programme in its ability to use multiple approaches to reach the objective becomes apparent. For example, LHCb will be able to measure α to about $5°$ precision using $B \to \rho \pi$ but would not be able to access the full information in the $\pi \pi$ and $\rho \rho$ channels, which is necessary to drive the uncertainty down to the $(1$–$2)°$ level of an SFF. Similarly, LHCb can certainly measure $\sin(2\beta)$ through mixing-induced CP violation in $B^0 \to J/\psi K_S^0$ decay to high accuracy (about 0.01) but will have less sensitivity to make the complementary measurements (e.g., in $J/\psi \pi^0$ and Dh^0) that help to ensure that the theoretical uncertainty is under control. LHCb plans to measure the angle γ with a precision of $(2$–$3)°$. An SFF is likely to be able to improve this precision to about $1°$. LHCb can make a precise measurement of the zero of the forward–backward asymmetry in $B^0 \to K^{*0} \mu^+ \mu^-$, but an SFF can also measure the inclusive channel $b \to s \ell^+ \ell^-$, which is theoretically a significantly cleaner observable [463].

The broad program of an SFF thus provides a very comprehensive set of measurements, extending what will already have been achieved by LHCb at that time. This will be of great importance for the study of flavor physics in the LHC era and beyond.

4.1.3 Phenomenological impact

The power of an SFF to observe NP and to determine the CKM parameters precisely is manifold. In the following, we present a few highlights of the phenomenological impact (for more detailed analyses, see [1044–1047, 1059]).

The measurements described in the previous section can be used to select a region in the $\bar{\rho}$–$\bar{\eta}$ plane. The numerical results in Table 57 indicate that a precision of a fraction of a percent can be reached, significantly improving the current situation and providing a generic test of the presence of NP at that level of precision.

There is also an impressive impact of an SFF on the parameters of the MSSM with generic squark mass matrices parameterized using the mass insertion (MI) approximation [97]. The analysis presented here is based on results and techniques developed in Refs. [104, 105, 107]. Figure 64 shows a simulation of how well the mass insertions (MIs), related to the off-diagonal entries of the squark mass matrices, could be reconstructed at an SFF. Figure 64 displays the allowed region in the $\text{Re}(\delta^d_{ij})_{AB}$–$\text{Im}(\delta^d_{ij})_{AB}$ plane with a value of $(\delta^d_{ij})_{AB}$ allowed from the present upper bound, $m_{\tilde{g}} = 1$ TeV and using the SFF measurements as constraints. The relevant constraints come from $\mathcal{B}(b \to s\gamma)$, $A_{CP}(b \to s\gamma)$, $\mathcal{B}(b \to s\ell^+\ell^-)$, $A_{CP}(b \to s\ell^+\ell^-)$, Δm_{B_s} and A^s_{SL}. It is apparent the key role of $A_{CP}(b \to s\gamma)$ together with the branching ratios of $b \to s\gamma$ and $b \to s\ell^+\ell^-$. The zero of the forward–backward asymmetry in $b \to s\ell^+\ell^-$, missing in the present analysis, is expected to give an additional strong constraint, further improving the already excellent extraction of $(\delta^d_{23})_{LR}$ shown in Fig. 64.

The search for FCNC transitions of charged leptons is one of the most promising directions to search for physics beyond the SM. In the last few years, neutrino physics has provided unambiguous indications about the nonconservation of lepton flavor, we therefore expect this phenomenon to occur also in the charged lepton sector.

Rare FCNC decays of the τ lepton are particularly interesting since the LFV sources involving the third generation are naturally the largest. In Fig. 65, we show the prediction for $\mathcal{B}(\tau \to \mu\gamma)$ within a SUSY $SO(10)$ framework for the accessible LHC SUSY parameter space $M_{1/2} \leq 1.5$ TeV, $m_0 \leq 5$ TeV and $\tan\beta = 40$ [1051]. Note that the measurement of $\mathcal{B}(\tau \to \mu\gamma)$ at an SFF can distinguish the scenario where LFV is governed by neutrino mixing matrix U_{PMNS} from the scenario where LFV is governed by the quark mixing matrix V_{CKM}.

In SUSY models, the squark and slepton mass matrices are determined by various SUSY-breaking parameters,

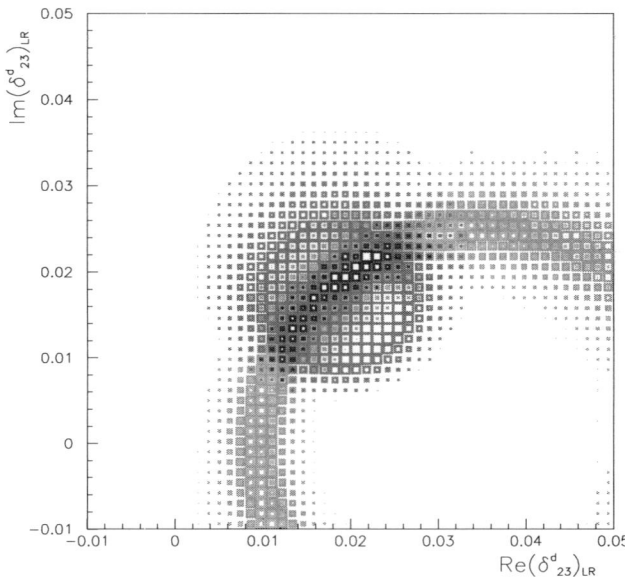

Fig. 64 (Color online) Density plot of the region in the $\text{Re}(\delta^d_{23})_{LR}$–$\text{Im}(\delta^d_{23})_{LR}$ plane for $m_{\tilde{q}} = m_{\tilde{g}} = 1$ TeV generated using SFF measurements. Different colours correspond to different constraints: $\mathcal{B}(B \to X_s \gamma)$ (*green*), $\mathcal{B}(B \to X_s \ell^+ \ell^-)$ (*cyan*), $A_{CP}(B \to X_s \gamma)$ (*magenta*), all together (*blue/black*). Central values of constraints corresponds to assuming $(\delta^d_{13})_{LL} = 0.028 e^{i\pi/4}$

Table 57 Uncertainties of the CKM parameters obtained from the SM fit using the experimental and theoretical information available today (left) and at the time of an SFF (right)

Parameter	SM fit today	SM fit at an SFF
$\bar{\rho}$	0.163 ± 0.028	± 0.0028
$\bar{\eta}$	0.344 ± 0.016	± 0.0024
α (°)	92.7 ± 4.2	± 0.45
β (°)	22.2 ± 0.9	± 0.17
γ (°)	64.6 ± 4.2	± 0.38

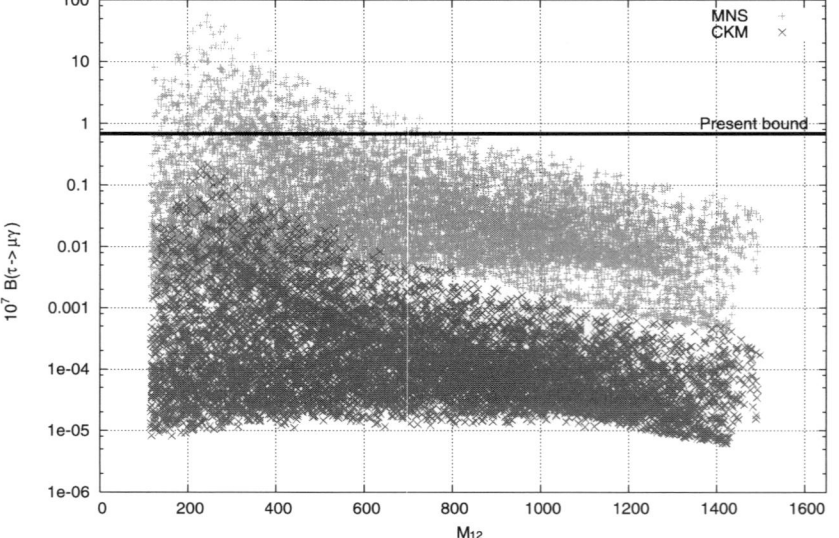

Fig. 65 $\mathcal{B}(\tau \to \mu\gamma)$ in units of 10^{-7} vs. the high-energy universal gaugino mass ($M_{1/2}$) within an $SO(10)$ framework [1051]. The plot is obtained by scanning the LHC accessible parameter space $m_0 \leq 5$ TeV for $\tan\beta = 40$. *Green* or *light* (*red* or *dark*) *points* correspond to the scenario where LFV is governed by the PMNS (CKM) mixing matrix. The *thick horizontal line* denotes the present experimental sensitivity. The expected SFF sensitivity is 2×10^{-9}

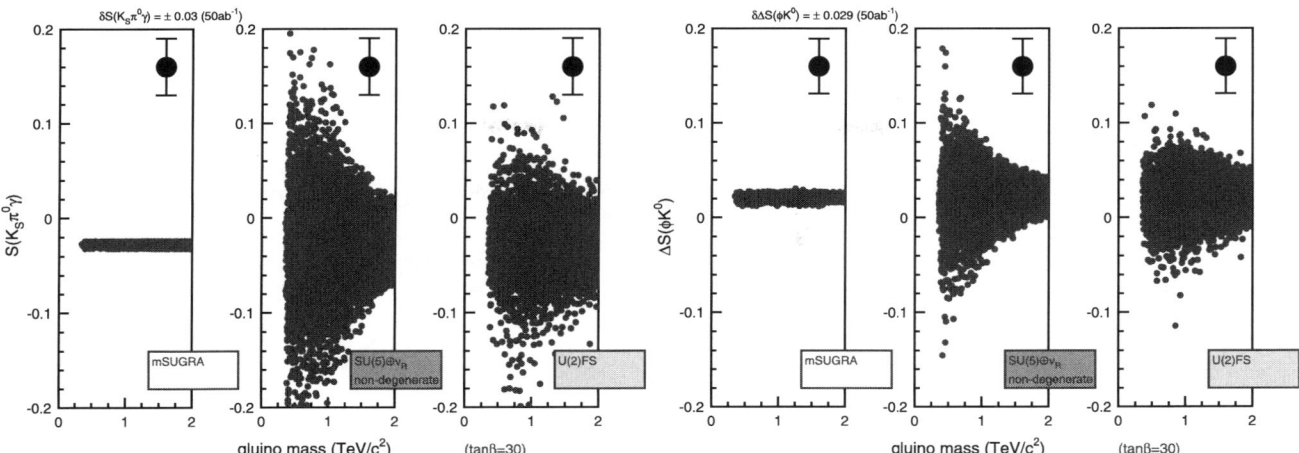

Fig. 66 Time-dependent asymmetry of $B^0 \to K_S^0 \pi^0 \gamma$ and the difference between the time-dependent asymmetries of $B^0 \to \phi K_S^0$ and $B^0 \to J/\psi K_S^0$ modes for three SUSY breaking scenarios: mSUGRA (*left*), $SU(5)$ SUSY GUT with right-handed neutrinos in nondegenerate case (*middle*), and MSSM with $U(2)$ flavor symmetry (*right*). The expected SFF sensitivities are also shown

and hence an SFF has the potential to study SUSY breaking scenarios through quark and lepton flavor signals. This will be particularly important when SUSY particles are found at the LHC, because flavor off-diagonal terms in these mass matrices could carry information on the origin of SUSY breaking and interactions at high-energy scales such as the GUT and the seesaw neutrino scales. Combined with the SUSY mass spectrum obtained at energy frontier experiments, it may be possible to clarify the whole structure of SUSY breaking. In order to illustrate the potential of an SFF to explore the SUSY breaking sector, three SUSY models are considered, and various flavor signals are compared. These are (i) the minimal supergravity model (mSUGRA), (ii) an $SU(5)$ SUSY GUT model with right-handed neutrinos, (iii) the MSSM with $U(2)$ flavor symmetry [1052, 1053]. Flavor signals in the $b \to s$ sector are shown in Fig. 66 for these three SUSY breaking scenarios. Scatter plots of the time-dependent asymmetry of $B \to K_S^0 \pi^0 \gamma$ and the difference between the time-dependent asymmetries of $B \to \phi K_S^0$ and $B \to J/\psi K_S^0$ modes are presented as a function of the gluino mass. Sizable deviations can be seen for $SU(5)$ SUSY GUT and $U(2)$ flavor symmetry cases even if the gluino mass is 1 TeV. The deviation is large enough to be identified at SFF. On the other hand, the deviations are much smaller for the mSUGRA case.

The correlation between $\mathcal{B}(\tau \to \mu\gamma)$ and $\mathcal{B}(\mu \to e\gamma)$ is shown in Fig. 67 for the nondegenerate $SU(5)$ SUSY GUT case. In this case, both processes can reach current upper bounds. It is thus possible that improvements in the $\mu \to e\gamma$ search at the MEG experiment and in the $\tau \to \mu\gamma$ search at an SFF lead to discoveries of muon and tau LFV processes, respectively.

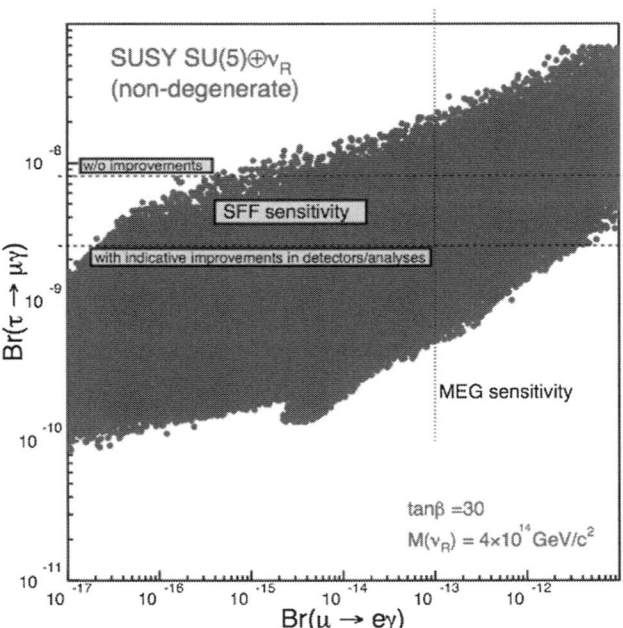

Fig. 67 Correlation between $\mathcal{B}(\tau \to \mu\gamma)$ and $\mathcal{B}(\mu \to e\gamma)$ for $SU(5)$ SUSY GUT with right-handed neutrinos in the nondegenerate case. Expected search limits at the SFF for $\mathcal{B}(\tau \to \mu\gamma)$ and for $\mathcal{B}(\mu \to e\gamma)$ from MEG are also shown

4.1.4 Summary

In conclusion, the physics case of an SFF collecting an integrated luminosity of (50–75) ab^{-1} is well established. Many NP sensitive measurements involving B and D mesons and τ leptons, unique to a SFF, can be performed with excellent sensitivity to new particles with masses up to $\sim$100 (or even $\sim$1000) TeV. The possibility to operate at the $\Upsilon(5S)$ resonance makes some measurements with B_s mesons also accessible, and options to run in the tau–charm threshold region and possibly with one or two polarized beams further broadens the physics reach. Flavor- and CP-violating couplings of new particles accessible at the LHC can be measured in most scenarios, even in the unfavorable cases assuming MFV. Together with the LHC, an SFF could be soon starting the project of reconstructing the NP Lagrangian. Admittedly, this daunting task would be difficult and take many years, but it provides an exciting objective for accelerator-based particle physics in the next decade and beyond.

4.2 Super$B$ proposal

The two asymmetric B factories, PEP-II [1054, 1055] and KEKB [1056], and their companion detectors, BaBar [1057] and Belle [1058], have produced a wealth of flavor physics results, subjecting the quark and lepton sectors of the SM to a series of stringent tests, all of which have been passed. With the much larger data sample that can be produced at a super B factory, qualitatively new studies will be possible, including searches for FCNCs, lepton-flavor violating processes and new sources of CP violation at sensitivities that could reveal physics beyond the SM. These studies will provide a uniquely important source of information about the details of the NP uncovered at hadron colliders in the coming decade [1059].

In light of this strong physics motivation, there has been a great deal of activity over the past six years aimed at designing an e^+e^- B factory that can produce samples of b, c and τ decays 50 to 100 times larger than will exist when the current B factory programs end.

Upgrades of PEP-II [1060] and KEKB [1061] to super B factories that accomplish this goal have been considered at SLAC and at KEK. These machines are extrapolations of the existing B factories with higher currents, more bunches and smaller β functions (1.5 to 3 mm). They also use a great deal of power (90 to 100 MW), and the high currents, approaching 10 A, pose significant challenges for detectors. To minimize the substantial wallplug power, the SuperPEP-II design doubled the current RF frequency to 958 MHz. In the case of SuperKEKB, a factor of two increase in luminosity is assumed for the use of crab crossing, which is currently being tested at KEKB, see Sect. 4.3.

SLAC has no current plans for an on-site accelerator-based high-energy physics program, so the SuperPEP-II proposal is moribund. The SuperKEKB proposal is considered as a future option of KEK. The problematic power consumption and background issues associated with the SLAC and KEK-based super B factory designs have now, however, motivated a new approach to super B factory design, using low emittance beams to produce a collider with a luminosity of 10^{36} but with reduced power consumption and lower backgrounds. This collider is called Super$B$. Design parameters of the existing colliders PEP-II and KEKB are compared with those of SuperPEP-II, SuperKEKB and Super$B$ in Table 58.

The Super B Conceptual Design Report [1045] describes a nascent international effort to construct a very high luminosity asymmetric e^+e^- flavor factory. The machine can use an existing tunnel or it could be built at a new site, such as the campus of the University of Rome "Tor Vergata", near the INFN National Laboratory of Frascati. The report was prepared by an international study group set up by the President of INFN at the end of 2005, with the charge of studying the physics motivation and the feasibility of constructing a SFF that would come into operation in the first

Table 58 Comparison of B factory and super B factory designs

	PEP-II	KEKB	SuperPEP-II	SuperKEKB	Super B
E_{LER} (GeV)	3.1	3.5	3.5	3.5	4
E_{HER} (GeV)	9	8	8	8	7
N_{part} ($\times 10^{10}$)	8	5.8	10	12	6
I_{LER} (A)	2.95	1.68	4.5	9.4	2.28
I_{HER} (A)	1.75	1.29	2.5	4.1	1.3
Wallplug power (MW)	22.5	45	~100	~90	17
Crossing angle (mrad)	0	±15	0	0	±17
Bunch length σ_z (mm)	11	6	1.7	3	7
σ_y^* (nm)	6900	2000	700	367	35
σ_x^* (µm)	160	110	58	42	5.7
β_y^* (mm)	11	6	1.5	3	0.3
Vertical beam–beam tune shift ξ_y	0.068	0.055	0.12	0.25	0.17
Luminosity (cm^{-2} s^{-1}) ($\times 10^{34}$)	1.1	1.6	70	80	100

half of the next decade with a peak luminosity in excess of 10^{36} cm^{-2} s^{-1} at the $\Upsilon(4S)$ resonance.

The key idea in the Super B design is the use of low emittance beams produced in an accelerator lattice derived from the ILC Damping Ring Design, together with a new collision region, again with roots in the ILC final focus design, but with important new concepts developed in this design effort. Remarkably, Super B produces this very large improvement in luminosity with circulating currents and wallplug power similar to those of the current B factories. There is clear synergy with ILC R&D; design efforts have already influenced one another, and many aspects of the ILC Damping Rings and Final Focus would be operationally tested at Super$B$ in which the bending magnets bear a treater burden in producing the needed damping. A comparison of the Super$B$ HER and LER rings with the ILC damping rings is given in Table 59.

There is quite a lot of siting flexibility in the Super B CDR design. Since the required damping times are produced by wigglers in straight sections, the radius of the ring can be varied (within limits, of course) to accommodate other sites and/or to optimize cost. Smaller radius designs are also being explored, in which the bending magnets bear a greater burden in producing the needed damping.

Employing concepts developed for the ILC damping rings and final focus in the design of the Super$B$ collider, one can produce a two-order-of-magnitude increase in luminosity with beam currents that are comparable to those in the existing asymmetric B factories. Background rates and radiation levels associated with the circulating currents are comparable to current values; luminosity-related backgrounds such as those due to radiative Bhabhas increase substantially. With careful design of the interaction region, including appropriate local shielding, and straightforward revisions of detector components, upgraded detectors based on BaBar or Belle are a good match to the machine environment: in this discussion, we use BaBar as a specific example. Required detector upgrades include: reduction of the radius of the beam pipe, allowing a first measurement of track position closer to the vertex and improving the vertex resolution (this allows the energy asymmetry of the collider to be reduced to 7 on 4 GeV); replacement of the drift chamber, as the current chamber will have exceeded its design lifetime; replacement of the endcap calorimeter, with faster crystals

Table 59 Parameters of the Super B HER and LER rings compared with the ILC damping rings

	LER	HER	ILC DR
Energy (GeV)	4	7	5
Luminosity (cm^{-2} s^{-1})		1×10^{36}	–
C (m)		2249	6695
Crossing angle (mrad)		2×17	–
Longitudinal polarization (%)	0	80	80
Wiggler field Bw (T)	1.00	0.83	1.67
L_{bend} (m) (Arc/FF)	0.45/0.75/5.4	5.4/5.4	3/6/–
Number of Bends (Arc/FF)	120/120/16	120/16	126/–
U_0 (MeV/turn)	1.9	3.3	8.7
Wiggler length: L_{tot}(m)	100	50	200
Damping time τ_s, τ_x (ms)	16/32	16/32	12.9/25.7
σ_z (mm)	6	6	9
ϵ_x (nm-rad)	1.6	1.6	0.8
ϵ_y (pm-rad)	4	4	2
σ_E(%)	0.084	0.09	0.13
Momentum compaction	1.8×10^{-4}	3.1×10^{-4}	4.2×10^{-4}
Synchrotron tune ν_s	0.011	0.02	0.067
V_{RF} (MV), N_{cavities}	6, 8	18, 24	24, 18
N_{part} ($\times 10^{10}$)	6.16	3.52	2.0
I_{beam} (A)	2.3	1.3	0.4
P_{beam} (MW)	4.4	4.3	3.5
f_{rf} (MHz)		476	650
N_{bunches}		1733	2625

having a smaller Molière radius, since there is a large increase in Bhabha electrons in this region.

Super B has two additional features: the capability of running at center-of-mass energies in the τ/charm threshold region and longitudinal polarization of the electron (high energy) beam. The luminosity in the 4 GeV region will be an order of magnitude below that in the $\Upsilon(4S)$ region, but even so, data-taking runs of only one month at each of the interesting energies ($\psi(3770)$, 4.03 GeV, τ threshold, etc.) would produce an order of magnitude more integrated luminosity than will exist at the conclusion of the BES-II program. The polarization scheme is discussed in some detail in the Super B CDR [1045]. The electron beam can be polarized at a level of 85%, making it possible to search for T violation in τ production due to the presence of an electric dipole moment or for CP violation in τ decay, which is not expected in the Standard Model.

The Super B design has been undertaken subject to two important constraints: (1) the lattice is closely related to the ILC Damping Ring lattice and (2) as many PEP-II components as possible have been incorporated into the design. A large number of PEP-II components can, in fact, be reused: The majority of the HER and LER magnets, the magnet power supplies, the RF system, the digital feedback system and many vacuum components. This will reduce the cost and engineering effort needed to bring the project to fruition.

The crabbed waist design employs a large "Piwinski angle" $\phi = \frac{\theta}{2}\frac{\sigma_z}{\sigma_x}$, where θ is the full geometric crossing angle of the beams at the interaction point. By producing the large Piwinski angle through the use of a large crossing angle and a very small horizontal beam size and having β_y comparable to the size of the beam overlap area, it is possible simultaneously to produce a very small beam spot, reduce the vertical tune shift and suppress vertical synchrobetatron resonances. However, new beam resonances then arise, which can be suppressed by using sextupoles in phase with the IP in the x plane and with a $\pi/2$ phase difference in the y plane. This is the crabbed waist transformation. These optical elements have an impact on the dynamic aperture of the lattice; studies carried out after the Super B CDR indicate that an adequate dynamic aperture can be achieved. The longer bunch length made possible by the new scheme has the further advantage of reducing the problems of higher-order mode heating, coherent synchrotron radiation and high power consumption. Beam sizes and particle densities are, however, in a regime where Touschek scattering is an important determinant of beam lifetime.

The Super B concept is a breakthrough in collider design. The invention of the "crabbed waist" final focus can, in fact, have impact even on the current generation of colliders. A test of the crabbed waist concept is planned to take place at Frascati in late 2007 or early 2008; a positive result of this test would be an important milestone as the Super B design progresses. The low emittance lattice, fundamental as well to the ILC damping ring design, allows high luminosity with modest power consumption and demands on the detector.

Since the circulating currents in Super B are comparable to those in the current B factories, an upgrade of one of the existing B factory detectors, BaBar or Belle is an excellent match to the Super B machine environment. As an example, we will describe the changes envisioned in an upgrade of BaBar, beginning with the components closest to the beamline.

Developments in silicon sensors and materials technology make it possible to improve the resolution of the silicon vertex tracker (SVT) and to reduce the diameter of the beam pipe. This allows reduction of the energy asymmetry of Super B to 7 on 4 GeV, saving on power costs and slightly improving solid angle coverage. The first layer of the SVT will initially be composed of striplets, with an upgrade to pixels in the highest luminosity regime. The main tracking chamber will still be a drift chamber, although with smaller cell size. The radiators of the DIRC particle identification system will be retained, but the readout system will be replaced with a version that occupies a smaller volume. The barrel CsI (Tl) electromagnetic calorimeter will also be retained, but the forward endcap will be replaced with LYSO (Ce) crystals, which are faster and more radiation-hard. A small backward region calorimeter will be added, mainly to serve as a veto in missing energy analyses. The superconducting coil and instrumented flux return (IFR) will be retained, with the flux return segmentation and thickness modified to improve muon identification efficiency. The instrumentation in the endcap regions of the IFR will be replaced with scintillator strips for higher rate capability. The basic architecture of the trigger and data acquisition system will be retained, but components must be upgraded to provide a much-increased bandwidth.

Super B [1062] is an extremely promising approach to producing the very high luminosity asymmetric B factory that is required to observe and explore the contributions of physics beyond the SM to heavy quark and τ decays. Its physics capabilities are complementary to those of an experiment such as LHCb at a hadron machine [1063].

INFN has formed an International Review Committee to critically examine the Super B Conceptual Design Report and give advice as to further steps, including submission of the CDR to the CERN Strategy Group, requests for funding to the Italian government and application for European Union funds.

Should the proposal process move forward, it is expected that the collider and detector projects will be realized as an international collaborative effort. Members of the Super B community will apply to their respective funding agencies for support, which will ultimately be recognized in Memoranda of Understanding. A cadre of accelerator experiments must be assembled to detail the design of Super B, while an international detector/physics collaboration is formed. The prospect of the reuse of substantial portions of PEP-II and BaBar raises the prospect of a major in-kind contribution from the US DOE and/or other agencies that contributed to BaBar construction; support of the project with other appropriate in-kind contributions is also conceivable. It is anticipated that the bulk of the US DOE contribution would be in kind, in the form of PEP-II components made available with the termination of the SLAC heavy flavor program. These include the HER and LER magnets, the RF and digital feedback systems, power supplies and vacuum components and the BaBar detector as the basis for an upgraded Super B detector.

The BaBar model of international collaboration, based on experience gained at CERN and other major laboratories in building and managing international collaborations over the past several decades, is expected to serve as a model for the Super B effort [1062]. The funding agencies of the participating countries will have a role, together with the host agency and host laboratory, in the management of the enterprise, as well as a fiscal role through an International Finance Committee and various review committees.

4.3 Accelerator design of SuperKEKB

The design of SuperKEKB has been developed since 2002 [1064]. The baseline design extends the same scheme as the present KEKB, as described below. The recently developed nano-beam scheme will be further studied as an option of SuperKEKB, while maintaining the baseline design for the time being. The possibility of an intermediate solution between these two schemes is not excluded a priori.

4.3.1 Baseline design of SuperKEKB

SuperKEKB is a natural extension of present KEKB. The baseline parameters of SuperKEKB are listed in Table 60.

The luminosity goal, 8×10^{35} cm^{-2} s^{-1}, is about 50 times higher than present KEKB. The gains of the luminosity will be achieved by higher currents ($\times 3$–$\times 6$), smaller focus parameter β_y^* ($\times 2$) and higher beam–beam parameter ξ_y ($\times 4.5$).

A higher stored current requires more RF sources and accelerating cavities. The baseline design adopts the same RF frequency, 509 MHz, as the present KEKB. The number of klystrons will be doubled, and the number of cavities will be increased by 50%. The total wall-plug power will be doubled. An option to adopt 1 GHz RF system to reduce the power is under consideration. The cavities will be modified for high current operation. The normal conducting accelerator with resonantly-coupled energy storage (ARES) cavity will have higher stored energy ratio of the storage cavity to the accelerating cavity. The superconducting cavity will have a new higher-order mode (HOM) absorber to dissipate 5 times more HOM power, namely 50 kW per cavity. The design of the RF system and the cavities has been basically done and prototyping is going on [1065–1069].

To store the high current, it is necessary to replace all existing beam pipes in both rings. In the positron ring, beam pipes with antechamber and special surface treatment such as TiN coating are required to suppress the electron cloud. The antechambers are necessary to store such high currents to absorb the power of the synchrotron radiation in both rings. Also all vacuum components such as bellows and gate valves must be replaced with low-impedance and high-current capable version. The small β_y^* requires shorter bunch length, which raises another reason to replace the beam pipes, otherwise the HOM loss and associated heating of the components will be crucial. The design of the beam pipes, of the bellows and of the gate valves for SuperKEKB has been done, and some prototypes were tested at present KEKB. There still remain a few R&D issues in beam collimators and coherent synchrotron radiation [1070–1075].

SuperKEKB will switch the charges of the beams from present KEKB to store positrons and electrons in the HER

Table 60 Parameters of SuperKEKB and present KEKB for the low (LER) and high (HER) energy rings

	SuperKEKB LER/HER	KEKB LER/HER	
Flavor	e^+/e^-	e^-/e^+	
Beam energy	3.5/8	3.5/8	GeV
Beam current	9.4/4.1	1.7/1.4	A
β_y^*/β_x^*	3/200	6/600	mm
Beam-beam ξ_y	~0.25	0.055	
Number of bunches/beam	5000	1400	
Horizontal emittance ε_x	6–12	18–24	nm
Bunch length σ_z	3	6	mm
Peak luminosity $\mathcal{L}$	8	0.17	10^{35} cm^{-2} s^{-1}
Wall-plug power	~100	45	MW

and the LER, respectively. The charge switch will relax the electron-cloud instability and reduce the amount of the positron production. For the charge switch, the injector linac will be upgraded with a C-band system, whose prototype has already been built and tested successfully. Also new ideas such as single-crystal target for the positron production have been already utilized to increase the intensity of the positrons [1076, 1077].

All existing magnets of KEKB will be reused in SuperKEKB, except the interaction region (IR), which must be renewed for smaller β^*. The final focusing superconducting quadrupole with compensation solenoid will be made stronger and their prototype has already been produced. Also the crossing angle will be increased from 22 to 30 mrad. A local chromaticity correction system, which is currently installed in the LER, will be added in the HER. Another issue with the smaller β^* is the aperture for the injected beam, especially for positrons. A new damping ring for positrons will be necessary in the injector linac to reduce the injection emittance and to increase the capture efficiency of the positrons [1078].

The boost in the beam–beam parameter ξ_y assumes the success of "crab crossing", which recovers an effective head-on collision under crossing angle by tilting each bunch by a half crossing angle. The crab cavities have been built and operated at KEKB since February 2007, basically showing the design performance in the voltage, Q-value, phase stability, etc. The associated tilt of the beam and the effective head-on collision have been confirmed in various observations including streak cameras. The resulting beam–beam parameter reached 0.086, which is higher than the geometrical gain by about 15%. Further studies are necessary to realize higher beam–beam parameters (>0.1) predicted by simulations for the present KEKB [1079–1084].

A number of beam instrumentations and controls will be upgraded at SuperKEKB, including beam position monitors, feedbacks, visible light and X-ray monitors, etc. Also utilities such as water cooling system will be reinforced [1085].

The current estimate of the total cost of the upgrade for SuperKEKB is about 300 M€ (1 € ~ 150 Y), excluding the salaries for the KEK employees in the accelerator group (about 90 FTE/year). If the upgrade of the RF system is deferred, the initial cost will be reduced to 200 M€.

One of the options to reduce the cost of the construction and electricity is to change the energy asymmetry from 8 + 3.5 GeV to 7 + 4 GeV. An early study has been done for the option resulting in a reduction by about 30 M€ in the construction and 12 MW in the electricity. Such a possibility will be investigated further.

This machine should have a flexibility to run at the charm threshold. The damping time and the emittance can be controlled by adding wigglers in the HER for that purpose. A polarized beam for the collision needs an intensive study on the implementation of spin rotators.

4.3.2 Studies for nano-beam scheme at KEK

The crab waist scheme is one of the most innovative features of the nano-beam Super$B$ design (Sect. 4.2 and [1045]). Simulation by K. Ohmi has shown that the crab waist scheme can improve the luminosity of present KEKB as powerfully as crab crossing with crab cavities. Actually crab waist can be even better than crab crossing, as it only needs conventional sextupole magnets whose construction and operation will be much easier than the state-of-the-art crab cavities. Special efforts have been made at KEK to develop such a design of the lattice to involve sextupole magnets at present KEKB.

This study of the lattice has shown that the dynamic aperture of the ring is drastically reduced by tuning on the crab sextupole magnets. These sextupoles are paired via I or $-I$ transformation, and the IP is located within the pair. If the transformation between the pair is completely linear, the nonlinearity of the first sextupole is completely absorbed by the second. This kind of cancellation has been successfully shown in existing machines including KEKB. In the case of the crab waist, however, there is the IP in the middle of the pair, and the nonlinearities around the IP violate the cancellation of the nonlinear terms of the sextupoles. At least two kinds of nonlinearity, the fringe field of the final focusing quadrupoles and the kinematical terms in the drift space around the IP, has been known to be inevitable, and either one of them is enough to degrade the dynamic aperture by 50%. As the fringe field and the kinematical terms are quite fundamental for the elements around the IP, it is not possible to remove them. The hope is to put several nonlinear magnets around the IP to cancel the nonlinearity at the IP.

The degradation of the dynamic aperture by the crab waist sextupoles will be also a serious problem for the future Super$B$. The dynamic aperture for a Super$B$ lattice was studied. The stable horizontal amplitude with the crab-sextupoles were dropped by 70% for the on-momentum particles and even worse for the off-momentum, synchrotron-oscillating particles. Again it has been known that the fringe field and the kinematical terms at the IP are the reasons of the reduction of the dynamic aperture.

One of the questions on the nano-beam scheme is that no strong-strong simulation has been done. Because of the relatively long bunch length, such a simulation will need more than 100 times more computer power than what is needed for usual schemes. Some preliminary efforts are going on for intermediate bunch length or with simplified models.

The nano-beam scheme can be also attractive even without the crab waist, because it has the potential to achieve 10^{36} cm^{-2} s^{-1} with smaller beam current. Therefore the KEKB team has decided to study the nano-beam scheme as an option of SuperKEKB, to make a flexible lattice and an IP design which is compatible with the nano-beam and also with the high-current scheme. Such a design study will identify fundamental and technical issues on the nano-beam scheme more specifically.

4.4 LHCb upgrade

4.4.1 Introduction

Flavor physics has played a major role in the formulation of the SM of particle physics. An example is the observation of CP violation which, in the SM, can be explained with three generations of quarks. However despite its success, the SM is seen as an effective low-energy theory because it cannot explain dark matter and the force hierarchy. The search for evidence of physics beyond the SM is the main goal of particle physics over the next decade.

The LHC at CERN will start operating in 2008 and will start to look for the Higgs boson and for NP particles which are expected in many models at the 1 TeV scale. However probing NP at the TeV scale is not restricted to direct searches at the high-energy frontier.

Flavor physics also has excellent potential to probe NP. In the SM, FCNCs are suppressed as these only occur through loop diagrams. Hence these decays are very sensitive to NP contributions which, in principle, could contribute with magnitude $\mathcal{O}(1)$ to these virtual quantum loops. The NP flavor sector could also exhibit CP violation and be very different from what is observed in the SM. In fact, the existing experimental limits from the flavor physics point to either a suppression of the couplings also for NP or an even higher NP mass scale.

LHCb is a dedicated heavy-flavor physics experiment designed to make precision measurements of CP violation and of rare decays of B hadrons at the LHC [1086]. LHCb will start taking data in 2008 and plans to record an integrated luminosity of ~ 0.5 fb^{-1} in the first physics run. During the following five years, LHCb expects to accumulate a data sample of ~ 10 fb^{-1}. This will put LHCb in an excellent position to probe physics beyond the SM. The expected performance is summarized in Sect. 4.4.2.

During this first phase of LHC operations, particle physics will reach a branch point. Either new physics beyond the SM will have been discovered at the general purpose detectors (ATLAS and CMS) and LHCb or new physics will be at a higher mass scale. In both scenarios, we will then almost certainly require a substantial increase in sensitivities to flavor observables, either to study the flavor structure of the newly discovered particles or to probe NP through loop processes at even higher mass scales.

The LHCb detector is optimized to operate at a luminosity of 2 to 5×10^{32} cm^{-2} s^{-1}, which is a factor of 20 to 50 below the LHC design luminosity. The LHC accelerator will reach its design luminosity of 10^{34} cm^{-2} s^{-1} after a few years of operation. The LHC machine optics allows LHCb to focus the beams in order to run at a luminosity of up to 50% of the LHC luminosity. To profit from the higher peak luminosities that are available at the LHC, the LHCb experiment is proposing an upgrade to extend its physics programme. The plan to operate the LHCb detector at ten times the design luminosity, i.e. at 2×10^{33} cm^{-2} s^{-1}, is described in Sect. 4.4.3. The LHCb upgrade would allow the LHCb experiment to probe NP in the flavor sector at unprecedented sensitivities.

Initial studies of the physics reach of the proposed LHCb upgrade are discussed in Sect. 4.4.4. To profit from these higher luminosities, the LHCb experiment requires an upgrade such that the detectors and triggers are able to cope with these larger luminosities. This is described in Sect. 4.4.5. A summary and conclusions are given in Sect. 4.4.6.

4.4.2 LHCb physics programme—the first five years

The large cross section of 500 µb for $b\bar{b}$-quark production in pp collisions at 14 TeV centre-of-mass energy will allow the LHCb experiment to collect much larger data samples of B mesons than previously available. The expected performance for measurements with LHCb has been determined by a full simulation [1048]. Many of these results have been described in detail in Sect. 3 of this report. We expect exciting results from the LHCb experiments over the next five years. Here we summarize some of the anticipated highlights.

In the SM, FCNC $b \to s$ transitions are suppressed as these only occur through loop diagrams. Of particular interest is the decay $B_s^0 \to \mu^+\mu^-$, which is very rare. The SM branching ratio $\mathcal{B}(B_s^0 \to \mu^+\mu^-)$ is calculated at $(3.86 \pm 0.15) \times 10^{-9}$ (see (128)) [27]. Physics beyond the SM can enhance this branching ratio considerably. For example, in the CMSSM [571], the branching ratio increases as $\tan^6 \beta$, where $\tan \beta$ is the ratio of the Higgs vacuum expectation values. The current limits from CDF and D0 are about a factor 20 above the SM prediction. Using their good invariant mass resolution $\sigma(M_{\mu\mu}) \approx 20$ MeV and low trigger threshold on the transverse momentum $p_T \geq 1$ GeV, LHCb will be able to probe the full CMSSM parameter space. With 10 fb^{-1} of data, LHCb expects to discover $B_s^0 \to \mu^+\mu^-$ with 5σ significance at the SM level [598].

Another major goal is to probe the weak phase ϕ_s of B_s^0 mixing. This is another excellent NP probe as the SM prediction for ϕ_s is very small: $\phi_s = -2\lambda^2 \eta \approx -0.035$, where λ and η are Wolfenstein parameters of the CKM matrix [1087]. Currently there are no strong constraints on ϕ_s available, and large CP violation in B_s^0 mixing is allowed [676, 678, 712, 714, 715]. The LHCb experiment expects to collect 131k $B_s^0 \to J/\psi\phi$ decays with a 2 fb^{-1} data sample. The corresponding precision on ϕ_s is estimated to be $\sigma(\phi_s) \approx 0.023$ [686]. A value of ϕ_s of $\mathcal{O}(0.1)$ or larger could be observed by LHCb. This would be a clear signal for NMFV beyond the SM [10].

LHCb will perform measurements of the CKM angle γ using two interfering diagrams in neutral and charged $B \to DK$ decays as well as $B_s^0 \to D_s^\mp K^\pm$ decays. The interference arises due to decays which are common to D^0 and $\bar{D}^0$ mesons such as $D^0(\bar{D}^0) \to K_S^0 \pi^+ \pi^-$ (Dalitz decay [637]) and $D^0(\bar{D}^0) \to K^\mp \pi^\pm, K^+ K^-$ (ADS and GLW [629, 635]), or through B_s mixing. The expected γ sensitivities for 2 fb^{-1} of LHCb data are estimated at $\sigma(\gamma) \sim 7°$–$15°$. When combining these measurements LHCb expects to achieve a precision $\sigma(\gamma) \sim 2.5°$ in a 10 fb^{-1} data sample [1048]. This will improve substantially the γ measurements from the B factories, which currently have an uncertainty of about 30° [389].

In Table 61, we show expected LHCb signal yields, background to signal ratios and sensitivity to physics observables based on a 2 fbinv data sample.

4.4.3 LHCb luminosity upgrade

After the first five years of operation of the LHCb experiment, the LHC will hopefully provide answers to some of the open questions of particle physics and, very possible, produce a few new puzzles. To be able to make progress in determining the flavor structure of new physics beyond the SM or probing higher mass scales, it is very likely that the required precision for several flavor physics observables will need to be improved substantially. It is also expected that the precision of many LHCb physics results will remain limited by the statistical error of the collected data. The following questions arise: What is the scientific case for collecting even larger data samples? Is LHCb exploiting the full potential for B physics at hadron colliders? Note that LHCb is the only dedicated heavy flavor experiment approved to run after 2010. In the remainder of this report, we will try to answer these questions.

The LHCb experiment has commenced studying the feasibility of upgrading the detector such that it can operate at a luminosity $\mathcal{L} \sim 2 \times 10^{33}$ cm^{-2} s^{-1}, which is ten times larger than the design luminosity [1089]. This upgrade would allow LHCb to collect a data sample of about 100 fb^{-1} during five years of running. This increased luminosity is achievable by decreasing the amplitude function β^* at the LHCb interaction point. The LHCb upgrade does not require the

Table 61 Expected signal yields S, signal to background ratios B/S and sensitivities for 2 fb^{-1} of data. The parameter s_0 is the zero point in the forward–backward asymmetry A_{FB}, and A_{CP} is the asymmetry in direct CP violation

	Decay	Yield S	B/S	Precision
γ	$B_s^0 \to D_s^\mp K^\pm$	6.2k	<0.18	$\sigma(\gamma) \sim 10°$
	$B^0 \to \pi^+ \pi^-$	36k	0.46	$\sigma(\gamma) \sim 5°$
	$B_s^0 \to K^+ K^-$	36k	<0.06	
	$B^0 \to D^0(K^-\pi^+, K^+\pi^-)K^{*0}$	3.4k, 0.6k	<1.0, <2.8	$\sigma(\gamma) \sim 6°$–$10°$
	$B^0 \to D^0(K^+K^-, \pi^+\pi^-)K^{*0}$	0.7k, 0.2k	<1.4, <5	
	$B^- \to D^0(K^-\pi^+, K^+\pi^-)K^-$	56k, 410	0.6, 2.0	$\sigma(\gamma) \sim 6°$–$10°$
	$B^- \to D^0(K^+K^-/\pi^+\pi^-)K^-$	5.8k, 2.0k	1.0, 3.6	
	$B^- \to D^0(K_S^0 \pi^+\pi^-)K^-$	5k	<0.2–1	$\sigma(\gamma) \sim 15°$
α	$B^0 \to \pi^+\pi^-\pi^0$	14k	<0.8	$\sigma(\alpha) \sim 8.5°$
	$B^{+,0} \to \rho^+\rho^0, \rho^+\rho^-, \rho^0\rho^0$	7k, 2k, 1.2k	1, <5, <5	
β	$B^0 \to J/\psi K_S^0$	333k	1.1	$\sigma(\sin 2\beta) \sim 0.015$
Δm_s	$B_s^0 \to D_s^- \pi^+$	140k	<0.4	$\sigma(\Delta m_s) \sim 0.007$ ps^{-1}
ϕ_s	$B_s^0 \to J/\psi\phi$	131k	0.12	$\sigma(\phi_s) \sim 0.023$ rad
	$B_s^0 \to \phi\phi$	3.1k	<0.8	$\sigma(\phi_s) \sim 0.11$ rad
Rare	$B_s^0 \to \mu^+\mu^-$	20	<6.2	
Decays	$B^0 \to K^{*0}\mu^+\mu^-$	7.2k	0.5	$\sigma(s_0) \sim 0.46$ GeV^2
	$B^0 \to K^{*0}\gamma$	68k	0.6	$\sigma(A_{\text{CP}}) \sim 0.01$
	$B_s^0 \to \phi\gamma$	11.5k	<0.55	

planned LHC luminosity upgrade (Super-LHC) as the LHC design luminosity is 10^{34} cm^{-2} s^{-1}, although it could operate at Super-LHC. Thus an upgrade of LHCb could be implemented as early as 2014.

As the number of interactions per beam crossing will increase to $n \sim 4$, this will require improvements to the LHCb sub-detectors and trigger. A major component of the LHCb upgrade will be the addition of a first-level detached vertex trigger, which will use information from the tracking detectors [1090, 1091]. This trigger has the potential of increasing the trigger efficiencies for decays into hadronic final states by at least a factor of two. The implementation of this detached vertex trigger will require large modifications to the detector read-out electronics which will be discussed in Sect. 4.4.5.

4.4.4 Physics with the LHCb upgrade

A 100 fb^{-1} data sample would allow one to improve the sensitivity of LHCb to unprecedented levels such that NP can be probed at the 1% level. Here we present estimates for a few selected channels. These are based on the following assumptions, which have yet to be demonstrated: maintaining trigger and reconstruction efficiencies at high luminosity running and making use of a detached vertex trigger to double the trigger efficiency for hadronic modes. Systematic errors are only treated in a very simplified way. Hence the quoted sensitivities have very large uncertainties and should be treated with caution. However, these estimates are extremely useful to motivate simulation studies for validating these assumptions. In addition, as soon as LHCb will start taking data, the simulations for low luminosity running can be verified with data.

NP can be probed by studying FCNCs in hadronic $b \to s$ transitions. One approach is to compare the time-dependent CP asymmetry in a hadronic penguin loop decay with a decay based on a tree diagram when both decays have the same weak phase. In hadronic FCNC transitions, unknown massive particles could make a sizable contribution to the $b \to s$ penguin loop, whereas tree decays are generally insensitive to NP. The B factories measure the CP asymmetry $\sin 2\beta^{\text{eff}}$ in the penguin decay $B^0 \to \phi K_S^0$. A value for $\sin 2\beta^{\text{eff}}$, which is different from $\sin 2\beta$ measured in $B^0 \to J/\psi K_S^0$, would signal physics beyond the SM. Within the current available precision, all $\sin 2\beta^{\text{eff}}$ measurements are in reasonable agreement with the SM, but most central values are lower than expected. For example, we find for the decay $B^0 \to \phi K_S^0$ that $\Delta S(\phi K_S^0) = \sin 2\beta^{\text{eff}} - \sin 2\beta = 0.29 \pm 0.17$ [1088].

This approach can also be applied to B_s^0 mesons which will be exploited by LHCb. Within the SM, the weak mixing phase ϕ_s is expected to be almost the same when comparing the time-dependent CP asymmetry of the hadronic penguin decay $B_s^0 \to \phi\phi$ with the tree decay $B_s^0 \to J/\psi\phi$. Due to a cancellation of the B_s^0 mixing and decay phase, the SM prediction for the sine-term, $S(\phi\phi)$, in the time-dependent asymmetry of $B_s^0 \to \phi\phi$ is very close to zero [834]. Thus any measurement of $S(\phi\phi) \neq 0$ would be a clear signal for NP and definitively rule out MFV [10]. From a full simulation LHCb expects to collect 3100 $B_s^0 \to \phi\phi$ events in 2 fb^{-1} of data with a background to signal ratio $B/S < 0.8$ at 90% C.L. [835]. The $S(\phi\phi)$ sensitivity has been studied using a toy Monte Carlo, taking resolutions and acceptance from the full simulation. After about 5 years LHCb expects to have accumulated a data sample of 10 fb^{-1} and will measure $S(\phi\phi)$ with a precision of $\sigma(S(\phi\phi)) = 0.05$ [835]. This precision is expected to be statistically limited, since systematic errors are likely much lower.

The LHCb upgrade will substantially improve the measurement of $S(\phi\phi)$, since this is a hadronic decay mode which will benefit most from the first-level detached vertex trigger. Scaling the sensitivity up to a data sample of 100 fb^{-1}, we estimate a precision of $\sigma(S(\phi\phi)) \sim 0.01$ to 0.02 rad. This sensitivity presents an exciting NP probe at the percent level which will arguably be (one of) the most precise time-dependent CP study in $b \to s$ transitions.

In a similar study, LHCb investigated the $b \to s$ penguin decay $B_d^0 \to \phi K_S^0$. A yield of 920 events is expected in 2 fb^{-1} of integrated luminosity, and the background to signal ratio is $0.3 < B/S < 1.1$. The sensitivity for the time-dependent CP violating asymmetry $\sin 2\beta^{\text{eff}}$ is estimated to be 0.10 in a 10 fb^{-1} data sample [836]. This is a hadronic decay which will also profit from a first-level detached vertex trigger. With 100 fb^{-1} of integrated luminosity, LHCb upgrade will allow to improve the $\sin 2\beta^{\text{eff}}$ sensitivity for $B_d^0 \to \phi K_S^0$ to ~ 0.025 to 0.035.

Using the tree decay $B_s^0 \to J/\psi\phi$, LHCb will also probe NP in the CP violation of B_s^0 mixing. With a 10 fb^{-1} data sample, the weak phase ϕ_s will be determined with a precision of 0.01 [1048]. This corresponds to $\sim 3.5\sigma$ significance for the SM expectation of ϕ_s for which the theoretical uncertainty is very precise ($\mathcal{O}(0.1\%)$). This precision is expected to be still statistically limited. A significantly larger data-set would allow LHCb to search for NP in B meson mixing at an unprecedented level. An upgrade of LHCb has the potential to measure the SM value of ϕ_s with $\sim 10\sigma$ significance ($\sigma(\phi_s) \sim 0.003$) in $B_s^0 \to J/\psi\phi$ decays. To control systematic errors at this level will be very challenging.

In the SM, the angle γ can be determined very precisely with tree decays, which are theoretically very clean. When combining all γ measurements in $B \to DK$ and $B_s^0 \to D_s^\mp K^\pm$ (including systematics), LHCb will constrain the value of γ to about 2.5°. However, it will not be possible to push below the desired 1° precision. Therefore, a very precise determination of γ in tree decays is an important objective of the LHCb upgrade physics programme. The expected yields in 100 fb^{-1} of data are very large: Examples are 620k $B_s^0 \to D_s^\mp K^\pm$, 500k $B \to D(K_S^0 \pi^+ \pi^-)K$ and 5600k $B \to D(K\pi)K$ events, respectively. All these γ modes will benefit greatly from an improved first-level trigger strategy that does not rely solely on high transverse momentum hadrons. Simple statistical extrapolations show that several individual modes will give a potential statistical uncertainty close to 1°. Systematic uncertainties will be very important. However, these uncertainties are largely uncorrelated amongst the modes and, in many cases, can be measured in control samples. Therefore, a global determination to below 1° of the tree level unitarity triangle will be possible [1049]. This will act as a standard candle to be compared to all loop determinations of the unitarity triangle parameters.

The very rare decay $B_s^0 \to \mu^+\mu^-$ is a key to many extensions beyond the SM. With a 100 fb^{-1} data sample, LHCb upgrade would be able to make a precision measurement of the branching ratio $\mathcal{B}(B_s^0 \to \mu^+\mu^-)$ to about ~5% at the SM level. This will allow LHCb upgrade to either measure precisely the flavor properties of new SUSY particles discovered at the LHC or to put very stringent constraints on all SUSY models in the large $\tan\beta$ regime [571].

The LHCb upgrade should also aim to observe the even rarer decay $B_d^0 \to \mu^+\mu^-$ which has an SM branching ratio of $(1.06 \pm 0.04) \times 10^{-10}$ (see (131)). The ratio $\mathcal{B}(B_d^0 \to \mu^+\mu^-)/\mathcal{B}(B_s^0 \to \mu^+\mu^-)$ is sensitive to new physics beyond the SM and will allow to distinguish between different models. This search will be extremely challenging as it requires an excellent understanding of the detector to reduce the muon fake rate due to backgrounds from hadronic two body modes to an acceptable level.

LHCb will exploit the semileptonic decay $B \to K^{*0}\mu^+\mu^-$, which is sensitive to new physics in the small $\tan\beta$ range. Using a full simulation LHCb expects to collect 7200 $B \to K^{*0}\mu^+\mu^-$ per 2 fb^{-1} [508]. In addition to the forward–backward asymmetry, A_{FB}, these large data samples will allow LHCb to measure the differential decay rates in the di-muon mass squared, q^2, and the angular distributions and to probe NP through the transversity amplitude $A_T^{(2)}$ and the K^{*0} longitudinal polarization [476]. In the theoretically favored region of $1 < q^2 < 6$ GeV$^2/c^4$, the resolution in $A_T^{(2)}$ is estimated at 0.16 with 10 fb^{-1} of integrated luminosity [510]. While this data sample might provide a hint of NP, a ten-fold increase in statistics will allow one to probe new physics at the few percent level and cover a large region of the MSSM parameter space. With a 100 fb^{-1} data sample, LHCb upgrade expects to collect 360k $B \to K^{*0}\mu^+\mu^-$ events. The corresponding precision for $A_T^{(2)}$ is estimated to be 0.05 to 0.06.

There are several other channels which have a large potential for probing NP with a 100 fb^{-1} data sample. An excellent example is $B_s^0 \to \phi\gamma$, which is sensitive to the photon polarization and right-handed currents [407]. Using a full simulation, LHCb expects a yield of 11.5k $B_s^0 \to \phi\gamma$ events in 2 fb^{-1} of data with a background to signal ratio <0.55 at 90% C.L. [454]. The sensitivity of this decay to NP arising in right-handed currents is under study. LHCb upgrade would also be able to search for NP by studying the decays $B_s \to \phi\mu^+\mu^-$ and $B \to \pi(\rho)\mu^+\mu^-$.

The very large charm sample would allow LHCb upgrade to search for NP in D^0 mixing and CP violation in charm decays. The expected statistical sensitivity on the parameters x'^2, y' and y_{CP} are 2×10^{-5}, 2.8×10^{-4} and 1.5×10^{-4}, respectively (Table 47). An LHCb upgrade could also probe lepton flavor violation in the decay mode $\tau \to \mu^+\mu^-\mu^+$ with an estimated sensitivity of 2.4×10^{-9} [1092].

The SM as well as SUSY or extra dimension models can be augmented by additional gauge sectors [1093–1095]. This is a very general consequence of string theories [1096–1098]. These gauge sectors can only be excited by high-energy collisions. An example is the "hidden valley" sector. The manifestations of many of these models could be new v-flavored particles with a long lifetime [1093]. These can decay to a pair of b and $\bar{b}$ quarks that produce jets in the detector. An example is the Higgs decay process $H \to \pi_v^0 \pi_v^0$ followed by $\pi_v^0 \to b\bar{b}$. LHCb is designed to detect b-flavored hadrons and thus in a good position to detect decays of long-lived new particles. The LHCb vertex detector (VELO) is ~1 m long making it possible to measure these decays. LHCb upgrade will increase the sensitivity to much lower production cross section for these processes.

In Table 62, we present a summary of the expected sensitivities for selected key measurements, discussed above and that could be performed with an upgrade of the LHCb experiment. These sensitivities will exceed the range for probing NP from LHCb and B factories considerably, and they will also improve upon the precision of SM parameters.

We now compare the physics potential of the LHCb upgrade collecting a 100 fb^{-1} data sample with that of an SFF

Table 62 Expected sensitivity for LHCb upgrade with an integrated luminosity of 100 fb^{-1}. A factor two of improvement for the L0 hadron trigger and systematic error estimates are shown as a range

Observable	LHCb upgrade sensitivity
$S(B_s \to \phi\phi)$	0.01–0.02
$S(B_d \to \phi K_S^0)$	0.025–0.035
$\phi_s (J/\psi \phi)$	0.003
$\sin(2\beta) (J/\psi K_S^0)$	0.003–0.010
$\gamma (B \to D^{(*)} K^{(*)})$	<1°
$\gamma (B_s \to D_s K)$	(1–2)°
$\mathcal{B}(B_s \to \mu^+\mu^-)$	(5–10)%
$\mathcal{B}(B_d \to \mu^+\mu^-)$	3σ
$A_T^{(2)} (B \to K^{*0}\mu^+\mu^-)$	0.05–0.06
$A_{FB}(B \to K^{*0}\mu^+\mu^-) s_0$	0.07 GeV^2

based on a 50 to 75 ab^{-1} data sample, which is discussed in Sect. 4.1.

The strengths of the two proposals are surprisingly complementary. For example, the cleaner environment of an e^+e^- collider allows the SFF to make inclusive measurements of $b \to s\gamma$, of the CKM matrix element V_{ub} and of rare decays with missing energy such as $B^+ \to \ell^+\nu$. However, LHCb upgrade is unique in its potential to exploit the physics of B_s^0 mesons, especially in B_s^0 oscillations. A key motivation for LHCb upgrade is the ability to probe new physics in hadronic $b \to s$ penguin transitions by measuring the time-dependent CP asymmetry in the decay $B_s^0 \to \phi\phi$ with a precision of 0.01 to 0.02. The SFF will make complementary measurements by studying the time-dependent CP asymmetries of $b \to s$ transitions in several B_d^0 decays.

The LHCb upgrade will be able to measure CP violation in the interference of mixing and decay in both B_s^0 and B_d^0 mesons. This will allow LHCb to probe NP simultaneously in FCNC with $B_d^0 \to J/\psi K_S^0$ and $B_s^0 \to J/\psi \phi$ (tree) and $B_d^0 \to \phi K_S^0$ and $B_s^0 \to \phi\phi$ (hadronic $b \to s$ penguin) to the unprecedented level of $\sim 1\%$.

The LHCb upgrade will probe NP contributions to right-handed currents by measuring the time-dependent CP asymmetry in the decay $B_s^0 \to \phi\gamma$. The SFF will make complementary measurements and exploit their better reconstruction efficiencies for decays with several neutral particles in the final state to measure the photon polarization of $B_d^0 \to K_S^0 \pi^0 \gamma$.

In channels where both approaches are possible, the sensitivities are often comparable. LHCb upgrade usually will have larger statistics, but systematic errors in the hadronic environment will be more difficult to control. Both LHCb upgrade and SFF propose to measure $\sin 2\beta$ to 0.01 and the UT angle γ with 1° precision.

A SFF can measure the zero of the forward–backward asymmetry in the inclusive channel $b \to s\ell^+\ell^-$, but the LHCb upgrade will collect a substantially larger sample of 360k $B_d^0 \to K^{*0} \mu^+ \mu^-$ decays compared to 11k at an SFF. This will enable LHCb to measure the asymmetry $A_T^{(2)}$ to $\sim 5\%$. Only the LHCb upgrade will be able to measure the $B_s^0 \to \mu^+ \mu^-$ branching ratio to $\sim 5\%$. This will either help to determine the flavor structure of new particles discovered at the LHC or will severely constrain the corresponding model parameters.

4.4.5 LHCb detector and trigger upgrade

We start out by presenting the limitations of the LHCb detector and trigger which prevent LHCb from operating the detectors at higher luminosity. At the design luminosity of 2×10^{32} cm^{-2} s^{-1}, the visible cross section is 63 mb which corresponds to about 10 MHz of bunch crossings with at least one visible interaction. Note that increasing the luminosity from 2 to 10×10^{32} cm^{-2} s^{-1} will only increase the number of interactions by a factor of two, since the number of bunch crossings with visible interactions increases from 10 to 26 MHz.

The LHCb experiment has a two-level trigger system. The Level-0 trigger (L0) is implemented in hardware, and the Higher Level Trigger (HLT) is running on a large CPU farm. The L0 trigger operates at 40 MHz. The purpose of L0 is to reduce this rate to 1.1 MHz, which is the maximum at which all LHCb detectors can be read-out by the front-end electronics. The L0 trigger selects objects (hadron h, e, and γ) with high transverse energy, $E_T^{h,e,\gamma}$, in the electromagnetic and hadronic calorimeters and the two highest transverse momentum (p_T^μ) muons in the muon system. At the nominal luminosity of 2×10^{32} cm^{-2} s^{-1}, the typical trigger thresholds are $E_T^h \geq 3.5$ GeV, $E_T^{e,\gamma} \geq 2.5$ GeV and $p_T^\mu \geq 1$ GeV. Events with multiple interactions are vetoed.

Simulations show that the L0 muon trigger efficiency for reconstructible events at the design luminosity of 2×10^{32} cm^{-2} s^{-1} is around 90% and that the output rate raises almost linearly with luminosity up to 5×10^{32} cm^{-2} s^{-1}. For larger luminosities, the loss in efficiency is minor. At the design luminosity, the muon trigger uses about 15% of the L0 bandwith. However, the L0 hadron trigger has a lower performance. The efficiencies of this trigger for hadronic decays are only about 40% at the design luminosity, whereas the L0 hadron trigger uses about 70% of the L0 bandwidth. At higher peak luminosity, the rate of visible pp interaction increases, which requires an increase in the threshold. The corresponding loss in efficiency results in an almost constant yield for the hadron trigger [1090].

This illustrates that the existing trigger does not scale with luminosity, in particular the hadronic trigger will not allow operating the LHCb experiment at ten times the design luminosity. The total trigger efficiency including the HLT for hadronic B decays is expected to be 25 to 30% [1048]. The goal of the LHCb upgrade should also be to improve the hadron trigger efficiency by at least a factor two.

We have commenced initial studies which investigate how to upgrade the LHCb detector and triggers such that the experiment can operate at luminosities $\mathcal{L} \sim 2 \times 10^{33}$ cm^{-2} s^{-1}. These show that the only way to achieve this is to measure both the momentum and the impact parameter of charged B decay products simultaneously. The present front-end architecture is not compatible with this requirement. The vertex and tracking detectors are read-out at a maximum rate of 1.1 MHz, thus this information is not available to the L0 trigger.

Hence the LHCb upgrade has opted for a front-end electronics which will read-out all LHCb sub-detectors at the full bunch crossing rate of 40 MHz of the LHC. Data will be transmitted over optical fibres to an off detector interface board which is read out by the DAQ. This has clear advantages as it would allow the implementation of a L0 displaced vertex trigger in a CPU farm. In fact all trigger decisions would be software-based which allows flexibility.

A initial study for the 40 MHz trigger uses $B_s^0 \to D_s^\mp K^\pm$ decays simulated at a luminosity of 6×10^{32} cm^{-2} s^{-1}. Events with large numbers of interactions are employed to simulate larger effective luminosities up to 2×10^{33} cm^{-2} s^{-1}. Assuming enough CPU power to process an event rate of 5 MHz, we obtain a trigger efficiency of 66% for this channel. The requirements are a transverse energy $E_T > 3$ GeV from the L0 hadron trigger which has an efficiency of 76% for signal combined with a matched track that has a transverse momentum $p_T > 2$ GeV/c and an impact parameter $\delta > 50\mu$m. In this combined trigger, the minimum bias rate does not depend strongly on the luminosity, and the triggered event yield scales linearly with the luminosity. In addition, the total trigger efficiency is 60% larger when compared with the existing baseline.

However, this approach requires a replacement of the front-end electronics for all sub-detectors, with the exception of the muon chambers which are already read out at 40 MHz. Replacing the front-end electronics will require new sensors for several sub-systems. Besides the VELO silicon sensors, the silicon sensors of the tracking stations will need to be replaced. The sensors close to the beam will suffer from a ten-fold increase in radiation, and hence more radiation hard sensors will be required. The RICH photon detectors have encapsulated front-end electronics and need to be replaced entirely.

The vertex detector (VELO) silicon sensors undergo radiation damage, and it is expected that these will need to be replaced when 6 to 8 fb^{-1} of luminosity has been collected [1099]. However the channel occupancy in the VELO is ~ 1% at design luminosity. When increasing the luminosity by a factor of ten to 2×10^{33} cm^{-2} s^{-1}, the occupancy only increases to ~ 3%, and the corresponding efficiency loss is small.

A preliminary study of the performance of the electromagnetic calorimeter (ECAL) at high luminosity shows only a small degradation for the selection efficiency of the decay $B_s^0 \to \phi\gamma$. It might be necessary to upgrade the inner section of ECAL to improve its granularity and energy resolution. The increased radiation level of irradiation leads to a degradation of the energy resolution and will require that half the inner ECAL section will need to be replaced after 3 years of operation at 2×10^{33} cm^{-2} s^{-1}.

R&D efforts have started on technologies for radiation-hard vertex detectors that will be able to operate in the LHC radiation environments at LHCb upgrade luminosities. The detector sensors will need to be able to operate at radiation doses of about 10^{15} 1 MeV equivalent neutrons/cm^2. Initial studies of Czochralski and n-on-p sensors irradiated up to 4.5×10^{14} 24 GeV protons/cm^2 are promising and show that the charge collection efficiencies saturate at acceptable bias voltages [1099]. Pixel sensors are very radiation hard, and R&D on this technology has started.

Two different vertex-detector geometries are envisaged. One is to shorten the strips, the other is to use pixels. Removing the RF foil that separates the VELO sensors from the primary beam-pipe vacuum would reduce the radiation length before the first measurement by 3% and improve the proper time resolution of B-meson decays.

4.4.6 Summary and conclusions

The LHC will open a new window for discovering NP. The LHCb experiment will probe NP with precision studies of flavor observables, whereas the general purpose detectors ATLAS and CMS aim to directly observe new particles. Both approaches are required to study the mass hierarchy and the couplings of the NP. LHCb will collect an integrated luminosity of about 10 fb^{-1} during its first five years. Very likely the LHC results will show that a significantly better sensitivity will be required for both, the direct and indirect approaches. Here we present a proposal to upgrade the LHCb detectors to be able to operate at ten times the design luminosity, i.e. at 2×10^{33} cm^{-2} s^{-1}, and to collect a data sample of 100 fb^{-1} with an improved detector. Initial sensitivities for physics with LHCb upgrade are presented. These show that LHCb upgrade has the potential to probe new physics at unprecedented levels that is mainly complementary to the proposed SFF. The upgraded LHCb experiment will include a first-level detached vertex trigger for which a new front-end architecture must be designed. A more radiation hard vertex detector is required to cope with the increased radiation doses.

5 Assessments[30]

In Sect. 1, we briefly introduced several NP scenarios and discussed their impact on FCNC and CP-violating processes. Then, in Sect. 3, we considered several benchmark channels that are particularly sensitive to NP, discussing the present status and future developments. The aim of this section is to summarize the present status of NP flavor scenarios, to identify possible patterns of NP signals and to describe the first attempts that have been made during this workshop to connect constraints on NP (and possible NP signals) in flavor and high-energy physics. The first two items are discussed in Sect. 5.1, the last one is presented in Sect. 5.3.

---

[30]Section coordinators: S. Heinemeyer, F. Parodi, L. Silvestrini.

5.1 New-physics patterns and correlations

The past decade has witnessed enormous progress in the field of flavor physics: B factories have studied flavor and CP violation in B_d–$\bar{B}_d$ mixing and in an impressive number of B decays; the Tevatron has produced the first results on B_s–$\bar{B}_s$ mixing and has studied several BRs and CP asymmetries in B and B_s decays; very recently, B-factories have established the first evidence of D–$\bar{D}$ mixing. This flourishing of experimental results has been accompanied by several remarkable improvements on the theory side, both in perturbative and nonperturbative computations. Let us just mention the NNLO calculation of $BR(b \to s\gamma)$, the proof of factorization in nonleptonic B decays in the infinite mass limit and the first unquenched results on B physics from lattice QCD.

Thanks to these experimental and theoretical achievements, we now have a rather precise idea of the flavor structure of viable NP extensions of the SM. The general picture emerging from the generalized Unitarity Triangle analysis performed in [7, 9, 210] and from the very recent data on D–$\bar{D}$ mixing [933, 950, 957, 1100] is that no new sources of CP violation of $\mathcal{O}(1)$ are observed in B_d, K and D mixing amplitudes. However, the possibility of NP CP-violating effects in B_s mixing is still open. Concerning $\Delta F = 1$ processes, the situation is quite different. In particular, large NP contributions to $s \to dg$, $b \to dg$ and $b \to sg$ transitions are not at all excluded. Sizable NP effects in $s \to dZ$, $b \to dZ$ and $b \to sZ$ vertices are also possible, although the available experimental data excludes order-of-magnitude enhancements. Finally, FC Higgs interactions generated by NP can still give large enhancements of scalar vertices, although the upper bounds on $B_s \to \mu^+\mu^-$ are getting tighter and tighter.

To summarize, we can say that, although the idea of MFV is phenomenologically appealing [10, 12, 82, 84, 190, 891, 1050], an equally possible alternative is that NP is contributing more to $\Delta F = 1$ transitions than to $\Delta F = 2$ ones. Within the class of $\Delta F = 1$ transitions, (chromo-)magnetic and scalar vertices are peculiar, since they require a chirality flip to take place, which leads to a down-type quark mass suppression within the SM. On the other hand, NP models can weaken this suppression if they contain additional heavy fermions and/or additional sources of chiral mixing. In this case, they can lead to spectacular enhancements for the coefficients of (chromo-)magnetic and scalar operators. Furthermore, if the relevant new particles are colored, they can naturally give a strong enhancement of chromomagnetic operators, while magnetic operators might be only marginally modified. The electric dipole moment of the neutron puts strong constraints on new sources of CP violation in chirality-flipping flavor-conserving operators involving light quarks, but this does not necessarily imply the suppression of flavor-violating operators, especially those involving b quarks. Therefore, assuming that NP is sizable in several $\Delta F = 1$ processes is perfectly legitimate given the present information available on flavor physics.

Thus, we can identify at least three classes of viable weakly-interacting NP extensions of the SM:[31]

1. Models with exact MFV.
2. Models with small ($\mathcal{O}(10\%)$) departures from MFV.
3. Models with enhanced scalar or chromomagnetic $\Delta F = 1$ vertices and a suitable suppression of NP contributions to $\Delta F = 2$ processes.

In models belonging to the third class, we expect sizable NP effects in B physics. From a theoretical point of view, a crucial observation is the strong breaking of the SM $SU(3)^5$ flavor symmetry by the top quark Yukawa coupling. This breaking necessarily propagates in the NP sector, so that in general it is very difficult to suppress NP contributions to CP violation in b decays, and these NP contributions could be naturally larger in $b \to s$ transitions than in $b \to d$ ones. This is indeed the case in several flavor models (see for example Ref. [1101]).

Another interesting argument is the connection between quark and lepton flavor violation in grand unified models [110, 1102–1104]. The idea is very simple: the large flavor mixing present in the neutrino sector, if mainly generated by Yukawa couplings, should be shared by right-handed down-type quarks that sit in the same $SU(5)$ multiplet with left-handed leptons. Once again, one expects in this case large NP contributions to $b \to s$ transitions.

5.2 Correlations between FCNC processes

On general grounds, it is difficult to establish correlations between FCNC processes without specifying not only the NP flavor structure but also the details of the NP model. However, there is a notable exception given by models of Constrained Minimal Flavor Violation (see Sect. 1 for the definition of this class of MFV models). While correlating $\Delta F = 1$ to $\Delta F = 2$ processes is not possible without specifying the details of the model, in the case of CMFV, there are several interesting correlations between FCNC processes. In CMFV, all NP effects can be reabsorbed in a redefinition of the top-mediated contribution to FCNC amplitudes. Thus, all processes that involve the same top-mediated amplitude are exactly correlated. This has interesting phenomenological consequences, allowing for stringent tests of CMFV by looking at correlated observables [10, 12, 190, 893, 1105].

It is enough to go from CMFV to MFV to destroy many of these correlations: for example, in MFV models with two

---

[31] Strongly-interacting NP most probably lies beyond the reach of direct searches at the LHC and so will not be discussed here [9].

Higgs doublets at large $\tan\beta$, it is in general not possible to connect K, B and B_s decays in a model-independent way. However, interesting correlations remain present also at large $\tan\beta$. For example, the enhancement of $B_s \to \mu^+\mu^-$ corresponds in general to a depletion of Δm_s [30] (actually, both features might be phenomenologically acceptable [32]).

Of course, within a specific model, it is in general possible to correlate $\Delta F = 1$ and $\Delta F = 2$ processes and to fully exploit the constraining power of flavor physics. The most popular example is given by the minimal supergravity models, where one can combine not only all the information from flavor physics but also the available lower bounds on SUSY particles and the constraints from electroweak physics, dark matter and cosmology [1106–1122, 1151–1155]. Interesting correlations between FCNC processes are also present in the CMSSM if one considers more general SUSY spectra than minimal supergravity [86, 1050].

Even allowing for new sources of flavor and CP violation to be present, correlations remain present between the several flavor observables generically affected by the same NP flavor-violating parameter. An interesting example is given by SUSY models with enhanced chromomagnetic $b \to s$ vertices (see e.g. [107]).

Another general class of NP models in which interesting correlations between FCNC processes can be established is given by SUSY-GUTs. Grand unification implies the equality of soft SUSY breaking terms at the GUT scale. Thus, any new source of flavor and CP violation present in squark masses must also be present in slepton masses, leading to a correlation between squark and slepton FCNC processes [69]. An extensive discussion of these correlations has been carried out in [70]. As an example, we present in Fig. 68 (from [70]) the constraints on $(\delta_{13}^d)_{RR}$ (defined in Sect. 1.3.5) from hadronic constraints only (upper left), leptonic constraints only (upper right), all constraints (lower left) and all constraints with improved leptonic bounds (lower right). In this interesting case, hadronic and leptonic bounds have comparable strengths. Exploiting the GUT correlation, it is possible to combine them to obtain a much tighter constraint on $(\delta_{13}^d)_{RR}$.

5.3 Connection to high-energy physics

Recent low-energy data from flavor physics experiments showed relatively good agreement with the SM prediction (taking into account the theory uncertainties). This imposes strong constraints on any NP scenario. In view of the new results and the new bounds on physics beyond the SM, the demand for scenarios that could be used for studies at ATLAS or CMS (or more generally for setting up the infrastructure for future studies once ATLAS and CMS have collected their first data) was issued. These scenarios should be in agreement with all existing B and K physics data and possibly show interesting signatures at the LHC experiments.

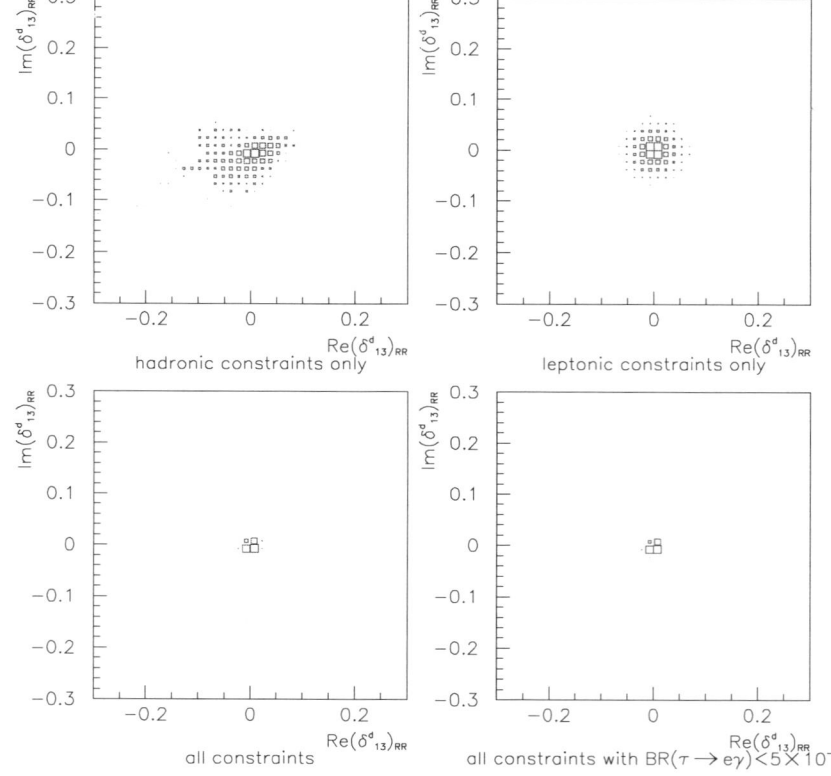

Fig. 68 Allowed region in the $\text{Re}(\delta_{13}^d)_{RR}$–$\text{Im}(\delta_{13}^d)_{RR}$ plane using hadronic constraints only (*upper left*), leptonic constraints only (*upper right*), all constraints (*lower left*) and all constraints with improved leptonic bounds (*lower right*)

In this respect, the question which parameter choices are useful as a benchmark scenario depends on the purpose of the actual investigation. If one is interested, for instance, in setting exclusion limits on the SUSY parameter space from the nonobservation of SUSY signals at the experiments performed up to now, it is useful to use a benchmark scenario which gives rise to "conservative" exclusion bounds. An example for a benchmark scenario of this kind is the $m_h^{\max}$-scenario [1123, 1124] used for the Higgs search at LEP [1125] and the Tevatron [1126, 1127]. Another purpose for using benchmark scenarios is to study "typical" experimental signatures of e.g. SUSY models and to investigate the experimental sensitivities and the achievable experimental precisions for these cases. For this application, it seems reasonable to choose "typical" parameters (a notion which is of course hard to define) of certain SUSY-breaking scenarios (see e.g. the "Snowmass Points and Slopes" [1129]). In this context, it can also be useful to consider "pathological" regions of parameter space or "worst-case" scenarios.

In the perspective of future improvements on B and K physics data, it is also worth to consider the possibility of a *positive* signal of NP selected by some low-energy observable. In this perspective, it is useful to consider benchmark scenarios with well-defined low-energy signatures, such as the MFV scenario with large $\tan\beta$ discussed in [32], or models with small flavor-breaking structures departing from the minimal structure of the constrained MSSM. These cases are particularly useful to explore the capability of future flavor-physics measurements in constraining a limited set of the SUSY parameter space, both separately and in conjunction with future ATLAS/CMS data.

A related issue concerning the definition of appropriate scenarios is whether a benchmark scenario chosen for investigating physics at ATLAS and CMS should be compatible with additional information from other experiments (beyond B and K physics). This refers in particular to constraints from cosmology or the measurement of the anomalous magnetic moment of the muon, $(g-2)_\mu$ [1128]. On the one hand, applying constraints of this kind gives rise to "more realistic" benchmark scenarios (see e.g. [1129]). On the other hand, one relies in this way on further assumptions (and has to take account of experimental and theoretical uncertainties related to these additional constraints), and it could eventually turn out that one has narrowed down the range of possibilities too much by applying these constraints. This applies in particular if slight modifications of the model under investigation have a minor impact on collider phenomenology but could significantly alter the bounds from cosmology and low-energy experiments. For instance, the presence of a small amount of R-parity violation in a SUSY model would strongly affect the constraints from dark matter relic abundance, while leaving the phenomenology at high-energy colliders essentially unchanged. Thus we restrict ourselves to scenarios which are compatible with flavor physics, with existing lower bounds on new particles (e.g. the bound on the lightest MSSM Higgs boson [1125, 1130]) and with other electroweak precision data, see Ref. [1131] and references therein.

The general procedure of setting up new scenarios follows the steps:

1. Identify the models of interest.
2. Identify within these models the regions of the parameter space that are compatible with the existing constraints from flavor physics, electroweak precision physics and direct bounds.
3. Identify specific sub-regions which could be selected by future improvements on flavor physics.
4. Study the most interesting points in view of their high-energy phenomenology that can be explored at ATLAS and CMS.
5. Set up the infrastructure for the analysis of (possible) data that will be collected at ATLAS and CMS to test the new high-energy results against existing low-energy data.

Concerning the first step, the model(s) which exhibited most interest during this workshop are the MSSM with (N)MFV. Consequently, in the following, we concentrate on this class of SUSY models.

Within the second and third step, it is desirable to connect different codes (e.g. working in the (N)MFV MSSM, see Sect. 1.5.1) to each other. Especially interesting is the combination of codes that provide the evaluation of (low-energy) flavor observables and others that deal with high-energy (high-p_T) calculations for the same set of parameters. This combination would allow one to test the ((N)MFV MSSM) parameter space with the results from flavor experiments as well as from high-energy experiments such as ATLAS or CMS.

A relatively simple approach for the combination of different codes is their implementation as sub-routines, called by a "master code" (see Sects. 5.3.3, 1.5.2). This master code takes care of the correct definition of the input parameters for the various subroutines. Concerning the last step, the application and use of the master code would change once experimental data showing a deviation from the SM predictions is available. This can come either from the ongoing flavor experiments or latest (hopefully) from ATLAS and CMS. If such a "signal" appears at the LHC, it has to be determined to which model and to which parameters within a model it can correspond. Instead of checking parameter points (to be investigated experimentally) for their agreement with experimental data, now a scan over a chosen model could be performed. Using the master code with its subroutines, each scan point can be tested against the "signal", and preferred parameter regions can be obtained using a χ^2 evaluation. It is obvious that the number of evaluated observables has to be as large as possible, i.e. the number of subroutines (implemented codes) should be as big as possible.

5.3.1 The first approach: prediction of b-physics observables from SUSY measurements

The first approach was followed in collaboration with ATLAS.

An LHC experiment will hopefully be able to measure a significant number of SUSY parameters based on the direct measurement of SUSY decays. The experimental potential in this field has been studied in detail for various benchmark points. Based on these studies, a possible approach is to focus on specific models for which many SUSY parameters can be measured at the LHC and to try to answer the following questions:

1. How precisely can b-physics variables be predicted using measured SUSY parameters?
2. Vice versa: can we use b-physics measurements to constrain badly measured SUSY parameters?
3. Is the precision of the measurements on the two sides adequate to rule out minimal flavor violation and/or to constrain flavor violation in the squark sector?

We will show in the following the application of this approach, especially of question (1), to a point of the MSSM space which was adopted as a benchmark point by the Supersymmetry Parameter Analysis (SPA) group [1132]. This model is defined in terms of the parameters of the mSUGRA model ($m_0 = 70$ GeV, $m_{1/2} = 250$ GeV, $A_0 = -300$ GeV, $\tan\beta = 10$, $\mu > 0$). This is a modification of the point SPS1a, essentially achieved by lowering m_0 from 100 to 70 GeV, originally defined in [1129] to take into account more recent results on dark matter density.

The values of the sparticle masses at tree level, computed with the program ISASUSY 7.71 [1133], are given in Table 63. Constraints on the sparticles masses can be obtained from measurements of the kinematics of the SUSY cascade decays [1134–1136]. This program has been carried out recently for the SPS1a model point [1137], assuming the performance of the ATLAS detector. The resulting constraints allow the measurement of the masses of $\tilde{\chi}_1^0$, $\tilde{\chi}_2^0$, $\tilde{\chi}_4^0$, $\tilde{g}$, $\tilde{q}_L$, $\tilde{q}_R$, $\tilde{b}_1$, $\tilde{b}_2$ $\tilde{\ell}_R$ $\tilde{\ell}_L$, $\tilde{\tau}_1$, where $\tilde{q}_L$ and $\tilde{q}_R$ are the average of the masses of the squarks of the first two generations. All these masses should be measurable with an uncertainties of a few percent for an integrated luminosity of 300 fb^{-1}. The estimated uncertainties will be used as an input to this study.

For the stop sector, a detailed study is available [1139], always performed in the framework of the ATLAS Collaboration. This analysis studies the tb invariant mass distribution in SUSY events. This distribution, shown in the left panel of Fig. 69, shows the characteristic kinematic edge which can be expressed as a function of the masses. Two

Table 63 Masses of the sparticles in the considered model as calculated at tree level with ISAJET 7.71 [1133]

Sparticle	Mass [GeV]	Sparticle	Mass [GeV]
$\tilde{\chi}_1^0$	97.2	$\tilde{\chi}_2^0$	180.1
$\tilde{\chi}_3^0$	398.4	$\tilde{\chi}_4^0$	413.8
$\tilde{\ell}_L$	189.4	$\tilde{\ell}_R$	124.1
$\tilde{\tau}_1$	107.7	$\tilde{\tau}_2$	194.2
$\tilde{t}_1$	347.3	$\tilde{t}_2$	562.3
$\tilde{u}_L$	533.3	$\tilde{g}$	607.0
h	116.8	A	424.6

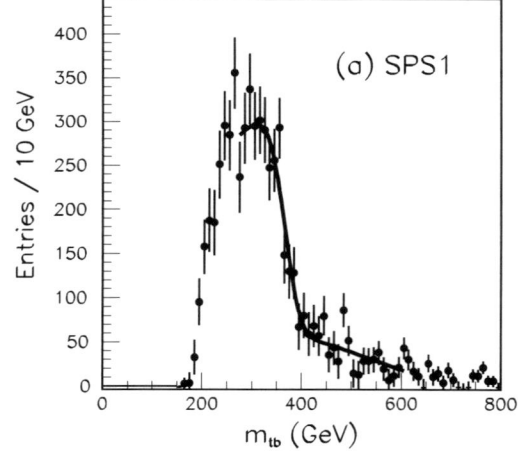
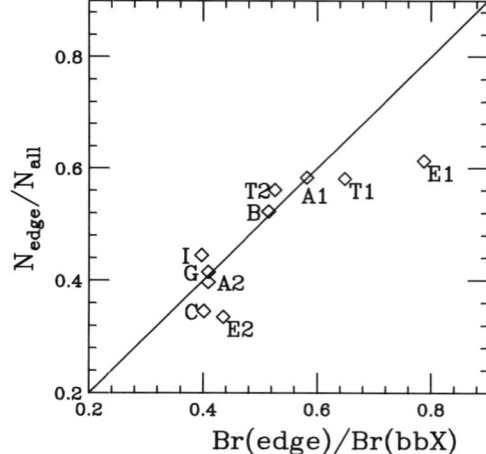

Fig. 69 *Left*: m_{tb} distribution for model point SPS1a. *Right*: relationship between $N_{\text{edge}}/N_{\text{all}}$ and $BR(\text{edge})/BR(bbX)$ for different model points as described in [1139]. Both figures from [1139]

main SUSY decay chains yield a tb final state signature:

$$\tilde{g} \to \tilde{t}_1 t \to tb\tilde{\chi}_1^\pm \qquad (231)$$

and

$$\tilde{g} \to \tilde{b}_1 b \to tb\tilde{\chi}_1^\pm. \qquad (232)$$

Therefore the position of the end-point in the tb mass distribution (M_{tb}^{fit}) will measure the average of the edges for the two decays weighted by the relative BR, which yields a constraint on a number of MSSM parameters:

$$M_{tb}^{\text{fit}} = f(m_{\tilde{t}_1}, m_{\tilde{b}_1}, m_{\tilde{g}}, m_{\tilde{\chi}_1^\pm}, \theta_{\tilde{t}}, \theta_{\tilde{b}}).$$

From the height of the observed kinematic distribution one can also measure the ratio of events in the tb mass distribution to all SUSY events with a b pair in the final state, $N_{\text{edge}}/N_{\text{all}}$. This observable is well correlated, as shown in the right panel of Fig. 69, with the quantity $BR(\text{edge})/BR(\tilde{g} \to bbX)$, where $BR(\text{edge})$ is the sum of the BR's for the decays (231) and (232) above. Finally direct searches in the SUSY Higgs sector yield additional constraints on the MSSM soft parameters.

The next step is the extraction of the soft SUSY-breaking parameters from the measured sparticle masses and branching ratios. We use a Monte Carlo technique relying on the generation of simulated experiments sampling the probability density functions of the measured observables. We proceed in the following way:

1. An 'experiment' is defined as a set of measurements, each of which is generated by picking a value from a Gaussian distribution with mean given by the central value calculated from the input parameters of the considered model and width given by the estimated statistical + systematic uncertainty of each measurement.
2. For each experiment, we extract the constraints on the MSSM model as we will describe in the following.

As a result of this calculation, we obtain a set of MSSM models, each of which is the "best" estimate for a given Monte Carlo experiment of the model generating the observed measurement pattern. For each of these models, the b-physics observables can be calculated.

Three groups of soft SUSY-breaking parameters are relevant for the prediction of b-physics observables:

– the parameters of the neutralino mixing matrix, M_1, M_2, μ, $\tan\beta$;
– m_A, the mass of the pseudoscalar Higgs, defining (together with $\tan\beta$) the Higgs sector at tree level;
– the masses and mixing angles of third generation squarks $\tilde{t}$ and $\tilde{b}$.

For the first two, a detailed discussion is given in [1140], which we will briefly summarize here.

In the SPA point, only the mass of three neutralinos (1, 2 and 4) can be measured. The three masses give a strong constraint on M_1, M_2, μ but have little sensitivity to $\tan\beta$. Therefore we use a fixed input value for $\tan\beta$ and calculate the values of M_1, M_2, μ from numerical inversion of the neutralino mixing matrix. We will then study 'a posteriori' the dependence on $\tan\beta$. The resultant uncertainty on M_1, M_2, μ is $\sim$(5–6) GeV, corresponding to the uncertainty on neutralino masses. By varying $\tan\beta$ in the range $3 < \tan\beta < 30$, the calculated values vary by less than 5 GeV.

Information on $\tan\beta$ and m_A can in principle be extracted from the study of the Higgs sector. The ATLAS potential for discovery is shown in Fig. 70 from [1135]. The light Higgs boson h can be discovered over the whole parameter space, but the measurement of its mass only provides somewhat loose constraints depending on the knowledge of the parameters of the stop sector. Much stronger constraints would be provided by the measurement of the mass and production cross-section of one or more of the heavy Higgs bosons. For the model under consideration with $\tan\beta = 10$ and $m_A \sim 425$ GeV, heavy Higgs bosons cannot be discovered at the LHC in their SM decay modes. Moreover, the heavy Higgs bosons cannot be produced in chargino–neutralino cascade decays, because the decays are kinematically closed. The only possibility would be the detection of $A/H \to \tilde{\chi}_2^0 \tilde{\chi}_2^0 \to 4\ell\ell$. Unfortunately the rate is very small, $\sim$40 events/experiment for 300 fb^{-1} before experimental cuts. A very detailed background study would be needed to assess the detectability of this signal.

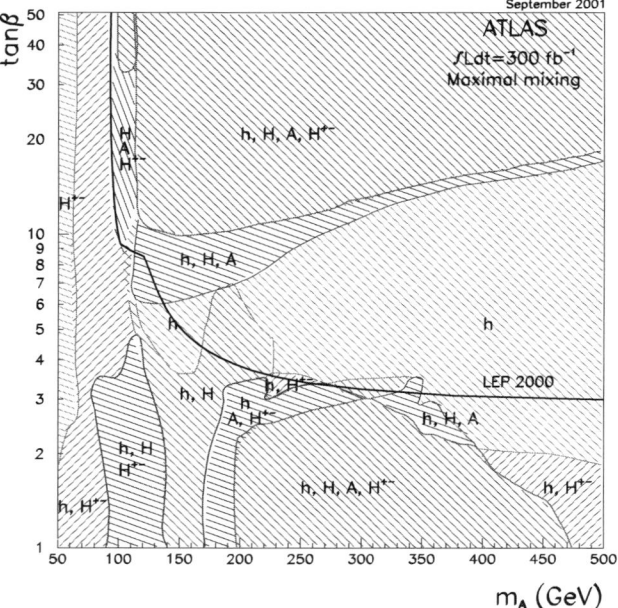

Fig. 70 Reach of the ATLAS experiment in the m_A–$\tan\beta$ plane for an integrated luminosity of 300 fb^{-1}. For each region in the plane, the detectable Higgs bosons are marked

We can now turn to the extraction of parameters of the stop-sbottom sector. The sector is defined by 5 soft SUSY-breaking parameters: $m(Q_3)$, the mass of the left-handed third generation doublet; $m(t_R)$ and $m(b_R)$, the masses of the stop and sbottom right-handed singlets; A_t and A_b, the stop and sbottom trilinear couplings. More convenient mixing variables would be $\theta_{\tilde{b}}$ and $\theta_{\tilde{t}}$, the left–right sbottom and stop mixing angles. For the considered point, 5 measurements will be available at the LHC:

- $m_{\tilde{b}_1}, m_{\tilde{b}_2}, BR(\tilde{g} \to b\tilde{b}_2 \to bb\tilde{\chi}_2^0)/BR(\tilde{g} \to b\tilde{b}_1 \to bb\tilde{\chi}_2^0)$ $(BR(\tilde{b}))$ [1137];
- $M_{tb}^{\text{fit}}, BR(\text{edge})/BR(\tilde{g} \to bbX)$ $(BR(\tilde{t}))$ [1139].

The assumed experimental errors on these variables are given in Table 64.

It is therefore possible to solve the available constraints for $m_{\tilde{t}_1}, \theta_{\tilde{b}}, \theta_{\tilde{t}}$, as discussed in [1141]. In [1141], the parameters of the gaugino matrix were assumed to be measured with infinite precision at the ILC, and the errors on the parameters in the stop sector were estimated by mapping the region in the $\theta_{\tilde{t}}$–$m_{\tilde{t}_1}$ plane compatible within the estimated errors with the nominal values of the five observables.

We incorporate the LHC uncertainties on the measurement of M_1, M_2, μ, and we use the technique of building Monte Carlo experiments described above.

The strategy is to scan the three-dimensional space $m_{\tilde{t}_1}, \theta_{\tilde{b}}, \theta_{\tilde{t}}$ and to find the point in space which reproduces the measured values of $M_{tb}, BR(\tilde{t}), BR(\tilde{b})$. For fixed $m_{\tilde{t}_1}$, the measurement of the position in the $\theta_{\tilde{b}}$–$\theta_{\tilde{t}}$ plane is given by combining the crossing of the line corresponding to the measured value of $BR(\tilde{b})$ with the line corresponding to the measured values of $BR(\tilde{b})$. We show in Fig. 71 respectively the band constrained by $\pm 1\sigma$ around the input values of $BR(\tilde{b})$ and $BR(\tilde{t})$ when all the other MSSM parameters are kept fixed. Because of the rather loose constraints on $BR(\tilde{b})$ and the low statistics in the $\tilde{b}_2$ peak, the region where the two bands cross, which roughly represents the allowed region in the plane, extends from the region around the input value ($\theta_{\tilde{t}} = 0.933, \theta_{\tilde{b}} = 0.42$) with a very low tail towards the region of high $\theta_{\tilde{b}}$ and low $\theta_{\tilde{t}}$.

Table 64 Assumed uncertainties for the LHC measurements in stop-bottom sector. The assumed statistics is 300 fb^{-1}. The only systematic error considered is the jet energy scale error on the mass/end point measurements

Variable	Value	Error
$m_{\tilde{g}} - m_{\tilde{b}_1}$	128.7 GeV	1.6 GeV
$m_{\tilde{g}} - m_{\tilde{b}_2}$	86.9 GeV	2.5 GeV
$BR(\tilde{b})$	0.70	0.05
$BR(\tilde{t})$	0.21	0.08
M_{tb}	411.3 GeV	5.4 GeV

The results of the scan are shown in Fig. 72. In the left plot, we show the distribution of the measured $m_{\tilde{t}_1}$ values for the considered ensemble of MC experiments. The RMS of the distribution is ~ 17 GeV, corresponding to a $\sim 5\%$ uncertainty on the light stop mass. The measured values in the $\theta_{\tilde{t}}$ versus $\theta_{\tilde{b}}$ plane are shown in the plot on the right of Fig. 72. As expected from the discussion above, a significant number of experiments yield a high value of $\theta_{\tilde{b}}$ and a low value of $\theta_{\tilde{t}}$.

The conclusions on the MSSM parameter measurement for the SPA model point under the assumption of no FCNC effects from sfermion mixing matrices are thus:

- neutralino/chargino mixing matrices fixed with $\sim 5\%$ if the value of $\tan\beta$ is known;
- slepton sector well constrained, including stau mixing angle;
- masses of first two generations squarks (L & R) and of gluino measured at $\sim(5$–$10)\%$ level;
- enough constraints to fix the 5 parameters of the stop/sbottom sector. For fixed $\tan\beta$, uncertainty of $\sim 5\%$ on stop mass, long tails in the measurement of $\theta_{\tilde{b}}$ and $\theta_{\tilde{t}}$;
- weak constraints on $\tan\beta$ and m_A.

We can now, based on the expected precision for the measurement of MSSM parameters, estimate how precisely observables in the b-sector can be predicted. We focus on two variables:

- $BR(B_s \to \mu\mu)$;
- $BR(B \to X_s\gamma)$.

Two public programs micrOMEGAs 1.3.6 [1142] and IS-ARED [1133] allow the evaluation of these two variables from an input set of MSSM parameters. Both programs work in the MFV framework and are based on the most recent NLO calculations. The results from micrOMEGAs 1.3.6 were used for the present exercise.

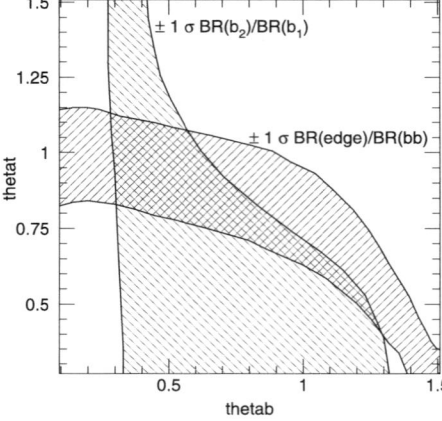

Fig. 71 Allowed 1σ bands on the $\theta_{\tilde{b}}$–$\theta_{\tilde{t}}$ plane respectively for the measurement of $BR(\tan\beta)$ (*downwards hatching*) and of $BR(\tilde{t})$ (*upwards hatching*)

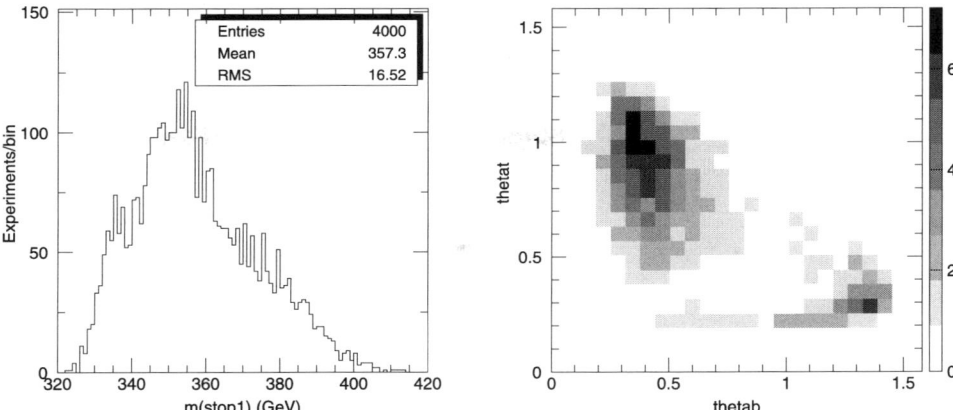

Fig. 72 *Left*: distribution of the calculated $\tilde{t}_1$ mass for an ensemble of Monte Carlo experiments at the LHC. *Right*: distribution of the calculated $\theta_{\tilde{t}}$ versus $\theta_{\tilde{b}}$ for an ensemble of Monte Carlo experiments. The assumed statistics is 300 fb^{-1}

The study is done in different steps. We first perform scans in the parameter space to evaluate the sensitivity of the two observables to the key parameters. Thereafter, based on the method of Monte Carlo experiments described above, we evaluate the expected value of $BR(B_s \to \mu\mu)$ and $BR(B \to X_s\gamma)$ for each Monte Carlo experiment. The spread of the obtained distributions is taken as the experimental uncertainty of the observables. Since m_A and $\tan\beta$ are badly constrained by the LHC measurements, this is done keeping m_A and $\tan\beta$ fixed.

The dependence of $BR(B_s \to \mu\mu)$ on m_A, $\tan\beta$ is shown in the left panel of Fig. 73. Since $BR(B_s \to \mu\mu) \propto \tan^6\beta/m_A^4$, this measurement has a strong constraining power on $\tan\beta$ if $\tan\beta \gtrsim 15$. For lower values of $\tan\beta \sim$, the effect becomes too small, and SUSY is indistinguishable from the SM. The present limits from the Tevatron experiments only eliminate a small region of the parameter space with small m_A and large $\tan\beta$. The expected 90% bound from ATLAS, 6.6×10^{-9} for 30 fb^{-1} [1143], would allow us to exclude a region in the m_A–$\tan\beta$ plane similar to the one excluded by nondiscovery of $H/A \to \tau\tau$. For higher $\tan\beta$, the measurement of a deviation from the SM would provide a nice cross-check with $\tan\beta$ as measured from H/A production.

The value of $BR(B \to X_s\gamma)$ in the m_A–$\tan\beta$ plane is shown in the right panel of Fig. 73. The present world average for $BR(B \to X_s\gamma)$ [502], $(3.3 \pm 0.4) \times 10^{-4}$, would select a narrow band in the m_A–$\tan\beta$ plane, thus providing essentially no bound on m_A and a strong constraint on the allowed $\tan\beta$ range, in the MFV hypothesis.

We show in Fig. 74 the values of $BR(B_s \to \mu\mu)$ and $BR(B \to X_s\gamma)$ in the $m_{\tilde{t}_1}$–$\theta_{\tilde{t}}$ plane with the other parameters fixed (see Fig. 75 below for an analysis of the effect of their uncertainty). The variation of $BR(B_s \to \mu\mu)$ over the considered space is moderate. The present experimental error on the measurement of $BR(B \to X_s\gamma)$ already defines a very small slice in the $m_{\tilde{t}_1}$–$\theta_{\tilde{t}}$ plane. For fixed $\theta_{\tilde{t}}$, the dependence on $m_{\tilde{t}_1}$ is not very strong. We therefore conclude that a precise measurement of $\theta_{\tilde{t}}$ is the key ingredient for the prediction of $BR(B \to X_s\gamma)$ from the LHC SUSY data.

As a next step, we verify that the experimental uncertainty on the two considered observables is indeed dominated by the measurement of m_A, $\tan\beta$, $m_{\tilde{t}_1}$ and $\theta_{\tilde{t}}$. To this effect, we calculate $BR(B_s \to \mu\mu)$ and $BR(B \to X_s\gamma)$ for all the Monte Carlo experiments, letting all of the MSSM parameters fluctuate according to the experimental error, except the four parameters mentioned above. The result is shown in Fig. 75. In these conditions, the uncertainty is small, 0.3% on the prediction of $BR(B_s \to \mu\mu)$ and 1% for the prediction of $BR(B \to X_s\gamma)$. These parametric uncertainties do not include the theoretical uncertainties in the calculation of the two observables.

Finally, we can evaluate how precisely we can predict the b-physics observables by varying all the MSSM parameters, according to the expected measurement precision at the LHC for the SPA point, except m_A and $\tan\beta$, which are kept fixed. The results are shown in Fig. 76. We observe a $\sim 5\%$ uncertainty on the prediction for $BR(B_s \to \mu\mu)$ and a $\sim 15\%$ uncertainty on the prediction for $BR(B \to X_s\gamma)$. For both observables, one can roughly observe two populations corresponding to the regions in the $\theta_{\tilde{b}}$–$\theta_{\tilde{t}}$ plane observed in Fig. 72. The experiments in the tail of mismeasured $\theta_{\tilde{t}}$ and $\theta_{\tilde{b}}$ contribute respectively to the region of high values of $BR(B_s \to \mu\mu)$, and to the bump for low values of $BR(B \to X_s\gamma)$.

We have thus shown that for the considered model, good enough measurements of MSSM parameters are possible at the LHC to provide predictions for $BR(B \to X_s\gamma)$, $BR(B_s \to \mu\mu)$ as a function of the two unconstrained variables, m_A and $\tan\beta$.

Once the LHC data are available, one can imagine different scenarios, e.g.:

– $A/H \to \tau\tau$ is observed, and $\tan\beta$ and m_A measured: at this point, a consistency check would be possible among the $\tan\beta$ constraints provided by the Higgs measurement and the one provided by the b-physics observ-

Fig. 73 *Left*: curves of equal value for $BR(B_s \to \mu\mu)$ in the m_A–$\tan\beta$ plane. *Right*: curves of equal value for $BR(B \to X_s\gamma)$. The MSSM parameters are as defined for the SPA point, and the calculations are performed using MicrOMEGAs

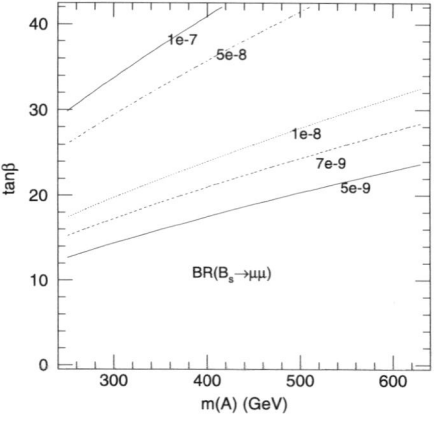

Fig. 74 *Left*: curves of equal value for $BR(B_s \to \mu\mu)$ in the $m_{\tilde{t}_1}$–$\theta_{\tilde{t}}$ plane. *Right*: curves of equal value for $BR(B \to X_s\gamma)$ in the $m_{\tilde{t}_1}$–$\theta_{\tilde{t}}$ plane. The MSSM parameters are as defined for the SPA point, and the calculations are performed using MicrOMEGAs

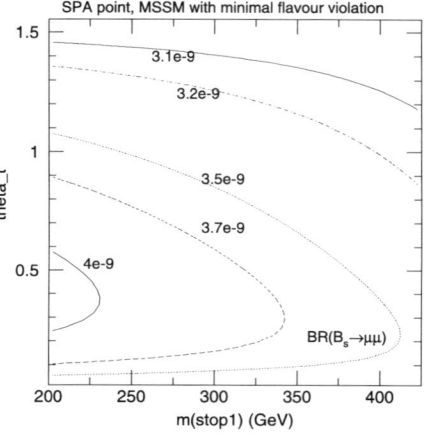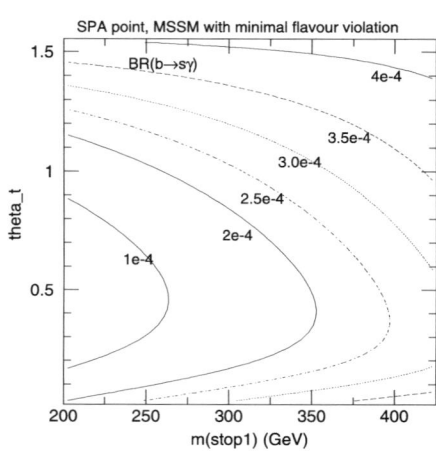

Fig. 75 Distribution of the predictions $BR(B_s \to \mu\mu)$ (*left*) and $BR(B \to X_s\gamma)$ (*right*) for an ensemble of LHC experiments when m_A, $\tan\beta$, $m_{\tilde{t}_1}$, $\theta_{\tilde{t}}$, $\theta_{\tilde{b}}$ are kept fixed at the nominal values, and all the remaining MSSM parameters are smeared according to the expected measurement uncertainty

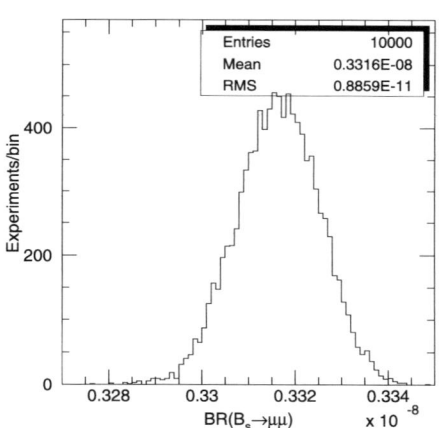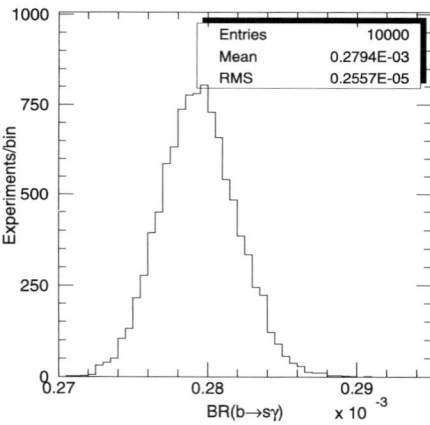

ables calculated in the MFV scheme. A significant disagreement, once all the experimental and statistical uncertainties are evaluated, would indicate the presence of flavor violation in the squark sector.

– $\tan\beta$ is not constrained by high-p_T searches:
a signal for nonminimal flavor violation could still be provided by the inconsistency of the $\tan\beta$ regions constrained by respectively $m(h)$, $BR(B \to X_s\gamma)$, and $BR(B_s \to \mu\mu)$. In case of consistency, the results could be taken as a measurement of the $\tan\beta$ parameter.

Relevant questions at this point are: what are the precisions required on the MSSM, on the b-physics measurements and on the theoretical calculations to be able to claim a signal for flavor-changing terms in the squark mass matrices?

In case the measurements are consistent with MFV, what additional constraints on the flavor violation sector can be

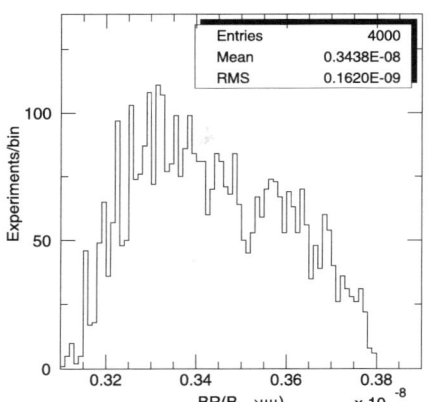
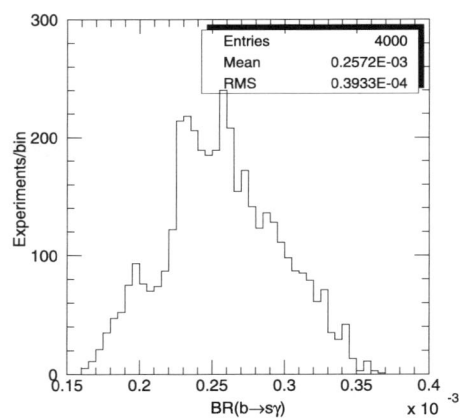

Fig. 76 Distribution of the predictions $BR(B_s \to \mu\mu)$ (*left*) and $BR(B \to X_s\gamma)$ (*right*) for an ensemble of LHC experiments when m_A, $\tan\beta$, are kept fixed at the nominal values, and all the remaining MSSM parameters, including the ones defining the stop sector, are smeared according to the expected measurement uncertainty

extracted by combining MSSM studies and b-physics measurements?

Various analyses are available in the literature [107, 113], based on assessing present allowed regions of nondiagonal elements in the super-CKM matrix, parametrized in terms of $(\delta_{23}^d)_{AB}$, where AB can be RR, LL, RL, LR. Bounds on δ are normally given for some special choice of soft SUSY-breaking parameters, e.g., $m(\tilde{q}) = m_{\tilde{g}} = \mu = -A_u$ for different choices of $m(\tilde{q})$. Additional variables are also considered, such as ΔM_B, $BR(B \to X_s \ell^+\ell^-)$, $A_{\mathrm{CP}}(B \to X_s\gamma)$.

Based on the study presented here, it would be interesting to repeat these analyses but for the parameters of a specific SUSY point, incorporating the expected experimental errors on the SUSY parameters. As a result of these studies, one could get guidance on which are the MSSM measurements crucial to discover flavor violation, thus pointing the way for the investigation of SUSY models in high-p_T physics.

5.3.2 The second approach: SUSY measurements in b-physics favored parameter spaces

A second, somewhat complementary, approach was followed in collaboration with CMS physicists.

5.3.2.1 B-physics favored parameter space

The model under investigation is the MSSM, in the first step with MFV, and possibly in a later stage also with NMFV. The compatibility with flavor physics was taken into account following [32], where the MSSM parameter space was analysed under the assumption of heavy scalar quarks and leptons and of large $\tan\beta$. The range of SUSY parameters has been restricted to the values listed in Table 65. Here $\tan\beta$ is the ratio of the two vacuum expectation values, M_A denotes the mass of the CP-odd Higgs boson, μ is the Higgs mixing parameter, $M_{\tilde{q},\tilde{l}}$ are the diagonal soft SUSY-breaking parameters in the scalar quark and scalar lepton sectors, respectively. All the trilinear couplings are set to be equal to A_t (the trilinear Higgs–stop coupling), while $m_{\tilde{g}}$, M_2 and M_1 are the gluino mass and the soft SUSY-breaking parameters in the

Table 65 Selected ranges and "best values" of the SUSY parameters for the "CMS analysis" in the MFV MSSM (following [32]): $\tan\beta$ is the ratio of the two vacuum expectation values, M_A denotes the mass of the CP-odd Higgs boson, μ is the Higgs mixing parameter, $M_{\tilde{q},\tilde{l}}$ are the diagonal soft SUSY-breaking parameters in the scalar quark and scalar lepton sectors, respectively; A_t is the trilinear Higgs–stop coupling, where all trilinear couplings are set equal; $m_{\tilde{g}}$, M_2 and M_1 are the gluino mass and the soft SUSY-breaking parameters in the gaugino sector. All parameters are assumed to be real

Parameter	Range	"Best" value(s)
$\tan\beta$	30–50	40
M_A [GeV]	300–1000	300, 500, 800, 1000
A_t [GeV]	-2000–(-1000)	$-1000, -2000$
μ [GeV]	500–1000	500, 1000
$M_{\tilde{q}}$ [GeV]	>1000	1000, 2000
$M_{\tilde{l}}$	$1/2\,M_{\tilde{q}}$	
$M_{\tilde{g}}$	$M_{\tilde{q}}$	
M_2 [GeV]		300, 500
M_1	$1/2\,M_2$	

chargino/neutralino sector. All parameters are assumed to be real. The upper part of Table 65 is the more relevant parameters, while the lower part has a smaller impact on the flavor physics phenomenology.

The ranges in [32] are generally compatible with the existing low-energy constrains. However, one expects to be able to select narrow sub-regions by more precise measurements of specific B-physics observables, such as $BR(B \to \tau\nu)$ or $BR(B_s \to \mu^+\mu^-)$. The "best" values denote specific points for which a more detailed investigation of the high-energy signatures at CMS has been performed.

5.3.2.2 Experimental analysis

The strategy followed by CMS physicists is to apply an already understood search analysis to the sample of MSSM points that are consistent with flavor constraints as described above. The starting point is [1144], in which CMS studied the production and decay of SUSY particles via inclusive final states including muons, high-p_T jets, and large missing transverse energy. In that work, a fully simulated and reconstructed low mass (LM1) Constrained MSSM (CMSSM) point was taken as the benchmark for selection optimization and study of systematic effects. Even though the study was performed within the context of CMSSM, the method is not specific to the CMSSM framework and should apply equally well in other contexts including, i.e. also in the general MSSM.

The response of the CMS detector to incident particles was simulated using a GEANT4-based framework [1145], known as the Object-oriented Simulation for CMS Analysis and Reconstruction (OSCAR) [1146]. The inclusion of pile-up and the reconstruction of analysis objects (muons, jets, etc) from hits in the detector was performed by a software framework known as the Object-oriented Reconstruction for CMS Analysis (ORCA) [1146]. In addition, a stand-alone fast simulation, known as the CMS FAst MOnte Carlo Simulation (FAMOS) framework [1146], was used to facilitate simulations involving CMSSM parameter scans. The fast simulation FAMOS has been shown to adequately represent the full CMS simulation [1144]. In both the full and fast simulations, hits from minimum bias events are superimposed on the main simulated event to reproduce the pile-up conditions expected for a luminosity of 2×10^{33} cm^{-2} s^{-1}.

Because the work presented in [1144] is an inclusive study of signatures involving at least one muon accompanied by multiple jets and large $\not{E}_T$, several SM processes contribute as sources of background and had to be taken into account. Accordingly, the main backgrounds studied in [1144] correspond to QCD dijet (2.8 million events with $0 < \hat{p}_T < 4$ TeV/c), top ($t\bar{t}$) production (3.3 million events), electroweak single-boson production (4.4 million events with $0 < \hat{p}_T < 4.4$ TeV/c) and electroweak dibosons production (1.2 million events). All backgrounds used were fully simulated and reconstructed.

The method employed in [1144] is to search for an excess in the number of selected events, compared with the number of events predicted from the SM. A Genetic Algorithm (GARCON [1147]) was used for the optimization of cuts to select the LM1 CMSSM point and results in: $E_T^{\text{miss}} > 130$ GeV, $E_T^{j1} > 440$ GeV, $E_T^{j2} > 440$ GeV, $|\eta^{j1}| < 1.9$, $|\eta^{j2}| < 1.5$, $|\eta^{j3}| < 3$, $\cos[\Delta\phi(j1,j2)] < 0.2$, $-0.95 < \cos[\Delta\phi(\not{E}_T, j1)] < 0.3$, $\cos[\Delta\phi(\not{E}_T, j2)] < 0.85$. Assuming 10 fb^{-1} of collected data, this set of cuts would expect to select a total of 2.5 background events from the SM and 311 signal events from the CMSSM LM1 benchmark signal point [1144].

In order to extend the work presented in [1144] to the context of the MSSM parameter space suggested by flavor considerations as described above, several points within the ranges of the MSSM parameters listed in Table 65 were sampled and simulated using the CMS fast simulation FAMOS. (The Pythia parameters used to generate each MSSM point may be found in [1144].) In the CMS exercise, the same set of selection cuts presented above, is directly applied (i.e., not reoptimized) to each simulated MSSM point. Finally, the number of selected events from each simulated MSSM point is tallied and compared with the expected number of standard model background events ($N_B = 2.5$).

It has been shown that the analysis method also works for this "new" part of the MSSM parameter space. Clearly, an optimization could enhance the analysis power. More detailed results will be presented elsewhere.

5.3.3 The "master code": multi-parameter fit to electroweak and low-energy observables

A first attempt to develop a "master code" as described above (see also Sect. 1.5.2) has been started in the course of this workshop in collaboration with physicists from CMS [208].

Based on flavor physics computer code from [32] and the more high-energy observable oriented computer code FeynHiggs [199–201], a first version of a "master code" has been developed. This "master code" combines calculations from both low-energy and electroweak observables in one common code. Great care has been taken to ensure that both sets of calculations are steered with a consistent set of input parameters. The current version of the "master code" is restricted to applications in the MSSM parameter space assuming MFV. Table 66 shows the observables which are currently considered in the "master code".

However, in the future, it is foreseen to significantly extend the "master code" by including other calculations both for different NP models as well as additional observables (e.g. cosmology constraints), see [1122] for the latest updates and developments. With the help of the "master code", it will eventually be possible to test model points from the low-energy side (via flavor and electroweak observables) and from the high-energy side (via the measurements of ATLAS/CMS). Thus a model point can be tested with *all* existing data.

Using the "master code" as a foundation, an additional code layer containing a χ^2 fit [1150] has been added to determine the consistency of a given set of MSSM parameters with the constraints defined in Table 66. Other studies of this kind using today's data can been performed in [1151–1155]. Studies using the anticipated data from the LHC and the ILC are carried out and documented in [1156, 1157].

Using the "master code", we will present a few showcases for a global χ^2 fit using a *simplified* version of the MSSM. The fit considers the following parameters: M_A (the CP-odd Higgs boson mass), $\tan\beta$ (the ratio of the two vacuum expectation values), $M_{\tilde{q},\tilde{l}}$ (a common diagonal soft SUSY-breaking parameter for squarks and sleptons, respectively), A (a common trilinear Higgs-sfermion coupling), μ

Table 66 List of available constraints in the "master code". The shown values and errors represent the current best understanding of these constraints. Smaller errors for M_W^{SUSY} and $\sin^2\theta_W^{SUSY}$ are possible using a dedicated code [1148, 1149], which is, however, so far not included in the "master code" (see, however, [1122])

Observable	Source	Constraint	Theo. error
$R_{BR_{b\to s\gamma}} = BR_{b\to s\gamma}^{SUSY}/BR_{b\to s\gamma}^{SM}$	[32]	1.127 ± 0.12	0.1
$R_{\Delta M_s} = \Delta M_s^{SUSY}/\Delta M_s^{SM}$	[32]	0.8 ± 0.2	0.1
$BR_{b\to\mu\mu}$	[32]	$< 8.0 \times 10^{-8}$	2×10^{-9}
$R_{BR_{b\to\tau\nu}} = BR_{b\to\tau\nu}^{SUSY}/BR_{b\to\tau\nu}^{SM}$	[32]	1.125 ± 0.52	0.1
$\Delta a_\mu = a_\mu^{SUSY} - a_\mu^{SM}$	FeynHiggs	$(27.6 \pm 8.4) \times 10^{-10}$	2.0×10^{-10}
M_W^{SUSY}	FeynHiggs	(80.398 ± 0.025) GeV	0.020 GeV
$\sin^2\theta_W^{SUSY}$	FeynHiggs	0.23153 ± 0.00016	0.00016
$M_h^{light}(SUSY)$	FeynHiggs	> 114.4 GeV	3.0 GeV

(the Higgs mixing parameter), M_1 and M_2 (the soft SUSY-breaking parameters in the chargino/neutralino sector) and $m_{\tilde{g}} = M_3$ (the gluino mass). All parameters are assumed to be real. Some further simplifying restrictions are applied: For the parameter μ, we require $|\mu| > M_2$. This ad-hoc ansatz is fully sufficient for our illustrative studies, but in the future, it will be replaced with a more sophisticated treatment of the parameters and of the experimentally excluded phase space regions (e.g. sparticle mass limits, etc.). In addition, the ansatz assumes $M_{\tilde{l}} = a_{\tilde{q},\tilde{l}} \times M_{\tilde{q}}$ as well as fixed values for M_1, M_2 and M_3. The initially assumed values of $a_{\tilde{q},\tilde{l}} = 0.5$, $M_2 = 200$ GeV, $M_3 = 300$ GeV and $M_1 = M_2/2$ are later varied within reasonable ranges to evaluate the systematic impact of the assumption on the final results.

The χ^2 is defined as

$$\chi^2 = \sum_i^{N_{const.}} \frac{(\text{Const.}_i - \text{Pred.}_i(\text{MSSM}))^2}{\Delta\text{Const.}^2 + \Delta\text{Pred.}^2}, \quad (233)$$

where Const.$_i$ represents the measured values (constraints), and Pred.$_i$ defines the MSSM parameter-dependent predictions of a given constraint. These predictions are obtained from the "master code". They depend on SM parameters like m_t, m_b and α_s. Some of these parameters still exhibit significant uncertainties which need to be taken into account in the fit procedure. In a simple χ^2 approach, it is straightforward to include these parametric uncertainties as fit parameters with penalty constraints. For our study, the uncertainty of the top quark mass was found to be by far the dominating parametric uncertainty. The required minimization of the χ^2 is carried out by the well-known and very reliable fit package Minuit [1150].

In the following section, we present some illustrative showcases that utilize this global χ^2 fit to extract quantitative results. However, these studies are mainly meant to demonstrate the potential and usefulness of "external" constraints for the interpretation of forthcoming discoveries and for the corresponding model parameter extraction.

5.3.3.1 Scan in the lightest Higgs-boson mass M_h One of the most important predictions of the MSSM is the existence of a light neutral Higgs boson with $M_h \leq 135$ GeV [199, 200]. This upper limit, together with the lower limit obtained at LEP, $M_h^{direct} \geq 114.4$ GeV [1125, 1130][32] represents a tight constraint on the remaining allowed parameter space of the MSSM. In the MSSM (with the simplifications explained above), M_h depends mainly on the average squark mass $M_{\tilde{q}}$, the Higgs mixing parameter μ, the trilinear Higgs-squark coupling A and $\tan\beta$. However, these parameters are also important for the predictions of low-energy and electroweak observables in the MSSM. Therefore, a global fit using the constraints listed in Table 66 not only allows a consistent extraction of the important MSSM parameters but will also provide a prediction for the most probable light Higgs boson mass M_h in the MSSM. A convenient way to illustrate the sensitivity of these parameters to M_h is a scan of the preferred parameter space as a function of this variable. For this procedure, the global χ^2 fit is performed repeatedly each time with a different value for the M_h constraint. Therefore, the extracted set of MSSM parameters for each individual fit corresponds to the preferred parameter space for a given value of M_h. While all M_h scan values below the lower limit of $M_h^{direct} > 114.4$ GeV are already excluded by experiment, it is nevertheless interesting to see the results of the M_h scan over the entire parameter space (i.e. also for M_h values $\lesssim 115$ GeV). For that reason, the lower M_h limit from the direct search at LEP has not been included in the χ^2 fit.

5.3.3.2 M_h scan using today's (pre-LHC) constraint values and errors Figure 77 shows the results of the M_h scan using the constraint values listed in Table 66. Since these values represent today's best knowledge of these observables, this result provides a first estimate of how low-energy and electroweak measurements constrain the MSSM parameter space. In the following, we will refer to this scan result as *today's M_h scan*.

It is important to note that the $M_h \approx [110, 125]$ GeV region seems to be preferred by the χ^2 scan. On the one hand,

---
[32] It is possible that the current lower limit could be even further improved before the LHC will start data taking in 2008 by the currently running Tevatron experiments CDF and D0.

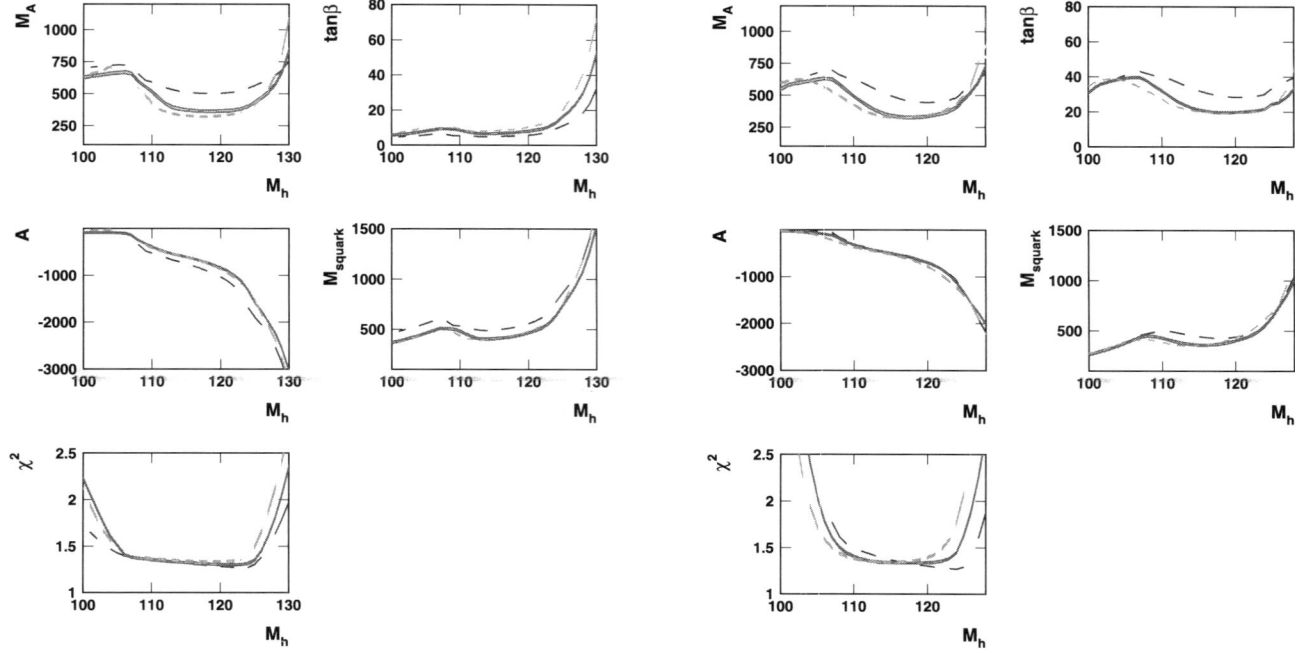

Fig. 77 (Color online) This figure shows the result of the extracted MSSM fit parameters and the corresponding χ^2 distribution (*lower right plot* in each case) for the two scan scenarios: *today's M_h scan* (*left five plots*) and *2009-EW-LowE M_h scan* (*right five plots*). Each *plot* shows three scan results, where the *full-red curve* corresponds to the default assumptions of $M_2 = 200$ GeV, $M_3 = 300$ GeV and $a_{\tilde{q},\tilde{l}} = 0.5$. The *blue-dashed line* (*large dash*) changes $a_{\tilde{q},\tilde{l}} = 0.33$ with respect to the default setting, while the *green-dashed line* (*small dash*) modifies $M_2 = 300$ GeV, $M_3 = 500$ GeV with respect to the default setting

all M_h values in this distinguished region of minimal χ^2 are almost equally likely. On the other hand, values outside this window (i.e. <110 GeV or >125 GeV) are clearly disfavored by the low-energy and electroweak constraints. This is an interesting observation suggesting that today's low-energy and electroweak data prefer a light MSSM Higgs boson with a mass significantly higher than the most probable value for the SM Higgs boson. For comparison, the current preferred value from the general electroweak fit is $M_h^{SM} \approx 80$ GeV [1158–1160].

In order to qualitatively estimate the systematic impact of the assumed parameter values ($M_2 = 200$ GeV, $M_3 = 300$ GeV and $a_{\tilde{q},\tilde{l}} = 0.5$) on the scan results, a variation of the parameter values within reasonable ranges has been carried out. Figure 77 shows the results of two of these cross checks: the blue-dashed line corresponds to the parameter setting $M_2 = 200$ GeV, $M_3 = 300$ GeV and $a_{\tilde{q},\tilde{l}} = 0.33$, while the green-dashed line uses $M_2 = 300$ GeV, $M_3 = 500$ GeV and $a_{\tilde{q},\tilde{l}} = 0.5$. The observed variation is rather small indicating that the general conclusions are not strongly affected by the assumed parameter setting of these quantities. In particular the preferred minimal χ^2 region of M_h remains almost unchanged.

The overall χ^2 minimum of *today's M_h scan* is at $M_h \approx 123$ GeV, and the preferred values of the important MSSM parameters are $M_A \approx 400$ GeV, $\tan\beta \approx 10$, $A \approx -1000$ GeV and $M_{\tilde{q}} \approx 500$ GeV. These values are qualitatively compatible with the range of "allowed" MSSM parameter space reported in Sect. 5.3.2. The fact that *today's M_h scan* prefers somewhat lower values for $\tan\beta$ and $M_{\tilde{q}}$ is mainly explained by the change in the experimental Belle result of $R_{BR_{b \to \tau\nu}}$ from 0.7 ± 0.3 to 1.125 ± 0.52 [326]. Using 0.7 ± 0.3 instead of the other more recent (corrected) value yields $\tan\beta \approx 30$, $M_{\tilde{q}} \approx 700$ GeV and $A \approx -1500$ GeV but does not change the general conclusion of the results (e.g. the preferred M_h range remains the same).

Figure 78 shows a comparison of the predicted constraint values and their corresponding measurements obtained from *today's M_h scan*. The measurements and their errors are also listed in Table 66. In general, good agreement between prediction and measurement is observed in the preferred minimal χ^2 region of $M_h \approx [110, 125]$ GeV. The fact that the χ^2 scan prefers a prediction of $R_{\Delta M_s}$ very close to unity is explained by (1) the already rather tight limit on $BR(B_s \to \mu^+\mu^-) < 8 \times 10^{-8}$ and (2) the large value of $R_{BR_{b \to \tau\nu}}$. Both constraints prefer low values of $\tan\beta$ and thus result in a prediction of $R_{\Delta M_s} \approx 1$. However, today's experimental value is still within one sigma compatible with this prediction.

Another interesting observation is the prediction of $BR(B_s \to \mu^+\mu^-)$. Although the constraint used for this quantity allows values up to $BR(B_s \to \mu^+\mu^-) < 8 \times 10^{-8}$, the scan predicts (in the interesting M_h region) an almost

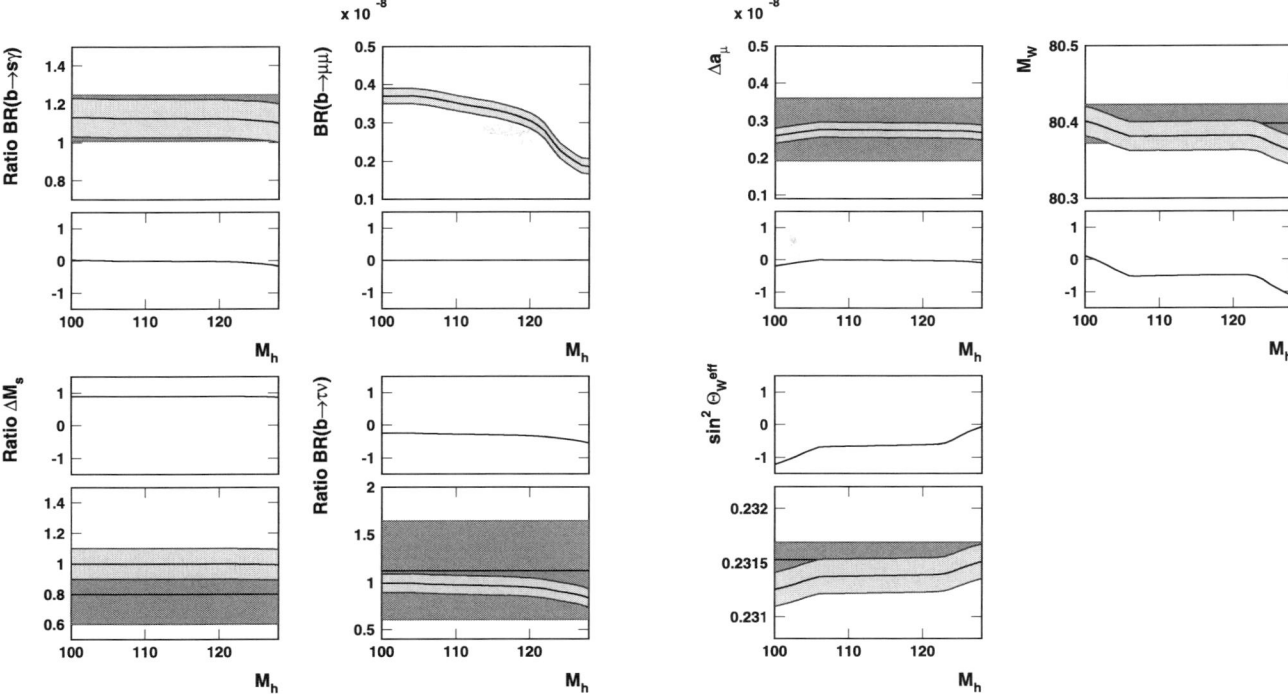

Fig. 78 This figure shows a comparison of the predicted constraint values (*yellow/light shaded band*) and their corresponding measurements (constant *green/dark shaded band*) obtained from *today's* M_h scan. All *plots* show a comparison of prediction versus measurement (*plots with bands*) as well as their corresponding pull contributions $\frac{\text{Const.}_i - \text{Pred.}_i (\text{MSSM})}{\sqrt{\Delta\text{Const.}^2 + \Delta\text{Pred.}^2}}$ to the overall χ^2

constant value of $BR(B_s \to \mu^+\mu^-) \approx (3.0\text{--}4.0) \times 10^{-9}$. This is an interesting observation, because this value coincides well with the SM prediction of $BR(B_s \to \mu^+\mu^-)^{\text{SM}} \approx 3.5 \times 10^{-9}$. This might suggest that the current low-energy and electroweak data prefer a value of $BR(B_s \to \mu^+\mu^-)$ close to its SM prediction. It will be interesting to see whether the soon forthcoming combined result of $R_{BR_{b \to \tau\nu}}$ from BaBar and Belle will confirm this trend. If this is the case, spectacular effects from new (MSSM) physics contributions seem rather unlikely for $B_s \to \mu^+\mu^-$.

5.3.3.3 Interpretation of potential LHC discoveries The LHC will start collecting physics data in 2008. For that reason, the first results are not expected before early 2009. In the meantime, however, it is likely that most of the considered low-energy and electroweak constraints will further improve. Therefore, in 2009, it will be possible to even more strongly restrict the allowed MSSM parameter space. Table 67 lists the assumed constraint values that might be achieved by this time period. The assumed values and errors are only chosen for illustrative purposes. The sole intention of this study is to demonstrate the potential of low-energy and electroweak data to constrain the parameter space of NP and to eventually provide guidance for the interpretation of potential NP discoveries at the LHC. Figure 77 (five

Table 67 Assumed constraint values and errors for the *2009-EW-LowE* scenario

Observable	Constraint	Theo. error
$R_{BR_{b \to s\gamma}}$	1.127 ± 0.1	0.1
$R_{\Delta M_s}$	0.8 ± 0.2	0.1
$BR_{b \to \mu\mu}$	$(3.5 \pm 0.35) \times 10^{-8}$	2×10^{-9}
$R_{BR_{b \to \tau\nu}}$	0.8 ± 0.2	0.1
Δa_μ	$(27.6 \pm 8.4) \times 10^{-10}$	2.0×10^{-10}
M_W^{SUSY}	(80.392 ± 0.020) GeV	0.020 GeV
$\sin^2 \theta_W^{\text{SUSY}}$	0.23153 ± 0.00016	0.00016
$M_h^{\text{light}}(\text{SUSY})$	> 114.4 GeV	3.0 GeV

plots on the right) shows the results of the χ^2 scan using the constraints listed in Table 67. In the following, we refer to these results as *2009-EW-LowE M_h scan*. Similar to the results from the *today's M_h scan*, the general results and conclusions of this study are largely unaffected by the variation of the assumed values for M_2, M_3 and $a_{\tilde{q},\tilde{l}}$. As shown in Fig. 77, the χ^2-preferred M_h region becomes even more pronounced. Hence, the allowed MSSM parameters space is further reduced. In particular this information will become very useful in the case of LHC discoveries and their corresponding interpretation. In order to illustrate this property, we define a few hypothetical scenarios:

- *2009-EW-LowE*:
  This scenario includes only the observables listed in Table 67. The overall χ^2 minima for this scenario is achieved for $M_A \approx 350$ GeV, $\tan\beta \approx 22$, $\mu \approx 5$ GeV, $A \approx -450$ GeV and $M_{\tilde{q}} \approx 350$ GeV. The corresponding prediction of the light MSSM Higgs boson mass is $M_h \approx 115$ GeV.
- *LHC-$M_{\tilde{q}}$*:
  This scenario includes *2009-EW-LowE* and additionally assumes that the relevant squark mass[33] $M_{\tilde{q}}$ is known at the level of 10%. To be consistent with *2009-EW-LowE*, we therefore define: $M_{\tilde{q}} = 350 \pm 35$ GeV.
- *LHC-$M_{\tilde{q}}$-M_A*:
  This scenario includes *LHC-$M_{\tilde{q}}$* and additionally assumes that the mass of $M_{AH^\pm}$ is known to 10%. To be consistent with *2009-EW-LowE*, we therefore define: $M_A = 355 \pm 35$ GeV.
- *LHC-$M_{\tilde{q}}$-M_A-M_h*:
  This scenario includes *LHC-$M_{\tilde{q}}$-M_A* and additionally assumes that the mass of M_h is measured with a 3 GeV error. To be consistent with *2009-EW-LowE*, we therefore define: $M_h = 115 \pm 3$ GeV.

Figure 79 shows the results of the M_h scan for the scenario *2009-EW-LowE* and the scenario *LHC-$M_{\tilde{q}}$-M_A*. As expected, the χ^2 allowed region of M_h is reduced to a small window by including the additional information of $M_A = 355 \pm 35$ GeV and $M_{\tilde{q}} = 350 \pm 35$ GeV. This information can, for example, be utilized to test the consistency of a discovered light Higgs boson candidate with:

(a) other discoveries of MSSM particle candidates (in our case squark and heavy Higgs candidates);
(b) low-energy and electroweak constraints.

Assuming that a light Higgs boson candidate has been observed and that its mass is measured with an error of $\Delta M_h = \pm 3$ GeV, Fig. 80 shows the $\Delta\chi^2$ distributions for the scenario *2009-EW-LowE* (green small-dashed line), *LHC-$M_{\tilde{q}}$* (blue large-dashed line) and *LHC-$M_{\tilde{q}}$-M_A* (red full line).

As defined above, all scenarios correspond to one MSSM parameter set that has a χ^2 minimum for $M_h \approx 115$ GeV, see Fig. 81. The $\Delta\chi^2$, and therefore also the exclusion limits, are defined with respect to this MSSM parameter set. For the most constraining scenario, all masses above ≈ 130 GeV are excluded at 95% C.L. Therefore, in this hypothetical case, M_h must be below 130 GeV in order to be compatible with the other observed LHC discoveries as

---
[33]For example, this could be achieved by a determination of the stop mass. In particular this mass is important for the determination of the lightest Higgs boson mass M_h in the MSSM.

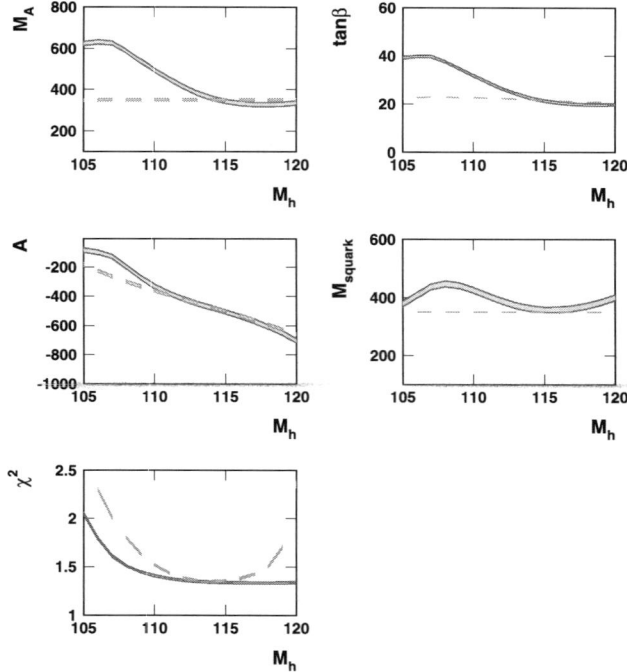

Fig. 79 This figure shows the result of the extracted MSSM fit parameter and the corresponding χ^2 distribution for the two scan scenarios: *2009-EW-LowE* M_h scan (*full-red curve*) and *LHC-$M_{\tilde{q}}$-M_A-M_h* scan (*green-dashed curve*)

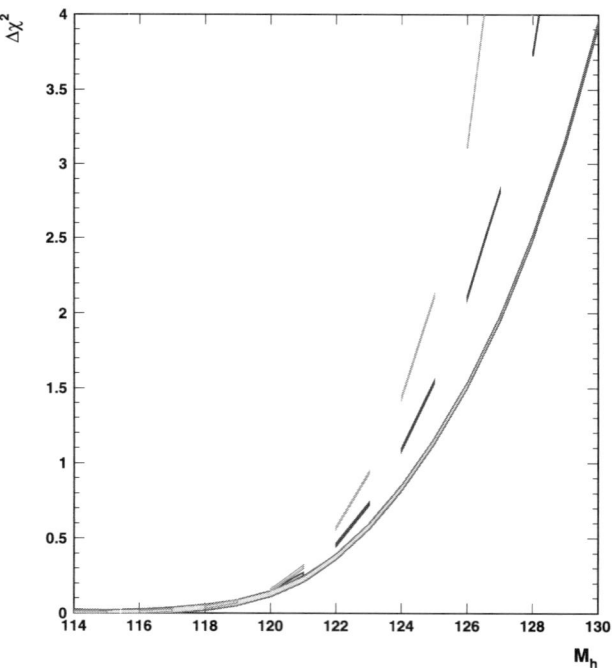

Fig. 80 $\Delta\chi^2$ distribution for scenario *LHC-$M_{\tilde{q}}$-M_A* testing the hypothesis that a discovered light Higgs boson candidate with a mass error of: $\Delta M_h = 3$ GeV (*red curve*), 2 GeV (*blue/dark dashed curve*) and 1 GeV (*green/light dashed curve*) is compatible with the MSSM

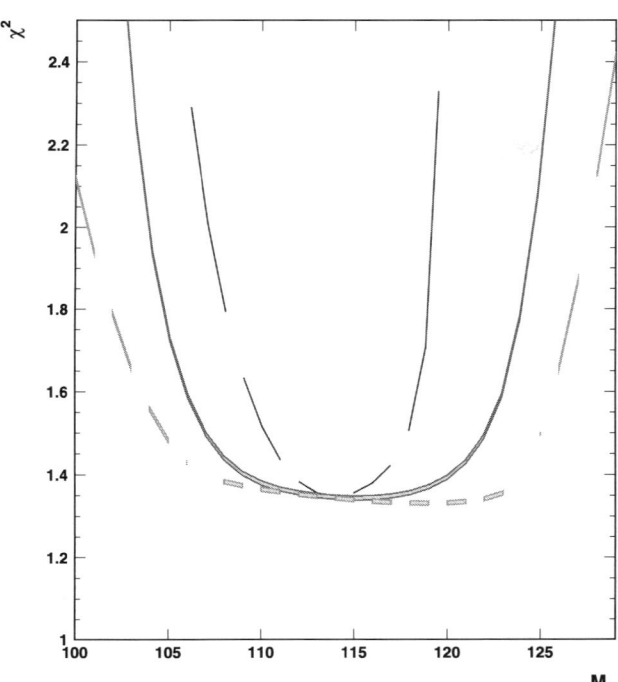

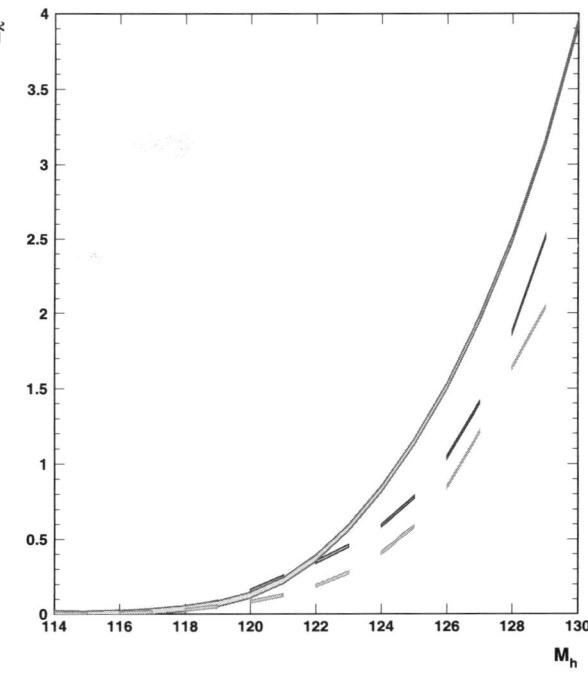

Fig. 81 χ^2 distribution as a function of M_h for the three scenarios: *2009-EW-LowE* M_h *scan* (*full-red curve*), *today's* M_h *scan* (*green-small-dashed curve*) and *LHC-$M_{\tilde{q}}$-M_A-M_h scan* (*blue-large-dashed curve*)

Fig. 82 $\Delta\chi^2$ distribution for scenario *LHC-$M_{\tilde{q}}$-M_A* (*red curve*), *LHC-$M_{\tilde{q}}$* (*blue/dark dashed curve*) and *2009-EW-LowE* (*green/light dashed curve*). All curves are evaluated with an assumed error of $\Delta M_h = 3$ GeV

well as with the low-energy and electroweak constraints. A discovery of a lightest Higgs boson with a mass above 130 GeV would rule out the MSSM at 95% C.L. It is clear that the exclusion limit depends on the assumed error for M_h. For scenario *LHC-$M_{\tilde{q}}$-M_A*, Fig. 82 compares the results for $\Delta M_h = \pm 3$, $\Delta M_h = \pm 2$ and $\Delta M_h = \pm 1$. With an assumed error of 2 GeV, the 95% C.L. exclusion limit would be around $M_h \approx 128$ GeV, while for a 1 GeV error, it would be as stringent as $M_h \approx 126$ GeV.

Therefore, together with the discoveries of a stop candidate and a heavy Higgs candidate, the consistency of a measured light Higgs candidate within the MSSM hypothesis can be tested. It should be noted that without the use of low-energy and electroweak constraints, this consistency test would be much weaker. For example, the three LHC discoveries alone will not significantly constrain the important MSSM parameters $\tan\beta$ and A. This feature is clearly demonstrated in Fig. 83. Without the inclusion of the low-energy and electroweak constraints, the parameters $\tan\beta$ and A are much less determined. Thus, the overall sensitivity of the consistency test is significantly worse.

Another way to illustrate the potential of external constraints for the interpretation of NP discoveries and the eventual extraction of the model parameters is shown in Fig. 84, which displays the $\Delta\chi^2 = 1$ contours of the four different scenarios for various parameter combinations. Although

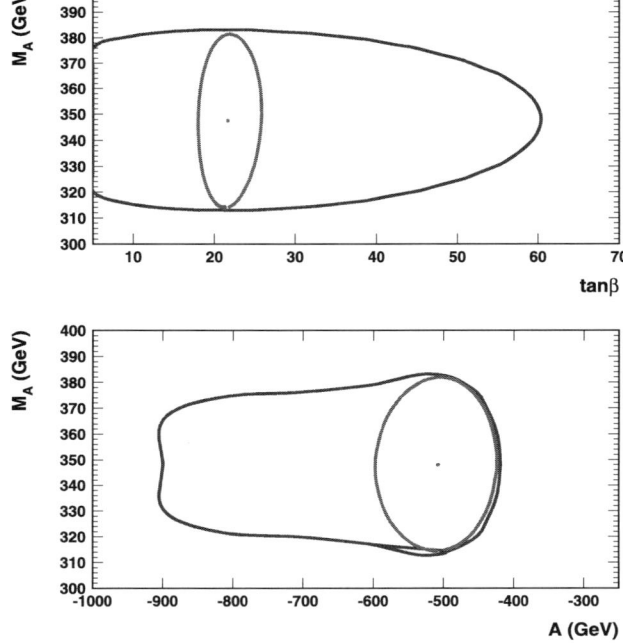

Fig. 83 The *red/lighter contour* corresponds to scenario *LHC-$M_{\tilde{q}}$-M_A-M_h* that includes the low-energy and electroweak constraints, while the *blue/darker contour* makes the same assumptions about the assumed LHC discoveries but does not include any external constraints

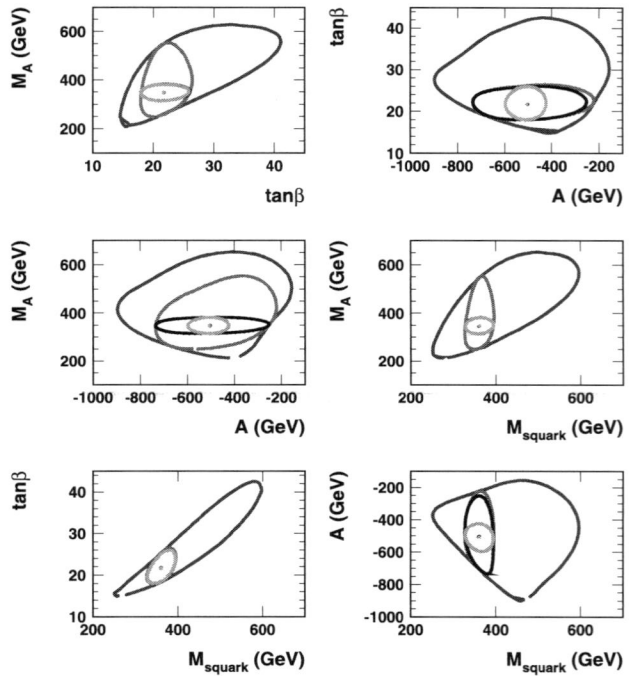

Fig. 84 This figure shows the $\Delta\chi^2 = 1$ contours of the four scenarios: *2009-EW-LowE* (*blue/outermost contour*), *LHC-$M_{\tilde{q}}$* (*red/lighter contour*), *LHC-$M_{\tilde{q}}$-M_A* (*black contour*) and *LHC-$M_{\tilde{q}}$-M_A-M_h* (*green/innermost contour*). Sometimes the inner contours partially (or even completely) overlap. In this case, the *green/innermost contour* covers the *black*, or the *black* covers the *red/lighter* one

2009-EW-LowE (blue contour) only utilizes indirect constraints (i.e. no direct measurement of NP quantities), the MSSM parameter space is already rather restricted. Adding $M_{\tilde{q}} = (350 \pm 35)$ GeV (red contour) in particular helps to further constrain $\tan\beta$ and to some extent also M_A, while measuring also the heavy (black contour) and also the light Higgs boson mass (green contour) will restrict the allowed range for A rather significantly. Also here the use of the external low-energy and electroweak constraints is essential to determine the important MSSM parameters $\tan\beta$ and A.

5.3.3.4 Outlook In order to fully exploit this interesting potential, it will be important to extend the "master code" by adding additional calculations such as extra low-energy observables, as well as, potentially, constraints from cosmology data (see [1122]). This will eventually yield an important tool for the comprehensive interpretation of future NP discoveries.

5.4 Discrimination between new physics scenarios

At present, the SM gives a fully consistent description of all experimental data in the flavor sector, apart from a few, not yet statistically significant deviations. This means that flavor physics can at present only rule out models that produce too large deviations from the SM; in practice, this means giving an upper bound on new sources of flavor and CP violation for a fixed NP scale or giving a lower bound on the NP scale for fixed values of the NP flavor parameters. As discussed in Sect. 5.1, this gives us hints on the flavor structure of NP models with new particles up to the TeV range. However, to fully exploit the constraining power of flavor physics, additional (external) information on the spectrum of new particles must be provided. First examples of the combination of flavor and high-p_T information have been presented in Sect. 5.3, and there is increasing activity in this direction.

Acknowledgements This work was supported in part by the National Science Foundation, the Natural Sciences and Engineering Council of Canada and the DFG cluster of excellence Origin and Structure of the Universe (www.universe-cluster.de), Germany. J.L.R. was supported by the United States Department of Energy through Grant No. DE FG02 90ER40560. A.A.P. was supported in part by the U.S. National Science Foundation CAREER Award PHY-0547794 and by the U.S. Department of Energy under Contract DE-FG02-96ER41005.

The authors of this report thank all further participants in this workshop series for their presentations during the five workshop meetings, for contributing to the discussions and for useful comments.

References

1. R. Fleischer, G. Isidori, J. Matias, J. High Energy Phys. **0305**, 053 (2003). arXiv:hep-ph/0302229
2. J.M. Soares, L. Wolfenstein, Phys. Rev. D **47**, 1021 (1993)
3. N.G. Deshpande, B. Dutta, S. Oh, Phys. Rev. Lett. **77**, 4499 (1996). arXiv:hep-ph/9608231
4. J.P. Silva, L. Wolfenstein, Phys. Rev. D **55**, 5331 (1997). arXiv:hep-ph/9610208
5. A.G. Cohen, D.B. Kaplan, F. Lepeintre, A.E. Nelson, Phys. Rev. Lett. **78**, 2300 (1997). arXiv:hep-ph/9610252
6. Y. Grossman, Y. Nir, M.P. Worah, Phys. Lett. B **407**, 307 (1997). arXiv:hep-ph/9704287
7. M. Bona et al. (UTfit Collaboration), Phys. Rev. Lett. **97**, 151803 (2006). arXiv:hep-ph/0605213
8. J. Charles et al. (CKMfitter Group), Eur. Phys. J. C **41**, 1 (2005). arXiv:hep-ph/0406184
9. M. Bona et al. (UTfit Collaboration), arXiv:0707.0636 [hep-ph]
10. G. D'Ambrosio, G.F. Giudice, G. Isidori, A. Strumia, Nucl. Phys. B **645**, 155 (2002). arXiv:hep-ph/0207036
11. R.S. Chivukula, H. Georgi, Phys. Lett. B **188**, 99 (1987)
12. A.J. Buras, P. Gambino, M. Gorbahn, S. Jager, L. Silvestrini, Phys. Lett. B **500**, 161 (2001). arXiv:hep-ph/0007085
13. A. Ali, D. London, Eur. Phys. J. C **9**, 687 (1999). arXiv:hep-ph/9903535
14. A. Bartl, T. Gajdosik, E. Lunghi, A. Masiero, W. Porod, H. Stremnitzer, O. Vives, Phys. Rev. D **64**, 076009 (2001). arXiv:hep-ph/0103324
15. S. Laplace, Z. Ligeti, Y. Nir, G. Perez, Phys. Rev. D **65**, 094040 (2002). arXiv:hep-ph/0202010
16. A.J. Buras, Acta Phys. Pol. B **34**, 5615 (2003). arXiv:hep-ph/0310208
17. A.J. Buras, A. Poschenrieder, M. Spranger, A. Weiler, Nucl. Phys. B **678**, 455 (2004). arXiv:hep-ph/0306158
18. K. Agashe, M. Papucci, G. Perez, D. Pirjol, arXiv:hep-ph/0509117
19. T. Feldmann, T. Mannel, J. High Energy Phys. **0702**, 067 (2007). arXiv:hep-ph/0611095

20. L.J. Hall, R. Rattazzi, U. Sarid, Phys. Rev. D **50**, 7048 (1994). arXiv:hep-ph/9306309
21. T. Blazek, S. Raby, S. Pokorski, Phys. Rev. D **52**, 4151 (1995). arXiv:hep-ph/9504364
22. M. Carena, D. Garcia, U. Nierste, C.E.M. Wagner, Nucl. Phys. B **577**, 88 (2000). arXiv:hep-ph/9912516
23. G. Degrassi, P. Gambino, G.F. Giudice, J. High Energy Phys. **0012**, 009 (2000). arXiv:hep-ph/0009337
24. M. Carena, D. Garcia, U. Nierste, C.E.M. Wagner, Phys. Lett. B **499**, 141 (2001). arXiv:hep-ph/0010003
25. C. Hamzaoui, M. Pospelov, M. Toharia, Phys. Rev. D **59**, 095005 (1999). arXiv:hep-ph/9807350
26. S.R. Choudhury, N. Gaur, Phys. Lett. B **451**, 86 (1999). arXiv:hep-ph/9810307
27. K.S. Babu, C.F. Kolda, Phys. Rev. Lett. **84**, 228 (2000). arXiv:hep-ph/9909476
28. G. Isidori, A. Retico, J. High Energy Phys. **0111**, 001 (2001). arXiv:hep-ph/0110121
29. A.J. Buras, P.H. Chankowski, J. Rosiek, L. Slawianowska, Nucl. Phys. B **659**, 3 (2003). arXiv:hep-ph/0210145
30. A.J. Buras, P.H. Chankowski, J. Rosiek, L. Slawianowska, Phys. Lett. B **546**, 96 (2002). arXiv:hep-ph/0207241
31. W.S. Hou, Phys. Rev. D **48**, 2342 (1993)
32. G. Isidori, P. Paradisi, Phys. Lett. B **639**, 499 (2006). arXiv:hep-ph/0605012
33. E. Lunghi, W. Porod, O. Vives, Phys. Rev. D **74**, 075003 (2006). arXiv:hep-ph/0605177
34. G. Isidori, F. Mescia, P. Paradisi, D. Temes, arXiv:hep-ph/0703035
35. B. Grinstein, V. Cirigliano, G. Isidori, M.B. Wise, Nucl. Phys. B **763**, 35 (2007). arXiv:hep-ph/0608123
36. V. Cirigliano, B. Grinstein, G. Isidori, M.B. Wise, Nucl. Phys. B **728**, 121 (2005). arXiv:hep-ph/0507001
37. M. Grassi (MEG Collaboration), Nucl. Phys. Proc. Suppl. **149**, 369 (2005)
38. L.J. Hall, L. Randall, Phys. Rev. Lett. **65**, 2939 (1990)
39. M. Dine, A.E. Nelson, Phys. Rev. D **48**, 1277 (1993). arXiv:hep-ph/9303230
40. M. Dine, A.E. Nelson, Y. Shirman, Phys. Rev. D **51**, 1362 (1995). arXiv:hep-ph/9408384
41. G.F. Giudice, R. Rattazzi, Phys. Rep. **322**, 419 (1999). arXiv:hep-ph/9801271
42. W. Altmannshofer, A.J. Buras, D. Guadagnoli, arXiv:hep-ph/0703200
43. J.R. Ellis, D.V. Nanopoulos, Phys. Lett. B **110**, 44 (1982)
44. R. Barbieri, R. Gatto, Phys. Lett. B **110**, 211 (1982)
45. M. Dine, W. Fischler, M. Srednicki, Nucl. Phys. B **189**, 575 (1981)
46. S. Dimopoulos, S. Raby, Nucl. Phys. B **192**, 353 (1981)
47. M. Dine, W. Fischler, Phys. Lett. B **110**, 227 (1982)
48. M. Dine, M. Srednicki, Nucl. Phys. B **202**, 238 (1982)
49. M. Dine, W. Fischler, Nucl. Phys. B **204**, 346 (1982)
50. L. Alvarez-Gaume, M. Claudson, M.B. Wise, Nucl. Phys. B **207**, 96 (1982)
51. C.R. Nappi, B.A. Ovrut, Phys. Lett. B **113**, 175 (1982)
52. S. Dimopoulos, S. Raby, Nucl. Phys. B **219**, 479 (1983)
53. M. Dine, A.E. Nelson, Y. Nir, Y. Shirman, Phys. Rev. D **53**, 2658 (1996). arXiv:hep-ph/9507378
54. E. Poppitz, S.P. Trivedi, Phys. Rev. D **55**, 5508 (1997). arXiv:hep-ph/9609529
55. N. Arkani-Hamed, J. March-Russell, H. Murayama, Nucl. Phys. B **509**, 3 (1998). arXiv:hep-ph/9701286
56. H. Murayama, Phys. Rev. Lett. **79**, 18 (1997). arXiv:hep-ph/9705271
57. S. Dimopoulos, G.R. Dvali, R. Rattazzi, G.F. Giudice, Nucl. Phys. B **510**, 12 (1998). arXiv:hep-ph/9705307
58. S. Dimopoulos, G.R. Dvali, R. Rattazzi, Phys. Lett. B **413**, 336 (1997). arXiv:hep-ph/9707537
59. M.A. Luty, Phys. Lett. B **414**, 71 (1997). arXiv:hep-ph/9706554
60. T. Hotta, K.I. Izawa, T. Yanagida, Phys. Rev. D **55**, 415 (1997). arXiv:hep-ph/9606203
61. L. Randall, Nucl. Phys. B **495**, 37 (1997). arXiv:hep-ph/9612426
62. Y. Shadmi, Phys. Lett. B **405**, 99 (1997). arXiv:hep-ph/9703312
63. N. Haba, N. Maru, T. Matsuoka, Phys. Rev. D **56**, 4207 (1997). arXiv:hep-ph/9703250
64. C. Csaki, L. Randall, W. Skiba, Phys. Rev. D **57**, 383 (1998). arXiv:hep-ph/9707386
65. Y. Shirman, Phys. Lett. B **417**, 281 (1998). arXiv:hep-ph/9709383
66. P. Langacker, N. Polonsky, J. Wang, Phys. Rev. D **60**, 115005 (1999). arXiv:hep-ph/9905252
67. K.S. Babu, Y. Mimura, arXiv:hep-ph/0101046
68. A. Delgado, G.F. Giudice, P. Slavich, arXiv:0706.3873 [hep-ph]
69. M. Ciuchini, A. Masiero, L. Silvestrini, S.K. Vempati, O. Vives, Phys. Rev. Lett. **92**, 071801 (2004). arXiv:hep-ph/0307191
70. M. Ciuchini, A. Masiero, P. Paradisi, L. Silvestrini, S.K. Vempati, O. Vives, arXiv:hep-ph/0702144
71. M. Albrecht, W. Altmannshofer, A.J. Buras, D. Guadagnoli, D.M. Straub, arXiv:0707.3954 [hep-ph]
72. S. Weinberg, Phys. Rev. D **26**, 287 (1982)
73. N. Sakai, T. Yanagida, Nucl. Phys. B **197**, 533 (1982)
74. A.Y. Smirnov, F. Vissani, Phys. Lett. B **380**, 317 (1996). arXiv:hep-ph/9601387
75. C.S. Aulakh, R.N. Mohapatra, Phys. Lett. B **119**, 136 (1982)
76. L.J. Hall, M. Suzuki, Nucl. Phys. B **231**, 419 (1984)
77. I.H. Lee, Nucl. Phys. B **246**, 120 (1984)
78. J.R. Ellis, G. Gelmini, C. Jarlskog, G.G. Ross, J.W.F. Valle, Phys. Lett. B **150**, 142 (1985)
79. L.E. Ibanez, G.G. Ross, Nucl. Phys. B **368**, 3 (1992)
80. V.D. Barger, G.F. Giudice, T. Han, Phys. Rev. D **40**, 2987 (1989)
81. K. Enqvist, A. Masiero, A. Riotto, Nucl. Phys. B **373**, 95 (1992)
82. M. Misiak, S. Pokorski, J. Rosiek, Adv. Ser. Direct. High Energy Phys. **15**, 795 (1998). arXiv:hep-ph/9703442
83. M. Ciuchini, G. Degrassi, P. Gambino, G.F. Giudice, Nucl. Phys. B **527**, 21 (1998). arXiv:hep-ph/9710335
84. M. Ciuchini, G. Degrassi, P. Gambino, G.F. Giudice, Nucl. Phys. B **534**, 3 (1998). arXiv:hep-ph/9806308
85. A. Freitas, E. Gasser, U. Haisch, arXiv:hep-ph/0702267
86. A.J. Buras, P. Gambino, M. Gorbahn, S. Jager, L. Silvestrini, Nucl. Phys. B **592**, 55 (2001). arXiv:hep-ph/0007313
87. R. Barbieri, L.J. Hall, A. Strumia, Nucl. Phys. B **449**, 437 (1995). arXiv:hep-ph/9504373
88. L. Silvestrini, arXiv:0705.1624 [hep-ph]
89. M. Dugan, B. Grinstein, L.J. Hall, Nucl. Phys. B **255**, 413 (1985)
90. S. Dimopoulos, S.D. Thomas, Nucl. Phys. B **465**, 23 (1996). arXiv:hep-ph/9510220
91. M.J. Duncan, J. Trampetic, Phys. Lett. B **134**, 439 (1984)
92. E. Franco, M.L. Mangano, Phys. Lett. B **135**, 445 (1984)
93. J.M. Gerard, W. Grimus, A. Raychaudhuri, G. Zoupanos, Phys. Lett. B **140**, 349 (1984)
94. J.M. Gerard, W. Grimus, A. Masiero, D.V. Nanopoulos, A. Raychaudhuri, Phys. Lett. B **141**, 79 (1984)
95. J.M. Gerard, W. Grimus, A. Masiero, D.V. Nanopoulos, A. Raychaudhuri, Nucl. Phys. B **253**, 93 (1985)
96. P. Langacker, B. Sathiapalan, Phys. Lett. B **144**, 401 (1984)
97. L.J. Hall, V.A. Kostelecky, S. Raby, Nucl. Phys. B **267**, 415 (1986)

98. M.J. Duncan, Nucl. Phys. B **221**, 285 (1983)
99. J.F. Donoghue, H.P. Nilles, D. Wyler, Phys. Lett. B **128**, 55 (1983)
100. A. Bouquet, J. Kaplan, C.A. Savoy, Phys. Lett. B **148**, 69 (1984)
101. F. Gabbiani, A. Masiero, Nucl. Phys. B **322**, 235 (1989)
102. J.S. Hagelin, S. Kelley, T. Tanaka, Nucl. Phys. B **415**, 293 (1994)
103. E. Gabrielli, A. Masiero, L. Silvestrini, Phys. Lett. B **374**, 80 (1996). arXiv:hep-ph/9509379
104. F. Gabbiani, E. Gabrielli, A. Masiero, L. Silvestrini, Nucl. Phys. B **477**, 321 (1996). arXiv:hep-ph/9604387
105. D. Becirevic et al., Nucl. Phys. B **634**, 105 (2002). arXiv:hep-ph/0112303
106. M. Ciuchini et al., J. High Energy Phys. **9810**, 008 (1998). arXiv:hep-ph/9808328
107. M. Ciuchini, E. Franco, A. Masiero, L. Silvestrini, Phys. Rev. D **67**, 075016 (2003) [Erratum: Phys. Rev. D **68**, 079901 (2003)]. arXiv:hep-ph/0212397
108. T. Besmer, C. Greub, T. Hurth, Nucl. Phys. B **609**, 359 (2001). arXiv:hep-ph/0105292
109. G.L. Kane, P. Ko, H.b. Wang, C. Kolda, J.h. Park, L.T. Wang, Phys. Rev. D **70**, 035015 (2004). arXiv:hep-ph/0212092
110. R. Harnik, D.T. Larson, H. Murayama, A. Pierce, Phys. Rev. D **69**, 094024 (2004). arXiv:hep-ph/0212180
111. K.I. Okumura, L. Roszkowski, J. High Energy Phys. **0310**, 024 (2003). arXiv:hep-ph/0308102
112. J. Foster, K.I. Okumura, L. Roszkowski, Phys. Lett. B **609**, 102 (2005). arXiv:hep-ph/0410323
113. J. Foster, K.I. Okumura, L. Roszkowski, J. High Energy Phys. **0508**, 094 (2005). arXiv:hep-ph/0506146
114. J. Foster, K.I. Okumura, L. Roszkowski, J. High Energy Phys. **0603**, 044 (2006). arXiv:hep-ph/0510422
115. J. Foster, K. Okumura, L. Roszkowski, Phys. Lett. B **641**, 452 (2006). arXiv:hep-ph/0604121
116. M. Ciuchini, L. Silvestrini, Phys. Rev. Lett. **97**, 021803 (2006). arXiv:hep-ph/0603114
117. M. Ciuchini, E. Franco, V. Lubicz, G. Martinelli, I. Scimemi, L. Silvestrini, Nucl. Phys. B **523**, 501 (1998). arXiv:hep-ph/9711402
118. M. Ciuchini, E. Franco, D. Guadagnoli, V. Lubicz, V. Porretti, L. Silvestrini, J. High Energy Phys. **0609**, 013 (2006). arXiv:hep-ph/0606197
119. W.M. Yao et al. (Particle Data Group), J. Phys. G **33**, 1 (2006)
120. M. Bona et al. (UTfit Collaboration), J. High Energy Phys. **0610**, 081 (2006). arXiv:hep-ph/0606167
121. S. Chen et al. (CLEO Collaboration), Phys. Rev. Lett. **87**, 251807 (2001). arXiv:hep-ex/0108032
122. P. Koppenburg et al. (Belle Collaboration), Phys. Rev. Lett. **93**, 061803 (2004). arXiv:hep-ex/0403004
123. B. Aubert et al. (BaBar Collaboration), Phys. Rev. D **72**, 052004 (2005). arXiv:hep-ex/0508004
124. B. Aubert et al. (BaBar Collaboration), Phys. Rev. Lett. **93**, 081802 (2004). arXiv:hep-ex/0404006
125. M. Iwasaki et al. (Belle Collaboration), Phys. Rev. D **72**, 092005 (2005). arXiv:hep-ex/0503044
126. A. Abulencia et al. (CDF Collaboration), arXiv:hep-ex/0609040
127. D. Becirevic, V. Gimenez, G. Martinelli, M. Papinutto, J. Reyes, J. High Energy Phys. **0204**, 025 (2002). arXiv:hep-lat/0110091
128. C.R. Allton et al., Phys. Lett. B **453**, 30 (1999). arXiv:hep-lat/9806016
129. R. Babich, N. Garron, C. Hoelbling, J. Howard, L. Lellouch, C. Rebbi, Phys. Rev. D **74**, 073009 (2006). arXiv:hep-lat/0605016
130. Y. Nakamura et al. (CP-PACS Collaboration), arXiv:hep-lat/0610075
131. A.J. Buras, G. Colangelo, G. Isidori, A. Romanino, L. Silvestrini, Nucl. Phys. B **566**, 3 (2000). arXiv:hep-ph/9908371
132. M. Ciuchini et al., J. High Energy Phys. **0107**, 013 (2001). arXiv:hep-ph/0012308
133. N. Arkani-Hamed, A.G. Cohen, H. Georgi, Phys. Lett. B **513**, 232 (2001). arXiv:hep-ph/0105239
134. D.E. Kaplan, M. Schmaltz, J. High Energy Phys. **0310**, 039 (2003). arXiv:hep-ph/0302049
135. M. Schmaltz, J. High Energy Phys. **0408**, 056 (2004). arXiv:hep-ph/0407143
136. N. Arkani-Hamed, A.G. Cohen, E. Katz, A.E. Nelson, T. Gregoire, J.G. Wacker, J. High Energy Phys. **0208**, 021 (2002). arXiv:hep-ph/0206020
137. N. Arkani-Hamed, A.G. Cohen, E. Katz, A.E. Nelson, J. High Energy Phys. **0207**, 034 (2002). arXiv:hep-ph/0206021
138. T. Han, H.E. Logan, B. McElrath, L.T. Wang, Phys. Rev. D **67**, 095004 (2003). arXiv:hep-ph/0301040
139. C. Csaki, J. Hubisz, G.D. Kribs, P. Meade, J. Terning, Phys. Rev. D **67**, 115002 (2003). arXiv:hep-ph/0211124
140. A.J. Buras, A. Poschenrieder, S. Uhlig, Nucl. Phys. B **716**, 173 (2005). arXiv:hep-ph/0410309
141. A.J. Buras, A. Poschenrieder, S. Uhlig, arXiv:hep-ph/0501230
142. A.J. Buras, A. Poschenrieder, S. Uhlig, W.A. Bardeen, J. High Energy Phys. **0611**, 062 (2006). arXiv:hep-ph/0607189
143. S.R. Choudhury, N. Gaur, A. Goyal, N. Mahajan, Phys. Lett. B **601**, 164 (2004). arXiv:hep-ph/0407050
144. J.Y. Lee, J. High Energy Phys. **0412**, 065 (2004). arXiv:hep-ph/0408362
145. S. Fajfer, S. Prelovsek, Phys. Rev. D **73**, 054026 (2006). arXiv:hep-ph/0511048
146. W.J. Huo, S.H. Zhu, Phys. Rev. D **68**, 097301 (2003). arXiv:hep-ph/0306029
147. H.C. Cheng, I. Low, J. High Energy Phys. **0309**, 051 (2003). arXiv:hep-ph/0308199
148. H.C. Cheng, I. Low, J. High Energy Phys. **0408**, 061 (2004). arXiv:hep-ph/0405243
149. J. Hubisz, P. Meade, A. Noble, M. Perelstein, J. High Energy Phys. **0601**, 135 (2006). arXiv:hep-ph/0506042
150. I. Low, J. High Energy Phys. **0410**, 067 (2004). arXiv:hep-ph/0409025
151. N. Cabibbo, Phys. Rev. Lett. **10**, 531 (1963)
152. M. Kobayashi, T. Maskawa, Prog. Theor. Phys. **49**, 652 (1973)
153. J. Hubisz, S.J. Lee, G. Paz, J. High Energy Phys. **0606**, 041 (2006). arXiv:hep-ph/0512169
154. B. Pontecorvo, Sov. Phys. JETP **6**, 429 (1957) [Zh. Eksp. Teor. Fiz. **33**, 549 (1957)]
155. B. Pontecorvo, Sov. Phys. JETP **7**, 172 (1958) [Zh. Eksp. Teor. Fiz. **34**, 247 (1957)]
156. Z. Maki, M. Nakagawa, S. Sakata, Prog. Theor. Phys. **28**, 870 (1962)
157. M. Blanke, A.J. Buras, A. Poschenrieder, S. Recksiegel, C. Tarantino, S. Uhlig, A. Weiler, Phys. Lett. B **646**, 253 (2007). arXiv:hep-ph/0609284
158. M. Blanke, A.J. Buras, A. Poschenrieder, S. Recksiegel, C. Tarantino, S. Uhlig, A. Weiler, J. High Energy Phys. **0701**, 066 (2007). arXiv:hep-ph/0610298
159. M. Blanke, A.J. Buras, A. Poschenrieder, C. Tarantino, S. Uhlig, A. Weiler, J. High Energy Phys. **0612**, 003 (2006). arXiv:hep-ph/0605214
160. S.R. Choudhury, A.S. Cornell, A. Deandrea, N. Gaur, A. Goyal, arXiv:hep-ph/0612327
161. M. Blanke, A.J. Buras, B. Duling, A. Poschenrieder, C. Tarantino, arXiv:hep-ph/0702136
162. D. Lucchesi (CDF and D0 Collaborations), Nucl. Phys. Proc. Suppl. **163**, 165 (2007)
163. C. Promberger, S. Schatt, F. Schwab, arXiv:hep-ph/0702169

164. S. Eidelman et al. (Particle Data Group), Phys. Lett. B **592**, 1 (2004)
165. P. Paradisi, J. High Energy Phys. **0510**, 006 (2005). arXiv:hep-ph/0505046
166. P. Paradisi, J. High Energy Phys. **0602**, 050 (2006). arXiv:hep-ph/0508054
167. P. Paradisi, J. High Energy Phys. **0608**, 047 (2006). arXiv:hep-ph/0601100
168. T. Kaluza, Sitzungsber. Preuss. Akad. Wiss. Berlin (Math. Phys.) **1921**, 966 (1921)
169. O. Klein, Z. Phys. **37**, 895 (1926) [Surv. High Energy Phys. **5**, 241 (1986)]
170. I. Antoniadis, Phys. Lett. B **246**, 377 (1990)
171. J.D. Lykken, Phys. Rev. D **54**, 3693 (1996). arXiv:hep-th/9603133
172. E. Witten, Nucl. Phys. B **471**, 135 (1996). arXiv:hep-th/9602070
173. P. Horava, E. Witten, Nucl. Phys. B **475**, 94 (1996). arXiv:hep-th/9603142
174. P. Horava, E. Witten, Nucl. Phys. B **460**, 506 (1996). arXiv:hep-th/9510209
175. E. Caceres, V.S. Kaplunovsky, I.M. Mandelberg, Nucl. Phys. B **493**, 73 (1997). arXiv:hep-th/9606036
176. N. Arkani-Hamed, S. Dimopoulos, G.R. Dvali, Phys. Lett. B **429**, 263 (1998). arXiv:hep-ph/9803315
177. I. Antoniadis, N. Arkani-Hamed, S. Dimopoulos, G.R. Dvali, Phys. Lett. B **436**, 257 (1998). arXiv:hep-ph/9804398
178. N. Arkani-Hamed, S. Dimopoulos, G.R. Dvali, Phys. Rev. D **59**, 086004 (1999). arXiv:hep-ph/9807344
179. L. Randall, R. Sundrum, Phys. Rev. Lett. **83**, 3370 (1999). arXiv:hep-ph/9905221
180. T. Appelquist, H.C. Cheng, B.A. Dobrescu, Phys. Rev. D **64**, 035002 (2001). arXiv:hep-ph/0012100
181. A.J. Buras, M. Spranger, A. Weiler, Nucl. Phys. B **660**, 225 (2003). arXiv:hep-ph/0212143
182. P. Colangelo, F. De Fazio, R. Ferrandes, T.N. Pham, Phys. Rev. D **73**, 115006 (2006). arXiv:hep-ph/0604029
183. P. Colangelo, F. De Fazio, R. Ferrandes, T.N. Pham, Phys. Rev. D **74**, 115006 (2006). arXiv:hep-ph/0610044
184. T. Inami, C.S. Lim, Prog. Theor. Phys. **65**, 297 (1981) [Erratum: Prog. Theor. Phys. **65**, 1772 (1981)]
185. P. Skands et al., J. High Energy Phys. **0407**, 036 (2004). arXiv:hep-ph/0311123
186. B.C. Allanach et al., arXiv:0801.0045 [hep-ph]
187. L. Silvestrini, Int. J. Mod. Phys. A **21**, 1738 (2006). arXiv:hep-ph/0510077
188. P. Gambino, U. Haisch, M. Misiak, Phys. Rev. Lett. **94**, 061803 (2005). arXiv:hep-ph/0410155
189. C. Bobeth, A.J. Buras, T. Ewerth, Nucl. Phys. B **713**, 522 (2005). arXiv:hep-ph/0409293
190. C. Bobeth, M. Bona, A.J. Buras, T. Ewerth, M. Pierini, L. Silvestrini, A. Weiler, Nucl. Phys. B **726**, 252 (2005). arXiv:hep-ph/0505110
191. G. Degrassi, P. Gambino, P. Slavich, arXiv:0712.3265 [hep-ph]
192. A.J. Buras, T. Ewerth, S. Jager, J. Rosiek, Nucl. Phys. B **714**, 103 (2005). arXiv:hep-ph/0408142
193. S. Bejar, J. Guasch, J. Sola, Nucl. Phys. B **600**, 21 (2001). arXiv:hep-ph/0011091
194. S. Bejar, J. Guasch, J. Sola, in *Proceedings of the 5th International Symposium on Radiative Corrections (RADCOR 2000)*, ed. by H.E. Haber. arXiv:hep-ph/0101294
195. S. Bejar, J. Guasch, J. Sola, Nucl. Phys. B **675**, 270 (2003). arXiv:hep-ph/0307144
196. S. Bejar, F. Dilme, J. Guasch, J. Sola, J. High Energy Phys. **0408**, 018 (2004). arXiv:hep-ph/0402188
197. S. Bejar, J. Guasch, J. Sola, J. High Energy Phys. **0510**, 113 (2005). arXiv:hep-ph/0508043
198. S. Bejar, J. Guasch, J. Sola, Nucl. Phys. Proc. Suppl. **157**, 147 (2006). arXiv:hep-ph/0601191
199. S. Heinemeyer, W. Hollik, G. Weiglein, Eur. Phys. J. C **9**, 343 (1999). arXiv:hep-ph/9812472
200. G. Degrassi, S. Heinemeyer, W. Hollik, P. Slavich, G. Weiglein, Eur. Phys. J. C **28**, 133 (2003). arXiv:hep-ph/0212020
201. S. Heinemeyer, W. Hollik, F. Merz, S. Penaranda, Eur. Phys. J. C **37**, 481 (2004). arXiv:hep-ph/0403228
202. T. Hahn, M. Perez-Victoria, Comput. Phys. Commun. **118**, 153 (1999). arXiv:hep-ph/9807565
203. T. Hahn, Comput. Phys. Commun. **140**, 418 (2001). arXiv:hep-ph/0012260
204. T. Hahn, W. Hollik, J.I. Illana, S. Penaranda, arXiv:hep-ph/0512315
205. T. Hahn, arXiv:hep-ph/0605049
206. B.C. Allanach, Comput. Phys. Commun. **143**, 305 (2002). arXiv:hep-ph/0104145
207. W. Porod, Comput. Phys. Commun. **153**, 275 (2003). arXiv:hep-ph/0301101
208. O. Buchmüller, private communication
209. M. Bona et al. (UTfit Collaboration), J. High Energy Phys. **0507**, 028 (2005). arXiv:hep-ph/0501199
210. M. Bona et al. (UTfit Collaboration), J. High Energy Phys. **0603**, 080 (2006). arXiv:hep-ph/0509219
211. M. Bona et al. (UTfit Collaboration), arXiv:hep-ph/0701204
212. R. Brun, F. Rademakers, Nucl. Instrum. Methods A **389**, 81 (1997)
213. ckmfitter.in2p3.fr
214. G. Buchalla, A.J. Buras, M.E. Lautenbacher, Rev. Mod. Phys. **68**, 1125 (1996). arXiv:hep-ph/9512380
215. M. Beneke, G. Buchalla, M. Neubert, C.T. Sachrajda, Phys. Rev. Lett. **83**, 1914 (1999). arXiv:hep-ph/9905312
216. M. Beneke, G. Buchalla, M. Neubert, C.T. Sachrajda, Nucl. Phys. B **591**, 313 (2000). arXiv:hep-ph/0006124
217. C.W. Bauer, S. Fleming, D. Pirjol, I.W. Stewart, Phys. Rev. D **63**, 114020 (2001). arXiv:hep-ph/0011336
218. C.W. Bauer, D. Pirjol, I.W. Stewart, Phys. Rev. D **65**, 054022 (2002). arXiv:hep-ph/0109045
219. J. Chay, C. Kim, Phys. Rev. D **68**, 071502 (2003). arXiv:hep-ph/0301055
220. J. Chay, C. Kim, Nucl. Phys. B **680**, 302 (2004). arXiv:hep-ph/0301262
221. C.W. Bauer, D. Pirjol, I.Z. Rothstein, I.W. Stewart, Phys. Rev. D **70**, 054015 (2004). arXiv:hep-ph/0401188
222. Y.Y. Keum, H.N. Li, A.I. Sanda, Phys. Rev. D **63**, 054008 (2001). arXiv:hep-ph/0004173
223. H.N. Li, S. Mishima, A.I. Sanda, Phys. Rev. D **72**, 114005 (2005). arXiv:hep-ph/0508041
224. M. Beneke, G. Buchalla, M. Neubert, C.T. Sachrajda, Nucl. Phys. B **606**, 245 (2001). arXiv:hep-ph/0104110
225. R.J. Hill, T. Becher, S.J. Lee, M. Neubert, J. High Energy Phys. **0407**, 081 (2004). arXiv:hep-ph/0404217
226. T. Becher, R.J. Hill, J. High Energy Phys. **0410**, 055 (2004). arXiv:hep-ph/0408344
227. G.G. Kirilin, arXiv:hep-ph/0508235
228. M. Beneke, D. Yang, Nucl. Phys. B **736**, 34 (2006). arXiv:hep-ph/0508250
229. M. Beneke, S. Jäger, Nucl. Phys. B **751**, 160 (2006). arXiv:hep-ph/0512351
230. P. Colangelo, A. Khodjamirian, arXiv:hep-ph/0010175
231. P. Ball, R. Zwicky, Phys. Rev. D **71**, 014015 (2005). arXiv:hep-ph/0406232
232. P. Ball, R. Zwicky, Phys. Rev. D **71**, 014029 (2005). arXiv:hep-ph/0412079
233. P. Ball, R. Zwicky, Phys. Lett. B **633**, 289 (2006). arXiv:hep-ph/0510338

234. F. De Fazio, T. Feldmann, T. Hurth, Nucl. Phys. B **733**, 1 (2006). arXiv:hep-ph/0504088
235. J. Chay, C. Kim, A.K. Leibovich, Phys. Lett. B **628**, 57 (2005). arXiv:hep-ph/0508157
236. A. Khodjamirian, T. Mannel, N. Offen, Phys. Lett. B **620**, 52 (2005). arXiv:hep-ph/0504091
237. S.J. Lee, M. Neubert, Phys. Rev. D **72**, 094028 (2005). arXiv:hep-ph/0509350
238. V.M. Braun, D.Y. Ivanov, G.P. Korchemsky, Phys. Rev. D **69**, 034014 (2004). arXiv:hep-ph/0309330
239. M. Beneke, M. Neubert, Nucl. Phys. B **651**, 225 (2003). arXiv:hep-ph/0210085
240. M. Beneke, M. Neubert, Nucl. Phys. B **675**, 333 (2003). arXiv:hep-ph/0308039
241. A.R. Williamson, J. Zupan, Phys. Rev. D **74**, 014003 (2006). arXiv:hep-ph/0601214
242. M. Beneke, G. Buchalla, M. Neubert, C.T. Sachrajda, Phys. Rev. D **72**, 098501 (2005). arXiv:hep-ph/0411171
243. A. Khodjamirian, T. Mannel, M. Melcher, B. Melic, Phys. Rev. D **72**, 094012 (2005). arXiv:hep-ph/0509049
244. A. Khodjamirian, arXiv:hep-ph/0607347
245. D. Zeppenfeld, Z. Phys. C **8**, 77 (1981)
246. M. Gronau, D. London, Phys. Rev. Lett. **65**, 3381 (1990)
247. Y. Nir, H.R. Quinn, Phys. Rev. Lett. **67**, 541 (1991)
248. J. Charles et al., arXiv:hep-ph/0607246
249. E. Baracchini, G. Isidori, Phys. Lett. B **633**, 309 (2006). arXiv:hep-ph/0508071
250. A. Khodjamirian, T. Mannel, M. Melcher, Phys. Rev. D **68**, 114007 (2003). arXiv:hep-ph/0308297
251. M. Beneke, T. Feldmann, Nucl. Phys. B **685**, 249 (2004). arXiv:hep-ph/0311335
252. N. Kivel, J. High Energy Phys. **0705**, 019 (2007). arXiv:hep-ph/0608291
253. V. Pilipp, Nucl. Phys. B **794**, 154 (2008). arXiv:0709.3214 [hep-ph]
254. V. Pilipp, arXiv:0709.0497 [hep-ph]
255. G. Bell, Nucl. Phys. B **795**, 1 (2008). arXiv:0705.3127 [hep-ph]
256. G. Bell, PhD thesis, LMU München, 2006. arXiv:0705.3133 [hep-ph]
257. M. Beneke, S. Jäger, Nucl. Phys. B **768**, 51 (2007). arXiv:hep-ph/0610322
258. B. Aubert et al. (BaBar Collaboration), arXiv:hep-ex/0608003
259. Y. Chao et al. (Belle Collaboration), PoS **HEP2005**, 256 (2006)
260. P. Golonka, Z. Was, Eur. Phys. J. C **45**, 97 (2006). arXiv:hep-ph/0506026
261. V.L. Chernyak, A.R. Zhitnitsky, Phys. Rep. **112**, 173 (1984)
262. V.M. Braun, arXiv:hep-ph/0608231
263. P. Ball, V.M. Braun, N. Kivel, Nucl. Phys. B **649**, 263 (2003). arXiv:hep-ph/0207307
264. A.G. Grozin, Int. J. Mod. Phys. A **20**, 7451 (2005). arXiv:hep-ph/0506226
265. P. Ball, R. Zwicky, arXiv:hep-ph/0609037
266. P. Ball, R. Zwicky, J. High Energy Phys. **0604**, 046 (2006). arXiv:hep-ph/0603232
267. V.M. Braun et al., Phys. Rev. D **68**, 054501 (2003). arXiv:hep-lat/0306006
268. D. Becirevic, V. Lubicz, F. Mescia, C. Tarantino, J. High Energy Phys. **0305**, 007 (2003). arXiv:hep-lat/0301020
269. P. Ball, V.M. Braun, Phys. Rev. D **54**, 2182 (1996). arXiv:hep-ph/9602323
270. P. Ball, G.W. Jones, R. Zwicky, Phys. Rev. D **75**, 054004 (2007). arXiv:hep-ph/0612081
271. P. Ball, M. Boglione, Phys. Rev. D **68**, 094006 (2003). arXiv:hep-ph/0307337
272. A. Khodjamirian, T. Mannel, M. Melcher, Phys. Rev. D **70**, 094002 (2004). arXiv:hep-ph/0407226
273. V.M. Braun, A. Lenz, Phys. Rev. D **70**, 074020 (2004). arXiv:hep-ph/0407282
274. P. Ball, R. Zwicky, J. High Energy Phys. **0602**, 034 (2006). arXiv:hep-ph/0601086
275. P.A. Boyle et al. (UKQCD Collaboration), Phys. Lett. B **641**, 67 (2006). arXiv:hep-lat/0607018
276. V.M. Braun et al., arXiv:hep-lat/0606012
277. P. Ball, V.M. Braun, A. Lenz, J. High Energy Phys. **0605**, 004 (2006). arXiv:hep-ph/0603063
278. P. Ball, G.W. Jones, arXiv:hep-ph/0702100
279. P. Ball, A.N. Talbot, J. High Energy Phys. **0506**, 063 (2005). arXiv:hep-ph/0502115
280. D. Becirevic, A.B. Kaidalov, Phys. Lett. B **478**, 417 (2000). arXiv:hep-ph/9904490
281. P. Ball, R. Zwicky, Phys. Lett. B **625**, 225 (2005). arXiv:hep-ph/0507076
282. S.W. Bosch, G. Buchalla, J. High Energy Phys. **0501**, 035 (2005). arXiv:hep-ph/0408231
283. M. Beneke, T. Feldmann, D. Seidel, Eur. Phys. J. C **41**, 173 (2005). arXiv:hep-ph/0412400
284. P. Ball, R. Zwicky, arXiv:hep-ph/0608009
285. R.N. Faustov, V.O. Galkin, Z. Phys. C **66**, 119 (1995)
286. D. Melikhov, Phys. Rev. D **56**, 7089 (1997). arXiv:hep-ph/9706417
287. W. Lucha, D. Melikhov, S. Simula, Phys. Rev. D **74**, 054004 (2006). arXiv:hep-ph/0606281
288. S. Godfrey, N. Isgur, Phys. Rev. D **32**, 189 (1985)
289. D. Ebert, V.O. Galkin, R.N. Faustov, Phys. Rev. D **57**, 5663 (1998). [Erratum: Phys. Rev. D **59**, 019902 (1999)]. arXiv:hep-ph/9712318
290. M. Wirbel, B. Stech, M. Bauer, Z. Phys. C **29**, 637 (1985)
291. N. Isgur, D. Scora, B. Grinstein, M.B. Wise, Phys. Rev. D **39**, 799 (1989)
292. D. Scora, N. Isgur, Phys. Rev. D **52**, 2783 (1995). arXiv:hep-ph/9503486
293. D. Melikhov, Phys. Rev. D **53**, 2460 (1996). arXiv:hep-ph/9509268
294. D. Melikhov, M. Beyer, Phys. Lett. B **452**, 121 (1999). arXiv:hep-ph/9901261
295. D. Melikhov, B. Stech, Phys. Rev. D **62**, 014006 (2000). arXiv:hep-ph/0001113
296. M. Beyer, D. Melikhov, Phys. Lett. B **436**, 344 (1998). arXiv:hep-ph/9807223
297. M. Beyer, D. Melikhov, N. Nikitin, B. Stech, Phys. Rev. D **64**, 094006 (2001). arXiv:hep-ph/0106203
298. F. Krüger, D. Melikhov, Phys. Rev. D **67**, 034002 (2003). arXiv:hep-ph/0208256
299. D. Melikhov, N. Nikitin, Phys. Rev. D **70**, 114028 (2004). arXiv:hep-ph/0410146
300. M.A. Ivanov, J.G. Körner, S.G. Kovalenko, C.D. Roberts, Phys. Rev. D **76**, 034018 (2007). arXiv:nucl-th/0703094
301. M. Okamoto, PoS **LAT2005**, 013 (2006). arXiv:hep-lat/0510113
302. T. Onogi, PoS **LAT2006**, 017 (2006). arXiv:hep-lat/0610115
303. M. Della Morte, PoS **LAT2007**, 008 (2007). arXiv:0711.3160
304. C. Dawson, PoS **LAT2005**, 007 (2006).
305. W. Lee, PoS **LAT2006**, 015 (2006). arXiv:hep-lat/0610058
306. A. Jüttner, PoS **LAT2007**, 014 (2007). arXiv:0711.1239
307. C. Allton et al. (RBC/UKQCD Collaboration), arXiv:hep-lat/0701013
308. S.R. Beane, P.F. Bedaque, K. Orginos, M.J. Savage, arXiv:hep-lat/0606023
309. C. Bernard et al. (MILC Collaboration), arXiv:hep-lat/0609053
310. E. Follana, C.T.H. Davies, G.P. Lepage, J. Shigemitsu, arXiv:0706.1726 [hep-lat]
311. A. Jüttner, J. Rolf (ALPHA Collaboration), Phys. Lett. B **560**, 59 (2003). arXiv:hep-lat/0302016

312. G.M. de Divitiis et al., Nucl. Phys. B **672**, 372 (2003). arXiv:hep-lat/0307005
313. C. Aubin et al., Phys. Rev. Lett. **95**, 122002 (2005). arXiv:hep-lat/0506030
314. H.W. Lin, S. Ohta, A. Soni, N. Yamada, Phys. Rev. D **74**, 114506 (2006). arXiv:hep-lat/0607035
315. M. Della Morte et al. (ALPHA Collaboration), Phys. Lett. B **581**, 93 (2004) [Erratum: Phys. Lett. B **612**, 313 (2005)]. arXiv:hep-lat/0307021
316. D. Guazzini, R. Sommer, N. Tantalo, arXiv:hep-lat/0609065
317. A. Ali Khan et al. (CP-PACS Collaboration), Phys. Rev. D **64**, 054504 (2001). arXiv:hep-lat/0103020
318. C. Bernard et al. (MILC Collaboration), Phys. Rev. D **66**, 094501 (2002). arXiv:hep-lat/0206016
319. M. Wingate et al., Phys. Rev. Lett. **92**, 162001 (2004). arXiv:hep-ph/0311130
320. S. Aoki et al. (JLQCD Collaboration), Phys. Rev. Lett. **91**, 212001 (2003). arXiv:hep-ph/0307039
321. A. Gray et al. (HPQCD Collaboration), Phys. Rev. Lett. **95**, 212001 (2005). arXiv:hep-lat/0507015
322. V. Gadiyak, O. Loktik, Phys. Rev. D **72**, 114504 (2005). arXiv:hep-lat/0509075
323. M. Artuso et al. (CLEO Collaboration), Phys. Rev. Lett. **95**, 251801 (2005). arXiv:hep-ex/0508057
324. M. Artuso et al. (CLEO Collaboration), arXiv:hep-ex/0607074
325. B. Aubert et al. (BaBar Collaboration), arXiv:hep-ex/0607094
326. K. Ikado et al., Phys. Rev. Lett. **97**, 251802 (2006). arXiv:hep-ex/0604018
327. D. Becirevic et al., Nucl. Phys. B **705**, 339 (2005). arXiv:hep-ph/0403217
328. N. Tsutsui et al. (JLQCD Collaboration), PoS **LAT2005**, 357 (2006). arXiv:hep-lat/0510068
329. C. Dawson et al., Phys. Rev. D **74**, 114502 (2006). arXiv:hep-ph/0607162
330. D.J. Antonio et al., arXiv:hep-lat/0610080
331. M. Okamoto (Fermilab Lattice Collaboration), arXiv:hep-lat/0412044
332. C. Aubin et al. (Fermilab Lattice Collaboration), Phys. Rev. Lett. **94**, 011601 (2005). arXiv:hep-ph/0408306
333. M. Okamoto et al., Nucl. Phys. Proc. Suppl. **140**, 461 (2005). arXiv:hep-lat/0409116
334. E. Dalgic et al., Phys. Rev. D **73**, 074502 (2006) [Erratum, to appear]. arXiv:hep-lat/0601021
335. G.M. de Divitiis, E. Molinaro, R. Petronzio, N. Tantalo, arXiv:0707.0582 [hep-lat]
336. K.C. Bowler et al. (UKQCD Collaboration), Phys. Lett. B **486**, 111 (2000). arXiv:hep-lat/9911011
337. A. Abada et al., Nucl. Phys. B **619**, 565 (2001). arXiv:hep-lat/0011065
338. A.X. El-Khadra et al., Phys. Rev. D **64**, 014502 (2001). arXiv:hep-ph/0101023
339. S. Aoki et al. (JLQCD Collaboration), Phys. Rev. D **64**, 114505 (2001). arXiv:hep-lat/0106024
340. M. Ablikim et al. (BES Collaboration), Phys. Lett. B **597**, 39 (2004). arXiv:hep-ex/0406028
341. J.M. Link et al. (FOCUS Collaboration), Phys. Lett. B **607**, 233 (2005). arXiv:hep-ex/0410037
342. A. Ali Khan et al. (CP-PACS Collaboration), Phys. Rev. D **64**, 114506 (2001). arXiv:hep-lat/0105020
343. T.A. DeGrand (MILC Collaboration), Phys. Rev. D **69**, 014504 (2004). arXiv:hep-lat/0309026
344. D. Becirevic et al., Eur. Phys. J. C **37**, 315 (2004). arXiv:hep-lat/0407004
345. Y. Aoki et al., Phys. Rev. D **73**, 094507 (2006). arXiv:hep-lat/0508011
346. P. Dimopoulos et al. (ALPHA Collaboration), Nucl. Phys. B **749**, 69 (2006). arXiv:hep-lat/0601002
347. J.M. Flynn, F. Mescia, A.S.B. Tariq (UKQCD Collaboration), J. High Energy Phys. **0411**, 049 (2004). arXiv:hep-lat/0406013
348. Y. Aoki et al., Phys. Rev. D **72**, 114505 (2005). arXiv:hep-lat/0411006
349. E. Gamiz et al. (HPQCD Collaboration), Phys. Rev. D **73**, 114502 (2006). arXiv:hep-lat/0603023
350. D.J. Antonio et al., arXiv:hep-ph/0702042
351. D. Becirevic et al., PoS **LAT2005**, 218 (2006). arXiv:hep-lat/0509165
352. E. Dalgic et al., arXiv:hep-lat/0610104
353. A. Donini et al., Phys. Lett. B **470**, 233 (1999). arXiv:hep-lat/9910017
354. D. Becirevic, G. Villadoro, Phys. Rev. D **70**, 094036 (2004). arXiv:hep-lat/0408029
355. D. Becirevic et al. (SPQcdR Collaboration), Phys. Lett. B **501**, 98 (2001). arXiv:hep-ph/0010349
356. H. Neuberger, Phys. Lett. B **417**, 141 (1998). arXiv:hep-lat/9707022
357. H. Neuberger, Phys. Lett. B **427**, 353 (1998). arXiv:hep-lat/9801031
358. P.H. Ginsparg, K.G. Wilson, Phys. Rev. D **25**, 2649 (1982)
359. M. Lüscher, Phys. Lett. B **428**, 342 (1998). arXiv:hep-lat/9802011
360. C. Gattringer et al. (BGR Collaboration), Nucl. Phys. B **677**, 3 (2004). arXiv:hep-lat/0307013
361. R. Sommer, Nucl. Phys. B **411**, 839 (1994). arXiv:hep-lat/9310022
362. J. Gasser, H. Leutwyler, Ann. Phys. **158**, 142 (1984)
363. J. Gasser, H. Leutwyler, Nucl. Phys. B **250**, 465 (1985)
364. C. Aubin, C. Bernard, Phys. Rev. D **68**, 034014 (2003). arXiv:hep-lat/0304014
365. C. Aubin, C. Bernard, Phys. Rev. D **68**, 074011 (2003). arXiv:hep-lat/0306026
366. C.W. Bernard, M.F.L. Golterman, Phys. Rev. D **49**, 486 (1994). arXiv:hep-lat/9306005
367. S.R. Sharpe, Phys. Rev. D **56**, 7052 (1997) [Erratum: Phys. Rev. D **62**, 099901 (2000)]. arXiv:hep-lat/9707018
368. S.R. Sharpe, N. Shoresh, Phys. Rev. D **62**, 094503 (2000). arXiv:hep-lat/0006017
369. S. Dürr, PoS **LAT2005**, 021 (2006). arXiv:hep-lat/0509026
370. S.R. Sharpe, PoS **LAT2006**, 022 (2006). arXiv:hep-lat/0610094
371. M. Creutz, PoS **LAT2007**, 007 (2007). arXiv:0708.1295
372. A.S. Kronfeld, PoS **LAT2007**, 016 (2007). arXiv:0711.0699
373. M.A. Clark, PoS **LAT2006**, 004 (2006). arXiv:hep-lat/0610048
374. J.M. Flynn, J. Nieves, Phys. Rev. D **75**, 013008 (2007). arXiv:hep-ph/0607258
375. S. Bertolini, F. Borzumati, A. Masiero, Phys. Rev. Lett. **59**, 180 (1987)
376. M. Misiak et al., Phys. Rev. Lett. **98**, 022002 (2007). arXiv:hep-ph/0609232
377. C. Bobeth, M. Misiak, J. Urban, Nucl. Phys. B **574**, 291 (2000). arXiv:hep-ph/9910220
378. M. Misiak, M. Steinhauser, Nucl. Phys. B **683**, 277 (2004). arXiv:hep-ph/0401041
379. M. Gorbahn, U. Haisch, Nucl. Phys. B **713**, 291 (2005). arXiv:hep-ph/0411071
380. M. Gorbahn, U. Haisch, M. Misiak, Phys. Rev. Lett. **95**, 102004 (2005). arXiv:hep-ph/0504194
381. M. Czakon, U. Haisch, M. Misiak, J. High Energy Phys. **0703**, 008 (2007). arXiv:hep-ph/0612329
382. K. Melnikov, A. Mitov, Phys. Lett. B **620**, 69 (2005). arXiv:hep-ph/0505097
383. I. Blokland, et al., Phys. Rev. D **72**, 033014 (2005). arXiv:hep-ph/0506055
384. H.M. Asatrian et al., Nucl. Phys. B **749**, 325 (2006). arXiv:hep-ph/0605009

385. H.M. Asatrian et al., Nucl. Phys. B **762**, 212 (2007). arXiv:hep-ph/0607316
386. H.M. Asatrian et al., arXiv:hep-ph/0611123
387. K. Bieri, C. Greub, M. Steinhauser, Phys. Rev. D **67**, 114019 (2003). arXiv:hep-ph/0302051
388. M. Misiak, M. Steinhauser, Nucl. Phys. B **764**, 62 (2007). arXiv:hep-ph/0609241
389. E. Barberio et al. (Heavy Flavor Averaging Group (HFAG) Collaboration), arXiv:0704.3575 [hep-ex]
390. A.F. Falk, M.E. Luke, M.J. Savage, Phys. Rev. D **49**, 3367 (1994). arXiv:hep-ph/9308288
391. I.I. Bigi et al., arXiv:hep-ph/9212227
392. G. Buchalla, G. Isidori, S.J. Rey, Nucl. Phys. B **511**, 594 (1998). arXiv:hep-ph/9705253
393. M.B. Voloshin, Phys. Lett. B **397**, 275 (1997). arXiv:hep-ph/9612483
394. A. Khodjamirian et al., Phys. Lett. B **402**, 167 (1997). arXiv:hep-ph/9702318
395. Z. Ligeti, L. Randall, M.B. Wise, Phys. Lett. B **402**, 178 (1997). arXiv:hep-ph/9702322
396. A.K. Grant et al., Phys. Rev. D **56**, 3151 (1997). arXiv:hep-ph/9702380
397. S.J. Lee, M. Neubert, G. Paz, arXiv:hep-ph/0609224
398. T. Becher, M. Neubert, Phys. Rev. Lett. **98**, 022003 (2007). arXiv:hep-ph/0610067
399. J.R. Andersen, E. Gardi, J. High Energy Phys. **0701**, 029 (2007). arXiv:hep-ph/0609250
400. I. Bigi, N. Uraltsev, Int. J. Mod. Phys. A **17**, 4709 (2002). arXiv:hep-ph/0202175
401. A. Kapustin, Z. Ligeti, H.D. Politzer, Phys. Lett. B **357**, 653 (1995). arXiv:hep-ph/9507248
402. Z. Ligeti et al., Phys. Rev. D **60**, 034019 (1999). arXiv:hep-ph/9903305
403. A. Ali, A.Y. Parkhomenko, Eur. Phys. J. C **23**, 89 (2002). arXiv:hep-ph/0105302
404. A. Ali, E. Lunghi, A.Y. Parkhomenko, Phys. Lett. B **595**, 323 (2004). arXiv:hep-ph/0405075
405. S.W. Bosch, G. Buchalla, Nucl. Phys. B **621**, 459 (2002). arXiv:hep-ph/0106081
406. T. Becher, R.J. Hill, M. Neubert, Phys. Rev. D **72**, 094017 (2005). arXiv:hep-ph/0503263
407. D. Atwood, M. Gronau, A. Soni, Phys. Rev. Lett. **79**, 185 (1997). arXiv:hep-ph/9704272
408. B. Grinstein, Y. Grossman, Z. Ligeti, D. Pirjol, Phys. Rev. D **71**, 011504 (2005). arXiv:hep-ph/0412019
409. B. Grinstein, D. Pirjol, Phys. Rev. D **73**, 014013 (2006). arXiv:hep-ph/0510104
410. A.L. Kagan, M. Neubert, Phys. Lett. B **539**, 227 (2002). arXiv:hep-ph/0110078
411. C. Sachrajda, T. Vladikas, private communication
412. R. Zwicky, in preparation
413. A. Khodjamirian, Nucl. Phys. B **605**, 558 (2001). arXiv:hep-ph/0012271
414. P. Ciafaloni, A. Romanino, A. Strumia, Nucl. Phys. B **524**, 361 (1998). arXiv:hep-ph/9710312
415. F.M. Borzumati, C. Greub, Phys. Rev. D **58**, 074004 (1998). arXiv:hep-ph/9802391
416. S. Bertolini, F. Borzumati, A. Masiero, G. Ridolfi, Nucl. Phys. B **353**, 591 (1991)
417. R. Barbieri, G.F. Giudice, Phys. Lett. B **309**, 86 (1993). arXiv:hep-ph/9303270
418. N. Oshimo, Nucl. Phys. B **404**, 20 (1993)
419. M.A. Diaz, Phys. Lett. B **304**, 278 (1993). arXiv:hep-ph/9303280
420. Y. Okada, Phys. Lett. B **315**, 119 (1993). arXiv:hep-ph/9307249
421. R. Garisto, J.N. Ng, Phys. Lett. B **315**, 372 (1993). arXiv:hep-ph/9307301
422. F. Borzumati, Z. Phys. C **63**, 291 (1994). arXiv:hep-ph/9310212
423. V.D. Barger et al., Phys. Rev. D **51**, 2438 (1995). arXiv:hep-ph/9407273
424. F. Borzumati, C. Greub, T. Hurth, D. Wyler, Phys. Rev. D **62**, 075005 (2000). arXiv:hep-ph/9911245
425. C. Bobeth, M. Misiak, J. Urban, Nucl. Phys. B **567**, 153 (2000). arXiv:hep-ph/9904413
426. F. Borzumati, C. Greub, Y. Yamada, Phys. Rev. D **69**, 055005 (2004). arXiv:hep-ph/0311151
427. G. Degrassi, P. Gambino, P. Slavich, Phys. Lett. B **635**, 335 (2006). arXiv:hep-ph/0601135
428. M.E. Gomez et al., Phys. Rev. D **74**, 015015 (2006). arXiv:hep-ph/0601163
429. K.i. Okumura, L. Roszkowski, Phys. Rev. Lett. **92**, 161801 (2004). arXiv:hep-ph/0208101
430. K. Agashe, N.G. Deshpande, G.H. Wu, Phys. Lett. B **514**, 309 (2001). arXiv:hep-ph/0105084
431. C.S. Kim, J.D. Kim, J.h. Song, Phys. Rev. D **67**, 015001 (2003). arXiv:hep-ph/0204002
432. K. Agashe, G. Perez, A. Soni, Phys. Rev. Lett. **93**, 201804 (2004). arXiv:hep-ph/0406101
433. K. Agashe, G. Perez, A. Soni, Phys. Rev. D **71**, 016002 (2005). arXiv:hep-ph/0408134
434. A. Ali, E. Lunghi, C. Greub, G. Hiller, Phys. Rev. D **66**, 034002 (2002). arXiv:hep-ph/0112300
435. G. Hiller, F. Kruger, Phys. Rev. D **69**, 074020 (2004). arXiv:hep-ph/0310219
436. G. Belanger, F. Boudjema, A. Pukhov, A. Semenov, Comput. Phys. Commun. **174**, 577 (2006). arXiv:hep-ph/0405253
437. A. Djouadi, J.L. Kneur, G. Moultaka, arXiv:hep-ph/0211331
438. S. Heinemeyer, W. Hollik, G. Weiglein, Comput. Phys. Commun. **124**, 76 (2000). arXiv:hep-ph/9812320
439. K. Abe et al. (Belle Collaboration), Phys. Lett. B **511**, 151 (2001). arXiv:hep-ex/0103042
440. M. Nakao et al. (Belle Collaboration), Phys. Rev. D **69**, 112001 (2004). arXiv:hep-ex/0402042
441. S. Nishida et al. (Belle Collaboration), Phys. Rev. Lett. **93**, 031803 (2004). arXiv:hep-ex/0308038
442. Y. Ushiroda et al. (Belle Collaboration), Phys. Rev. D **74**, 111104 (2006). arXiv:hep-ex/0608017
443. Y. Ushiroda et al., Phys. Rev. Lett. **94**, 231601 (2005). arXiv:hep-ex/0503008
444. B. Aubert et al. (BaBar Collaboration), Phys. Rev. D **72**, 052004 (2005). arXiv:hep-ex/0508004
445. B. Aubert et al. (BaBar Collaboration), Phys. Rev. Lett. **97**, 171803 (2006). arXiv:hep-ex/0607071
446. B. Aubert et al. (BaBar Collaboration), arXiv:0708.1652 [hep-ex]
447. B. Aubert et al. (BaBar Collaboration), Phys. Rev. Lett. **96**, 221801 (2006). arXiv:hep-ex/0601046
448. B. Aubert et al. (BaBar Collaboration), Phys. Rev. D **73**, 012006 (2006). arXiv:hep-ex/0509040
449. B. Aubert et al. (BaBar Collaboration), Phys. Rev. Lett. **93**, 021804 (2004). arXiv:hep-ex/0403035
450. B. Aubert et al. (BaBar Collaboration), Phys. Rev. D **70**, 112006 (2004). arXiv:hep-ex/0407003
451. B. Aubert et al. (BaBar Collaboration), Phys. Rev. D **75**, 051102 (2007). arXiv:hep-ex/0611037
452. B. Aubert et al. (BaBar Collaboration), Phys. Rev. D **72**, 051103 (2005). arXiv:hep-ex/0507038
453. I. Belyaev (LHCb Collaboration), Acta Phys. Pol. B **38**, 905 (2007), LHCb note LHCb-2005-001,2005
454. L. Shchutska, I. Belyaev, A. Golutvin, LHCb Public note LHCb-2007-030, 2007

455. S. Viret, F. Ohlsson-Malek, M. Smizanska ATLAS-PHYS-PUB-2005-006, 2004
456. F. Legger, LHCb Public note LHCb-2006-012, 2006
457. F. Legger, LHCb Public note LHCb-2006-013, 2006
458. G. Hiller et al., arXiv:hep-ph/0702191
459. H.H. Asatryan, H.M. Asatrian, C. Greub, M. Walker, Phys. Rev. D **65**, 074004 (2002). arXiv:hep-ph/0109140
460. H.H. Asatrian, H.M. Asatrian, C. Greub, M. Walker, Phys. Lett. B **507**, 162 (2001). arXiv:hep-ph/0103087
461. A. Ghinculov, T. Hurth, G. Isidori, Y.P. Yao, Nucl. Phys. B **648**, 254 (2003). arXiv:hep-ph/0208088
462. H.M. Asatrian, K. Bieri, C. Greub, A. Hovhannisyan, Phys. Rev. D **66**, 094013 (2002). arXiv:hep-ph/0209006
463. A. Ghinculov, T. Hurth, G. Isidori, Y.P. Yao, Nucl. Phys. B **685**, 351 (2004). arXiv:hep-ph/0312128
464. P. Gambino, M. Gorbahn, U. Haisch, Nucl. Phys. B **673**, 238 (2003). arXiv:hep-ph/0306079
465. C. Bobeth, P. Gambino, M. Gorbahn, U. Haisch, J. High Energy Phys. **0404**, 071 (2004). arXiv:hep-ph/0312090
466. G. Buchalla, G. Isidori, Nucl. Phys. B **525**, 333 (1998). arXiv:hep-ph/9801456
467. A. Ali, G. Hiller, L.T. Handoko, T. Morozumi, Phys. Rev. D **55**, 4105 (1997). arXiv:hep-ph/9609449
468. C.W. Bauer, C.N. Burrell, Phys. Lett. B **469**, 248 (1999). arXiv:hep-ph/9907517
469. C.W. Bauer, C.N. Burrell, Phys. Rev. D **62**, 114028 (2000). arXiv:hep-ph/9911404
470. Z. Ligeti, F.J. Tackmann, Phys. Lett. B **653**, 404 (2007). arXiv:0707.1694 [hep-ph]
471. H.H. Asatryan, H.M. Asatrian, C. Greub, M. Walker, Phys. Rev. D **66**, 034009 (2002). arXiv:hep-ph/0204341
472. H.M. Asatrian, H.H. Asatryan, A. Hovhannisyan, V. Poghosyan, Mod. Phys. Lett. A **19**, 603 (2004). arXiv:hep-ph/0311187
473. T. Huber, E. Lunghi, M. Misiak, D. Wyler, Nucl. Phys. B **740**, 105 (2006). arXiv:hep-ph/0512066
474. T. Huber, T. Hurth, E. Lunghi, arXiv:0712.3009 [hep-ph]
475. K.S.M. Lee, Z. Ligeti, I.W. Stewart, F.J. Tackmann, Phys. Rev. D **75**, 034016 (2007). arXiv:hep-ph/0612156
476. F. Krüger, J. Matias, Phys. Rev. D **71**, 094009 (2005). arXiv:hep-ph/0502060
477. E. Lunghi, J. Matias, J. High Energy Phys. **0704**, 058 (2007). arXiv:hep-ph/0612166
478. M. Beneke, T. Feldmann, D. Seidel, Nucl. Phys. B **612**, 25 (2001). arXiv:hep-ph/0106067
479. A. Ali, G. Kramer, G.h. Zhu, Eur. Phys. J. C **47**, 625 (2006). arXiv:hep-ph/0601034
480. B. Grinstein, D. Pirjol, Phys. Rev. D **70**, 114005 (2004). arXiv:hep-ph/0404250
481. G. Burdman, Phys. Rev. D **57**, 4254 (1998). arXiv:hep-ph/9710550
482. A. Ali, P. Ball, L.T. Handoko, G. Hiller, Phys. Rev. D **61**, 074024 (2000). arXiv:hep-ph/9910221
483. P. Ball, V.M. Braun, Phys. Rev. D **58**, 094016 (1998). arXiv:hep-ph/9805422
484. B. Grinstein, D. Pirjol, Phys. Rev. D **73**, 094027 (2006). arXiv:hep-ph/0505155
485. G. Hiller, A. Kagan, Phys. Rev. D **65**, 074038 (2002). arXiv:hep-ph/0108074
486. C.H. Chen, C.Q. Geng, Phys. Rev. D **64**, 074001 (2001). arXiv:hep-ph/0106193
487. T.M. Aliev, A. Ozpineci, M. Savci, Phys. Rev. D **67**, 035007 (2003). arXiv:hep-ph/0211447
488. T.M. Aliev, A. Ozpineci, M. Savci, Nucl. Phys. B **649**, 168 (2003). arXiv:hep-ph/0202120
489. A.K. Giri, R. Mohanta, J. Phys. G **31**, 1559 (2005)
490. G. Turan, J. Phys. G **31**, 525 (2005)
491. T.M. Aliev, M. Savci, J. Phys. G **26**, 997 (2000) arXiv:hep-ph/9906473
492. G. Turan, J. High Energy Phys. **0505**, 008 (2005)
493. A.K. Giri, R. Mohanta, Eur. Phys. J. C **45**, 151 (2006). arXiv:hep-ph/0510171
494. C.H. Chen, C.Q. Geng, J.N. Ng, Phys. Rev. D **65**, 091502 (2002). arXiv:hep-ph/0202103
495. T.M. Aliev, M. Savci, J. High Energy Phys. **0605**, 001 (2006). arXiv:hep-ph/0507324
496. M.B. Voloshin, Phys. Lett. B **397**, 275 (1997)
497. J.W. Chen, G. Rupak, M.J. Savage, Phys. Lett. B **410**, 285 (1997)
498. F. Krüger, L.M. Sehgal, Phys. Lett. B **380**, 199 (1996)
499. K. Abe et al. (Belle Collaboration), Belle-CONF-0415. arXiv:hep-ex/0410006
500. A. Ishikawa et al., Phys. Rev. Lett. **96**, 251801 (2006). arXiv:hep-ex/0603018
501. B. Aubert et al. (BaBar Collaboration), Phys. Rev. D **73**, 092001 (2006). arXiv:hep-ex/0604007
502. E. Barberio et al. (Heavy Flavor Averaging Group (HFAG)), arXiv:hep-ex/0603003
503. Q.S. Yan, C.S. Huang, W. Liao, S.H. Zhu, Phys. Rev. D **62**, 094023 (2000). arXiv:hep-ph/0004262
504. K.S.M. Lee, I.W. Stewart, Phys. Rev. D **74**, 014005 (2006). arXiv:hep-ph/0511334
505. K.S.M. Lee, Z. Ligeti, I.W. Stewart, F.J. Tackmann, Phys. Rev. D **74**, 011501 (2006). arXiv:hep-ph/0512191
506. A. Ishikawa, in *6th Workshop on a Higher Luminosity B Factory*, KEK, Tsukuba, Japan, 16–18 November 2004
507. J. Hewett et al., SLAC-R-709. arXiv:hep-ph/0503261
508. J. Dickens, V. Gibson, C. Lazzeroni, M. Patel, LHCb Public note, LHCb-2007-038
509. J. Dickens, V. Gibson, C. Lazzeroni, M. Patel, LHCb Public note, LHCb-2007-039
510. U. Egede, LHCb Public note, LHCb-2007-057
511. A. Rimoldi, A. Dell'Acqua, eConf **C0303241**, TUMT001 (2003). arXiv:physics/0306086
512. M. Smizanska, CERN ATLAS Communication Note, ATL-COM-PHYS-2003-038
513. D.J. Lange, Nucl. Instrum. Methods A **462**, 152 (2001)
514. M. Smizanska, J. Catmore, CERN ATLAS Communication Note, ATL-COM-PHYS-2004-041
515. D. Melikhov, N. Nikitin, S. Simula, Phys. Rev. D **57**, 6814 (1998). arXiv:hep-ph/9711362
516. T. Sjostrand, L. Lonnblad, S. Mrenna, P. Skands, arXiv:hep-ph/0308153
517. ATLAS Collaboration, CERN-LHCC-97-16
518. ATLAS Collaboration, CERN-LHCC-2003-022
519. T. Lagouri, N. Kanaya, A. Krasznahorkay, CERN ATLAS Communication Note, ATL-COM-PHYS-2006-007
520. J. Marriner, CDF Internal Note 2724, 1994
521. A. Policicchio, G. Crosetti, CERN ATLAS Communication Note, ATL-COM-PHYS-2007-005
522. G. Hiller, eConf **C030603**, MAR02 (2003). arXiv:hep-ph/0308180
523. Y. Dincer, L.M. Sehgal, Phys. Lett. B **521**, 7 (2001). arXiv:hep-ph/0108144
524. A. Dedes, J.R. Ellis, M. Raidal, Phys. Lett. B **549**, 159 (2002). arXiv:hep-ph/0209207
525. T. Fujihara, S.K. Kang, C.S. Kim, D. Kimura, T. Morozumi, Phys. Rev. D **73**, 074011 (2006). arXiv:hep-ph/0512010
526. G. Buchalla, G. Hiller, G. Isidori, Phys. Rev. D **63**, 014015 (2001). arXiv:hep-ph/0006136
527. T. Becher, S. Braig, M. Neubert, A.L. Kagan, Phys. Lett. B **540**, 278 (2002). arXiv:hep-ph/0205274

528. C. Bobeth, T. Ewerth, F. Kruger, J. Urban, Phys. Rev. D **64**, 074014 (2001). arXiv:hep-ph/0104284
529. C. Bobeth, T. Ewerth, F. Kruger, J. Urban, Phys. Rev. D **66**, 074021 (2002). arXiv:hep-ph/0204225
530. C.S. Huang, W. Liao, Phys. Lett. B **525**, 107 (2002). arXiv:hep-ph/0011089
531. A. Dedes, A. Pilaftsis, Phys. Rev. D **67**, 015012 (2003). arXiv:hep-ph/0209306
532. P.H. Chankowski, J. Kalinowski, Z. Was, M. Worek, Nucl. Phys. B **713**, 555 (2005). arXiv:hep-ph/0412253
533. K. Abe et al. (Belle Collaboration), arXiv:hep-ex/0508009
534. T.M. Aliev, C.S. Kim, Y.G. Kim, Phys. Rev. D **62**, 014026 (2000). arXiv:hep-ph/9910501
535. C. Bobeth, G. Hiller, G. Piranishvili, J. High Energy Phys. **0712**, 040 (2007). arXiv:0709.4174 [hep-ph]
536. D.A. Demir, K.A. Olive, M.B. Voloshin, Phys. Rev. D **66**, 034015 (2002). arXiv:hep-ph/0204119
537. S.R. Choudhury, A.S. Cornell, N. Gaur, G.C. Joshi, Phys. Rev. D **69**, 054018 (2004). arXiv:hep-ph/0307276
538. A.S. Cornell, N. Gaur, S.K. Singh, arXiv:hep-ph/0505136
539. T. Feldmann, J. Matias, J. High Energy Phys. **0301**, 074 (2003). arXiv:hep-ph/0212158
540. G. Hiller, Phys. Rev. D **70**, 034018 (2004). arXiv:hep-ph/0404220
541. R.E. Marshak, Riazuddin, C.P. Ryan, *Theory of Weak Interactions in Particle Physics* (Wiley, New York, 1969)
542. Y. Grossman, Z. Ligeti, E. Nardi, Nucl. Phys. B **465**, 369 (1996). [Erratum: Nucl. Phys. B **480**, 756 (1996)]. arXiv:hep-ph/9510378
543. B. Aubert et al. (BaBar Collaboration), arXiv:hep-ex/0608019
544. A. Masiero, P. Paradisi, R. Petronzio, Phys. Rev. D **74**, 011701 (2006). arXiv:hep-ph/0511289
545. P. Colangelo, F. De Fazio, P. Santorelli, E. Scrimieri, Phys. Lett. B **395**, 339 (1997). arXiv:hep-ph/9610297
546. D. Melikhov, N. Nikitin, S. Simula, Phys. Lett. B **428**, 171 (1998). arXiv:hep-ph/9803269
547. A.F. Falk, B. Grinstein, Nucl. Phys. B **416**, 771 (1994). arXiv:hep-ph/9306310
548. T.M. Aliev, C.S. Kim, Phys. Rev. D **58**, 013003 (1998). arXiv:hep-ph/9710428
549. C.S. Kim, Y.G. Kim, T. Morozumi, Phys. Rev. D **60**, 094007 (1999). arXiv:hep-ph/9905528
550. Z. Ligeti, M.B. Wise, Phys. Rev. D **53**, 4937 (1996). arXiv:hep-ph/9512225
551. K. Abe et al. (Belle Collaboration), arXiv:hep-ex/0507034
552. K. Abe et al. (Belle Collaboration), arXiv:hep-ex/0608047
553. B. Aubert et al. (BaBar Collaboration), Phys. Rev. Lett. **94**, 101801 (2005). arXiv:hep-ex/0411061
554. N. Satoyama et al. (Belle Collaboration), arXiv:hep-ex/0611045
555. BaBar Collaboration, B. Aubert et al., arXiv:hep-ex/0607110
556. G. Buchalla, A.J. Buras, Nucl. Phys. B **400**, 225 (1993)
557. M. Misiak, J. Urban, Phys. Lett. B **451**, 161 (1999). arXiv:hep-ph/9901278
558. G. Buchalla, A.J. Buras, Nucl. Phys. B **548**, 309 (1999). arXiv:hep-ph/9901288
559. A.J. Buras, Phys. Lett. B **566**, 115 (2003). arXiv:hep-ph/0303060
560. J.F. Gunion, H.E. Haber, G.L. Kane, S. Dawson, arXiv:hep-ph/9302272
561. H.E. Logan, U. Nierste, Nucl. Phys. B **586**, 39 (2000). arXiv:hep-ph/0004139
562. P.H. Chankowski, L. Slawianowska, Phys. Rev. D **63**, 054012 (2001). arXiv:hep-ph/0008046
563. G. Isidori, A. Retico, J. High Energy Phys. **0209**, 063 (2002). arXiv:hep-ph/0208159
564. A. Dedes, H.K. Dreiner, U. Nierste, Phys. Rev. Lett. **87**, 251804 (2001). arXiv:hep-ph/0108037
565. H.P. Nilles, Phys. Lett. B **115**, 193 (1982)
566. H.P. Nilles, Nucl. Phys. B **217**, 366 (1983)
567. A. Chamseddine, R. Arnowitt, P. Nath, Phys. Rev. Lett. **49**, 970 (1982)
568. R. Barbieri, S. Ferrara, C. Savoy, Phys. Lett. B **119**, 343 (1982)
569. L. Hall, J. Lykken, S. Weinberg, Phys. Rev. D **27**, 2359 (1983)
570. S.K. Soni, H.A. Weldon, Phys. Lett. B **126**, 215 (1983)
571. J.R. Ellis, K.A. Olive, V.C. Spanos, Phys. Lett. B **624**, 47 (2005). arXiv:hep-ph/0504196
572. J.R. Ellis, K.A. Olive, Y. Santoso, V.C. Spanos, J. High Energy Phys. **0605**, 063 (2006). arXiv:hep-ph/0603136
573. R. Dermisek, S. Raby, L. Roszkowski, R. Ruiz De Austri, J. High Energy Phys. **0304**, 037 (2003). arXiv:hep-ph/0304101
574. R. Dermisek, S. Raby, L. Roszkowski, R. Ruiz de Austri, J. High Energy Phys. **0509**, 029 (2005). arXiv:hep-ph/0507233
575. A.G. Akeroyd, S. Recksiegel, J. Phys. G **29**, 2311 (2003). arXiv:hep-ph/0306037
576. G. Valencia, S. Willenbrock, Phys. Rev. D **50**, 6843 (1994). arXiv:hep-ph/9409201
577. B. Aubert et al. (BaBar Collaboration), Phys. Rev. Lett. **94**, 221803 (2005). arXiv:hep-ex/0408096
578. B. Aubert et al. (BaBar Collaboration), Phys. Rev. Lett. **96**, 241802 (2006). arXiv:hep-ex/0511015
579. D0 Collaboration, D0-NOTE-CONF-5344-2007
580. CDF-Run II Collaboration, CDF-PUBLIC-NOTE-8176
581. R.P. Bernhard, arXiv:hep-ex/0605065
582. A. Abulencia et al. (CDF Collaboration), Phys. Rev. Lett. **95**, 221805 (2005) [Erratum: Phys. Rev. Lett. **95**, 249905 (2005)]. arXiv:hep-ex/0508036
583. V.M. Abazov et al. (D0 Collaboration), Phys. Rev. Lett. **94**, 071802 (2005). arXiv:hep-ex/0410039
584. D. Acosta et al. (CDF Collaboration), Phys. Rev. Lett. **93**, 032001 (2004). arXiv:hep-ex/0403032
585. F. Abe et al. (CDF Collaboration), Phys. Rev. D **57**, 3811 (1998)
586. M. Acciarri et al. (L3 Collaboration), Phys. Lett. B **391**, 474 (1997)
587. M.C. Chang et al. (Belle Collaboration), Phys. Rev. D **68**, 111101 (2003). arXiv:hep-ex/0309069
588. T. Bergfeld et al. (CLEO Collaboration), Phys. Rev. D **62**, 091102 (2000). arXiv:hep-ex/0007042
589. B. Abbott et al. (D0 Collaboration), Phys. Lett. B **423**, 419 (1998). arXiv:hep-ex/9801027
590. C. Albajar et al. (UA1 Collaboration), Phys. Lett. B **262**, 163 (1991)
591. H. Albrecht et al. (ARGUS Collaboration), Phys. Lett. B **199**, 451 (1987)
592. P. Avery et al. (CLEO Collaboration), Phys. Lett. B **183**, 429 (1987)
593. R. Antunes-Nobrega et al. (LHCb Collaboration), CERN-LHCC-2003-031
594. N. Panikashvili (ATLAS Collaboration), Nucl. Phys. Proc. Suppl. **156**, 129 (2006)
595. CMS Collaboration, CERN-LHCC-2002-026
596. N. Nikitin et al., Nucl. Phys. Proc. Suppl. **156**, 119 (2006)
597. C. Eggel, U. Langenegger, A. Starodumov, CMS-CR-2006-071
598. D. Martinez, J.A. Hernando, F. Teubert, LHCb Public note, CERN-LHCb-2007-033
599. A. Nikitenko, A. Starodumov, N. Stepanov, arXiv:hep-ph/9907256
600. ATLAS Collaboration, CERN-LHCC-99-15, ATLAS-TDR-15
601. N. Nikitin, S. Sivoklokov, M. Smizanska, D. Tlisov, K. Toms, ATL-COM-PHYS-2006-086
602. N. Nikitin, P. Reznicek, S. Sivoklokov, M. Smizanska, K. Toms, Nucl. Phys. Proc. Suppl. **163**, 147 (2007)

603. H. Boos, T. Mannel, J. Reuter, Phys. Rev. D **70**, 036006 (2004). arXiv:hep-ph/0403085
604. M. Ciuchini, M. Pierini, L. Silvestrini, Phys. Rev. Lett. **95**, 221804 (2005). arXiv:hep-ph/0507290
605. M. Gronau, J. Zupan, Phys. Rev. D **71**, 074017 (2005)
606. B. Aubert et al. (BaBar Collaboration), Phys. Rev. Lett. **87**, 091801 (2001). arXiv:hep-ex/0107013
607. K. Abe et al. (Belle Collaboration), Phys. Rev. Lett. **87**, 091802 (2001). arXiv:hep-ex/0107061
608. B. Aubert et al. (BaBar Collaboration), arXiv:hep-ex/0607107
609. K.-F. Chen et al. (Belle Collaboration), arXiv:hep-ex/0608039
610. B. Aubert et al. (BaBar Collaboration), Phys. Rev. Lett. **87**, 241801 (2001). arXiv:hep-ex/0107049
611. I. Dunietz, H.R. Quinn, A. Snyder, W. Toki, H.J. Lipkin, Phys. Rev. D **43**, 2193 (1991)
612. A. Bondar, T. Gershon, P. Krokovny, Phys. Lett. B **624**, 1 (2005). arXiv:hep-ph/0503174
613. B. Aubert et al. (BaBar Collaboration), Phys. Rev. D **71**, 032005 (2005). arXiv:hep-ex/0411016
614. R. Itoh et al. (Belle Collaboration), Phys. Rev. Lett. **95**, 091601 (2005). arXiv:hep-ex/0504030
615. B. Aubert et al. (BaBar Collaboration), arXiv:hep-ex/0607105
616. P. Krokovny et al. (Belle Collaboration), Phys. Rev. Lett. **97**, 081801 (2006)
617. Y. Grossman, H.R. Quinn, Phys. Rev. D **58**, 017504 (1998). arXiv:hep-ph/9712306
618. B. Aubert et al. (BaBar Collaboration), Phys. Rev. Lett. **95**, 041805 (2005). arXiv:hep-ex/0503049
619. K. Abe et al. (Belle Collaboration), arXiv:hep-ex/0507039
620. B. Aubert et al. (BaBar Collaboration), arXiv:hep-ex/0607106
621. K. Abe (Belle Collaboration), arXiv:hep-ex/0608035
622. B. Aubert et al. (BaBar Collaboration), arXiv:hep-ex/0607097
623. A.E. Snyder, H.R. Quinn, Phys. Rev. D **48**, 2139 (1993)
624. B. Aubert et al. (BaBar Collaboration), arXiv:hep-ex/0608002
625. K. Abe et al. (Belle Collaboration), arXiv:hep-ex/0609003
626. B. Aubert et al. (BaBar Collaboration), arXiv:hep-ex/0408099
627. I.I.Y. Bigi, A.I. Sanda, Phys. Lett. B **211**, 213 (1988)
628. A.B. Carter, A.I. Sanda, Phys. Rev. Lett. **45**, 952 (1980)
629. M. Gronau, Phys. Lett. B **557**, 198 (2003). arXiv:hep-ph/0211282
630. D. Atwood, A. Soni, Phys. Rev. D **71**, 013007 (2005). arXiv:hep-ph/0312100
631. Y. Grossman, A. Soffer, J. Zupan, Phys. Rev. D **72**, 031501 (2005)
632. M. Gronau, D. London, Phys. Lett. B **253**, 483 (1991)
633. M. Gronau, D. Wyler, Phys. Lett. B **265**, 172 (1991)
634. M. Gronau, Phys. Rev. D **58**, 037301 (1998). arXiv:hep-ph/9802315
635. D. Atwood, I. Dunietz, A. Soni, Phys. Rev. Lett. **78**, 3257 (1997). arXiv:hep-ph/9612433
636. D. Atwood, I. Dunietz, A. Soni, Phys. Rev. D **63**, 036005 (2001). arXiv:hep-ph/0008090
637. A. Giri, Y. Grossman, A. Soffer, J. Zupan, Phys. Rev. D **68**, 054018 (2003). arXiv:hep-ph/0303187
638. A. Bondar, in *Proceedings of BINP Special Analysis Meeting on Dalitz*, 24–26 September 2002, unpublished
639. A. Poluektov et al. (Belle Collaboration), Phys. Rev. D **73**, 112009 (2006)
640. B. Aubert et al. (BaBar Collaboration), arXiv:hep-ex/0607104
641. B. Aubert et al. (BaBar Collaboration), Phys. Rev. Lett. **95**, 121802 (2005)
642. H. Muramatsu et al. (CLEO Collaboration), Phys. Rev. Lett. **89**, 251802 (2002) [Erratum: Phys. Rev. Lett. **90**, 059901 (2003)]. arXiv:hep-ex/0207067
643. G.J. Feldman, R.D. Cousins, Phys. Rev. D **57**, 3873 (1998). arXiv:physics/9711021
644. D. Atwood, A. Soni, Phys. Rev. D **68**, 033009 (2003). arXiv:hep-ph/0206045
645. D. Atwood, A. Soni, Phys. Rev. D **68**, 033003 (2003)
646. D. Asner, talk given at the 3rd Workshop on the CKM Unitarity Triangle (CKM2005). ckm2005.ucsd.edu/WG/WG5/thu3/Asner-WG5-S4.pdf
647. B. Kayser, D. London, Phys. Rev. D **61**, 116013 (2000). arXiv:hep-ph/9909561
648. A. Bondar, A. Poluektov, Eur. Phys. J. C **47**, 347 (2006)
649. O. Long, M. Baak, R.N. Cahn, D. Kirkby, Phys. Rev. D **68**, 034010 (2003). arXiv:hep-ex/0303030
650. B. Aubert et al. (BaBar Collaboration), Phys. Rev. D **71**, 112003 (2005). arXiv:hep-ex/0504035
651. B. Aubert et al. (BaBar Collaboration), Phys. Rev. D **73**, 111101 (2006). arXiv:hep-ex/0602049
652. F.J. Ronga et al. (Belle Collaboration), Phys. Rev. D **73**, 092003 (2006). arXiv:hep-ex/0604013
653. I. Dunietz, Phys. Lett. B **427**, 179 (1998). arXiv:hep-ph/9712401
654. O. Deschamps, F. Machefert, S. Monteil, P. Perret, P. Robbe, A. Robert, M.-H. Schune, LHCb Public note, LHCb-2007-046
655. P.F. Harrison, H.R. Quinn (eds.), *The BaBar Physics Book*, SLAC-R-504 (1998), Sect. 6.5
656. M. Patel, LHCb Public note, LHCb-2006-066
657. M. Patel, LHCb Public note, LHCb-2007-043
658. V. Gibson, C. Lazzeroni, J. Libby, LHCb Public note, LHCb-2007-048
659. A. Powell, LHCb Public note, LHCb-2007-004
660. K. Akiba, M. Gandelman, Decays, LHCb Public note, HCb-2007-050
661. J. Rademacker, G. Wilkinson, Phys. Lett. B **647**, 400 (2007). arXiv:hep-ex/0611272
662. S. Cohen, M. Merk, E. Rodrigues, LHCb Public note, LHCb-2007-041
663. R. Fleischer, Nucl. Phys. B **671**, 459 (2003). arXiv:hep-ph/0304027
664. V.V. Gligorov, LHCb Public note, LHCb-2007-044
665. G. Wilkinson, LHCb Public note, LHCb-2005-036
666. S. Amato, M. Gandelman, C. Gobel, L. de Paula, LHCb Public note, LHCb-2007-045
667. V. Abazov (D0 Collaboration), arXiv:hep-ex/0701007
668. V.M. Abazov et al. (D0 Collaboration), Phys. Rev. D **74**, 092001 (2006). arXiv:hep-ex/0609014
669. D. Buskulic et al. (ALEPH Collaboration), Phys. Lett. B **377**, 205 (1996)
670. F. Abe et al. (CDF Collaboration), Phys. Rev. D **59**, 032004 (1999). arXiv:hep-ex/9808003
671. P. Abreu et al. (DELPHI Collaboration), Eur. Phys. J. C **16**, 555 (2000). arXiv:hep-ex/0107077
672. K. Ackerstaff et al. (OPAL Collaboration), Phys. Lett. B **426**, 161 (1998). arXiv:hep-ex/9802002
673. V.M. Abazov et al. (D0 Collaboration), Phys. Rev. Lett. **97**, 241801 (2006). arXiv:hep-ex/0604046
674. A.S. Dighe, I. Dunietz, R. Fleischer, Eur. Phys. J. C **6**, 647 (1999). arXiv:hep-ph/9804253
675. D. Acosta et al. (CDF Collaboration), Phys. Rev. Lett. **94**, 101803 (2005). arXiv:hep-ex/0412057
676. V.M. Abazov et al. (D0 Collaboration), Phys. Rev. Lett. **98**, 121801 (2007). arXiv:hep-ex/0701012
677. A. Abulencia et al. (CDF-Run II Collaboration), Phys. Rev. Lett. **97**, 062003 (2006) [AIP Conf. Proc. **870**, 116 (2006)]. arXiv:hep-ex/0606027
678. Y. Grossman, Y. Nir, G. Raz, Phys. Rev. Lett. **97**, 151801 (2006). arXiv:hep-ph/0605028
679. V.M. Abazov et al. (D0 Collaboration), Phys. Rev. D **76**, 057101 (2007). arXiv:hep-ex/0702030

680. J. Borel, L. Nicolas, O. Schneider, J. Van Hunen, LHCb Public note, LHCb-2007-017
681. P. Vankov, G. Raven, LHCb Public note, LHCb-2007-055
682. G. Balbi, V.M. Vagnoni, S. Vecchi, LHCb Public note, LHCb-2007-032
683. G. Balbi, S. Vecchi, LHCb Public note, LHCb-2007-056
684. M. Calvi, O. Leroy, M. Musy, LHCb Public note, LHCb-2007-058
685. L. Fernández, EPFL Ph. D. thesis 3613, CERN-THESIS-2006-042
686. L. Fernández, in *Physics at LHC*, Cracow, Poland, LHCb Public note, LHCb-2006-047. Acta Phys. Pol. B **38**, 931 (2007)
687. D. Volyanskyy, J. van Tilburg, LHCb Public note, LHCb-2006-027
688. S. Jimenez-Otero, EPFL PhD thesis 3779, CERN-THESIS-2007-051
689. U. Nierste, arXiv:hep-ph/0406300
690. O. Leroy, F. Muheim, S. Poss, Y. Xie, LHCb Public note, LHCb-2007-029
691. N. Brook, N. Cottingham, R.W. Lambert, F. Muheim, J. Rademacker, P. Szczypka, Y. Xie, LHCb Public note, LHCb-2007-054
692. V.V. Gligorov, J. Rademacker, LHCb Public note, LHCb-2007-053
693. J. Rademacker, Nucl. Instrum. Methods **570**, 525 (2007). arXiv:hep-ex/0502042
694. V. Ciulli et al., CMS NOTE-2006/090
695. CMS Collaboration, The TriDAS project technical design report, vol. 2: Data acquisition and high-level trigger, CERN/LHCC 2002-26, CMS TDR 6.2, 2002
696. CMS Collaboration, CMS physics technical design report, vol. I: Detector performance and software, CERN/LHCC 2006-001, CMS TDR 8.1, 2006
697. I. Dunietz, R. Fleischer, U. Nierste, Phys. Rev. D **63**, 114015 (2001). arXiv:hep-ph/0012219
698. S. Dambach, U. Langenegger, A. Starodumov, Nucl. Instrum. Methods A **569**, 824 (2006). arXiv:hep-ph/0607294
699. A. Starodumov, Z. Xie, B_s decay vertex resolution, CMS NOTE 1997/85
700. ATLAS Collaboration, ATLAS: Detector and physics performance technical design report, vol. 1, CERN-LHCC-99-14
701. A.J. Buras, M. Jamin, P.H. Weisz, Nucl. Phys. B **347**, 491 (1990)
702. M. Beneke, G. Buchalla, C. Greub, A. Lenz, U. Nierste, Phys. Lett. B **459**, 631 (1999). arXiv:hep-ph/9808385
703. M. Beneke, G. Buchalla, A. Lenz, U. Nierste, Phys. Lett. B **576**, 173 (2003). arXiv:hep-ph/0307344
704. M. Ciuchini, E. Franco, V. Lubicz, F. Mescia, C. Tarantino, J. High Energy Phys. **0308**, 031 (2003). arXiv:hep-ph/0308029
705. V. Gimenez, J. Reyes, Nucl. Phys. Proc. Suppl. **94**, 350 (2001). arXiv:hep-lat/0010048
706. S. Hashimoto, K.I. Ishikawa, T. Onogi, M. Sakamoto, N. Tsutsui, N. Yamada, Phys. Rev. D **62**, 114502 (2000). arXiv:hep-lat/0004022
707. S. Aoki et al. (JLQCD Collaboration), Phys. Rev. D **67**, 014506 (2003). arXiv:hep-lat/0208038
708. D. Becirevic, D. Meloni, A. Retico, V. Gimenez, V. Lubicz, G. Martinelli, Eur. Phys. J. C **18**, 157 (2000). arXiv:hep-ph/0006135
709. L. Lellouch, C.J. Lin (UKQCD Collaboration), Phys. Rev. D **64**, 094501 (2001). arXiv:hep-ph/0011086
710. N. Yamada et al. (JLQCD Collaboration), Nucl. Phys. Proc. Suppl. **106**, 397 (2002). arXiv:hep-lat/0110087
711. M. Beneke, G. Buchalla, I. Dunietz, Phys. Rev. D **54**, 4419 (1996). arXiv:hep-ph/9605259
712. A. Lenz, U. Nierste, J. High Energy Phys. **0706**, 072 (2007). arXiv:hep-ph/0612167
713. C. Tarantino, arXiv:hep-ph/0702235
714. Z. Ligeti, M. Papucci, G. Perez, Phys. Rev. Lett. **97**, 101801 (2006). arXiv:hep-ph/0604112
715. P. Ball, R. Fleischer, Eur. Phys. J. C **48**, 413 (2006). arXiv:hep-ph/0604249
716. M. Rescigno, private communication
717. R. Fleischer, Phys. Lett. B **365**, 399 (1996). arXiv:hep-ph/9509204
718. Y. Grossman, M.P. Worah, Phys. Lett. B **395**, 241 (1997). arXiv:hep-ph/9612269
719. R. Fleischer, Int. J. Mod. Phys. A **12**, 2459 (1997). arXiv:hep-ph/9612446
720. M. Ciuchini, E. Franco, G. Martinelli, A. Masiero, L. Silvestrini, Phys. Rev. Lett. **79**, 978 (1997). arXiv:hep-ph/9704274
721. D. London, A. Soni, Phys. Lett. B **407**, 61 (1997). arXiv:hep-ph/9704277
722. G. Buchalla, G. Hiller, Y. Nir, G. Raz, J. High Energy Phys. **0509**, 074 (2005). arXiv:hep-ph/0503151
723. M. Beneke, Phys. Lett. B **620**, 143 (2005). arXiv:hep-ph/0505075
724. H.Y. Cheng, C.K. Chua, A. Soni, Phys. Rev. D **72**, 014006 (2005). arXiv:hep-ph/0502235
725. H.N. Li, S. Mishima, Phys. Rev. D **74**, 094020 (2006). arXiv:hep-ph/0608277
726. H.Y. Cheng, C.K. Chua, A. Soni, Phys. Rev. D **72**, 094003 (2005). arXiv:hep-ph/0506268
727. H.J. Lipkin, Y. Nir, H.R. Quinn, A. Snyder, Phys. Rev. D **44**, 1454 (1991)
728. T. Gershon, M. Hazumi, Phys. Lett. B **596**, 163 (2004). arXiv:hep-ph/0402097
729. Y. Grossman, Z. Ligeti, Y. Nir, H. Quinn, Phys. Rev. D **68**, 015004 (2003). arXiv:hep-ph/0303171
730. K. Abe et al. (Belle Collaboration), arXiv:hep-ex/0208030
731. M. Gronau, J.L. Rosner, Phys. Lett. B **564**, 90 (2003). arXiv:hep-ph/0304178
732. K. Abe et al. (Belle Collaboration), arXiv:hep-ex/0609006
733. M. Gronau, J.L. Rosner, Phys. Rev. D **72**, 094031 (2005). arXiv:hep-ph/0509155
734. B. Aubert et al. (BaBar Collaboration), arXiv:hep-ex/0507016
735. B. Aubert et al. (BaBar Collaboration), arXiv:hep-ex/0607108
736. G. Engelhard, G. Raz, Phys. Rev. D **72**, 114017 (2005). arXiv:hep-ph/0508046
737. G. Engelhard, Y. Nir, G. Raz, Phys. Rev. D **72**, 075013 (2005). arXiv:hep-ph/0505194
738. C.K. Chua, in *The Proceedings of 4th Flavor Physics and CP Violation Conference (FPCP 2006)*, Vancouver, BC, Canada, 9–12 April 2006, p. 008. arXiv:hep-ph/0605301
739. B. Aubert et al. (BaBar Collaboration0, arXiv:hep-ex/0607112
740. M. Ciuchini, M. Pierini, L. Silvestrini, Phys. Rev. D **74**, 051301 (2006). arXiv:hep-ph/0601233
741. M. Ciuchini, M. Pierini, L. Silvestrini, arXiv:hep-ph/0602207
742. M. Gronau, D. Pirjol, A. Soni, J. Zupan, arXiv:hep-ph/0608243
743. B. Grinstein, R.F. Lebed, Phys. Rev. D **53**, 6344 (1996). arXiv:hep-ph/9602218
744. M.J. Savage, M.B. Wise, Phys. Rev. D **39**, 3346 (1989) [Erratum: Phys. Rev. D **40**, 3127 (1989)]
745. L.L. Chau, H.Y. Cheng, W.K. Sze, H. Yao, B. Tseng, Phys. Rev. D **43**, 2176 (1991) [Erratum: Phys. Rev. D **58**, 019902 (1998)]
746. M. Gronau, O.F. Hernandez, D. London, J.L. Rosner, Phys. Rev. D **50**, 4529 (1994). arXiv:hep-ph/9404283
747. M. Gronau, O.F. Hernandez, D. London, J.L. Rosner, Phys. Rev. D **52**, 6356 (1995). arXiv:hep-ph/9504326
748. M. Gronau, O.F. Hernandez, D. London, J.L. Rosner, Phys. Rev. D **52**, 6374 (1995). arXiv:hep-ph/9504327

749. A.S. Dighe, M. Gronau, J.L. Rosner, Phys. Lett. B **367**, 357 (1996) [Erratum: Phys. Lett. B **377**, 325 (1996)]. arXiv:hep-ph/9509428
750. A.S. Dighe, M. Gronau, J.L. Rosner, Phys. Rev. D **57**, 1783 (1998). arXiv:hep-ph/9709223
751. C.W. Bauer, D. Pirjol, Phys. Lett. B **604**, 183 (2004). arXiv:hep-ph/0408161
752. C.W. Chiang, J.L. Rosner, Phys. Rev. D **65**, 074035 (2002) [Erratum: Phys. Rev. D **68**, 039902 (2003)]. arXiv:hep-ph/0112285
753. C.W. Chiang, M. Gronau, J.L. Rosner, Phys. Rev. D **68**, 074012 (2003). arXiv:hep-ph/0306021
754. C.W. Chiang, M. Gronau, Z. Luo, J.L. Rosner, D.A. Suprun, Phys. Rev. D **69**, 034001 (2004). arXiv:hep-ph/0307395
755. C.W. Chiang, M. Gronau, J.L. Rosner, D.A. Suprun, Phys. Rev. D **70**, 034020 (2004). arXiv:hep-ph/0404073
756. C.W. Chiang, Y.F. Zhou, arXiv:hep-ph/0609128
757. Y.F. Zhou, Y.L. Wu, J.N. Ng, C.Q. Geng, Phys. Rev. D **63**, 054011 (2001). arXiv:hep-ph/0006225
758. X.G. He, Y.K. Hsiao, J.Q. Shi, Y.L. Wu, Y.F. Zhou, Phys. Rev. D **64**, 034002 (2001). arXiv:hep-ph/0011337
759. Y.L. Wu, Y.F. Zhou, Eur. Phys. J. direct C **5**, 014 (2003) [Eur. Phys. J. C **32S1**, 179 (2004)]. arXiv:hep-ph/0210367
760. Y.L. Wu, Y.F. Zhou, Phys. Rev. D **71**, 021701 (2005). arXiv:hep-ph/0409221
761. Y.L. Wu, Y.F. Zhou, Phys. Rev. D **72**, 034037 (2005). arXiv:hep-ph/0503077
762. M. Neubert, J.L. Rosner, Phys. Lett. B **441**, 403 (1998). arXiv:hep-ph/9808493
763. M. Neubert, J.L. Rosner, Phys. Rev. Lett. **81**, 5076 (1998). arXiv:hep-ph/9809311
764. M. Neubert, J. High Energy Phys. **9902**, 014 (1999). arXiv:hep-ph/9812396
765. M. Gronau, D. Pirjol, T.M. Yan, Phys. Rev. D **60**, 034021 (1999) [Erratum: Phys. Rev. D **69**, 119901 (2004)]. arXiv:hep-ph/9810482
766. C.W. Bauer, I.Z. Rothstein, I.W. Stewart, Phys. Rev. D **74**, 034010 (2006). arXiv:hep-ph/0510241
767. M. Gronau, J.L. Rosner, arXiv:hep-ph/0610227
768. X.Q. Hao, X.G. He, X.Q. Li, arXiv:hep-ph/0609264
769. M. Gronau, J.L. Rosner, Phys. Rev. D **58**, 113005 (1998). arXiv:hep-ph/9806348
770. A.J. Buras, R. Fleischer, S. Recksiegel, F. Schwab, Eur. Phys. J. C **45**, 701 (2006). arXiv:hep-ph/0512032
771. R. Fleischer, Phys. Lett. B **459**, 306 (1999). arXiv:hep-ph/9903456
772. R. Fleischer, Eur. Phys. J. C **52**, 267 (2007). arXiv:0705.1121 [hep-ph]
773. R. Fleischer, J. Matias, Phys. Rev. D **61**, 074004 (2000). arXiv:hep-ph/9906274
774. R. Fleischer, J. Matias, Phys. Rev. D **66**, 054009 (2002). arXiv:hep-ph/0204101
775. D. London, J. Matias, J. Virto, Phys. Rev. D **71**, 014024 (2005). arXiv:hep-ph/0410011
776. M. Gronau, Phys. Lett. B **492**, 297 (2000). arXiv:hep-ph/0008292
777. D. Tonelli (CDF Collaboration), PoS **HEP2005**, 258 (2006). arXiv:hep-ex/0512024
778. R. Fleischer, S. Recksiegel, F. Schwab, Eur. Phys. J. C **51**, 55 (2007). arXiv:hep-ph/0702275
779. A.J. Buras, R. Fleischer, S. Recksiegel, F. Schwab, Acta Phys. Pol. B **36**, 2015 (2005). arXiv:hep-ph/0410407
780. A.J. Buras, R. Fleischer, S. Recksiegel, F. Schwab, Nucl. Phys. B **697**, 133 (2004). arXiv:hep-ph/0402112
781. A.J. Buras, R. Fleischer, S. Recksiegel, F. Schwab, Phys. Rev. Lett. **92**, 101804 (2004). arXiv:hep-ph/0312259
782. A. Soni, D.A. Suprun, arXiv:hep-ph/0609089
783. A. Soni, D.A. Suprun, Phys. Lett. B **635**, 330 (2006). arXiv:hep-ph/0511012
784. M. Gronau, J.L. Rosner, J. Zupan, Phys. Rev. D **74**, 093003 (2006). arXiv:hep-ph/0608085
785. M. Gronau, J.L. Rosner, J. Zupan, Phys. Lett. B **596**, 107 (2004). arXiv:hep-ph/0403287
786. M. Gronau, Y. Grossman, J.L. Rosner, Phys. Lett. B **579**, 331 (2004). arXiv:hep-ph/0310020
787. M. Gronau, J.L. Rosner, Phys. Rev. D **71**, 074019 (2005). arXiv:hep-ph/0503131
788. G. Raz, arXiv:hep-ph/0509125
789. M. Gronau, J.L. Rosner, Phys. Rev. D **59**, 113002 (1999). arXiv:hep-ph/9809384
790. H.J. Lipkin, Phys. Lett. B **445**, 403 (1999). arXiv:hep-ph/9810351
791. M. Gronau, J.L. Rosner, Phys. Rev. D **65**, 013004 (2002) [Erratum: Phys. Rev. D **65**, 079901 (2002)]. arXiv:hep-ph/0109238
792. M. Gronau, Y. Grossman, G. Raz, J.L. Rosner, Phys. Lett. B **635**, 207 (2006). arXiv:hep-ph/0601129
793. M. Gronau, Phys. Lett. B **627**, 82 (2005). arXiv:hep-ph/0508047
794. S. Baek, D. London, J. Matias, J. Virto, J. High Energy Phys. **0602**, 027 (2006). arXiv:hep-ph/0511295
795. A.S. Safir, J. High Energy Phys. **0409**, 053 (2004). arXiv:hep-ph/0407015
796. S. Descotes-Genon, J. Matias, J. Virto, Phys. Rev. Lett. **97**, 061801 (2006). arXiv:hep-ph/0603239
797. S. Descotes-Genon, J. Matias, J. Virto, Phys. Rev. D **76**, 074005 (2007). arXiv:0705.0477 [hep-ph]
798. S. Baek, D. London, J. Matias, J. Virto, arXiv:hep-ph/0610109
799. A.J. Buras, L. Silvestrini, Nucl. Phys. B **569**, 3 (2000). arXiv:hep-ph/9812392
800. M. Ciuchini, E. Franco, G. Martinelli, L. Silvestrini, Nucl. Phys. B **501**, 271 (1997). arXiv:hep-ph/9703353
801. M. Ciuchini, E. Franco, G. Martinelli, M. Pierini, L. Silvestrini, Phys. Lett. B **515**, 33 (2001). arXiv:hep-ph/0104126
802. M. Ciuchini, M. Pierini, L. Silvestrini, arXiv:hep-ph/0703137
803. M. Ciuchini et al., in preparation
804. R. Barbieri, A. Strumia, Nucl. Phys. B **508**, 3 (1997). arXiv:hep-ph/9704402
805. A.L. Kagan, M. Neubert, Phys. Rev. D **58**, 094012 (1998). arXiv:hep-ph/9803368
806. S.A. Abel, W.N. Cottingham, I.B. Whittingham, Phys. Rev. D **58**, 073006 (1998). arXiv:hep-ph/9803401
807. A. Kagan, arXiv:hep-ph/9806266
808. R. Fleischer, T. Mannel, Phys. Lett. B **511**, 240 (2001). arXiv:hep-ph/0103121
809. E. Lunghi, D. Wyler, Phys. Lett. B **521**, 320 (2001). arXiv:hep-ph/0109149
810. M.B. Causse, arXiv:hep-ph/0207070
811. G. Hiller, Phys. Rev. D **66**, 071502 (2002). arXiv:hep-ph/0207356
812. S. Khalil, E. Kou, Phys. Rev. D **67**, 055009 (2003). arXiv:hep-ph/0212023
813. S. Baek, Phys. Rev. D **67**, 096004 (2003). arXiv:hep-ph/0301269
814. K. Agashe, C.D. Carone, Phys. Rev. D **68**, 035017 (2003). arXiv:hep-ph/0304229
815. J.F. Cheng, C.S. Huang, X.h. Wu, Phys. Lett. B **585**, 287 (2004). arXiv:hep-ph/0306086
816. D. Chakraverty, E. Gabrielli, K. Huitu, S. Khalil, Phys. Rev. D **68**, 095004 (2003). arXiv:hep-ph/0306076
817. S. Khalil, E. Kou, Phys. Rev. Lett. **91**, 241602 (2003). arXiv:hep-ph/0303214

818. S. Khalil, E. Kou, in *The Proceedings of 2nd Workshop on the CKM Unitarity Triangle*, Durham, England, 5–9 April 2003, p. WG305. arXiv:hep-ph/0307024
819. J.F. Cheng, C.S. Huang, X.H. Wu, Nucl. Phys. B **701**, 54 (2004). arXiv:hep-ph/0404055
820. S. Khalil, E. Kou, Phys. Rev. D **71**, 114016 (2005). arXiv:hep-ph/0407284
821. E. Gabrielli, K. Huitu, S. Khalil, Nucl. Phys. B **710**, 139 (2005). arXiv:hep-ph/0407291
822. S. Khalil, Mod. Phys. Lett. A **19**, 2745 (2004) [Afr. J. Math. Phys. **1**, 101 (2004)]. arXiv:hep-ph/0411151
823. S. Khalil, Phys. Rev. D **72**, 035007 (2005). arXiv:hep-ph/0505151
824. M. Ciuchini et al., in preparation
825. M. Ciuchini, Nucl. Phys. Proc. Suppl. B **109**, 307 (2002). arXiv:hep-ph/0112133
826. A.L. Kagan, arXiv:hep-ph/0407076
827. M. Endo, S. Mishima, M. Yamaguchi, Phys. Lett. B **609**, 95 (2005). arXiv:hep-ph/0409245
828. J.A. Casas, S. Dimopoulos, Phys. Lett. B **387**, 107 (1996). arXiv:hep-ph/9606237
829. B. Aubert et al. (BaBar Collaboration), arXiv:hep-ex/0508017
830. B. Aubert et al. (BaBar Collaboration), arXiv:hep-ex/0607101
831. B. Aubert et al. (BaBar Collaboration), arXiv:hep-ex/0607096
832. B. Aubert et al. (BaBar Collaboration), Phys. Rev. Lett. **98**, 031801 (2007). arXiv:hep-ex/0609052
833. B. Aubert et al. (BaBar Collaboration), Phys. Rev. Lett. **97**, 171805 (2006). arXiv:hep-ex/0608036
834. M. Raidal, Phys. Rev. Lett. **89**, 231803 (2002). arXiv:hep-ph/0208091
835. S. Amato, J. McCarron, F. Muheim, B. Souza de Paula, Y. Xie, LHCb Public note, LHCb-2007-047
836. Y. Xie, LHCb Public note, LHCb-2007-130
837. A.G. Akeroyd et al., Physics at super-B factory, to be posted on hep-ex
838. M. Hazumi, Contribution to ICHEP 2006
839. S. Hashimoto et al., Letter of intent for KEK superB factory, KEK-REPORT-2004-4
840. B. Aubert et al. (BaBar Collaboration), Phys. Rev. D **74**, 051106 (2006). arXiv:hep-ex/0607063
841. BaBar Collaboration, Phys. Rev. D **73**, 071102 (2006)
842. B. Aubert et al. (BaBar Collaboration), Phys. Rev. Lett. **98**, 051802 (2007). arXiv:hep-ex/0607109
843. Belle Collaboration, arXiv:hep-ex/0608034
844. B. Aubert et al. (BaBar Collaboration), arXiv:hep-ex/0607098
845. Belle Collaboration, Phys. Rev. Lett. **96**, 171801 (2006)
846. BaBar Collaboration, arXiv:hep-ex/0610073
847. Belle Collaboration, Phys. Rev. Lett. **94**, 221804 (2005)
848. BaBar Collaboration, arXiv:hep-ex/0607057
849. Belle Collaboration, arXiv:hep-ex/0505039
850. A. Datta, D. London, Int. J. Mod. Phys. A **19**, 2505 (2004)
851. D. London, N. Sinha, R. Sinha, Phys. Rev. D **69**, 114013 (2004)
852. C.-D. Lu, Y.-L. Shen, W. Wang, arXiv:hep-ph/0606092
853. BaBar Collaboration, Phys. Rev. D **74**, 031104 (2006)
854. A. Abulencia et al. (CDF Collaboration), Phys. Rev. Lett. **97**, 211802 (2006). arXiv:hep-ex/0607021
855. A. Carbone, N. Nardulli, S. Pennazzi, A. Sarti, V. Vagnoni, LHCb Public note, CERN-LHCb-2007-059
856. O. Leitner, Z.J. Ajaltouni, Nucl. Phys. Proc. Suppl. **174**, 169 (2007). arXiv:hep-ph/0610189
857. G. Buchalla, A.J. Buras, Phys. Rev. D **54**, 6782 (1996). arXiv:hep-ph/9607447
858. A.J. Buras, F. Schwab, S. Uhlig, arXiv:hep-ph/0405132
859. G. Isidori, in *The Proceedings of 4th Flavor Physics and CP Violation Conference (FPCP 2006)*, Vancouver, BC, Canada, 9–12 April 2006, p. 035. arXiv:hep-ph/0606047
860. D. Bryman, A.J. Buras, G. Isidori, L. Littenberg, Int. J. Mod. Phys. A **21**, 487 (2006). arXiv:hep-ph/0505171
861. W.J. Marciano, Z. Parsa, Phys. Rev. D **53**, 1 (1996)
862. F. Mescia, C. Smith, Phys. Rev. D **76**, 034017 (2007). arXiv:0705.2025 [hep-ph]
863. A.J. Buras, M. Gorbahn, U. Haisch, U. Nierste, Phys. Rev. Lett. **95**, 261805 (2005). arXiv:hep-ph/0508165
864. A.J. Buras, M. Gorbahn, U. Haisch, U. Nierste, J. High Energy Phys. **0611**, 002 (2006). arXiv:hep-ph/0603079
865. G. Isidori, F. Mescia, C. Smith, Nucl. Phys. B **718**, 319 (2005). arXiv:hep-ph/0503107
866. A.J. Buras, A. Romanino, L. Silvestrini, Nucl. Phys. B **520**, 3 (1998). arXiv:hep-ph/9712398
867. V.V. Anisimovsky et al. (E949 Collaboration), Phys. Rev. Lett. **93**, 031801 (2004). arXiv:hep-ex/0403036
868. S. Adler et al. (E787 Collaboration), Phys. Rev. Lett. **88**, 041803 (2002). arXiv:hep-ex/0111091
869. S. Adler et al. (E787 Collaboration), Phys. Rev. Lett. **79**, 2204 (1997). arXiv:hep-ex/9708031
870. Y. Grossman, Y. Nir, Phys. Lett. B **398**, 163 (1997). arXiv:hep-ph/9701313
871. G. Buchalla, G. D'Ambrosio, G. Isidori, Nucl. Phys. B **672**, 387 (2003). arXiv:hep-ph/0308008
872. G. Isidori, C. Smith, R. Unterdorfer, Eur. Phys. J. C **36**, 57 (2004). arXiv:hep-ph/0404127
873. U. Haisch, in *Proceedings of KAON International Conference*, Laboratori Nazionali di Frascati dell'INFN, Rome, Italy, 21–25 May 2007. arXiv:0707.3098 [hep-ph]
874. C. Tarantino, in *Proceedings of KAON International Conference*, Laboratori Nazionali di Frascati dell'INFN, Rome, Italy, 21–25 May 2007. arXiv:0706.3436 [hep-ph]
875. C. Smith, in *Proceedings of KAON International Conference*, Laboratori Nazionali di Frascati dell'INFN, Rome, Italy, 21–25 May 2007. arXiv:0707.2309 [hep-ph]
876. J.K. Ahn et al. (E391a Collaboration), arXiv:0712.4164 [hep-ex]
877. A. Alavi-Harati et al. (KTeV Collaboration), Phys. Rev. Lett. **93**, 021805 (2004). arXiv:hep-ex/0309072
878. A. Alavi-Harati et al. (KTeV Collaboration), Phys. Rev. Lett. **84**, 5279 (2000). arXiv:hep-ex/0001006
879. G. Isidori, F. Mescia, P. Paradisi, C. Smith, S. Trine, J. High Energy Phys. **0608**, 064 (2006). arXiv:hep-ph/0604074
880. G. Buchalla, G. Isidori, Phys. Lett. B **440**, 170 (1998). arXiv:hep-ph/9806501
881. G. Isidori, G. Martinelli, P. Turchetti, Phys. Lett. B **633**, 75 (2006). arXiv:hep-lat/0506026
882. A.J. Buras, M.E. Lautenbacher, M. Misiak, M. Munz, Nucl. Phys. B **423**, 349 (1994). arXiv:hep-ph/9402347
883. G. D'Ambrosio, G. Ecker, G. Isidori, J. Portoles, J. High Energy Phys. **9808**, 004 (1998). arXiv:hep-ph/9808289
884. S. Friot, D. Greynat, E. De Rafael, Phys. Lett. B **595**, 301 (2004). arXiv:hep-ph/0404136
885. J.R. Batley et al. (NA48/1 Collaboration), Phys. Lett. B **576**, 43 (2003). arXiv:hep-ex/0309075
886. J.R. Batley et al. (NA48/1 Collaboration), Phys. Lett. B **599**, 197 (2004). arXiv:hep-ex/0409011
887. F. Mescia, C. Smith, S. Trine, J. High Energy Phys. **0608**, 088 (2006). arXiv:hep-ph/0606081
888. M. Gorbahn, U. Haisch, Phys. Rev. Lett. **97**, 122002 (2006). arXiv:hep-ph/0605203
889. G. Isidori, R. Unterdorfer, J. High Energy Phys. **0401**, 009 (2004). arXiv:hep-ph/0311084
890. J.M. Gerard, C. Smith, S. Trine, Nucl. Phys. B **730**, 1 (2005). arXiv:hep-ph/0508189
891. M. Blanke, A.J. Buras, D. Guadagnoli, C. Tarantino, J. High Energy Phys. **0610**, 003 (2006). arXiv:hep-ph/0604057

892. G. Buchalla, A.J. Buras, Phys. Lett. B **333**, 221 (1994). arXiv:hep-ph/9405259
893. A.J. Buras, R. Fleischer, Phys. Rev. D **64**, 115010 (2001). arXiv:hep-ph/0104238
894. Y. Nir, M.P. Worah, Phys. Lett. B **423**, 319 (1998). arXiv:hep-ph/9711215
895. G. Colangelo, G. Isidori, J. High Energy Phys. **9809**, 009 (1998). arXiv:hep-ph/9808487
896. G. Isidori, P. Paradisi, Phys. Rev. D **73**, 055017 (2006). arXiv:hep-ph/0601094
897. T. Moroi, Phys. Lett. B **493**, 366 (2000). arXiv:hep-ph/0007328
898. D. Chang, A. Masiero, H. Murayama, Phys. Rev. D **67**, 075013 (2003). arXiv:hep-ph/0205111
899. P.L. Cho, M. Misiak, D. Wyler, Phys. Rev. D **54**, 3329 (1996). arXiv:hep-ph/9601360
900. S. Davidson, D.C. Bailey, B.A. Campbell, Z. Phys. C **61**, 613 (1994). arXiv:hep-ph/9309310
901. R. Barbier et al., arXiv:hep-ph/9810232
902. Y. Grossman, G. Isidori, H. Murayama, Phys. Lett. B **588**, 74 (2004). arXiv:hep-ph/0311353
903. A. Deandrea, J. Welzel, M. Oertel, J. High Energy Phys. **0410**, 038 (2004). arXiv:hep-ph/0407216
904. N.G. Deshpande, D.K. Ghosh, X.G. He, Phys. Rev. D **70**, 093003 (2004). arXiv:hep-ph/0407021
905. G. Anelli et al., CERN-SPSC-2005-013, SPSC-P-326, 11/06/2005
906. G. Unal (NA48 Collaboration), Frascati Phys. Ser. **21**, 361 (2001). arXiv:hep-ex/0012011
907. http://j-parc.jp/index-e.html
908. http://j-parc.jp/NuclPart/PAC_for_NuclPart.html
909. The J-PARC proposals, including the kaon experiments. Available from http://j-parc.jp/NuclPart/Proposal_0606_e.html
910. I.I.Y. Bigi, Y.L. Dokshitzer, V.A. Khoze, J.H. Kuhn, P.M. Zerwas, Phys. Lett. B **181**, 157 (1986)
911. G. Burdman, E. Golowich, J. Hewett, S. Pakvasa, Phys. Rev. D **66**, 014009 (2002)
912. S. Bianco, F.L. Fabbri, D. Benson, I. Bigi, Riv. Nuovo Cimento **26**(N7), 1 (2003)
913. G. Burdman, I. Shipsey, Annu. Rev. Nucl. Part. Sci. **53**, 431 (2003)
914. A.A. Petrov, eConf **C030603**, MEC05 (2003). arXiv:hep-ph/0311371
915. H.N. Nelson, in *Proceedings of the 19th International Symposium on Photon and Lepton Interactions at High Energy LP99*, ed. by J.A. Jaros, M.E. Peskin. arXiv:hep-ex/9908021
916. A. Datta, D. Kumbhakar, Z. Phys. C **27**, 515 (1985)
917. A.A. Petrov, Phys. Rev. D **56**, 1685 (1997)
918. K. Abe et al. (Belle Collaboration), Phys. Rev. D **72**, 071101 (2005)
919. C. Cawlfield et al.(CLEO Collaboration), Phys. Rev. D **71**, 077101 (2005)
920. B. Aubert et al. (BaBar Collaboration), Phys. Rev. D **70**, 091102 (2004)
921. M. Hosack (FOCUS Collaboration), Fermilab-Thesis-2002-25 (2002)
922. E.M. Aitala et al. (E791 Collaboration), Phys. Rev. Lett. **77**, 2384 (1996)
923. I.I.Y. Bigi, in *Proceedings of the XXXIII International Conference on High Energy Physics*, Berkeley, CA, USA, 1986, SLAC-PUB-4074
924. G. Blaylock, A. Seiden, Y. Nir, Phys. Lett. B **355**, 555 (1995)
925. S. Bergmann, Y. Grossman, Z. Ligeti, Y. Nir, A.A. Petrov, Phys. Lett. B **486**, 418 (2000)
926. L. Zhang et al. (Belle Collaboration), Phys. Rev. Lett. **96**, 151801 (2006)
927. J. Li et al. (Belle Collaboration), Phys. Rev. Lett. **94**, 071801 (2005)
928. J.M. Link et al. (FOCUS Collaboration), Phys. Lett. B **618**, 23 (2005)
929. B. Aubert et al. (BaBar Collaboration), Phys. Rev. Lett. **91**, 171801 (2003)
930. R. Godang et al. (CLEO Collaboration), Phys. Rev. Lett. **84**, 5038 (2000)
931. E.M. Aitala et al. (E791 Collaboration), Phys. Rev. D **57**, 13 (1998)
932. R. Barate et al. (ALEPH Collaboration), Phys. Lett. B **436**, 211 (1998)
933. B. Aubert et al. (BaBar Collaboration), Phys. Rev. Lett. **98**, 211802 (2007)
934. A. Abulencia et al. (CDF Collaboration), Phys. Rev. D **74**, 031109 (2006)
935. P. Spradlin, G. Wilkinson, F. Xing, LHCb Public note, LHCb-2007-049
936. W.M. Yao et al. (Particle Data Group), J. Phys. G **33**, 1 (2006), review by D. Asner on pp. 713–716
937. E.M. Aitala et al. (E791 Collaboration), Phys. Lett. B **403**, 377 (1997)
938. J.M. Link et al. (FOCUS Collaboration), Phys. Lett. B **491**, 232 (2000) [Erratum: Phys. Lett. B **495**, 443 (2000)]
939. B. Aubert et al. (BaBar Collaboration), Phys. Rev. D **71**, 091101 (2005)
940. S. Malvezzi, AIP Conf. Proc. **549**, 569 (2002)
941. S. Kopp et al. (CLEO Collaboration), Phys. Rev. D **63**, 092001 (2001)
942. G. Brandenburg et al. (CLEO Collaboration), Phys. Rev. Lett. **87**, 071802 (2001)
943. D.M. Asner et al. (CLEO Collaboration), Phys. Rev. D **70**, 091101 (2004)
944. D. Cronin-Hennessy et al. (CLEO Collaboration), Phys. Rev. D **72**, 031102 (2005)
945. I.I.Y. Bigi, A.I. Sanda, Camb. Monogr. Part. Phys. Nucl. Phys. Cosmol. **9**, 1 (2000)
946. X.C. Tian et al. (Belle Collaboration), Phys. Rev. Lett. **95**, 231801 (2005)
947. S.A. Dytman et al. (CLEO Collaboration), Phys. Rev. D **64**, 111101 (2001)
948. B. Aubert et al. (BaBar Collaboration), Phys. Rev. Lett. **97**, 221803 (2006)
949. D.M. Asner et al. (CLEO Collaboration), Phys. Rev. D **72**, 012001 (2005)
950. K. Abe et al. (Belle Collaboration), Phys. Rev. Lett. **99**, 131803 (2007)
951. B. Aubert et al. (BaBar Collaboration), Phys. Rev. Lett. **91**, 121801 (2003)
952. K. Abe et al. (Belle Collaboration), arXiv:hep-ex/0308034
953. K. Abe et al. (Belle Collaboration), Phys. Rev. Lett. **88**, 162001 (2002)
954. S.E. Csorna et al. (CLEO Collaboration), Phys. Rev. D **65**, 092001 (2002)
955. J.M. Link et al. (FOCUS Collaboration), Phys. Lett. B **485**, 62 (2000)
956. E.M. Aitala et al. (E791 Collaboration), Phys. Rev. Lett. **83**, 32 (1999)
957. M. Staric et al. (Belle Collaboration), Phys. Rev. Lett. **98**, 211803 (2007)
958. I.I.Y. Bigi, in *Tau Charm Factory Workshop*, Stanford, CA, 23–27 May 1989, UND-HEP-89-BIG01
959. D. Atwood, A.A. Petrov, Phys. Rev. D **71**, 054032 (2005)
960. D.M. Asner, W.M. Sun, Phys. Rev. D **73**, 034024 (2006)
961. D.M. Asner et al. (CLEO Collaboration), Int. J. Mod. Phys. A **21**, 5456 (2006)
962. A.F. Falk, Y. Grossman, Z. Ligeti, A.A. Petrov, Phys. Rev. D **65**, 054034 (2002)

963. A.F. Falk, Y. Grossman, Z. Ligeti, Y. Nir, A.A. Petrov, Phys. Rev. D **69**, 114021 (2004)
964. I.I.Y. Bigi, N.G. Uraltsev, Nucl. Phys. B **592**, 92 (2001)
965. E. Golowich, S. Pakvasa, A.A. Petrov, arXiv:hep-ph/0610039
966. K. Abe et al. (Belle Collaboration), Phys. Rev. D **73**, 051106 (2006)
967. M. Saigo et al. (Belle Collaboration), Phys. Rev. Lett. **94**, 091601 (2005)
968. F. Buccella, M. Lusignoli, G. Miele, A. Pugliese, P. Santorelli, Phys. Rev. D **51**, 3478 (1995)
969. J.M. Link et al. (FOCUS Collaboration), Phys. Lett. B **622**, 239 (2005)
970. L.M. Sehgal, M. Wanninger, Phys. Rev. D **46**, 1035 (1992) [Erratum: Phys. Rev. D **46**, 5209 (1992)]
971. D. Acosta et al. (CDF Collaboration), Phys. Rev. Lett. **94**, 122001 (2005)
972. G. Burdman, E. Golowich, J.L. Hewett, S. Pakvasa, Phys. Rev. D **52**, 6383 (1995)
973. C. Greub, T. Hurth, M. Misiak, D. Wyler, Phys. Lett. B **382**, 415 (1996)
974. Q. Ho-Kim, X.Y. Pham, Phys. Rev. D **61**, 013008 (2000)
975. S. Prelovsek, D. Wyler, Phys. Lett. B **500**, 304 (2001)
976. S. Fajfer, S. Prelovsek, P. Singer, Phys. Rev. D **64**, 098502 (2001)
977. S. Fajfer, P. Singer, J. Zupan, Eur. Phys. J. C **27**, 201 (2003)
978. Q. He et al. (CLEO Collaboration), Phys. Rev. Lett. **95**, 221802 (2005)
979. B. Aubert (BaBar Collaboration), arXiv:hep-ex/0607051
980. S. Fajfer, S. Prelovsek, arXiv:hep-ph/0610032
981. S. Prelovsek, PhD thesis. arXiv:hep-ph/0010106
982. S. Fajfer, S. Prelovsek, P. Singer, D. Wyler, Phys. Lett. B **487**, 81 (2000)
983. S. Fajfer, S. Prelovsek, P. Singer, Eur. Phys. J. C **6**, 471 (1999)
984. S. Fajfer, S. Prelovsek, P. Singer, Phys. Rev. D **58**, 094038 (1998)
985. T.E. Coan et al. (CLEO Collaboration), Phys. Rev. Lett. **90**, 101801 (2003)
986. J.M. Link et al. (FOCUS Collaboration), Phys. Lett. B **572**, 21 (2003)
987. A. Korn (CDF Collaboration), arXiv:hep-ex/0305054
988. E.M. Aitala et al. (E791 Collaboration), Phys. Lett. B **462**, 401 (1999)
989. K. Kodama et al. (E653 Collaboration), Phys. Lett. B **345**, 85 (1995)
990. A. Freyberger et al. (CLEO Collaboration), Phys. Rev. Lett. **76**, 3065 (1996) [Erratum: Phys. Rev. Lett. **77**, 2147 (1996)]
991. P.L. Frabetti et al. (E687 Collaboration), Phys. Lett. B **398**, 239 (1997)
992. M. Artuso, Int. J, Mod. Phys. A **21**, 1697 (2006) arXiv:hep-ex/0510052
993. D. Asner, arXiv:hep-ex/0410014 (expanded version of review in Eidelman et al., Phys. Lett. B **592**, 1 (2004), pp. 664–667)
994. Y. Grossman, Z. Ligeti, A. Soffer, Phys. Rev. D **67**, 071301 (2003)
995. J.L. Rosner, D.A. Suprun, Phys. Rev. D **68**, 054010 (2003)
996. C. Cawlfield et al. (CLEO Collaboration), Phys. Rev. D **74**, 031108 (2006)
997. A. Bondar, in *The Proceedings of the BINP Special Analysis Meeting on Dalitz Analysis*, 24–26 September, 2002, unpublished
998. A. Bondar, A. Poluektov, in *Proceedings of the CKM 2006 Workshop*
999. B. Grinstein, Phys. Rev. Lett. **71**, 3067 (1993)
1000. D. Silverman, H. Yao, Phys. Rev. D **38**, 214 (1988)
1001. A.G. Akeroyd, Prog. Theor. Phys. **111**, 295 (2004). arXiv:hep-ph/0308260
1002. J.L. Hewett, arXiv:hep-ph/9505246
1003. G. Bonvicini et al. (CLEO Collaboration), Phys. Rev. D **70**, 112004 (2004)
1004. P. Rubin et al. (CLEO Collaboration), Phys. Rev. D **73**, 112005 (2006)
1005. M. Chadha et al. (CLEO Collaboration), Phys. Rev. D **58**, 032002 (1998)
1006. Yu. Alexandrov et al. (BEATRICE Collaboration), Phys. Lett. B **478**, 31 (2000)
1007. A. Heister et al. (ALEPH Collaboration), Phys. Lett. B **528**, 1 (2002)
1008. G. Abbiendi et al. (OPAL Collaboration), Phys. Lett. B **516**, 236 (2001)
1009. M. Acciarri et al. (L3 Collaboration), Phys. Lett. B **396**, 327 (1997)
1010. T.W. Chiu, T.H. Hsieh, J.Y. Lee, P.H. Liu, H.J. Chang, Phys. Lett. B **624**, 31 (2005)
1011. D. Becirevic, P. Boucaud, J.P. Leroy, V. Lubicz, G. Martinelli, F. Mescia, F. Rapuano, Phys. Rev. D **60**, 074501 (1999)
1012. J. Bordes, J. Penarrocha, K. Schilcher, J. High Energy Phys. **0511**, 014 (2005)
1013. S. Narison, arXiv:hep-ph/0202200
1014. D. Ebert, R.N. Faustov, V.O. Galkin, Phys. Lett. B **635**, 93 (2006)
1015. G. Cvetic, C.S. Kim, G.L. Wang, W. Namgung, Phys. Lett. B **596**, 84 (2004)
1016. Z.G. Wang, W.M. Yang, S.L. Wan, Nucl. Phys. A **744**, 156 (2004)
1017. L.A.M. Salcedo, J.P.B. de Melo, D. Hadjmichef, T. Frederico, Braz. J. Phys. **34**, 297 (2004)
1018. J.F. Amundson, J.L. Rosner, M.A. Kelly, N. Horwitz, S.L. Stone, Phys. Rev. D **47**, 3059 (1993)
1019. N.E. Adam et al. (CLEO Collaboration), Phys. Rev. Lett. **97**, 251801 (2006)
1020. M. Ablikim et al. (BES Collaboration), Phys. Lett. B **608**, 24 (2005)
1021. T.E. Coan et al. (CLEO Collaboration), Phys. Rev. Lett. **95**, 181802 (2005)
1022. G.S. Huang et al. (CLEO Collaboration), Phys. Rev. Lett. **95**, 181801 (2005)
1023. Y. Gao, Talk given at ICHEP 2006, Moscow, Russia, August 2006
1024. J.M. Link et al. (FOCUS Collaboration), Phys. Lett. B **535**, 43 (2002)
1025. M.R. Shepherd et al., Phys. Rev. D **74**, 052001 (2006)
1026. S. Fajfer, J. Kamenik, AIP Conf. Proc. **806**, 203 (2006)
1027. G.S. Huang et al. (CLEO Collaboration), Phys. Rev. D **94**, 011802 (2005)
1028. K. Abe et al. (Belle Collaboration), arXiv:hep-ex/0510003
1029. B. Aubert et al. (BaBar Collaboration), arXiv:hep-ex/0607077
1030. S. Stone, in *Heavy Flavours*, ed. by A.J. Buras, M. Lindner (World Scientific, Singapore, 1992)
1031. R.J. Hill, Phys. Rev. D **73**, 014012 (2006)
1032. S. Fajfer, J. Kamenik, Phys. Rev. D **72**, 034029 (2005)
1033. S. Fajfer, J. Kamenik, N. Kosnik, Phys. Rev. D **74**, 034027 (2006)
1034. Q. He et al. (CLEO Collaboration), Phys. Rev. Lett. **95**, 121801 (2005) [Erratum: Phys. Rev. Lett. **96**, 199903 (2006)]
1035. N. Adam et al. (CLEO Collaboration), arXiv:hep-ex/0607079
1036. B.Y. Blok, M.A. Shifman, Sov. J. Nucl. Phys. **45**, 301 (1987) [Yad. Fiz. **45**, 478 (1987)]
1037. H.Y. Cheng, Phys. Lett. B **335**, 428 (1994)
1038. J.L. Rosner, Phys. Rev. D **60**, 114026 (1999)
1039. C.W. Chiang, J.L. Rosner, Phys. Rev. D **65**, 054007 (2002)
1040. C.W. Chiang, Z. Luo, J.L. Rosner, Phys. Rev. D **67**, 014001 (2003)

1041. Y.L. Wu, M. Zhong, Y.F. Zhou, Eur. Phys. J. C **42**, 391 (2005)
1042. X.Y. Wu, X.G. Yin, D.B. Chen, Y.Q. Guo, Y. Zeng, Mod. Phys. Lett. A **19**, 1623 (2004)
1043. S. Ritt (MEG Collaboration), Nucl. Phys. Proc. Suppl. **162**, 279 (2006)
1044. T. Browder, M. Ciuchini, T. Gershon, M. Hazumi, T. Hurth, Y. Okada, A. Stocchi, arXiv:0710.3799 [hep-ph]
1045. M. Bona et al., INFN/AE-07/2, SLAC-R-856, LAL 07-15. arXiv:0709.0451 [hep-ex]
1046. A.G. Akeroyd et al. (SuperKEKB Physics Working Group), arXiv:hep-ex/0406071
1047. M. Hazumi et al. (SuperKEKB Physics Working Group), in preparation
1048. O. Schneider, in *1st LHCb Collaboration Upgrade Workshop*, January 2007. Available from http://indico.cern.ch/conferenceDisplay.py?confId=8351
1049. G. Wilkinson, in *1st LHCb Collaboration Upgrade Workshop*, January 2007. Available from http://indico.cern.ch/conferenceDisplay.py?confId=8351
1050. E. Gabrielli, G.F. Giudice, B **433**, 3 (1995). [Erratum: Nucl. Phys. B **507**, 549 (1997)]. arXiv:hep-lat/9407029
1051. L. Calibbi, A. Faccia, A. Masiero, S.K. Vempati, Phys. Rev. D **74**, 116002 (2006). arXiv:hep-ph/0605139
1052. T. Goto, Y. Okada, Y. Shimizu, T. Shindou, M. Tanaka, Phys. Rev. D **70**, 035012 (2004). arXiv:hep-ph/0306093
1053. T. Goto, Y. Okada, T. Shindou, M. Tanaka, Phys. Rev. D **77**, 095010 (2008). arXiv:0711.2935 [hep-ph]
1054. PEP-II Conceptual Design Report, SLAC-372, LBL-PUB-5303, CALT-68-1715, UCRL-ID-106426, UC-IIRPA-91-01 (1991)
1055. J. Seeman et al., in *European Particle Accelerator Conference (EPAC 06)*, Edinburgh, Scotland, 26–30 June 2006, SLAC-PUB-12023
1056. S. Kurokawa, E. Kikutani, Nucl. Instrum. Methods A **499**, 1 (2003), and other papers that volume
1057. B. Aubert et al. (BaBar Collaboration), Nucl. Instrum. Methods A **479**, 1 (2002)
1058. A. Abashian, et al. (Belle Collaboration), Nucl. Instrum. Methods A **479**, 117 (2002)
1059. J. Hewett, D. Hitlin (eds.), arXiv:hep-ph/0503261, and references therein
1060. J. Seeman et al., in *Proceedings of the 9th European Particle Accelerator Conference (EPAC 2004)*, Lucerne, 2004
1061. S. Hashimoto (ed.), KEK-REPORT-2004-4,2004
1062. The URL for the Super*B* web site is http://www.pi.infn.it/SuperB/
1063. N. Harnew et al., Nucl. Instrum. Methods A **408**, 137 (1998), also see updates in these proceedings
1064. S. Hashimoto (ed.), KEK-REPORT-2004-4, June 2004
1065. K. Akai, in *10th European Particle Accelerator Conference (EPAC06)*, Edinburgh, UK, 26–30 June 2006, pp. 19–23, KEK-PREPRINT-2006-020, KEK-PREPRINT-2006-20
1066. T. Abe, T. Kageyama, K. Akai, K. Ebihara, H. Sakai, Y. Takeuchi, KEK-PREPRINT-2005-78, KEK, Tsukuba, November 2005. Phys. Rev. Spec. Top. Accel.Beams **9**, 062002 (2006)
1067. T. Kageyama, in *Particles and Nuclei*, Santa Fe, 2005. AIP Conf. Proc. **842**, 1064 (2006)
1068. Y. Takeuchi, T. Abe, T. Kageyama, H. Sakai, in *Particle Accelerator Conference*, Knoxville, May 2005, p. 1195, KEK-PREPRINT-2005-53, PAC-2005-WPAT010
1069. T. Abe, T. Kageyama, Z. Kabeya, T. Kawasumi, T. Nakamura, K. Tsujimoto, K. Tajiri, in *Particle Accelerator Conference*, Knoxville, May 2005, p. 1051, PAC-2005-TPPT007
1070. Y. Suetsugu, K. Kanazawa, K. Shibata, H. Hisamatsu Nucl. Instrum. Methods A **556**, 399 (2006)
1071. Y. Suetsugu et al., Nucl. Instrum. Methods A **554**, 92 (2005)
1072. Y. Suetsugu, K.I. Kanazawa, N. Ohuchi, K. Shibata, M. Shirai, in *Particle Accelerator Conference*, Knoxville, May 2005, p. 3203, PAC-2005-RPPE052
1073. Y. Suetsugu, H. Hisamatsu, K.I. Kanazawa, N. Ohuchi, K. Shibata, M. Shirai, in *Particle Accelerator Conference*, Knoxville, May 2005, p. 3256, PAC-2005-RPPE053
1074. K.I. Kanazawa, H. Fukuma, H. Hisamatsu, Y. Suetsugu, in *Particle Accelerator Conference*, Knoxville, May 2005, p. 1054, PAC-2005-FPAP007
1075. J.W. Flanagan, K. Ohmi, H. Fukuma, S. Hiramatsu, M. Tobiyama, E. Perevedentsev, Phys. Rev. Lett. **94**, 054801 (2005)
1076. T. Suwada et al., KEK-PREPRINT-2006-56, November 2006. Phys. Rev. Spec. Top. Accel. Beams **10**, 073501 (2007)
1077. T. Kamitani et al., in *Particle Accelerator Conference*, Knoxville, May 2005, p. 1233, PAC-2005-TPPT011
1078. N. Ohuchi, Y. Funakoshi, H. Koiso, K. Oide, K. Tsuchiya, in *Particle Accelerator Conference*, Knoxville, May 2005, p. 2470, PAC-2005-MPPT037
1079. K. Ohmi, K. Oide, Phys. Rev. Spec. Top. Accel. Beams **10**, 014401 (2007)
1080. K. Ohmi, K. Oide, E. Perevedentsev, in *10th European Particle Accelerator Conference (EPAC06)*, Edinburgh, UK, June 2006, pp. 616–618
1081. K. Ohmi, Y. Funakoshi, S. Hiramatsu, K. Oide, M. Tobiyama in *10th European Particle Accelerator Conference (EPAC06)*, Edinburgh, UK, June 2006, pp. 619–621
1082. K. Ohmi, M. Tawada, Y. Funakoshi, in *Particle Accelerator Conference*, Knoxville, May 2005, p. 925, KEK-PREPRINT-2005-42, PAC-2005-TPPP004
1083. K. Akai, Y. Morita, in *Particle Accelerator Conference*, Knoxville, May 2005, p. 1129, KEK-PREPRINT-2005-26, PAC-2005-TPPT008
1084. T. Abe et al., arXiv:0706.3248 [physics.ins-det]
1085. M. Tobiyama, J.W. Flanagan, H. Fukuma, S. Kurokawa, K. Ohmi, S.S. Win, Phys. Rev. ST Accel. Beams **9**, 012801 (2006)
1086. R. Antunes Nobrega et al., LHCb TDR 9, CERN/LHCC/2003-30, 2003
1087. I.I.Y. Bigi, A.I. Sanda, Nucl. Phys. B **193**, 85 (1981)
1088. Heavy Flavor Averaging Group, updated Summer 2007. http://www.slac.stanford.edu/xorg/hfag/
1089. F. Muheim, Nucl. Phys. Proc. Suppl. **170**, 317 (2007)
1090. H. Dijkstra, The LHCb upgrade, in *FPCP07*, Bled, Slovenia, May 2007
1091. C. Parkes, The LHCb detector, in *EPS07*, Manchester, UK, July 2007
1092. V. Obraztsov, in *BEACH2006*, Lancaster, UK, July 2006
1093. M.J. Strasser, K.M. Zurek, arXiv:hep-ph/0604261
1094. M.J. Strasser, K.M. Zurek, arXiv:hep-ph/0605193
1095. M.J. Strassler, arXiv:hep-ph/0607160
1096. M. Cvetic, P. Langacker, G. Shiu, Phys. Rev. D **66**, 066004 (2002). arXiv:hep-ph/0205252
1097. N. Arkani-Hamed, S. Dimopoulos, S. Kachru, arXiv:hep-th/0501082
1098. V. Barger, P. Langacker, G. Shaughnessy, New J. Phys. **9**, 333 (2007). arXiv:hep-ph/0702001
1099. C. Parkes, Nucl. Instrum. Methods A **569**, 115 (2006)
1100. M. Ciuchini, E. Franco, D. Guadagnoli, V. Lubicz, M. Pierini, V. Porretti, L. Silvestrini, arXiv:hep-ph/0703204
1101. A. Masiero, M. Piai, A. Romanino, L. Silvestrini, Phys. Rev. D **64**, 075005 (2001). arXiv:hep-ph/0104101
1102. S. Baek, T. Goto, Y. Okada, K.i. Okumura, Phys. Rev. D **63**, 051701 (2001). arXiv:hep-ph/0002141
1103. J. Hisano, Y. Shimizu, Phys. Lett. B **565**, 183 (2003). arXiv:hep-ph/0303071
1104. C.S. Huang, T.J. Li, W. Liao, Nucl. Phys. B **673**, 331 (2003). arXiv:hep-ph/0304130

1105. A.J. Buras, R. Buras, Phys. Lett. B **501**, 223 (2001). arXiv:hep-ph/0008273
1106. W. de Boer, A. Dabelstein, W. Hollik, W. Mosle, U. Schwickerath, Z. Phys. C **75**, 627 (1997). arXiv:hep-ph/9607286
1107. W. de Boer, A. Dabelstein, W. Hollik, W. Mosle, U. Schwickerath, arXiv:hep-ph/9609209
1108. G.C. Cho, K. Hagiwara, Nucl. Phys. B **574**, 623 (2000). arXiv:hep-ph/9912260
1109. Y.M. Cho, I.P. Neupane, Int. J. Mod. Phys. A **18**, 2703 (2003). arXiv:hep-th/0112227
1110. J. Erler, D.M. Pierce, Nucl. Phys. B **526**, 53 (1998). arXiv:hep-ph/9801238
1111. G. Altarelli, F. Caravaglios, G.F. Giudice, P. Gambino, G. Ridolfi, J. High Energy Phys. **0106**, 018 (2001). arXiv:hep-ph/0106029
1112. A. Djouadi, M. Drees, J.L. Kneur, J. High Energy Phys. **0108**, 055 (2001). arXiv:hep-ph/0107316
1113. W. de Boer, M. Huber, C. Sander, D.I. Kazakov, Phys. Lett. B **515**, 283 (2001)
1114. W. de Boer, C. Sander, Phys. Lett. B **585**, 276 (2004). arXiv:hep-ph/0307049
1115. G. Belanger, F. Boudjema, A. Cottrant, A. Pukhov, A. Semenov, Nucl. Phys. B **706**, 411 (2005). arXiv:hep-ph/0407218
1116. J.R. Ellis, K.A. Olive, Y. Santoso, V.C. Spanos, Phys. Rev. D **69**, 095004 (2004). arXiv:hep-ph/0310356
1117. J.R. Ellis, D.V. Nanopoulos, K.A. Olive, Y. Santoso, Phys. Lett. B **633**, 583 (2006). arXiv:hep-ph/0509331
1118. J.R. Ellis, S. Heinemeyer, K.A. Olive, A.M. Weber, G. Weiglein, arXiv:0706.0652 [hep-ph]
1119. E.A. Baltz, P. Gondolo, J. High Energy Phys. **0410**, 052 (2004). arXiv:hep-ph/0407039
1120. B.C. Allanach, Phys. Lett. B **635**, 123 (2006). arXiv:hep-ph/0601089
1121. B.C. Allanach, K. Cranmer, C.G. Lester, A.M. Weber, arXiv:0705.0487 [hep-ph]
1122. O. Buchmueller et al., arXiv:0707.3447 [hep-ph]
1123. M. Carena, S. Heinemeyer, C.E.M. Wagner, G. Weiglein, Eur. Phys. J. C **26**, 601 (2003). arXiv:hep-ph/0202167
1124. M. Carena, S. Heinemeyer, C.E.M. Wagner, G. Weiglein, Eur. Phys. J. C **45**, 797 (2006). arXiv:hep-ph/0511023
1125. S. Schael et al. (LEP Working Group for Higgs boson searches), Eur. Phys. J. C **47**, 547 (2006). arXiv:hep-ex/0602042
1126. A. Abulencia et al. (CDF Collaboration), Phys. Rev. Lett. **96**, 011802 (2006). arXiv:hep-ex/0508051
1127. V.M. Abazov et al. (D0 Collaboration), Phys. Rev. Lett. **95**, 151801 (2005). arXiv:hep-ex/0504018
1128. G.W. Bennett et al. (Muon G-2 Collaboration), Phys. Rev. D **73**, 072003 (2006). arXiv:hep-ex/0602035
1129. B.C. Allanach et al., Eur. Phys. J. C **25**, 113 (2002). arXiv:hep-ph/0202233
1130. R. Barate et al. (LEP Working Group for Higgs boson searches), Phys. Lett. B **565**, 61 (2003). arXiv:hep-ex/0306033
1131. S. Heinemeyer, W. Hollik, G. Weiglein, Phys. Rep. **425**, 265 (2006). arXiv:hep-ph/0412214
1132. J.A. Aguilar-Saavedra et al., Eur. Phys. J. C **46**, 43 (2006). arXiv:hep-ph/0511344
1133. H. Baer, F.E. Paige, S.D. Protopopescu, X. Tata, arXiv:hep-ph/0312045
1134. H. Bachacou, I. Hinchliffe, F.E. Paige, Phys. Rev. D **62**, 015009 (2000). arXiv:hep-ph/9907518
1135. ATLAS Collaboration, ATLAS detector and physics performance technical design report, CERN/LHCC 99-14/15, 1999
1136. B.C. Allanach, C.G. Lester, M.A. Parker, B.R. Webber, J. High Energy Phys. **0009**, 004 (2000). arXiv:hep-ph/0007009
1137. B.K. Gjelsten, E. Lytken, D.J. Miller, P. Osland, G. Polesello, ATLAS internal note ATL-PHYS-2004-007, 2004, published in [1138]
1138. G. Weiglein et al. (LHC/LC Study Group), Phys. Rep. **426**, 47 (2006). arXiv:hep-ph/0410364
1139. J. Hisano, K. Kawagoe, M.M. Nojiri, Phys. Rev. D **68**, 035007 (2003). arXiv:hep-ph/0304214
1140. M.M. Nojiri, G. Polesello, D.R. Tovey, J. High Energy Phys. **0603**, 063 (2006). arXiv:hep-ph/0512204
1141. J. Hisano, K. Kawagoe, M.M. Nojiri, Determination of stop and sbottom sector by LHC and ILC, published in [1142]
1142. G. Belanger, F. Boudjema, A. Pukhov, A. Semenov, Comput. Phys. Commun. **149**, 103 (2002). arXiv:hep-ph/0112278
1143. See contribution by XY, these proceedings
1144. D. Acosta et al., CMS NOTE 2006/134, 2006
1145. J. Allison et al., IEEE Trans. Nucl. Sci. **53**(1), 270–278 (2006)
1146. CMS Collaboration, CERN/LHCC 2006-001, 2006
1147. S. Abdullin et al., arXiv:hep-ph/0605143
1148. S. Heinemeyer, W. Hollik, D. Stockinger, A.M. Weber, G. Weiglein, J. High Energy Phys. **0608**, 052 (2006). arXiv:hep-ph/0604147
1149. S. Heinemeyer, W. Hollik, A.M. Weber, G. Weiglein, arXiv:0710.2972 [hep-ph]
1150. seal.web.cern.ch/seal/snapshot/work-packages/mathlibs/minuit
1151. J.R. Ellis, S. Heinemeyer, K.A. Olive, G. Weiglein, J. High Energy Phys. **0502**, 013 (2005). arXiv:hep-ph/0411216
1152. B.C. Allanach, C.G. Lester, Phys. Rev. D **73**, 015013 (2006). arXiv:hep-ph/0507283
1153. R.R. de Austri, R. Trotta, L. Roszkowski, J. High Energy Phys. **0605**, 002 (2006). arXiv:hep-ph/0602028
1154. J.R. Ellis, S. Heinemeyer, K.A. Olive, G. Weiglein, J. High Energy Phys. **0605**, 005 (2006). arXiv:hep-ph/0602220
1155. B.C. Allanach, C.G. Lester, A.M. Weber, J. High Energy Phys. **0612**, 065 (2006). arXiv:hep-ph/0609295
1156. P. Bechtle, K. Desch, P. Wienemann, Comput. Phys. Commun. **174**, 47 (2006). arXiv:hep-ph/0412012
1157. R. Lafaye, T. Plehn, D. Zerwas, arXiv:hep-ph/0404282
1158. ALEPH Collaboration, Phys. Rep. **427**, 257 (2006). arXiv:hep-ex/0509008
1159. ALEPH Collaboration, arXiv:hep-ex/0511027
1160. M. Grünewald, private communication